Atherosclerosis Disease Management

Jasjit S. Suri • Chirinjeev Kathuria
Filippo Molinari
Editors

Atherosclerosis Disease Management

Springer

Editors
Jasjit S. Suri
Biomedical Technologies, Inc.
Denver, Colorado
USA
and
Idaho State University (Affiliated)
Pocatello, Idaho
USA
jsuri@comcast.net

Filippo Molinari
BioLab
Department of Electronics
Politecnico di Torino
Torino, Italy
filippo.molinari@polito.it

Chirinjeev Kathuria
Planet Space, Inc.
Chicago, Illinois
USA

ISBN 978-1-4419-7221-7 e-ISBN 978-1-4419-7222-4
DOI 10.1007/978-1-4419-7222-4
Springer New York Dordrecht Heidelberg London

Library of Congress Control Number: 2010937645

Printed on acid-free paper

Springer is part of Springer Science+Business Media (www.springer.com)

Contents

Contributors

Gudmundur Johann Arason
Department of Immunology, Faculty of Medicine, University of Iceland, Sturlugötu 7, 101, Reykjavík, Iceland

Karol P. Budohoski
Acedemic Neurosurgery Unit, University of Cambridge, Cambridge, UK

Stéphane G. Carlier
Columbia University Medical Center, New York, New York, USA

Bernard Chiu
Imaging Research Laboratories, Graduate Program in Biomedical Engineering, Robarts Research Institute, London, ON, Canada

Wouter Driessen
David H. Koch Center, Anderson Cancer Center, University of Texas, Houston, TX, USA

Mario Eandi
Istituto di Farmacologia, Università degli Studi, Torino, Italy

Michaela Egger
Imaging Research Laboratories, Department of Medical Biophysics, Robarts Research Institute, University of Western Ontario, London, ON, Canada

Ayman El-Baz
Bioimaging Laboratory, University of Louisville, Louisville, KY, USA

Mohamed Abou El-Ghar
Urology and Nephrology Department, University of Mansoura, Mansoura, Egypt

Ahmed Elnakib
Bioimaging Laboratory, University of Louisville, Louisville, KY, USA

Stanislav Emelianov
Department of Biomedical Engineering, The University of Texas at Austin, Austin, TX, USA

J. M. A. van Engelshoven
Department of Radiology, Maastricht University Medical Center, Cardiovascular Research Institute Maastricht, Maastricht, The Netherlands

Lorena Esposito
Department of Neurology, Klinikum Rechts der Isar, Technische Universitaet Muenchen, Ismaningerstr. 22, 81675, Muenchen, Germany

Robert Falk
Director, Medical Imaging Division, Jewish Hospital, Louisville, KY, USA

Mostafa Fatemi
Ultrasound Research Lab, Department of Physiology and Biomedical Engineering, Mayo Clinic College of Medicine, 200 First Street SW, Rochester, MN 55905, USA

Aaron Fenster
Imaging Research Laboratories, Department of Medical Imaging, Department of Medical Biophysics, Graduate Program in Biomedical Engineering, Robarts Research Institute, University of Western Ontario, London, ON, Canada

Emanuele Ferrero
Vascular and Encovascular Surgery Unit, Mauriziano Umberto I Hospital, Turin, Italy

Michelangelo Ferri
Vascular and Endovascular Surgery Unit, Mauriziano Umberto I Hospital, Turin, Italy

Regina Feurer
Department of Neurology, Klinikum Rechts der Isar, Technische Universitaet Muenchen, Ismaningerstr. 22, 81675, Muenchen, Germany

Hao Gao PhD candidate in Biomechanics
Brunel Institute for Bioengineering, Brunel University, Uxbridge, UK

Sara Giordano
Neurology Division, Gradenigo Hospital, Torino, Italy

James F. Greenleaf Ph.D
Ultrasound Research Lab, Department of Physiology and Biomedical Engineering, Mayo Clinic College of Medicine, 200 First Street SW, Rochester, MN 55905, USA

Jonathan H. Gillard MD, FRCR
University Department of Radiology, University of Cambridge, Cambridge, UK

Georgy Gimel'farb
Department of Computer Science, University of Auckland, Auckland, New Zealand

Pierangela Giustetto
Neurology Division, Gradenigo Hospital, Torino, Italy

H. H. G. Hansen
Clinical Physics Laboratory, Department of Pediatrics, Radboud University Nijmegen Medical Center, Nijmegen, The Netherlands

Hofstra L
Department of Cardiology, Maastricht University Medical Center, Cardiovascular Research Institute Maastricht, Maastricht, The Netherlands

T. Idzenga
Clinical Physics Laboratory, Department of Pediatrics, Radboud University Nijmegen Medical Center, Nijmegen, The Netherlands

Birgit Kantor
Cardiovascular Diseases Division, Internal Medicine Department, Mayo Clinic, 200 First Street SW, Rochester, MN 55905, USA

Andrei Karpiouk
Department of Biomedical Engineering, The University of Texas at Austin, Austin, TX, USA

Jose Katz
Department of Medicine, Columbia University, New York, NY 10033, USA

Patrick Kee
6431 Fannin, MSB 1.247, Houston, TX 77030, USA
patrick.kee@uth.tmc.edu

Peter J. Kirkpatrick
Acedemic Neurosurgery Unit, University of Cambridge, Cambridge, UK

M. E. Kooi
Department of Radiology, Maastricht University Medical Center, Cardiovascular Research Institute Maastricht, Maastricht, The Netherlands

C. L. de Korte
Clinical Physics Laboratory, Department of Pediatrics, Radboud University Nijmegen Medical Center, Nijmegen, The Netherlands

Amin Katouzian
Heffner Biomedical Imaging Lab, Biomedical Eng. Dep., Columbia University, 1210 Amsterdam Ave., 373 Eng. Terrace, New York, NY 10027, USA

Girish J. Kotwal
InflaMed Inc, Louisville, KY, USA
Sullivan University College of Pharmacy, Louisville, KY, USA

Adam Krasinski
Imaging Research Laboratories, Department of Medical Biophysics, Robarts Research Institute, University of Western Ontario, London, ON, Canada

R. M. Kwee
Department of Radiology, Maastricht University Medical Center,
Cardiovascular Research Institute Maastricht, Maastricht, The Netherlands

Andrew F. Laine
Biomedical Engineering Department, Columbia University,
1210 Amsterdam Avenue, New York, NY, USA

William Liboni
Neurology Division, Gradenigo Hospital, Torino, Italy

John C. Lieske
Division of Nephrology, Department of Internal Medicine, Hypertension,
Laboratory Medicine, and Pathology, Stabile 703, Mayo Clinic,
200 First Street SW, Rochester, MN 55905, USA

Christos P. Loizou
Department of Computer Science, School of Sciences, Intercollege,
92 Ayias Phylaxeos Street, P. O. Box 51604, CY-3507 Limassol, Cyprus

Quan Long senior lecturer
Biomedical Engineering, Brunel University, London, UK

Andrea Marsico
Head of the Anatomo-Pathology Division of the Koelliker Hospital, Torino, Italy
and
Adjunct Professor at the University of Torino, Torino, Italy
and
Senior Consultant in Histo-Cytopathology, Polyclinic of Monza, Italy

W. H. Mess
Department of Clinical Neurophysiology, Maastricht University Medical Center,
Cardiovascular Research Institute Maastricht, Maastricht, The Netherlands

Virginia M. Miller
Departments of Surgery and Physiology and Biomedical Engineering,
Mayo Clinic, 4-62 Medical Science Building, 200 First Street SW,
Rochester, MN 55905, USA

Filippo Molinari
Biolab – Dipartimento di Elettronica, Politecnico di Torino,
Corso Duca degli Abruzzi, 24, 10129, Torino, Italy

Franco Nessi
Vascular and Encovascular Surgery Unit, Mauriziano Umberto I Hospital,
Turin, Italy

R. J. van Oostenbrugge
Department of Neurology, Maastricht University Medical Center,
Cardiovascular Research Institute Maastricht, Maastricht, The Netherlands

Constantinos S. Pattichis
Department of Computer Science, University of Cyprus, Kallipoleos 75, P.O. Box 20537, CY-1678 Nicosia, Cyprus

Grace Parraga
Imaging Research Laboratories, Department of Medical Imaging, Department of Medical Biophysics, Graduate Program in Biomedical Engineering, Robarts Research Institute, University of Western Ontario, London, ON, Canada

Enrica Pavanelli
Neurology Division, Gradenigo Hospital, Torino, Italy

Cristina Pislaru
Ultrasound Research Lab, Department of Physiology and Biomedical Engineering, Mayo Clinic College of Medicine, 200 First Street SW, Rochester, MN 55905, USA

Holger Poppert
Department of Neurology, Klinikum Rechts der Isar, Technische Universitaet Muenchen, Ismaningerstr. 22, 81675 Muenchen, Germany

Nicoletta Rebaudengo
Neurology Division, Gradenigo Hospital, Torino, Italy

Tobias Saam
Standort Innenstadt Klinikum, Institut für Klinische Radiologie, Universität Muenchen, Vaillant-Einheit Maistrasse 11, Muenchen, Germany

Luca Saba
Department of Radiology, Azienda Ospedaliero Universitaria (A.O.U.), di Cagliari – Polo di Monserrato, Monserrato (Cagliari) 09045, Italy

Dirk Sander
Neurologische Klinik Medical Park Loipl, Thanngasse 15, 83483 Bischofswiesen, Germany

Rakesh Sharma
Department of Medicine, Columbia University, New York, NY 10033, USA; Center of Nanobiotechnology, Florida State University and Tallahassee Memorial Hospital, Tallahassee, FL 32304, USA; Innovations And Solutions Inc, 3945 West Pensacola Street, Tallahassee, FL 32304, USA

Maria K. Schwartz
Allergic Diseases Research, Guggenheim 4, Mayo Clinic, 200 First Street SW, Rochester, MN 55905, USA

J. David Spence
Imaging Research Laboratories, Stroke Prevention & Atherosclerosis Research Centre, Robarts Research Institute, London, ON, Canada

Jimmy Su
Department of Biomedical Engineering, The University of Texas at Austin, Austin, TX, USA

Jasjit S. Suri
Biomedical Technologies Inc, Denver, CO, USA;
Idaho State University, Pocatello, ID, USA;
Eigen Inc, Grass Valley, CA, USA

Tjun Y Tang
University Department of Radiology, University of Cambridge, Cambridge, UK

Perla Thorbjornsdottir
Department of Immunology, Landspitali University Hospital, LSH Hringbraut (hus 14), 101 Reykjavik, Iceland

Gudmundur Thorgeirsson
Department of Medicine, Landspitali University Hospital,
LSH Hringbraut (hus 14), 101 Reykjavik, Iceland;
Faculty of Medicine, University of Iceland, Sturlugötu 7, 101 Reykjavík, Iceland

Rikin A. Trivedi
Box 166, Department of Neurosurgery, Addenbrooke's Hospital,
Hills Road, CB2 0QQ Cambridge, UK

Andrea Viazzo
Vascular and Encovascular Surgery Unit, Mauriziano Umberto I Hospital, Turin, Italy

Bo Wang
Department of Biomedical Engineering, The University of Texas at Austin, Austin, TX, USA

J. E. Wildberger
Department of Radiology, Maastricht University Medical Center,
Cardiovascular Research Institute Maastricht, Maastricht, The Netherlands

Doug Yeager
Department of Biomedical Engineering, The University of Texas at Austin, Austin, TX, USA

Victoria EL Young
University Department of Radiology, University of Cambridge, Cambridge, UK

Guang Zeng
Department of Electrical and Computer Engineering, Clemson University, Clemson SC, USA; Mayo Clinic, Rochester, MN, USA

Fuxing Zhang
Research scientist at School of Medicine at University of Colorado, Denver, CO, USA

Part I
Histology, Pathologies and Associated Risks

Chapter 1
Introduction to the Pathology of Carotid Atherosclerosis: Histologic Classification and Imaging Correlation

Naima Carter-Monroe, Saami K. Yazdani, Elena Ladich, Frank D. Kolodgie, and Renu Virmani

Abstract Understanding the natural history of carotid atherosclerosis is essential in the management of patients at risk for stroke. Atherosclerotic plaque at the carotid bifurcation is the underlying cause of the majority of ischemic strokes and the degree of carotid stenosis is strongly associated with stroke risk in symptomatic patients. Pathologic studies comparing symptomatic and asymptomatic carotid plaques have demonstrated that specific plaque characteristics are associated with ischemic brain injury and the mechanisms underlying plaque instability in the carotid circulation are similar to those in the coronary circulation. This chapter will focus on the morphologic classification of carotid atherosclerosis based on a modification of the AHA classification system (with a comparison to atherosclerosis in the coronary vasculature) and will consider morphologic differences between carotid plaques in asymptomatic vs. symptomatic patients. In addition, we provide brief overview of the burgeoning number of imaging modalities used in the characterization of carotid plaques, as they compare to histologic studies.

Keywords Atherosclerosis • Fibroatheroma • Thin-cap fibroatheroma • Plaque rupture • Plaque erosion • Carotid • Endarterectomy • Plaque morphology • Inflammation • Magnetic resonance imaging • Angiography • Doppler ultrasound

1.1 Introduction

Despite advances in diagnostic and therapeutic interventions aimed at eradicating the scourge of cardiovascular disease, in the year 2006 alone, one out of every six deaths was due to coronary artery disease, with a total mortality of 425,425 persons in the US population. For the same year, in approximately 1 out of every 8.6 death certificates, or a total of 282,754 deaths, heart failure was recorded as an underlying cause of death or a precipitating factor. Current projections on cardiac-related disease

R. Virmani (✉)
CVPath Institute, Inc., 19 Firstfield, Road, Gaithersburg 20878, MD, USA
e-mail: rvirmani@cvpath.org

Jasjit S. Suri et al. (eds.), *Atherosclerosis Disease Management*,
DOI 10.1007/978-1-4419-7222-4_1, © Springer Science+Business Media, LLC 2011

in the US estimate that 785,000 people will have a new coronary event, 470,000 will have recurrent disease, and 195,000 will have a silent first myocardial infarction for 2010 [1].

As the third leading cause of death in the USA, stroke proves to be just as devastating given that in 1 year approximately 795,000 people will suffer a new or recurrent stroke. Of these cases, approximately 500,000 are first attacks and 200,000 recurrent attacks. In 2006, stroke contributed to approximately 1 in 18 deaths in the USA [1]. Ischemic stroke accounts for the largest number of new strokes (88%) followed by intracerebral hemorrhage (9%) and subarachnoid hemorrhage (3%) [2]. Atherosclerotic plaque at the carotid bifurcation is the underlying cause of the majority of ischemic strokes and the degree of carotid stenosis is strongly associated with stroke risk in symptomatic patients [3]. However, the degree of stenosis does not always predict those patients who will develop vulnerable lesions as low-grade lesions may also result in cerebrovascular events. Pathologic studies comparing symptomatic and asymptomatic carotid plaques have demonstrated that specific plaque characteristics are associated with ischemic brain injury and the mechanisms underlying plaque instability in the carotid circulation are similar to those in the coronary circulation [4, 5]. In fact, plaque morphology is considered an additional independent risk factor for cerebral infarction.

Before launching into a discussion of the pathological aspects of atherosclerotic disease of the carotid, the rich history of the medical assessment of atherosclerosis and evolution of pathological evaluation will be presented. The pathology and natural history of atherosclerotic carotid disease in light of our current knowledge of coronary atherosclerosis will follow. While the precise sequence of events leading to carotid plaque vulnerability is as yet unknown, certain early lesions and more advanced progressive lesions have been characterized and will be presented according to a modified classification scheme originally devised for the coronary circulation. In addition, the screening and current medical imaging modalities to assess carotid atherosclerosis and correlation with histologic findings will be discussed.

1.2 Atherosclerosis: A Historical Perspective

Atherosclerosis is an "ancient disease" with a fascinating history, beginning with its characterization in medical works of ancient Egyptians, Greek, and Romans (both atherosclerosis and cardiovascular disease in general). Roman Emperor Hadrian (76–138 AD) according to accounts by classical historian Dio Cassius (recorded 80 years after Hadrian's death), died from congestive heart failure secondary to hypertension and coronary atherosclerosis [6]. This fascinating history leads up to a duel of ideas between Rudolf Virchow and Carl von Rokitansky in the middle of the nineteenth century. Both observed cellular inflammatory changes in atherosclerotic lesions of the vessels they examined. Rokitansky held that these inflammatory changes were secondary in nature. Virchow, however, postulated that inflammation played a primary role in the process of atherogenesis [7].

Conventional wisdom has cast atherosclerosis to be a disease of modern man secondary to modern diet and stress despite the historical evidence outlined above and (more extensively) in other texts. However, paleopathology paints different picture, with findings of atherosclerotic lesions in mummies [8]. Microscopic examination of preserved vessels extracted from the mummified remains of the ancients showed evidence of atheroma, lipid deposition, medical calcification. Radiological exam revealed calcification of aorta and other large vessels. Allam et al. utilized whole-body, six-slice computed X-ray tomographic imaging (CT) to visualize calcium hydroxyapatite in vessel walls on 22 mummies kept at the Egyptian National Museum of Antiquities in Cairo, Egypt. Presence of calcium hydroxyapatite in a clearly defined artery upon CT imaging considered diagnostic for atherosclerosis (based on current convention) [9] and calcification along an artery's probable course considered "probable atherosclerosis." In these mummies, who lived between 1981 BCE and 334 CE, CT imaging found definite evidence of atherosclerosis in the form of calcium hydroxyapatite deposition in 5 of 16 mummies (30%), and probable atherosclerosis in 4 of 16 (25%). Calcification was more prevalent in those mummies who died at age 45 years or older (87%) as opposed to those dying before age of 45 (25%) [10].

1.3 Introduction to Carotid Artery Atherosclerosis

1.3.1 Pathologic Evaluation of the Carotid Endarterectomy Specimen

Carotid endarterectomy (CEA) has become the principal technique for cerebral revascularization in symptomatic and asymptomatic patient with extracranial carotid occlusive disease. CEA has become the most commonly performed vascular operation with an estimated 117,000 procedures performed annually in the USA. While the precise sequence of events leading to carotid plaque vulnerability is as yet unknown, certain early lesions and more advanced progressive lesions have been characterized and will be presented according to a modified classification scheme originally devised for the coronary circulation. It is in the interest of the pathologist to evaluate the endarterectomy specimen optimally, as only a detailed histologic examination of the carotid plaque specimen may demonstrate the underlying plaque morphology responsible for the disease, especially in symptomatic lesions.

Most surgeons remove the carotid plaques from the carotid artery bifurcation along with 10–15 mm of the internal and, if necessary, the external carotid artery. In all cases, the fixed specimens should be X-rayed to allow not only the identification of calcification but also provide information as to the extent of the luminal narrowing. Since most specimens are calcified, there is a necessity for most specimens to be decalcified in EDTA before histologic studies (Fig. 1.1). After decalcification, the specimen is cut transversely at 3–4 mm intervals beginning at the bifurcation. The entire specimen should be evaluated, as the culprit lesion

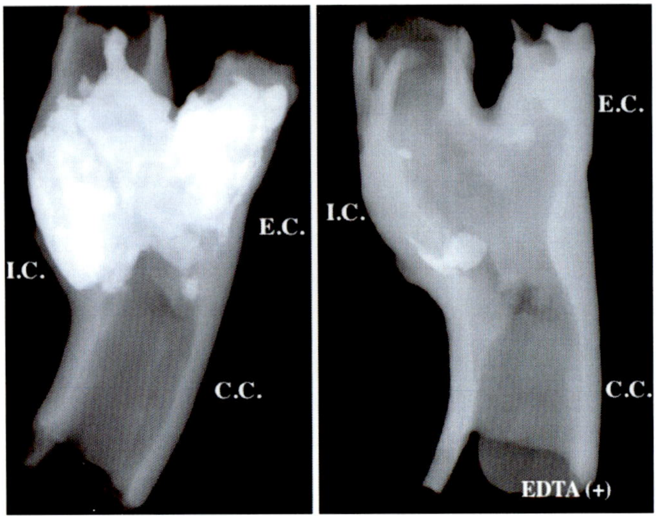

Fig. 1.1 Radiograph of a carotid endarterectomy specimen with extensive calcification in the internal and external carotid artery, beginning at the bifurcation site (*left*). The same specimen in A after 96 h of decalcification in ethylenediaminetetracetate (EDTA) (*right*). Note the severe narrowing of the lumen (*arrow*)

may not be limited to the most severely narrowed segment. Carotid plaque types share similarities with those found in the coronary circulation and may be classified according to AHA guidelines or by the simplified classification scheme described below [11].

1.3.2 Localization of Plaque at the Carotid Bifurcation

The earliest pathologic studies described the occurrence of atherosclerosis near branch ostia, bifurcations and bends, suggesting that flow dynamics play an important role in its induction. Atherosclerotic plaque tends to occur at regions where flow velocity and shear stress are reduced. It has been demonstrated that blood flow is disturbed at the carotid bifurcation where it departs from a laminar unidirectional pattern. The greatest atherosclerotic plaque accumulation typically occurs on the outer wall of the proximal segment and the sinus of the internal carotid artery, in the region of the lowest wall shear stress (Fig. 1.2). Plaque thickness is the least on the flow divider side at the junction of the internal and external carotid arteries where wall stress is the highest [12]. Thus, the unique geometrical configuration and flow properties of the carotid bifurcation contribute to the formation of atherosclerotic plaque, which may lead to critical carotid stenosis. However, plaque complications, regardless of the degree of the stenosis, are frequently the critical determinant of clinical consequences. At the carotid bifurcation, hemodynamic

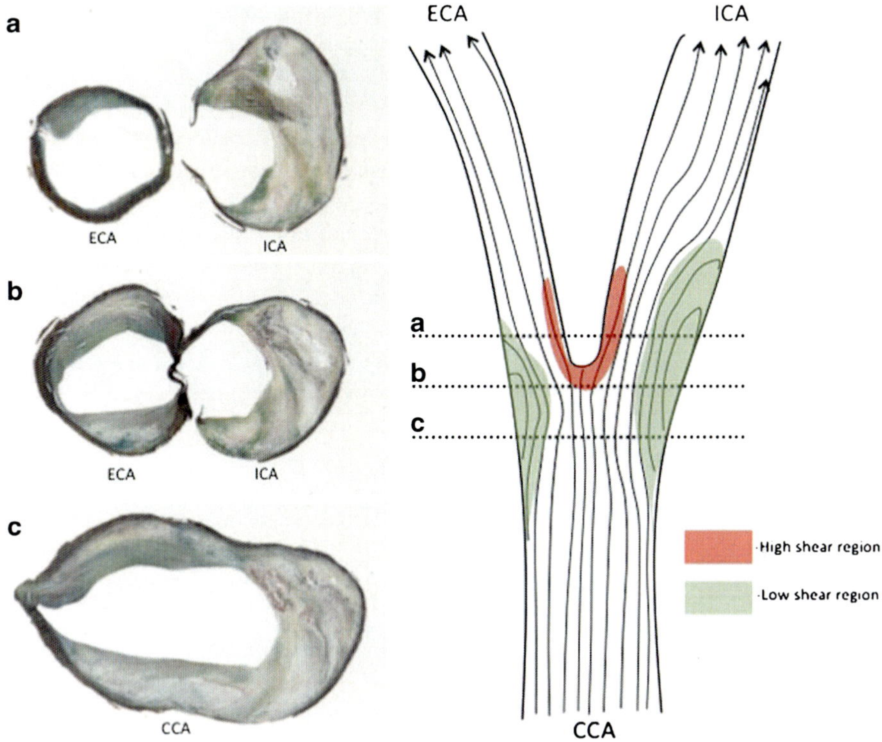

Fig. 1.2 Atherosclerotic disease at the carotid bifurcation. Plaque formation typically develops at the lateral walls of the bifurcation, as blood tends to separate and form low regions of shear stress. At the carina, flow remains parallel to the vessel wall. (**a**–**c**) demonstrate typical neointimal growth observed at the common carotid artery (CCA), internal carotid artery (ICA), and external carotid artery (CCA). It can be observed that within the carina (high shear regions), minimal neointima is developed

conditions may affect both the development and consequences of potentially catastrophic plaque complications.

1.4 Classification of Atherosclerotic Disease

1.4.1 The AHA Classification Scheme

The earliest classification system for atherosclerotic disease consisted of only two categories – the "fatty streak" and the atheromatous plaque. Considered as the precursor lesion to the atheromatous plaque, the fatty streak was defined as a lesion consisting of smooth muscle cells, lipid laden macrophages, and other inflammatory cells embedded within a proteoglycan–collagen matrix. The atheromatous plaque represented a continuation from the fatty streak stage, as a raised lesion with a lipid-rich necrotic core and an overlying fibrous cap. Within this necrotic core, varying amounts of cholesterol and cholesterol esters are deposited [13].

In a series of three reports, the AHA classification scheme was introduced using a numerical classification to stratify the various forms of coronary lesions [14–16]. This scheme was more sophisticated and focused on linear progression of human atherosclerotic disease progressing from unaffected normal intima (and adaptive intimal changes/thickening), to pre-atherosclerotic intimal lesions (Types II, III) to advanced disease (IV, V, VI). In brief, the first category or the Type I lesions represented the very initial changes, with only an increase in intimal macrophages and appearance of the foam cell – macrophages filled with lipid droplets. Type II lesions are grossly identifiable as "the fatty streak" layers of foam cells and lipid droplets interspersed within layers of intimal smooth muscle cells. Type III lesions are considered intermediate lesions (a bridge between Type II and Type IV), characterized by pools of extracellular lipid [16]. The atheroma as the first of the advanced lesions, falls within the Type IV category, and is characterized by a larger, confluent, and more disruptive lipid core. Next in the sequence is the fibroatheroma, or Type V lesion, in which the lipid core remains sequestered from the lumen by layers of fibrous connective tissue, with (Type Va) or without (Vb) calcification. Some variants of the Type V lesion have minimal lipid deposition (Vc). The Type VI lesion extends the Type V lesion to include plaques with fissure, hematoma, and/or thrombus formation [15].

This scheme assumes that the "atheroma" is a stable lesion, following Virchow's deduction that the "atheroma," is a fatty mass encapsulated within a fibrous cap much like purulent material in an abscess is encapsulated within a capsule [17]. This capsule must be disrupted in order for the thrombogenic core to gain exposure to the vascular lumen and cause initiation of the coagulation cascade. It is based on this paradigm, that the concept of plaque rupture as the critical event leading to atherosclerotic death has been accepted [18]. In one autopsy-based study, evidence of plaque rupture associated with thrombosis was identified in 73% of cases, plaque fissure with intraplaque fibrin deposition and hemorrhage seen in 8% of cases, and 19% with no evidence of thrombi [19].

1.4.2 Limitations of the AHA Classification

Over time and with observation of more lesions, many have noted limitations to the AHA classification. Specifically, one limitation entails the lack of direct, experimental human or animal studies to prospectively model the progression of atherosclerotic disease. Animal models rarely progress beyond Type IV, the atheroma, which is considered to be the most stable of the advanced lesions. This is not the case in humans, where clinically evident lesions fall in the type V and VI categories, and type IV lesions are usually clinically silent except in cases of severe lipidemia in which the atheromatous core can become occlusive because of increase in size alone [20].

A second limitation involves the analysis of human arteries, primarily from autopsy material. Several studies involving the analysis of autopsy derived human coronary specimens have shown exceptions to the classification rules of the AHA system, including a study by van der Wal et al. [21] involving a series of 20 patients undergoing sudden cardiac death with plaque rupture seen in 60% of the coronary lesions. The remaining 40% of lesions showed "superficial erosion" – a diagnostic category not addressed in the AHA schema. In approximately half of the cases of "superficial erosion," a fibrous cap heavily infiltrated by macrophages and T-lymphocyte and overlying a necrotic core was identified. The second series of studies evaluated coronary vessels from greater than 200 cases of sudden coronary death [22–26]. "Sudden coronary death" is defined as an unexpected death witnessed within 6 h of the onset of symptoms or death of a person known to be in stable condition <24 h before death [25]. Surprisingly, only one-third of the lesions in this series could be classified as plaque rupture, and 35% of lesions with thrombi failed to show a rupture site. And many of these lesions did not show significant inflammation.

We have proposed modifications to the AHA classification to address the aforementioned issues, mainly to the classification of the "intermediate" and "advanced" categories of plaque morphology [11]. This modified system includes seven categories as detailed in Table 1.1 including intimal xanthoma, intimal thickening, pathological intimal thickening, fibrous cap atheroma (fibroatheroma), thin cap fibrous cap atheroma, calcified nodule, and fibrocalcific plaque. Although both the AHA and Modified AHA classification systems explicitly refer to coronary artery atherosclerosis, observation shows that there are sufficient similarities between atherosclerotic lesions in the carotid and femoral vasculature in order to extent this classification system to those vascular beds. Figure 1.3 provides various examples of the lesion types according to the modified classification.

Table 1.1 Modified classification based on morphologic description

Early nonsymptomatic carotid disease
Diffuse intimal thickening
Intimal xanthoma
Intermediate lesion
Pathologic intimal thickening
Progression of atherosclerosis leading to plaque enlargement
Plaque hemorrhage (+/– calcification)
Thin cap fibroatheroma (+/– calcification)
Lesions with thrombi
Plaque rupture with luminal thrombus
Plaque rupture with ulceration
Plaque rupture with organizing thrombus
Plaque erosion
Calcified nodule
Stable atherosclerotic plaque
Healed rupture/erosion
Fibrocalcific plaque
Total occlusion

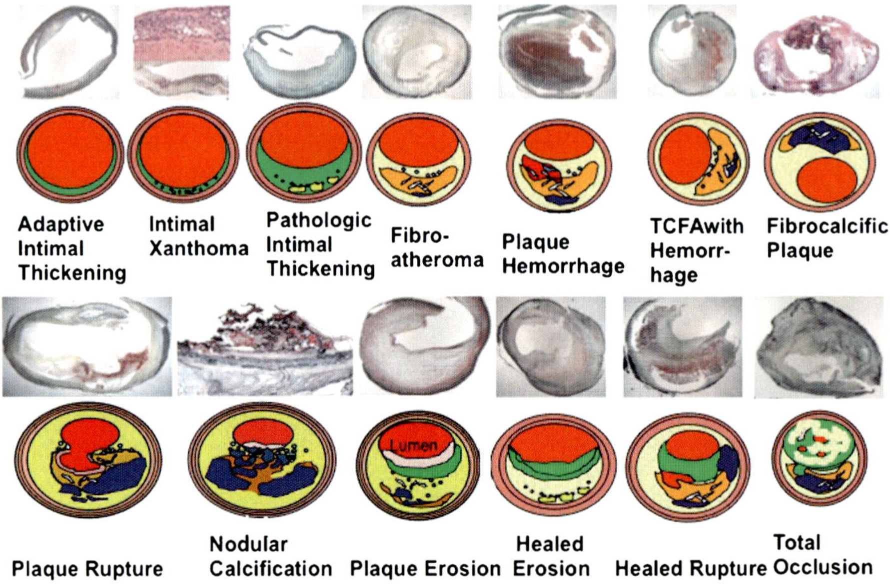

Fig. 1.3 The progression of atherosclerotic disease in the carotid. As shown in this series of cartoons and corresponding microscopic images, carotid atherosclerotic disease progresses from early, pre-atherosclerotic adaptive lesions to the more advanced symptomatic lesions such as the thin cap fibroatheroma (TCFA), plaque rupture, and total occlusion, chiefly as a result of the evolution of lipid pools emerging at the stage of pathological intimal thickening (PIT) into the necrotic core of the fibroatheroma and other advanced lesions

1.4.3 Pathologic Features of Atherosclerosis and Modifications to the AHA Classification

1.4.3.1 Early, Asymptomatic Lesions

Intimal Thickening and Intimal Xanthoma

Both intimal thickening and intimal xanthomas are considered the earliest, prelesional stage of the disease. "Intimal xanthoma" replaces the type I "fatty streak" or "initial lesion" in the AHA classification and is characterized by focal accumulations of lipid laden macrophages noticed in the arterial walls of the very young and known to regress with time. Adaptive intimal thickening replaces the Type II "intimal lesion" or "intimal thickening" characterized by smooth muscle cells and proteoglycan matrix with variable amounts of lipid and absent to minimal infiltrating inflammatory cells. In the carotid intimal thickening and plaque formation have been demonstrated to predominately occur at the outer wall of the proximal segment and at the sinus of the internal carotid artery. Both regions experience the lowest wall shear stress in the carotid and share the distinction of being areas of maximal plaque burden in cases of advanced atherosclerotic disease. As postulated for coronary vessels, it is the intimal mass lesion that serves as the most likely precursor of advanced atherosclerotic lesions [27].

Pathological Intimal Thickening

These lesions mark the transition from the early pre-atherosclerotic lesions (the "intimal mass" or "intimal xanthoma") to the more advanced lesions (i.e., the fibroatheroma) discussed below. Both we and the authors of the AHA classification scheme agree that the majority of human atheromatous lesions originate as preexisting intimal masses, and not from the intimal xanthoma seen in juvenile patients [28]. When these lesions progress to the pre-atheromatous or "pathological" stage, they are characterized by acellular regions located within the deeper intimal layers (close to the media) filled with proteoglycans and extracellular lipid pools. In this lesion we begin to notice the presence of inflammatory cells, as macrophages and T-lymphocyte aggregate toward the luminal side of the intima at the periphery of the lipid pools [11, 16]. The lipid pools most often arise from areas of adaptive intimal thickening, AHA Type III lesions fit into this category and are commonly observed in the coronary, carotid, and ileofemoral arteries.

1.4.3.2 Advanced Symptomatic Lesions

Fibrous Cap Atheroma

This category encompasses plaques categorized as AHA Type IV and V lesions and includes those lesions with a "fibrous cap" overlying a lipid core [11, 15]. This fibrous cap consists of smooth muscle cells embedded in a proteoglycan matrix with infiltration by variable numbers of macrophages and/or T-lymphocyte. The underlying lipid core is composed of variable amounts of extracellular lipid, necrotic debris, and cholesterol crystals often surrounded by macrophages (Fig. 1.4).

Progression of Atherosclerosis Leading to Plaque Enlargement

Intraplaque Hemorrhage

Intraplaque hemorrhage is common in advanced coronary atherosclerotic disease. It is believed to arise from the disruption of thin-walled microvessels (vasa vasorum) that are lined by discontinuous epithelium without supporting smooth muscle cells. Several investigators have suggested that intraplaque hemorrhage and rupture of the fibrous cap are associated with an increased density of microvessels [24, 29]. In the carotid circulation, the incidence of intraplaque hemorrhage has been reported as higher in symptomatic patients (84% vs. 56% of asymptomatic) [4]. Several studies in fact have cited intraplaque hemorrhage as an important process

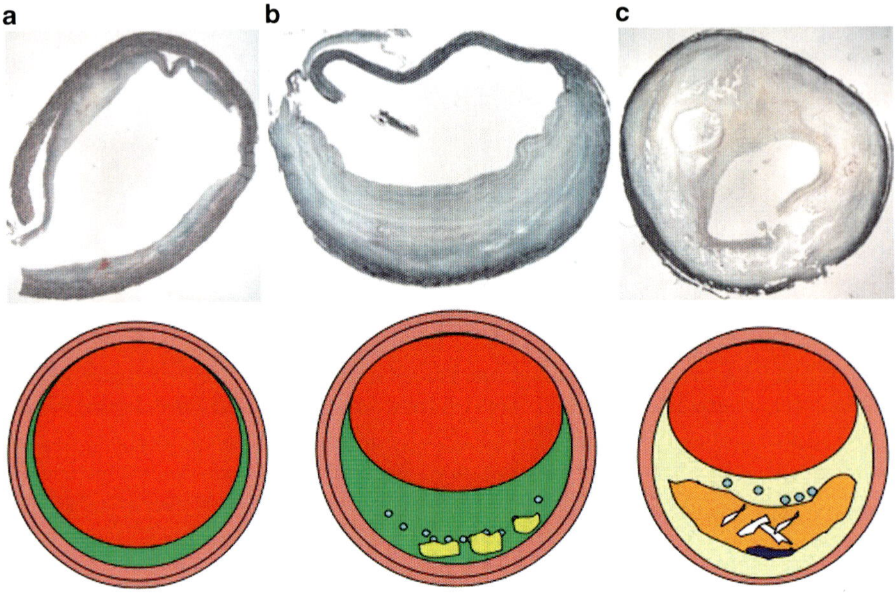

Fig. 1.4 A focus review of the progression of carotid atherosclerosis. (**a**) Adaptive intimal thickening (AIT) in a microscopic section from a carotid artery showing mild increase in smooth muscle cells and proteoglycan matrix (*green area* in cartoon image). (**b**) An intermediate lesion, pathological intimal thickening (PIT), in which one begins to see smooth muscle cells loss, extracellular lipid pools (*yellow areas* in cartoon image), and macrophage infiltration (represented by *blue circles*). (**c**) A well-developed fibroatheroma illustrating the classic necrotic core (*orange area* in cartoon) composed of cholesterol clefts (*white areas*) and necrotic debris with the overlying fibrous cap, often infiltrated by macrophages (*blue circles*)

associated with carotid plaque progression and the development of neurologic symptoms suggesting that hemorrhage may be related to disruption of the plaque or may lead to critical stenosis [30–33]. Plaque vascularity has been shown to correlate with intraplaque hemorrhage and the presence of symptomatic carotid disease [29]. These new blood vessels could play an active role in the metabolic activity of the plaque and ultimately control the processes that govern plaque progression. In addition, fibrin is a common finding in mature atherosclerotic lesions and most likely represents chronic hemorrhage within the plaque.

Thin Cap Fibrous Atheroma (Vulnerable Plaque)

This category expounds upon the fibrous cap atheroma to include those cases not included explicitly in the AHA classification with a quantifiably thinner fibrous cap, defined as a thickness <65 μm, and a relatively large necrotic core, often representing approximately 25% of the plaque area [26]. Studies have shown that the "thin cap" experiences loss of both extracellular matrix and smooth muscle cells, often accompanied by hemorrhage, calcification and abundant vasa vasorum [23, 34].

It is this thinning of the fibrous cap that leading to fissures and ruptures that results in total fibrous cap disruption in the coronary [11, 35], carotid [4], aortic [36], and femoral arteries. And it is this disruption of the fibrous cap, exposing the highly thrombogenic substances of the underlying necrotic core to the lumen, that is one factor responsible for luminal thrombosis. Given that 75% of thrombi in patients experiencing sudden coronary death are secondary to plaque rupture, early recognition and treatment of the thin cap fibroatheroma is of the utmost importance in the fight against premature death secondary to coronary atherosclerotic disease [35].

In the carotid artery, our laboratory has measured a mean vulnerable cap thickness of 72 ± 24 μm. Therefore, we have defined carotid vulnerable plaque thickness as less than 120 μm. Another recent study has defined carotid vulnerable plaque thickness as less than 165 μm based on a mean (±SD) cap thickness of 70 ± 47 μm [37]. Carotid plaques follow a similar pattern of disruption with fibrous cap thinning and infiltration of macrophages (Fig. 1.5). In a recent study, 47% of carotid ruptured plaque occurred in arterial segments with less than 70% luminal narrowing. Furthermore, a high prevalence of vulnerable plaques occurred in segments not significantly narrowed (80% of cases) [37]. These data suggest that the culprit

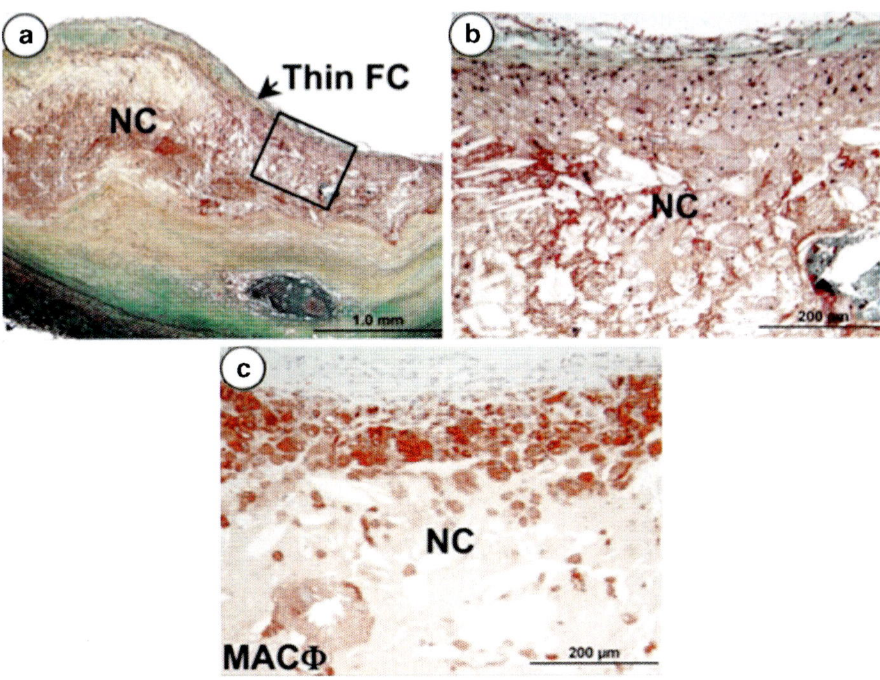

Fig. 1.5 Thin-cap fibroatheroma "vulnerable" plaque in the carotid artery. (**a**) A Movat Pentachrome-stained image of a thin cap fibroatheroma consisting of a relatively large necrotic core (NC) covered by a thin fibrous cap (FC). (**b**) Demonstrates a high-power image demonstrating infiltration of foamy macrophages in the fibrous cap. Macrophages (MACΦ) can be more clearly seen on oil-red-o staining in (**c**) (modified from [85])

lesions and their precursors occur more commonly in less severely narrowed vessels. Moreover, the data highlight the important tenet that plaques may progress to a substantial size before significant luminal stenosis occurs.

Lesions with Thrombi

This category includes lesions shown by observation to predispose to luminal thrombosis. As will be discussed further, these lesions are not mutually exclusive and thus can co-exist in the vascular bed and even in the same plaque.

Plaque Rupture with Luminal Thrombus/Organizing Thrombus

"Plaque rupture" is a descriptive term for phenomenon in which the fibrous cap becomes disrupted and an overlying luminal thrombus is in continuity with the underlying necrotic core. Ruptures typically have an enlarged necrotic core and the area of fibrous cap disruption shows both loss of smooth muscle cells and infiltration by macrophages and lymphocytes (Fig. 1.6). An acute thrombus is

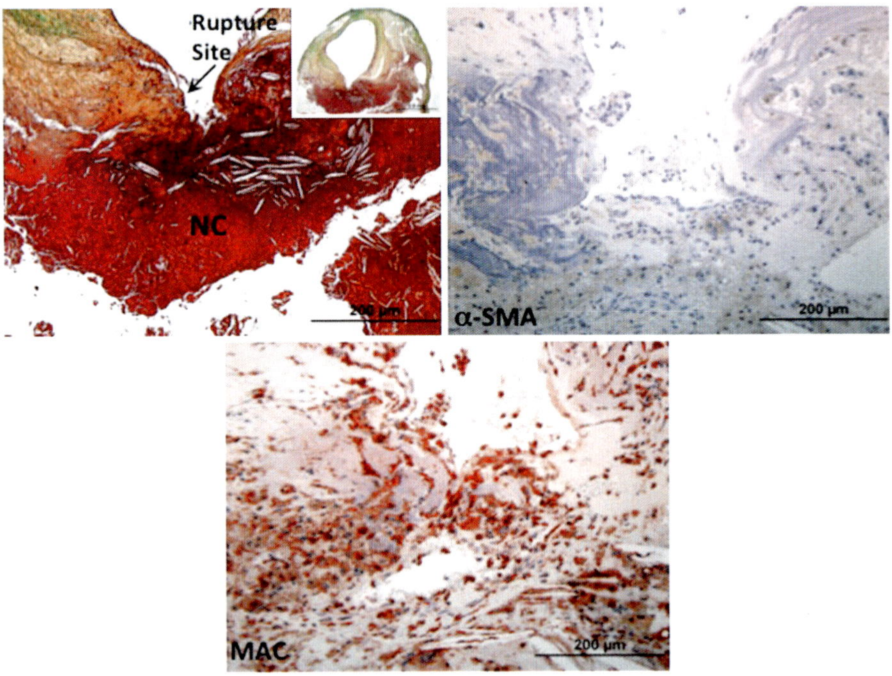

Fig. 1.6 Plaque rupture with ulceration in the carotid artery. (**a**) A Movat Pentachrome-stained image demonstrates a disrupted fibrous cap (*arrow*) with a relatively large necrotic core (NC). (**b**) shows a lack of actin-positive smooth muscle cells (ASMA) in the region of rupture. (**c**) demonstrates an abundance of CD-68 positive macrophages (MACΦ) at site of rupture (modified from [85])

characterized by platelet aggregates with few red blood cells and scattering of acute inflammatory cells. Over time, the thrombus may become organized, a process which involves infiltration of endothelial and smooth muscle cells and neovascularization. In cases of sudden coronary death, at least 75–80% of patients dying suddenly show the presence of acute or organized thrombi, while the rest demonstrate "critical" (≥75%) cross-sectional area luminal narrowing [38]. While plaque rupture with luminal thrombus is considered to be the major etiology of stroke, thrombi occupying large portions of the lumen in the carotid are unusual [4]. Spagnoli et al. identified thrombotically active plaques in 74% of patients with ipsilateral stroke [5].

Plaque Rupture with Ulceration

Most investigators agree that plaque rupture with ulceration is the dominant mechanism that leads to thrombus formation with subsequent embolization and cerebral ischemic events [4, 31]. Because of the differing hemodynamic properties of the carotid vs. coronary circulation, ulceration is a more common phenomenon in the carotid artery where sheer stress is higher compared to the coronary circulation. Ulceration is defined by an excavated necrotic core with a discontinuous fibrous cap. Thrombus, if present, is found lying in the excavated crater.

Plaque Erosion

Although plaque erosions account for approximately 30–35% of cases of thrombotic sudden coronary death, plaque erosion is an infrequent cause of thrombosis in carotid atherosclerotic disease [4, 5]. It has been proposed that the rarity of plaque erosions may be related to the higher flow in the carotid location vs. the coronary circulation. It is believed that erosion is the result of vasospasm and loss of endothelial cells. Because the carotid artery is a large vessel, it is not surprising that erosions are very infrequently observed in the carotid atherosclerosis and is the least frequent cause of carotid thrombosis.

Calcified Nodule

The "calcified nodule" represents the least frequent cause of luminal thrombus accounting for 2–5% of coronary thrombi [38]. This term refers to a lesion with fibrous cap disruption and thrombi associated with eruptive dense calcified nodules. The plaque is heavily calcified consisting of calcified plates and a surrounding area of fibrosis in the presence or absence of a necrotic core (Fig. 1.7). The luminal region of the plaque shows the presence of breaks in the calcified plate, sometimes even bone formation, and interspersed fibrin with a disrupted surface fibrous cap. Although still infrequent in the carotid location, it is more frequently observed in carotid plaque ruptures vs. coronary accounting for 6–7% of thrombi (RV unpublished data).

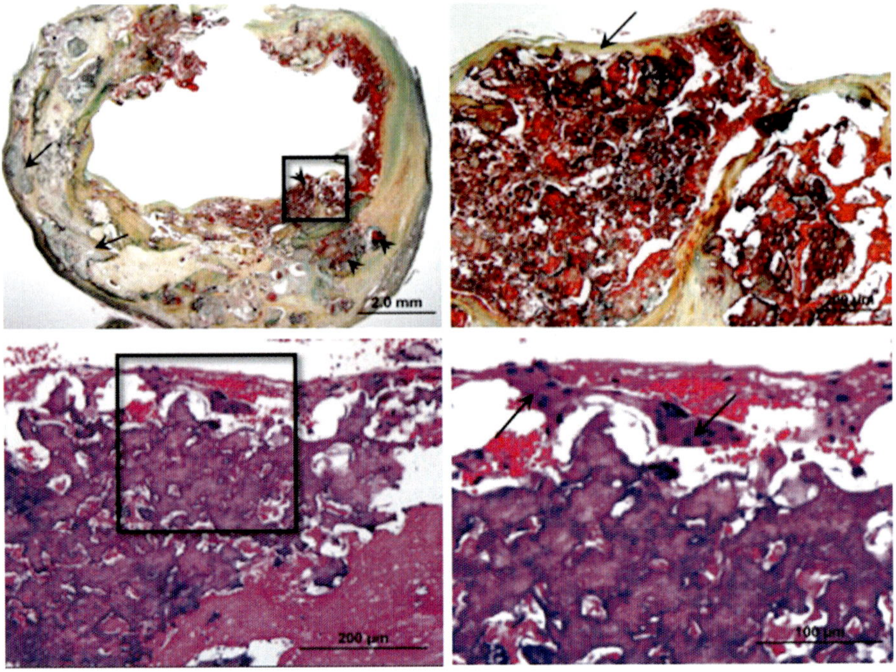

Fig. 1.7 Calcific nodule with luminal thrombus in the carotid artery. (**a**) A Movat Pentachrome-stained image demonstrates extensive calcification by large plates and multiple smaller nodules (*black arrows*). In (**b**) a high-power images demonstrates a thin fibrous cap (*black arrow*) over the region of nodular calcification. (**c**, **d**) Show high-power H&E images of the luminal surface of a nodular calcification with luminal layering of platelet/fibrin thrombus and multinucleated osteoclasts (*arrows*)

1.4.3.3 Stable Atherosclerotic Plaque

Healed Rupture/Erosion

Healed lesions define a third category of atherosclerotic disease. These consist of healed plaque ruptures (HPRs), erosions and total occlusions. Multiple HPRs are also described in the carotid arteries and similar to the coronary circulation the degree of luminal narrowing may be related to the layering of multiple healed repair sites. In a recent study, it was demonstrated that healed ruptures were present in 13.9% of stroke patients, 11.5% of TIA patients, and 16.6% of asymptomatic patients [37]. While it has been shown that in coronary artery disease progressive narrowing occurs because of thrombosis, thrombus does not typically occupy a large portion of the carotid lumen and may explain why the prevalence of multiple HPRS appears to be somewhat less frequent in the carotid artery.

Fibrocalcific Plaques

These lesions are characterized by thick fibrous plaques overlying extensive accumulation of calcium in the intima close to the media. This form of plaque is normally seen in patients with stable angina. Coronary calcification correlates highly with plaque burden but its effect on plaque instability is less evident. Those that are >75% narrowed likely represent burnt-out lesions. Since necrotic core is usually minimal to absent in these plaques, this lesion is not considered a true fibroatheroma. However, it is possible that the fibrocalcific lesion is the end stage of a process of atheromatous plaque rupture and/or erosion with healing and calcification.

In carotid plaques, calcification is more likely to begin at the surface, resulting in eruption of calcified nodules. Also, asymptomatic carotid plaques are, in the majority of cases, fibrocalcific plaques.

Chronic Total Occlusion

Chronic total occlusions may demonstrate varied histology depending on the age of the lesion. Older lesions demonstrate luminal obstruction characterized by dense collagen and/or proteoglycan with interspersed capillaries, arterioles, smooth muscle cells, and inflammatory cells. These lesions may also show earlier phases of organizing thrombi containing fibrin, red blood cells, and granulation tissue. Total occlusions often demonstrate shrinkage of the artery, perhaps due to the effect of collagen within plaque and/or adventitia. This is not as common a lesion in the carotid location as in the coronary arteries, which is likely the effect of high flow causing thrombus to embolize.

1.4.4 Carotid vs. Coronary Disease: Differences in Plaque Morphology

Despite the many similarities demonstrated in plaque morphology between the carotid and coronary circulation, there are several unique features of carotid plaque morphology related to the high flow rates and the shear forces caused by the bifurcation of the common carotid artery into the internal and external carotids. One of the most important is the ulcerated plaque, which is rare in the coronary artery circulation but relatively common in the carotid and other elastic arteries. While ulceration is associated with thrombotic lesions in symptomatic patients, thrombus is not always present at the ulcerative site, a phenomenon most likely related to embolic mechanisms in the carotid circulation.

Plaque hemorrhage in the carotid artery is much more frequent than in the coronary arteries and may be related to high flow rates and pressures in the lumen and the vasa vasorum. The maximum frequency of hemorrhage is observed in arteries with 50–75% cross-sectional area luminal narrowing [24]. We have reported in coronary plaques that intraplaque hemorrhage is responsible for necrotic core enlargement and excessive foamy macrophages in the fibrous caps [39]. Red blood cell membranes are the richest source of free cholesterol as compared to any other cell in the body. The free cholesterol in the necrotic core is believed to arise from apoptotic cell death of foamy macrophages. However, we have shown that the free cholesterol found in fibroatheromas, thin cap fibroatheromas, and plaque ruptures is also derived from erythrocytes that become trapped in the necrotic core when intraplaque hemorrhages occur. Takaya et al. reported that patients with carotid intraplaque hemorrhage at 18 months follow-up had larger necrotic cores as well as accelerated plaque progression as compared to patients without intraplaque hemorrhage [33].

The frequency of calcification is similar in coronary and carotid arteries, with maximum calcification seen in carotid arteries that are narrowed in greater than 70% cross-sectional area. However, the frequency of calcified nodules is slightly higher in carotid disease (approximately 6–7%) as compared to 2–5% in coronary artery disease. Finally, plaque erosion, which accounts for 30–35% of coronary thrombosis, is a distinctly rare entity in the carotid artery.

1.5 Risk Factors Contributing to Symptomatic Carotid Disease and Correlation with Plaque Morphology

The correlation of risk factors with stroke is complicated by the multiple etiologic categories of stroke, including carotid atherothrombosis, aortic arch plaque embolization, thromboembolism for ischemic strokes, and hypertensive hemorrhagic strokes. Patients with carotid disease present with a spectrum of risk factors similar to coronary disease, including hypertension, and atherogenic and thrombotic factors. Although hypertension is by far the most important risk factor for the development of strokes, other risks include impaired cardiac function, diabetes, nonvalvular atrial fibrillation, migraine, and family history, and risk factors recognized as modifiable include cigarette smoking, low level of physical activity, and obesity [40, 41]. The incidence of stroke increases in proportion to both systolic and diastolic blood pressure and is elevated in Blacks, who have higher rates of hypertension [42]. The relative risk of ischemic stroke in diabetic patients ranges from 1.8- to 6-fold greater in case-control studies. More recently, attention has focused on inflammatory markers of atherosclerosis. High C-reactive protein (CRP) levels have been shown to be a predictor of risk of future cardiovascular events. Similarly, independent of other cardiovascular risk factors, elevated plasma CRP levels significantly predict the risk of future ischemic stroke and TIA in the asymptomatic elderly population [43]. High CRP at hospital discharge is a predictor of future cardiovascular events and death in patients admitted with ischemic stroke.

Serum lipids have long been associated with coronary artery disease, but not with cerebrovascular disease. However, clinical trials using β-hydroxy-β-methylglutaryl-CoA (HMG-CoA) reductase inhibitors (statins) have shown a reduction of stroke risk in patients with coronary artery disease and elevated cholesterol levels. It has been shown that lipid-lowering therapy selectively depletes the lipid cores in carotid plaques. Zhao et al. analyzed CEA specimens from patients treated for 10 years with lipid-lowering agents in the Familial Atherosclerosis Treatment Study (FATS). This study demonstrated that the lipid core was significantly smaller in treated patients, although the extent of calcification was greater than nontreated controls, and the fibrous tissue content was the same [44].

Smoking, another independent risk factor for stroke, is associated with an increased arterial wall stiffness, increase in fibrinogen levels, increased platelet aggregation and hematocrit, and decreased HDL-cholesterol [42]. Hypercoagulable states associated with the development of stroke include antiphospholipid syndrome, factor V Leiden, prothrombin 20210 mutation, protein C and S deficiency and high fibrinogen levels. Nonfasting total homocysteine levels are an independent risk factor for incidence of stroke in elderly persons [26].

Several studies have correlated plaque morphology to risk factors in the carotid and coronary circulation. Spagnoli et al. have shown that the fibrous carotid plaque correlated with aging and diabetes, the granulomatous plaque with hypertensive females, and the foam cell rich xanthomatous plaque exhibiting extensive alcianophilia with hypercholesterolemia. In smokers, plaques were frequently complicated by mural thrombosis [26]. Mauriello et al. studying CEA specimens showed that patients with the highest tertile of fibrinogen (>407 mg/dl) had a high incidence of thrombosis (67%) compared with plaques of subjects with the lower and middle tertile (22 and 29%, $p=0.002$ and $p=0.009$, respectively) [45]. Plaque rupture was significantly associated with high fibrinogen level (54%, $p=0.003$). Multivariate analysis revealed that hyperfibrinogenemia was an independent predictor of fibrous cap thickness (inverse correlation), macrophage foam cell infiltration of the cap, and thrombosis. When accounting for the other risk factors, hyperfibrinogenemia remained an independent predictor of carotid thrombosis [45]. It is becoming increasingly evident that more studies correlating plaque morphology with risk factors are needed to further improve our understanding of carotid disease and target risk factor modification as more detailed assessment of plaque composition is possible with improved imaging.

1.6 Comparison of Carotid Plaque Histology from Symptomatic and Asymptomatic Patients

In general, few pathologic studies have correlated carotid and aortic plaque morphology with cerebral findings, and as a result, the mechanisms by which carotid atherosclerosis results in cerebrovascular symptoms are less well understood than those linking coronary disease and myocardial symptoms. Overall, most studies demonstrated that the pathology of symptomatic plaques is similar to that of culprit coronary plaques. Furthermore, some of these studies have demonstrated that thrombus triggered by plaque rupture is one of the major determinants of ischemic stroke in patients affected by carotid atherosclerotic disease [5]. The majority of ischemic strokes appear to result from embolization from an atherosclerotic plaque or acute occlusion of the carotid artery and propagation of the thrombus distally rather than static occlusion [46]. While recent reports highlight significant differences in the frequency of plaque rupture between symptomatic and asymptomatic patients, other factors have also been associated with ischemic stroke. These include surface irregularity, plaque vascularity, ulceration, fibrous cap thinning, and infiltration of the fibrous cap by macrophages and T cells [4, 46–48].

Previously, we reviewed 44 CEA specimens (from 25 asymptomatic and 19 symptomatic patients). The asymptomatic and symptomatic patients had similar mean percent stenosis (77% vs. 74%). Thirty-three patients were men and 11 were women, with a mean age of 74 years for asymptomatic patients and 70 years for symptomatic patients. Patients were considered symptomatic if they had experienced stroke, transient ischemic attack (TIA), or amaurosis fugax ipsilateral to the carotid lesion being studied. Other risk factors, including hypertension, diabetes mellitus, coronary artery disease, smoking history, serum cholesterol, and triglyceride levels were similar between groups. Each plaque was evaluated for the presence of a necrotic core, calcification, microscopic ulceration, plaque rupture, intraplaque hemorrhage, thrombus, infiltration of smooth muscle cells, fibrous cap thinning, infiltration of the fibrous cap with foam cells and intraplaque fibrin. The study showed that symptomatic carotid artery disease is more frequently associated with plaque rupture (74%) than is asymptomatic disease (32%) suggesting critical differences in plaque morphology between patients with symptomatic and asymptomatic disease. In addition, fibrous cap thinning was noted in 95% of symptomatic patients and in 48% of asymptomatic plaques ($p=0.003$). Infiltration of the fibrous cap with foam cells was also significantly more common in the symptomatic plaques (84% vs. 44% of asymptomatic plaques, $p=0.006$). Intraplaque fibrin was seen in 100% of symptomatic plaques vs. 68% of asymptomatic plaques; $p=0.008$ [4].

Bassiouny et al. performed a study of CEA specimens comparing symptomatic high-grade stenosis lesions and asymptomatic autopsy specimens without high-grade carotid artery stenosis. They showed that high-grade carotid stenotic plaques were associated with a significantly higher incidence of ulceration (53%), thrombosis (49%), and lumen irregularity (78%) compared to nonstenotic asymptomatic plaques (6, 0, and 17%, respectively; $p<0.01$). Although these features were more prominent in symptomatic patients, they were also present in 80% of the stenotic bifurcations, and did not distinguish between symptomatic endarterectomy and asymptomatic autopsy lesions [49].

In a subsequent report from the same group, disruption or ulceration of the fibrous cap was more common in the symptomatic than asymptomatic plaques. In addition, the number of macrophages infiltrating the fibrous cap was three times greater in the symptomatic plaques compared with the asymptomatic plaques ($1{,}114 \pm 1{,}104$ vs. 385 ± 622, respectively, $p<0.009$) [47].

A summation analysis published in 2000 by Golledge et al. summarized the findings of several pathology studies that had evaluated carotid plaque histology from symptomatic and asymptomatic patients. Studies included for review were limited to those demonstrating similar stenosis severity between the asymptomatic and symptomatic patients [46]. The results demonstrated that plaque rupture or ulceration is more common in symptomatic vs. asymptomatic patients (48% vs. 31%, $p<0.001$). Both luminal thrombus (40% vs. 35%) and intraplaque hemorrhage (48% vs. 50%) appeared to be similar for symptomatic and asymptomatic patients. In addition, most studies showed that the fibrous cap is thinner and inflammation is more common, with a greater number of macrophages and T cells detected in the cap of symptomatic plaques.

In a recent study performed by Spagnoli et al. 269 carotid plaques were analyzed (96 plaques from patients with ipsilateral major stroke, 91 plaques from patients with TIA, and 82 plaques from asymptomatic patients). The study demonstrated that thrombosis associated with plaque rupture is one of the major determinants of ischemic stroke in patients affected by carotid atherosclerotic disease [5]. There was no difference in degree of stenosis between the groups. In patients with ipsilateral major stroke, a thrombotically active plaque was observed in 74% (96 patients) of plaques. In this patient group, acute thrombus was associated with cap rupture in 90.1% of 71 thrombosed plaques and 9.9% with luminal surface erosion. In contrast, 35.2% of patients with TIA and 14.6% of asymptomatic patients were found to have thrombotically active plaque. Erosion was approximately twice as common in patients with TIA compared to stroke.

Cap inflammation consisted of numerous inflammatory cells composed predominantly of monocytes, macrophages, and T-lymphocyte. The amount of inflammation in stroke patients was almost twice the amount as that in patients with TIA. In addition, the severity of clinical events correlated significantly with the degree of inflammation in ruptured plaques suggesting that inflammatory cells residing in the carotid plaque may contribute to cell and tissue injury in ischemic brain disease.

In the largest histology study of symptomatic carotid lesions (526 carotid plaques in which symptomatic lesions included stroke ($n=159$) and TIA ($n=367$)), the histological features related to the nature and timing of presenting symptoms were described [48]. The study demonstrated a high prevalence of cap rupture (58.1%), intraplaque hemorrhage (64.6%), and marked plaque inflammation (66.8%) in the symptomatic plaques. The overall prevalence of histology features was similar for patients with stroke, cerebral TIA, and amaurosis fugax (a subset of TIA). Dense plaque macrophage infiltration was strongly associated with both cap rupture and time since stroke, suggesting possible causal links between plaque inflammation and plaque instability. There was a negative association between plaque macrophages and time since stroke with a continued decline in plaque removed up to 180 days after stroke. Although evidence of previous intraplaque hemorrhage was strongly associated with cap rupture, the association with time was weak. This study demonstrated that macrophage infiltration is greatest in those symptomatic plaques with rupture. Combined with previous findings of this group, demonstrating a very strong association between cap rupture on histology and surface ulceration on angiography, and between angiographic ulceration and risk of subsequent stroke, it appears that plaques with dense infiltration by macrophages are associated with a higher risk of stroke.

1.7 Imaging Modalities for Carotid Arteries and Histopathological Correlation

It is well known that symptomatic patients with severe stenosis (≥70%) benefit from CEA [50], and it is therefore vital to image and accurately measure the site of severe narrowing. However, pathological assessment of carotid plaques has also demonstrated that risk of emboli is not only associated with the size of the lesion but also the type of lesion, i.e., plaque morphology [37]. The ability to image and identify the "vulnerable plaque" and especially those with hemorrhage will be vital in early diagnosis, prevention, and treatment of stroke and neurological side effects. Below is a brief review of the current imaging modalities used to detect carotid artery disease. More detail in imaging of carotid plaques can be found in recent reviews [51–53].

1.7.1 Digital Subtraction Angiography

The digital subtraction angiogram (DSA) represents the traditional and gold standard in assessing the severity of luminal narrowing in the carotid artery. The clinician can accurately observe and measure the luminal diameter along the length of the artery in a 2-D plane, moreover, flow characteristics such as separation and eddies can be visualized. In estimating stenosis rates at the most severe narrowing, a reference diameter of a healthy distal internal carotid artery is obtained [54]. There are several disadvantages associated with angiograms, since the angiogram visualizes the vessel lumen, while the biology of plaque is occurring in the vessel wall. In addition, plaque features such as thin fibrous cap atheromas (vulnerable plaques) cannot be identified. Moreover, the technique is invasive, so there is an associated labor, time, expense, and risk associated with the catheterization procedures.

1.7.2 Doppler Ultrasound

Doppler ultrasound provides an inexpensive and fast noninvasive technique to assess carotid lesions [55]. This technique can provide local diameter and wall thickness measurements, as well as flow characteristics including wave forms, gradient, and peak velocities. The accuracy of such measurements is highly dependent on the operator [56], Doppler angles [57], and artifacts due to highly calcified plaques, which are generally observed in highly stenosed carotid lesions. Stenosis rates are assessed by comparing the flow measurements of the site of the stenosis to the common carotid regions. Although Doppler ultrasound provides quantitative data at the site of severe narrowing, an overview of the vascular anatomy and surrounding structures is not obtained. Therefore, ultrasound has been suggested at least in early nonstenotic lesions to be superior for risk assessment as compared to DSA. Digital subtraction angiography still remains the method of choice to rule out obstructive carotid disease in asymptomatic and symptomatic individuals in most centers; however, it is being slowly replaced by newer noninvasive technologies, and since the peak of digital subtraction angiography in the 1980s, it is on the decline [51].

1.7.3 CT Angiography

Although angiography and ultrasound remain the gold standards, the use of CT in assessing carotid artery stenosis has steadily increased. Multi-slice helical CT scanners enable fast and accurate vessel imaging with minimal discomfort by peripheral injection of a contrast agent. With evolution of CT technology, high-resolution images of the luminal volume of arteries can be imaged from the aortic arch to the circle of Willis. Luminal assessment of locations of severe stenosis can be obscured with extensive plaque calcification, especially when circumferential. But overall, the sensitivity and specificity of CT angiography for carotid artery stenosis is similar to digital subtraction angiography [58–60]. Limitations of CT angiography include ionization radiation dose and ballooning artifacts of heavy calcification. Recently, feasibility and the evaluation of carotid plaque composition assessment using dual-source CT as compared to histopathologic specimens after CEA has been demonstrated, with excellent correlation (100%) for calcification and 85% for the detection of low-density fatty plaques [61]. Most centers today consider noninvasive testing, alone or in combination, to be sufficiently accurate to replace digital subtraction angiography for the routine assessment of carotid disease.

1.7.4 Magnetic Resonance Imaging

The development of high-resolution MRI has emerged as one of the most promising techniques [33] fibrous tissue, calcium, and lipid can be identified using signal intensity variation from four different weightings [44]. High-resolution MRI not only allows differentiation of lipid core but also is able to assess presence of plaque hemorrhage and provide an estimate of the age of hemorrhage present [33, 62] (see Figs. 1.8 and 1.9). A recent study by Oikawa et al. has demonstrated a high level of agreement between characterization of carotid plaque, with each layer of the plaque

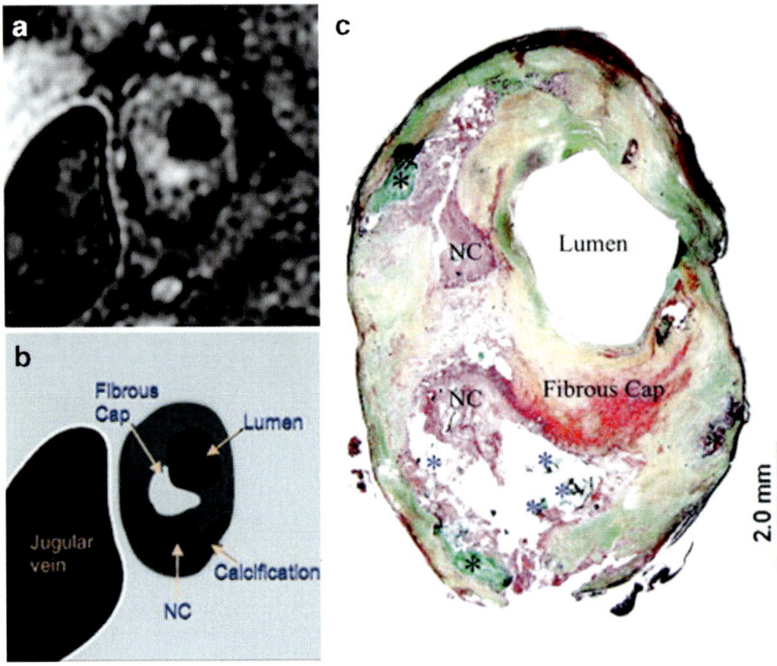

Fig. 1.8 Characterization of atheroma composition using high-resolution MRI. (**a**) In vivo contrast-enhanced BBMRI of the distal right common carotid artery of a 71-year-old man with asymptomatic carotid stenosis. Enhancing tissue highlights the relatively thick fibrous cap against the lumen and necrotic core (NC) as outlined in the schematic (**b**). The composition is confirmed in the endarterectomy section stained by Movat at the corresponding location (**c**). The specimen demonstrates a hemorrhagic NC containing calcifications (*blue asterisk*) with more dense calcifications (*black asterisk*) along its outer wall. Dark patches within the core seen on the MRI likely represent these focal calcifications. Note, part of the necrotic core and calcified areas have artifactually fallen off the section during cutting and staining (reprinted from [68])

measured as percentage thickness of the arterial wall, using routine histology and MRI (see Fig. 1.10) [63]. However, it is not possible to measure the precise thickness of the fibrous cap as these measures exceed the resolution of the MRI (200–500 μm) [64]. High-resolution MRI has been utilized to assess the effect of lipid-lowering treatment on carotid plaque composition [44] and has been verified in ex vivo models [65]. Furthermore, determination of plaque components of moderate-grade carotid stenosis using MRI techniques (T1-weighted coronal three-dimensional volume acquisition with high sensitivity for methemoglobin) has excellent accuracy for the detection of intraplaque hemorrhage [33]. Preliminary work by Takaya et al. showed that the presence of thin cap or ruptured caps, intraplaque hemorrhage, or large lipid-rich necrotic core was associated with subsequent neurological events at a mean interval of 38 months [66]. Others have shown that plaques of symptomatic patients are more likely to have features of vulnerable or ruptured characteristics on MRI than those from asymptomatic patients when the stenosis is similar [67, 68].

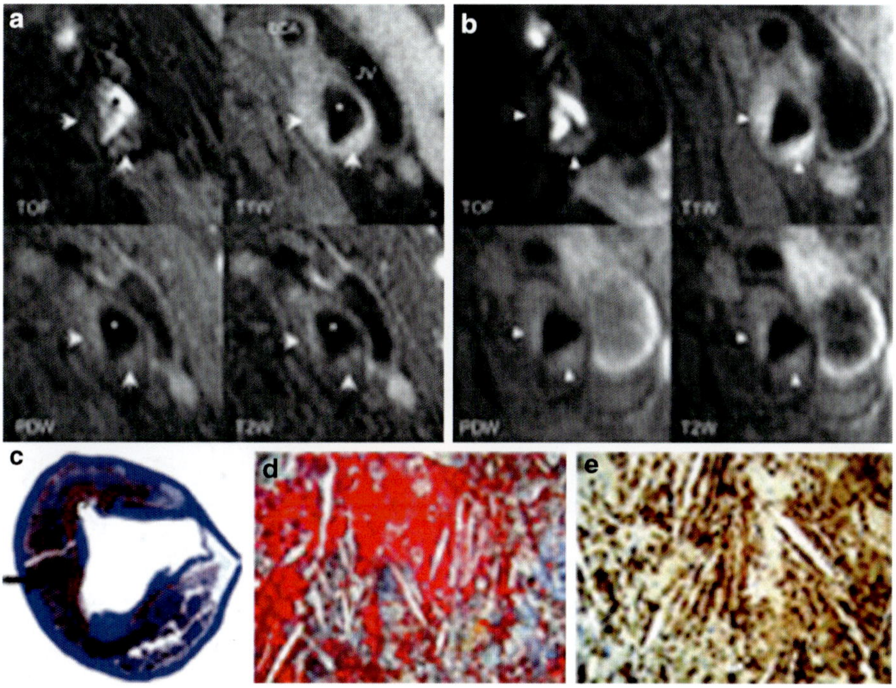

Fig. 1.9 (**a**) Signal intensity of type II hemorrhage at baseline examination. Type II hemorrhage is identified by hyperintense signals on TOF, T1W, PDW, and T2W images of left internal carotid artery (*arrow*). *Asterisks* show location of lumen. (**b**) Images from 18-month follow-up scan showed a similar signal intensity pattern in the same regions (*arrowheads*). (**c**) Matched Mallory's trichrome-stained section from excised CEA specimen. (**d**) High-power (×400) field taken from region (*arrow* in **c**) deep within necrotic core showing hemorrhagic debris and cholesterol clefts. (**e**) Glycophorin A immunostaining of the same region (×400) in an adjacent section shows extensive staining of hemorrhagic debris indicating the presence of erythrocyte membranes. JV indicates jugular vein; ECA, external carotid artery (reprinted from Takaya et al. [33] *with permission*)

MRI has also been utilized to demonstrate ethnically based differences in carotid plaque morphology, with Chinese patients demonstrating larger necrotic cores and lesser degrees of calcification than US counterparts [69]. MRI imaging in the near future can potentially be combined with other modalities, including PET, to provide both structural and metabolic information about the plaque [70].

1.7.5 Role of Inflammation in Imaging

Time has vindicated Virchow's original observation that cellular inflammatory response is a contributing factor in the evolution of atherosclerotic disease. As early as 1815, Joseph Hodgson postulated that inflammation was the underlying cause of

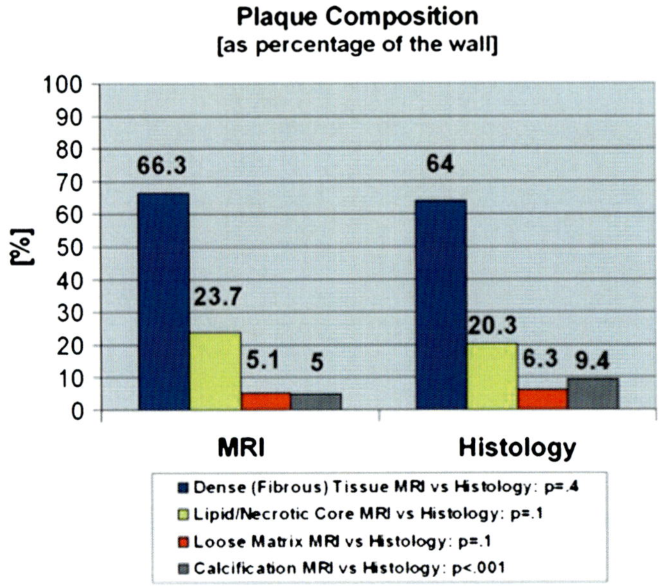

Fig. 1.10 Plaque composition calculated as the percentage of the vessel wall area, calculated per artery, and then averaged across all arteries for magnetic resonance imaging (MRI) and histology (reprinted from Oikawa et al. [63] *with permission*)

atherosclerosis and that the disease process occurred in the intima [71]. However, by the mid twentieth-century, the theory of atherosclerosis as an inflammatory disease had become prematurely obsolete and conspicuously absent from the literature [72]. Aschoff in 1907 and Anitschkow a few years later began the demise of the atherosclerosis as an inflammatory disease theory and ushered in the era of the lipid hypothesis as the underlying cause of atherosclerosis [73]. Aschoff identified of cholesterol esters in the "fatty" material in diseased vessels and Anitschkow demonstrated the induction of arterial lipid deposits in rabbits fed a high cholesterol diet [74, 75]. Russell Ross, revived the concept of atherosclerosis as an inflammatory disease by the introduction of the response to injury theory and by the promotion of role of lipid oxidation as a cause of atherosclerosis [76].

Current research has shown that beginning at the earliest stage of pre-atherosclerotic lesions, namely, the fatty streak, oxidized low-density lipoprotein (LDL) deposited in the intima acts as a chemoattractant to monocyte-derived macrophages [77]. In addition, these monocytes express receptors for chemoattractant substances such as monocyte chemotactic protein-1 (MCP-1) and for adhesion molecules such as vascular cell adhesion molecule (VCAM-1) and intercellular adhesion molecule-1 (ICAM-1) [78]. Using these receptors, monocytes are able to migrate and localize to the subintimal layers where they begin to express scavenger receptors such as CD68, SRAI/II, and FcRIII [79]. The inflammatory cascade continues with the elaboration of numerous cytokines by foam cells in the neointima and the secretion of proteolytic enzymes, such as capthesins, and metalloproteinases by activated macrophages.

It is at this point we observe the intersection of molecular and imaging research. Several of the aforementioned inflammatory mediators have been the focus of experimental studies aimed at developing novel biomarkers to aid in molecular imaging of plaques, including the successful radiolabeling of LDL [80], autologous leukocytes [81], MCP-1, VCAM-1, and metallinoproteinases [78]. Unfortunately, the clinical utility of utilizing these markers for identification of active inflammation in atherosclerotic plaques has not been established. More promising results have been achieved with the successful imaging of active macrophages by ^{18}F-labeled FDG [82] and the labeling of dying macrophages by annexin-45 [83, 84].

1.8 Conclusions

The morphology of atherosclerotic plaque types in the carotid vascular bed share many similarities with histologic descriptions of atherosclerosis in the coronary arteries, which ultimately aids in the classification of carotid disease. Similar to coronary atherosclerosis, thrombosis triggered by plaque rupture is one of the major determinants of ischemic stroke. And as the field of cardiovascular pathology evolves, so too does the ability of imaging modalities, MRI in particular, to aid in the early diagnosis of carotid atherosclerosis in vivo. Surveillance and early diagnosis of vulnerable lesions will prove to be of the utmost importance in future efforts to tailor therapeutic interventions in patients at risk for cerebrovascular events and effectively stratify patients into meaningful risk categories.

References

1. Lloyd-Jones D, Adams RJ, Brown TM, Carnethon M, Dai S, De Simone G, Ferguson TB, Ford E, Furie K, Gillespie C, Go A, Greenlund K, Haase N, Hailpern S, Ho PM, Howard V, Kissela B, Kittner S, Lackland D, Lisabeth L, Marelli A, McDermott MM, Meigs J, Mozaffarian D, Mussolino M, Nichol G, Roger V, Rosamond W, Sacco R, Sorlie P, Stafford R, Thom T, Wasserthiel-Smoller S, Wong ND, Wylie-Rosett J. Heart disease and stroke statistics – 2010 update. A report from the American Heart Association. *Circulation* 2010;121(7):948–54.
2. Thom T, Haase N, Rosamond W, Howard VJ, Rumsfeld J, Manolio T, Zheng ZJ, Flegal K, O'Donnell C, Kittner S, Lloyd-Jones D, Goff DC, Jr., Hong Y, Adams R, Friday G, Furie K, Gorelick P, Kissela B, Marler J, Meigs J, Roger V, Sidney S, Sorlie P, Steinberger J, Wasserthiel-Smoller S, Wilson M, Wolf P. Heart disease and stroke statistics – 2006 update: a report from the American Heart Association Statistics Committee and Stroke Statistics Subcommittee. *Circulation* 2006;113:e85–151.
3. North American Symptomatic Carotid Endarterectomy Trial Collaborators. Beneficial effect of carotid endarterectomy in symptomatic patients with high-grade carotid stenosis. *N Engl J Med* 1991;325:445–53.
4. Carr S, Farb A, Pearce WH, Virmani R, Yao JS. Atherosclerotic plaque rupture in symptomatic carotid artery stenosis. *J Vasc Surg* 1996;23:755–65; discussion 765–6.
5. Spagnoli LG, Mauriello A, Sangiorgi G, Fratoni S, Bonanno E, Schwartz RS, Piepgras DG, Pistolese R, Ippoliti A, Holmes DR, Jr. Extracranial thrombotically active carotid plaque as a risk factor for ischemic stroke. *JAMA* 2004;292:1845–52.

6. Petrakis N. Diagonal earlobe creases, type A behavior and the death of Emporer Hadrian (medicine in perspective). *West J Med* 1980;132:87–91.
7. Mayerl C, Lukasser M, Sedivy R, Niederegger H, Seiler R, Wick G. Atherosclerosis research from past to present – on the track of two pathologists with opposing views, Carl von Rokitansky and Rudolf Virchow. *Virchows Arch* 2006;449:96–103.
8. Magee R. Arterial disease in antiquity. *Med J Aust* 1998;169:663–6.
9. Budoff MJ, Achenbach S, Blumenthal RS, Carr JJ, Goldin JG, Greenland P, Guerci AD, Lima JA, Rader DJ, Rubin GD, Shaw LJ, Wiegers SE. Assessment of coronary artery disease by cardiac computed tomography: a scientific statement from the American Heart Association Committee on Cardiovascular Imaging and Intervention, Council on Cardiovascular Radiology and Intervention, and Committee on Cardiac Imaging, Council on Clinical Cardiology. *Circulation* 2006;114:1761–91.
10. Allam AH, Thompson RC, Wann LS, Miyamoto MI, Thomas GS. Computed tomographic assessment of atherosclerosis in ancient mummies. *JAMA* 2009;302:2091–4.
11. Virmani R, Kolodgie FD, Burke AP, Farb A, Schwartz SM. Lessons from sudden coronary death: a comprehensive morphological classification scheme for atherosclerotic lesions. *Arterioscler Thromb Vasc Biol* 2000;20:1262–75.
12. Glagov S, Zarins C, Giddens DP, Ku DN. Hemodynamics and atherosclerosis. Insights and perspectives gained from studies of human arteries. *Arch Pathol Lab Med* 1988;112:1018–31.
13. Virmani R, Ladich ER, Burke AP, Kolodgie FD. Histopathology of carotid atherosclerotic disease. *Neurosurgery* 2006;59:S219–27; discussion S3–13.
14. Stary HC, Blankenhorn DH, Chandler AB, Glagov S, Insull W, Jr., Richardson M, Rosenfeld ME, Schaffer SA, Schwartz CJ, Wagner WD, et al. A definition of the intima of human arteries and of its atherosclerosis-prone regions. A report from the Committee on Vascular Lesions of the Council on Arteriosclerosis, American Heart Association. *Arterioscler Thromb* 1992;12:120–34.
15. Stary HC, Chandler AB, Dinsmore RE, Fuster V, Glagov S, Insull W, Jr., Rosenfeld ME, Schwartz CJ, Wagner WD, Wissler RW. A definition of advanced types of atherosclerotic lesions and a histological classification of atherosclerosis. A report from the Committee on Vascular Lesions of the Council on Arteriosclerosis, American Heart Association. *Arterioscler Thromb Vasc Biol* 1995;15:1512–31.
16. Stary HC, Chandler AB, Glagov S, Guyton JR, Insull W, Jr., Rosenfeld ME, Schaffer SA, Schwartz CJ, Wagner WD, Wissler RW. A definition of initial, fatty streak, and intermediate lesions of atherosclerosis. A report from the Committee on Vascular Lesions of the Council on Arteriosclerosis, American Heart Association. *Circulation* 1994;89:2462–78.
17. Virchow R. Cellular Pathology Based on Physiological and Pathological Histology. Birmingham, AL: Classics of Medicine Library, 1858.
18. Ross R. Atherosclerosis – an inflammatory disease. *N Engl J Med* 1999;340:115–26.
19. Davies MJ. Anatomic features in victims of sudden coronary death. Coronary artery pathology. *Circulation* 1992;85:I19–24.
20. Caligiuri G, Levy B, Pernow J, Thoren P, Hansson GK. Myocardial infarction mediated by endothelin receptor signaling in hypercholesterolemic mice. *Proc Natl Acad Sci U S A* 1999;96:6920–4.
21. van der Wal AC, Becker AE, van der Loos CM, Das PK. Site of intimal rupture or erosion of thrombosed coronary atherosclerotic plaques is characterized by an inflammatory process irrespective of the dominant plaque morphology. *Circulation* 1994;89:36–44.
22. Farb A, Burke AP, Tang AL, Liang TY, Mannan P, Smialek J, Virmani R. Coronary plaque erosion without rupture into a lipid core. A frequent cause of coronary thrombosis in sudden coronary death. *Circulation* 1996;93:1354–63.
23. Burke AP, Farb A, Malcom GT, Liang Y, Smialek J, Virmani R. Effect of risk factors on the mechanism of acute thrombosis and sudden coronary death in women. *Circulation* 1998;97:2110–6.
24. Burke AP, Farb A, Malcom GT, Liang Y, Smialek JE, Virmani R. Plaque rupture and sudden death related to exertion in men with coronary artery disease. *JAMA* 1999;281:921–6.

25. Farb A, Tang AL, Burke AP, Sessums L, Liang Y, Virmani R. Sudden coronary death. Frequency of active coronary lesions, inactive coronary lesions, and myocardial infarction. *Circulation* 1995;92:1701–9.
26. Burke AP, Farb A, Malcom GT, Liang YH, Smialek J, Virmani R. Coronary risk factors and plaque morphology in men with coronary disease who died suddenly. *N Engl J Med* 1997;336:1276–82.
27. Ladich EL, Virmani R, Kolodgie FD, Burke AP. Natural History of Carotid Atherosclerosis. London, UK: Informa UK Ltd, 2008.
28. Velican C. A dissecting view on the role of the fatty streak in the pathogenesis of human atherosclerosis: culprit or bystander? *Med Interne* 1981;19:321–37.
29. Mofidi R, Crotty TB, McCarthy P, Sheehan SJ, Mehigan D, Keaveny TV. Association between plaque instability, angiogenesis and symptomatic carotid occlusive disease. *Br J Surg* 2001;88:945–50.
30. Bornstein NM, Krajewski A, Lewis AJ, Norris JW. Clinical significance of carotid plaque hemorrhage. *Arch Neurol* 1990;47:958–9.
31. Avril G, Batt M, Guidoin R, Marois M, Hassen-Khodja R, Daune B, Gagliardi JM, Le Bas P. Carotid endarterectomy plaques: correlations of clinical and anatomic findings. *Ann Vasc Surg* 1991;5:50–4.
32. Imparato AM. The carotid bifurcation plaque – a model for the study of atherosclerosis. *J Vasc Surg* 1986;3:249–55.
33. Takaya N, Yuan C, Chu B, Saam T, Polissar NL, Jarvik GP, Isaac C, McDonough J, Natiello C, Small R, Ferguson MS, Hatsukami TS. Presence of intraplaque hemorrhage stimulates progression of carotid atherosclerotic plaques: a high-resolution magnetic resonance imaging study. *Circulation* 2005;111:2768–75.
34. Fishbein MC, Siegel RJ. How big are coronary atherosclerotic plaques that rupture? *Circulation* 1996;94:2662–6.
35. Davies MJ, Thomas AC. Plaque fissuring – the cause of acute myocardial infarction, sudden ischaemic death, and crescendo angina. *Br Heart J* 1985;53:363–73.
36. Lendon CL, Davies MJ, Born GV, Richardson PD. Atherosclerotic plaque caps are locally weakened when macrophages density is increased. *Atherosclerosis* 1991;87:87–90.
37. Mauriello A, Sangiorgi GM, Virmani R, Trimarchi S, Holmes DR, Jr., Kolodgie FD, Piepgras DG, Piperno G, Liotti D, Narula J, Righini P, Ippoliti A, Spagnoli LG. A pathobiologic link between risk factors profile and morphological markers of carotid instability. *Atherosclerosis* 2009;208:572–80.
38. Virmani R, Burke AP, Farb A, Kolodgie FD. Pathology of the vulnerable plaque. *J Am Coll Cardiol* 2006;47:C13–8.
39. Kolodgie FD, Gold HK, Burke AP, Fowler DR, Kruth HS, Weber DK, Farb A, Guerrero LJ, Hayase M, Kutys R, Narula J, Finn AV, Virmani R. Intraplaque hemorrhage and progression of coronary atheroma. *N Engl J Med* 2003;349:2316–25.
40. Wolf PA. Prevention of stroke. *Lancet* 1998;352(Suppl 3):SIII15–8.
41. Wolf PA, Grotta JC. Cerebrovascular disease. *Circulation* 2000;102:IV75–80.
42. Goldstein LB, Adams R, Becker K, Furberg CD, Gorelick PB, Hademenos G, Hill M, Howard G, Howard VJ, Jacobs B, Levine SR, Mosca L, Sacco RL, Sherman DG, Wolf PA, del Zoppo GJ. Primary prevention of ischemic stroke: a statement for healthcare professionals from the Stroke Council of the American Heart Association. *Circulation* 2001;103:163–82.
43. Rost NS, Wolf PA, Kase CS, Kelly-Hayes M, Silbershatz H, Massaro JM, D'Agostino RB, Franzblau C, Wilson PW. Plasma concentration of C-reactive protein and risk of ischemic stroke and transient ischemic attack: the Framingham study. *Stroke* 2001;32:2575–9.
44. Zhao XQ, Yuan C, Hatsukami TS, Frechette EH, Kang XJ, Maravilla KR, Brown BG. Effects of prolonged intensive lipid-lowering therapy on the characteristics of carotid atherosclerotic plaques in vivo by MRI: a case-control study. *Arterioscler Thromb Vasc Biol* 2001;21:1623–9.
45. Mauriello A, Sangiorgi G, Palmieri G, Virmani R, Holmes DR, Jr., Schwartz RS, Pistolese R, Ippoliti A, Spagnoli LG. Hyperfibrinogenemia is associated with specific histocytological

composition and complications of atherosclerotic carotid plaques in patients affected by transient ischemic attacks. *Circulation* 2000;101:744–50.
46. Golledge J, Greenhalgh RM, Davies AH. The symptomatic carotid plaque. *Stroke* 2000;31:774–81.
47. Bassiouny HS, Sakaguchi Y, Mikucki SA, McKinsey JF, Piano G, Gewertz BL, Glagov S. Juxtalumenal location of plaque necrosis and neoformation in symptomatic carotid stenosis. *J Vasc Surg* 1997;26:585–94.
48. Redgrave JN, Lovett JK, Gallagher PJ, Rothwell PM. Histological assessment of 526 symptomatic carotid plaques in relation to the nature and timing of ischemic symptoms: the Oxford plaque study. *Circulation* 2006;113:2320–8.
49. Bassiouny HS, Davis H, Massawa N, Gewertz BL, Glagov S, Zarins CK. Critical carotid stenoses: morphologic and chemical similarity between symptomatic and asymptomatic plaques. *J Vasc Surg* 1989;9:202–12.
50. European Carotid Surgery Trialists' Collaborative Group. MRC European Carotid Surgery Trial: interim results for symptomatic patients with severe (70–99%) or with mild (0–29%) carotid stenosis. *Lancet* 1991;337:1235–43.
51. JM UK-I, Young V, Gillard JH. Carotid-artery imaging in the diagnosis and management of patients at risk of stroke. *Lancet Neurol* 2009;8:569–80.
52. Hermus L, van Dam GM, Zeebregts CJ. Advanced Carotid Plaque Imaging. *Eur J Vasc Endovasc Surg* 2009;39:125–33.
53. Nighoghossian N, Derex L, Douek P. The vulnerable carotid artery plaque: current imaging methods and new perspectives. *Stroke* 2005;36:2764–72.
54. Barnett HJ, Taylor DW, Eliasziw M, Fox AJ, Ferguson GG, Haynes RB, Rankin RN, Clagett GP, Hachinski VC, Sackett DL, Thorpe KE, Meldrum HE, Spence JD. Benefit of carotid endarterectomy in patients with symptomatic moderate or severe stenosis. North American Symptomatic Carotid Endarterectomy Trial Collaborators. *N Engl J Med* 1998;339:1415–25.
55. Weinberger J, Ramos L, Ambrose JA, Fuster V. Morphologic and dynamic changes of atherosclerotic plaque at the carotid artery bifurcation: sequential imaging by real time B-mode ultrasonography. *J Am Coll Cardiol* 1988;12:1515–21.
56. Mikkonen RH, Kreula JM, Virkkunen PJ. Reproducibility of Doppler ultrasound measurements. *Acta Radiol* 1996;37:545–50.
57. Grant EG, Benson CB, Moneta GL, Alexandrov AV, Baker JD, Bluth EI, Carroll BA, Eliasziw M, Gocke J, Hertzberg BS, Katarick S, Needleman L, Pellerito J, Polak JF, Rholl KS, Wooster DL, Zierler E. Carotid artery stenosis: grayscale and Doppler ultrasound diagnosis – Society of Radiologists in Ultrasound consensus conference. *Ultrasound Q* 2003;19:190–8.
58. Moll R, Dinkel HP. Value of the CT angiography in the diagnosis of common carotid artery bifurcation disease: CT angiography versus digital subtraction angiography and color flow Doppler. *Eur J Radiol* 2001;39:155–62.
59. Anderson GB, Ashforth R, Steinke DE, Ferdinandy R, Findlay JM. CT angiography for the detection and characterization of carotid artery bifurcation disease. *Stroke* 2000;31:2168–74.
60. Josephson SA, Bryant SO, Mak HK, Johnston SC, Dillon WP, Smith WS. Evaluation of carotid stenosis using CT angiography in the initial evaluation of stroke and TIA. *Neurology* 2004;63:457–60.
61. Das M, Braunschweig T, Muhlenbruch G, Mahnken AH, Krings T, Langer S, Koeppel T, Jacobs M, Gunther RW, Mommertz G. Carotid plaque analysis: comparison of dual-source computed tomography (CT) findings and histopathological correlation. *Eur J Vasc Endovasc Surg* 2009;38:14–9.
62. Chu B, Kampschulte A, Ferguson MS, Kerwin WS, Yarnykh VL, O'Brien KD, Polissar NL, Hatsukami TS, Yuan C. Hemorrhage in the atherosclerotic carotid plaque: a high-resolution MRI study. *Stroke* 2004;35:1079–84.
63. Oikawa M, Ota H, Takaya N, Miller Z, Hatsukami TS, Yuan C. Carotid magnetic resonance imaging. A window to study atherosclerosis and identify high-risk plaques. *Circ J* 2009;73:1765–73.

64. Redgrave JN, Gallagher P, Lovett JK, Rothwell PM. Critical cap thickness and rupture in symptomatic carotid plaques: the oxford plaque study. *Stroke* 2008;39:1722–9.
65. Shinnar M, Fallon JT, Wehrli S, Levin M, Dalmacy D, Fayad ZA, Badimon JJ, Harrington M, Harrington E, Fuster V. The diagnostic accuracy of ex vivo MRI for human atherosclerotic plaque characterization. *Arterioscler Thromb Vasc Biol* 1999;19:2756–61.
66. Takaya N, Yuan C, Chu B, Saam T, Underhill H, Cai J, Tran N, Polissar NL, Isaac C, Ferguson MS, Garden GA, Cramer SC, Maravilla KR, Hashimoto B, Hatsukami TS. Association between carotid plaque characteristics and subsequent ischemic cerebrovascular events: a prospective assessment with MRI – initial results. *Stroke* 2006;37:818–23.
67. JM UK-I, Tang TY, Patterson A, Graves MJ, Howarth S, Li ZY, Trivedi R, Bowden D, Kirkpatrick PJ, Gaunt ME, Warburton EA, Antoun NM, Gillard JH. Characterisation of carotid atheroma in symptomatic and asymptomatic patients using high resolution MRI. *J Neurol Neurosurg Psychiatry* 2008;79:905–12.
68. Wasserman BA, Wityk RJ, Trout HH, III, Virmani R. Low-grade carotid stenosis: looking beyond the lumen with MRI. *Stroke* 2005;36:2504–13.
69. Saam T, Cai JM, Cai YQ, An NY, Kampschulte A, Xu D, Kerwin WS, Takaya N, Polissar NL, Hatsukami TS, Yuan C. Carotid plaque composition differs between ethno-racial groups: an MRI pilot study comparing mainland Chinese and American Caucasian patients. *Arterioscler Thromb Vasc Biol* 2005;25:611–6.
70. Davies JR, Rudd JH, Fryer TD, Graves MJ, Clark JC, Kirkpatrick PJ, Gillard JH, Warburton EA, Weissberg PL. Identification of culprit lesions after transient ischemic attack by combined 18F fluorodeoxyglucose positron-emission tomography and high-resolution magnetic resonance imaging. *Stroke* 2005;36:2642–7.
71. Acierno L. Atherosclerosis (Arteriosclerosis). The History of Cardiology. New York: Partheno Publishing Group, 1994.
72. Saphir O, Charles C. Thomas. Inflammatory factors in arteriosclerosis. In: Blumenthal HT, ed. *Cowdry's Arteriosclerosis,* Springfield, IL 1967:415.
73. Madjid M, Aboshady I, Casscells SW, Virmani R, Willerson JT. Introduction: history of atherosclerosis and vunerable plaque research. In: Virmani R, Narula J, Leon MB, Willerson JT, eds. The Vunerable Atherosclerosis Plaque: Strategies for Diagnosis and Management. Malden, MA: Blackwell Futura, 2007:3–17.
74. Aschoff L. Verh Dtsch Pathol Gesellsch. 1907:106.
75. Anitschkow N, Chalatow S. On experimental cholesterin steatosis and its significance in the orgin of some pathologic processes. *Centrbl Allg Pathol Pathol Anat* 1913;24:1–9.
76. Ross R. The pathogenesis of atherosclerosis: a perspective for the 1990s. *Nature* 1993;362:801–9.
77. Napoli C, D'Armiento FP, Mancini FP, Postiglione A, Witztum JL, Palumbo G, Palinski W. Fatty streak formation occurs in human fetal aortas and is greatly enhanced by maternal hypercholesterolemia. Intimal accumulation of low density lipoprotein and its oxidation precede monocyte recruitment into early atherosclerotic lesions. *J Clin Invest* 1997;100:2680–90.
78. Narula J, Garg P, Achenbach S, Motoyama S, Virmani R, Strauss HW. Arithmetic of vulnerable plaques for noninvasive imaging. *Nat Clin Pract Cardiovasc Med* 2008;5(Suppl 2): S2–10.
79. Tahara N, Imaizumi T, Virmani R, Narula J. Clinical feasibility of molecular imaging of plaque inflammation in atherosclerosis. *J Nucl Med* 2009;50:331–4.
80. Lees AM, Lees RS, Schoen FJ, Isaacsohn JL, Fischman AJ, McKusick KA, Strauss HW. Imaging human atherosclerosis with 99mTc-labeled low density lipoproteins. *Arteriosclerosis* 1988;8:461–70.
81. Virgolini I, Muller C, Fitscha P, Chiba P, Sinzinger H. Radiolabelling autologous monocytes with 111-indium-oxine for reinjection in patients with atherosclerosis. *Prog Clin Biol Res* 1990;355:271–80.
82. Rudd JH, Warburton EA, Fryer TD, Jones HA, Clark JC, Antoun N, Johnstrom P, Davenport AP, Kirkpatrick PJ, Arch BN, Pickard JD, Weissberg PL. Imaging atherosclerotic

plaque inflammation with [18F]-fluorodeoxyglucose positron emission tomography. *Circulation* 2002;105:2708–11.

83. Kolodgie FD, Petrov A, Virmani R, Narula N, Verjans JW, Weber DK, Hartung D, Steinmetz N, Vanderheyden JL, Vannan MA, Gold HK, Reutelingsperger CP, Hofstra L, Narula J. Targeting of apoptotic macrophages and experimental atheroma with radiolabeled annexin V: a technique with potential for noninvasive imaging of vulnerable plaque. *Circulation* 2003;108:3134–9.
84. Kietselaer BL, Reutelingsperger CP, Heidendal GA, Daemen MJ, Mess WH, Hofstra L, Narula J. Noninvasive detection of plaque instability with use of radiolabeled annexin A5 in patients with carotid-artery atherosclerosis. *N Engl J Med* 2004;350:1472–3.
85. Chakfe N., Durand B., Kretz JG. New Technologies in Vascular Biomaterials: Fundamentals About Stents II. Strasbourg, France: EUROPROT, 2007.

Biographies

Naima Carter-Monroe, MD is currently a staff pathologist at CVPath Institute, Inc. She received her medical degree from Howard University College, completed her residency in Anatomic and Clinical Pathology at the University of Maryland Medical Center, and recently completed a fellowship in Health Sciences Informatics at the Johns Hopkins University College of Medicine. Her research interests include the pathology of atherosclerosis and cardiac valves, and the use of informatics tools in the practice of pathology.

Saami K. Yazdani, Ph.D. is currently a research scientist at CVPath Institute, Inc., where his primary interest involves the pathologic assessment of catheter-based clinical devices and the development of in vitro flow systems for the evaluation of coronary stents. He received doctorate of philosophy in Biomedical Engineering at the Wake Forest University School of Medicine and has extensive research experience in the fields of tissue engineering and in vitro evaluation of cardiovascular stents.

Elena R. Ladich, MD is currently the Chief of Anatomic Cardiovascular Pathology at CVPath Institute, Inc. in Gaithersburg, MD. She completed both her medical training and residency in Anatomic and Clinical Pathology at Georgetown University School of Medicine. Dr. Ladich has trained in both environmental and toxicologic pathology and cardiovascular pathology at the Armed Forces Institute of Pathology in Washington DC. Her academic interests include the natural history of atherosclerosis and cardiac valves.

Frank D. Kolodgie, PhD is currently Associate Director at CVPath Institute, Inc. Dr. Kolodgie has held faculty appointments at the University of Maryland School of Medicine and the Hahnemann University Medical School. He received both a masters and doctorate of philosophy from the University of Maryland at Baltimore in pathology. For twenty years his research has focused on the molecular pathology of human coronary atherosclerosis, including the role of inflammation, apoptosis and neoangiogenesis in the progression of atherosclerotic disease.

Renu Virmani, MD, President and Medical Director at CVPath Institute, Inc., has over 30 years of experience in the field of cardiovascular pathology and has held faculty appointments at Georgetown University, University of Maryland School of Medicine, the Uniformed Services University of Health Sciences, and Vanderbilt University. Prior to founding CVPath Institute, she served chair of the department of Cardiovascular Pathology at the Armed Forces Institute of Pathology for 20 years. Her research interests include the natural history of atherosclerosis, which lead her to develop the Modified AHA classification for atherosclerosis. In addition, she has been at the forefront of the pathologic evaluation the body's physiologic response to coronary stents.

Chapter 2
Cardiovascular Risk in Subjects with Carotid Pathologies

Fulvio Orzan, Matteo Anselmino, and Margherita Cannillo

Abstract Although atherosclerotic disease often becomes clinically evident only in one particular vascular distribution during a person's lifespan, it is an ubiquitous process, that can be detected in all arteries, depending on the diagnostic test adopted and the threshold chosen. Disease of the carotid arteries is frequent and the present chapter will focus on the correlation of this pathology with coronary heart disease.

Keywords Carotid disease • Coronary heart disease • Cardiovascular risk • Intima-media thickness • Cardiac events

Disease of the carotid arteries is frequent. Among the 5,201 subjects aged 65 years or older enrolled in the Cardiovascular Health Study, detectable carotid stenosis was present in 75% of men and 62% of women, although the prevalence of ≥50% stenosis was low: 7% in men and 5% in women [38].

Although atherosclerotic disease often becomes clinically evident only in one particular vascular distribution during a person's lifespan, it is an ubiquitous process, that can be detected in all arteries, depending on the diagnostic test adopted and the threshold chosen [48].

We shall review this problem focusing on the following points:

1. Carotid disease is an index of diffuse atherosclerosis, including coronary heart disease (CHD).
2. For this reason, cardiovascular risk and CHD are to be reckoned within patients with (a) symptomatic cerebrovascular disease (CVD), that is transient ischemic attack (TIA) or stroke, (b) asymptomatic CVD, (c) candidates for carotid endarterectomy.

F. Orzan (✉)
Division of Cardiology, Department of Internal Medicine, University of Turin, Corso A.M. Dogliotti, 14, 10126 Turin, Italy
e-mail: fulvio.orzan@unito.it

Jasjit S. Suri et al. (eds.), *Atherosclerosis Disease Management*,
DOI 10.1007/978-1-4419-7222-4_2, © Springer Science+Business Media, LLC 2011

3. Controversies about need for CHD testing in patients with CVD and measures to be taken in case of positive testing.
4. Role of the intima–media thickness (IMT) of the carotid artery in detecting occult coronary artery disease (CAD).

2.1 Carotid Disease is an Index of Diffuse Atherosclerosis, Including Coronary Heart Disease

Disease that is manifest at one site increases the chances that it will be found at other vascular beds as well.

Pathological studies support this concept. In 200 consecutive medicolegal autopsies Mathur et al. found a significant correlation between the coronary and intracranial arterial beds. Coronary atherosclerosis appeared to develop first, about 20 years earlier than cerebral atherosclerosis [36].

Another study collected 1,042 complete sets of cerebral arteries (intracranial and extracranial), coronary arteries, and aortas from autopsied persons 10–69 years of age [51]. On a group basis there was an association between the amount of raised lesions in one artery and that in another. However, on an individual basis there was only a slight degree of association between extent of raised lesions in either aorta or coronary arteries and that in the carotid arteries and there appeared to be no correlation between either aorta or coronary arteries and the other cerebral arteries. The conclusion was that "it seems impossible to predict in individual cases the amount of atherosclerosis in the cerebral arteries from the amount of lesions in the aorta and the coronary arteries."

Again, from 803 consecutive autopsies of neurologic patients, Gongora-Rivera et al. [21] reported a prevalence of coronary stenoses ≥50% in 37.5% of 341 patients with stroke. Pathologic evidence of myocardial infarction was found in 40.8%. Two-thirds of cases of myocardial infarction were clinically silent.

In a prospective series of 506 patients with extracranial CVD, coronary arteriography revealed significant CAD (≥70% stenosis) in 37% of patients with clinical evidence of ischemic heart disease, compared with 16% of those without [31].

The REACH registry included 67,888 patients from North America, Latin America, Western and Eastern Europe, Middle East, Asia, Australia, and Japan. These were outpatients aged 45 or older, with established clinical CAD, CVD, or peripheral vascular disease, or with at least three atherosclerotic risk factors, defined as follows: diabetes mellitus, high blood pressure, hypercholesterolemia, current cigarette smoking, ankle-brachial index of less than 0.9, asymptomatic carotid stenosis (12,389 patients). Polyvascular disease (clinical, symptomatic disease coexistent in two or three territories) was recognized in a significant proportion, 15.9% [3].

The age of a patient under scrutiny is important, as there is a well-recognized tendency for symptomatic CVD to "lag" 10–15 years behind CAD [15, 36]. In a geriatric population one is not surprised at finding that CAD, CVD, and peripheral vascular disease are frequently coexistent [37]. However, in terms of primary prevention, the age threshold must be kept low and the carotid artery abnormalities should be weighed together with the traditional cardiovascular risk factors (hypertension, diabetes, cholesterol, and smoking).

Subclinical atherosclerosis was examined by means of IMT and coronary calcium calcification (CAC) in more than 4,000 participants of the CARDIA and MESA studies, aged 32–50 years. The degree of both carotid IMT and CAC, was significantly higher in subjects with a high calculated lifetime cardiovascular risk, and showed a higher rate of progression [2].

In conclusion, patients CAD must be evaluated within the context of diffuse, polyvascular atherosclerosis. Indeed symptomatic carotid disease or >50% obstruction of a carotid artery is now rightly considered a coronary risk equivalent by the National Cholesterol Education Program NCEP [24].

What are then the practical consequences of the association? It depends on the clinical scenario in which we encounter our CVD patient.

2.2 Cardiovascular Risk and CHD in Patients with Specific Conditions

Cardiovascular risk and CHD are to be reckoned within patients with (1) asymptomatic CVD, (2) symptomatic CVD, and (3) candidates for carotid endarterectomy.

2.2.1 Asymptomatic CVD

CAD is relatively rare in asymptomatic CVD: its prevalence is estimated to vary from 2 to 8% [20]. Nevertheless, when present, it importantly influences the prognosis.

In patients with asymptomatic carotid stenosis and no history of CAD, who have diabetes, or peripheral vascular disease, the risk of cardiac events is similar to that of patients with a history of CAD [9]

In the medical therapy arm of the randomized ACAS study (medical therapy vs. carotid endarterectomy for ≥60% asymptomatic stenosis), there was a rate of 3.8 deaths per 100 person-years, of which half were caused by myocardial infarction or other cardiac diseases [15].

2.2.2 Patients with TIA/Stroke

After a stroke, short-term (<90 days) mortality due to cardiac causes ranges between 2 and 5% [1]. In the VISTA database, 161 of 864 patients (19%), experienced at least one serious cardiac adverse event (defined as a nonfatal episode of ventricular tachycardia, ventricular fibrillation, myocardial infarction, pulmonary edema/moderate-severe cardiac failure, or cardiac death) within the 12 weeks following an ischemic stroke. At baseline, 44.7% had a diagnosis of ischemic heart disease, and 28.5% had suffered from congestive heart failure [42].

The long-term mortality from cardiac causes also is substantial, and sometimes it is even superior to that caused by recurrent stroke [1].

Already in 1982 from the Framingham study it was known that cardiovascular disease (congestive heart failure, CHD, and hypertension) was the leading cause of death for the long-term survivors of a group of 394 stroke victims (57% of whom had an atherothrombotic brain infarction) [45].

Six hundred seventy-five patients enrolled by the Oxfordshire Community Stroke Project were prospectively followed up for up to 6.5 years. After a stroke, the average annual risk of death was 9.1%, 2.3-fold the risk in people from the general population. After the first year, cardiovascular disease became the most common cause of death [10].

In the Cardiovascular Health Study, among 546 subjects with first ischemic stroke, deaths, recurrent strokes, and CHD events were identified over 3.2-year follow-up. After the first year, the stroke and CHD rates were similar (5.2 and 4.6%, respectively). Lacunar strokes had the lowest mortality (11.9%) and recurrence rates (4.3%) [26].

A meta-analysis published by Touzè et al. included 39 studies with 65,996 patients, and a mean follow-up of 3.5 years after a TIA and an ischemic stroke. The annual risk of a myocardial infarction was 2.2%, 1.1% for fatal myocardial infarction [52].

In the Nomas study 655 patients aged >40 years were followed after a stroke. Median follow-up in survivors was 4 years. There were 86 vascular deaths, including myocardial infarction ($n=17$), congestive heart failure ($n=8$), sudden cardiac death ($n=15$), fatal strokes ($n=39$), and other ($n=7$). The 30-day mortality rate was 5.3%. The annual risk of myocardial infarction or vascular death among 30-day survivors was 2% [12]. The risk was strongly associated with the etiology of the stroke, being highest after an embolic type, and lowest after a lacunar type, precisely as had been found in the Cardiovascular Health Study [26].

The prevalence of cardiac disease at entry in patients who have suffered from a TIA/stroke is estimated to be 20–30% [1] but how many have no signs/symptoms of cardiac disease?

Several small studies have shown that patients with TIA and stroke have a high prevalence of asymptomatic CHD.

In a 6-month prospective study of 232 patients with cerebral ischemia, 100 (43%) had a prior history of heart disease. In 132 patients without prior heart disease, 47 (36%) were found to have cardiac disease: 6 atrial fibrillation, 31 cardiomegaly or left ventricular hypertrophy, 9 ischemic heart disease and 1 left bundle branch block [18].

After a TIA or a mild stroke, of 34 patients without clinical evidence of heart disease, 14 had abnormal cardiac scans [44].

The exercise test was positive in 26% of 140 patients with cerebral ischemia and without symptoms or electrocardiographic signs of ischemic heart disease [13].

In 33 patients with no history of CAD, 11 (33%) presented reversible ischemic defects by thallium-201 scintigraphy scan [35].

Chimowitz et al. studied 69 patients with TIA or stroke and without overt CAD with a cardiac stress test (adenosine or dipyridamole thallium myocardial perfusion imaging, exercise thallium myocardial perfusion imaging, or exercise ECG). The frequency of abnormal stress tests was 50% (15 of 30) in patients with large-artery cerebrovascular disease vs. 23% (9 of 39) in patients with other causes of brain ischemia [8].

Coronary artery stenosis was observed by computed tomography in 25.4% of stroke patients without previous clinical evidence of CHD [49].

The Stroke Council and the Council on Clinical Cardiology of the American Heart Association/American Stroke Association estimate that 20–40% of stroke patients may have abnormal test for silent cardiac ischemia [1].

In conclusion, patients with CVD, both asymptomatic, and after a TIA/stroke have an increased risk of myocardial infarction and cardiac death.

Patients with an ischemic stroke (except possibly those with a lacunar stroke) and those with a 50% carotid artery obstruction must be considered to have a coronary risk equivalent [7].

Patients who have suffered from a large-vessel atherosclerotic stroke should be considered for cardiac screening [1].

2.2.3 *Candidates for Carotid Endarterectomy*

Of 335 consecutive patients who underwent carotid endarterectomy between 1969 and 1973 at the Cleveland Clinic, fatal myocardial infarction accounted for 60% of early deaths within 30 days of operation and occurred in 1.8% of the entire series. Among the patients who survived operation, the 5-year mortality rate was 27%, and the 11-year mortality rate was 48%. Myocardial infarction caused 37% of the deaths that occurred within 5 years after operation and 38% of the deaths that have occurred within 11 years. Improvement in actuarial survival ($p<0.05$) and reduction in the late mortality rate ($p<0.01$) were statistically significant for the subset of patients with suspected CAD who had aortocoronary bypass graft procedures [30].

At the Mayo Clinic 177 patients who underwent carotid endarterectomy were stratified as to the presence ($n=64$) or absence ($n=93$) of overt CAD or prior myocardial revascularization ($n=20$) at the time of endarterectomy. 8-year relative survival was 89% in those without and 75% in those with overt CAD. The cumulative incidence of a cardiac event at 8 years after carotid endarterectomy was greater in those with overt CAD than in those without (61 vs. 25%, $p<0.0001$). In multivariate analysis, untreated CAD and diabetes were the only independent predictors of subsequent cardiac events [43].

In the NASCET study, 2,985 patients were randomized either to carotid endarterectomy or to medical therapy. In the first month after surgery the rate of cardiovascular complications was 8.1 vs. 1.2% in those allocated to medical treatment. Only a history of myocardial infarction or angina and a history of hypertension were statistically significant risk factors for medical complications [41].

The Swedish vascular registry reported a substantial reduction in long-term survival for 631 asymptomatic patients who underwent carotid endarterectomy. Among the predictors of decreased longevity were diabetes, cardiac disease, and previous vascular surgery. Patients with a cardiovascular event after carotid endarterectomy demonstrated a statistically significant decreased 5-year survival (2.2% among those alive vs. 11.4% among those dead; $p<0.001$) [29].

Given this background, at the Cleveland Clinic coronary arteriography before carotid endarterectomy has been the policy since 1978. In a prospective series of 506 patients (32–94 years old, mean 65) with extracranial CVD and previous neurologic symptoms ($N=288$) or asymptomatic carotid bruits ($N=218$), CAD (≥70% stenosis) was documented in 37% of patients with clinical evidence of ischemic heart disease, compared with 16% of those without. Severe inoperable coronary disease was especially common (14%) among diabetics [31].

Among the 1,662 patients of the ACAS study (carotid endarterectomy for asymptomatic carotid artery stenosis), 8.4% had a positive history for CAD, defined by the presence of angina, myocardial infarction, previous CABG, or an abnormal electrocardiogram [15].

These observations are confirmed by the aforementioned NASCET study, in which the risk of combined outcome of severe myocardial infarction and cardiac death was evaluated according to the baseline history of ischemic heart disease (angina, myocardial infarction) and of risk factors (age ≥75 years, history of diabetes, history of hypertension, smoking in past year, left ventricular hypertrophy on ECG, myocardial infarction on ECG, or creatinine >115 μmol/L). With history of ischemic heart disease at entry (1,124 patients), the 5-year risk of combined outcome of severe myocardial infarction and cardiac death was 16.5%. Without history at entry (1,691 patients), risk was 6.7%. The 5-year risk of severe myocardial infarction or cardiac death increased to 33.9% for patients with ≥4 risk factors (age ≥75 years, history of diabetes, history of hypertension, smoking in past year, left ventricular hypertrophy on ECG, myocardial infarction on ECG, or creatinine >115 μmol/L) plus a history of ischemic heart disease and to 23.5% for those without history of ischemic heart disease [17].

In addition, patients with EF of 35% or less are at increased risk for perioperative cardiac complications and reduced overall survival following carotid surgery [27].

Urbinati et al. [53] stratified 172 patients with symptomatic carotid stenosis 70–99% into four groups: (1) no coronary symptoms, no ischemia; (2) coronary ischemia (by exercise ECG and/or Thallium scanning) without symptoms; (3) unable to exercise; (4) cardiac ischemic symptoms. Kaplan–Meier estimated curves of survival free from fatal and nonfatal coronary events were 97, 51, 49, and 59%, respectively. The rate of events was higher for patients with cardiac ischemia, whether with or without symptoms ($p<0.001$, group 1 vs. groups 2 and 3; $p<0.01$, group 1 vs. group 4).

In an effort to identify patients at risk, Ombrellaro et al. tested 174 carotid endarterectomy patients with preoperative dipyridamole myocardial scintigraphy. Preoperative histories of myocardial infarction and chest pain were significant independent predictors of adverse cardiac outcomes ($p<0.05$) while scintigraphy was not [40].

Landesberg et al. [32] reported an observational study of 255 carotid endarterectomy candidates who underwent Thallium scanning. Those with moderate to severe reversible defects or with multiple reversible defects were referred for coronary arteriography and subsequent coronary revascularization. Patients who underwent coronary revascularization had a survival rate similar to that of patients with a normal or mildly abnormal thallium scan. Their survival was better than that of patients with a significantly abnormal scan, who did not receive a coronary revascularization. It is noteworthy, however, that 47% of the entire group had a history of ischemic heart disease.

2.3 Controversies About Need for CHD Testing in Patients with CVD and Measures to Be Taken in Case of Positive Testing

A general start of how to deal with cardiovascular risk in candidates for carotid endarterectomy is to follow the ACC/AHA 2007 guidelines on perioperative cardiac evaluation and care for noncardiac surgery, in which carotid endarterectomy is classified as bearing an intermediate cardiac risk (1–5%) [16]. In the guidelines cardiovascular risk evaluation is based on the Revised Cardiac Index [33] and six steps are suggested:

1. *Emergency surgery?* ⇒ Vital signs, ECG, Blood chemistry ⇒ *Proceed to surgery.*
2. Nonemergent surgery ⇒ *Active cardiac conditions present*? (listed below) ⇒ *Cardiac evaluation* is recommended:
   (a) Unstable angina (CCS class III/IV stable, if sedentary)
   (b) Decompensated heart failure (NYHA functional class IV; worsening or new-onset heart failure)
   (c) Significant arrhythmias
      - High-grade atrioventricular block
      - Mobitz II atrioventricular block
      - Third-degree atrioventricular heart block
      - Symptomatic ventricular arrhythmias
      - Supraventricular arrhythmias (including atrial fibrillation) with uncontrolled ventricular rate (HR greater than 100 bpm at rest)
      - Symptomatic bradycardia
      - Newly recognized ventricular tachycardia
   (d) Severe valvular disease
      - Severe aortic stenosis
      - Symptomatic mitral stenosis
3. Nonemergent surgery, no active cardiac conditions present, *Low risk surgery*? ⇒ *Proceed to surgery.*
4. Intermediate risk surgery? (carotid endarterectomy) ⇒ *Functional capacity* (4 METS, “climbing 1 flight of stairs”).

5. Low functional capacity? Assess clinical risk factors:

   (a) Ischemic heart disease, defined as: class I//II angina, previous MI, positive cardiac ischemia test, ECG with abnormal Q waves
   (b) History of previous congestive heart failure, presence of peripheral edema, bilateral rales, S3, or X-ray with pulmonary vascular redistribution
   (c) Preoperative insulin treatment for diabetes mellitus
   (d) Preoperative creatinine greater than 2 mg/dL

No clinical risk factors ⇒ *Proceed to surgery.*
Risk Factor(s) present: consider testing *if it will change management.*
To summarize:

1. In a nonemergent condition a minimum work-up should include: History, Physical Examination, ECG, Blood and urinalysis.
2. An active cardiac condition calls for cardiac consultation.
3. In the absence of active cardiac conditions, an assessment of functional capacity is recommended.
4. The presence of risk factors does not necessarily call for cardiac testing; this is "considered" only if the result will change management.
5. History taking is instrumental (active cardiac conditions, clinical risk factors).

Unfortunately point 3 is ambiguous: functional capacity is either assessed rather imprecisely on clinical grounds (much as the NYHA classification), or it requires cardiac testing. This leads to the ambiguity of point 4: testing should change management only if we have reasonable evidence that action taken following the result of the test will prevent adverse events (myocardial infarction, serious arrhythmias, acute heart failure) or death. Searching for CAD in individuals with CVD is relevant if the prevalence of CAD in a population is high and if there is a demonstrated benefit in treating such patients [5, 52]. Such crucial information is at the moment lacking or conflicting, and the reader is referred to the exchange by Gregoratos and Brett for an exhaustive review of these points [23].

Although addressed at vascular surgery, not specifically at carotid artery disease, Kertai has produced a thorough review about whether and how to screen such patients, and how best to treat them [28].

Finally an attempt is also made at defining the "high-risk cardiac patient" in relation to carotid endarterectomy. In the SAPPHIRE study it has been defined as "Clinically significant cardiac disease: congestive heart failure, abnormal stress test, or need for open-heart surgery" [57].

The recent Controversies in Carotid Artery Revascularization have proposed the following definition: congestive heart failure (NYHA class III–IV); unstable angina (CCS class III/IV); myocardial infarction in the last 30 days; severe CAD (left main, ≥2 vessel disease); left ventricular ejection fraction ≤30%; heart surgery planned in the next 30 days [55].

The merit of different clinical indices of cardiac risk, and their weaknesses are well explained and discussed by Devereaux et al. [11].

The optimal management of an asymptomatic CVD patient whose cardiac testing turns positive is controversial. In fact, revascularization is not recommended in asymptomatic patients by the guidelines [16].

In patients with stable CAD, prophylactic coronary revascularization before high-risk noncardiac surgery does not confer any beneficial effects, when compared with optimized medical management, in terms of perioperative mortality, myocardial infarction, long-term mortality, or adverse cardiac events [56].

Even in high-risk patients undergoing major vascular surgery preoperative coronary revascularization was not associated with improved postoperative or long-term outcome compared with the best medical treatment [47].

2.4 Intima–media Thickness of the Carotid Artery and Cardiac Risk

2.4.1 Correlation with Coronary Angiography

In 1994 Geroulakos et al. examined with high-resolution B-mode ultrasound 75 patients who underwent coronary angiography for assessment of chest pain and 40 normal controls matched for age and sex. The IMT of the common carotid artery for the controls was 0.71 ± 0.16 mm and for the patients 0.91 ± 0.18 mm ($p<0.005$). In patients with normal coronary angiogram the IMT was 0.73 ± 0.1 mm. In the group with one-vessel disease it increased to 0.91 ± 0.17 mm ($p<0.05$), in the group with two-vessel disease it was 0.96 ± 0.17 mm ($p<0.01$), and in the group with three-vessel disease it was 0.99 ± 0.21 mm ($p<0.01$). There was a significant linear trend between IMT and the number of involved vessels ($p<0.0001$, $r=0.44$) [19].

Kafetzakis et al. reported in their study of 184 patients submitted to coronary angiography and to duplex ultrasonography of the carotid, femoral, and popliteal arteries. IMT of the carotid and femoral arteries were independent predictive factors of obstructive CAD [25].

2.4.2 Correlation with Cardiac Events

The association of preexisting CHD, CVD, and peripheral vascular disease with carotid and popliteal IMT (measured by B-mode ultrasound) was assessed in 13,870 subjects enrolled in the Atherosclerosis Risk in Communities (ARIC) Study. A substantially greater arterial wall thickness was observed in middle-aged adults with prevalent cardiovascular disease. Both carotid and popliteal arterial IMT were related to clinically manifest cardiovascular disease affecting distant vascular beds, such as the cerebral, peripheral, and coronary artery vascular beds [6].

In the Rotterdam study, the risk of myocardial infarction increased 43% per standard deviation increase (0.163 mm) in common carotid IMT. The risk was particularly increased in subjects with an IMT in the upper quintile of the distribution (0.908 mm) relative to the reference category (0.75 mm), but the association between IMT and risk of myocardial infarction did not show a clearly linear pattern. Those who developed myocardial infarction had at baseline a higher prevalence of hypercholesterolemia and diabetes, and had more frequently suffered already from a heart attack, so that adjustment for cardiovascular risk factors attenuated the magnitude of the associations and their statistical significance [4].

The combined measure of IMT was significantly associated with the risk of myocardial infarction in 5,858 participants to the Cardiovascular Health Study, whose carotid arteries were evaluated with high-resolution B-mode ultrasonography [39].

A meta-analysis in 2007 confirmed that carotid IMT is a strong predictor of future vascular events, both stroke and myocardial infarction [34].

However, after reviewing 18 studies, Wald and Beswick [54] concluded that neither carotid plaque nor IMT is sufficiently discriminatory between affected and unaffected individuals to be a worthwhile screening test for CHD. While it is true that the relative risk is higher for those with a very high IMT compared with a very low IMT, this only confirms that an association is present. Because the groups being compared are mutually exclusive, and most people in the middle of the distribution are not considered in the analysis, the value of the ultrasound exam as a screening test is low, in the range of that of serum cholesterol or diastolic blood pressure.

Needless to say, the quality of the ultrasound study must be kept high [14] in case of therapeutic decisions. For IMT measurement, it should meet the criteria meticulously described by the American Society of Echocardiography, which has produced a specific protocol [50] (Table 2.1).

Table 2.1 Highlights

Coronary artery disease must be evaluated within the context of diffuse, polyvascular atherosclerosis
Patients with cerebrovascular disease, both asymptomatic, and after a TIA/stroke present an increased risk of myocardial infarction and cardiac death
Symptomatic carotid disease or >50% obstruction of a carotid artery is a coronary risk equivalent
Patients who have suffered from a large-vessel atherosclerotic stroke should be considered for cardiac screening
Carotid endarterectomy is an intermediate cardiac risk surgery:
1. In a nonemergent condition a minimum work-up should include: History, Physical Examination, ECG, Blood, and urinalysis
2. An active cardiac condition calls for cardiac consultation
3. In the absence of active cardiac conditions, an assessment of functional capacity is recommended
4. The presence of risk factors does not necessarily call for cardiac testing, that should be considered only if the result will change management
5. History taking is instrumental (active cardiac conditions, clinical risk factors)

2.5 Conclusions

In summary, the increased attention to cardiac (coronary) conditions in patients with carotid disease is wholly justified both for subjects with asymptomatic disease, patients after a TIA or stroke, and patients before a planned carotid endarterectomy (and afterwards).

This increased awareness should translate in more stringent primary and secondary prevention of the atherosclerotic disease [22, 24, 46].

Whether one should screen for CAD the patient with carotid disease who has no cardiac symptoms, but significant risk factors, the test to be adopted, and which therapy should be subsequently implemented are issues that await the results of controlled clinical trials.

References

1. Adams RJ, Chimowitz MI, Alpert JS, Awad IA, Cerqueria MD, Fayad P, Taubert KA. Coronary risk evaluation in patients with transient ischemic attack and ischemic stroke. A scientific statement for healthcare professionals from the Stroke Council and the Council on Clinical Cardiology of the American Heart Association/American Stroke Association. Circulation 2003;108:1278–1290
2. Berry JD, Liu K, Folsom AR, Lewis CE, Carr JJ, Polak JF, Shea S, Sidney S, O'Leary DH, Chan C, Lloyd-Jones DM. Prevalence and progression of subclinical atherosclerosis in younger adults with low short-term but high lifetime estimated risk for cardiovascular disease: the Coronary Artery Risk Development in Young Adults Study and Multi-Ethnic Study of Atherosclerosis. Circulation 2009;119:382–389
3. Bhatt DL, Steg PG, Ohman EM, Hirsch AT, Ikeda Y, Mas J-L, Goto S, Liau C-S, Richard AJ, Roether J, Wilson PWF. International prevalence, recognition, and treatment of cardiovascular risk factors in outpatients with atherothrombosis. JAMA 2006;295:180–189
4. Bots ML, Hoes AW, Koudstaal PJ, et al. Common carotid intima- media thickness and risk of stroke and myocardial infarction (The Rotterdam Study). Circulation 1997;96:1432–1437
5. Brett AS. Coronary assessment before noncardiac surgery. Curent strategies are flawed. Circulation 2008;117:3145–3151
6. Burke GL, Evans GW, Riley WA, Sharrett AR, Howard G, Barnes RW, Rosamond W, Crow RS, Rautaharju PM, Heiss G. Arterial wall thickness is associated with prevalent cardiovascular disease in middle-aged adults. The Atherosclerosis Risk in Communities (ARIC) Study Stroke. 1995 Mar;26(3):386–91
7. Chaturvedi S. Should stroke be considered both a brain attack and a heart attack? Stroke 2007;38:1713–1714
8. Chimowitz MI, Poole RM, Starling MR, Schwaiger M, Gross MD. Frequency and severity of asymptomatic coronary disease in patients with different causes of stroke. Stroke 1997;28:941–945
9. Chimowitz MI, Weiss DG, Cohen SL, et al. Cardiac prognosis of patients with carotid stenosis and no history of coronary artery disease: Veterans Affairs Cooperative Study Group 167. Stroke 1994;25:759–765
10. Dennis MS, Burn JP, Sandercock PA, Bamford JM, Wade DT, Warlow CP.Long-term survival after first-ever stroke: the Oxfordshire Community Stroke Project. Stroke 1993;24:796–800
11. Devereaux PJ, Goldman L, Yusuf S, Gilbert K, Leslie K, Guyatt GH. Surveillance and prevention of major perioperative ischemic cardiac events in patients undergoing noncardiac surgery: a review. CMAJ 2005;173:779–788

12. Dhamoon M, Tai W, Boden-Albala B, Runek T, Paik MC, Sacco RL, Elkind MSV. Risk of myocardial infarction or vascular death after first ischemic stroke: the Northern Manhattan study. Stroke 2007;38:1752–1758
13. Di Pasquale G, Pinelli G, Grazi P, et al. Incidence of silent myocardial ischaemia in patients with cerebral ischaemia. Eur Heart J 1988;9(suppl N):104–107
14. Elgersma OE, van Leeuwen MS, Meijer R, Eikelboom BC, van der Graaf Y. Lumen reduction measurements of the internal carotid artery before and after Levovist enhancement: reproducibility and agreement with angiography. J Ultrasound Med 1999;18:191–201
15. Endarterectomy for asymptomatic carotid artery stenosis. Executive Committee for the Asymptomatic Carotid Atherosclerosis Study. JAMA 1995;273:1421–1428
16. Fleisher LA, Beckman JA, Brown KA, Calkins H, Chaikoff E, Fleischmann KE, Freeman WK, Froelich JB, Kasper EK, Kersten JR, Riegel B, Robb JF. ACC/AHA 2007 guidelines on perioperative cardiac evaluation and care for noncardiac surgery: a report of the American College of Cardiology/American Heart Association Task Force on Practice Guidelines. Circulation 2007;116:1971–1996
17. Gates PC, Eliasziw M, Algra A, Barnett HJ, Gunton RW; North American Symptomatic Carotid Endarterectomy Trial Group. Identifying patients with symptomatic carotid artery disease at high and low risk of severe myocardial infarction and cardiac death. Stroke 2002;33:2413–2416
18. Gates P, Peppard R, Kempster P, Harris A, Pierce M. Clinically unsuspected cardiac disease in patients with cerebral ischaemia. Clin Exp Neurol 1987;23:75–80
19. Geroulakos G, O'Gorman DJ, Kalodiki E, et al. The carotid intima-media thickness as a marker of the presence of severe symptomatic coronary artery disease. Eur Heart J 1994;15:781–785
20. Goldstein LB, Adams R, Alberts MJ, Appel LJ, Brass LM, Bushnell CD, Culebras A, DeGraba TJ, Gorelick PB, Guyton JR, Hart RG, Howard G, Kelly-Hayes M, Nixon JV, Sacco RL. Primary prevention of ischemic stroke. A guideline from the American Heart Association/American Stroke Association Stroke Council: cosponsored by the Atherosclerotic Peripheral Vascular Disease Interdisciplinary Working Group; Cardiovascular Nursing Council; Clinical Cardiology Council; Nutrition, Physical Activity, and Metabolism Council; and the Quality of Care and Outcomes Research Interdisciplinary Working Group: The American Academy of Neurology affirms the value of this guideline. Stroke 2006;37:1583
21. Gongora-Rivera F, Labreuche J, Jaramillo A, Steg PG, Hauw JJ, Amarenco P. Autopsy prevalence of coronary atherosclerosis in patients with fatal stroke. Stroke 2007;38:1203–1210
22. Goldstein LB, Adams R, Alberts MJ, Appel LJ, Brass LM, Bushnell CD, Culebras A, DeGraba TJ, Gorelick PB, Guyton JR, Hart RG, Howard G, Kelly-Hayes M, Nixon JV, Sacco RL. Primary prevention of ischemic stroke: a guideline from the American Heart Association/American Stroke Association Stroke Council: cosponsored by the Atherosclerotic Peripheral Vascular Disease Interdisciplinary Working Group; Cardiovascular Nursing Council; Clinical Cardiology Council; Nutrition, Physical Activity, and Metabolism Council; and the Quality of Care and Outcomes Research Interdisciplinary Working Group. Circulation 2006;113:e873–e923
23. Gregoratos G. Current guideline-based preoperative evaluation provides the best management of patients undergoing noncardiac surgery. Circulation 2008;117;3134–3144
24. Grundy SM, Cleeman JI, Merz CN, Brewer HB Jr, Clark LT, Hunninghake DB, Pasternak RC, Smith SC Jr, Stone NJ; National Heart, Lung, and Blood Institute, American College of Cardiology Foundation, American Heart Association. Implications of recent clinical trials for the national cholesterol education program adult treatment panel III guidelines. Circulation 2004;110:227–239
25. Kafetzakis A, Kochiadakis G, Laliotis A, Peteinarakis I, Touloupakis E, Igoumenidis N, Katsamouris A. Association of subclinical wall changes of carotid, femoral, and popliteal arteries with obstructive coronary artery disease in patients undergoing coronary angiography. Chest 2005;128:2538–2543
26. Kaplan RC, Tirschwell DL, Longstreth WT Jr, Manolio TA, Heckbert SR, Lefkowitz D, El-Saed A, Psaty BM. Vascular events, mortality, and preventive therapy following ischemic stroke in the elderly. Neurology 2005;65:835–842

27. Kazmers A, Cerqueira MD, Zierler RE. The role of preoperative radionuclide left ventricular ejection fraction for risk assessment in carotid surgery. Arch Surg 1988;123:416–419
28. Kertai MD. Preoperative coronary revascularization in high-risk patients undergoing vascular surgery: a core review. Anesth Analg 2008;106:751–758
29. Kragsterman B, Bjorck M, Lindback J, Bergqvist D, Parsson H. Long-term survival after carotid endarterectomy for asymptomatic stenosis. Stroke 2006;37:2886–2891
30. Hertzer NR, Lees CD. Fatal myocardial infarction following carotid endarterectomy: three hundred thirty-five patients followed 6–11 years after operation. Ann Surg 1981;194:212–218
31. Hertzer R, Young JR, Beven EG, Graor RA, O'Hara PJ, Ruschhaupt WF, deWolfe VG, Maljovec LC. Coronary angiography in 506 patients with extracranial cerebrovascular disease. Arch Intern Med 1985;145:849–852
32. Landesberg G, Wolf Y, Schechter D, Mosseri M, Weissman C, Anner H, Chisin R, Luria MH, Kovalski N, Bocher M, Erel J, Berlatzky Y. Preoperative thallium scanning, selective coronary revascularization, and long-term survival after carotid endarterectomy. Stroke 1998;29:2541–2548
33. Lee TH, Marcantonio ER, Mangione CM, Thomas Eric J, Polanczyk CA, Cook EF, Sugarbaker DJ, Donaldson MC, Poss R, Ho KKL, Ludwig LE, Pedan A, Goldman L. Derivation and prospective validation of a simple index for prediction of cardiac risk of major noncardiac surgery. Circulation 1999;100:1043–1049
34. Lorenz MW, Markus HS, Bots ML, Rosvall M, Sitzer M. Prediction of clinical cardiovascular events with carotid intima-media thickness. Circulation 2007;115:459–467
35. Love BB, Grover-McKay M, Biller J, et al. Coronary artery disease and cardiac events with asymptomatic and symptomatic cerebrovascular disease. Stroke 1992;23:939–945
36. Mathur KS, Kashyap SK, Kumar V. Correlation of the extent and severity of atherosclerosis in the coronary and cerebral arteries. Circulation 1963;27:929–934
37. Ness J, Aronow WS. Prevalence of coexistence of coronary artery disease, ischemic stroke, and peripheral arterial disease in older persons, mean age 80 years, in an academic hospital-based geriatrics practice. J Am Geriatr Soc 1999;47:1255–1256
38. O'Leary DH, Polak JF, Kronmal RA, Kittner SJ, Bond MG, Wolfson SK, Bommer W, Price TR, Gardin JM, Savage PJ. Distribution and correlates of sonographically detected carotid artery disease in the cardiovascular health study. The CHS Collaborative Research Group. Stroke 1992;23:1752–1760
39. O'Leary DH, Polak JF, Kronmal RA, Manolio TA, Burke GL, Wolfson SK. Carotid-artery intima and media thickness as a risk factor for myocardial infarction and stroke in older adults. N Engl J Med 1999;340:14–22
40. Ombrellaro MP, Dieter RA III, Freeman M, Stevens SL, Goldman MH. Role of dipyridamole myocardial scintigraphy in carotid artery surgery. J Am Coll Surg 1995;181(5):451–458
41. Paciaroni M, Eliasziw ML, Kappelle, LJ, et al. Medical complications associated with carotid endarterectomy. Stroke 1999;30:1759–1763
42. Prosser J, MacGregor L, Lees KR, Diener H-C, Hacke W, Davis S; VISTA Investigators. Predictors of early cardiac morbidity and mortality after ischemic stroke. Stroke 2007;38:2295–2302
43. Rihal CS, Gersh BJ, Whisnant JP, Rooke TW, Sundt TM Jr, O'Fallon WM, Ballard DJ. Influence of coronary heart disease on morbidity and mortality after carotid endarterectomy: a population-based study in Olmsted County, Minnesota (1970–1988). J Am Coll Cardiol 1992;19:1254–1260
44. Rokey R, Rolak LA, Harati Y, Kutka N, Verani MS. Coronary artery disease in patients with cerebrovascular disease: a prospective study. Ann Neurol 1984;16:50–53
45. Sacco RL, Wolf PA, Kannel WB, McNamara PM. Survival and recurrence following stroke: the Framingham Study. Stroke 1982;13:290–295
46. Sacco RL, Adams R, Albers G, Alberts MJ, Benavente O, Furie K, Goldstein LB, Gorelick P, Halperin J, Harbaugh R, Johnston SC, Katzan I, Kelly-Hayes M, Kenton EJ, Marks M, Schwamm LH, Tomsick T. Guidelines for prevention of stroke in patients with ischemic stroke or transient ischemic attack: a statement for healthcare professionals from the American

Heart Association/American Stroke Association Council on Stroke: co-sponsored by the Council on Cardiovascular Radiology and Intervention: the American Academy of Neurology affirms the value of this guideline. Stroke 2006;37:577–617

47. Schouten O, van Kuijk JP, Flu WJ, Winkel TA, Welten GM, Boersma E, Verhagen HJ, Bax JJ, Poldermans D; DECREASE Study Group. Long-term outcome of prophylactic coronary revascularization in cardiac high-risk patients undergoing major vascular surgery (from the randomized DECREASE-V Pilot Study). Am J Cardiol 2009;103:897–901
48. Shah AM, Banerjee T, Mukherjee D. Coronary, peripheral and cerebrovascular disease: a complex relationship. Herz 2008;33:475–480
49. Seo WK, Yong HS, Koh SB, Suh SI, Kim JH, Yu SW, Lee JY. Correlation of coronary artery atherosclerosis with atherosclerosis of the intracranial cerebral artery and the extracranial carotid artery. Eur Neurol 2008;59(6):292–298
50. Stein JH, Korcarz CE, Hurst RT, Lonn E, Kendall CB, Mohler ER, Najjar SS, Rembold CM, Post WS; American Society of Echocardiography Carotid Intima-Media Thickness Task Force. Use of carotid ultrasound to identify subclinical vascular disease and evaluate cardiovascular disease risk: a consensus statement from the American Society of Echocardiography Carotid Intima-Media Thickness Task Force. Endorsed by the Society for Vascular Medicine. J Am Soc Echocardiogr 2008;21:93–111
51. Solberg LA, McGarry PA, Moossy J, Strong JP, Tejada C, Löken AC. Severity of atherosclerosis in cerebral arteries, coronary arteries, and aortas. Ann N Y Acad Sci 1968;149:956–973
52. Touzé E, Varenne O, Chatellier G, Peyrard S, Rothwell PM, Mas JL. Risk of myocardial infarction and vascular death after transient ischemic attack and ischemic stroke: a systematic review and meta-analysis. Stroke 2005 36:2748–2755
53. Urbinati S, Di Pasquale G, Andreoli A, Lusa AM, Carini G, Grazi P, Labanti G, Passarelli P, Corbelli C, Pinelli G. Preoperative noninvasive coronary risk stratification in candidates for carotid endarterectomy. Stroke 1994;25:2022–2027
54. Wald DS, Beswick JP. Carotid ultrasound screening for coronary heart disease: results based on a meta-analysis of 18 studies and 44861 subjects. J Med Screen 2009;16:147–154
55. White CJ, Beckman JA, Cambria RP, Comerota AJ, Gray WA, Hobson RW II, Iyer SS; Writing Group 5. Atherosclerotic Peripheral Vascular Disease Symposium II: controversies in carotid artery revascularization. Circulation 2008;118:2852–2859
56. Wong EY, Lawrence HP, Wong DT. The effects of prophylactic coronary revascularization or medical management on patient outcomes after noncardiac surgery – a meta-analysis. Can J Anaesth 2007;54:705–717
57. Yadav JS, Wholey MH, Kuntz RE, Fayad P, Katzen BT, Mishkel GJ, Bajwa TK, Whitlow P, Strickman NE, Jaff MR, Popma JJ, Snead DB, Cutlip DE, Firth BG, Ouriel K; Stenting and Angioplasty with Protection in Patients at High Risk for Endarterectomy Investigators. Protected carotid-artery stenting versus endarterectomy in high-risk patients. N Engl J Med 2004;351:1493–1501

Biographies

Fulvio Orzan, MD, PhD, is currently Assistant Professor on faculty of the Department of Internal Medicine and Cardiology of the University of Torino (Torino, Italy). He is also on faculty of the Cardiology Unit of the S. Giovanni Battista "Molinette" Hospital of Torino. His research interests are related to the investigation and treatment of cardiac and cardiovascular pathologies.

Matteo Anselmino (12-03-1978). During his PhD at Karolinska Institutet, Stockholm, Sweden he worked on the "Euro Heart Survey on Diabetes and the Heart" focusing on treatment patterns in patients with coronary artery disease with and without abnormal glucose regulation. To date he is Assistant Professor at the Cardiology Division of the Internal Medicine Department of the University of Turin, Italy committed to clinical and research experience at the Cardiac Pacing and Electrophysiology Unit.

Margherita Cannillo, MD, PhD, obtained the Italian Laurea in Medicine and Surgery from the University of Torino (Torino, Italy) in 2008. She is now specializing at the Cardiology School of the University of Torino. Her principal research interests are related to the morpho-functional assessment of the right ventricle in subjects with septal defects.

Chapter 3
Neurological Evaluation and Management of Patients with Atherosclerotic Disease

William Liboni, Enrica Pavanelli, Nicoletta Rebaudengo, Filippo Molinari, and Jasjit S. Suri

Abstract The management of the atherosclerotic patient with neurological symptoms is extremely important in clinical practice. Symptoms (usually transient ischemic attacks or minor strokes) are one of the principal elements on the basis of which the surgical decision is made. We started an experimentation aimed at validating the use of contrast-enhanced ultrasonography in the clinical management of the neurological patient.

Keywords Carotid plaque • Atherosclerosis • Neurological symptoms • NASCET • ECST • Surgery • Patient management

3.1 Stroke as Social Problem

The ischemic or hemorrhagic stroke is, among the epidemiological emergencies, one of the most serious clinical and welfare problems all around the world. In developed countries, in fact, stroke represents the leading cause of permanent invalidity and the second cause of dementia [1]. Also, it is the third cause of global deaths, causing about 10–12% of all the deceases. Recent statistics from the World Health Organization place stroke as the second cause of death starting from 2,040.

Considering the average statistics of Europe, North America, and Japan, stroke has an annual incidence in the range 1.68–2.97 new cases every 1,000 citizens. About 80% of the new cases is caused by an ischemic stroke, whereas the remaining 20% is constituted by cerebral and subarachnoidal hemorrhages [2]. The total number of annual strokes is constituted by new cases for the 80% and by relapses for the 20%. Therefore, the current forecasts will not be smoothed by new interventions; stroke will afflict the health of billions of people.

F. Molinari (✉)
Biolab, Department of Electronics, Politecnico di Torino, Torino, Italy
e-mail: filippo.molinari@polito.it

Jasjit S. Suri et al. (eds.), *Atherosclerosis Disease Management*,
DOI 10.1007/978-1-4419-7222-4_3, © Springer Science+Business Media, LLC 2011

One of the factors facilitating the increase of stroke incidence is the average aging of the world population. Stroke risk, in fact, increases with age: practically, it doubles every 10 years from 45 years on. The risk of relapses varies from 10 to 15% in the first year after the stroke onset and from 4 to 9% in the first 5 years. The maximal incidence is observed around 85 years. Therefore, 75% of the strokes occur after 65 years of age and only 5% occur before 45 years of age. Considering the population of elderly people (range 65–85 years), stroke prevalence is about 7%, with a weak higher probability in men (7.4%) than in women (6.6%).

Among the stroke patients, 20–30% die within 3 months from the stroke attack; 40–50% are permanently impaired and lose autonomy, whereas about 10% show a relapse within 12 months. The social costs connected to stroke are, therefore, extremely high. If the forecasts on the increasing incidence of stroke are correct, in near future there will be few economies that will be able to sustain the augmented stroke costs.

3.2 Stroke and Atherosclerosis Early Prevention and Management

Secondary prevention of the principal causes of cerebrovascular pathologies is therefore becoming a major problem. Clearly, prevention is crucial in the subjects that show neurological symptoms. However, it is fundamental also in the population of patients without symptoms, but with a peripheral vascular cardiac pathology or with a high risk of cardiovascular accidents.

Instrumental diagnosis for assessing and documenting the progression of the atherosclerotic pathology offers several different approaches. Each technique has different sensibility and specificity performance in relation to the investigated organ:

- Cardiac district
- Peripheral vascular district
- Encephalic district

Among the diagnostic systems used in stroke and atherosclerosis diagnosis, the most important are as follows:

- *Computer Tomography (CT).* Basically, CT is used in precise and accurate estimation of the stenosis degree of arteries. Early signs of atherosclerosis can be quantified by computing the stenosis state of the major body arteries. Clearly, CT is optimal for second diagnosis and surgical planning, but it cannot be considered the most important tool for first diagnosis. Modern CT scanners can effectively reconstruct the vascular bed with a very high degree of fidelity (angio-CT). Major drawback of the CT methodology is the use of dangerous radiations and the need for a contrast agent.
- *Magnetic Resonance Imaging (MRI).* Several studies all around the world demonstrated the excellent performance of MRI in artery characterization and plaque analysis. Specifically, the focal study of plaques is becoming the standard for

obtaining a "*virtual histology*" indication. Magnetic fields and radiofrequencies are less dangerous than X-rays; therefore, MRI is preferable to CT when patient safety is a major issue. However, MRI has very high associated costs: experienced personnel and top instrumentation are necessary for obtaining optimal diagnostic performance. The MRI device, also, has very high associated management costs.

- *Ultrasounds (US)*. Undoubtedly, the US equipment is low cost, portable, and with low management expenses. If CT and MRI require specialized centers, US can be performed at the patient's bedside. This is a major advantage when prevention is a major issue, since US can be widely diffused on the territory without the need for the patients to go to a specific center. US are mechanical radiations propagating mechanical energy exclusively. Therefore, their associated risk is very low. However, US are operator dependent and prone to subjective interpretations and errors. This is a very important issue, since the minor costs and the higher versatility should not induce in forgetting the lower accuracy of the US methodology.

The need for a screening in secondary prevention recently pushed the need for accurate, repeatable, and relatively cheap techniques. Therefore, the performance of the different techniques have been compared and carefully evaluated worldwide. Several studies, in fact, delineated the diagnostic performance in presence of different symptoms and surgical indication of the patients. Well-established and accepted indications came from the NASCET study [3], which showed that the symptomatic patient with carotid stenosis must be surgically treated [4]. Currently, the clinician still can decide for the best therapeutic strategy in the presence of an asymptomatic patient with stenosis degree higher than 70%. If stenosis is lower than 70%, it is very debated to indicate for surgical intervention: if surgical treatment is recommended, this should be done in a service with very good skills and performance, that is, with a complication probability lower than 1.5% (as recommended by ACAS [3] and ACST) [5]. The use of stents is currently generally accepted and indicated, with the exception of symptomatic patients with very closed stenosis and contraindications to surgery [6].

The clinical experience made after the publication of the NASCET study showed that by evaluating only the degree of stenosis poses some problems:

(a) Stenosis is not the only pathogenic element for stroke.
(b) Some patients may be totally asymptomatic even in the presence of a high stenosis degree.
(c) The nature of the artery wall alterations is an important issue that correlates with clinical symptoms onset.

Therefore, the knowledge of the carotid wall compromission and of the atheromasic plaque composition is a discriminant factor that is independent on the entity of the narrowing of the artery lumen [7]. The above-mentioned clinical analysis techniques, even if with different invasiveness, risk index, and associated costs, may provide informative and crucial contents about the atheroma (and beside the estimation of the stenosis degree).

Ultrasounds possess a wider applicability with respect to radiological techniques. Also, they can be applied to a wider number of citizens by screening protocols. Therefore, ultrasounds could be potentially used both for serial evaluations than for first diagnosis issues.

In the following, we describe a study we conducted on the management of the atherosclerotic patient with neurological symptoms, describing the clinical trial we built in our Institution (the Neurology Division of the Gradenigo Hospital of Torino, Italy) and the clinical examinations we made.

3.3 Objectives and End Points

The first objective of this experience was the internal and external validation of the ultrasound methodology in the instrumental diagnosis of the atheromasic plaque of the carotids. This objective is very important since it is documented that ultrasound-based instrumental measurements are strongly correlated to ground truth (i.e., the *IMT* ultrasound based measurement is nowadays a clinical routine that is accepted and used in almost all the ultrasound vascular divisions, cardiological divisions, and neurological divisions [8]). However, if geometrical measurements have been validated and standardized, the use of ultrasounds in plaque analysis is a recent topic. In many studies, the invasive variant of the ultrasound analysis has been used (IVUS: Intra-Vascular Ultrasound). It has been shown that IVUS is potentially very effective in plaque characterization. However, invasiveness is a major problem. Therefore, one of the aims of this experience was to test traditional B-Mode imaging as a potential tool for plaque imaging.

The second objective was the verification of the sensibility and specificity of plaque characterization by ultrasounds. We aimed at correlating the results of the ultrasound-based plaque characterization to histology.

The third and last objective was the comparison of the routinely angio-CT and angio-MRI plaque characterization techniques with the ultrasound examination.

The primary end point of our study was the identification and characterization of the plaque, as expression of the atherosclerotic pathology and located at the bifurcation of the carotid artery, by using instrumental ultrasound diagnostic systems, coupled to an integrated processing of the images, leading to specific sensibility and specificity.

- Plaque characterization and identification of predictive factors.
- Correlation between plaque and critical event in symptomatic patients.
- Correlation between plaque characterization made by ultrasounds and MRI imaging.
- Correlation between plaque characterization made by ultrasounds and histology.

Correlated to this point, we focused on the added value of the ultrasound contrast-agent examination and of the paramagnetic MRI contrast agent in the definition of the plaque criticity.

Secondary endpoints were as follows:

- A study on the inter- and intra-operators variability in the interpretation of the neurological data.
- Operator compliance with respect to the ultrasound methodology and contribution to automation and signal processing, in order to enhance the objectivity of the diagnostic process and smoothing the intrinsical subjectivity of the ultrasound methodology.

The overall study design was observational/experimental. In our clinical experimentation, we used both ultrasound contrast agent (second generation contrast agent, made of gaseous microbubbles) and MRI contrast agent (paramagnetic gadolinium-based chemical agent with intravasal confinement).

Most of the data collected during this experimentation constituted the basis for other correlated research studies, which are reported in others chapters of this book (i.e., the contrast-enhanced ultrasound technique for plaque characterization and the metabolomic profiling of atherosclerotic patients). In the following sections of this chapter, we illustrate the clinical protocol we proposed and some sample results of our experimentation.

3.4 Experimental Setup

The experimental setup herein described was carried at the Neurology Service of the Gradenigo Hospital from June 2008 to December 2009.[1] The protocol focused on the patients basin of a neurology service, thus it considered mainly subjects suffering from atherosclerosis and that showed neurological symptoms.

3.4.1 Patients Inclusion Criteria

We selected four classes of patients:

(a) *Symptomatic patients with surgical indication*. Patients with a recent history of transient ischemic attack (TIA), minor stroke, or non-invalidating stroke. The patients had either a surgical carotid plaque (stenosis greater than 50% if measured according to NASCET or 70% if according to ECST [9]), or a borderline but complicated plaque (according to the gray-scale median five-classes classification scheme [10]), with consensus to possible interventional/surgical procedure.

[1]The present study was partially granted by the CRT Foundation of Torino, Italy, under the research project "Which diagnosis for which therapy in welfare conservation", aimed at optimizing the clinical treatment of disease with large social impact.

(b) *Symptomatic patients without surgical indication*. Patients with a recent history of TIA, minor stroke, or non-invalidating stroke. The patients had a noncritical and nonsurgical atheromasic plaque (stenosis degree lower than 50% NASCET or lower than 70% ECST).
(c) *Asymptomatic patients with surgical indication*. Patients without neurological symptoms, but with peripheral arteriopathy, ischemic cardiopathy, or other risk factors (i.e., diabetes, hypertension, hypercholesterolemia). Patients had either a surgical carotid plaque (stenosis greater than 60% if measured according to NASCET or 80% if according to ECST), or a borderline but complicated plaque (according to the gray-scale median five-classes classification scheme [10]), with consensus to possible interventional/surgical procedure.
(d) *Asymptomatic patients without surgical indication*. Patients without a recent history of neurological symptoms, but with peripheral arteriopathy, ischemic cardiopathy, or other risk factors (i.e., diabetes, hypertension, hypercholesterolemia). The patients had a noncritical and nonsurgical atheromasic plaque (stenosis degree lower than 60% NASCET or lower than 80% ECST).

In classes *c* and *d*, we increased the critical value of the stenosis degree of 10% with respect to standard guidelines, considering 60% for the NASCET criterion and 80% for the ECST one.

3.4.2 Exclusion Criteria

We excluded from the tests:

- The subjects who did not accept the experimental protocol in full and did not sign the informed consent.
- The subjects with severe pathologies carrying physical and psychological impairments.
- The subjects with altered consciousness status were discarded.
- The subjects in acute phase or with urgent surgical intervention needed within 48 h.
- The subjects affected by pulmonary pathologies.
- The subjects with positive anamnesis for recent cardiac events.
- The subject with renal insufficiency.
- The subjects with allergy to milk and its derivates.
- The subjects with contraindication to the administration of the contrast agents.

3.4.3 Ethical Issues

The patient recruitment followed standard guidelines. Data were anonymized and all the exams were carried out in a blind fashion. The corresponding keys for coupling data to the specific patient were known only by the person responsible for the project. All the safety issues were considered in order to guarantee privacy, to ensure the storing and backup of the dataset, and to enable the safe and encrypted data transmission. Informed consensus was obtained for each subject participating in the study. The present protocol received the approval from the Ethical Committee of the Gradenigo Hospital.

3.4.4 Functional and Instrumental Clinical Examinations

3.4.4.1 Clinical Examination

All the patients underwent

- Neurological visit
- Angio-surgical visit
- Cardiological visit

3.4.4.2 Laboratory and Hematochemical Exams

First level hematochemical exams: (1) complete hemochrome with leucocitary formula; (2) glicemy; (3) creatinine; (4) total cholesterol; (5) LDL and HDL cholesterol; (6) AST; (7) ALT; (8) ALP; (9) GGT; (10) apoprotein A; (11) apoprotein B; (12) TSH; (13) FT4; (14) VES; (15) PCR; (16) anti-citrulline antibody; (17) anti-beta 2GPI antibody; (18) homocysteine; (19) B12 vitamin; (20) folates; (21) PT; (22) PTT; (23) fibrinogen; (24) antithrombin III; (25) coagulation C protein; (26) S protein; (27) APC resistance; (28) microalbuminuria.

Second level hematochemical exams: (1) FT3; (2) LDH; (3) QPE; (4) 677-MTHFR mutation; (5) prothrombin polymorphism; (6) V factor mutation; (7) lactic acid; (8) ASLO; (9) Alpha-1 acid glicoprotein; (10) ANA; (11) IC; (12) cryoglobuline; (13) anti-borrelia antibodies; (14) CEA; (15) Ca 19.9; (16) alpha fetoprotein; (17) ferritine; (18) NSE; (19) Ca 125; (20) Ca 15.3; (21) PSA; (22) breath test.

3.4.4.3 Instrumental Examinations

Ultrasound examination

- EcoColorDoppler with high-definition (7.5–12 MHz) linear surface probe.
- B-Mode 2-D contrast-enhanced study of the carotid wall and/or plaque.

From a methodological point of view, the basal examination is conducted before and after the injection of contrast agent, at the level of the carotid bifurcation, common carotid, and first tract of the internal carotid, with axial and longitudinal projections. The intima-media thickness (IMT) was measured. We also noted and measured pathological modifications of the artery wall echogenicity, amplitude modifications of the vessel lumen and of the blood flow velocity, and echogenicity modifications after the contrast agent injection [11, 12].

MRI examination

All the exams were conducted in 1.5 T static magnetic field, with field gradients higher than 36 mT/m. The MRI investigation documented possible encephalic lesions due to vascular impairment, by using the SE-TSE technique, with T1-T2 images, Flair, and diffusion images.

Plaque was studied with a dedicated surface coil, with black/bright blood imaging, in basal conditions and after paramagnetic contrast agent administration.

Finally, we performed a 3-D angio-MRI extracranial reconstruction in basal conditions and after contrast-agent injection, with evaluation of vessel sections in axial projections, by maximum intensity projections.

3.5 Sample Instrumental Data

In this section, we show some samples of the acquired data from our protocol. Traditional and routinely examinations (such as MRI) are coupled to new ultrasound processing strategies for a better representation and understanding of the atheromasic disease. As we said, most of the data coming from this study were used in specific contributions mainly dedicated to the ultrasound carotid wall segmentation and plaque characterization.

Figure 3.1 reports the angio-MRI examination carried out on a 69-year-old male patient who was diagnosed an carotid plaque. The maximum intensity projection reconstruction of the supra-aortic vessels documented the presence of a closed

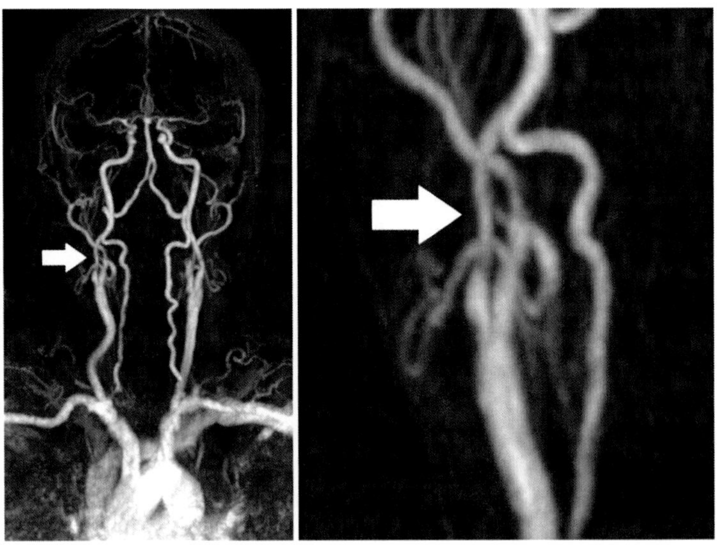

Fig. 3.1 Angio-MRI of the supra-aortic vessels. The *white arrows* indicate a significant narrowing of the internal carotid artery lumen as consequence of a atheromasic plaque. The *left image* is the complete MRI reconstruction and the *right image* is the zoomed version of the right artery branch

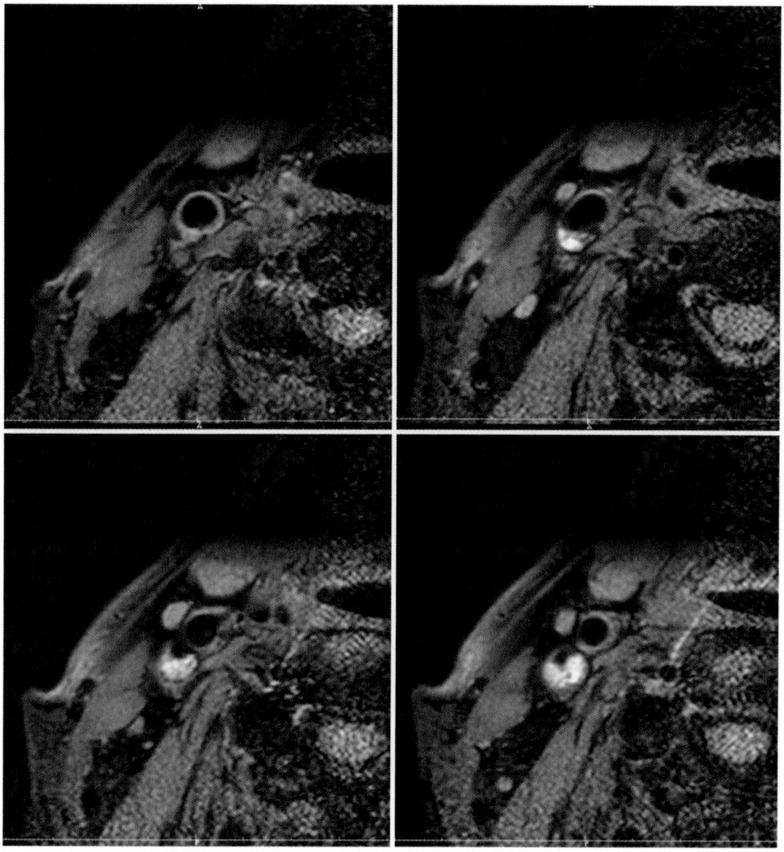

Fig. 3.2 T1-weighted carotid images acquired with dedicated coil. The plaque is visible in the carotid bulb, with hyper intensity in T1 and inhomogeneous aspect

stenosis in the right internal carotid artery corresponding to 50% following NASCET criterion. The plaque had a length approximately equal to 2 cm. There were no other significant alterations of the vessels in the extracranial and intracranial circulation.

The study at the level of the cranial basis with dedicated surface coil evidenced hyper intensity in T1, located at the carotid bulb (Fig. 3.2). The appearance of the plaque is inhomogeneous and irregular. The T2 representation confirms the inhomogeneous aspect of the plaque, with less defined hyper intense representation and hypo intense areas within (Fig. 3.3). Such representation could be indicative of a degenerative component with calcific content and vascularization. Such plaque, even if within the stenosis limit of the NASCET guidelines, can be thought of as complicated and, possibly, at high risk for the patient. In fact, the mixed presence of calcifications and vascular components could reveal the presence of a mixed tissue with hembolic behavior. To further confirm the plaque characterization, we also reported the plaque representation in bright-blood technique (Fig. 3.4).

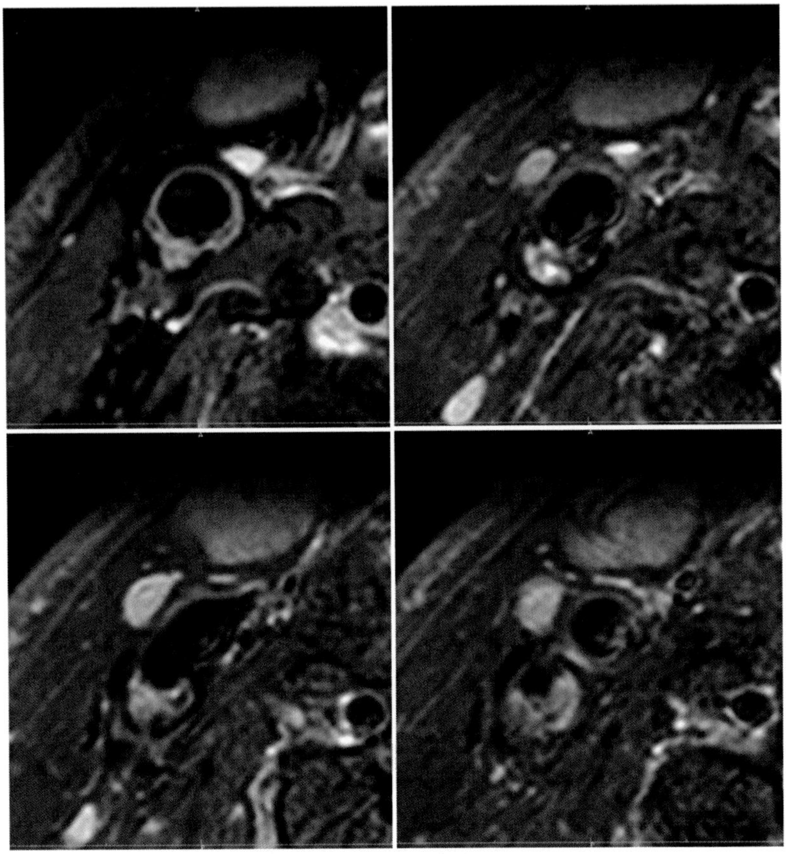

Fig. 3.3 T2-weighted representation of the same plaque of Figs. 3.1 and 3.2. T2 images reveal the inhomogeneous appearance of the plaque protruding into the internal carotid artery

Figure 3.5 reports the sonographic appearance of the plaque. The left panel reports the traditional B-Mode representation of the plaque. The shadow cone artifact reveals the calcium content of the plaque. The right panel of Fig. 3.5 reports the ColorDoppler image of the plaque. The blood flow velocity at the carotid bulb level demonstrates the effect of the critical stenosis. The ultrasound examination was completed by the injection of 1.5 ml of contrast agent and by a subsequent imaging of the tissutal enhancement obtained. The images were recorded about 40–50 s after the contrast agent injection, to avoid the early arterial phase and to let the contrast agent diffuse into the microcirculation.

The patient underwent carotid endarterectomy 3 weeks after examination. The specimen of the carotid was sent to histology for reporting. Figure 3.6 is comprehensive of the ultrasound examination and of the histological characterization. Figure 3.6a reports the traditional B-Mode image before contrast agent injection. Figure 3.6b shows the contrast-enhanced ultrasound image after the administration of 1.5 ml of contrast agent. Figure 3.6c shows the H&E 95× histology specimen. This panel reports the central slice of the plaque, in order to make the results comparable to longitudinal

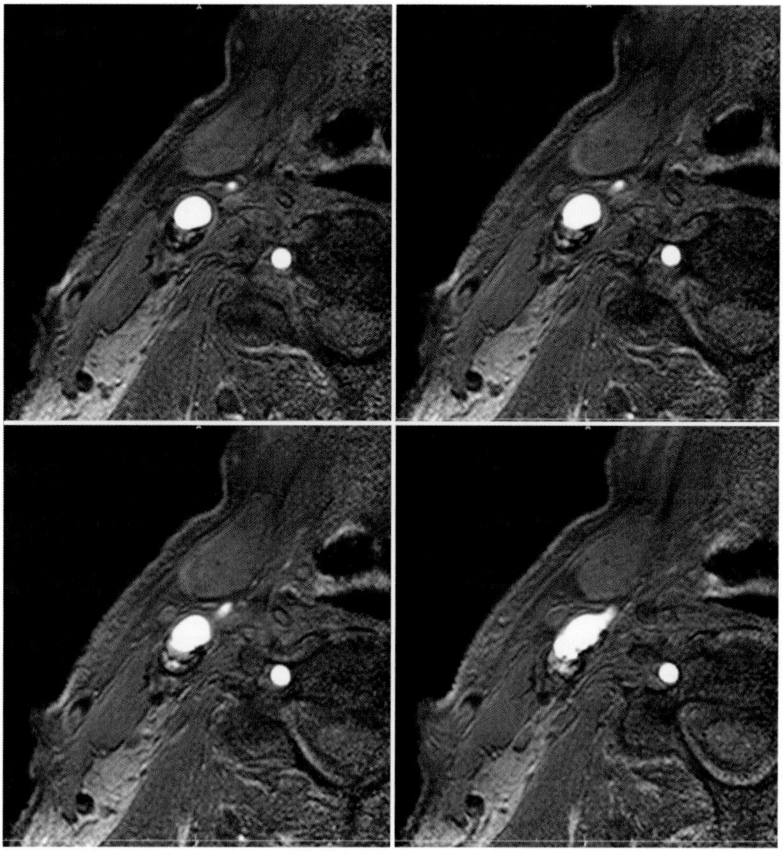

Fig. 3.4 Bright-blood MRI characterization of the plaque in Figs. 3.1–3.3

ultrasound images. It is possible to observe a diffused blue/violet color that traces the diffuse calcium content of the plaque. The white spots represent lipids. Therefore, this plaque was clearly mixed and complicated, since it coupled a clacific component to a soft and unstable one. Finally, Fig. 3.6d reports the contrast-enhanced processed image. The details of the image processing strategy are reported in Chap. 8. The plaque is clearly calcific with diffuse calcium crystals in intima and media.

3.6 Results and Impact on the Management of the Patient with Neurological Symptoms

Presently, the management of the atherosclerotic patient is not standardized in all centers. As we already mentioned, there are two groups of patients (i.e., symptomatic without surgical indication and asymptomatic with surgical indication) who are not always treated the same way in all the clinical divisions. This gray zone is where this study concentrated most of the efforts.

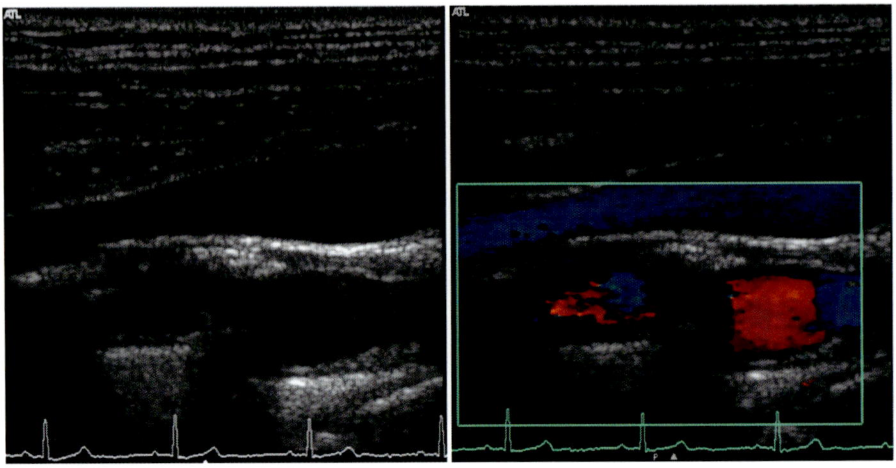

Fig. 3.5 *Left panel:* sonographic appearance of the plaque represented by traditional B-Mode imaging. The presence of a calcific component is revealed by the shadow cone artifact of the plaque. *Right panel:* ColorDoppler imaging showing the perturbed blood flow velocity in correspondence of the carotid bulb

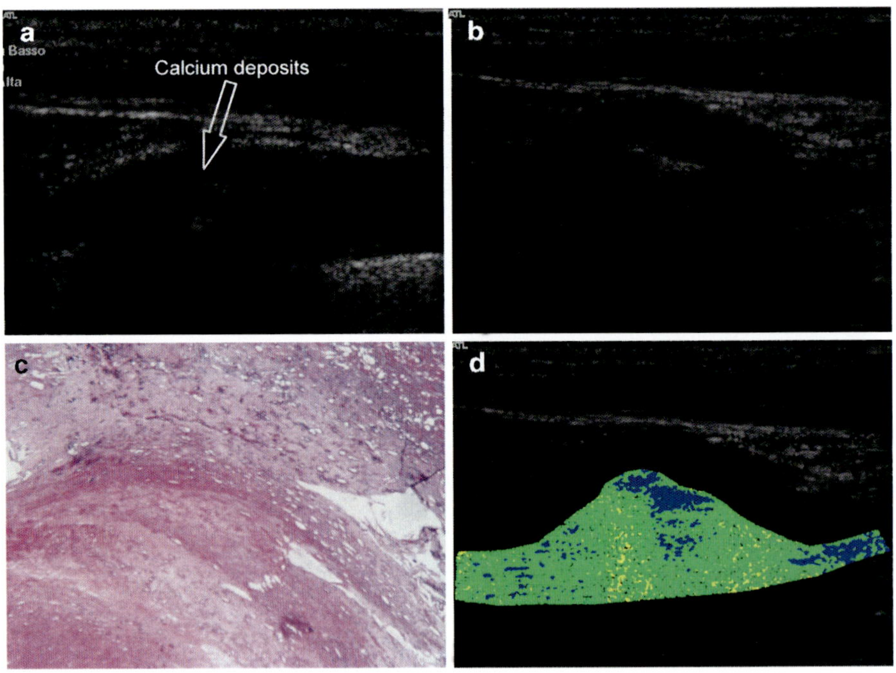

Fig. 3.6 (**a**) Traditional B-Mode image before contrast agent injection. (**b**) ceUS image after the administration of 1.5 ml of contrast agent. (**c**) H&E ×95 histology of the specimen. Central slice of the plaque. (**d**) ceUS processed image. The plaque is clearly calcific with diffuse calcium crystals in intima and media

The first objective of this work was the internal and external validation of the ultrasound methodology in the instrumental diagnosis of the atheromasic plaque of the carotids. Our first results showed that, by using contrast-enhanced ultrasound imaging, it is possible to effectively characterize plaque composition. From a clinical management point of view, this result is extremely important for three reasons.

- First, atherosclerotic patients, being either symptomatic or not, routinely undergo ultrasound examinations of the supra-aortic vessels. Hence, if the ultrasound methodology proved effective in describing plaque composition, further time consuming and costly examinations could be avoided. In this historical moment, where welfare and public health cover a big percentage of the national expenses of countries, the optimization of the clinical management protocols is of extreme importance. Such importance is increased by the fact that the atherosclerotic pathology afflicts a large number of citizens. Thus, there is a big scope of improvement for public health. Currently, the only criteria for deciding the surgical intervention are the presence of symptoms and the stenosis degree. Plaque composition covers a lower importance and is a further decision element in complicated or borderline cases. By fostering our experience in this field, we aim at providing more data to the vascular surgeon in order to optimize the surgical planning and treatment.
- Second, MRI is unsuitable to monitoring purposes and patients' follow up. Re-stenosis, cerebrovascular risks, and cardiological/neurological complications, however, indicate the need for more versatile diagnostic instruments. In a near future, it might be possible to assess changes in the arterial wall tissues by using ultrasounds. Again, this would improve the clinical management of the patients and reduce the overall associated costs.
- Third, as drug therapies are becoming more and more effective and aggressive, ultrasounds could constitute the basis for monitoring and evaluating the effects of pharmacological therapies.

Our second objective was the verification of the sensibility and specificity of plaque characterization by ultrasounds. Results are better described in Chap. 8, but some major observations are reported in the following. After endarterectomy, of 20 plaques, 12 were found to be soft and ulcerated at the histologic analysis, while 8 were calcified. A blind plaque assessment made on the basis of the contrast-enhanced ultrasound images before surgical treatment found ten soft plaques and ten calcified. Overall, sensitivity of plaque characterization was about 100% and specificity about 80%. Clearly, this performance evaluation is qualitative, subjective, and obtained on relatively small sample, but it provides encouraging results. From a diagnostic point of view, in fact, a first ultrasound-based plaque evaluation could be very important to establish the subsequent treatment course.

Presently, in the Neurological Department of the Gradenigo Hospital of Torino (Italy), this experimentation is ongoing and is presently used as clinical pipeline. A strict cooperation with the Vascular Surgery of the "Umberto I" Hospital of Torino and with the Histology and Pathology Department of the Koelliker Hospital of Torino is enabling pretreatment diagnosis, vascular intervention, and histological reporting of the specimens. This team will provide consolidated results about importance of this methodology in the management of atherosclerosis patients with neurological symptoms.

References

1. V.P. Schepers, M. Ketelaar, I.G. van de Port, J.M. Visser-Meily, and E. Lindeman, Comparing contents of functional outcome measures in stroke rehabilitation using the International Classification of Functioning, Disabil Health Disabil Rehabil, 29(3), (2007), 221–30.
2. V.P. Schepers, M. Ketelaar, A.J. Visser-Meily, V. de Groot, J.W. Twisk, and E. Lindeman, Functional recovery differs between ischaemic and haemorrhagic stroke patients, J Rehabil Med, 40(6), (2008), 487–9.
3. M. Fisher, A. Martin, M. Cosgrove, and J.W. Norris, The NASCET-ACAS plaque project. North American Symptomatic Carotid Endarterectomy Trial. Asymptomatic Carotid Atherosclerosis Study, Stroke, 24(12 Suppl), (1993), I24–5; discussion I31–2.
4. H.J. Barnett and H.E. Meldrum, Carotid endarterectomy: a neurotherapeutic advance, Arch Neurol, 57(1), (2000), 40–5.
5. P.M. Rothwell and C.P. Warlow, Interpretation of operative risks of individual surgeons. European Carotid Surgery Trialists' Collaborative Group, Lancet, 353(9161), (1999), 1325.
6. P.M. Rothwell and C.P. Warlow, Prediction of benefit from carotid endarterectomy in individual patients: a risk-modelling study. European Carotid Surgery Trialists' Collaborative Group, Lancet, 353(9170), (1999), 2105–10.
7. P.M. Rothwell, R. Gibson, and C.P. Warlow, Interrelation between plaque surface morphology and degree of stenosis on carotid angiograms and the risk of ischemic stroke in patients with symptomatic carotid stenosis. On behalf of the European Carotid Surgery Trialists' Collaborative Group, Stroke, 31(3), (2000), 615–21.
8. P.M. Rothwell, R.J. Gibson, J. Slattery, and C.P. Warlow, Prognostic value and reproducibility of measurements of carotid stenosis. A comparison of three methods on 1001 angiograms. European Carotid Surgery Trialists' Collaborative Group, Stroke, 25(12), (1994), 2440–4.
9. L. Saba and G. Mallarini, A comparison between NASCET and ECST methods in the study of carotids Evaluation using Multi-Detector-Row CT angiography, Eur J Radiol, (2009), (epub ahead of print).
10. I. Mayor, S. Momjian, P. Lalive, and R. Sztajzel, Carotid plaque: comparison between visual and grey-scale median analysis, Ultrasound Med Biol, 29(7), (2003), 961–6.
11. F. Molinari, S. Delsanto, P. Giustetto, W. Liboni, S. Badalamenti, and J.S. Suri, User-independent plaque segmentation and accurate intima-media thickness measurement of carotid artery wall using ultrasound. In, Advances in diagnostic and therapeutic ultrasound imaging, J.S. Suri, C. Kathuria, R.F. Chang, et al., eds. pp. 111–140 (Artech House, Norwood, MA, 2008).
12. F. Molinari, W. Liboni, E. Pavanelli, P. Giustetto, S. Badalamenti, and J.S. Suri, Accurate and automatic carotid plaque characterization in contrast enhanced 2-d ultrasound images, Conf Proc IEEE Eng Med Biol Soc, 2007, (2007), 335–8.

Biographies

Dr. William Liboni received the Italian Laurea in Medicine in 1969 from the Università degli Studi di Torino and then specialized in radiology and nuclear medicine in 1975 and 1993, respectively. Since 1994 he directed the Neurology Division at the Gradenigo Hospital of Torino, Italy. His research focuses on the functional assessment of neurologically impaired subjects and on the non-invasive monitoring of chronic pathologies.

Dr. Enrica Pavanelli received the Italian Laurea in Medicine in 1993 from the Università degli Studi di Torino, Italy. In 1997 she specialized in neurology. Since 1997 she is on faculty of the Neurology Division of the Gradenigo Hospital of Torino, Italy. She is dedicated to cerebrovascular diseases, Parkinson diseases and ultrasound diagnosis.

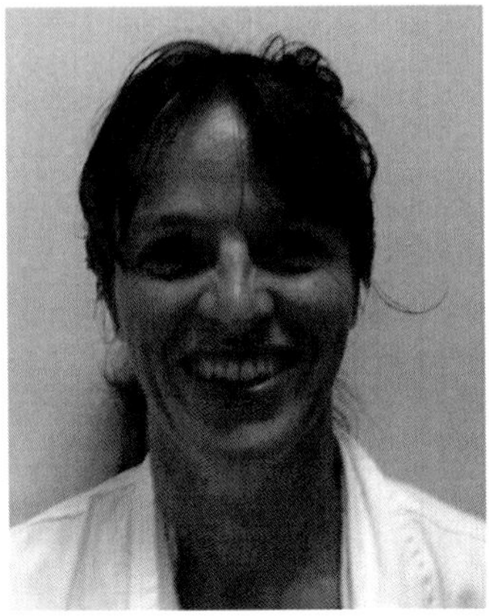

Dr. Nicoletta Rebaudengo received the Italian Laurea in Medicine and then specialized in Neurology from the Università degli Studi di Torino (Italy) in 1989 and 1993, respectively. From 1991 to 1993 she was post-doctoral fellow at Fidia Georgetown Research Foundation at Georgetown University of Washington D.C. Since 1993 she has been working at the Gradenigo Hospital of Torino as neurologist. In particular, she is dedicated to cephalea studies and ultrasound diagnosis.

Dr. Filippo Molinari received the Italian Laurea and the Ph.D. in electrical engineering from the Politecnico di Torino, Torino, Italy, in 1997 and 2000, respectively. He is leader in ultrasound imaging focused towards tissue characterization, vascular quantification for diagnostics and therapeutics. Currently, he is Assistant Professor at Politecnico di Torino, Italy – Department of Electronics.

Dr. Jasjit S. Suri is an innovator, scientist, a visionary, an industrialist and an internationally known world leader in Biomedical Engineering. Dr. Suri has spent over 20 years in the field of biomedical engineering/devices and its management. He received his Doctrate from University of Washington, Seattle and Business Management Sciences from Weatherhead, Case Western Reserve University, Cleveland, Ohio. Dr. Suri was crowned with President's Gold medal in 1980 and the Fellow of American Institute of Medical and Biological Engineering for his outstanding contributions.

Chapter 4
Pathology of Atherosclerotic Disease

Andrea Marsico

Abstract The term atherosclerosis is derived from the Greek words "atheroma" and "sclerosis" which mean, respectively, "mush" and "hardening." These two concepts, seemingly antithetical, immediately make the idea of what processes anatomical substrate of the disease: hardening of the vascular wall and lumen reduction of the same vessel as a result of the formation of a plaque whose content is at least partially and initially soft consistency. This chapter will introduce, with the help of pictorial samples and specimens, the pathology analysis of carotid plaques. Specimen pictures as well as fatty streak diagrams will help comprehensive discussion about plaque morphology, composition, and appearance.

Keywords Atherosclerosis • Plaque • Carotid artery

The term atherosclerosis is derived from the Greek words "atheroma" and "sclerosis" which mean, respectively, "mush" and "hardening." These two concepts, seemingly antithetical, immediately make the idea of what processes anatomical substrate of the disease: hardening of the vascular wall and lumen reduction of the same vessel as a result of the formation of a plaque whose content is at least partially and initially soft consistency.

Atherosclerosis is part of a more general framework called "arteriosclerosis," whose etymology shows the reduced elasticity of the arterial vascular wall. Arteriosclerosis, in addition to atherosclerosis, includes other diseases such as calcific sclerosis of Mönckeberg and the arteriolosclerosis.

These three diseases recognize target districts: Sclerosis of Mönckeberg mainly affects small and medium caliber vessels in the muscles of the limbs, the

A. Marsico (✉)
Head of the Anatomo-Pathology Division of the Koelliker Hospital, Torino, Italy
and
Adjunct Professor at the University of Torino, Torino, Italy
and
Senior Consultant in Histo-Cytopathology, Polyclinic of Monza, Italy
e-mail: dott.marsico@gmail.com

Jasjit S. Suri et al. (eds.), *Atherosclerosis Disease Management*,
DOI 10.1007/978-1-4419-7222-4_4, © Springer Science+Business Media, LLC 2011

arteriolosclerosis predominantly affects small vessels (especially the kidney). Atherosclerosis is rather a systemic process and affects both arteries of elastic type of large-sized vessels (aorta, iliac) and light arteries with no elastic component (muscle vessels, coronary arteries).

4.1 Epidemiology of Atherosclerosis

Atherosclerosis is an extremely common disease, whose prevalence is inversely proportional to the degree of richness of the study population: low in African and Asian countries, reaches its peak in the USA and Europe (Western). The countries of the Mediterranean (including Italy) are placed in an intermediate position.

Atherosclerosis is the leading cause of death in industrialized countries and is expected to become the leading cause of death even in developing countries by 2025.

4.2 Risk Factors

Atherosclerotic disease is commonly considered a multifactorial disease. As often happens in human disease, the simultaneous presence of more risk factors had a cumulative effect in respect of the risk, increasing it exponentially.

Risk factors can be summarized into two groups: modifiable and nonmodifiable:

Nonmodifiable risk factors:

- Gender: male sex has a higher incidence (than women of childbearing age).
- Age: although atherosclerotic processes begin in childhood, increase their complications from the fourth decade of life.
- Heredity: familiar hypercholesterolemia, diabetes, and hypertension.
- Hyperhomocysteinemia: increased blood concentration of homocysteine (hereditary), which causes endotgelial damages on an oxidative basis.

Modifiable risk factors (are modifiable by drug treatment, dietary or lifestyle changes):

- Hyperlipemia and hypercholesterolemia: a key component of the plaque (atheroma) is made from cholesterol. The data cholesterol "in itself" is not indicative of an increased risk of developing atherosclerotic disease. Cholesterol, although chemically an alcohol, is not soluble in aqueous medium made from human plasma and must be transported through proteins that act as carriers. Cholesterol carrier protein in the sense "centrifugal" or to the suburbs are represented by LDL and VLDL (low and very low density lipoproteins) and cholesterol carrier protein in the sense "centripetal" or from the periphery toward the hepatic catabolism, are known as HDL (high density lipoprotein). Without this clarification, you can understand how HDL component took the slang name of "good cholesterol" actually having the ability to "drain" the main component of the plaque to the excretion of biliary type.

- Hypertension: pathogenetic whose action is likely to be linked to action "lying" on the vessel walls than a direct action, which increases the permeability of the vessel against the LDL.
- Cigarette smoke: the mechanism of action is not fully known, but is probably due to oxidative stress of endothelial wall.
- Diabetes mellitus: in this case there is an increased vascular permeability to LDL-related nonenzymatic process of glycosilation lipoproteins themselves, in turn induced by glucose.

4.3 Normal Anatomy of the Arterial Vascular Wall

To understand the pathogenesis of atherosclerosis and its complications, it is essential to know the normal anatomy of the artery walls (Fig. 4.1).

- As you can see from the table in Fig. 4.1, the wall of the artery (in this schematic seen in frontal section) consists of concentric structures called adventitia, media, intima, and endothelium (from outside to inside) respectively.
- Tunica adventitia is essentially a fibromuscular-type structure.
- Tunica media consists of layers of muscle cells intermingled with elastic fibers called "lamellar units." The percentage of muscle with elastic component varies across layers. The largest percentage of elastic fibers is found in the aorta and, in general, in large caliber vessels; the arteries of the limbs contain, however, a minimum.
- Intimate tunica (also called intima tunica) is a thin elastic layer upon which the endothelium is in direct contact with the blood stream. Contrary to what may seem, the endothelium is not simply a "wallpaper" lining the vessel lumen; it is instead a real "body organ" with endocrine functions related to the maintenance of vascular tone. The endothelium produces substances such as nitric monoxide (NO - a powerful vasodilator), prostacyclin (which is another vasodilator),

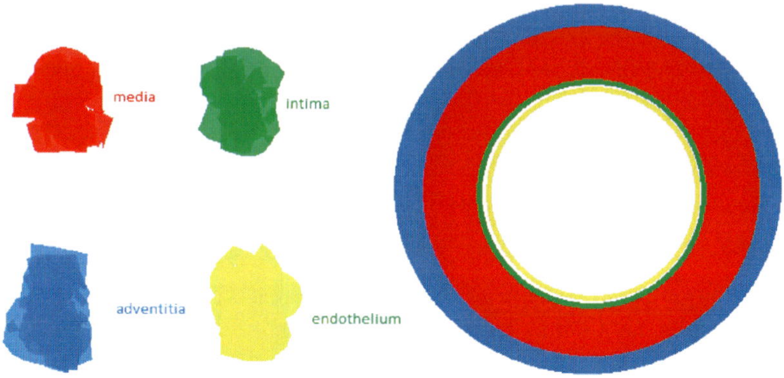

Fig. 4.1 Normal vessel anatomy diagram

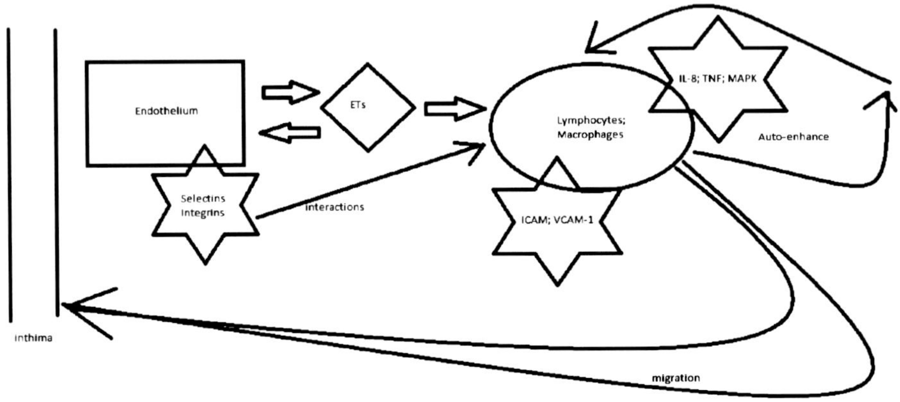

Fig. 4.2 Endothelial damage drawing

and "endothelial factors" (ET - which have vasoconstriction function). Hence, the intima layer participates to the homeostasis maintainence. When damaged, the endothelium excretes ETs. ET1 induces chemotaxis of inflammatory factors and the expression of selectins and integrins and adhesion molecules (ICAM-1, VCAM-1), as well as of immunoglobulins by lymphocytes and macrophages themselves. All these molecules facilitate the adhesion of inflammatory cells on the endothelial surface and their migration in the subendothelial and intimal layers.

The family of ETs have also action as a mitotic stimulation of smooth muscle cells (Fig. 4.2).

4.4 Pathology

The morphological features of atherosclerotic disease have been well codified and universally accepted in the work appeared in 1995, in *Circulation* [32].

The American Heart Association recognizes six stages of atherosclerosis, according to Fig. 4.3.

Within this classification, we can also recognize a further staging between primary lesions and complicated lesions.

4.5 Elementary Atheromasic Injuries

- Initial injury: is the presence of isolated foamy macrophages.
- Atherosclerotic fatty streak: It appears from the early years of life, especially in the thoracic-abdominal aorta. The name derives from the appearance of the typical lesions: small yellowish "drops" (spots) sometimes confluent to form strips.

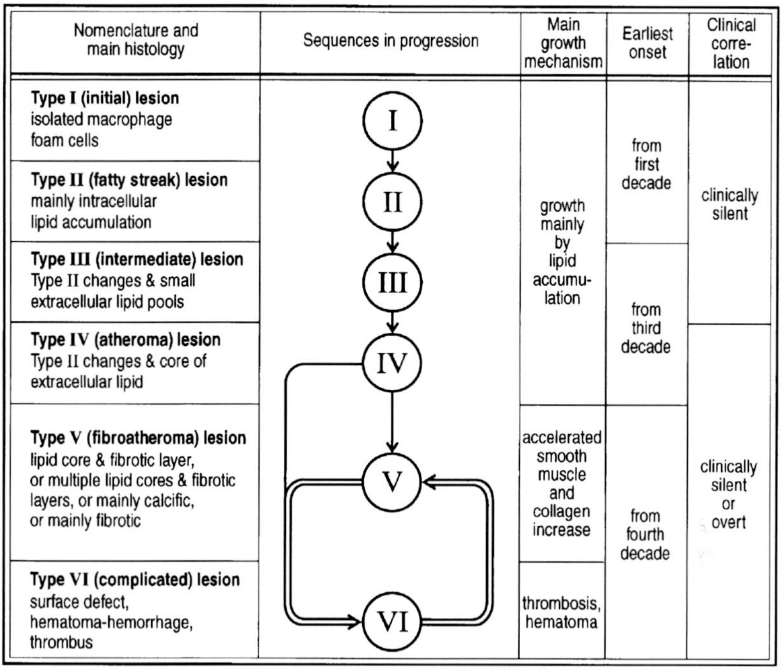

Fig. 4.3 Classification of the six types of artery wall lesion, as proposed by Stary HC et al. in 1995 [32]

Histologically, these lesions take the form of fat-free lesions or are contained inside foamy histiocytes (macrophages). Besides the macrophages, occasionally also lymphocytes can also be found.

- Intimal hyperplasia: Characterized by a concentric stenosis (not eccentric as that of the striae) formed by the intimal proliferation of smooth muscle cells belonging to the medial and migration into the subintimal after endothelial injury (resulting in release of ETS) (Figs. 4.4–4.6).

Factors released from the damaged endothelium determine a myocyte mitotic activity and also induce the synthesis of matrix proteins (colleagenes, glicolsamminoglicani). Later, due to the effect of factors released from the damaged endothelium, especially TGF, the myocytes migrated in the intima become foamy histiocytes. This process transforms the histology of the plaque from "fibrous plaque" (myocytes and protein material) to fibroatheromasic plaque (myocytes transformed into macrophages). This sequence is not universally accepted in the literature: some authors do not consider the fibroatheromasic plaque as a result of erosive endothelial damage, but rather as a result of a functional impairment. Conversely, the fibrous plaque was found after interventional maneuvers on the vascular wall (e.g., percutaneous angioplasty) with obvious micro-loss of continuity of the endothelium. This particular condition, we can call it "iatrogenic," can be called "accelerated atherosclerosis" because its onset time is significantly reduced compared to traditional forms (Fig. 4.7).

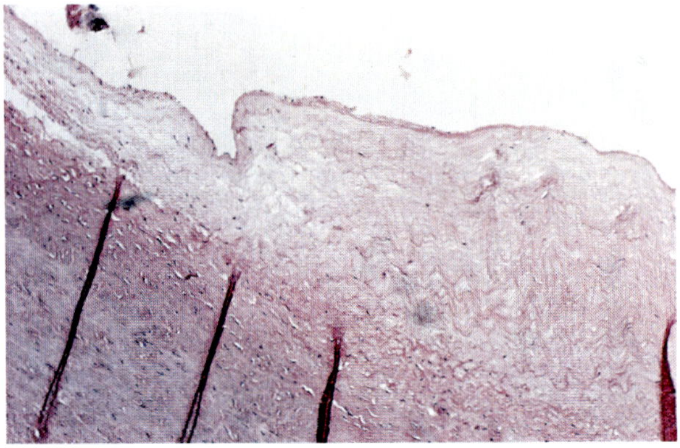

Fig. 4.4 Intimal hyperplasia (courtesy of Professor Roberto Navone, University of Turin)

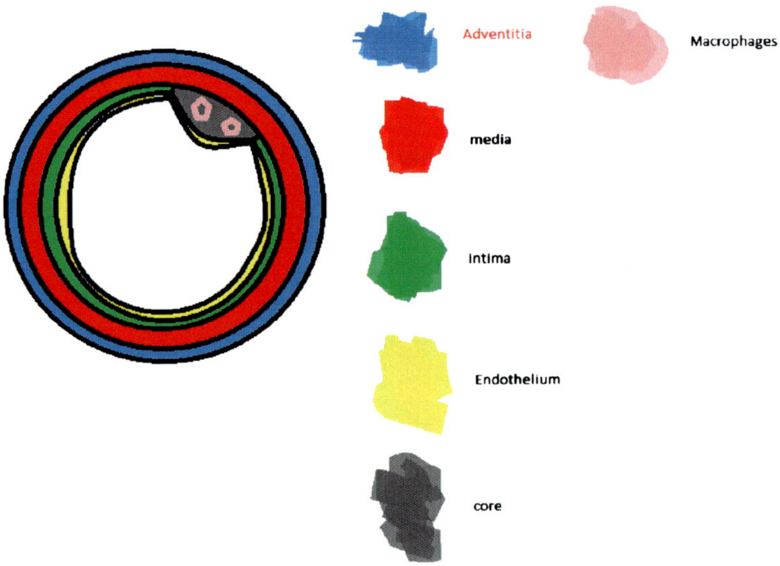

Fig. 4.5 Fatty streak diagram relative to the intimal hyperplasia

- Fibrous plaques, as just alluded, may be understood as the consequence of a prolonged intimal hyperplasia in which the myocytes migrated within the tunica intima media have undertaken a synthesizing pathway. From the macroscopic point of view, the section of the vessel wall shows eccentric thickening and zonal, whitish, and hard consistency. There may be phenomena of calcification, whose origin still remains obscure. Hypotheses have been advanced concerning the precipitation which is direct within the atheromatous gruel. Calcification is mediated by cytokines synthesized within the plaque, with a mechanism that is very similar to that of bone mineralization (Figs. 4.8 and 4.9).
- Fibroatherosclerotic plaque: that is the most common form of "atherosclerotic disease."

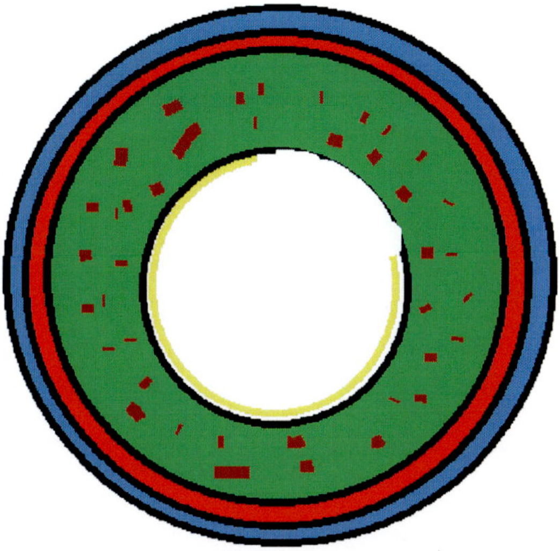

Fig. 4.6 Intimal hyperplasia diagram, note the endothelial damage

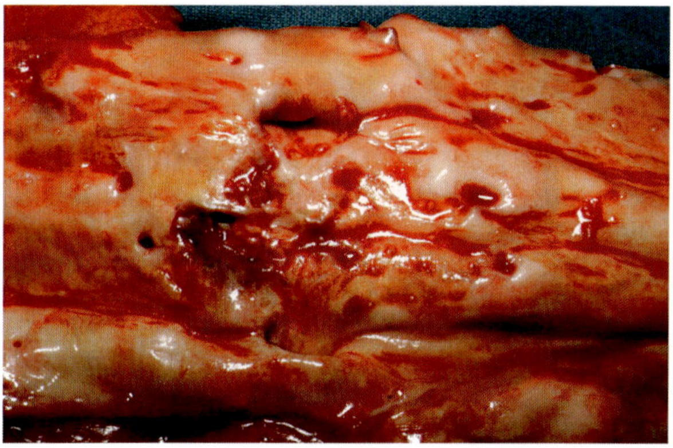

Fig. 4.7 Ulceration in a plaque (courtesy of Professor Roberto Navone, University of Turin)

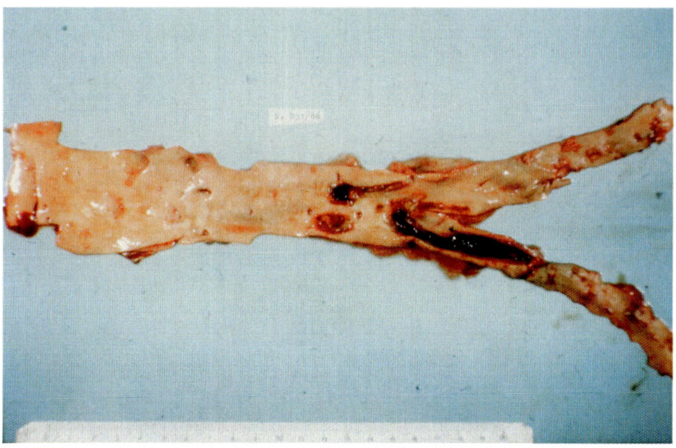

Fig. 4.8 Thrombosis and ulceration in a plaque (courtesy of Professor Roberto Navone, University of Turin)

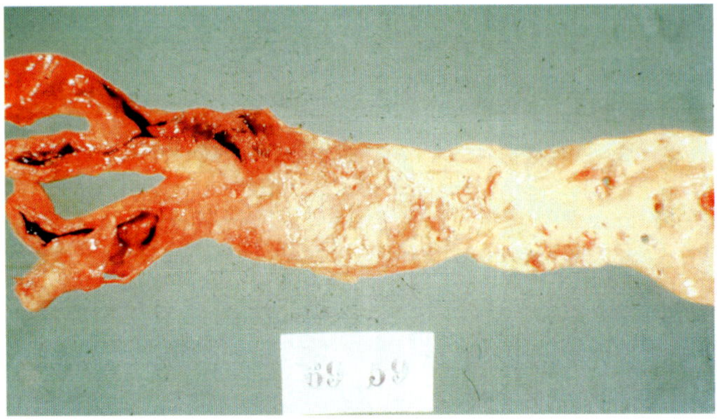

Fig. 4.9 Another example of thrombosis and ulceration in a plaque (courtesy of Professor Roberto Navone, University of Turin)

Fig. 4.10 Fibroatherosclerotic plaque diagram

Macroscopically in the form of hard plaque, white or yellowish anatomically, as shown in the diagram below, we find the following:

- "Core" or the central part varies in texture from soft to poltaceous as in the case of classic atheroma.
- "Fibrous cap," which defines and limits the atheroma and is of great prognostic significance, as we will see later.
- "Shoulder plaque," or the most lateral part of the capsule itself, between the plaque and the core. In the shoulder, we find newly formed vessels, foamy and multinucleated histiocytes, and chronic inflammatory elements (Fig. 4.10).

Anatomically, fibroatherosclerotic plaque occurs with greater frequency in the abdominal aorta (which is also the site of dangerous aneurysms), in the coronary and carotid arteries, and in the polygon of Willis.

4.6 *Relationship between* fibrous capsule and prognostic significance of the lesions

If the plaque has a rich fibrous component and, conversely, the atheromasic component is little or absent, the lesion will have little tendency to run into complications, and therefore is called "stable plaque."

The opposite condition, a plate full of poltaceous atheroma with a thin capsule will more frequently encounter complications and is called "vulnerable plaque" (Figs. 4.11–4.13).

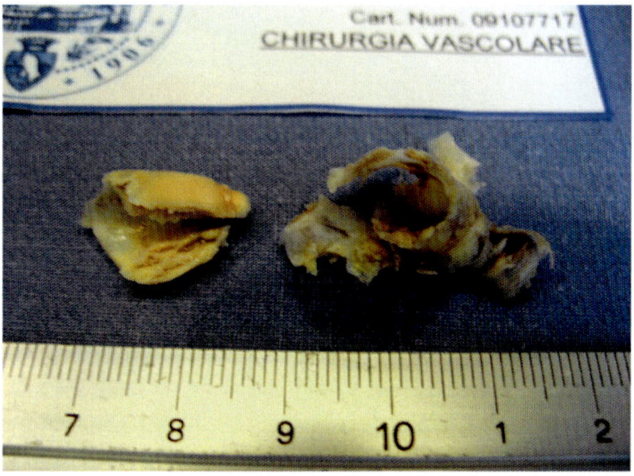

Fig. 4.11 Large and complicated vulnerable plaque at the carotid

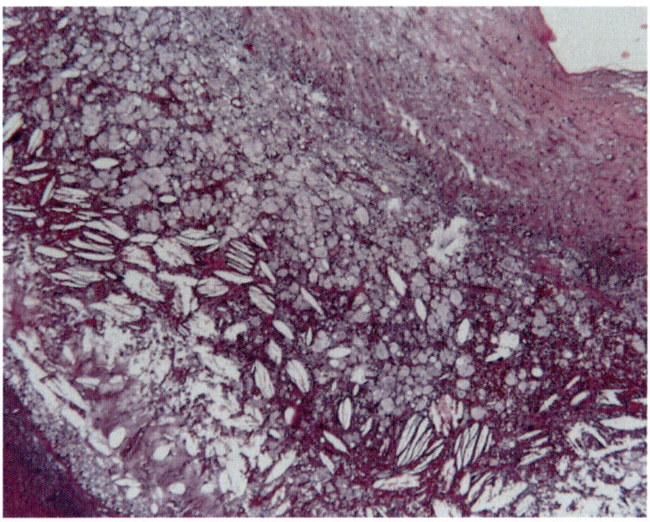

Fig. 4.12 "Shoulder" of a carotid fibroatheromasic plaque: note the abundance of foamy histiocytes and the presence of cholesterol esters in crystalline form, as well as chronic inflammation

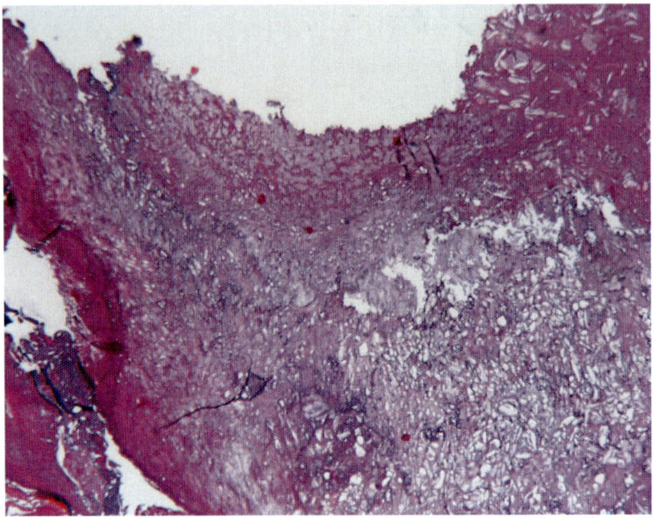

Fig. 4.13 Border area between the shoulder and core of an unstable and complicated lesion of the carotid. You may notice the presence of cholesterol esters crystalline form, rare histiocytes, the presence of massive calcified sclerosis, and the ulceration of the endothelium (*top*) due to the disappearance of the fibrous capsule

The mode of progression in the evolution of the plaque in the process from "stable plaque" to "vulnerable plaque" is unknown; however, an active role of the hyperlypoproteinemia both in physical sense (accumulation of lipids) and in activating the proliferation of smooth muscle cells has been assumed. There are also physical phenomena in the homeostatic adaptation, suffering from atherosclerotic vascular segment; these phenomena are named as follows:

- Active remodeling: intervening until the vascular wall is extensible, so it increases the caliber of the vessel lumen
- Passive remodeling: occurring when vasodilation is no longer possible. It determines a change in the wall composition and a consequent reduction of the vessel lumen.

4.7 Complicated Atheromasic Injuries

4.7.1 Calcification

It is a fundamental phenomenon in plaque progression by providing a suitable substrate introduction of acute events such as cracking and rupture of plaque.

The vessel hit by calcified plaque becomes obviously hard and inextensible, crumbly in consistency, and the transition zones between calcified areas and fibroatheromasic areas become a "minor resistance locus" for a break.

The biochemical mechanism that leads to calcification is currently not known.

4.7.2 Ulcers and Plaque Rupture

Ulcers and plaque rupture are events that are triggered by hydrostatic pressure on the sides of the vessel wall and, of course, the atheromatous plaque. In the "loci of minor resistance" described above, solutions of continuous hood sizes can form from small ulcers to the cleft of the vessel.

Obviously, these lesions are not static but dynamic in their evolution. "Primum movens" in the event of acute plaque is the exposure of the interior of the plaque itself to the blood stream, resulting in release of biochemical mediators that can cause a further narrowing of the lumen through a phenomenon of vasospasm (mechanism involved, e.g., in acute myocardial infarction). If the size of continuous hood is larger, it can penetrate more blood resulting in a phenomenon of obstruction of the lumen, but also of hemorrhage and thrombosis within the lipid core (this can be both endoluminal and intralaminar) with consequent exposure of the blood clotting pathway to the atheroma gruel. If the crack reaches sufficient size, it is possible that the vessel would meet the phenomena of bulging wall with the formation of aneurysmal dilation, potentially fatal, as the base for vascular rupture and internal bleeding (this phenomenon is more frequent in the abdominal aorta). The sharp and sudden penetration of blood in the solution of continuous hood causes the "rupture of the aneurysm," while wearing out the real wall is related to the disorganization of the anatomical structure of the wall made by the presence of atheromatous plaque. Finally, formation of a "floating bridge" consisting of a part of fibrous cap detached from the continuous solution, floating within the lumen, with the possibility of sudden obstructions and penetrations of blood in the plaque, with a fluid regulatory mechanism called "valve" occurs (Figs. 4.14 and 4.15).

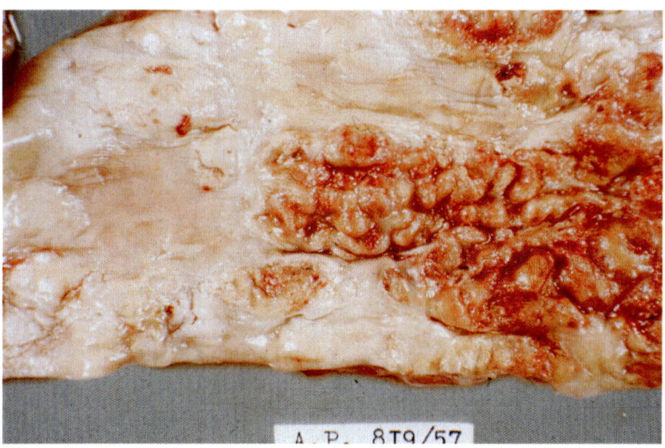

Fig. 4.14 Ulcerated and complicated plaque of abdominal aorta (courtesy of Professor Roberto Navone, University of Turin)

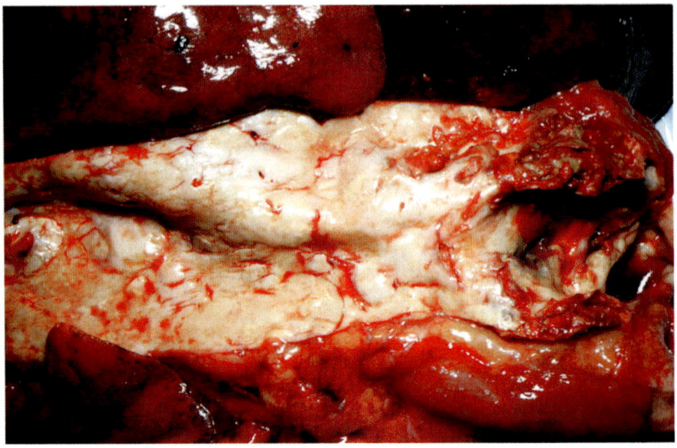

Fig. 4.15 Another sample of ulcerated and complicated plaque of abdominal aorta (courtesy of Professor Roberto Navone, University of Turin)

4.8 Etiopathogenetic Theories

Atherosclerotic disease is multifactorial, and the mechanisms that cause this disease are still not completely known.

In principle, two mechanisms have been postulated that would imply the formation of plaque: they affect the integrity or not of the endothelium.

Erosive endothelial damage: Probably related to more severe forms, in which plaque rupture occurs. Consists in the ulcer, or the loss, even focal part of the endothelial lining of the vascular wall with consequent exposure of intimal layer. This erosion can be of mechanical nature (e.g., iatrogenic: catheterization), of physic nature (always iatrogenic as a result of radiotherapy), and of biochemical-cellular nature. In the last variant, in which no physical mechanisms come into play, erosion appears to be related to endothelial release of proteolytic enzymes by both inflammatory elements (macrophages, lymphocytes) and the endothelium itself (in the parts facing the ulceration).

At this point, there is an enzymatic chain (i.e., formed by pro-enzymes and enzymes, all with proteinasic activity) that is able to strip the endothelium that facilitates the migration and proliferation of smooth muscle cells within intimal–subintimal layer. In this chain enzyme, substances such as plasminogen, plasmin, and the chymase would all be involved in a complex game of adjustments that also involves cytokines, proto oncogenes, growth factors and steroids.

Non erosive endothelial damage: Occurs where the fibroatheromatous plaque is dominated by intact endothelium. It is the most mysterious condition now, although experimental evidence has demonstrated the existence.

Some mechanisms underlying the origin of the intact endothelium plaque have been hypothesized:

- "Impairmement" on endothelial metabolic activities
- Changes in hemodynamics (in particular the loss of laminar flow as happens at the bifurcation)
- Metabolic disorders such as hypercholesterolemia

Obviously each of these mechanisms is likely to present multiple levels of interaction with others.

Activity of microorganisms: In this case, actions of many agents have been postulated. The main suspect is the *Chlamidia pneumnoniae*, especially in combination with cigarette smoke. The mechanism seems associated with the capacity to stimulate the uptake of lipoproteins by macrophages after phagocytosis of the bacteria, and by bacterial lipopolysaccharide and heat shock protein, to promote the transformation of macrophages into foam cells. Other agents suspected of having a role in pathogenesis are Chlamidia species, *Helicobacter pylori*, and Cytomegalovirus.

Autoimmune diseases: From epidemiological point of view, patients with SLE have a fivefold increased risk of developing acute coronary event. Also progression is more rapid in patients with SLE. The mechanism appears to be related to oxidation of lipoproteins that may have a cross reaction against epitopes of antibodies to phospholipids typical of SLE.

References

1. Raja BS, Susma AM, et al. Pathogenesis of atherosclerosis: a multifactorial process. Exp Clin Cardiol. Spring 2002;7(1):40–53.
2. Gallo, D'amati. Anatomia Patologica: la sistematica. UTET. 2007;264–277.
3. Kolodgie FD, Gold HK, Burke AP, et al. Intraplaque hemorrhage and progression of coronary atheroma. N Engl J Med. Dec 11 2003;349(24):2316–25.
4. Ross R. Atherosclerosis – an inflammatory disease. N Engl J Med. Jan 14 1999;340(2):115–26. [Medline].
5. Farmer JA, Gotto AM. Dyslipidemia and other risk factors for coronary artery disease. In: Braunwald E, ed. Heart Disease: A Textbook of Cardiovascular Medicine. 5th ed. Philadelphia: WB Saunders; 1997:1126–60.
6. Bakhru A, Erlinger TP. Smoking cessation and cardiovascular disease risk factors: results from the Third National Health and Nutrition Examination Survey. PLoS Med. Jun 2005;2(6):e160. [Medline].
7. Streppel MT, Ocké MC, Boshuizen HC, Kok FJ, Kromhout D. Long-term wine consumption is related to cardiovascular mortality and life expectancy independently of moderate alcohol intake: the Zutphen Study. J Epidemiol Community Health. Jul 2009;63(7):534–40. [Medline].
8. Djoussé L, Lee IM, Buring JE, Gaziano JM. Alcohol consumption and risk of cardiovascular disease and death in women: potential mediating mechanisms. Circulation. Jul 21 2009;120(3):237–44. [Medline].
9. [Best Evidence] Sinha R, Cross AJ, Graubard BI, Leitzmann MF, Schatzkin A. Meat intake and mortality: a prospective study of over half a million people. Arch Intern Med. Mar 23 2009;169(6):562–71. [Medline].
10. [Best Evidence] Ferdowsian HR, Barnard ND. Effects of plant-based diets on plasma lipids. Am J Cardiol. Oct 1 2009;104(7):947–56. [Medline].
11. Anderson TJ. Assessment and treatment of endothelial dysfunction in humans. J Am Coll Cardiol. Sep 1999;34(3):631–8. [Medline].

12. Arbustini E, Dal Bello B, Morbini P. Plaque erosion is a major substrate for coronary thrombosis in acute myocardial infarction. Heart. Sep 1999;82(3):269–72. [Medline].
13. Bazzano LA. Folic acid supplementation and cardiovascular disease: the state of the art. Am J Med Sci. Jul 2009;338(1):48–9. [Medline].
14. Burke AP, Farb A, Malcom GT. Coronary risk factors and plaque morphology in men with coronary disease who died suddenly. N Engl J Med. May 1 1997;336(18):1276–82. [Medline].
15. Davies MJ. Stability and instability: two faces of coronary atherosclerosis. The Paul Dudley White Lecture 1995. Circulation. Oct 15 1996;94(8):2013–20. [Medline].
16. Davies MJ. The pathophysiology of acute coronary syndromes. Heart. Mar 2000;83(3):361–6. [Medline].
17. Diaz MN, Frei B, Vita JA. Antioxidants and atherosclerotic heart disease. N Engl J Med. Aug 7 1997;337(6):408–16. [Medline].
18. Fuster V., Lewis A. Conner Memorial Lecture. Mechanisms leading to myocardial infarction: insights from studies of vascular biology. Circulation. Oct 1994;90(4):2126–46. [Medline].
19. Gimbrone MA Jr, Nagel T, Topper JN. Biomechanical activation: an emerging paradigm in endothelial adhesion biology. J Clin Invest. Dec 1 1997;100(11 Suppl):S61–5. [Medline].
20. Glagov S, Weisenberg E, Zarins CK. Compensatory enlargement of human atherosclerotic coronary arteries. N Engl J Med. May 28 1987;316(22):1371–5. [Medline].
21. Hiro T, Fujii T, Yoshitake S. Longitudinal visualization of spontaneous coronary plaque rupture by 3D intravascular ultrasound. Circulation. Mar 28 2000;101(12):E114–5. [Medline].
22. Ibañez B, Badimon JJ, Garcia MJ. Diagnosis of atherosclerosis by imaging. Am J Med. Jan 2009;122(1):S15–25. [Medline]. [Full Test].
23. Khandanpour N, Loke YK, Meyer FJ, Jennings B, Armon MP. Homocysteine and peripheral arterial disease: systematic review and meta-analysis. Eur J Vasc Endovasc Surg. Jun 26 2009;38(3):316–22. [Medline].
24. Labarthe DR. Cardiovascular diseases: a global public health challenge. In: Labarthe DR, ed. Epidemiology and Prevention of Cardiovascular Diseases. Gaithersburg, MD: Aspen Publishers; 1998:3–16.
25. Libby P. Changing concepts of atherogenesis. J Intern Med. Mar 2000;247(3):349–58. [Medline].
26. Libby P. Atherosclerosis. In: Fauci A, et al., eds. Harrison's Principles of Internal Medicine. 14th ed. New York: McGraw-Hill; 1998:1345–52.
27. Libby P. Molecular bases of the acute coronary syndromes. Circulation. Jun 1 1995;91(11): 2844–50. [Medline].
28. Ridker P, Libby P. Nontraditional coronary risk factors and vascular biology: the frontiers of preventive cardiology. J Investig Med. Oct 1998;46(8):338–50. [Medline].
29. Ross R. The pathogenesis of atherosclerosis. In: Braunwald E, ed. Heart Disease: A Textbook of Cardiovascular Medicine. Philadelphia: WB Saunders; 1997:1105–25.
30. Salonen JT, Salonen R. Ultrasound B-mode imaging in observational studies of atherosclerotic progression. Circulation. Mar 1993;87(3 Suppl):II56–65. [Medline].
31. Selwyn AP, Kinlay S, Libby P. Atherogenic lipids, vascular dysfunction, and clinical signs of ischemic heart disease. Circulation. Jan 7 1997;95(1):5–7. [Medline].
32. Stary HC, Chandler AB, Dinsmore RE. A definition of advanced types of atherosclerotic lesions and a histological classification of atherosclerosis. A report from the Committee on Vascular Lesions of the Council on Arteriosclerosis, American Heart Association. Circulation. Sep 1 1995;92(5):1355–74. [Medline].
33. Topper JN, Gimbrone MA Jr. Blood flow and vascular gene expression: fluid shear stress as a modulator of endothelial phenotype. Mol Med Today. Jan 1999;5(1):40–6. [Medline].
34. Weissberg PL. Atherogenesis: current understanding of the causes of atheroma. Heart. Feb 2000;83(2):247–52. [Medline].

Biography

Andrea Marsico, MD, PhD, is a Contract professor at the University of Turin. He is currently Head of pathology division of the KOELLIKER Hospital (Turin, Italy) and Senior consultant in histo-cytopathology at the Health Group "Policlinico di Monza" (Vercelli, Italy). His research interests include the validation of new diagnostic techniques in oncology and atherosclerosis.

Chapter 5
Stress Analysis on Carotid Atherosclerotic Plaques by Fluid Structure Interaction

Hao Gao and Quan Long

Abstract Biomechanical factors have been considered to be important triggers in plaque rupture. Plaque stress analysis has been used for providing critical information in assessing rupture risk for specific plaques. The research combined plaque MR imaging, computational modeling, on plaque stress analysis presented in this chapter will advance our understandings of plaque rupture, and may establish new procedures for patient diagnose, management, and treatment in the future.

Keywords Atherosclerosis • Carotid plaque • Stress analysis • Fluid interaction • MRI

Abbreviation

TIA	transient ischemic attack
MRI	magnetic resonance imaging
FEM	finite element method
FSI	fluid structure interaction
TOF	time of flight
ImT2W_FatSat	intermediate T_2 weighted with fat saturation
T2W	T_2 weighted
STIR	short T_1 inversion recovery
CCA	common carotid artery
ICA	internal carotid artery
ECA	external carotid artery
CFD	computational fluid dynamics
WSS	wall shear stress
WSS_*t*max	maximum wall shear stress in the whole cardiac cycle
VWTS	Von Mises stress

Q. Long (✉)
Brunel University, Uxbridge, Middlesex UB8 3PH, UK
e-mail: quan.long@brunel.ac.uk

Jasjit S. Suri et al. (eds.), *Atherosclerosis Disease Management*,
DOI 10.1007/978-1-4419-7222-4_5, © Springer Science+Business Media, LLC 2011

VWTS_*t*max	maximum VWTS in a cycle
VWTS_*s*max	spatial maximum VMTS at a time
rcVWTS	relative cyclic VWTS
FPS	first-principle stress
min_FT	minimum fibrous cap thickness
max_SD	maximum degree of stenosis

5.1 Background

Rupture of carotid plaques is a major cause of cerebrovascular thrombotic events such as transient ischemic attacks (TIA) and stroke [31]. Despite years of research on the subject, the exact mechanisms of plaque rupture still remain unclear. Plaques that are prone to rupture may often be clinically silent until the time of rupture. Much research has been devoted in recent years to find rupture risk factors. It is believed that a large lipid pool, a thin fibrous cap, and a high content of inflammatory cells are generally contributing to a vulnerable plaque [24].

Plaque rupture can be considered as a mechanical failure. Finite element studies have shown that mechanical stress concentrations often occur in the plaque shoulders (or boundaries of a lipid pool region) and thinner fibrous cap regions where plaques most often rupture [7]. Recent developments in high-resolution multispectral MRI have enabled the visualization of plaque components in vivo [5, 16, 46]. It provides more realistic plaque geometries for stress analysis [11, 40]. Patient-specific plaque stress analysis has been used to compare the stress level between symptomatic and asymptomatic patients [20] or multicases [12] to provide better patient management. Indexes based on stress analysis also have been proposed for plaque vulnerability assessment based on MRI data [49]. Therefore, plaque stress analysis can provide important information on the understanding of plaque rupture mechanism and may eventually provide risk assessment for individual plaque vulnerability.

5.1.1 Stress-Related Plaque Rupture Hypothesis

Several rupture mechanisms have been proposed in recent years which can normally be classified into two categories. One is biological abnormalities and the other is biomechanical factors. The biomechanical-related plaque rupture hypothesis can be summarized as follows.

(a) *Local maximum stress*: the local stress concentration in the fibrous cap region, which is higher than the critical value fibrous cap can sustain, is considered to be the main reason of rupture in light of biomechanical factors [7, 36]. A critical stress value of 300 kPa has been proposed by Cheng et al. [7], which will induce plaque rupture based on their 2D stress analysis on ruptured and stable atherosclerotic lesions.

(b) *Fatigue failure*: Bank et al. [3] hypothesized that mechanical fatigue on fibrous cap caused by pulsatile blood flow may be an important factor causing plaque rupture.

(c) *Extremely high wall shear stress*: Extremely high wall shear stress has been linked to plaque rupture recently [14, 15] based on the findings that plaque ulceration occurs at the high wall shear stress locations.

(d) *Debonding effect by microcalcification*: The hypothesis for vulnerable plaque rupture due to stress-induced debonding around cellular microcalcifications in thin fibrous caps was proposed [43] based on the fact that cellular-level micro-calcifications in a thin cap can cause local stress concentrations that lead to interfacial debonding.

(e) *Other hypothesis* includes injury due to turbulent flow in the stenosis [21], rupture of the vasa vasorum [4], etc.

Compared with the others, the local maximum stress hypothesis is the most wildly accepted one, which is also the one that we are working on. Several studies have been conducted to correlate the high stress regions with rupture sites to provide evidence for plaque rupture caused by extreme stresses. Cheng et al. [7] studied the stress concentration locations compared with the rupture sites based on 2D histological plaque samples; they found that the circumferential stress concentrations have a good correlation with the rupture sites. However, plaque rupture may not always occur at the regions of highest stress. Lee et al. [19] used in vitro balloon angioplasty to cause plaque rupture after the intravascular ultrasound imaging of the plaques. In combination with structure analysis, they found that 82% fractures occurred in the regions with high circumferential stress. Ohayon et al. [26, 27] found that the peak circumferential stress areas correlated well with plaque rupture sites by comparing the plaque rupture locations on postangioplasty intravascular ultrasound images. Tang et al. [35, 37] suggested that a local increase in stress/strain could be a cause of plaque rupture, and it may be used for plaque rupture risk assessment. However, the local maximum stress hypothesis of plaque rupture has not been clearly verified in vivo, due to difficulties such as (1) there is no technique that can directly measure wall stress in the plaque in vivo, (2) the geometry of a specific plaque at pre- and postrupture status is practically unachievable. Nonetheless, to work on any stress-related hypothesis, obtaining accurate stress distribution on the specific plaque is essential.

5.1.2 Stress Studies on Plaques

Since there is no way to accurately measure the stress in plaques, computational modeling has been extensively applied to obtain the stress in plaque regions, especially finite element method (FEM). The realistic plaque geometry obtained from medical images including regions of multicomponents such as lipid pool, thin

Table 5.1 The main tasks in plaque stress analysis models

Tasks in plaque stress analysis	
Plaque components to be included in a model	1. Arterial wall geometry (multilayers) 2. Fibrous cap structure (shape, size, components, thickness) 3. Lipid core (shape, size, components) 4. Calcification (size, location) 5. Hemorrhage 6. Inflammation activities (fibrous cap weakening/erosion) 7. Angiogenesis
Plaque geometry model complexity	1. Idealized/realistic model 2. 2D/3D model 3. Imaging source for plaque reconstruction 4. In vivo/ex vivo model 5. The accuracy/reproducibility of plaque reconstruction
Loading and boundary conditions	1. Pulsating blood pressure 2. Wall shear stress 3. Axial/circumferential residual stress/strain 4. Motion caused by surrounding tissue
Material properties	1. Existed experimental data for plaque components 2. Proper material model (isotropic/anisotropic, linear/nonlinear) 3. Patient-specific/general material model 4. Stiffness variation (heterogeneous material property)
Simulation procedure complexity	1. Plane stress/strain calculation (2D) 2. 3D structure-only analysis 3. 3D fluid-only analysis 4. 2D fluid–structure interaction analysis 5. 3D fluid–structure interaction analysis

fibrous cap, and even microcalcifications, can be incorporated into the stress analysis by FEM. Finite element models of plaque stress analysis have been developed from 2D to 3D, from idealized models to patient-specific plaques, and from structure analysis only to fluid structure interaction (FSI) analysis. Recently, FSI has gained great popularity for plaque stress analysis due to its ability to provide solutions on both blood flow patterns and wall stress [10, 12, 36, 37]. There are normally five main tasks involved in building a plaque mechanical analysis model which include: (a) acquisition of plaque components, (b) reconstruction of plaque geometry; (c) application of mechanical load and dynamic boundary conditions; (d) definition of suitable material model; and (e) FEM simulation and postprocessing. Table 5.1 lists breakdowns of each task for different level of simulations. The following section provides a brief review of the simulation technique developments.

5.1.2.1 2D Versus 3D Structure-Only Stress Analysis

Since 1990s, 2D structure stress analysis has been extensively applied to plaque models for predicting stress distributions from idealized models to patient-specific models. The morphological features of plaques could have significant effects on plaque stress distribution. A decreasing fibrous cap thickness can considerably increase the peak circumferential stress, while increase in the stenosis severity will decrease the peak stress, suggesting that plaque rupture may not associate with a high degree of stenosis [22]. Based on a 2D stress analysis study comparing the simulated stress values for ruptured and nonruptured plaques, Cheng et al. [7] proposed a rupture critical stress with a value of 300 kPa. Recent 2D modeling by Li [20] based on in vivo MRI data showed that maximal predicted plaque stresses in symptomatic patients were higher than those in asymptomatic patients, indicating the possibility that plaques with higher stresses may be more prone to be symptomatic and to rupture in the near future.

Compared to the 3D plaque stress analysis, 2D models have the advantage of simpler procedures on building models, have fewer requirements of simulation skills and computational power, and are also much faster, which allows the simulation to be done for many more cases. The main disadvantage, however, is that the result can be unrealistic, because of the assumptions of plane strain/stress, simplified material models, or distorted geometry. The maximum and minimum stress values from histological models could be 28 and 69% higher than those from MRI-based models, respectively, according to Tang's study [35]. Therefore, realistic 3D plaque models are needed for accurate stress analysis in plaque regions.

With 3D models, the exact plaque geometries can be modeled, such as the multilayer model of arterial wall, irregularities of plaque shape, etc. Moreover, without the assumptions of plane stress/strain, the predicted stress would be much closer to the realities. Holzapfel et al. [17] studied diseased vessels by introducing a multilayer anisotropic 3D model incorporating the histological structure of the arterial wall. They compared their 3D models with the existing biomechanical models, characterized by isotropic material response, assumption of plane strain, etc. Their results showed that those assumptions used in 2D modeling may result a wider range of stress value distribution between the maximum and minimum stress comparing with a 3D model. It indicates that the simplifications, such as linear material model, 2D plaque geometry, need to be well justified when interpreting stress results with plaque rupture.

5.1.2.2 3D Structure Analysis Only Versus Fluid Structure Interaction Stress Analysis

In addition to the differences of modeling dimensions, simulation procedures adopted in the modeling form another source of simulation variations. In the physiological situations, plaques are under pulsatile blood pressure loading. The pressure acting on the luminal wall is different spatially, especially in the narrow lumen region where the blood flow is accelerated and pressure will drop according to Bernoulli's equation. Therefore, the uniform pressure loading on luminal wall of the 3D FE model, which is normally adopted in structure analysis only simulation, cannot predict the realistic stress distribution. In order to accurately model the plaque stress distribution, one must consider the hemodynamic loadings in the plaque region, and the corresponding wall dynamics under the pulsatile loading caused by blood flow. Therefore, the coupled 3D blood flow simulation in the plaque region is required as one part of plaque stress analysis. In other words, fluid structure interaction (FSI) simulation is required in order to produce the realistic results.

FSI simulation has been applied to arterial wall study since the last decade. Zhao et al. [50] first introduced MRI-based FSI models to predict wall shear stress and wall stress patterns in healthy subjects and tried to define regions of low wall shear stress and high wall stress and analyze their correlations. Tang et al. [34] has developed a 3D FSI model from 3D ex vivo MR images of a human atherosclerotic carotid artery to identify critical flow and stress/strain conditions, which may be related to plaque rupture. In the following years, Tang and colleagues did a series of studies on stress analysis with atherosclerotic plaques from ex vivo to in vivo states [35]. Their results showed that large lipid pools and thin fibrous caps are associated with both extreme maximum and minimum stress/strain levels. Large cyclic stress and strain resulted from pulsating pressure were identified, which may lead to plaque rupture due to fatigue.

Furthermore, efforts have been made to use plaque stresses for assessing and predicting plaque vulnerability [38]. The local maximal stress hypothesis and a stress-based computational plaque vulnerability index were proposed to assess plaque vulnerability by Tang et al. [36], in which a critical stress in the fibrous cap was chosen to determine the vulnerability index. Their results showed 85% agreement with the assessment given by histo-pathological analysis. A similar study on the assessment of plaque vulnerability by MRI and computational biomechanics has also been performed by Zheng et al. [49], the stresses/strains were calculated in the plaque sections, an index including entire plaque area, lipid-cap thickness, and stress components was proposed.

Stress analysis by 3D FSI on multiple patients with plaques reconstructed from in vivo MRI has been developed in our group. The predicted stress factors have been linked with plaque rupture risk assessment. The predicted stress with reliable plaque stress models will advance understanding of plaque rupture mechanism and be helpful for a better diagnosis and treatment of patients with plaques. In this chapter, we concentrate on the stress analysis with carotid plaques reconstructed from in vivo MRI, including (1) carotid plaque geometry reconstruction from in vivo MRI; (2) stress analysis with four patients with different carotid arterial plaque morphology; (3) study on patients with recent TIA to investigate the correlation between extreme stress conditions and plaque rupture sites; and (4) impacts of plaque components' morphology variations to stress distributions, and some other issues are also discussed briefly, such as plaque geometry reconstruction reproducibility, material model variations, etc.

5.2 Methodology

5.2.1 Carotid Plaque Reconstructions from In Vivo MRI

5.2.1.1 MR Imaging Acquisition

Patients with carotid arterial plaques, recruited from a specialist neurovascular clinic, were chosen for the study. Multispectral MRI scans were performed on the subjects to provide plaque morphology. The protocol was approved by the local ethics committee, and written informed consent was obtained from each patient before the study.

Multicontrast imaging was conducted using a 1.5 T MRI system (GE Diagnostic Imaging, Waukesha, WI) and a four-channel phased-array neck coil (PACC, Machnet BV, Elde, The Netherlands). Movement artifacts were minimized using a dedicated vacuum-based head restraint system (VAC-LOK Cushion, Oncology Systems Limited, UK) to fix the head and neck in a comfortable position and allow close apposition of the surface coils. After an initial coronal localizer sequence, axial 2D time-of-flight (TOF) MR angiography was performed to identify the location of the carotid bifurcation and the region of maximal stenosis on each side. Axial images were acquired through the common carotid artery (CCA) 12 mm below the carotid bifurcation to a point 12 mm distal to the extent of the stenosis identified on the TOF sequence to cover a large range of carotid bifurcation. The following 2D, ECG-gated, blood-suppressed, fast spin echo pulse sequences were used in the plaque region: intermediate T_2 weighted (ImT2W_FatSat: TR/TE: 2*RR/46) with fat saturation; T_2 weighted (T2W: TR/TE: 2*RR/100); short T_1 inversion recovery (STIR: TR/TE/TI: 2*RR/46/150). The pixel size was $0.39\times0.39\times3$ mm in all cases. The field of view was 10 cm and matrix size 256×256. These images were used to delineate the various plaque components such as the fibrous cap, the lipid core.

5.2.1.2 3D Carotid Bifurcation and Arterial Plaque Reconstruction

Plaque Components Segmentation

The artery and plaque geometries were obtained from the multispectral MR scans. An in-house program developed in Matlab was used to facilitate the segmentation of lipid core, arterial wall, and lumen, which have different signal characteristics when imaged using the multispectral protocol. The plaque region was identified and reconstructed based on T2W, ImT2W_FatSat, and STIR images (when applicable), and the healthy arterial part was reconstructed based on TOF images; the detailed image segmentation and 3D reconstruction procedures described below as in Fig. 5.1 were for one subject. The whole procedure was explained for the subject.

The lumen area in T2W images appears with very low intensity, forming a dark region compared with the arterial wall. As shown in Fig. 5.2, the arterial wall can be defined by the band region around the lumen. The lumen boundary can be easily detected by conventional image segmentation methods such as region growing or active contour. The outer wall boundaries sometimes have very low contrast with the surrounding tissue, especially when they are attached to the vein at regions of healthy wall which makes the segmentation difficult. In this situation, a fixed thickness of the local arterial wall of 2–4 image pixels (0.8–1.6 mm) was assigned. With the information of the clearly defined luminal wall, the outer wall locations could be obtained indirectly.

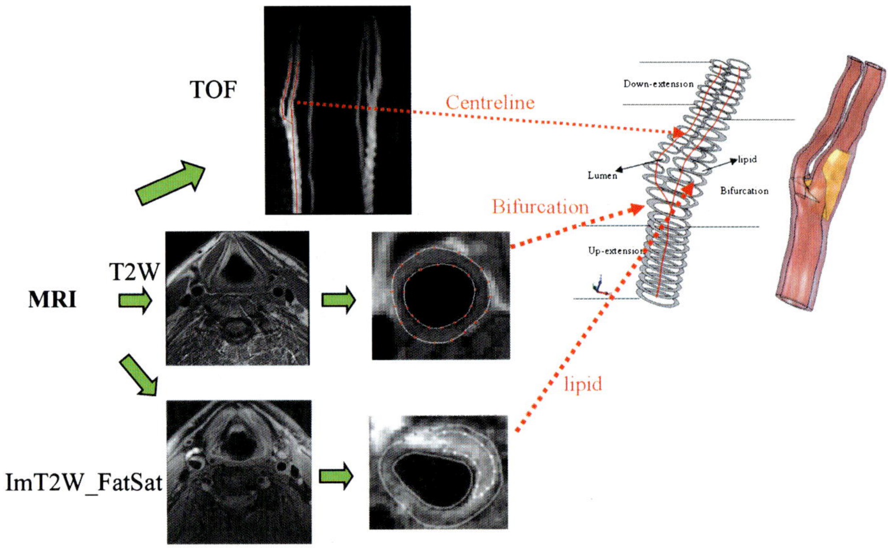

Fig. 5.1 General procedure of 3D plaque geometry reconstruction

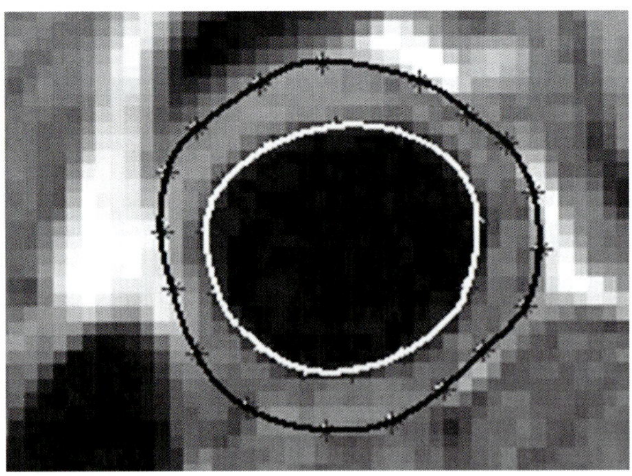

Fig. 5.2 Segmentation of the T2-weighted images for lumen and arterial outer boundary

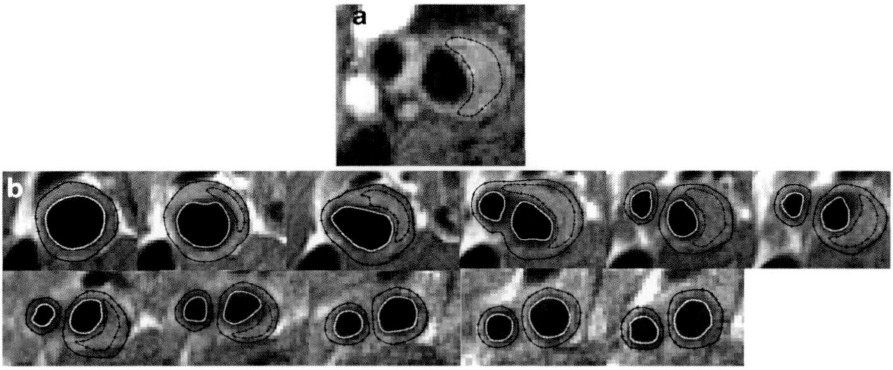

Fig. 5.3 (**a**) The segmentation of the lipid core region; (**b**) the combination of segmentation of arterial wall and lipid core

ImT2W_FatSat images were mainly used for lipid core segmentation. In total, seven 2D images were available for the subject for definition of the lipid core. Regions of lipid core, fibrous cap, and lumen appeared slightly darker, brighter, and black, respectively [32, 39, 41]. STIR images could also provide further help on segmentations. In STIR images, the area with very high signal adjacent to the lumen was classified as fibrous cap, areas of low signal intensity below the fibrous cap were classified as lipid core. Combining the artery geometry from T2W images, the lipid core could be segmented, as shown in Fig. 5.3a. Because of the relative thin arterial wall and also the need for segmentation of the lipid region in the arterial wall, one pixel of segmentation error might impose great uncertainties in the stress analysis. A manual segmentation method was used to define all plaque components for a better control of plaque structure segmentation quality.

Since the arterial inner walls, outer walls, and lipid core boundary were delineated by T2W, ImT2W-FatSat, and STIR images, respectively, it was necessary to perform a registration procedure between images obtained from T2W, ImT2W-FatSat, and STIR sequences. A linear transformation method [9] was employed in this study. Figure 5.3b shows an example of the segmentation of flow lumen, arterial wall, and lipid core by the combination and coregistration of T2W, ImT2W-FatSat, and STIR images.

3D Geometry Reconstruction of Carotid Bifurcation

Reconstruction of the arterial regions beyond the plaque segment was based on the TOF images. The center points of the lumen contour provided by TOF and T2W images were used to construct the center lines for CCA, internal carotid artery (ICA), and external carotid artery (ECA), shown in Fig. 5.4a. The 2D slices from different MRI series could be aligned along the center lines, with axial coordinates assigned. For the healthy regions upstream and downstream of the plaque, only luminal surface information was available from the 2D TOF scan. A constant wall thickness was assigned for these regions to construct the arterial wall. The value of the thickness was set to be the same as the arterial wall at the exit planes of the bifurcation model.

After 2D segmentation and registration, the region boundary points were imported into SolidWorks™ 2006 for each slice. Complete contours were generated using a B-spline interpolation. Because the in vivo MR-measured plaque was subject to physiological pressure loading, while stress simulation normally starts from zero loading condition, a proper shrinkage procedure should be applied from the geometry obtained from in vivo MRI data to represent physiological loading condition. This had been done during the geometry reconstruction. The shrinkage procedure was: (a) 10% shrinkage in the axial direction from reconstructed geometry based on existing literatures and (b) 10% lumen area shrinkage (about 5% shrinkage in diameter) determined by trial and error to give best match with in vivo geometry under mean pressure loading (100 mmHg). The calculation of lumen area shrinkage rate was as follows: (a) one cross section in CCA was chosen, the arterial wall was reconstructed for 2D structure analysis. (b) The shrinkage rates varied from 0 to 20% were applied to the reconstructed 2D arterial wall. In the shrinkage procedure, the vessel cross-section area was kept the same (incompressible). A pressure of $(p_s+p_d)/2$, where p_s and p_d represent systolic and diastolic pressure, respectively, was applied to the lumen. (c) The deformed arterial wall with different initial lumen shrinkage rates was compared to in vivo morphology, the rate which had the best match with the in vivo shape and size was chosen as the best fit lumen shrinkage rate. In addition, the outer wall was also reduced in proportion to match the arterial wall volume in the in vivo state.

After the shrinkage procedure, 3D reconstruction was carried out by the loft method as shown in Fig. 5.4b. Finally, the complete plaque model contained two parts: (a) lipid core (indicated in Fig. 5.4) and (b) arterial wall region. The arterial wall region between the lumen and lipid core was treated as the fibrous cap, which was part of arterial wall in our models, therefore the fibrous cap was indirectly identified by the relative lipid region location and luminal walls.

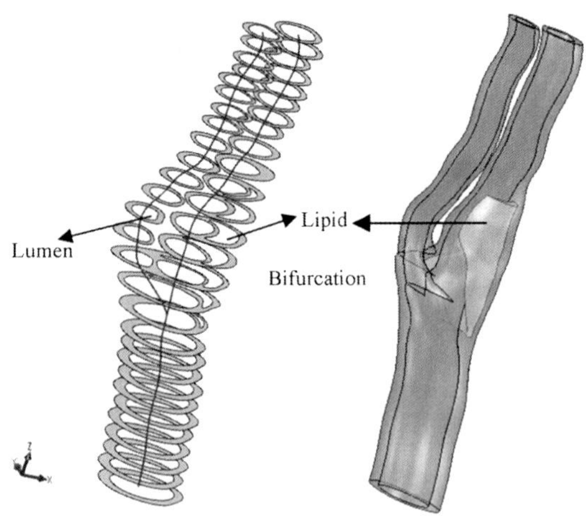

Fig. 5.4 (**a**) 2D contours of subject 1; (**b**) 3D geometry reconstruction of subject 1

5.2.2 FSI Simulation and Boundary Conditions

The carotid arterial wall was assumed to be nonlinear, isotropic, and incompressible, with the five-parameter Monney–Rivilin model [1] being used to describe its material properties. The strain energy function is given by:

$$W = C_{10}(I_1 - 3) + C_{01}(I_2 - 3) + C_{20}(I_1 - 3)^2 + C_{11}(I_1 - 3)(I_2 - 3) \\ + C_{02}(I_2 - 3)^2 + \frac{1}{d}(J - 1)^2. \tag{1}$$

The material constants were $C_{10} = 50.445\,\text{kPa}$, $C_{01} = 30.491\,\text{kPa}$, $C_{20} = 40\,\text{kPa}$, $C_{11} = 120\,\text{kPa}$, $C_{02} = 10\,\text{kPa}$, and d=1.44e^{-7} according to the published experimental results [45]. The lipid core was assumed to be very soft with a 1 kPa Young's modulus and a 0.49 Poisson ratio [8]. The interface between lipid core and arterial wall was treated as a so-called "always bonded" contact; the pure penalty method was employed for the contact connection. Computational nodes at the outlet plane of ICA and ECA were fixed in all directions, and an axial prestretch of 11% based on the shrunk geometry was applied at the inlet plane of CCA for the structure analysis. The structure model was meshed with an unstructured mesh consisting of nearly 100,000 10-node 3D tetra elements.

The fluid domain was meshed in ICEM CFD11.0 with a much finer grid of around one million 3D-tetra cells. Blood was treated as an incompressible, Newtonian fluid with a viscosity of 4×10^{-3} Pa s and a density of 1,067 kg m^{-3}. The flow was assumed to be laminar fluid. Transient simulations were carried out with time-dependent pressure at the inlet of the CCA and mass flow rates at the ICA and ECA. Based on the phase-contrast cine images of the patient, volumetric flow rate could be calculated at the inflow plane of CCA and outflow plane of ICA. The flow rate at ECA was derived from the flow rate difference between CCA and ICA. The cross-sectional area changing curve on the CCA inlet plane during a cycle measured by phase-contrast cine images were used to generate a pressure–time curve in CCA and the pressure value in CCA was rescaled according to the pressure–time curve with a range of 80–110 mmHg. Figure 5.5 shows the boundary conditions for the patient.

The FSI simulation requires the CFD code to pass the values of normal force (pressure) and fluid shear force to the structure analysis solver, while the arterial wall displacements obtained from structure simulation are fed back to the CFD for remeshing. For each time step, data are exchanged in a loop of staggered iterations, which guarantee convergence in both fluid and structure domains. The inner surface of the carotid arterial wall and the corresponding fluid boundary were defined as the fluid–structure interface. There are two ways for FSI realizations: (1) one-way FSI and (2) two-way FSI (fully coupled FSI). The main difference between them is that in one-way FSI, the results are only transferred in one direction, either from CFD to structure analysis or from structure analysis to CFD, while in two-way FSI, the result from both simulation domains will be fed back to the other domain to provide fully coupled simulation. The general FSI procedure set up for plaque stress analysis is shown in Fig. 5.6. In this

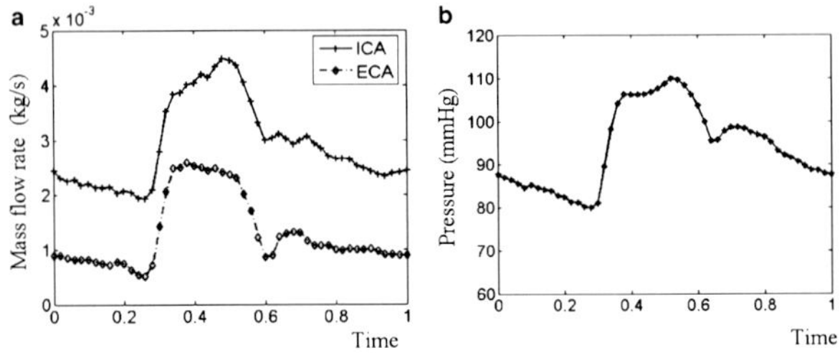

Fig. 5.5 Boundary conditions for blood flow domain. (**a**) Mass flow rate for ICA and ECA; (**b**) pressure profile for CCA

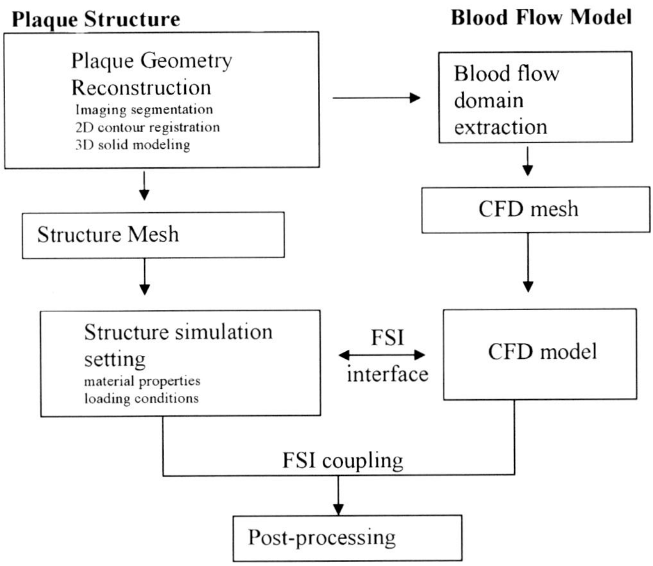

Fig. 5.6 Illustration of a general FSI procedure for carotid arterial plaque stress analysis

study, one-way FSI indicates that the pressure loads are transferred from CFD to the structure analysis to provide realistic loading condition. Mesh density sensitivity analyses were performed on both fluid and solid domains until differences between solutions from two consecutive meshes were negligible (less than 2% of the absolute value). The comparisons of the solutions for this purpose included Von Mises stress at 30 selected points in different regions. Since the simulation started from a time point at mid-diastole with milder changed flow rates, the periodic results could be reached after one cycle.

For postprocessing, parameters had been defined as *fluid flow parameters* and *solid domain parameters*.

Fluid flow parameters: The maximum wall shear stress in the whole cardiac cycle (WSS_*t*max) was used to represent fluid domain results. The definition is:

$$\mathrm{WSS\_}t\max(x,y,z)=\max(|\vec{\tau}_w(x,y,z,t)|),\quad t\subset(0,T),$$

where $\vec{\tau}_w(x,y,z,t)$ is the instantaneous WSS vector, T is the cardiac cycle, t is the time, and $|\vec{\tau}_w(x,y,z,t)|$ is the WSS value at time t.

Solid domain parameters: Since the stress in the arterial wall is a second-order tensor which has six components, the Von Mises stress form, named VWTS, was chosen to represent the wall tensile stress distribution and level in the diseased plaque. In the study, the maximum VWTS in a cycle (VWTS_*t*max), spatial maximum VMTS at a time (VWTS_*s*max), and relative cyclic VWTS (rcVWTS) at each computational node were used to represent the plaque stress distributions. The mathematical definitions are:

$$\mathrm{VWTS}(x,y,z,t)=\sqrt{\frac{(\sigma_1-\sigma_2)^2+(\sigma_2-\sigma_3)^2+(\sigma_3-\sigma_1)^2}{2}}(x,y,z,t),\quad t\subset(0,T),$$

$$\mathrm{VWTS\_}t\max(x,y,z)=\max(\mathrm{VWTS}(x,y,z,t)),\quad t\subset(0,T),$$

$$\mathrm{VWTS\_}s\max(t)=\max(\mathrm{VWTS}(x,y,z,t)),\quad t\subset(0,T),$$

$$\mathrm{rcVWTS}(x,y,z)=\frac{(\mathrm{VWTS\_}t\max(x,y,z)-\mathrm{VWTS\_}t\min(x,y,z))}{\mathrm{VWTS\_}t\mathrm{mean}(x,y,z)},\quad t\subset(0,T),$$

where σ_1, σ_2, and σ_3 are the first, second, and third principle stresses, respectively, for location (x, y, z) at time t, VWTS_*t*max, VWTS_*t*min, and VWTS_*t*mean represent maximum, minimum, and average VWTS, respectively, in a cardiac cycle in a specific region.

5.3 Simulation Result on Different Applications

In this section, the FSI procedure described above was applied for carotid plaque stress analysis, including (1) stress analysis with multiple patients having carotid plaques; (2) stress analysis with patients who suffered TIA recently; and (3) the impacts of fibrous cap thickness and lipid core size on plaque stress distributions.

5.3.1 Stress Analyses with Multiple Patients

Fully coupled FSI simulation was performed on four subjects with varied severity of carotid plaques (S1, S2, S3, and S4). The geometries are shown in Fig. 5.7. Briefly, S1 had low stenosis degree but a large lipid core, indicating a positive remodeling; S2 and S3 had higher degrees of stenosis with a large lipid core, indicating

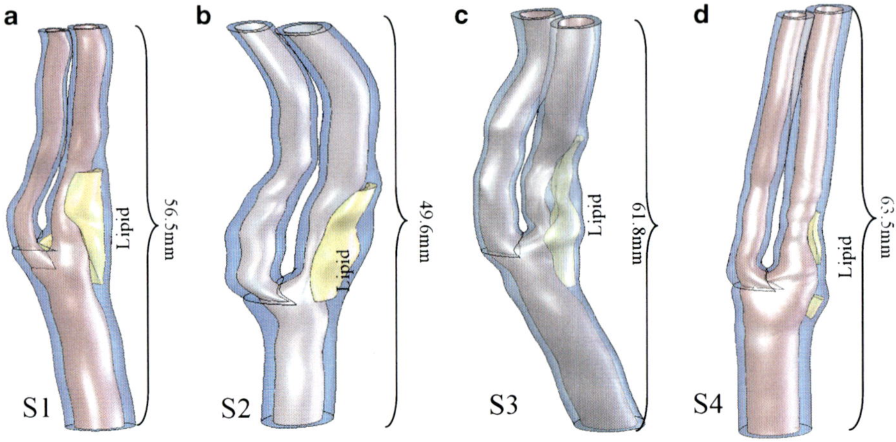

Fig. 5.7 3D plaque geometry reconstructions for four subjects, S1, S2, S3, and S4, respectively

negative remodeling; S4 had low stenosis degree with two small lipid cores. The flow boundary conditions for the four subjects were assumed to be the same, shown in Fig. 5.5.

5.3.1.1 Fluid Domain Results

The pressure and wall shear stress distributions for S1 and S2 at systole phase are shown in Fig. 5.8. The pressure in plaque regions drops slightly from upstream of the plaques. This pressure drop for S1 is around 300 Pa, which becomes higher in S2, about 450 Pa due to the higher stenosis degree. WSS is generally higher in plaque throats. Due to the high degree of stenosis, S2 has the highest intra plaque WSS among all subjects. The distributions of $WSS\_t\max$ for the four subjects are shown in Fig. 5.9. Because of the minor area reduction, WSS values in S1 and S4 are much lower than other cases. A large poststenosis recirculation zone can be found immediately distal to the stenotic region in S2 and S3.

5.3.1.2 Wall Tensile Stress

Figure 5.10a shows the VWTS distribution for S1 in the lumen surface at systole phase. A local high stress region can be found downstream of plaque, right on the fibrous cap, indicated by "a1." The stresses at three locations for a cardiac cycle were extracted shown in Fig. 5.10a indicated by "a1," "a2," and "a3." They are selected as a1 in the location which experiences local maximum VWTS; a2 and a3 locate at the fibrous cap center and upstream boundary, respectively. By examining the whole cardiac cycle, the region downstream of plaque always experiences higher VWTS than other plaque regions for S1, the stress curves have similar trends

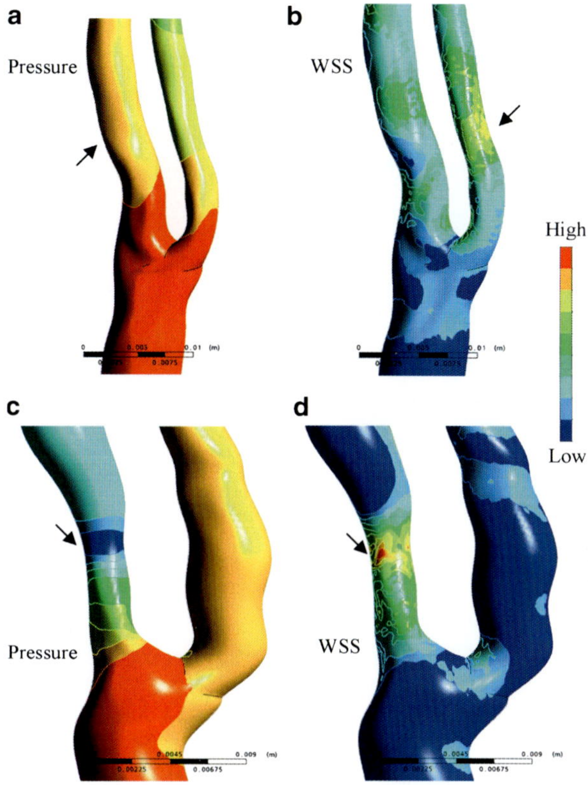

Fig. 5.8 Pressure, WSS distributions for the subjects S1, S2, only showing the stenotic region. Locations of maximum WSS and minimum pressure are indicated by *arrows*. (**a**) Pressure distribution for S1, maximum value: 14.5 kPa; (**b**) WSS distribution for S1, maximum value: 13.5 Pa; (**c**) pressure distribution for S2, maximum value: 14.5 kPa; (**d**) WSS distribution for S2, maximum value: 21.8 Pa

to the imposed pressure boundary condition of pressure–time curve in CCA, as shown in Fig. 5.5. Figure 5.10b shows the maximum principle stress, which has similar distributions to VWTS. Figure 5.10c shows the equivalent strain at the lumen surface, again similar distributions can be found as VWTS distribution.

Figure 5.11 shows detailed stress distributions for the four subjects at peak systole. The general stress distribution contours are presented for the whole plaque with longitudinal cutting plane. VWTS distributions on the two chosen transversal planes in the plaque regions are showed with the local high stress concentrations indicated by the arrows. In general, VWTS is higher at the luminal wall, lower at the arterial outer wall, and lowest in the lipid region. Compared to the other subjects, S4 has more uniform distributions of VWTS along the circumferential direction due to its small lipid core. For the sections with a thin fibrous cap, the stress concentration regions appear at one or both edges of the lipid core (or plaque shoulders).

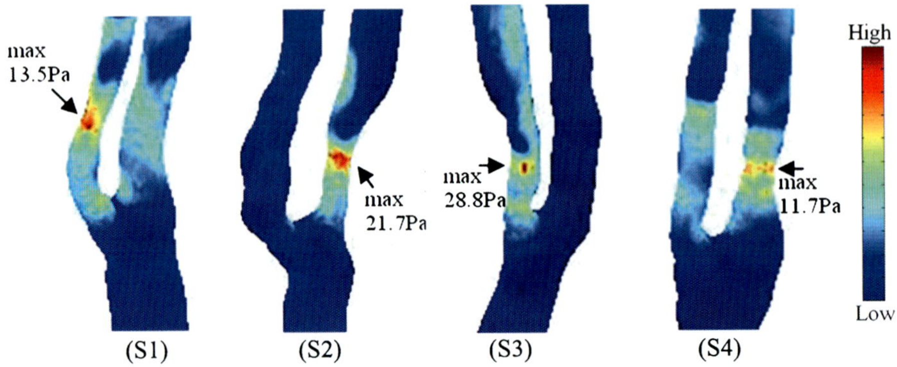

Fig. 5.9 WSS_*t*max distributions for the subjects S1, S2, S3, and S4, only showing the stenotic region. The maximum locations are indicated by *arrows*

5.3.1.3 Wall Tensile Stress in the Fibrous Cap

The fibrous cap thickness was extracted for each subject, defined as the shortest distance between the fibrous cap surface (lumen side) and lipid region, shown in Fig. 5.12. Figure 5.13 provides more details of the distributions of temporal maximum VWTS (VWTS_*t*max in the third column) within a cycle, and relative VWTS variation (rcVWTS in the fourth column) on the fibrous cap at the luminal side during a cycle. The fibrous cap thickness distribution for each plaque is shown in the second column of Fig. 5.13. Although the MR image spatial resolution is 0.39 mm, the 3D surface interpolation during model reconstruction can still produce fibrous cap regions with a thickness less than 0.39 mm. The minimum fibrous cap thickness is 0.18, 0.29, 0.23, and 0.34 mm for S1–S4, respectively.

Generally, a thin fibrous cap is associated with a high VWTS [8]. This trend has been followed in cases of S1, S2, and S3. For example, in the first column of Fig. 5.13, for case S1, four thin fibrous cap regions can be identified as marked by R1, R2, R3, and R4, with the thinnest region at R1. The corresponding stress distributions for S1 are shown in the second column. Four local high stress regions marked by R1, R2, R3, and R4 can be found at areas possessing a thin fibrous cap. A similar trend could be found in S2 and S3, with high stress appearing at upstream left and right side for S2, upstream left and downstream right for S3. They are all related to thin fibrous caps. However, the stress concentration region in S4 seems not to follow the trend very well due to the small lipid cores and large fibrous cap thickness. With regard to the VWTS_*t*max distribution for all cases, stress concentrations are more likely to occur at the boundaries of the fibrous cap, which represent the shoulder regions of the plaque at the transversal plane. The rcVWTS maps in the third column of Fig. 5.13 provide information about the relative VWTS variations during a cycle. It is found that high rcVWTS regions negatively correlate with high VWTS zones in S1, S2, and S3, with high rcVWTS regions close to the middle in the circumferential and longitudinal directions of the fibrous cap.

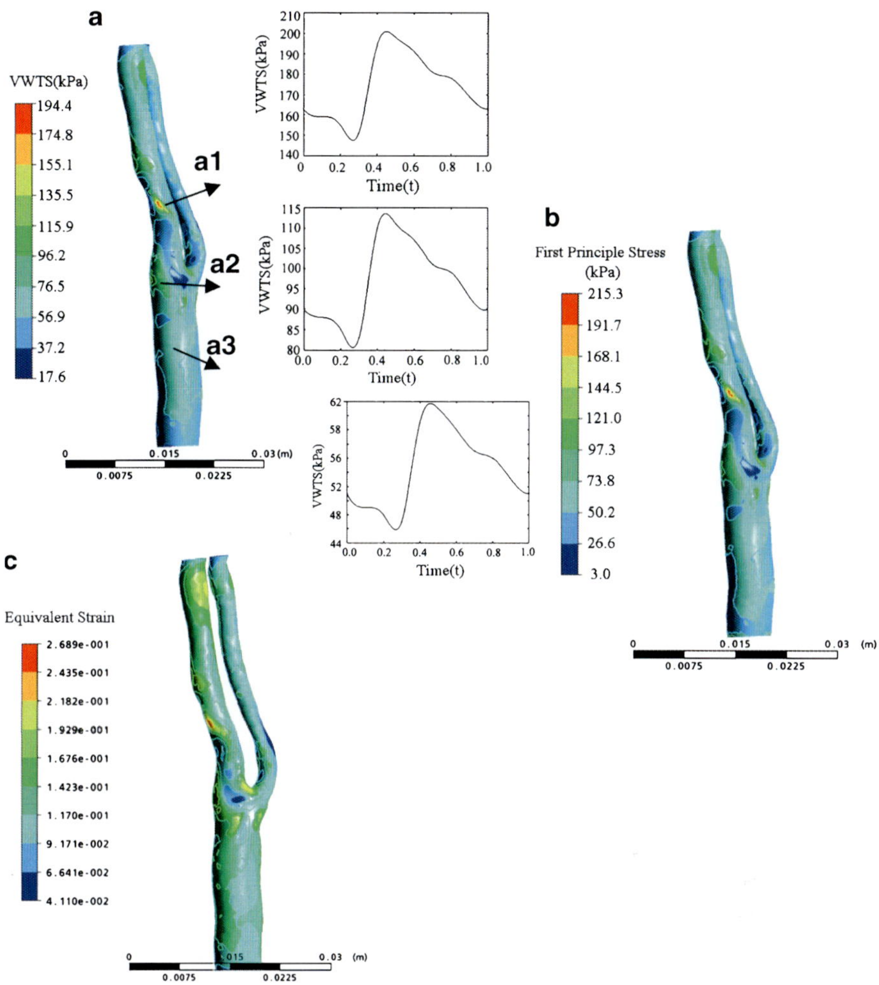

Fig. 5.10 Stress/strain distribution for subject 1 in the lumen surface at systole phase. (**a**) VWTS; (**b**) first principle stress; (**c**) equivalent strain

5.3.1.4 Impact of Plaque Morphology to the Stress Distributions

Figure 5.14 presents several stress and geometry factors for all subjects in the plaque region. They are the spatial maximum/mean value of WSS_*t*max, VWTS_*t*max, and rcVWTS in the defined fibrous cap surface, the minimum fibrous cap thickness (min_FT), and maximum degree of stenosis (max_SD) in the plaque region. From Fig. 5.14, the stress factors derived from WSS are highly correlated with the degree of stenosis; a higher degree of stenosis will produce higher values of maximum WSS and mean WSS. However, the maximum VWTS_*t*max values do not appear to be dependent on the degree of stenosis; they are more

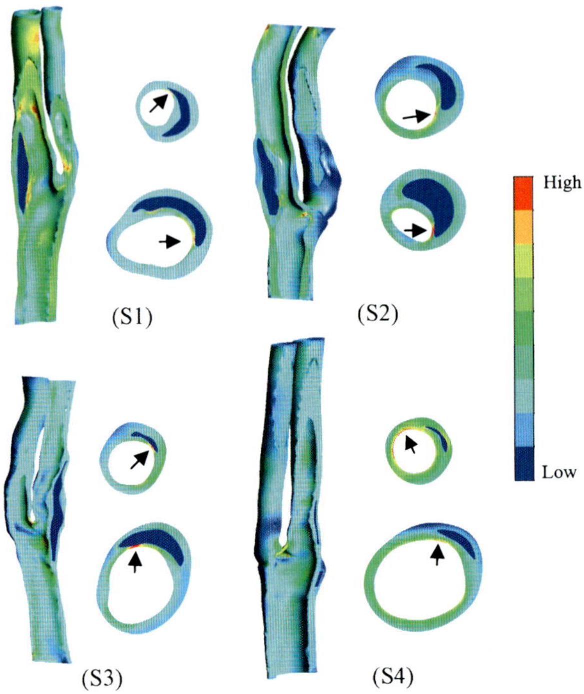

Fig. 5.11 Stress distributions for four subjects at peak systole. It shows general stress distributions in the whole geometry with cutting windows (*longitudinal view*); and VWTS distributions on two transversal sections at plaque regions

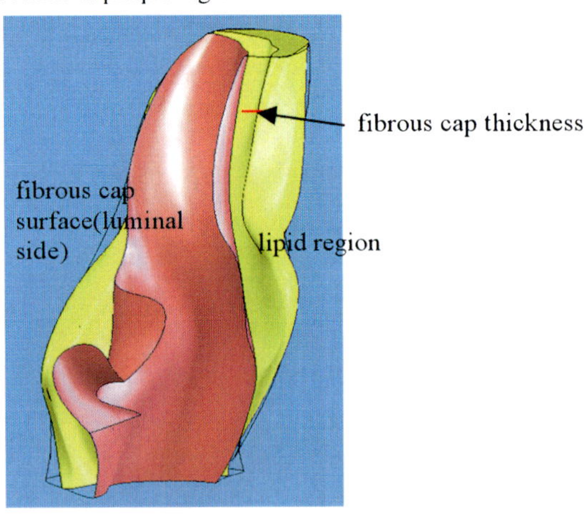

Fig. 5.12 Fibrous cap thickness definition

closely related to the fibrous cap thickness (min_FT). The rcVWTS values are similar in all subjects, with a higher level in S2 and S3. When combining minimum fibrous cap thickness with VWTS factors, S1 possesses both the highest local maximum stress value, and very thin fibrous cap, indicating a higher rupture risk than the

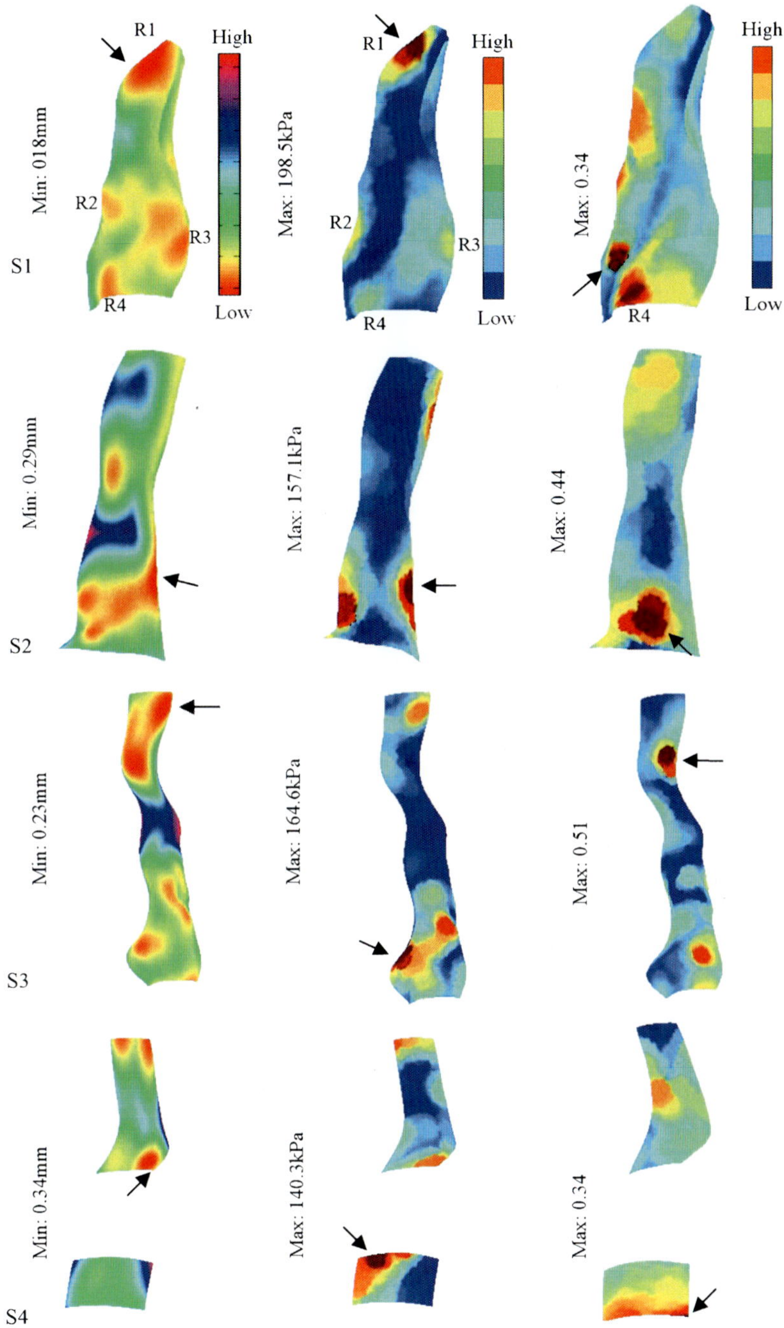

Fig. 5.13 Fibrous cap thickness and stress distribution on the fibrous cap for the four subjects. The *first column* represents the fibrous cap thickness; the *second column* is VWTS_*t*max distribution in the fibrous cap, the maximum VWTS_*t*max locations are indicated by *arrows*; the *third column* is rcVWTS distribution, the maximum rcVWTS locations are indicated by *arrows*

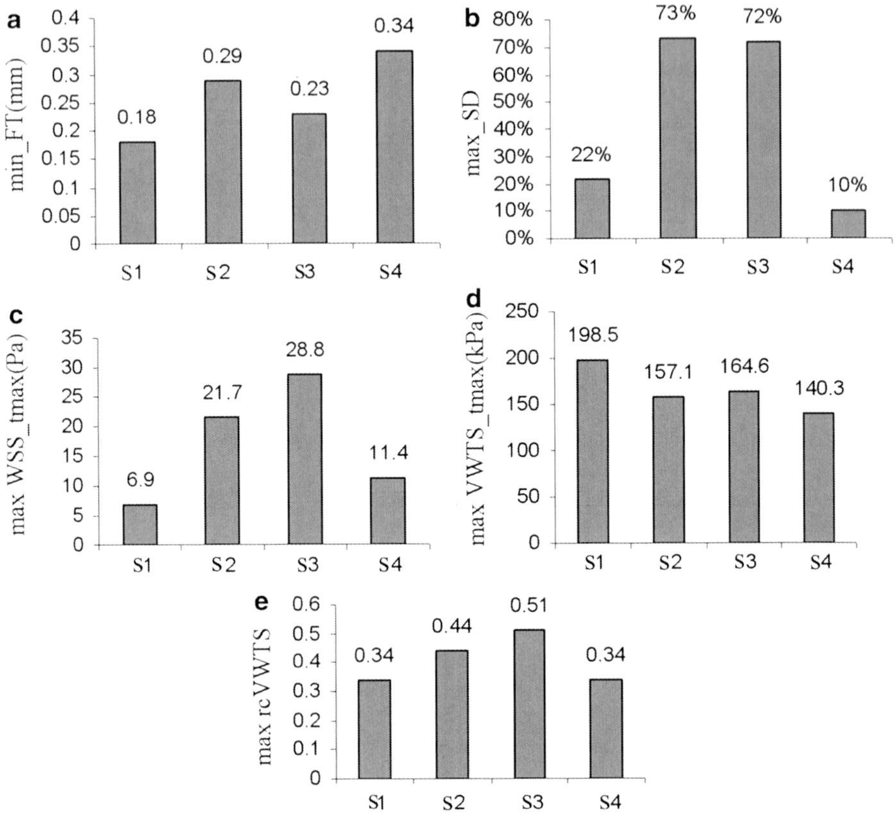

Fig. 5.14 Stress and morphological factors for S1, S2, S3, and S4 with actual values presented on the top of each bar. (**a**) Minimum fibrous cap thickness (min_FT); (**b**) maximum degree of stenosis (max_SD); (**c**) temporal maximum wall shear stress (WSS_*t*max); (**d**) maximum value of temporal maximum wall tensile stress over one cardiac cycle (max VWTS_*t*max), (**e**) maximum relative cyclic wall tensile stress (max rcVWTS)

other subjects. In summary, the plaque stability order from high risk to low risk could be S1, S3, S2, and S4 based on the FSI stress analysis.

5.3.2 *Stress Analysis with TIA Patients*

It is believed that both plaque morphology and the biomechanical environment of the plaque influence plaque vulnerability. However, the direct evidence between stress and plaque rupture is still not available. If plaque geometry is available at both pre- and postrupture status, it would be possible to provide some proofs of the plaque rupture hypothesis. However, it is generally difficult to obtain plaque morphology before a rupture. MRI scan of the patient who suffered TIA recently can be performed which will provide the critical plaque morphology information of the rupture sites. If the preruptured plaque geometry can be reconstructed with some assumptions, combined with the stress analysis, it will provide a possible way to study the plaque rupture with stress distributions.

Three patients, who suffered from a recent TIA, were recruited from a specialist neurovascular clinic, and underwent carotid MRI within 72 h. The study procedure was reviewed and approved by the regional research ethics committee. All patients gave written informed consent. During plaque geometry reconstruction, special attention was paid to the segmentations of lipid core and fibrous cap. The ruptured fibrous cap was deemed present if there was a clear defect, discontinuity, or ulceration within the fibrous cap [16, 47]. To reconstruct preruptured plaque geometry, a constant thickness of 0.39 mm (one pixel) was assumed for the fibrous cap at rupture region where the fibrous cap signal was missing in images. Geometries are shown in Fig. 5.15 (named TIA1, TIA2, and TIA3). One-way FSI was applied to the stress analysis to reduce the simulation time.

First-principle stress (FPS), which represents the highest stretching stress in the wall, was used here to represent the wall stress distribution for the three subjects at the systolic cardiac phase. Figure 5.16 shows the stress results for all three subjects. Figure 5.16 (1a, 2a, 3a) shows FPS distributions with a longitudinal cutting plane. Figure 5.16 (1b, 2b, 3c) presents FPS distributions on different cross sections from CCA to ICA, covering the whole plaque region. FPS concentration regions could be found in the fibrous cap, especially when the fibrous cap is thin, indicated by arrows. Figure 5.16 (1c) shows FPS stress distributions in the lumen surface for TIA1, the local maximum stress regions can also be found in the plaque region, indicated by the arrow, which is at the plaque rupture region (indicated in Fig. 5.15). The similar results can be found in TIA2 and TIA3, the local maximum stress concentrations are found around the rupture sites.

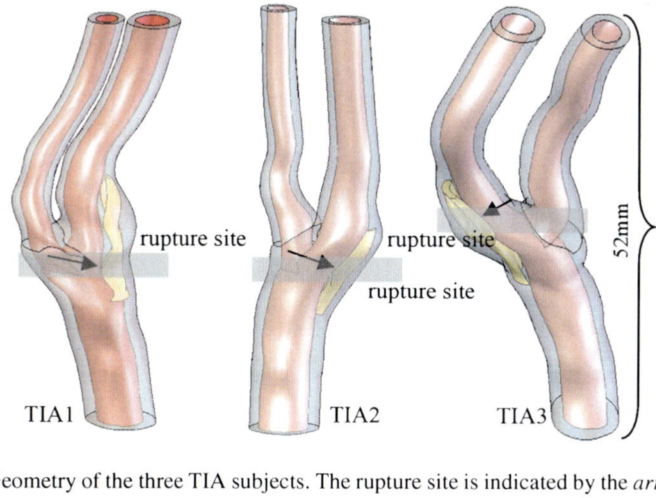

Fig. 5.15 Geometry of the three TIA subjects. The rupture site is indicated by the *arrow with gray band*. The total model length is 52 mm for every subject

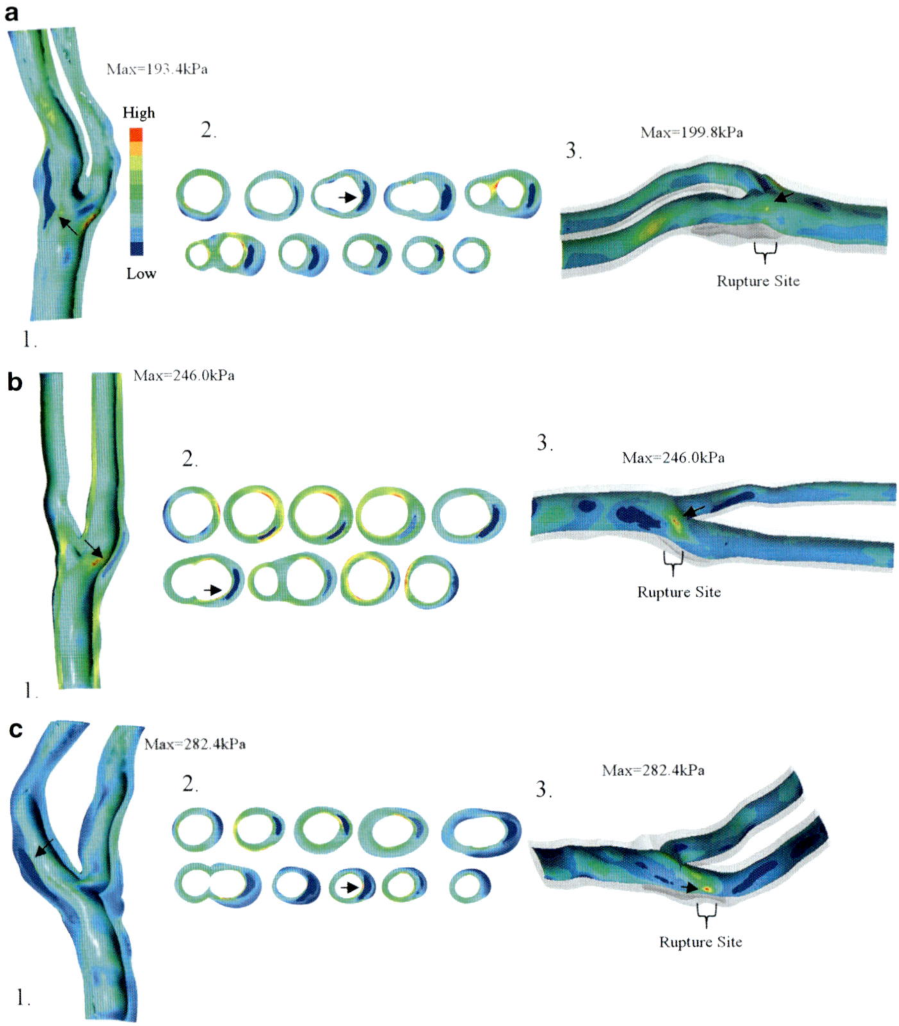

Fig. 5.16 General stress distribution at a systole phase for the three TIA cases. (a) TIA1; (b) TIA2; (c) TIA3

5.3.3 Effects of Lipid Core Volume and Fibrous Cap Thickness on Stress Distribution

Plaque structure features and components have been considered to be important factors in plaque vulnerability. It is known that fibrous cap thickness and lipid core have considerable effects on predicted stress distributions and levels. With multipatient-specific plaque models, stress analysis could provide certain information regarding plaque rupture risk. However, it is not easy to change one plaque feature, and keep others the same to study individual parameter in plaque stress analysis on realistic geometry. Furthermore, the parameter studies on lipid core size and fibrous cap thickness are mainly based on 2D stress analysis so far. Therefore, a realistic multicomponent arterial plaque geometry (shown in Fig. 5.17a) was chosen as a baseline plaque model. Based on this, a total of 13 carotid bifurcation models were manipulated (shown in Fig. 5.17b) with varied combinations of fibrous cap thickness and lipid core volume but the same degree of stenosis. The detailed description of the rule on generating the simulation model geometry is shown in Fig. 5.17. The summary of the studied cases can be found in Table 5.2. One-way FSI simulations were performed on the cases to investigate the impacts of the specific combination of fibrous cap thickness and lipid core volume to the stress distribution.

Figure 5.18a presents the $VWTS\_s\max$ curves in one cardiac cycle for cases 1, 2, 3, and 4, which have the same minimum fibrous cap thickness of 0.15 mm. The $VWTS\_s\max$ value is generally higher than 300 kPa for cases 1, 2, 3 in peak systole. For case 4, the $VWTS\_s\max$ value is close to 300 kPa. Among the cases, the $VWTS\_s\max$ value decreases gradually for the cases with smaller lipid core. Figure 5.18b presents the $VWTS\_s\max$ distribution for cases 2, 5, 8, and 11, which have the same lipid core volume. $VWTS\_s\max$ decreases dramatically with the increased fibrous cap thickness. The $VWTS\_s\max$ value only exceeds 300 kPa in case 2. The $VWTS\_s\max$ curve and values are almost the same for cases 8 and 11 with the peak value below 200 kPa.

By examining $VWTS\_s\max$ values at the fibrous cap region for all 13 cases in a cycle, it can be found that the $VWTS\_s\max$ value is much more sensitive to the

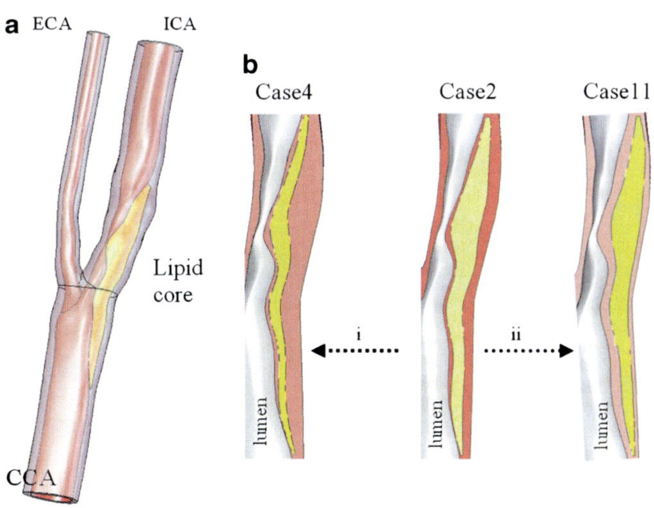

Fig. 5.17 Diagrams of arterial model reconstruction and manipulation of simulation cases. (**a**) The reconstructed geometry for case 2 (base model). (**b**) Schematic reconstruction of simulation cases: (i) fibrous cap thickness remains constant and lipid core volume is reduced proportionally, from case 2 to case 4; (ii) lipid core volume remains constant, and the fibrous cap thickness is increased proportionally, from case 2 to case 11

Table 5.2 Descriptions of the plaque geometry of the simulation cases constructed in the study

	Fibrous cap thickness (mm)			
Lipid core size (%)	0.15	0.35	0.58	0.64
26.46	Case 1			
22.4	Case 2	Case 5	Case 8	Case 11
18.15	Case 3	Case 6	Case 9	Case 12
12.23	Case 4	Case 7	Case 10	Case 13

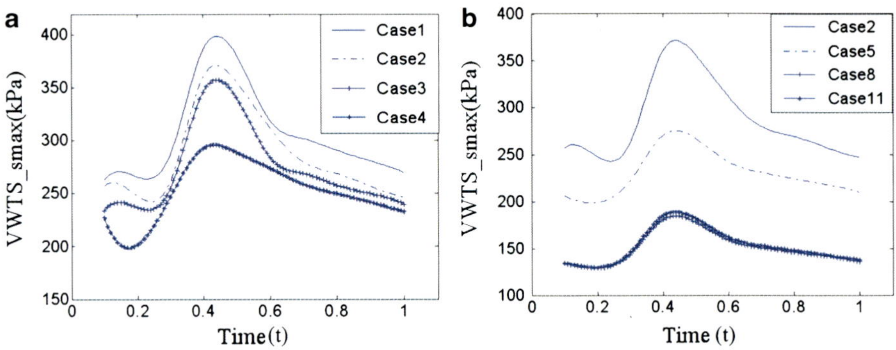

Fig. 5.18 VWTS_smax distributions in one cycle in the selected cases. (**a**) Cases have the same fibrous cap thickness of 0.15 mm; (**b**) cases have the same lipid core size with volume ratio of 22.4%

changes in fibrous cap thickness than in lipid core volume. For models with the same fibrous cap thickness, the maximum stress is nearly in the same range. However, for the same lipid core size, the stress changes greatly from thinner fibrous cap to a thicker one. If 300 kPa can be used as a critical rupture stress value [7], the simulation cases could be separated and classified as vulnerable plaques and stable plaques.

5.4 Modeling Procedure Uncertainties Analysis

5.4.1 Geometry Reconstruction Reproducibility

Because the uncertainties in MR imaging and 3D reconstruction can significantly influence the following stress analysis results, it is necessary to assess the uncertainty of 3D plaque geometry reconstruction. Three investigators (IV1, IV2, IV3) were invited to segment the MRI images to define the lumen wall, arterial outer wall, and lipid core boundary independently following the above protocols on the data for case S1 from Sect. 5.2.1.2.1 by using the same in-house program. Three parameters were defined to quantify the interoperator segmentation results. They were (a) arterial wall thickness, which is the distance between lumen and arterial

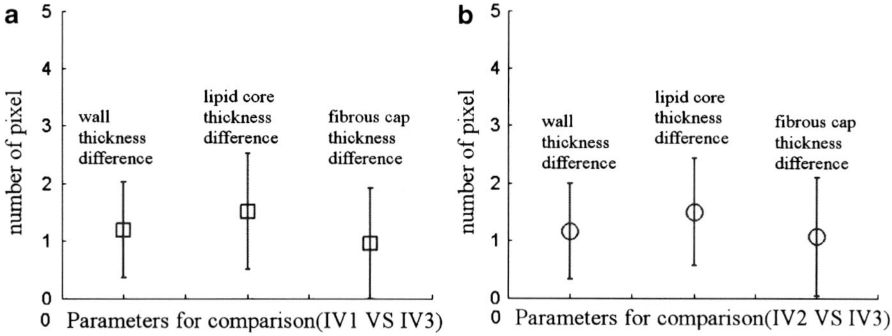

Fig. 5.19 Plot of means ± standard deviation of interobserver disagreement for wall thickness, lipid core thickness, and fibrous cap thickness on subject 1, (**a**) parameters for comparison between IV1 and IV3; (**b**) parameters for comparison between IV2 and IV3

outer wall; (b) lipid core thickness or the distance between the lipid core boundary along the same radius direction; and (c) fibrous cap thickness.

The results show that estimation of wall thickness is highly consistent between the different investigators (linear correlation coefficients >0.9 with $P<0.05$), however, correlations of lipid core size and fibrous cap thickness were much poorer with correlation coefficients of about 0.5 ($P<0.05$). Figure 5.19 shows the differences in calculated parameters between IV1 and IV3, and IV2 and IV3, of all images in the plaque region of case S1. From these analyses, interoperator reproducibility for the arterial wall reconstruction based on MR images was high, the disagreement among the operators was around one pixel. The reproducibility was lower for the segmentation of lipid core and fibrous cap. Therefore, much attention and effort need to be paid to the accurate segmentation of lipid core and fibrous cap structure.

5.4.2 *Variation of Material Model Definition*

From Williamson's study [44], the stress analysis may be used with confidence because the uncertainties in material properties cause relatively small errors in the stress prediction. In this chapter, two plaque components have been considered in the plaque region: arterial wall (including fibrous cap) and lipid region. Based on the baseline model of S1, material properties for arterial wall and lipid region were changed from −100% softer to 100% stiffer (one material was changed, while the other one was kept the same as the baseline model). Static simulations were performed with different material properties. Local maximum VWTS at fibrous cap increases by about 8% with a 100% increase in vessel wall stiffness, and decreased by about 5% with a 100% decrease in vessel wall stiffness. For the variation of lipid core stiffness, it led to less than 10% changes for local maximum VWTS at fibrous cap. The results had confirmed the finding from Williamson's study [44]. This does not mean that patient-specific material models are not important in plaque stress analysis. Comparing the plaque morphology between one plaque and another, the material property variation may play a less significant role in determining the plaque stress level.

5.4.3 Axial Stretch

It has been well accepted that the arteries experience significant axial stretch in vivo. Due to lack of methods in measuring the axial stretch in vivo, the data regarding axial stretch in human carotids are scant. In order to quantify the effects caused by axial stretch in the study, simulations with different axial stretch were carried out on S1 to investigate their influence. The simulations were performed with different axial stretch imposed in the CCA plane with 0, 10, 20, and 30%. Maximum VWTS in the whole geometry and plaque region was summarized in Table 5.3. The results indicate that the axial prestretch could have great influence on stress distribution in plaque region. A large axial prestretch can cause an extreme stress condition at the healthy part of artery, and the maximum stress values are also sensitive to the varied axial prestretch rate.

5.4.4 Residual Stress

A biological tissue does not become stress free when all external loads are removed. Previous studies have shown that such residual stress/strain tends to maintain the integrity of vessel structure and make the stress distribution more uniform throughout the arterial wall. However, the effects of residual stress on plaque stability have not been well studied. Ohayon's study [28] showed the residual stress/strain in plaques is not negligible and may dramatically affect the physiological peak stress in a thin fibrous cap. The study from our group showed that the circumferential residual stress has limited effects on the actual plaque stress distribution under normal physiological loadings [29]. Those controversial results indicate that more efforts need to be made to understand the effect of residual stress and strain on plaque stability.

Table 5.3 The comparison of stress results with different axial prestretch

Simulations	Axial stretch (%)	Global Max VWTS (kPa)	Higher than Sim1	Local Max VWTS in Plaque (kPa)	Higher than Sim1 (%)
Simulation 1	0	167	–	167	–
Simulation 2	10	195	17%	195	17
Simulation 3	20	264	58%	242	45
Simulation 4	30	660	295%	484	189

5.5 Discussions and Conclusion

Recent advances in MRI technology have greatly enhanced plaque morphology classification and delineation with high resolution. With anatomically realistic plaque geometry, FSI simulation is able to provide detailed stress analysis for plaques, which is believed to be helpful for plaque rupture risk assessment.

For MRI-based plaque stress analysis, 3D in vivo image resolution and accurate plaque components reconstruction are still big concerns for reliable stress prediction. In the study, MRI special resolution is 0.39 mm, which may be enough for presentation purpose, but it is not adequate to build a stress analysis model, especially when defining a very thin fibrous cap. Studies in our group showed that the uncertainties of the geometry reconstruction on different plaque components have varied impacts on the stress analysis [11]. The overestimation of a near luminal wall lipid region can induce higher stress concentrations and generally higher stress levels in the plaque regions. The impacts on the stress distribution in the plaque region caused by over/underestimation of healthy arterial wall are much less and insignificant compared with the variations caused by the over/underestimation of the lipid region. The increased sensitivity of lipid region segmentation accuracy to the stress analysis would impose certain difficulties when interpreting the plaque stress analysis results. Recently, high-resolution, contrast-enhanced MRI has been used for the analysis of the atherosclerotic plaque. This can provide more information regarding the components of the plaque [48] and could be useful for the precise reconstruction of plaque geometry and, finally, a more reliable stress prediction.

WSS was believed to play an important role in the plaque initiation and development [13]. Recently Groen et al. [15] proposed that the high WSS in the fibrous cap could rupture the vulnerable plaque. WSS is normally high in the plaque region, which may contribute to the instability of the thin fibrous cap, especially in the long term. From earlier studies [2], WSS with a level above 40 Pa is able to cause structure change or even damage in the endothelial cells in the lumen side. In our study with FSI simulations, WSS is usually less than the critical value (40 Pa). It was proposed that high WSS could induce thinning of the fibrous cap and destabilizing plaques [15]. Up until now, the direct link between plaque rupture and high wall shear stress has not been established.

Generally, the extremely high stress value occurs at the thinnest fibrous cap thickness among the studied subjects. When looking at individual plaques for S1, S2, S3, and S4, stress concentration regions are confined to the thinner fibrous cap regions, with high stress values located at the shoulder regions. These are the major rupture risk regions responsible for 60% of plaque ruptures according to Shah [33]. On the other hand, during the long development period of a plaque, with constant remodeling, a constantly high stress region in the fibrous cap may be associated with a strengthened collagen structure. As a result, rupture critical stress would be increased in the region. It may partially explain the fact that not all plaque ruptures occur at the shoulder regions. The pulsating nature of arterial pressure and the consequent cyclic arterial wall stress/strain not only modulate the cellular function of endothelial cells [18], but also cause fatigue damage in the fibrous cap [42]. The introduction of rcVWTS in the study aims to assess the VWTS oscillation level on the fibrous cap. A higher rcVWTS represents a high variation of VWTS in a cycle, which indicates a high probability of fatigue damage in the region. The results show that high rcVWTS regions occur in the central region of the fibrous cap, rather than the thinnest region as in S3, which agrees well with Lovett's result in which 16.9% of ruptures occur in the middle of the plaque in symptomatic patients [23]. However, histological evidence of plaque fatigue is needed to support correlation of the rcVWTS value with rupture probability.

Based on the stress analysis, plaque rupture risk assessments among the four subjects (S1, S2, S3, and S4) were made. The results show that higher stress (indicating a high rupture risk) is associated with a thinner fibrous cap and a large lipid region. In terms of morphology features, the lipid volume size from large to small for the four subjects is S1, S3, S2, S4 and the fibrous cap thickness from thin to thick is S1, S3, S2, S4, while the degree of stenosis from high to low is S2, S3, S4, S1. The predicted stress level on the fibrous cap from high to low is S1, S3, S2, S4, which followed the similar trend from morphological analysis. It also indicates that varied morphological features of the plaque components and the induced stress factors could provide more information of the rupture risk, rather than the degree of stenosis [25]. A higher degree stenosis usually causes a more disturbed flow around the plaque, a greater pressure drop through the stenosis, and a higher wall shear stress in throats. All of those effects may contribute to the plaque instability. However, the stress level in S1 is higher than S2 and S3, which had higher stenosis degrees than the one in S1. Therefore, the rupture risk could be underestimated if only based on the stenosis degree. It must be stated here that the rupture risk assessment cannot be concluded with only local stress data. Rupture of a fibrous cap is a very complex process [6] which is promoted by not only the mechanical factors but also other abnormalities in tissue and cells such as elevated inflammatory activity, degraded collagen structures, and so on [30]. Comprehensive plaque vulnerability assessment should involve a combination of systemic markers, morphological features, and biomechanical factors.

Challenges. The main challenges of obtaining a realistic plaque stress analysis are (a) noninvasive methods for accurately determining plaque components; (b) material properties; (c) pathological information regarding inflammation and fibrous cap erosion; and (d) realistic boundary conditions for stress analysis. The development of automatic or less labor-intensive plaque geometry reconstruction methods are also necessary if the procedure is going to be used in routine clinical practice. Multiscale stress analysis from cellular level to the tissue level may be another important area in term of simulation technical development in order to provide more insights into stress-induced plaque rupture.

5.6 Conclusion

Developments in high-resolution multispectral MRI have allowed plaque components to be visualized in vivo, providing more realistic plaque geometries for stress analysis. The progress in numerical modeling in stress analysis has made fluid structure interaction analysis with patient-specific plaque models possible. Extreme stress distributions in the plaque region can be predicted, which may be used for plaque rupture risk assessment. The combination of plaque MR imaging analysis, computational modeling, and clinical study/validation would advance our understandings of plaque rupture and establish new procedures for patient diagnose, management, and treatment.

Acknowledgment This project is supported by the British Heart Foundation (FS/06/048). The authors like to thank Dr. ZY Li, Dr. M Graves, MD JH Gillard from Department of Radiology, Cambridge University, for their contributions to all MR images, and collaborations in the project.

References

1. ANSYS™ 11.0 Documentation.
2. Aarts P, Steendijk P, Sizma JJ, Heethaar RM(1986) Fluid shear as a possible mechanism for platelet diffusivity in flowing blood. Journal of Biomechanics, 19(10):799–805.
3. Bank AJ, Versluis A, Dodge SM, Douglas WH(2000) Atherosclerotic plaque rupture: A fatigue process? Medical Hypotheses. 55(6):480–484.
4. Barger AC, Beeuwkee R, Lainey LL, Silverman KJ(1984) Hypothesis: vasa vasorum and neovascularization of coronary arteries. A possible role in the pathophyiology of atherosclerosis. The New England Journal of Medicine. 310:175–177.
5. Cai, JM, Hatsukami TS, Ferguson MS, Small R, Polissar NL, Yuan C(2002) Classification of human carotid atherosclerotic lesions with in vivo multicontrast magnetic resonance imaging. Circulation. 106:1368–1373.
6. Casscells W, Naghavi M, Willerson JT(2003) Vulnerable atherosclerotic plaque: A multifocal disease. Circulation. 107 (16):2072–2075.
7. Cheng GC, Loree HM, Kamm RD, Fishbein MC, Lee RT(1993) Distribution of circumferential stress in ruptured and stable atherosclerotic lesions. A structural analysis with histopathological correlation. Circulaiton. 87(4):1179–1187.
8. Finet G, Ohayon J, Rioufol G(2004) Biomechanical interaction between cap thickness, Lipid Core composition and blood pressure in vulnerable coronary plaque: Impact on stability or instability. Coronary Artery Disease. 15(1):13–20.
9. Goshtasby AA(2005) 2D and 3D image registration for medical, remote sensing and industrial applications. Wiley Interscience, pp107–140.
10. Gao H, Long Q(2008) Effects of varied lipid core volume and fibrous cap thickness on stress distribution in carotid arterial plaques by 3D one-way FSI simulations. Journal of Biomechanics. 41(14):3053–3059.
11. Gao H, Long Q, Graves M, Gillard JH, Li ZY(2009a) Study of reproducibility of human arterial plaque reconstruction and its effects on stress analysis based on multispectral in vivo magnetic resonance imaging. Journal of Magnetic resonance imaging. 30(1):85–93.
12. Gao H, Long Q, Graves M, Gillard JH, Li ZY(2009b) Carotid arterial plaque stress analysis using fluid-structure interactive simulation based on in-vivo magnetic resonance images of four patients. Journal of Biomechanics. 42:1416–1423.
13. Gibson CM, Diaz L, Kandarpa K, Sacks FM, Pasternak RC, Sandor T, Feldman CL, Stone PH(1993) Relation of vessel wall shear stress to atherosclerosis progression in human coronary arteries. Arteriosclerosis and Thrombosis. 13(2):310–315.
14. Groen HC, Gijsen FJH, Van Der Lugt A, Ferguson MS, Hatsukami TS, Van Der Steen AFW, Yuan C, Wentzel JJ(2007) Plaque rupture in the carotid artery is localized at the high shear stress region: A case report. Stroke. 38(8): 2379–2381.

15. Groen HC, Gijsen FJH, Van Der Lugt A, Ferguson MS, Hatsukami TS, Van Der Steen AFW, Yuan C, Wentzel JJ(2008) High shear stress influences plaque vulnerability. Netherlands Heart Journal. 16(7–8):280–283.
16. Hatsukami TS, Ross R, Polissar NL, Yuan C(2000) Visualization of Fibrous cap thickness and rupture in human atherosclerotic carotid plaque in vivo with high-resolution magnetic resonance imaging. Circulation. 102:959–964.
17. Holzapfel GA, Stadler M, Schulze-Bauer CA(2002) A layer-specific three dimensional model for the simulation of balloon angioplasty using magnetic resonance imaging and mechanical testing. Annals of Biomedical Engineering. 30(6):753–767.
18. Kaissis JD, Liais CD, Sumpio BE(2004) Effects of cyclic strain on vascular cells. Endothelium. 11:17–28.
19. Lee RT, Loree HM, Cheng GC, Lieberman EH, Jaramillo N, Schoen FJ(1993) Computational structure analysis based on intravascular ultrasound imaging before in vitro angioplasty: prediction of plaque fracture locations. Journal of American College of Cardiology. 21(3):777–782.
20. Li ZY, Howarth SPS, Tang T, Graves MJ, U-King-Im J, Trivedi RA, Kirkpatrick PJ, Gillard JH(2007) Structural analysis and magnetic resonance imaging predict plaque vulnerability: A study comparing symptomatic and asymptomatic individuals. Journal of Vascular Surgery. 45(4):768–775.
21. Loree HM, Kamm RD, Atkinson CM, Lee RT(1991) Turbulent pressure fluctuations on surface of model vascular stenosis. American Journal of Physiology. Heart and Circulatory Physiology. 261:H644–H650.
22. Loree HM, Kamm RD, Stringfellow RG, Lee RT(1992) Effects of fibrous cap thickness on peak circumferential stress in model atherosclerotic vessels. Circulation Research. 71(4): 850–858.
23. Lovett JK, Rothwell PM(2003) Site of carotid plaque ulceration in relation to direction of blood flow: An angiographic and pathological study. Cerebrovascular Diseases. 16:369–375.
24. Naghavi M, Libby P, et al.(2003) From vulnerable plaque to vulnerable patient: a call for new definition and risk assessment strategies: Part I. Circulation. 108:1664–1672.
25. Nighoghossian N, Derex L, Douek P(2005) The Vulnerable Carotid Artery Plaque : Current Imaging Methods and New Perspectives. Stroke. 36:2764–2772.
26. Ohayon J, Teppaz P, Finet G, Rioufol G(2001) In-vivo prediction of human coronary plaque rupture locations using intravascular ultrasound and the finite element method. Coronary Artery Disease. 12(8):655–663.
27. Ohayon J, Finet G, Treyve F, Rioufol G, Dubreuil O(2005) A three dimensional finite element analysis of stress distribution in a coronary atherosclerotic plaque : in vivo prediction of plaque rupture location. Biomechanics Applied to Computer Assisted Surgery. 37:225–241.
28. Ohayon J, Dubreuil O, Tracqui P, Le Floch S, et al.(2007) Influence of residual stress/strain on the biomechanical stability of vulnerable coronary plaques: Poetntial impact for evaluating the risk of plaque rupture. American Journal of Physiology-Heart and Circulatory Physiology. 293(3):H1987–H1996.
29. Pocaterra M, Gao H, Das S, Pinelli M, Long Q(2009) Circumferential residual stress distribution and its influence in a diseased carotid artery. Proceedings of 2009 ASME summer bioengineering conference, Lake Tahoe, CA, June 17–21.
30. Richardson PD(2002). Biomechanics of plaque rupture: Progress, problems, and new frontiers. Annals of Biomedical Engineering. 30: 524–536.
31. Rosamond W, Flegal K, et al.(2007) Heart disease and stroke statistics – 2007 update, Circulation.115:69–171.
32. Saam T, Ferguson MS, Yarnykh VL, Takaya N, Xu D, Polissar NL, Hatsukami TS, Yuan C.(2005) Quantitative evaluation of carotid plaque composition by in vivo MRI. *Arteriosclerosis, Thrombosis, and Vascular Biology*. 25:234–239.
33. Shah PK(2003) Mechanisms of plaque vulnerability and rupture. Journal of the American College of Cardiology. 41(4):S15–S22.
34. Tang D, Yang C, Zheng J, Woodard PK, Sicard GA, Saffitz JE, Kobayashi S, Yuan C(2004a) 3D computational mechanical analysis for human atherosclerotic plaques using MRI-

based models with fluid-structure interactions. Lecture Notes in Computer Science. 3217 (1 PART 2):328–336.
35. Tang D, Yang C, Zheng J, Woodard PK, Sicard GA, Saffitz JE, Yuan C(2004b) 3D mri-based multicomponent fis models for atherosclerotic plaques. Annals of Biomedical Engineering. 32(7):947–960.
36. Tang D, Yang C, Zheng J, Woodard PK, Saffitz JE, Peteruccelli JD, Sicard GA, Yuan C(2005a) Local maximal stress hypothesis and computational plaque vulnerability index for atherosclerotic plaque assessment. Annals of Biomedical Engineering. 33(12):1789–1801.
37. Tang D, Yang C, Zheng J, Woodard PK, Saffitz JE, Sicard GA, Pilgram TK, Yuan C(2005b) Quantifying effects of plaque structure and material properties on stress distributions in human atherosclerotic plaques using 3D FSI models. Journal of Biomechanical Engineering. 127(7):1185–1194.
38. Tang D, Teng Z, Canton G, Hatsukami TS, Dong L, Huang X, Yuan C(2009) Local critical stress correlates better than global maximum stress with plaque morphological features linked to atherosclerotic plaque vulnerability: An in vivo multi-patient study. Biomedical Engineering Online. 8:15.
39. Toussaint J, Southern JF, Fuster V, et al.(1995) T2-Weighted contrast for NMR charaxterization of human atherosclerosis. Arteriosclerosis, Thrombosis, and Vascular Biology. 15:1533–1542.
40. Touze E, Toussaint J-F, Coste J, Schmitt E, Bonneville F, Vandermarcq P, Louis J; Oppenheim C(2007). Reproducibility of high-resolution MRI for the identification and the quantification of carotid atherosclerotic plaque components. Stroke. 38:1812–1819.
41. Trivedi RA, Mallawarachi C, U-king-Im JM, et al.(2006) Identifying Inflamed carotid plaques using in Vivo USPIO-enhanced MR imaging to label plaque macrophages. Arteriosclerosis, Thrombosis, and Vascular Biology. 26:1601–1606.
42. Versluis A, Bank AJ, Douglas WH(2006) Fatigue and plaque rupture in myocardial infarction. Journal of the Biomechanics. 39:339–347.
43. Vengrenyuk Y, Carlier S, Xanthos S, Cardoso L, Ganatos P, Virmnani R, Einav S, Gilchrist L, Weinbaum S(2006) A hypothesis for vulnerable plaque rupture due to stress-induced debonding around cellular microcalcificaitons in thin fibrous cap. Proceedings of the National Academy of Sciences of the United States of America. 103(40):14678–14683.
44. Williamson SD, Lam Y, Younis HF, Huang H, Patel S, Kaazempur-Mofrad MR, Kamm RD(2003) On the sensitivity of wall stresses in diseased arteries to variable material properties. Journal of Biomechanical Engineering. 125(1):147–155.
45. Yang C, Tang D, Yuan C, Hatsukami TS, Zheng J, Woodard PK(2007) In vivo/ex vivo MRI-based 3D non-Newtonian FSI models for human atherosclerotic plaques compared with fluid/wall-only models. Computer Modeling in Engineering and Sciences. 19(3):233–245.
46. Yuan C, Kerwin WS, Ferguson MS, Polissar N, Zhang S, Cai J, Hatsukami TS(2002a) Contrast-enhanced high resolution MRI for atherosclerotic carotid artery tissue charactrization. Journal of magnetic Resonance Imaging. 15:62–67.
47. Yuan C, Zhang SX, Polissar NL, Echelard D, et al.(2002b) Identification of fibrous cap rupture with magnetic resonance imaging is highly associated with recent transient ischemic attack or stroke. Circulation. 105:181–185.
48. Yuan C, Kerwin WS, Ferguson MS, Polissar N, Zhang S, Cai J, Hatsukami TS(2005) Contrast-enhanced high resolution MRi for atherosclerotic carotid artery tissue characterization. Journal of Magnetic Resonance Imaging. 15:62–67.
49. Zheng J, El Naqa I, Rowold EF, Pilgram KT, Woodard KO, Saffitz EJ, Tang D(2005) Quantitative assessment of coronary artery plaque vulnerability by high resolution magnetic resonance imaging and computational biomechanics: a pilot study ex vivo. Magnetic Resonance in Medicine. 54:1360–1368.
50. Zhao SZ, Ariff B, Long Q, Thom SA, Hughes AD, Xu XY(2002) Inter-individual variations in wall shear stress and mechanical stress distribution at the carotid artery bifurcation of healthy humans. Journal of Biomechanics. 35:1367–1377.

Biographies

Mr. Hao Gao, PhD candidate in Biomechanics at Brunel Institute for Bioengineering, Brunel University, Uxbridge, UK, sponsored by British Heart Foundation (FS/06/048). His major research interests are computational simulation of biomedical-related problems, including bio-fluid and -structure dynamics by the applications of various numerical methods.

Dr. Q. Long, received his first degree in Biomechanics in Huazhong University of Science and Technology, China in 1985 and gained his PhD degree in Mechanical Engineering in City University, London, in 1999. He is now a senior lecturer of Biomedical Engineering in Brunel University, London, UK. He has been working in Biomedical Engineering for 25 years. His main expertises are in numerical simulation (including CFD and FEM) in biomechanical problems. In recent years, his main focus of research is on the studies of biomechanical factors on atherosclerosis development and human venerable plaque rupture.

Part II
Ultrasound Imaging

Chapter 6
Methods in Atherosclerotic Plaque Characterization Using Intravascular Ultrasound Images and Backscattered Signals

Amin Katouzian, Stéphane G. Carlier, and Andrew F. Laine

Abstract We will review existing supervised as well as unsupervised image- and spectrumderived algorithms in the context of atherosclerotic plaque characterization and detection of vulnerable plaques. We will further elaborate more on challenges involved in characterization of plaques from tissue preparation, data collection, and registration toward classification.

Keywords Atherosclerosis • Plaque characterization • IVUS

6.1 Introduction

Often, patients with chest pain and high-risk factors associated with cardiovascular disease (i.e., elevated cholesterol) develop atherosclerotic plaques that can either cause stable coronary artery stenosis leading to angina pectoris during exercise or lead to acute coronary or vascular events such as myocardial infarction or stroke when they rupture [1]. Most cases of myocardial infarction and stroke occur when a thrombus is formed on a previously stable plaque that ruptures [2]. Persons at risk typically have no premonitory symptoms, and angiographic studies of coronary arteries in patients with nonfatal acute coronary syndromes (ACS) showed that most such events were due to rapid progression of mild, hemodynamically insignificant lesions [3–5]. These patients could undergo percutaneous coronary intervention (PCI) procedure also known as coronary angioplasty. During such a therapeutic procedure, interventional cardiologist diagnoses the site of an occlusion or narrowing using angiogram (X-ray along with radio-opaque contrast injected in the coronary arteries), which offers two-dimensional (2D) representations of a three-dimensional (3D) structure (the coronary artery tree with atherosclerotic plaques). This lack of

A.F. Laine (✉)
Biomedical Engineering Department, Columbia University, 351 Engineering Terrace MC-8904, 1210 Amsterdam Avenue, New York, NY 10027, USA
e-mail: laine@columbia.edu

Jasjit S. Suri et al. (eds.), *Atherosclerosis Disease Management*,
DOI 10.1007/978-1-4419-7222-4_6,

adequate geometrical and pathological information has motivated experts to take advantage of alternative 3D imaging modalities. For example, an interventional cardiologist may deploy an IVUS catheter to take cross-sectional images of arterial wall. Pathological and more recent IVUS studies of ruptured plaques showed that underlying lesions leading to acute coronary syndrome actually present a large plaque burden, often without much lumen compromise, a positive remodeling, and a thin, macrophage-rich cap that covers a large necrotic core [6–13]. The ultimate goal is to detect plaques at risk and treat them selectively before they rupture in order to prevent mortality and morbidity [14]. For this purpose, researchers have developed various tissue characterization algorithms through different imaging modalities to detect vulnerable plaques [15–19] in coronary arteries. Among them, intravascular ultrasound (IVUS) has been the most favored and clinically approved imaging modality, since it is relatively inexpensive and provides adequate spatial resolution along with sufficient penetration, while other comparable imaging techniques, such as near infrared (NIR) and optical coherence tomography (OCT) with excellent resolution, lack adequate penetration. In this chapter, we review existing IVUS-based tissue characterization techniques in the context of pathohistological atherosclerotic studies along with associated challenges from tissue preparation toward automatic classification.

6.2 IVUS Data Collection Specification

6.2.1 In Vivo Acquisition

Generally, an IVUS catheter is advanced in the left or right coronary artery down and possibly in some side branches on a guide wire coming out of a guiding catheter inserted in the femoral artery, Fig. 6.1. Acquisition of cross-sectional ultrasound images of right coronary arteries (RCA), left anterior descending (LAD), and left circumflex (LCX) coronary arteries can be performed with rotating single-element or phased array transducers. The catheter pullback speed varies between 0.5 and 1 mm/s and the frame rate can be set to 30 or 60 frames/s. The radiofrequency (RF) data are continuous time, real-valued, and band-limited signals and will be digitized at periodic time intervals. The digitizer's sampling frequency specifies number of samples per each A-line. For example, Boston Scientific's (Fremont, CA) iLab and Volcano's (Rancho Cordova, CA) s5™ scanners equipped with digitizer boards that sample IVUS signals at the rate of 400 and 200 MHz, respectively. The digitized data will be saved into a hard disk for post-processing and future study. In order to construct a typical IVUS image, the envelope of each RF signal will be computed by the corresponding analytical signal followed by decimation and interpolation in axial and radial directions, respectively. A logarithmic compression can also be used to enhance the image quality followed by quantization. Then, the resulting grayscale image is transformed to Cartesian coordinates to generate an IVUS frame. Further image processing techniques may be implemented

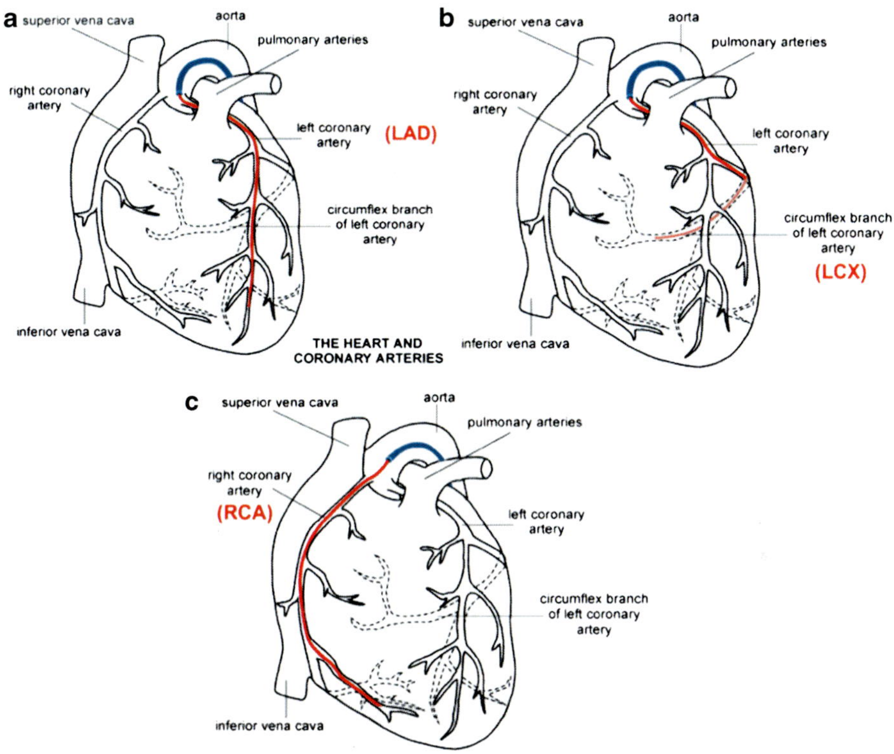

Fig. 6.1 The heart and coronary arteries. Illustration of advanced schematic catheter (*red*) into LAD (**a**), LCX (**b**), and RCA (**c**)

to enhance the image quality for better clinical quantification. Figure 6.2 illustrates the schematic cross-sectional anatomy of arterial wall and atherosclerotic plaque along with two distinct cross-sectional grayscale IVUS images acquired with 64-element phased array 20 MHz as well as a single-element rotating 45-MHz VOLCANO transducers. Comparing IVUS probes, higher center frequency improves spatial resolution, preferred by most of interventional cardiologists, at the cost of more variability among extracted spectrum-derived features [20]. We will discuss this in more detail in the next sections.

6.2.2 In Vitro Setup and Specimen Preparation

An atherosclerotic tissue characterization algorithm developed using in vitro data will work on in vivo data to the extent that the tissue signatures remain similar. As long as differences found in tissue signatures between in vitro and in vivo imaging

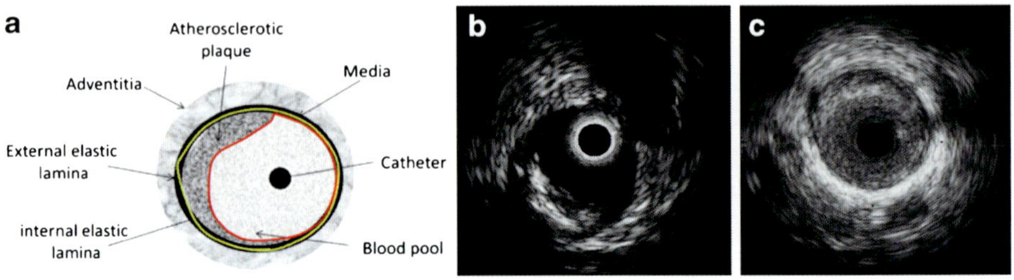

Fig. 6.2 Schematic cross-sectional anatomy of arterial wall and atherosclerotic plaque, the lumen border (*red*) and vessel wall border (*yellow*) (**a**), two distinct cross-sectional grayscale IVUS images acquired with 64-elements phased array 20 MHz (**b**), and single-element rotating 45 MHz (**c**) VOLCANO transducers

situations are systematic, it may still be possible to empirically retune an in vitro-trained algorithm and make it work in vivo. In order to reduce such variations and develop reliable algorithm for in vivo classification, researchers dissect arteries from human hearts collected from autopsy and transplant surgery within 24-h post-mortem and explantation, respectively. Nevertheless, animal models also have been suggested and used for the characterization of atherosclerotic plaques [21].

Since the overall justification of in vivo real-time plaque characterization is made by interventional cardiologists through use of classified tissues, it is indispensable to train the classifier with the most reliable features. In order to collect a training data set, the regions of interest (ROIs) on the arteries are marked and relative cross-sectional histology images obtained. The IVUS-histology frame alignment plays a crucial role because the IVUS frames are labeled through the interpretation of the corresponding histology images. Subsequently, the signals are assigned to labeled tissues and relative features are extracted. The IVUS-histology matching problem becomes more challenging due to the (1) curvature of arteries, especially in the LCX, (2) registration between an IVUS image and its histology since the IVUS imaging plane and the slicing plane of microtome are somewhat different, and (3) shrinkage of the arteries after formalin fixation. Here, we describe two methodologies to prepare specimen and obtain the most precise one-to-one matched IVUS-histology cross-sections.

6.2.2.1 Local Marking of ROIs

In this methodology, as proposed in [22], the arteries are not dissected as the whole heart is examined. Electrocauterization of small distal arteries is performed when necessary to avoid any leakage. A cannula is fixed in the ostium of the left main coronary artery and a circulating system consisting of phosphate-buffered saline (PBS) is used to insure a constant pressure (100 mmHg) and flow in order to maintain arteries physiologically opened at 37°C. The IVUS catheter is introduced and advanced on a 0.014" guide wire and manual pullback is performed to search for sites of significant plaques. One of the main challenges in this study is to mark ROIs as precisely as possible to get the best match of IVUS-histology cross-sections. In order to do so, a fluoroscopic X-ray system is deployed to visualize the tip of the IVUS catheter. After stabilizing of catheter at the site of interest, two needles were implanted under fluoroscopic guidance into surrounding fat so that they crossed above the tip of the IVUS catheter. Thereafter, a third needle is passed through the crossing point to mark it by a suture, Fig. 6.3. The first two needles are then removed. This procedure is repeated for 3–5 different ROIs per artery and the corresponding RF signals are acquired as described previously while complete automatic pullback is taken. Sutures are used to locate the ROIs for further histology processing and slicing hearts are fixed cannulating both coronary arteries to recirculate 10% buffered formaldehyde under 100 mmHg for 3 h. In the next step, the arteries are prepared for staining after decalcification.

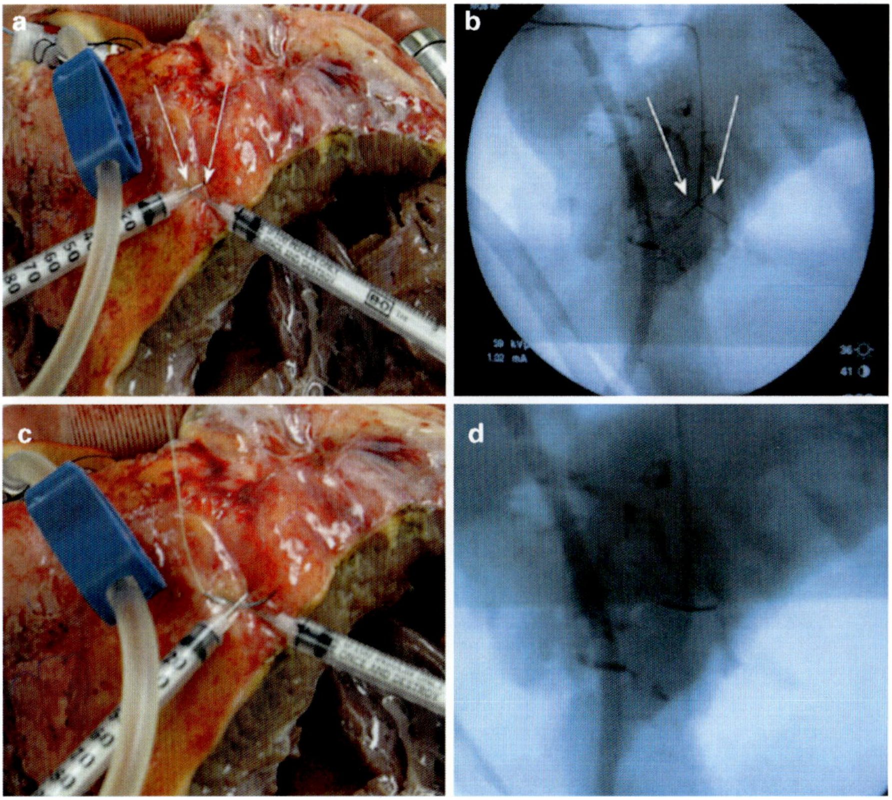

Fig. 6.3 Marking ROIs on LAD of a transplant heart; two needles (*pointed with white arrows*) are placed into surrounding fat (**a**) at the tip of the IVUS catheter visualized under fluoroscopic guidance (**b**). The third needle is passed through the crossed point and a suture is done to mark the ROI (**c**). The first two needles are removed and the procedure is repeated for each ROI (**d**)

Blocks of 2–3 mm are cut from distal locations of the sutures. The proximal end is marked with ink. Histology slices are taken from distal to proximal after arteries are decalcified and embedded in paraffin. In this methodology, for each block, three serial sections are taken at 500-μm intervals, which correspond to 30 frames in IVUS acquisition, Fig. 6.4. The first two sections are stained with hematoxylin and eosin (H&E) and Movat Pentachrome, and the last section could be kept unstained for future additional staining (e.g., Sirius Red). Hematoxylin and eosin are among the most commonly used stains in histopathology. Hematoxylin turns nuclei blue; eosin turns the cytoplasm (mostly composed of proteins arginine and lysine) pink. Another stain, the Russell–Movat pentachrome, colors cytoplasm in red, elastic fibers in black, collagen and reticulum fibers in yellow to greenish, and proteoglycans in blue. Clear areas might represent water, carbohydrate, lipid, gas or decalcified areas. Sirius red, on the other hand, has been used exclusively for collagen staining in cardiovascular histopathology. Collagen fibers are stained red and can be distinguished morphologically from other tissue components in the plaque. Taking a complete automatic pullback (or a short pullback along each ROI) and sectioning three slices at intervals of 500 μm allows finding the best possible match between ROI's cross-sections and corresponding histology images. Besides reflection of the needle, natural markers such as side branches and small calcified inclusions could be used to improve the IVUS-histology

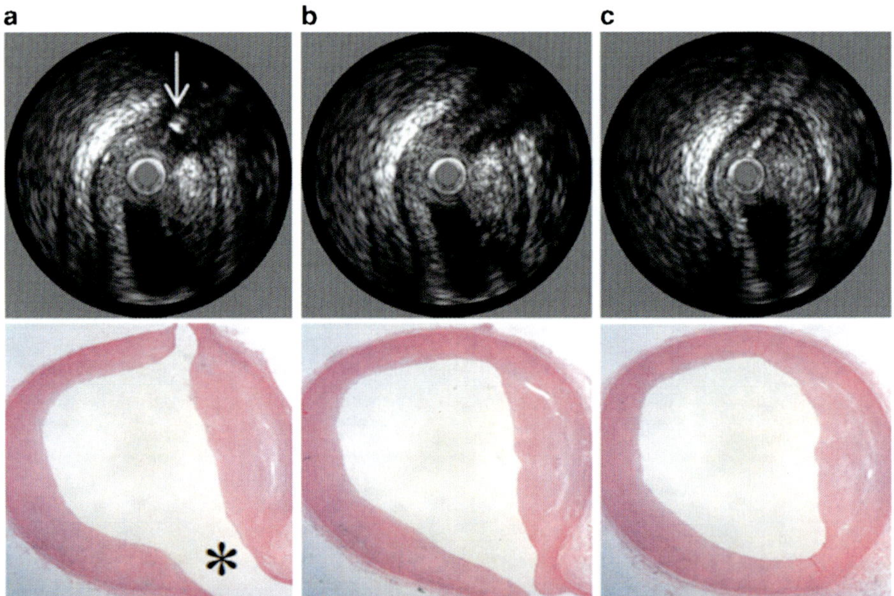

Fig. 6.4 Taking three sections at each ROI with 500-μm intervals, which corresponds to 30 frames. Marked ROI with reflection of the needle (*white arrow*) and side branch (*asterisk*) with corresponding histology (**a**). Two IVUS frames with 500-μm intervals and corresponding histology sections (**b, c**)

frames correlation accuracy. Figure 6.4a shows a marked ROI with needle and a side branch as well as corresponding histology image.

6.2.2.2 Systematic Marking of ROIs

This methodology has been introduced and developed by Jennifer Lisauskas at Infraredx (Burlington, MA) as part of tissue characterization project through near infrared (NIR) signals. In this methodology, the arteries are dissected from heart, placed in tissue cage fixture, Fig. 6.5a, and attached to a circulating fluid flow system, Fig. 6.5b. Average length of the arterial segments attached to the fluid system could be 50 mm. The arterial segments are perfused with saline at body temperature (37°C) with pulsatile flow (60 bpm, 135 mL/min) at physiologic pressure (80–120 mmHg). Then, an IVUS catheter is advanced on a 0.014" guide wire and a complete automatic pullback is performed from the distal to the proximal side. The same procedure is repeated using human blood. Prior to the experiment, the blood is agitated by hand for approximately 1 min to mix and then stirred at low speed with a magnetic stir bar using a hot/stir plate until the experiments begun. The RF data are collected as described previously.

After imaging, the arteries are pressure fixed with 10% buffered formaldehyde followed by decalcification. The histology blocks are prepared every 2 mm (corresponding to 120 frames of the IVUS pullback) using the sidebars. All blocks are embedded in paraffin and sectioned for histological staining. Two 5-μm thick histologic cross-sections are stained with H&E and Russell–Movat pentachrome. The transplant arteries included a third histologic cross-section, which remained unstained for subsequent staining (e.g., Sirius Red).

This methodology has two main advantages. First, the orientation of artery is not changed throughout the whole procedure. Therefore, more reliable IVUS-histology pairs could be obtained and the number of cross-sections of interest (CSIs) per vessel is significantly increased (average of 25 regions) compared to the traditional methods (3–5 regions). Secondly, it allows us to investigate the effect of flowing that is expected to introduce signal changes (i.e., attenuation) and may alter classification results.

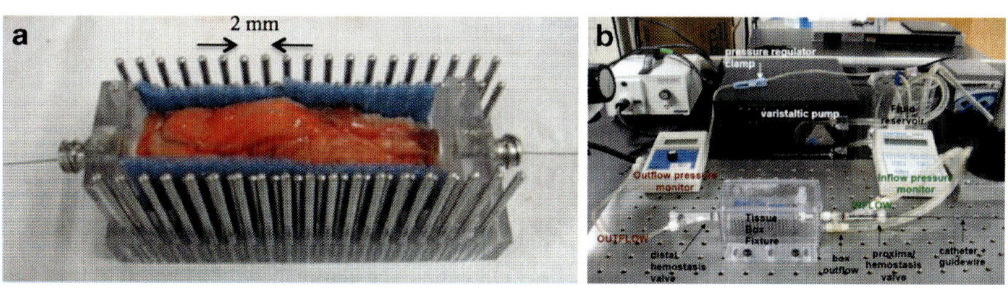

Fig. 6.5 Tissue cage fixture (**a**), in vitro experiment setup (**b**) developed by Jennifer Lisauskas at Infraredx (Burlington, MA)

6.3 New Intravascular Ultrasound Methods for Atherosclerotic Plaque Characterization

6.3.1 Spectral and RF-Based Approaches

Postprocessing of the backscattered RF IVUS signal has been developed in order to better characterize plaque composition. It has been shown that the ultrasound RF-signal analysis can provide quantitative information on tissue microstructures [23, 24]. The basic tenet of ultrasound tissue characterization is that different tissue types imprint their own "signature" on the backscattered echo received by the transducer. Hence, spectral analysis of the ultrasound RF signals has been extensively studied and employed in different applications [25–28]. This is because the tissues of interest generate signals that are stochastic (i.e., their internal constituents exhibit randomness in size, position, orientation, etc.) and the spectra represent the ensemble properties of the scatterers. If the spatial autocorrelation function describing these factors has a known form, then the theoretical scattering models can be used with spectral data to estimate two physical scatterer properties: the effective sizes of constituent scatterers and their "acoustic concentration" [27, 29]. It is a statistically observable fact that on averaging a large number of spectra obtained from homogenous areas of tissue in carefully controlled in vitro experiments, different types of tissue give rise to recognizably different spectra, Fig. 6.6. In most recent methods of tissue characterization, the spectrum is summarized using a few numbers (e.g., seven parameters in [30]) to capture its basic shape, so-called spectral signature.

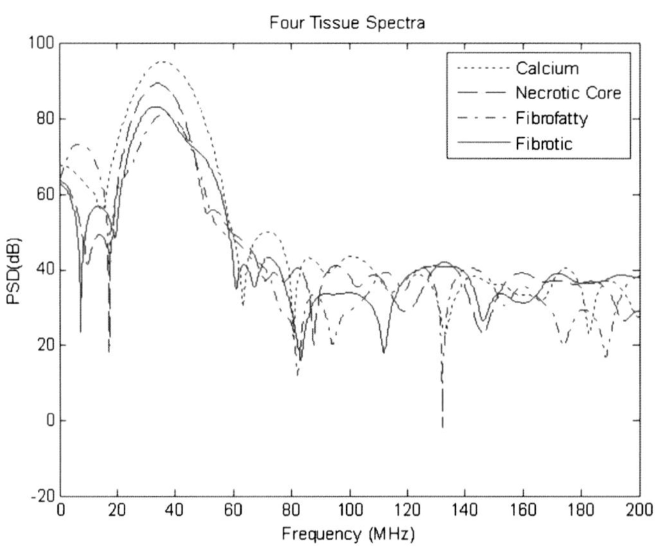

Fig. 6.6 Dissimilarity among four tissue spectra, calcium, necrotic core, fibrofatty, and fibrotic measured in sample size of 64 collected from one cross-section of interest (CSI)

6.3.1.1 IVUS-Virtual Histology

In 1983, Lizzi et al. presented a sliding-window analysis method to identify tumors in the eye and liver by using two spectral signatures extracted from calibrated tissue spectra [31]. Those parameters were the slope of the regression line fitted to the mid-band portion of calibrated spectrum and intercept of that regression line at zero frequency. They demonstrated that the slope and intercept are indicative of the scatterer size and concentration. Nair et al. [30] extended Lizzi's approach and introduced the first commercially available IVUS-based atherosclerotic tissue characterization algorithm known as virtual histology (VH). This methodology has now been implemented in the Volcano (Rancho Cordova, CA) IVUS clinical scanners that offer near real-time tissue characterization in vivo using a 20-MHz phased array transducer.

This approach is based on seven spectral parameters (intercept, slope, mid-band-fit (MBF), and minimum and maximum powers and their corresponding frequencies) extracted from tissue spectra, which were normalized through the spectrum of an acrylic reflector. The normalization is necessary in order to eliminate transducer impulse response prior to feature extraction. Basically, a short ultrasonic pulse is transmitted toward perfect reflector and the spectrum of the reflected signal is considered as an estimate of the transducer frequency response. This is not a feasible approach for clinical real-time plaque characterization but has been performed for in vitro experiments. The same group deployed blind deconvolution to prevail this problem [32]. They also added integrated backscatter (IB) coefficient as an additional feature [33] and later showed that autoregressive (AR) models were superior to the Fourier technique [34]. Figure 6.7 demonstrates an arbitrary tissue spectrum, averaged Plexiglas spectrum fitted with Gaussian model, normalized spectrum, and extracted features through linear regression fit. Finally, they employed classification tree (CT) to differentiate plaque compositions. Figure 6.8 shows an IVUS CSI, corresponding histology image, and VH result. In 15 patients with versus 15 patients without ACS undergoing directional coronary atherectomy, histopathology of the retrieved atherectomy specimens was compared with that of IVUS-virtual histology (IVUS-VH). Predictive accuracy of IVUS-VH was 87% for fibrous tissue, 87% for fibrofatty tissue, 88% for necrotic core, and 97% for dense calcium regions. Figure 6.9 illustrates four grayscale IVUS images acquired in vivo with 20-MHz phased array transducer and corresponding VH results. The dark green, light green, red, white, and gray represent fibrotic, fibrofatty, necrotic core, calcium, and media, respectively.

6.3.1.2 IVUS-Integrated Backscatter

Kawasaki et al. introduced an alternative tissue classification scheme based solely on another RF-derived parameter: the integrated backscatter (IB). In this approach, RF signals arewindowed and the IB value is computed in frequency domain as follows:

$$\frac{1}{f_{\mathrm{BW}_{\max}} - f_{\mathrm{BW}_{\min}}} \int_{f_{\mathrm{BW}_{\min}}}^{f_{\mathrm{BW}_{\min}}} \mathrm{PSD}(f)\,df, \tag{6.1}$$

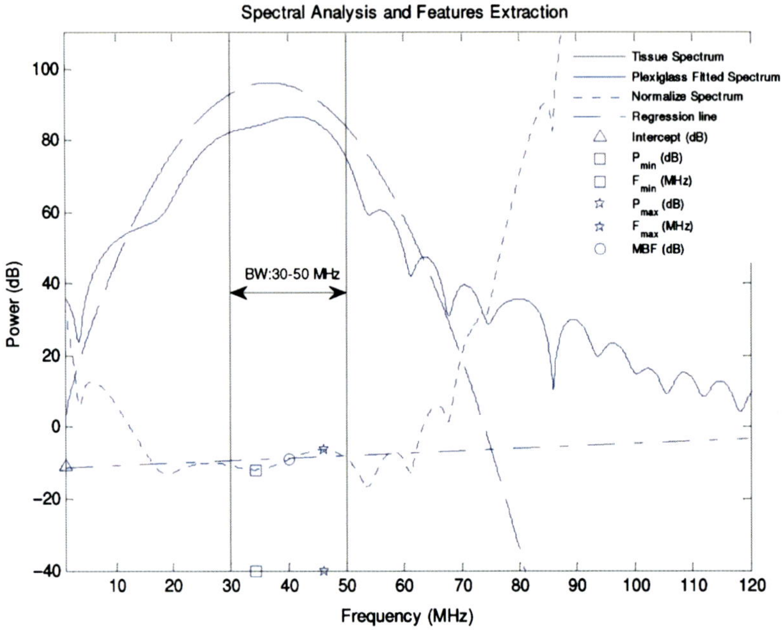

Fig. 6.7 Linear regression fit to normalized tissue's spectrum and extracted features; intercept, slope, mid-band fit (MBF), and minimum and maximum powers and their corresponding frequencies. Slopes (decibels per megahertz) of regression line and integrated backscatter coefficient have not been shown. The corresponding IVUS data have been collected with 40-MHz unfocused transducer

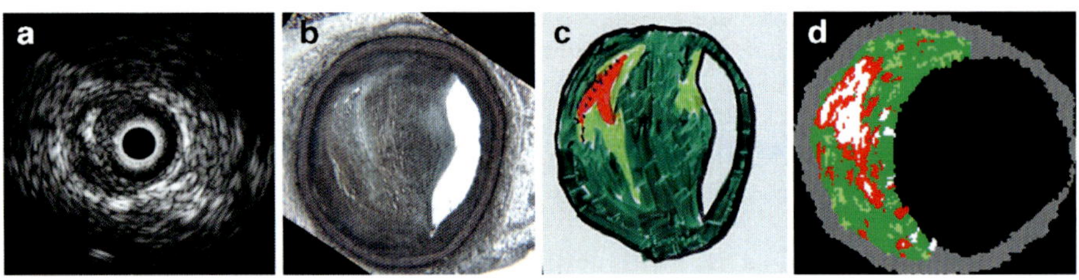

Fig. 6.8 Original IVUS grayscale image acquired in vitro (**a**), histological validation showing a fibroatheroma with a thick fibrotic cap (**b**), pathologist's pencil sketch showing the four major tissue types present in the lesion using the IVUS-VH color-coded classification: *dark green codes* for fibrotic tissue; *light green*, fibrofatty tissue; *red*, necrotic core. Microcalcifications are here marked with *small dark points* (**c**), IVUS-VH coding with calcium here coded in *white* (**d**)

where PSD is estimated power spectral density and $f_{\mathrm{BW}_{\min}}$ *and* $f_{\mathrm{BW}_{\mathrm{mix}}}$ are the minimum and maximum frequency in the specified bandwidth, respectively. This system is presently distributed only in Japan (YD Co. Ltd, Tokyo) and uses the IVUS catheter from Boston Scientific (Fremont, CA) based on a 40-MHz single rotating crystal [35, 36]. Color-coded maps of plaques are constructed by use of these IB data, which reflects the plaque histology of autopsied coronary arteries. IVUS-integrated

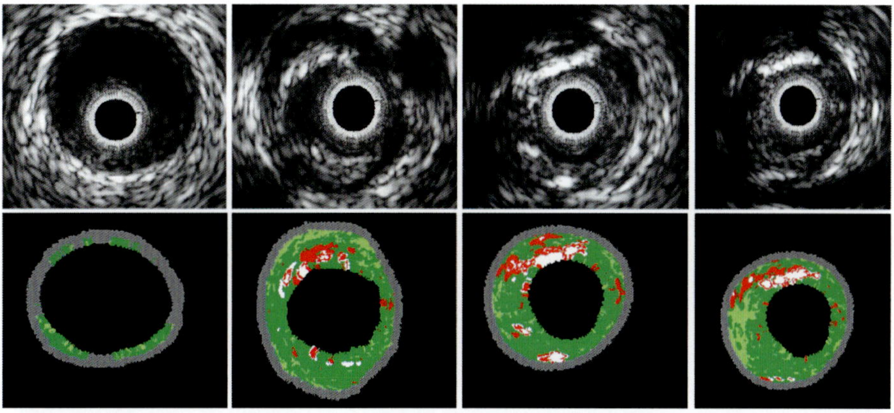

Fig. 6.9 IVUS-VH in vivo results. Four IVUS grayscale images (*top row*) acquired with 20 MHz 64-element phased array transducer from distal (*left*) to proximal (*right*) and their corresponding VH results (*bottom row*)

backscatter (IVUS-IB) values have been divided into five categories to construct color-coded maps: thrombus, intimal hyperplasia or lipid core, fibrous tissue, mixed lesions, and calcification, Fig. 6.10. Comparisons of IVUS-IB with histopathology demonstrated a high sensitivity for characterizing calcification, fibrosis, and lipid pool (100, 94, and 84%, respectively) and a high specificity (99, 84, and 97%, respectively). The tissue IB values were calibrated by subtracting the IB values from the IB value of stainless steel placed at a distance of 1.5 mm from the catheter. In the ex vivo studies, each site of each tissue characteristic was placed at a distance of ~1.5 mm from the catheter. They further validated their IVUS-IB in vivo results with corresponding angioscopy images. The same group recently compared and reported the overall agreement between IVUS-IB and IVUS-VH in the tissue characterization of plaques from the same coronary arterial cross-sections and quantified their results with corresponding histopathological images [37], Fig. 6.11.

6.3.1.3 IVUS Elastography

The principle of elastography is to study the response of tissue to mechanical excitation (e.g., compression). Several groups have deployed elasticity imaging in order to assess the local mechanical properties of atherosclerotic plaque or vessel wall using different techniques such as envelope-based elastography [38], spectral-based strain imaging [39], phase sensitive speckle tracking [40], and cross-correlation-based technique between RF signals in consecutive frames [41]. De Korte et al. [41] computed the relative local displacements between IVUS images acquired at two different levels of intravascular pressure with a 30-MHz single element mechanically rotating transducer. These displacements were estimated from the time shift between gated radiofrequency echo signals using cross-correlation algorithms with interpolation around the peak. Then, the strain information was presented in an independent

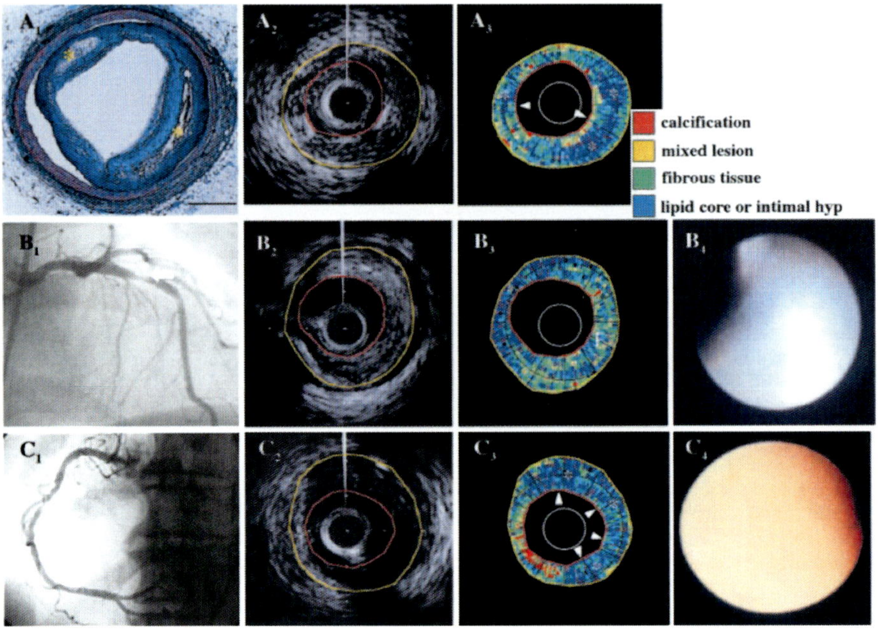

Fig. 6.10 Color-coded maps of coronary arterial plaques constructed by IB-IVUS and histology, coronary angiography, or angioscopy. (**a**) Autopsy study of coronary arterial plaque. A_1, Histological finding with fibrosis, mixed lesion, calcification, and large (*right*) and small (*left*) lipid cores (*asterisk*) stained with Masson's trichrome. *Bar* 1 mm. A_2, Conventional IVUS image of same segment as A_1. A_3, Color-coded map of intima of A_1 constructed by IB-IVUS. (**b**) In vivo study of coronary arterial plaque. B_1, Angiography of left coronary artery. *Arrow* indicates a segment with 60% diameter stenosis. B_2, Conventional IVUS image of same segment as shown by *arrow* in B_1. B_3, Color-coded map of intima of B_1 constructed by IB-IVUS. B_4, Angioscopic finding of plaque at right in B_3. Note that white plaque is related to fibrous tissue. (**c**) In vivo study of coronary arterial plaque. C_1, Angiography of right coronary artery. *Arrow* indicates a segment with 40% diameter stenosis. C_2, Conventional IVUS image of same segment as shown by *arrow* in C_1. C_3, Color-coded map of intima in C_1 constructed by IB-IVUS. Note large lipid core (*asterisk*) with fibrous cap (*arrowhead*). C_4, Angioscopic finding of plaque at top in C_3. Note that *yellow plaque* is related to thin fibrous cap. *Red and yellow contours* represent traced lumen and vessel wall borders, respectively

complimentary 2D image of the strain called elastogram. Authors demonstrated the feasibility of this technique for intravascular purposes using phantom studies and studies on human arteries both in vitro and in vivo. The results were further validated with conventional IVUS images, corresponding histology images, and compression modulus values. Figure 6.12 illustrates two in vitro IVUS grayscale images, resulting elastogram, and corresponding histopathological images. The results demonstrate the capability of intravascular elastography to characterize different plaque types and classify tissues based on their mean strain values to fibrous, fibrofatty and fatty. Additional IVUS elastography (IVE) validation has been reported by Maurice et al. [42] that employed the Lagrangian speckle model estimator (LSME) along with the scaling factor estimator (SFE) to compute the radial strain elastograms.

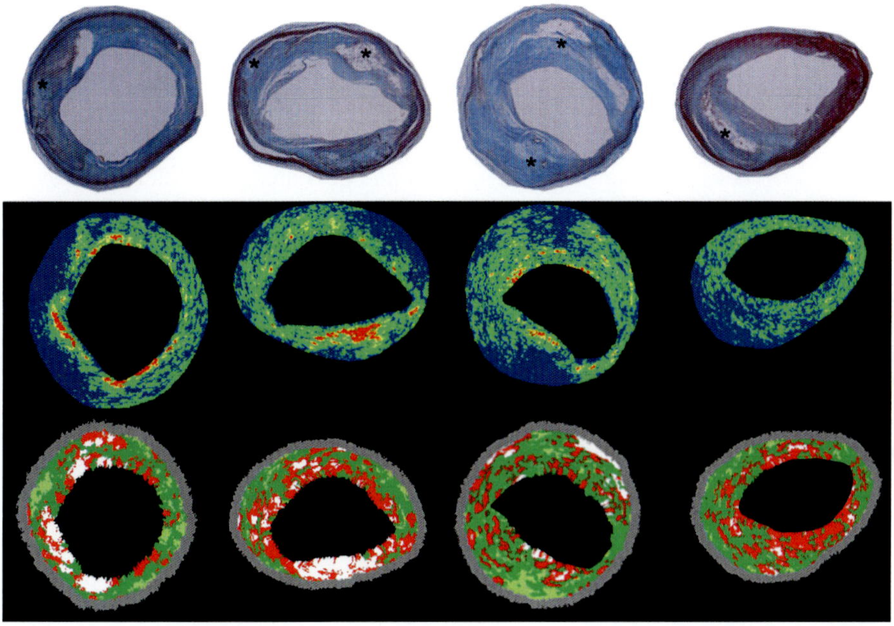

Fig. 6.11 Comparison between IVUS-IB results with corresponding IVUS-VH. Histological image (Masson's trichrome staining) (*top row*). *Asterisk* depicts lipid pool region, corresponding color-coded maps constructed by the IVUS-IB technique (*middle row*), *red*, *yellow*, *green*, and *blue colors* represent calcification, dense fibrosis, fibrosis and lipid pool, respectively. Corresponding IVUS-VH images (*bottom row*)

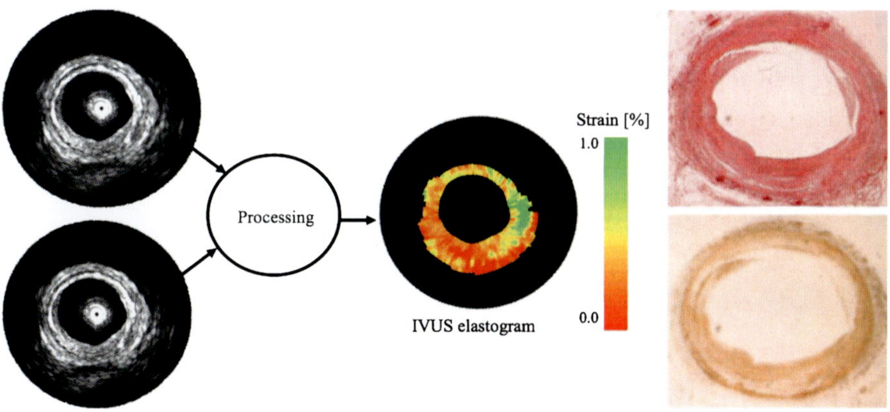

Fig. 6.12 Two in vitro IVUS grayscale images acquired at two different levels of pressures (85 and 90 mmHg) and resulting elastogram along with corresponding histopathological images, Picro-sirus red (**a**), Alpha-actin (**b**). The IVUS image reveals an eccentric plaque between 2 and 11 o'clock. The elastogram shows that the plaque can be divided into two parts: a low-strain part (0.2%) between 4 and 11 o'clock and a high-strain part (1%) between 2 and 4 o'clock, both compared with the moderate strain (0.5%) in the normal vessel wall. The histology reveals that the region between 4 and 11 o'clock is fibrous material and the region between 2 and 4 o'clock is fatty material

6.3.2 Texture-Based Approaches

IVUS findings have shown that the sonographic differences yielded the visual discrimination among plaque constituents [43, 44]. In other words, variations of intensities are attributed to the repetitive tissue microstructure patterns. These have motivated researchers to develop texture-based algorithms on IVUS images to differentiate tissue types [45–47]. One of the main differences between texture-derived atherosclerotic plaque characterization algorithms with their spectrum-based counterparts is that no RF signal is needed. This can be advantageous since usually RF signals are not accessible. On the other hand, the appearance of images may vary depends on selected parameters during acquisition (i.e., time gain compensation), normalization, or reconstruction (i.e., nonlinear transformation) that makes the problem of interest more challenging.

6.3.2.1 IVUS-Prognosis Histology

It has been shown that the multiscale approach for texture discrimination is the most compatible analysis to human and mammalian vision processing systems due to its conservation of energy in both spatial and frequency domains [48, 49]. Beck et al. [50] employed spatial-frequency-localized expansions and their generalization to 2D to discern the textural patterns on constructed images from backscattered IVUS signals, while the geometrically oriented decompositions were provided at this dimension. They presented an effective texture-derived atherosclerotic tissue characterization algorithm using overcomplete discrete wavelet packet frames (DWPF) and showed that orthonormal wavelet basis functions reliably translate ultrasonic information of tissue microstructures into visual textures that could be used as features for classification. Unlike discrete wavelet transform (DWT) and discrete wavelet packet transform (DWPT), the decompositions are translation invariant in DWPF. In other words, no decimation occurs between levels, Fig. 6.13. The multichannel wavelet schematic in

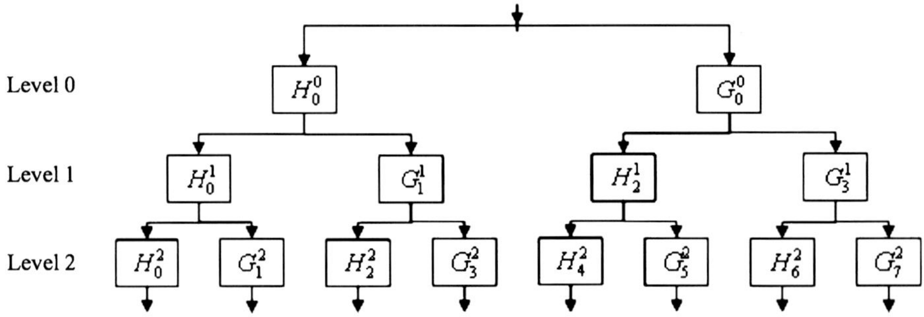

Fig. 6.13 Tree structure for discrete wavelet packet frames (DWPF), which behaves like a filter bank so selection of filters at the first level becomes crucial and their characteristics may impact results

Fig. 6.13 behaves like a filter bank. Symmetry, frequency response, and boundary accuracy are the most crucial factors in selection of the highpass, G, and the lowpass, H, filters that directly impact the classification results. Hence, the authors selected Lemarie–Battle [51] wavelets that are symmetric (have linear phase response) and quadrature mirror filters (QMF), Fig. 6.14. For each IVUS image, a separable tensor product is used, which leads to orientation selectivity in decomposition tree. The envelope of features is extracted at the bottom of decomposition tree and a k-means classifier is deployed to classify tissue types into calcified, fibrotic, fibrolipidic and no tissues. They deployed scoring approach to quantify the results and asked an independent histologist to score each PH image by eye-balling. For higher accuracy, each histology image was divided into four or more distinguishable regions that separated homogeneous from heterogeneous regions. The accuracy of characterization was evaluated for each region separately and averaged for composite rating for each tissue type. The overall classification performance was evaluated on 83 cross-sections of interest and reported to be 87.75, 90.87 and 99.70% for fibrotic, fibrolipidic, and calcified regions, respectively. Figure 6.15 demonstrates two IVUS grayscale images, corresponding histology images, and generated PH images. Authors concluded that the extracted textural features were perfectly suited for classification and capturing characteristics of atherosclerotic plaque components with fairly high correlation to their histology image correspondences. This would resolve one of the main limitations of the IVUS, which is discrimination between fibrous and fatty tissues [52, 53]. Due to unsupervised classification of presented technique, the algorithm currently does not detect the necrotic core directly. However, the vulnerability of plaque can be deduced by contrasting the lipid pool regions on PH images in a similar way as Okubo et al. has reported [37].

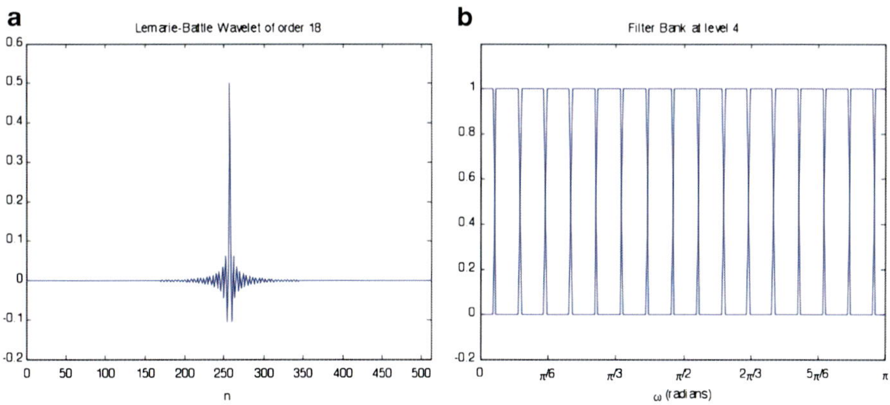

Fig. 6.14 Lemarie–Battle filter of order 18 (**a**), constructed filter bank at level 4 (**b**)

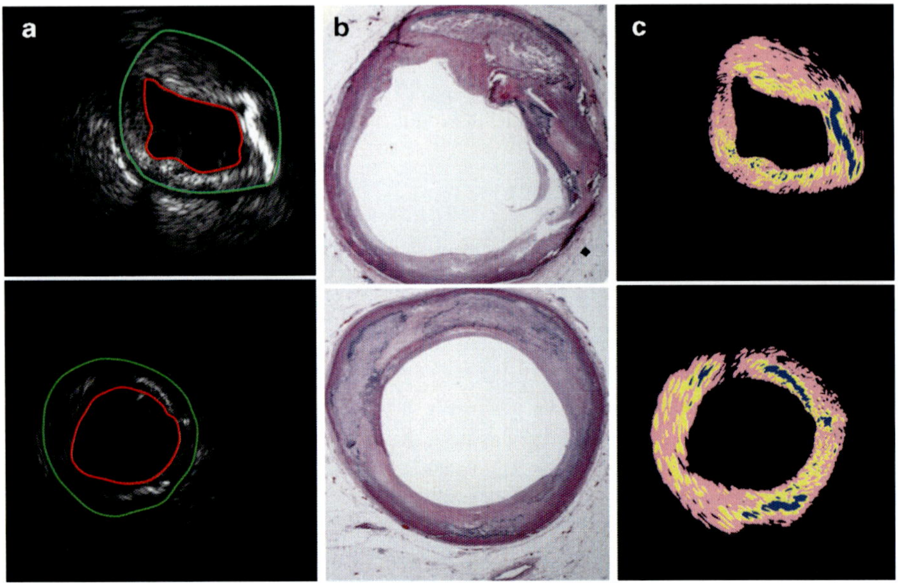

Fig. 6.15 Manually traced vessel wall (*green*) and lumen (*red*) borders, on IVUS B-mode image (**a**), corresponding H&E histology images of tow cross-sections of interest (**b**), resulting IVUS-PH images generated by the algorithm (*blue*, *yellow*, and *pink colors* represent calcified, fibrotic and fibrolipidic components) (**c**)

6.3.2.2 IVUS-Error Correcting Output Codes

Jeremias et al. [54] introduced a new technique to characterize intravascular ultrasound tissues based on different types of features, such as radial frequency, texture-based features, and combined features. They presented a subclass approach of error correcting output codes (ECOC) that splits the tissue classes into different subsets according to the applied base classifier. To deal with the classification of multiple tissues, the use of robust multiclass learning techniques is required. In this sense, ECOC shows to robustly combine binary classifiers to solve multiclass problems. Complex IVUS data sets containing overlapping data are solved by splitting the original set of classes into subclasses and embedding the binary problems in a problem-dependent ECOC design.

In the proposed technique, three types of features are employed. The first ones obtained from RF signals by computing the energy of A-line and the energy of windowed spectrum. The second ones extracted based on texture-based features from reconstructed images including cooccurrence Matrix [55], local binary patterns [56], and Gabor filters [57, 58]. Additionally, taking into account that highly nonechogenic plaques produce significant shade in the radial direction of the vessel, a complementary feature that represents the presence of shading in the image, was included in the feature set. Finally, the slope-based features proposed by Nair et al. [30].

Authors used the RF signals and their reconstructed images from a set of ten different patients. To generate the data set on texture features, the intersection between segmented images has been mapped into a feature vector. Then, all the features collected were categorized by patient and each of the three possible plaque types. The image features were extracted by using the previous texture descriptors: cooccurrence matrix, local binary patterns, and Gabor filters. Those features were calculated for each pixel and gathered in a feature vector of 68 dimensions. An example of a manual and automatic texture-based segmentation for the same sample is shown in Fig. 6.16. To generate the data set of RF features, the RF signals were acquired with single element rotating 40-MHz Boston Scientific catheters using a 12-bit acquisition card with a sampling rate of 200 MHz. To analyze the RF signals, the sliding window was composed of 64 samples of depth and 12 radial A-lines, and the displacement was fixed in 16 samples and four A-lines. The power spectrum of the window ranges from 0 to 100 m/s and it was sampled by 100 points. Then, it was complemented with two energy measures yielding a 102 feature vector. We also consider a third data set that concatenates the descriptors from the previous RF and texture-based features, obtaining a feature vector of length 170 features.

Authors compared their technique with the state-of-the-art ECOC coding designs: one-versus-one [59], one-versus-all [60], dense random [61], sparse random [61], and discriminant ECOC (DECOC) [62]. Three different base classifiers were applied over each ECOC configuration: nearest mean classifier (NMC) with the classification decision using the Euclidean distance between the mean of the classes, discrete AdaBoost with 40 iterations of decision stumps [63], and Fisher linear discriminant analysis (FLDA). The method automatically characterizes different tissues, showing performance improvements over the state-of-the-art ECOC

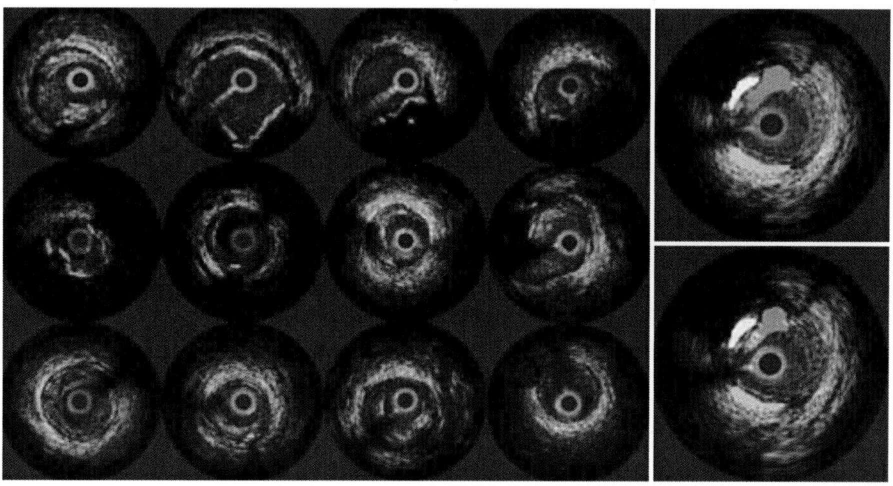

Fig. 6.16 *Left*: IVUS data set samples. *Right*: (*top*) segmentation by a physician and (*down*) Automatic classification with texture-based features. The *white area* corresponds to calcium, the *light gray area* to fibrosis, and the *dark gray area* to soft plaque

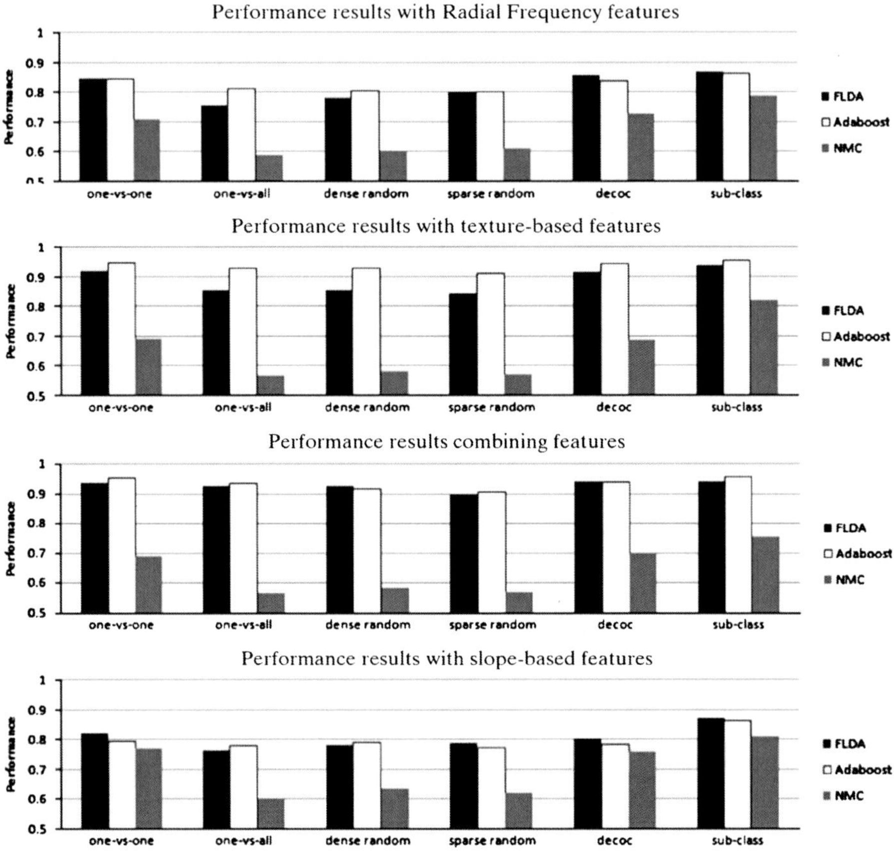

Fig. 6.17 Performance results for different sets of features, ECOC designs, and base classifiers on the IVUS data set

techniques for FLDA, Discrete AdaBoost, and NMC. In particular, the results show higher performance when using texture-based features compared with RF signals and slope-based features, and slights improvements when the sets of features are combined, Fig. 6.17.

6.3.2.3 IVUS-Image-Based Histology (IVUS-IBH)

Daugman [64] developed a technique for the characterization of atherosclerotic plaques via textural processing of IVUS images. In this approach, similar to IVUS-prognosis histology (IVUS-PH), the generated tissue color maps can be provided for every IVUS frames acquired during pullback, while in IVUS-VH methodology the VH images are generated for every other 30 frames due to ECG-gated protocol, Fig. 6.18. This would resolve the poor longitudinal resolution in current IVUS-VH

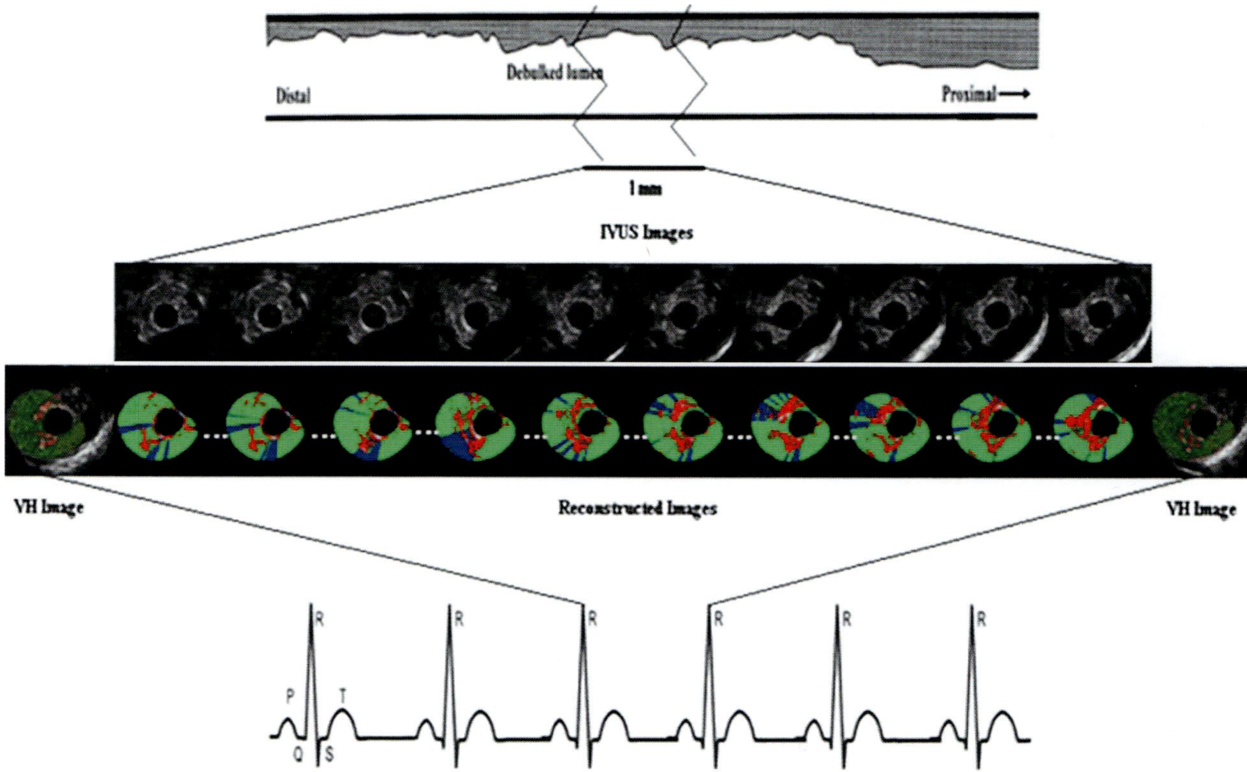

Fig. 6.18 Illustration of enhancement of the longitudinal resolution of atherosclerosis plaque composition characterization of grayscale IVUS using IVUS-IBH method compared to IVUS-VH. In IVUS-IBH approach, the tissue color maps are generated for every frames, similar to IVUS-PH algorithm, while in IVUS-VH algorithm the VH images are provided for every other 30 frames due to ECG gating protocol

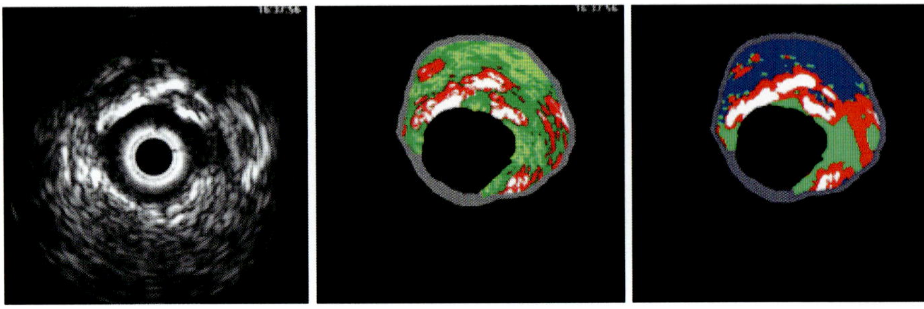

Fig. 6.19 An in vivo acquired IVUS grayscale image (**a**), corresponding VH image (**b**), generated tissue color map through IVUS-IBH approach (**c**). The *red*, *white*, *green*, and *blue colors* represent necrotic core, calcified, fibrofatty, and acoustic shadowing regions, respectively

technique. What makes this approach different from existing technique is the detection of acoustic shadowing behind the arc of calcified plaques. These shadow regions, which exist and displayed in the plaque area of some IVUS images appears as echo-soft; so, when treated within other parts of plaque area, although they are mostly calcium and necrotic core tissues, normally should be classified to the lipid or fibrofatty classes [65, 66].

Acoustic shadow regions displayed in IVUS grayscale images cannot represent any useful information for plaque component analysis [67], however, IVUS-VH images classify tissues behind the arc of calcified regions in shadow areas. In proposed technique, "shadow regions" is detected through two thresholding steps: one thresholding procedure for detecting the high-intensity regions and another one to identify the existence of the low-intensity regions behind the regions detected in the previous step and mark them as "shadow region." In order to characterize the rest of plaque regions, two feature extraction methods, local binary pattern (LBP) [68], and the run-length method [69] were combined. In vivo and ex vivo results were validated with corresponding IVUS-VH images as gold standard and the overall accuracy for all tissue types found to be 73%. Sensitivities (specificities) were reported 80% (86%) for dense calcium, 80% (92%) for fibrofatty tissues, and 60% (81%) for necrotic core. Figure 6.19 shows an in vivo acquired IVUS frame and generated tissue color map through IBH approach along with corresponding VH image for comparison.

6.4 Challenges Associated with Atherosclerotic Tissue Characterization Algorithms

In this section, we consider a number of important factors in the characterization of atherosclerotic plaques that need to be studied more comprehensively and precisely to make existing algorithms more reliable so interventional cardiologist can make their decisions more confidently. In coronary artery plaque characterization, the extracted features for each tissue type are so stringent particularly for closely related tissues such as fibrotic and fibrolipidic in which any perturbation like change of pressure or flowing blood may lead to misclassification of one tissue for another. Although these factors play a vital role in the characterization of atherosclerotic plaques, they have not been studied as inclusive as they should have been to date.

6.4.1 Variability of Tissue Signatures

The principal challenge in building a tissue characterization system is to develop a proper definition for tissue signatures that maintains their similarity within each tissue type and distinction between tissue types. This is indeed a challenge since the spectral/RF-derived tissue signatures corresponding to a single tissue type can, in general, be shown to vary across different cross-sectional slices even within the same vessel particularly when high frequency transducer (40 MHz up) is used [20]. Figure 6.20 shows the variations among slope and intercept extracted from 92, 73, 299, and 1234 ROIs of necrotic core, calcified, fibrolipid, and fibrotic signals, respectively, in three distinct frames with a ten-frame interval. Each ROI contained 64 Hamming-windowed samples. For each ROI, the signals were passed through a linear phase band-pass Butterworth filter of order ten with cutoff frequencies of $f_{c_{min}} = 10\text{MHz}$ and $f_{c_{mix}} = 80\text{MHz}$. Then, five adjacent line spectra were computed using a 4,096-point FFT. In order to get a smoother spectrum, a median filter of size 5×3 was slided on the calculated spectra. The source of such variations can be related to: (1) image formation, such as small changes in the angle of incidence of the ultrasound beam or variations in the geometric configuration of scatterers, which could affect features such as IB [70]; (2) genuine changes in physical characteristics within the particular tissue type; and (3) variation in transducer properties that confound the recovery of tissue type from tissue signatures. For instance, the transducer center frequency appears to be fairly consistent around a frequency, that is, off from the nominal frequency (e.g., 40 MHz), Table 6.1. This may not directly impact any extracted

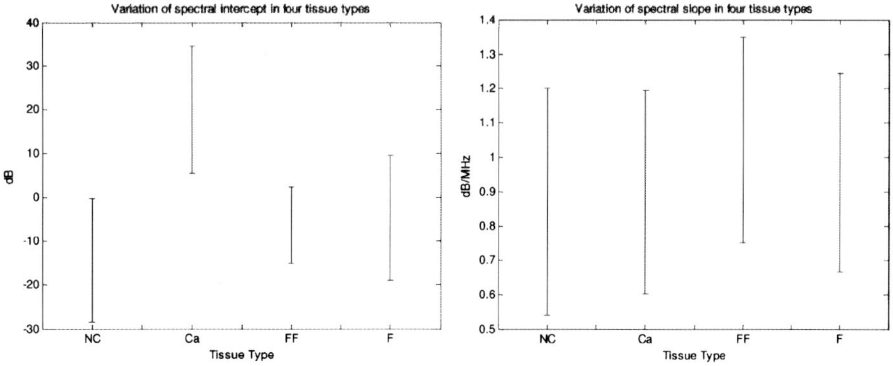

Fig. 6.20 Variations of two spectral features, intercept (**a**), and slope (**b**), extracted from four tissue types; *NC* necrotic core, *Ca* calcium, *FF* fibrofatty, *F* fibrotic. The data were collected using 40-MHz single element transducer

Table 6.1 Statistical variations of transducers' spectral parameters

	Transducer 1	Transducer 2	Transducer 3
Center frequency (MHz)	34.48 ± 0.14	41.93 ± 0.31	36.46 ± 0.69
Bandwidth at −6 dB (% of center frequency)	29.31 ± 1.64	37.34 ± 2.28	32.32 ± 2.65
Bandwidth at −12 dB (% of center frequency)	59.22 ± 2.82	63.68 ± 3.39	66.79 ± 5.36

spectral feature but becomes crucial when the bandwidth is measured in order to extract features within a specific bandwidth. Perturbations in extracted features are also increased if a broader bandwidth is chosen. These variations may alter the classification results. The texture-derived algorithms untangle this limitation and can be deployed independently of the transducer's center frequency and bandwidth.

6.4.2 Consistency Among PH Images in Adjacent Frames

The best way to investigate the reliability of any atherosclerotic tissue characterization algorithm is to examine the consistency among generated tissue color maps in adjacent frames. We know that the local properties of lesions are not changed very much in successive frames due to rich resolution in pullback direction. In other words, the plaque characteristics are gradually changed during pullback and abrupt changes among successive tissue color map images are not expected. Figure 6.21 shows the grayscale in-vivo acquired IVUS images (40 MHz single element Boston Scientific transducer) and the corresponding changes in plaque constituents in IVUS-PH images for five consecutive frames.

6.4.3 Effects of Change of Pressure

During catheterization procedure, the coronary artery is dilated and relaxed due to pressure change. The effects of change of pressure on extracted features, especially those derived form normalized spectrum, are unknown and have not been investigated and reported yet. In fact any reliable tissue characterization algorithm must be as less sensitive as possible to any change including pressure and its consistency should be validated. For this reason, we performed an experiment and changed the pressure while acquiring the IVUS images at a single cross-section of interest (no pullback). Figure 6.22 displays the effects of change of pressure on the grayscale IVUS images and constructed PH images at three different levels of pressure. Although the sizes of pathological structures are changed, the global interpretations on the PH images remain unvaried, keeping in mind that some of the variations are due to the contractions/expansions of the plaque in 3D volume.

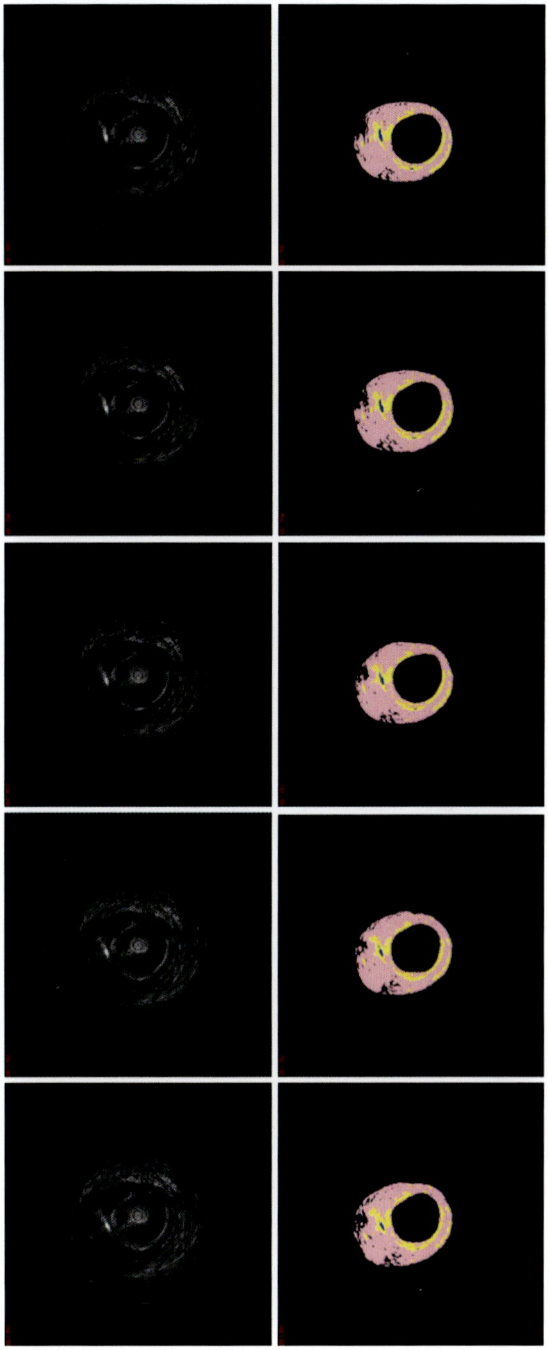

Fig. 6.21 Five consecutive in vivo IVUS grayscale images acquired with 40-MHz single element Boston Scientific transducer (*top row*) and corresponding generated IVUS-PH images (*bottom row*). The results show consistency among classified tissues in consecutive PH images

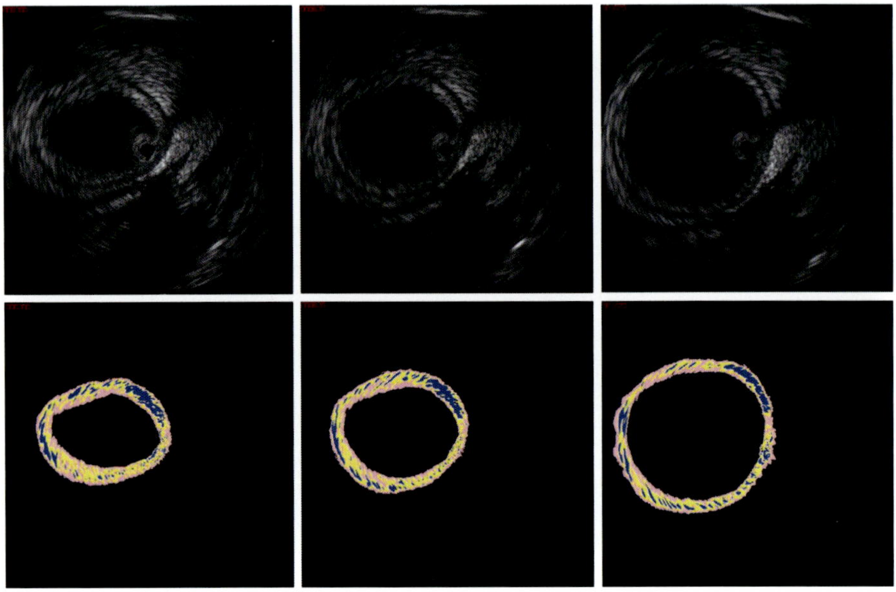

Fig. 6.22 Effects of change of pressure on generated PH images. In vitro grayscale IVUS images acquired at three different levels of pressures for a particular cross section of interest (no pullback) (*top row*), and corresponding constructed PH images (*bottom row*)

6.4.4 Effects of Flowing Blood

It has not been verified by any group to date whether in vitro trained classifier can be used for in vivo classification or not. In general, the supervised classifier is trained using in vitro-derived features deploying PBS, while the effects of flowing blood on in vivo constructed tissue color maps remained unknown. For this purpose, we acquired the IVUS signals in PBS and human blood solution for each segment. Figure 6.23 demonstrates the IVUS grayscale images of a cross-section of interest acquired using circulating Saline as well as human blood and corresponding constructed PH images along with H&E histology image. As seen, imaging of the plaque in blood introduces some differences although the overall interpretation seems to be very similar. However, not much is known about the nature of these differences, other than that imaging plaque through blood (which is difficult to duplicate in vitro) can be expected to introduce some differences and also that tissue fixation through formalin (which situation is absent in vivo) might alter the tissue response to ultrasound. It is, therefore, crucial that we explore the nature of these dissimilarities (i.e., the attenuation of signals in blood, blood flow rate, etc.), since it impacts the in vivo classification accuracy using an in vitro trained classifier. As long as the differences found in the tissue signatures between the in vitro and in vivo imaging situations are systematic, it may still be possible to empirically retune an in vitro-trained algorithm to reliably perform in vivo. Since the initial angular position of the transducer element during

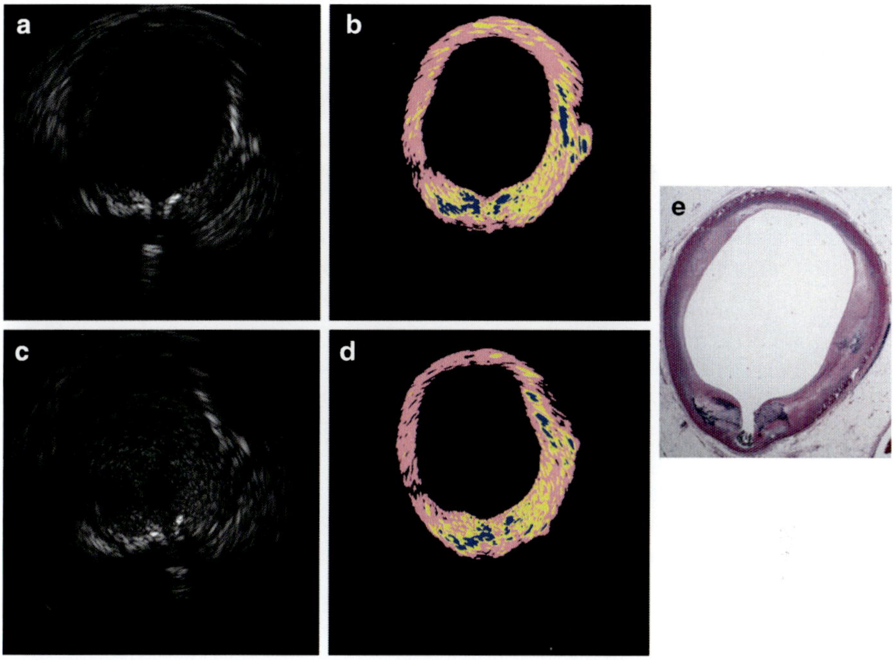

Fig. 6.23 IVUS grayscale image of a cross section of interest using circulating saline (**a**) and corresponding constructed PH image (**b**), IVUS grayscale image of the same cross section acquired with circulating human blood (**c**) and corresponding constructed PH image (**d**), H&E histology image of the same cross section (**e**)

acquisition for two cross section of interest was different, the IVUS grayscale images were slightly rotated.

6.4.5 Interpretation of Histological Images, Labeling of IVUS Frames, and Sufficiency of Data Sets

Since the overall justification of in vivo real-time plaque characterization is made by interventional cardiologists through the use of classified tissues, it is indispensable to train the classifier with the most reliable features. The labeling of IVUS images and validation of tissue color maps are compared with corresponding histology images. Although the histology image is the gold standard, its interpretation can be subjective. Consequently, experts may categorize tissues differently, and as a result, different training data sets may lead to different classification. For example, one can simply categorize tissues as fibrotic, lipid pool, necrotic core and calcified, or one may differentiate between the levels of presence of fatty materials and add fibrolipid to compromise between fibrotic and lipid. Basically, in order to collect a training data set, the regions of interest in the arteries are marked and relative cross-sectional histology images obtained. Hence, the IVUS-histology frame alignment plays a crucial role in this study because the IVUS frames are labeled through the interpretation of the corresponding histology images. Subsequently, the signals are assigned to labeled tissues and relative features are extracted. The IVUS-histology matching problem becomes more challenging due to the: (1) curvature of arteries, especially in the LCX; (2) registration between an IVUS image and its relative histology since the IVUS imaging plane and the slicing plane of the microtome are somewhat different; and (3) shrinkage of the arteries after formalin fixation.

When assessing the accuracy of a characterization system, one must use a cross-validation approach in which different subsets of the database are, in turn, set aside for testing against an algorithm trained on remaining data. It is important that the test subsets be drawn from separate distributions completely absent in the training data. For IVUS tissue characterization, this translates to drawing test data from data not represented in the training data. Note that this condition is not met by creating a test set by simply drawing a random set of signatures from the entire database. If such procedures that estimate accuracy are not followed, it might be estimated overly optimistically, especially for small training set sizes. Once the proper method of assessing accuracy is employed, the estimate turns out to be much lower consistent with what is encountered in clinical evaluations of currently existing algorithms. In fact, increasing the number of ROIs will help to assemble more comprehensive signature database.

6.4.6 Classification of Tissues Behind the Arc of Calcified Plaques

We previously pointed to the unreliability of tissue classification in the regions behind the arc of calcium [67] and raised the question that whether classified tissues in these regions are dependable or not. We concluded that the acoustic shadowing in the grayscale IVUS images, caused by attenuation of signals in the dense-calcified tissues, limited lesion assessment in these regions. We examined cross sections of interest that contained calcified plaques and confirmed that not enough features existed throughout the regions behind the arc of calcified region extended to the vessel wall. Figure 6.24 illustrates a cross section of interest with calcified plaque,

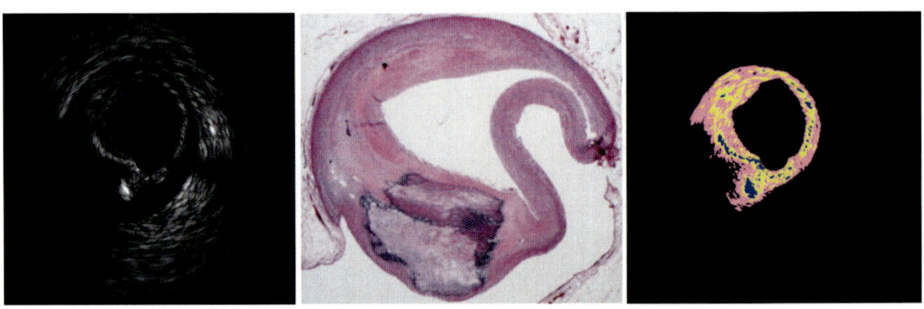

Fig. 6.24 An in vitro acquired IVUS grayscale image (**a**), corresponding H&E histology image (**b**), corresponding constructed PH image (**c**). The arc of calcified plaque appears from 6 o'clock and extended toward 9 o'clock

in which the corresponding region was partially labeled as background or lack of tissue. The results clearly show that there is not much informative features behind arc of calcified regions, from 6 to 9 o'clock, to classify any tissue.

Subsequently, Mintz and Missel [71] mention that "highly calcified lesions are an anatomical limitation to IVUS assessment. However, it is important to differentiate cross-sectional area analysis from volumetric analysis. For grayscale IVUS [image], an arc of 90° makes IVUS measurements of plaque inaccurate owing to extensive shadowing caused by signal attenuation. This is true for a single cross-sectional frame at the minimum lumen area (MLA) site, it but may be not the case for a cross-section 2 or 3 mm away." In fact, newly developed algorithms [20, 54, 64] have taken this limitation into account and either mark tissues in these regions with "no tissue" [20] or detect them automatically and do not classify them [64].

6.5 Summary and Conclusion

There has never been a better time for IVUS catheters. They recently have been enabled to differentiate colors, and this is more than just an esthetic upgrade such as that for television 50 years ago: the additional information shown is extracted from the spectral content of the RF IVUS signal and appears to be related to the atherosclerotic plaque composition. In this chapter, we also reviewed existing texture-based atherosclerotic plaque characterization algorithms as an alternative to spectrum-derived techniques without any one-to-one comparison. In fact discussion on advantages and disadvantages of each presented technique is beyond the scope of this book and it needs further investigation as well as validation.

References

1. M. Naghavi, P. Libby, E. Falk, *et al.*, "From vulnerable plaque to vulnerable patient: a call for new definitions and risk assessment strategies – Part I," *Circulation*, 108, 1664–1672 (2003).
2. J. Mackay, G. Mensah, World Health Organization and US Centers for Disease Control and Prevention, "The atlas of heart disease and stroke," Geneva: WHO (2004).
3. J. A. Ambrose, M. A. Tannenbaum, D. Alexopoulos, *et al.*, "Angiographic progression of coronary artery disease and the development of myocardial infarction," *J. Am. Coll. Cardiol.*, 12(1), 56–62 (1988).
4. D. Hackett, G. Davies, A. Maseri, "Pre-existing coronary stenoses in patients with first myocardial infarction are not necessarily severe," *J. Eur. Heart*, 9(12), 1317–23 (1988).
5. D. Giroud, J. M. Li, P. Urban, *et al.*, "Relation of the site of acute myocardial infarction to the most severe coronary arterial stenosis at prior angiography," *J. Am. Cardiol.*, 69(8), 729–732 (1992).
6. P. D. Richardson, M. J. Davies, G. V. R. Born, "Influence of plaque configuration and stress distribution on fissuring of coronary atherosclerotic plaques," *Lancet*, 21, 941–944 (1989).
7. J. H. Qiao, M. C. Fishbein, "The severity of coronary atherosclerosis at sites of plaque rupture with occlusive thrombosis," *J. Am. Coll. Cardiol.*, 17(5), 1138–1142 (1991).
8. A. P. Burke, A. Farb, G. T. Malcom, *et al.*, "Coronary risk factors and plaque morphology in men with coronary disease who died suddenly," *N. Engl. J. Med.*, 336(18), 1276–1282 (1997).

9. R. Virmani, F. D. Kolodgie, A. P. Burke, *et al.*, "Lessons from sudden coronary death: a comprehensive morphological classification scheme for atherosclerotic lesions," *Arterioscler. Thromb. Vasc. Biol.*, 20(5), 1262–1275 (2000).
10. E. Falk, "Pathogenesis of atherosclerosis," *J. Am. Coll. Cardiol.*, 47(8 Suppl), C7–C12 (2006).
11. P. Schoenhagen, K. M. Ziada, S. R. Kapadia, *et al.*, "Extent and direction of arterial remodeling in stable versus unstable coronary syndromes: an intravascular ultrasound study," *Circulation*, 101(6), 598–603 (2000).
12. A. Maehara, G. S. Mintz, A. B. Bui, *et al.*, "Morphologic and angiographic features of coronary plaque rupture detected by intravascular ultrasound," *J. Am. Coll. Cardiol.*, 40(5), 904–910 (2002).
13. K. Fujii, G. S. Mintz, S. G. Carlier, *et al.*, "Intravascular ultrasound profile analysis of ruptured coronary plaques," *Am. J. Cardiol.*, 98(4), 429–435 (2006).
14. J. E. Muller, A. Tawakol, S. Kathiresan, J. Narula, "New opportunities for identification and reduction of coronary risk: treatment of vulnerable patients, arteries, and plaques," *J. Am. Coll. Cardiol.*, 47, C2–C6 (2006).
15. W. C. Little, "Angiographic assessment of the culprit coronary artery lesion before acute myocardial infarction," *Am. J. Cardiol.*, 66, 44G–47G (1990).
16. J. E. Muller, G. H. Tofler, "Triggering and hourly variation of onset of arterial thrombosis," *Ann. Epidemiol.*, 2, 393–405 (1992).
17. E. Falk, "Why do plaques rupture?" *Circulation*, 86, III30–III42 (1992).
18. P. Libby, "Molecular bases of the acute coronary syndromes," *Circulation*, 91, 2844–2850 (1995).
19. R. Virmani, A. P. Burke, F. D. Kolodgie, A. Farb, "Pathology of the thin-cap fibroatheroma: a type of vulnerable plaque," *J. Inteven. Cardiol.*, 16(3), 267–272 (2003).
20. A. Katouzian, S. Sathyanarayana, B. Baseri, E. E. Konofagou, S. G. Carlier, "Challenges in atherosclerotic plaque characterization with intravascular ultrasound (IVUS): from data collection to classification," *IEEE Trans. Inf. Technol. Biomed.*, 12(3), 315–327 (2008).
21. J. F. Granada, D. W. Bradley, H. K. Win, C. L. Alviar, A. Builes, E. I. Lev, R. Barrios, D. G. Schulz, A. E. Raizner, G. L. Kaluza, "*In vivo* plaque characterization using intravascular ultrasound-virtual histology in a porcine model of complex coronary lesions," *Arterioscler. Thromb. Vasc. Biol.*, 27(2), 387–393 (2007).
22. A. Katouzian, S. Sathyanarayana, W. Li, T. Thomas, S. G. Carlier, "Challenges in tissue characterization from backscattered intravascular ultrasound signals," Proceeding of SPIE (2006).
23. L. Fellingham, F. Sommer, "Ultrasonic characterization of tissue structure in the *in vivo* human liver and spleen," *IEEE Trans. Sonics Ultrason.*, SU-31(4), 418–428 (1984).
24. L. Landini, L. Verrazzani, "Spectral characterization of tissue microstructure by ultrasound:Astochastic approach," *IEEE Trans. Ultrason. Ferroelectr. Freq. Control*, 37(5), 448–456 (1990).
25. W. C. A. Pereira, S. L. Bridal, A. Coron, P. Laugier, "Singular spectrum analysis applied to backscattered ultrasound signals from *in vitro* human cancellous bone specimens," *IEEE Trans. Ultrason. Ferroelectr. Freq. Control*, 51(3), 302–312 (2004).
26. J. Y. David, S. A. Jones, D. P. Giddens, "Modern spectral analysis techniques for blood flow velocity and spectral measurements with pulsed doppler ultrasound," *IEEE Trans. Biomed. Eng.*, 38(6), 589–596 (1991).
27. F. L. Lizzi, M. Ostromogilsky, E. J. Feleppa, M. C. Rorke, and M. M. Yaremko, "Relationship of ultrasonic spectral parameters to features of tissue microstructure," *IEEE Trans. Ultrason. Ferroelectr. Freq. Control*, UFFC-34(3), 319–329 (1986).
28. F. Ossant, M. Lebertre, L. Pourcelot, F. Patat, "Ultrasonic characterization of maturation of fetal lung microstructure: an animal study," *Ultrasound Med. Biol.*, 27(2), 157–169 (2001).
29. F. L. Lizzi, M. A. Laviola, and D. J. Coleman, "Tissue signature characterization utilizing frequency domain analysis," in *Proc. IEEE Ultrason. Symp.*, 714–719 (1976).
30. A. Nair, B. D. Kuban, N. Obuchowski, D. G. Vince, "Assessing spectral algorithms to predict atherosclerotic plaque composition with normalized and raw intravascular ultrasound data," *Ultrasound Med. Biol.*, 27(10), 1319–1331 (2001).

31. F. L. Lizzi, M. Greenebaum, E. J. Feleppa, M. Elbaum, D. J. Coleman, "Theoretical framework for spectrum analysis in ultrasonic tissue characterization," *J. Acoust. Soc. Am.*, 73(4), 1366–1373 (1983).
32. A. Nair, D. G. Vince, D. Calvetti, "Blind data calibration of intravascular ultrasound data for automated tissue characterization," *Proc. IEEE Ultrason. Symp.*, 2, 1126–1129 (2004).
33. A. Nair, B. D. Kuban, M. Tuzcu, P. Schoenhagen, S. E. Nissen, D. G. Vince, "Coronary plaque classification with intravascular ultrasound radiofrequency data analysis," *Circulation*, 106(17), 2200–2206 (2002).
34. A. Nair, D. Calvetti, D. G. Vince, "Regularized autoregressive analysis of intravascular ultrasound backscatter: improvement in spatial accuracy of tissue maps," *IEEE Trans. Ultrason. Ferroelectr. Freq. Control*, 51(4), 420–431 (2004).
35. M. Kawasaki, H. Takatsu, T. Noda, *et al.*, "Noninvasive quantitative tissue characterization and two-dimensional color-coded map of human atherosclerotic lesions using ultrasound integrated backscatter: comparison between histology and integrated backscatter images," *J. Am. Coll. Cardiol.*, 38(2), 486–92 (2002).
36. M. Kawasaki, H. Takatsu, T. Noda, *et al.*, "*In vivo* quantitative tissue characterization of human coronary arterial plaques by use of integrated backscatter intravascular ultrasound and comparison with angioscopic findings," *Circulation*,105(21), 2487–2492 (2002).
37. M. Okubo, M. Kawasaki, Y. Ishihara, *et al.*, "Tissue characterization of coronary plaques: comparison of integrated backscatter intravascular ultrasound with virtual histology intravascular ultrasound," *Circulation Jpn.*, 72, 1631–1639 (2008).
38. L. K. Ryan, F. S. Foster, "Ultrasonic measurement of differential displacement and strain in vascular model," *Ultrason. Imaging*, 19, 19–38 (1997).
39. H. E. Talhmai, L. S. Wilson, M. L. Neale, "Spectral tissue strain: a new technique for imaging tissue strain using intravascular ultrasound," *Ultrasound Med. Biol.*, 20, 759–772 (1994).
40. B. M. Shapo, J. R. Crowe, A. R. Skovoroda, M. Eberle, N. A. Cohn, M. O'Donnell, "Displacement and strain imaging of coronary arteries with intraluminal ultrasound," *IEEE Trans. UFFC*, 43, 234–246 (1996).
41. C. L. de Korte, A. F. W. van der Steen, E. I. Céspedes, G. Pasterkamp, S. G. Carlier, F. Mastik, A. H. Schoneveld, P. W. Serruys, N. Bom, "Characterization of plaque components and vulnerability with intravascular ultrasound elastography," *Physics Med. Biol.*, 45, 1465–1475 (2000).
42. R. L. Maurice, J. Fromagea, M. R. Cardinal, M. Doyley, E. de Muinck, J. Robb, G. Cloutier, "Characterization of atherosclerotic plaques and mural thrombi with intravascular ultrasound elastography: a potetial method evaluated in an aortic rabbit model and a human coronary artery," *IEEE Trans. Inf. Technol. Biomed.*, 12(3), 290–298 (2008).
43. J. M. Tobis, J. Mallery, D. Mahon, K. Lehmann, P. Zalesky, J. Griffith, J. Gessert, M. Moriuchi, M. McRae, M. L. Dwyer, "Intravascular ultrasound imaging of human coronary arteries *in vivo*. Analysis of tissue characterization with comparison to *in vitro* histological specimens," *Circulation*, 83, 913–926 (1991).
44. K. Gad, M. B. Leon, "Characterization of atherosclerotic lesions by intravascular ultrasound: possible role in unstable coronary syndromes and in interventional theraputics procedure," *Am. J. Cardiol.*, 68, 85B–91B (1991).
45. X. Zhang, C. R. McKay, M. Sonka, "Tissue characterization in intravascular ultrasound images," *IEEE Trans. Med. Imaging*, 17(6), 889–899 (1998).
46. K. L. Caballero, O. Pujol, J. Barajas, J. Mauri, P. Radeva, "Assessing *in-vivo* IVUS tissue classification accuracy between normalized image reconstruction and RF analysis," *MICCAI*, 82–89 (2006).
47. E. Brunenberg, O. Pujol, B. H. Romeny, P. Radeva, "Automatic IVUS segmentation of atherosclerotic plaque with stop & go snake," *Med. Image Comput. Comput. Assist. Interv.*, 9(2), 9–16 (2006).
48. R. De Valois, K. De Valois, "Spatial vision," New York: Oxford University Press (1988).
49. J. Beck, A. Sutter, R. Ivry, "Spatial frequency channels and perceptual grouping in texture segregation," *Comput. Vis. Graph. Image Process.*, 37, 299–325 (1987).

50. A. Katouzian, B. Baseri, E. E. Konofagou, A. F. Laine, "Texture-driven coronary artery plaque characterization using wavelet packet signatures," *IEEE International Symposium on Biomedical Imaging: From Nano to Macro* (ISBI) (2008).
51. S. G. Mallat, "A theory of multiresolution signal decomposition: The wavelet representation," *IEEE Trans. Pattern Anal. Mach. Intell.*, 11(7), 674–693 (1989).
52. T. Hiro, C. Y. Leung, S. De Guzman, V. J. Caiozzo, A. R. Fervid, H. Karimi, R. H. Helfant, J. M. Tobis, "Are soft echoes really soft? Intravascular ultrasound assessment of mechanical properties in human atherosclerotic tissue," *Am. Heart J.*, 133, 1–7 (1997).
53. A. Jeremias, M. L. Kolz, T. S. Ikonen, J. F. Gummert, A. Oshima, M. Hayase, Y. Honda, N. Komiyama, G. J. Berry, R. E. Morris, P. G. Yock, P. J. Fitzgerald, "Feasibility of *in vivo* intravascular ultrasound tissue characterization in the detection of early vascular transplant rejection," *Circulation*, 100, 2127–2130 (1999).
54. S. Escalera, O. Pujol, J. Mauri, P. Radeva, "Intravascular ultrasound tissue characterization with sub-class error-correcting output codes," *J. Sign. Process. Syst.*, 55(1–3), 35–47 (2009).
55. P. Ohanian, R. Dubes, "Performance evaluation for four classes of textural features," *Pattern Recognit.*, 25, 819–833 (1992).
56. T. Ojala, M. Pietikainen, T. Maenpaa, "Multiresolution gray-scale and rotation invariant texture classification with local binary patterns," *IEEE Trans. Pattern Anal. Mach. Intell.*, 24, 971–987 (2002).
57. J. G. Daugman, "Uncertainty relation for resolution in space, spatial frequency, and orientation optimized by two-dimensional visual cortical filters," *J. Opt. Soc. Am.*, 2(A), 1160–1169 (1985).
58. A. C. Bovik, M. Clark, W. S. Geisler, "Multichannel texture analysis using localized spatial filters," *IEEE Trans. Pattern Anal. Mach. Intell.*, 12(1), 55–73 (1990).
59. T. Hastie, R. Tibshirani, "Classification by pairwise grouping," *NIPS*, 26, 451–471 (1998).
60. N. J. Nilsson, "Learning machines," New York: McGraw-Hill (1965).
61. E. Allwein, R. Schapire, Y. Singer, "Reducing multiclass to binary: a unifying approach for margin classifiers," *J. Mach. Learn. Res.*, 1, 113–141 (2002).
62. O. Pujol, P. Radeva, J. Vitrià, "Discriminant ECOC: a heuristic method for application dependent design of error correcting output codes," *IEEE Trans. Pattern Anal. Mach. Intell., 28*, 1007–1012 (2006).
63. J. Friedman, T. Hastie, R. Tibshirani, "Additive logistic regression: a statistical view of boosting," *Ann. Stat.*, 38, 337–374 (1998).
64. A. Taki, A. Roodaki, O. Pauly, S. Setarehdan, G. Unal, N. Navab, "A new method for characterization of coronary plaque composition via ivus images," *IEEE International Symposium on Biomedical Imaging: From Nano to Macro* (ISBI) (2009).
65. N. Bruining, S. Verheye, M. Knaapen, "Three-dimensional and quantitative analysis of atherosclerotic plaque composition by automated differential echogenicity," *Catheter. Cardiovasc. Interv.*, 70(7), 968–978 (2007).
66. G. S. Mintz, S. E. Nissen, W. D. Anderson, "American college of cardiology clinical expert consensus document on standards for acquisition, measurement and reporting of intravascular ultrasound studies (IVUS): a report of the American college of cardiology task force on clinical expert consensus documents," *J. Am. Coll. Cardiol.*, 37, 1478–1492 (2001).
67. K. Tanaka, S. Carlier, A. Katouzian, G. Mintz, "Characterization of the intravascular ultrasound radiofrequency signal within regions of acoustic shadowing behind calcium," *J. Am. Coll. Cardiol.*, 49(9 Suppl B), 29B (2007).
68. T. Ojala, M. Pietikäinen, T. Mäenpää, "Multiresolution gray-scale and rotation invariant texture classification with local binary patterns," *IEEE Trans. Pattern Anal. Mach. Intell.*, 24(7), 971–987 (2002).
69. X. Tang, "Texture information in run length matrices," *IEEE Trans. Image Process.*, 7(11), 1602–1609 (1998).
70. E. Picano, L. Landdini, A. Distante, *et al.*, "Angle dependence of ultrasonic backscatter in arterial tissues: a study *in vitro*," *Circulation*, 72, 572–576 (1985).
71. G. S. Mintz, E. Missel, "What is behind the calcium? The relationship between calcium and necrotic core on virtual histology analyses: reply," *Letters to the Editor, European Heart Journal*, 30, 125–126 (2009).

Biographies

Amin Katouzian (S'05) received the B.S. and M.S. degrees in Electrical Engineering from Sistan & Baluchestan University (Iran) and Fairleigh Dickinson University (NJ, USA), respectively, and M.Phil. degree in Biomedical Engineering from Columbia University (NY, USA). He is currently a Ph.D. candidate in the Heffner Biomedical Engineering Lab (HBIL) at Department of Biomedical Engineering in Columbia University. His research interests are in medical image analysis, Statistical random process and pattern recognition, Multidimensional feature extraction and classification, Cue extraction, Multiscale wavelet analysis, Neural Networks, and Audio/Speech signal processing and recognition.

Stéphane G. Carlier received his MD degree from Université Libre de Bruxelles (Belgium) in 1996 and defended his PhD in Biomedical Engineering at Erasmus University, Rotterdam (Netherlands) in 2001. He is an interventional cardiologist working presently at the University Hospital (UZ) Brussel and at the St Maria Hospital in Halle, Belgium. He held previously the position of Assistant Professor of Clinical Medicine at Columbia University, New York and was also the Director of the Intravascular Imaging & Physiology Corelab of the Cardiovascular Research Foundation. He has an extensive record of international publications covering a variety of aspects in academic cardiology. In particular, he contributed significantly to the topic of intracoronary imaging. He serves on several reviewing boards of medical and bioengineering journals. His research interests include intravascular imaging, cardiac and vascular dynamics, physiology and signal processing. He is a member of the European Association of Percutaneous Cardiovascular Interventions (EAPCI) of the European Society of Cardiology.

Andrew F. Laine received his D.Sc. degree from Washington University (St. Louis) School of Engineering and Applied Science in Computer Science, in 1989 and BS degree from Cornell University (Ithaca). He is Director of the Heffner Biomedical Imaging Laboratory in the Department of Biomedical Engineering at Columbia University in New York City and is Professor of Biomedical Engineering and Radiology (Physics). He served as Chair of Technical Committee (TC-BIIP) on Biomedical Imaging and Image Processing for the EMBS (2006–2009); and on the IEEE ISBI (International Symposium on Biomedical Imaging) steering committee since 2006 and Program Chair for the IEEE EMBS annual conference in 2006 (New York City, USA), 2011 (Boston, USA), Program Co-Chair for IEEE ISBI in 2008 (Paris, France); Vice President of Publications for IEEE EMBS since 2008. His research interests include quantitative image analysis; cardiac functional imaging: ultrasound and MRI, retinal imaging, intravascular imaging and biosignal processing. He is also Fellow of IEEE and AIMBE.

Chapter 7
Despeckle Filtering of Ultrasound Images*

Christos P. Loizou and Constantinos S. Pattichis

Abstract It is well known that speckle is a multiplicative noise that degrades the visual evaluation in ultrasound imaging. This necessitates the need for robust despeckling techniques for both routine clinical practice and tele-consultation. The recent advancements in ultrasound instrumentation and portable ultrasound devices necessitate the need of more robust despeckling techniques for enhanced ultrasound medical imaging for both routine clinical practice and tele-consultation. The objective of this chapter is to introduce the theoretical background of a number of despeckle filtering techniques and to carry out a comparative evaluation of despeckle filtering based on texture analysis, image quality evaluation metrics, and visual evaluation by medical experts, on ultrasound images of the carotid artery bifurcation. In this chapter, a total of ten despeckle filters are presented based on local statistics, median filtering, pixel homogeneity, geometric filtering, homomorphic filtering, anisotropic diffusion, nonlinear coherence diffusion, and wavelet filtering. Our results suggest that the first-order statistics filter *DsFlsmv* gave the best performance, followed by the geometric filter *DsFgf4d* and the homogeneous mask area filter *DsFlsminsc*. These filters improved the class separation between the asymptomatic and the symptomatic classes based on the statistics of the extracted texture features, gave only a marginal improvement in the classification success rate, and improved the visual assessment carried out by the two experts. Most importantly, a despeckle filtering and evaluation protocol is proposed based on texture analysis, image quality evaluation metrics, and visual evaluation by experts. In conclusion, the proper selection of a despeckle filter is very important in the enhancement of ultrasonic imaging of the carotid artery. Further work is needed to evaluate at a larger scale and in clinical practice the performance of the proposed despeckle filters in the automated segmentation, texture analysis, and classification of carotid ultrasound imaging.

*To be published in the book *Atherosclerosis Disease Management*

C.P. Loizou (✉)
School of Sciences, Department of Computer Science, Intercollege, 92 Ayias Phylaxeos Street, Limassol 3507, Cyprus
e-mail: loizou.c@lim.Intercollege.ac.cy

Jasjit S. Suri et al. (eds.), *Atherosclerosis Disease Management*,
DOI 10.1007/978-1-4419-7222-4_7,

Keywords Speckle • Despeckle filtering • Ultrasound imaging • Texture analysis • Carotid artery

7.1 Introduction

The use of ultrasound in the diagnosis and assessment of arterial disease is well established because of its noninvasive nature, its low cost, and the continuing improvements in image quality [1]. Speckle, a form of locally correlated multiplicative noise, corrupts medical ultrasound imaging making visual observation difficult [2, 3]. The presence of speckle noise in ultrasound images has been documented since the early 1970s where researchers, such as Burckhardt [2], Wagner [3], and Goodman [4], described the fundamentals and the statistical properties of the speckle noise. Speckle is not truly a noise in the typical engineering sense, since its texture often carries useful information about the image being viewed. It is the primary factor, which limits the contrast resolution in diagnostic ultrasound imaging, thereby limiting the detectability of small, low contrast lesions and making the ultrasound images generally difficult for the nonspecialist to interpret [2–6]. Due to the speckle presence, ultrasound experts with sufficient experience may not often draw useful conclusions from the images [6]. Speckle noise also limits the effective application of image processing and analysis algorithms (i.e., edge detection and segmentation) and display in 2D and volume rendering in 3D. Therefore, speckle is most often considered a dominant source of noise in ultrasound imaging and should be filtered out [2, 5, 6] without affecting important features of the image. The objective of this chapter was to carry out a comparative evaluation of despeckle filtering techniques based on texture analysis, image quality evaluation metrics as well as visual assessment by the experts on 440 ultrasound images of the carotid artery bifurcation. Results of this study were also published in [7–9]. Furthermore, to introduce the theoretical background (equations), examples of the Matlab™ code for a number of despeckle filters are given. Finally, at the end of this book, an appendix provides details about the despeckle filtering Matlab™ toolbox which can be download at http://www.medinfo.cs.ucy.ac.cy/index.php/matlab-software.

The wide spread of mobile and portable telemedicine ultrasound scanning instruments also necessitates the need for better image-processing techniques, in order to offer a clearer image to the medical practitioner. This makes the use of efficient despeckle filtering a very important task. Early attempts to suppress speckle noise were implemented by averaging of uncorrelated images of the same tissue recorded under different spatial positions [5, 10, 11]. While these methods are effective for speckle reduction, they require multiple images of the same object to be obtained [12]. Speckle reducing filters originated from the synthetic aperture radar (SAR) community [10]. These filters have then later been applied to ultrasound imaging since the early 1980s [13]. Filters that are used widely in both SAR and ultrasound imaging include the Frost [14], Lee [10, 15, 16], and Kuan [12, 17].

Table 7.1 summarizes the despeckle filtering techniques that are investigated in this study, grouped under the following categories: local statistics, median filtering, homogeneity, geometric, homomorphic, anisotropic diffusion, and wavelet filtering. Furthermore, in Table 7.1, the main investigators, the methodology used, and the corresponding filter names are given. These filters are briefly introduced in this section, and presented in greater detail in Sect. 7.2.

Some of the local statistic filters are the Lee [10, 15, 16], the Frost [14], and the Kuan [12, 17]. The Lee and Kuan filters have the same structure, whereas the Kuan is a generalization of the Lee filter. Both filters form the output image by

Table 7.1 An overview of despeckle filtering techniques

Speckle reduction technique	Method	Investigator	Filter name
Linear filtering	Moving window utilizing local statistics	[7–17]	*DsFlsmv*
		[9–15]	
	(a) mean (m), variance (σ^2)	[7, 9, 34]	
	(b) mean, variance, third and fourth moments (higher statistical moments), and entropy	[2–16]	*DsFlsmvsk1d, DsFlsmvsk2d*
	(c) Homogeneous mask area filters		*DsFlsminsc*
	(d) *DsFwiener* filtering		*DsFwiener*
Nonlinear filtering	Median filtering	[7, 35]	*DsFmedian*
	Linear scaling of the gray level values	[7, 9, 48]	*DsFls* *DsFca DsFlecasort*
	Based on the most homogeneous neighborhood around each pixel	[7, 9, 18]	*DsFhomog*
	Nonlinear iterative algorithm (geometric filtering)	[7, 9, 11]	*DsFgf4d*
	The image is logarithmically transformed, the fast Fourier transform (FFT) is computed, denoised, the inverse FFT is computed, and finally exponentially transformed back	[2, 7, 19, 20]	*DsFhomo*
Diffusion filtering	Nonlinear filtering technique for simultaneously performing contrast enhancement and noise reduction	[2, 5, 7, 9, 13, 14, 21–25]	*DsFad*
		[5, 7, 9]	
	Exponential damp kernel filters utilizing diffusion	[7, 9, 26]	
		[7, 9, 26]	
	Speckle reducing anisotropic diffusion based on the coefficient of variation		*DsFsrad*
	Coherence enhancing diffusion		*DsFnldif*
Wavelet filtering	Only the useful wavelet coefficients are utilized	[9, 16, 27–31, 37–39]	*DsFwaveltc*

Source: [7], © 2005.

computing the central pixel intensity inside a filter-moving window, which is calculated from the average intensity values of the pixels and a coefficient of variation inside the moving window. Kuan considered a multiplicative speckle model and designed a linear filter, based on the minimum-mean-square error (MMSE) criterion that has optimal performance when the histogram of the image intensity is Gaussian distributed. The Lee [10] filter is a particular case of the Kuan filter based on a linear approximation made for the multiplicative noise model. The Frost [14] makes a balance between the averaging and the all-pass filters. It was designed as an adaptive Wiener filter that assumed an autoregressive exponential model for the image.

In the homogeneity group, the filtering is based on the most homogeneous neighborhood around each image pixel [9, 18]. Geometric filters [11] are based on nonlinear iterative algorithms, which increment or decrement the pixel values in a neighborhood based upon their relative values. The method of homomorphic filtering [19, 20] is similar to the logarithmic point operations used in histogram improvement, where dominant bright pixels are de-emphasized. In the homomorphic filtering, the fast Fourier transform (FFT) of the image is calculated, then denoised, and then the inverse FFT is calculated.

Some other despeckle filtering methods, such as anisotropic diffusion [2, 21–25], speckle reducing anisotropic diffusion [5], and coherence anisotropic diffusion [26] presented recently in the literature, are nonlinear filtering techniques for simultaneously performing contrast enhancement and noise reduction by utilizing the coefficient of variation [5]. Furthermore, in the wavelet category, filters for suppressing the speckle noise were documented. These filters are making use of a realistic distribution of the wavelet coefficients [2, 16, 27–32] where only the useful wavelet coefficients are utilized. Different wavelet shrinkage approaches were investigated usually based on Donoho's work [31].

The majority of speckle reduction techniques have certain limitations that can be briefly summarized as follows:

1. They are sensitive to the size and shape of the window. The use of different window sizes greatly affects the quality of the processed images. If the window is too large over smoothing will occur, subtle details of the image will be lost in the filtering process and edges will be blurred. On the other hand, a small window will decrease the smoothing capability of the filter and will not reduce the speckle noise thus making the filter not effective.
2. Some of the despeckle methods based on window approaches require thresholds to be used in the filtering process, which have to be estimated empirically. The inappropriate choice of a threshold may lead to average filtering and noisy boundaries thus leaving the sharp features unfiltered [7, 11, 15].
3. Most of the existing despeckle filters do not enhance the edges but they only inhibit smoothing near the edges. When an edge is contained in the filtering window, the coefficient of variation will be high and smoothing will be inhibited. Therefore, speckle in the neighborhood of an edge will remain after filtering. They are not directional in the sense that in the presence of an edge, all smoothing

is precluded. Instead of inhibiting smoothing in directions perpendicular to the edge, smoothing in directions parallel to the edge is allowed.
4. Different evaluation criteria for evaluating the performance of despeckle filtering are used by different studies. Although most of the studies use quantitative criteria like the mean-square error (MSE) and speckle index (C), there are additional quantitative criteria, such as texture analysis and classification, image quality evaluation metrics, and visual assessment by experts, that could be investigated.

To the best of our knowledge, there is only one study which investigated despeckle filtering on ultrasound images of the carotid artery and proposed speckle reducing anisotropic diffusion as the most appropriate method [5]. This technique was compared with the Frost [14], Lee [15], and the homomorphic filtering [20] and documented that anisotropic diffusion performed better.

In this study, we compare the performance of ten despeckle filters on 440 ultrasound images of the carotid artery bifurcation. The performance of these filters was evaluated using texture analysis, the kNN classifier, image quality evaluation metrics, and visual evaluation by two experts. The results of our study show that despeckle filtering improves the class separation between asymptomatic and symptomatic ultrasound images of the carotid artery.

In the following section, a brief overview of despeckle filtering techniques is presented. In Sect. 7.3, the methodology is presented, covering the material, recording of ultrasound images, texture and statistical analysis, the kNN classifier, image quality evaluation metrics, and the experiment carried out for visual evaluation is described. Sects. 7.4 and 7.5 present the results and discussion, respectively. Finally, Sect. 7.6 presents the summary and future directions, where a despeckling filtering and evaluation protocol is also proposed. A despeckle filtering toolbox proposed in [9], which includes the MATLAB code for despeckling, image quality evaluation, and texture analysis, is also available to download at http://www.medinfo.cs.ucy.ac.cy.

7.2 Despeckle Filtering

In order to be able to derive an efficient despeckle filter, a speckle noise model is needed. The speckle noise model may be approximated as multiplicative, if the envelope signal which is received at the output of the beamformer of the ultrasound imaging system, is captured before logarithmic compression, and may be defined as:

$$y_{i,j} = x_{i,j} n_{i,j} + a_{i,j}, \tag{7.1}$$

where $y_{i,j}$ represents the noisy pixel in the middle of the moving window, $x_{i,j}$ represents the noise-free pixel, $n_{i,j}$ and $a_{i,j}$ represent the multiplicative and additive noise, respectively, and i, j are the indices of the spatial locations that belong to the 2D space of real numbers, $i, j \in \Re^2$. Logarithmic compression is applied to the envelope-detected echo signal in order to fit it in the display range [26, 33]. It has been shown that the logarithmic compression affects the speckle noise statistics in such a way that the local mean becomes proportional to the local variance rather than the standard deviation [26, 28, 30, 33] [see also (7.8)]. More specifically, logarithmic compression affects the high-intensity tail of the Rayleigh and Rician probability density function (PDF) more than the low intensity part. As a result, the speckle noise becomes very close to white Gaussian noise corresponding to the uncompressed Rayleigh signal [33]. Since the effect of additive noise is considerably smaller compared with that of multiplicative noise, (7.1) may be written as:

$$y_{i,j} \approx x_{i,j} n_{i,j}. \tag{7.2}$$

Thus, the logarithmic compression transforms the model in (7.2) into the classical signal in additive noise form as:

$$\log(y_{i,j}) = \log(x_{i,j}) + \log(n_{i,j}), \tag{7.3a}$$

$$g_{i,j} = f_{i,j} + nl_{i,j}. \tag{7.3b}$$

For the rest of the paper, the term $\log(y_{i,j})$, which is the observed pixel on the ultrasound image display after logarithmic compression, is denoted as $g_{i,j}$, and the terms $\log(x_{i,j})$ and $\log(n_{i,j})$, which are the noise-free pixel and noise component after logarithmic compression, as $f_{i,j}$ and $nl_{i,j}$, respectively [see (7.3b)].

7.2.1 *Local Statistics Filtering*

Most of the techniques for speckle reduction filtering in the literature use local statistics. Their working principle may be described by a weighted average calculation using subregion statistics to estimate statistical measures over different pixel windows varying from 3×3 up to 15×15. All these techniques assume that the speckle noise model has a multiplicative form as given in (7.2) [7–16, 26, 28].

7.2.1.1 First-Order Statistics Filtering (*DsFlsmv, DsFwiener*)

The filters utilizing the first-order statistics such as the variance and the mean of the neighborhood may be described with the model as in (7.3). Hence, the algorithms in this class may be traced back to the following equation [5, 7–17]:

$$f_{i,j} = \overline{g} + k_{i,j}(g_{i,j} - \overline{g}), \tag{7.4}$$

where $f_{i,j}$ is the estimated noise-free pixel value, $g_{i,j}$ is the noisy pixel value in the moving window, $\overline{g}$ is the local mean value of an $N_1 \times N_2$ region surrounding and including pixel $g_{i,j}$, $k_{i,j}$ is a weighting factor with $k \in [0..1]$ and i, j as the pixel coordinates. The factor $k_{i,j}$ is a function of the local statistics in a moving window. It can be found in the literature [9, 10, 12, 15], and may be derived in different forms, that:

$$k_{i,j} = \frac{1 - \overline{g}^2 \sigma^2}{\sigma^2 (1 + \sigma_n^2)}, \tag{7.5}$$

$$k_{i,j} = \frac{\sigma^2}{\overline{g}^2 \sigma_n^2 + \sigma^2}, \tag{7.6}$$

$$k_{i,j} = \frac{\sigma^2 - \sigma_n^2}{\sigma^2}. \tag{7.7}$$

The values σ^2 and σ_n^2 represent the variance in the moving window and the variance of noise in the whole image, respectively. The noise variance may be calculated for the logarithmically compressed image, by computing the average noise variance over a number of windows with dimensions considerably larger than the filtering window. In each window, the noise variance is computed as:

$$\sigma_n^2 = \sum_{i=1}^{p} \frac{\sigma_p^2}{\overline{g}_p}, \tag{7.8}$$

where σ_p^2 and $\overline{g}_p$ are the variance and mean of the noise in the selected windows, respectively, and p is the index covering all windows in the whole image [26, 27, 33]. If the value of $k_{i,j}$ is 1 (in edge areas), this will result to an unchanged pixel, whereas a value of 0 (in uniform areas) replaces the actual pixel by the local average, $\overline{g}$, over a small region of interest [see (7.4)]. In this study, the filter *DsFlsmv* uses (7.5). The filter *DsFwiener* uses a pixel-wise adaptive Wiener method [2–6, 14] implemented as given in (7.4), with the weighting factor $k_{i,j}$ as given in (7.7). For both despeckle filters *DsFlsmv* and *wiener*, the moving window size was 5×5. For running the despeckle filter *DsFlsmv*, see the example given in the Appendix.

7.2.1.2 Homogeneous Mask Area Filtering (*DsFlsminsc*)

The *DsFlsminsc* is a 2D filter operating in a 5×5 pixel neighborhood by searching for the most homogenous neighborhood area around each pixel, using a 3×3 subset window [9, 34]. The middle pixel of the 5×5 pixel neighborhood is substituted with the average gray level of the 3×3 pixel mask with the smallest speckle index, C, where C for log-compressed images is given by:

$$C = \frac{\sigma_s^2}{\overline{g}_s}, \tag{7.9}$$

where σ_s^2 and $\overline{g}_s$ represent the variance and mean of the 3×3 window, respectively. The window with the smallest C is the most homogenous semiwindow, which presumably does not contain any edge. The filter is applied iteratively until the gray levels of almost all pixels in the image do not change.

7.2.2 Median Filtering (DsFmedian)

The filter *DsFmedian* [35] is a simple nonlinear operator that replaces the middle pixel in the window with the median value of its neighbors. The moving window for the *DsFmedian* filter was 7×7 pixels.

7.2.3 Maximum Homogeneity Over a Pixel Neighborhood Filtering (DsFhomog)

The *DsFhomog* filter is based on an estimation of the most homogeneous neighborhood around each image pixel [36]. The filter takes into consideration only pixels that belong to the processed neighborhood (7×7 pixels) using (7.10), under the assumption that the observed area is homogeneous. The output image is then given by:

$$f_{i,j} = \left(c_{i,j}g_{i,j}\right) / \sum_{i,j} c_{i,j}, \text{ with } c_{i,j} = 1 \text{ if} (1-2\sigma_n)\,\overline{g} \leq g_{i,j} \leq (1+2\sigma_n)\,\overline{g}, \quad c_{i,j} = 0 \text{ otherwise.} \tag{7.10}$$

The *DsFhomog* filter does not require any parameters or thresholds to be tuned, thus making the filter suitable for automatic interpretation.

7.2.4 Geometric Filtering (DsFgf4d)

The concept of the geometric filtering is that speckle appears in the image as narrow walls and valleys. The geometric filter, through iterative repetition, gradually tears down the narrow walls (bright edges) and fills up the narrow valleys (dark edges), thus smearing the weak edges that need to be preserved.

The *DsFgf4d* filter [11] investigated in this study uses a nonlinear noise reduction technique. It compares the intensity of the central pixel in a 3×3 neighborhood with those of its eight neighbors and, based upon the neighborhood pixel intensities, it increments or decrements the intensity of the central pixel such that it becomes more representative of its surroundings. The operation of the geometric filter *DsFgf4d* may be described with Fig. 7.1 and has the following form:

1. *Select direction and assign pixel values*
   Select the direction be NS and the corresponding three consecutive pixels be a, b, c (see Fig. 7.1a, b).

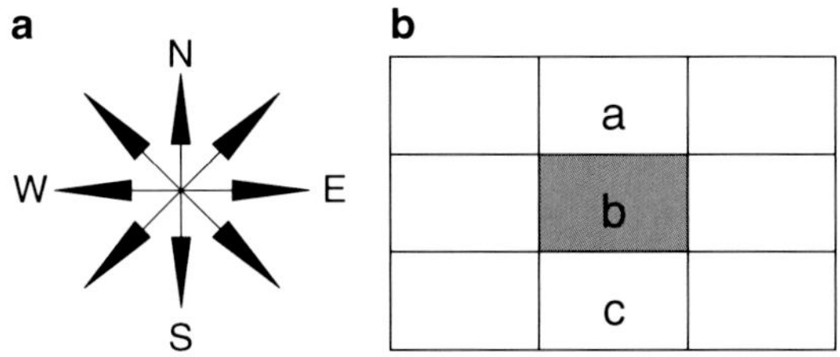

Fig. 7.1 (**a**) Directions of implementation of the *DsFgf4d* geometric filter, (**b**) pixels selected for the NS direction (intensity of central pixel b is adjusted based on the values of intensities of pixels a, b, and c). Source: [9], © 2008

2. *Carry out central pixel adjustments*
   Do the following intensity adjustments (see Fig. 7.1b)
   if $a \geq b+2$ then $b=b+1$
   if $a>b$ and $b \leq c$ then $b=b+1$,
   if $c>b$ and $b \leq a$ then $b=b+1$,
   if $c \leq b+2$ then $b=b+1$,
   if $a \leq b-2$ then $b=b-1$,
   if $a<b$ and $b \geq c$ then $b=b-1$,
   if $c<b$ and $b \geq a$ then $b=b-1$
   if $c \leq b-2$ then $b=b-1$.
3. Repeat steps 1 and 2 for directions west-east (WE), west-north to south-east (WN-SE), and north-east to west-south (NE to WS) (see Fig. 7.1a).

7.2.5 Homomorphic Filtering (*DsFhomo*)

The *DsFhomo* filter performs homomorphic filtering for image enhancement by calculating the FFT of the logarithmic-compressed image, applying a denoising homomorphic filter function $H(.)$, and then performing the inverse FFT of the image [19, 20]. The homomorphic filter function $H(.)$ maybe constructed either using a band-pass Butterworth or a high-boost Butterworth filter. In this study, a high-boost Butterworth filter was used with the homomorphic function [19]:

$$H_{u,v} = \gamma_{\mathrm{L}} + \frac{\gamma_{\mathrm{H}}}{1+(D_0 / D_{u,v})^2}, \tag{7.11a}$$

with

$$D_{u,v} = \sqrt{\left(u-\frac{N}{2}\right)^2 + \left(v-\frac{N}{2}\right)^2}, \tag{7.11b}$$

where $D_0 = 1.8$ is the cut-off frequency of the filter; $\gamma_{\mathrm{L}} = 0.4$ and $\gamma_{\mathrm{H}} = 0.6$ are the gains for the low and high frequencies, respectively; u and v are the spatial coordinates of the frequency-transformed image; and N the dimensions of the image in the u and v space.

This form of filtering sharpens the features and flattens the speckle variations in an image.

7.2.6 Diffusion Filtering

Diffusion filters remove noise from an image by modifying the image via solving a partial differential equation (PDE). The smoothing is carried out depending on the image edges and their directions. Anisotropic diffusion is an efficient nonlinear technique for simultaneously performing contrast enhancement and noise reduction. It smoothes homogeneous image regions but retains image edges [5, 24, 25] without requiring any information from the image power spectrum. It may thus directly be applied to logarithmic-compressed images. Consider applying the isotropic diffusion equation given by $\mathrm{d}g_{i,j,t}/\mathrm{d}t = \mathrm{div}(d\nabla g)$ using the original noisy image, $g_{i,j,t=0}$, as the initial condition, where $g_{i,j,t=0}$ is an image in the continuous domain, i, j specifies spatial position, t is an artificial time parameter, d is the diffusion constant, and ∇g is the image gradient. Modifying the image according to this linear isotropic diffusion equation is equivalent to filtering the image with a Gaussian filter. In this section, we will present conventional anisotropic diffusion (*DsFad*) and coherent nonlinear anisotropic diffusion (*DsFnldif*).

7.2.6.1 Anisotropic Diffusion Filtering (*DsFad*)

Perona and Malik [25] replaced the classical isotropic diffusion equation, as described above, by the introduction of a function $d_{i,j,t} = f(|\nabla g|)$ that smoothes the original image while trying to preserve brightness discontinuities with:

$$\frac{\mathrm{d}g_{i,j,t}}{\mathrm{d}t} = \mathrm{div}\left[d_{i,j,t}\nabla g_{i,j,t}\right] = \left[\frac{\mathrm{d}}{\mathrm{d}i}d_{i,j,t}\frac{\mathrm{d}}{\mathrm{d}i}g_{i,j,t}\right] + \left[\frac{\mathrm{d}}{\mathrm{d}j}d_{i,j,t}\frac{\mathrm{d}}{\mathrm{d}j}g_{i,j,t}\right], \quad (7.12\mathrm{a})$$

where $|\nabla g|$ is the gradient magnitude and $d(|\nabla g|)$ is an edge stopping function, which is chosen to satisfy $d \to 0$ when $|\nabla g| \to \infty$ so that the diffusion is stopped across edges. This function, called the diffusion coefficient, $d(|\nabla g|)$, which is a monotonically decreasing function of the gradient magnitude, $|\nabla g|$, yields intraregion smoothing not interregion smoothing [21, 22, 24, 25] by impeding diffusion at image edges. It increases smoothing parallel to the edge and stops smoothing perpendicular to the edge, as the highest gradient values are perpendicular to the edge and dilated across edges. The choice of $d(|\nabla g|)$ can greatly affect the extent to which discontinuities are preserved. For example, if $d(|\nabla g|)$ is constant at all locations, then smoothing progresses in an isotropic manner. If $d(|\nabla g|)$ is allowed to vary according to the local image gradient, then we have anisotropic diffusion.

A basic anisotropic PDE is given in (7.12a). Two different diffusion coefficients were proposed in [25] and also derived in [24]. The diffusion coefficient suggested were

$$d(|\nabla g|) = \frac{1}{1+(|\nabla g_{i,j}|/K)^2}, \tag{7.12b}$$

where K, in (7.12b), is a positive gradient threshold parameter, known as diffusion or flow constant [24]. In our study, the diffusion coefficient in (7.12b) was used as it was found to perform better in our images of carotid artery.

A discrete formulation of the anisotropic diffusion in (7.12a) is [2, 24, 25]

$$\frac{\mathrm{d}g_{i,j}}{\mathrm{d}t} = \frac{\lambda}{|\eta_s|}\begin{Bmatrix} d_{i+1,j,t}\left[g_{i+1,j} - g_{i,j}\right] + d_{i-1,j,t}\left[g_{i-1,j} - g_{i,j}\right] + \\ d_{i,j+1,t}\left[g_{i,j+1} - g_{i,j}\right] + d_{i,j-1,t}\left[g_{i,j-1} - g_{i,j}\right] \end{Bmatrix}, \tag{7.13a}$$

where the new pixel gray value $f_{i,j}$ at location i, j is

$$f_{i,j} = g_{i,j} + \frac{1}{4}\frac{\mathrm{d}g_{i,j}}{\mathrm{d}t}, \tag{7.13b}$$

where $d_{i+1,j,t}$, $d_{i-1,j,t}$, $d_{i,j+1,t}$, and $d_{i,j-1,t}$ are the diffusion coefficients for the west, east, north, and south pixel directions, in a four pixel neighborhood, around the pixel i, j where diffusion is computed, respectively. The coefficient of variation leads to the largest diffusion where the nearest-neighbor difference is large (the largest edge), while the smallest diffusion is calculated where the nearest-neighbor difference is small (the weakest edge). The constant $\lambda \in \Re^+$ is a scalar that determines the rate of diffusion, η_s represents the spatial neighborhood of pixel i, j, and $|\eta_s|$ is the number of neighbors (usually four except at the image boundaries). Perona and Malik [25] linearly approximated the directional derivative in a particular direction as $\nabla g_{i,j} = g_{i+1,j} - g_{i,j}$ (for the east direction of the central pixel i, j). Modifying the image according to the above equation in (7.13), which is a linear isotropic diffusion equation, is equivalent to filtering the image with a Gaussian filter. The parameters for the anisotropic diffusion filter used in this study were $\lambda = 0.25$, $\eta_s = 8$, and the parameter $K = 30$, which was used for the calculation of the edge stopping function $d(|\nabla g|)$ in (7.12b).

7.2.6.2 Coherent Nonlinear Anisotropic Diffusion Filtering (*DsFnldif*)

The applicability of the *DsFad* filter (7.12) is restricted to smoothing with edge enhancement, where $|\nabla g|$ has higher magnitude at edges. In general, the function $d(|\nabla g|)$ in (7.12) can be put into a tensor form that measures local coherence of structures such that the diffusion process becomes more directional in both the gradient and the contour directions, which represent the directions of maximum and minimum variations, respectively. Therefore, the *DsFnldif* filter will take the form:

$$\frac{\mathrm{d}g_{i,j,t}}{\mathrm{d}t} = \mathrm{div}\left[D\nabla g\right] \tag{7.14a}$$

where $D \in \Re^{2x2}$ is a symmetric positive semidefinite diffusion tensor representing the required diffusion in both gradient and contour directions and, hence, enhancing coherent structures as well as edges. The design of D as well as the derivation of the coherent nonlinear anisotropic diffusion model may be found in [26] and isgiven as:

$$D = (\omega_1 \, \omega_2) \begin{pmatrix} \lambda_1 & 0 \\ 0 & \lambda_2 \end{pmatrix} \begin{pmatrix} \omega_1^{\mathrm{T}} \\ \omega_2^{\mathrm{T}} \end{pmatrix} \tag{7.14b}$$

with

$$\lambda_1 = \begin{cases} \alpha\left(1 - \dfrac{(\mu_1 - \mu_2)^2}{s^2}\right)^{\mathrm{if}(\lambda_1 - \lambda_2)^2 \le s^2} \\ 0, \qquad\qquad\qquad\qquad \mathrm{else} \end{cases} \tag{7.14c}$$

$$\lambda_2 = \alpha.$$

where the eigenvectors ω_1 and ω_2 and the eigenvalues λ_1 and λ_2 correspond to the directions of maximum and minimum variations and the strength of these variations, respectively. The flow at each point is affected by the local coherence, which is measured by $(\mu_1 - \mu_2)$ in (7.14c).

The parameters used in this study for the *nldif* filter were $s^2 = 2$ and $\alpha = 0.9$, which were used for the calculation of the diffusion tensor D, and the parameter step size $m = 0.2$, which defined the number of diffusion steps performed. The local coherence is close to zero in very noisy regions and diffusion must become isotropic ($\mu_1 = \mu_2 = \alpha = 0.9$), whereas in regions with lower speckle noise the local coherence must correspond to $(\mu_1 - \mu_2)^2 > s^2$ [26].

7.2.7 *Wavelet Filtering (DsFwaveltc)*

Speckle reduction filtering in the wavelet domain, used in this study, is based on the idea of the Daubenchies Symlet wavelet and on soft-thresholding denoising. It was first proposed by Donoho [31] and also investigated by [27, 28, 37–40]. The Symmlets family of wavelets, although not perfectly symmetrical, were designed to have the least asymmetry and highest number of vanishing moments for a given compact support [31]. The *DsFwaveltc* filter implemented in this study is described as follows:

1. Estimate the variance of the speckle noise σ_n^2 from the logarithmic-transformed noisy image with (7.8).
2. Compute the discrete wavelet transform (DWT) using the Symlet wavelet for two scales.
3. For each sub-band

(a) Compute a threshold [29, 31]

$$T = \begin{cases} (T_{\max} - \alpha(j-1))\sigma_n, & \text{if } T_{\max} - \alpha(j-1) > T_{\min}, \\ T_{\min}\sigma_n, & \text{else,} \end{cases} \tag{7.15}$$

where α is a decreasing factor between two consecutive levels, $T_{\max}$ is a maximum factor for σ_n, and $T_{\min}$ is a minimum factor. The threshold T is primarily calculated using σ_n and a decreasing factor $T_{\max} - \alpha(j-1)$.

(b) Apply the thresholding procedure in (a) on the wavelet coefficients in step 2.

4. Invert the multiscale decomposition to reconstruct the despeckled image f.

7.3 Methodology

7.3.1 *Material*

A total of 440 ultrasound images of the carotid artery bifurcation, 220 asymptomatic and 220 symptomatic, were investigated in this study. Asymptomatic images were recorded from patients at risk of atherosclerosis in the absence of clinical symptoms, whereas symptomatic images were recorded from patients at risk of atherosclerosis who have already developed clinical symptoms, such as a stroke episode.

7.3.2 *Recording of Ultrasound Images*

In this study, ultrasound images of the carotid artery bifurcation were acquired using the ATL HDI-3000 ultrasound scanner. The ATL HDI-3000 ultrasound scanner is equipped with 64 elements fine pitch high-resolution, 38-mm broadband array, a multielement ultrasound scan head with an operating frequency range of 4–7 MHz, an acoustic aperture of 10 mm $\times$ 8 mm, and a transmission focal range of 0.8–11 cm [41]. In this work, all images were recorded as they are displayed in the ultrasound monitor after logarithmic compression. The images were recorded digitally on a magneto optical drive, with a resolution of 768 $\times$ 756 pixels with 256 gray levels. The image resolution was 16.66 pixels/mm.

7.3.3 *Despeckle Filtering*

Ten despeckle filters were investigated as presented in Sect. 7.2 and were applied on the 440 logarithmically compressed ultrasound images.

7.3.4 Texture Analysis

Texture provides useful information for the characterization of atherosclerotic plaque [42]. In this study, a total of 56 different texture features were extracted both from the original and the despeckled images as follows [42, 43]:

1. *Statistical features* (*SF*): (1) Mean, (2) Median, (3) Variance (σ^2), (4) Skewness (σ^3), (5) Kurtosis (σ^4), and (6) Speckle index (σ / m).
2. *Spatial gray level dependence matrices* (*SGLDM*) as proposed by Haralick et al. [43]: (1) Angular second moment (ASM), (2) Contrast, (3) Correlation, (4) Sum of squares: variance, (5) Inverse difference moment (IDM), (6) Sum average, (7) Sum variance, (8) Sum entropy, (9) Entropy, (10) Difference variance, (11) Difference entropy, (12) and (13) Information measures of correlation. Each feature was computed using a distance of 1 pixel. Also, for each feature, the mean values and the range of values were computed and were used as two different feature sets.
3. *Gray level difference statistics* (*GLDS*) [44]: (1) Contrast, (2) ASM, (3) Entropy, and (4) Mean.
4. *Neighborhood gray tone difference matrix* (*NGTDM*) [45]: (1) Coarseness, (2) Contrast, (3) Business, (4) Complexity, and (5) Strength.
5. *Statistical feature matrix* (*SFM*) [46]: (1) Coarseness, (2) Contrast, (3) Periodicity, and (4) Roughness.
6. *Laws texture energy measures* (*TEM*) [46]: For the laws TEM extraction, vectors of length $l = 7$, $L = (1,6,15,20,15,6,1)$, $E = (-1,-4,-5,0,5,4,1)$ and $S = (-1,-2,1,4,1-2,-1)$ were used, where L performs local averaging, E acts as an edge detector, and S acts as a spot detector. The following TEM features were extracted: (1) LL – texture energy (TE) from LL kernel, (2) EE – TE from EE kernel, (3) SS – TE from SS kernel, (4) LE – average TE from LE and EL kernels, (5) ES – average TE from ES and SE kernels, and (6) LS – average TE from LS and SL kernels.
7. *Fractal dimension texture analysis* (*FDTA*) [46]: Hurst coefficient, $H^{(k)}$ for resolutions $k = 1, 2, 3, 4$.
8. *Fourier power spectrum* (*FPS*) [46]: (1) Radial sum and (2) Angular sum.

7.3.5 Distance Measures

In order to identify the most discriminant features separating asymptomatic and symptomatic ultrasound images before and after despeckle filtering, the following distance measure was computed for each feature [42]:

$$\text{dis}_{zc} = \frac{|m_{za} - m_{zs}|}{\sqrt{\sigma_{za}^2 + \sigma_{zs}^2}}, \tag{7.16}$$

where z is the feature index, c if o indicates the original image set and if f indicates the despeckled image set, m_{za} and m_{zs} are the mean values, and σ_{za} and σ_{zs}

are the standard deviations of the asymptomatic and symptomatic classes, respectively. The most discriminant features are the ones with the highest distance values [42]. If the distance after despeckle filtering is increased, i.e.,

$$\mathrm{dis}_{zf} > \mathrm{dis}_{zo} \tag{7.17}$$

then it can be derived that the classes may be better separated.

For each feature, a percentage distance was computed as

$$\mathrm{feat\_dis}_z = (\mathrm{dis}_{zf} - \mathrm{dis}_{zo}) \times 100. \tag{7.18}$$

For each feature set, a score distance was computed as

$$\mathrm{Score\_Dis} = \frac{1}{N}\sum_{z=1}^{N}(\mathrm{dis}_{zf} - \mathrm{dis}_{zo}) \times 100, \tag{7.19}$$

where N is the number of features in the feature set. It should be noted that for all features, a larger feature distance shows improvement.

7.3.6 *Univariate Statistical Analysis*

The Wilcoxon rank sum test was used in order to detect if for each texture feature a significant (S) difference or not (NS) exists between the original and the despeckled images at $p<0.05$.

7.3.7 *kNN Classifier*

The statistical k-nearest-neighbor (kNN) classifier using the Euclidean distance with $k=7$ was also used to classify a plaque image as asymptomatic or symptomatic [42]. The leave-one-out method was used for evaluating the performance of the classifier, where each case is evaluated in relation to the rest of the cases. This procedure is characterized by no bias concerning the possible training and evaluation bootstrap sets. The kNN classifier was chosen because it is simple to implement and computationally very efficient. This is highly desired due to the many feature sets and filters tested [46].

7.3.8 *Image Quality Evaluation Metrics*

Differences between the original $g_{i,j}$ and the despeckled $f_{i,j}$ images were evaluated using image quality evaluation metrics. The following measures, which are easy to compute and have clear physical meaning, were computed [8, 9]:

1. The MSE

$$\mathrm{MSE} = \frac{1}{MN}\sum_{i=1}^{M}\sum_{j=1}^{N}(g_{i,j} - f_{i,j})^2, \tag{7.20}$$

which measures the quality change between the original and processed image in an $M \times N$ window [47].

2. The root MSE (RMSE), which is the square root of the squared error averaged over an $M \times N$ window [48]

$$\mathrm{RMSE} = \sqrt{\frac{1}{MN}\sum_{i=1}^{M}\sum_{j=1}^{N}(g_{i,j} - f_{i,j})^2}. \tag{7.21}$$

3. The error summation in the form of the Minkowski metric, which is the norm of the dissimilarity between the original and the despeckled images [49]:

$$\mathrm{Err} = \left(\frac{1}{MN}\sum_{i=1}^{M}\sum_{j=1}^{N}\left|g_{i,j} - f_{i,j}\right|^{\beta}\right)^{1/\beta} \tag{7.22}$$

computed for $\beta = 3$ (Err_3) and $\beta = 4$ (Err_4). For $\beta = 2$, the RMSE is computed as in (7.21), whereas for $\beta = 1$ the RMSE is computed as the absolute difference and for $\beta = \infty$ the maximum difference measure.

4. The geometric average error (GAE) is computed as follows [50]. The GAE is approaching zero, if there is a very good transformation (small differences) between the original and the despeckled image, and high vice versa. This measure is also used for tele-ultrasound, when transmitting ultrasound images and is defined as:

$$\mathrm{GAE} = \left(\prod_{i=1}^{M}\prod_{j=1}^{N}\sqrt{g_{i,j} - f_{i,j}}\right)^{1/MN}. \tag{7.23}$$

5. The SNR is given by [51]:

$$\mathrm{SNR} = 10\log_{10}\frac{\sum_{i=1}^{M}\sum_{j=1}^{N}(g_{i,j}^2 + f_{i,j}^2)}{\sum_{i=1}^{M}\sum_{j=1}^{N}(g_{i,j} - f_{i,j})^2}. \tag{7.24}$$

6. The peak SNR (PSNR) is computed using [51]:

$$\mathrm{PSNR} = -10\log_{10}\frac{\mathrm{MSE}}{g^2_{\max^2}} \tag{7.25}$$

where $g^2_{\max}$ is the maximum intensity in the unfiltered image. The PSNR is higher for a better-transformed image and lower for a poorly transformed image. It measures image fidelity, i.e., how closely the despeckled image resembles the original image.

7. The mathematically defined universal quality index [52] models any distortion as a combination of three different factors, which are: loss of correlation, luminance distortion, and contrast distortion and is derived as:

$$Q = \frac{\sigma_{gf}}{\sigma_f \sigma_g} \cdot \frac{2\overline{f}\overline{g}}{(\overline{f})^2 + (\overline{g})^2} \cdot \frac{2\sigma_f \sigma_g}{\sigma_f^2 + \sigma_g^2}, \quad -1 < Q < 1, \tag{7.26}$$

where $\overline{g}$ and $\overline{f}$ represent the mean of the original and despeckled values with their standard deviations σ_g and σ_f of the original and despeckled values of the analysis window, and σ_{gf} represents the covariance between the original and despeckled windows. Q is computed for a sliding window of size 8×8 without overlapping. Its highest value is 1 if $g_{i,j} = f_{i,j}$, while its lowest value is –1 if $f_{i,j} = 2\overline{g} - g_{i,j}$.

8. The structural similarity index (SSIN) between two images [49], which is a generalization of (7.26), is given by:

$$\text{SSIN} = \frac{(2\overline{g}\overline{f} + c_1)(2\sigma_{gf} + c_2)}{(\overline{g}^2 + \overline{f}^2 + c_1)(\sigma_g^2 + \sigma_f^2 + c_2)}, \quad -1 < \text{SSIN} < 1, \tag{7.27}$$

where $c_1 = 0.01dr$ and $c_2 = 0.03dr$ with $dr = 255$ representing the dynamic range of the ultrasound images. The range of values for the SSIN lies between –1 for a bad and 1 for a good similarity between the original and despeckled images, respectively. It is computed, similarly to the Q measure, for a sliding window of size 8×8 without overlapping.

It is noted that a new image of quality metric based on natural scene statistics and mutual information between the original and the filtered images has recently been proposed by Sheikh et al. [53]. This metric will be investigated in a future study.

7.3.9 *Visual Evaluation by Experts*

Visual evaluation can be broadly categorized as the ability of an expert to extract useful anatomical information from an ultrasound image. The visual evaluation varies of course from expert to expert and is subject to the observer's variability [54]. The visual evaluation, in this study, was carried out according to the ITU-R recommendations with the double stimulus continuous quality scale (DSCQS) procedure [50]. A total of 100 ultrasound images of the carotid artery bifurcation (50 asymptomatic and 50 symptomatic) were evaluated visually by two vascular experts, a cardiovascular surgeon and a neurovascular specialist before and after despeckle filtering. For each case, the original and the despeckled images (despeckled with filters *DsFlsmv*, *DsFlsminsc*, *DsFmedian*, *DsFwiener*, *DsFhomog*, *DsFgf4d*, *DsFhomo*, *DsFad*, *DsFnldif*, and *DsFwaveltc*) were presented to the two experts without labeling at random. The experts were asked to assign a score in the one to five scale corresponding to low and high subjective visual perception criteria.

Five was given to an image with the best visual perception. Therefore, the maximum score for a filter is 500, if the expert assigned the score of 5 for all the 100 images. For each filter, the score was divided by five to be expressed in percentage format. The experts were allowed to give equal scores to more than one image in each case. For each class and for each filter, the average score was computed.

The two vascular experts evaluated the area around the distal common carotid, 2–3 cm before the bifurcation, and the bifurcation. It is known that measurements taken from the far wall of the carotid artery are more accurate than those taken from the near wall [55]. Furthermore, the experts were examining the image in the lumen area, in order to identify the existence of a plaque or not.

7.4 Results

In this section, we present the results of the ten despeckle filters described in Sect. 7.2, applied on 220 asymptomatic and 220 symptomatic ultrasound images of the carotid artery bifurcation. A total of 56 texture features were computed, and the most discriminant ones are presented. Furthermore, the performance of these filters is investigated for discriminating between asymptomatic and symptomatic images using the statistical kNN classifier. Moreover, nine different image quality evaluation metrics were computed as well as visual evaluation scores carried out by two experts.

7.4.1 Evaluation of Despeckle Filtering on a Symptomatic Ultrasound Image and a Cardiac Image

Figure 7.2 shows an ultrasound image of the carotid together with the despeckled images. The best visual results as assessed by the two experts were obtained by the filters *DsFlsmv* and *DsFlsminsc*, whereas the filters *DsFgf4d*, *DsFad*, and *DsFnldif* also showed good visual results but smoothed the image considerably and thus edges and subtle details may be lost. Filters that showed a blurring effect are the *DsFmedian*, *DsFwiener*, *DsFhomog*, and *DsFwaveltc*. Filters *DsFwiener*, *DsFhomog*, and *DsFwaveltc* showed poorer visual results.

Figure 7.3 shows two original ultrasound cardiac images in (a) and (e) and the despeckled images in (b), (c), (d) and (f), (g), (h) with filters *DsFsrad*, *DsFlsmv*, and *DsFgf4d*, respectively. The moving sliding window applied to all images was 7×7 pixels. The best visual results as assessed by the two experts were obtained by the filters *DsFlsmv* (see Fig. 7.3b, f) after 4 iterations, the filter *DsFsrad* (see Fig. 7.3d, h) after 50 iterations and a coefficient of variation 0.025, and the *DsFgf4d* (see Fig. 7.3c, g) after 4 iterations. The rest of the filters also showed good visual results but smoothed the image loosing subtle details, affecting also the edges.

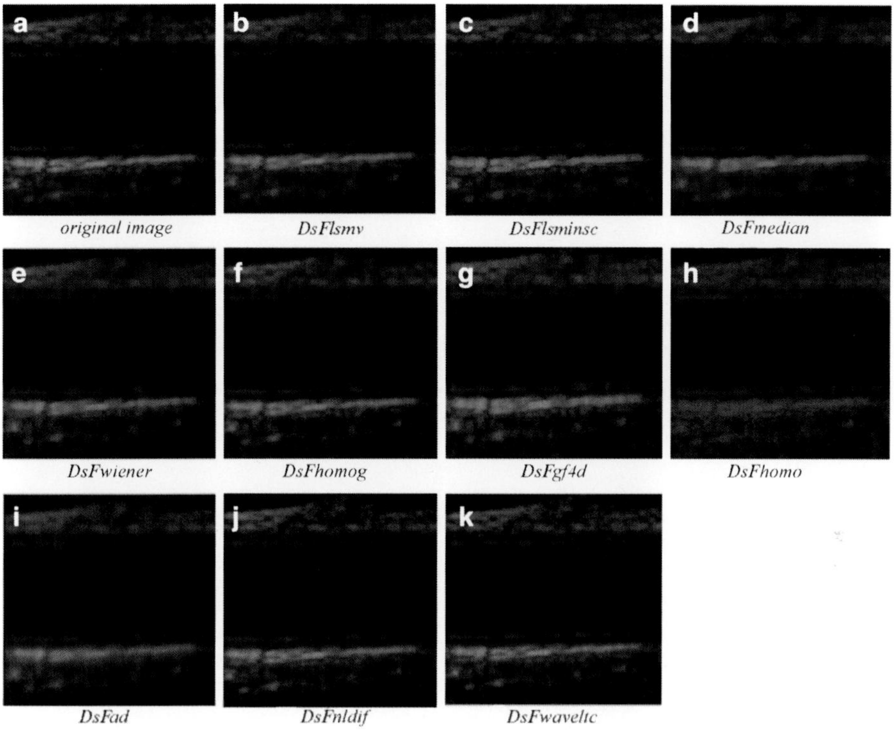

Fig. 7.2 Original ultrasound image of the carotid artery (2–3 cm proximal to bifurcation) given in (**a**), and the despeckled filtered images given in (**b**–**k**). Source: [7], © 2005

7.4.2 *Texture Analysis: Distance Measures, Table 7.2*

Despeckle filtering and texture analysis were carried out on 440 ultrasound images of the carotid. Table 7.2 tabulates the results of feat_dis_z (7.18), and Score_Dis (7.19) for SF, SGLDM range of values, and NGTDM feature sets for the ten despeckle filters. The results of these feature sets are only presented, since were the ones with best performance. The filters are categorized as local statistics, median, maximum homogeneity (HF), geometric (GF), homomorphic (HM), diffusion, and wavelet filters, as introduced in Sects. 7.1 and 7.2. Also the number of iterations (Nr. of It.) for each filter is given, which was selected based on C and on the visual evaluation of the two experts. When C was minimally changing then the filtering process was stopped. The bolded values represent the values that showed an improvement after despeckle filtering compared with the original. The last row in each subtable shows the Score_Dis for all features, where the highest value indicates the best filter in the subtable. In addition, a total score distance Score_Dis_T was computed for all feature sets shown in the last row of Table 7.2. Some of the despeckle filters, shown in Table 7.2, changing the number of texture features by increasing the distance between the two classes (positive values in Table 7.2), and

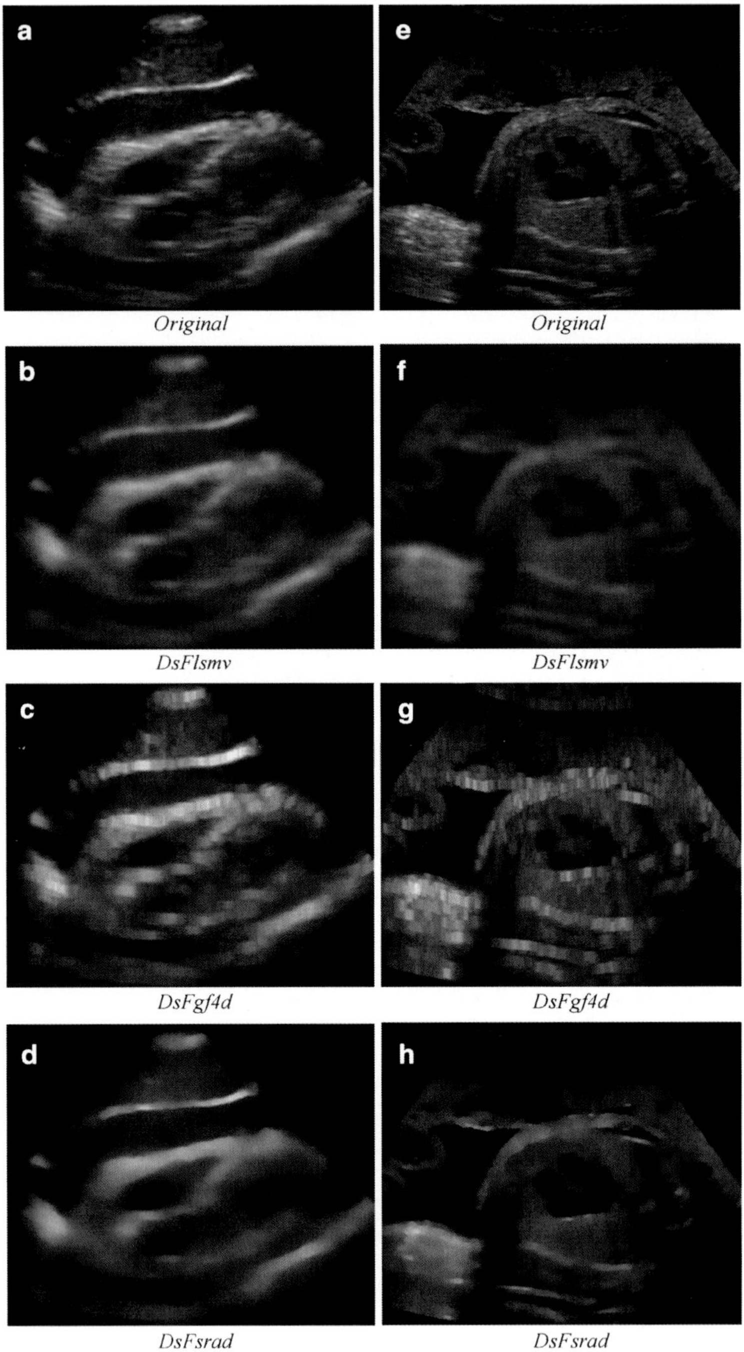

Fig. 7.3 (**a**) and (**e**) Original cardiac ultrasound images and the despeckled images with the filters *DsFlsmv*, *DsFgf4d*, and *DsFsrad*, given in the *left* and *right columns* for the *first and second images*, respectively. Source: [9], © 2008

therefore making the identification and separation between asymptomatic and symptomatic plaques more feasible. A positive feature distance shows improvement after despeckle filtering, whereas a negative shows deterioration.

In the first part of Table 7.2, the results of the SF features are presented, where the best $Score\_Dis$ is given for the filter *DsFhomo* followed by the *DsFlsminsc*, *DsFlsmv*, *DsFhomog*, *DsFnldif*, *DsFwaveltc*, *DsFmedian*, and *DsFwiener*, with the worst $Score\_Dis$ given by *gf4d*. All filters reduced the speckle index *C*. Almost all filters reduced significantly the variance σ^2 and the kurtosis σ^3 of the histogram, as it is seen from the bolded values in the first part of Table 7.2.

In the second part of the Table 7.2, the results of the SGLDM range of values for the features set are tabulated. The filters with the highest $Score\_Dis$ in the SGLDM range of values for the features set are *DsFhomo*, *DsFlsminsc*, *DsFmedian*, *DsFad*, and *DsFhomog*, whereas all the other filters (*DsFnldif*, *DsFwiener*, *DsFwaveltc*, *DsFgf4d*, and *DsFlsmv*) are presenting a negative $Score\_Dis$. Texture features, which improved in most of the filters, are the contrast, correlation, sum of squares variance (SOSV), sum average, and sum variance.

In the third part of Table 7.2, for the NGTDM feature set, almost all filters showed an improvement in $Score\_Dis$. Best filters in the NGTDM feature set were the *DsFhomo*, *DsFlsminsc*, *DsFhomog*, and *DsFlsmv*. Texture features improved at most were the completion, coarseness, and contrast. The completion of the image was increased by all filters.

Finally, in the last row of Table 7.2, the total score distance $Score\_Dis\_T$ for all feature sets is shown, where best values were obtained by the filters *DsFhomo*, *DsFlsminsc*, *DsFmedian*, *DsFlsmv*, *DsFhomog*, and *DsFad*.

7.4.3 Texture Analysis: Univariate Statistical Analysis, Table 7.3

Table 7.3 shows the results of the rank sum test, which was performed on the SGLDM range of values for the features set of Table 7.2 for the ten despeckle filters. The test was performed to check if significant differences exist between the features computed on the 440 original and the 440 despeckled images. Filters that resulted with the most significant number of features after despeckle filtering, as shown in the column "Score" of Table 7.3, were the following: *DsFlsmv*, *DsFgf4d*, *DsFlsminsc*, and *DsFnldif*. The rest of the filters gave a lower number of significantly different features. Features that showed a significant difference after filtering were the IDM, ASM, sum of entropy, contrast, correlation, SOSV, and sum variance ($\sum \mathrm{Var}$). These features were mostly affected after despeckle filtering and they were significantly different.

7.4.4 Texture Analysis: kNN Classifier, Table 7.4

Table 7.4 shows the percentage of correct classifications score for the kNN classifier with $k=7$ for classifying a subject as asymptomatic or symptomatic. The classifier

Table 7.2 Feature distance (7.18) and Score_Dis (7.19) For SF, SGLDM range of values, and Ngtdm texture feature sets between asymptomatic and symptomatic carotid plaque ultrasound images

	Local statistics				HF	GF	HM	Diffusion		Wavelet
Feature	*DsFlsmv*	*DsFlsminsc*	*DsFwiener*	*DsFmedian*	*DsFhomog*	*DsFgf4d*	*DsFhomo*	*DsFad*	*DsFnldif*	*DsFwaveltc*
Number of iteration	4	1	2	2	1	3	2	20	5	5
SF – statistical features										
Mean	**14**	**22**	**19**	**4**	**11**	**3**	**164**	**18**	**5**	**15**
Median	−5	−17	−26	−5	−5	−15	**110**	−29	−6	−15
σ^2	**18**	**38**	**18**	**7**	**13**	−2	**140**	**9**	**7**	**18**
σ^3	**12**	**16**	**5**	**9**	**7**	−0.1	**149**	**17**	**7**	**8**
σ^4	−12	−14	−7	−6	−4	−3	**117**	−21	**6**	−9
C	0.4	0.3	0.3	0.4	0.3	0.4	0.08	0.3	0.4	0.3
$Score\_dis$	27	45	9	9	22	−17	680	−6	19	17
SGLDM range of values – spatial gray level dependence matrix										
ASM	−21	−0.5	−29	2	−4	−8	−47	−25	−17	−20
Contrast	**47**	**107**	**14**	**64**	**32**	−3	**165**	**104**	**13**	**22**
Correlation	**12**	**59**	**15**	**24**	−5	**2**	**10**	**54**	−4	−4
SOSV	**9**	**40**	**18**	**10**	**16**	−2	**101**	**9**	**8**	**20**
IDM	−50	−11	−48	2	−29	−8	**94**	−54	−34	−43
SAV	**17**	**24**	**23**	**7**	**15**	**3**	**169**	**22**	**6**	**18**
$\sum Var$	**19**	**38**	**18**	**9**	**15**	−2	**90**	**9**	**8**	**20**
$\sum Entr$	−34	−14	−49	**3**	−19	−4	−11	−47	−30	−36

Score _ dis	−1	243	−38	121	21	−22	571	72	−50	−23
Coarseness	**30**	**87**	**4**	**9**	−16	−7	**72**	−36	−37	−33
Contrast	**7**	−0.3	−9	**8**	**0.4**	−4	**105**	**5**	−27	−15
Busyness	**17**	**26**	−30	**8**	**1**	−4	**48**	−14	−39	**8**
Completion	**64**	**151**	**21**	**53**	**80**	**2**	**150**	**63**	**18**	**27**

Bolded values show improvement after despeckle filtering

ASM angular second moment, *SOSV* sum of squares variance, *IDM* inverse difference moment, *SAV* sum average, $\sum$ Var sum variance, *HF* homogeneity, *GF* geometric, *HM* homomorphic

Source: [7], © 2005

Table 7.3 Wilcoxon rank sum test for the SGLDM range of values for the texture features applied on the 440 ultrasound images of carotid plaque before and after despeckle filtering

	Local statistics			Median	HF	GF	HM	Diffusion		Wavelet	Score
Feature	*DsFlsmv*	*DsFlsminsc*	*DsFwiener*	*DsFmedian*	*DsFhomog*	*DsFgf4d*	*DsFhomo*	*DsFad*	*DsFnldif*	*DsFwaveltc*	
ASM	S	S	NS	NS	S	S	NS	S	S	S	7
Contrast	S	NS	NS	NS	NS	S	NS	NS	S	NS	3
Correlation	S	S	NS	NS	NS	S	NS	NS	NS	NS	3
SOSV	S	NS	NS	NS	NS	S	NS	NS	NS	NS	2
IDM	S	S	NS	S	S	S	S	NS	S	S	8
SAV	NS	NS	NS	NS	NS	NS	NS	NS	NS	NS	0
$\sum$Var	S	S	NS	NS	NS	NS	NS	NS	NS	NS	2
$\sum$Entropy	S	S	NS	NS	NS	S	NS	NS	S	S	5
Score	7	5	0	1	2	6	1	1	4	3	

The test with S and NS shows the features that are and are not significantly different after filtering at $p<0.05$, respectivelyScore illustrates the number of Ss
ASM angular second moment, *SOSV* sum of squares variance, *IDM* inverse difference moment, *SAV* sum average, $\sum$Var sum variance, *HM* homomorphic, *HF* homogeneity, *GF* geometric
Source: [7], © 2005

Table 7.4 Percentage of correct classifications score for the kNN classifier with $k=7$ for the original and the filtered image sets

	Number of		Local statistics			Median	HF	GF	HM	Diffusion		Wavelet
Feature set	features	Original	*DsFlsmv*	*DsFlsminsc*	*DsFwiener*	*DsFmedian*	*DsFhomog*	*DsFgf4d*	*DsFhomo*	*DsFad*	*DsFnldif*	*DsFwaveltc*
SF	5	59	**62**	**61**	**61**	57	**63**	59	**65**	**60**	52	**61**
SGLDMm	13	65	63	64	62	63	**69**	**67**	**68**	61	**66**	63
SGLDMr	13	70	66	**72**	64	66	65	70	69	64	65	65
GLDS	4	64	63	**66**	61	**69**	64	**66**	**72**	59	58	62
NGTDM	5	64	63	**68**	60	**69**	63	**65**	57	60	61	62
SFM	4	62	62	60	62	58	55	**65**	**68**	59	56	55
TEM	6	59	**68**	52	**60**	59	**66**	**60**	**65**	53	**60**	**60**
FDTA	4	64	63	**66**	53	**68**	53	62	**73**	55	54	62
FPS	2	59	54	**64**	59	58	59	59	59	52	48	55
Average		63	63	**64**	60	63	62	**64**	**66**	58	58	61

Bolded values indicate improvement after despeckling
SF statistical features, *SGLDMm* spatial gray level dependence matrix mean values, *SGLDMr* spatial gray level dependence matrix range of values, *GLDS* gray level difference statistics, *NGTDM* neighborhood gray tone difference matrix, *SFM* statistical feature matrix, *TEM* laws texture energy measures, *FDTA* fractal dimension texture analysis, *FPS* Fourier power spectrum, *HF* homogeneity, *GF* Geometric, *HM* homomorphic
Source: [7], © 2005

was evaluated using the leave-one-out method [46] on 220 asymptomatic and 220 symptomatic images of the original and despeckled images. The percentage of correct classifications score is given for the following feature sets: statistical features (SF), spatial gray level dependence matrix mean values (SGLDMm), spatial gray level dependence matrix range of values (SGLDMr), gray level difference statistics (GLDS), neighborhood gray tone difference matrix (NGTDM), statistical feature matrix (SFM), laws texture energy measures (TEM), fractal dimension texture analysis (FDTA), and Fourier power spectrum (FPS). Filters that showed an improvement in classifications success score compared with that of the original image set were, in average (last row of Table 7.4), the filter *DsFhomo* (3%), *gf4d* (1%), and *DsFlsminsc* (1%).

Feature sets that mostly benefited by the despeckle filtering were the SF, GLDS, NGTDM, and TEM when counting the number of cases that the correct classifications score was improved. Less improvement was observed for the feature sets FDTA, SFM, SGLDMm, FPS, and SGLDMr. For the feature set SGLDMr better results are given for the *DsFlsminsc* filter with an improvement of 2%. This is the only filter that showed an improvement for this class of features. For the feature set TEM, the filter *DsFlsmv* shows the best improvement with 9%, whereas for the FPS feature set the filter *DsFlsminsc* shows the best improvement with 5%. The filter *DsFlsminsc* showed improvement in the GLDS and NGTDM feature sets, whereas the filter *DsFlsmv* showed improvement for the feature sets SF and TEM.

7.4.5 Image Quality Evaluation Metrics, Table 7.5

Table 7.5 tabulates the image quality evaluation metrics presented in Sect. 7.3.8, for the 220 asymptomatic and 220 symptomatic ultrasound images between the original and the despeckled images, respectively. Best values were obtained for the *DsFnldif*, *DsFlsmv*, and *DsFwaveltc* with lower MSE, RMSE, Err3, and Err4 and higher SNR and PSNR. The GAE was 0.00 for all cases, and this can be attributed to the fact that the information between the original and the despeckled images remains unchanged. Best values for the universal quality index (Q) and the SSIN were obtained for the filters *DsFlsmv* and *DsFnldif*.

7.4.6 Visual Evaluation by Experts, Table 7.6

Table 7.6 shows the results of the visual evaluation of the original and despeckled images made by two experts, a cardiovascular surgeon and a neurovascular specialist. The last row of Table 7.6 presents the overall average percentage (%) score assigned by both experts for each filter.

For the cardiovascular surgeon, the average score showed that the best despeckle filter is the *DsFlsmv* with a score of 62%, followed by *DsFgf4d*, *DsFmedian*,

Table 7.5 Image quality evaluation metrics computed for the 220 asymptomatic and 220 symptomatic images

	Local statistics			Median	HF	GF	HM	Diffusion		Wavelet
Feature set	*DsFlsmv*	*DsFlsminsc*	*DsFwiener*	*DsFmedian*	*DsFhomog*	*DsFgf4d*	*DsFhomo*	*DsFad*	*DsFnldif*	*DsFwaveltc*
Asymptomatic images										
MSE	13	86	19	131	42	182	758	132	8	11
RMSE	3	9	4	10	6	13	27	11	2	3
Err3	7	17	5	25	14	25	38	21	5	4
Err4	11	26	7	41	24	40	49	32	10	5
GAE	0	0	0	0	0	0	0	0	0	0
SNR	25	17	23	16	21	14	5	14	28	25
PSNR	39	29	36	29	34	27	20	28	41	39
Q	0.83	0.78	0.74	0.84	0.92	0.77	0.28	0.68	0.93	0.65
SSIN	0.97	0.88	0.92	0.94	0.97	0.88	0.43	0.87	0.97	0.9
Symptomatic images										
MSE	33	374	44	169	110	557	1452	374	8	23
RMSE	5	19	6	13	10	23	37	19	3	5
Err3	10	33	9	25	20	43	51	31	5	6
Err4	16	47	11	38	30	63	64	43	7	8
GAE	0	0	0	0	0	0	0	0	0	0
SNR	24	13	22	16	17	12	5	12	29	25
PSNR	34	23	33	26	28	21	17	23	39	36
Q	0.82	0.77	0.7	0.79	0.87	0.75	0.24	0.63	0.87	0.49
SSIN	0.97	0.85	0.89	0.81	0.94	0.85	0.28	0.81	0.97	0.87

MSE mean-square error, *RMSE* randomized mean-square error, *Err3, Err4* minowski metrics, *GAE* geometric average error, *SNR* signal-to-noise ratio, *PSNR* peak signal-to-noise ratio, *Q* universal quality index, *SSIN* structural similarity index
Source: [8], © 2006

Table 7.6 Percentage scoring of visual evaluation of the original and despeckled images [50 asymptomatic (A) and 50 symptomatic (S)] by the experts

			Local statistics		Median	HF	GF	HM	Diffusion	Wavelet
Experts	A/S	Original	*DsFlsmv*	*DsFlsminsc*	*DsFmedian*	*DsFhomog*	*DsFgf4d*	*DsFhomo*	*DsFnldif*	*DsFwaveltc*
Cardiovascular	A	33	75	33	43	47	61	19	43	32
surgeon	S	48	49	18	57	43	42	20	33	22
Average %		41	62	26	50	45	52	19	38	27
Neurovascular	A	70	76	73	74	63	79	23	52	29
specialist	S	66	67	63	58	45	65	55	41	28
Average %		68	71	68	66	54	72	39	47	28
Overall average %		54	67	47	58	50	62	29	43	28

HF homogeneity, *GF* geometric, *HM* homomorphic
Source: [7], © 2005

DsFhomog, and original with scores of 52, 50, 45, and 41%, respectively. For the neurovascular specialist, the average score showed that the best filter is the *DsFgf4d* with a score of 72%, followed by *DsFlsmv*, original, *DsFlsminsc*, and *DsFmedian* with scores of 71, 68, 68, and 66% respectively. The overall average percent score shows that the highest score was given to the filter *DsFlsmv* (67%), followed by *DsFgf4d* (62%), *DsFmedian* (58%), and original (54%). It should be emphasized that the despeckle filter *DsFlsmv* is the only filter that was graded with a higher score than the original by both experts for the asymptomatic and symptomatic image sets.

We may observe a difference in the scorings between the two vascular specialists and this is because the cardiovascular surgeon is primarily interested in the plaque composition and texture evaluation, whereas the neurovascular specialist is interested to evaluate the degree of stenosis and the lumen diameter in order to identify the plaque contour. Filters *DsFlsmv* and *DsFgf4d* were identified as the best despeckle filters by both specialists as they improved visual perception with overall average scores of 67 and 62%, respectively. The filters *DsFwaveltc* and *DsFhomo* were scored by both specialists with the lowest overall average scores of 28 and 29%, respectively.

By examining the visual results of Fig. 7.2, the statistical results of Tables 7.2–7.5, and the visual evaluation of Table 7.6, we can conclude that the best filters are *DsFlsmv* and *DsFgf4d*, which may be used for both plaque composition enhancement and plaque texture analysis, whereas the filters *DsFlsmv*, *DsFgf4d*, and *DsFlsminsc* are more appropriate to identify the degree of stenosis and therefore may be used when the primary interest is to outline the plaque borders.

7.4.7 Intima–Media Complex and Plaque Segmentation

In a recent study [56], we developed and evaluated a snake segmentation method for detecting the intima–media complex (IMC) in ultrasound imaging of the carotid artery after normalization and speckle reduction filtering. It was shown that the application of normalization and speckle reduction filtering prior to segmentation improves both the manual and the automated IMC segmentation results. It should be noted that the despeckle filter *DsFlsmv*, with a moving sliding window of 5×5, was iteratively applied four times, on the area around the IMC and not to the whole image, on all images prior to the IMC segmentation.

Figure 7.4 shows a longitudinal ultrasound image of the CCA with the manual delineations from the two experts (Fig. 7.4b, c), the automatic initial contour estimation (Fig. 7.4d), and the Williams and Shah snakes segmentation results for the cases of no preprocessing (NP) (Fig. 7.4e), despeckled (DS) (Fig. 7.4f), normalized (N) (Fig. 7.4g), and normalized despeckled (NDS) (Fig. 7.4h). The detected IMT_{mean}, IMT_{max}, and IMT_{min} values are shown with a double, single, and dashed line boxes, respectively. The results in Fig. 7.4 showed that the IMT was detected in all snakes segmentation measurements but with variations between experts and methods.

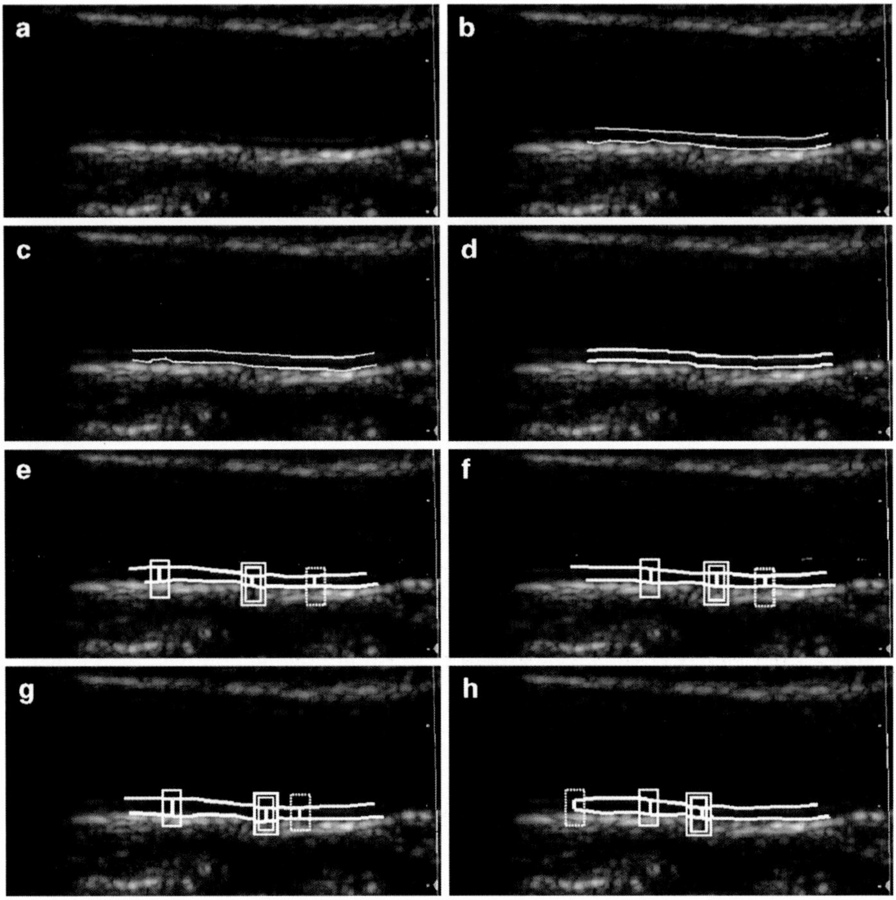

Fig. 7.4 (**a**) Original longitudinal ultrasound image of the carotid artery, (**b**) manual delineation from Expert 1, (**c**) manual delineation from Expert 2, (**d**) initial contour estimation, and the segmentation results of the IMT for (**e**) no preprocessing (NP), (**f**) despeckled (DS), (**g**) normalized (N), and (**h**) normalized despeckled (NDS) images. The detected IMT_{mean}, IMT_{max}, and IMT_{min} are shown with a *double, single, and dashed line boxes* respectively. Source: [9], © 2008

In another study [60], we proposed and evaluated an integrated plaque segmentation system based on normalization, speckle reduction filtering, and snakes segmentation. Four different snake segmentation methods were investigated in [60], namely: (1) the Williams and Shah, (2) the Balloon, (3) the Lai and Chin, and (4) the gradient vector flow (GVF). These were applied on 80 plaque ultrasound images of the CCA. The comparison of the four different plaque snakes segmentation methods showed that the Lai and Chin segmentation method gave slightly better results, although these results were not statistically significant when compared with the other three snakes segmentation methods. It was also shown that the application of normalization and speckle reduction filtering prior to segmentation improves both the manual and the automated plaque segmentation results.

Figure 7.5 illustrates an original longitudinal ultrasound B-Mode image of a carotid plaque with a manual delineation made by the expert in (a), and the results of the William and Shah segmentation in (b), the Balloon segmentation in (c), the Lai and Chin segmentation in (d), and the GVF segmentation in (e). Figure 7.5f shows the superimposition of the segmentation contours computed in Fig. 7.5b–e. As illustrated in Fig. 7.5f, both the manual and the snakes segmentation contours are visually very similar. It should be noted that the despeckle filter *DsFlsmv*, with a moving sliding window of 5×5, was iteratively applied four times on all images prior to segmentation.

7.5 Discussion

Despeckle filtering is an important operation in the enhancement of ultrasound images of the carotid artery, both in the case of texture analysis and in the case of image quality evaluation and of visual evaluation by the experts. In this study, a total of ten despeckle filters were comparatively evaluated on 440 ultrasound images of the carotid artery bifurcation, and the validation results are summarized in Table 7.7.

As given in Table 7.7, filters *DsFlsmv*, *DsFlsminsc*, and *DsFhomo* improved the class separation between the asymptomatic and the symptomatic classes (see also Table 7.2). Filters *DsFlsmv*, *DsFlsminsc*, and *DsFgf4d* gave a high number of significantly different features (see Table 7.3). Filters *DsFlsminsc*, *DsFgf4d*, and *DsFhomo* gave only a marginal improvement in the percentage of correct classifications success rate (see Table 7.4). Moreover, filters *DsFlsmv*, *DsFnldif*, and *DsFwaveltc* gave better image quality evaluation results (see Table 7.5). Filters *DsFlsmv* and *DsFgf4d* improved the visual assessment carried out by the experts (see Table 7.6). It is clearly shown that filter *DsFlsmv* gave the best performance, followed by the filters *DsFlsminsc* and *gf4d* (see Table 7.7). Filters *DsFlsmv* or *DsFgf4d* can be used for despeckling of asymptomatic images where the expert is interested mainly in the plaque composition and texture analysis. Filters *DsFlsmv* or *DsFgf4d* or *DsFlsmnsc* can be used for despeckling of symptomatic images where the expert is interested in identifying the degree of stenosis and the plaque borders. Filters *DsFhomo*, *DsFnldif*, and *DsFwaveltc* gave poor performance.

Filter *DsFlsmv* gave very good performance with respect to: (1) preserving the mean and the median as well as decreasing the variance and the speckle index of the image, (2) increasing the distance of the texture features between the asymptomatic and the symptomatic classes, (3) significantly changing the SGLDM range of values for the texture features after filtering based on the Wilcoxon rank sum test, (4) marginally improving the classification success rate of the kNN classifier for the classification of asymptomatic and symptomatic images in the cases of SF, SMF, and TEM feature sets, and (5) improving the image quality. The *DsFlsmv* filter, which is a simple filter, is based on local image statistics. It was first introduced in [10, 15, 16] by Jong-Sen Lee and co-workers, and it was tested on a few SAR images with satisfactory results. It was also used for SAR imaging in [14] and image restoration in [17], again with satisfactory results.

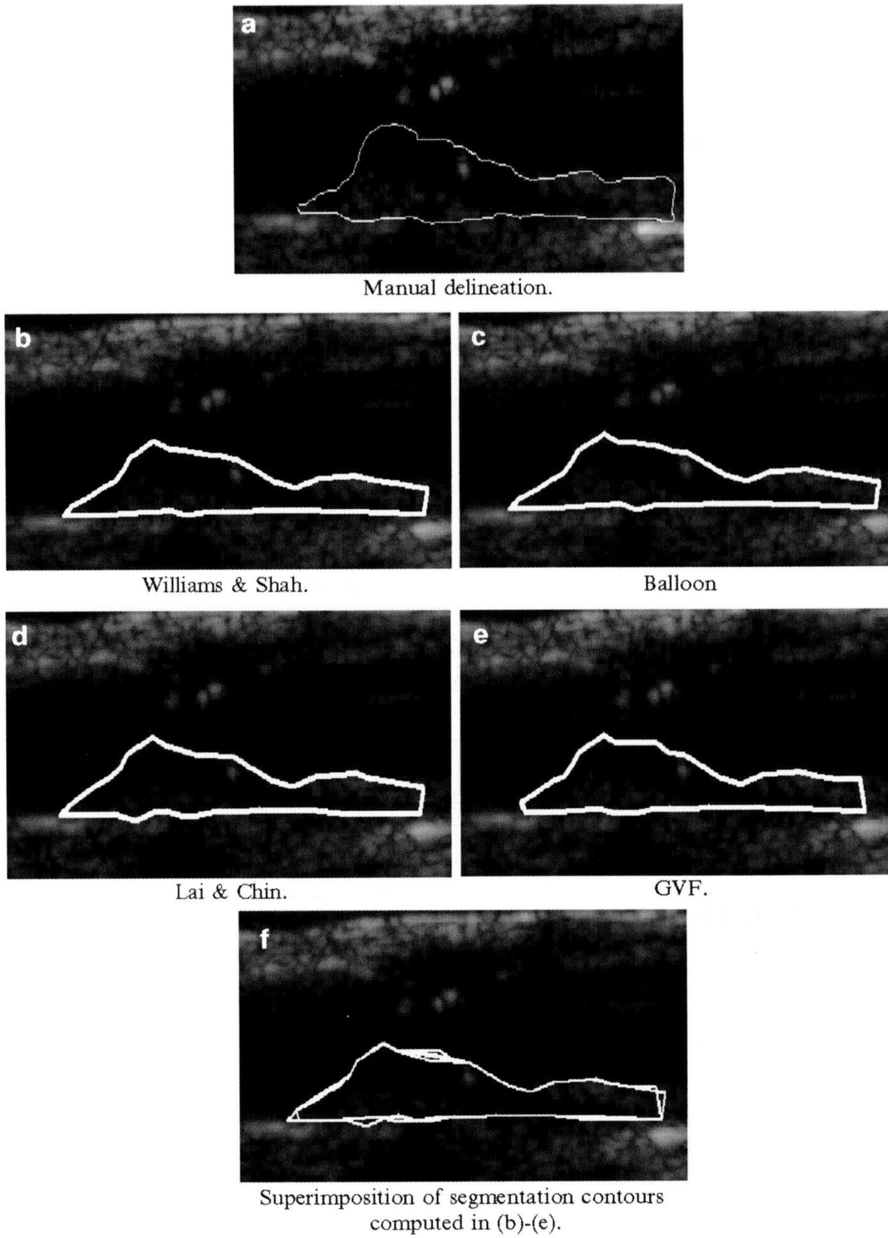

Fig. 7.5 Plaque segmentation results on a longitudinal ultrasound B-Mode image of the carotid artery: (**a**) manual, (**b**) Williams and Shah, (**c**) Balloon, (**d**), Lai and Chin, (**e**) GVF snake, and (**f**) superimposition of segmentation contours computed in (**b**–**e**). Source: [9], © IEEE 2008

Table 7.7 Summary findings of Despeckle filtering in ultrasound imaging of the carotid artery

Despeckle filter	Statistical and texture features (Table 7.2)	Statistical analysis (Table 7.3)	kNN classifier (Table 7.4)	Image quality evaluation (Table 7.5)	Optical Perception evaluation (Table 7.6)
Local statistics					
DsFlsmv	✓	✓		✓	✓
DsFlsminsc	✓	✓	✓		
Geometric filtering					
DsFgf4d		✓	✓		✓
Homomorphic filtering					
DsFhomo	✓		✓		
Diffusion filtering					
DsFnldif				✓	
Wavelet filtering					
DsFwaveltc				✓	

Source: [9], © 2008

Filter *DsFlsminsc* gave the best performance with respect to: (1) preserving the mean as well as decreasing the variance and the speckle index and increasing the contrast of the image, (2) increasing the distance of the texture features between the asymptomatic and the symptomatic classes, (3) significantly changing the SGLDM texture features after filtering based on the Wilcoxon rank sum test, (4) improving the classification success rate of the kNN classifier for the classification of asymptomatic and symptomatic images in the cases of SF, SGLDMr, GLDS, NGTDM, FDTA, and FPS feature sets. Filter *DsFlsminsc* was originally introduced by Nagao in [34] and was tested on an artificial and a SAR image with satisfactory performance. In this study, the filter was modified by using the speckle index instead of the variance value for each subwindow [as described in Sect. 7.2.1.2 (7.9)].

Filter *DsFgf4d* gave very good performance with respect to: (1) decreasing the speckle index, (2) marginally increasing the distance of the texture features between the asymptomatic and the symptomatic classes, (3) significantly changing the SGLDM range of values for the texture features after filtering based on the Wilcoxon rank sum test, (4) improving the classification success rate of the kNN classifier for the classification of asymptomatic and symptomatic images in the cases of SGLDMm, GLDS, NGTDM, SFM, and TEM feature sets. The geometric filter *DsFgf4d* was introduced by Crimmins [11], and was tested visually on a few SAR images with satisfactory results.

Filters used for speckle reduction in ultrasound imaging by other investigators include: *DsFmedian* [35], *DsFwiener* [14], *DsFhomog* [9, 18], *DsFhomo* [19, 20], *DsFad* [5], and *DsFwaveltc* [31]. However, these filters were evaluated on a small number of images, and their performance was tested based only on the mean, median, standard deviation, and speckle index of the image before and after filtering.

The *DsFmedian* and the *DsFwiener* filters were originally used by many researchers for suppressing the additive and later for suppressing the multiplicative noise in different types of images [2–10, 14, 35]. The results of this study showed that the *DsFwiener* and *DsFmedian* filters were not able to remove the speckle noise and produced blurred edges in the filtered image (see Fig. 7.2). In this study, the *DsFmedian* filter performed poorer as shown in Tables 7.2–7.4.

The *DsFhomog* [9, 18]and *DsFhomo* [2, 19, 20] filters were recently used by some researchers for speckle reduction but our results in Tables 7.2, 7.3, and 7.5 and the visual evaluation of the experts in Table 7.6 showed poor performance especially for the *DsFhomo* filter.

Anisotropic diffusion is an efficient nonlinear technique for simultaneously performing contrast enhancement and noise reduction. It smoothes homogeneous image regions but retains image edges [25]. Anisotropic diffusion filters usually require many iteration steps compared with the local statistic filters. In a recent study [5], speckle reducing anisotropic diffusion filtering was proposed as the most appropriate filter for ultrasound images of the carotid artery. However, in this study, *DsFad*, as shown in Tables 7.2–7.6, performed poorer compared with *DsFlsmv*, *DsFgf4d*, and *DsFlsminsc*.

Furthermore, wavelet filtering proposed by Donoho in [31] was investigated for suppressing the speckle noise in SAR images [16, 37], real-world images [27], and ultrasound images [28] with favorable results. In this study, it is shown that the *DsFwaveltc* filter gave poorer performance for removing the speckle noise from the ultrasound images of the carotid artery (Tables 7.2 and 7.3).

It was shown in [58–60] that the preprocessing of ultrasound images of the carotid artery with normalization and speckle reduction, followed by the snakes initialization and the Williams and Shah segmentation algorithm, can be used successfully in the measurement of IMT complementing the manual measurements. The comparison of the four different plaque snakes segmentation methods, proposed in our recent study [60] for the segmentation of the atherosclerotic carotid plaque from ultrasound images, showed that the Lai and Chin snakes segmentation method gave slightly better results although these results were not statistically significant when compared with the other three snakes segmentation methods (Williams and Shah, Balloon, and GVF).

In conclusion, despeckle filtering is an important operation in the enhancement of ultrasonic imaging of the carotid artery. In this study, it was shown that simple filters based on local statistics (*DsFlsmv* and *DsFlsminsc*) and geometric filtering (*DsFgf4d*) could be used successfully for the processing of these images. In this context, despeckle filtering can be used as a preprocessing step for the automated segmentation of the IMT [56] and the carotid plaque, followed by the carotid plaque texture analysis and classification. This field has also been investigated by our group [9, 57]. Our findings show promising results, however, further work is required to evaluate the performance of the suggested despeckle filters at a larger scale as well as their impact on clinical practice. In addition, the usefulness of the proposed despeckle filters in portable ultrasound systems and in wireless telemedicine systems still has to be investigated.

7.6 Summary and Future Directions

7.6.1 Summary

Despeckle filtering applications have been a rapidly emerging research area in recent years. The basic principles, the theoretical background, and the algorithmic steps of a representative set of despeckle filters were covered in this book. Moreover, selected representative applications of image despeckling covering a variety of ultrasound image-processing tasks are presented. Most importantly, a despeckle filtering and evaluation protocol are documented in Table 7.8. The source code of the algorithms discussed in this book has been made available on the web, thus enabling researchers to more easily exploit the application of despeckle filtering in their problems under investigation.

A total of ten different despeckle filters were documented in this book based on linear filtering, nonlinear filtering, diffusion filtering, and wavelet filtering. We have evaluated despeckle filtering on 440 (220 asymptomatic and 220 symptomatic) ultrasound images of the carotid artery bifurcation, based on the visual evaluation by two medical experts, the texture analysis measures, and the image quality evaluation metrics. A linear despeckle filter based on local statistics (*DsFlsmv*) improved the class separation between the asymptomatic and the symptomatic classes, gave only a marginal improvement in the percentage of correct classifications success rate based on the texture analysis and the kNN classifier, and

Table 7.8 Despeckle filtering and evaluation protocol

1	*Recording of ultrasound images*: Ultrasound images are acquired by ultrasound equipment and stored for further image processing. Regions of interest (ROIs) could be selected for further processing
2	*Normalize the image*: The stored images may be retrieved and a normalized procedure may be applied (as described, for example, in Sect. 7.3.2)
3	*Apply despeckle filtering*: Select the set of filters to apply despeckling together with their corresponding parameters (like moving window size, iterations, and other)
4	*Texture features analysis*: After despeckle filtering, the user may select ROIs (i.e., the plaque or the area around the IMT) and extract texture features. Distance metrics between the original and the despeckled images may be computed (as well as between different classes of images if applicable)
5	*Compute image quality evaluation metrics*: On the selected ROIs compute image quality evaluation metrics between the original noisy and the despeckled images
6	*Visual quality evaluation by experts*: The original and/or despeckled images may be visually evaluated by experts
7	*Select the most appropriate despeckle filter/filters:* Based on steps 3–6 construct a performance evaluation table and select the most appropriate filter(s) for the problem under investigation

Source: [9], © 2008

improved the visual assessment by the experts. It was also found that the *DsFlsmv* despeckle filter can be used for despeckling asymptomatic images where the expert is interested mainly in the plaque composition and texture analysis, whereas a geometric despeckle filter (*DsFgf4d*) can be used for despeckling of symptomatic images where the expert is interested in identifying the degree of stenosis and the plaque borders.

The results of this study suggest that the first-order statistics despeckle filter *DsFlsmv* may be applied on ultrasound images to improve the visual perception and automatic image analysis. Furthermore, despeckle filtering was investigated as a preprocessing step for the automated segmentation of the IMT [56], media and intima layers [59], the carotid plaque [60], followed by the carotid plaque texture analysis and classification (as documented in the above paragraph). In a recent study [58], where the texture characteristics of the media and intima layer of the CCA was investigated, we have found significant differences among texture features extracted from the IL, ML, and IMC from different age groups. Furthermore, for some texture features, we found that they follow trends that correlate with patient's age. For example, the gray-scale median (GSM) of the ML falls linearly with increasing MLT and with increasing age. Our findings suggest that ultrasound image texture analysis of the media layer has potential as an assessment biomarker for the risk of stroke.

Despeckle filters *DsFlsmv*, *DsFlsminsc*, and *DsFgf4d* gave the best performance for the segmentation tasks. It was shown in [8, 58] that when normalization and speckle reduction filtering is applied on ultrasound images of the carotid artery prior to IMT segmentation, the automated segmentation measurements are closer to the manual measurements. This field has also been investigated by our group [9, 57]. Our findings showed promising results; however, further work is required to evaluate the performance of the suggested despeckle filters at a larger scale as well as their impact on clinical practice. In addition, the usefulness of the proposed despeckle filters in portable ultrasound systems and in wireless telemedicine systems still has to be investigated.

For those readers whose principal need is to use existing image despeckle filtering technologies and apply them on different types of images, there is no simple answer regarding which specific filtering algorithm should be selected without a significant understanding of both the filtering fundamentals and the application environment under investigation. A number of issues would need to be addressed. These include the availability of images to be processed/analyzed, the required level of filtering, the application scope (general purpose or application specific), the application goal (for extracting features from the image or for visual enhancement), the allowable computational complexity, the allowable implementation complexity, and the computational requirements (e.g., real time or offline). We believe that a good understanding of the contents of this book can help the readers make the right choice of selecting the most appropriate filter for the application under development. Furthermore, the despeckle filtering evaluation protocol documented in Table 7.8 could also be exploited.

7.6.2 Future Directions

The despeckle filtering algorithms and the measures for image quality evaluation introduced in this chapter can also be generalized and applied to other image- and video-processing applications. Only a small number of filtering algorithms and image quality evaluation metrics were investigated in this chapter, and numerous extensions and improvements can be envisaged.

In general, the development of despeckle filtering algorithms for image despeckling is a well-investigated field and many researchers have been involved in this subject, but there is still not an appropriate method proposed, which will satisfy both the visual and the automated interpretation of image processing and analysis tasks. Most importantly, more comparative studies of despeckle filtering are necessary, where different filters could be evaluated by multiple experts as well as based on image quality and evaluation metrics as proposed in this chapter.

In addition, the issue of video despeckling is still in its infancy, although it is noted that the proposed methodology and filtering algorithms documented in this chapter may be also investigated in video sequences (by frame filtering). There are many issues related to video despeckle filtering that remain to be solved. In general, the development of a multiplicative model based on video sequences is required, since most of the models developed for video filtering were for additive noise [61, 62]. Furthermore, the utilization of the motion details by using motion estimation, in order to estimate pixels that need to be filtered in the neighboring frames should also be utilized as also proposed in [63].

Despeckle filtering may also be applied in the preprocessing of ultrasound images for other organs, including the detection of hyperechoic or hypoechoic lesions in the kidney, liver, spleen, thyroid, kidney, breast, and other. It may be particularly effective when combined with harmonic imaging, since both can increase tissue contrast. Speckle reduction can also be extremely valuable when attempting to fuse ultrasound with computed tomography (CT), MRI, positron emission tomography (PET), or optical coherence tomography (OCT) images. For example, when a lesion is suspected on a CT scan but it is not clearly visible, ultrasound despeckle filtering can be applied in order to accentuate subtle borders that may be masked by speckle.

Ultrasound imaging instrumentation linked with imaging hardware and software technology has been rapidly advancing in the last two decades. Although these advanced imaging devices produce higher-quality images and video, the need still exists for better image- and video-processing techniques including despeckle filtering. Toward this direction, it is anticipated that the effective use of despeckle filtering (by exploiting the filters and algorithms documented in this book) will greatly help in producing images with higher quality. These images that would be not only easier to visualize and to extract useful information, but would also enable the development of more robust image preprocessing and segmentation algorithms, minimizing routine manual image analysis and facilitating more accurate automated measurements of both industrially and clinically relevant parameters.

	LINEAR FILTERING: LINEAR SCALING FILTER (DSFLSMV)
1	Load the image for filtering
2	Specify the region of interest to be filtered, the moving window size (nhood) and the number of iterations (n)
3	Compute the noise variance σ_n^2 with (7.5) for the whole image
4	Starting from the left upper corner of the image, compute for each moving window the coefficient $k_{i,j}$ in (7.5)
5	Compute $f_{i,j}$ in (7.4) and replace the noisy middle point in each moving window $g_{i,j}$, with the new computed value $f_{i,j}$
6	Repeat steps 4 and 5 for the whole image by sliding the moving window from left to right
7	Repeat steps 3 to 6 for n iterations
8	Compute the image quality evaluation metrics and the texture features for the original and the despeckled images
9	Display the original and despeckled images, the image quality and evaluation metrics, and the texture features

Fig. 7.6 Linear filtering: linear scaling filter (*DsFlsmv*)

7.8 Appendix: An Example of Running the Despeckle Filtering Toolbox

The Despeckle Filtering toolbox [9] can be downloaded at: http://www.medinfo.cs.ucy.ac.cy/index.php/matlab-software

An example of using the toolbox with the linear scaling filter (*Dsflsmv*) depicting the steps to be followed is given in Fig. 7.6, whereas the corresponding Matlab code can be found in the toolbox.

References

1. Lamont D, Parker L, White M, Unwin N, Bennett SMA, Cohen M, Richardson D, Dickinson HO, Adamson A, Alberti KGMM, Graft AW. Risk of cardiovascular disease measured by carotid intima-media thickness at age 49–51: life course study. BMJ 2000:**320**:273–8
2. Burckhardt CB. Speckle in ultrasound B-mode scans. IEEE Trans Sonics Ultrason 1978:**SU-25**:1:1–6
3. Wagner RF, Smith SW, Sandrik JM, Lopez H. Statistics of speckle in ultrasound B-scans. IEEE Trans Sonics Ultrason 1983:**30**:156–63
4. Goodman JW. Some fundamental properties of speckle. J Opt Soc Am 1976:**66**:11:1145–9
5. Yongjian Y, Acton ST. Speckle reducing anisotropic diffusion. IEEE Trans Image Process 2002:**11**:11:1260–70
6. Prager RW, Gee AH, Treece GM, Berman L. Speckle detection in ultrasound images using first order statistics. Department of Engineering, University of Cambridge, Cambridge, GUED/F-INFENG/TR, 2002:1–17

7. Loizou CP, Pattichis CS, Christodoulou CI, Istepanian RSH, Pantziaris M, Nicolaides AN. Despeckle filtering in ultrasound imaging of the carotid artery. IEEE Trans Ultrason Ferroelectr Freq Control 2005:**52**:2:1653–69
8. Loizou CP, Pattichis CS, Pantziaris MS, Tyllis T, Nicolaides AN. Quantitative quality evaluation of ultrasound imaging in the carotid artery. Med Biol Eng Comput 2006:**44**:5:414–26
9. Loizou CP, Pattichis CS. Despeckle filtering algorithms and software for ultrasound imaging. Synthesis lectures on algorithms and software for engineering, Ed. Morgan & Claypool Publishers, San Rafael, CA, 2008
10. Lee JS. Speckle analysis and smoothing of synthetic aperture radar images. Comput Graph Image Process 1981:**17**:24–32
11. Busse L, Crimmins TR, Fienup JR. A model based approach to improve the performance of the geometric filtering speckle reduction algorithm. IEEE Ultrason Symp 1995:**2**:1353–6
12. Kuan DT, Sawchuk AA, Strand TC, Chavel P. Adaptive restoration of images with speckle. IEEE Trans Acous Speech Signal Process 1989:**ASSP-35**:373–83
13. Insana M, Hall TJ, Gledndon GC, Posental SJ. Progress in quantitative ultrasonic imaging. SPIE Med Imaging III, Image Format 1989:**1090**:2–9
14. Frost VS, Stiles JA, Shanmungan KS, Holtzman JC. A model for radar images and its application for adaptive digital filtering of multiplicative noise. IEEE Trans Pattern Anal Mach Intell 1982:**4**:2:157–65
15. Lee JS. Digital image enhancement and noise filtering by using local statistics. IEEE Trans Pattern Anal Mach Intell 1980:**PAMI-2**:2:165–8
16. Lee JS. Refined filtering of image noise using local statistics. Comput Graph Image Process 1981:**15**:380–9
17. Kuan DT, Sawchuk AA. Adaptive noise smoothing filter for images with signal dependent noise. IEEE Trans Pattern Anal Mach Intell 1985:**PAMI-7**:2:165–77
18. Christodoulou CI, Loizou CP, Pattichis C.S, Pantziaris M, Kyriakou E, Pattichis MS, Schizas CN, Nicolaides AN. Despeckle filtering in ultrasound imaging of the carotid artery. Proc 2nd Joint Eng Med Biol/Biomed Eng Soc (EMBS/BMES), Houston, TX, 2002:1027–8
19. Solbo S, Eltoft T. Homomorphic wavelet based-statistical despeckling of SAR images. IEEE Trans Geosci Remote Sens 2004:**42**:4:711–21
20. Saniie J, Wang T, Bilgutay N. Analysis of homomorphic processing for ultrasonic grain signal characterization. IEEE Trans Ultrason Ferroelectr Freq Control 1989:**3**:365–75
21. Jin S, Wang Y, Hiller J. An adaptive non-linear diffusion algorithm for filtering medical images. IEEE Trans Inf Technol Biomed 2000:**4**:4:298–305
22. Weickert J, Romery B., Viergever M. Efficient and reliable schemes for nonlinear diffusion filtering. IEEE Trans Image Process 1998:**7**:398–410
23. Rougon N, Preteux F. Controlled anisotropic diffusion. Conf Nonlinear Image Processing VI, IS&T/SPIE Symposium on Electron Imaging, Science and Technology, San Jose, CA, 1995:1–12
24. Black M, Sapiro G, Marimont D, Heeger D. Robust anisotropic diffusion. IEEE Trans Image Process 1998:**7**:3:421–32
25. Perona P, Malik J. Scale-space and edge detection using anisotropic diffusion. IEEE Trans Pattern Anal Mach Intell 1990:**12**:7:629–39
26. Abd-Elmoniem K, Youssef A-B, Kadah Y. Real-time speckle reduction and coherence enhancement in ultrasound imaging via nonlinear anisotropic diffusion. IEEE Trans Biomed Eng 2002:**49**:9:997–1014
27. Zhong S, Cherkassky V. Image denoising using wavelet thresholding and model selection. Proc IEEE Int Conf Image Processing, Vancouver, 2000:1–4
28. Achim A, Bezerianos A, Tsakalides P. Novel Bayesian multiscale method for speckle removal in medical ultrasound images. IEEE Trans Med Imaging 2001:**20**:8:772–83
29. Zong X, Laine A. Geiser E. Speckle reduction and contrast enhancement of echocardiograms via multiscale nonlinear processing. IEEE Trans Med Imaging 1998:**17**:4:532–40
30. Hao X, Gao S, Gao X. A novel multiscale nonlinear thresholding method for ultrasonic speckle suppressing. IEEE Trans Med Imaging 1999:**18**:9:787–94
31. Donoho DL. Denoising by soft thresholding. IEEE Trans Inf Theory 1995:**41**:613–27

32. Wink AM, Roerdink JBTM. Denoising functional MR images: a comparison of wavelet denoising and Gaussian smoothing. IEEE Trans Med Imaging 2004:**23**:3:374–87
33. Dutt V. Statistical analysis of ultrasound echo envelope. PhD dissertation, Mayo Graduate School, Rochester, MN, 1995
34. Nagao M, Matsuyama T. Edge preserving smoothing. Comput Graph Image Process 1979:**9**:394–407
35. Huang T, Yang G, Tang G. A fast two-dimensional median filtering algorithm. IEEE Trans Acous Speech Signal Process 1979:**27**:1:13–8
36. Ali SM, Burge RE. New automatic techniques for smoothing and segmenting SAR images. Signal Processing 1988:**14**:335–46
37. Medeiros FNS, Mascarenhas NDA, Marques RCP, Laprano CM. Edge preserving wavelet speckle filtering. 5th IEEE Southwest Symposium on Image Analysis and Interpretation, Santa Fe, NM, 2002:281–85
38. Moulin P. Multiscale image decomposition and wavelets. In Hand Image and Video Processing, Ed. by A. Bovik, Academic Press, New York, 2000:289–300
39. Scheunders P. Wavelet thresholding of multivalued images. IEEE Trans Image Process 2004:**13**:4:475–83
40. Gupta S, Chauhan RC, Sexana SC. Wavelet-based statistical approach for speckle reduction in medical ultrasound images. Med Biol Eng Comput 2004:**42**:189–92
41. A Philips Medical System Company. Comparison of image clarity, SonoCT real-time compound imaging versus conventional 2D ultrasound imaging. Report, ATL Ultrasound, 2001
42. Christodoulou CI, Pattichis CS, Pantziaris M, Nicolaides AN. Texture-based classification of atherosclerotic carotid plaques. IEEE Trans Med Imaging 2003:**22**:7:902–12
43. Haralick RM, Shanmugam K, Dinstein I. Texture features for image classification. IEEE Trans Syst Man Cybern 1973:**SMC-3**:610–21
44. Weszka JS, Dyer CR, Rosenfield A. A comparative study of texture measures for terrain classification. IEEE Trans Syst Man Cybern 1976:**SMC-6**:269–85
45. Amadasun M, King R. Textural features corresponding to textural properties. IEEE Trans Syst Man Cybern 1989:**19**:5:1264–74
46. Wu CM, Chen YC, Hsieh K-S. Texture features for classification of ultrasonic images. IEEE Trans Med Imaging 1992:**11**:141–52
47. Chen TJ, Chuang KS, Wu J, Chen SC, Hwang IM, Jan ML. A novel image quality index using Moran I statistics. Phys Med Biol 2003:**48**:131–37
48. Gonzalez R, Woods R. Digital image processing, 2nd Edition, Prentice Hall Inc., Upper Saddle River, 2002:419–20
49. Wang Z, Bovik A, Sheikh H, Simoncelli E. Image quality assessment: from error measurement to structural similarity. IEEE Trans Image Process 2004:**13**:4:600–12
50. Winkler S. Vision models and quality metrics for image processing applications. PhD dissertation, University of Lausanne, Lausanne, 2000
51. Sakrison D. On the role of observer and a distortion measure in image transmission. IEEE Trans Commun 1977:**25**:1251–67
52. Wang Z, Bovik A. A universal quality index. IEEE Signal Process Lett 2002:**9**:3:81–4
53. Sheikh HR, Bovik AC, de Veciana G. An information fidelity criterion for image quality assessment using natural scene statistics. IEEE Trans Image Process 2005:**14**:12:2117–28
54. Krupinski E, Kundel H, Judy P, Nodine C. The medical image perception society, key issues for image perception research. Radiology 1998:**209**:611–2
55. Elatrozy T, Nicolaides AN, Tegos T, Zarka A, Griffin M, Sabetai M. The effect of B-mode ultrasonic image standardization of the echodensity of symptomatic and asymptomatic carotid bifurcation plaque. Int Angiol 1998:**17**:3:179–86
56. Loizou CP, Pattichis CS, Pantziaris MS, Tyllis T, Nicolaides AN. Snakes based segmentation of the common carotid artery intima media. Med Biol Eng Comput 2007:**45**:35–49
57. Pattichis CS, Kyriakou E, Christodoulou CI, Pattichis MS, Loizou CP, Pantziaris M, Nicolaides AN. Cardiovascular: ultrasound imaging in vascular cases. In Wiley Encyclopaedia of Biomed Engineering, Ed. by M. Akay, Wiley, Hoboken, 2006:1–12

58. Loizou CP, Pantziaris M, Pattichis MS, Kyriakou E, Pattichis CS. Ultrasound image texture analysis of the intima and media layers of the common carotid artery and its correlation with age and gender. Comput Med Image Graph 2009:**33**:4:317–24
59. Loizou CP, Pattichis CS, Nicolaides AN, Pantziaris M. Manual and automated media and intima thickness measurements of the common carotid artery. IEEE Trans Ultrason Ferroelectr Freq Control 2009:**56**:5:983–94
60. Loizou CP, Pattichis CS, Pantziaris M, Nicolaides AN. An integrated system for the segmentation of atherosclerotic carotid plaque. IEEE Trans Inf Technol Biomed 2007:**11**:5:661–7
61. Jung J, H, Hong K, Yang S. Noise reduction using variance characteristics in noisy image sequence. Int Conf Consumer Electronics, Lasvegas, 2005:213–4
62. Bertalmio M, Caselles V, Pardo A. Movie denoising by average of warped lines. IEEE Trans Image Process 2007:**16**:9:233–47
63. Zlokoliza V. Advanced nonlinear methods for video denoising. PhD dissertation, Ghent University, Ghent, 2006

Biographies

Christos P. Loizou was born in Cyprus in October 1962 and received his BSc degree in Electrical Engineering, the Dipl.-Ing. (MSc) degree in Computer Science and Telecommunications from the University of Kaisserslautern, Kaisserslautern, Germany, and the PhD degree from the Department of Computer Science, Kingston University, London, UK on ultrasound image analysis of the carotid artery in 1990 and 2005, respectively. He is currently an Assistant Professor in the Department of Computer Science at Intercollege in Cyprus. His research interests include medical imaging, in investigating the risk of stroke, and the multiple sclerosis disease.

Constantinos S. Pattichis was born in Cyprus on January 30, 1959 and received his Diploma as Technician Engineer from the Higher Technical Institute in Cyprus in 1979, the BSc in Electrical Engineering from the University of New Brunswick, Canada, in 1983, the MSc in Biomedical Engineering from the University of Texas at Austin, USA, in 1984, the MSc in Neurology from the University of Newcastle Upon Tyne, UK, in 1991, and the PhD in Electronic Engineering from the University of London, UK, in 1992. He is currently a Professor in the Department of Computer Science of the University of Cyprus. His research interests include ehealth, medical imaging, biosignal analysis, and intelligent systems.

Chapter 8
Use of Ultrasound Contrast Agents in Plaque Characterization

Filippo Molinari, William Liboni, Pierangela Giustetto, Enrica Pavanelli, Sara Giordano, and Jasjit S. Suri

Abstract Plaque characterization is gaining increasing importance owing to the need of a better determination of the intervention strategies in atherosclerotic patients. Currently, asymptomatic patients are treated by drugs administrations, whereas symptomatic patients are further investigated for assessing their suitability to surgical intervention. Guidelines and international consensuses state that the degree of stenosis and the symptoms are the only criteria for the selection of the surgical intervention. However, some problems still remain. First, intervention could be stenting or endarterectomy. Second, a considerable percentage of patients present restenosis after intervention.

Due to these limitations, the characterization of the inner content of a plaque seems a suitable way to gain a deeper comprehension of the patient's status and, therefore, for the onset of a more appropriate therapeutic line. MRI and CT imaging are the most used and accepted methods for noninvasive plaque characterization. If CT is the election methodology for plaques with important calcifications, MRI proved optimal tissue discrimination. Hence, by using CT and MRI, it is possible to differentiate soft and hard plaques. CT, however, makes use of ionizing radiations. Conversely, MRI imaging requires the use of contrast agents that could be little tolerated by some patients. Both CT and MRI are also expensive techniques and cannot be used in follow-up or monitoring protocols.

In this chapter, an ultrasound-based methodology for the characterization of carotid plaques is shown. This technique requires the injection of a small volume (about 1.5 ml) of contrast agent and the acquisition of postcontrast images. The rationale of this technique is that poorly perfused tissues (like lipids) show a lower contrast enhancement with respect to highly perfused tissues (such as the fibrous and the muscular tissue).

The technique consists of two steps. First, the image is automatically segmented by a completely user-independent algorithm. Then, the portion of the wall corresponding to the plaque is analyzed and intensity is assigned to a specific tissue.

F. Molinari (✉)
Biolab, Department of Electronics, Politecnico di Torino, Torino, Italy
e-mail: filippo.molinari@polito.it

Jasjit S. Suri et al. (eds.), *Atherosclerosis Disease Management*,
DOI 10.1007/978-1-4419-7222-4_8,

In this study, performance evaluation was done against histology. Twenty plaque specimens were sent to pathology for reporting. Qualitative correlation of the histology report and of the contrast-enhanced ultrasound analysis was performed. Results showed that this methodology is effective in differentiating soft by hard plaques. The plaque components that can be effectively identified are hemorrhagic blood, lipids, fibrous/muscular tissue, and calcium.

Despite the need for a deeper investigation and a quantitative evaluation of the results, this methodology showed encouraging results. This analysis architecture is under validation in a Neurology Division and is aimed at being used for follow-up of the patients and quantification of the drug therapy effects.

Keywords Contrast-enhanced ultrasound imaging • Carotid artery • Histology • Plaque characterization

8.1 Introduction

The atherosclerotic process refers to the degeneration of the arterial wall and the deposition of lipid and other blood-borne material within the arterial wall of almost all vascular territories [1]. The atherosclerotic process that takes place in the carotid arteries has been widely studied in last 20 years. Carotid wall lesions, in fact, have been correlated to different pathologies such as coronary and cerebrovascular diseases [2], poststroke cognitive impairments [3], platelet aggregability [4], aortic valve damage [5], and diabetes [6]. Large multicenter and cohort studies have been devoted to the screening of carotid diseases as predictors of severe pathologies over population [7–10].

Once born from the aortic arc, the carotid arteries (CA) follow the neck axis, at a depth of about 2–4 cm. This anatomical positioning and the relatively big diameter make the ultrasound examination of the CAs very simple and effective. Clinically, the echographic and echoDoppler analysis of the CAs is the routine examination for first diagnosis and follow-up.

The most widely used indicator of cardiovascular and cerebrovascular risk is the carotid artery intima-media thickness (IMT) [11]. The IMT, as risk indicator, possesses many advantages:

(i) It is an early marker of atherosclerosis progression;
(ii) Its measure is highly repeatable;
(iii) It can be measured noninvasively;
(iv) It can be used to quantify pathology monitoring and/or drug therapy efficacy.

In clinical practice, IMT is usually measured by ultrasound imaging. Beside IMT, recent studies have demonstrated the need for a deeper and more precise characterization of the CA wall. Some authors proposed the characterization of the CA wall echogenicity as a complement to IMT [12–14]. If IMT measurement is usually done manually by an expert sonographist during the examination, further characterization of the artery wall status require advanced, complex, and automated image processing techniques. Moreover, several research groups from all around the world are starting studies devoted to the noninvasive analysis of carotid plaque composition [13, 15–17].

Arterial plaque is probably the most evident final effect of the atherosclerotic process. Several studies demonstrated that a careful characterization of the plaque is extremely important in the management of the atherosclerotic disease [18–21]. In recent years, most of the techniques for plaque characterization relied on in vitro or in ex vivo samples [21, 22]. However, in vitro characterizations are unsuitable to goals like plaque evolution monitoring and quantification of drug therapy.

In the following section, we briefly summarize the possible techniques for plaque characterization.

- *Computer Tomography (CT) plaque imaging.* CT imaging is widely used for plaque measurement and characterization. Multidetector CT has recently emerged for measuring luminal stenosis, coronary calcium, and even the extent of noncalcified coronary plaque volume [1]. Essentially, CT imaging is very effective in visualizing fibrous and hard plaques, with high calcium content. Soft tissues being poorly represented in CT imaging, soft and unstable plaques can be poorly characterized. By administrating a radiology contrast agent, the precise quantification of the vessel stenosis and plaque volume is possible in noncalcified plaques too.
- *Magnetic Resonance Imaging (MRI).* Nowadays, MRI plays a fundamental role in arterial plaque assessment. Like CT, MRI imaging enables the precise and easy quantification of the plaque volume and of the luminal stenosis degree. Some recent studies have shown that MRI imaging (and also CT imaging) have very good agreement with digital subtraction angiography, which remains the historical gold standard method to assess the degree of lumen stenosis and the basis of the NASCET [23] and ECST [24] criteria. However, MRI imaging showed great characterization capabilities. MRI is very effective in tissue discrimination; hence, both calcified and noncalcified plaques can be represented with a very good detail level [25–27]. Saam et al. [28] proposed MRI imaging as the golden standard for noninvasive arterial plaque characterization. In their review, they summarized the status of MRI with regard to depiction of the criteria that define vulnerable plaques by using existing MR techniques. They placed the accent on the word "vulnerable," thus introducing a further evaluation criterion that surpasses the traditional one of internal lumen stenosis. Finally, MRI has been widely used to characterize the CA wall and possible CA plaques, both in traditional MRI [29, 30] and contrast-enhanced imaging [31, 32]. Major advantages of the MRI over ultrasounds are the high signal-to-noise ratio, the possibility of 3-D volumetric scans, and the independence of the imaging technique on the operator. Unfortunately, the imaging sequence (i.e., white-blood or black-blood sequences) and the magnetic field intensity have a big impact on the image quality and on the possibility of tissue characterization. Moreover, MRI contrast agents

are less tolerated by patients and at higher risk of adverse effects than ultrasound agents. MRI is also an expensive technique that requires postprocessing and reporting; hence, it is not suited for real-time examinations and for monitoring purposes.

- *Intra-Vascular Ultrasound (IVUS) imaging*. Since 2002 IVUS was used for the accurate characterization of artery plaques. Brathwaite et al. [33] developed an automatic classification algorithm to differentiate between four common lesion types in atherosclerotic arteries: calcific, fibro-calcific, fibrous, and fibro-fatty. Recently, by using IVUS, Nasu et al. [34] experimented the possibility of quantifying the effect of drug therapy on plaque progression/regression. IVUS methodology was the first to be indicated as a possible way to obtain "*virtual histology*," i.e., an indirect tissue characterization ideally performed at a histological level.

Beside the well-consolidated above-mentioned imaging techniques, there are some other growing methodologies that are now used at experimental level. Vibroacoustic imaging and elastography are ultrasound based imaging modalities that also point towards "virtual histology." Here, focus is made on the different dynamic responses to incident radiation made by tissues with different mechanical properties. A major advantage of these techniques with respect to traditional ultrasound imaging is the signal-to-noise ratio (SNR). The application in CA assessment is still under development, even though pilot studies reported extremely positive and encouraging results [35, 36].

Nevertheless, ultrasound imaging offers interesting potentialities, and it is currently a technique showing great expansion. This chapter describes an ultrasound-based technique for the characterization of carotid plaques. By using custom-developed software, we compare the ultrasound-based plaque characterization to histology, considering a population of patients who underwent surgical plaque removal.

8.2 Basics of Plaque Characterization by Ultrasounds

The ultrasound examination of the CA has some important advantages:

(i) It is a real-time methodology;
(ii) It is a relatively low-cost technique;
(iii) It does not require complex instrumentation;
(iv) It can be done repeatedly at very low risk (ultrasounds are nonionizing radiations);
(v) It can be done bedside or as homecare assistance. However, ultrasound images are quite difficult to process due to their intrinsic nature.

If compared to tomography techniques such as CT and MRI, ultrasounds possess the advantages of being fast, practical, cheaper, and suited to follow-up. Hence, even though presently MRI and multislice CT are the best emerging techniques for precise plaque characterization, ultrasound-based plaque analysis has scope of implementation in the relatively high percentage of patients who need follow-up.

Ultrasounds already proved effective in IMT measurement and plaque segmentation. Many studies demonstrated that ultrasound-measured quantities are validated and in good agreement with manual measurements [7, 37–50]. Our research team has a wide background in the specific field of carotid vascular ultrasound: since 2005, we have developed automated techniques for the CA wall segmentation [40, 41], plaque segmentation [51, 52], CA automated location [53, 54], and accurate IMT measurement [55]. Our goal was to foster the ultrasound potentialities to plaque virtual histology.

First of all, it is very difficult to perform virtual histology by using ultrasounds. The scattered ultrasound echo is function of a difference in the acoustic impedance between two tissues, and not of tissue density. Therefore, it is impossible to directly link tissues to echogenicity (i.e., the same tissue may be imaged with different gray levels when its neighboring tissues change).

Second, ultrasound images suffer from different noise sources and artifacts. Speckle noise is perhaps the most important disturbance [56, 57]. It is generated by the multiple scatters and absorption generated by tissues. Its impact on image quality is very detrimental: in ultrasound imaging, a tissue is never homogeneous but appears as a random distribution of gray tones of different intensity. If the trained human operator can easily cope with speckle noise, this disturbance still represents a major challenge to many automated image processing techniques. Calcium deposits can originate other image artifacts by absorbing most of the incident radiation and dropping a shadow cone underneath them. Finally, multiple reverberations and backscattering may be originated by dense structures or blood clots.

By summarizing, characterization techniques based on traditional 2D or 3D ultrasound imaging still suffers from performance problems and lack of tissue definition. In a pilot study, our team showed the possibility of CA plaque characterization by using contrast-enhanced ultrasound imaging [13]. The rationale was that high-frequency and high-resolution ultrasound scans performed after the injection of a contrast agent could be useful for differentiating poorly perfused tissues from highly perfused tissues in the CA wall. Results presented were relative to a relatively small population of patients, and there was no correlation with other imaging techniques or with histology.

In this chapter, we show the qualitative results of a histology-based validation, along with the latest improvements in our image processing strategy.

8.3 Experimental Protocol and Patients Selection

We enrolled in this study 20 nonconsecutive patients of the Neurology Division of the Gradenigo Hospital of Torino (Italy). The patients' mean age was 63.1 ± 12.4 years, with a range of 54–88 years. Eighteen patients were males. All the patients were referred to the Neurology Division of the Gradenigo Hospital due to neurological symptoms: 12 of them had transient ischemic attacks, six had previous history of embolic events, and two had both. None of the patients developed a stroke.

Subjects with cardiac (i.e., the presence of a patent foramen ovale) or pulmonary emboli origin were preliminary-screened and discarded. Patients were nonconsecutive since we had to select subjects indicated for surgical intervention. All the patients showed one carotid plaque: 15 were distal (i.e., located on the far carotid wall) and five proximal (i.e., located on the near carotid wall). Twelve plaques were located in the internal CA or very close to the bulb, eight on the common tract of the CA.

The subjects underwent the following testing protocol:

- Traditional ultrasound examination of the carotid arteries and of the supraaortic vessels. A reference B-Mode image of the plaque was acquired.
- A second B-Mode image, acquired at the same instant of the cardiac cycle of the previous one, was recorded after about 40 s from the injection of 1.5 ml of a contrast agent (SonoVue, Bracco, Italy). After this time, the contrast agent was mainly present in the microcirculation, thus giving enhancement of the tissues. Figure 8.1 shows an example of a B-Mode image of a plaque acquired before (Fig. 8.1a) and after (Fig. 8.1b) the injection of the contrast agent. The white arrow indicates the plaque.

The patients signed an informed consent prior to being submitted to the analysis, in accordance to the local rules. The testing protocol was approved by the Ethical Committee of the Gradenigo Hospital (Torino, Italy).

After testing, the selected patients underwent carotid endarterectomy (CEA). Plaque was surgically removed and sent to histology for reporting. All pathological specimens were transected and stained with hematoxylin and eosin (H&E). Histological reporting was used as ground truth (GT).

A neurologist (W.L.) performed the carotid scans (with and without contrast agent) by means of an ATL HDI-5000 ultrasound device equipped with a 10 MHz

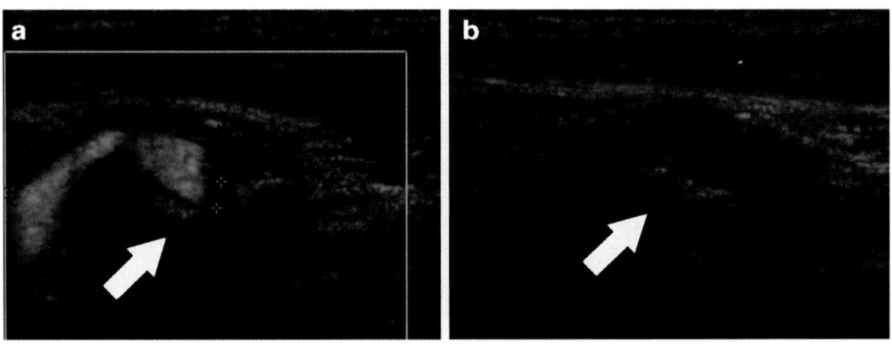

Fig. 8.1 B-Mode representation of a mixed distal plaque of the CA. (**a**) Traditional B-Mode image acquired at the beginning of the examination. The *shaded gray color* inside the lumen reflects the color Doppler imaging. The manually placed markers were used to estimate the lumen narrowing. (**b**) B-Mode contrast-enhanced image. The *white arrows* indicate the plaque

linear probe. All the images were downloaded to a computer in DICOM format, discretized on 8 bits, and represented by means of a linear gray scale. Four experts (a neurologist, a cardiologist, and two radiology technicians) manually segmented all the images and traced the profiles of the lumen–intima (LI) and media–adventitia (MA) interfaces, following the plaque profile. The average segmentations were used to measure the segmentation error of our automatic algorithm.

8.4 Ultrasound Images Segmentation Strategy

We processed our image dataset by an automated technique we developed in 2005 [41] and refined in 2008 [13, 52]. We named such technique as *CULEX* (*C*ompletely *U*ser-independent *L*ayers *EX*traction). *CULEX* consists of two steps:

1. Automated tracings of the near (AD_N) and far (AD_F) adventitia profiles;
2. Snake-based segmentation of the lumen–intima and media–adventitia boundaries starting from the detected far adventitia.

The detailed structure of *CULEX* has already been described in previously published works [41, 52]. In the following, we summarize the structure of *CULEX*.

The core of the algorithm is the pixel analysis based on local statistics. Carotid appearance can be seen as a mixture model with different characteristics: (1) pixels belonging to the CCA lumen are dark (i.e., of low-intensity value) and surrounded by a dark homogeneous neighborhood; (2) the pixels belonging to the CCA wall are bright (i.e., of high-intensity value) and surrounded by a relatively homogeneous neighborhood; (3) all the remaining pixels should have average intensity with an inhomogeneous neighborhood. By computing the mean value and the standard deviation of a 10×10 square neighborhood of each pixel, we can separate the lumen pixels from the others by placing proper thresholds on the neighborhood mean intensity and standard deviation values. Empirically, we found that the better discrimination for the lumen pixels was intensity lower than 0.08 and standard deviation lower than 0.14 on a normalized (0,1) scale [41, 53].

The image is then processed columnwise. The deepest local intensity maximum is marked as far adventitia candidate. The algorithm then decreases the row index (i.e., moves towards the near wall) and searches for a lumen pixel. Lumen is defined as the first local intensity minimum with neighborhood mean and standard deviation lower than 0.08 and 0.14, respectively. By further decreasing the row index, the algorithm marks the position of the near adventitia wall (AD_N) as the first local intensity maximum, i.e., found just after the end of the lumen. Output of this first step is the tracing of AD_N, of AD_F, and of the artery lumen (L). The artery lumen is a continuous line passing between AD_N and AD_F, but not necessarily in the centerline of the vessel.

The image is then considered columnwise, and only the pixels comprised between AD_N and AD_F are processed. A gradient enhancing the vertical transitions is used and the two higher intensity maxima of the gradient are considered as first guess of the far wall LI and MA transitions. The intensity criterion we used was relative to the intensity profile of the specific column: we only considered maxima whose intensity was above the 90th-percentile of the intensities distribution of the column pixels. The sequence of LI and MA points is then refined using the snake-based algorithm [58], where the elasticity and tension parameters: $\alpha(s) = 0.1$ and $\beta(s) = 0.01$. The same procedure was applied to the points comprised between the lumen profile L and the AD_N.

Figure 8.2 sketches the *CULEX* strategy: left panel depicts the near (AD_N), the far (AD_F) adventitia, and the lumen (L); right panel reports a representative *CULEX* segmentation of the near and far CA wall. The legend of the profiles acronym is reported by the caption of Fig. 8.2.

Extensive characterization of the *CULEX* algorithm can be found in a study by Delsanto et al. [41]. In other studies *CULEX* was benchmarked against other segmentation strategies [55, 59]. On the overall, we showed that the LI segmentation error was about 1.2 ± 0.9 pixels for the near wall and 0.9 ± 0.8 for the far wall. The MA segmentation error was about 1.9 ± 1.6 pixels for the near wall and 0.5 ± 0.7 for the far wall. By considering an axial resolution approximately equal to 62.5 μm/pixel, we obtained segmentation errors equal to about 75 μm for the proximal LI, 56 μm for the distal LI, 118 μm for the proximal MA, and 31 μm for the distal MA. The performance metric we used was the Euclidean distance of the computer-generated profiles from human tracings.

CULEX proved effective in plaque imaging too. The segmentation strategy is completely automated, so there is no need for an a priori knowledge of the image type. When tested on plaques, we measured the performance by considering the percentage of misclassified pixels (with respect to GT). Calcified and hard plaques were segmented with an average error of about $8\% \pm 5\%$. When working on soft and anechoic plaques, the segmentation error increased up to $12\% \pm 6\%$.

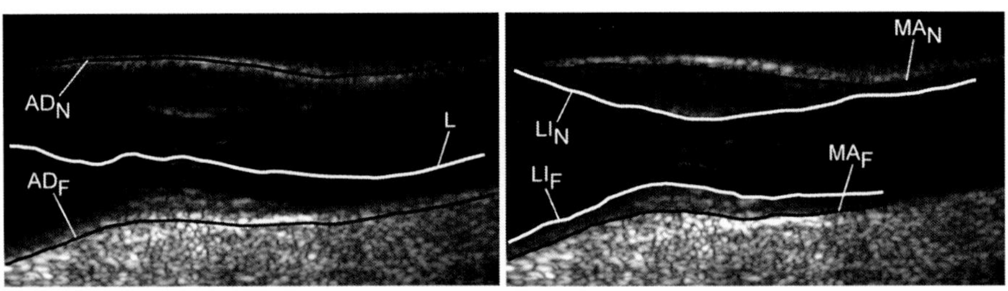

Fig. 8.2 Example of *CULEX* segmentation of a CA B-Mode ultrasound image. *Left panel*: original image with automated tracings of the near adventitia (AD_N), far adventitia (AD_F), and the lumen (L). Adventitia profiles are traced in *black*, lumen in *white*. *Right panel*: automated lumen–intima tracings for the near (LI_N) and far (LI_F) wall depicted in *white*; tracings of the media–adventitia layers of the near (MA_N) and far (MA_F) wall depicted in *black*

8.5 Plaque Characterization in Contrast: Enhanced Ultrasound Images

We used contrast–enhanced ultrasound (ceUS) B-Mode imaging for plaque characterization. Contrast-enhanced imaging has amplitude and spatial resolution that is very high, and thus, it can effectively represent the wall microcirculation. Due to lack of spatial resolution, the representation of such circulation is impossible by using Power- or Color-Doppler approaches. In fact, most of the ultrasound scanners significantly loose spatial detail when used in compound imaging.

When performing virtual histology by using an imaging technique, the principal components of the tissues must be differentiated. In CA plaques, the four most important components are: the lipids, the muscular and the fibrous tissues, calcium, and blood hemorrhages that are typical of ulcerated plaques. Therefore, aim of the ceUS image analysis is the differentiation of such plaque components.

The hypothesis of this approach is that different tissues should have different levels of blood perfusion. Hence, the ultrasound contrast agent being strictly non-diffusive, it is expected that once the initial phase of enhancement of the vessels' lumen has ended, the contrast agent caused enhancement to the wall tissues. Therefore, we expected that wall tissues had enhancement linked to their characteristics:

- The lipidic tissue should be poorly perfused and, therefore, remain with a dark appearance on the ceUS image;
- The fibrous and muscular tissues should have higher perfusion and, thus, acquire a brighter appearance on the ceUS image.

Blood hemorrhages are expected to have a dark appearance. In fact, once the initial enhancement phase in the macrocirculation has terminated, it is not expected that hemorrhages get enhancement by the microcirculation. Being blood essentially anechoic, thrombus is represented by black.

Calcium deposits are unperfused. Hence, their appearance should not change between traditional B-Mode and ceUS representation. Hence, we assumed the calcium as bright.

Following the previously explained hypothesis, we adopted this processing strategy for the ceUS images:

1. ceUS plaque images were first segmented by using *CULEX* to derive the LI and MA profiles of the near and far wall. This step, as described in Sect. 8.4, is completely automated.
2. The image was intensity normalized, so that the vessel lumen corresponded to 0 and the adventitia wall to about 190 (in a range 0–255). Then, all the pixels constituting the plaque were color-coded according to the following correspondence:

   (a) Hemorrhagic blood: gray level in the range 0–4 (color red)
   (b) Lipids: gray level in the range 8–26 (color yellow)

   (c) Muscle and fibrous tissue: gray level in the range 41–76 and 112–196 (color green);
   (d) Calcium: gray level in the range 211–255 (color blue).

3. The automatically segmented plaque was then color-coded, and the results were compared to histology. In this study, we did not use any correlation technique between ceUS images and histology images. An expert pathologist, who analyzed all the specimens of this study, visually correlated the histologic report with the ceUS images. The pathologist was blinded with respect to the ceUS image; he first fixed, analyzed, and reported all the specimens, and only at the end, he correlated results with ceUS data.

The values correspondences between tissues and gray levels in the ceUS images were derived by the studies from Lal et al. [60, 61]. They showed that correlation of B-mode ultrasonographic morphology with histological characteristics of atherosclerotic carotid plaques remained ill-defined since the classification of plaques with recently reported measures of plaque echogenicity and heterogeneity had been unsatisfactory. The gray-scale ranges they used were based on the analysis of control subjects. This helped define the amount of intraplaque hemorrhage, lipid, fibromuscular tissue, and calcium within carotid plaque images.

In their studies, Lal et al. used traditional B-Mode imaging. The innovation of our approach relies on the use of ceUS. In fact, by using harmonic imaging, the spatial resolution of the image increases. This, intrinsically, enables a better localization of the different tissues within the plaque. Then, since the ultrasound contrast agent remains within the vessels, specific enhancement is expected in the plaque region. Hence, we believe that ceUS is an optimal choice for the selective enhancement of the SNR, leading to a better overall characterization.

The contrast agent we used has microbubbles of diameter ranging between 5 μm and about 100 μm. Working with at a ultrasound frequency of 10 MHz, we could detect only groups of microbubbles, being the wavelength of the used ultrasound beam sensibly larger than the diameters of the microbubbles. Hence, one of the first limitations of this representation is that, by using this methodology, only tissues perfused by vessels of 100–200 μm of diameters can be clearly observed. Smaller vessels are likely to produce a diffuse gray enhancement on the image, but it would be difficult to localize it in a precise point of the image due to lack of spatial resolution.

Figure 8.3 reports an example of ceUS image enhancement after the injection of the contrast agent. Figure 8.3a is relative to a soft and anechoic distal CA plaque (indicated by the white arrow). Its soft content makes the plaque mostly invisible. Figure 8.3b shows the same plaque appearance 40 s after the injection of 1.5 ml of contrast agent. The plaque now is visible and its texture can be appreciated and analyzed. It can be noted that the plaque enhancement is different at different points: the left portion of the plaque seems darker, whereas the right half seems brighter.

Figure 8.4 reports the *CULEX* segmentation of the plaque in Fig. 8.3. Figure 8.4a demonstrates that soft plaques cannot be effectively recognized and segmented due to poor representation. Figure 8.4b shows the *CULEX* performance

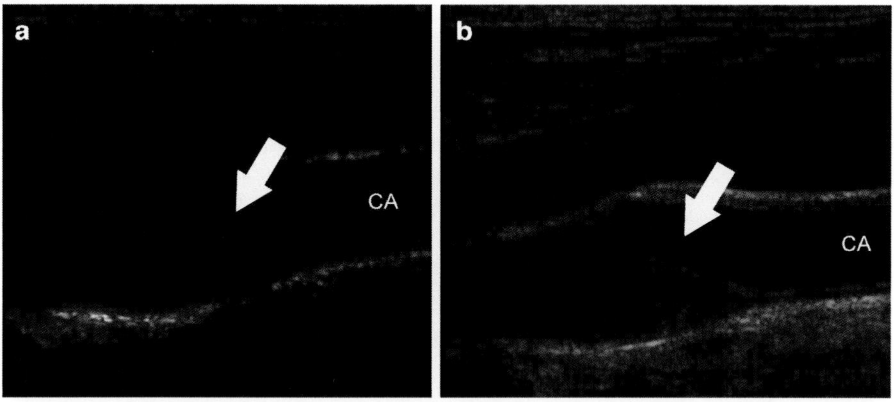

Fig. 8.3 Example of ceUS image enhancement. (**a**) Original B-Mode image of a soft distal plaque. The *white arrow* indicates the presence of the plaque that is mostly anechoic. (**b**) ceUS image of the same plaque obtained after the injection of 1.5 ml of contrast agent. The image was acquired 40 s after the contrast agent injection. The plaque (*white arrow*) is now clearly observable

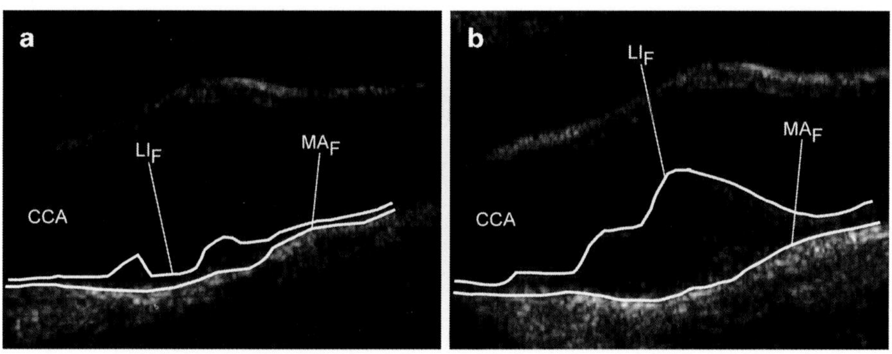

Fig. 8.4 *CULEX* segmentation of the anechoic plaque of Fig. 8.3. (**a**) *CULEX* segmentation of the original B-Mode image. Performance is poor and plaque cannot be segmented properly due to poor representation. (**b**) ceUS image better represents plaque, and segmentation is very accurate. The percentage of misclassified pixels for this image was equal to 5%

on ceUS images. Segmentation was very accurate, and the percent of misclassified pixel was as low as 5%.

Figure 8.5 reports the comparison between the *CULEX* segmentation (white lines) and the manual tracings of the experts (red line for the LI and blue for the MA). It can be noticed that *CULEX* automated segmentation on ceUS images is very accurate and effective. Automated segmentation constitutes the basis for the automated plaque characterization procedure.

Finally, Fig. 8.6 reports the color-coded image after the application of our characterization procedure. It can be observed that only the pixels within the LI and MA

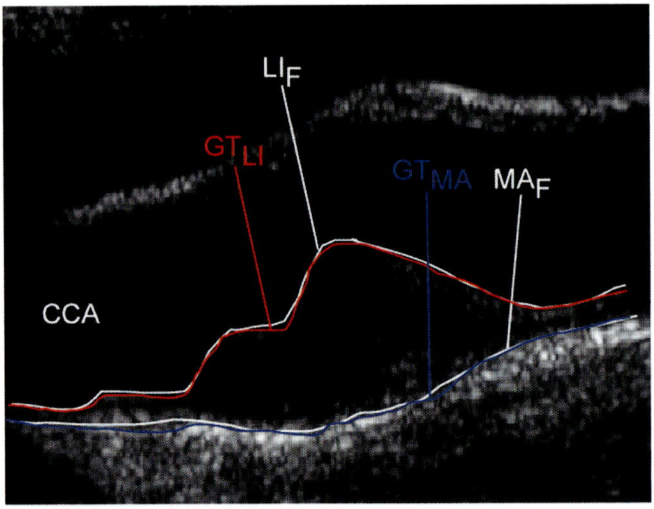

Fig. 8.5 Manual segmentation (GT) and *CULEX* segmentation performed on a ceUS image of an anechoic plaque

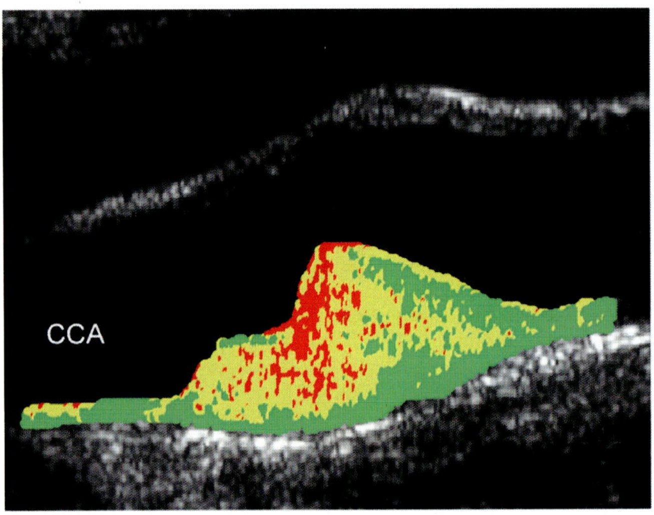

Fig. 8.6 ceUS image of the plaque in Fig. 8.3 after analysis and tissue characterization. Only the pixels of the plaque have been colored. As expected, the *left half* of the plaque is predominantly soft and with a large percentage of hemorrhagic blood. The *right half* is more compact and predominantly muscular or fibrous. There is no calcium in the plaque. (© IEEE 2007 - Reproduced with permission from [13])

boundaries have been processed. As expected, the plaque under analysis was clearly soft and anechoic and the predominant component of the plaque itself is lipidic tissue (yellow). The left half of the plaque seems softer than the right one: it is characterized by a high content of hemorrhagic blood (red) as well as a great amount of lipids. The right half is more stable, being dominated by muscular and fibrous tissue (green). Calcium is absent. On the overall, this plaque is mixed, with a predominant soft composition that makes it a plaque at high risk of rupture.

8.6 ceUS Plaque Characterization and Histology

All the surgically removed specimens were analyzed by H&E images. The same laboratory and the same operator reported all the specimens. This ensured maximal reliability and reproducibility of the results.

In this section, we present some results we obtained. At the end of the section, concluding remarks about the correlation between histology and ceUS plaque characterization is provided.

8.6.1 Plaque with Calcium Deposits

In this example, we show the specimen obtained from a 73-year-old male patient. The plaque caused a critical stenosis of the right internal CA. The patient was symptomatic and the plaque was firstly diagnosed during a cardiological visit. Symptoms were blood pressure increase and subjective instability, but without cognitive or senso-motor deficits. The degree of stenosis was 80% (NASCET criterium).

After CEA, the specimen was sliced and colored according to the experimental protocol (Sect. 8.3). Figure 8.7 reports the plaque specimen (a and b) along with the H&E images of two slices (c and d). The magnification of the histological H&E images is 95×. The diffuse violet color indicates calcium. It can be observed that calcium is present in a diffused way. Figure 8.7c and d are relative to the central portion of the plaque specimen. Calcium is clearly visible in proximity of the CA lumen (location in the intima and media layers).

Figure 8.8 reports the ceUS characterization of the plaque in Fig. 8.7. Figure 8.7a is relative to the original B-Mode image. Calcium is visible due to the shadow cone it projects on the image. The calcium deposit is indicated by the arrow in Fig. 8.8a. Figure 8.8b depicts the ceUS image, where it is possible to observe the better representation of the plaque and of part of the underneath tissues, which are almost invisible in Fig. 8.8a. Figure 8.8c is the H&E image of the central slice of the plaque, roughly corresponding to the central vertical column of Fig. 8.8b. Finally, Fig. 8.8d reports the color-coded ceUS plaque characterization.

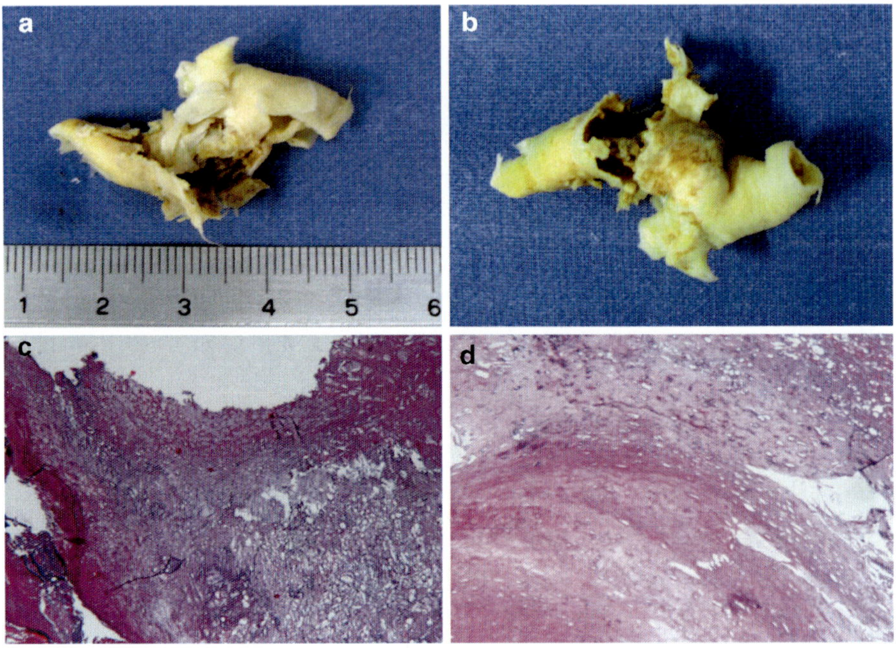

Fig. 8.7 Specimen of a CA hard plaque with high calcium content. (**a**, **b**) Specimen after surgical removal and before histological analysis. (**c**, **d**) H&E ×95 images. The violet diffuse color depicts calcium

The calcium deposits have been correctly detected by the ceUS characterization. Also, lipids within the CA wall are present and correctly reported. In the H&E image (Fig. 8.8c) lipids are depicted as white spots. Overall, this plaque is a mostly calcific, with predominance of muscular/fibrous tissue and small lipid content. From the comparison of Fig. 8.8c and d, it is possible to observe a good correlation of the ceUS characterization with histology.

Artifacts are the major problem in ceUS analysis of hard plaques with overt calcifications. Calcium absorbs most of the ultrasound radiation, projecting a shadow cone that precludes the observation of the tissues underneath. ceUS representation improves the image quality, since the echoes from soft tissues are enhanced. However, when in presence of highly calcified plaques, this methodology could not be applicable. It must be said that heavily calcified plaques are anyway easy to diagnose by traditional ultrasound imaging, they being characterized by a specific pattern: calcium is represented by a bright white with a black shadow cone underneath. Therefore, heavily calcified plaques could be better analyzed and imaged by using CT or MRI techniques. This methodology finds it utility in the analysis of mixed plaques, when the composition is crucial. Specifically, it has been shown that the analysis of calcium content could provide important information for the appropriate selection of the surgical intervention strategy [62, 63].

8.6.2 *Soft Unstable Plaque*

This plaque was removed by CEA from a 82-years-old male patient who was referred to the Neurology Division due to recurrent TIA episodes. The plaque was located in the bulb of the left CA, with protrusion in the first tract of the common CA. Surface was smooth and regular, with concentric development. The stenosis degree was measured in 75% following ECST criterion, and 65% following NASCET. The MRI examination confirmed the presence of a 25-mm mixed plaque, with fibrous component in the deeper layers and soft unstable subintima component. Calcium areas were almost absent.

Figure 8.9 reports the histological specimen and the H&E images. The plaque dimensions are reported by Fig. 8.9a and b. Figure 8.9c and d show the overt lipid content of the plaque, depicted by white cholesterol "needles" with subintima location. The "foamy" aspect of the plaque in Fig. 8.9d is due to the presence of macrophages, indicating an ongoing inflammatory process.

Figure 8.10 reports the original B-Mode image (Fig. 8.10a), the ceUS image (b), and the H&E 95× specimen (c). Figure 8.10d reports the color-coded plaque

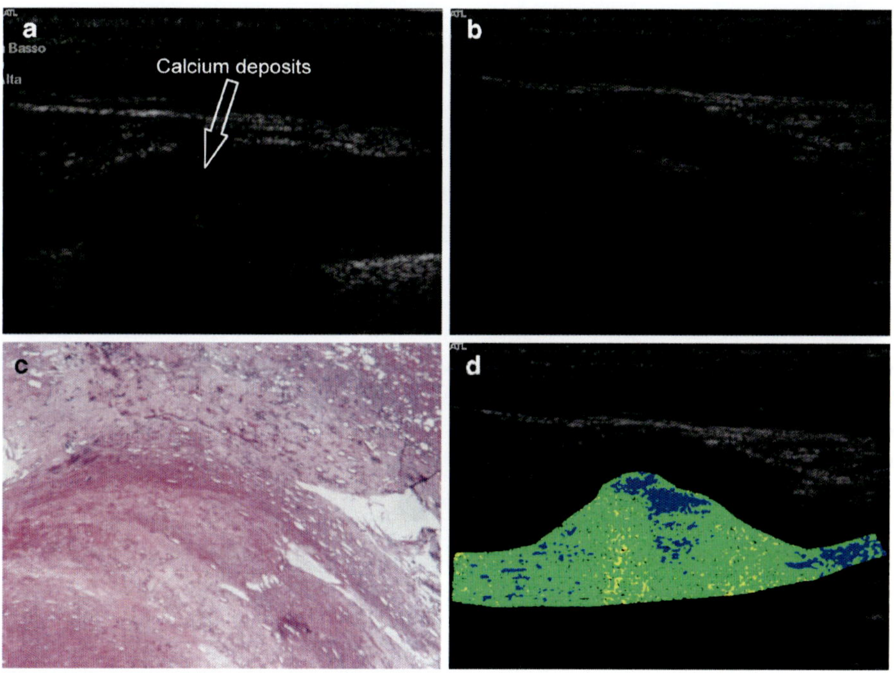

Fig. 8.8 (**a**) Traditional B-Mode image before contrast agent injection. (**b**) ceUS image after the administration of 1.5 ml of contrast agent. (**c**) H&E ×95 histology of the specimen. Central slice of the plaque. (**d**) ceUS processed image. The plaque is clearly calcific with diffuse calcium crystals in intima and media

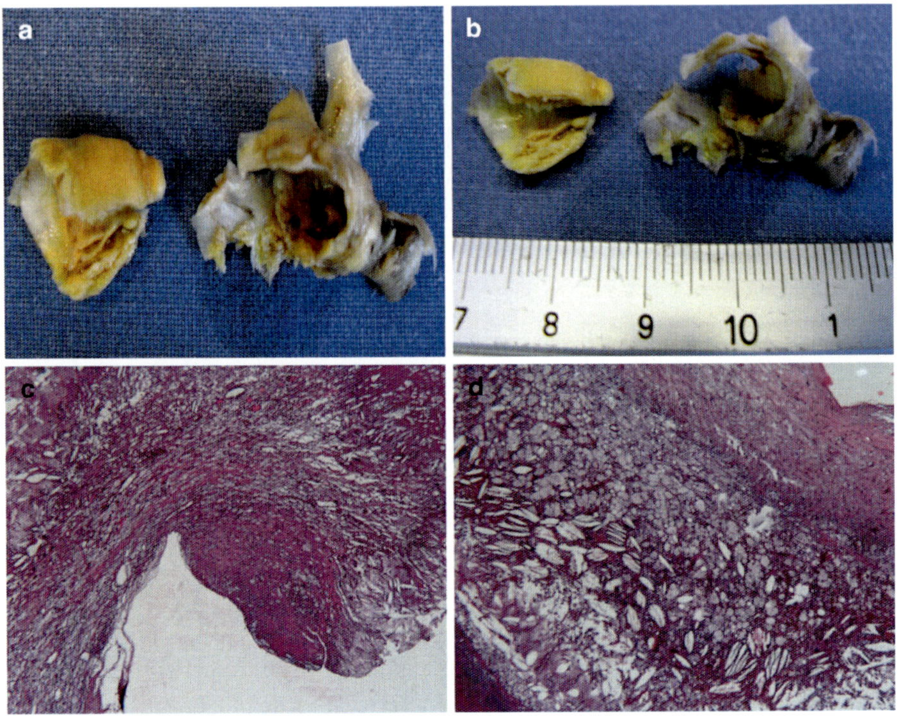

Fig. 8.9 Specimen of a CA soft plaque with prevalent lipid content. (**a**, **b**) Specimen after surgical removal and before histological analysis. (**c**, **d**) H&E ×95 images. The *white spots* depicts lipids. Specifically, the needle-shaped spots are cholesterol deposits. The "foamy" aspect of the image in proximity of the intima layer reveals the inflammatory process with macrophages

characterization. The lipidic content is clearly represented, with essentially a subintimal location, as reported by histology. The red hemorrhages present in proximity of the intima layer are compatible with the ulcerated aspect of the plaque (Fig. 8.9c) and the inflammatory process (Figs. 8.9d or 8.10c). The deeper layers of the plaque are coded by green, thus indicating a fibrous tissue. Again, this characterization is in accordance with histology.

Our characterization procedure proved effective in presence of mixed and soft plaques. By coding the ceUS images, it was possible to discriminate the lipidic and the fibrous content, thus giving important information about the plaque stability. In our testing database, all the mixed soft plaques were correctly identified by this ceUS characterization strategy. Concordance with histology was very high.

From a technical point of view, soft plaques do not pose any problem for their automated segmentation and characterization. Soft plaques are usually well represented by conventional ultrasound imaging. ceUS improves the representation and the contrast resolution, by better separating soft from hard tissues (i.e., fat from fibrous/muscular tissues).

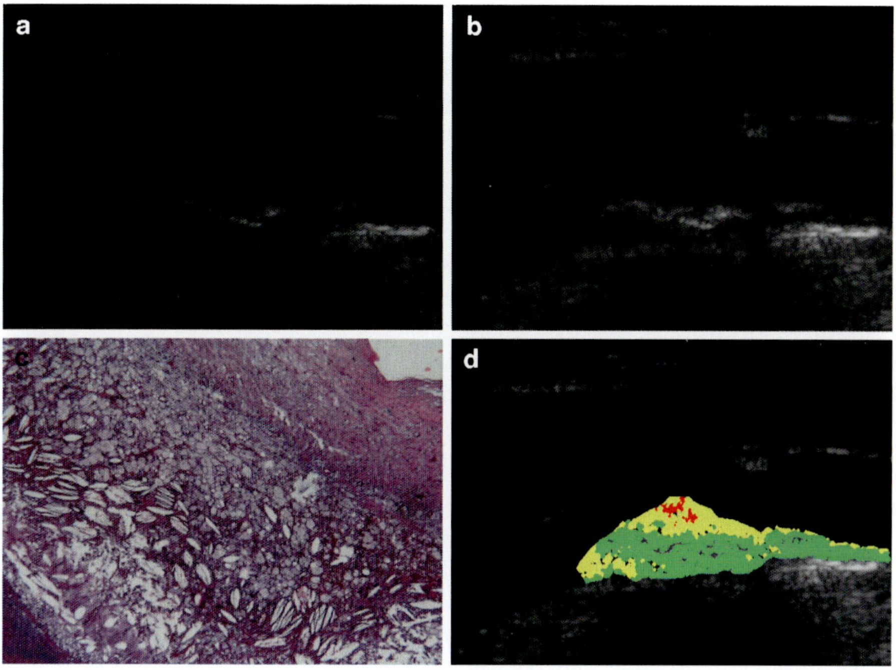

Fig. 8.10 (**a**) Traditional B-Mode image before contrast agent injection. (**b**) ceUS image after the administration of 1.5 ml of contrast agent. (**c**) H&E ×95 histology of the specimen. Central slice of the plaque. (**d**) ceUS processed image. The plaque reveals a moderate hemorrhagic blood content located in proximity of the intima layer. This is consistent with the appearance of the intima in (**c**) and with the overt presence of macrophages, indicating an ongoing inflammatory process. The plaque is mainly lipidic, with a deep fibrous/muscular tissue

8.7 Discussions, Limitations, and Future Perspectives

We proposed a completely automated methodology for plaque characterization in contrast-enhanced ultrasound imaging. Results were validated against histology. A selected population of patients with CA plaques was analyzed, and a high level of concordance was observed. Therefore, ceUS plaque characterization seems a promising technique for the noninvasive assessment of artery plaques.

The herein proposed characterization methodology is completely automated. This is particularly important, since it ensures the possibility of processing relatively large databases without the need for any user interaction. Also, the absence of user interaction ensures standardization of the procedure and the results, as it is required in large multicenter studies.

Automated segmentation of the *CULEX* algorithm tested on ceUS images was satisfactory. The segmentation errors were compatible with the goal of plaque characterization. The segmentation of plaqued CAs is more challenging than that of healthy CAs or of CAs with increased IMT (but without plaques). However, segmentation percent errors lower than 10% ensure that the plaque is correctly recognized and traced.

As already mentioned, the segmentation becomes more problematic when the plaque is highly calcific. This is because the higher the calcium content, the higher the artifacts in the image. In some conditions, plaque segmentation may be impossible. Since plaque segmentation is the basis of our automated technique, highly calcified plaques may be impossible to analyze.

Soft plaques pose minor challenges to our procedure. Usually, no artifacts are present in such images; hence, soft plaques are segmented with low errors.

Our procedure was implemented in MATLAB environment and tested on a dual 2.5 GHz Apple PowerPC, equipped with 8 GB of RAM. The average time required for plaque characterization was 56.3 ± 2.8 s, subdivided in about 49 s for segmentation and 7 s for ceUS characterization. Segmentation is still a costly step that cannot be implemented in real time. This is because *CULEX* adopts a strategy based on local statistics, gradients, and snake iterations, thus requiring about 50 s for reaching final segmentation.

This study has some limitations that we discuss here.

- First, validation was only qualitative. An expert pathologist (the same that obtained the H&E images) visually scored the ceUS color-coded images and correlated it to the histological report. The indication of "good," "average," or "poor" correlation, therefore, was subjective. Nevertheless, we obtained a "good" correlation in 18 images out of 20 and an "average" correlation in two images. These images that obtained a lower scoring were relative to highly calcified plaques that were poorly segmented. Therefore, some calcium deposits in the intima layers were not correctly tracked and correlation to histology resulted scarce. Despite the need for a wider validation and a quantitative correlation score, we believe that contrast-enhanced ultrasound imaging could have a future in noninvasive plaque characterization.
- The second limitation was the use of 2D imaging. Correlation with histology was crucial since 2D imaging depicted a single longitudinal view of the plaque. Also, 2D ceUS images are longitudinal, whereas H&E images are transverse. From a practical point of view, in fact, it is very difficult to obtain longitudinal slices from the plaque specimens. This makes the correlation between the two imaging techniques complicated. The authors are now working towards a 3D ultrasound imaging technique with ceUS characterization to compare the different plaque views (ceUS and histology) by adopting superior registration techniques.
- The third limitation and scope of improvement of this work is the selection of the gray levels correspondence values to tissues. We adopted the same levels that Lal et al. derived after validation on a large dataset of normal CA images. However, as we showed, ceUS images have intensity levels that are clearly different from conventional ultrasound images. Therefore, redefinition or tuning of the gray levels could be required. However, this step would require a larger database and a numerical evaluation of the results. Once the 3-D methodology is ready, we will introduce this optimization step in our processing pipeline.
- Finally, the overall procedure is still lengthy and not suitable to online processing. The DICOM image must be transferred from the ultrasound scanner to a computer, then it must be segmented and, finally, analyzed. Scope of improvement is certainly the segmentation technique, which should be suitable for online processing without detriment of the segmentation performance.

The herein proposed methodology, however, has some major advantages and interesting points.

- The ceUS-based methodology is completely automated. This, as we already mentioned, makes the overall technique suitable to large and standardized studies.
- The segmentation performance of *CULEX* technique was independent on the scanner used to acquire the image [53]. This makes the technique extremely robust and applicable in virtually any center with suitable equipment and practice in contrast-enhanced vascular imaging.
- Noninvasivity is certainly one of the major advantages of this technique. Undoubtedly, MRI and CT are election techniques for measuring stenosis, plaque volume, and for plaque characterization. However, such techniques have the disadvantage of being expensive and impracticable for monitoring. Ultrasound examinations, besides being safe for the patient, can be performed at bedside as well as in homecare situations. This makes ultrasounds the election technique for plaque monitoring. Therefore, this characterization methodology could find its application in the follow-up of treated patients. Surgically treated patients, in fact, are exposed to the risk of restenosis. In this case, the coupling between ultrasounds and ceUS plaque characterization could be very helpful for early visualization of possible restenosis signs. Conversely, patients that undergo pharmacological treatment would benefit from accurate plaque characterization if changes occur. Progression/regression of the plaque is now merely assessed by relying on the stenosis degree, usually measured by considering the prestenosis and the post-stenosis blood flow velocities. In future, we believe that plaque composition could also play an important role in the selection of the most proper drug therapy.

Presently, this technique is under extensive validation and in the clinical pipeline of the Neurology Division of the Gradenigo Hospital of Torino, Italy.

References

1. J.J. Badimon, B. Ibanez, and G. Cimmino, Genesis and dynamics of atherosclerotic lesions: implications for early detection, Cerebrovasc Dis, 27 Suppl 1, (2009), 38–47.
2. S.H. Johnsen and E.B. Mathiesen, Carotid plaque compared with intima-media thickness as a predictor of coronary and cerebrovascular disease, Curr Cardiol Rep, 11(1), (2009), 21–7.
3. Y.H. Lee and S.J. Yeh, Correlation of common carotid artery intima media thickness, intracranial arterial stenosis and post-stroke cognitive impairment, Acta Neurol Taiwan, 16(4), (2007), 207–13.
4. Y. Fusegawa, H. Hashizume, T. Okumura, Y. Deguchi, Y. Shina, Y. Ikari, and T. Tanabe, Hypertensive patients with carotid artery plaque exhibit increased platelet aggregability Thromb Res, 117(6), (2006), 615–22.
5. M. Soylu, A.D. Demir, O. Ozdemir, Y. Uzun, and S. Korkmaz, Relationship between plaque morphology of carotid artery and aortic valve calcification Angiology, 54(6), (2003), 637–40.
6. L.A. Lange, D.W. Bowden, C.D. Langefeld, L.E. Wagenknecht, J.J. Carr, S.S. Rich, W.A. Riley, and B.I. Freedman, Heritability of carotid artery intima-medial thickness in type 2 diabetes Stroke, 33(7), (2002), 1876–81.
7. P.J. Touboul, R. Hernandez-Hernandez, S. Kucukoglu, K.S. Woo, E. Vicaut, J. Labreuche, C. Migom, H. Silva, and R. Vinueza, Carotid artery intima media thickness, plaque and Framingham

cardiovascular score in Asia, Africa/Middle East and Latin America: the PARC-AALA study Int J Cardiovasc Imaging, 23(5), (2007), 557–67.

8. A.H. Thakore, C.Y. Guo, M.G. Larson, D. Corey, T.J. Wang, R.S. Vasan, R.B. D'Agostino, Sr., I. Lipinska, J.F. Keaney, Jr., E.J. Benjamin, and C.J. O'Donnell, Association of multiple inflammatory markers with carotid intimal medial thickness and stenosis (from the Framingham Heart Study) Am J Cardiol, 99(11), (2007), 1598–602.
9. R. Elosua, L.A. Cupples, C.S. Fox, J.F. Polak, R.A. D'Agostino, Sr., P.A. Wolf, C.J. O'Donnell, and J.M. Ordovas, Association between well-characterized lipoprotein-related genetic variants and carotid intimal medial thickness and stenosis: The Framingham Heart Study Atherosclerosis, 189(1), (2006), 222–8.
10. P.J. Touboul, J. Labreuche, E. Vicaut, and P. Amarenco, Carotid intima-media thickness, plaques, and Framingham risk score as independent determinants of stroke risk Stroke, 36(8), (2005), 1741–5.
11. A. Simon, J. Gariepy, G. Chironi, J.L. Megnien, and J. Levenson, Intima-media thickness: a new tool for diagnosis and treatment of cardiovascular risk J Hypertens, 20(2), (2002), 159–69.
12. H. Bliman, Could the sonographic characterization of carotid plaque improve the surrogacy of intima-media thickness for vascular outcomes? Int J Cardiol, (2009), (epub ahead of print).
13. F. Molinari, W. Liboni, E. Pavanelli, P. Giustetto, S. Badalamenti, and J.S. Suri, Accurate and automatic carotid plaque characterization in contrast enhanced 2-d ultrasound images Conf Proc IEEE Eng Med Biol Soc, 2007, (2007), 335–8.
14. C.P. Loizou, M. Pantziaris, M.S. Pattichis, E. Kyriacou, and C.S. Pattichis, Ultrasound image texture analysis of the intima and media layers of the common carotid artery and its correlation with age and gender Comput Med Imaging Graph, 33(4), (2009), 317–24.
15. H. Shi, T. Varghese, C.C. Mitchell, M. McCormick, R.J. Dempsey, and M.A. Kliewer, In vivo attenuation and equivalent scatterer size parameters for atherosclerotic carotid plaque: Preliminary results, Ultrasonics, 49(8), (2009), 779–85.
16. M. Ogita, H. Funayama, T. Nakamura, K. Sakakura, Y. Sugawara, N. Kubo, J. Ako, S.E. Ishikawa, and S. Momomura, Plaque characterization of non-culprit lesions by virtual histology intravascular ultrasound in diabetic patients: Impact of renal function J Cardiol, 54(1), (2009), 59–65.
17. E. Vicenzini, M.F. Giannoni, F. Benedetti-Valentini, and G.L. Lenzi, Imaging of carotid plaque angiogenesis Cerebrovasc Dis, 27 Suppl 2, (2009), 48–54.
18. B.G. Rubin, Impact of plaque characterization on carotid interventions Perspect Vasc Surg Endovasc Ther, 18(4), (2006), 312–5.
19. J. Qian, A. Maehara, G.S. Mintz, M.P. Margolis, A. Lerman, J. Rogers, S. Banai, S. Kazziha, C. Castellanos, L. Dani, M. Fahy, G.W. Stone, and M.B. Leon, Impact of gender and age on in vivo virtual histology-intravascular ultrasound imaging plaque characterization (from the global Virtual Histology Intravascular Ultrasound [VH-IVUS] registry) Am J Cardiol, 103(9), (2009), 1210–4.
20. P. Schoenhagen, M. Barreto, and S.S. Halliburton, Quantitative plaque characterization with coronary CT angiography (CTA): current challenges and future application in atherosclerosis trials and clinical risk assessment Int J Cardiovasc Imaging, 24(3), (2008), 313–6.
21. T.T. de Weert, M. Ouhlous, P.E. Zondervan, J.M. Hendriks, D.W. Dippel, M.R. van Sambeek, and A. van der Lugt, In vitro characterization of atherosclerotic carotid plaque with multidetector computed tomography and histopathological correlation Eur Radiol, 15(9), (2005), 1906–14.
22. M.S. Nguyen, O. Salvado, D. Roy, G. Steyer, M.E. Stone, R.D. Hoffman, and D.L. Wilson, Ex vivo characterization of human atherosclerotic iliac plaque components using cryo-imaging J Microsc, 232(3), (2008), 432–41.
23. M. Fisher, A. Martin, M. Cosgrove, and J.W. Norris, The NASCET-ACAS plaque project. North American Symptomatic Carotid Endarterectomy Trial. Asymptomatic Carotid Atherosclerosis Study Stroke, 24(12 Suppl), (1993), I24–5; discussion I31–2.
24. L. Saba and G. Mallarini, A comparison between NASCET and ECST methods in the study of carotids Evaluation using Multi-Detector-Row CT angiography Eur J Radiol, (2009), (epub ahead of print).
25. M. Egger, A. Krasinski, B.K. Rutt, A. Fenster, and G. Parraga, Comparison of B-mode ultrasound, 3-dimensional ultrasound, and magnetic resonance imaging measurements of carotid atherosclerosis J Ultrasound Med, 27(9), (2008), 1321–34.

26. U.K.-I. JM, V. Young, and J.H. Gillard, Carotid-artery imaging in the diagnosis and management of patients at risk of stroke Lancet Neurol, 8(6), (2009), 569–80.
27. N. Nighoghossian, L. Derex, and P. Douek, The vulnerable carotid artery plaque: current imaging methods and new perspectives Stroke, 36(12), (2005), 2764–72.
28. T. Saam, T.S. Hatsukami, N. Takaya, B. Chu, H. Underhill, W.S. Kerwin, J. Cai, M.S. Ferguson, and C. Yuan, The vulnerable, or high-risk, atherosclerotic plaque: noninvasive MR imaging for characterization and assessment Radiology, 244(1), (2007), 64–77.
29. K. Yoshida, O. Narumi, M. Chin, K. Inoue, T. Tabuchi, K. Oda, M. Nagayama, N. Egawa, M. Hojo, Y. Goto, Y. Watanabe, and S. Yamagata, Characterization of carotid atherosclerosis and detection of soft plaque with use of black-blood MR imaging AJNR Am J Neuroradiol, 29(5), (2008), 868–74.
30. D.P. Hinton-Yates, R.C. Cury, L.L. Wald, G.C. Wiggins, B. Keil, R. Seethmaraju, D. Gangadharamurthy, C.S. Ogilvy, G. Dai, S.L. Houser, J.R. Stone, and K.L. Furie, 3.0 T plaque imaging Top Magn Reson Imaging, 18(5), (2007), 389–400.
31. B.A. Wasserman, W.I. Smith, H.H. Trout, 3rd, R.O. Cannon, 3rd, R.S. Balaban, and A.E. Arai, Carotid artery atherosclerosis: in vivo morphologic characterization with gadolinium-enhanced double-oblique MR imaging initial results Radiology, 223(2), (2002), 566–73.
32. C. Yuan, W.S. Kerwin, M.S. Ferguson, N. Polissar, S. Zhang, J. Cai, and T.S. Hatsukami, Contrast-enhanced high resolution MRI for atherosclerotic carotid artery tissue characterization J Magn Reson Imaging, 15(1), (2002), 62–7.
33. P. Brathwaite, A. Nagaraj, B. Kane, D.D. McPherson, and E.L. Dove. Automatic classification and differentiation of atherosclerotic lesions in swine using IVUS and texture features. Proc. of Computers in Cardiology, (2002), 109–112.
34. K. Nasu, E. Tsuchikane, O. Katoh, N. Tanaka, M. Kimura, M. Ehara, Y. Kinoshita, T. Matsubara, H. Matsuo, K. Asakura, Y. Asakura, M. Terashima, T. Takayama, J. Honye, A. Hirayama, S. Saito, and T. Suzuki, Effect of fluvastatin on progression of coronary atherosclerotic plaque evaluated by virtual histology intravascular ultrasound JACC Cardiovasc Interv, 2(7), (2009), 689–96.
35. H. Ribbers, R.G. Lopata, S. Holewijn, G. Pasterkamp, J.D. Blankensteijn, and C.L. de Korte, Noninvasive two-dimensional strain imaging of arteries: validation in phantoms and preliminary experience in carotid arteries in vivo Ultrasound Med Biol, 33(4), (2007), 530–40.
36. M. Fatemi and J.F. Greenleaf, Probing the dynamics of tissue at low frequencies with the radiation force of ultrasound Phys Med Biol, 45(6), (2000), 1449–64.
37. D.C. Cheng, A. Schmidt-Trucksass, K.S. Cheng, and H. Burkhardt, Using snakes to detect the intimal and adventitial layers of the common carotid artery wall in sonographic images Comput Methods Programs Biomed, 67(1), (2002), 27–37.
38. C.I. Christodoulou, C.S. Pattichis, M. Pantziaris, and A. Nicolaides, Texture-based classification of atherosclerotic carotid plaques IEEE Trans Med Imaging, 22(7), (2003), 902–12.
39. E. de Groot, S.I. van Leuven, R. Duivenvoorden, M.C. Meuwese, F. Akdim, M.L. Bots, and J.J. Kastelein, Measurement of carotid intima-media thickness to assess progression and regression of atherosclerosis Nat Clin Pract Cardiovasc Med, 5(5), (2008), 280–8.
40. S. Delsanto, F. Molinari, P. Giustetto, W. Liboni, and S. Badalamenti, CULEX-completely user-independent layers extraction: ultrasonic carotid artery images segmentation Conf Proc IEEE Eng Med Biol Soc, 6, (2005), 6468–71.
41. S. Delsanto, F. Molinari, P. Giustetto, W. Liboni, S. Badalamenti, and J.S. Suri, Characterization of a completely user-independent algorithm for carotid artery segmentation in 2-D ultrasound images IEEE Trans Instrum Meas, 56(4), (2007), 1265–1274.
42. F. Destrempes, J. Meunier, M.F. Giroux, G. Soulez, and G. Cloutier, Segmentation in ultrasonic B-mode images of healthy carotid arteries using mixtures of Nakagami distributions and stochastic optimization IEEE Trans Med Imaging, 28(2), (2009), 215–29.
43. F. Faita, V. Gemignani, E. Bianchini, C. Giannarelli, L. Ghiadoni, and M. Demi, Real-time measurement system for evaluation of the carotid intima-media thickness with a robust edge operator J Ultrasound Med, 27(9), (2008), 1353–61.

44. S. Golemati, J. Stoitsis, E.G. Sifakis, T. Balkizas, and K.S. Nikita, Using the Hough transform to segment ultrasound images of longitudinal and transverse sections of the carotid artery Ultrasound Med Biol, 33(12), (2007), 1918–32.
45. M.A. Gutierrez, P.E. Pilon, S.G. Lage, L. Kopel, R.T. Carvalho, and S.S. Furuie, Automatic measurement of carotid diameter and wall thickness in ultrasound images Comput Cardiol, 29, (2002), 359–362.
46. E. Kyriacou, M.S. Pattichis, C.I. Christodoulou, C.S. Pattichis, S. Kakkos, and A. Nicolaides, Ultrasound imaging in the analysis of carotid plaque morphology for the assessment of stroke. In, Plaque characterization using multimodality imaging: pixel to molecular, J.S. Suri, C. Yuan, D.L. Wilson, and S. Laxminarayan, eds. pp. 241–275 (IOS Press, Amsterdam, 2005).
47. Q. Liang, I. Wendelhag, J. Wikstrand, and T. Gustavsson, A multiscale dynamic programming procedure for boundary detection in ultrasonic artery images IEEE Trans Med Imaging, 19(2), (2000), 127–42.
48. C. Liguori, A. Paolillo, and A. Pietrosanto, An automatic measurement system for the evaluation of carotid intima-media thickness IEEE Trans Instrum Meas, 50(6), (2001), 1684–1691.
49. C.P. Loizou, C.S. Pattichis, M. Pantziaris, T. Tyllis, and A. Nicolaides, Snakes based segmentation of the common carotid artery intima media Med Biol Eng Comput, 45(1), (2007), 35–49.
50. P. Pignoli and T. Longo, Evaluation of atherosclerosis with B-mode ultrasound imaging J Nucl Med Allied Sci, 32(3), (1988), 166–73.
51. S. Delsanto, F. Molinari, W. Liboni, P. Giustetto, S. Badalamenti, and J.S. Suri, User-independent plaque characterization and accurate IMT measurement of carotid artery wall using ultrasound Conf Proc IEEE Eng Med Biol Soc, 1, (2006), 2404–7.
52. F. Molinari, S. Delsanto, P. Giustetto, W. Liboni, S. Badalamenti, and J.S. Suri, User-independent plaque segmentation and accurate intima-media thickness measurement of carotid artery wall using, ultrasound. In, Advances in diagnostic and therapeutic ultrasound imaging, J.S. Suri, C. Kathuria, R.F. Chang, et al., eds. pp. 111–140 (Artech House, Norwood, MA, 2008).
53. F. Molinari, W. Liboni, P. Giustetto, S. Badalamenti, and J.S. Suri, Automatic computer-based tracings (ACT) in longitudinal 2-D ultrasound images using different scanners J Mech Med Biol, 9(4), (2009), 481–505.
54. F. Molinari, G. Zeng, and J.S. Suri, An integrated approach to computer-based automated tracing and its validation for 200 common carotid arterial wall ultrasound images: a new technique. J Ultrasound Med, 29(3), (2010), 399–418.
55. F. Molinari, G. Zeng, and J.S. Suri, Intima-media thickness: setting a standard for completely automated method for ultrasound, IEEE Transaction on Ultrasonics Ferroelectrics and Frequency Control, 57(5), (2010), 1112–1124
56. O.V. Michailovich and A. Tannenbaum, Despeckling of medical ultrasound images, IEEE Trans Ultrason Ferroelectr Freq Control, 53(1), (2006), 64–78.
57. C.P. Loizou, C.S. Pattichis, C.I. Christodoulou, R.S. Istepanian, M. Pantziaris, and A. Nicolaides, Comparative evaluation of despeckle filtering in ultrasound imaging of the carotid artery, IEEE Trans Ultrason Ferroelectr Freq Control, 52(10), (2005), 1653–69.
58. D.J. Williams and M. Shah, A Fast Algorithm for Active Contours and Curvature Estimation, Cvgip-Image Understanding, 55(1), (1992), 14–26.
59. F. Molinari, G. Zeng, and J.S. Suri, Greedy Technique and Its Validation for Fusion of Two Segmentation Paradigms Leads to an Accurate Intima-Media Thickness Measure in Plaque Carotid Arterial Ultrasound, Journal for Vascular Ultrasound, 34(2), (2010), 63–73.
60. B.K. Lal, R.W. Hobson, 2nd, M. Hameed, P.J. Pappas, F.T. Padberg, Jr., Z. Jamil, and W.N. Duran, Noninvasive identification of the unstable carotid plaque, Ann Vasc Surg, 20(2), (2006), 167–74.
61. B.K. Lal, R.W. Hobson, 2nd, P.J. Pappas, R. Kubicka, M. Hameed, E.Y. Chakhtoura, Z. Jamil, F.T. Padberg, Jr., P.B. Haser, and W.N. Duran, Pixel distribution analysis of B-mode ultrasound scan images predicts histologic features of atherosclerotic carotid plaques, J Vasc Surg, 35(6), (2002), 1210–7.
62. H. Katano and K. Yamada, Analysis of calcium in carotid plaques with Agatston scores for appropriate selection of surgical intervention, Stroke, 38(11), (2007), 3040–4.
63. Y. Niwa, H. Katano, and K. Yamada, Calcification in carotid atheromatous plaque: delineation by 3D-CT angiography, compared with pathological findings, Neurol Res, 26(7), (2004), 778–84.

Biographies

Dr. Filippo Molinari received the Italian Laurea and the Ph.D. in electrical engineering from the Politecnico di Torino, Torino, Italy, in 1997 and 2000, respectively. He is leader in ultrasound imaging focused towards tissue characterization, vascular quantification for diagnostics and therapeutics. Currently, he is Assistant Professor at Politecnico di Torino, Italy – Department of Electronics. Dr. Filippo Molinari received the Italian Laurea and the Ph.D. in electrical engineering from the Politecnico di Torino, Torino, Italy, in 1997 and 2000, respectively. He is leader in ultrasound imaging focused towards tissue characterization, vascular quantification for diagnostics and therapeutics. Currently, he is Assistant Professor at Politecnico di Torino, Italy – Department of Electronics.

Dr. William Liboni received the Italian Laurea in Medicine in 1969 from the Università degli Studi di Torino and then specialized in radiology and nuclear medicine in 1975 and 1993, respectively. Since 1994 he directed the Neurology Division at the Gradenigo Hospital of Torino, Italy. His research focuses on the functional assessment of neurologically impaired subjects and on the non-invasive monitoring of chronic pathologies.

Dr. Pierangela Giustetto received the Italian Laurea in Radiology techniques for medical images and radiotherapy from University di Torino, Italy,in 1987. In 2006, she received the Italian first level Laurea in Physics from the University of Torino, Italy. She is now completing her M.S. degree in Physics working in the field of advanced ultrasound technologies and interactions with human tissues and nanomolecules.

Dr. Enrica Pavanelli received the Italian Laurea in Medicine in 1993 from the Università degli Studi di Torino, Italy. In 1997 she specialized in neurology. Since 1997 she is on faculty of the Neurology Division of the Gradenigo Hospital of Torino, Italy. She is dedicated to cerebrovascular diseases, Parkinson diseases and ultrasound diagnosis.

Dr. Sara Giordano graduated in Biomedical Engineering at the Politecnico di Torino, Torino, Italy. She was researcher under grant at the Laboratory for Engineering of Neuromuscular System of the Politecnico di Torino and then she took a stage in telemedicine at TelecomItalia Labs. Since 2008 she has been working at the Gradenigo Hospital of Torino as clinical engineer, in charge of medical devices management and projects in the neurological field.

Dr. Jasjit S. Suri is an innovator, scientist, a visionary, an industrialist and an internationally known world leader in Biomedical Engineering. Dr. Suri has spent over 20 years in the field of biomedical engineering/devices and its management. He received his Doctrate from University of Washington, Seattle and Business Management Sciences from Weatherhead, Case Western Reserve University, Cleveland, Ohio. Dr. Suri was crowned with President's Gold medal in 1980 and the Fellow of American Institute of Medical and Biological Engineering for his outstanding contributions.

Chapter 9
An Integrated Approach to Computer-Based Automated Tracing and IMT Measurement for Carotid Artery Longitudinal Ultrasound Images

Filippo Molinari, Guang Zeng, and Jasjit S. Suri

Abstract In this chapter, we describe a novel algorithm called CALEXia (*C*ompletely *A*utomated *L*ayers *EX*traction technique based on *integrated approach*) for the automated localization and segmentation of the carotid artery wall in longitudinal 2D B-Mode ultrasound images.

CALEXia consists of two parts: (1) a module for the automated localization of the carotid artery in the longitudinal image frame and (2) a module for the automated segmentation of the far carotid wall and for the intima–media thickness (IMT) measurement. CALEXia exploits the image information by combining feature extraction, fitting, and fuzzy classification.

We tested CALEXia on a database consisting of 200 images. Manual segmentations traced by experts were considered as ground truth. IMT was measured as a distance between the lumen–intima (LI) and the media–adventitia (MA) computer-generated boundaries. The distance between boundaries was computed by using the polyline distance metric (PDM), which provides a robust and reliable measure of the distance between two profiles with different number of points. Similarly, the system performance in terms of segmentation error was computed by relying on the PDM.

CALEXia correctly identified the position of the carotid artery in 190 images out of 200 (failure rate of 5%). Possible error sources were the presence of blood backscattering and either the jugular vein or a plaque protruding in the artery lumen. The overall system segmentation errors were equal to 1.22 ± 1.30 pixels (0.076 ± 0.081 mm) for the LI and 0.45 ± 0.47 pixels (0.028 ± 0.029 mm) for the MA interface. The IMT measurement error was equal to 0.76 ± 0.69 pixels (0.048 ± 0.043 mm).

Despite a bias in the LI interface detection, CALEXia proved effective in automated carotid location and segmentation. Its modular architecture could constitute a general basis for the development of automated carotid images processing and IMT measurement algorithms.

F. Molinari (✉)
Biolab – Dipartimento di Elettronica, Politecnico di Torino, Corso Duca degli Abruzzi, 24, 10129 Torino, Italy
e-mail: filippo.molinari@polito.it

Jasjit S. Suri et al. (eds.), *Atherosclerosis Disease Management*,
DOI 10.1007/978-1-4419-7222-4_9,

Keywords Ultrasound imaging • Carotid artery • Automated tracings • Geometric features • Integration • Performance evaluation • Validation

9.1 Introduction

According to the recent data from the World Health Organization, cardiovascular diseases (CVDs) represent the third leading cause of death in Western countries [1]. CVDs include coronary artery disease, cerebrovascular disease, peripheral artery disease, heart failure, and hypertension. Beside a correct lifestyle, prevention is the key to reduce the impact of CVDs on the total number of global deaths.

Ultrasound examination is one of the most used diagnostic tools to assess CVDs. Ultrasounds offer advantages in clinical practice such as: they are nonionizing radiations, they allow a safe and relatively quick examination of the patient, they have no dangerous biological effects, and the ultrasound equipment is relatively low cost if compared with other imaging devices. Unfortunately, ultrasounds are operator dependent. Moreover, the ultrasound images are usually quite noisy and require training to be correctly interpreted.

The most threatening pathology associated to CDVs is atherosclerosis. Atherosclerosis is a progressive degeneration that causes reduction of the artery lumen and thickening of the artery wall. The atherosclerotic process has been correlated to the increased risk of CVDs such as stroke and heart attack [2]. Therefore, the monitoring and quantification of the atherosclerotic process is of fundamental importance in clinical practice [3]. The intima–media thickness (IMT) of the common carotid artery (CCA) is the most widely used marker to monitor the atherosclerotic process and assessment of CVD risk [4]. Numerical analysis of the CCA images requires ad hoc computer programs. In the literature, several algorithms for the automated segmentation of the CCA ultrasound images have been proposed [5–8]. However, almost all the techniques require a certain degree of user interaction that precludes real automation. User interaction results in final measurements that can be biased by the operator choices; also, the automatic processing of large databases is impossible.

The segmentation process is conceptually made of two distinct parts:

1. Recognition of the carotid artery and tracing of near and far adventitia layers in the 2D B-Mode ultrasound image.
2. Tracing of the boundaries between lumen–intima and media–adventitia.

The above-mentioned techniques have automated step 2, but user interaction is usually needed to perform step 1. Figure 9.1a represents the anatomical course of the CCA along the neck. Figure 9.1b is a longitudinal B-Mode ultrasound representation of a tract of a healthy CCA. Several factors may complicate the automatic detection of the CCA and tracings of adventitia layers in the image.

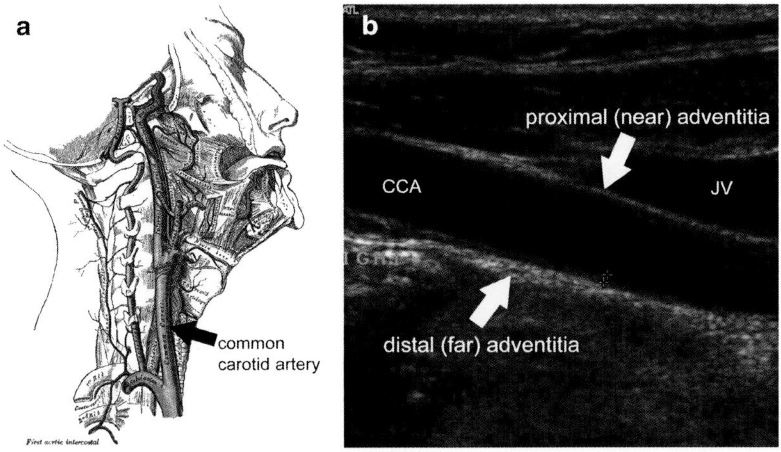

Fig. 9.1 (**a**) Anatomical view of the common carotid artery (CCA) course along the neck. (**b**) B-Mode longitudinal representation. The CCA appears as a dark stripe (the lumen) surrounded by two echogenic lines (near and far adventitia layers). JV jugular vein (© AIUM 2010 - Reproduced with permission from [20])

- First of all, the CCA can be easily confounded with the jugular vein, since their echographic appearance is very similar.
- Also, the morphological aspect of the CCA changes due to individual variability (the artery can be straight, curved, or even coiled) or due to different insonation angles.
- Noise represents another major problem: in many practical conditions, it may be more visible in a specific region of the image, thus making the recognition of the CCA particularly critical in that image portion.
- The image artifacts (particularly, blood backscattering) affect longitudinal B-Mode images.

The CCA can be thought as a region of dark (the lumen) comprised between two bright continuous stripes (the adventitia layers of the CCA wall). Therefore, locating the CCA means tracing the adventitia (AD) boundaries.

In the literature, there were few techniques dealing with this topic (step 1).

1. In 2007, Golemati et al. proposed the use of the Hough transform for locating the CCA in longitudinal and transverse images [9]. This approach had the advantage of being suitable to longitudinal and transverse B-Mode images. However, being the Hough transform essentially a geometrical operator, it was effective only in processing images where the CCA appeared as straight and horizontal.
2. Rossi et al. [10] recently presented an efficient technique for the automatic recognition of the carotid artery in longitudinal images. The maxima of the column envelopes marked the position of the adventitia layers. A template-matching algorithm processed each envelope in order to find the center of the carotid lumen, which was the point that comprised the two intensity peaks given by the

artery walls. This technique proved effective in the presence of fibrous plaques and noise. However, this method was not independent of the ultrasound image scanner.
3. In 2009, the authors developed a technique for the automatic recognition of the CCA in longitudinal images [11]. We showed that our technique was effective in recognizing the CCA in normal and pathologic images. We also demonstrated that performances were independent on the ultrasound scanner used to acquire the images. The CCA was detected by automatically tracing the near and far adventitia profiles.

This contribution was the latest improvement of a previously developed technique for the IMT measurement of the CCA. We named this technique as CULEXsa, which stands for *C*ompletely *U*ser-independent *L*ayers *EX*traction algorithm based on *signal analysis* [12–14]. CULEXsa was an innovative methodology, but still had a scope of improvement in three areas, which would improve the overall performance. First, the distances between the automatically traced profiles and the human tracings were close to 8 pixels. This caused an IMT measurement error equal to about 40 μm. Most of the best-performing algorithms allows for IMT measurement error of the order of 10 μm [7, 8, 15], with top performance of about 1 μm [16]. Second, noise could still represent a problem for the automatic recognition of the CCA, causing automatic tracings to be distant from the adventitia layers (this problem was found in about 5% of the analyzed images) and precluding the CCA segmentation and IMT measurement. Also, noise resulted in a failure rate of about 15% of the images. Third, the computational cost of the algorithm precluded real-time implementation. The final segmentation step of CULEXsa was based on active parametric contours (*snakes*) and, therefore, was of difficult tuning in the presence of noise. Improvement of performance was difficult, given the difficulty of driving the snake in the presence of critical images [5, 8].

Therefore, we propose a totally new and superior architecture. The algorithm herein presented has been developed with three aims:

1. Improving the overall performance in terms of CCA detection and robustness to noise.
2. Testing a new solution for IMT measurement.
3. Lightening the computational burden.

The new algorithm is a combination of feature extraction, fitting, classification, and fuzzy clustering. It is completely user independent; hence, we called it as CALEXia, which stands for *C*ompletely *A*utomated *L*ayers *EX*traction technique based on *integrated approach*.

The chapter is organized as follows: Sect. 9.2 describes the architecture and segmentation strategy of the algorithm based on integrated approach, Sect. 9.3 describes the metric we used to measure the performance and IMT, Sect. 9.4 describes the results obtained in terms of CCA recognition and automated IMT estimation, and Sect. 9.5 discusses strengths and weaknesses of the new approach.

9.2 CALEXia Architecture

9.2.1 Automatic Recognition of the Common Carotid Artery

CALEXia is based on an initial assumption that the artery should occupy the whole image frame width. With reference to Fig. 9.1b, we developed our technique by considering the artery structure as a region with low intensity (the lumen) surrounded by two bright stripes at each side (the adventitia layers of the carotid arterial walls). Therefore, if we consider the vertical gradient of the image, the arterial wall can be thought as a sequence of contiguous pixels that are local intensity maxima in the vertical direction.

The algorithm adopts a column-wise approach for processing the ultrasound image. For each column, the algorithm searches for seed points, which are high-intensity local maxima surrounded by low-intensity local minima. Seed points have a high probability of belonging to high-intensity structures (and, therefore, to the adventitia layers). The algorithm then traces the line segments by connecting these seed points. Such line segments should form the adventitial layer of the CCA. To overcome the problem of false positives (i.e., traced lines that do not correspond to the adventitia layers, but to other structures of the image) and of over-segmentation (i.e., very short line segments that need to be connected to form a continuous adventitia profile), line segments are first combined, fitted, and connected; then, a classification procedure discards all the line segments that have a low probability of including as a vessel lumen. The novelty of the system lies in the development of a multistage well-connected integrated approach based on the concept of geometric feature extraction whose coefficients are estimated using learning-based discrimination procedure. Discrimination procedure is modeled as a linear combination of geometric features, both for seed determination and line fitting protocols. The weights of the discriminators are derived by the perceptron applied to the true line segments and false positives of the randomly selected training set. The perceptron is a neural network commonly employed in classification procedures. Basically, it is a linear classifier that, by combining the input data with coefficients (called weights), maps its input in a binary domain of zeros and ones. We used the perceptron to discriminate true positives (mapped to one) to false positives (mapped to zero). Like other neural networks, the perceptron requires training to effectively perform discrimination. To train our perceptrons, we used the 15-images subset described in Sect. 9.3.1.

9.2.1.1 Selection of Seed Points

The concept of seed point selection is based on a linear discriminator that bins the true seed points and false-negative seed points separately. The linear discriminator model is designed based on two key properties of seed points: height of the seed candidate and breadth of the seed candidate. The design of the discriminator parameters is based on a training set randomly selected from the given population.

The whole idea is to find the possible seed points that are above a certain threshold and are likely to be the edges of the lumen, but not inside the lumen.

For implementation, first, we attenuate the speckle noise by convolving the Gaussian low-pass filter with known size, mean, and standard deviation. We took convolving window to be 9×9 pixels with mean and standard deviation to be 0 and 20 pixels, respectively. The image is then considered column-wise (i.e., seed points are searched along vertical lines). These points are the local intensity maxima in the vertical direction.

To reduce the computational cost of the procedure, we decimated the number of columns of the image. The detected local maxima are candidate seed points.

For each seed candidate, we calculate two properties:

- *Intensity e*, which corresponds to the height of the seed candidate on the vertical intensity profile
- *Breadth b*, which corresponds to the distance between the two neighboring local minima that are on the opposite sides of the seed candidate

To distinguish true seed points from local maxima due to background noise, we use a 3×1 linear discriminator **v** applied to the vector of seed candidates with property represented as: $\mathbf{p} = [b\ e\ 1]$. The definition of the discriminator **v** was based on the 15-images training data set. Currently, this number proved to be stable; however, further discussions can be seen on this choice. We plotted the distribution of the true seed points (Fig. 9.2a) and of the false positives (Fig. 9.2b). The black line shows the decision boundary, which was computed by a linear perceptron [17]. Our experiments yield the linear discriminator as: $\mathbf{v} = [0.3\ 74.9\ {-82.4}]^{\mathrm{T}}$. The values of the linear discriminator are summarized in Table 9.1. This linear discriminator was kept same for all the 200 images of the validation database. The criteria for seed point selection was based on the threshold u_{T}, mathematically laid out to follow the equality:

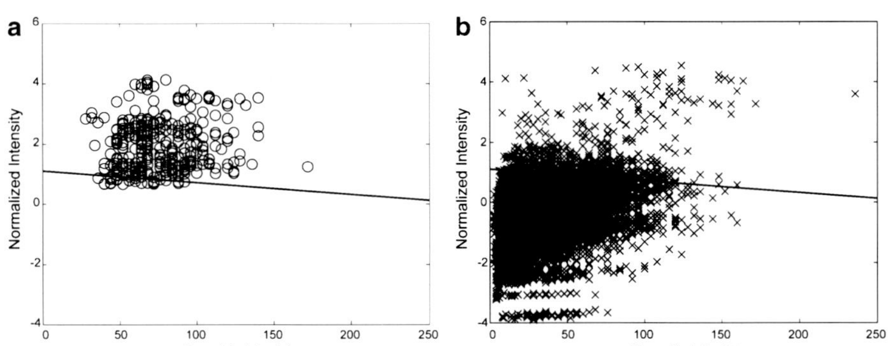

Fig. 9.2 Distribution of the true seed points [*circles – left panel*, (**a**)] and of the false positives [*crosses – right panel*, (**b**)] on the 15 images of the testing database. The *black lines* represent the decision region of the linear discriminator. The *horizontal axis* represents the breadth and the *vertical axis* the normalized intensity of the points (normalization was computed as the difference of each pixel intensity with respect to the mean value and divided by the standard deviation of the image intensity value) (© AIUM 2010 - Reproduced with permission from [20])

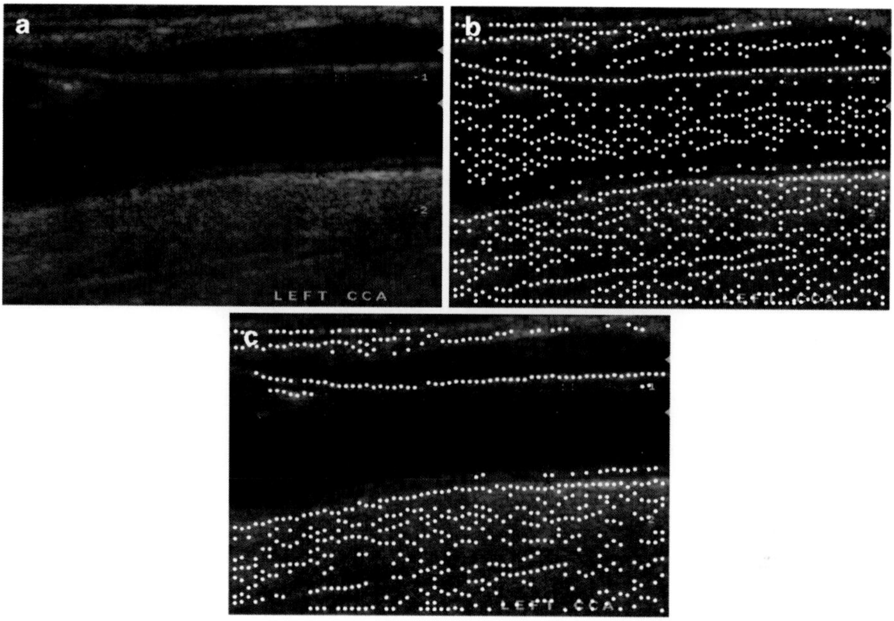

Fig. 9.3 Detected seed points in a sample ultrasound image. (**a**) Original B-Mode image. (**b**) Detected local maxima. (**c**) Selected seed points based on linear discriminator model (© AIUM 2010 - Reproduced with permission from [20])

Table 9.1 Numerical values of the linear classifiers used in the design of CALEXia (© AIUM 2010 - Reproduced with permission from [20])

Discriminator	Weights of the linear discriminator
$\mathbf{v}$ – detection of the seed points	$[0.3\ 74.9\ -82.4]^T$
$\mathbf{u}_c$ – validation of the line segments	$[8.5\ 8.8\ -16.6]^T$
$\mathbf{u}_d$ – connectedness of the line segments	$[1.8\ 2.8\ -1.2]^T$

The values were derived using a training set of 15 images

$\mathbf{p}\cdot\mathbf{v} > u_T$ (here $\cdot$ denotes the dot product between the two vectors). Here u_T is equal to zero.

The process of detecting the seed points is sketched in Fig. 9.3: the leftmost panel depicts the original image (Fig. 9.3a), the central panel shows the detected local maxima for all the columns of the image (Fig. 9.3b), and the right panel reports the selected seed points (Fig. 9.3c) using linear discriminator.

9.2.1.2 Fitting of Line Segments and Its Tracing

This stage incorporates the iterative line segment formation, once the seed points are estimated from the resample signal envelope using the linear discrimination model. The algorithm in this stage consists of a seed point selection at random, along with its closest neighbor seed point, and a line segment is fit to this pair. The segment is then extended and adjusted by iteratively incorporating nearby seed points whose feature of *spread* and *residue* is smaller than the predefined thresholds (definition of *spread* and *residue* is shown below). The process of adding line points is continues until no more line points are found in the current neighborhood or until the next candidate has already been added to another line.

Although the foregoing procedure quickly fits line segments to seed points, it suffers from over-segmentation caused by the presence of noise (particularly, of speckle noise). In addition, to improve the performance of the procedure in the presence of a curved artery, we need to join the line segments that belong to a possibly curved artery wall, whose orientation might change. On the other hand, since this procedure is based solely upon the coherence of the locations of local maxima in the image intensity function, it may trace any bright background noise (thus, detecting false adventitia layers). Therefore, we adopted a multistep approach that: (1) *combines* over-segmented line segments; (2) *connects* line segments belonging to a possibly curved artery wall; and (3) removes the incorrect line tracings that are caused by noise. The remaining portion of this subsection is dedicated on *combinability* and *connectedness*.

For each fitted line segment s_i, four geometric features are measured:

(a) *Support*, defined as the number of seed points contained by the segment s_i (indicated by $g_1(s_i)$);
(b) *Residue*, defined as the mean squared error of the seed points with respect to their perpendicular distance from s_i (indicated by $g_2(s_i)$);
(c) *Spread*, defined as the shortest path length connecting the seed points divided by the number of seed points s_i contains (indicated by $g_3(s_i)$);
(d) *Width stability*, defined as the percentage of points along the fitted line segments s_i whose width (perpendicular distance to the nearest intensity edge) is within some tolerance of the estimated width of the adventitial layers (indicated by $g_4(s_i)$).

We associate an energy function U_D to each segment s_i by linearly combining the above-defined first three features:

$$U_\mathrm{D}(s_i)=\sum_{i=1}^{3}\omega_i g_i(s_i). \tag{9.1}$$

First, pairs of line segments are tested for *combinability*. Two line segments s_i and s_j are combined into a single segment s_c if the intensity energy of the combined segment is lower than the sum of the two line segment energies, i.e., if:

$$U_\mathrm{D}(s_\mathrm{c})\le U_\mathrm{D}(s_i)+U_\mathrm{D}(s_j). \tag{9.2}$$

In (9.1), the support g_1 is the same for both sides of the formula (i.e., for the single line segments and for the resulting combined segment s_c). Fundamentally, this condition ensures the combination of line segments that are close to each other and well aligned.

To remove the false positives (i.e., the traced line segments that do not constitute adventitia layers), a subsequent *validation* step is performed. Let s_i be the detected line segment and let D_valid be the predicate that is true if a carotid artery wall exists in the ultrasound image at the location of s_i. Let D'_valid be the complement of D_valid.

According to Bayes' decision rule, we consider the line segment as detected from the CCA wall if its posteriori probability $P(D_{\text{valid}} \mid s_i)$ is greater than that of its complement $P(D'_{\text{valid}} \mid s_i)$, thus if:

$$\frac{P(D_{\text{valid}} \mid s_i)}{P(D'_{\text{valid}} \mid s_i)} = \exp\{\Psi\,(D_{\text{valid}} \mid s_i)\} > 1. \tag{9.3}$$

For computational reason, we use the *support* $g_1(s_i)$ and the *width stability* $g_4(s_i)$ and thus $\Psi\,(D_{\text{valid}} \mid s_i)$ is mathematically defined according to the following formula:

$$\Psi\,(D_{\text{valid}} \mid s_i) = \omega_1 g_1(s_i) + \omega_4 g_4(s_i). \tag{9.4}$$

The function $\psi\,(D_{\text{valid}} \mid s_i)$ is determined by the linear discriminator $\mathbf{u}_{c,}$ where the weights ω_1 and ω_4 were determined by applying the linear discriminator to the training data, with a similar procedure to that described in Sect. 9.2.1.1. The numerical values we obtained by the training set are reported in Table 9.1. Figure 9.4 reports the discriminator values derived by a perceptron applied to true segments and false positives of the training set. As shown in Fig. 9.4, the linear discriminator value of a valid line segment is positive, while that of an invalid line segment is negative. Therefore, a positive sign is used in (9.3) to make sure that the posteriori probability $P(D_{\text{valid}} \mid s_i)$ of a valid line segment is larger than its complement $P(D'_{\text{valid}} \mid s_i)$.

Combinability and *validation* of line segments were performed in order to fulfill the above-described first step: reduce over-segmentation by combining line segments. In order to perform the second step (i.e., connecting line segments possibly belonging to a curved vessel), we introduce the *connectedness* function. First, we adopt the definition of attraction region proposed by Stoica et al. [18]; we define an attraction region A_i of a segment s_i as a set of points such that $\delta \in A_i$ if and only if $\|\delta_i - p_i\| < \tau_i$ or $\|\delta_i - q_i\| < \tau_i$, where p_i and q_i are the two end points of s_i and τ_i is equal

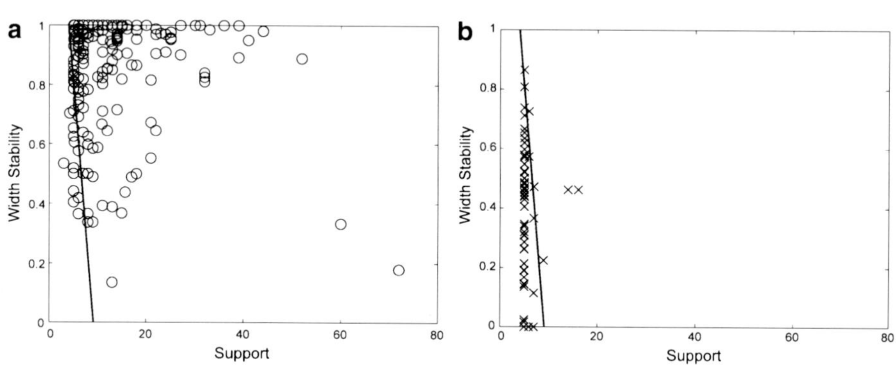

Fig. 9.4 Linear discriminator applied to the line segments. The *left panel* [*circles* – (**a**)] shows the true segments; the *right panel* [*crosses* – (**b**)] depicts the false positives. The *black lines* represent the discriminator line. *Horizontal axis* reports the support of each line segment, *vertical axis* its width stability (expressed in percentage) (© AIUM 2010 - Reproduced with permission from [20])

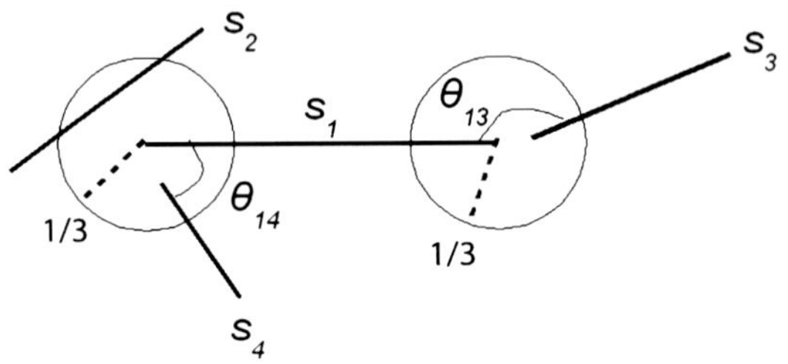

Fig. 9.5 A segments s_1 and its attraction region represented by the *circles* of one-third radius. Segments s_3 and s_4 interact with s_1, whereas s_2 does not (© AIUM 2010 - Reproduced with permission from [20])

to one-third the length of the segment s_i. Two segments s_i and s_j are considered to have an attraction interaction with one another if $p_i \in A_j$ or $q_i \in A_j$ or $p_j \in A_i$ or $q_i \in A_i$. With respect to this definition, line segments can be connected at both of its end points, either of its end points, or they can be unconnected. Figure 9.5 shows an example of four line segments: the circles represent the attraction region of the segment s_1. Segments s_3 and s_4 interact with s_1 while s_2 does not.

For any two attracted segments s_i and s_j, two geometric features are measured:

(a) *Proximity*, that is the minimum Euclidean distance between the end points of the two segments (indicated by $h_1(s_i, s_j)$;
(b) *Alignment*, that measures the curvature of the line segments as: $\left|(\pi - \theta_{ij})/\pi\right|$, where θ_{ij} is the angle between s_i and s_j (indicated by $h_2(s_i, s_j)$).

With these definitions and the weights ω_1' and ω_2', we formulate the intersection energy U_1 of two attracted line segments s_i and s_j as the following linear combination:

$$U_1(s_i, s_j) = \sum_{j=1}^{2} \omega_j' h_j(s_i, s_j). \tag{9.5}$$

This interaction function $U_1(s_i, s_j)$ penalizes overlapped and neighboring segments that are not well aligned.

Let D_{conn} be the predicate that is true if the two attracted line segments s_i and s_j belong to a curved vessel and let D'_{conn} be its complement. We declare two attracted line segments s_i and s_j as *connectable* if the posteriori validation probability $P(D_{\text{conn}} | s_i, s_j)$ is greater than that of its complement $P(D'_{\text{conn}} | s_i, s_j)$, thus if:

$$\frac{P(D_{\text{conn}} | s_i, s_j)}{P(D'_{\text{conn}} | s_i, s_j)} = \exp\left\{-\Psi(D_{\text{conn}} | s_i, s_j)\right\} > 1, \tag{9.6}$$

where, the function $\Psi(D_{\text{conn}} | s_i, s_j)$ is based on *proximity* $h_1(s_i, s_j)$ and *alignment* $h_2(s_i, s_j)$:

$$\Psi(D_{\text{conn}} | s_i, s_j) = \omega_1' h_1(s_i, s_j) + \omega_2' h_2(s_i, s_j). \tag{9.7}$$

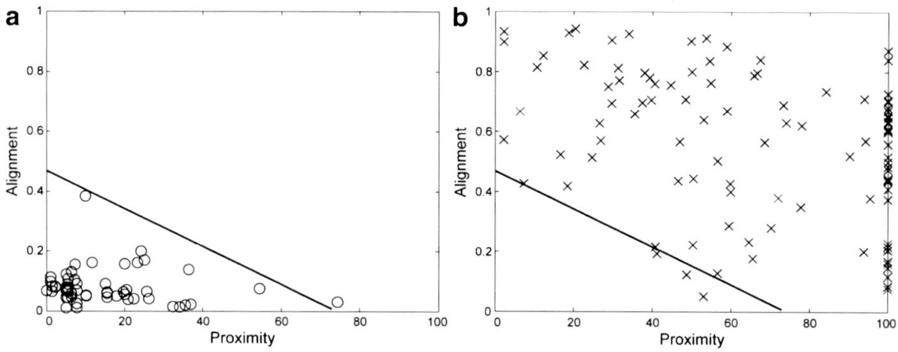

Fig. 9.6 Linear discriminator applied to the line segments. The *left panel* [*circles* – (**a**)] shows the true segments; the *right panel* [*crosses* – (**b**)] depicts the false positives. The *black lines* represent the discriminator line. *Horizontal axis* reports the proximity of each line segment, *vertical axis* its alignment (© AIUM 2010 - Reproduced with permission from [20])

The weights ω_1' and ω_2' are determined by applying a linear discriminator $\mathbf{u}_d$ to the training data. The numerical values we obtained by the training set are reported in Table 9.1. Figure 9.6 reports the discriminator values derived by a perceptron applied to true segments and false positives of the training set. The linear discriminator value of two connectable line segments is negative. Therefore, a negative sign is used in (9.6) to make sure that the posteriori probability $P\left(D_{\text{conn}} \middle| s_i, s_j\right)$ of two connectable line segments is larger than its complement.

Figure 9.7 summarizes the steps of fitting applied to the traced line segments. Figure 9.7a depicts the line segments obtained by the region of growing technique applied to the seed points; it is evident for the problem of over-segmentation and of false positives (i.e., line tracings that do not correspond to adventitia layers). Figure 9.7b reports the line segments after *combination*, where nearby and almost aligned line segments have been merged. Figure 9.7c shows the remaining line segments after validation, whereas Fig. 9.7d depicts the final result after the *connectedness* testing. It can be noticed that in Fig. 9.7d, line segments tracing the adventitial wall in proximity of the curvature (indicated in Fig. 9.7c by the white arrow A on the proximal wall and the arrow B on the distal wall) have been merged together (white arrows A′ and B′ in Fig. 9.7d). This ensures the possibility of following the curved wall profiles.

Figure 9.7d reveals that the line fitting method traces line segments that still over-segment in the image. In fact, line fitting method segments any feature that has a constant intensity in a given direction of the ultrasound image. This is because intensity and breadth are the features of seed points, which are then paired to form line segments. Such features represent false-positive detections.

Due to the bifurcation and crossover of linear features, we take an extra precautionary step. This is because sometimes segments from different structures may be misconnected. To avoid the problem of misjoining for any segments s_i with m (where $m > 1$) interacted line segments at the same end point, we connected it to the segments s_j which produce the minimum interaction energy $U_I\,(s_i, s_j)$:

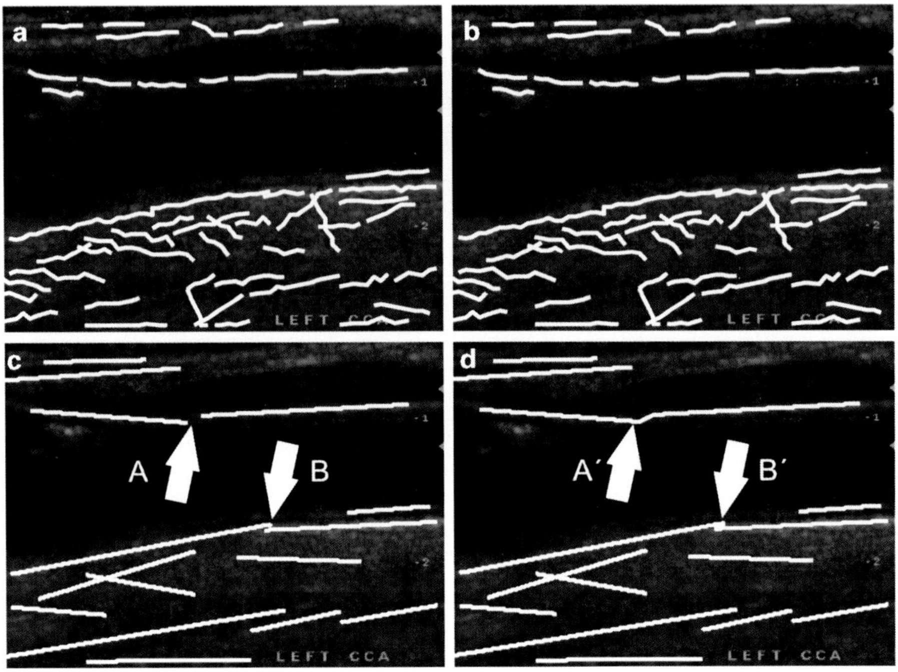

Fig. 9.7 Detected line segments in a sample image. (**a**) Line segments after fitting. (**b**) Combined line segments. (**c**) Valid line segments. (**d**) Connected line segments. The *white arrows A and B* in (**c**) indicate a couple of valid but disconnected line segments on the proximal (**a**) and distal (**b**) wall. The *white arrows A′ and B′* in (**d**) show the connection of the line segments (© AIUM 2010 - Reproduced with permission from [20])

$$\left\{s_i, s_j\right\} = \underset{l=1,\ldots m}{\operatorname{argmin}} \left\{U_{\mathrm{I}}\left(s_i, s_l\right)\right\} = \underset{l=1,\ldots m}{\operatorname{argmin}} \left\{\psi\left(D_{\mathrm{conn}} \left| s_i, s_l\right.\right)\right\}. \tag{9.8}$$

As shown in Fig. 9.8, although the two segments s_2 and s_3 interact with s_1 at the same end point, and both the two pairs (s_1, s_2) and (s_1, s_3) can be connected, we only connect (s_1, s_3) because U_{I} (s_1, s_3) = 0.38 is lower than U_{I} (s_1, s_2) = 1.04.

9.2.1.3 Line Segments Recognition and Classification

Once the line segments are traced, we must ensure that these line segments are applicable to correct the region along the CCA. Thus, we require a methodology that can classify the line segments in lumen vs. nonlumen region. This classification step must consider each pair of line segments and test if the two line segments can be part of the artery lumen. The structure of this classification step is represented in Fig. 9.9. Figure 9.9a reports an image with three line segments, indicated by SE_1, SE_2, and SE_3. Clearly, SE_1 and SE_2 are the true positive line segments (i.e., the tracings corresponding to the adventitia layers), whereas SE_3 is a false-positive detection.

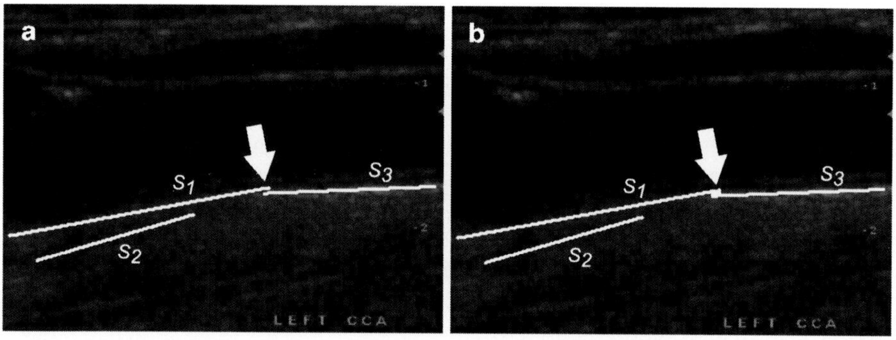

Fig. 9.8 An example of connecting line segments with multi-interaction: (**a**) line segments with multi-interaction and (**b**) connected line segments. The *white arrow* indicates line segments s_1 and s_3 that are disconnected in (**a**) (*left panel*) and connected in (**b**) (*right panel*) (© AIUM 2010 - Reproduced with permission from [20])

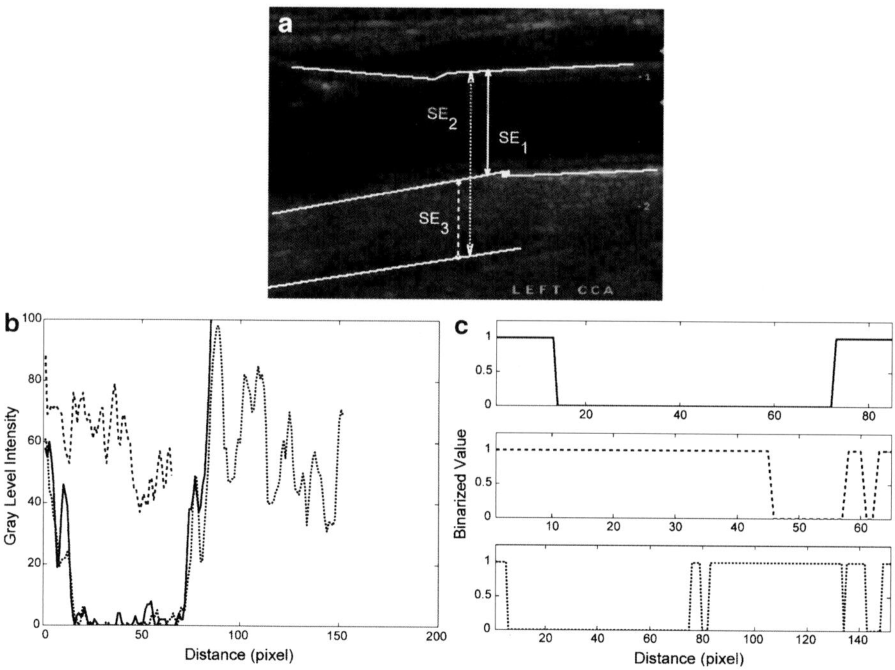

Fig. 9.9 Procedure for the selection of the correct pair of signal envelopes. (**a**) Among the three profiles, SE_1 (*continuous line*) contains the artery structure ($R_{lumen} = 70\%$); SE_2 (*dotted line*) contains the artery structure plus unwanted artifacts ($R_{lumen} = 46\%$); SE_3 (*dashed line*) contains only artifacts ($R_{lumen} = 20\%$). (**b**) Intensity profile of each signal envelope pair. (**c**) the binarized signal envelope for the estimation of R_{lumen} (© AIUM 2010 - Reproduced with permission from [20])

The three segments are considered in pairs. For each pair of segments, we plotted the intensity profile (Fig. 9.9b) by averaging the gray levels along the columns. In Fig. 9.9b, the continuous profile (corresponding to the pair SE_1–SE_2) is characterized by a large interval in which the intensity is close to zero (i.e., it comprises the vessel lumen). The other profiles (dashed and dotted lines in Fig. 9.9b) never reaches low values, since either they come from segments that do not contain the lumen (dashed line, pair SE_1–SE_3) or they contain the lumen but also other echogenic structures (dotted line, pair SE_2–SE_3). To assess if a given pair can be considered as containing only the lumen, we binarize the intensity profiles using Otsu's criterion [19]. Figure 9.9c reports the three intensity profiles after binarization. The threshold derived from Otsu's criterion is used to detect pixels belonging to the lumen. If a pixel is comprised between the segment pairs and has intensity lower than the threshold, it is counted as lumen pixel. We defined as R_{lumen} the percentage of points belonging to the lumen for each segment pair. Finally, the segment pair that has the highest R_{lumen} among all the pairs is considered as correct adventitia tracings. In the example of Fig. 9.9, this procedure leads to the consideration of the pair SE_1–SE_2 (continuous lines) as the final adventitia layer tracings. The final far (distal) adventitia tracing and the near (proximal) adventitia tracing are denoted as AD_F and AD_N, respectively.

9.2.2 IMT Measurement Strategy

The innovative segmentation strategy consists of five steps [21].

1. To attenuate the effect of noise, we low-pass filter the image by using a 5×5 pixels Gaussian kernel with zero mean and standard deviation equal to one.
2. We identify a rectangular region of interest (ROI) starting from the traced far adventitia profile (indicated by AD_F in Fig. 9.10a). The width of the ROI is taken equal to the support of the AD_F tracing. The upper horizontal limit of the ROI is calculated by moving the uppermost point of the AD_F to 20 pixels (1.25 mm) in the vertical direction (i.e., toward the CCA lumen). The lower limit of the ROI is calculated by shifting the deepest point of the profile to 10 pixels (0.625 mm) downwards (i.e., toward the bottom of the image). We chose these shift values to ensure that the ROI comprised a portion of the CCA lumen and the tissues underneath the CCA. Moreover, sometimes a fixed size lower and upper half ROI will reach the near wall, causing clustering problem in the procedure explained in next step. By using the uppermost and deepest point, we extract a ROI containing lumen and far wall only. An example of the resultant ROI is reported in Fig. 9.10b (white dashed rectangle).
3. We processed only the pixels in the ROI. The ROI is considered column-wise. For each column, we extract the signal envelope corresponding to the vertical intensity profile. With reference to Fig. 9.10c, the white dashed line indicates a column of the ROI and the black line in Fig. 9.10d represents the signal envelope. We considered only the pixels comprised between the row-index 0 and the row-index corresponding to the position of the adventitial layer (in Fig. 9.10d, the AD_F position is indicated by the vertical black dashed line. The corresponding

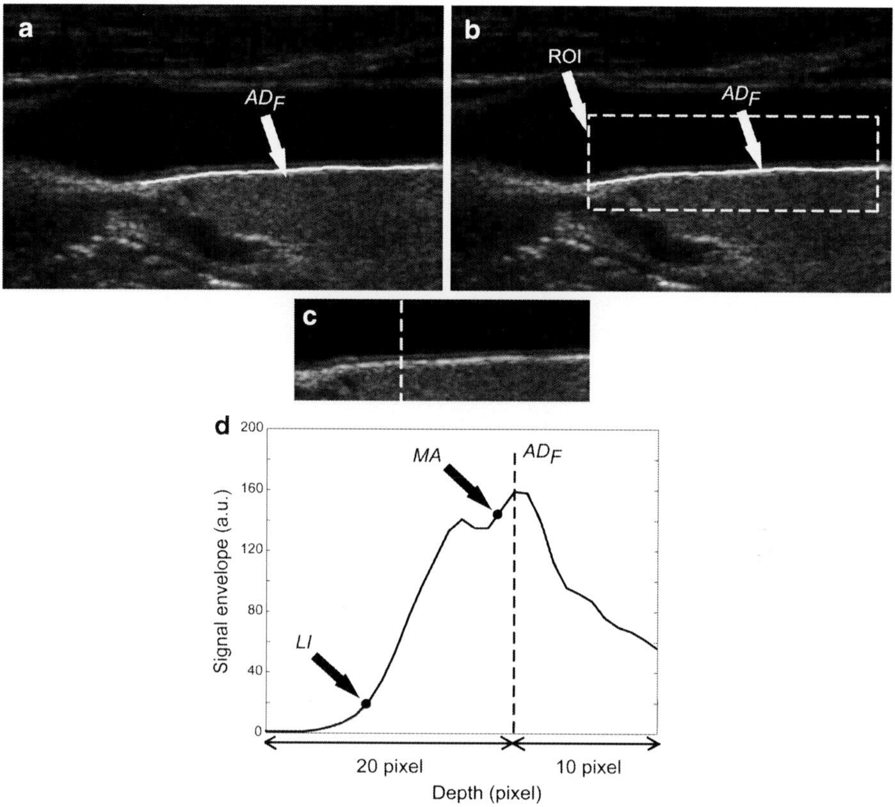

Fig. 9.10 Schematic representation of segmentation performed by CALEXia. (**a**) Original image with far adventitia tracing (AD_F). The AD_F represent the starting point of the segmentation procedure. (**b**) Rectangular ROI automatically derived by the AD_F tracing. The ROI width is same as AD_F tracing. The upper limit of the ROI is 20 pixels (1.25 mm) above the uppermost point of the AD_F profile. The lower limit of the ROI is 40 pixels (2.5 mm) below the deepest point of the AD_F tracing. (**c**) The previously extracted ROI is the only portion of the image that is segmented. The pixels are considered column-wise (the *white dashed line* indicates one column of the ROI). (**d**) Signal envelope of the intensity profile calculated along the *white dashed line* in panel (**c**). The *horizontal axis* reports the index row (i.e., the depth of the image. Depth increases toward the *right end of the graph*). The *vertical axis* reports the signal envelope amplitude in arbitrary units. The *vertical dashed line* marks the position of the far adventitia. The *black circles* represent the boundaries between lumen–intima (LI) and media–adventitia (MA) detected by the fuzzy *K-means* classifier (© IEEE 2010 - Reproduced with permission from [21])

row-index is equal to 37). In some images, because of the curvature of CCA walls, a portion of the near wall AD_N can be included in the ROI. To avoid the misclassification caused by the near wall AD_N inside the ROI, the extracted signal envelope is trimmed based on the position of the uppermost point of the AD_F and the deepest point of the AD_N.

4. The points of the signal envelope comprised between 0 and AD_F were clustered by using a fuzzy *K-means* classifier. We fixed the number of classes equal to three, ideally: artery lumen, intima–media complex, and adventitia layer. The classifier

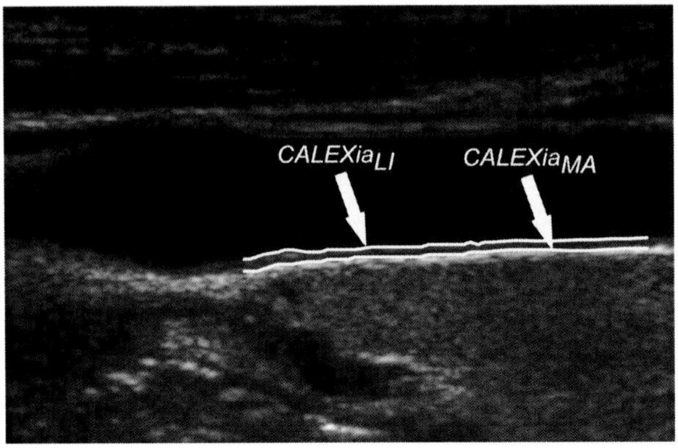

Fig. 9.11 Image segmentation performed by CALEXia. The image is same as in Fig. 9.10a. CALEXia$_{LI}$ represents the lumen–intima tracing. CALEXia$_{MA}$ represents the media–adventitia tracing (© IEEE 2010 - Reproduced with permission from [21])

used a Euclidean distance metric. Initialization was done by feeding the gray level intensity values along the signal envelope as input. Artery lumen, intima and media layers, and adventitial layer can be easily identified by sorting the center value of each class. The pixels at the boundaries between the clusters were considered as the markers of the LI and MA interfaces. The black dots in Fig. 9.10d are overlaid to the signal envelope indicate the position of the detected LI and MA interfaces.

5. Steps 3 and 4 are repeated for all the columns of the ROI. If in a given column the classifier failed to find three contiguous regions, then we discarded that column. The sequence of the identified LI and MA markers constitute the final segmentation of the CCA wall. The final profiles were regularized by a B-Spline. The result of the segmentation process for the image in Fig. 9.10a is reported in Fig. 9.11.

9.3 Design of Performance Metric

In this section, we give an overview of the image database we used to test our novel algorithm and describe the metric used to assess performance.

9.3.1 Image Database

The image database consisted of 200 B-mode 2D ultrasound longitudinal images of the common tract of the carotid artery. All the images were transferred to a computer via a DICOM communication port in log-compressed 8-bit grayscale. We set an axial resolution equal to 62.5 μm/pixel for each image.

The images were taken from 130 subjects: 50 normal subjects and 80 patients suffering from atherosclerosis who were recruited by the Neurology division of the

Gradenigo Hospital (Torino, Italy), where all the ultrasound examinations were conducted. The subjects' age ranged from 25 to 83 years (mean: 48.3 years; standard deviation: 9.9 years). Seventy subjects were males. All the patients were clinically evaluated before being included in the study and all the subjects signed an informed consent before the ultrasound examination. The study received the approval by the Institutional Committee of the Gradenigo Hospital.

Three expert operators (two technicians and a neurologist) independently traced the adventitial layers of the CCA in all the images. Manual tracings were performed by a graphical user interface that was previously developed by the authors [14]. We considered the average profile of the human tracings as ground truth (GT).

We also used another small set of 15 images for training purposes. This sample included eight normal carotids and seven carotids with increased IMT. The images were randomly selected. These 15 images were not included in the validation procedure.

9.3.2 *Polyline Distance Metric and Performance Metric*

In this section, we will illustrate the performance metric we chose to evaluate our technique. In previous studies [11, 12, 14, 20], we adopted the mean absolute error as performance metric. However, we found that in some conditions, it could give biased results. In the following, we will show the definition of the segmentation errors and of the IMT measurement error in terms of the polyline distance metric (PDM) between boundaries.

9.3.2.1 Polyline Distance Metric

We developed a measure similar to the approach as developed by Suri et al. [22]. This polyline distance measure seems to be a robust and reliable indicator of the distance between the two given boundaries, which truly represents the distances between boundary shapes along the artery. The basic idea is to measure the distance of each vertex of a boundary to the segments of the other boundary. The measured distance becomes very robust because it is independent of the number of points of each boundary. In the following, we briefly present the mathematical details about the computation of the polyline distance measure.

Consider two boundaries B_1 and B_2 as depicted in Fig. 9.12. We can define the distance $d(v,s)$ between a vertex v and a segment s. Let us consider the vertex $v=(x_0,y_0)$ on the boundary B_1 and the segment s formed by the endpoints $v_1=(x_1,y_1)$ and $v_2=(x_2,y_2)$ of B_2. We can define the polyline distance $d(v,s)$ between the vertex v and a polyline segment s, mathematically represented as:

$$d(v,s)=\begin{cases} d_{\perp}, & 0\le\lambda\le 1, \\ \min\{d_1,d_2\}, & \lambda<0,\lambda>1, \end{cases} \tag{9.9}$$

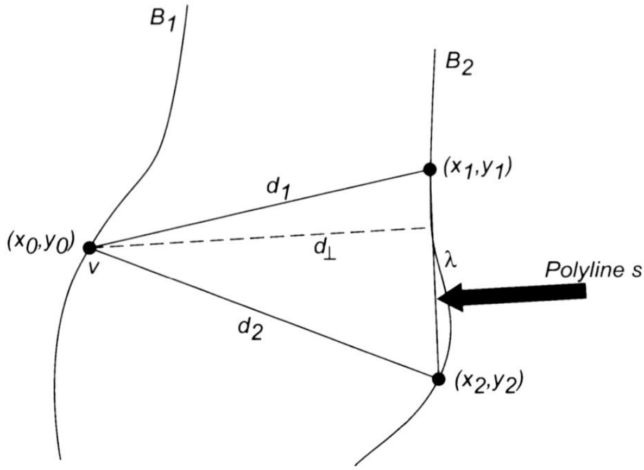

Fig. 9.12 Schematic representation of the polyline distance measure. The distance between vertex v and segment s is represented by $d_\perp$. If the intersection between $d_\perp$ and s falls close to the segment endpoints, the polyline distance is the minimum between d_1 and d_2

where

$$d_1 = \sqrt{(x_0 - x_1)^2 + (y_0 - y_1)^2}, \tag{9.10}$$

$$d_2 = \sqrt{(x_0 - x_2)^2 + (y_0 - y_2)^2}, \tag{9.11}$$

$$\lambda = \frac{(y_2 - y_1)(y_0 - y_1) + (x_2 - x_1)(x_0 - x_1)}{(x_2 - x_1)^2 + (y_2 - y_1)^2}, \tag{9.12}$$

$$d_\perp = \frac{(y_2 - y_1)(x_1 - x_0) + (x_2 - x_1)(y_0 - y_1)}{\sqrt{(x_2 - x_1)^2 + (y_2 - y_1)^2}}, \tag{9.13}$$

being:

- d_1 and d_2 are the Euclidean distances between the vertex v and the endpoints of segment s
- λ is the distance along the vector of the segment s
- $d_\perp$ is the perpendicular distance between v and the segment s

The polyline distance from vertex v to the boundary B_2 can be defined as $d(v, B_2) = \min_{s \in B_2} \{d(v,s)\}$. The distance between the vertexes of B_1 and the segments of B_2 is defined as the sum of the distances from the vertexes of B_1 to the closest segment of B_2:

$$d(B_1, B_2) = \sum_{v \in B_1} d(v, B_2). \tag{9.14}$$

Similarly, it is possible to calculate $d(B_2, B_1)$ (i.e., the distance between the vertices of B_2 and the closest segment of B_1) by simply swapping the boundaries. The polyline distance between boundaries is defined as:

$$D(B_1, B_2) = \frac{d(B_1, B_2) + d(B_2, B_1)}{(\text{No. of vertices of } B_1 + \text{No. of vertices of } B_2)}. \tag{9.15}$$

This measure $D(B_1, B_2)$ is very important since it reflects the average distance between the two computer-generated boundaries and the corresponding ground truth profiles.

9.3.2.2 Mean System Error

We defined the mean system error corresponding to CALEXia boundaries as the mean value of the error between these corresponding boundaries and the GT. Let N be the number of images in the database. This measure is mathematically given as:

$$\delta_{\text{LI}}^{\text{CALEXia}} = \frac{1}{N} \sum_{\text{images}} D(\text{CALEXia}_{\text{LI}}, \text{GT}_{\text{LI}}), \tag{9.16}$$

$$\delta_{\text{MA}}^{\text{CALEXia}} = \frac{1}{N} \sum_{\text{images}} D(\text{CALEXia}_{\text{MA}}, \text{GT}_{\text{MA}}), \tag{9.17}$$

where $\delta_{\text{LI}}^{\text{CALEXia}}$ and $\delta_{\text{MA}}^{\text{CALEXia}}$ are CALEXia mean system errors on LI and MA interfaces, respectively; $D(\text{CALEXia}_{\text{LI}}, \text{GT}_{\text{LI}})$ and $D(\text{CALEXia}_{\text{MA}}, \text{GT}_{\text{MA}})$ represent the polyline distance between CALEXia tracings and ground truth on a given image.

9.3.2.3 IMT Metric

The IMT value was calculated as the polyline distance between the LI and the MA boundary. This was then compared with the IMT for the GT.

$$\text{IMT}_{\text{CALEXia}} = D(\text{CALEXia}_{\text{LI}}, \text{CALEXia}_{\text{MA}}), \tag{9.18}$$

$$\text{IMT}_{\text{GT}} = D(\text{GT}_{\text{LI}}, \text{GT}_{\text{MA}}). \tag{9.19}$$

In (9.18) and (9.19), $\text{IMT}_{\text{CALEXia}}$ and IMT_{GT} are the IMT measures for CALEXia and GT, respectively. The mean IMT measurement error is defined as

$$\Omega_{\mathrm{IMT}}^{\mathrm{CALEXia}} = \frac{1}{N}\sum_{\mathrm{images}} \left|\mathrm{IMT}_{\mathrm{CALEXia}} - \mathrm{IMT}_{\mathrm{GT}}\right|. \tag{9.20}$$

9.4 Performance Evaluation and Benchmarking

9.4.1 Automated Tracings of the Carotid Artery

CALEXia proved very effective in automatically locating the CCA by tracing the near and far adventitia layer profiles (called $\mathrm{CALEXia}_N$ and $\mathrm{CALEXia}_F$, respectively). Figure 9.13 reports sample CALEXia performances on three different images. The left column of Fig. 9.13 represents the original B-Mode images; the right column depicts the CALEXia tracings of far and near layers with ultrasound image in the background. Figure 9.13a shows normal CCA that is not horizontally oriented in the frame; Fig. 9.13b shows how the CALEXia tracings correctly overlapped on the adventitial wall. Figure 9.13c documents the case of an artery that is not horizontal in the frame and presents a curvature. Figure 9.13d demonstrates the capability of CALEXia to follow curved walls while avoiding the noise due to backscattering. Figure 9.13e is relative to a plagued vessel: the white arrow in Fig. 9.13e indicates a fibrous plaque on the near wall. Figure 9.13f shows the correct performance of the CALEXia technique even in plagued arteries.

Figure 9.14 reports the comparison between CALEXia and GT tracings in three different cases. The GT tracings were obtained as the average of the three human operators' tracings. The white continuous lines in Fig. 9.14a, b represent the CALEXia proximal (near) adventitia, the white dotted lines represent the GT (indicated by GT_N in Fig. 9.14a, b). The black continuous lines in Fig. 9.14a, b represent the CALEXia distal (far) adventitia, the black dotted lines represent the GT (indicated by GT_F in Fig. 9.14a, b).

Since computer-generated boundaries and manually traced profiles may not have the same support, we first computed the common support between $\mathrm{CALEXia}_N$ and $\mathrm{CALEXia}_F$ and the corresponding GT boundaries. According to Rossi et al. [10], we considered the carotid artery as correctly located in the image if the distance between the computer-generated boundaries of the adventitia layers and the corresponding GT boundaries was lower than 1.5 mm (corresponding to 24 pixels). The distance between the computer-generated boundaries and the GT boundaries was calculated by using the PDM. We found that in ten images this distance was greater than 1.5 mm. Therefore, CALEXia failed in automatically locating the CCA in 5% of the images. Such images were removed from the database and they were further analyzed to detect the failure causes.

By using the absolute distance metric, we obtained slightly worse results: 16 images showed an absolute distance of the computer-generated boundaries from the GT boundaries higher than 1.5 mm (corresponding to 8% of the images). On the near wall, the average PDM between $\mathrm{CALEXia}_N$ and GT_N was equal to 8.3 ± 7.6

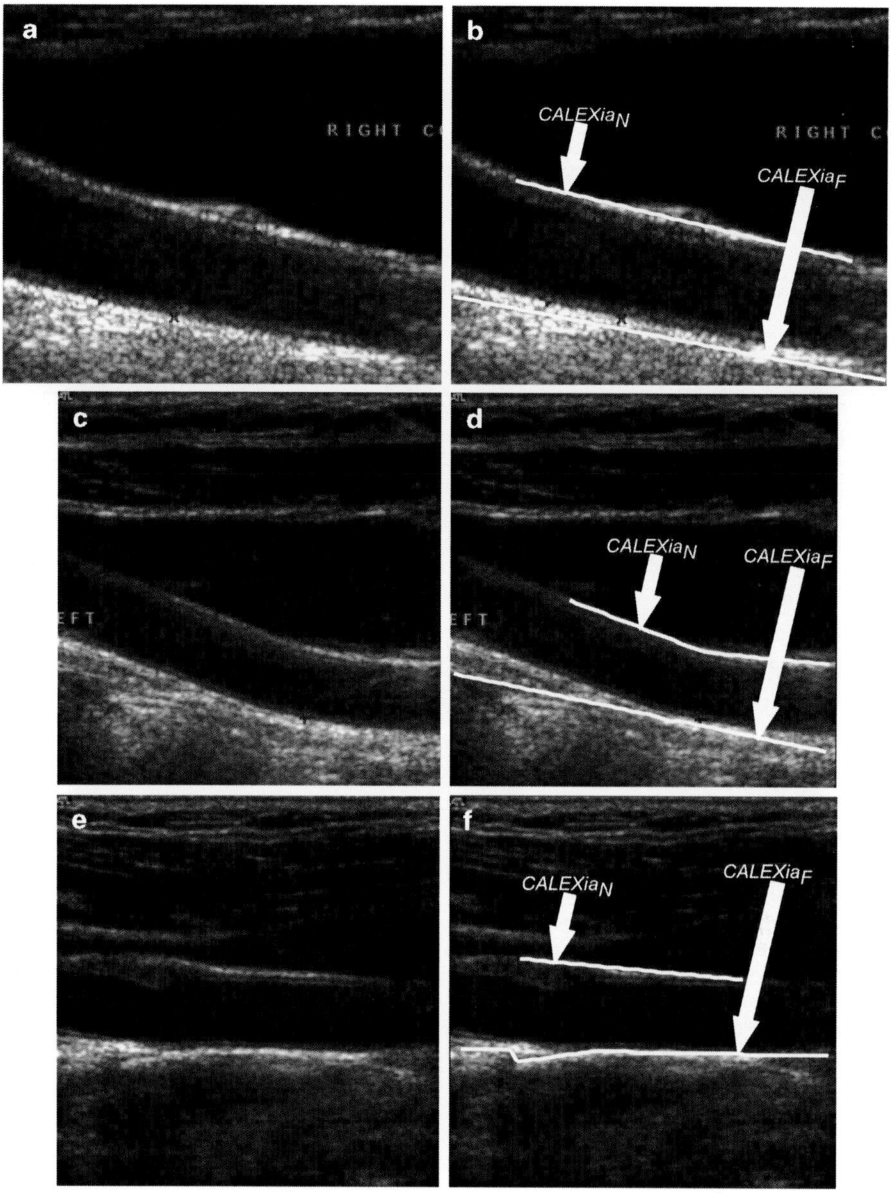

Fig. 9.13 Samples of CALEXia performance in CCA detection and adventitia automated tracings. *Left column* shows the original images; *right column* shows the CALEXia tracings overlaid on the ultrasound images. (**a**, **b**) Healthy CCA that is not horizontal. (**c**, **d**) Nonhorizontal CCA with curvature. (**e**, **f**) Plagued vessel with a fibrous plaque [*white arrow* in (**e**)] on the near wall. CALEXia shows adequate performances, even in the presence of blood backscattering [*white arrows* in (**a**) and (**c**)] (© AIUM 2010 - Reproduced with permission from [20])

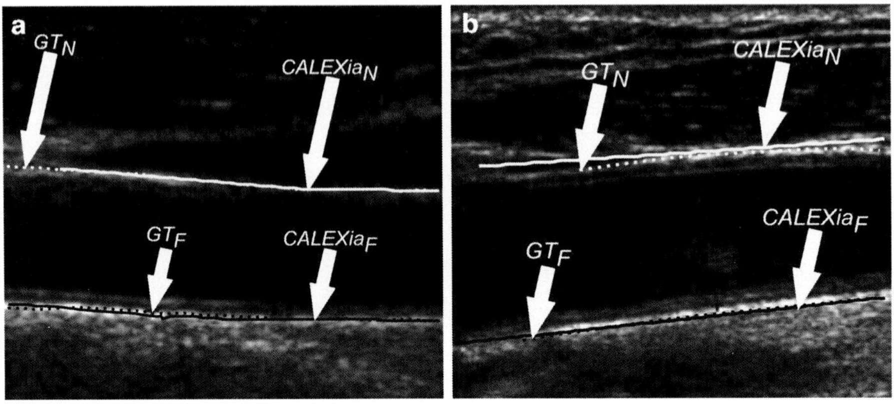

Fig. 9.14 Samples of CALEXia tracings compared with human tracings (GT). Proximal (near) adventitia tracings of CALEXia are depicted by *white continuous lines*, whereas *white dotted lines* indicate the GT near adventitia (GT_N in the figures). Distal (far) adventitia tracings of CALEXia are depicted by *black continuous lines*, whereas *black dotted lines* indicate the GT far adventitia (GT_F in the figures) (© AIUM 2010 - Reproduced with permission from [20])

pixels, corresponding to 0.52 ± 0.48 mm. On the far wall, the performance was equal to 6.6 ± 8.3 pixels, corresponding to 0.41 ± 0.52 mm. Results demonstrate that CALEXia is very accurate in automatically tracing the near and far adventitia profile, thus localizing the CCA in the image frame. This result is of great importance in the optic of an automated procedure for the CCA wall segmentation. This first module of CALEXia, in fact, can be thought of as a general localization procedure to be used in the development of automated techniques. Section 9.6 discusses the possible problems arising in locating the CCA when using the CALEXia architecture.

9.4.2 Carotid Wall Segmentation and IMT Measurement

In this section, we will show the performance of CALEXia in terms of LI and MA segmentation and IMT measurement. As described in Sect. 9.3.2, performance was evaluated by using the PDM.

Figure 9.15 reports a sample of CALEXia tracings overlaid to ground truth (GT). CALEXia tracings are represented by dashed lines, GT by continuous lines. Figure 9.15a reports LI tracings (white lines) and Fig. 9.15b MA tracings (black lines). Again, to avoid biased results due to different support of the profiles, we first computed the common support between the LI boundaries of CALEXia and GT, and that of the MA profiles. The mean distance between CALEXia and GT profiles for the LI boundary, averaged over the 200 images, that we called $\delta_{LI}^{CALEXia}$ in (9.16), was equal to 1.22 ± 1.30 pixels, corresponding to 0.076 ± 0.081 mm. The

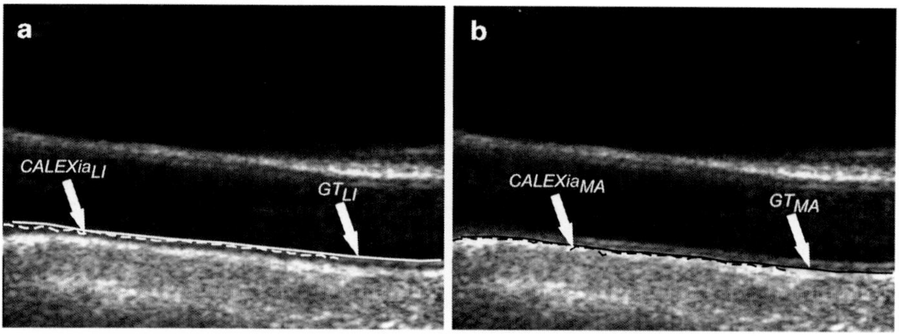

Fig. 9.15 Comparative samples of CALEXia (*dashed lines*) segmentation compared with GT (*continuous lines*). (**a**) LI tracings (*white lines*); (**b**) MA tracings (*black lines*)

analogous mean system error for the MA boundary, called $\delta_{MA}^{CALEXia}$ and defined by (9.17), was equal to 0.45 ± 0.47 pixels, corresponding to 0.028 ± 0.029 mm.

A direct comparison of these results with previously published performance is not straightforward. In fact, in the literature, the absolute mean distance is usually used as performance metric [5, 8, 23]. However, in the following, we will provide a brief comparison of the herein obtained results with respect to the most performing algorithms published so far.

- The LI segmentation error was slightly higher than most of the previously published techniques. Cheng et al. [5] obtained LI segmentation errors equal to 0.062 ± 0.060 mm, adopting a snake-based segmentation. They used the mean squared error metric. Destrempes et al. proposed a segmentation technique based on the modeling of the carotid wall by a mixture of Nakagami distributions [23]. They reached segmentation errors as low as 0.021 ± 0.030 mm. Other snake-based methodologies [12] showed segmentation errors equal to 0.059 ± 0.065 mm. The reason for this lower performance of CALEXia with respect to other approaches resides in the final fuzzy *K-means* classifier. Mainly, LI segmentation is made difficult by superimposed noise (usually in the form of blood backscattering). In Sect. 9.5, we provide more detailed discussion about the possible error sources.
- The MA segmentation error was one of the best results ever obtained. Only Destrempes et al. obtained a lower segmentation error (equal to 0.016 ± 0.007 mm) [23]. Again, the result cannot be directly compared due to different evaluation metrics. Anyway, most of the snake-based approaches do not reach performance better than about 0.040 ± 0.050 mm. This excellent result has been made possible by the superior architecture of CALEXia, which integrates many image features before performing segmentation. On the MA layer, particularly, CALEXia could effectively exploit the intensity difference between the media and the adventitia layers while maintaining a coherent boundary shape (imposed by the feature extraction and classification). As a result, the CALEXia MA tracings were very closed to GT tracings.

The IMT measurement error $\Omega_{\mathrm{IMT}}^{\mathrm{CALEXia}}$, defined by (9.20), was equal to 0.76±0.69 pixels, corresponding to 0.048±0.043 mm. This result is in line with the performance of most user-dependent techniques and outperforms some other algorithms based on active contours [6] or snakes [8, 12]. It must be said that the best-performing algorithms for the IMT measurement present errors of the order of 10 μm, which can be reduced to about 1 μm in special conditions [16]. The limit of CALEXia in IMT measurement is given by the average performance on the LI tracing. The next section provides a comprehensive discussion about merits, limits, and possible error sources.

9.5 CALEXia Merits, Problems, and Perspectives

We have developed a novel technique for automatic computer-based tracing of CCA in longitudinal B-Mode ultrasound images, its segmentation, and IMT measurement. The new technique consists of an integrated approach combining geometric feature extraction, line fitting, and fuzzy classification. We characterized CALEXia performance by using the PDM, which ensures a stable and reliable result independent of the number of points of the boundaries. In this section, we discuss the positive and negative aspects of CALEXia and its future perspectives as a computer diagnostic tool.

CALEXia offered satisfactory performance when used on a database of real images. In the following, we summarize its advantages.

- *User independence*. CALEXia is a completely user-independent tool. This ensures that CALEXia can be used to process large sets of images without the need for any human interaction. Most of the best-performing techniques that have been developed to aid CCA wall segmentation and IMT measurement require a certain degree of user interaction. Such interaction precludes real automation and makes the result dependent on the operator choices.
- *Suitability to different carotid morphologies*. There is variability in the appearance of the CCA in the longitudinal B-Mode image frame. Most of the CCAs appear as straight and horizontal in the image, but some others may not be horizontally placed. Some vessels are curved or even kinked. Our strategy of automated CCA location enables the processing of almost any kind of CCA morphology. This characteristic increases the versatility of CALEXia. Also, sonographers are not required to scan the CCA in a predefined and standard projection, but they can choose the best projection according to each subject and to the specific artery. CALEXia proved effective in recognizing the CCA also in plagued vessels.
- *Low MA segmentation error*. As described in Sect. 9.4.2, CALEXia showed a very low MA segmentation error when compared with ground truth. The $\delta_{\mathrm{MA}}^{\mathrm{CALEXia}}$ error was lower than 0.5 pixel, allowing for an error of about 28 μm. This result is the best among the user-independent algorithms and better of most of the user-dependent approaches.

- *Real-time implementation*. We implemented and tested CALEXia on a dual-2.5 GHz PowerPc computer equipped by 8 GB of RAM. CALEXia was implemented in a MATLAB (The MathWorks, Natick, MA, USA) framework. The average processing time was equal to 3.1 ± 0.7 s, thus CALEXia enabled a real-time segmentation and IMT measurement. This result was made possible due to decimation of the image columns. When detecting seed points (Sect. 9.2.1.1), the image is first decimated and then the intensity profile of each column is processed. Decimation enables a strong reduction of the computational cost while preserving the correct morphology of the traced boundaries. Moreover, once trained, the linear classifiers used to perform fitting and feature extraction are very rapid and computationally light. CALEXia was developed with the precise aim of being used as a real-time support to the sonographer.

CALEXia failed in recognizing the CCA in ten images over the 200 we tested. We spotted two particular conditions that caused nonperfect adventitial tracings by CALEXia.

1. The first condition is represented by "multilayered" images. An example of such a kind can be seen in Fig. 9.16. Figure 9.16a reports the original ultrasound image: the jugular vein structure (indicated by JV) is parallel to the CCA and placed exactly above it. The image assumes a "multilayered" aspect since we can observe a repeated pattern of lumen walls (i.e., dark stripe – bright lines). The deepest pattern is formed by the far adventitia, the CCA lumen, and the near CCA adventitia that overlaps with the distal (far) border of the jugular vein (white arrow in Fig. 9.16a). The second pattern is made by the near CCA adventitia overlapped to the distal (far) JV border, the JV lumen, and the near JV border. The CALEXia algorithm traces line segments in correspondence to the three borders (far CCA wall, near CCA wall, and near JV wall), but it fails to exactly find the final classification of the correct line

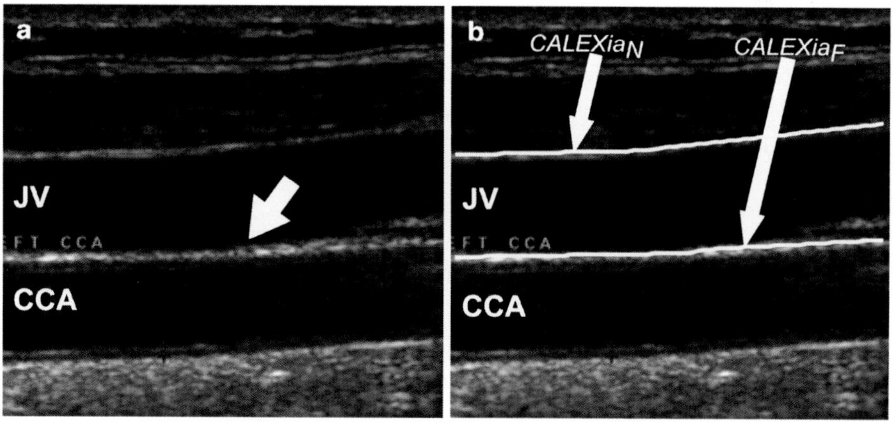

Fig. 9.16 (**a**) Example of "multilayered" structure with the jugular vein (JV) overlapped to the common carotid artery (CCA). (**b**) CALEXia fails in detecting the correct pair of line segments corresponding to the artery walls and segments the JV (© AIUM 2010 - Reproduced with permission from [20])

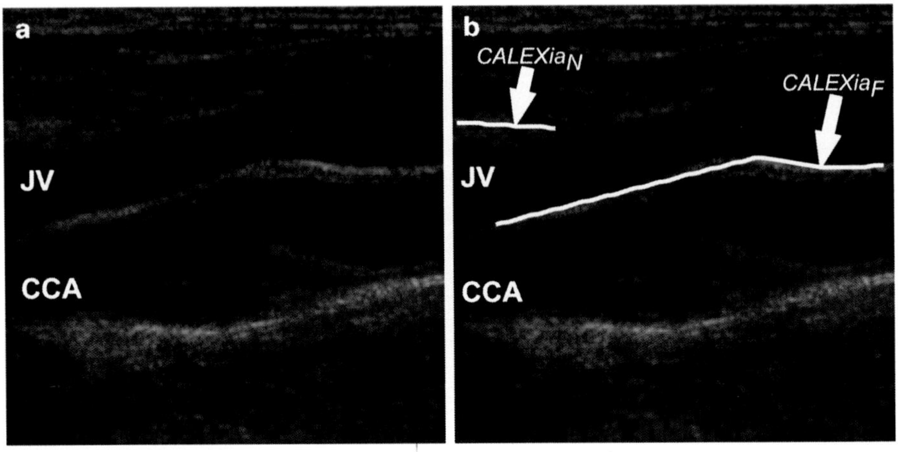

Fig. 9.17 Example of incorrect CALEXia segmentation due to a deeper plagued vessel. (**a**) The distal (far) plaque is quite big and protruding into the common carotid artery (CCA) lumen. Also, the jugular vein (JV) overlaps the CCA. (**b**) CALEXia fails in detecting the correct pair of line segments corresponding to the artery walls and segments a small portion of the JV (© AIUM 2010 - Reproduced with permission from [20])

pairs, since the JV lumen appears as darker than the lumen of the CCA. Blood backscattering that is present in the CCA lumen (but not in the JV lumen) makes the CALEXia algorithm to select the line segments shown in Fig. 9.16b. Therefore, the algorithm recognizes the JV instead of the CCA.

2. Deeper plaques protruding in the artery lumen represent the second possible cause of nonperfect CALEXia segmentation. Figure 9.17a depicts an echolucent type-II soft plaque on the distal (far) border of the CCA. Again, CALEXia incorrectly assigns the line segments to the carotid adventitial layers due to the fact that the CCA lumen is not black and homogeneous. The plaque significantly perturbs the average intensity level of the CCA lumen. Therefore, the final line segment assignment recognizes the JV instead of the CCA. The final tracings of the CALEXia method is reported in Fig. 9.17b.

We did not find any other cause of possible malfunctioning of the CALEXia algorithm.

On the LI interface, the segmentation error of CALEXia ($\delta_{\mathrm{LI}}^{\mathrm{CALEXia}}$) is greater than that of most of other techniques. The segmentation of this interface is most critical to CALEXia. This is due to the fact that the fuzzy *K-means* classifier, which we use to mark the transitions between lumen and intima (Fig. 9.10d), underestimates the LI position. The CALEXia average error on the LI interface is the only case in which we experienced segmentation errors greater than one pixel. To test possible improvement on the LI interface tracing, we experimented different classifiers. In some cases, a better estimation of the LI position caused a detriment in the estimation of the MA position. The paper reports the optimal performance achieved. Also, some efficient classifiers require the selection of a threshold or critical value. Being complete automation of a fundamental requirement of CALEXia, we preferred to rely on the automation ensured by *K-means*.

There are two factors limiting the CALEXia performance: (1) the underestimation of the LI interface (as discussed above) and (2) the segmentation of image regions characterized by a very low signal-to-noise (SNR) ratio. The solution of the two problems is not straightforward. Our aim is to improve the segmentation performances without losing the versatility of the algorithm. In fact, this is a generalized technique that, making use of a superior integrated architecture, is effective in processing almost any kind of image. The robustness to noise is good, since only 5% of the randomly selected images were incorrectly processed. Our methodology is effective in coping with different anatomies, since it can also process carotids that are not horizontal in the image, as well as curved arteries.

There are three scopes of improvement of this new methodology.

1. To estimate the location of carotid artery wall, line segments are fitted into the selected seed points. A series of geometric-based and intensity-based features are measured for each fitted line segment. *Spread* and *residue* are used to combine over-segmented line segments if they are near each other and well aligned. *Width stability* is used to validate line segments with small width variation. *Proximity* and *alignment* can help to penalize segments that are connected but not well aligned. Two linear discriminators are learned from the training set for connecting and validating line segments. Their true positive rates are 98 and 92%, respectively. From a theoretical point of view, the linear discriminator we designed for removing bright background noise could be more deeply investigated to reduce false-negative decisions, especially for carotid artery walls affected by high-backscattering noise. A strong linear discriminator including features that correspond to the stability of gray level intensity distribution may help to solve this problem.
2. We tested the effect of different training sets on the performance of the linear discriminators. We divided the 200-images database in sets of 15 images each that we used to train our classifiers (thus resulting in 13 subsets or trail runs). Then, we calculated the performances of the 13 trail runs in terms of numbers of correctly processed images, LI and MA segmentation errors, and IMT measurement error for the remaining images of the database. No training data set produced significant differences in the CALEXia performance and the results were stable. The overall performance of the system (averaged on the 13 trail runs) were:

   (a) Number of correctly processed images equal to 5%.
   (b) LI segmentation error equal to 1.18 ± 1.09 pixels; MA segmentation error equal to 0.029 ± 0.030 pixels.
   (c) IMT estimation error equal to 0.75 ± 0.69 pixels.

   Therefore, performance was statistically similar for all the different trained trials. Future work may be done to further optimize the size of the training sample, in order to better cope with the variability of the ultrasound longitudinal carotid images that can be encountered in clinical practice.

3. IMT performance is usually dependent on the region of the image that is chosen for segmentation. It was demonstrated that a snake-based segmentation procedure could be very effective in aiding the human operator in measuring the intima and media thicknesses once the operator had selected the image region in which to perform IMT measurement [7, 8]. CALEXia, like other automated techniques, segment the entire image and, therefore, estimate IMT also in regions of the image affected by a high level of noise. This is a bias in the IMT final measurement. A scope of improvement would be an intelligent strategy to a priori select the image region in which the noise is lower and the IMT measurement more reliable.

9.6 Conclusions

We developed a completely user-independent algorithm for automated tracing and the layer extraction of the carotid artery wall in ultrasound images. This novel technique (CALEXia), which represents a further step in carotid automated ultrasound image processing, was used to measure the IMT. The data analysis of the segmentation and IMT measurement errors showed that the tracing of the LI boundary may have a scope of improvement. The underestimation of the IMT value depends on the quality of the LI tracing. Despite little LI challenge, CALEXia out-performed the most of the previously developed techniques in segmenting the MA interface. CALEXia may represent a generalized and standard methodology toward completely automated and accurate IMT measurement. The authors are working toward further improvement of the segmentation performances and the adoption of CALEXia in routine clinical environment.

Acknowledgements The Authors would like to express their gratitude to Dr. William Liboni, MD (Neurology Service, Gradenigo Hospital, Torino, Italy) for providing the data and for plaque classification.

References

1. W.H. Organization. Cardiovascular disease. Available from: http://www.who.int/cardiovascular_diseases/en/.
2. S.H. Johnsen and E.B. Mathiesen, Carotid plaque compared with intima-media thickness as a predictor of coronary and cerebrovascular disease, Curr Cardiol Rep, 11(1), (2009), 21–7.
3. P.R. Hunziker, C. Imsand, D. Keller, N. Hess, V. Barbosa, F. Nietlispach, N. Liel-Cohen, A.E. Weyman, M. Pfisterer, and P. Buser, Bedside quantification of atherosclerosis severity for cardiovascular risk stratification: a prospective cohort study, J Am Coll Cardiol, 39(4), (2002), 702–9.
4. P. Pignoli and T. Longo, Evaluation of atherosclerosis with B-mode ultrasound imaging, J Nucl Med Allied Sci, 32(3), (1988), 166–73.
5. D.C. Cheng, A. Schmidt-Trucksass, K.S. Cheng, and H. Burkhardt, Using snakes to detect the intimal and adventitial layers of the common carotid artery wall in sonographic images, Comput Methods Programs Biomed, 67(1), (2002), 27–37.

6. M.A. Gutierrez, P.E. Pilon, S.G. Lage, L. Kopel, R.T. Carvalho, and S.S. Furuie, Automatic measurement of carotid diameter and wall thickness in ultrasound images, Comput Cardiol, 29, (2002), 359–62.
7. C.P. Loizou, C.S. Pattichis, A.N. Nicolaides, and M. Pantziaris, Manual and automated media and intima thickness measurements of the common carotid artery, IEEE Trans Ultrason Ferroelectr Freq Control, 56(5), (2009), 983–94.
8. C.P. Loizou, C.S. Pattichis, M. Pantziaris, T. Tyllis, and A. Nicolaides, Snakes based segmentation of the common carotid artery intima media, Med Biol Eng Comput, 45(1), (2007), 35–49.
9. S. Golemati, J. Stoitsis, E.G. Sifakis, T. Balkizas, and K.S. Nikita, Using the Hough transform to segment ultrasound images of longitudinal and transverse sections of the carotid artery, Ultrasound Med Biol, 33(12), (2007), 1918–32.
10. A.C. Rossi, P.J. Brands, and A.P. Hoeks, Automatic recognition of the common carotid artery in longitudinal ultrasound B-mode scans, Med Image Anal, 12(6), (2008), 653–65.
11. F. Molinari, W. Liboni, P. Giustetto, S. Badalamenti, and J.S. Suri, Automatic computer-based tracings (ACT) in longitudinal 2-D ultrasound images using different scanners, Journal of Mechanics in Medicine and Biology, 9(4), (2009), 481–505.
12. S. Delsanto, F. Molinari, P. Giustetto, W. Liboni, S. Badalamenti, and J.S. Suri, Characterization of a completely user-independent algorithm for carotid artery segmentation in 2-D ultrasound images, IEEE Trans Instrum Meas, 56(4), (2007), 1265–74.
13. S. Delsanto, F. Molinari, W. Liboni, P. Giustetto, S. Badalamenti, and J.S. Suri, User-independent plaque characterization and accurate IMT measurement of carotid artery wall using ultrasound, Conf Proc IEEE Eng Med Biol Soc, 1, (2006), 2404–7.
14. F. Molinari, S. Delsanto, P. Giustetto, W. Liboni, S. Badalamenti, and J.S. Suri, "User-independent plaque segmentation and accurate intima-media thickness measurement of carotid artery wall using ultrasound," Advances in diagnostic and therapeutic ultrasound imaging, J.S. Suri, C. Kathuria, R.F. Chang et al., eds., Norwood, MA: Artech House, (2008), pp. 111–140.
15. J.H. Stein, C.E. Korcarz, M.E. Mays, P.S. Douglas, M. Palta, H. Zhang, T. Lecaire, D. Paine, D. Gustafson, and L. Fan, A semiautomated ultrasound border detection program that facilitates clinical measurement of ultrasound carotid intima-media thickness, J Am Soc Echocardiogr, 18(3), (2005), 244–51.
16. F. Faita, V. Gemignani, E. Bianchini, C. Giannarelli, L. Ghiadoni, and M. Demi, Real-time measurement system for evaluation of the carotid intima-media thickness with a robust edge operator, J Ultrasound Med, 27(9), (2008), 1353–61.
17. F. Rosenblatt, The perceptron: a probabilistic model for information storage and organization in the brain, Psychol Rev, 65(6), (1958), 386–408.
18. R. Stoica, X. Descombes, and J. Zerubia, A Gibbs point process for road extraction from remotely sensed images, Int J Comput Vis, 57, (2004), 121–36.
19. N. Otsu, A threshold selection method from gray-level histograms, IEEE Trans Syst Man Cybern, 9(1), (1979), 62–6.
20. F. Molinari, G. Zeng, and J.S. Suri, An integrated approach to computer-based automated tracing and its validation for 200 common carotid arterial wall ultrasound images: A new technique, J Ultras Med, 29, (2010), 399–418.
21. F. Molinari, G. Zeng, and J.S. Suri, Intima-media thickness: setting a standard for completely automated method for ultrasound, IEEE Transaction on Ultrasonics Ferroelectrics and Frequency Control, 57(5), (2010), 1112–1124.
22. J.S. Suri, R.M. Haralick, and F.H. Sheehan, Greedy algorithm for error correction in automatically produced boundaries from low contrast ventriculograms, Pattern Anal Appl, 3(1), (2000), 39–60.
23. F. Destrempes, J. Meunier, M.F. Giroux, G. Soulez, and G. Cloutier, Segmentation in ultrasonic B-mode images of healthy carotid arteries using mixtures of Nakagami distributions and stochastic optimization, IEEE Trans Med Imaging, 28(2), (2009), 215–29.

Biographies

Dr. Filippo Molinari received the Italian Laurea and the Ph.D. in electrical engineering from the Politecnico di Torino, Torino, Italy, in 1997 and 2000, respectively. He is leader in ultrasound imaging focused towards tissue characterization, vascular quantification for diagnostics and therapeutics. Currently, he is Assistant Professor at Politecnico di Torino, Italy – Department of Electronics.

Dr. Guang Zeng received the B.S. degree from Xiangtan University, China in 1998. He received the M.S. degree in 2005 and the Ph.D. degree in 2008 from Clemson University, SC, USA, both in Electrical Engineering. He is currently working in the Aging and Dementia Imaging Research Laboratory, Mayo Clinic, Rochester, MN. His research interests include biomedical image processing, pattern recognition and computer vision.

Dr. Jasjit S. Suri is an innovator, scientist, a visionary, an industrialist and an internationally known world leader in Biomedical Engineering. Dr. Suri has spent over 20 years in the field of biomedical engineering/devices and its management. He received his Doctrate from University of Washington, Seattle and Business Management Sciences from Weatherhead, Case Western Reserve University, Cleveland, Ohio. Dr. Suri was crowned with President's Gold medal in 1980 and the Fellow of American Institute of Medical and Biological Engineering for his outstanding contributions.

Chapter 10
Inter-Greedy Technique for Fusion of Different Segmentation Strategies Leading to High-Performance Carotid IMT Measurement in Ultrasound Images

Filippo Molinari, Guang Zeng, and Jasjit S. Suri

Abstract User-based estimation of intima–media thickness (*IMT*) of carotid arteries leads to subjectivity in its decision support systems, while being used as a cardiovascular risk marker. During automated computer-based decision support, we developed segmentation strategies that follow three main courses of our contributions: (a) signal processing approach, combined with snakes and fuzzy *K-means* (*CULEXsa*), (b) integrated approach based on seed and line detection, followed by probability-based connectivity and classification (*CALEXia*), and (c) morphological approach, with watershed transform and fitting (*WS*). These grayscale segmentation algorithms yielding carotid wall boundaries has certain bias along with their own merits. We have recently developed a fusion technique, which combines two carotid wall boundaries using ground truth (*GT*) as an ideal marker and is helpful in removing bias.

Here, we have extended this fusion concept by taking the merits of these multiple boundaries, the so-called Inter-Greedy (*IG*) approach. Further, we estimate IMT from these fused boundaries from multiple sources. Starting from the technique with the overall least system error (the snake-based one), we iteratively swapped the vertices of the profiles until we minimized its overall distance with respect to ground truth. The fusion boundary was the Inter-Greedy boundary. We used the polyline distance metric for performance evaluation and error minimization. We ran the segmentation protocol over the database of 200 carotid longitudinal B-mode ultrasound images and compared the performance of all the four techniques (CALEXia, CULEXsa, WS, and IG).

The mean error of Inter-Greedy technique yielded 0.32 ± 0.44 pixel (20.0 ± 27.5 μm) for the *LI* boundary (a $33.3 \pm 5.6\%$ improvement over initial best performing technique) and 0.21 ± 0.34 pixel (13.1 ± 21.3 μm) for *MA* boundary (a $32.3 \pm 6.7\%$ improvement). IMT measurement error for Greedy method was 0.74 ± 0.75 pixel (46.3 ± 46.9 μm), a $43.5 \pm 2.4\%$ improvement.

F. Molinari (✉)
Biolab, Department of Electronics, Politecnico di Torino, Corso Duca degli Abruzzi, 24, 10129 Torino, Italy
e-mail: filippo.molinari@polito.it

Jasjit S. Suri et al. (eds.), *Atherosclerosis Disease Management*,
DOI 10.1007/978-1-4419-7222-4_10, © Springer Science+Business Media, LLC 2011

Keywords Ultrasound imaging • Carotid artery • Automated tracings • Geometric features • Integration • Greedy method • Performance evaluation • Validation • Benchmarking

10.1 Introduction

The atherosclerotic process is strictly linked to the increased risk of cardiovascular and cerebrovascular disease. Atherosclerosis refers to the deposit of lipids in the artery wall. The increase of the artery walls' thickness causes a reduction of the lumen with possible vascular problems. The most widely used and validated marker of atherosclerosis is the intima–media thickness (*IMT*) of the carotid artery. Large multicenter studies have shown that computer-measured IMT is a risk marker for cardiovascular disorders [1, 2] and that it has predictive value for incident myocardial infarction [3].

The IMT measurement is usually performed by relying on ultrasound images. Manual measurement of the IMT by expert sonographers is highly reliable, but time-consuming. Also, the results depend on the experts, with their associated subjectivity. Therefore, manual segmentation is impractical in large studies where standardization of the protocol is required and where the number of patients is very high. Computer-based IMT measurement has been growing in importance since the first studies by Pignoli et al., who showed feasibility of a computer IMT measurement in B-Mode ultrasound images [4]. Since then, algorithms have moved towards user independence, real-time computation, and high performance.

Conceptually, the artery wall is formed by three concentric layers (the innermost called *intima*, the middle called *media*, and the outermost called *adventitia*), and measuring the IMT means finding the boundaries between the artery lumen and the intima (the so-called lumen–intima boundary – *LI*) and between the media and adventitia layers (the media–adventitia boundary – *MA*). The Euclidean or orthogonal distance between the LI and the MA boundaries is considered as the estimation of the IMT.

Noise has been a challenging obstacle limiting the IMT measurement capabilities of computer algorithms. Ultrasounds produce images with lower spatial resolution and signal-to-noise ratio than other imaging modalities (angiography, multislice computer tomography, and magnetic resonance imaging). However, since ultrasound devices are less expensive, portable, real-time and use no ionizing radiations, hence, IMT is usually measured on B-Mode ultrasound images in clinical practice. As a result, there is a growing interest towards high-performance automated ultrasound techniques.

Since 2005 [5], our research team has been developing completely automated segmentation algorithms and IMT measurement strategies. Complete automation was a key requisite of our strategies: we aimed at developing techniques that could be easily used for large studies as well as on already-acquired datasets without the need for user interaction. We developed a snake-based segmentation strategy (called *CULEXsa*, Completely User-independent Layer EXtraction algorithm based on signal approach) in 2005–2007 [5–7]. CULEXsa proved very effective in carotid wall segmentation and IMT measurement but had scope for improvement in the number of images that could be independently processed. We showed that in about 8% of the images, CULEXsa failed in automatically tracing the LI and MA boundaries due to noise. Also, it was difficult to adapt the snake parameters to the different CCA morphologies.

In 2009, we proposed another automated algorithm that was based on a different approach. We called it *CALEXia*, from Completely Automated Layers EXtraction based on integrated approach [8]. As this technique was based on an integration of feature extraction, fitting, and classification, CALEXia was complementary to CULEXsa. We showed that CALEXia outperformed CULEXsa in MA tracing accuracy. Also, CALEXia was more versatile compared to CULEXsa and performed correctly in 95% of the tested images (against 92% of CULEXsa) [9]. However, IMT measurement performance was still inadequate for clinical use.

We combined the CALEXia and CULEXsa traced profiles and obtained an improvement of about 12.5% on LI tracing accuracy and 16.1% on MA. IMT measurement error dropped to about 3.6%, leading to a measurement error of about 83 μm. Such values are compatible with clinical use, but still higher than some proposed user-dependent approaches [10–12].

Recently, we tested the effectiveness of morphological approaches (i.e., the watershed transform – *WS*) for automated CCA location and wall segmentation in ultrasound imaging [13]. Our rationale was to introduce other complementary techniques into our processing pipeline, to achieve optimal segmentation and IMT measurement performance.

These complementary techniques led us to adopt a Greedy Approach, to improve IMT measurement. The image processing algorithms add a bias, which is difficult to remove as they are very small in nature. We, thus, have to depend upon the point (x,y) locations, which are represented by these boundaries, obtained using image processing methods. One way to remove the bias is to use Ground Truth information as the gold standard. The bias effect can be removed by regressing the boundaries obtained using image processing methods and comparing them with that of Ground Truth. This bias can come from multiple sources of multiple image processing methods. In this chapter, we apply an Inter-Greedy (*IG*) approach to the LI and MA boundaries traced by the three image segmentation techniques (namely, CALEXia, CULEXsa, and WS), to remove their bias errors. These three automated techniques were complementary and used different foundations for segmentation. Therefore, the goal of this research was to calibrate IMT measurement by fusing the different output of image processing approaches. Performance validation was done against manual tracings on a large dataset of 200 images, acquired by multiple scanners. We combined iteratively the three automatically produced boundaries to decrease the overall system distance from Ground Truth. LI and MA boundary calibration was obtained by substituting points among the different techniques so that the obtained boundary was closer to Ground Truth than the initial ones (Fig. 10.1).

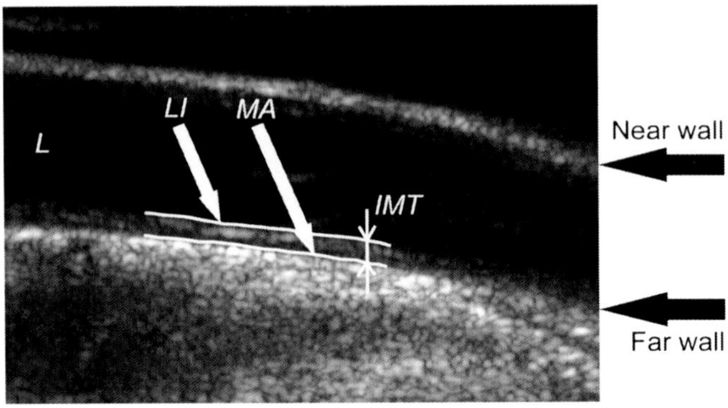

Fig. 10.1 Schematic representation of the lumen–intima (LI) and media–adventitia (MA) interfaces in longitudinal carotid B-Mode images. The *black arrows* on the *right* indicate the near and fall carotid walls. The intima–media thickness (IMT) is calculated as the distance between LI and MA (© Springer 2010 - doi: 10.1007/s10916-010-9507-y - Reproduced from [22])

The chapter is organized as follows: Sect. 10.2 describes the CULEXsa architecture, Sect. 10.3 the CALEXia, and Sect. 10.4 the watershed-based segmentation strategy. Section 10.5 illustrates the Inter-Greedy approach, whereas Sect. 10.6 discusses the performance metrics we used to compare results to Ground Truth (constituted by manual tracings). The results of the segmentation performance comparison among techniques is reported in Sect. 10.7, while Sect. 10.8 shows the segmentation error in each boundary vertex. The IMT accuracy is discussed in Sect. 10.9. Conclusions of this work are traced in Sect. 10.10.

10.2 Architecture of CULEXsa

CULEXsa is a completely user-independent algorithm, which consists of two steps, for extracting the layers of the artery wall:

1. Automated tracings of the near (AD_N) and far (AD_F) adventitia profiles.
2. Snake-based segmentation of the LI and MA boundaries starting from the detected far adventitia.

The detailed structure of CULEXsa has already been described in previously published works [5, 10]. In the following section, we summarize the structure of CULEXsa.

The core of the algorithm is the pixel analysis based on local statistics. Carotid appearance can be seen as a mixture model with different characteristics: (1) pixels belonging to the CCA lumen are dark (i.e., of low intensity value) and surrounded by a dark homogeneous neighborhood; (2) pixels belonging to the CCA wall are bright (i.e., of high intensity value) and surrounded by a relatively homogeneous neighborhood; (3) all the remaining pixels have average intensity with an inhomogeneous neighborhood. By computing the mean value and the standard deviation of a 10×10 square neighborhood of each pixel, we can separate the lumen pixels from the others by placing proper thresholds on the neighborhood mean intensity and standard deviation values. Empirically, we found that the better discrimination for the lumen pixels was intensity lower than 0.08 and standard deviation lower than 0.14 on a normalized (0,1) scale [5, 14].

The image is then processed columnwise. The deepest local intensity maximum is marked as far adventitia candidate. The algorithm then decreases the row index (i.e., moves towards the near wall) and searches for a lumen pixel. Lumen is defined as the first local intensity minimum with neighborhood mean and standard deviation lower than 0.08 and 0.14, respectively. By further decreasing the row index, the algorithm marks the position of the near adventitia wall (AD_N) as the first local intensity maximum, i.e., found just after the end of the lumen.

CULEXsa segments only the far wall. The image is considered columnwise, and only the pixels comprised between AD_N and AD_F are processed. A gradient enhancing the vertical transitions is used, and the two higher intensity maxima of the gradient are considered as the first guess of LI and MA transitions. The intensity criterion we used was relative to the intensity profile of the specific column: we only considered maxima whose intensity was above the 90th-percentile of the intensities distribution of the column pixels. The sequence of LI and MA points is then refined using the snake-based algorithm [17], where the elasticity and tension parameters are: $\alpha(s) = 0.1$ and $\beta(s) = 0.01$.

Recently, we demonstrated that CULEXsa is effective in detecting the CCA in normal as well as pathological images and that its performances are independent of the ultrasound OEM scanner used for data acquisition [14]. Figure 10.2 sketches the CULEXsa strategy: Fig. 10.2a (left panel) depicts the near (AD_N), the far (AD_F) adventitia, and the lumen (*L*); Fig. 10.2b (right panel) reports a representative CULEXsa segmentation of the far wall of the CCA.

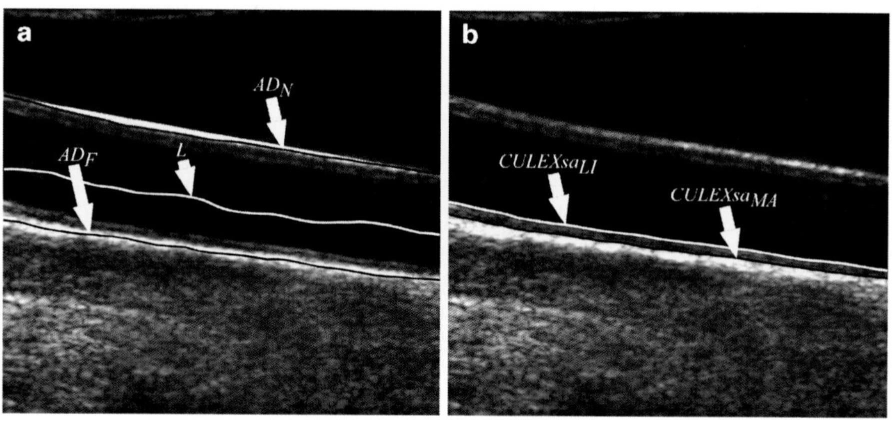

Fig. 10.2 (**a**) Automated location of the common carotid artery made by CULEXsa: the algorithms traces the profiles of the near (AD_N) and far (AD_F) adventitia layers and of the lumen (L). (**b**) Image segmentation performed by CULEXsa: $CULEXsa_{LI}$ represents the lumen–intima tracing; $CULEXsa_{MA}$ represents the media–adventitia tracing

The principal limitations of CULEXsa are a low signal-to-noise ratio (mainly due to blood backscattering) and the presence of ultrasound image artifacts. Noise (particularly, speckle noise) has a detrimental effect on local statistics. In fact, in the presence of noise, the lumen vessel might not be dark and homogeneous as we modeled it. Therefore, local statistics may be unable to correctly locate the pixels belonging to the lumen. As a consequence, CULEXsa may fail in detecting the AD_N and AD_F candidates, thus precluding the correct wall segmentation. Image artifacts are very common in ultrasound imaging. The most usual ones are the shadow cone (projected by calcium deposits in the vessel wall) and the multiple echoes or reverberations. These artifacts alter the intensity levels of the carotid wall, thus introducing errors in the detection of the LI and MA.

10.3 Architecture of CALEXia

CALEXia is a completely automated procedure for carotid layers extraction that, unlike CULEXsa, is based on an integrated approach consisting of feature extraction, fitting, and classification [15]. CALEXia consists of two parts:

1. A module that automatically locates the CCA in the image.
2. A segmentation procedure that automatically traces the LI and the MA contours of the distal (far) wall, once CCA has been localized.

Conceptually, in step 1, CALEXia exploits the image information to automatically detect the near and far adventitia. The local intensity maxima are detected for each column located on the CCA using a linear discriminator. These points are called "seed points." Seed points are then linked to form line segments. An intelligent procedure removes short or false line segments and joins close and aligned segments. This procedure avoids oversegmentation of the artery wall. Line segment classification extracts the line pair that contains the artery lumen in between.

Once CCA has been detected, the image is scanned columnwise. The intensity profile of each column is processed by using a fuzzy *K-means* classifier, which is initialized by the intensity values. The classifier assigns pixel to three clusters: (1) lumen, (2) the intima and media layers (which are not distinguishable in ultrasound images), and (3) the adventitia layer. The points at the transitions between the three clusters are taken as the LI and MA boundaries' markers. The sequence of the transition points of the entire column determines the segmentation of the LI and MA interfaces. Figure 10.3 reports an example of CALEXia automated tracings of the CCA (left figure) and of far wall segmentation (right figure).

CALEXia performances are limited by the following problems: (1) low signal-to-noise ratio and (2) low contrast at the interface between the artery lumen and intima layer. The effect of noise in CALEXia is the incorrect detection of the near and far adventitia. The line segments constituting the adventitial tracings are generated starting from the seed points. However, a high level of noise determines bad discrimination performances. False positives may be interpreted as true seed

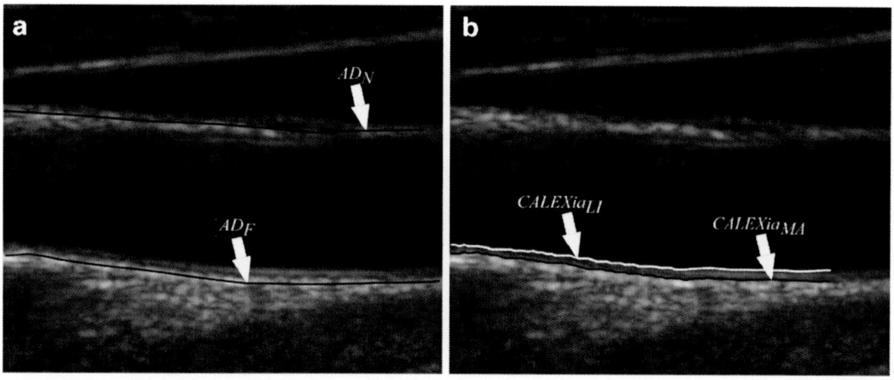

Fig. 10.3 (**a**) Automated location of the common carotid artery made by CALEXia: the algorithms traces the profiles of the near (AD_N) and far (AD_F) adventitia layers. (**b**) Image segmentation performed by CALEXia: $CALEXia_{LI}$ represents the lumen–intima tracing; $CALEXia_{MA}$ represents the media–adventitia tracing

points, thus originating incorrect line segments. In some images, the classification procedure cannot reject such line segments, and the CCA identification becomes problematic.

In some images, the interface between lumen and intima is characterized by a very smooth transition. This is especially evident in images characterized by blood backscattering (i.e., abnormal ultrasound echoes scattered by blood particles that make the lumen appear as gray instead of black). In such condition, the fuzzy *K-means* classifier may incorrectly detect the transitions between lumen and intima. In some cases, in fact, our strategy of classifying pixels in three clusters is not realistic. Again, inaccurate LI and MA tracings preclude the possibility of a precise IMT measurement.

10.4 Architecture Based on Morphological Approach of Watershed Transform

We used a marker-based watershed transform to perform image segmentation. Markers are the initial flooding points so that each marker starts at the watershed basins. We adopted morphological operators (erosion and reconstruction) to individuate the foreground markers. Figure 10.4 represents the overall watershed segmentation strategy. The original image (Fig. 10.4a) was first eroded by using a 12-pixel disk-shaped structuring element (Fig. 10.4b). This step ensured attenuation of noisy and small portions of the image. The image was then reconstructed against the original one (Fig. 10.4c), and further threshold was applied using Otsu's method (Fig. 10.4d). The white areas resulting from the threshold process were used to perform the marker-based watershed segmentation. The result for the original image in Fig. 10.4a is shown in Fig. 10.4e, where each color represents a watershed basin.

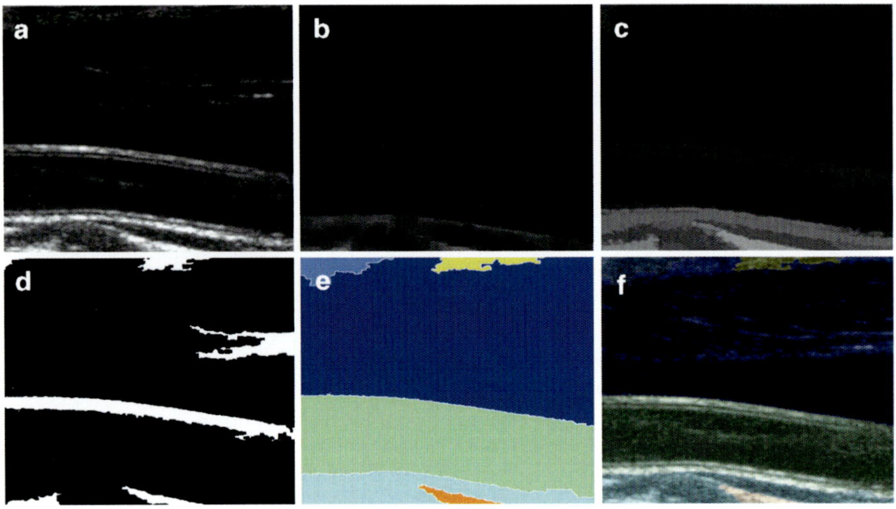

Fig. 10.4 Processing strategy of the WS technique. (**a**) Original B-Mode image. (**b**) Image eroded by using a 12-pixel disk-shaped structuring element. (**c**) Image reconstructed against original. (**d**) Segmentation made by Otsu's criterion. (**e**) Watershed segmentation of the original image. (**f**) Original image with basins overlapped and color-coded (© Springer 2010 - doi: 10.1007/s10916-010-9507-y - Reproduced from [24])

Figure 10.4f reports the basins overlaid to the original image. It can be noticed that the CCA is recognized in a unique watershed region that comprises the artery lumen and the walls.

From among the basins obtained by the watershed transform, we had to automatically select the one that comprised the CCA. We developed a methodology that specifically searches for the artery lumen, which is a dark and relatively homogeneous zone. A Canny edge detector was applied to each watershed-segmented region. Grayscale signal envelopes were extracted columnwise from between the detected edges in the original image. Along the extracted signal envelope, we determined whether points belonged to the lumen region by applying a predefined threshold value T_{lumen} (which was calculated by using Otsu's criterion). For each signal envelope, a column lumen region ratio ε was then calculated as the ratio of the number of classified lumen points to the length of the signal envelope. Only when ε was higher than a predefined threshold value ε_T, we saved the corresponding column for the following lumen region ratio calculation. We used a ε_T value equal to 60%. Finally, the lumen region ratio of each watershed-segmented region was simply calculated as $R^i_{\text{lumen}} = S_V / S_T$, where S_V is the number of valid signal envelopes and S_T is the total number of signal envelopes.

Conceptually, if a watershed region contained only the artery lumen, the value of R^i_{lumen} would be equal to 1; if it contained only tissue, its value would be ideally 0. Then, for each watershed region i, we calculated a geometric and an intensity-based feature: (1) H_i, the average region height (defined as the average distance between

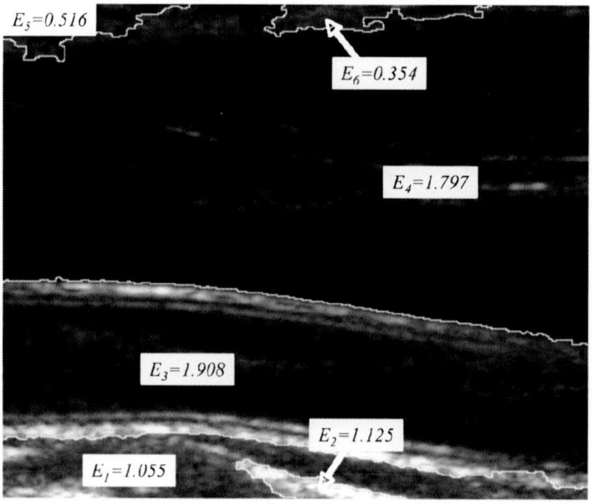

Fig. 10.5 Energy value of the detected basins. The basin with the higher energy is the one with the higher probability of containing the carotid artery structure (© Springer 2010 - doi: 10.1007/s10916-010-9507-y - Reproduced from [24])

the upper and bottom boundaries of that region), (2) I_i, the average region intensity. The three features were then combined to form the score E_i, defined as:

$$E_i = R^i_{\text{lumen}} \times \frac{255 - I_i}{255} \times \exp\left(\frac{H_i}{H}\right) \quad (10.1)$$

where H is the image height. The region containing the CCA was selected as the one with the greater E_i value. Figure 10.5 reports the E_i values for the image in Fig. 10.4. It can be observed that the region relative to the CCA is characterized by a greater score that the others ($E_5 = 1.908$).

We hypothesized that the watershed basin contained the entire CCA so that the upper boundary of the basin was in correspondence to the CCA near wall, and the bottom boundary was in correspondence to the CCA far wall. We traced an ROI starting from the bottom of the basin, which had the same width of the basin and had a height equal to 40 pixels. We chose this height after pilot tests. We found that 40 pixels is the optimal ROI height that comprises the artery lumen and the far wall while avoiding the near wall. Since we worked with images having an axial resolution of 62.5 μm, 50 pixels correspond to 3.125 mm. This value ensured an ROI that comprised the entire far wall and a portion of the artery lumen. The ROI was then considered columnwise. We processed the profile of each column by using a fuzzy *K-means* classifier, which was initialized by the intensity values. The classifier assigned the pixels to three clusters: (1) artery lumen (i.e., low intensity pixels), (2) intima and media layers (i.e., average intensity pixels), and (3) adventitia layer (i.e., high intensity pixels). The point at the transition between clusters (1) and (2) was considered as the marker of the LI interface, whereas the point at the transition

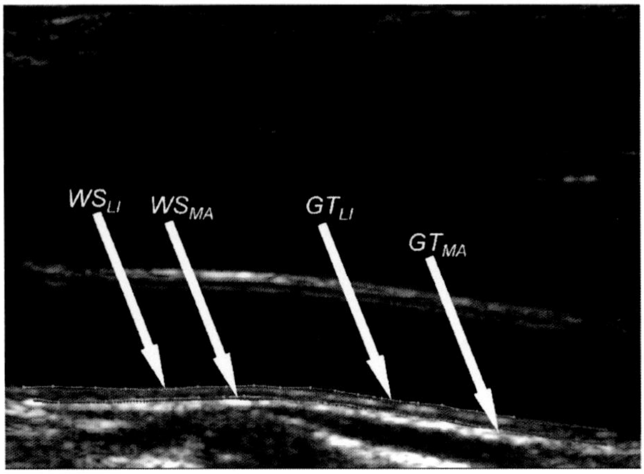

Fig. 10.6 Sample segmentation of WS technique compared to Ground Truth. WS_{LI} and WS_{MA} represent the lumen–intima and media–adventitia tracings respectively, whereas GT_{LI} and GT_{MA} represent the Ground Truth (© Springer 2010 - doi: 10.1007/s10916-010-9507-y - Reproduced from [24])

between the clusters (2) and (3) was the marker of the MA interface. All the columns in which the classifier could not find the three clusters were discarded. The sequence of the LI and MA markers constituted the final profiles. Figure 10.6 reports a sample of automatically traced LI and MA boundaries.

10.5 Inter-Greedy Approach for Fusion of Multiple Image Processing Boundaries

The Greedy Algorithm (*GA*) is an iterative technique that searches the global optimum of a problem by a series of local optimizations. Greedy algorithms were extensively used in segmentation procedures for error correction and performance optimization [19, 20]. The basic idea is to fuse two given boundaries in such a way that the fused boundary is closer to the ideal boundary (or *GT*) compared to the two given boundaries.

The need for GA in our IMT measurement framework was felt when we observed that different techniques had complementary features. We adopted a Greedy implementation, similar to that of Suri et al. [19], that is very well described using the "ball–basket" method. With reference to Fig. 10.7, we explain the greedy technique we used. Let us consider the LI interface. We start with the two boundaries, namely, B_1 and B_2 that represent the LI automated tracings of two techniques. The ideal boundary is represented by GT_{LI}. Let us suppose that B_1 has P_1 number of points and B_2 has P_2 number of points. This starting condition is reported by the upper panel of Fig. 10.7.

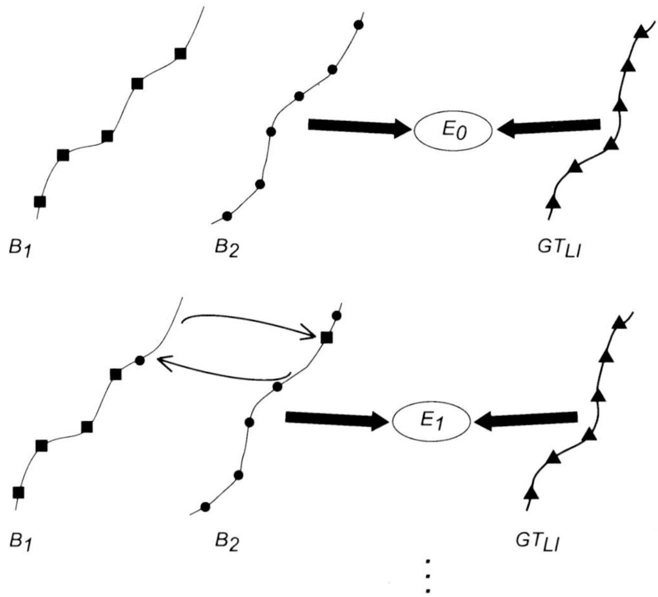

Fig. 10.7 Schematic representation of the greedy algorithm used to calibrate the segmentation boundaries. *Upper panel*: Considering an image in the database, we start with the B_1 (*black squares*) and the B_2 (*black circles*) lumen–intima boundaries. The initial error computed against ground truth (*triangles* – GT) is E_0. *Lower panel*: The first vertex of B_2 and B_1 is swapped. The new greedy boundary produces the error E_1. Only if the errors are lower than the initial error the B_2 points are swapped with B_1, otherwise they are discarded (© Springer 2010 - doi: 10.1007/s10916-010-9507-y - Reproduced from [24])

Figure 10.7 represents the first iteration of the greedy approach substitution cycle. We swap the first vertex of B_2 with the closest one of B_1 and compute the system error (i.e., the distance of this new boundary with respect to the ideal boundary, GT). This error is indicated by E_1 in Fig. 10.7 (lower panel). We iterate the cycle in Fig. 10.7 for every point of the B_2 boundary, obtaining P_2 values of system error. If the minimum of the P_2 values of system error is lower than the initial system error between B_1 and GT_{LI} (indicated by E_0 in Fig.10.7), the corresponding vertex of B_2 is substituted to the closest vertex of B_1. At this point, the procedure iterates until there are no more vertex swaps that decrease the system error or until all the P_2 vertices of B_2 have been inserted into B_1. The final boundary is the new greedy GA_{LI}. The GA_{LI} final boundary is interpolated by a bicubic spline, to smooth out any minor oscillations in the boundary profile. The same structure was used for the MA profile.

We applied this algorithm twice, since we merged the three boundaries coming from CULEXsa, CALEXia, and WS iteratively, obtaining Inter-Greedy boundaries. Having three starting techniques, we tried the following combinations:

1. First, we combined CULEXsa and CALEXia and obtained the greedy boundary GA_1, and then we combined GA_1 with WS to obtain the final Inter-Greedy boundary called IG_1.

2. First, we combined CULEXsa and WS and obtained the greedy boundary GA_2, and then we combined GA_2 with CALEXia to obtain the final Inter-Greedy boundary called IG_2.
3. First, we combined WS and CALEXia and obtained the greedy boundary GA_3, and then we combined GA_3 with CULEXsa to obtain the final Inter-Greedy boundary called IG_3.

We obtained the Inter-Greedy boundaries for all the 200 images of the database, thus obtaining 200 greedy LI and 200 greedy MA boundaries. We then evaluated the segmentation performances for all the methods: CULEXsa, CALEXia, WS, and IG_1, IG_2, and IG_3.

10.6 Performance Evaluation Metrics and Image Dataset

We developed a graphical user interface for three expert operators (a physician, a technician, and a neurologist), which was then used independently for segmenting all the 200 images of the testing database [16]. We considered GT as the average profile of the human tracings.

We defined the mean system error corresponding to CULEXsa, CALEXia, WS, and the three IG boundaries as the mean value of the error between the corresponding boundaries and the GT. These three measures were defined as the polyline distance between the computer-traced boundaries and the GT averaged on the 200 images of the database. Hence, we calculated six mean system errors, one for each boundary.

The polyline distance metric is a robust metric to define the distance between two boundaries. The basic idea is to measure the distance of each vertex of a boundary to the segments of the other boundary. Briefly, consider two boundaries B_1 and B_2. Let us consider the vertex $v = (x_0, y_0)$ on the boundary B_1 and the segment s formed by the end points $v_1 = (x_1, y_1)$ and $v_2 = (x_2, y_2)$ of B_2. The polyline distance between the vertex v and a polyline segment s can be defined as the perpendicular distance $d_\perp$ if such distance falls on the segment s, otherwise as the minimum between the Euclidean distances d_1 and d_2.

The polyline distance from vertex v to the boundary B_2 can be defined as the minimum distance between v and the segments of B_2. The distance between the vertexes of B_1 to the segments of B_2 is then defined as the sum of the distances from the vertexes of B_1 to the closest segment of B_2. Let us call this distance as $d(B_1, B_2)$. Similarly, it is possible to calculate the distance between the vertices of B_2 to the closest segment of B_1 (let us call this distance as $d(B_2, B_1)$). The polyline distance between boundaries is defined as:

$$D(B_1, B_2) = \frac{d(B_1, B_2) + d(B_2, B_1)}{(\text{Number of vertices of } B_1 + \#\text{ of vertices of } B_2)} \tag{10.2}$$

This measure $D(B_1, B_2)$ is very important since it reflects the average distance between two computer-generated boundaries and the corresponding ground truth profiles. Full details about polyline implementation and use for IMT measurement can be found in the works by Suri et al. [19] and Molinari et al. [21].

The IMT value was calculated as the polyline distance between the LI and the MA boundary, for each of the six techniques. This was then compared to the IMT for the GT. Specifically, the IMT value that we calculated by using one of the three techniques was compared to the corresponding IMT value calculated by GT. The difference was the IMT measurement error. The mean IMT measurement error was then computed by averaging all the IMT measurement errors on the 200 images of the database.

B-Mode images of the common tract of the carotid artery were acquired using three different scanners:

- ATL-HDI5000 (Philips Medical Systems, The Netherlands), equipped with a 50-mm linear probe (code L12-5), which worked in the frequency range of 5–12 MHz.
- LOGIQ7 (GE Medical Systems Ultrasound, UK), equipped with a 44-mm linear probe (code 9L), which worked in the frequency range of 2.5–8 MHz.
- MyLab70 [EsaoteItalia (Biosound), Genova, Italy], equipped with a 50-mm linear probe (code LA523), which worked in the frequency range of 4–13 MHz.

The use of multiple scanners was justified by the need of showing performance in a real clinical scenario, where different operators using different machines and settings may have acquired images. We resample all the images at a standard density of 16 pixels/mm [14, 15]. Axial resolution on the digitized image was thus equal to 0.0625 mm. All the images were coded to 256 grayscale level (8 bits). Figure 10.1 reports an example of B-Mode longitudinal CCA image with the black arrows indicating the near and far wall (right side of the image) and the LI and MA interfaces for the far wall (human tracings).

Our testing database consisted of 200 images acquired from 150 consecutive patients (age: 50–83 years; mean ± standard deviation = 69 ± 16 years) of the Neurology Division of the Gradenigo Hospital, Torino, Italy. Ninety subjects were males. All the subjects were symptomatic and were referred to the Neurology Division either for neurological or cardiovascular disorders. All the subjects were instructed about the study and signed and informed consent prior to being submitted to the ultrasound examination. The Institutional Committee of the Gradenigo Hospital approved this study.

10.7 Segmentation Performance

In the following section, we illustrate the segmentation performance of CALEXia, CULEXsa, WS and further compare their performances to Inter-Greedy technique. Out of 200 images, CALEXia could not correctly identify the CCA in ten images

Table 10.1 Mean absolute errors and standard deviations calculated on the image database for the three segmentation techniques (CALEXia, CULEXsa, WS) and the three Inter-Greedy boundaries (© Springer 2010-doi: 10.1007/s10916-010-9507-y - Reproduced from [24])

Techniques	Dimension	Lumen–intima (LI)	Media–adventitia (MA)	Intima–media thickness (IMT)
CALEXia	pixel	3.35 ± 13.26	2.97 ± 14.60	3.04 ± 2.30
	mm	0.209 ± 0.829	0.186 ± 0.913	0.190 ± 0.144
CULEXsa	pixel	0.48 ± 1.03	0.31 ± 0.58	1.31 ± 1.09
	mm	0.030 ± 0.064	0.019 ± 0.036	0.082 ± 0.068
WS	pixel	9.18 ± 28.06	9.46 ± 27.49	3.18 ± 3.23
	mm	0.573 ± 1.753	0.591 ± 1.718	0.199 ± 0.202
IG_1	pixel	0.32 ± 0.44	0.21 ± 0.34	0.74 ± 0.75
	mm	0.020 ± 0.028	0.013 ± 0.021	0.046 ± 0.047
IG_2	pixel	0.35 ± 0.55	0.26 ± 0.41	0.88 ± 0.87
	mm	0.022 ± 0.034	0.016 ± 0.026	0.055 ± 0.054
IG_3	pixel	0.41 ± 0.53	0.30 ± 0.48	0.96 ± 0.89
	mm	0.026 ± 0.033	0.019 ± 0.030	0.060 ± 0.056

The number of images is equal to 176. The third column reports the mean absolute error for the LI interface, the fourth for MA interface. The fifth column reports the IMT calculation error. The values are expressed both in pixel and mm (the axial resolution is 0.0625 mm/pixel)

(5% of the images), CULEXsa could not process 12 images out of 200 (6% of the images), and WS could not process two images out of 200 (1% of the images). Therefore, we removed 24 images from the database and tested our segmentation algorithms on $N = 176$ images.

Table 10.1 summarizes the system errors obtained by the three segmentation techniques (CULEXsa, CALEXia, and WS) along with the three Inter-Greedy boundaries (IG_1, IG_2, and IG_3, defined as per Sect. 10.5). The techniques are reported in the rows, while the column shows the morphology description: LI, MA, and IMT. The last three rows of Table 10.1 report the performance of the three Inter-Greedy combinations. Clearly, IG_1 was the combination offering best performance. It can be noticed that all the three Inter-Greedy combinations reduced the segmentation error and the IMT estimation bias when compared to the best performing segmentation technique (i.e., CULEXsa, reported by the second row of Table 10.1). Note that IG_1 offered the best performance, even there was no statistically significant difference among the performance of the three Inter-Greedy combinations (Student's *t*-test, $p > 0.05$).

Therefore, we discuss the performance of the three segmentation techniques in comparison with those of the best performing Inter-Greedy combination, i.e., IG_1. In the following sections, the term Inter-Greedy specifically refers to the combination IG_1.

10.7.1 Performance Evaluation of Segmentation Techniques for Lumen–Intima

The LI mean system errors are reported by the third column of Table 10.1. Using PDM, CULEXsa showed a mean error equal to 0.48 ± 1.03 pixel (30.0 ± 64.4 μm) and was considered as the starting greedy boundary for all the images. After fusion of CULEXsa and CALEXia, we obtain the Greedy Boundary – GA_l, followed by the fusion with WS boundary to obtain final Inter-Greedy boundary called IG_l. The resulting IG_l boundary showed a reduced mean error of 0.32 ± 0.44 pixel (20.0 ± 27.5 μm). The segmentation performance of IG_l improved by 33.3 ± 5.6% with respect to CULEXsa.

Figure 10.8 shows an example of segmentation improvement obtained by Inter-Greedy algorithm. The upper row of Fig. 10.8 (left panel) reports the original segmentation of the LI interface performed by CULEXsa (dashed white line – $CULEXsa_{LI}$). We took CULEXsa as the starting point because it represented the least PDM system error. The continuous white line represents ground truth (GT_{LI}). The automated segmentation is very accurate even though not fully perfect. The right panel (upper row) reports the zoomed portion of the rectangular window of

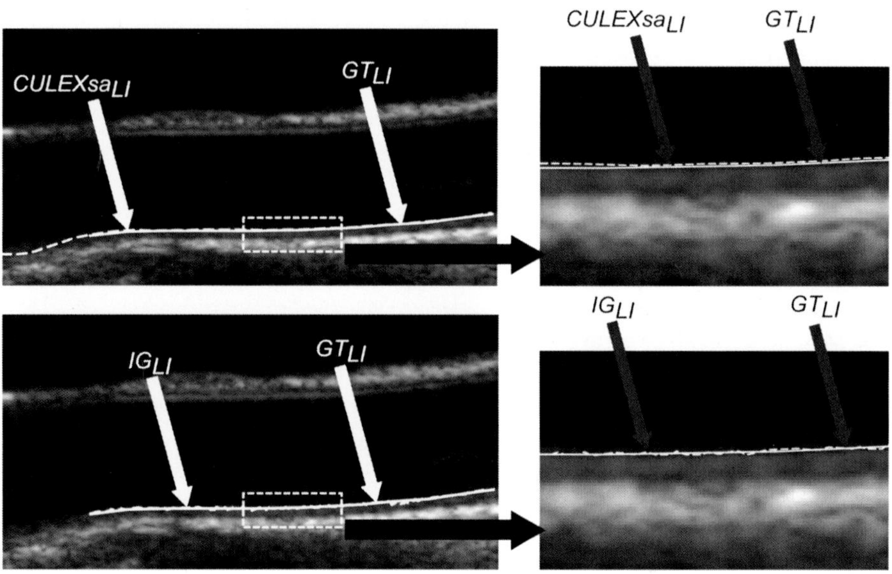

Fig. 10.8 Example of Inter-Greedy calibration of lumen–intima boundary. *Upper row*: Original image with ground truth (*white continuous line* – GT_{LI}) and CULEXsa boundary (*dashed white line* – $CULEXsa_{LI}$) overlapped (*left panel*). The *white rectangle* indicates a nonperfect CULEXsa segmentation. The CULEXsa boundary deviates from GT and segmentation is not accurate (*right panel*). *Bottom row*: Original image with ground truth (*white continuous line* – GT_{LI}) and Inter-Greedy boundary (*dashed white line* – IG_{LI}) overlapped (*left panel*). Zoomed view of the area in the *rectangle*: the greedy boundary is almost overlapped to GT (*right panel*) (© Springer 2010 - doi: 10.1007/s10916-010-9507-y - Reproduced from [24])

the left panel. This windows shows a portion of the image in which the automated tracing $CULEXsa_{LI}$ substantially derives from the ideal boundary. It can be noticed that CULEXsa tracing is neatly inside lumen region with respect to the Ground Truth. The bottom row (left panel) shows the final Inter-Greedy boundary (IG_{LI}) overlapped on the ground truth (GT_{LI}). The right panel of the bottom row, which is the same zoomed portion of the corresponding panel of the upper row, shows that IG_{LI} is now almost bumper-to-bumper with respect to GT_{LI}. For this specific image, the LI segmentation error decreased by 11% with respect to CULEXsa.

CULEXsa is superior to both CALEXia and WS for segmenting LI. However, CULEXsa segmentation has snakes methodology embedded in it. Such deformable models are limited by the fact that it is very difficult to precisely tune elasticity and rigidity parameters in each image. We used the parameters based on the previous works [5, 13], which suited our dataset. These values were similar to those found by other researchers (even though they used slightly different snake implementations) [9]. However, performance may vary from image to image, due to speckle noise and different carotid morphologies.

The Inter-Greedy approach may enable the correction of the snake boundary (CULEXsa) by integrating with vertices taken from CALEXia and WS, which in turn are based on totally different segmentation architectures. CALEXia and WS adopt a fuzzy *K-means* classifier to detect the LI interface. This strategy is globally less robust to noise but has the advantage of being very accurate where noise is low and contrast is high. Section 10.3 reports a thorough discussion about the characteristics of the CALEXia and WS vertices that were placed into the CULEXsa boundary vertices to produce the Inter-Greedy boundary.

10.7.2 Performance Evaluation of Segmentation Techniques for Media–Adventitia

The MA mean system errors are reported by the fourth column of Table 10.1. CULEXsa was considered as the starting greedy boundary as its overall system error measured by PDM was equal to 0.31 ± 0.58 pixel (with respect to Ground Truth). CULEXsa was then combined, first with CALEXia (which showed a PDM distance from GT equal to 2.97 ± 14.60 pixel) and second with WS (the distance of WS tracings from GT was 9.46 ± 27.49 pixel). The greedy boundary IG_{I} showed a mean segmentation error equal to 0.21 ± 0.34 pixel (13.1 ± 21.3 μm), which is significantly lower than that of the other techniques. The performance improvement obtained by greedy boundary compared to CALEXia was equal to 32.3 ± 6.7%.

Figure 10.9 depicts a comparison between CULEXsa segmentation of the MA interface and the improvement obtained by using the Inter-Greedy approach. The upper row compares $CULEXsa_{MA}$ to GT_{MA}, whereas the bottom row compares IG_{MA} and GT_{MA}. The left panel (upper row) of Fig. 10.9 reports the original segmentation of the MA interface performed by CULEXsa (dashed black line). The

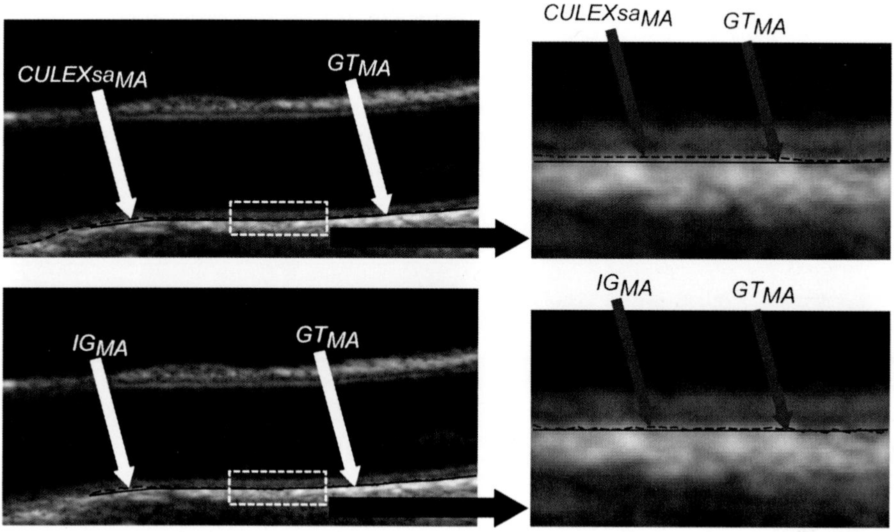

Fig. 10.9 Example of Inter-Greedy calibration of media–adventitia boundary. *Upper row*: Original image with ground truth (*black continuous line* – GT_{MA}) and CULEXsa boundary (*dashed black line* – $CULEXsa_{MA}$) overlapped (*left panel*). The *white rectangle* indicates a nonperfect CULEXsa segmentation. The CULEXsa boundary deviates from GT and segmentation is not accurate (*right panel*). *Bottom row*: Original image with ground truth (*black continuous line* – GT_{MA}) and Inter-Greedy boundary (*dashed white line* – IG_{MA}) overlapped (*left panel*). Zoomed view of the area in the *rectangle*: the greedy boundary is almost overlapped to GT (*right panel*) (© Springer 2010 - doi: 10.1007/s10916-010-9507-y - Reproduced from [24])

continuous black line represents ground truth (GT_{MA}). The right panel (upper row) reports the zoomed version of the small rectangular region of the left panel. This shows a portion of the image in which the automated tracing $CULEXsa_{MA}$ substantially derives from the ideal boundary due to an incorrect detection of the MA transition. The left panel of the bottom row shows the final Inter-Greedy boundary (IG_{MA}) overlapped with respect to ground truth (GT_{MA}). In the right panel (bottom row), which is the same zoomed portion of the corresponding right panel of the upper row, it can be observed that IG_{MA} now is almost overlapped to GT_{MA}. For this specific image, the MA segmentation error decreased by 10.7% with respect to CULEXsa.

Figures 10.10–10.12 report samples of Inter-Greedy performance. We presented nine ultrasound images selected from our testing dataset. In Fig. 10.10, the white dashed rectangle indicates the portion of the original image that is presented in Fig. 10.12 as zoomed view. Figure 10.11 reports the same images of Fig. 10.10 with Inter-Greedy (dashed lines) and Ground Truth (continuous lines) overlapped. Figure 10.12 reports same images of Fig. 10.10. The white dashed line represents the IG_{LI}, the white continuous line the GT_{LI}; the black dashed line the IG_{MA}, and the black continuous line the GT_{MA}. The Inter-Greedy tracings are almost overlapped to Ground Truth, demonstrating the performance of Inter-Greedy in boundaries calibration.

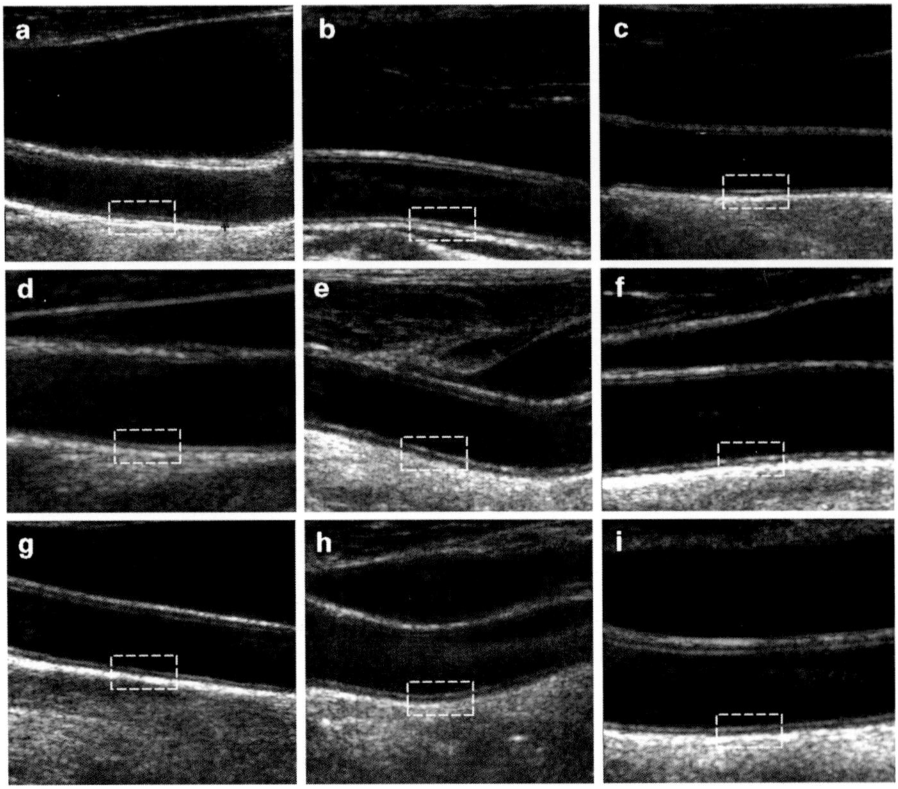

Fig. 10.10 Sample images selected from the testing dataset. The *white dashed rectangle* indicates the image portions that are shown in zoomed version in Fig. 10.12 (© Springer 2010 - doi: 10.1007/s10916-010-9507-y - Reproduced from [24])

10.7.3 Performance of Greedy with Respect to CALEXia, CULEXsa, and WS

Using the Inter-Greedy algorithm, we obtained LI and MA boundaries that were closer to ideal boundaries with respect to the original three series of tracings. The overall number of vertices that were substituted between CULEXsa and CALEXia in the first round of the Inter-Greedy algorithm was equal to 2270, whereas the number of points swapped between the intermediate greedy boundary (GA_1) and WS was equal to 1548. On MA boundaries, the number of vertices swapped between CULEXsa and CALEXia were equal to 2144, whereas the number of points swapped between the intermediate greedy boundary (GA_1) and WS were 1478. The total number of vertices that constituted profiles was 14,239 for LI and 14,312 for MA. Therefore, overall, the Inter-Greedy algorithm substituted about 26.8 and 25.3% of the vertices in the LI and MA boundaries, respectively.

We analyzed the characteristics of the swapped vertices. We adopted the signal-to-noise ratio (*SNR*) definition proposed by Masden et al. [18] and Suri et al. [19].

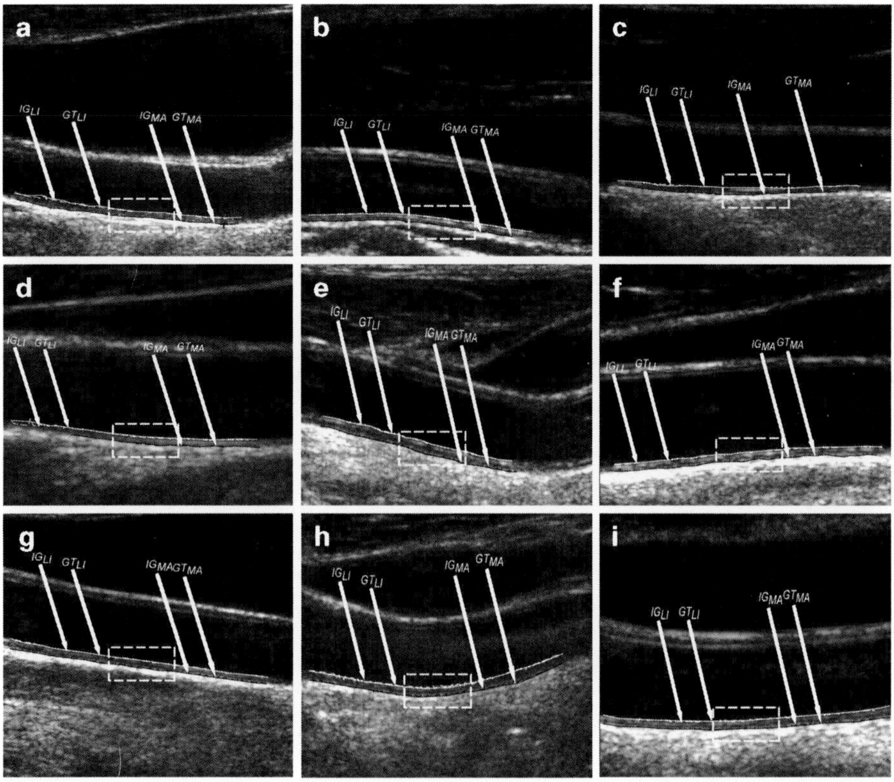

Fig. 10.11 Sample images selected from the testing dataset with computer generated Inter-Greedy tracings (*dashed lines*) and Ground Truth (*continuous lines*) overlapped. The *white dashed rectangle* indicates the image portions that are shown in zoomed version in Fig. 10.12 (© Springer 2010 - doi: 10.1007/s10916-010-9507-y - Reproduced from [24])

We considered all the vertices of the four boundaries (i.e., CULEXsa, CALEXia, WS, and IG_l). We estimated the SNR value in correspondence to each vertex. We calculated the mean intensity of the artery wall at the boundary vertex and subtracted the mean value of the noise computed in the artery lumen. The result was divided by the standard deviation of the artery lumen pixels for that vertex location. The rationale is that high SNR values can be obtained only in the presence of black and homogeneous lumen coupled to bright wall.

The average SNR on all the vertices was 3.3 ± 0.5 for LI and 7.9 ± 1.0 for MA. The average SNR value on the swapped vertices was 3.9 ± 0.8 for LI and 9.1 ± 1.2 for MA. Therefore, the average SNR in correspondence of the swapped vertices was about $18 \pm 3.7\%$ higher than the average SNR boundary (for LI) and $15 \pm 3.8\%$ (for MA).

This result confirmed our observations about the greedy characteristics. The best performing initial automated technique (CULEXsa) is very effective in coping with

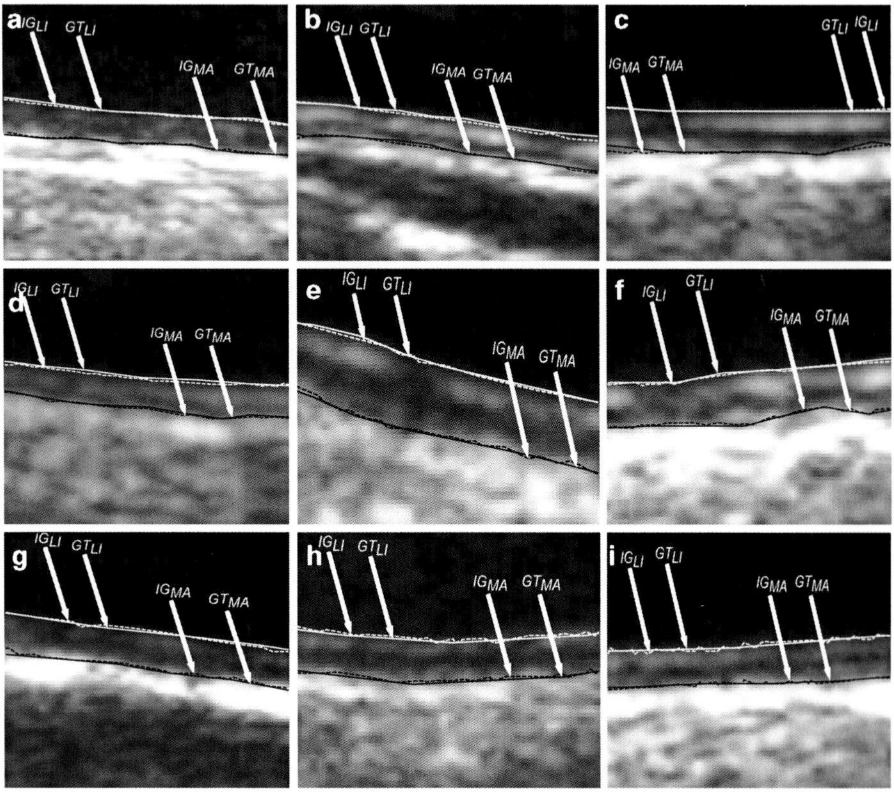

Fig. 10.12 Inter-Greedy boundaries shown in the *zoomed rectangular window* for the corresponding images shown in Fig. 10.10. *Solid lines* are computer estimated while *dashed lines* are Ground Truth. The *white dashed line* represents IG_{LI}, the *white continuous line* GT_{LI}, the *black dashed line* IG_{MA}, and the *black continuous line* GT_{MA} (© Springer 2010 - doi: 10.1007/s10916-010-9507-y - Reproduced from [24])

overall noise. In fact, rigidity and tension of the snake prevents final segmentation to derive significantly from the CCA wall layers. However, the lower the SNR, the higher must be the rigidity of the snake. A high rigidity values in snakes oversmooths the final LI and MA profiles, causing local suboptimal segmentations and, consequently, IMT estimation. CALEXia and WS, based on different architecture, cannot provide an initial global segmentation as accurate as CULEXsa, but their final *K-means* classifier allows for optimal local choice of the LI and MA interfaces when SNR is high. Hence, optimal Inter-Greedy boundary is obtained by preferring CALEXia and WS points when SNR is high, and CULEXsa points when SNR is low. As we have already discussed, we can conclude that the Inter-Greedy algorithm is very efficient in exploiting better segmentation solution to enhance overall performance.

We implemented this greedy algorithm in a Matlab framework (The MathWorks, Natick, MA) running on a dual 2.5-GHz PowerPC with 8 MB of RAM. The average computational time for getting the greedy boundary was equal to 117.2 ± 1.6 s per fusion in one image. Hence, an image was processed in less than two minutes (considering the CULEXsa, CALEXia, and WS profiles already available).

10.8 Error per Vertex for Different Segmentation Techniques

Figure 10.13 reports the system error per vertex (EPV) computed on the LI boundary for CALEXia, CULEXsa, WS, and IG *cases*. These are represented by circles, squares, diamonds, and triangles. Figure 10.13 confirms that the average greedy performance is superior to that of any single segmentation techniques, such as CALEXia, CULEXsa, and WS.

Figure 10.14 reports the system error per vertex computed on the MA boundary for CALEXia, CULEXsa, WS, and IG. The circles, squares, diamonds, and triangles represent EPV for CALEXia, CULEXsa, WS, and IG algorithms, respectively. The greedy boundary has the best performance when compared to the three initial boundaries in the entire MA tracing.

These results confirm the observation that the greedy algorithm essentially substitutes the boundary vertices on the basis of the segmentation performance of the techniques (i.e., on the basis of the local SNR). In fact, the system error per vertex

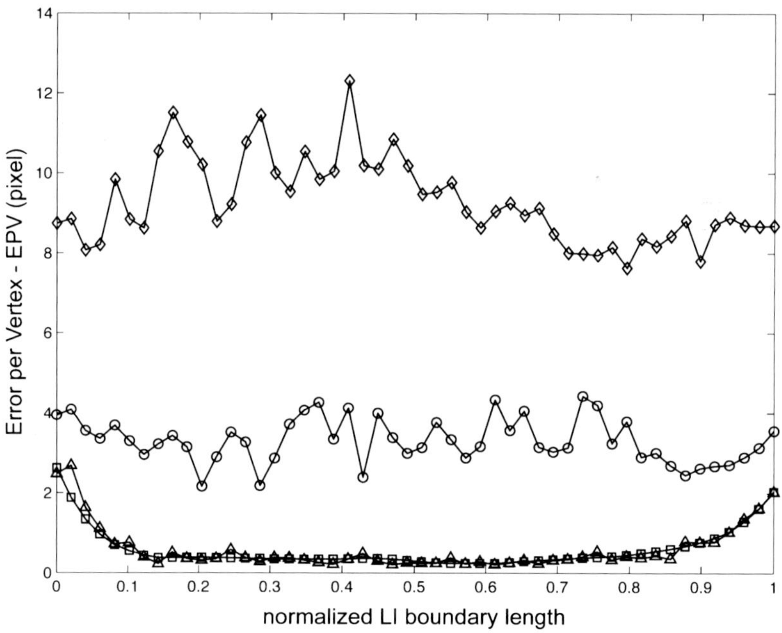

Fig. 10.13 Representation of CALEXia, CULEXsa, WS, and Inter-*Greedy* mean system error per vertex. All the lumen–intima boundaries were normalized between 0 and 1. The *graph* shows that *Inter-Greedy* boundary (*triangles*) offers the better performance than CALEXia (*circles*), CULEXsa (*squares*), and WS (*diamonds*) (© Springer 2010 - doi: 10.1007/s10916-010-9507-y - Reproduced from [24])

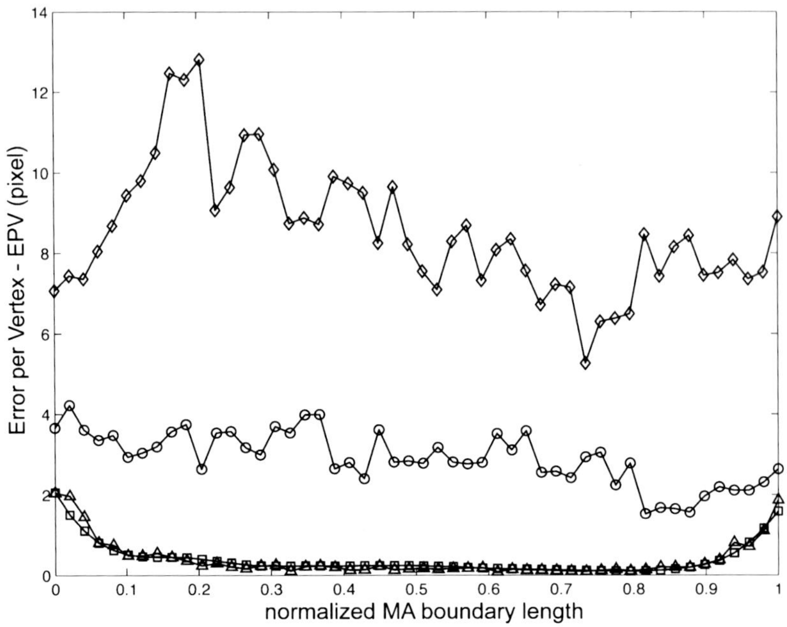

Fig. 10.14 Representation of CALEXia, CULEXsa, WS, and Inter-*Greedy* mean system error per vertex. All the media–adventitia boundaries were normalized between 0 and 1. The *graph* shows that *Inter-Greedy* boundary (*triangles*) offers the better performance than CALEXia (*circles*), CULEXsa (*squares*), and WS (*diamonds*) (© Springer 2010 - doi: 10.1007/s10916-010-9507-y - Reproduced from [24])

is almost flat if we exclude the border effect observable at the beginning (from 0.0 to 0.1) and end (from 0.9 to 1) of the normalized profiles. This means that, on an average, it is not possible to find a relation between the substituted vertex and its geometrical position along the boundary.

10.9 Intima–Media Thickness Measurement Performance of CALEXia, CULEXsa, WS, and Inter-Greedy Algorithms

The last column of Table 10.1 reports the mean IMT measurement errors for the CALEXia (third column), CULEXsa (fourth column), WS (fifth column), and IG (sixth column) boundaries. The Inter-Greedy approach led to a reduction in the IMT measurement error as big as $43.5 \pm 2.4\%$. The mean IMT measurement error was lower than 1 pixel. This is very important as subpixel errors ensure a highly reliable IMT measurement.

The Inter-Greedy mean IMT measurement error (46.3 μm) is the best performance obtained by completely automated techniques. This performance is inferior only to some very specifically tuned and semiautomated techniques. Faita et al., in fact, developed the best performing IMT measurement technique, in 2008. They declared an IMT measurement error of about 10 μm. However, their methodology required user interaction. Also, there was no discussion towards the effect of noise on the algorithm performance.

There are some considerations that must be made. First, to the best of our knowledge, CULEXsa, CALEXia, and WS are the only completely user-independent techniques that have so far proven robust in CCA segmentation and IMT measurement. Most of the techniques, in fact, require human interaction to locate the carotid artery in the image. This precludes real automation. Our results were obtained on a database of 200 randomly selected images acquired by multiple scanners. Therefore, our techniques proved effective in presence of noise, vessel pathology, and different carotid morphologic appearances.

Second, we measured the IMT values and the IMT measurement errors by relying on the polyline distance between boundaries. Usually, the IMT performance is estimated by the standard distance metrics (Euclidean distance, minimum distance between vertices of different boundaries, or Hausdorff distance). The standard distance metrics, however, are sensible to the number of points of the boundaries and to their geometrical location. Therefore, a direct comparison of our results with previously published works is not straightforward.

This consideration leads to the third and principal difference in our methodologies compared to others: our IMT results are biased because they ensemble IMT values estimated in high SNR points and IMT values estimated in low SNR points. When human operators start the segmentation by manually tracing the region of interest in which to perform IMT calculation, they are implicitly selecting the image region with higher SNR. With our automated procedure, this discrimination is difficult. The added value of using the Inter-Greedy approach relies in the fact that low-SNR IMT estimates can be replaced by IMT estimates coming from vertices with higher SNR (i.e., where segmentation and measurement performances are better), thus improving IMT measurement.

From a clinical point of view, IMT is considered normal if lower than 0.9 mm (900 μm). Therefore, measurement errors of about 46 μm (about 5% of normal value) must be considered as suitable for clinical use. The interoperator variability in IMT measurement can be as high as 25 μm. Hence, we believe that the Inter-Greedy approach could be very useful for aiding and supporting reliable and accurate IMT measurement.

10.10 Conclusions

We have presented the Inter-Greedy technique applied to the common carotid artery segmentation and IMT measurement. We started with three automatically generated boundaries and obtained an Inter-Greedy boundary by iterative combination of the initial three boundaries. The Inter-Greedy technique significantly improved the segmentation performances by 33% on the lumen–intima and 32% on the media–adventitia interface. The IMT measurement error was about 44% lower compared to CULEXsa with mean error of about 46 μm.

The Inter-Greedy method improved the segmentation and measurement performance by optimizing the carotid wall interface detection in spite of varying levels of SNR. When SNR is high, fuzzy-based segmentation techniques offer better performance, while snake maintains smoothness of the traced profiles when SNR is low and, therefore, reduces sensitivity to noise.

References

1. P. J. Touboul, M. G. Hennerici, S. Meairs *et al.*, "Mannheim carotid intima-media thickness consensus (2004-2006). An update on behalf of the Advisory Board of the 3rd and 4th Watching the Risk Symposium, 13th and 15th European Stroke Conferences, Mannheim, Germany, 2004, and Brussels, Belgium, 2006," *Cerebrovasc Dis*, vol. 23, no. 1, pp. 75–80, 2007.
2. P. J. Touboul, M. G. Hennerici, S. Meairs *et al.*, "Mannheim intima-media thickness consensus," *Cerebrovasc Dis*, vol. 18, no. 4, pp. 346–9, 2004.
3. I. M. van der Meer, M. L. Bots, A. Hofman *et al.*, "Predictive value of noninvasive measures of atherosclerosis for incident myocardial infarction: the Rotterdam Study," *Circulation*, vol. 109, no. 9, pp. 1089–94, Mar 9, 2004.
4. P. Pignoli and T. Longo, "Evaluation of atherosclerosis with B-mode ultrasound imaging," *J Nucl Med Allied Sci*, vol. 32, no. 3, pp. 166–73, Jul–Sep, 1988.
5. S. Delsanto, F. Molinari, P. Giustetto *et al.*, "CULEX-completely user-independent layers extraction: ultrasonic carotid artery images segmentation," *Conf Proc IEEE Eng Med Biol Soc*, vol. 6, pp. 6468–71, 2005.
6. S. Delsanto, F. Molinari, P. Giustetto *et al.*, "Characterization of a completely user-independent algorithm for carotid artery segmentation in 2-D ultrasound images," *IEEE Trans Instrum Meas*, vol. 56, no. 4, pp. 1265–74, 2007.
7. S. Delsanto, F. Molinari, W. Liboni *et al.*, "User-independent plaque characterization and accurate IMT measurement of carotid artery wall using ultrasound," *Conf Proc IEEE Eng Med Biol Soc*, vol. 1, pp. 2404–7, 2006.
8. F. Molinari, G. Zeng, and J. S. Suri, "An integrated approach to computer-based automated tracing and its validation for 200 common carotid arterial wall ultrasound images: a new technique," *J Ultras Med*, vol. 29, pp. 399–418, 2010.
9. F. Molinari, G. Zeng, and J. S. Suri, "Intima-media thickness: setting a standard for completely automated method for ultrasound," *IEEE Trans Ultrason Ferroelectr Freq Control*, vol. 57, no. 5, pp. 1112–24, 2010.
10. F. Faita, V. Gemignani, E. Bianchini *et al.*, "Real-time measurement system for evaluation of the carotid intima-media thickness with a robust edge operator," *J Ultrasound Med*, vol. 27, no. 9, pp. 1353–61, Sep, 2008.
11. J. H. Stein, C. E. Korcarz, M. E. Mays *et al.*, "A semiautomated ultrasound border detection program that facilitates clinical measurement of ultrasound carotid intima-media thickness," *J Am Soc Echocardiogr*, vol. 18, no. 3, pp. 244–51, Mar, 2005.
12. Q. Liang, I. Wendelhag, J. Wikstrand *et al.*, "A multiscale dynamic programming procedure for boundary detection in ultrasonic artery images," *IEEE Trans Med Imaging*, vol. 19, no. 2, pp. 127–42, Feb, 2000.
13. F. Molinari, G. Zeng, and J. S. Suri, "Carotid wall segmentation and imt measurement in longitudinal ultrasound images using morphological approach," in International Symposium on Biomedical Imaging, Rotterdam, The Netherlands, 2010.

14. E. Kyriacou, M. S. Pattichis, C. I. Christodoulou *et al.*, "Ultrasound imaging in the analysis of carotid plaque morphology for the assessment of stroke," *Plaque imaging: pixel to molecular level*, J. S. Suri, C. Yuan, D. L. Wilson *et al.*, eds., pp. 241–75, Amsterdam: IOS Press, 2005.
15. C. P. Loizou, C. S. Pattichis, M. Pantziaris *et al.*, "Snakes based segmentation of the common carotid artery intima media," *Med Biol Eng Comput*, vol. 45, no. 1, pp. 35–49, Jan, 2007.
16. F. Molinari, S. Delsanto, P. Giustetto *et al.*, "User-independent plaque segmentation and accurate intima-media thickness measurement of carotid artery wall using ultrasound," *Advances in diagnostic and therapeutic ultrasound imaging*, J. S. Suri, C. Kathuria, R. F. Chang *et al.*, eds., pp. 111–40, Norwood, MA: Artech House, 2008.
17. D. J. Williams and M. Shah, "A fast algorithm for active contours and curvature estimation," *CVGIP Image Underst*, vol. 55, no. 1, pp. 14–26, Jan, 1992.
18. E. Masden, J. Zagzebski, J. Frank, and J. Rownd, Automated system and method for testing resolution of ultrasound scanners. 1996: US Patent 5, 574, 212
19. J. S. Suri, R. M. Haralick, and F. H. Sheehan, "Greedy algorithm for error correction in automatically produced boundaries from low contrast ventriculograms," *Pattern Anal Appl*, vol. 3, no. 1, pp. 39–60, 2000.
20. J. J. Verbeek, N. Vlassis, and B. Krose, "Efficient greedy learning of gaussian mixture models," *Neural Comput*, vol. 15, no. 2, pp. 469–85, Feb, 2003.
21. F. Molinari, G. Zeng, and J. S. Suri, " Greedy technique and its validation for fusion of two segmentation paradigms leading to an accurate IMT measure in plaqued carotid arterial ultrasound," *J Vasc Ultrasound*, vol. 34, no. 2, pp. 63–73, 2010.
22. F. Molinari, G. Zeng, and J. Suri, Inter-Greedy Technique for Fusion of Different Segmentation Strategies Leading to High-Performance Carotid IMT Measurement in Ultrasound Images. *Journal of Medical Systems*: p. 1-15. (in press).

Biographies

Dr. Filippo Molinari received the Italian Laurea and the Ph.D. in electrical engineering from the Politecnico di Torino, Torino, Italy, in 1997 and 2000, respectively. He is leader in ultrasound imaging focused towards tissue characterization, vascular quantification for diagnostics and therapeutics. Currently, he is Assistant Professor at Politecnico di Torino, Italy – Department of Electronics.

Dr. Guang Zeng received the B.S. degree from Xiangtan University, China in 1998. He received the M.S. degree in 2005 and the Ph.D. degree in 2008 from Clemson University, SC, USA, both in Electrical Engineering. He is currently working in the Aging and Dementia Imaging Research Laboratory, Mayo Clinic, Rochester, MN. His research interests include biomedical image processing, pattern recognition and computer vision.

Dr. Jasjit S. Suri is an innovator, scientist, a visionary, an industrialist and an internationally known world leader in Biomedical Engineering. Dr. Suri has spent over 20 years in the field of biomedical engineering/devices and its management. He received his Doctrate from University of Washington, Seattle and Business Management Sciences from Weatherhead, Case Western Reserve University, Cleveland, Ohio. Dr. Suri was crowned with President's Gold medal in 1980 and the Fellow of American Institute of Medical and Biological Engineering for his outstanding contributions.

Chapter 11
Techniques and Challenges in Intima–Media Thickness Measurement for Carotid Ultrasound Images: A Review

Filippo Molinari, Guang Zeng, and Jasjit S. Suri

Abstract Last 10 years have witnessed the growth of many computer applications for the segmentation of the vessel wall in ultrasound imaging. Epidemiological studies showed that the thickness of the major arteries is an early and effective marker of onset of cardiovascular diseases. Ultrasound imaging, being fast, cheap, reliable, and safe, became a standard vascular assessment methodology.

This review is an attempt to discuss the most performing methodologies that have been developed so far to perform computer segmentation and intima–media thickness (IMT) measurement of the carotid arteries in ultrasound images. First we will give the rationale and the clinical relevance of computer measurements in clinical practice, then we will discuss the challenges that one has to face when approaching the segmentation of ultrasound vascular images.

The core of the paper is the presentation, discussion, benchmarking, and evaluation of different segmentation techniques, including edge detection, active contours, dynamic programming, local statistics, Hough transform, statistical modeling, integrated approach, and three-dimensional techniques. Also, we will discuss and compare the different performance metrics that have been proposed and used to perform validation. Best performing user-dependent techniques show an average IMT measurement of about 1 μm when compared to human tracings (Faita et al. J Ultras Med, 2008), whereas completely automated techniques show errors of about 10 μm.

The review ends with a discussion about the current standards in carotid wall segmentation and in an overview of the future perspectives, which may include the adoption of advanced and intelligent strategies to let the computer technique measure the IMT in the image portion where measurement is more reliable.

Keywords Ultrasound imaging • Carotid artery • Segmentation • Intima–media thickness • Atherosclerosis

---

F. Molinari (✉)
Biolab, Dipartimento di Elettronica, Politecnico di Torino, Corso Duca degli, Abruzzi, 24, Torino 10129, Italy
e-mail: filippo.molinari@polito.it

Jasjit S. Suri et al. (eds.), *Atherosclerosis Disease Management*,
DOI 10.1007/978-1-4419-7222-4_11,

11.1 Introduction

11.1.1 Rationale and Applications

According to recent data from the World Health Organization, cardiovascular diseases (CVDs) represent the third leading cause of death in Western countries [1]. CVDs include coronary artery disease, cerebrovascular disease, peripheral artery disease, heart failure, and hypertension. Beside a correct lifestyle, prevention is the key to reduce the impact of CVDs on the total number of global deaths. Therefore, the individuation of early markers of increased risk of CVDs is of paramount importance in clinical practice.

The earliest manifestation of the possible onset of a CVD is atherosclerosis. The atherosclerotic process refers to the degeneration of the arterial wall and the deposition of lipids and other blood-borne material within the arterial wall of almost all vascular territories [2–4]. Recent studies showed that the lipid deposit might be the consequence of an inflammatory process that takes place into the artery wall. The studies linking atherosclerosis and inflammation that were published in the last 10 years are more than 6,000 (see the following reviews among others: Antoniades et al. [5], Laufer et al. [6], Montecucco et al. [7], and Kovanen [8]).

The atherosclerotic process that takes place in the carotid arteries has been widely studied in last 20 years. Again, a thorough review of recent literature showed more than 22,000 contributions published in last 10 years. In the following, we selected only few relevant papers dealing with specific clinical investigations. Carotid wall lesions have been correlated to different pathologies such as coronary and cerebrovascular diseases [9], poststroke cognitive impairments [10], platelet aggregation [11], aortic valve damage [12], and diabetes [13]. Large multicentric and cohort studies have been devoted to the screening of carotid diseases as predictors of severe pathologies over population [14–17].

Once born from the aortic arc, the carotid arteries (CA) follow the neck axis, at a depth of about 2–4 cm. Figure 11.1 (left panel) depicts the anatomy of the carotid artery and its course along the neck; Fig. 11.1 (right panel) reports the supra-aortic arterial circulation, with the course of the common carotid artery and its bifurcation that originates from the internal and external carotid arteries. This anatomical positioning and the relatively big diameter make the ultrasound examination of the CAs very simple and effective. Clinically, the echographic and echoDoppler analysis of the CAs is the routine examination for first diagnosis and follow-up [18].

The most widely used indicator of cardiovascular and cerebrovascular risk is the carotid artery intima–media thickness (IMT) [19]. The IMT, as risk indicator, possesses many advantages [20]:

1. It is an early marker of atherosclerosis progression.
2. Its measure is highly repeatable.
3. It can be measured noninvasively.
4. It can be used to quantify pathology monitoring and/or drug therapy efficacy.

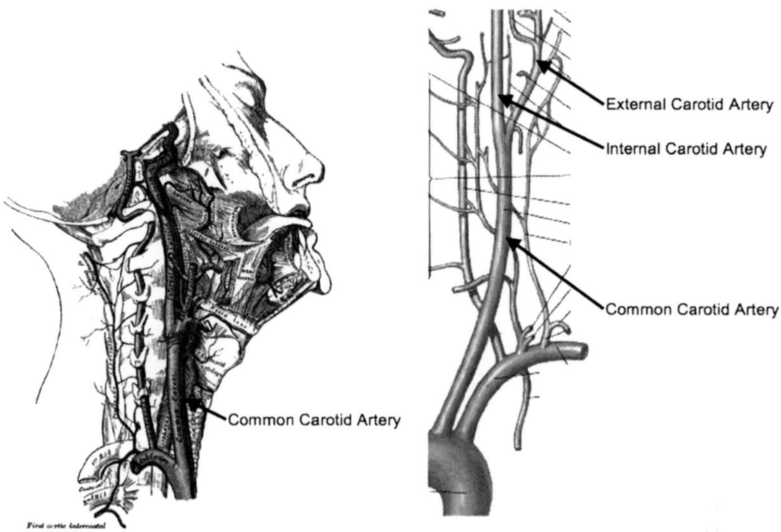

Fig. 11.1 *Left panel*: Anatomical sketch of the course of the common carotid artery along the neck. *Right panel*: overview of the supra-aortic circulation, with the aortic arc on the bottom and the course of the common, internal, and external carotid arteries (illustration by Henry Gray—"Anatomy of the human body", 1918. Copyright expired. Reproduced with permission from Bartleby.com. The original image can be found at the address: http://www.bartleby.com/107/146.html#i513)

Ultrasound examinations is perhaps the most used diagnostic tools to assess CVDs. Ultrasounds offers several advantages in clinical practice such as: (1) they are nonionizing radiations, (2) they allow a safe and relatively quick examination of the patient, (3) they have no dangerous biological effects, and (4) the ultrasound equipment is relatively low-cost if compared to other imaging devices. Unfortunately, ultrasounds are operator dependent. Moreover, the ultrasound images are usually quite noisy and require training to be correctly interpreted.

Ultrasound assessment of carotids, therefore, constituted the basis for many clinical studies.

11.1.2 Clinical Importance of Vessel Wall Segmentation

Segmentation of the CCA wall is the most important challenge in computer-aided clinical applications. Figure 11.2 sketches the structure of the CCA wall. The upper picture represents the transversal view (i.e., a plane that is orthogonal to the artery axis) of the CCA wall; the lower picture depicts the longitudinal view (i.e., a plane that is parallel to the artery axis). The artery wall consists of three layers: the innermost is called intima, the middle media, and the outermost adventitia. The intima

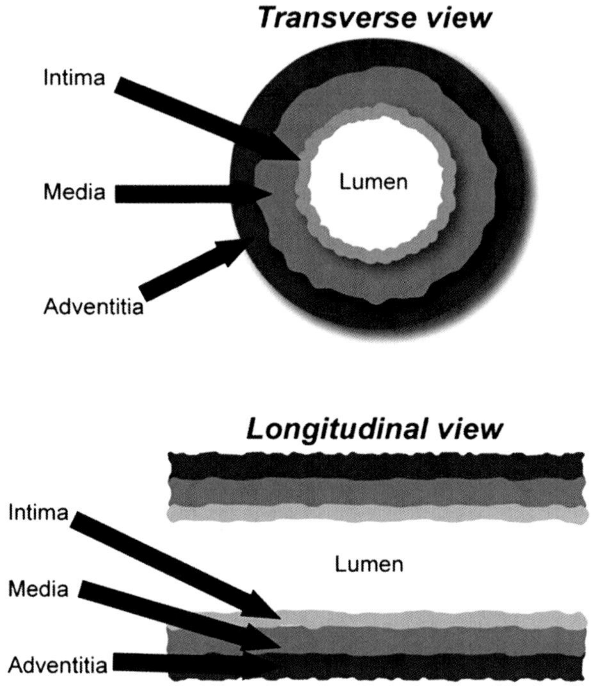

Fig. 11.2 *Upper panel*: Schematic representation of the transverse appearance of an artery. The structure is almost circular and the lumen of the vessel (*black* in ultrasound images) appears surrounded by a circular structure of different colors, each color representing a layer of the artery wall (from inside to outside: intima, media, and adventitia layer). *Bottom panel*: Schematic representation of an artery in longitudinal projection. The lumen of the vessel is surrounded by bright stripes originated by the artery wall layers. The structure is symmetrical, since the intima, media, and adventitia layer of the far wall is represented in inverted order on the near wall

layer is mainly constituted by a pavement endothelium, with polygonal, oval, or fusiform cells, followed by a subendothelial layer of connective tissue covered by a layer of elastic fibers, which represent the part of the wall that is in direct contact with blood. The middle coat (media) is mainly composed by muscular cells, with a transverse fibers orientation. The adventitia layer is composed by bundles of connective tissue and elastic fibers.

If we specifically refer to the CCA, segmentation of the artery wall means automatically tracing the profiles of the most important interfaces: the lumen–intima (LI) and the media–adventitia (MA) interface. When the clinical examination aims at measuring the IMT, the longitudinal view is preferred. This is due to the fact that the horizontal position of the artery layers is orthogonal to the propagation direction of the ultrasound beam; therefore, the ultrasound echoes are enhanced. In longitudinal projection, in fact, the structure of the carotid wall can be easily visualized. Conversely, in a transverse projection, the CCA appears as a dark circle where only the upper and bottom part of the wall is at focus. Usually, the lateral parts are poorly

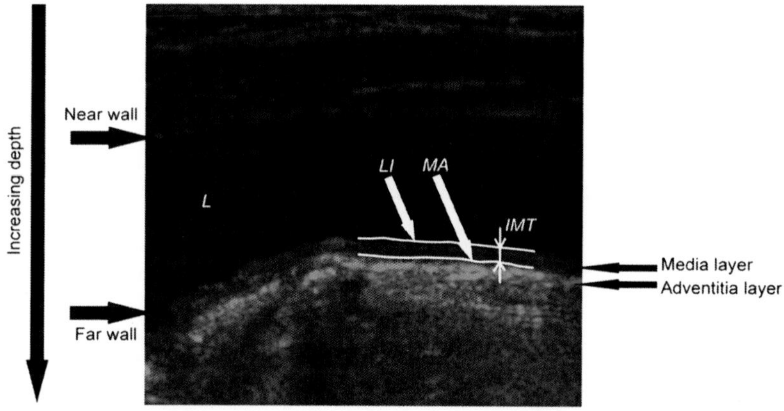

Fig. 11.3 Sample of echographic appearance of a common carotid artery in longitudinal projection. Depth increases from top to bottom. The *black stripe* corresponding to the artery lumen (L) is bounded upwards by the near wall and downwards by the far wall. In B-Mode longitudinal images the adventitia layer appears as bright and the media layer as a *dark grey*. Usually, the intima layer cannot be distinguished by the media. The IMT is calculated as the distance between the lumen–intima (LI) and the media–adventitia (MA) interfaces

represented due to longitudinal incidence with respect to the radiation. Hence, accurate measurement is more difficult.

Figure 11.3 depicts an ultrasound longitudinal B-Mode image of a healthy CCA. Depth is increasing in vertical from top to bottom. In the longitudinal view, the CCA is seen as a dark region (the lumen, indicated by L in Fig. 11.3) comprised between the near and far walls (black arrows on the left of the image). The intima layer is poorly represented in carotid images. The intima layer originates a thin and low-intensity echo, but its structure is then fused with the media layers because of poor difference in the acoustic impedance of the two adjacent layers [21]. The media layer is usually represented by a dark gray, whereas the adventitia layer is highly echogenic and it appears as a bright gray (black arrows on the right of the image). The white arrows point at the LI and MA interfaces. The IMT is the distance between the LI and the MA boundaries. In this figure, the LI and MA profiles were manually traced by an expert; IMT resulted in 15 pixels, corresponding to 0.937 mm.

Computer methods for segmenting the ultrasound CCA images mainly aim at tracing the LI and MA boundaries. The IMT, in fact, can be computed by measuring the distance of the LI from the MA boundary. Similarly, the media layer thickness can be estimated starting from the CCA walls segmentation [22]. By segmenting a cineloop of B-Mode images, it is also possible to calculate the media–media distance (i.e., the distance between the near and far media layers) and the carotid diameter [23]. Tortoli et al. [24] showed that it is possible to estimate the arterial stiffness by measuring the change in the artery diameter during the cardiac cycle. In 2007, Hermans et al. [25] showed that carotid IMT is increased in dialysis patients and that an important parameter to assess the overall status of the patient is the inhomogeneity of the IMT itself. Again, wall segmentation is the key to perform instrumental measurements of clinical use.

This review will focus on the techniques that have been developed to perform CCA wall segmentation, both in two-dimensional (2-D) and three-dimensional (3-D) images.

11.1.3 Relationship of Computer Measurements and CVD

The relationship between the artery IMT and the augmented risk of CVD disease is well known and documented. Several multicentric studies from all around the world (among others, see: Latin American Cities [26], Asia/Africa and Middle East [14], Europe [27], Japan [28], USA [29, 30], and China [31]) confirmed the importance of IMT monitoring as an early marker of CVD risk. Even though at different extent, the studies confirmed that the increase of the IMT value above 0.9–1 mm is indicative of a significant increase of CVD risk.

A high correlation between CVD and computer measurements can be observed in studies regarding plagued vessels [9, 32, 33]. IMT and atherosclerosis lesions were correlated to the telomere length in studies regarding aging [34].

Warwick et al. [35] showed that the degree of carotid artery stenosis, the length of the carotid artery plaque, the diameter of the carotid artery, and the blood hematocrit all independently significantly affect the required pump perfusion pressure to maintain adequate cerebral perfusion during cardiopulmonary bypass.

Plaque analysis is of fundamental importance to establish the need for surgical treatment. Presently, surgical intervention is established on the basis of the stenosis degree of a vessel. Even though stenosis is usually assessed by relying on volumetric data (i.e., CT or MRI) [36–38], the importance of the ultrasound assessment of the artery plaque is growing [39–41].

Computer methods for measuring CCA features, therefore, must be solid and robust. From a measurement point of view, this implies that accuracy, repeatability, robustness to noise, applicability to diverse pathologies, independence on human operator, independence on ultrasound OEM scanner, and measurement uncertainty characterization are all mandatory characteristics.

11.1.4 Monitoring of Carotid Wall Evolution

Ultrasound radiations exclusively propagate mechanical energy. This is a major advantage of the ultrasound methodology with respect to other radiological techniques for vascular assessment (e.g., CT and angiography). MRI does not make use of dangerous radiations, but it is quite expensive and requires, in some scanning sequences, the injection of a contrast agent that could be not well tolerated by patients [20]. Hence, ultrasounds are the best choice for patients' follow-up and for monitoring of disease progression/regression.

Many recent studies demonstrated the importance of a precise and proper follow-up of carotid IMT change when patients are subjected to drug therapy [42–44]. This enforces the need for robust computer approaches to CCA segmentation. Loizou et al. [22, 45] recently assessed the need for accurate media and intima thickness measurements in aging, showing that elderly people present altered media and intima ultrasound texture when compared to younger people. This study, though preliminary and based on a relatively small sample, reveals the potentialities of ultrasound examinations and computer analysis in patient follow-up.

Another major topic in ultrasound monitoring is the follow-up of plaques. Several studies were devoted to the computerized analysis of the plaque composition in ultrasound images [46–50]. Ultrasound-based "virtual histology" is now emerging as one of the election method to assess plaque composition [40]. Some studies performed ultrasound virtual histology by using intravascular ultrasound (IVUS) imaging [51–53], elastography [54, 55], or contrast-enhanced ultrasound imaging [56, 57]. In 2007, Vicenzini et al. [57] pointed out the strict relation between vasa-vasorum (i.e., the small vasculature that carries blood and nutrients to the major vessels wall) and plaque composition. Adventitial vasa-vasorum were scanned by using duplex ultrasound contrast-enhanced imaging. Results showed that fatty and fibrous plaques were rich of vasa-vasorum, whereas calcified, necrotic, and hemorrhagic plaques were not. Similarly, Molinari et al. [56] analyzed the postcontrast brightness of different plaques, with the purpose of characterizing plaque composition on the basis of the intensity level (i.e., of the contrast agent local concentration). These studies can be considered as preliminary, since a methodological validation against histology is still missing. Currently, validation is performed against the scoring given by expert sonographers or vascular surgeons.

Nevertheless, accurate plaque characterization and follow-up will constitute a key methodology in clinical practice. This goal requires ad-hoc ultrasound scanning protocols (usually with contrast agent injection) and specific processing algorithms.

11.2 Challenges in Carotid Wall Segmentation

It is well known that noise represents perhaps the most prominent problem in ultrasound imaging. Speckle noise was documented since 1979 [58]. It reduces the overall quality of the image by creating a "pixelated" effect that is detrimental both to human perception and to numerical processing algorithms.

Ultrasound images also have overall quality that is dependent on the scanner used and on its settings. In fact, being ultrasound imaging a human-dependent scanning technique, different operators may render the same object in different fashions.

Specifically referring to CCA images, another complication is given by the high variability in normal vessel morphology and in vessel appearance under pathology. In this section, we will discuss the major challenges encountered when developing an ultrasound CCA images segmentation algorithm.

11.2.1 Biological Variability in Normal and Pathology

Healthy CA's (i.e., arteries without plaque and with IMT lower than 1 mm) are usually represented horizontally in an image frame. This projection ensures the possibility of good artery focusing, while avoiding excessive attenuation given by neck muscles. However, the horizontal projection is not always feasible. Some CA's may be curved or even kinked. Some others may have a course that is not parallel to the neck axis. A complete dissertation of possible carotids anatomical variations was given by Lucev et al. [59] in his doctoral studies.

The morphology of the carotid changes when there is a presence of pathology or if the walls are plagued. Plaque is an alteration of the medial wall layer. From a geometrical point of view, it causes a hill-shaped carotid wall. In longitudinal images, therefore, the sonographer must modify the insonation plane in order to ensure an optimal representation of the plaque. Again, the overall CA appearance in the image may not be horizontal and straight.

This is a crucial point for most of the so far developed segmentation techniques. Several algorithms, in fact, offer good segmentation performances when in presence of horizontal, regular, and straight vessels; some others are very robust against CA morphology and appearance.

From a methodological point of view, if the technique aims at the computer-aided IMT measurement and if it requires a human interaction, then the appearance of the CA in the image frame is not crucial. The operator, in fact, can select the portion of the image where IMT measurement can be optimal in terms of CA positioning, noise, and geometry. On the contrary, if the algorithm is devoted to automated CA wall segmentation and IMT measurement, then morphology is crucial. On the overall, the techniques that are completely based on active contours or morphological operators are the most sensitive to CA appearance; the techniques based on feature extraction or local operators are less sensitive.

In 2008, Rossi et al. [60] proposed an automated algorithm for the CA detection in the image frame. This approach was based on modeling of the ultrasound radiofrequency (RF) signal and on its interpretation. The authors showed that it was possible to automatically detect the CA in the ultrasound image frame without any user interaction, with a very low failure rate. Clearly, however, this methodology is not applicable on already acquired images. Also, it requires an ad-hoc ultrasound scanner giving access to the raw RF signal.

Molinari et al. [61] proposed a user-independent technique for the automatic CA localization and adventitia layers tracings in ultrasound images. They tested their technique on different CA morphology (healthy and plagued vessels), orientation (horizontal and inclined), and appearance (straight and curved vessels). Also, their methodology showed same performances on images acquired by different ultrasound OEM scanners. This result, of high practical importance, introduces the issue of instrumental variability in CA ultrasound imaging.

11.2.2 Instrumental Variability

Ultrasound imaging is a real-time methodology in which the trained operator has full control of the device. Specifically, parameters like, focusing, scanning depth, pulse repetition frequency, gain, grayscale appearance, and time-gain compensation can be adjusted in real time. International best practice guidelines recommend adjusting the scanner parameters to the specific patient. Therefore, instrumental variability causes differences in ultrasound images acquired by different operators/ scanners.

Most of the so far proposed segmentation and IMT measurement techniques are not independent of the scanner used. To the very best of our knowledge, only few techniques specifically addressed the issue of scanner independence [61–64]. Loizou et al. proposed normalization of the images [62]. They suggested a linear grayscale remapping so that artery lumen should have intensities between 0 and 5, and adventitia between 180 and 190. Such values were proposed for an ultrasound image coded on 8 bits, hence on intensity values ranging from 0 (black) to 255 (white). Normalization attenuates the differences in gain settings made by the operators. The effect of using different scanners is also reduced. However, normalization requires human interaction, since the operator has to choose a lumen region and an adventitial region for intensity remapping. Thus, this approach is effective only in user-dependent frameworks.

Molinari et al. [61] showed that their local statistics algorithm was effective in separating lumen points from tissue points in images acquired with different scanners. This approach was completely automated and did not require any user interaction. Thus, this methodology is a viable solution when developing completely automated techniques.

Most of the other herein presented techniques (Sect. 3) are scanner dependent. In this case, retraining, tuning, or parameters optimization is required when the image characteristics alter.

11.2.3 Noise Sources

As we already mentioned, speckle noise is the most important noise source in ultrasound imaging. Loizou et al. [62, 65] presented a very detailed review on speckle noise in carotid longitudinal images. They studied different despeckling strategies: local statistics, median filtering, pixel homogeneity, geometric filtering, homomorphic filtering, anisotropic diffusion, nonlinear coherence diffusion, and wavelet filtering [65]. They showed that the first-order statistics filter (named as lsmv by the authors) gave the best performance in the specific case of carotid imaging. This filter is defined by the following equation:

$$J_{x,y} = \bar{I} + k_{x,y}(I_{x,y} - \bar{I}),$$

where $I_{x,y}$ is the intensity of the noisy pixel, $\bar{I}$ is the mean intensity of a $N \times M$ pixel neighborhood and $k_{x,y}$ is a local statistic measure. The noise-free pixel is indicated by $J_{x,y}$. Loizou et al. [65] mathematically defined $k_{x,y} = \dfrac{\sigma_I^2}{\bar{I}^2\sigma_I^2 + \sigma_n^2}$, where σ_I^2 epresents the variance of the pixels in the neighborhood, and σ_n^2 the variance of the noise in the cropped image. An optimal neighborhood size was between 5×5 and 7×7, depending on the intensity of speckle noise.

Zhang et al. [66] proposed a Laplacian pyramid-based nonlinear diffusion (LPND) filter in 2007. With this method, speckle was removed by nonlinear diffusion filtering of bandpass ultrasound images in Laplacian pyramid domain. For nonlinear diffusion in each pyramid layer, a gradient threshold was automatically determined by a variation of median absolute deviation (MAD) estimator. The performance of the proposed LPND method was compared with that of other speckle reduction methods [including the speckle reducing anisotropic diffusion (SRAD) and nonlinear coherent diffusion (NCD)]. In simulation and phantom studies, an average gain of 1.55 and 1.34 dB in contrast-to-noise ratio was obtained compared to SRAD and NCD, respectively. The performance of this filter was satisfactory in carotid imaging, since it effectively crushed noise while preserving the wall edges. However, due to lack of extensive validation on carotid images and computational burden, results were indicated as preliminary by the authors.

Another major noise source in carotid imaging is represented by ultrasound artifacts. The most common are (1) blood backscattering and (2) shadow cones.

1. Blood backscattering is caused by red cells rouleaux in the artery stream. Cells tend to aggregate and to follow the dominant streamlines in the artery lumen [67]. Blood clot or macro-aggregates are hyper-echogenic. The consequence is that artery lumen appears brighter than it should (ideally, the artery lumen should appear black). This artifact may pose problems to the algorithms that model the carotid artery as a black stripe (the lumen) surrounded by bright lines (the walls). In general, blood backscattering poses a challenge in tracking the transition between lumen and artery wall. Therefore, its main impact is on the LI boundary estimation. Normalization (as discussed in Sect. 2.2) is a good turnaround to attenuate the problem of backscattering.
2. Calcium may deposit within the artery wall as a consequence of atherosclerosis and wall degeneration. Being characterized by very high acoustic impedance, at the interface between tissue and calcium the reflection coefficient is about 1. This means that no ultrasound radiation can propagate beyond a calcium deposit. Calcium drops a shadow cone on the ultrasound image. Often, this problem

arises in plaque imaging, even though calcium may also be present in diseased but non-plagued arteries. The shadow cones affecting a portion of the image preclude segmentation (given the absence of echoes underneath). Some algorithms may suffer from such discontinuous image appearance and end with unreliable segmentations. Local statistics-based and distribution-based algorithms are the most sensitive to this disturbance while active contour-based and feature-based algorithms are less sensitive.

11.3 Computer Methods in Carotid Wall Segmentation

In the following, we describe the most performing techniques for carotid wall segmentation in ultrasound imaging. This is split into 2-D- and 3-D-based segmentation methods. Table 11.1 summarizes the techniques that are presented here. For each methodology, we describe principles, performance, advantages, and limitations. Particular attention will also be paid to the automation degree of each technique. Some methodologies, in fact, lack complete automation and require user interaction in order to perform segmentation. Also, some techniques can interact with the operator. This is done in order to ensure the possibility of correcting evident local segmentation failures.

11.3.1 Edge Tracking and Gradient-Based Techniques

The first approach to perform carotid wall segmentation was based on edge detection [68, 69]. Pignoli et al. [68] were the first to introduce computer methods in clinics with the aim of aiding IMT measurement. The observation of the ultrasound appearance of the CCA drove their approach. Basically, as depicted by Figs. 11.2 and 11.3, the CCA can be thought of as a dark region (the carotid lumen) surrounded by bright stripes (the near and far walls). Hence, Pignoli et al. measured the distal wall IMT by considering the intensity profile of a section of the image when moving from the centre of the vessel to the borders. Figure 11.4 sketches this idea: Fig. 11.4a reports the original B-Mode image; Fig. 11.4b shows the image after speckle removal (in this example we applied an iterative despeckling filter based on local statistics, as suggested by Loizou et al. [65]), Fig. 11.4c reports the intensity profile relative to the white dashed line in Fig. 11.4b. The adventitia layers (near – AD_N and far – AD_F) are clearly identifiable on the intensity profile graph, along with the lumen of the artery (L). Figure 11.4d reports the three points (AD_N, AD_F, and L) on the original image.

The adventitia layer is usually very bright, being formed by dense and fibrous tissue. Therefore, it is relatively easy to mark, on the intensity profile of a region of the far CCA wall, the transitions between the lumen and the artery wall (LI) and then the transition between the media and the adventitia layer (MA). The distance between the two transition points constitutes the IMT estimate.

Table 11.1 Summary of the most performing techniques for IMT measurement and carotid wall segmentation

Author	Methodology	Year	Performance metric	IMT measurement error (μm)	Far wall segmentation: LI interface (μm)	Far wall segmentation: MA interface (μm)	Near wall segmentation: LI interface (μm)	Near wall segmentation: MA interface (μm)	Complete automation	User correction
Touboul	Edge detection	1992	MAD	–	–	–	–	–	No	Yes
Liguori	Edge detection	2001	MAD	–	–	–	–	–	No	No
Stein	Edge detection	2005	MAD	12.0 ± 6.0	–	–	–	–	No	Yes
Faita	Edge detection	2009	MAD	10.0 ± 38.0	–	–	–	–	No	Yes
Liang	Dynamic programming	2000	MAD	42.0 ± 20.0	–	–	–	–	Yes	Yes
Gutierrez	Active contours	2002	MAD	90.0 ± 60.0	–	–	–	–	No	No
Destrempes	Nakagami modeling	2009	MAD + HD	–	21.0 ± 13.0	0.16 ± 7.0	–	–	No	No
Goleitiati	Hough transform	2007	XTB	–	–	–	–	–	Yes	No
Cheng	Snakes	2002	MSE	–	62.3 ± 60.5	38.4 ± 68.3	–	–	No	No
Loizou	Snakes	2007	MAD + HD	50.0 ± 25.0	–	–	–	–	No	Yes
Delsanto	Snakes	2006	MAD	63.0 ± 49.1	59.4 ± 65.0	48.1 ± 50.0	–	–	Yes	No
Molinari	Snakes	2008	MAD	35.0 ± 32.0	56.3 ± 50	50.0 ± 43.8	75.0 ± 56.3	131.3 ± 118.8	Yes	No
Molinari	Snakes	2009	PDM	10.0 ± 10.0	35.0 ± 32.0	37.0 ± 29.0	–	–	Yes	No
Molinari	Integrated approach	2009	MAD	54.0 ± 35.0	91.0 ± 93.0	25.0 ± 55.0	–	–	Yes	No

First column reports the author along with numbered reference. Second column reports measurement/segmentation methodology. Third column is publication year. Fourth column reports the performance metrics used to measure IMT or to assess segmentation performance. Fifth column is IMT, sixth is far wall (LI and MA), and seventh is near wall (LI and MA) segmentation performance. Eighth column indicates if the technique has implemented real automation, thus if no user interaction is needed. Ninth column reports the possibility of human correction of evident incorrect performances (used in interactive measurement frameworks) (© Elsevier 2010 - Reproduced by Molinari et al., Computer Methods and Programs in Biomedicine, 2010)

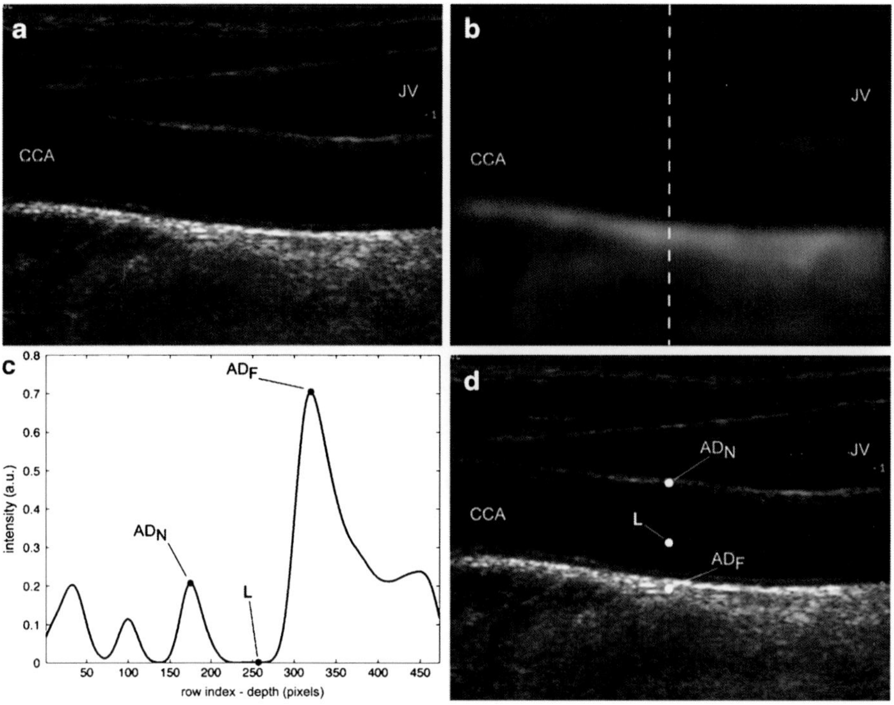

Fig. 11.4 Identification of the wall points in ultrasound longitudinal carotid imaging. (**a**) Original B-Mode image (*CCA*, Common Carotid Artery; *JV*, Jugular Vein). (**b**) Despeckled and low-passed image (see text for details about filtering). The *dashed white line* indicates a column of the image. (**c**) Intensity profile of the image column indicated by the *white dashed line* in (**c**). The horizontal axis reports the pixel row index (i.e., the depth); the vertical axis reports the intensity. The deepest local intensity maximum indicates the position of the far adventitia layer (AD_F), which is followed by a dark region (the artery lumen – L). The subsequent local intensity maximum represents the near adventitia layer (AD_N). (**d**) Detected wall points overlaid on the original image (figure taken from [61]. DOI No: 10.1142/S0219519409003115 - © World Scientific Publishing Company, http://www.worldscinet.com/jmmb/jmmb.shtml)

The same structure was adopted by Touboul et al. [69] that used a IMT measurement technique based on edge detection in several multicentric clinical and epidemiological studies [14, 27].

Neither of the above-referenced studies provided a thorough characterization of the algorithm performance. Rather, they tested the agreement between manual IMT estimations made by experienced sonographers with respect to IMT estimates made by using their computer-assisted method. They both showed a good correlation between the two sets of measures, with correlation coefficients ranging from about 70 to 97%.

In 2001, Liguori et al. [70] proposed a segmentation technique based on edge detection that made use of image gradients. They considered the artery as horizontally placed in the B-Mode image. For each column of the image they calculated the

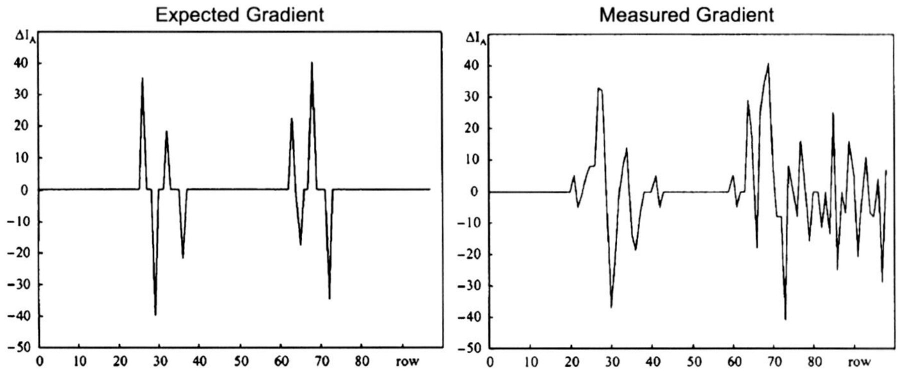

Fig. 11.5 *Left panel*: Theoretical representation of the expected gradient for a longitudinal ultrasound image of CCA. The gradient is relative to the intensity profile of a column of the image (see Fig. 11.4 for further details) The first set of transitions (*left panel*, from 20 to 40 on the horizontal axis) are relative to the near wall, the second (from 60 to 80) to the far wall. *Right panel*: Measured gradient. Gradients are usually noisy, therefore to correctly locate the layers on the gradient, intelligent or statistical approaches might be required (© IEEE-reproduced from [70])

gradient of the intensity profile. The ideal gradient is reported by Fig. 11.5 (left panel): it is assumed that all the pixels of the artery lumen are black and that the carotid wall layers originate the gradient transitions. The first set of transitions (left panel, from 20 to 40 on the horizontal axis) are relative to the near wall, the second (from 60 to 80) to the far wall. Experimentally, however, due to noise, the measured intensity gradient is different from the theoretical one. An example is reported by Fig. 11.5 (right panel). Liguori et al. adopted a statistical thresholding to reduce noise before computing the image gradient, in order to facilitate edge detection. Even though this methodology was not well characterized in terms of segmentation and IMT measurement performance, the authors provided a very exhaustive and complete metrological characterization. Complete details about metrological characterization of segmentation performance will be provided in Sect. 5; however, the authors demonstrated that the IMT measurement uncertainty for their approach was equal to 20 μm, which can be considered as a reasonable uncertainty in clinical applications. The technique proposed by Liguori et al. lacks full automation, since the user is required to manually select the portion of the image in which to perform IMT measurement. However, the authors proved the possibility of segmenting the near wall along with plagued walls, provided that the plaque had a smooth and regular surface.

In 2005, Stein et al. [71] proposed a gradient-based methodology for measuring the IMT of the carotid distal wall. The focus of their paper was about reproducibility and accuracy. They tested on 300 carotid segments. An expert and a novice operator performed the 300 measurements with and without the assistance of the computer method. The authors proved that computer-aided IMT measurement were faster, more reproducible, accurate, and independent on the operator skill. The IMT measurement error was 12.0 ± 6.0 μm. This method did not perform a real CCA wall segmentation, but rather was a computer system to aid and improve IMT estimation. No discussion was made about noise robustness.

The most performing and innovative gradient-based approach is represented by the recent work by Faita et al. [72]. The gradient-based segmentation, as already noted by Liguori et al. [70], mainly suffers from the problem of superimposed noise, which precludes a proper individuation of the LI and MA transitions. In this study, the gradient performance is improved by the use of a first-order absolute moment edge operator (FOAM) and a pattern recognition approach. The FOAM operator is defined as:

$$e(x,y) = \frac{1}{A_\Theta} \iint_\Theta \left| f(x,y) - f\left(x - \tau_x, y - \tau_y\right) \right| \mathrm{d}\tau_x \mathrm{d}\tau \tag{11.1}$$

where the original image $f(x,y)$ is evaluated on a circular domain Θ of area equal to A_Θ. Basically, this operator computes the mean dispersion of the values the image assumes in the Θ domain with respect to the value that it assumes in the central point of the domain. When (1) is computed in correspondence of a neat color transition, $e(x,y)$ is maximum. The algorithm then searches the maxima corresponding to the LI and MA transitions on the $e(x,y)$ profile (if one of the two maxima cannot be found, that intensity profile is discarded). The overall performance of this methodology was very high: IMT measurement error was equal to 10.0 ± 38.0 μm. Moreover, FOAM operator and intelligent procedures to discover maxima ensured a good robustness to noise. This technique is real time; hence, it is suited to clinical application. Limitations of such approach are the difficulty of processing curved vessels or vessel that does not appear as horizontal in the image. Also, full automation is precluded by the need for a manual selection of a region-of-interest (ROI) in the image that contains the carotid artery.

11.3.2 Dynamic Programming Techniques

Dynamic programming techniques were introduced starting from early 1990s to reduce the variability in ultrasound measurements [73–76]. Evidence was that manual measurements were subjected to high variability introduced by the operator skills and by overt drifts over time [77, 78]. In 1997, Wendelhag et al. [74] first introduced a dynamic programming procedure for the automatic detection of echo interfaces. This technique combined multiple measurements of echo intensity, intensity gradient, and boundary continuity. The system also included optional interactive modification by the human operator. An initial set of ultrasound images was manually segmented by expert operators and served as reference (ground truth). The estimated values of the three boundary features (echo intensity, intensity gradient, and boundary continuity) were linearly combined in order to create a cost function. Each image point was associated with a specific cost that in turn correlated with the likelihood of that point being located at the echo interface. Dynamic programming was used in order to reduce the computational cost given by the need of inspecting all the points of the image. To find the optimal weights, a training of the system was made, considering ground truth as reference. The IMT measurement error of this technique was better than 40 ± 36 μm. The authors demonstrated that the use of such computerized approach diminished the difference in the segmentation performance of differently trained operators. A major advantage of this methodology was complete automation: the dynamic programming approach allowed for analyzing the cost function for every pixel in the image, thus avoiding the need for any human action. Also, this technique permitted the human correction of evident erroneously segmented points. The major limitation relies in the need for training of the system. The three boundary features (echo intensity, intensity gradient, and boundary continuity) are dependent on the ultrasound scanner used and on the scanner settings. Therefore, retraining is required when scanner is changed. Also, no evidence of noise robustness was given, but, again, the three features of the cost function are all highly noise dependent.

In 2000, Liang et al. [79] proposed a dynamic programming technique based on multiscale analysis. Basically, this technique can be thought of as the evolution of the previous one by Wendelhag et al. [74]. Dynamic programming was applied in an iterative fashion, starting from the image in a coarse representation and then applying step-by-step refinement iteration. Hence, the global position of the artery in the image was estimated on a coarse scale (thus reducing computational cost), while precise position of the wall layers was estimated in a fine scale (thus ensuring accurate performance). This improved technique reduced the computational burden. The segmentation performances remained unchanged: the IMT measurement error was equal to 42.0 ± 20.0 μm. Advantages and disadvantages of this technique were same as for the one by Wendelhag et al. [74].

11.3.3 Active Contours (Snakes)-Based Segmentation

Active parametric contours, also called snakes, have been widely used in segmentation of medical images. The snake can be thought as a set of vertices connected by line segments, which can evolve under the action of different forces. At least six different contributions adopted snakes to perform carotid segmentation [63, 64, 80–83]. The popularity of the snake-based segmentation methodology relies in the fact that the artery LI and MA boundaries can be well distinguished in the image. Therefore, a straight deformable model could be able to adapt to such interfaces after a proper tuning of its parameters. Most of the studies adopted the traditional formulation of a snake as proposed by Williams and Shah [84]. From a geometrical point of view, a 2-D snake is a parametric contour represented by $\upsilon(s) = [x(s),\ y(s)]$, where $(x, y) \in \mathrm{R}^2$ denotes the spatial coordinates of an image and $s \in [0, 1]$ represents the parametric domain. The snake adapts itself by a dynamic process that minimizes a global energy function $E_{\text{snake}}(\upsilon)$ defined as follows:

$$E_{\text{snake}}(\upsilon) = E_{\text{int}}(\upsilon) + E_{\text{ext}}(\upsilon) \tag{11.2}$$

where $E_{int}(\upsilon)$ is the internal energy of the snake and $E_{ext}(\upsilon)$ is an external driving energy. Usually, for a general application, $E_{int}(\upsilon)$ depends on some constraints that are imposed to the snake and can be formulated as:

$$E_{int}(\upsilon) = \int_0^1 \{\alpha(s)|\upsilon'(s)|^2 + \beta(s)|\upsilon''(s)|^2\} \, ds \tag{11.3}$$

being $\alpha(s)$ and $\beta(s)$ the snake elasticity and rigidity, respectively. The internal energy prevents the snake from twisting or bending in an excessive way, so that the morphology of the features of the image can be better preserved. $E_{ext}(\upsilon)$ typically depends on relevant features of the image (i.e., borders, lines, points, etc.). The snake points should move toward the features of the image while remaining constrained by the internal forces. In most of the snake-based algorithms, $E_{ext}(\upsilon)$ is modeled by using the local image gradients. The snake reaches its equilibrium condition when the forces in (2) are balanced. Therefore, snakes require a fine-tuning of their parameters in order to be able to reach the equilibrium condition between forces in correspondence of the image relevant features. We can summarize the problems affecting the use of snakes in the following points:

- Need for optimization of the parameters
- Dependence on the initialization of the snake points
- Dependence on the number of points constituting the snake
- Sensitivity to noise

In the following, we describe the relevant characteristics of the best performing snake-based techniques for CCA segmentation and IMT measurement.

In 2002, Gutierrez et al. [80] proposed an automated technique for carotid IMT and lumen diameter estimation based on active contours and multiresolution analysis. They modeled the contour of each wall layer using the active contours by Lobregt and Virgever [85]. The active contour vertices could move being subjected to three different forces: internal, external, and damping (or viscous). The internal force was taken proportional to the contour curvature. The external force was the local magnitude of the image gradient. The damping force was proportional to the velocity of each contour vertex and in opposite direction. The three forces were linearly combined by three weighting factors. The authors did not provide any information about the weights optimization and about the iteration conditions. Overall IMT measurement performance was 90.0 ± 60.0 μm. Validation was done by comparison with human measurements. The novelty element of this technique was the use of a damping force that helped to smooth and stabilize the contour evolution. It is not possible to assess the robustness of the technique, since a discussion about noise is missing.

Cheng et al. [81] proposed a snake-based technique for CCA segmentation in 2002. They observed that traditional snakes could show poor segmentation performance in the region comprised between the intima and the adventitia layers. Specifically, being the intima and the adventitia bright features, a snake could remain trapped in-between for it could be attracted by the adventitia (i.e., downwards), but also by intima (i.e., upwards). Figure 11.6 sketches a sample of this

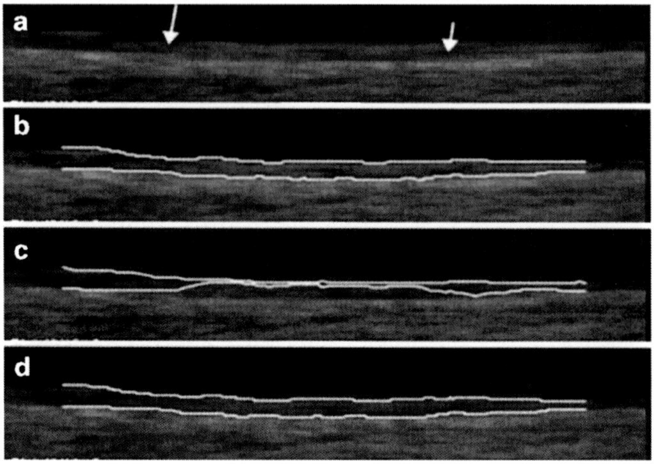

Fig. 11.6 Sample of possible snake erratic behavior in segmenting the LI and MA layers. (**a**) Original image of the far wall. The *white arrows* indicate dishomogeneities of the media layer. (**b**) Final segmentation of Cheng's snake [81]. (**c**) Erratic behavior of a traditional snake (ziplock snake): the lower snake collapses and the MA profile is not correctly traced. (**d**) Ground truth obtained by average of manual segmentations (reproduced from [81])

erratic behavior [81]. Figure 11.6a shows a portion of distal CCA wall; the white arrows indicate the presence of noise in the artery wall, between the intima and adventitia layers. Figure 11.6c shows the behavior of a traditional snake: the noise attracts the snake toward intima and adventitia cannot be correctly segmented. To overcome such limitation, the authors redefined the external energy as:

$$E_{\text{ext}}(\upsilon) = -\int_{\Omega} g_0 \left(\left|f_M * I\right| + D * I\right) \mathrm{d}s = -\int_{\Omega} F(\upsilon)\,\mathrm{d}s, \qquad 4$$

where I is the original ultrasound image, f_{M} represents the Macleod operator [86], and g_0 is the gravity constant (the symbol * denotes the bidimensional convolution). The matrix D is defined as:

$$D = A\begin{bmatrix} 0 & -1 & 0 \\ 0 & 0 & 0 \\ 0 & 1 & 0 \end{bmatrix},$$

where A is a positive weighting factor. Note that the central column of the matrix D is a vertical gradient that has a negative weight upwards (−1) and a positive weight downwards (+1). Therefore, this operator prevents possible trapping of the snake in-between the intima and adventitia layers, since now there is a repulsion force from intima and an attraction force toward adventitia. The behavior of such snake is represented by Fig. 11.6b where it is possible to observe the correct segmentation of the adventitia layer (Fig. 11.6d represent the manual segmentation – ground truth).

The authors validated their technique vs. human tracings. They used the mean squared error (MSE) of the distance between the computer-generated boundaries and the ground truth as performance metric. The MSE for the LI interface was equal to 62.3 ± 60.5 μm, whereas for the MA it was 38.4 ± 68.3 μm. Since the value of some parameters were not given, it is difficult to understand the behavior of this snake in presence of arteries with nonstraight appearance. Robustness to noise seemed adequate, but a characterization was not provided. Also, the issue of initialization was not addressed.

A complete and thorough summary of the snake performances in carotid segmentation was given in 2007 by Loizou et al. [63]. They addressed the problem of image preprocessing standardization in order to overcome some snake limitations. Specifically, they addressed the problems of noise and of parameters tuning. They observed that most of the CCA segmentation techniques were lacking speckle noise reduction. However, due to blood rouleaux in vessels, speckle noise and blood backscattering can be very high [62]. The authors proposed preliminary image intensity normalization followed by despeckling. Brightness normalization helped controlling the external energies in correspondence with the layer boundaries. Considering the typical ultrasound image discretized on 8 bits (i.e., on integers from 0 to 255), the authors proposed to scale image intensities so that the median of the blood was between 0 and 5, while the median of the adventitia layer was between 180 and 190. With this normalization, the optimal snake parameters were $\alpha(s) = 0.6$ and $\beta(s) = 0.4$. The external force was weighted by a parameter $\gamma(s) = 2$. Despeckling further helped to reduce snake sensibility to image noise. The IMT measurement error was 50.0 ± 25.0 μm, validated against human tracings. The authors also proposed an initialization technique that proved robust in presence of images with different noise level. However, manual interaction was still required to place the initial seed points of the snake in the artery lumen.

11.3.4 Local Statistics and Snakes

In 2007, Delsanto et al. [82, 87] proposed a combined approach of local statistics and snake-based segmentation to perform IMT measurement. The goal was to develop a completely user-independent segmentation strategy. The technique they proposed was conceptually made of two distinct parts: (1) a module aiming at locating the carotid artery in the image frame and (2) a segmentation and IMT measurement module. Local statistics were the basis of module 1; snakes of module 2.

To automatically locate the CCA in the image frame, the authors proposed to cluster into a bidimensional histogram the mean values and standard deviation values of the image pixels 10×10 neighborhood. The lumen pixels being ideally black and surrounded by other black pixels, the basic idea was that points located in the artery lumen should have very low neighborhood intensity mean and standard deviation values. Therefore, the automatic detection of lumen points was made possible in an automated fashion. Figure 11.7 depicts the procedure used to automatically locate the lumen points: Fig. 11.7a represents the original B-Mode ultrasound image; Fig. 11.7b represents the despeckled image; Fig. 11.7c the bidimensional histogram of the mean and standard deviation values of the 10×10 neighborhood of each pixel. The grey region corresponds to the region that was supposed to contain essentially lumen points. The same pixels falling into the grey region in Fig. 11.7c are depicted by a gray color in Fig. 11.7d. It is possible to observe that automatic individuation of the lumen vessel is possible.

Delsanto et al. then considered the image column-wise. With reference to Fig. 11.4, the intensity profile of each column was analyzed in order to find the transitions corresponding to the near and far adventitia layers (Fig. 11.4c). Valid adventitia points should have been separated by some lumen pixels (i.e., pixels falling in the corresponding gray region of the bidimensional histogram, as in Fig. 11.7c). This procedure led to the tracing of the adventitia layers.

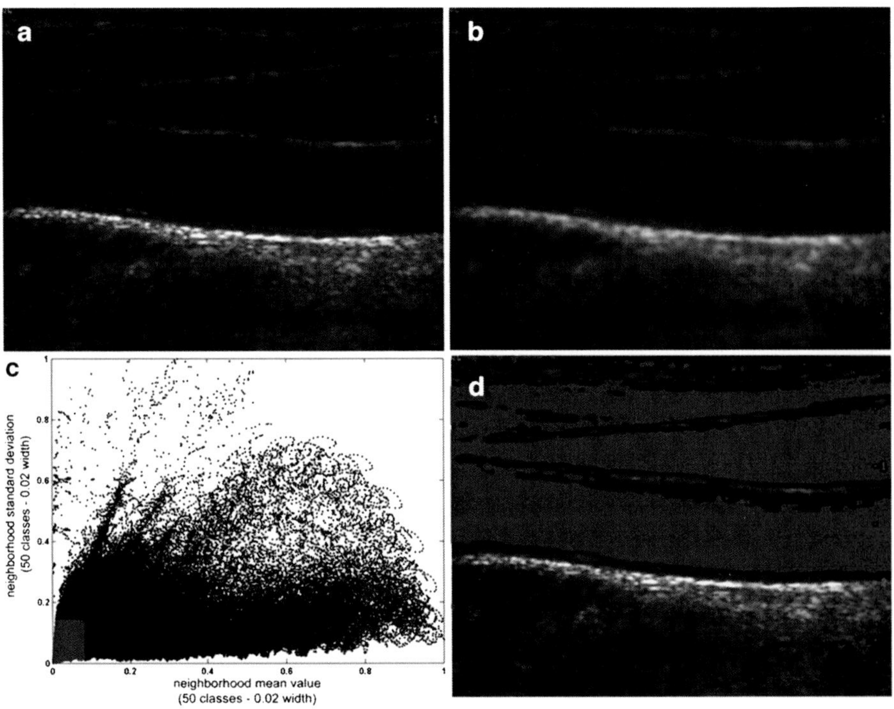

Fig. 11.7 Automated procedure for lumen points detection based on local statistics as proposed by Delsanto et al. [82]. (**a**) Original B-Mode image of a healthy CCA. (**b**) Despeckled image. (**c**) Bidimensional representation of the mean (*horizontal*) and standard deviation (*vertical*) of a 10×10 pixel neighborhood. The *grey* region represents the portion of the histogram in which it is supposed that a lumen point should fall, being it *black* and surrounded by other *black* pixels. (**d**) All the points falling into the *grey* region of (**c**) have been overlaid in *dark grey* on the original image. It can be seen that essentially only points belonging to the lumen have been detected (© IEEE-reproduced from [82])

After the user-independent localization of the artery, the segmentation was based on an initial gradient-based contour, followed by a snake regularization and refinement of the LI and MA profiles. The snake formulation was

$$E = \int_0^1 \frac{1}{2}\alpha(s)\left|\upsilon'(s)\right|^2 + \gamma(s)E_{\text{ext}}\left(\upsilon(s)\right)ds \tag{11.5}$$

with the optimal snake parameters $\alpha(s)$ and $\gamma(s)$ equal to 0.1 and 0.01, respectively. The LI and MA segmentation performance of this approach were equal to 59.4 ± 65.0 and 48.1 ± 50.0 μm, respectively, whereas the IMT measurement error was equal to 63.0 ± 49.1 μm. Validation was made against human tracings and the mean absolute error was considered as distance metric. This strategy was completely user-independent and allowed for the segmentation of the CCA wall and the accurate IMT measurement. However, the authors declared that noise could preclude the proper CCA localization (and, consequently, segmentation) in about 10% of the cases. Nevertheless, this strategy could cope with almost any kind of CCA projection, since the authors could segment also curved vessels or arteries with nonhorizontal appearance.

In 2008, the same research group proposed a further improvement of the technique that could segment also the near wall as well as plagued vessels (Molinari et al. [83, 88]). The processing strategy for the CCA location was identical to previous version. Once detected, the profiles of the near and far adventitia defined a ROI. A fuzzy K-means classifier was inserted in the segmentation module in order to cluster the ROI pixels. The authors considered three possible clusters: (a) lumen, (b) intima and media layers, (c) adventitia layer. The border between regions a and b was considered as the initial guess of the LI boundary; the border between regions b and c as the MA. Then, snakes were used to refine the LI and MA boundaries. Using the same performance metrics of Delsanto et al. [82] the authors obtained a LI segmentation error of 56.3 ± 50 and 75.0 ± 56.3 μm for the far and near wall, respectively. The MA segmentation error for the far and near wall was equal to 50.0 ± 43.8 and 131.3 ± 118.8 μm, respectively. The IMT measurement error was reduced to 35.0 ± 32.0 μm.

This technique proved also robust in segmenting plaques. On stable and fibrous plaques (with echogenic appearance), the average number of misclassified pixels was equal to $8 \pm 5\%$; on unstable and soft plaques (with echolucent appearance) the misclassification error increased to $12 \pm 4\%$. This result demonstrated the versatility of the strategy. The authors admitted a residual 8% of images that could not be automatically processed due to low signal-to-noise ratio or to failure in the ROI detection. They also pointed out the need for a better delineation of echolucent plaques.

In 2009, Molinari et al. [61] characterized their technique by testing over a 300 images database acquired with three different OEM ultrasound scanners. They proved that segmentation performance were independent on the scanner used. This result enforces the robustness of this local statistic and snake approach in terms of standard artery detection and segmentation. The last version of this technique showed far wall LI and MA segmentation errors equal to 35.0 ± 32.0 and

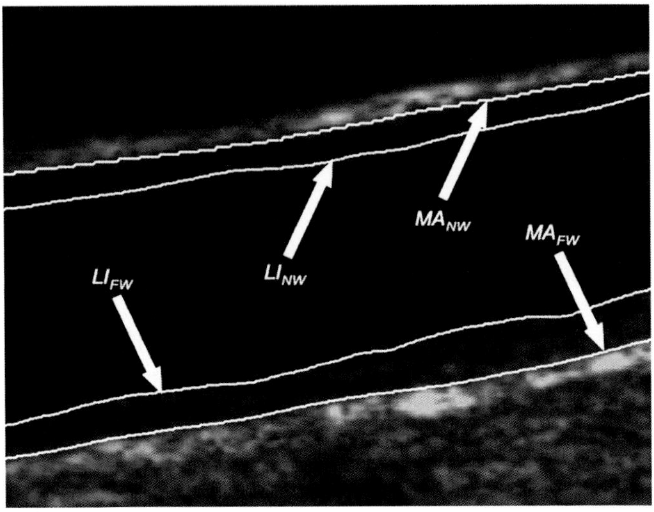

Fig. 11.8 Completely user-independent segmentation of the near and far carotid wall (technique proposed by Molinari et al. [83, 88]). LI_{NW} lumen–intima near wall interface, LI_{FW} lumen–intima far wall interface, MA_{NW} media–adventitia near wall interface, MA_{FW} media–adventitia far wall interface

37.0 ± 29.0 μm, respectively. The IMT measurement error dropped to 10.0 ± 10.0 μm. Figure 11.8 reports a sample of completely user-independent segmentation of the near and far carotid wall obtained by means of this approach.

11.3.5 Nakagami Modeling

This approach adopted the modeling of the intensity of a small image ROI to perform segmentation. The rationale was that if the ROI comprised the IMT complex, then its appearance should be easily modeled by brightness and speckle noise. The Nakagami distribution, in fact, proved effective in modeling the radiofrequency (RF) scattered ultrasound signal [89–91]. Destrempes et al. [92] (2009) proposed a segmentation strategy based on Nakagami mixture modeling and stochastic optimization. They considered small vertical ROIs (stripes) containing the IMT complex and analyzed the RF signal. A sample of the considered geometry is reported by the upper panel of Fig. 11.9. The authors modeled the vertical RF signals as mixture of three Nakagami distributions, with the following assumptions:

1. The lumen corresponded locally to the distribution with lower mean.
2. The IMT corresponded locally to the mixture.
3. The adventitia corresponded locally to the distribution with higher mean.

The lower panel of Fig. 11.9 sketches the assignment of the four profiles to the lumen, intima–media, and adventitia boundaries. In a first step, they computed the maximum a posteriori estimator of the proposed model, using the expectation maximization algorithm. The optimal segmentation, based on the estimated distributions as well as a statistical prior for a generic disease-free IMT (i.e., IMT lower than 70 μm and with small variations along the farcarotid wall), was computed using a variant of the exploration/selection algorithm. Convergence of this algorithm to the optimal solution was assured asymptotically and was independent of the initial solution. This technique proved very performing in CCA segmentation: using the mean absolute distance (MAD) metric the authors obtained LI and MA segmentation errors equal to 21.0 ± 13.0 and 0.16 ± 7.0 μm, respectively. The authors validated their methodology on about 30 sequences of ultrasound B-Mode images. The assumption 3 on the IMT value makes the technique little applicable in processing a generic pathology image. Also, the authors admitted that their

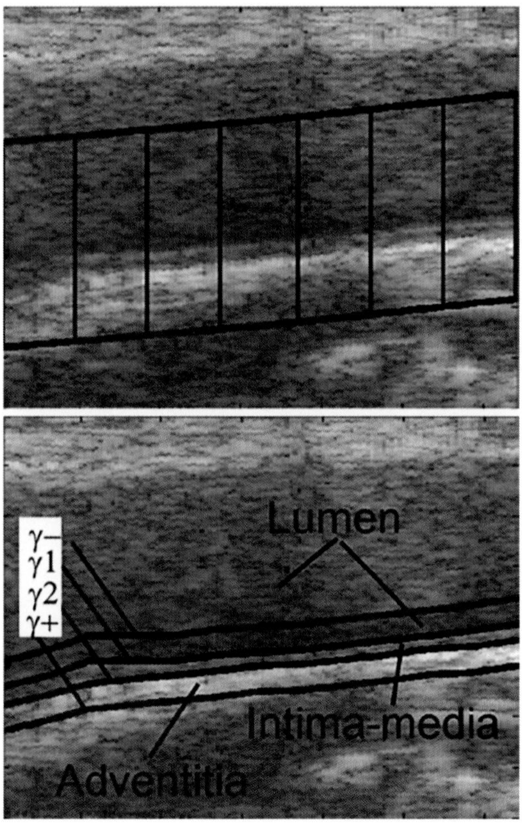

Fig. 11.9 *Upper panel*: Representation of the ROIs selected in dynamic processing strategies. Each ROI is vertical and contains the lumen, the intima–media structure, and the adventitia layer. *Bottom panel*: Assignment of the four boundaries (three distributions) to the structural elements in the ROIs (© IEEE 2009 - reproduced from [79])

method was well suited to a semiautomatic context that requires minimal manual initialization. No characterization on noise robustness was provided.

11.3.6 Hough Transform

Between 2004 and 2009, Golemati et al. [93–96] developed a segmentation algorithm based on the Hough transform. The Hough transform [97] is widely used in computer vision and image processing algorithms, since it allows for an easy and reliable detection of mathematically defined shapes in the image frame. The Hough transform is mostly used to detect lines and circles. The authors exploited the properties of the Hough transform and developed a technique that could automatically segment both longitudinal than transverse B-Mode images. The appearance of the CCA layers is, in fact, straight in longitudinal projections and circular in transverse projection (see the scheme in Fig. 11.2). The dominant lines corresponding to the LI and MA boundaries in longitudinal images were independently traced by using the Hough transform; their distance was taken as IMT estimation. In transverse images, the circles corresponding to the LI and MA boundaries were extracted. An example of CCA recognition and segmentation is reported by Fig. 11.10. The upper row represents the CCA in longitudinal projection, along with the near and far LI boundaries in different cardiac cycles. The lower row represents a CCA in transverse projection, with the white circle delineating the LI boundary in different cardiac cycles. The authors demonstrated that their methodology could also process weak plagued vessels. Clearly,

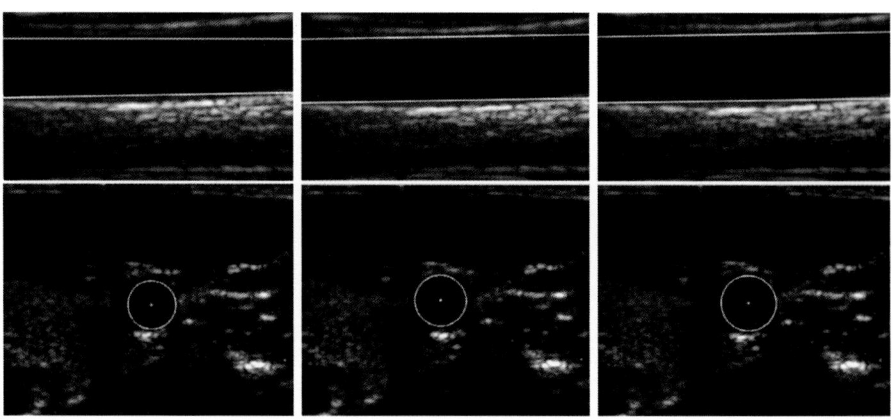

Fig. 11.10 *Upper row*: Samples of CCA recognition in longitudinal B-Mode images using the Hough transform. The *white lines* represent the detected lumen–intima interfaces in different cardiac cycle instants. *Bottom row*: Samples of CCA recognition in transverse images. The *circles* represent the Hough transform detection of the lumen–intima interfaces in different cardiac cycle instants (© Elsevier 2007 - reproduced from [94])

the limitation of such approach is the need for a straight and horizontal vessel in the image. If the CCA were straight but not horizontal, the Hough transform would not recognize the correct dominant lines. Validation of this technique was done against human tracings by using the cross-table: the authors computed the number of pixel correctly segmented (true positives and true negatives) and those incorrectly traced (false positives or negatives). Sensitivity was 96 and 82% in longitudinal and transverse projections, respectively; specificity was 96% in both the projections.

11.3.7 Integrated Approach

Molinari et al. developed a generalized architecture for vessel wall segmentation in 2009 [64]. Their technique consisted of two parts:

1. A module for the automatic location of the CCA in the image.
2. A segmentation procedure that automatically traces the LI and the MA contours of the distal (far) wall once CCA has been localized.

Conceptually, in step 1, the technique exploits the image information in order to automatically detect the near and far adventitia. The local intensity maxima of each column are processed by a linear discriminator to detect which are located on the CCA wall. These points are called "seed points." Figure 11.11 summarizes the functioning of this methodology. Figure 11.11a represents the original B-Mode image; Fig. 11.11b the seed points overlaid to the image. It can be noticed that the linear discriminator allows for seed points exclusively located on the carotid walls. Seed points are then linked to form line segments. An intelligent procedure removes short or false line segments and joins close and aligned segments. This procedure avoids oversegmentation of the artery wall. Figure 11.11c shows the three line segments (called LS_1, LS_2, and LS_3) obtained by seed points of Fig. 11.11b. Figure 11.11d shows the final tracings of the CCA wall after joining of, LS_2, and LS_3. Once CCA has been detected, the image is scanned column-wise: the intensity profile of each column is processed by using a fuzzy K-means classifier, which is initialized by the intensity values. The classifier assigns pixel to three clusters that are (1) lumen, (2) the intima and media layers, and (3) the adventitia layer. The points at the transitions between the three clusters are taken as the LI and MA boundaries markers.

This technique has the advantage of being completely user independent. Its integrated approach exploits the morphological features of the artery wall allowing for segmentation in more than 95% of the cases. The authors validated against human tracings, using the MAD as performance metric. The segmentation of the MA boundary was very accurate (25.0 ± 55.0 μm), but the LI was average (91.0 ± 93.0 μm). As a consequence, the IMT measurement error was higher than most of previously cited techniques (54.0 ± 35.0 μm).

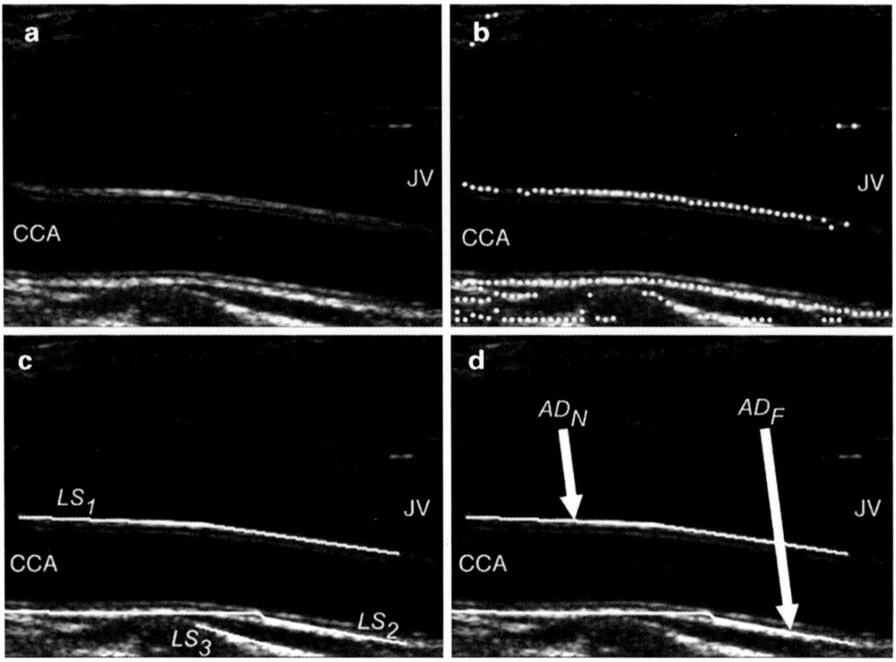

Fig. 11.11 Schematic representation of the integrated approach by Molinari et al. [64]. (**a**) Original image of the CCA (*JV* jugular vein). (**b**) Seed points detected by the linear discriminator. Note that seed points are featured on the vessel wall and in the tissue, the other local intensity maxima have been discarded. (**c**) Line segments obtained by connecting the seed points and discarding fragmented and misplaced lines. The three line segments (LS_1, LS_2, and LS_3) are considered in pair to find the two that comprise the artery lumen. (**d**) Final segmentation of the near (AD_N) and far (AD_F) adventitia layers (© AIUM 2010 - reproduced from [64])

11.3.8 3-D Segmentation Methods

Segmentation of 3-D vascular images grew essentially from about 2000 to nowadays. The development of segmentation techniques followed the rapid expansion of the 3-D ultrasound methodology in clinical practice. Fenster et al. recently discussed the need for a mechanic scanning in order to produce reliable vascular 3-D images [98]. They pointed out that manual scanning might preclude a proper vessel visualization and segmentation due to distorted morphology. To this purpose, they developed a probe holder equipped by a motor. This device could move the ultrasound probe along the neck, parallel to the neck axis, with a constant velocity, in order to be able to mimic the sonographer movement but in a reproducible and controlled fashion. Also, 3-D imaging requires a fast scanning device, to avoid movement artifacts due to patient motion, respiration, and cardiac contraction [98].

3-D vessel volumes can be observed by volume rendering or by representation in three orthogonal planes (i.e., axial, coronal, and sagittal). The rendering methodologies are usually taken by volumetric imaging techniques (i.e., MRI and CT) and are out of the scope of this review. 3-D carotid images are usually segmented using the vessel wall volume technique, which is typical of the MRI. The 3-D volume is processed slice-by-slice and each slice is segmented independently. Slices are transverse to the vessel axis. Most of the proposed techniques adopted active contours (snakes) [98–100]. Specifically, in the approach by Zahalka et al. [100] the user had to place a single seed point in the lumen of the artery and then the procedure automatically segmented the transverse image. An example of transverse image segmentation and 3-D carotid morphology and plaque reconstruction is given in Fig. 11.12. The authors validated their methodology by using a phantom with specific material simulating a plagued vessel [101].

The 3-D segmentation methodology was, however, mainly devoted to plaque analysis and characterization [102–106]. Specific aims were the assessment of plaque change after drug treatments [47, 103, 107]. In plagued arteries, it must be said that the 3-D methodology has some advantages over the traditional 2-D B-Mode representation:

1. 3-D imaging can represent the entire vessel (and plaque) volume and not only of a single projection or insonation plane as in 2-D.
2. 3-D imaging is optimal for volumetric studies, which are precluded in conventional 2-D imaging.
3. Some patients' morphologies are a limiting factor for 2-D imaging. For example, if a plaque is rich in calcium, it will drop a shadow cone on the 2-D image that would preclude a proper artery representation. In some cases, the insonation angle is critical due to the morphology of the vessels or of the neck. 3-D imaging is almost independent on such factors since the vessel is scanned from several different insonation planes.

Fenster et al. pointed out that the number and thickness of the slices influence plaque volume estimation [98]. Thicker slices ensure a lower number of slices and, therefore, a lower computational time. However, it was demonstrated that slices thicker than 3.0 mm would lead to an underestimation of the plaque volume. Therefore, the optimal 3-D ultrasound scan for plaque volume measurement should have a slice thickness between 1.0 and 3.0 mm.

11.3.9 IVUS Techniques

When performing an IVUS exam, a small ultrasound probe mounted on a catheter is inserted into the vessel so that the artery wall is imaged from the inside. This approach enhances the wall appearance due to the absence of interposed tissues between the probe and the wall. This methodology is invasive and usually applied to critical patients or in presurgical examinations.

The IMT can be measured also by IVUS images. However, we did not find any computer-based approach to the IMT measurement from IVUS images. Most of the

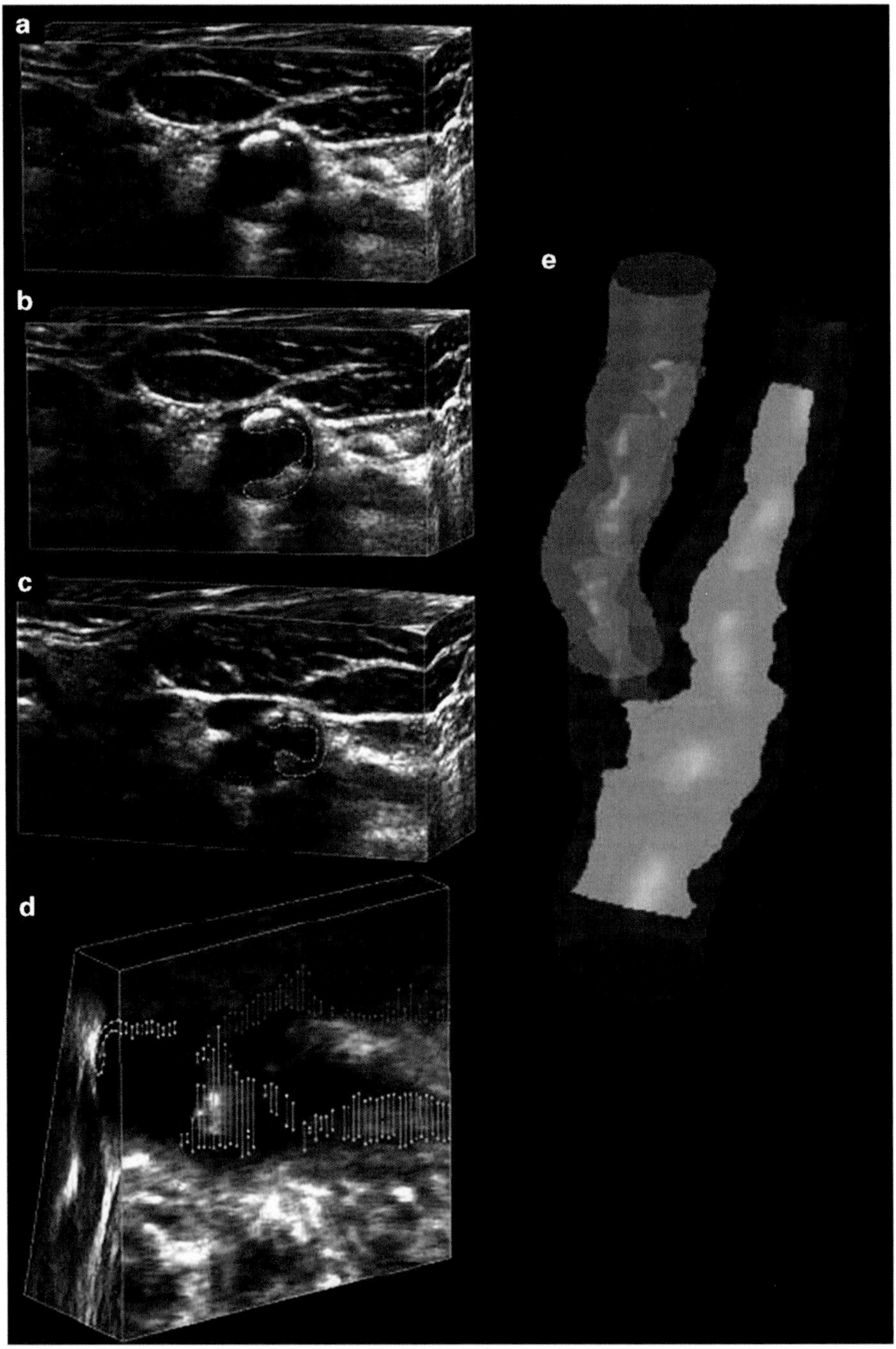

Fig. 11.12 Sample representation of 3-D carotid artery segmentation and volumetric reconstruction. The example refers to a plagued vessel. The slice-by-slice representation (**a**–**c**) is used to perform segmentation. (**d**) shows 3-D longitudinal reconstruction of the transverse segmentations. (**e**) Reports the final volume rendering (reproduced from [98])

studies are qualitative [108–110] or comparative of the IVUS performance with respect to other radiological techniques in atherosclerosis assessment [111].

3-D plaque imaging greatly benefited also from IVUS. Sanz-Requena et al. [112] proposed a semiautomated method for plaque segmentation and measurement in 3-D IVUS imaging. Their approach was based on snakes; particularly, they used gradient vector flow snakes that showed better performance in following the morphology of complicated plaques. This algorithm, besides segmenting the artery wall, offered the possibility of several calculations, including: plaque length, stenosis severity, and plaque volume. Plaque volume was estimated with an average error of about 5 mm^3, approximately equal to 20% of the nominal plaque volume. Validation was made against human tracings.

IVUS can be thought of as a potentially interesting application field for computer-based algorithms. It appears as a very powerful methodology in tissue characterization. Hence, if nowadays only preliminary studies can be found related to arterial wall IVUS characterization [112], we expect a growth of such technique in the carotid assessment scenario.

11.4 A Discussion on Correlation with Human Tracings

As we discussed in Sect. 3, most of the studies related to IMT measurement were validated against human tracings. In some studies, the number of human operators was not clearly declared [20, 81, 92], some other studies validated against two or more operators [45, 63, 68, 69, 72]. In the following, we will show the most used performance metrics to validate IMT measurements and computer traced boundaries (CTB).

11.4.1 Mean Absolute Distance

Given a CTB and a reference human-traced boundary (call it GT – ground truth), the MAD metrics can be defined as:

$$\varepsilon = \frac{1}{N}\sum_{y=1}^{N}\left|\mathrm{CTB}(y) - \mathrm{GT}(y)\right|, \tag{11.6}$$

where N is the number of points constituting the two boundaries, and y is the index spanning the columns of the image. This metrics, in fact, can be used when the two profiles have the same number of points. Usually, boundaries are interpolated so that there is a boundary point for every column of the image. This metrics have been used in several studies (see Table 11.1). Particularly, this metric is effective when the artery is straight and horizontal in the image. If the vessel wall is curved or inclined, then the MAD overestimates the distance between from GT.

If applied to the computation of IMT measurement error, the MAD becomes:

$$\varepsilon^{\mathrm{IMT}} = \frac{1}{P}\sum_{i=1}^{P}\left|\mathrm{IMT}_i^{\mathrm{CTB}} - \mathrm{IMT}_i^{\mathrm{GT}}\right|, \tag{11.7}$$

where P is the number of processed images; $\mathrm{IMT}_i^{\mathrm{CTB}}$ the computer-measured IMT for image i-th; and $\mathrm{IMT}_i^{\mathrm{GT}}$ the manually measured IMT on the same i-th image.

11.4.2 Hausdorff Distance

The Hausdorff distance (HD) between two boundaries is a measure of the longest distance that one has to travel if moving from a given point on a boundary and going to the other boundary. In other words, given the boundaries B_1 and B_2, one has to calculate the Euclidean distances of each vertex of B_1 from the vertices of B_2. Let's indicate with d_{12} the minimum distance of the most distant vertex of B_1 from the vertices of B_2. Similarly, let's define d_{21} the minimum distance of the most distant vertex of B_2 from the vertices of B_1. Figure 11.13 reports a reference for the computation of d_{12} and d_{21}. The HD can be mathematically defined as follows:

$$\mathrm{HD} = \max\{d_{12}, d_{21}\}. \tag{11.8}$$

The major limitation of HD is that only distances between vertices are measured. Therefore, this measure is significant when the two boundaries have almost the same number of vertices. Also, if the spatial sampling of the vertices of B_1 is very different from that of B_2, the HD may lead to unreliable results.

HD was used by Destrempes et al. [92] to measure IMT (distance between LI and MA boundaries) and to validate results (distance between computer tracings and GT).

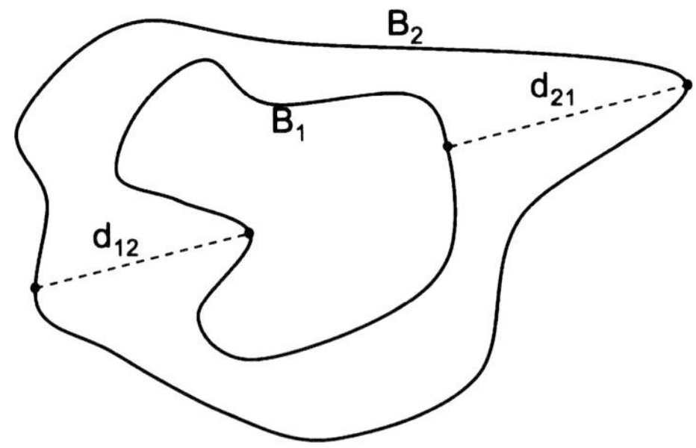

Fig. 11.13 Schematic representation of the Hausdorff distance (HD). The distance d_{12} is the shortest distance separating the most distant point of boundary B_1 from the closer point of boundary B_2. The distance d_{21} is the shortest distance separating the most distant point of B_2 from the closer point of B_1. The HD is the maximum between d_{12} and d_{21}

11.4.3 Polyline Distance Metric

In some conditions, the previously described performance metrics do not truly represent the distance between two boundaries. In 2000, Suri et al. [113] proposed to use the polyline distance, since this measure seems to be a more robust and reliable indicator of the distance between two given boundaries, which truly represents the distances between boundary shapes along the artery. The basic idea is to measure the distance of each vertex of a boundary to the segments of the other boundary. The measured distance becomes very robust because it is independent on the number of points of each boundary.

Consider two boundaries B_1 and B_2 as depicted in Fig. 11.14. We can define the distance $d(v,s)$ between a vertex v and a segment s. Let's consider the vertex $v = (x_0, y_0)$ on the boundary B_1 and the segment s formed by the endpoints $v_1 = (x_1, y_1)$ and $v_2 = (x_2, y_2)$ of B_2. We can define the polyline distance $d(v,s)$ between the vertex v and a polyline segment s, mathematically represented as follows:

$$d(v,s) = \begin{cases} d_{\perp}, & 0 \le \lambda \le 1, \\ \min\{d_1, d_2\}, & \lambda < 0, \lambda > 1, \end{cases} \tag{11.9}$$

where

$$d_1 = \sqrt{(x_0 - x_1)^2 + (y_0 - y_1)^2}, \tag{11.10}$$

$$d_2 = \sqrt{(x_0 - x_2)^2 + (y_0 - y_2)^2}, \tag{11.11}$$

$$\lambda = \frac{(y_2 - y_1)(y_0 - y_1) + (x_2 - x_1)(x_0 - x_1)}{(x_2 - x_1)^2 + (y_2 - y_1)^2}, \tag{11.12}$$

$$d_{\perp} = \frac{(y_2 - y_1)(x_1 - x_0) + (x_2 - x_1)(y_0 - y_1)}{\sqrt{(x_2 - x_1)^2 + (y_2 - y_1)^2}}, \tag{11.13}$$

being:

- d_1 and d_2 are the Euclidean distances between the vertex v and the endpoints of segment s.
- λ is the distance along the vector of the segment s.
- $d_{\perp}$ is the perpendicular distance between v and the segment s.

The polyline distance from vertex v to the boundary B_2 can be defined as $d(v, B_2) = \min_{s \in B_2} \{d(v,s)\}$. The distance between the vertexes of B_1 to the segments of B_2 is defined as the sum of the distances from the vertexes of B_1 to the closest segment of B_2:

$$d(B_1, B_2) = \sum_{v \in B_1} d(v, B_2) \tag{11.14}$$

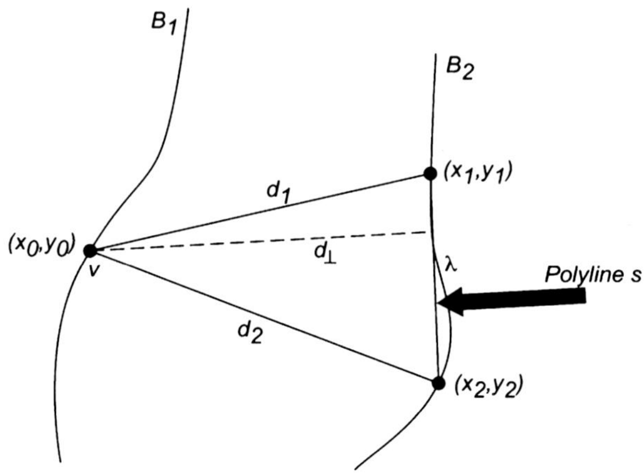

Fig. 11.14 Schematic representation of the polyline distance metric (PDM) between two boundaries B_1 and B_2. In the example, the PDM is computed between the vertex (x_0, y_0) of boundary B_1 and the polyline s of boundary B_2

Similarly, it is possible to calculate $d(B_2, B_1)$ (i.e., the distance between the vertices of B_2 to the closest segment of B_1) by simply swapping the boundaries. The polyline distance between boundaries is the defined as follows:

$$D(B_1, B_2) = \frac{d(B_1, B_2) + d(B_2, B_1)}{(\text{No. of vertices of } B_1 + \text{No. of vertices of } B_2)}. \tag{11.15}$$

This measure $D(B_1, B_2)$ is very important since it reflects the average distance between two computer-generated boundaries and the corresponding ground truth profiles.

11.4.4 Percent Statistic Test

The percent statistic test was first introduced by Chalana et al. [114] and then modified by Alberola-Lòpez et al. [115]. Basically, it is used to test if the computer-generated boundaries differ from manual tracings as much as the manual tracings differ from one another. This test is very important, since many authors pointed out the nontrascurable difference between human tracings made by different experts [22, 63, 68, 72, 78, 82, 83, 87]. The basic idea is that if the computer-generated boundary behaves like a human-generated boundary, it must have the same probability of falling within the inter-observer range as the manual segmentation. Let D_m be the maximum distance between any two tracings (i.e., $D_m = \max_{i,j} \{D_{ij}\}$ for $i \neq j$). A tracing falls in the inter-observer range if the distances separating it from the others tracings are all lower than D_m. Assuming the human tracings as independent and identically distributed, the probability p of the computer-generated boundary falling into the inter-observer range is equal to $p = \frac{n-1}{n+1}$, being $n+1$ the total number of tracings (i.e., n human tracing and 1 computer-generated boundary. In our study $n=3$) and $n-1$ the number of contours minus the two contours with distance D_m. Defining X_j as the event "the computer-generated tracing lies within the inter-observer range for the j-th image," then X_j is a random variable with Bernoulli distribution of parameters p and $q=1-p$. Having a database consisting of N images, the variable $Z = \sum_{j=1}^{N} \frac{X_j}{N}$ can be considered as normally distributed with mean value equal to p and standard deviation equal to pq/N. As we are trying to determine whether the computer-generated tracing falls outside the inter-observer range more often than the human tracings, we seek the one-sided confidence interval for the variable Z, i.e., the θ value for which $P(p-Z > \theta) = \alpha$, where $1-\alpha$ is the significance level. It was shown that $\theta = \sqrt{pq/N}\, z_{1-\alpha}$ [115], where $z_{1-\alpha}$ is the value of a normal standard variable leaving and area equal to $1-\alpha$ to its right. Therefore, the acceptance region for this test is where the critical value Z_0 is greater $p-\theta$

In some cited studies, this test was used to show that computer-generated LI and MA boundaries were to be considered as human tracings [82, 83, 87].

11.4.5 Manual and Computer-Measured IMT

When processing large sets of ultrasound images, the user comes up with two sets of IMT measurements: the computer-measured and the manual IMT values. Comparison between the two sets was done using the following criteria:

- *Correlation*: The two sets were correlated and the Pearson's R coefficient is used to give an estimate of the measurement agreement. This measure is statistically efficient but is unable to detect possible bias in computer measurements.
- *Bland-Altman plot*: The means of the automated and manual measurements are plotted with respect to the differences (see [63, 72] among others). This method is very effective in pointing out possible biases in IMT estimation. The importance of this comparison is very high for carotid IMT measurements, since all the so far developed techniques showed a negative IMT measurement bias. This means that computer methods underestimate IMT. As previously discussed, this could potentially be a problem in clinical practice, where an increase in IMT is index of augmented cardiovascular risk or of incipient atherosclerosis.

Delsanto et al. [82] preferred to give full characterization of the IMT estimation errors by representing the error histograms. We believe that this representation could be potentially useful in assessing the IMT measurement performance, since it shows the statistical error distribution.

Cheng et al. [81] were the only ones who used the MSE for performance evaluation.

11.5 Discussion and Future Perspectives

Accurate carotid wall segmentation in ultrasound imaging is the key to perform a precise IMT measurement. From a functional point of view, the segmentation strategies that have been proposed so far essentially are divided into two groups:

1. Techniques for the computer-aided IMT measurement in an integrated framework.
2. Completely user-independent techniques for CA wall segmentation.

The two groups can be seen as complementary: while the algorithms of group 1 are mainly devoted to the IMT measurement under the supervision of a human operator, the key point of the techniques in group 2 is complete automation and robustness over large image databases.

On the overall, the techniques belonging to group 1 offer better IMT measurement performances than those of group 2. This is due to the fact that, under human supervision, IMT is usually measured in a portion of the image where noise is low and there are no artifacts. The interaction with the user is certainly beneficial in terms of measurement performance.

To test the influence of the user interaction on the IMT measurement performance, we implemented and ran on the same image database two techniques: the first belonging to the group 1 (i.e., interactive techniques for accurate IMT measurement) and the second belonging to group 2 (i.e., completely user-independent technique). We choose the snake-based segmentation strategy proposed by Loizou et al. for group 1 [63] (we call it Tech1 in the following) and the completely automated technique based on feature extraction, fitting, and classification as proposed by Molinari et al. [64] for group 2 (which we call Tech2 in the following). A trained sonographer manually selected a rectangular ROI placed on the CA distal wall, in a region of the image free from blood backscattering, excessive noise, and artifacts. The length of the ROI was fixed to 9 mm (i.e., 144 pixels) for all the images. Then we ran the IMT measurement procedure in that region only with the two techniques. We tested the technique on a database consisting of 200 randomly selected longitudinal images containing both healthy and diseased vessels. We discarded images with plagued vessels. We used the MAD metrics to measure IMT and to calculate the measurement error. Tech1 showed an IMT measurement error equal to 20.0 ± 12.1 μm; Tech2 showed 20.6 ± 14.4 μm. By omitting user interaction, Tech2, which is completely automated, showed a measurement error that increased to 53.5 ± 32.3 μm. Therefore, user interaction is still crucial to obtain optimal IMT measurement performances.

One of the possible development scenarios for the techniques of group 2 is the adoption of an intelligent strategy for reducing the image to a ROI in which the IMT could be measured with optimal performances. Neural networks, fuzzy logic, trained classifiers, could all help in selecting the best image ROIs where to perform IMT measurement. However, the insertion of such strategy would increase the computational cost of the technique, which may preclude real-time IMT measurement.

A major advantage of the techniques belonging to group 2 is versatility: being developed in order to process large amounts of data, they are robust with respect to image characteristics (i.e., gray scale, gains, contrast, ...) and to vessel morphology (i.e., straight, curved, inclined, plagued vessels can all be processed). On the contrary, most of the techniques of group 1 offer very powerful performance but on image with selected characteristics. Future emerging techniques should, therefore, integrate versatility and performance, while keeping the processing time as short as possible. The most prominent future developments of CA segmentation in this direction may be based on morphological transforms and trained systems. The basic idea should be to merge the performance of techniques 1 with the versatility of techniques 2. Morphological transforms (like the watershed transform) may be of help in selecting the image regions with highest contrast and lower noise, while training systems may be used to correct local segmentation defects on the basis of previously acquired experience.

The direct comparison of IMT measurement performance, as discussed in Sect. 3, is not straightforward. A confounding factor is given by the axial resolution of the ultrasound scanner used. Most of the authors worked with images interpolated at the same resolution, typically comprised between 10 and 16 pixels/mm, [22, 61, 63, 64, 70, 82, 83, 88, 96, 116], even though many other authors did not provide clear data. The error conversion from pixel to millimeter could then give misleading results. As a general overview, given the resolution of commercially available scanners, we believe that an axial resolution better than 70 μm is optimal to perform IMT measurement.

The best performing technique so far developed is the one from Faita et al. [72]. Their best performance could reach 1 μm of IMT measurement error. Also, the use of the FOAM operator ensured robustness to noise and applicability in several practical conditions. Computational cost of this methodology was relatively low and anyway suitable to clinical use.

11.6 Conclusions

Computer algorithms for the processing of carotid ultrasound images have continuously grown in number and performance in last 10 years. Starting from early techniques mainly based on edge-detection approaches, complex architecture have been developed and are now available to perform user-independent ultrasound images segmentation. Active contours, dynamic programming, geometrical and modeling approaches, local statistics-based techniques, and integrated approaches have been presented to segment the carotid wall and trace the boundaries of the lumen–intima and media–adventitia interfaces.

Beside accurate IMT measurement, segmentation of the carotid artery is crucial for assessing the progression of the atherosclerotic disease. Monitoring of the carotid wall dimension and morphology may be used to clinically evaluate the effects of drug therapies or follow-up subjects with high cardiovascular risk.

Major challenges that will be faced in future will be the merging between high measurement performance and automation. Ultrasound vascular images present high variability caused by anatomy, ultrasound equipment, and operator skills. Characterization and validation studies will be required in order to carefully assess the effect of such variability on segmentation performance.

Standardization should be introduced in the validation process and performance metric adoption. We suggest to use the polyline distance metric (PDM) measure as performance metric: being independent of the number of points of a contour and of points location, PDM appears as the optimal measure for IMT measurement and performance evaluation (i.e., distance from ground truth).

This chapter is an extended and modified version of the work by Molinari et al. [117].

References

1. W.H. Organization. Cardiovascular disease. Available from: http://www.who.int/cardiovascular_diseases/en/
2. J.J. Badimon, B. Ibanez, and G. Cimmino, Genesis and dynamics of atherosclerotic lesions: implications for early detection, Cerebrovasc Dis, 27(Suppl 1), (2009), 38–47.
3. M. Walter, Interrelationships among HDL metabolism, aging, and atherosclerosis, Arterioscler Thromb Vasc Biol, 29(9), (2009), 1244–50.
4. A.M. Kampoli, D. Tousoulis, C. Antoniades, G. Siasos, and C. Stefanadis, Biomarkers of premature atherosclerosis, Trends Mol Med, 15(7), (2009), 323–32.
5. C. Antoniades, C. Bakogiannis, D. Tousoulis, A.S. Antonopoulos, and C. Stefanadis, The CD40/CD40 ligand system: linking inflammation with atherothrombosis, J Am Coll Cardiol, 54(8), (2009), 669–77.
6. E.M. Laufer, M.H. Winkens, J. Narula, and L. Hofstra, Molecular imaging of macrophage cell death for the assessment of plaque vulnerability, Arterioscler Thromb Vasc Biol, 29(7), (2009), 1031–8.
7. F. Montecucco and F. Mach, Update on statin-mediated anti-inflammatory activities in atherosclerosis, Semin Immunopathol, 31(1), (2009), 127–42.
8. P.T. Kovanen, Mast cells in atherogenesis: actions and reactions, Curr Atheroscler Rep, 11(3), (2009), 214–9.
9. S.H. Johnsen and E.B. Mathiesen, Carotid plaque compared with intima-media thickness as a predictor of coronary and cerebrovascular disease, Curr Cardiol Rep, 11(1), (2009), 21–7.
10. Y.H. Lee and S.J. Yeh, Correlation of common carotid artery intima media thickness, intracranial arterial stenosis and post-stroke cognitive impairment, Acta Neurol Taiwan, 16(4), (2007), 207–13.
11. Y. Fusegawa, H. Hashizume, T. Okumura, Y. Deguchi, Y. Shina, Y. Ikari, and T. Tanabe, Hypertensive patients with carotid artery plaque exhibit increased platelet aggregability, Thromb Res, 117(6), (2006), 615–22.
12. M. Soylu, A.D. Demir, O. Ozdemir, Y. Uzun, and S. Korkmaz, Relationship between plaque morphology of carotid artery and aortic valve calcification, Angiology, 54(6), (2003), 637–40.
13. L.A. Lange, D.W. Bowden, C.D. Langefeld, L.E. Wagenknecht, J.J. Carr, S.S. Rich, W.A. Riley, and B.I. Freedman, Heritability of carotid artery intima-medial thickness in type 2 diabetes, Stroke, 33(7), (2002), 1876–81.
14. P.J. Touboul, R. Hernandez-Hernandez, S. Kucukoglu, K.S. Woo, E. Vicaut, J. Labreuche, C. Migom, H. Silva, and R. Vinueza, Carotid artery intima media thickness, plaque and

Framingham cardiovascular score in Asia, Africa/Middle East and Latin America: the PARC-AALA study, Int J Cardiovasc Imaging, 23(5), (2007), 557–67.

15. A.H. Thakore, C.Y. Guo, M.G. Larson, D. Corey, T.J. Wang, R.S. Vasan, R.B. D'Agostino, Sr., I. Lipinska, J.F. Keaney, Jr., E.J. Benjamin, and C.J. O'Donnell, Association of multiple inflammatory markers with carotid intimal medial thickness and stenosis (from the Framingham Heart Study), Am J Cardiol, 99(11), (2007), 1598–602.
16. R. Elosua, L.A. Cupples, C.S. Fox, J.F. Polak, R.A. D'Agostino, Sr., P.A. Wolf, C.J. O'Donnell, and J.M. Ordovas, Association between well-characterized lipoprotein-related genetic variants and carotid intimal medial thickness and stenosis: the Framingham Heart Study, Atherosclerosis, 189(1), (2006), 222–8.
17. P.J. Touboul, J. Labreuche, E. Vicaut, and P. Amarenco, Carotid intima-media thickness, plaques, and Framingham risk score as independent determinants of stroke risk, Stroke, 36(8), (2005), 1741–5.
18. D. Gaitini and M. Soudack, Diagnosing carotid stenosis by Doppler sonography: state of the art, J Ultrasound Med, 24(8), (2005), 1127–36.
19. A. Simon, J. Gariepy, G. Chironi, J.L. Megnien, and J. Levenson, Intima-media thickness: a new tool for diagnosis and treatment of cardiovascular risk, J Hypertens, 20(2), (2002), 159–69.
20. E. de Groot, S.I. van Leuven, R. Duivenvoorden, M.C. Meuwese, F. Akdim, M.L. Bots, and J.J. Kastelein, Measurement of carotid intima-media thickness to assess progression and regression of atherosclerosis, Nat Clin Pract Cardiovasc Med, 5(5), (2008), 280–8.
21. G.S. Mintz, S.E. Nissen, W.D. Anderson, S.R. Bailey, R. Erbel, P.J. Fitzgerald, F.J. Pinto, K. Rosenfield, R.J. Siegel, E.M. Tuzcu, and P.G. Yock, American College of Cardiology Clinical Expert Consensus Document on Standards for Acquisition, Measurement and Reporting of Intravascular Ultrasound Studies (IVUS). A report of the American College of Cardiology Task Force on Clinical Expert Consensus Documents, J Am Coll Cardiol, 37(5), (2001), 1478–92.
22. C.P. Loizou, C.S. Pattichis, A.N. Nicolaides, and M. Pantziaris, Manual and automated media and intima thickness measurements of the common carotid artery, IEEE Trans Ultrason Ferroelectr Freq Control, 56(5), (2009), 983–94.
23. J.K. Balasundaram and R.S. Banu, A non-invasive study of alterations of the carotid artery with age using ultrasound images, Med Biol Eng Comput, 44(9), (2006), 767–72.
24. P. Tortoli, R. Bettarini, F. Guidi, F. Andreuccetti, and D. Righi, A simplified approach for real-time detection of arterial wall velocity and distension, IEEE Trans Ultrason Ferroelectr Freq Control, 48(4), (2001), 1005–12.
25. M.M. Hermans, J.P. Kooman, V. Brandenburg, M. Ketteler, J.G. Damoiseaux, J.W. Cohen Tervaert, I. Ferreira, P.L. Rensma, U. Gladziwa, A.A. Kroon, A.P. Hoeks, C.D. Stehouwer, and K.M. Leunissen, Spatial inhomogeneity of common carotid artery intima-media is increased in dialysis patients, Nephrol Dial Transplant, 22(4), (2007), 1205–12.
26. H. Schargrodsky, R. Hernandez-Hernandez, B.M. Champagne, H. Silva, R. Vinueza, L.C. Silva Aycaguer, P.J. Touboul, C.P. Boissonnet, J. Escobedo, F. Pellegrini, A. Macchia, and E. Wilson, CARMELA: assessment of cardiovascular risk in seven Latin American cities, Am J Med, 121(1), (2008), 58–65.
27. P.J. Touboul, M.G. Hennerici, S. Meairs, H. Adams, P. Amarenco, N. Bornstein, L. Csiba, M. Desvarieux, S. Ebrahim, M. Fatar, R. Hernandez Hernandez, M. Jaff, S. Kownator, P. Prati, T. Rundek, M. Sitzer, U. Schminke, J.C. Tardif, A. Taylor, E. Vicaut, K.S. Woo, F. Zannad, and M. Zureik, Mannheim carotid intima-media thickness consensus (2004–2006). An update on behalf of the Advisory Board of the 3rd and 4th Watching the Risk Symposium, 13th and 15th European Stroke Conferences, Mannheim, Germany, 2004, and Brussels, Belgium, 2006, Cerebrovasc Dis, 23(1), (2007), 75–80.
28. H. Watanabe, K. Yamane, G. Egusa, and N. Kohno, Influence of westernization of lifestyle on the progression of IMT in Japanese, J Atheroscler Thromb, 11(6), (2004), 330–4.
29. M.J. Roman, T.Z. Naqvi, J.M. Gardin, M. Gerhard-Herman, M. Jaff, and E. Mohler, American society of echocardiography report. Clinical application of noninvasive vascular ultrasound in

cardiovascular risk stratification: a report from the American Society of Echocardiography and the Society for Vascular Medicine and Biology, Vasc Med, 11(3), (2006), 201–11.
30. A.R. Bhuiyan, S.R. Srinivasan, W. Chen, T.K. Paul, and G.S. Berenson, Correlates of vascular structure and function measures in asymptomatic young adults: the Bogalusa Heart Study, Atherosclerosis, 189(1), (2006), 1–7.
31. L. Liu, F. Zhao, Y. Yang, L.T. Qi, B.W. Zhang, F. Chen, D. Ciren, B. Zheng, S.Y. Wang, Y. Huo, and L.S. Liu, The clinical significance of carotid intima-media thickness in cardiovascular diseases: a survey in Beijing, J Hum Hypertens, 22(4), (2008), 259–65.
32. J.M. U-King-Im, V. Young, and J.H. Gillard, Carotid-artery imaging in the diagnosis and management of patients at risk of stroke, Lancet Neurol, 8(6), (2009), 569–80.
33. J. Roquer, T. Segura, J. Serena, and J. Castillo, Endothelial dysfunction, vascular disease and stroke: the ARTICO study, Cerebrovasc Dis, 27(Suppl 1), (2009), 25–37.
34. T. De Meyer, E.R. Rietzschel, M.L. De Buyzere, M.R. Langlois, D. De Bacquer, P. Segers, P. Van Damme, G.G. De Backer, P. Van Oostveldt, W. Van Criekinge, T.C. Gillebert, and S. Bekaert, Systemic telomere length and preclinical atherosclerosis: the Asklepios Study, Eur Heart J, 30(24), (2009), 3074–81.
35. R. Warwick, P. Sastry, E. Fontaine, and M. Poullis, Carotid artery diameter, plaque morphology, and hematocrit, in addition to percentage stenosis, predict reduced cerebral perfusion pressure during cardiopulmonary bypass: a mathematical model, J Extra Corpor Technol, 41(2), (2009), 92–6.
36. North American Symptomatic Carotid Endarterectomy Trial Collaborators, Beneficial effect of carotid endarterectomy in symptomatic patients with high-grade carotid stenosis, N Engl J Med, 325(7), (1991), 445–53.
37. M. Fisher, A. Martin, M. Cosgrove, and J.W. Norris, The NASCET-ACAS plaque project. North American Symptomatic Carotid Endarterectomy Trial. Asymptomatic Carotid Atherosclerosis Study, Stroke, 24(12 Suppl), (1993), I24–5; discussion I31–2.
38. L. Saba and G. Mallarini, A comparison between NASCET and ECST methods in the study of carotids Evaluation using Multi-Detector-Row CT angiography, Eur J Radiol, (2009).
39. E. Kyriacou, M.S. Pattichis, C. Christodoulou, C.S. Pattichis, S. Kakkos, and A. Nicolaides, Ultrasound imaging in the analysis of carotid plaque morphology for the assessment of stroke, In: J.S. Suri, C. Yuan, D.L. Wilson, S. Laxminarayan (eds), Plaque characterization using multimodality imaging: pixel to molecular, pp. 241–275 (2005, IOS Press, Amsterdam).
40. B.J. Schiro and M.H. Wholey, The expanding indications for virtual histology intravascular ultrasound for plaque analysis prior to carotid stenting, J Cardiovasc Surg (Torino), 49(6), (2008), 729–36.
41. N. Nighoghossian, L. Derex, and P. Douek, The vulnerable carotid artery plaque: current imaging methods and new perspectives, Stroke, 36(12), (2005), 2764–72.
42. G. Riccioni, L.A. Bazzano, T. Bucciarelli, B. Mancini, E. di Ilio, and N. D'Orazio, Rosuvastatin reduces intima-media thickness in hypercholesterolemic subjects with asymptomatic carotid artery disease: the Asymptomatic Carotid Atherosclerotic Disease in Manfredonia (ACADIM) Study, Expert Opin Pharmacother, 9(14), (2008), 2403–8.
43. J.J. Kastelein, E. de Groot, and R. Sankatsing, Atherosclerosis measured by B-mode ultrasonography: effect of statin therapy on disease progression, Am J Med, 116(Suppl 6A), (2004), 31S–6.
44. J.J. Kastelein, A. Wiegman, and E. de Groot, Surrogate markers of atherosclerosis: impact of statins, Atheroscler Suppl, 4(1), (2003), 31–6.
45. C.P. Loizou, M. Pantziaris, M.S. Pattichis, E. Kyriacou, and C.S. Pattichis, Ultrasound image texture analysis of the intima and media layers of the common carotid artery and its correlation with age and gender, Comput Med Imaging Graph, 33(4), (2009), 317–24.
46. C.I. Christodoulou, C.S. Pattichis, M. Pantziaris, and A. Nicolaides, Texture-based classification of atherosclerotic carotid plaques, IEEE Trans Med Imaging, 22(7), (2003), 902–12.
47. C.D. Ainsworth, C.C. Blake, A. Tamayo, V. Beletsky, A. Fenster, and J.D. Spence, 3D ultrasound measurement of change in carotid plaque volume: a tool for rapid evaluation of new therapies, Stroke, 36(9), (2005), 1904–9.

48. R.A. Baldewsing, J.A. Schaar, F. Mastik, C.W.J. Oomens, and A.F.W. van der Steen, Assessment of vulnerable plaque composition by matching the deformation of a parametric plaque model to measured plaque deformation, IEEE Trans Med Imaging, 24(4), (2005), 514–28.
49. C.P. Loizou, C.S. Pattichis, M. Pantziaris, and A. Nicolaides, An integrated system for the segmentation of atherosclerotic carotid plaque, IEEE Trans Inf Technol Biomed, 11(6), (2007), 661–7.
50. S.G. Mougiakakou, S. Golemati, I. Gousias, A.N. Nicolaides, and K.S. Nikita, Computer-aided diagnosis of carotid atherosclerosis based on ultrasound image statistics, laws' texture and neural networks, Ultrasound Med Biol, 33(1), (2007), 26–36.
51. K. Nasu, E. Tsuchikane, O. Katoh, N. Tanaka, M. Kimura, M. Ehara, Y. Kinoshita, T. Matsubara, H. Matsuo, K. Asakura, Y. Asakura, M. Terashima, T. Takayama, J. Honye, A. Hirayama, S. Saito, and T. Suzuki, Effect of fluvastatin on progression of coronary atherosclerotic plaque evaluated by virtual histology intravascular ultrasound, JACC Cardiovasc Interv, 2(7), (2009), 689–96.
52. E.M. Tuzcu and P. Schoenhagen, Atherosclerosis imaging: intravascular ultrasound, Drugs, 64(Suppl 2), (2004), 1–7.
53. S.E. Nissen, Rationale for a postintervention continuum of care: insights from intravascular ultrasound, Am J Cardiol, 86(4B), (2000), 12H–7H.
54. C.L. de Korte and A.F. van der Steen, Intravascular ultrasound elastography: an overview, Ultrasonics, 40(1–8), (2002), 859–65.
55. C.L. de Korte, J.A. Schaar, F. Mastik, P.W. Serruys, and A.F. van der Steen, Intravascular elastography: from bench to bedside, J Interv Cardiol, 16(3), (2003), 253–9.
56. F. Molinari, W. Liboni, E. Pavanelli, P. Giustetto, S. Badalamenti, and J.S. Suri, Accurate and automatic carotid plaque characterization in contrast enhanced 2-d ultrasound images, Conf Proc IEEE Eng Med Biol Soc, 2007, (2007), 335–8.
57. E. Vicenzini, M.F. Giannoni, F. Puccinelli, M.C. Ricciardi, M. Altieri, V. Di Piero, B. Gossetti, F.B. Valentini, and G.L. Lenzi, Detection of carotid adventitial vasa vasorum and plaque vascularization with ultrasound cadence contrast pulse sequencing technique and echo-contrast agent, Stroke, 38(10), (2007), 2841–3.
58. J.G. Abbott and F.L. Thurstone, Acoustic speckle: theory and experimental analysis, Ultrason Imaging, 1(4), (1979), 303–24.
59. N. Lucev, D. Bobinac, I. Maric, and I. Drescik, Variations of the great arteries in the carotid triangle, Otolaryngol Head Neck Surg, 122(4), (2000), 590–1.
60. A.C. Rossi, P.J. Brands, and A.P. Hoeks, Automatic recognition of the common carotid artery in longitudinal ultrasound B-mode scans, Med Image Anal, 12(6), (2008), 653–65.
61. F. Molinari, W. Liboni, P. Giustetto, S. Badalamenti, and J.S. Suri, Automatic computer-based tracings (ACT) in longitudinal 2-D ultrasound images using different scanners, J Mech Med Biol, 9(4), (2009), 481–505.
62. C.P. Loizou, C.S. Pattichis, M. Pantziaris, T. Tyllis, and A. Nicolaides, Quality evaluation of ultrasound imaging in the carotid artery based on normalization and speckle reduction filtering, Med Biol Eng Comput, 44(5), (2006), 414–26.
63. C.P. Loizou, C.S. Pattichis, M. Pantziaris, T. Tyllis, and A. Nicolaides, Snakes based segmentation of the common carotid artery intima media, Med Biol Eng Comput, 45(1), (2007), 35–49.
64. F. Molinari, G. Zeng, and J.S. Suri, An integrated approach to computer-based automated tracing and its validation for 200 common carotid arterial wall ultrasound images: a new technique, J Ultrasound Med, 29, (2010), 399–418.
65. C.P. Loizou, C.S. Pattichis, C.I. Christodoulou, R.S.H. Istepanian, M. Pantziaris, and A. Nicolaides, Comparative evaluation of despeckle filtering in ultrasound imaging of the carotid artery, IEEE Trans Ultrason Ferroelectr Freq Control, 52(10), (2005), 1653–1669.
66. Z. Fan, Y. Yang Mo, K. Liang Mong, and K. Yongmin, Nonlinear diffusion in laplacian pyramid domain for ultrasonic speckle reduction, IEEE Trans Med Imaging, 26(2), (2007), 200–11.
67. J.F. Stoltz and M. Donner, New trends in clinical hemorheology: an introduction to the concept of the hemorheological profile, Schweiz Med Wochenschr Suppl, 43, (1991), 41–9.
68. P. Pignoli and T. Longo, Evaluation of atherosclerosis with B-mode ultrasound imaging, J Nucl Med Allied Sci, 32(3), (1988), 166–73.

69. P.J. Touboul, P. Prati, P.Y. Scarabin, V. Adrai, E. Thibout, and P. Ducimetiere, Use of monitoring software to improve the measurement of carotid wall thickness by B-mode imaging, J Hypertens Suppl, 10(5), (1992), S37–41.
70. C. Liguori, A. Paolillo, and A. Pietrosanto, An automatic measurement system for the evaluation of carotid intima-media thickness, IEEE Trans Instrum Meas, 50(6), (2001), 1684–91.
71. J.H. Stein, C.E. Korcarz, M.E. Mays, P.S. Douglas, M. Palta, H. Zhang, T. Lecaire, D. Paine, D. Gustafson, and L. Fan, A semiautomated ultrasound border detection program that facilitates clinical measurement of ultrasound carotid intima-media thickness, J Am Soc Echocardiogr, 18(3), (2005), 244–51.
72. F. Faita, V. Gemignani, E. Bianchini, C. Giannarelli, L. Ghiadoni, and M. Demi, Real-time measurement system for evaluation of the carotid intima-media thickness with a robust edge operator, J Ultrasound Med, 27(9), (2008), 1353–61.
73. C. Schmidt and I. Wendelhag, How can the variability in ultrasound measurement of intima-media thickness be reduced? Studies of interobserver variability in carotid and femoral arteries, Clin Physiol, 19(1), (1999), 45–55.
74. I. Wendelhag, Q. Liang, T. Gustavsson, and J. Wikstrand, A new automated computerized analyzing system simplifies readings and reduces the variability in ultrasound measurement of intima-media thickness, Stroke, 28(11), (1997), 2195–200.
75. I. Wendelhag, O. Wiklund, and J. Wikstrand, Arterial wall thickness in familial hypercholesterolemia. Ultrasound measurement of intima-media thickness in the common carotid artery, Arterioscler Thromb, 12(1), (1992), 70–7.
76. I. Wendelhag, T. Gustavsson, M. Suurkula, G. Berglund, and J. Wikstrand, Ultrasound measurement of wall thickness in the carotid artery: fundamental principles and description of a computerized analysing system, Clin Physiol, 11(6), (1991), 565–77.
77. I. Wendelhag, O. Wiklund, and J. Wikstrand, On quantifying plaque size and intima-media thickness in carotid and femoral arteries. Comments on results from a prospective ultrasound study in patients with familial hypercholesterolemia, Arterioscler Thromb Vasc Biol, 16(7), (1996), 843–50.
78. C.D. Furberg, R.P. Byington, and T.E. Craven, Lessons learned from clinical trials with ultrasound end-points, J Intern Med, 236(5), (1994), 575–80.
79. Q. Liang, I. Wendelhag, J. Wikstrand, and T. Gustavsson, A multiscale dynamic programming procedure for boundary detection in ultrasonic artery images, IEEE Trans Med Imaging, 19(2), (2000), 127–42.
80. M.A. Gutierrez, P.E. Pilon, S.G. Lage, L. Kopel, R.T. Carvalho, and S.S. Furuie, Automatic measurement of carotid diameter and wall thickness in ultrasound images, Comput Cardiol, 29, (2002), 359–62.
81. D.C. Cheng, A. Schmidt-Trucksass, K.S. Cheng, and H. Burkhardt, Using snakes to detect the intimal and adventitial layers of the common carotid artery wall in sonographic images, Comput Methods Programs Biomed, 67(1), (2002), 27–37.
82. S. Delsanto, F. Molinari, P. Giustetto, W. Liboni, S. Badalamenti, and J.S. Suri, Characterization of a completely user-independent algorithm for carotid artery segmentation in 2-D ultrasound images, IEEE Trans Instrum Meas, 56(4), (2007), 1265–74.
83. F. Molinari, S. Delsanto, P. Giustetto, W. Liboni, S. Badalamenti, and J.S. Suri, User-independent plaque segmentation and accurate intima-media thickness measurement of carotid artery wall using ultrasound, In: J.S. Suri, R.-F. Chang, C. Kathuria, F. Molinari, A. Fenster (eds), Advances in diagnostic and therapeutic ultrasound imaging, pp. 111–140 (2008, Artech House, Norwood, MA).
84. D.J. Williams and M. Shah, A fast algorithm for active contours and curvature estimation, Cvgip: Image Understanding, 55(1), (1992), 14–26.
85. S. Lobregt and M.A. Viergever, A discrete dynamic contour model, IEEE Trans Med Imaging, 14(1), (1995), 12–24.
86. I.D.G. Macleod, Quantitative study of the orientation bias of some edge detector schemes – comment, IEEE Trans Pattern Anal Mach Intell, 1(4), (1979), 408–9.

87. S. Delsanto, F. Molinari, P. Giustetto, W. Liboni, and S. Badalamenti, CULEX-completely user-independent layers extraction: ultrasonic carotid artery images segmentation, Conf Proc IEEE Eng Med Biol Soc, 6, (2005), 6468–71.
88. S. Delsanto, F. Molinari, W. Liboni, P. Giustetto, S. Badalamenti, and J.S. Suri, User-independent plaque characterization and accurate IMT measurement of carotid artery wall using ultrasound, Conf Proc IEEE Eng Med Biol Soc, 1, (2006), 2404–7.
89. P.M. Shankar, V.A. Dumane, T. George, C.W. Piccoli, J.M. Reid, F. Forsberg, and B.B. Goldberg, Classification of breast masses in ultrasonic B scans using Nakagami and K distributions, Phys Med Biol, 48(14), (2003), 2229–40.
90. P.M. Shankar, Estimation of the Nakagami parameter from log-compressed ultrasonic back-scattered envelopes, J Acoust Soc Am, 114(1), (2003), 70–2.
91. P.M. Shankar, A compound scattering pdf for the ultrasonic echo envelope and its relationship to K and Nakagami distributions, IEEE Trans Ultrason Ferroelectr Freq Control, 50(3), (2003), 339–43.
92. F. Destrempes, J. Meunier, M.F. Giroux, G. Soulez, and G. Cloutier, Segmentation in ultrasonic B-mode images of healthy carotid arteries using mixtures of Nakagami distributions and stochastic optimization, IEEE Trans Med Imaging, 28(2), (2009), 215–29.
93. S. Golemati, J. Stoitsis, T. Balkizas, and K. Nikita, Comparison of B-mode, M-mode and Hough transform methods for measurement of arterial diastolic and systolic diameters, Conf Proc IEEE Eng Med Biol Soc, 2(1), (2005), 1758–61.
94. S. Golemati, J. Stoitsis, E.G. Sifakis, T. Balkizas, and K.S. Nikita, Using the Hough transform to segment ultrasound images of longitudinal and transverse sections of the carotid artery, Ultrasound Med Biol, 33(12), (2007), 1918–32.
95. S. Golemati, T.J. Tegos, A. Sassano, K.S. Nikita, and A.N. Nicolaides, Echogenicity of B-mode sonographic images of the carotid artery: work in progress, J Ultrasound Med, 23(5), (2004), 659–69.
96. J. Stoitsis, S. Golemati, S. Kendros, and K.S. Nikita, Automated detection of the carotid artery wall in B-mode ultrasound images using active contours initialized by the Hough transform, Conf Proc IEEE Eng Med Biol Soc, 2008, (2008), 3146–9.
97. P.V.C. Hough, Method and means for recognizing complex patterns, U. Patent, Editor. 1962.
98. A. Fenster, G. Parraga, A. Landry, B. Chiu, M. Egger, and J.D. Spence, 3-D US Imaging of the Carotid Arteries, in Advances in diagnostic and therapeutic ultrasound imaging, J.S. Suri, et al., Editors. 2008, Artech House: Boston, 67–92.
99. F. Mao, J. Gill, D. Downey, and A. Fenster, Segmentation of carotid artery in ultrasound images: method development and evaluation technique, Med Phys, 27(8), (2000), 1961–70.
100. A. Zahalka and A. Fenster, An automated segmentation method for three-dimensional carotid ultrasound images, Phys Med Biol, 46(4), (2001), 1321–42.
101. A. Landry and A. Fenster, Theoretical and experimental quantification of carotid plaque volume measurements made by three-dimensional ultrasound using test phantoms, Med Phys, 29(10), (2002), 2319–27.
102. U. Schminke, L. Hilker, L. Motsch, B. Griewing, and C. Kessler, Volumetric assessment of plaque progression with 3-dimensional ultrasonography under statin therapy, J Neuroimaging, 12(3), (2002), 245–51.
103. B. Chiu, M. Egger, J.D. Spence, G. Parraga, and A. Fenster, Quantification of carotid vessel wall and plaque thickness change using 3D ultrasound images, Med Phys, 35(8), (2008), 3691–710.
104. M. Egger, A. Krasinski, B.K. Rutt, A. Fenster, and G. Parraga, Comparison of B-mode ultrasound, 3-dimensional ultrasound, and magnetic resonance imaging measurements of carotid atherosclerosis, J Ultrasound Med, 27(9), (2008), 1321–34.
105. J.C.R. Seabra, L.M. Pedro, J. Fernandes e Fernandes, and J.M. Sanches, A 3-D ultrasound-based framework to characterize the echo morphology of carotid plaques, IEEE Trans Biomed Eng, 56(5), (2009), 1442–53.
106. A. Fenster, C. Blake, I. Gyacskov, A. Landry, and J.D. Spence, 3D ultrasound analysis of carotid plaque volume and surface morphology, Ultrasonics, 44(Suppl 1), (2006), e153–7.

107. A. Krasinski, B. Chiu, J.D. Spence, A. Fenster, and G. Parraga, Three-dimensional ultrasound quantification of intensive statin treatment of carotid atherosclerosis, Ultrasound Med Biol, 35(11), (2009), 1763–72.
108. M. Amato, P. Montorsi, A. Ravani, E. Oldani, S. Galli, P.M. Ravagnani, E. Tremoli, and D. Baldassarre, Carotid intima-media thickness by B-mode ultrasound as surrogate of coronary atherosclerosis: correlation with quantitative coronary angiography and coronary intravascular ultrasound findings, Eur Heart J, 28(17), (2007), 2094–101.
109. T. Ogata, M. Yasaka, M. Yamagishi, O. Seguchi, K. Nagatsuka, and K. Minematsu, Atherosclerosis found on carotid ultrasonography is associated with atherosclerosis on coronary intravascular ultrasonography, J Ultrasound Med, 24(4), (2005), 469–74.
110. T. Wakeyama, H. Ogawa, H. Iida, A. Takaki, T. Iwami, M. Mochizuki, and T. Tanaka, Intima-media thickening of the radial artery after transradial intervention. An intravascular ultrasound study, J Am Coll Cardiol, 41(7), (2003), 1109–14.
111. T. Kume, T. Akasaka, T. Kawamoto, N. Watanabe, E. Toyota, Y. Neishi, R. Sukmawan, Y. Sadahira, and K. Yoshida, Assessment of coronary intima-media thickness by optical coherence tomography: comparison with intravascular ultrasound, Circ J, 69(8), (2005), 903–7.
112. R. Sanz-Requena, D. Moratal, D.R. Garcia-Sanchez, V. Bodi, J.J. Rieta, and J.M. Sanchis, Automatic segmentation and 3D reconstruction of intravascular ultrasound images for a fast preliminar evaluation of vessel pathologies, Comput Med Imaging Graph, 31(2), (2007), 71–80.
113. J.S. Suri, R.M. Haralick, and F.H. Sheehan, Greedy algorithm for error correction in automatically produced boundaries from low contrast ventriculograms, Pattern Anal Appl, 3(1), (2000), 39–60.
114. V. Chalana and Y. Kim, A methodology for evaluation of boundary detection algorithms on medical images, IEEE Trans Med Imaging, 16(5), (1997), 642–52.
115. C. Alberola-Lopez, M. Martin-Fernandez, and J. Ruiz-Alzola, Comments on: a methodology for evaluation of boundary detection algorithms on medical images, IEEE Trans Med Imaging, 23(5), (2004), 658–60.
116. J. Stoitsis, S. Golemati, and K.S. Nikita, a modular software system to assist interpretation of medical images—application to vascular ultrasound images, IEEE Trans Instrum Meas, 55(6), (2006), 1944–52.
117. F. Molinari, G. Zeng, and J.S. Suri, A state of the art review on intima-media thickness (IMT) measurement and wall segmentation techniques for carotid ultrasound, Computer Methods and Programs in Biomedicne, (2010 (in press)).

Biographies

Dr. Filippo Molinari received the Italian Laurea and the Ph.D. in electrical engineering from the Politecnico di Torino, Torino, Italy, in 1997 and 2000, respectively. He is leader in ultrasound imaging focused towards tissue characterization, vascular quantification for diagnostics and therapeutics. Currently, he is Assistant Professor at Politecnico di Torino, Italy – Department of Electronics.

Dr. Guang Zeng received the B.S. degree from Xiangtan University, China in 1998. He received the M.S. degree in 2005 and the Ph.D. degree in 2008 from Clemson University, SC, USA, both in Electrical Engineering. He is currently working in the Aging and Dementia Imaging Research Laboratory, Mayo Clinic, Rochester, MN. His research interests include biomedical image processing, pattern recognition and computer vision.

Dr. Jasjit S. Suri is an innovator, scientist, a visionary, an industrialist and an internationally known world leader in Biomedical Engineering. Dr. Suri has spent over 20 years in the field of biomedical engineering/devices and its management. He received his Doctrate from University of Washington, Seattle and Business Management Sciences from Weatherhead, Case Western Reserve University, Cleveland, Ohio. Dr. Suri was crowned with President's Gold medal in 1980 and the Fellow of American Institute of Medical and Biological Engineering for his outstanding contributions.

Chapter 12
3D Carotid Ultrasound Imaging

Grace Parraga, Aaron Fenster, Adam Krasinski, Bernard Chiu, Micaela Egger, and J. David Spence

Abstract Ultrasound (US) phenotypes of carotid atherosclerosis include intima–media thickness (IMT), total plaque area (TPA), total plaque volume (TPV), and Doppler ultrasound-based measurements of stenosis. Doppler US is a well-established screening tool in the assessment of stenosis severity. However, Doppler flow-velocity-based measurements do not provide information on plaque morphology, plaque vulnerability, or composition. The measurement of IMT from B-mode US images is a widely used US phenotype of atherosclerosis and has been regarded as a surrogate measurement of atherosclerosis as it correlates with vascular outcomes. Although the measurement of IMT has been validated in many studies, it is clear that many distinct biological pathways and mechanisms may be reflected by the measurement. More recently, TPA and TPV have emerged as useful US phenotypes of carotid atherosclerosis that measure plaque burden in 2D and 3D, respectively. Total plaque area has been shown to be a stronger predictor of coronary events than IMT [Spence et al., Stroke 33:2916–2922, 2002; Johnsen et al., Stroke 38(11):2873–2880, 2007].

In order to overcome some limitations and accelerate the translation of 3D US measurements of carotid atherosclerosis to clinical research and clinical practice, semiautomated methods of measurement and measurements that are derived from biological components of carotid disease with readily distinguishable US boundaries (enabling multiple observers to be trained in shorter time periods and with decreased interobserver variability) are required. This has stimulated the development and validation of a new 3D US measurement of

---

A. Fenster (✉)
Imaging Research Laboratories, Robarts Research Institute, 100 Perth Drive, P.O. Box 5015, London, ON, Canada, N6A 5K8
and
Department of Medical Imaging, University of Western Ontario, London, ON, Canada
and
Graduate Program in Biomedical Engineering, University of Western Ontario, London, ON, Canada
and
Department of Medical Biophysics, University of Western Ontario, London, ON, Canada
e-mail: afenster@imaging.robarts.ca

Jasjit S. Suri et al. (eds.), *Atherosclerosis Disease Management*,
DOI 10.1007/978-1-4419-7222-4_12,

carotid atherosclerosis – *vessel wall volume* (VWV), which is a measurement of vessel wall thickness and plaque within the common carotid artery, the internal and external carotid branches. This measurement can be more easily semiautomated, and observers can be trained to measure VWV in shorter durations and with greater reliability.

Our objective is to demonstrate that 3D US is a viable technique for quantifying the progression and regression of carotid atherosclerosis.

Keywords 3D ultrasound • Carotid atherosclerosis • Atherosclesotic plaque • Intima media thickness

12.1 Introduction

Ultrasound (US) phenotypes of carotid atherosclerosis include intima–media thickness (IMT) [1], total plaque area (TPA) [2], total plaque volume (TPV) [3–6], and Doppler ultrasound-based measurements of stenosis [7, 8]. Doppler US is a well-established screening tool in the assessment of stenosis severity [7–11]. However, Doppler flow-velocity-based measurements do not provide information on plaque morphology, plaque vulnerability, or composition. The measurement of IMT from B-mode US images is a widely used US phenotype of atherosclerosis and has been regarded as a surrogate measurement of atherosclerosis as it correlates with vascular outcomes [12–14]. Although the measurement of IMT has been validated in many studies, it is clear that many distinct biological pathways and mechanisms may be reflected by the measurement. For example, IMT may represent hypertensive medial hypertrophy [15, 16], compensatory intimal thickening due to mechanical forces of blood flow [17, 18], or the initial "fatty streak" stage of atherosclerosis that involves accumulation of macrophage foam cells in the artery wall [19]. The intima and media thickness fluctuate over time in response to a variety of factors, which may not necessarily be related to atherosclerotic plaque formation and progression. More recently, TPA [2] and TPV [3, 4, 20–24] have emerged as useful US phenotypes of carotid atherosclerosis that measure plaque burden in 2D and 3D, respectively. Total plaque area has been shown to be a stronger predictor of coronary events than IMT [51, 52]. Earlier work demonstrated that TPV in particular can be used to measure changes in plaque burden [3, 4, 20, 25, 26] and evaluate the effects of statin therapy [27, 28]. While the measurement of TPV provides valuable quantitative information about global plaque burden, it does not identify the locations in the vessel where volumetric changes are occurring. Furthermore, the measurement of TPV from 3D US images requires trained observers who are expert in 3D US image interpretation and in distinguishing vessel wall from plaque in 3D US images. Limitations of this approach include image interpretation and measurement differences within and between observers, long training times for observers, and long durations to perform manual segmentations. In order to overcome some of these limitations and accelerate the translation of 3D US measurements of carotid atherosclerosis to clinical research and clinical practice, semiautomated methods of measurement and measurements that are derived from biological components of carotid disease with readily distinguishable US boundaries (enabling multiple observers to be trained in shorter time periods and with decreased interobserver variability) are required [29]. This has stimulated the development and validation of a new 3D US measurement of carotid atherosclerosis – *vessel wall volume* (VWV), which is a measurement of vessel wall thickness and plaque within the common carotid artery, the internal and external carotid branches. This measurement can be more easily semiautomated, and observers can be trained to measure VWV in shorter durations and with greater reliability.

Unlike the measurement of 3D US TPV, which requires observers to distinguish plaque–lumen and plaque–outer vessel wall boundaries, the measurement of 3D US VWV requires an observer to manually outline the lumen–intima/plaque and media–adventitia boundaries – similar to the measurement of IMT. These boundaries are more straightforward to interpret than plaque–lumen and wall boundaries in 3D US images. In addition, VWV boundaries measurements are more regular and circular, which may simplify the development of semiautomated segmentation techniques. In this chapter, we review the method used to acquire 3D carotid ultrasound images and discuss its use in the measurement of TPV and VWV. Our objective is to demonstrate the utility of these approaches and demonstrate that 3D US is a viable technique for quantifying the progression and regression of carotid atherosclerosis.

12.2 3D Carotid Ultrasound Scanning Technique

In this section, we review the method used for acquiring 3D carotid US images and the tools required for visualizing and measuring TPA and VWV. For further information about the technical and computational aspects of 3D US, readers may refer to recent review articles and books on the subject [30–37].

Because images of the carotid arteries require at least a 4 cm of scanning length, real-time 3D (i.e., 4D) systems cannot be used effectively. Thus, all 3D US systems that are currently used to acquire images of the carotid arteries are conventional US transducers that produce 2D US images. Since the use of conventional US transducer must be moved over the carotid arteries to collect all the required 2D images necessary to reconstruct the 3D US image, a method to track the position and orientation of the transducer must be used. Over the past decade, two methods have been developed to image the carotid arteries: mechanical linear scanners and magnetically tracked freehand scanners. We have used the mechanical scanning approach, which is summarized here.

12.2.1 Mechanical Linear 3D Carotid Ultrasound Imaging

Linear scanners use a motorized mechanism to translate the transducer linearly along the neck of the patient as shown in Fig. 12.1. Transverse 2D images of the

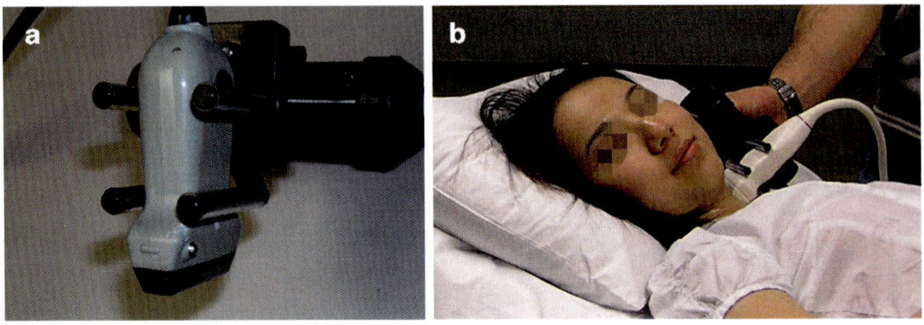

Fig. 12.1 (**a**) Photograph of a mechanical linear scanning mechanism used to acquire 3D carotid US images. The transducer is translated along the arteries, while conventional 2D US images are acquired by a computer and reconstructed into a 3D image in real time. (**b**) Photograph of the system being used to scan the carotid arteries

carotid arteries are acquired at regular spatial intervals as the transducer is moved over the carotid arteries. Each image in the set of acquired 2D images is spaced equally so that all images are parallel to each other, making 3D reconstruction easy and possible in real time. The length of the scan depends on the length of the mechanical scanning mechanism and can be made to be at least 4–6 cm. The resolution of the image in the 3D scanning direction (i.e., along the artery) depends on the elevational resolution of the transducer as well as the spacing between the acquired images. It can be optimized by varying the translating speed and sampling interval to match the sampling rate to the frame rate of the ultrasound machine and to match the sampling interval to half (or smaller) the elevational resolution of the transducer [38]. Typically, we acquire 2D US images every 0.2 mm. If the 2D US images are acquired at 30 frames per second, a 4-cm length will require 200 2D US images, which can be collected in 6.7 s without cardiac gating.

The simple predefined geometry of the acquired 2D US images allows the development of a simple algorithm to reconstruct a 3D image [38]. Thus, using this approach, a 3D image can be reconstructed as the 2D images are being acquired and immediate viewing of the 3D carotid image after scanning is possible to determine if additional 3D scans are necessary. This specific advantage of immediate review of 3D images after a scan significantly shortens the examination time and reduces digital storage requirements, as unnecessary 3D images are not required to be stored.

Because the 3D carotid ultrasound image is produced from a series of conventional 2D images, the resolution in the 3D image will not be isotropic. In the direction parallel to the acquired 2D US image planes, the resolution of the reconstructed 3D image will be equal to the original 2D images; however, in the direction of the 3D scan along the arteries, the resolution of the reconstructed 3D image will depend on the elevational resolution of the transducer and the interslice spacing [38]. Since the elevational resolution is worse than the inplane resolution of the 2D US images, the resolution of the 3D US image will be the poorest in the 3D scanning direction (i.e., elevation). Therefore, a transducer with good elevational resolution should be used to obtain optimal results.

Although the 3D mechanical scanning approach requires a mechanical mechanism to be held by the operator (Fig. 12.1), it offers three advantages: short imaging times, high-quality 3D images, and fast reconstruction times. However, bulkiness and weight of the mechanism sometimes make it inconvenient to use. Linear scanning has been successfully implemented in many vascular imaging applications using B-mode and color Doppler images of the carotid arteries [7, 8, 11, 39, 40], vascular test phantoms [9, 10, 41], and power Doppler images [7–11, 39, 40] and studies of carotid atherosclerosis [42–44]. An example of a mechanical scanning mechanism and its use is shown in Fig. 12.1, and two examples of linearly scanned 3D ultrasound images of carotid arteries with complex plaques are shown in Fig. 12.2.

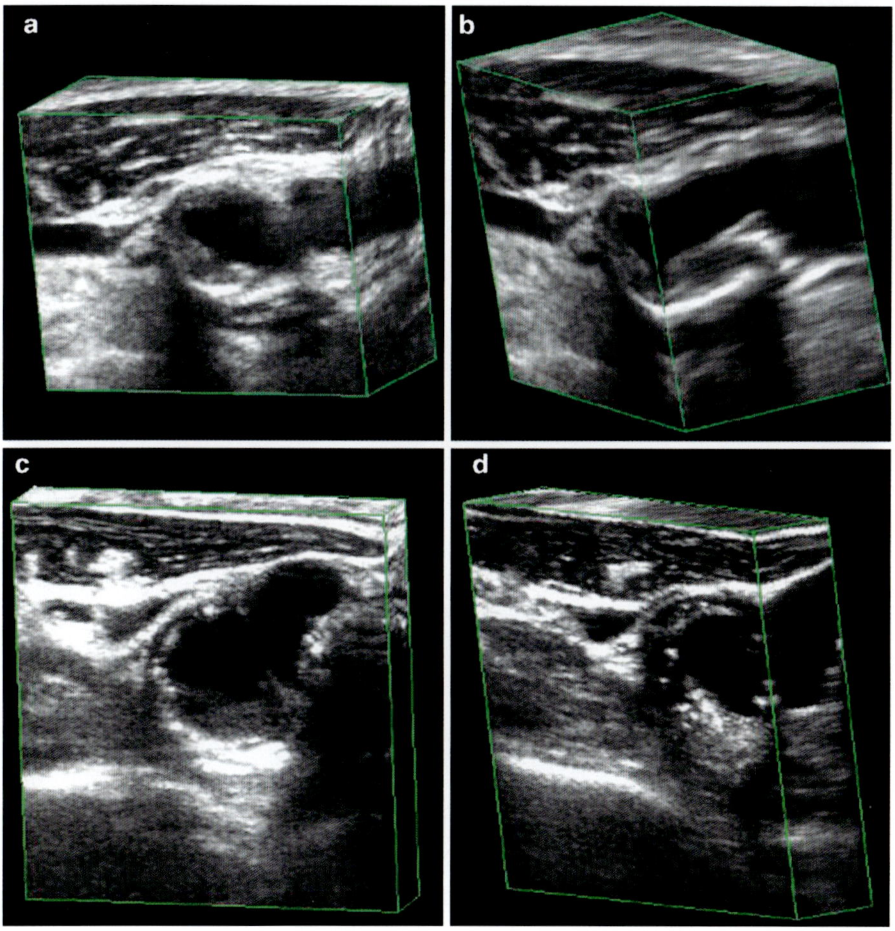

Fig. 12.2 Two 3D carotid ultrasound views of two different patients with complex and ulcerated carotid plaques. For each patient, a transverse (**a**) and (**c**) and longitudinal (**b**) and (**d**) views are shown side by side

12.2.2 3D Carotid Ultrasound Image Reconstruction

The 3D reconstruction procedure involves placing the acquired 2D US images in their correct location within the volume, generating a 3D image. The gray scale values of any voxel not sampled by the 2D US images are then calculated by interpolating between the adjacent 2D US images. As a result, all 2D image information is preserved, allowing viewing of the original 2D planes as well as any other views. Current inexpensive desktop computers are now sufficient for 3D reconstructions to occur while the 2D images are being acquired (i.e., in real time). Thus, it is possible to view the complete 3D carotid US image immediately after the acquisition of the 2D US images has been completed.

12.2.3 Viewing of 3D Carotid Ultrasound Images

Many 3D viewing methods have been developed over the past decade. The method we use most commonly is the *cube view* approach, which is based on multiplanar rendering using texture mapping. In this technique, a 3D US image is displayed as a polyhedron, and the appropriate US image for each plane is "painted" on the face of the cube (texture mapped). Users can rotate the polyhedron to obtain the desired orientation of the 3D US image as well as move any of the surfaces (i.e., by slicing of the 3D image parallel or obliquely) to the original, while the appropriate data is texture-mapped in real time onto the new face. As a result, users always have 3D image-based cues, which relate the plane being manipulated to the rest of the anatomy. These visual cues allow users to efficiently identify the desired structures [30, 31, 35, 36]. Examples of this approach are shown in Figs. 12.2 and 12.3.

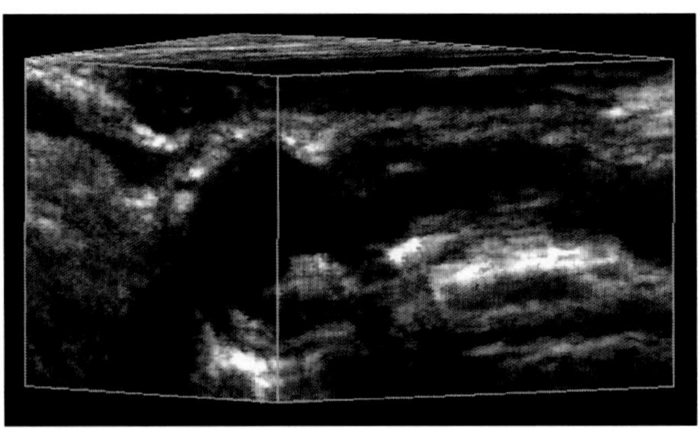

Fig. 12.3 An example of a 3D carotid US image of a patient with carotid atherosclerosis

12.3 Quantification of Carotid Atherosclerosis

12.3.1 Total Plaque Volume

Carotid atherosclerosis can be quantified using two of the following manual planimetry techniques: TPV and VWV techniques. When using the TPV technique, each 3D carotid US image is 'sliced' transverse to the vessel axis, starting from one end of the plaque using an interslice distance (ISD) of 1.0 mm. Using software developed in our laboratory, the plaque is contoured in each cross-sectional image using a cross-haired cursor (see Fig. 12.4). As the contours are manually outlined, the visualization software calculates the area of the contours automatically. Sequential areas enclosed by the contours are averaged and multiplied by the ISD to calculate the incremental volume. A summation of incremental volumes provides a measure of the TPV. After measuring a complete plaque volume, the 3D US image can be viewed in multiple orientations to verify that the entire plaque volume is outlined by the set of contours. A typical plaque requires about 10–30 slices, resulting in approximately 15 min of manual segmentation time.

12.3.2 Vessel Wall Volume

The VWV technique is commonly used for analyzing MR images and is an alternative method for quantifying atherosclerosis in the carotid arteries. Measurement of the VWV proceeds in a similar way to the measurement of TPV. Each 3D carotid US 3D image is 'sliced' transverse to the vessel axis, starting from one end of the 3D US image using an interslice distance (ISD) of 1.0 mm. In this approach, the lumen (blood–intima boundary) and the vessel wall (media–adventitia boundary) are segmented in each slice. The area inside the lumen boundary is subtracted from the area inside the vessel wall boundary to give the vessel wall area. Sequential areas are averaged and multiplied by the ISD to give the incremental VWV. The summation of incremental volumes provides a measure of the total VWV (Figs. 12.5 and 12.6).

12.4 3D Carotid US Studies

12.4.1 Monitoring Carotid Atherosclerosis Regression

A variety of carotid atherosclerosis measurement tools have been developed and used for monitoring of patients at risk of stroke, such as blood pressure and serum cholesterol levels. Here, we summarize the use of 3D carotid US to monitor

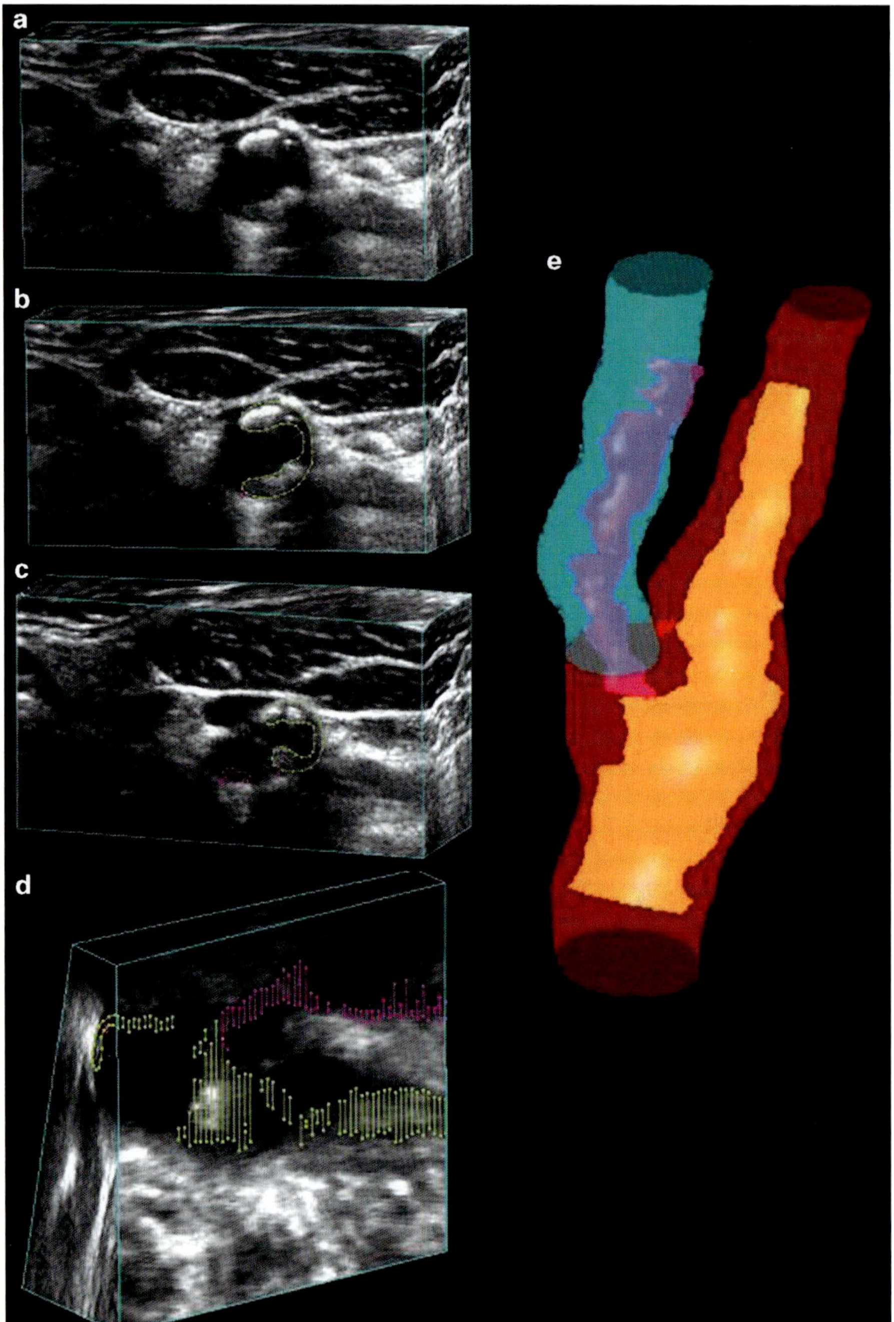

Fig. 12.4 Steps used in measurement of *total plaque volume* from 3D US images. (**a**) First, the 3D image is "sliced" to obtain a transverse view. (**b** and **c**) Using a mouse-driven cross-haired cursor, the plaque is outlined in successive image "slices" until all the plaques have been traversed. (**d**) The vessel can be sliced to reveal a longitudinal view with the outlines of the plaques. (**e**) After outlining all the plaques, the total volume can be calculated, and a mesh fitted to provides a view of the plaque surface together with the boundary of the vessel

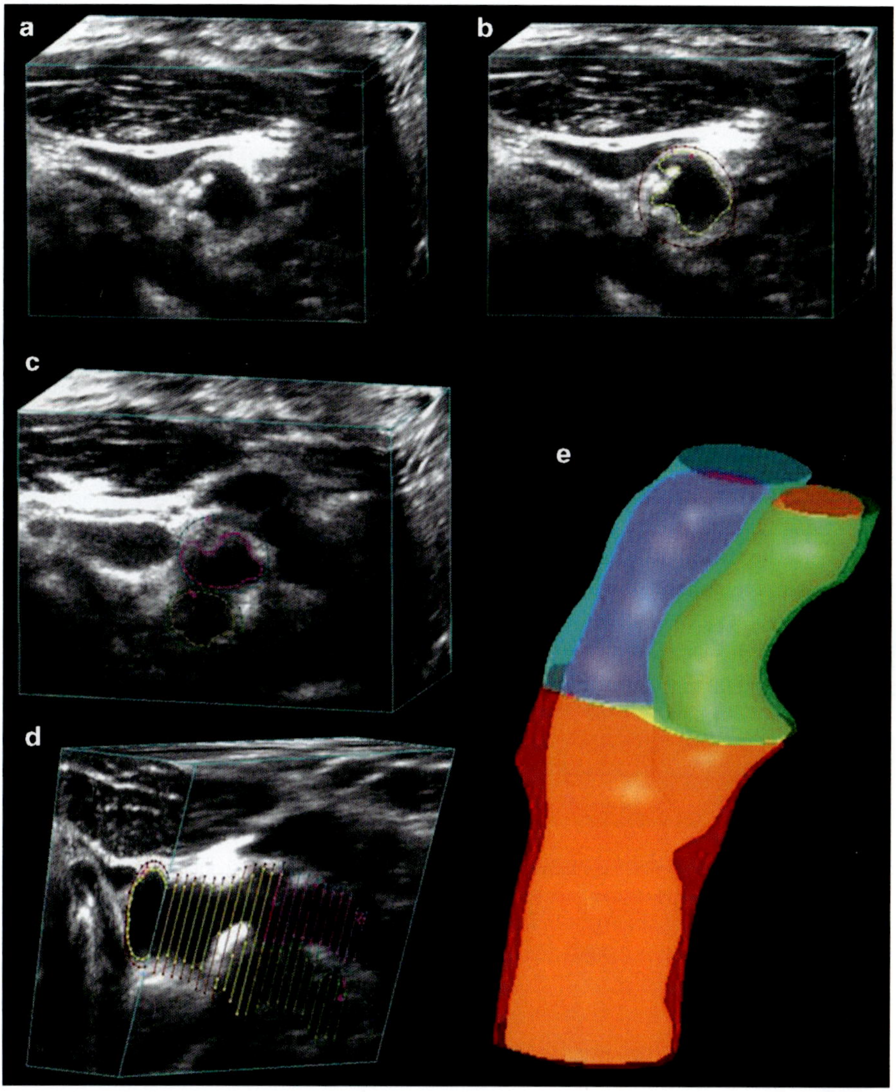

Fig. 12.5 Steps used in measurement of *vessel wall plus plaque* volume from 3D ultrasound images. (**a**) First, the 3D image is "sliced" to obtain a transverse view. (**b** and **c**) Using a mouse-driven cross-haired cursor, the vessel boundary and the lumen boundary plaque are outlined separately in successive image "slices" until all the slices have been traversed (typically 1.5 cm above and below the carotid bifurcation). (**d**) The vessel can be sliced to reveal a longitudinal view with the outlines and correct any errors. (**e**) After outlining has been completed, the *vessel wall plus plaque* volume can be calculated, and a mesh fitted to provides a view of the vessel and the lumen boundaries. Each branch of the carotid artery has been colored differently

response of carotid atherosclerosis to intensive statin treatment of carotid atherosclerosis. In order to visualize changes in the carotid artery and remodelling that occurs during intensive statin treatment and to try to exploit the inherent advantages of 3D imaging, we applied a novel 3D US measurement and method,

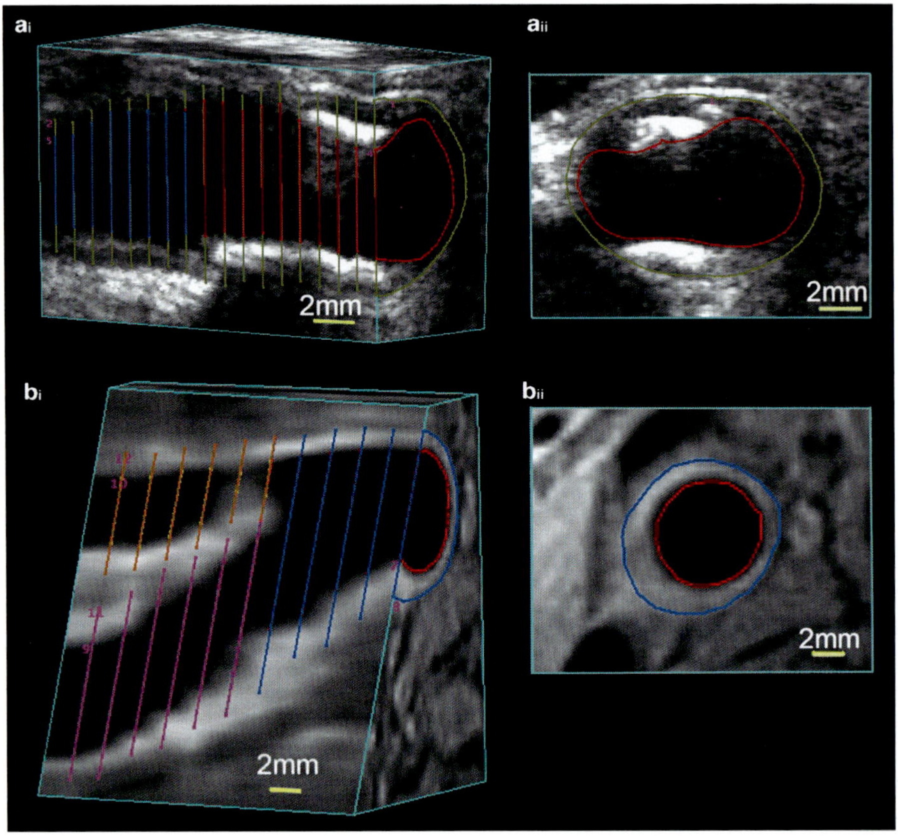

Fig. 12.6 3D views of VWV measurements. The 3D image is "sliced" to obtain a transverse view (Aii and Bii). Using a mouse-driven cross-haired cursor, the plaque is outlined in successive image "slices" until all the plaques have been traversed. The vessel can be sliced to reveal a longitudinal view with the outlines of the plaques (Ai and Bi). After outlining all the plaques, the total volume can be calculated, and a mesh fitted to provides a view of the plaque surface together with the boundary of the vessel

previously developed in our laboratory [3, 4, 28, 42, 43], that analyzes successive carotid artery vessel wall and lumen segmentation outlines to provide measurements of TPA, VWV and carotid VWV thickness and thickness difference maps.

12.4.1.1 TPA Measurements of Intensive Statin Treatment of Carotid Atherosclerosis

Fifty patients with asymptomatic carotid stenosis >60% as defined by carotid Doppler flow velocities were enrolled in this study [28]. Patients with a previous history of angina and myocardial infarction were excluded for safety reasons, but because carotid stenosis is associated with a high risk of cardiac events, we did not wish to expose patients to a long duration of placebo therapy. Patients with atrial fibrillation were also excluded due to the increase of cardiac motion, which alters significantly the carotid wall motion. All subjects gave consent to a protocol approved by the University of Western Ontario Standing Board of Human Research Ethics and were randomized to placebo vs. atorvastatin 80 mg daily for a duration of treatment of 3 months.

The subjects were imaged by 3D carotid US at baseline and 3 months later while recumbent on a gurney with their upper torso inclined approximately 15°. Both carotids were scanned over a scan distance of 4 cm, with the bifurcation located as closely as possible to the center of the volume. The best 3D carotid images of each carotid side was selected based on the position of the bifurcation in the 3D image and image qualities such as shadowing, cardiac motion etc, generating the selection of four images per patient at each time point. Measurements of TPV were made using manual planimetry as described above.

From the initial cohort of 50 subjects, baseline and 3-month TPV measurements were obtained in 38 cases. Some patients dropped out (refused to return for repeat measurement), one died, and some had their images excluded for technical reasons. Characteristics of the subjects were analyzed at baseline and showed that there were no significant differences in risk factors between the two treatment groups. Analysis of the results (see Fig. 12.7) of the TPV measurements showed that baseline plaque volume (mean ± SD) was 722.0 ± 473.7 mm^3 for the placebo group and 689.5 ± 410 mm^3 for the atorvastatin group ($p=0.83$); 3-month plaque volumes were 738.8 ± 494.7 mm^3 on placebo, 599.3 ± 355.2 mm^3 on atorvastatin ($p=0.34$). Over 3 months, plaque volume increased on placebo by 16.8 ± 74.1 mm^3, while on atorvastatin there was significant regression of plaque volume, by −90.3 ± 85.1 mm^3 ($p<0.0001$). The operator variability (SD) was found to be 53 mm^3.

12.4.1.2 VWV Measurements of Intensive Statin Treatment of Carotid Atherosclerosis

The same images used to measure TPV were also used to measure VWV. A single observer blinded to subject identity, treatment, and time point performed manual segmentation of all carotid volumes. Manual planimetry was used to measure VWV as described above. Briefly, to identify the media–adventitia boundary, all 3D US images were viewed simultaneously in the longitudinal and axial views. This allowed the observer to identify the characteristic double line pattern in the longitudinal view, which represented the intima–lumen and media–adventitia boundaries [45]. The carotid bifurcation was used as a point of reference to reduce interscan measurement variability, and the bifurcation was first identified within both follow-up and baseline 3D US images so that all measurements could be initialized with the same 2D image slice. The CCA was segmented for a maximum distance of 15 mm proximal to the bifurcation, and the internal and external carotid branches (ICA and ECA, respectively) were segmented 10 mm distal from the bifurcation. The enclosed areas were used to calculate VWV. Percent atheroma volume (PAV), which is a measure of

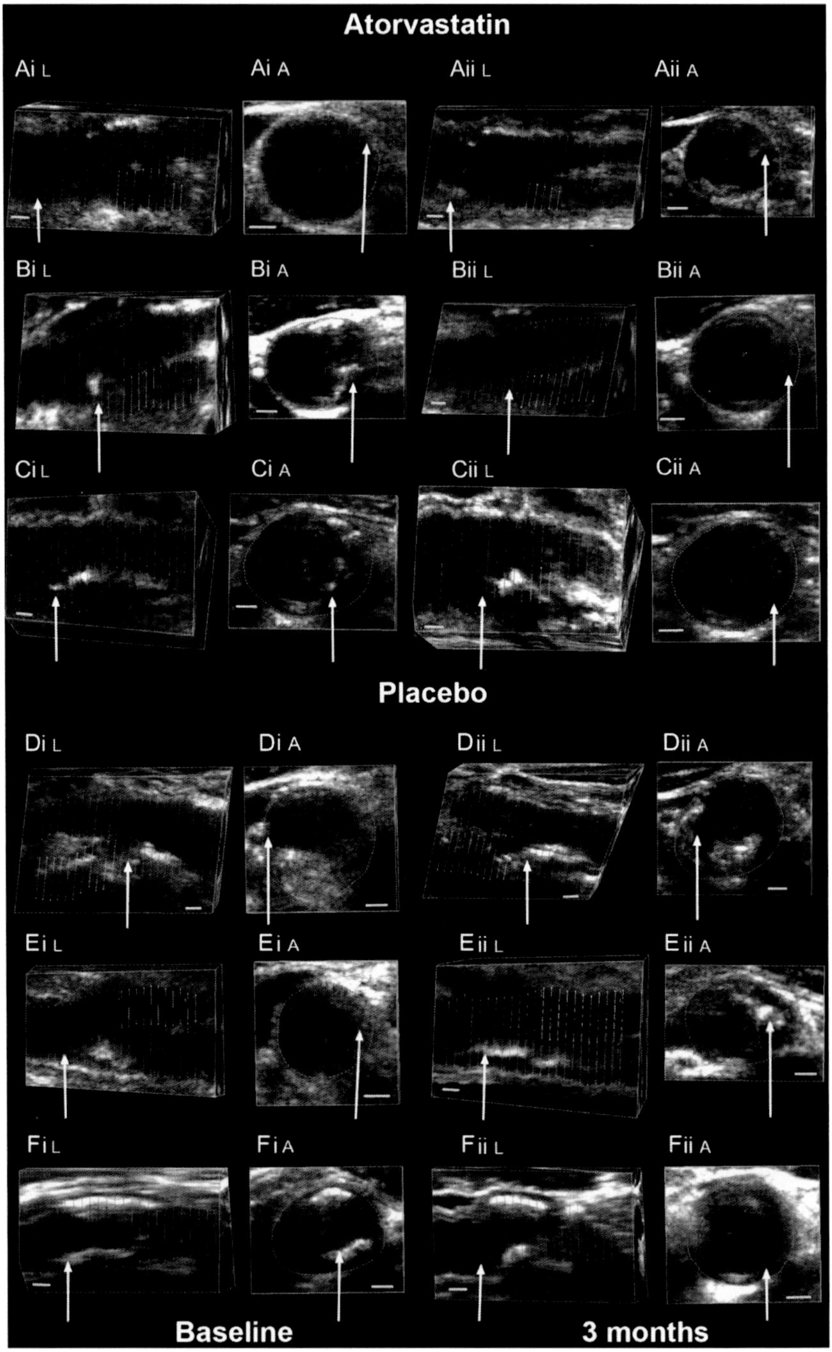

Fig. 12.7 Transverse and longitudinal 3D ultrasound of carotid atherosclerosis both longitudinal (L) and axial (A) views are shown. *Arrows* indicate regions corresponding to regions of interest in Fig. 12.9. All scale bars represent 2 mm. (**a**) A particular atorvastatin subject shown with large

atherosclerotic lesion burden previously developed for use in intravascular ultrasound (IVUS) studies [46], was also derived from the manually segmented contours. Percent atheroma volume (PAV) was calculated as the ratio of 3D US VWV and the entire outer wall volume.

Some of the results of the analysis of VWV and PAV for both treatment groups are shown in Fig. 12.7 and Table 12.1. Baseline VWV (mean ± SD) was 1330 ± 300 mm^3 for the atorvastatin treatment group and 1510±450 mm^3 for the placebo group, and the difference between treatment groups was not significant ($p=0.19$). After 3 months, subjects in the atorvastatin treatment group demonstrated a mean VWV change (mean ± SD) of −30 ± 110 mm^3, whereas for subjects in the placebo treatment group a mean VWV increase of 70 ± 140 mm^3 and this difference between groups was statistically significant ($p<0.05$). Ultrasound volumes demonstrating changes in selected subjects are seen in Fig. 12.7. The change in PAV (mean ± SD) was 0.2 ± 3.2%, for the atorvastatin treatment group and 1.9 ± 3.8% for subjects in the placebo treatment group and this difference was not significantly different ($p>0.05$) (Table 12.1). In addition, a repeated measures ANOVA demonstrated a significant interaction of time and treatment for VWV ($p<0.05$), but not for PAV. However, for PAV, ANOVA also detected a significant treatment effect ($p=0.045$).

12.4.1.3 Generation of 3D and 2D Carotid Maps

VWV segmentation allows for the generation of vessel wall and plaque thickness maps [47, 48] and plaque thickness difference maps. Briefly, as shown in Fig. 12.8

Table 12.1 Three-dimensional ultrasound atherosclerosis measurements

	Atorvastatin treatment group $n=16$	Placebo treatment group $n=19$	Difference between treatment groups
VWV (±SD) mm^3			
Baseline	1330 (300)	1510 (450)	
Follow-up	1300 (250)	1580 (490)	
Δ VWV (±SD) mm^3	−30 (110)	+70 (140)	$p=0.03$
Δ VWV (±SD)%	−1.4 (7.7)	4.9 (10.3)	$p=0.05$
PAV (±SD)%			
Baseline	53.8 (8.7)	47.4 (7.5)	
Follow-up	54.0 (8.6)	49.3 (7.7)	
Δ PAV (±SD)%	+0.2 (3.2)	+1.9 (3.8)	$p=0.17$

Fig. 12.7 (continued) positive change (increase) in VWV between baseline (i) and follow-up (ii). (**b**) A representative atorvastatin subject with mean negative change (decrease) in VWV between baseline (i) and follow-up (ii) (**c**) A particular atorvastatin subject with large negative change (decrease) in VWV between baseline (i) and follow-up (ii) (**d**) A particular placebo subject with large positive change (increase) in VWV between baseline (i) and follow-up (ii). (**e**) A representative placebo subject with mean positive change (increase) in VWV between baseline (i) and follow-up (ii) (**f**) A particular placebo subject with large negative change (decrease) in VWV between baseline (i) and follow-up (ii)

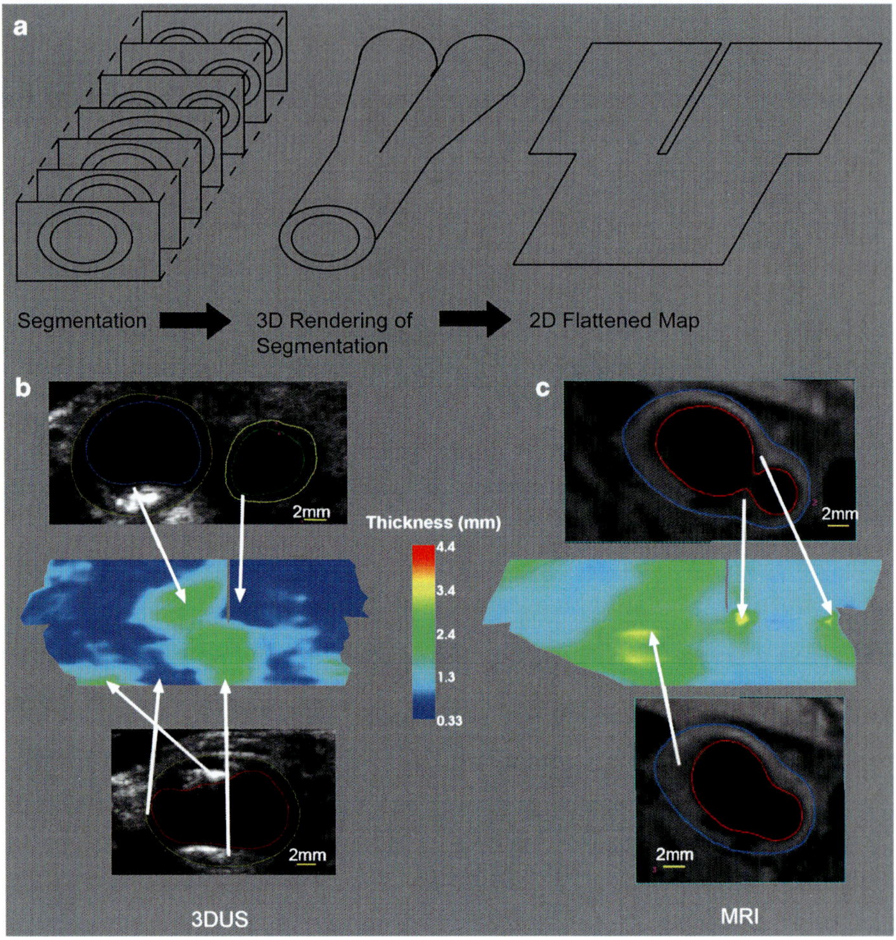

Fig. 12.8 Figure showing the generation of 2D and 3D carotid thickness maps. (**a**) Schematic diagram showing the flattening process of the 3D thickness maps generated from the measurements of VWV. (**b**) Sections "sliced" from the 3D carotid images and their correspondence to the 2D flattened maps

and previously described [47], for each carotid artery, a 3D carotid thickness map was generated by establishing corresponding points of the vessel wall and lumen segmentation surfaces with the resultant thickness of the vessel wall and plaque considered to be the distance between each pair of corresponding points. To map the 3D thickness map onto a 2D plane, the carotid map was bisected and flattened using arc-preserving [48] and area-preserving [49] algorithms. To generate carotid thickness difference maps [42], carotid maps from baseline and 3-month follow-up were registered [47] and digitally subtracted.

The carotid artery thickness difference maps shown in Fig. 12.9 demonstrate localized spatial vessel wall and plaque thickness changes over 3 months for 12

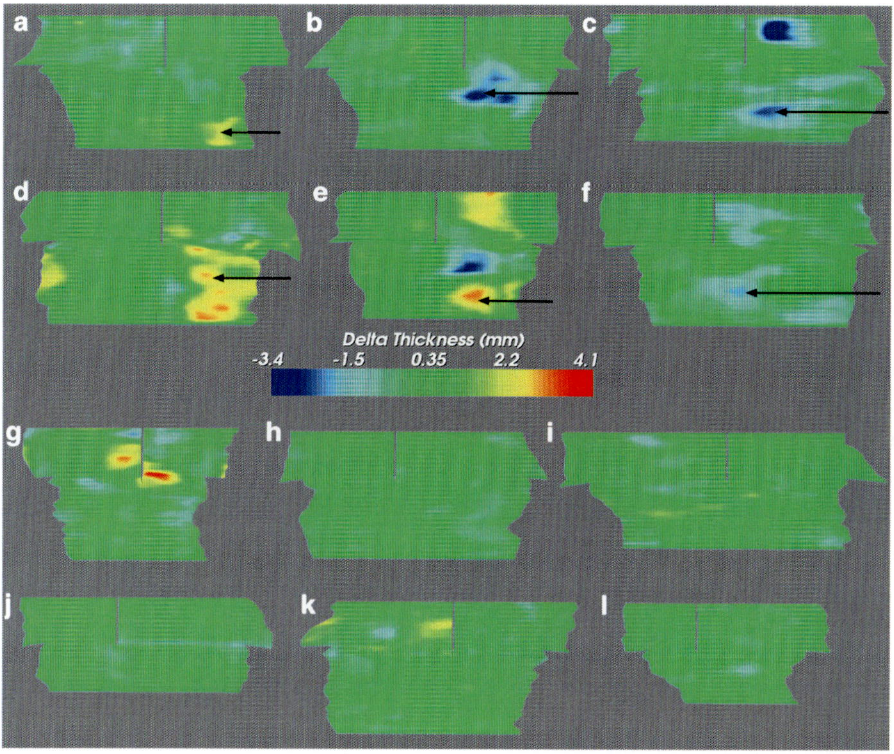

Fig. 12.9 Three-dimensional ultrasound thickness difference maps. Regions of interest on maps correspond to regions of interest demonstrated in Fig. 12.8. (**a**) Carotid thickness difference map for a representative subject from the atorvastatin treatment group with largest positive change (increase) in VWV measured over a 3-month follow-up period. (**b**) Carotid thickness difference map for a representative subject from the atorvastatin treatment group with mean VWV change measured over a 3-month follow-up period. (**c**) Carotid thickness difference map for a representative subject from the atorvastatin treatment group with greatest negative change (decrease) in VWV measured over a 3-month follow-up period. (**d**) Carotid thickness difference map for a representative subject from the placebo treatment groups with largest positive change (increase) in VWV measured over a 3-month follow-up period. (**e**) Carotid thickness difference map for a representative subject from the placebo treatment groups with mean change in VWV measured over a 3-month follow-up period. (**f**) Carotid thickness difference map for a representative subject from the placebo treatment group with greatest negative change (decrease) in VWV measured over a 3-month follow-up period. (**g–l**) Carotid thickness difference maps for six additional representative subjects from the atorvastatin (**g–i**) and placebo (**j–l**) groups

representative subjects from the atorvastatin treatment (Fig. 12.8a–c, g–i) and placebo (Fig. 12.8d–f, h–i) groups.

The resultant quantification of plaque and wall thickness is represented by color maps (see color bar in Fig. 12.9), normalized for all maps according to the measured minimum and maximum thickness changes observed for all 12 subjects. Representative maps are provided for two subjects with the largest VWV increases in the atorvastatin and placebo groups (Fig. 12.8a, d), two subjects representing the average amount of VWV change measured (Fig. 12.8b, e), and two subjects with the largest VWV decreases in both the atorvastatin (Fig. 12.8c) and placebo group (Fig. 12.8f). The maps show spatial changes in vessel wall and plaque thickness which, for subjects in the atorvastatin treatment group shown in Fig. 12.8a–c, can be observed as regions of decreased thickness in the common and internal carotid artery for two subjects (Fig. 12.4b, c) and a region of increased thickness in the CCA for the single subject in that treatment group that showed the corresponding greatest increase in VWV. Thickness difference maps for two representative patients in the placebo treatment group provided in Fig. 12.8d, e show large increases in vessel wall thickness within the CCA and ICA. For the single subject in the placebo treatment group that showed the greatest overall decrease in VWV, there was a corresponding small focal area of decreased vessel wall and plaque thickness in the common carotid artery.

12.4.1.4 Mapping Spatial and Temporal Changes in Carotid Atherosclerosis from 3D Images

The analysis of 3D carotid US images of changes in carotid plaque and vessel wall has the potential to provide quantitative and dynamic measures of the volumetric and spatial changes. Thus, we developed a method to analyze successive carotid artery wall and lumen segmentation outlines from 3D carotid US images and display these as spatial 2D maps of vessel thickness [48, 50]. To demonstrate this technique, we show results from two studies.

1. Three subjects with carotid plaque area greater than 0.5 cm^2 (moderate atherosclerosis subjects) who were scanned twice in 2 weeks to assess the variability of the 3D US measurements due to image variability, scanning parameter changes, sonographer changes, and observer variability.
2. Five subjects with carotid stenosis (carotid stenosis subjects) were scanned twice, once at baseline and once after 12 weeks of statin or placebo therapy, with three subjects having received 80 mg atorvastatin daily and two subjects receiving placebo treatment (discussed in Sect. 12.4.1.3 above).

Manual planimetry as described above was used to segment the vessel wall and lumen boundaries in the 3D carotid US images and generate VWV measurements. The carotid bifurcation was used as a point of reference [28] to reduce interscan measurement variability. Thus, the bifurcation was first identified and marked within the 3D US image so that all measurements could be initialized on the same 2D image slice. The boundaries were segmented along the 30-mm vessel axis, which was placed parallel to the longitudinal axis of the common carotid artery and centered on the carotid bifurcation. As described above, the segmentation proceeded on 3D US image slices with an interslice distance (ISD) of 1 mm. The common carotid artery was segmented proximally at a distance of 15 mm, and the internal and external carotid branches were each segmented distally at a distance of 10 mm from the bifurcation. The average area enclosed by two sequential contours was multiplied by the ISD to give the volume between adjacent slices, and the

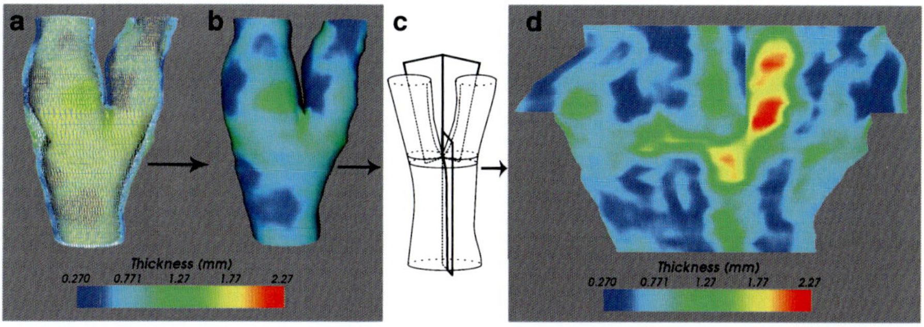

Fig. 12.10 (**a**) Five repetitions of manual image segmentation provide mean vessel wall and lumen outlines with the distance between each pair of corresponding points on the vessel wall and lumen surfaces, providing thickness of vessel wall and plaque. Once the correspondence points have been established between reconstructed vessel and lumen surfaces, (**b**) a 3D image rendered thickness map is generated (**c**), with the carotid artery opened along the planes (**d**) into a flattened thickness 2D map

interslice volumes were summed up to determine the total volume of each contour. VWV was calculated by subtracting the volume enclosed by the lumen contours from the volume enclosed by the vessel contours.

The technique used to generate the 3D and 2D maps of the carotid VWV was described in detail by Chiu et al. [48, 50] and is only summarized here. Briefly, as shown in Figs. 12.8 and 12.10, mean vessel wall and lumen surfaces (shown in Fig. 12.10a) are reconstructed from the five repeated segmentations of each carotid image [43]. The 3D carotid thickness map shown in Fig. 12.10b is generated by establishing corresponding points on the vessel wall and lumen surfaces with the resultant thickness of the vessel wall and plaque calculated to be the distance between each pair of corresponding points [50]. The 3D carotid map is then "cut" (as shown in Fig. 12.10c) to generate a flattened 2D thickness map (Fig. 12.10d) to map the 3D wall and plaque thickness maps shown in Fig. 12.10b onto a 2D plane [48]. The flattened 2D thickness map in Fig. 12.10d shows the spatial distribution of wall and plaque thickness within the artery and provides a continuous wall and plaque surface [48, 50] to facilitate the visualization and qualitative assessment of carotid artery wall and plaque thickness distribution.

In addition to generating thickness 2D maps, thickness difference 2D maps can also be generated by subtracting the 2D thickness map at baseline (scan) from the 2D thickness map obtained at a later time. The resultant thickness difference 2D map provides a continuous surface in which to examine location-specific changes in wall and plaque thickness [48, 50].

The utility of the 2D thickness change maps is shown in Figs. 12.11 and 12.12. Figure 12.11 shows "Slices" of 3D carotid US images obtained at baseline and 12 weeks later of three subjects. Figure 12.11a shows 3D image "slices" of a subject treated with atorvastatin at baseline and 12 weeks later, Fig. 12.11b shows 3D image "slices" of a subject treated with placebo at baseline and 12 weeks later, and Fig. 12.10c shows 3D image "slices" of a subject with moderate atherosclerosis who was scanned twice in 2 weeks.

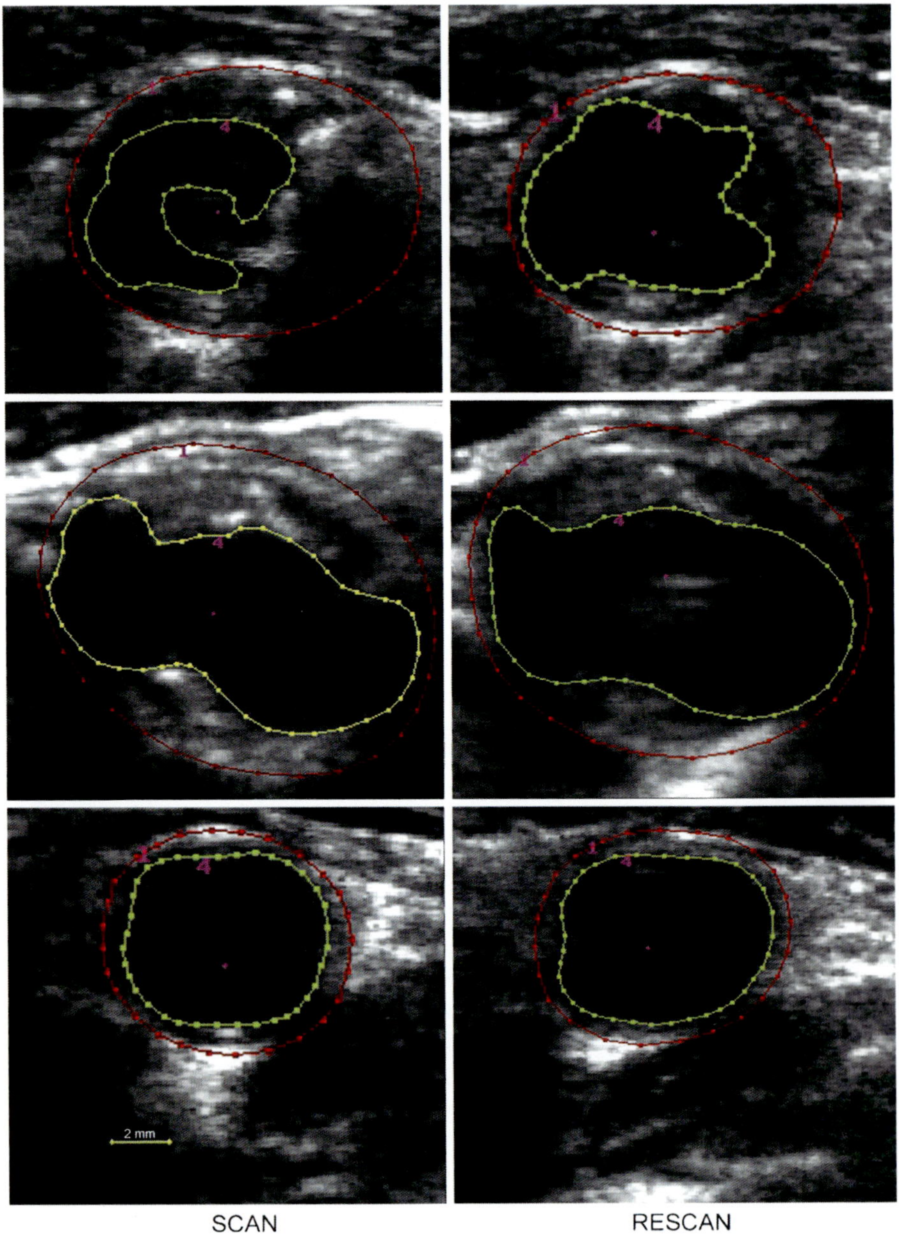

Fig. 12.11 (**a**) Carotid stenosis subject baseline scan (i) and 12 weeks later (ii) treated with atorvastatin. (**b**) Carotid stenosis subject baseline scan (i) and 12 weeks later (ii) treated with placebo. (**c**) Moderate atherosclerosis subject scan baseline scan (i) and 2 weeks later (ii) no treatment

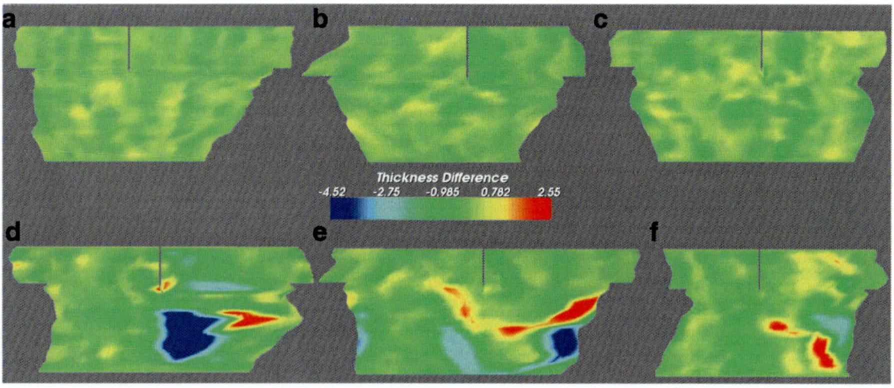

Fig. 12.12 (**a** and **b**) Moderate atherosclerosis subject difference between baseline scan and scan 2 weeks later. (**c**) Carotid stenosis subject treated with placebo – difference between baseline scan and a 3-months follow-up. (**d–f**) Carotid stenosis subjects treated with atorvastatin, 2D difference maps between baseline scan and a 3-month follow-up

Scan–rescan 2D thickness difference maps are shown in Fig. 12.12 for six subjects: (1) two moderate atherosclerosis subjects (Fig. 12.12a, b) scanned twice in 2 weeks, (2) three carotid atherosclerosis subjects treated with atorvastatin (Fig. 12.12d–f), scanned at baseline and 12 weeks later, and (3) a carotid atherosclerosis subject treated with placebo (Fig. 12.12c) and scanned at baseline and 12 weeks later. These six carotid 2D thickness difference maps allow qualitative examination of plaque and wall changes in these subjects over the different scan–rescan periods. Plaque and wall thickness differences between scan and rescan are color-coded with the color scale provided for the six topology difference maps, and the color scale is normalized according to the range of thickness changes observed for statin-treated subject (shown in Fig. 12.12e with a change in VWV at rescan of 280 ± 60 mm^3 VWV in 12 weeks).

Carotid 2D thickness difference maps for a carotid stenosis subject (Fig. 12.12c) scanned after 12 weeks of treatment with placebo and for two moderate atherosclerosis subjects scanned twice within 2 weeks (Fig. 12.12a, b) indicate no change (green = 0 mm thickness difference). However, thickness difference maps for all three carotid stenosis subjects treated with atorvastatin do show plaque and wall thickness changes in the common carotid artery with plaque and wall thickness changes ranging from −4.5 mm to +2.5 mm.

12.5 Discussion

3D ultrasound has already demonstrated clear advantages in obstetrics, cardiology, and image guidance of interventional procedures. Current 3D US technology is sufficiently advanced to allow real-time 3D imaging using 2D array transducers and near real-time 3D imaging with mechanically manipulated 1D transducers.

The current focus related to 3D US imaging is the establishment of the utility of 3D US in clinical applications and improved image analysis techniques allowing quantitative measurements in an efficient manner. Improved software tools for image analysis are promising to make 3D US a routine tool on ultrasound machines.

Although we have shown the potential of 3D carotid US imaging for monitoring carotid atherosclerosis progression and regression, this modality requires further development for it to become a routine tool for quantifying carotid disease and its changes. Chief among the required changes is the development of automated or semiautomated carotid plaque, lumen, and vessel wall segmentation. Manual segmentation has been demonstrated to have low intraoperator variability when used by a trained operator. However, the manual segmentation approach is slow and tedious. Thus, improved methods to segment the structures in the 3D carotid images would accelerate the use of this technology. However, before automated or semiautomated techniques are used routinely, they must be verified thoroughly. Use of manual segmentations by a trained observer can be used as the surrogate "gold" standard for analysis of accuracy, and repeated segmentations can be used for analysis of variability.

In addition to segmentation developments, analysis of changes in plaque texture may provide information on changes in plaque composition. Validation of these developments is more difficult and would require comparisons to MR plaque composition results or comparisons to histological sections of carotid vessel specimens.

Acknowledgements The authors acknowledge the financial support of the Canadian Institutes for Health Research and the Ontario Research Fund. Fenster holds a Canada Research Chair and acknowledges the support of the Canada Research Chair Program.

References

1. Barth JD. Carotid intima media thickness and beyond. Curr Drug Targets Cardiovasc Hematol Disord 2004; 4(2):129–145.
2. Spence JD, Hegele RA. Non-invasive assessment of atherosclerosis risk. Curr Drug Targets Cardiovasc Hematol Disord 2004; 4(2):125–128.
3. Landry A, Spence JD, Fenster A. Measurement of carotid plaque volume by 3-dimensional ultrasound. Stroke 2004; 35(4):864–869.
4. Landry A, Spence JD, Fenster A. Quantification of carotid plaque volume measurements using 3D ultrasound imaging. Ultrasound Med Biol 2005; 31(6):751–762.
5. Steinke W, Hennerici M. Three-dimensioal ultrasound imaging of carotid artery plaques. J Cardiovasc Technol 1989; 8:15–22.
6. Hatsukami TS, Thackray BD, Primozich JF et al. Echolucent regions in carotid plaque: preliminary analysis comparing three-dimensional histologic reconstructions to sonographic findings. Ultrasound Med Biol 1994; 20(8):743–749.
7. Pretorius DH, Nelson TR, Jaffe JS. 3-dimensional sonographic analysis based on color flow Doppler and gray scale image data: a preliminary report. J Ultrasound Med 1992; 11(5):225–232.
8. Picot PA, Rickey DW, Mitchell R, Rankin RN, Fenster A. Three-Dimensional Colour Doppler Imaging of the Carotid Artery. SPIE Proceedings: Image Capture, Formatting and Display 1444, 206–213, 1991.

9. Guo Z, Fenster A. Three-dimensional power Doppler imaging: a phantom study to quantify vessel stenosis. Ultrasound Med Biol 1996; 22(8):1059–1069.
10. Dabrowski W, Dunmore-Buyze J, Cardinal HN, Fenster A. A real vessel phantom for flow imaging: 3-D Doppler ultrasound of steady flow. Ultrasound Med Biol 2001; 27(1):135–141.
11. Picot PA, Rickey DW, Mitchell R, Rankin RN, Fenster A. Three-dimensional colour Doppler imaging. Ultrasound Med Biol 1993; 19(2):95–104.
12. Barnett P, Spence J, Manuck S, Jennings S. Psychological stress and the progression of carotid artery disease. J Hypertens 1997; 15(1):49–55.
13. Baldassarre D, Amato M, Bondioli A, Sirtori CR, Tremoli E. Carotid artery intima-media thickness measured by ultrasonography in normal clinical practice correlates well with atherosclerosis risk factors. Stroke 2000; 31(10):2426–2430.
14. Baldassarre D, Amato M, Pustina L et al. Measurement of carotid artery intima-media thickness in dyslipidemic patients increases the power of traditional risk factors to predict cardiovascular events. Atherosclerosis 2007; 191:403–408.
15. Owens GK. Control of hypertrophic versus hyperplastic growth of vascular smooth muscle cells. Am J Physiol 1989; 257(6 Pt 2):H1755–H1765.
16. Spence JD. Ultrasound measurement of carotid plaque as a surrogate outcome for coronary artery disease. Am J Cardiol 2002; 89(4A):10B–15B.
17. Stary HC, Blankenhorn DH, Chandler AB et al. A definition of the intima of human arteries and of its atherosclerosis-prone regions. A report from the Committee on Vascular Lesions of the Council on Arteriosclerosis, American Heart Association. Arterioscler Thromb 1992; 12(1):120–134.
18. Hennerici M, Baezner H, Daffertshofer M. Ultrasound and arterial wall disease. Cerebrovasc Dis 2004; 17(Suppl 1):19–33.
19. Stary HC, Chandler AB, Glagov S et al. A definition of initial, fatty streak, and intermediate lesions of atherosclerosis. A report from the Committee on Vascular Lesions of the Council on Arteriosclerosis, American Heart Association. Arterioscler Thromb 1994; 14(5):840–856.
20. Landry A, Fenster A. Theoretical and experimental quantification of carotid plaque volume measurements made by three-dimensional ultrasound using test phantoms. Med Phys 2002; 29(10):2319–2327.
21. Delcker A, Diener HC. Quantification of atherosclerotic plaques in carotid arteries by three-dimensional ultrasound. Br J Radiol 1994; 67(7999):672–678.
22. Delcker A, Diener HC. 3D ultrasound measurement of atherosclerotic plaque volume in carotid arteries. Bildgebung 1994; 61(2):116–121.
23. Delcker A, Tegeler C. Influence of ECG-triggered data acquisition on reliability for carotid plaque volume measurements with a magnetic sensor three-dimensional ultrasound system. Ultrasound Med Biol 1998; 24(4):601–605.
24. Palombo C, Kozakova M, Morizzo C et al. Ultrafast three-dimensional ultrasound: application to carotid artery imaging. Stroke 1998; 29(8):1631–1637.
25. Delcker A, Diener HC, Wilhelm H. Influence of vascular risk factors for atherosclerotic carotid artery plaque progression. Stroke 1995; 26(11):2016–2022.
26. Schminke U, Motsch L, Griewing B, Gaull M, Kessler C. Three-dimensional power-mode ultrasound for quantification of the progression of carotid artery atherosclerosis. J Neurol 2000; 247(2):106–111.
27. Zhao XQ, Yuan C, Hatsukami TS et al. Effects of prolonged intensive lipid-lowering therapy on the characteristics of carotid atherosclerotic plaques in vivo by MRI: A case-control study. Arterioscler Thromb Vasc Biol 2001; 21(10):1623–1629.
28. Ainsworth CD, Blake CC, Tamayo A, Be[illegible]tsky V, Fenster A, Spence JD. 3D ultrasound measurement of change in carotid plaque [illegible]e: a tool for rapid evaluation of new therapies. Stroke 2005; 36(9):1904–1909.
29. Zahalka A, Fenster A. An automated segmentation method for three-dimensional carotid ultrasound images. Phys Med Biol 2001; 46(4):1321–1342.
30. Fenster A, Downey DB, Cardinal HN. Three-dimensional ultrasound imaging. Phys Med Biol 2001; 46(5):R67–R99.
31. Nelson TR, Downey DB, Pretorius DH, Fenster A. Three-Dimensional Ultrasound. Philadelphia, PA: Lippincott Williams & Wilkins, 1999.

32. Nelson TR, Pretorius DH. Three-dimensional ultrasound imaging. Ultrasound Med Biol 1998; 24(9):1243–1270.
33. Baba K, Jurkovic D. Three-Dimensional Ultrasound in Obstetrics and Gynecology. Pearl River, NY: Parthenon Publishing Group, 1997.
34. Downey DB, Fenster A. Three-dimensional ultrasound: a maturing technology. Ultrasound Q 1998; 14(1):25–40.
35. Fenster A, Downey DB. Three-dimensional ultrasound imaging. In: Beutel J, Kundel H, Van Metter R, editors. Handbook of Medical Imaging, Volume 1, Physics and Psychophysics. Bellingham, WA: SPIE Press, 2000: 433–509.
36. Fenster A, Downey DB. Three-dimensional ultrasound imaging. Annu Rev Biomed Eng 2000; 2:457–475.
37. Fenster A, Downey DB. Basic principles and applications of 3-D ultrasound imaging. In: Stergiopoulos S, editor. An Advanced Signal Processing Handbook. Boca Raton, FL: CRC Press, 2001: 14–34.
38. Smith W, Fenster A. Statistical analysis of decorrelation-based transducer tracking for three-dimensional ultrasound. Med Phys 2003; 30(7):1580–1591.
39. Downey DB, Fenster A. Vascular imaging with a three-dimensional power Doppler system. AJR Am J Roentgenol 1995; 165(3):665–668.
40. Fenster A, Tong S, Sherebrin S, Downey DB, Rankin RN. Three-Dimensional Ultrasound Imaging. Proceedings of SPIE Physics of Medical Imaging 2432, 176–184, 1995.
41. Hughes SW, D'Arcy TJ, Maxwell DJ et al. Volume estimation from multiplanar 2D ultrasound images using a remote electromagnetic position and orientation sensor. Ultrasound Med Biol 1996; 22(5):561–572.
42. Egger M, Chiu B, Spence JD, Fenster A, Parraga G. Mapping spatial and temporal changes in carotid atherosclerosis from three-dimensional ultrasound images. Ultrasound Med Biol 2008; 34(1):64–72.
43. Egger M, Spence JD, Fenster A, Parraga G. Validation of 3D ultrasound vessel wall volume: an imaging phenotype of carotid atherosclerosis. Ultrasound Med Biol 2007; 33(6):905–914.
44. Egger M, Krasinski A, Rutt BK, Fenster A, Parraga G. Comparison of B-mode ultrasound, 3-dimensional ultrasound and magnetic resonance imaging measurements of carotid atherosclerosis. J Ultrasound Med 2008; 27(9):1321–1334.
45. Pignoli P, Tremoli E, Poli A, Oreste P, Paoletti R. Intimal plus medial thickness of the arterial wall: a direct measurement with ultrasound imaging. Circulation 1986; 74(6):1399–1406.
46. Nissen SE, Tsunoda T, Tuzcu EM et al. Effect of recombinant ApoA-I Milano on coronary atherosclerosis in patients with acute coronary syndromes: a randomized controlled trial. JAMA 2003; 290(17):2292–2300.
47. Chiu B, Egger M, Spence D, Parraga G, Fenster A. Quantification of carotid vessel wall and plaque thickness change using 3D ultrasound images. Med Phys 2008; 35:3691–3710.
48. Chiu B, Egger M, Spence JD, Parraga G, Fenster A. Quantification of progression and regression of carotid vessel atherosclerosis using 3D ultrasound images. Conf Proc IEEE Eng Med Biol Soc 2006; 1:3819–3822.
59. Chiu B, Egger M, Spence DJ, Parraga G, Fenster A. Area-preserving flattening maps of 3D ultrasound carotid arteries images. Med Image Anal 2008; 12(6):676–688.
50. Chiu B, Egger M, Spence JD, Parraga G, Fenster A. Quantification of carotid vessel atherosclerosis. Proc SPIE 6143, 2006.
51. Spence JD, Eliasziw M, DiCicco M, Hackam [illegible]G, Galil R, Lohmann T. Carotid plaque area: a tool for targeting and evaluating vascular p[illegible]tive therapy. Stroke 2002; 33:2916–2922.
52. Johnsen SH, Mathiesen EB, Joakimsen O, Sten[illegible] E, Wilsgaard T, Løchen ML, Njølstad I, Arnesen E. Carotid atherosclerosis is a stronger predictor of myocardial infarction in women than in men: a 6-year follow-up study of 6226 persons: the Tromsø study. Stroke 2007; 38(11):2873–2880.

Biographies

Dr. Parraga's laboratory is focused on understanding the underlying mechanisms of carotid atherosclerosis using novel imaging technologies as a tool in patient-based biomedical engineering, biophysics and computational research. An important goal of the lab is to identify and validate radiological intermediate endpoints as potential therapeutic targets for use in clinical trials using 3D Ultrasound methods as well as quantitative image processing software tools. Dr. Parraga completed her PhD in Biochemistry at the University of Washington in Seattle Washington and after completing post-doctoral studies at the Biozentrum, University of Basel, (Basel, Switzerland), she joined F. Hoffman La Roche AG as a Scientist in Pharmaceutical Research and Development. In 2004, she returned to academic research within the Imaging Research Laboratories, Robarts Research Institute. Dr. Parraga's work has been published in high impact journals including Science, the Proceedings of the National Academy of Sciences, Circulation, Radiology and the American Journal of Respiratory and Critical Care Medicine. She also currently serves as a member of the National Institutes of Health (NIH) Biomedical Imaging Technologies study panel, the Canadian Institutes of Health Research (CIHR) operating grant peer-review panel and co-Chairs a National Science and Engineering Research Council (NSERC) peer-review panel. Her lab is currently funded by CIHR Operating, New Investigator and Team grants and the Ontario Institute for Cancer Research.

Dr. Fenster received his PhD degree in 1976 from the Department of Medical Biophysics of the University of Toronto. His first academic appointment was at the Department of Radiology and Medical Biophysics of the University of Toronto from 1979 to 1987, and the Director of the Radiological Research laboratories of the Department of Radiology. In 1987 he moved to London and became a Scientist and founding Director of the Imaging Research Laboratories at the Robarts Research Institute and Professor at The University of Western Ontario (UWO) in Radiology and Medical Biophysics. He also was the founder and current Associate Director of the graduate Program in Biomedical Engineering. He is the Chair of the basic Science Division of the Department of Medical Imaging and the Director of the Biomedical Imaging Research Centre at UWO. Currently, he holds a Canada Research Chair-Tier 1 in Biomedical Engineering and he is the recipient of the 2007 Premier's Award for Innovative Leadership and UWO's 2008 Hellmuth Prize for achievement in research. Fenster's group has focused on the development of 3D ultrasound imaging with diagnostic and surgical/therapeutic cancer applications in humans as well as mouse research models. His team developed the world's firsts in 3D ultrasound imaging of the carotids and prostate, 3D ultrasound guided prostate cryosurgery and brachytherapy, 3D ultrasound guided prostate and breast biopsy for early diagnosis of cancer and 3D ultrasound images of mouse tumours and their vasculature. Fenster's research has resulted in 36 patents (26 issued and 10 pending) and the formation of two companies in London (Life Imaging Systems and Enhanced Vision Systems), with Fenster as a founding scientist. In addition, some of his patents have been licensed to 11 different companies, which have commercialized them for world-wide distribution.

Adam Krasinski, MSc received his undergraduate training in Biophysics at the University of Guelph (Guelph, Canada) and his MSc training in Medical Biophysics at The University of Western Ontario. He currently works as a research technician at Robarts Research Institute in collaboration with Drs. Fenster and Parraga.

Bernard Chiu received his Bachelor of Science degree in electrical engineering at the University of Calgary, Canada, in 2001, the Master of Applied Science degree in electrical and computer engineering at the University of Waterloo, Canada, in 2003, and the Ph.D. degree in biomedical engineering at the University of Western Ontario, Canada, in 2008. He is currently a senior fellow in Department of Radiology at the University of Washington, Seattle, WA, USA. His interests include medical image processing, quantification and visualization.

Micaela Egger, MSc received her undergraduate training at Queen's University (Kingston, Canada) and her MSc training in Medical Biophysics at The University of Western Ontario. She is currently pursuing an MD degree in Geneva Switzerland.

J. David Spence, BA, MD, MBA, FRCP(C), FAHA, FCAHS Stroke Prevention & Atherosclerosis Research Centre, Robarts Research Institute, London, Canada Professor Spence has focused on prevention of stroke throughout his career. He pioneered the measurement of 2-Dimensional carotid total plaque area beginning in 1990. Since 1994 he has collaborated with Dr. Aaron Fenster, and since 2004 with Dr. Grace Parraga, both at the Robarts Research Institute, in measurement of 3-Dimensional plaque volume and vessel wall volume. His research program focuses on measurement of atherosclerosis by ultrasound, for patient management, genetic research and for assessing effects of new therapies.

Part III
X-Rays, CT, and MR Clinical Imaging

Chapter 13
CT Imaging in the Carotid Artery

Luca Saba

Facts are the air of scientists. Without them you can never fly

Linus Pauling (1901–1994) Nobel Prize in Chemistry 1954 and for Peace 1962

Abstract In the study of carotid arteries, computed tomography angiography (CTA) allows to analyze, other than the luminal narrowing, the carotid plaque, the carotid wall, as well as the surrounding structures. Moreover by using CTA it is possible to have three-dimensional information with high spatial resolution. This chapter will detail the role of CTA for the assessment of cardiovascular pathology with an emphasis on the detection, analysis and characterization of carotid atherosclerosis.

Keywords Carodit arteries • Imaging • Computed tomography

13.1 General Introduction

Computed tomography (CT) can provide exquisite, rapid, high-resolution imaging of the body and in particular of the vascular system. The introduction of spiral CT, in the early 1990s, made it possible to cover body regions rapidly so that the vascular enhancement following intravenous injection could be captured during one scan by allowing the development of the so-called computed tomography angiography (CTA). The introduction of multi-detector row technology gave a tremendous boost to the CTA that is now considered by several authors a reference standard in advanced vascular imaging. CT technology has seen more rapid improvements than other radiological techniques: over the past decade the scanners performance doubled approximately every 2 years and the number of slices in CT doubled approximately

L. Saba (✉)
Department of Radiology, Azienda Ospedaliero Universitaria (A.O.U.), di Cagliari – Polo di Monserrato, s.s. 554, Monserrato 09045, Cagliari, Italy
e-mail: lucasaba@tiscali.it

Jasjit S. Suri et al. (eds.), *Atherosclerosis Disease Management*,
DOI 10.1007/978-1-4419-7222-4_13, © Springer Science+Business Media, LLC 2011

every 2.5 years. Advances in spatial and temporal resolution and image reconstruction software have helped in the evaluation of vascular artery pathology and, in particular, in coronary artery anatomy (and vessel patency) and in carotid artery plaque detection and characterization.

Over the years, CTA – together with Eco-Colour-Doppler and Magnetic Resonance Angiography – has taken over most diagnostic vascular procedures from invasive catheter angiography. In fact, CTA has several advantages and, in particular, is highly standardized which makes it a very fast and robust procedure that is the technique of choice in many vascular disease. In particular, in the study of carotid arteries, the use of CTA instead of catheter angiography allows to analyze other than the luminal narrowing the carotid plaque, the carotid wall, as well as the surrounding structures. Moreover by using CTA it is possible to have three-dimensional (3D) information with high spatial resolution.

This chapter will detail the role of CTA for the assessment of cardiovascular pathology with an emphasis on the detection, analysis and characterization of carotid atherosclerosis.

13.2 CT Principles

13.2.1 General Overview

CTA, in particular, multi-detector row CTA (MDCTA) is a powerful non-invasive imaging tool with a number of advantages over the others non-invasive imaging techniques. CT has experienced tremendous technological developments since its introduction more that 35 years ago. The first generation of CT scanners were developed in the 1970s [1, 2] and numerous innovations have improved the utility and application field of the CT, such as the introduction of helical systems that allowed the development of the "volumetric CT" concept. By using the first generation CT scanner the acquisition of a single image took several minutes whereas nowadays large volumes are covered in few seconds. Basically the CT is an X-ray tomographic technique in which an X-ray source rotates around an object and the X-ray passes through the object from various directions. The X-ray attenuation along each of the many paths through the object is analyzed by a mathematical image reconstruction (*inverse radon transformation*) to calculate the local attenuation coefficient at each point within the acquisition volume. After this phase the local attenuation coefficient are normalized to yield CT numbers for every point of the matrix. At the end of the process these CT numbers are converted into shades of grey. In the first and second generation scanner (sequential scanner) the scan volume was covered in a stepwise manner: after acquiring an axial section the table was moved by a certain amount and the next scan was performed. Obviously, this scanning technique was slow and suffered from discontinuities at the border between sections.

A major improvement in the CT technology was the incorporation of continuously rotating X-ray tubes with a single detector row longitudinally positioned in the gantry.

This was the introduction of *third-generation geometry* (spiral CT) resulting in fundamental and far reaching improvements of CT imaging and opening a spectrum of entirely new applications including CT angiography with new visualization techniques such as Maximum Intensity Projection (MIP), Multi-Planar-reconstruction (MPR), Curved Planar Reconstruction (CPR) and Volume Rendering Technique (VRT). With these scanners, an X-ray tube and a detector array rotate synchronously and continuously around the patient and it became possible to acquire a data volume in continuous fashion. The generation of the raw data was obtained during multiple rotations while the patient's table moved through the scan plane. This technique became the basis of CT angiography because it allowed to image rapidly and continuously a large volume so that transient phase of contrast enhancement could be captured within one scan. Single-detector row scanners use a single detector arc or detector ring that consists of parallel detector elements that are much smaller in the direction of the arc (*xy plane*) than in the *z*-direction. In multi-detector technology each element on the detector array is subdivided along the *z*-axis and these parallel detector rows can be electronically combined to yield between 4 and 256 separate sections per rotation depending on the type of the scanner.

However, by using the spiral CT technology the goal of isotropic imaging within one breath hold could not be reached because a compromise between scan range, longitudinal resolution and scan time should be accepted. In the 1998 it was proposed the first four-detector row scanner with a rotation time of 0.5 s. This scanner allowed to increase scanning performance by a factor of 6 compared with the previous single slice scanners rotating at 0.75 s. The most important clinical benefit was the possibility to scan a given scan range in a given time at a substantially improved spatial resolution. The introduction of 16-slice CT scanners in 2001 finally allowed to obtain true isotropic CT scanning with the potentiality to produce 1,000 images in less than 15 s. As compared to four-detector row scanners, performance of 32–64-detector row scanners has been increased more than 20 time owing to more detector rows and faster rotation speed.

A recent major improvement in the CT technology is the introduction of the dual source CT [3–6]. Isotropic voxels, high spatial and temporal resolution, use of fast contrast material injection rate and post-processing tools improved sensitivity and specificity of MDCTA [7, 8]. MDCTA allows full vascular imaging in few seconds from the aorta to the circle of Willis. For instance CT angiography of the carotid arteries and the circle of Willis with 16 section at 0.75 mm collimation, 0.5 s rotation time and pitch of 1.5 requires only 9 s for a scan range of 250 mm (with a table feed of 36 mm/s). Scanners with 64 or more detector rows offer further better performances: gantry rotation times decreased to 0.33 s, and section widths of 0.5 mm are available. So, examination of the whole length of the carotid arteries from the aortic arch to the circle of Willis (250 mm approximately scan range) requires only 4 s for a 64-detector row CT (64×0.6 mm, pitch of 1.33, 0,33-s rotation time). By providing this large coverage MDCTA can now evaluate besides the carotid bifurcation, other important atherosclerotic sites, like the aortic arch, the origin of the supra-aortic vessels, the carotid siphon and the vertebro-basilar arteries. Faster acquisitions provide a better image quality reducing artefacts (respiration, swallowing, pulsation of the aorta, short arteriovenous circulation). MDCT can provide lesser artefacts than single row helical CT with a faster table speed. Another advantage of MDCTA for atherosclerotic plaque evaluation is the increased in-plane resolution, the decreased slice thickness and the subsequent ability to obtain isotropic voxel so that more detailed analysis of atherosclerotic plaque morphology and luminal plaque surface may now be possible.

13.2.2 Scanning Parameters

To correctly apply CT angiography it is extremely important to understand the scanning parameters and it is possible to group them into three major categories: (1) acquisition parameters, (2) derivative parameters, and (3) reconstruction parameters. In the *acquisition parameters* there are: number of active detector rows, section collimation, rotation time, table feed per rotation, pitch factor, scan length, tube voltage, tube load. The *number of active detector rows* is one for the single detector row CT, four for the 4-MDCT detector row and so on. The *section collimation* is determined by the total width of the acquired volume in the centre of the scan field divided by the number of sections. The *rotation time* is the time it takes for the tub-detector unit to rotate one time around the patient. The *table feed* indicates how much the table moves during the rotation time. The *pitch* provides the relation between table feed and total width of the acquired volume according to this formula pitch = table feed/ (number of active detector rows × section collimation). The *scan length* may vary according to the clinical indications. In the *derivative parameters* there are *table speed* (that is determined between the ratio of table feed and rotation time) and *scan duration* (measured in second is determined by scan length/ table speed). In the *reconstruction parameters* there are field of view (FOV), Matrix size, reconstruction filter, section thickness and reconstruction interval. The *FOV* determines which part of the data will end up in the images. The *matrix size* of the reconstructed image usually is a matrix 512 × 512, but it is possible to use other matrix like 256 × 256, 768 × 768 or even 1,024 × 1,024 for the modern scanner. The matrix size works as a limiting factor for spatial resolution. Too small matrix can reduce spatial resolution; however, a very large matrix size cannot increase resolution beyond the system resolution. The *reconstruction filter* is necessary to reconstruct usable images from the projectional raw data.

CTA relies on thin section spiral imaging and optimum timing of contrast material injection to ensure proper enhancement of the target territory. Pre-contrast scan is not mandatory in the analysis of carotid arteries, but recent works demonstrate that may be helpful for the evaluation of the carotid plaque enhancement (CPE) and carotid wall enhancement [9, 10]

13.3 Basic Post-processing Techniques in CTA

The advent of 16, 64, 256 and over slice helical CT has provided the impetus for many changes in CT applications and implementation of study protocols reproduce. The impact has especially been felt with the ability to isotropic data sets which can be visualized as a true volume in a 3D world. The latest generation of CT systems allows for truly isotropic imaging in virtually any application. As a consequence, the distinction between transverse and in-plane resolution is gradually becoming a historical remnant and the traditional axial slice is loosing its clinical predominance. In fact, it is now replaced by interactive viewing manipulation of isotropic volume images. The routine availability of high-quality volume data sets opens the way for new visualization paradigms and novel evaluation techniques, the corresponding large number of images also poses a challenge to the viewing and reading environment of the radiologist. Image processing and 3D reconstruction of diagnostic images represents a necessary tool for depicting complex anatomical structures and understanding pathological changes in terms of both morphology and function.

To study carotid arteries and other vessels using MDCT, it is possible to use different techniques of post-processing in addition to axial scan. First introduced in the late 1980s, post-processing methods currently show an impressive visual capacity of representing anatomic structures, but they are not always adequate in detecting and visualizing all pathologic conditions. Therefore, they have to be used critically to obtain the best diagnostic efficacy. There are several types of image reconstructions available, and authors believe that the simultaneous use of all of these tools is time consuming and incorrect.

It is possible to classify the technique that display of 3D models into *projectional* and *perspective* methods. Projectional methods are those in which a 3D volume is projected into a bi-dimensional plane; in the perspective methods a 3D virtual world is displayed by means of techniques that aim to reproduce the perspective of the human eye looking at the physical world.

Projectional methods include CT image-reformatting approaches such as multi-planar reformations (MPR) in the different spatial planes. More specific projection techniques include MIP and minimum-intensity projection (MinIP). The reformatting process does not modify the CT data but uses them in off-axis views and displays the images in an orientation different from native acquisition. Surface and volume rendering use algorithms that generate 3D views of sectional two-dimensional (2D) data. Volume rendering displays the entire volume preserving the whole dynamic range of the image, whereas surface rendering is based on the extraction of an intermediate surface description of the relevant objects from the volume data.

Several reformatting techniques are being proposed in the last years but the most frequently used nowadays in the assessment of carotid arteries are these four: MIP, MPR, CPR and VR. All of these techniques show strengths and pitfalls since they are based upon post-processing procedures of the CT data.

In the *MIP* only the voxel with highest CT number is displayed depending on the voxel position along the projecting ray [11]; therefore, MIP enables the evaluation of each voxel along a line from the viewer's eye through the volume of data and to select the maximum voxel value, which is used as the displayed value. MIP represents the simulation of rays that are traced from the operator through the object to the display screen, and only the relative maximum value detected along each ray path into the selected slab is retained by the computer. MIP may generate high-quality DSA-like images providing an overview of the target vessel. Often, it is relevant to use (thin or thick) slab MIP [12] to limit the over-projection phenomenon, in order to avoid time-consuming editing to exclude ebony structures (Fig. 13.1). MIP has a number of related artefacts and shortcomings that must be taken into account to properly interpret the rendered image. The displayed pixel intensity represents only the structure with the highest intensity along the projected ray. A high-density structure such as calcifications will obscure information from intravascular contrast material (Fig. 13.2). This limitation can be partially overcome with use of non-linear transfer function or through volume editing to eliminate undesired data. MIP misses the visualization of depth, leading sometimes to misinterpretation of the anatomical relationships. In MIPs, differences in attenuation can be detected, thus plaque mural calcifications are distinguishable but, especially, an extensive mural calcification hampers the demonstration of vessel lumen. Intra-luminal pathology may be obscured if it is surrounded by enhanced blood flow; hence, the detection of intravascular thrombi, emboli or dissection is limited in MIP.

The use of the highest intensity value also increases the mean background intensity of the image, effectively selecting the "noisiest" voxels and thereby decreasing the visibility of vessels in enhancing structures such as the kidney and liver. MIP images are typically not displayed with surface shading or other depth cues, which can make assessment of 3D relationships difficult. Also, volume averaging (the effect of finite volume resolution) coupled with the MIP algorithm commonly leads to MIP artefacts. A normal small vessel passing obliquely through a volume may have a "string of beads" appearance, because it is only partially represented by voxels along its length. Sliding thin-slab MIPs are also useful for the demonstration of the complex vascular structures. Targeted MIP is useful to avoid artefacts from overlying structures.

MPR is a simple reformatting method in which the images results from the retrospectively reconstructed axial data and obtained image may be oriented in every spatial direction. 2D MPR is a widely used post-processing technique and with arbitrary chosen views, it is very useful in CT angiography (Fig. 13.3). With the use of 16 or higher number it is possible to generate isotropic images. Oblique sectioning, particularly when interactive, requires a visualization method for plane placement and orientation verification. One method uses selected orthogonal images with a line indicating the intersection of the oblique image with the orthogonal Images. MPR images show great utility in the quantitative lumen analysis and in axial image plane definition because they depict lumen shape.

CPR is a 2D post-processing method that shows the cross-sectional profile of a vessel along its length [13–15]. Often structures of interest may have curvilinear

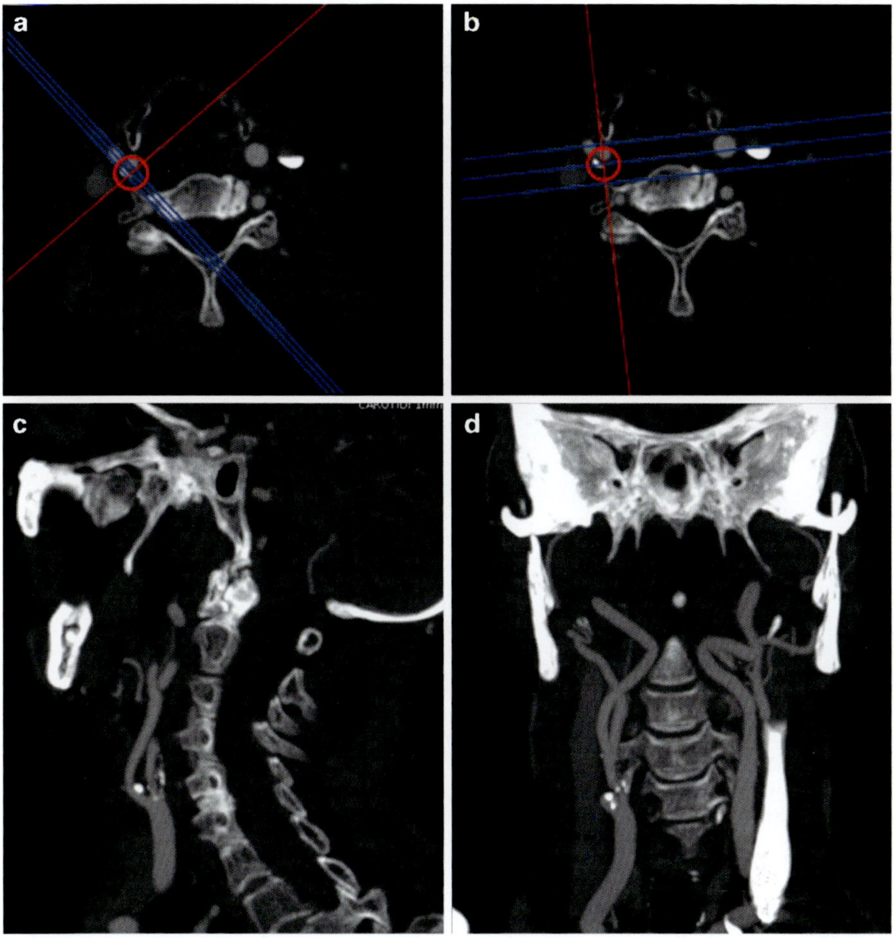

Fig. 13.1 MIP post-processed images. In (**a, b**) a thick slab (1 cm) MIP is visible. In (**a**) CTA axial image is visible, and the *blue* and *red* lines represent the MIP reconstruction direction and volume; in (**b**) the MIP is given. In (**c, d**) a thick slab (3 cm) MIP is visible. In (**a**) CTA axial image is visible, and the *blue* and *red* lines represent the MIP reconstruction direction and volume; in (**d**) the MIP post-processed image is given

morphology that orthogonal and oblique reformations cannot capture in a single 2D image. This restriction of planar sections can be overcome using curvilinear reformatting techniques. A trace along an arbitrary path on any orthogonal plane identifies a set of pixels in the orthogonal image that have a corresponding row of voxels through the volumetric data set. Each row of voxels for each pixel on the trace can be displayed as a line of a new image, which corresponds to the curved planar structure lying along the trace (Fig. 13.4). It is possible to consider Curved planar reformatting as a variation of MPR. However, by using the MPR there are some limitations. An important issue concerning the use of the CPR is that the vascular

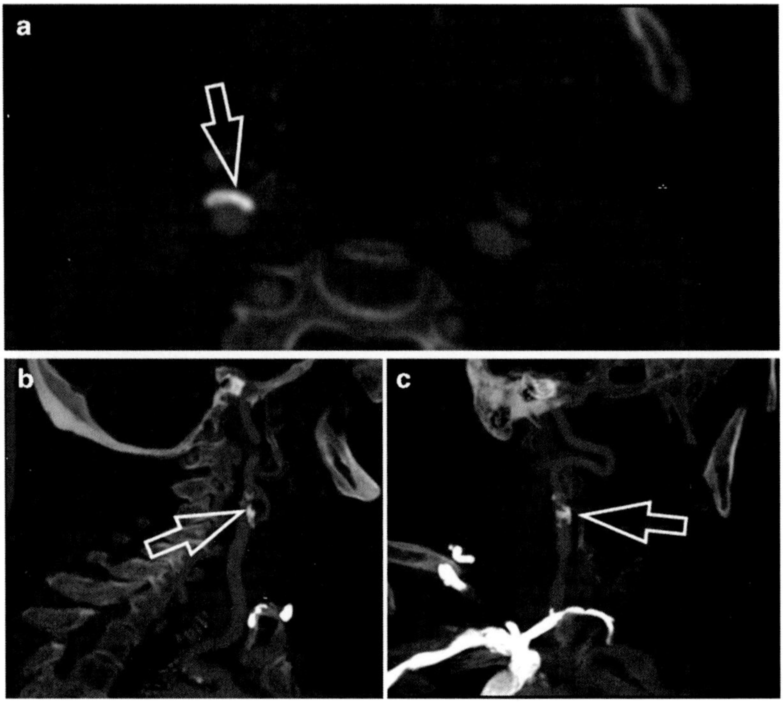

Fig. 13.2 MIP post-processed image. Presence of calcification may hamper MIP visualization of carotid images. In (**a**) CTA axial image is visible. MIP oriented according to a sagittal plane (**b**) allows to view the regular opacified lumen, whereas in the MIP oriented according to coronal plane the big calcification located in the right internal carotid artery hide the residual lumen and did not allow a correct quantification of the residual lumen

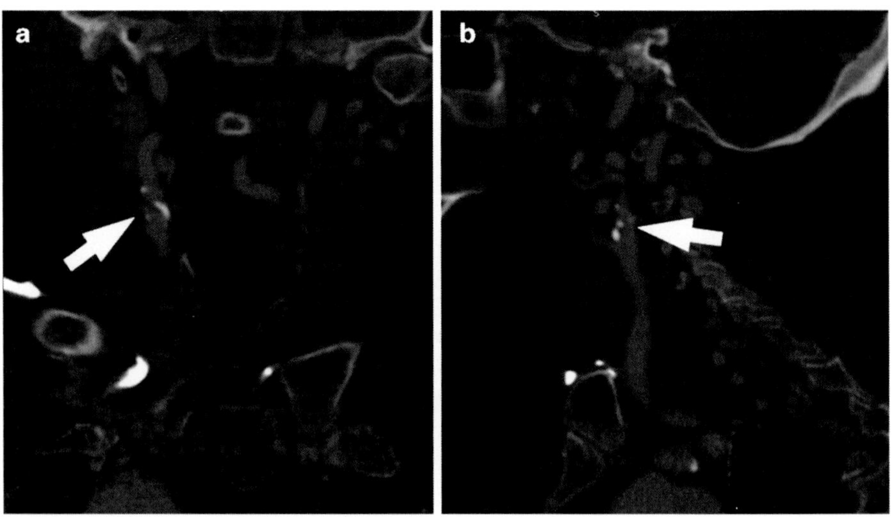

Fig. 13.3 MPR post-processed images according to coronal (**a**) and sagittal (**b**) planes

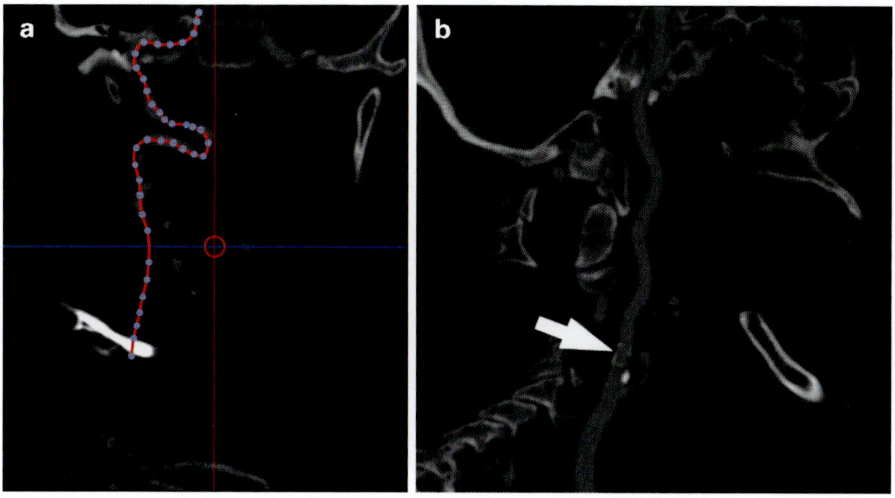

Fig. 13.4 CPR post-processing techniques. In (**a**) the trace along an arbitrary path that identifies a set of pixels in the orthogonal image that have a corresponding row of voxels through the volumetric data set is visible. Each row of voxels for each pixel on the trace can be displayed as a line of a new image, which corresponds to the curved planar structure lying along the trace (**b**)

images depend on the course of the selected curved plane; in fact, if the operator selects an inexact plane, it will generate an erroneous stenosis morphology and degree. In general, CPR is operator dependent and not suitable for measurement of diameters while the operator may not accurately point the centre of the vessel leading to distortions of anatomic structures. However, there is commercially available software with automatic assessment of the vessel using CPR, where the automatic assessment of the centre of the lumen facilitates automatic measurements of vessel cross-sectional dimensions in diameters and in area value.

VR represents another important and useful system to analyze carotid artery [16]. This technique incorporates all CT raw data into a resulting image producing high-quality 3D images (Fig. 13.5). The rendering technique is the computer algorithm used to transform conventional serial transaxial CT imaging data into simulated 3D images. Rendering methods can be divided into two classes: surface based (often using thresholding) techniques and volume-based (often using "percentage" classification) techniques. The type of rendering technique has great impact on the quality of the final images in any given 3D application. The purpose of volume visualization is to represent and analyze volume data in a realistic way by integrating a series of axial CT sections into a form that is easier to interpret than the original images. With the aid of advanced computers the VRT has developed into one of the most fascinating technique for 3D image display. VR technique uses data from all imaged voxels through volume data management. VR image formation for display consists of algorithms for transfer functions. In VR the voxels are assigned with an opacity value and this value can be chosen between total transparency and opacity. Maintenance of the data from all voxels in VR allows the editing of 3D display with

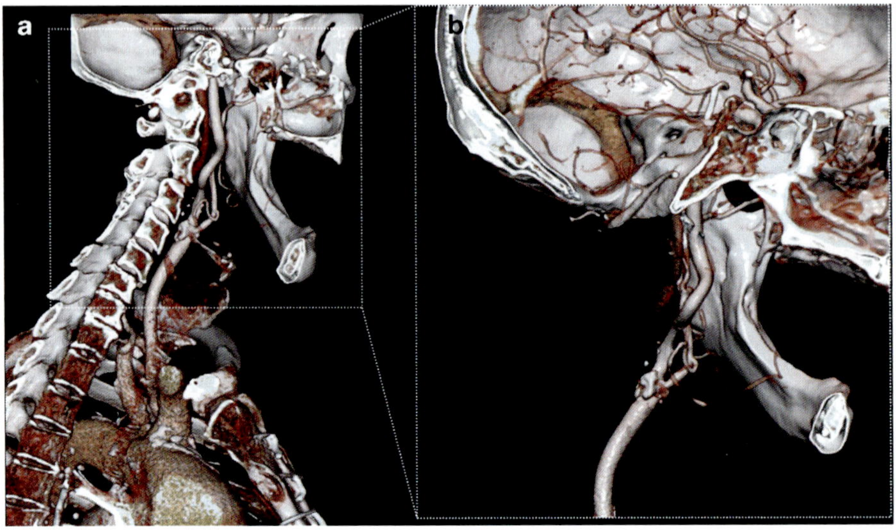

Fig. 13.5 VR post-processed images

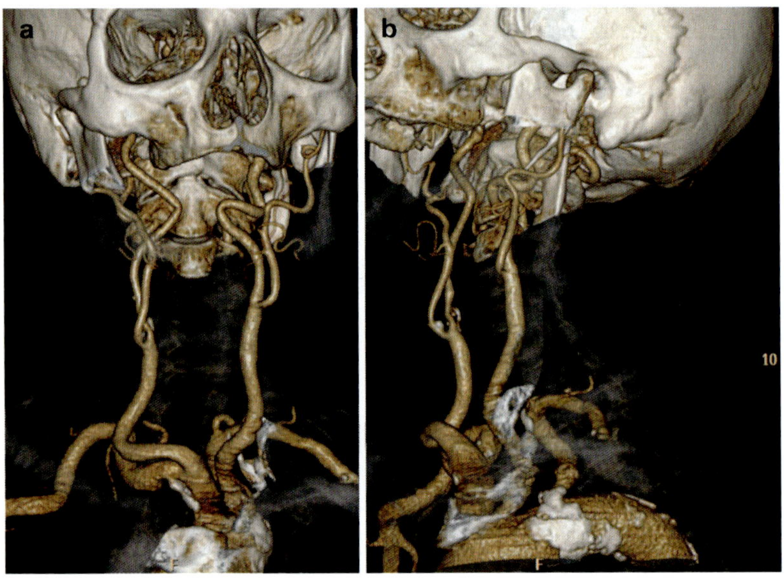

Fig. 13.6 In VR post-processed images, the aortic arch and the supra-aortic vessels are clearly visible

a clip plane in real time. Colouring of the structures having different opacity value for each anatomic structure is possible (Fig. 13.6). All 3D-rendering techniques represent a 3D volume of data in one or more 2D planes, conveying the spatial relationships inherent in the data with the use of visual depth cues. VR relies on mathematic formulas to determine for each pixel what portion of the data in memory should be displayed on the screen and how that portion should be weighted to best represent spatial relationships. The key to high-quality 3D imaging is the use of thin sections reconstructed at close inter-slice spacing. Specific scan protocols in regard to kVp and mAs will vary between different scanners but the selection of parameters must balance image quality with minimizing patient dose for the study. In cases where IV contrast is used (cases of CT angiography) the timing and delivery of contrast material relative to scan acquisition is critical.

Voxel selection is usually accomplished by projecting lines (rays) through the data sets that correspond to the pixel matrix of the desired 2D image. Differences in the images produced with various types of 3D rendering are the result of variations in how voxels are selected and weighted. To optimally display the anatomical structures, volume rendering enables modulation of window width and level, opacity, and percentage classification, and enables also the interactive change of perspective of 3D rendering in real time.

The most common method used to determine the percentage contents is probabilistic classification involving a trapezoidal approximation. This method for determining tissue-type percentages works well for CT data (as well as other types of data). For trapezoidal classification, each tissue type is assigned a nominal value range that, in theory, represents that tissue type exactly. A voxel with a signal within that nominal value range is considered to contain 100% of that tissue. Around this ideal nominal value range, another range is defined by choosing a high and low point representing attenuation values at which a voxel would contain none of the designated tissue. Voxels with signal intensities that lie between the 0% point and the corresponding 100% points are assigned a corresponding percentage between 0 and 100%. Thus, a voxel with signal intensity precisely halfway between the 0 and 100% points would be assigned 50% of that issue. A voxel with signal intensity three-fourths of the way towards the 100% point would be assigned 75% of that tissue. All values between the 0 and 100% points represent voxels in which volume averaging is present. This trapezoidal classification models closely the actual volume averaging in CT voxels.

Manual or semi-automatic editing is typically performed to "eliminate" an object such as an organ from surrounding structures; in particular, in the analysis of carotid arteries it is frequently used in the bone removal (Fig. 13.7) function that allows to delete bones that are superimposing with carotid arteries. Rendering parameters are applied to the full-volume data sets and affect the appearance of the image to be displayed. Volume rendering typically segments data on the basis of voxel attenuation. The window can be adjusted to standard settings; however, real-time rendering also permits the user to interactively alter the window setting and instantly see the changes. "Opacity" refers to the degree with which structures that appear close to the user obscure structures that are farther away. Low opacity values allow the user to "see through" structures. The use of percentage classification is a more advanced feature of volume rendering which distinguishes between different groups of voxels pertaining to different tissues in an optimal way. Each group is approximated by a trapezoid in the software that can be interactively manipulated

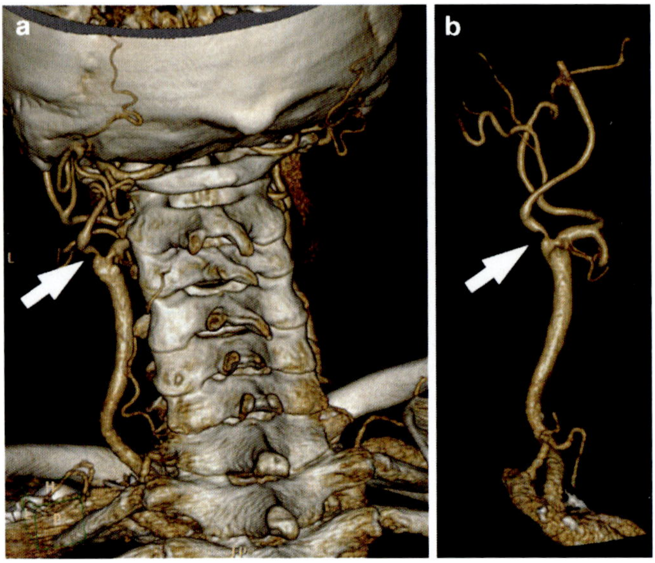

Fig. 13.7 Severe stenosis in the left infernal carotid artery demonstrated by VR (**a**). In (**b**) the left carotid without bones, after manual editing for bone removal, is visible

to alter the visibility of the tissue, and multiple trapezoidal distributions can be displayed simultaneously.

Image display is the final step of image processing and relates to the process by which a "virtual" 3D representation of a volume data set is "flattened" onto one or more 2D planes; this process is required because a 3D model must be represented with a computer monitor, which is a bi-dimensional device. The most commonly used display method is raycasting, which is a basic technique for displaying a volume of data in two dimensions. In this technique, an array of parallel lines (rays) are mathematically projected through the volume in alignment with each pixel within a desired 2D plane. A weighted sum of these voxel values encountered by the ray is calculated for each pixel and then scaled to the particular range of greyscale values in the display. By using advanced VR algorithms it is possible to distinguish mural calcifications from the residual lumen [17] and this algorithm is very useful in detecting ulcerated carotid plaques [7].

13.3.1 CTA Issues

In the common application of CT two major issues arise: the radiation dose and the risk of contrast material administration. Radiation exposure to the patient at CT and the consequent potential radiation hazard have recently gained consideration in both the scientific and public literature [18, 19]. The National Council on Radiation

Protection and Measurements recently reported that in the USA the per capita dose of radiation from medical imaging has increased by a factor of nearly six, since early 1980s [20, 21]. In biologic material exposed to X-rays the most common scenario is the creation of radicals that can interact with the nearby DNA to cause strand breaks or base damage [18]. In fact, experimental and epidemiologic evidence has linked exposure to low-dose, ionizing radiation with the development of solid cancers and leukaemia. Although most of the quantitative estimates of the radiation-induced cancer risk are derived from analyses of atomic bomb survivors, there are other supporting studies based on workers in nuclear industry that suggest an increased risk of cancer for those who were exposed to an average dose of approximately 20 mSv, which is a typical organ dose from a single CT scan for an adult [22, 23]. In fact, MDCT delivers a higher dose of radiation than many other radiological imaging modalities. For example, the effective dose of MDCTA is about three to six times that of conventional angiography [23]. It is fundamental to reduce radiation exposure. To obtain a reduction it is necessary to adjust the dose to the patient's weight and size [24]. However, authors agreed that radiation dose during MDCT should not be reduced at the cost of diagnostic performance [25].

Another important point is the potential reaction to contrast material. Nowadays non-ionic low-osmolarity contrast media have completely replaced intravenous ionic contrast media in most of the institutions. Several studies have examined adverse reactions to ionic and non-ionic contrast media and have consistently reported fewer adverse events associated with the use of non-ionic agents. However contrast material injection has a slight risk of causing an intolerance [26–28]. Reactions usually occur immediately and include itching, flushing or difficulty of swallowing or breathing but also nephropathy may be a serious consequence. An extremely interesting study of Callahan et al. found a strong and direct relationship between patient age and incidence of contrast material reactions in the paediatric population, suggesting increase in intravenous contrast material reaction with age through 18 years and then a decrease in the incidence. The risk of death is 1/50,000–1/500,000 [29]. Overall iodine is safe and has been used for many years in millions of CT exams. Iodine contrast increases the sensitivity of the MDCTA exams and it is worldwide accepted that the benefits of using iodine contrast typically outweighs the risks.

13.4 Carotid Artery Imaging

13.4.1 Imaging Diagnostic Flowchart

The choice of a specific imaging modality to assess the carotid artery depends on several parameters and depends largely on the clinical indication for imaging and the skills available in individual centres. In 2008, Raff and colleagues [30] proposed a diagnostic algorithm for the correct use of the imaging modalities according to the different clinical indication of the patients. They proposed that for patients with an high likelihood of vascular disease, CTA may represent an appropriate first exam. On the other hand, for screening of patients with a lower likelihood of neurovascular pathology US-ECD should be selected. If significant stenotic disease of the ICA is detected, CTA as well as MRA can be used to confirm the diagnosis and to accurately determine the precise degree of stenosis. For those asymptomatic patients scheduled for surgery as coronary artery bypass graft, abdominal aortic aneurysm and lower limb ischemia, US-ECD represent an accurate and cost-effective non-invasive screening tool [30, 31]. DSA is infrequently necessary and only in cases of severe multiple vessels disease, for which assessment of flow direction and collateral patterns may be important or when the image quality of non-invasive procedure is of limited value.

13.4.2 Imaging Techniques

Four major types of carotid imaging methods are currently available: DSA, US-ECD, CTA and MRA. DSA has been the gold standard of carotid imaging for the assessment of carotid stenosis and has been used in all the major randomized controlled trials of carotid endarterectomy (CEA), but this technique suffers from some important limitations (high cost, procedural risk) and non-invasive methods are increasingly used instead of DSA. Despite advances in catheter technology and expertise, DSA remains associated with a small but important risk of neurological complications. Such risk potentially reduces benefits gained by potential revascularization in the same procedure, and these concerns have generated substantial interest in non-invasive techniques for carotid imaging such as US-ECD, CTA, and MRA. In the following paragraphs will be discussed briefly potentialities, limitations and application of the most used imaging techniques with the exclusion of CTA that was already discussed in the first part of this chapter.

13.4.3 Digital Subtraction Angiography

DSA is a type of fluoroscopy technique used in interventional and diagnostic radiology. DSA is an invasive method that requires femoral artery puncture and intra-arterial injection of contrast medium, usually in the aortic arch or in the common carotid artery. It is possible to obtain images of the stenotic lumen with an high spatial resolution and an estimation of low dynamics, such as slow and delayed blood flow. DSA was considered in the past years as the gold standard for the study of carotid artery pathology because of its high spatial (50 μm) and temporal resolution (10 ms). This method allows to obtain optimal definition of the opacified lumen, as well as the presence of plaque characteristics as lumen irregularity or plaque ulcerations [32–34] (Fig. 13.8). DSA suffers of several limitations in fact it is invasive, labour intensive, expensive, time expensive and requires a period of bed

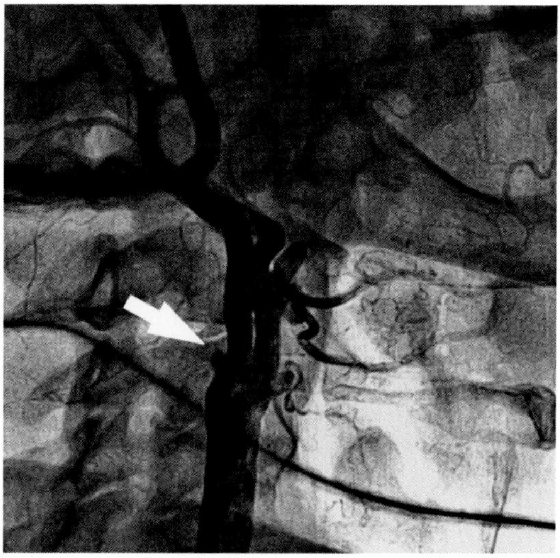

Fig. 13.8 DSA image of the right internal carotid artery. The *white arrow* indicates a small ulcer in the right ICA proximal to the point of maximum stenosis

rest [35]. The main concern of DSA is the risk of neurological complications: in fact the risk of permanent neurological complications is up to 0.9% and transient neurological complications up to 2% [35–39]. Another less important complication is the risk of groin haematoma that has been reported to be up to 8%, although these haematoma rarely cause significant morbidity or delay in hospital discharge. In the last 10 years the concept of vulnerable plaque is acquiring more importance [40]. Wasserman et al. [41] have suggested that in the carotid artery, it is time to look beyond the degree of stenosis and to analyze and characterize the plaque composition. The DSA is not capable to analyze the plaque characteristics, because it is only a "luminal methodology," but non-invasive technique like US-ECD, MRA and CTA can study the plaque composition and structure. For these reasons (application field, risk and costs), DSA should be used only in particular condition like flow-dynamic analysis and small collateralization visualization.

13.4.4 Ultrasound Echo Colour Doppler

US-ECD is globally accepted as the standard of care for the initial diagnosis of carotid artery bifurcation disease. This high-resolution, non-invasive technique is readily available, rapidly applicable, and can be performed at relatively low cost. In the USA, US-ECD may be the only diagnostic imaging modality performed before CEA. Therefore, the information obtained with US-ECD must be reliable and reproducible. The accuracy of carotid US hinges on practicing meticulous scanning technique and requires excellent technical equipment. It is important to remember that there are several pitfalls that may mislead the operator to falsely interpret colour and spectral Doppler findings. High frequency linear transducers (>7 MHz) are ideal for the assessment of plaque morphology and quantification of intima–media thickness (IMT), whereas lower frequency linear transducers (<7 MHz) are preferred for Doppler examinations [42]. In some particular conditions, like short muscular neck, the use of a linear transducer may not be possible so that a curved-array transducer should be selected.

Initial Doppler criteria were proposed by Fell et al. [43] where authors utilized measurements of the spectral waveform to predict ICA stenosis. Subsequently, alternative criteria were developed which have resulted in excellent sensitivity and specificity of US-ECD to determine high-grade stenosis [44]. Jahromi et al. [45] in a meta-analyses of published criteria for US-ECD and they observed sensitivities of 98% and specificities of 88% for detecting >50% ICA stenosis; and 94 and 90%, respectively, for detecting >70% ICA stenosis.

The extent, location and characteristics of atherosclerotic plaque in the common carotid artery, internal carotid artery and external carotid artery should be documented and the vessels should be imaged as completely as possible by applying a cephalic angulation of the transducer at the level of the mandible (Fig. 13.9). The Doppler analysis should be always performed and the velocity of

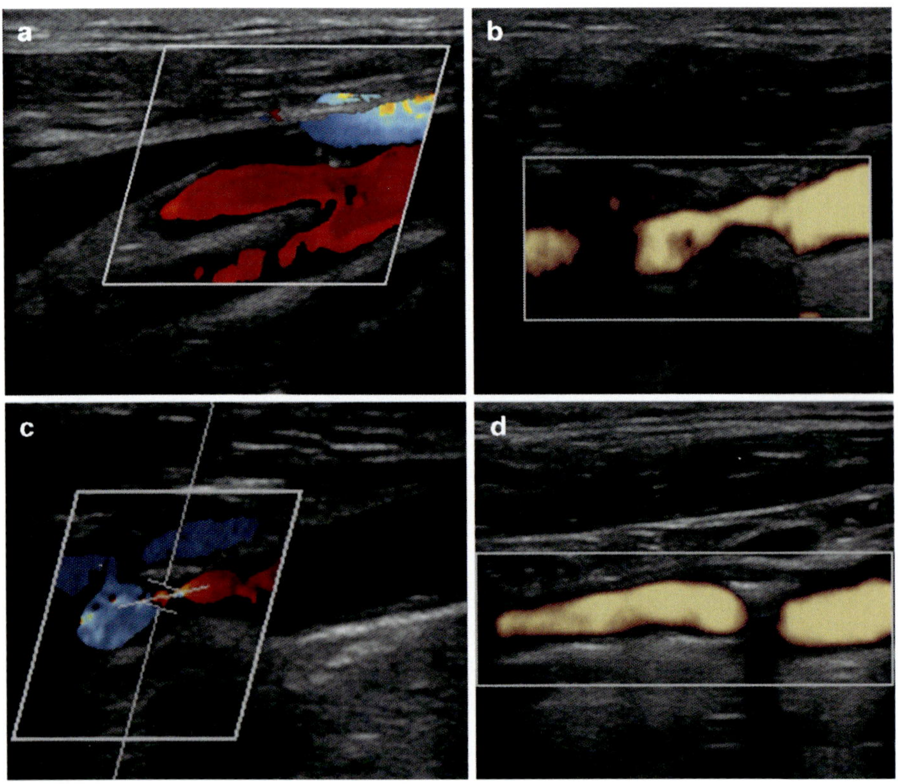

Fig. 13.9 Examples of different carotid plaque studies by using US-ECD

blood flow in the mid-CCA and in the proximal ICA should be measured [42]. Ultrasound images can be evaluated either visually or objectively by a computer-assisted grey-scale median (GSM) measurement. It was demonstrated that the visual evaluation of plaque echogenicity has only fair reproducibility [46] whereas objective characterization is more reliable and less observer dependent [47]. However, there is no consensus yet on which GSM threshold value is most sensitive to distinguish vulnerable from stable plaques because computer-assisted GSM measurement only assesses the median brightness of the entire plaque and regional instability, such as haemorrhage, may be present within a plaque even with a high GSM value. For these reasons other methods were proposed to analyze the plaque like the stratified grey-scale measurement of carotid plaque echogenicity, real-time compound ultrasonography or the pixel segmentation with tissue mapping [48–50].

Several ultrasound markers to identify high-risk patients have been reported in literature, including carotid stenosis evaluation, plaque echogenicity and irregularity. Hypo-echoic plaques are more likely to be symptomatic than hyper-echoic ones, because they contain more soft tissue (lipid and haemorrhage), while hyper-echoic plaques are primarily composed of fibrous tissue and calcifications [51–53]. Several studies analyzed the relationship between plaque echogenicity and symptoms. In particular, the Tromsø study followed up 223 patients with carotid stenosis between 35 and 99% and 215 controls for 3 years [54] and the authors observed that the relative risk (RR) of ipsilateral cerebrovascular events in the Hypo-echoic groups was 3.52 (95% CI, 1.0–12.4). The Cardiovascular Health Study [55] showed that asymptomatic elderly patients with a hypo-echoic plaque have a RR of ipsilateral ischemic stroke of 2.78 (95% CI, 1.4–5.7), independent of degree of stenosis and other cardiovascular risk factors. Nowadays, US-ECD represents an optimal choice as first-line exam of the carotid artery; however, its limitations in concordance between observers and in the identification of some plaque risk factors such as ulceration impose the use of a second line exam (CTA, MRA) before surgical or interventional procedures.

13.4.5 Magnetic Resonance Angiography

Several Studies have shown that MRA can be used to quantify carotid stenosis degree and accurately characterize the composition and morphology of human carotid atherosclerotic plaque. In particular, in the last decade, significant progress has been made towards the non-invasive detection of vulnerable atherosclerotic plaque using MRA [56, 57]. This imaging technique does not involve ionizing radiation, enables visualization of the vessel lumen [58, 59] and can be repeated serially to track progression or regression of the plaque.

MRA of the carotid arteries has gone through a long evolutionary period to become a routine imaging modality for evaluation of stenosis at many centres [60, 61]. Nowadays, in the carotid artery stenosis quantification the MRA sensitivity in demonstrating stenosis >70% is much better than that of US-ECD [62]. The first application of MR in carotid arteries started in mid-1980s [63, 64] and in the last years there was a significant improvement thanks to the increase of the magnetic field (from 0.3 to 3 Tesla and more), the development of coils dedicated to the carotid artery analysis (surface coils), the creation of advanced sequences for the data acquisition (Black Blood techniques, Time of Flight) and thanks to the use of contrast material that improve the vessel lumen visualization.

The first MRA method, phase-contrast MRA, was developed 30 years ago and was quickly followed by 2D and 3D time-of-flight (TOF) MRA. TOF MRA has been widely adopted for an array of clinical indications but is relatively insensitive to slow flow and is associated with long scan times and signal voids, all of which can lead to poor-quality imaging and over-estimation of stenosis. More recently, contrast-enhanced (CE) MRA has been introduced. CE MRA produces high-quality images in a very short period of time and may alleviate some of the drawbacks associated with TOF MRA.

MR angiography, however, is sensitive to artefacts caused by the slow and turbulent flow associated with high-grade stenosis. Of particular importance are the potential overcall of stenosis grade and the differentiation of high-grade stenosis from occlusion and the accurate diagnosis of moderate-grade (50–69%) stenosis for the patient to receive optimal management. Carotid MR angiographic studies tend to overestimate the degree of high stenoses [65]. Other than the morphologic techniques others have been developed to quantify the stenosis of the artery, for example MR flow quantification with the phase-contrast method enables non-invasive measurement of the volumetric flow rates and velocity curves at any portion of a vessel desired and could, therefore, provide additional information about the hemodynamics of a stenosis. The accuracy of the phase-contrast method has been validated in vitro and in vivo [66, 67]. The unparalleled sensitivity of MRI to soft tissue signal has been exploited to examine not only the indirect manifestation of atherosclerosis as a narrowing of the vascular lumen but to assess the plaque itself [68–72]. There have been extensive investigations directed at developing and validating MR methods that can essentially reproduce histological evaluation of plaque composition using in vivo methods. Several sequences can be used to study the plaque components but the single sequence that has been most widely used in characterization of plaque composition is the T2-weighted fast spin-echo sequence. On these images, the lipid core appears as a hypointense region, fibrous cap appears relatively hyperintense, and calcification appears as a very dark region. The other principal component in the atheroma that can be readily defined is the location of fresh intra-plaque haemorrhage consisting principally of meta-haemoglobin.

In addition to the intrinsic contrast that can be generated in different plaque components using multi-contrast MR methods, recent studies used MR to identify the presence and activity of specific molecules involved in plaque inflammation, in particular, using ultra-small super paramagnetic particles of iron oxide (USPIOs) [73–75]. USPIOs are iron oxide nano-particles stabilized with low molecular weight dextran with a mean diameter of 30 nm. These relatively small particles have a much larger half-life in blood than the conventional superparamagnetic iron oxide particles, with a mean diameter of 150 nm. Because of their long half-life in blood, USPIOs can be taken up by macrophages in the whole body. The animal studies indicate that the USPIOs are phagocytosed by macrophages in atherosclerotic plaques, which causes a signal decrease on MR images. Because a preponderance of macrophages is an important feature of a high-risk plaque, USPIO-enhanced MRI is a promising method for the in vivo differentiation between low- and high-risk plaques.

13.4.6 Other Imaging Modalities

Research interest has increasingly focused on inflammatory biomarkers as a means of predicting future risk of rupture. In fact, there is evidence that the inflammatory process dominated by macrophages within the carotid plaques increases the risk of rupture and subsequent thromboembolic events [76]. The inflammation precedes the calcification, and the present imaging techniques, for example angiography or contrast-enhanced CT, have only limited capacity to find small non-calcified plaques, and cannot detect inflammation within the plaques.

Nuclear medicine techniques have been developed in order to study the carotid plaque inflammation. Molecular nuclear medicine imaging has the potential to furnish functional information on cell biologic events which determine the risk of plaque rupture; moreover, besides their non-invasive nature, nuclear medicine techniques have the potential to evaluate important determinants of plaque vulnerability, taking into account specific cellular or biochemical changes that characterize these lesions.

Nuclear medicine images are based on the administration of a radionuclide tracer compound to the patient, and its subsequent detection by techniques such as single-photon emission computed tomography (SPECT) and positron emission tomography (PET).

Nuclear imaging techniques are very sensitive in detecting radioactive tracers targeted at carotid plaques. Disadvantages of nuclear imaging techniques are the lack of detailed anatomic information in the area of tracer uptake and exposure of the patient to ionizing radiation [77].

13.4.6.1 [18F]-Fluorodeoxyglucose Positron Emission Tomography

[18F]-Fluorodeoxyglucose positron emission tomography (FDG-PET) represents a promising method to study and characterize the carotid vulnerable plaque. PET imaging is based on the detection of gamma photons from the emission of positrons. Radionuclides used in PET scanning are typically isotopes with short half-lives such as ^{11}C (~20 min), ^{13}N (~similar 10 min), ^{15}O (~2 min) and ^{18}F (~110 min). PET images are derived from the detection of positron emitting radionuclides, labelled to biochemical and metabolic substrates, and fluorine-18 deoxyglucose (FDG) is the most employed radiotracer.

FDG is transferred by glucose transporter proteins on the cell surface into living cells where it is trapped after the first metabolic stage and is phosphorylated by hexokinase enzyme to FDG-6 phosphate that is not metabolized. High accumulation of FDG appears especially in those cells that have a need for high quantities of glucose for their energy metabolism, such as inflammatory cells [77] because the degree of cellular FDG uptake is related to the cellular metabolic rate and the number of glucose transporters. The inflammatory nature of atherosclerosis is now well recognized. From the initial phases of leukocyte recruitment, to eventual rupture of the vulnerable plaque, inflammatory mediators appear to play a central role in the pathogenesis of atherosclerosis, so that Falk defines atherosclerosis as a "multifocal, smoldering, immuno-inflammatory disease of medium-sized and large arteries fueled by lipids" [78].

The FDG ability to localize inflammatory cells may be useful for diagnosis of vascular diseases such as large-vessel arteritis and in animal models it is has also shown that FDG-PET can detect atherosclerosis-like lesions [79–81]. In fact, a significant characteristic of this technique is that FDG is well correlated with the level of macrophage infiltration in the lesions [82]. It has also been reported that the FDG-PET signal in plaques is reduced following a period of statin treatment [83]. On the best of our knowledge, the first data on FDG-PET imaging in human atherosclerotic carotid plaque inflammation was reported in 2002 by Rudd et al. [84]. In this study, eight patients who suffered a recent carotid territory TIA and had an internal carotid artery stenosis >70% were found to have a significantly increased FDG uptake into all eight symptomatic plaques compared to the six asymptomatic plaques on the contralateral side. In another study by Tawakol et al. [85] published in 2006, a group of 17 patients was studied and a significant correlation ($r=0.70$; $p<0.0001$) was found between FDG uptake and the macrophage staining from the corresponding histological sections. From these studies it was shown that FDG PET can provide in vivo a non-invasive measure of the severity of carotid plaque inflammation. Another potential application of FDG PET is the evaluation of plaque inflammation after drug therapy by using it as a surrogate marker of anti-atheroma drug efficacy: Tahara et al. [86] observed that in patients who underwent 3 months of simvastatin treatment there was a reduction of FDG uptake in the atherosclerotic plaques whereas dietary management alone did not.

13.4.6.2 Single-Photon Emission Computed Tomography

SPECT is similar to PET in detecting gamma radiation and typical radioisotopes used in SPECT include ^{99m}Technetium, ^{123}Iodine and ^{111m}Indium. In general, these radioactive substances have longer decay times than those used in PET and can also be labelled with compounds and then injected as pharmaceuticals. Technetium Tc-99m-labelled annexin A5 has been used for the specific targeting of vulnerable atherosclerotic lesions, in fact annexin A5 is a normally circulating protein which targets phosphatidylserine, a molecule expressed on the cell membrane during apoptotic cell death and apoptosis is an important indicator of atherosclerotic plaque vulnerability, and it occurs primarily in smooth muscle cells and macrophages. Kietselaer et al. [87] published the first clinical study with radio-labelled annexin A5 in humans and they observed that Tc-99 annexin A5 uptake was seen in the cervical region in the patients with recent TIA whereas no uptake was detectable in the other patients who had suffered a TIA more than 3 months before imaging and who had been treated with anti-platelet agents and high-dose statins. These patients underwent CEA and histological examination demonstrated that the positive Tc99m-annexin A5 uptake correlated with plaque macrophage content, whereas both patients with negative annexin imaging scans showed stable lesions with abundant smooth muscle cells and collagen, and negligible macrophage infiltration and no intra-plaque haemorrhage.

13.4.6.3 Scintigraphy

It is also possible to study carotid arteries by using scintigraphic procedures and, in particular, *Platelet scintigraphy*, *Interleukin-2 (IL2) scintigraphy* and *Low-density lipoprotein scintigraphy*. Scintigraphy with radio-labelled platelets is a technique that allows imaging of platelet-rich thrombi on atherosclerotic lesions. Manca et al. [88] correlated in vivo ^{111}In-labelled platelet scintigraphy to histology in a series of 22 patients and found a sensitivity and specificity of 89% and 100% for the detection of carotid plaque thrombosis. IL2 is a cytokine that acts by binding to its receptor (IL2R), expressed mainly on activated lymphocytes, but also on activated monocytes and macrophages, and smooth muscle cells. On the best of our knowledge, only one in vivo IL2 scintigraphy study has been performed in humans [89] and the authors observed that carotid plaques strongly correlated with the percentage and absolute number of IL2R-positive cells at histology. The third scintigraphic technique is the low-density lipoprotein scintigraphy, in fact the radio-labelled autologous low-density lipoproteins specifically accumulate within monocytes and macrophages [90]. Only few studies had been performed by using this technique but the results in the identification of carotid plaque inflammation are promising [91–93].

13.5 Carotid Artery Pathology and Stroke Risk

Atherosclerosis is a common systemic problem with multiple arterial sites affected. Some degree of carotid artery narrowing has been reported in up to 75% of men and 62% of women aged 65 years [94]. The first study which put in correlation carotid artery lesions with incidence of stroke is attributed to Savory who in 1856 [95] reported the case of a young woman affected by left mono ocular symptoms, right hemiplegy and dysesthesia. In this woman, the autoptic examination revealed an occlusion of the distal tract of the left internal carotid artery associated with a bilateral obstruction of sub-clavian artery. Since this first case report, many others have been so far described [96–98] which put in evidence that there was a deep correlation between carotid pathology and cerebral symptoms. Currently, this pathology is the third leading cause of severe disability in the Western World causing 4.5 millions of deaths [82, 83, 99]. Each year, in the USA, 795,000 people experience a new or recurrent stroke and in 2005 this pathology accounted for 1 in every 17 deaths [100]. Moreover, it is expected that the burden will increase greatly during the next 20 years [101]. Therefore, stroke represents a severe health problem and the identification of its risk factors is extremely important [102, 103]. To date, stroke pathogenesis is not completely understood, even if it may be shortly ascribed to two major process: (a) hypo-perfusion from high stenosis/occlusion of the vessel [104, 105] and (b) distal embolization [106, 107]. However, while it is likely that some strokes, associated with carotid artery disease, result from hypo-perfusion, the majority of such events appear be a consequence of embolization of an atherosclerotic plaque or acute occlusion of the carotid artery, with distal propagation of the thrombus. The risk of stroke in patients with carotid atherosclerosis is closely associated with the severity of luminal stenosis: for asymptomatic patients with less than 75% stenosis the yearly risk of stroke is less than 1%, but this risk increases to 5% for patients with stenosis greater 75%. Moreover, this risk is much higher in the symptomatic patients with a 10% in the first year with risk rising 35% over the next 5 years [108–111].

In the past years, three multi-centric randomized studies, NASCET (North American Symptomatic Carotid Endarterectomy Trial), ECST (European Carotid Surgery Trial) and ACAS (Asymptomatic Carotid AtheroSclerosis Group), provided cut-off values stenosis degree indicating possible benefits of CEA [112–114]. In particular, the pooled analysis of these three trials proved the benefits (70–99% NASCET stenosis with a risk reduction of 16%, $p<0.001$) of undergoing an endarterectomy for those patients with symptomatic high-grade stenosis (70–99%) after 5 years of follow-up [115]. For symptomatic patients with mild stenosis (50% or less in the NASCET) the risks during the surgical procedure outweighed the benefits, whereas in patients with a moderate stenosis (50–70% NASCET) there was a moderate benefit with an absolute risk reduction of 4.6% after 5 years of follow-up.

NASCET, ECST and ACAS evaluated the degree of stenosis as the percentage reduction in the linear diameter of the artery. Differences exist in the evaluation of stenosis degree between NASCET, ECST and CSI and values derived from these three classification methods are not equal if we consider the same carotid stenosis, so it is always important to specify the classification used. With the *NASCET criteria* the ratio between the lumen diameter at the stenosis and normal lumen diameter distal, where there is no stenosis, is calculated (Fig. 13.10).

With the *ECST criteria* the ratio between the lumen diameter at the stenosis and the total carotid diameter (including the plaque) is calculated. With the *CSI-index criteria* between the lumen diameter at the stenosis and normal lumen of the proximal common carotid artery first multiplied by 1.2 is calculated. In fact the CSI method uses a fixed conversion factor of 1.2 to estimate the carotid bulb size based on a "fixed anatomic relationship" between the CCA and the carotid bulb. In both NASCET and ECST, to quantify the degree of the stenosis, measurements have to be performed on a strictly perpendicular plane to the carotid axis. It was demonstrated by Saba et al. [116] that NASCET and ECST methods show a strength correlation

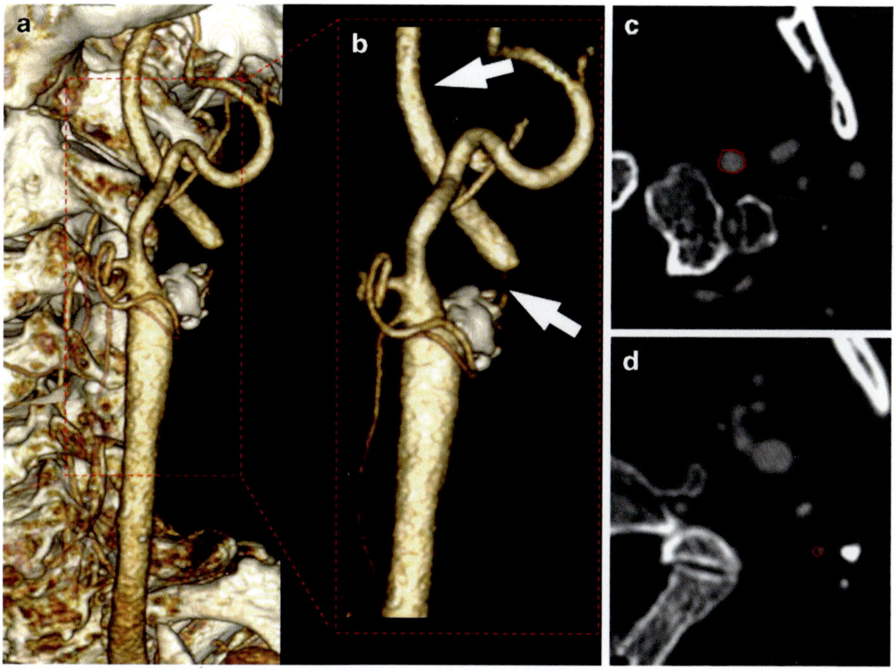

Fig. 13.10 NASCET method. The ratio between the lumen diameter at the stenosis and normal lumen diameter distal where there is no stenosis is calculated

and that inter-observer and intra-observer agreement values are high for both NASCET and ECST.

In the evaluation of carotid artery stenosis US-ECD, CTA and MRA are widely used. Sonography has come into widespread use for classification of carotid stenosis since 1975, when a quantitative examination protocol was developed that is now nearly universal. Nowadays US-ECD is commonly performed to screen patients with possible carotid artery disease, but the accuracy of US-ECD is moderate if our purpose is the assessment of stenosis degree [117]. In recent years, particularly in the USA, sonography was used as the only imaging diagnostic test for the patients selection for carotid endarterectomy, but the use of this technique as the unique test has been widely discussed [118–120]. In fact Eliasziw et al. [117] demonstrated that the only US-ECD use for the patient selection may determine some critical errors. The use of colour-flow Doppler can increase accuracy, in particular, in differentiating near occlusion from occlusion [121]. In a study which compared CTA to US-ECD with conventional angiography as the reference, CTA was found to yield higher sensitivity, specificity and predictive values than US-ECD in assessing high-grade stenosis and distinguishing it from complete occlusion.

Nowadays, MDCTA represents probably the best imaging methods to quantify the carotid artery stenosis with a sensitivity in the evaluation of stenosis degree that may be compared with angiography but with less risks [122–125] and CTA sensitivity for stenosis between 70 and 99%, in internal carotid artery with angiography as reference standard, can reach excellent results. Moreover, since almost a decade, workstations available are capable of performing high-quality 2D and 3D post-processed images.

MRA has become widely used as a non-invasive diagnostic imaging modality for carotid artery stenosis quantification. It avoids the radiation and iodinated contrast exposure associated with CTA. MRA sensitivity is quite good in determining stenosis degree. In a wide series of studies that assumed carotid angiography as gold standard, the median sensitivity for a high-grade lesion was 93% whereas the median specificity was 88% [126]. A recent meta-analysis of Debrey and colleagues, published in 2008 [127], indicated that MRA is highly accurate for the diagnosis of high-grade ICA stenosis and occlusion with both TOF and CE techniques, with CE MRA having the edge over TOF MRA, that MRA is of high accuracy for distinguishing occlusions from high-grade stenosis, particularly CE MRA and that both CE MRA and especially TOF MRA appear to be poor diagnostic tools for moderate ICA stenoses. 3D TOF provides superior resolution but its sensitivity in measuring the flow is lower compared with 2D TOF. CEMRA is a very promising technique and it is not impaired by slow-flow situations.

It was reported by some authors [128, 129], that both US-ECD and MRA had tendency to overestimate stenosis degree whereas CTA underestimates it. In a recent study, Saba et al. [130] demonstrated that that NASCET stenosis measured in MDCTA and PSV values measured by US-ECD have a good correlation: the use of a threshold of 283 cm/s allows obtaining good value of sensitivity and specificity (sensitivity and specificity : 75 and 88.6%, respectively). Probably the use of 3-Tesla systems will increase MRA performance in carotid stenosis quantification [131].

The methodology of carotid stenosis quantification is widely debated because NACET, ECTS and CSI-index are indirect ratio–percent methods. These trials imaged carotid arteries by using conventional angiography and methods of deriving percent stenosis ratios were adopted because standardized stenosis measurement were not consistent with film (in conventional angiography) and because when it was used with the digital angiography there were different degree of magnification and lack of millimetre calibration. With the introduction of MDCTA, thanks to its high spatial resolution (with isotropic voxel of 3 mm), new direct mm-method has been proposed by Bartlett and colleagues [29, 132, 133] in order to overcome limitations of the classical percent methods. It was well demonstrated that there is a linear relationship between millimetre carotid bulb stenosis diameter and NASCET. Threshold values of 1.4–2.2 mm can be used to evaluate moderate stenosis (50–69% according to NASCET) with a sensitivity of 75% and a specificity of 93.8%. A carotid diameter of 1.3 mm corresponds to 70% stenosis and this value was proposed by Bartlett as a threshold for severe carotid stenosis with a sensitivity of 88% and a specificity of 92%. Saba and colleagues [134] demonstrated that the simple millimetre measurement of stenosis can reliably predict NASCET-type, ECST-type and CSI-type percent stenosis. Similar correlation are present between ECST and mm-method and CSI-index and mm-method. The use of a direct mm-method for the quantification of the carotid arteries is an excellent option because it is possible to avoid cumbersome calculus and because direct mm-method allows to overcome the use of three different methods of measurement. The simple millimetre measurement of stenosis can reliably predict NASCET-type, ECST-type and CSI-type percent stenosis. Until now only studies performed by using CTA were published to quantify the carotid stenosis by using mm-methods but with the increase in spatial resolution (thanks to dedicated surface coils and 3-Tesla fields) it is reasonable to think that it will be possible to use even the MRA to perform this task.

At the present time the most powerful predictor of high-risk carotid atherosclerosis is the presence of recent focal neurological symptoms ipsilateral to a carotid stenosis [30, 135, 136] with a stroke risk very high in the first weeks after TIA or minor stroke and this risk declines rapidly thereafter. Moreover, it was well demonstrated by Wardlaw and colleagues [31] that the most cost-effective diagnostic strategies for carotid stenosis are those that offers surgery to a larger proportion of patients quickly after the warning TIA/minor stroke, in particular, those who suffer a 70–99% NASCET stenosis. For this reason there is a short time window for the stroke prevention necessitating a rapid identification of patients with substantial carotid high-risk plaque. A prompt and accurate diagnosis of carotid artery disease is critical when planning a therapeutical strategy.

13.6 From the Concept of Luminal Narrowing to the Carotid Vulnerable Plaque

Traditionally, the degree of luminal stenosis has been used as a measure of atherosclerotic disease severity. In 1988, however, angiographic studies on coronary arteries [137–140] demonstrated that moderate coronary artery stenosis may lead to acute myocardial infarction and subsequent histopathologic study results showed that plaque erosion and disruption were common morphologic features found in symptomatic lesions by indicating that *luminal narrowing* was not the sole predictor of myocardial infarction. Similar observations were observed later in the carotid arteries [41, 141, 142]. Even low-grade stenosis can result in cerebrovascular events, so it is fundamental to look beyond degree of stenosis and to characterize plaque morphology. A large, complicated, and potentially unstable carotid plaque may not compromise the lumen by an amount that would be considered a significant stenosis. This realization has created tremendous interest in the development of non-invasive methods that can help (a) quantify total plaque volume (disease burden) by using methods that do not depend on the size of the residual lumen and (b) reliably identify the morphologic features of the vulnerable plaque. Although still relatively new, success in this active area of research could have a tremendous impact on the diagnosis and management of cardiovascular disease [143–145]. For these reasons for the carotid arteries was introduced the concept of "vulnerable plaque" that refers to an atherosclerotic plaque that contains a large necrotic lipid core covered by a thin or disrupted fibrous cap and it is characterized by a higher tendency to rupture, resulting in embolization or thrombosis. Two main class of plaque determinant can be analyzed by using imaging: plaque luminal morphology (plaque irregularities and ulcerations), plaque components (fibrous cap, intra-plaque haemorrhage, thrombus, calcification). Another important parameter is the plaque eccentricity/remodelling.

The composition of the atherosclerotic plaque changes during the progression of atherosclerotic disease, and this plays an important role in the development of a vulnerable plaque. Until now, most data on the temporal change in the composition of the atherosclerotic plaque have been based on cross-sectional studies in which vessel specimens obtained during autopsy or operation (carotid endarterectomy) were evaluated. Atherosclerotic plaques were classified based on their composition and that classification was considered to reflect the temporal natural history of atherosclerotic disease.

The classical plaque classification according to the American Heart Association (AHA) comprises six stages. The early stages of plaque development according to the AHA (stages I–IV) follow a characteristic and uniform sequence, whereas the fate of the atheromatous plaque (stages V + VI) is less predictable. The *stage I* identifies the incorporation of scattered macrophages and foam cells within the arterial wall triggered by intralesional atherogenic lipoprotein. *Stage II* is represented by the development of fatty streaks, whereas *stage III* lesions already exhibit extracellular fats which break up cell–cell connections in smooth muscle layers. By definition, the atheroma (*stage IV*) then contains an actual fatty necrotic core. It is extremely important to underline that atheroma may not narrow the vascular lumen since it has been shown that affected vessels compensate for the increase in plaque volume through a widening of their external circumference rather than protrusion of plaque into the vascular lumen. Equally atheroma may be entirely asymptomatic; however, it is at this point of plaque evolution where symptoms may first occur. Further developments are less well understood: *stage V* lesions contain significant fibrous matter with *Va* lesions having a fatty core as well as a multilayered thick fibrous cap (fibroatheroma), *Vb* lesions are largely calcified while *Vc* lesions are predominantly fibrous. Complex *stage VI* lesions feature fissures or areas of internal haemorrhage or focal apposition thrombosis. The reparative process following thrombosis, comprising growth of fibroblasts, shrinkage of the thrombus and its incorporation into the vessel wall then does leave a narrowed lumen, particularly since complex lesions may undergo repeated cycles of rupture, thrombosis and remodelling.

Critically, this classification fails to identify the two crucial stages of plaque development, namely plaque the atheroma covered by no more than a thin fibrous cap as well as the erosion of the plaque surface. This has been addressed in a review by Virmani now suggesting seven categories of lesions [146]. Rupture of so-called vulnerable plaque is therefore the key event, both for the gradual vessel occlusion as well as the clinically more relevant scenario, occasionally even fatal accelerated plaque development. Today, a classification system describing five phases of plaque development more closely resembles the current thinking [147]: *phase 1*, comprising the histological stages I–III, constitutes early plaque development and is clinically silent. Advanced lesions (*phase 2*) can histologically be either stage IV (atheroma) or Va (thick cap) and carry the risk of rupture but may not be stenosing the vessel.

Phases 3 and *4* define acute phases with plaque rupture and subsequent thrombosis (3) or repeated thrombosis (4). Phase 3 may still be clinically silent while phase 4 through its increasing haemodynamic effect is always symptomatic. Finally, *phase 5* plaque is again a haemodynamically symptomatic lesion which actively remodels through the in-sprouting of vasa vasorum, dense dystrophic calcification (stage Vb) or fibrous scarring (stage Vc).

13.7 Plaque Luminal Morphology

CTA allows to analyze the plaque surface whereby a differentiation can be made between plaque irregularities and plaque ulceration. With microscopic evaluation of the carotid plaque it became clear that angiographic ulceration and irregularities were strongly associated with the presence of plaque rupture, plaque haemorrhage, a large lipid core size and less fibrous tissue, i.e. features that are all closely related with the concept of vulnerable plaque. De Weert and colleagues [148] demonstrated that MDCTA can assess atherosclerotic carotid plaque surface morphology with differentiation between smooth, irregular and ulcerated surfaces.

13.7.1 Smooth Surface

This condition indicate a regular luminal morphology at the level of the plaque without any sign of ulceration or irregularity. Usually the lumen profile should be plain (Fig. 13.11).

13.7.2 Plaque Irregularities

Irregularities in the surface of the plaque are frequently associated with the stroke/TIA presence: Rothwell and colleagues [106] demonstrated, by using angiography how plaque surface irregularity was an independent predictor of ipsilateral ischemic stroke and these observations were confirmed also by using MDCTA. A criteria proposed to identify a plaque morphology as irregular is the presence of pre- or post-stenotic dilatation and/or if the plaque surface morphology showed irregularities without any sign of ulceration (Fig. 13.11).

13.7.3 Plaque Ulcerations

Plaque ulceration has been defined as "an intimal defect larger than 1,000 μm in width, exposing the necrotic core of the atheromatous plaque [149] (Fig. 13.12). Ulcerated carotid plaques are associated with a higher risk of embolism and occlusion.

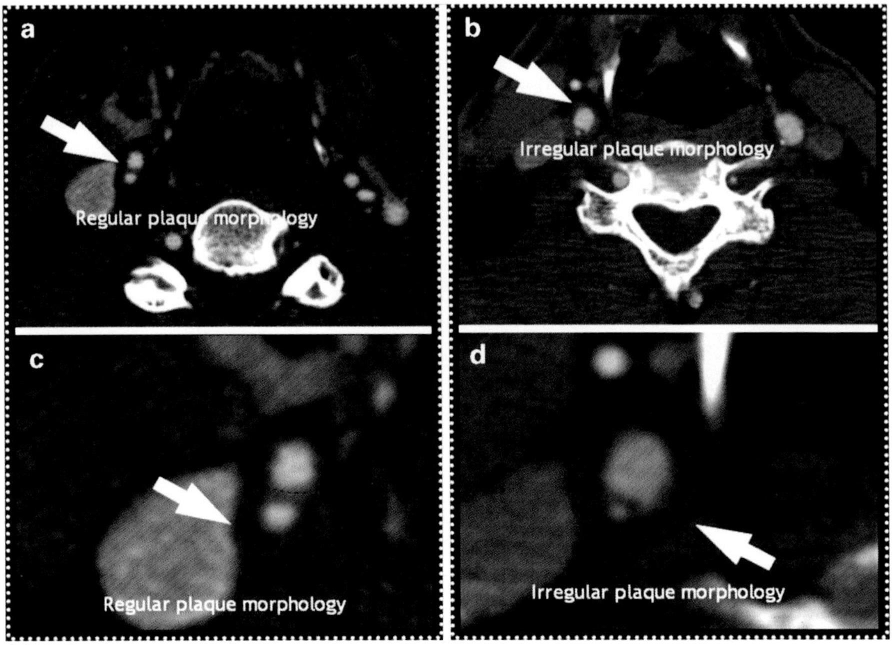

Fig. 13.11 CTA axial images. In (**a**, **b**) two different examples of regular plaque morphology are visible whereas in (**c**, **d**) two examples of irregular plaque morphology are visible

In the NASCET study, in the group of patients who received medical therapy, 30% of patients who had a severe carotid stenosis associated with an ulcerated plaque suffered an ischemic cerebral event within 2 years, while only 17% of patients with severe stenosis but no ulcerated plaques had an ischemic cerebral event within 2 years [112, 135, 150, 151]. Kim et al. [150] demonstrated that the presence of ulceration alone represents an important risk for neurological symptoms and a high-grade stenosis, combined with plaque ulceration, produces an *increased risk* of stroke. Authors demonstrated that ulceration is more frequent in symptomatic carotid artery than in asymptomatic carotid artery. In particular ulceration was found in 11% of the symptomatic carotid artery and in 40% of the carotid arteries with a moderate to severe degree of stenosis (30–99%).

It is demonstrated that hypercholesterolemia is positively and significantly associated with the presence of complicated plaques and this association may be explained by the atherogenic effect of lipoprotein(a) in the presence of high plasma LDL-C levels which increases lipid deposition in atherosclerotic plaque making the plaque more prone to the rupture.

DSA is not sufficiently reliable for detecting plaque ulceration with a 46% sensitivity and a 74% specificity [153–156]. In the 2007 Saba et al. [157] compared MDCTA and US-ECD in the detection of plaque ulceration by using surgery as gold standard and they observed that MDCTA, in particular when post-processing techniques were used [7], determined markedly higher results compared to

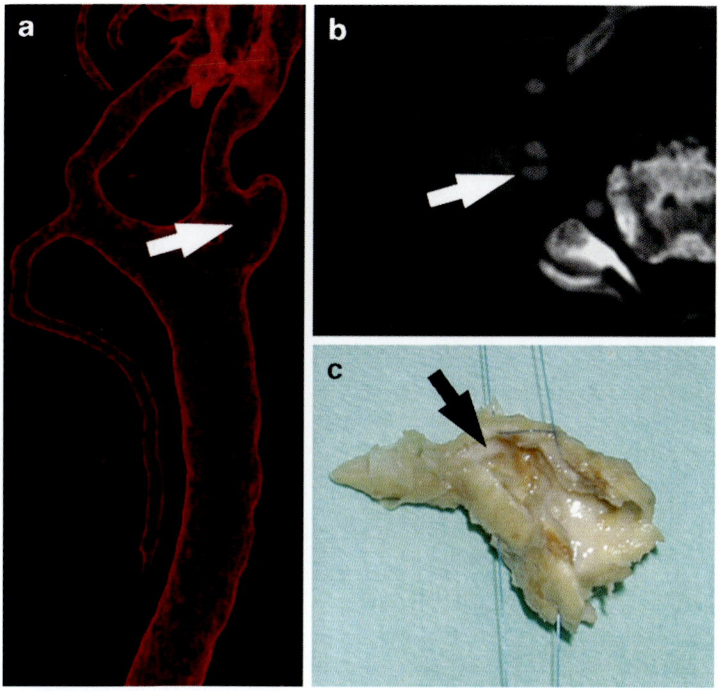

Fig. 13.12 Volume-rendered post-processed image (**a**) and axial scan (**b**) show a severe and irregular ulcerated plaque of the right internal carotid artery (*white arrows*). The ulcer (3 mm in depth) was confirmed after carotid endarterectomy (**c**, *black arrow*)

US-ECD (93% vs. 37.5%); these values are similar to data produced by other authors [158–160] and higher compared to Walker and colleagues who evaluated 165 CTA studies, compared them with endarterectomy specimens and reported a sensitivity of 60% and a specificity of 74%. MRA shows good efficacy in detecting plaque ulcerations [161].

A study of Redgrave published in 2006 of symptomatic carotid endarterectomy specimens from 526 consecutive patients with a stenosis degree of 75–90% found ulceration in 58% of cases. This is an extremely high value compared to the findings of MDCTA, MRA, US and DSA but can be explained with three reasons: (1) higher resolution of histology that enables the detection of even small ulceration; (2) the presence of thrombus formation on the location of a rupture that may fill the ruptured site, which may lead to a non-visualization with MDCTA; and (3) higher volume of calcifications in severe stenosis which may hamper accurate detection of small ulcerations by MDCTA.

Some authors state that calcified plaque could be a limitation in the use of the CT angiography for the ulcerated plaque detection [162, 163]. Plaque ulceration has been more frequently observed proximal to the point of maximum luminal stenosis (Fig. 13.13), which is exposed to higher wall shear stress. In the CTA

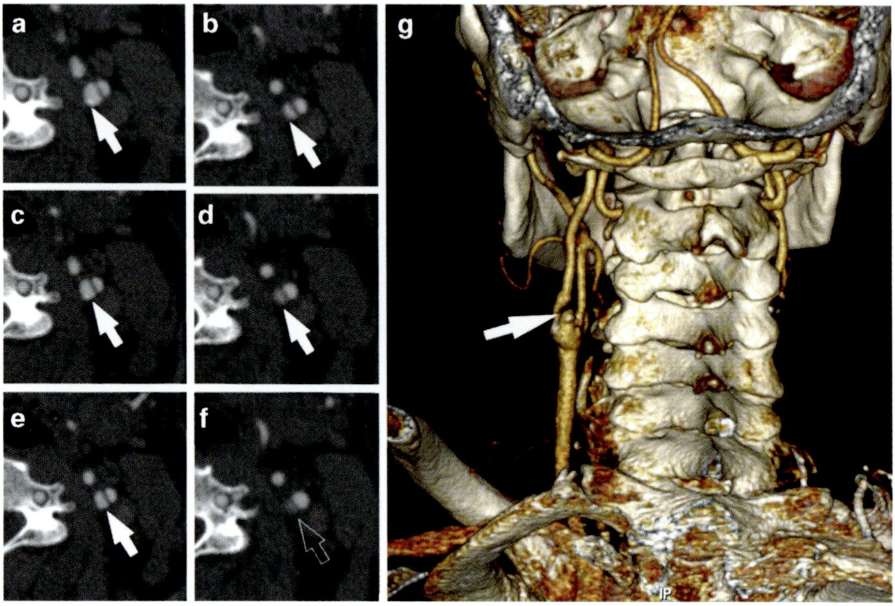

Fig. 13.13 Axial scan (**a–f**) and volume-rendered post-processed image (**g**) show a severe and irregular ulcer (*white arrow*) of the left internal carotid artery (*white arrows*). The ulcer (3 mm in depth) is proximal to the point of maximum stenosis, as clearly visible in the serial CTA images (**a–f**) and in the VR post-processed image (**g**)

analysis the most common criteria used to identify the presence of an atherosclerotic carotid plaque is the presence of extension of contrast material beyond the vascular lumen into the surrounding plaque.

Ulcerated plaques are usually categorized according to the classification described by Lovett et al. [164] where type 1 is an ulcer that points out perpendicular to the lumen; type 2 has a narrow neck and points out proximally and distally; type 3 has an ulcer neck proximally and points out distally whereas type 4 has an ulcer neck distally and points out proximally. Lovett et al [164] demonstrated that type 1 and type 3 are the most frequent type of rupture but at present there is no evidence that a specific type of plaque is associated with a different clinical behaviour.

Usually the localization of the ulcer can be described as proximal or distal according to the point of maximum stenosis and it is extremely interesting to observe that the ulcerated and irregular plaques are more frequently encountered with a higher degree of stenosis. This data was first demonstrated in the ECST study by using DSA (71% at the proximal site) and then by Saba and de Weert. It is interesting to underline that the higher frequency of ulceration in the proximal site observed in the carotid artery is similar in the coronary arteries; in fact an intracoronary ultrasound study found that 69% of the ulcerated ruptured plaques were proximal to the narrowest luminal point [40]. The fact that the proximal site shows a predilection for ulceration is in concordance with shear stress theories; in fact it is thought that high shear stress on the plaque surface, determined by the luminal narrowing, weakens the fibrous cap and determines the development of ulceration.

By using CTA for the detection of ulceration, the presence of calcification may represent a problem because big calcifications may hamper identification of ulceration, but this is a problem that CTA shares with other imaging modalities and in particular with ultrasound where the acoustic shadowing from calcifications obscures the presence of ulcerations.

Our group demonstrated that the type of the plaque (see next section) may influence the prevalence of plaque ulceration; in fact is was demonstrated that the fatty plaques are statistically associated with the presence of ulcerations compared to the mixed ones.

13.8 Analysis of the Different Types of Plaque

It is well demonstrated from histological studies that carotid plaques may be very different each other, and composition may change in the plaques. First studies at using CT to image the components of carotid atheroma have been described by Porsche et al. [165, 166]. In these studies, 55 patients undergoing carotid endarterectomy were imaged preoperatively using single slice spiral CT angiography. A total of 165 histological sections were available for analysis. Plaque density was measured (in Hounsfield units) on axial CT sections and the presence or absence of ulceration was noted. The CT images were co-registered with the appropriate histological sections. ANOVA testing revealed a statistically significant decrease in CT attenuation values as total lipid content of the plaque increased, but the standard deviation of these values was very high. The authors were unable to demonstrate any other histological factors that showed a statistically significant correlation with CT attenuation. With the development of new generation scanners and, in particular, 16 detector row and over new horizons opened for MDCT. In particular, MDCT allows evaluating carotid atherosclerosis with thinner slices (0.5–1.0 mm) and less volume averaging and therefore more detailed analysis of plaque composition may now became possible. The identification of the plaque components and in particular to identify the "type" of plaque represents a key problem because it is well demonstrated that a specific type of plaque (and therefore a consequent specific plaque histological composition) is associated with an increased or reduced risk of cerebrovascular symptoms. It is possible that even low-grade stenosis in the carotid arteries can lead to the development of cerebrovascular events, for this reason it may be important to look beyond the stenosis degree and determine plaque characteristics. The first author who identified the relationship between the presence of symptoms and plaque type was Schroder et al. [136] by analyzing coronary arteries and he differentiated plaque type into three categories *fatty*, *mixed* and *calcified* plaques. In this classification fatty (soft plaques) were considered as those plaques with a density value <50 HU, mixed plaques were those plaques with a density value between 50 and 119 HU and calcified plaques were those with a density >120 HU (Fig. 13.14).

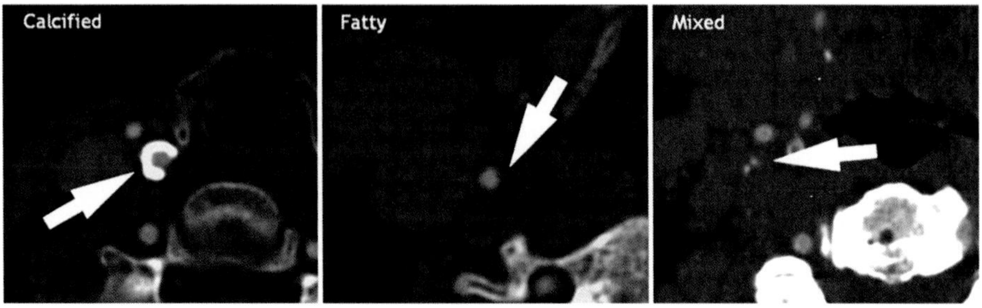

Fig. 13.14 CTA axial images that show the three different type of carotid artery plaques: calcified plaque (**a**), fatty plaque (**b**), and mixed plaque (**c**)

Other groups used this classification to study carotid arteries, and in particular Saba et al. proposed for the measure of HU attenuation the use of a circular or elliptical region of interest (ROI) cursor located in the predominant area of the plaque of at least 1 mm^2; those areas showing contamination by contrast medium or calcification non-contributory to the stenosis should be avoided. Moreover, regions of beam hardening in calcified areas were also avoided. This kind of method for the HU measurement of the plaque is subjective, but studies demonstrated an optimal inter-observer reproducibility. One of the major limitation of this kind of classification is that it represents simplification of the carotid plaque. In fact, it is possible that all the three components at are present the same time and their percentage may represent a significant role. For this reason this classification was modified by de Weert et al. who changed the HU threshold to consider a fatty as fatty, mixed or calcified by using automated computer software that can measure all the plaque by identifying the components by demonstrating that in vivo quantification of the volume of atherosclerotic carotid plaque and its components is possible with MDCTA. The authors used a free software (ImageJ™ plus PolyMeasure™ add on; Rasband, National Institute of Mental Health, Bethesda, MD, USA). In this classification fatty were considered those plaque with a density value <60 HU, mixed plaques were those plaque with a density value between 60 and 129 HU and calcified plaque were those with a density >130 HU (Fig. 13.15).

This method is extremely reliable in characterizing plaque elements as fatty, mixed or calcified but is time consuming (30–60 min): in fact it is necessary to manually draw the ROI for the plaque and the ROI for the opacified lumen for each slice where a plaque is visible. The inter-observer concordance is extremely good but the time necessary to complete. By using these automatic method some problems can arise; in particular, the difficulty in differentiation between a normal vessel wall and a slightly thickened (diseased) vessel wall, influences the assessment of the most proximal and distal image with atherosclerosis and thus the length of the atherosclerotic lesion. Because the plaque volume measurements include the original vessel wall, inclusion of additional images with normal vessel wall increases the amount of measured volumes considerably. The second issue is the manual outlining of the

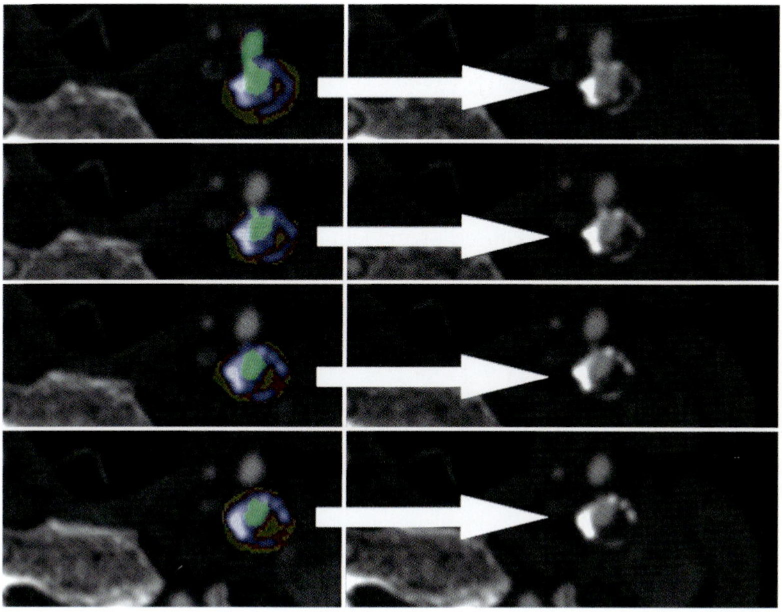

Fig. 13.15 Example of semi-automated classification of the plaque by using ImageJ™ plus PolyMeasure™ add on

outer border of the vessel wall, in fact some parts of the vessel wall can easily be differentiated from the surrounding tissue due to the low density of peri-arterial fat or the presence of calcifications at the outer border of the plaque [167]. However, other parts have the same density as the peri- and paravertebral and sternocleidomastoid muscle, which are frequently positioned along the artery. The erroneous manual inclusion of peri-arterial fat in the ROI leads to the classification of this fat as lipid in the plaque. An additional problem in the assessment of the outer border of the vessel wall is that the size of calcifications is influenced by differences in window-level setting. Because these calcifications are often located at the border of an atherosclerotic plaque, different window-level settings between observers will influence the assessment of the outer vessel wall between observers, and thereby introduce variability in the assessment of plaque volume and calcified volume [167]. Another problem is the differentiation of contrast-enhanced lumen from atherosclerotic plaque. In some plaque without calcifications at the inner border of the plaque the differentiation is automated and based on a threshold and the only variability is caused by a difference in the measurement of luminal attenuation, which was fortunately low. In case a calcification borders the lumen, a threshold-based approach would merge the lumen with the calcification. In such cases, manual drawing of the border between lumen and calcification was necessary which introduced a variability in plaque volume and calcium volume measurements [167].

In a previous work [168], to investigate the interpretation of hypodense regions within atherosclerotic plaque in MDCT images, each hypodense region was divided in to three different regions based on a range of HV (<0 HU, 0–30 HU, 30–60 HU). For each range of HV the number of hypodense regions in the MDCT images was assessed and the results were subsequently compared with the histological section. On histology, hypodense regions were true-positive for lipid core if the whole region fell within a lipid core area (i.e., lipid, haemorrhage or necrotic debris) and false-positive for lipid core if these regions included (besides lipid core) calcified or fibrous tissue. Hypodense regions along the vessel wall border were counted separately because they were caused by inadequate positioning of the outer contour with inclusion of peri-arterial fat in the analysis. The sequent analysis of hypodense regions showed that 54 hypodense regions were present in the HV range of <0 HU: 28 regions were true-positive for lipid core and 26 regions were located at the vessel wall borders. With a HV range of 0–30 HU, 118 hypodense regions were present: 49 regions were true-positive, 3 regions false-positive, and 66 regions were located at the vessel wall borders. With a HV range of 30–60 HU, 144 hypodense regions were present: 28 regions were true-positive, 95 were false-positive, and 21 regions were located at the vessel wall borders. This study showed that the HU measured in the centre of fibrous-rich regions and lipid core is significantly different and indicated that very hypodense regions (<30 HU) in the centre of atherosclerotic carotid plaque are associated with the presence of a lipid core (i.e., lipid, haemorrhage or necrotic debris). Quantification of fibrous tissue was also analyzed and this study demonstrated a strong correlation to histology. Because of the fact that fibrous tissue is a stabilizing feature of an atherosclerotic plaque the quantification could qualify fibrous tissue as an important predictor of plaque stability. It is important to underline that fatty plaques in CTA are associated with the presence of a lipid core. The lipid core, once thought to be an inert deposition of lipid is now known to be a highly biologically active area. The core itself is predominantly composed of foamy macrophages that are dead or dying with more viable cells usually found at the interface between the core and the fibrous cap, particularly at the shoulders of the plaque. As macrophages translocate across the intimal endothelium as a result of endothelial dysfunction, they take up oxidized LDL cholesterol which activates them and encourages them to continue to do this. Activated macrophages secrete proteolytic enzymes such as matrix metalloproteinases (e.g., MMP-7, MMP-9) which degrade proteins and collagens making up the extracellular matrix of the fibrous cap, so weakening it and ultimately increasing its risk of rupture. This hypothesis is confirmed in works that demonstrated how the fatty plaque, detected by CTA are more frequently associated with the development of cerebrovascular symptoms and with the presence of CT brain lesions. Plaque composition is considered an important feature of the so-called vulnerable plaque [169]. Histological studies that examined carotid endarterectomy specimens have shown that the composition of the plaque is influenced by cardiovascular risk factors: hypercholesterolemia and hyperfibrinogenemia were correlated with plaques rich in foam cells. Authors found that hypercholesterolemia was independently associated with the proportion of lipid and calcium in the plaque. In comparison to the histological studies, it was found that patients with hypercholesterolemia had a smaller proportion of lipid and a larger proportion of calcium in their symptomatic plaques. This might be explained by the large proportion of patients on lipid-lowering drugs. IVUS studies in coronary arteries have revealed that statins reduce the lipid content of atherosclerotic plaques. MRI studies on the carotid plaque have shown that prolonged lipid-lowering drug therapy is associated with markedly decreased lipid content.

13.9 Carotid Plaque Volume

It is current opinion that atherosclerotic plaque rupture, more than the luminal narrowing, plays an important role in acute events, like TIA and minor strokes. Accumulation of atherosclerotic plaque in the carotid artery may lead to positive remodelling in which the artery enlarges to preserve the luminal area. In addition, a certain amount of atherosclerosis should be present in the carotid bulb before it causes stenosis. Consequently, plaque volume is usually underestimated by the degree of stenosis. It is current thinking that plaque volume is a better descriptor of the severity of atherosclerotic disease than degree of stenosis, and some researchers [169] advocate evaluating the role of plaque volume as an additional parameter in the assessment of stroke risk and ultimately in treatment decision-making. Plaque volume and plaque composition may be parameters that help in a better risk prediction and selection of patients who could benefit from surgical or endovascular intervention. CT angiography has established itself as an accurate modality to assess the presence of atherosclerotic disease and to grade the severity of stenosis. In addition, MDCTA has the ability to identify and quantify different plaque components (lipid, fibrous tissue, and calcifications). Non-invasive in vivo assessment of atherosclerotic plaque volume and relative contribution of the different plaque components have important clinical implications and in particular it provides optimal parameters, together with the severity of stenosis, for cardiovascular risk assessment.

The Rotterdam group [169] lead by Professor Aad van der Lugt found a moderate association between the severity of luminal stenosis and plaque volume [in 20 out of 61 patients with a stenosis grade of 0%, atherosclerotic plaque was present (PV = 411 ± 472 mm^3). In 18 of the 20 patients with a stenosis grade of 1–29%, plaque was present with a PV of 543 ± 464 mm^3]. The correlation between plaque volume and severity of stenosis in the group of patients who had an atherosclerotic plaque ($n = 57$) was moderate with a *R*s of 0.66 ($P < 0.0001$)]. It is very interesting to observe that a relatively large proportion of the patients with cerebrovascular symptoms had atherosclerotic plaque with no luminal narrowing measured according to the NASCET criteria. It is possible to find several reasons because severity of stenosis is not a good reflection of the amount of atherosclerotic disease. Firstly, the configuration of a normal carotid bulbus and bifurcation allows the accumulation of extensive atherosclerotic plaque before a luminal stenosis becomes visible on angiography. Secondly, vessel remodelling plays an additional role since the discrepancy between the atherosclerotic plaque load and angiographic appearance of the lumen is also present in other vessel beds. Thirdly, extensive atherosclerotic disease can be present without stenosis of the lumen, and vice versa, stenosis can be caused by a focal accumulation of a small amount of atherosclerotic plaque. Other authors evaluated the carotid artery plaque volume [170] and observed that there is an excellent inter-obsever agreement in the quantification of the plaque volume by using CT with a correlation coefficient of 0.999.

Another interesting observation is that plaque composition changes with increasing plaque volume. According to the AHA criteria, which describe advanced atherosclerotic lesions in the carotid artery as containing more lipid and more calcium, we found an increase in the proportions of lipid and calcification with increasing plaque volume.

It was demonstrated that smoking was independently related to severity of stenosis and plaque volume, whereas the other classical risk factors did not show a statistically significant association.

13.10 Plaque Components

13.10.1 Fibrous Cap

The fibrous cap is a layer of fibrous connective tissue, which arise from the migration and proliferation of vascular smooth muscle cells and from matrix deposition and the normal fibrous cap contains macrophages and smooth muscle cells, whereas the fibrous cap of an atheroma is composed of smooth muscle cells, macrophages, foam cells, lymphocytes, elastin and collagen. In 1992, Loree et al. [171] published a study to determine the effect on the overall stress on the plaque when varying the overall degree of stenosis and thickness of fibrous cap. They hypothesized that the final link in the chain of plaque rupture was likely to be mechanical stress on a chronically weakened fibrous cap leading finally to rupture. Their simulations found that circumferential stress on the plaque was exquisitely linked to the thickness of the fibrous cap and degree of arterial remodelling (according to the law of Laplace); the thinner the cap, the greater the stress even at relatively low levels of stenosis. As observed in patients of different ages, and in animal models, FC formation appears to be a relatively late event in atherosclerosis, occurring as a result of progression from a macrophage-rich fatty streak lesion. The thickness of the FC can be extremely variable. In 2008, Redgrave et al. analyzed 526 carotid plaques and showed values ranging from 10 to 1,750 μm. The modified AHA classification of atherosclerotic lesions defines a coronary thin FC as one that is below 65 μm in thickness because this was the 95th percentile cap thickness at the point of rupture in a series of 41 fatally ruptured coronary plaques; however, the degree of cap thickness that renders a carotid plaque prone to rupture is still being debated. It is well known that a ruptured fibrous cap (which overlies the thrombogenic necrotic core) is an important component of the vulnerable coronary and carotid atherosclerotic plaque [172–174]. More evidence comes from cardiological post-mortem studies looking at culprit coronary lesions for sudden cardiac death. These showed that responsible lesions could be split broadly into three groups; one, the so-called fibroatheroma, with a very thin ruptured fibrous cap and extensive lipid core with associated intra-luminal thrombus formation; two, the erosion, with no evidence of plaque rupture or intra-plaque haemorrhage but a denuded fibrous cap with no overlying vascular endothelium and an associated overlying thrombus; three, a thrombus overlying a superficial calcified nodule. The fibroatheroma lesion with thin fibrous cap was by far the most common. It is also well demonstrated that a thin fibrous cap and its rupture are responsible for the majority of strokes distal to a carotid stenosis. Various papers have reported a histological comparison between asymptomatic and symptomatic plaques obtained at carotid endarterectomy. Golledge et al. has reviewed the findings of these studies and a rate of 48% (76/157) plaque ulceration/rupture in symptomatic patients was found opposed to a 31% (35/113) rate in asymptomatic patients. Further evidence of the risk posed by thin FCs is provided by the study of Zhi-Yong Li et al., in which an analysis in a carotid artery model was performed to evaluate the effect of FC thickness on plaque vulnerability. In this analysis the authors demonstrated that the risk of plaque rupture increased exponentially as FC thickness decreased. In 1997, Bassiouny et al. demonstrated that the minimum thickness of the FC is related to the risk of an ipsilateral neurological event. They found that the mean distance between the necrotic core and lumen was 500 μm in asymptomatic subjects and 270 μm in symptomatic subjects.

Recent studies demonstrated that high-resolution MRA is capable of distinguishing intact, thick fibrous caps from thin or ruptured caps in human carotid atherosclerosis in vivo. Thick fibrous caps appear as a juxtaluminal band of low signal in TOF MR images. In plaques with thin fibrous caps, this dark juxtaluminal band is absent. In plaques with fibrous cap rupture, the dark band is absent and there is a region of hyperintense signal adjacent to the lumen. In a study by Yuan et al. [175], it is stated that the presence of a thin or ruptured fibrous cap, in MRA images obtained from patients scheduled for carotid endarterectomy, is highly associated with recent TIA or stroke. In a study of Saba et al. [176] the presence of fissured fibrous cap is associated with cerebrovascular events ($p=0.0032$). The MRA is nowadays the best imaging technique to study the fibrous cap, its thickness and its fissuration [175–182]. Trivedi et al. [180] stated that the MRA accurately quantifies the relative thickness of fibrous cap by using the histological findings as reference standard, and Cai et al. [177] by using contrast material gadolinium demonstrated that the analysis of T1W pre- and post-contrast material allows to obtain accurate morphological measurement of the fibrous cap ($r=0.73$; $p=0.001$). US-ECD was sometimes used to study the fibrous cap but the results published in literature are sub-optimal: Sztajzel et al. studied 31 plaques derived from 28 patients undergoing carotid endarterectomy by using ultrasound and they obtained a sensitivity and specificity of 73 and 67% in quantification of the thickness of the fibrous cap. In Waki et al. it is shown that US can distinguish between thin and thick fibrous cap in correlation with histology.

The use of CT in assessing FCs has not been studied extensively (Fig. 13.16). The possibilities for state-of-the-art CT with respect to the visualization of the FC remain uncertain.

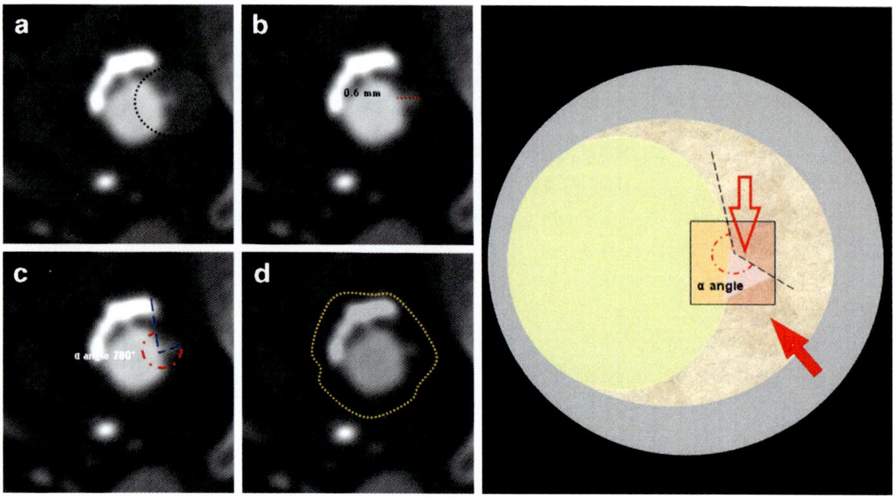

Fig. 13.16 Scheme of the criteria proposed by Saba et al. The four criteria are depicted as follows: (1) the presence of contrast material in a cleft located in the inner carotid surface (**a**), (2) dimension <1 mm in depth (**b**), (3) an angle of 230° or more with the lumen (**c**), and (4) the presence of atherosclerotic plaque into which the contrast material projects (**d**). The alpha angle is determined by the intersection of the carotid inner lumen profile and the adjacent profile of the "in plus" image

In the past some research has been conducted using spiral CT to determine plaque morphology, with mixed results [183, 184]. However, there is some evidence that MDCTA is suitable for determining plaque morphology, and both histological correlation [184] and inter-observer agreement [185] are quite good. These studies suggest that MDCTA could be a useful modality when MRI is less available. Saba et al. [186] proposed multi-detector-row CT angiography for the study of fissured FC. In this study 147 patients were analyzed and four criteria were proposed to identify a rupture of the FC: (1) the presence of a contrast material in a cleft located in the inner carotid surface, (2) whose dimension was less than 1 mm in depth with (3) an angle of 230° or more with the lumen, and (4) the presence of an atherosclerotic plaque into which the contrast material in a cleft projects. With these four criteria the inter-observer agreement obtained was good (0.781) and, moreover, it was observed that the presence of a ruptured FC is significantly associated with cerebrovascular symptoms (OR 3.86; $p=0.0032$). Recently, Wintermark et al. [43–45] demonstrated that it is possible to measure FC thickness. First, they used an automatic computer classifier algorithm for detection of plaque components. Thereafter, they quantitatively assessed the thickness of the FC with a good correlation with histology ($R^2=0.77$). It is important to note that the study by Wintermark [44] analyzed thick FCs (1.1 ± 1 mm in the histological analysis) and that we should test whether MDCT can analyze thinner FCs, because the threshold to consider a FC as thin is considered nowadays to be 200–250 μm.

13.10.2 Intra-plaque Haemorrhage

Intra-plaque haemorrhage is commonly observed in atherosclerotic plaques and some authors think that it can be due to rupture of plaque neovasculature [183]. Extravasation of red blood cells or iron accumulation in plaque may represent plaque instability and promote plaque progression [40, 184]. The mechanism that results in intraplaque haemorrhages is not completely understood: Paterson [185] has proposed that intra-plaque haemorrhage is secondary to rupture of the vasa vasorum while Constantinides and others [189, 187] have suggested that haemorrhage into a plaque occurs from cracks or fissures that originate at the luminal surface. Haemorrhages are more common when a stenosis exceeding 70% diameter reduction is present [188]. MRA can easily depict intra-plaque haemorrhage and it offers the possibility of distinguishing between acute and non-acute haemorrhages. In vivo MRA imaging is able to depict carotid intra-plaque haemorrhage with good sensitivity and moderate to good specificity. Moody et al. [189] showed that T1-weighted images of the carotid arteries can be used to accurately identify histologically confirmed complicated plaque with haemorrhage. Moreover, it is possible to distinguish the temporal phase of the haemorrhage according to different signal sequences.

In the MDCTA image, it is possible to visualize intra-plaque haemorrhage as a region where there is evidence of a low contrast material passage. In the axial images, intra-plaque haemorrhage is quite visible and when its dimensions exceed 1 mm it is possible to observe it by using the various image-reformatting techniques. Recently, Ajduk et al. [190] studied CTA carotid plaque characterizes with histological comparison and they demonstrated, by using ROC curve statistics, that the presence of HU value <31 HU are statistically associated with the presence of intra-plaque haemorrhage with a sensitivity of 100% and a specificity of 64%. However, the limited number of patients analyzed (31 subjects) did not allow to consider these results as completely significant. Another problem is that these data are discordant compared to those obtained by Wintermark et al. [6] where it was reported a overlap in CT Hounsfield densities between connective tissue and haemorrhage (mean HU value of haemorrhage is 97.5 ± 22 SD whereas mean HU value of connective tissue is 46.4 ± 19.9 SD) and it is clearly visible that how there is a big difference between Wintermark vs. Ajduk values in the intra-plaque haemorrhage values. On the basis of the current data and indication of the literature it is still debated whether it is possible to correctly detect intra-plaque haemorrhage with CT without the use of automated software classification analysis [191–193].

13.10.3 Plaque Thrombus

Plaque thrombus represents a significant risk factor for development of cereb rovascular events. It is possible to find thrombus in complicated plaques and unstable plaques. Presence of thrombus may produce distal embolization. Surface thrombus were visible in near 30% of the cases of ulcerated plaque, as reported by

Rothwell et al. [194] in their angiographic/pathologic analysis on 1,671 patients. In the past years the diagnosis of intra-luminal thrombus of the internal carotid artery was obtained by using angiography with anticoagulant therapy [195]. Nowadays, the best imaging method to detect the carotid thrombus is MRA. The detection of thrombus and the determination of its age on MR images are mainly related to the physical characteristics and visual appearance of thrombus [196–198]. Some authors proposed to use the water diffusion properties of atherothrombotic tissue to distinguish fresh thrombi from 1-week-old thrombi and occluding old thrombi on MR images [197]. CTA can also detect carotid thrombus [193].

13.10.4 Plaque Calcification

Intra-plaque calcium deposits in the coronary artery plaque is related to an increased morbidity and mortality [199–201]. However, the presence of calcium in the carotid artery plaque seems to be related to a lower risk cerebrovascular disease [54, 170, 202–206]. The detection of calcification is valuable for the estimation of the thromboembolic predisposition of the carotid atherosclerotic lesion, since plaque calcifications are thought to be associated with more stable plaques. It is likely that calcium confers stability in the atheromatous carotid plaque, resulting in protection against the biomechanical stress and the disruption. With CT angiography it is extremely easy to detect calcification in the carotid plaque (Fig. 13.17)

The mechanism by which calcium confers stability to plaques has both a mechanical and functional basis. In fact, it was demonstrated that calcification does not significantly impact the biomechanical stress on the fibrous cap, unlike

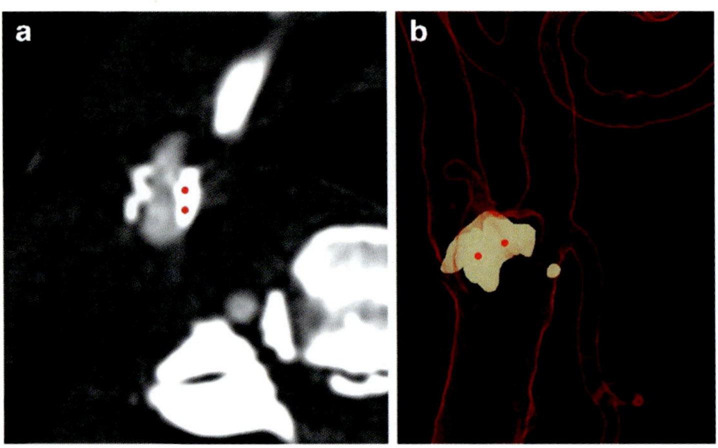

Fig. 13.17 Examples of calcified carotid artery plaque as visible in CTA axial image (**a**) and in the volume-rendered post-processed image (**b**)

lipid pools which increase the stress. The presence of intra-plaque calcium can be evaluated by using CTA, MRA or US-ECD, but the most sensitive technique in its detection, characterization and quantification remains CTA; in fact, the presence of calcification in atherosclerotic plaques can be easily detected with CT due to the high attenuation of the X-rays by calcium hydroxyapatite which leads to a high-density structure in the plaque. The first author that demonstrated that calcification (coronary calcifications) could be easily detected and quantified with Electronic Beam Computed Tomography was Agaston [207]. Later on this principle of calcium quantification was also used in the carotid arteries. Authors proposed to use a density of 130 HU in a ROI having an area >1 mm. Calcification in MDCTA can be detected without the presence of luminal contrast media; however, even in presence of contrast media the threshold value allows to easily differentiate calcium from contrast medium. One of the most important problems by using the Agaston score for coronary calcification is the significant measurement variability (up to 30%), and these data may be reasonable, the same for the carotid arteries even if the variability for carotid calcifications is not yet known. Severe calcifications may hamper correct quantification of lipid core.

13.11 Plaque Eccentricity and Remodelling

From the study on coronary arteries, it is known that plaque eccentricity is strongly associated with the acute coronary syndrome. Important and recent studies demonstrated that the same concept used for the coronary arteries can be applied for the carotid arteries. In fact, in 2008, Ohara and colleagues [208] demonstrated by using MDCTA that the presence of eccentric plaques was associated with a significantly increased incidence of ipsilateral events compared with patients with concentric stenosis. These findings are concordant with previous computational simulation that used carotid bifurcation models and where it was demonstrated that there were differences between concentric and eccentric stenosis with respect to the severity and distribution of wall shear stress as well as to the size of the post-stenotic recirculation zone [209]. Hardie and colleagues [210] studied the remodelling ratio and the eccentricity index of atherosclerotic carotid plaque by using the MDCTA and they observed that the expansive carotid remodelling was significantly greater in patients with cerebral ischemic symptoms than in asymptomatic patients by suggesting that the extent of expansive remodelling quantified by MDCTA may indicate underlying atherosclerotic plaque vulnerability.

The concept of remodelling indicates the morphological and ultra-structural variation of a plaque in time. Histological analysis of coronary plaques performed within 1 week after infarction showed morphological features of instability, while plaques taken later were histologically similar to those in patients with stable angina. Plaques changes and it is possible that some determinants of instability are transitory.

13.12 Carotid Plaque Enhancement

The presence of CPE in the carotid plaque is a "work-in-progress" topic. In fact, at the time of publishing of this book there is only one published paper in the literature by Romero et al. [10] and a congress paper presented by Saba et al. [9]. The work of Romero studied the presence of enhancement in the carotid wall and they observed that by using a threshold and contrast enhancement of 10 HU there is a positive association with the development of cerebrovascular symptoms. This result is consistent with previous histological studies that have demonstrated the presence of neovascularization of the adventitial vasa vasorum in symptomatic ICA walls. CTA-detected carotid wall enhancement may represent vasa vasorum neovascularization based on both its anatomic localization and rapid-enhancing time course. The authors underlined that a confounding factor in the interpretation of carotid wall enhancement is that dense calcification may obscure subtle wall enhancement.

Saba et al. [9] reported in the European Congress of Radiology of Wien in March 2010 that the MDCTA can measure the CPE with an optimal reproducibility between observers and demonstrated that the presence of plaque enhancement is strongly associated with the presence of cerebrovascular symptoms. Moreover, it was observed that fatty plaques show a bigger enhancement compared to mixed ones and that fatty plaques show more frequent enhancement compared to mixed ones. This study confirms that the plaque enhancement "ipso facto" cannot be considered a good marker of plaque instability (in fact, 50% of patients without symptoms showed contrast enhancement whereas the 84% of patients with cerebrovascular symptoms showed plaque instability), but it is possible that a possible solution is the measurement of the degree of enhancement. By using MDCT it is possible to reliably quantify HU variation in the tissues, and the ROC analysis indicated that 15 HU may represent a good threshold to consider CPE as symptomatic. The use of 15 HU as threshold allows to obtain a specificity of 83.33% and a sensitivity of 76.47%.

This data has important clinical implications because it may be used for the plaque risk stratification and identification of vulnerable plaque characteristics is of outmost importance because of the high risk to precipitate acute thrombotic occlusion. The observed plaque may be ascribed to the neovascularization within the carotid plaque. It has been demonstrated, by using different imaging techniques [9, 211, 212] That enhancement of carotid plaque with use of contrast material, correlates with histological density of neovessels within the carotid plaque and the cause of the enhancement is thought to be increase vascularity of the adventitial vasa vasorum feeding the plaque neovasculature (Fig. 13.18). The plaque neovascularization has been well established and confirmed in histological studies as a consistent feature of plaque in patients with cerebrovascular symptoms [213, 214] and researchers have indicated neovascularization as an important factor contributing to vulnerability of atherosclerotic plaque [25, 215]; in fact, plaque neovascularization was found to be more extensive in symptomatic and vulnerable plaques [213–215]. Work is ongoing regarding this difficult topic. Since it has been found that rupture of neointimal vessels and vasa vasorum can lead to plaque haemorrhage and thrombosis, some attempt

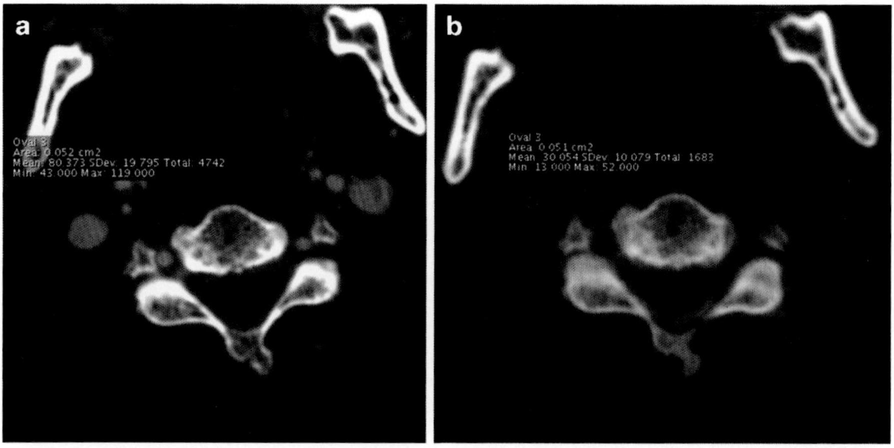

Fig. 13.18 MDCTA axial images after (**a**) and before (**b**) administration of contrast material

has been made to quantify and characterize the extent of neovessel formation using MR imaging. U-King-Im et al. showed in a small study that neovescularisation, as assessed by immunohistochemistry following endarterectomy, correlated reasonably well with late uptake of Gadolinium (Gd) into the plaque. They used a standard T1-weighted sequence, imaging 0, 8 and 12 min following the administration of Gd. Although a small sample size, plaques showing a greater degree of neovessel formation tended to show a greater late signal enhancement. More studies need to be performed in this area to determine whether or not plaque enhancement can be considered a viable clinical tool. Wasserman et al. have recently studied the wash-in kinetics of gadolinium into fibrous cap and lipid core of atheromatous plaque which may well correlate with neovessel density.

13.13 Other Imaging Concepts on Carotid

13.13.1 The Carotid Artery Wall Thickness

An increased common carotid artery intima–media thickness (CCA-IMT) measured by B-mode ultrasound occurs in an earlier phase of the atherosclerotic process and it is a powerful predictor of stroke risk and its severity [216–219]; for this reason, CCA-IMT is nowadays the most used surrogate marker of stroke [100]. The risk associated with a specific increase in IMT is age dependent, the risk being higher in younger individuals. In healthy adults, IMT ranges from 0.25 to 1.5 mm [220]. Epidemiological studies have reported associations between a range of cardiovascular risk factors (smoking, blood pressure, elevated blood cholesterol) and IMT [221–223]. Previous papers demonstrated that age is one of the most powerful determinants of IMT, with an increase from 0.01 to 0.02 mm per year. With the

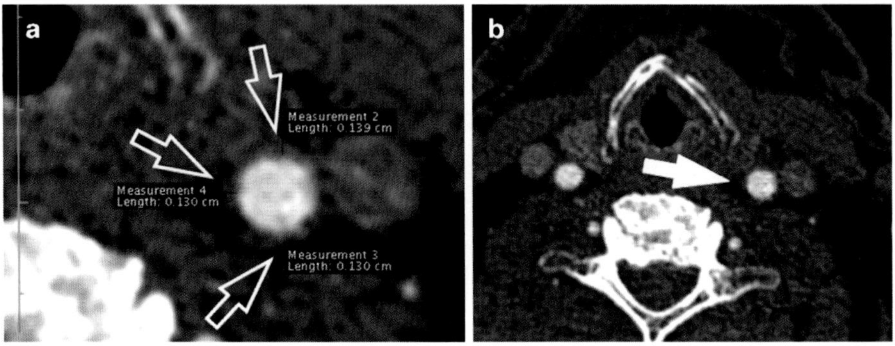

Fig. 13.19 CTA axial images indicating how to measure the carotid artery wall thickness

development of CT technologies studies have been made to use the MDCTA to quantify the carotid artery wall thickness (CAWT) [224] (Fig. 13.19) and authors demonstrated that MDCTA can easily measure the carotid wall with an optimal inter-observer agreement.

The use of MDCTA to evaluate CAWT is extremely useful; compared to the Sonography the IMT method suffers from low inter-observer and inter-method agreement, whereas the MDCTA offers optimal reproducibility. Moreover [224], it was clearly demonstrated that by using a threshold of 1 mm in the CAWT, there was an important statistical association between thick CAWT and stroke ($p<0.0001$); in fact, patients with ≥1 mm CAWT had stroke with an odds ratio of 8.16, in comparison with patients with <1 mm CAWT. Recently, it was demonstrated that CAWT and IMT had a strength correlation with a correlation coefficient, r, of 0.9855.

13.13.2 Automated Plaque Analysis

The carotid artery plaque is an extremely complex structure, characterized by numerous components. The presence of specific determinants (i.e. haemorrhage, fibrous tissue, lipid rich necrotic core, calcification), their percentage and their volume are important to quantify the vulnerability risk of a specific type of plaque. It is impossible for human being to quantitatively measure the volume of the plaque and the different percentages of plaque's components; therefore, there is the necessity for the use of computer software that can perform this task. The automatic analysis of the carotid atherosclerotic plaque is a challenging task, owing to the limited contrast between the plaque components and the surrounding soft tissue. These applications are extremely important because, for example, it is well demonstrated that the plaque volume is a better descriptor of the severity of atherosclerotic disease than the degree of stenosis [114]. It is possible to distinguish two types of categories: semi-automated software and full-automated software. In the semi-automated software the radiologist identifies the carotid artery plaque (its external and internal boundaries) and the software automatically classifies the plaque composition according to specific threshold identified by the operator. In fact, the most important and difficult step towards automatic atherosclerotic plaque quantification is the difficult task of detecting the outer vessel wall boundary, which encloses both the vessel lumen and the atherosclerotic plaque. Once the outer vessel wall is defined and the lumen is segmented it facilitates the further quantification of the plaque. By using this kind of software, the group of Aad van der Lugt demonstrated that it is possible to analyze the composition and volume of the carotid plaque [167–169] in patients who underwent MDCTA. Wintermark et al. [192] reported also that they could assess large intra-plaque haemorrhage with a kappa value of 0.712 between CTA and histology by using specific automatic software.

The full-automated software is more complex because it should identify the carotid plaque automatically. To the best of our knowledge, there is no commercial software of this kind yet, but several groups are working to create one.

13.14 Conclusion

Atherosclerosis of the carotid arteries remains a major cause of death in industrialized countries, and therefore, non-invasive, in vivo assessment of carotid plaque components is fundamental in therapy determination. Improvement in imaging techniques allows to study the unstable or vulnerable plaque with precision allowing a correct patient's risk stratification. Nowadays, MDCTA represents an advanced and reliable methodology to study carotid plaque characteristics. By using CTA it is possible to study carotid artery stenosis and plaque composition at the same time. The development of automated, computer-based, methods of plaque analysis will allow further precise stroke risk assessment.

Acknowledgments I would like to thank Professor Allan J Fox for his stimulating discussions and the precious support he gave to me in the study of carotids.

References

1. Prokop M. General principles of MDCT. Eur J Radiol 2003;45:S4–S10.
2. Walter D, De Man B, Iatrou M, Edic PM. Future generation CT imaging. Thorac Surg Clin 2004;14:135–149.
3. Prokop M. Multislice CT: technical principles and future trends. Eur Radiol 2003;13:M3–M13.
4. Fishman E. A brief overview of CT angiography. Suppl Appl Radiol 2003;32:9–11.
5. Das M, Braunschweig T, Muhlenbruch G, Mahnken AH, Krings T, Langer S, et al. Carotid plaque analysis: comparison of dual source computed tomography (CT) findings and histopathological correlation. Eur J Vasc Endovasc Surg 2009;38:14–19.
6. Wintermark M, Jawadi SS, Rapp JH, Tihan T, Thong E, Glidden DV. High resolution CT imaging of carotid artery atherosclerotic plaques. AJNR Am J Neuroradiol 2008;29:875–882.
7. Saba L, Caddeo G, Sanfilippo R, Montisci R, Mallarini G. Efficacy and sensitivity of axial scans and different reconstruction methods in the study of the ulcerated carotid plaque by using multi-detector-row CT angiography. Comparison with surgical results. AJNR Am J Neuroradiol 2007;28:716–723.

8. Rydberg J, Buckwalter KA, Caldemeyer KS, et al. Multisection CT: scanning techniques and clinical applications. Radiographics 2000;20:1787–1806.
9. Saba L, Sanfilippo R, Atzeni M, Ribuffo D, Montisci M, Mallarini G. Carotid plaque enhancement after contrast material administration and symptomaticity: an evaluation by using multi-detector-row CT angiography. Proceedings of the ECR 2010. Insights into Imaging 2010;1(suppl 1):225.
10. Romero JM, Babiarz LS, Forero NP, Murphy EK, Schaefer PW, Gonzalez RG, Lev MH. Arterial wall enhancement overlying carotid plaque on CT angiography correlates with symptoms in patients with high grade stenosis. Stroke 2009;40:1894–1896.
11. Prokop M, Shin HO, Schanz A, Schaefer-Prokop C. Use of maximum intensity projections in CT angiography: a basic review. Radiographics 1997;17:433–451.
12. Ertl-Wagner BB, Bruening R, Blume J, Hoffmann RT, Mueller-Schunk S, Snyder B, Reiser MF. Relative value of sliding-thin slab multiplar reformations and sliding-thin-slab-maximum intensity projections as reformatting techniques in multisection CT angiography of the cervicocranial vessels. AJNR Am J Neuroradiol 2006;27:107–113.
13. Rubin GD, Silverman SG. Helical (spiral) CT of the retroperitoneum. Radiol Clin North Am 1995;33:903–932.
14. Rankin SC. Spiral CT: vascular applications. Eur J Radiol 1998;28:18–29.
15. Rubin GD, Dake MD, Semba CP. Current status of three-dimensional spiral CT scanning for imaging the vasculature. Radiol Clin North Am 1995;33:51–70.
16. Calhoun PS, Kuszyk BS, Heath DG, Carley JC, Fishman EK. Three-dimensional volume rendering of spiral CT data: theory and method. Radiographics 1999;19:745–764.
17. de Monyè C, Weert TT, Zaalberg W, Cademartiri F, Siepman DAM, Dippel DWJ, van der Lugt A. Optimization of CT Angiography of the carotid artery with a 16-MDCT scanner: cranio-caudal scan direction reduces contrast material-related perivenous artifacts. AJR Am J Roentgenol 2006;186:1737–1745.
18. Brennen DJ, Hall EJ. Computed tomography – an increasing source of radiation exposure. N Engl J Med 2007;357:2277–2284.
19. Nickoloff E, Alderson P. Radiation exposure to patients from CT: reality public perception, and policy. AJR Am J Roentgenol 2001;177:285–287.
20. National Council on Radiation Protection and Measurements. Ionizing Radiation exposure of the population on the United States: recommendation of the National Council on Radiation Protection and Measurements. Report no 160. Bethesda, MD: NCRP, March 2009.
21. Mettler FA Jr, Thomasden BR, Bhargavan M, et al. Medical Radiation exposure in the U.S. in 2006: preliminary results. Health Phys 2008;95:502–507.
22. Cardis E, Vrijheid M, Blettner M. The 15-country collaborative study of cancer risk among radiation workers in the nuclear industry: estimates of radiation-related cancer risk. Radiat Res 2007;167:396–416.
23. Hunold P, Vogt FM, Schmermund A. Radiation exposure during cardiac CT: effective doses at multi-detector-row CT and electron-beam CT. Radiology 2003;226:145–152.
24. Flohr TG, Schaller S, Stierstorfer K, et al. Multi-detector row CT systems and image-reconstruction techniques. Radiology 2005;235:756–773.
25. Kalra MK, Maher MM, Toth TL. Strategies for CT radiation dose optimization. Radiology 2004;230:619–628.
26. Lasser EC. A coherent biochemical basis for increased reactivity to contrast material in allergic patients : a novel concept. AJR Am J Roentgenol 1987;149:1281–1285.
27. Cohan RH, Bullard MA, Ellis JH. Local reactions after injection of iodinated contrast material: detection, management and outcome. Acad Radiol 1997;4:711–718.
28. Cashaman JD, McCredie J, Henry DA. Intravenous contrast media: use and associated mortality. Med J Aust 1991;155:618–623.
29. Bartlett ES, Walters TD, Symons SP, Fox AJ. Quantification of carotid stenosis on CT Angiography. AJNR Am J Neuroradiol 2006;27:13–19.
30. Jaff MR, Goldmakher GV, LEV MH, Romero JM. Imaging of the carotid arteries: the role of duplex ultrasonography, magnetic resonance arteriography, and computerized tomographic arteriography. Vasc Med 2008;13:281–292.

31. Wardlaw JM, Chappell FM, Stevenson M, De Nigris E, Thomas S, Gillard J, et al. Accurate, pratical and cost-effective assessment of carotid stenosis in the UK. Health Technol Assess 2006;10:1–182.
32. Runge VM, Kirsch JE, Lee C. Constrast-enhanced MR angiography. Magn Reson Imaging 1993;3:233–239.
33. Comerota AJ, Katz ML, White JV, Grosh JD. The preoperative diagnosis of the ulcerated carotid atheroma. J Vasc Surg 1990;11:505–510.
34. Liapis CD, Paraskevas KI. Do carotid surface irregularities correlate with the development of cerebrovascular symptoms ? An analysis of the supporting studies, the opposing studies and the possible pathomechanism. Vascular 2006;14:88–92.
35. U-King-Im J, Hollingworth W, Trivedi RA. Contrast-enhanced MR angiography vs intra-arterial digital subtraction angiography for carotid artery imaging: activity-based cost analysis. Eur Radiol 2004;14:730–735.
36. Willinsky RA, Taylor Sm, TerBrugge K, Farb RI, Tomlinsong G, Montanera W. Neurologic complications of cerebral angiography: prospective analysis of 2,899 procedures and review of the literature. Radiology 2003;227:522–528.
37. Hankey HJ, Warlow CP, Molyneaux AJ. Complications of cerebral angiography for patients with mild carotid territory ischemia being considered for carotid endarterectomy. J Neurol Neurosurg Psychiatry 1990;53:542–548.
38. Berczi V, Randall M, Balamurugan R. Safety of arch aortography for assessment of carotid arteries. Eur J Vasc Endovasc Surg 2006;31:3–7.
39. Jonston DC, Chapman KM, Goldstein LB. Low rate of complications of cerebral angiography in routine clinical practice. Neurology 2001;57:2012–2014.
40. Naghavi M, Libby P, Falk E, et al. From Vulnerable claque to vulnerable patient: a call for new definitions and risk assessment strategies. Part I. Circulation 2003;108:1664–1672.
41. Wasserman BA, Wityk RJ, Trout HH, Virmani R. Looking beyond the lumen with MRI. Stroke 2005;36:2504–2513.
42. Tahmasebpour HR, Buckley AR, Cooperberg PL, Fix CH. Sonographic examination of the carotid arteries. Radiographics 2005;25:1561–1575.
43. Fell, G, Phillips, DJ, Chikos, PM, Harley, JD, Thiele, BL, Strandness, DE. Ultrasonic duplex scanning for disease of the carotid artery. Circulation 1981;64:1191–1195.
44. Sabeti, S, Schillinger, M, Mlekusch, W. Quantification of internal carotid artery stenosis with duplex US: comparative analysis of different flow velocity criteria. Radiology 2004;232:431–439.
45. Jahromi, AS, Cina, CS, Liu, Y, Clase, CM. Sensitivity and specificity of color duplex ultrasound measurement in the estimation of internal carotid artery stenosis: a systematic review and meta-analysis. J Vasc Surg 2005;41:962–972.
46. de Bray JM, Baud JM, Delanoy P. Reproducibility in ultrasonic characterization of carotid plaques. Cerebrovasc Dis 1998;8:273–277.
47. Sabetai MM, Tegos TJ, Nicolaides AN, Dhanjil S, Pare GJ, Stevens JM. Reproducibility of computer quantified carotid plaque echogenicity: can we overcome the subjectivity? Stroke 2000;31:2189–2196.
48. Sztajzel R, Momjian S, Momjian-Mayor I. Stratified gray-scale median analysis and color mapping of the carotid plaque: correlation with endarterectomy specimen histology of 28 patients. Stroke 2005;36:741–745.
49. Lal BK, Hobson RW II, Hameed M. Noninvasive identification of the unstable carotid plaque. Ann Vasc Surg 2006;20:167–174.
50. Kern R, Szabo K, Hennerici M, Meairs S. Characterization of carotid artery plaques using real-time compound B-mode ultrasound. Stroke 2004;35:870–875.
51. Sabetai MM, Tegos TJ, Nicolaides AN. Hemispheric symptoms and carotid plaque echomorphology. J Vasc Surg 2000;31:39–49.
52. Gronholdt ML, Wiebe BM, Laursen H, Nielsen TG, Schroeder TV, Sillesen H. Lipid-rich carotid artery plaques appear echolucent on ultrasound B-mode images and may be associated with intraplaque haemorrhage. Eur J Vasc Endovasc Surg 1997;14:439–445.

53. Lai BK, Hobson RW 2nd, Pappas PJ. Pixel distribution analysis of B-mode ultrasound scan images predicts histologic features of atherosclerotic carotid plaques. J Vasc Surg 2002;35:1210–1217.
54. Mathiesen EB, Bonaa KH, Joakimsen O. Echolucent plaques are associated with high risk of ischemic cerebrovascular events in carotid stenosis: the Tromso Study. Circulation 2001;103:2171–2175.
55. Polak JF, Shemanski L, O'Leary DH. Hypoechoic plaque at US of the carotid artery: an independent risk factor for incident stroke in adults aged 65 years or older. Cardiovascular Health Study. Radiology 1998;208:649–654.
56. Saam T, Hatsukami TS, Takaya N. The vulnerable, or high risk atherosclerotic plaque: noninvasive MR imaging for characterization and assessment. Radiology 2007;244:64–77.
57. Briley-Saebo KC, Mulder WJ, Mani V. Magnetic resonance imaging of vulnerable atherosclerotic plaques: current imaging strategies and molecular imaging probes. J Magn Reson Imaging 2007;27:460–479.
58. Choudhury RP, Fuster V, Badimon JJ, Fisher EA, Fayad ZA. MRI and characterization of atherosclerotic plaque: emerging applications and molecular imaging. Arterioscler Thromb Vasc Biol 2002;22:1065.
59. Yuan C, Mitsumori LM, Beach KW, Maravilla KR. Carotid atherosclerotic plauqe: non invasive MR characterization and identification of vulnerable lesions. Radiology 2001;221:285–299.
60. Rapp JH, Saloner D. Current status of carotid imaging by MRA. Cardiovasc Surg 2003;11:445–447.
61. Huston J III, Fain SB, Wald JT. Carotid artery: elliptic centric contrast-enhanced MR angiography compared with conventional angiography. Radiology 2001;218:138–143.
62. Kim WY, Danias PG, Stuber M. Coronary magnetic resonance angiography for the detection of coronary stenoses. N Engl J Med 2001;345:1863–1869.
63. Alvarez O, Edwards JH, Hyman RA. MR recognition of internal carotid artery occlusion. AJNR Am J Neuroradiol 1986;7:359–360.
64. Wehrli FW, Shimakawa A, Gullberg GT, MacFall JR. Time-of-flight MR flow imaging: selective saturation recovery with gradient refocusing. Radiology 1986;160:781–785.
65. Anderson CM, Saloner D, Lee RE. Assessment of carotid artery stenosis by MR angiography: comparison with x-ray angiography and color-coded Doppler ultrasound. AJNR AM J Neuroradiol 1992;13:989–1003.
66. Kondo C, Caputo CR, Semelka R, Foster E, Shimakawa A, Higgins CB. Right and left ventricular stroke volume measurements with velocity-encoded cine MR imaging: in vitro and in vivo validation. AJR AM J Roentgenol 1991;157:9–16.
67. Pelc LR, Pelc NJ, Rayhill SC. Arterial and venous blood flow: noninvasive quantisation with MR imaging. Radiology 1992;185:809–812.
68. Yuan C, Petty C, O'Brien KD, Hatsukami TS, Eary JF, Brown BG. In vitro and in situ magnetic resonance imaging signal features of atherosclerotic plaque-associated lipids. Arterioscler Thromb Vasc Biol 1997;17:1496–1503.
69. Kerwin WS, O'Brien KD, Ferguson MS, Polissar N, Hatsukami TS, Yuan C. Inflammation in carotid atherosclerotic plaque: a dynamic contrast-enhanced MR imaging study. Radiology 2006;241:459–468.
70. Rauch U, Osende JI, Fuster V, Badimon JJ, Fayad Z, Chesebro JH. Thrombus formation on atherosclerotic plaques: pathogenesis and clinical consequences. Ann Intern Med 2001;134:224–238 [Abstract/Free Full Text].
71. Fisher M, Blumenfeld AM, Smith TW. The importance of carotid artery plaque disruption and hemorrhage. Arch Neurol 1987;44:1086–1089.
72. Fleiner M, Kummer M, Mirlacher M, Sauter G, Cathomas G, Krapf R, Biedermann BC. Arterial neovascularization and inflammation in vulnerable patients: early and late signs of symptomatic atherosclerosis. Circulation 2004;110:2843–2850.
73. Ruehm SG, Corot C, Vogt P. Magnetic resonance imaging of atherosclerotic plaque with ultrasmall superparamagnetic particles of iron oxide in hyperlipidemic rabbits. Circulation 2001;103:415–422.

74. Schmitz SA, Coupland SE, Gust R. Superparamagnetic iron oxide-enhanced MRI of atherosclerotic plaques in Watanabe hereditable hyperlipidemic rabbits. Invest Radiol 2000;35:460–471.
75. Trivedi RA, Mallawarachi C, JM UK-I, Graves MJ, Horsley J, Goddard MJ, Brown A, Wang L, Kirkpatrick PJ, Brown J, Gillard JH. Identifying inflamed carotid plaques using in vivo uspio-enhanced MR imaging to label plaque macrophages. Arterioscler Thromb Vasc Biol 2006;26:1601–1606.
76. Libby P. Inflammation in atherosclerosis (review). Nature 2002;420:868–874.
77. Kapoor V, McCook BM, Torok FS. An introduction to PETCT imaging. Radiographics 2004;24:523–543.
78. Falk E. Pathogenesis of atherosclerosis. J Am Coll Cardiol 2006;47:C7–C12.
79. Bleeker-Rovers CP, Bredie SJ, van der Meer JW, Corstens FH, Oyen WJ. F-18-fluorodeoxyglucose positron emission tomography in diagnosis and follow-up of patients with different types of vasculitis. Neth J Med 2003;61:323–329.
80. Blockmans D, De Ceuninck L, Vanderschueren S, Knockaert D, Mortelmans L, Bobbaers H. Repetitive 18-fluorodeoxyglucose positron emission tomography in isolated polymyalgia rheumatica: a prospective study in 35 patients. Rheumatology (Oxford) 2007;46:672–677.
81. Lederman RJ, Raylman RR, Fisher SJ, Kison PV, San H, Nabel EG. Detection of atherosclerosis using a novel positron-sensitive probe and 18-fluorodeoxyglucose (FDG). Nucl Med Commun 2001;22:747–753.
82. Adelman SM: Economic impact. In McDowell FM report on the national survey of stroke. Stroke 1981;12:1.
83. Robins M, Baum H. National survey of stroke: incidence. Stroke 1981;12:1;45–57.
84. Rudd JH, Warburton EA, Fryer TD. Imaging atherosclerotic plaque inflammation with [18F]-fluorodeoxyglucose positron emission tomography. Circulation 2002;105:2708–2711.
85. Tawakol A, Migrino RQ, Bashian GG. In vivo 18Ffluorodeoxyglucose positron emission tomography imaging provides a noninvasive measure of carotid plaque inflammationin patients. J Am Coll Cardiol 2006;48:1818–1824.
86. Tahara N, Kai H, Ishibashi M, Nakahura H, Kaida H, Baba K, et al. Simvastatin attenuates plaque inflammation: evaluation by fluorodeoxyglucose positron emission tomography. J Am Coll Cardiol 2006;48:1825–1831.
87. Kietselaer BL, Hofstra L, Dumont EA. The role of labelled Annexin A5 in imaging of programmed cell death. From animal to clinical imaging. Q J Nucl Med 2003;47:349–361.
88. Manca G, Parenti G, Bellina R. ^{111}In platelet scintigraphy for the noninvasive detection of carotid plaque thrombosis. Stroke 2001;32:719–727.
89. Annovazzi A, Bonanno E, Arca M. 99mTc-interleukin-2 scintigraphy for the in vivo imaging of vulnerable atherosclerotic plaques. Eur J Nucl Med Mol Imaging 2006;33:117–126.
90. Iuliano L, Mauriello A, Sbarigia E, Spagnoli LG, Violi F. Radiolabeled native low-density lipoprotein injected into patients with carotid stenosis accumulates in macrophages of atherosclerotic plaque: effect of vitamin E supplementation. Circulation 2000;101:1249–1254.
91. Lees AM, Lees RS, Schoen FJ. Imaging human atherosclerosis with 99mTc-labeled low density lipoproteins. Arteriosclerosis 1988;8:461–470.
92. Iuliano L, Signore A, Vallabajosula S. Preparation and biodistribution of 99m technetium labelled oxidized LDL in man. Atherosclerosis 1996;126:131–141.
93. Virgolini I, O'Grady J, Lupattelli G. In vivo quantification of cholesterol content in human carotid arteries by quantitative gamma-camera imaging after injection of autologous low density lipoproteins (LDL). Int J Rad Appl Instrum B 1992;19:245–250.
94. O' Leary DH, Polak JF, Kronmal RA, Kittner SJ, Bond MG, Wolfson SK, et al. Distribution and correlated of Sonographically detected carotid artery disease in the cardiovascular health study. The CHS Collaborative research group. Stroke 1992;23:1752–1760.
95. Savory WS. Case of a young woman in whom the main arteries of both upper extremities and of the left side of the neck were throughout completely obliterated. Med Chir Trans 1856;39:205–219.

96. Gowers WR. On a case of simultaneous embolism of central retinal and middle cerebral arteries. Lancet 1875;2:794.
97. Chiari H. Uber des verhalten des teilungswinkels der carotis communis bei der endarteriitis chronica deformans. Verh Dtsch Ges Pathol 1905;9:326–330.
98. Guthrie LG, Mayou S. Right hemiplegia and atrophy of left optic nerve. Proc R Soc Med 1908;1:180.
99. Gibbs RGJ, Todd JC, Irvine C, Lawrensen R, Newson R, Greenhalgh RM, Davies AH. Relationship between the regional and national incidence of transient ischemic attack and stroke and performance of carotid endarterectomy. Eur J Vasc Surg 1998;16:47–52.
100. Lloyd-Jones D, Adams R, Carnethon M, De Simone G, Ferguson TB, Flegal K, et al. Heart disease and stroke statistics-2009 update: a report from the American Heart Association Statistics Committee and Stroke Statistics Subcommittee. Circulation 2009;119:480–486.
101. Donnan GA, Fisher M, Macleod M, Davis SM. Stroke. Lancet 2008;371:1612–1623.
102. Wolinsky FD, Bentler SE, Cook EA, Chrischilles EA, Liu L, Wright KB, Geweke JF, Obrizan M, Pavlik CE, Ohsfeldt RL, Jones MP, Wallace RB, Rosenthal GE. A 12-year prospective study of stroke risk in older Medicare beneficiaries. BMC Geriatr 2009;9:17.
103. Holman RR, Paul SK, Bethel MA, Neil HA, Matthews DR. Long-term follow-up after tight control of blood pressure in type 2 diabetes. N Engl J Med 2008;359:1565–1576.
104. Grego F, Antonello M, Lepidi S. Is controlateral carotid artery occlusion a risk factor for carotid endarterectomy? Ann Vasc Surg 2005;19:882–889.
105. Fox AJ, Eliasziw M, Rothwell PM. Identification, prognosis, and management of patients with carotid artery near occlusion. AJNR Am J Neuroradiol 2005;26:2086–2094.
106. Marcus HS, MacKinnon A. Asymptomatic embolization detected by Doppler ultrasound predicts stroke risk in symptomatic carotid artery stenosis. Stroke 2005;36:971–975.
107. McCann RL. Surgical management of carotid artery atherosclerotic disease. South Med J 1994;86:S23–S28.
108. Autret A, Pourcelot L, Saudeau D, Marchal C, Bertrand P, de Boisvilliers S. Stroke risk in patients with carotid stenosis. Lancet 1987;1:888–890.
109. Hertzer NR, Flanagan RA, Beven EG, O'Hara PJ. Surgical versus non operative treatment of asymptomatic carotid stenosis. 290 patients documented by intravenous angiography. Ann Surg 1986;204:163–171.
110. Dennis MS, Bamford JM, Sandercock PA, Warlow CP. A comparison of risk factors and prognosis for transient ischemic attacks and minor ischemic strokes. The Oxfordshire Community Stroke Project. Stroke 1989;20:1494–1499.
111. Meissner I, Wiebers DO, Whisnant JP, O'Fallon WM. The natural history of asymptomatic carotid artery occlusive lesions. JAMA 1987;258:2704–2707.
112. North American Symptomatic Carotid Endarterectomy Trial Collaborators. Beneficial effect of carotid endarterectomy in symptomatic patients high with grade stenosis. N Engl J Med 1991;325:445–453.
113. European Carotid Surgery Trialists' Collaborative Group. Randomised trial of endarterectomy for recently symptomatic carotid stenosis: final results of the MRC European Carotid Surgery Trial (ECST). Lancet 1998;351:1379–1387.
114. Asymptomatic Carotid Atherosclerosis Group. Endarterectomy for asymptomatic carotid artery stenosis. JAMA 1995;73:1421–1428.
115. Rothwell PM, Eliasziw M, Gutnikov SA. Analysis of pooled data from the randomized controlled trials of endarterectomy for symptomatic carotid stenosis. Lancet 2003;361:107–116.
116. Saba L, Mallarini G. A comparison between NASCET and ECST methods in the study of carotids. Evaluation using multi-detector-row CT angiography. European Journal of Radiology, PMID: 19464837.
117. Eliasziw M, Rankin RN, Fox AJ. Accuracy and prognostic consequences of ultrasonography in identifying severe carotid artery stenosis. Stroke 1995;26:1747–1752.
118. Goodson SF, Flanigan P, Bishara RA. Can carotid duplex scanning supplant arteriography in patients with focal carotid territory symptoms? J Vasc Surg 1987;5:551–557.

119. Polak JF, Kalina P, Donaldson MC. Carotid endarterectomy: preoperative evaluation of candidates with combined Doppler sonography and MR angiography: work in progress. Radiology 1993;186:333–338.
120. Dawson DL, Zierler RE, Strandness DE, The role of duplex scanning and arteriography before carotid endarterectomy: a prospective study. J Vasc Surg 1993;18:673–680.
121. Link J, Brossmann J, Grabener M, Mueller-Hulsbeck S, Steffens JC, Brinkmann G, Heller M. Spiral CT angiography and selective digital subtraction angiography of internal carotid artery stenosis. AJNR Am J Neuroradiol 1996;17:89–94.
122. Hirai T, Korogi Y, Ono K, Murata Y, Takahashi M, Suginohara K, Uemura S. Maximum stenosis of extracranial internal carotid artery: effect of luminal morphology on stenosis measurement by using CT angiography and conventional DSA. Radiology 2001;221:802–809.
123. Verhoek G, Costello P, Khoo EW, Wu R, Kat E, Fitridge RA. Carotid bifurcation CT angiography: assessment of interactive volume rendering. J Comput Assist Tomogr 1999;23:590–596.
124. Sameshima T, Futami S, Morita Y, Yokogami K, Miyahara S, Sameshima Y, et al. Clinical usefulness and problems with three-dimensional CT angiography for the evaluation of atherosclerotic stenosis of the carotid artery: comparison with conventional angiography, MRA and ultrasound sonography. Surg Neurol 1999;51:300–309.
125. Seemann MD, Englmeier K, Schuhmann DR, Minx C, Furst H, Heuck A, Reiser MF. Evaluation of the carotid and vertebral arteries: comparison of 3D SCTA and IA-DSA – work in progress. Eur Radiol 1999;9:105–112.
126. Yucel EK, Anderson CM, Edelman RR, et al. Magnetic resonance angiography. Update on applications for extracranial arteries. Circulation 1999;100:2284–2301.
127. Debrey SM, Yu H, Lynch JK, Lovbald KV, Wright VL, Janket SKD, et al. Diagnostic accuracy of magnetic resonance angiography for internal carotid artery disease: a systematic review and meta-analysis. Stroke 2008;39:2237–2248.
128. Nederkoorn PJ, Mali WPT, Eikelboom BC, et al. Preoperative diagnosis of carotid artery stenosis. Accuracy of non-invasive testing. Stroke 2002;33:2003–2008.
129. Magarelli N, Scarabino T, Simeone AL, Florio F, Carriero A, Salvolini U, Bonomo L. Carotid stenosis: a comparison between MR, and spiral CT angiography. Neuroradiology 1998;40:367–373.
130. Saba L, Sanfilippo R, Montisci R, Mallarini G. Correlation between US-PSV and MDCTA in the quantification of carotid artery stenosis. Eur J Radiol 2010;74:99–103.
131. Nael K, Villablanca JP, Pope WB, et al. Supraaortic arteries: contrast enhanced MR angiography at 3.0 T – highly accelerated parallel acquisition for improved spatial resolution over an extended field of view. Radiology 2007;42:600–609.
132. Bartlett ES, Symons SP, Fox AJ. Correlation of carotid stenosis diameter and cross-sectional areas with CT angiography. AJNR Am J Neuroradiol 2006;27:638–642.
133. Bartlett ES, Walters TD, Symons SP, Fox AJ. Carotid stenosis index revised with direct CT angiography measurement of carotid arteries to quantify carotid stenosis. Stroke 2007;38:286–291.
134. Saba L, Mallarini G. Comparison between percentage methods and mm-method in the quatification of carotid artery stenosis, Manchester, United Kingdom. Proceedings of the UKRC, 8–10 June 2009 Manchester, United Kingdom, British Journal of Radiology congress, Series p 66.
135. Barnett HJ, Taylor DW, Eliasziw M, Fox AJ, Ferguson GG, Haynes RB, et al. Benefit of carotid endarterectomy in patients with symptomatic moderate or severe stenosis. North American Symptomatic carotid Endarterectomy Trial Collaborators. N Engl J Med 1998;339:1415–1425.
136. Executive committee for the Asymptomatic Carotid Atherosclerosis Study. Endarterectomy for asymptomatic carotid artery stenosis. JAMA 1995;273:1421–1428.
137. Ambrose JA, Tannenbaum MA, Alexopoulos D. Angiographic progression of coronary artery disease and the development of myocardial infarction. J Am Coll Cardiol 1988;12:56–62.

138. Little WC, Constantinescu M, Applegate RJ. Can coronary angiography predict the site of a subsequent myocardial infarction in patients with mild-to-moderate coronary artery disease? Circulation 1988;78:1157–1166.
139. Fuster V, Stein B, Ambrose JA, Badimon L, Badimon JJ, Chesebro JH. Atherosclerotic plaque rupture and thrombosis: evolving concepts. Circulation 1990;82(suppl 3):II47–II59.
140. Libby P. The interface of atherosclerosis and thrombosis: basic mechanisms. Vasc Med 1998;3:225–229.
141. Lovett JK, Gallagher PJ, Hands LJ, Walton J, Rothwell PM. Histological correlates of carotid plaque surface morphology on lumen contrast imaging. Circulation 2004;110:2190–2197.
142. Lovett JK, Gallagher PJ, Rothwell PM. Reproducibility of histological assessment of carotid plaque: implications for studies of carotid imaging. Cerebrovasc Dis 2004;18:117–123.
143. Budinger T, Berson A, McVeigh E, et al. Magnetic resonance imaging of the cardiovascular system. J Cardiovasc Magn Reson 1999;1:53–58.
144. Falk E, Fernandez-Ortiz A. Role of thrombosis in atherosclerosis and its complications. Am J Cardiol 1995;75:3B–11B.
145. Kullo IJ, Edwards WD, Schwartz RS. Vulnerable plaque: pathobiology and clinical implications. Ann Intern Med 1998;129:1050–1060.
146. Virmani R, et al. Lessons from sudden coronary death: a comprehensive morphological classification scheme for atherosclerotic lesions. Arterioscler Thromb Vasc Biol 2000;20:1262–1275.
147. Fuster V, et al. Atherothrombosis and high-risk plaque. Part I: evolving concepts. J Am Coll Cardiol 2005;46:937–954.
148. de Weert TT, Cretier S, Groen HC, et al. Atherosclerotic plaque surface morphology in the carotid bifurcation assessed with multidetector computed tomography angiography. Stroke 2009;40:1334–1340.
149. Sitzer M, Muller W, Siebler M, Hort W, Kniemeyer H-W, Jancke L, Steinmetz H. Plaque ulceration and lumen thrombus are the main sources of cerebral microemboli in high-grade internal carotid artery stenosis. Stroke 1995;26:1231–1233.
150. Eliasziw M, Streifler JY, Fox AJ. Significance of plaque ulceration in symptomatic patients with high grade carotid stenosis: North American Symptomatic Carotid Endarterectomy Trial. Stroke 1994;25:304–308.
151. Morgenstern LB, Fox AJ, Sharpe BL, Eliasziw HJ, Grotta JC. The risks and benefits of carotid endarterectomy in patients with near occlusion of the carotid artery. North American Symptomatic Carotid Endarterectomy Trial (NASCET) group. Neurology 1997;48:911–915.
152. Kim DI, Lee SJ, Lee BB, Kim Yi, Chung CS, Seo DW, Lee KH, Ko YH, Kim DK, DO YS, Byun HS. The relationship between the angiographic findings and the clinical features of carotid artery plaque. Surg Today 2000;30:37–42.
153. Randoux B, Marro B, Koskas F. Carotid artery stenosis: prospective comparison of CT, three-dimensional gadolinium-enhanced MR, and conventional angiography. Radiology 2001;220:179–185.
154. Runge VM, Kirsch JE, Lee C. Contrast-enhanced MR angiography. J Magn Reson Imaging 1993;3:233–239.
155. Streifler JY, Eliaziw M, Fox AJ, Benavente OR, Hachiniski VC, Ferguson GG, Barnett HJM. Angiographic detection of carotid plaque ulceration: comparison with surgical observations in a multicenter study. Stroke 1994;25:1130–1132.
156. Walker LJ, Ismail Á, McMeekin W, Lambert D, Mendelow AD, Birchall D. Computed tomography angiography for the evaluation of carotid atherosclerotic plaque: correlation with histopathology of endarterectomy specimens. Stroke 2002;33:977–981.
157. Saba L, Caddeo G, Sanfilippo R, Montisci R, Mallarini G. CT and US in the study of ulcerated carotid plaque compared with surgical results. Advantages of multi-detector-row CT angiography. AJNR Am J Neuroradiol 2007;28:1061–1066.
158. Debernardi S, Martincich L, Lazzarò D. CT angiography in the assessment of carotid atherosclerotic disease: results of more than two years' experience. Radiol Med 2004;108:116–127.

159. Lammie GA, Wardlaw J, Allan P, Ruckley CV, Peek R, Signorini DF. What pathological components indicate carotid atheroma activity and can these be identified reliably using ultrasound? Eur J Ultrasound 2000;11:77–86.
160. Bluth EI, McVay LV, Merritt CR, Sullivan A. The identification of ulcerative plaque with high resolution duplex carotid scanning. J Ultrasound Med 1998;7:73–76.
161. Alvarez-Linera J, Benito-Leon J, Escribano J, Campollo J, Gesto R. Prospective evaluation of carotid artery stenosis: elliptic centric contrast-enhanced MR angiography and spiral CT angiography compared with digital subtraction angiography. AJNR Am J Neuroradiol 2003;24:1012–1019.
162. Cumming MJ, Morrow IA. Carotid artery stenosis: a prospective comparison of CT angiography and conventional angiography. AJR Am J Roentgenol 1994;163:517–523.
163. Schwartz RB, Jones KM, Chernoff DM, et al. Common carotid artery bifurcation: evaluation with spiral CT. Radiology 1992;185:513–519.
164. Lovett JK, Gallagher PJ, Hands LJ, et al. Histological correlates of carotid plaque surface morphology on lumen contrast imaging. Circulation 2004;110:2190–2197.
165. Porsche C, Walker L, Mendelow AD, et al. Assessment of vessel wall thickness in carotid atherosclerosis using spiral CT angiography. Eur J Vasc Endovasc Surg 2002;23: 437–440.
166. Porsche C, Walker L, Mendelow D, et al. Evaluation of cross-sectional luminal morphology in carotid atherosclerotic disease by use of spiral CT angiography. Stroke 2001;32:2511–2515.
167. de Weert TT, de Moynè C, Meijering E, Booij R, Niessen WJ, Dippel DW, van der Lugt A. Assessment of atherosclerotic carotid plaque volume with multidetector-row CT angiography. Int J Cardiovasc Imaging 2008;24:751–759.
168. de Weert TT, Ouholus M, Meijring E, Zondervan PE, Hendriks JM, van Sambeek MR, et al. In vivo characterization and quantification of atherosclerotic plaque components with multidetector computer tomography and histopathological correlation. Arterioscler Thromb Vasc Biol 2006;26:2366–2372.
169. Ouhlous M, Flach HZ, de Weert TT. Carotid plaque composition and cerebral infarction: MR imaging study. AJNR Am J Neuroradiol 2005;26:1044–1049.
170. Nandalur KR, Hardie HD, Raghavan P, Schipper MJ, Baskurt E, Kramer CM. Composition of the stable carotid plaque: insights from a multi-detector computed tomography study of plaque volume. Stroke 2007;38:935–940.
171. Loree HM, Kamm RD, Stringfellow RG, et al. Effects of fibrous cap thickness on peak circumferential stress in model atherosclerotic vessels. Circ Res 1992;71:850–858.
172. Li ZY, Howarth S, Tang T, Graves M, U-King-Im J, Gillard JH. Does calcium deposition play a role in the stability of atheroma ? Location may be the key. Cerebrovasc Dis 2007;24:452–459.
173. Virmani R, Burke, AP, Kolodgie FD, Farb A. Vulnerable plaque: the pathology of unstable coronary lesions. J Interv Cardiol 2002;15:439–446.
174. Falk E. Coronary thrombosis: pathogenesis and clinical manifestation. Am J Cardiol 1991;68:28B–35B.
175. Yuan C, Zhang SX, Polissar NL, Echelard D, Ortiz G, Davis JW, Ellington E, Ferguson MS, Hatsukami TS. Identification of fibrous cap rupture with magnetic resonance imaging is highly associated with recent transient ischemic attack or stroke. Circulation 2002;105:181–185.
176. Saba L, Mallarini G. Fissured fibrous cap of vulnerable carotid plaques and symptomaticity: are they correlated? Preliminary results by using MDCTA. Cerebrovasc Dis 2009;27:322–327.
177. Cai J, Hatsukami TS, Ferguson MS, et al. In vivo quantitative measurement of intact fibrous cap and lipid-rich necrotic core size in atherosclerotic carotid plaque: comparison of high-resolution, contrast-enhanced magnetic resonance imaging and histology. Circulation 2005;112:3437–3444.
178. Parker DL, Yuan C, Blatter DD. MR angiography by multiple thin slab 3D acquisition. Magn Reson Med 1991;17:434–451.

179. Mitsumori LM, Hatsukami TS, Ferguson MS, Kerwin WS, Cai J, Yuan C. In vivo accuracy of multisequence MR imaging for identifying unstable fibrous caps in advanced human carotid plaques. J Magn Reson Imaging 2003;17:410–420.
180. Trivedi RA, King-Im J, Graves MJ, et al. Multi-sequence in vivo MRI can quantify fibrous cap and lipid core components in human carotid atherosclerotic plaques. Eur J Vasc Endovasc Surg 2004;28:207–213.
181. Kramer CM, Cerilli LA, Hagspiel K, DiMaria JM, Epstein FH, Kern JA. Magnetic resonance imaging identifies the fibrous cap in atherosclerotic abdominal aortic aneurysm. Circulation 2004;109:1016–1021.
182. Wasserman BA, Smith WI, Trout HH III, Cannon RO III, Balaban RS, Arai AE. Carotid artery atherosclerosis: in vivo morphologic characterization with gadolinium enhanced double-oblique MR imaging-initial results. Radiology 2002;223:566–573.
183. Takaya N, Yuan C, Chu B, et al. Presence of intraplaque hemorrhage stimulates progression of carotid atherosclerotic plaques. Circulation 2005;111:2768–2775.
184. Kolodgie FD, Gold HK, Burke AP, et al. Intraplaque hemorrhage and progression of coronary atheroma. N Engl J Med 2003;349:2316–2325.
185. Paterson JC. Capillary rupture with intimal hemorrhage as a causative factor in coronary thrombosis. Arch Pathol 1938;25:474–487.
186. Constantinides P. Plaque fissuring in human coronary thrombosis. J Atheroscler Res 1966;6:1–17.
187. Davies MJ, Thomas AC. Plaque fissuring: the cause of acute myocardial infarction, sudden ischaemic death, and crescendo angina. Br Heart J 1985;53:363–373.
188. Beach KW, Hatsukami T, Detmer PR. Carotid artery intraplaque hemorrhage and stenting velocity. Stroke 1993;24;314–319.
189. Moody AR, Murphy RE, Morgan PS. Characterization of complicated carotid plaque with magnetic resonance direct thrombus imaging in patients with cerebral ischemia. Circulation 2003;107:3047–3052.
190. Ajduk M, Pavic L, Bulimbasic S, Sarlija M, Pavic P, et al. Multidetector-row computed tomography in evaluation of atherosclerotic carotid plaques complicated with intraplaque hemorrhage. Ann Vasc Surg 2009;23:186–193.
191. Oliver TB, Lammie GA, Wright AR. Atherosclerotic plaque at the carotid bifurcation: CT angiographic appearance with histopathologic correlation. AJNR AM J Neuroradiol 1999;20:897–901.
192. Wintermark M, Jawadi SS, Rapp JH, Tihan T, Tong E, Glidden DV, et al. High resolution CT imaging of carotid artery atherosclerotic plaques. AJNR AM J Neuroradiol 2008;29:875–882.
193. Saba L, Sanfilippo R, Pirisi R, Pascalis L, Montisci R, Mallarini G. Multi-detector row CT angiography in the study of atherosclerotic carotid artery. Neuroradiology 2007;49:623–637.
194. Rothwelll PM, Gibson R, Warlow CP. Interrelation between plaque surface morphology and degree of stenosis on carotid angiograms and the risk of ischemic stroke in patients with symptomatic carotid stenosis. Stroke 2000;31:615–621.
195. Pelz DM, Buchan A, Fox AJ, Barnett HJM, Vinuela F. Intraluminal thrombus of the internal carotid arteries: angiographic demonstration with anticoagulant theraphy alone. Radiology 1986;160:369–373.
196. Wentzel JJ, Aguiar SH, Fayad ZA. Vascular MRI in the diagnosis and therapy of the high risk atherosclerotic plaque. J Interv Cardiol 2003;16:129–142.
197. Toussaint JF, Southern JF, Fuster V, Kantor HL. Water diffusion properties of human atherosclerosis and thrombosis measured by pulse field gradient nuclear magnetic resonance. Arterioscler Thromb Vasc Biol 1997;17:542–546.
198. Bradley WG Jr. MR appearance of hemorrhagein the brain. Radiology 1993;189:15–26.
199. Sangiorgi G, Rumberger JA, Severson A. Arterial calcification and not lumen stenosis is highly correlated with atherosclerotic plaque burden in humans: a histologic study of 723 coronary artery segments using nondecalcifying methodology. J Am Coll Cardiol 1998;31:126–133.
200. Kondos GT, Hoff JA, Sevrukov A. Electronbeam tomography coronary artery calcium and cardiac events: a 37-month follow-up of 5635 initially asymptomatic low- to intermediate-risk adults. Circulation 2003;107:2571–2576.

201. Shaw LJ, Raggi P, Schisterman E, Berman DS, Callister TQ. Prognostic value of cardiac risk factors and coronary artery calcium screening for all-cause mortality. Radiology 2003;228:826–833.
202. Nandalur KR, Baskurt E, Hagspiel KD, Phillips CD, Kramer CM. Calcified Carotid atherosclerosic Plaque is associated less with ischemic symptoms than is noncalcified plaque on MDCT. AJR Am J Roentgenol 2005;184:295–298.
203. Saba L, Montisci R, Sanfilippo R, Mallarini G. Multidetector row CT of the brain and carotid artery: a correlative analysis. Clin Radiol 2009;64:767–778.
204. Fanning NF, Walters TD, Fox AJ, Symons SP. Association between calcification on the cervical carotid artery bifurcation and white matter ischemia. AJNR Am J Neuroradiol 2006;27:378–383.
205. Hunt JL, Fairman R, Mitchell ME, Carpenter JP, Golden M, Khalapyan T, Wolfe M, Neschis D, Milner R, Scoll B, Cusak A, Moheler ER III. Bone formation in carotid plaques. A clinicopathological study. Stroke 2002;33:1214–1219.
206. Jeziorska M, McCollum C, Wooley DE. Observations on bone formation and remodelling in advanced atherosclerotic lesions of human carotid arteries. Virchows Arch 1998;433:559–565.
207. Agaston AS, Janowitz WR, Hildner FJ. Quantification of coronary artery calcium using ultrafast computed tomography. J Am Coll Cardiol 1990;15:827–832.
208. Ohara T, Toyoda K, Otsubo R, Nagatsuka K, Kubota Y, Yasaka M, et al. Eccentric stenosis of the carotid artery associated with ipsilateral cerebrovascular events. AJNR Am J Neuroradiol 2008;29:1200–1203.
209. Tambasco M, Steinman DA. Path-dependent hemodynamics of the stenosed carotid bifurcation. Ann Biomed Eng 2003;31:1054–1065.
210. Hardie AD, Kramer CM, Raghavan P.The impact of expansive arterial remodelling on clinical presentation in carotid artery disease: a multi-detector-row CT angiography study. AJNR Am J Neuroradiol 2007;28:1067–1070.
211. Coli S, Magnoni M, Sangiorgi G, et al. Contrast enhanced ultrasound imaging of intraplaque neovascularization in carotid arteries: correlation with histology and plaque echogenicity. J Am Coll Cardiol 2008;52:223–230.
212. Xiong L, Deng YB, Zhu Y, Liu YN, Bi XJ. Correlation of carotid plaque neovascularization detected by using contrast-enhanced US with clinical symptoms. Radiology 2009;251:583–589.
213. Moreno PR, Purushothaman KR, Fuster V, et al. Plaque neovascularization is increased in ruptured atherosclerotic lesions of human aorta: implications for plaque vulnerability. Circulation 2004;110:2032–2038.
214. McCarthy MJ, Loftus IM, Thompson MM, et al. Angiogenesis and the atherosclerotic carotid plaque: an association between Symptomatology and plaque morphology. J Vasc Surg 1999;30:261–268.
215. Dunmore BJ, McCarthy MJ, Naylor AR, Brindle NP. Carotid plaque instability and ischemic symptoms are linked to immaturity of microvessels within plaques. J Vasc Surg 2007;45:155–159.
216. Lorenz MW, von Kegler S, Steinmetz H, Markus HS, Sitzer M. Carotid intima-media thickening indicates a higher vascular risk across a wide age range: prospective data from the Carotid Atherosclerosis Progression Study (CAPS). Stroke 2006;37(1):87–92.
217. O'Leary DH, Polak JF, Kronmal RA, Manolio TA, Burke GL, Wolfson SK. Carotid-artery intima and media thickness as a risk factor for myocardial infarction and stroke in older adults. Cardiovascular Health Study Collaborative Research Group. N Engl J Med 1999;340(1):14–22.
218. Coll B, Feinstein SB. Carotid intima-media thickness measurements: techniques and clinical relevance. Curr Atheroscler Rep 2008;10(5):444–450.
219. Heliopoulos I, Papaoiakim M, Tsivgoulis G, Chatzintounas T, Vadikolias K, Papanas N, Piperidou C. Common carotid intima media thickness as a marker of clinical severity in patients with symptomatic extracranial carotid artery stenosis. Clin Neurol Neurosurg 2009;111(3):246–250.
220. Veller MG, Fisher CM, Nicolaides AN. Measurement of the ultrasonic intima-media complex thickness in normal subjects. J Vasc Surg 1993;17:719–725.

221. Allan PL, Mowbray PI, Lee AJ, Fowkes FG. Relationship between carotid intima-media thickness and symptomatic and asymptomatic peripheral arterial disease: the Edinburgh Artery Study. Stroke 1997;28:348–353.
222. Howard G, Sharrett AR, Heiss G, Evans GW, Chambless LE, Riley WA, Burke GL, ARIC Investigators. Carotid artery intimal-medial thickness distribution in general populations as evaluated by B-mode ultrasound. Stroke 1993;24:1297–1304.
223. Salonen R, Salonen JT. Determinants of carotid intima-media thickness: a population-based ultrasonography study in eastern Finnish men. J Intern Med 1991;229:225–231.
224. Saba L, Sanfilippo R, Pascalis L, Montisci R, Caddeo G, Mallarini G. Carotid artery wall thickness and Ischemic symptoms evaluation using multi-detector row CT angiography. Eur Radiol 2008;18:1962–1971.

Biography

Dr Luca Saba currently works in the A.O.U. of Cagliari (Italy). Dr Saba research fields are focused on Multi-Detector-Row CT applications, Neuroradiology, and Diagnostic in Vascular Sciences. His works, as lead author, achieved several high impact factor, peer-reviewed, Journals as *American Journal of Neuroradiology, European Radiology, European Journal of Radiology, Acta Radiologica, Cardiovascular and Interventional Radiology, Journal of Computer Assisted Tomography, American Journal of Roentgenology, Neuroradiology, Clinical Radiology, Journal of Cardiovascular Surgery, Cerebrovascular Diseases.* Dr Saba presented more than 300 papers in National and International Congress (RSNA, ESGAR, ECR, ISR, AOCR, AINR, JRS, SIRM, AINR). He currently serves on the Editorial Board of several Radiological Journals.

Chapter 14
Fast, Accurate Unsupervised Segmentation of 3D Magnetic Resonance Angiography

Ayman El-Baz, Georgy Gimel'farb, Ahmed Elnakib, Robert Falk, and Mohamed Abou El-Ghar

Abstract Accurate automatic extraction of a 3D cerebrovascular system from images obtained by time-of-flight (TOF) or phase-contrast (PC) magnetic resonance angiography (MRA) is a challenging segmentation problem due to small size objects of interest (blood vessels) in each 2D MRA slice and complex surrounding anatomical structures, e.g. fat, bones, or gray and white brain matter. We show that due to a multimodal nature of MRA data, blood vessels can be accurately separated from background in each slice by a voxel-wise classification based on precisely identified probability models of voxel intensities. To identify the models, an empirical marginal probability distribution of intensities is closely approximated with a linear combination of discrete Gaussians (LCDG) with alternate signs by using our previous EM-based techniques for precise LCG-approximation adapted to deal with the LCDGs. High accuracy of the proposed approach is experimentally validated on 85 real MRA datasets (50 TOF and 35 PC) as well as synthetic MRA data for special 3D geometrical phantoms of known shapes.

Keywords Segmentation • Magnetic Resonance Angiography (MRA) • Blood vessels • Expectation-Maximization (EM) • Modified EM • Linear Combination of Discrete Gaussians (LCDG)

14.1 Introduction

Accurate segmentation of MRA images to extract a 3D cerebrovascular system is one of the most important problems in practical computer-assisted medical diagnostics. PC-MRA provides good suppression of background signals and quantifies blood flow velocity vectors for each voxel. Time-of-flight magnetic resonance angiography (TOF-MRA) is less quantitative, but it is fast and provides images with high contrast. Most popular present-day techniques for extracting blood

A. El-Baz (✉)
Bioimaging Laboratory, University of Louisville, Louisville, KY, USA
e-mail: aselba01@louisville.edu

Jasjit S. Suri et al. (eds.), *Atherosclerosis Disease Management*,
DOI 10.1007/978-1-4419-7222-4_14,

vessels from the MRA data are scale-space filtering, deformable models, statistical models, and hybrid methods.

Multiscale filtering [1–6] enhances curvilinear structures in 3D medical images by convolving an image with Gaussian filters at multiple scales. Eigenvalues of the Hessian at each voxel are analyzed to determine the local shapes of 3D structures (by the eigenvalues, voxels from a linear structure like a blood vessel differ from those for a planar structure, speckle noise, or unstructured component). The multiscale filter output forms a new enhanced image such that the curvilinear structures become brighter whereas other components, e.g. speckle noise and planar structures such as skin, are darker [1, 5, 6]. Such an image can be directly visualized [5] or thresholded [1] or segmented using a deformable model [6]. Alternatively, the obtained eigenvalues define a candidate set of voxels corresponding to centerlines of the vessels [2–4]. Multiscale filter responses at each of the candidates determine how likely that voxel belongs to a vessel of each particular diameter. The maximal response over all the diameters (scales) is assigned to each voxel, and a surface model of the entire vascular structure is reconstructed from the estimated centerlines and diameters. After segmenting the filtered MRA image by thresholding, anisotropic diffusion techniques are used to remove noise but preserve small vessels [3, 7, 8].

An alternative medial-axes-based multiscale approach assumes that the vessel centerlines are often the brightest and detects them as intensity ridges of the image [9]. The vessel's width is then determined by multiscale filter responses. This algorithm has been used in conjunction with 2D/3D registration to incorporate information from a pair of X-ray angiograms [10]. By involving differential geometry, the volumetric MRA image is treated as a hypersurface in a 4D space whose extrema of curvature correspond to the vessel centerlines [11].

Deformable model approaches to 3D vascular segmentation attempt to approximate the boundary surface of the blood vessels. An initial boundary is evolving to optimize a surface energy that depends on image gradients and surface smoothness [12]. Topologically adaptable surfaces make classical deformable models more efficient for segmenting intracranial vasculature [13]. Geodesic active contours implemented with level set techniques offer flexible topological adaptability to segment the MRA images [14], including more efficient adaptation to local geometric structures represented, e.g., by tensor eigenvalues [15]. Fast segmentation of blood vessel surfaces is obtained by inflating a 3D balloon with fast marching methods [16]. Two-step segmentation of a 3D vascular tree in [17] is first carried out locally in a small volume of interest. Then, a global topology is estimated to initialize a new volume of interest. A multiscale geometrical flow is proposed in [18] to segment the vascular tree.

Compared to the scale-space filtering, the deformable models produce much better experimental results but have a common drawback, namely, a manual initialization. In addition, both the groups are slow compared to statistical approaches.

Statistical extraction of a vascular tree is completely automatic, but its accuracy depends on underlying probability models. The MRA images are multimodal in that the signals (intensities or gray levels) in each region of interest (e.g., blood vessels, brain tissues, etc.) are associated with a particular dominant mode of the total marginal probability distribution of signals. To the best of our knowledge, adaptive statistical approaches for extracting blood vessels from the MRA images have been proposed so far only by Wilson and Noble [19] for the TOF-MRA data and Chung and Noble [20] for the PC-MRA data. The former approach represents the marginal data distribution with a mixture of two Gaussians and one uniform component for the stationary CSF, brain tissues, and arteries, respectively, whereas the latter approach replaces the Gaussians with the more adequate Rician distributions. To identify the mixture, i.e., to estimate all its parameters, a conventional EM algorithm is used in both the cases. It was called a "modified EM" in [19], after replacing gray levels in individual pixels considered by their initial EM scheme with a marginal gray level distribution. Actually, such a modification simply returns to what is in common use for decades in probability density estimation (see e.g., [21]), while the individual pixels appeared in their initial scheme only as an unduly verbatim replica of a general EM framework.

Different hybrid approaches attempt to combine the aforementioned three approaches. For instance, a region-based deformable contour for segmenting tubular structures is derived in [22], by combining signal statistics and shape information. A combination of a Gaussian statistical model with the maximum intensity projection (MIP) images acquired at three orthogonal directions [23] allows for extracting blood vessels iteratively from images acquired by rotational angiography. The MIP Z-buffer is segmented using a continuity criterion to generate candidate sets of "seed" voxels being then coupled with a global threshold to extract the whole tree using region growing techniques [24].

Cylinder matching [25, 26] detects vessels by minimizing the inertia moments of a cylinder and using prior knowledge about the intensity profiles in and at the edge of a vessel. A more generalized technique in [27] approximates the vessel's cross-section by a polygon. Continuity and orientation between the consecutive slices are used to calculate a locally optimal shape for the polygon with good accuracy. An octree decomposition of a velocity field PC-MRA image is proposed in [28] to find an optimal tessellation. Each block of the octree contains at most one feature defined by gray levels and orientation vectors.

An alternative approach in [29] extracts an initial shape of vessels by image thresholding. Then, a locally smooth surface is formed by region growing using binary morphological operations. A recursive hybrid segmentation framework in [30, 31] combines a prior Gibbs random field model, marching cubes, and deformable models. First, the Gibbs model is used to estimate object boundaries using region information from 2D slices. Then, the estimated boundaries and the marching cubes technique are used to construct a 3D mesh specifying the initial geometry of a deformable model. Finally, the deformable model fits to the data under the 3D image gradient forces.

The above overview shows the following limitations of the existing approaches:

1. Most of them presume only a single type of images (e.g. TOF- or PC-MRA).
2. Most of them require user interaction to initialize a vessel of interest.

3. Some deformable models assume the circular vessel cross-sections; this holds for healthy people, but not for patients with stenosis or aneurysm.
4. All but statistical approaches are computationally expensive.
5. Known statistical approaches use only predefined probability models that cannot fit all the cases because actual intensity distributions for blood vessels depend on the patient, scanner, and scanning parameters.

Below, we show that the fast and highly accurate statistical approach to extract blood vessels is obtained when the probability models of each region of interest in TOF or PC-MRA images are precisely identified rather than predefined as in [19, 20].

In our approach, the empirical gray level distribution for each MRA slice is closely approximated with an LCDG. Then, the latter is split into three individual LCDGs, one per a region of interest. These regions associated with the three dominant modes relate to darker bones and fat, gray brain tissues, and bright blood vessels, respectively. The identified models specify an intensity threshold for extracting blood vessels in that slice. Finally, a 3D connectivity filter is applied to the extracted voxels to select the desired cerebrovascular system.

As our multiple experiments show, more precise region models result in significantly better segmentation accuracy compared to other methods.

14.2 Slice-Wise Segmentation with the LCDG Models

We use the expected log-likelihood as a model identification criterion. Let $X = (X_s : s = 1,...,S)$ denote a 3D MRA image containing S coregistered 2D slices $X_s = (X_s(i,j) : (i,j) \in R; X_s(i,j) \in Q)$. Here, R and $Q = \{0,1,...,Q-1\}$ are a rectangular arithmetic lattice supporting the 3D image and a finite set of Q-ary intensities (gray levels), respectively. Let $F_s = f_s(q) : q \in Q; \sum_{q \in Q} f_s(q) = 1$ where q denotes the gray level be an empirical marginal probability distribution of gray levels for the MRA slice X_s.

In accordance with [32], each such slice is considered as a K-modal image with a known number K of the dominant modes related to the regions of interest (in our particular case, $K=3$). To segment the slice by separating the modes, we have to estimate from F_s individual probability distributions of signals associated with each mode. In contrast to a conventional mixture of Gaussians, one per region [21], or a slightly more flexible mixtures involving other simple distributions, one per region, as e.g. in [19, 20], we will closely approximate F_s with a linear combination of discrete Gaussians (LCDG). Then, the LCDG for the whole image is partitioned into the like submodels relating to each dominant mode.

The DG is defined as the probability distribution $\Psi_\theta = (\psi(q|\theta) : q \in Q)$ on Q of gray levels such that each probability $\psi(q|\theta)$ relates to the cumulative Gaussian probability function $\Phi_\theta(q)$ as follows (here, θ is a shorthand notation $\theta = (\mu, \sigma^2)$ for the mean, μ, and variance, σ^2):

$$\psi(q|\theta) = \begin{cases} \Phi_\theta(0.5) \text{ for } q = 0, \\ \Phi_\theta(q+0.5) - \Phi_\theta(q-0.5) \text{ for } q = 1,\ldots,Q-2, \\ 1 - \Phi_\theta(Q-1.5) \text{ for } q = Q-1. \end{cases}$$

The LCDG with C_p positive and C_n negative components such that $C_\mathrm{p} \geq K$:

$$p_{w,\Theta}(q) = \sum_{r=1}^{C_\mathrm{p}} w_{\mathrm{p},r}\psi(q|\theta_{\mathrm{p},r}) - \sum_{l=1}^{C_\mathrm{n}} w_{\mathrm{n},l}\psi(q|\theta_{\mathrm{n},l}) \tag{14.1}$$

has obvious restrictions on its weights $w = [w_{\mathrm{p},.}, w_{\mathrm{n},.}]$, namely, all the weights are nonnegative and

$$\sum_{r=1}^{C_\mathrm{p}} w_{\mathrm{p},r} - \sum_{l=1}^{C_\mathrm{n}} w_{\mathrm{n},l} = 1. \tag{14.2}$$

Generally, the true probabilities are nonnegative: $p_{w,\Theta}(q) \geq 0$ for all $q \in Q$. Therefore, the probability distributions comprise only a proper subset of all the LCDGs in (14.1) which may have negative components $p_{w,\Theta}(q) < 0$ for some $q \in Q$.

Our goal is to find a K-modal probability model that closely approximates the unknown marginal gray level distribution. Given F_s, the Bayesian estimate F of the latter is as follows [21]: $f(q) = (|R|f_s(q)+1)/(|R|+Q)$, and the desired model has to maximize the expected log-likelihood of the statistically independent empirical data by the model parameters:

$$L(w,\Theta) = \sum_{q\in Q} f(q)\log p_w\Theta(q). \tag{14.3}$$

For simplicity, we do not restrict the identification procedure to only the true probability distributions but also check the validity of the restrictions during the procedure itself. The Bayesian probability estimate F with no zero or unit values in (14.3) ensures that a sufficiently large vicinity of each component $f(q)$ complies to the restrictions.

To precisely identify the LCDG-model including the numbers of its positive and negative components, we adapt to the LCDGs our EM-based techniques introduced in [32] for identification of a probability density with a continuous LCG model. For completeness, the adapted algorithms are outlined in Appendix.

The entire segmentation algorithm is as follows.

1. For each successive MRA slice $X_s : s = 1,\ldots,S$,
(a) Collect the marginal empirical probability distribution $F_s = f_s(q) : q \in Q$ of gray levels.
(b) Find an initial LCDG-model that closely approximates F_s by using an initializing algorithm in "Sequential EM-based initialization" of Appendix to estimate the numbers $C_\mathrm{p} - K$, C_n and parameters w, θ (weights, means, and variances) of the positive and negative DGs.

(c) Refine the LCDG-model with the fixed C_p and C_n by adjusting all other m parameters with a modified EM algorithm in "Modified EM algorithm for refining LCDGs" of Appendix.
(d) Split the final LCDG-model into K submodels, one per each dominant mode, by minimizing the expected errors of misclassification and select the LCDG-submodel with the largest mean value (i.e., the submodel corresponding to the brightest pixels) as the model of the desired blood vessels.
(e) Extract the blood vessels voxels in this MRA slice using the intensity threshold t separating best their LCDG-submodel from the background ones.
2. Eliminate artifacts from the whole set of the extracted voxels using a connectivity filter that selects the largest connected tree structure built by a 3D volume growing algorithm [33].[1]

The main goal of the whole procedure is to find the threshold for each MRA slice that extracts the brighter blood vessels from their darker background in such a way that the vessels boundaries are accurately separated from the surrounding structures with sometimes almost the same brightness along these boundaries.

The initialization at Step 1b always produces the LCDG with the nonnegative starting probabilities

$p_{w,\Theta}(q)$. While the refinement at Step 1c increases the likelihood, the probabilities continue to be nonnegative. In our experiments below, the opposite situations have never been met.

14.3 Experimental Results

Experiments in extracting blood vessels have been conducted with the 3D TOFMRA and PC-MRA images of the following spatial resolution and size acquired with the Picker 1.5T Edge MRI scanner:

	Resolution (mm)	Size of each data set (voxels)
TOF-MRA	$0.43 \times 0.43 \times 1.0$	$512 \times 512 \times 93$
PC-MRA	$0.43 \times 0.43 \times 1.0$	$512 \times 512 \times 123$

Both the image types are three-modal ($K=3$) with the aforementioned signal classes of dark gray bones and fat, gray brain tissues, and light-gray blood vessels. Typical 2D TOF- and PC-MRA slices and their three-class dominant Gaussian mixtures P_3 approximating the estimated marginal distributions F are shown in Fig. 14.1.

[1] Step 2 is necessary due to MRA sensitivity to tissues with short T1 responses (e.g. subcutaneous fat) that may obscure the blood vessels in the segmented volume.

Both the image types are three-modal ($K=3$) with the aforementioned signal classes of dark gray bones and fat, gray brain tissues, and light-gray blood vessels. Typical 2D TOF- and PC-MRA slices and their 3-class dominant Gaussian mixtures P_3 approximating the estimated marginal distributions F are shown in Fig. 14.1.

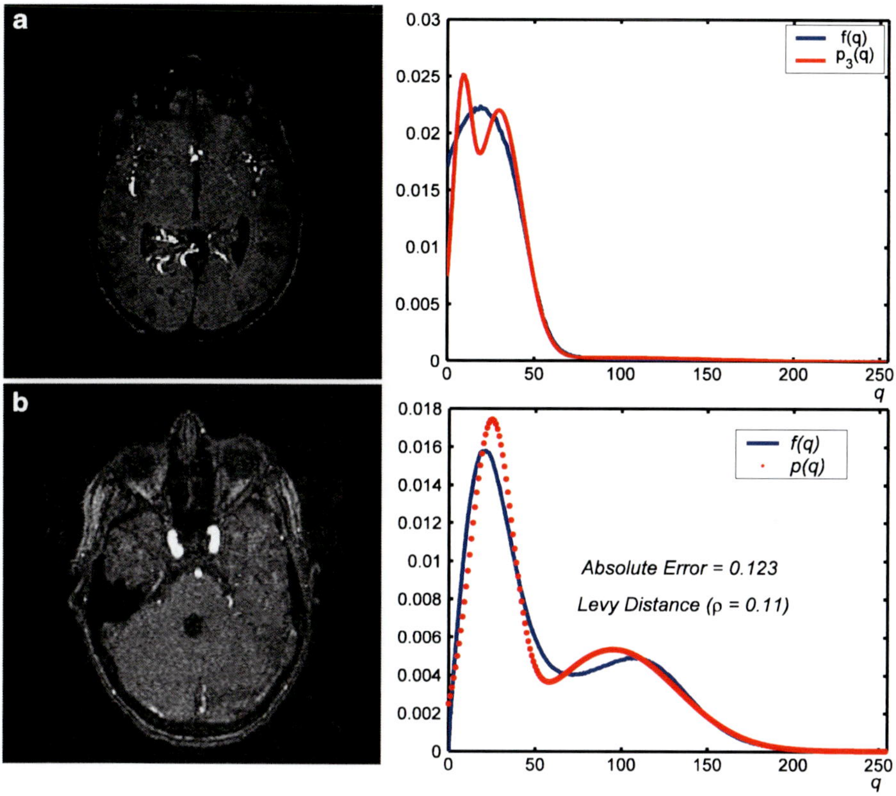

Fig. 14.1 Typical TOF-MRA (**a**) and PC-MRA (**b**) slices with the dominant Gaussian mixtures $P_3 = (p_3(q) : q \in Q)$ laid over the distributions $F = (f(q) : q \in Q$

14.3.1 Segmentation of Natural TOF- and PC-MRA Images

Figure 14.2 illustrates how Step 1b of our algorithm builds an initial LCDG model for the TOF- and PC-MRA images in Fig. 14.1. Absolute deviations $|f(q) - p_3(q)|$ are scaled up to make the unit sums of the positive or absolute negative deviations for $q = 0, \ldots, Q-1$. The minimum approximation errors are obtained in both the cases with the six-component Gaussian mixtures. Each initial LCDG model can be split, if necessary, into the three LCDG-submodels for each signal class.

Figure 14.3 presents the final LCDG-models after Step 1c of our algorithm and shows successive changes of the log-likelihood during the refining EM process. For the TOF- and PC-MRA images, first nine EM iterations increase the log-likelihood from −5.7 to −5.2 and −5.5 to −4.4, respectively. The final LCDG submodels of each class (Step 1d) suggest that thresholds $t = 192$ and $t = 73$ separate blood vessels from the TOF- and PC-MRA images, respectively, with the minimum expected misclassification error. To highlight the advantages of our approach, Fig. 14.4 shows the approximation of the distributions F for the TOF-and PC-MRA images

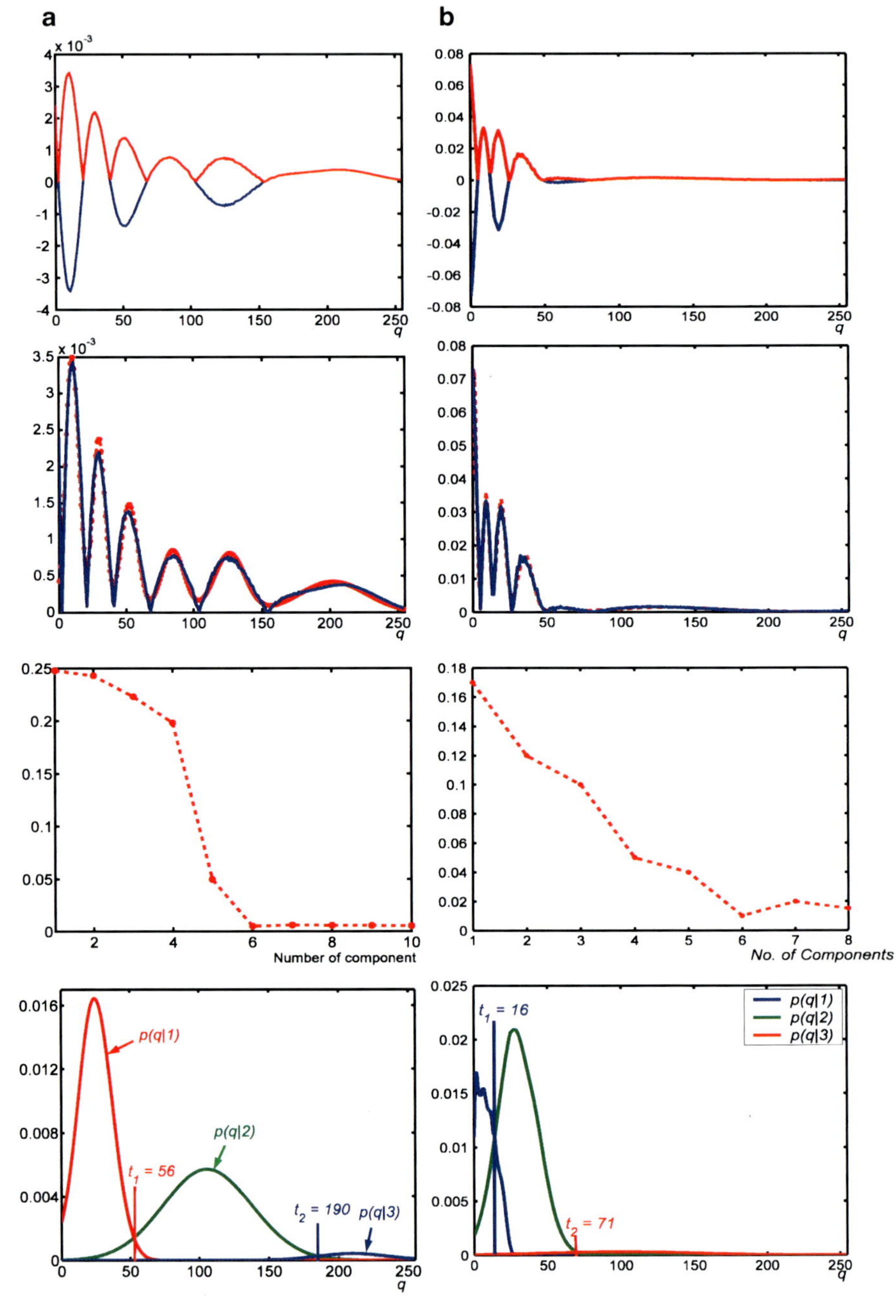

Fig. 14.2 TOF- (**a**) and PC-MRA (**b**): from *left* to *right* – alternating and absolute deviations between F and P_3; the mixture model of the absolute deviations; the absolute approximation error in function of the number of Gaussians approximating the scaled-up absolute deviations, and the initial LCDG-submodels of each signal class

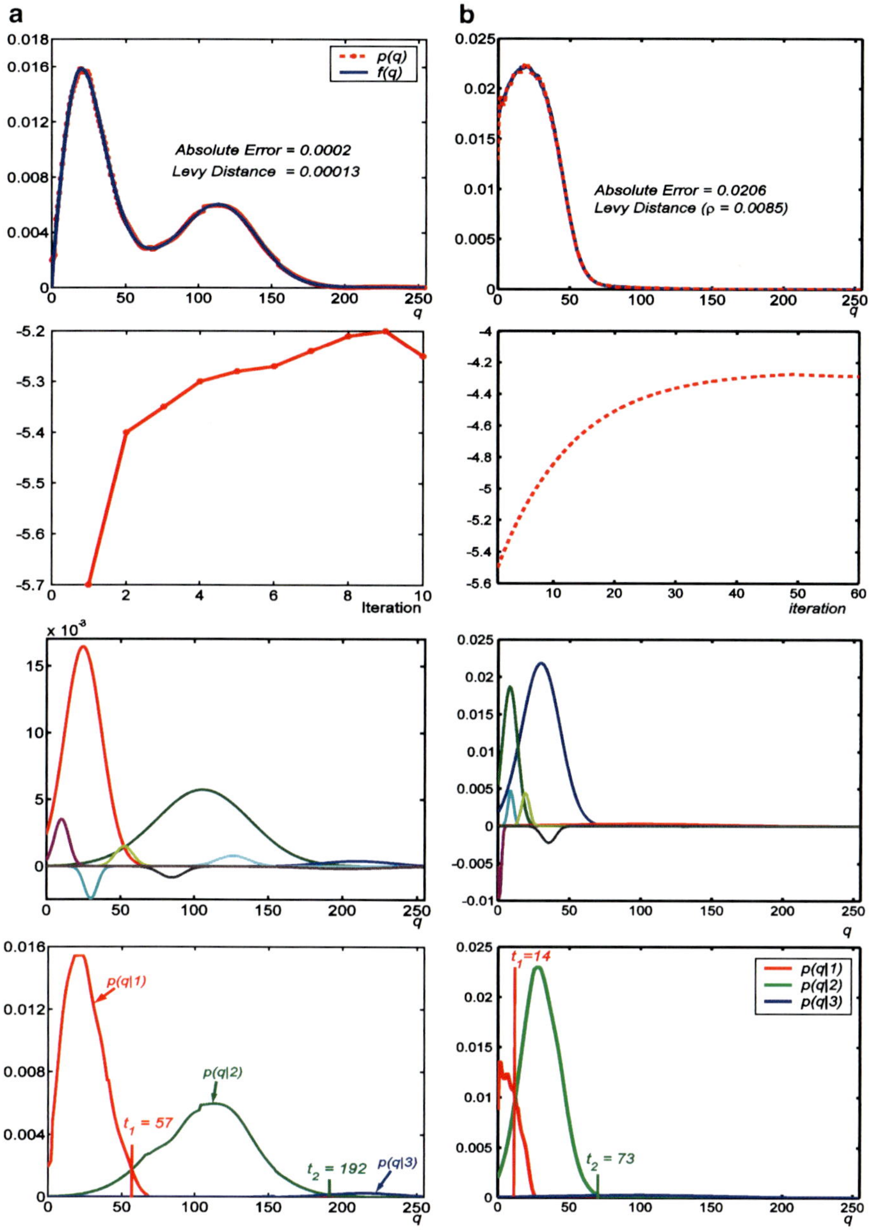

Fig. 14.3 TOF- (**a**) and PC-MRA (**b**): from *left* to *right* – the final 3-class LCDG-model laid over the distribution F; the log-likelihood changes for the refining EM-iterations; the DGs for the final model, and the final LCDG submodels of each class

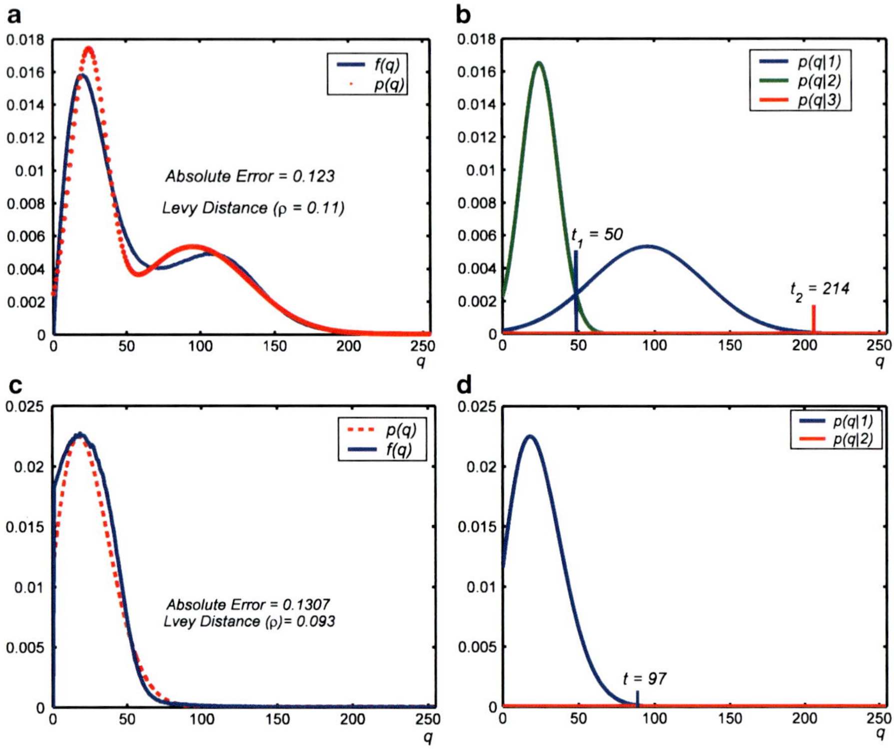

Fig. 14.4 The Wilson–Noble's [19] (**a**, **b**) and Chung–Noble's [20] (**c**, **d**) models: the estimated distribution (**a**, **c**) and the class submodels (**b**, **d**)

in Fig. 14.1 with the three-component Wilson–Noble's [19] and Chung–Noble's [20] mixtures, respectively. Our approach provides considerably higher approximation quality in terms of the Levy interdistribution distance [34] and the total absolute difference between two distributions:

		Levy distance	Absolute difference
TOF-MRA	Our approachWilson–Noble [19]	0.000130.110	0.000200.123
PC-MRA	Our approachChung–Noble [20]	0.00260.110	0.00850.093

Comparisons of the above approaches on 50 natural TOF-MRA and 35 PCMRA data sets confirm our more precise model yields much higher segmentation accuracy. Typically, higher separation thresholds of the Wilson–Noble's or Chung–Noble's approaches (e.g. $t=214$ versus our $t=192$ for the TOF-MRA and $t=97$ versus our $t=73$ for the PC-MRA in the above examples) miss some blood vessels. For example, the test results in Figs. 14.5 and 14.6 after applying the connectivity filter (Step 2) to our and Wilson–Noble's or Chung–Noble's segmentation, respectively, show these latter segmentations fail to detect sizeable parts of the brain vascular trees assigned by expert radiologists to the actual trees and extracted by

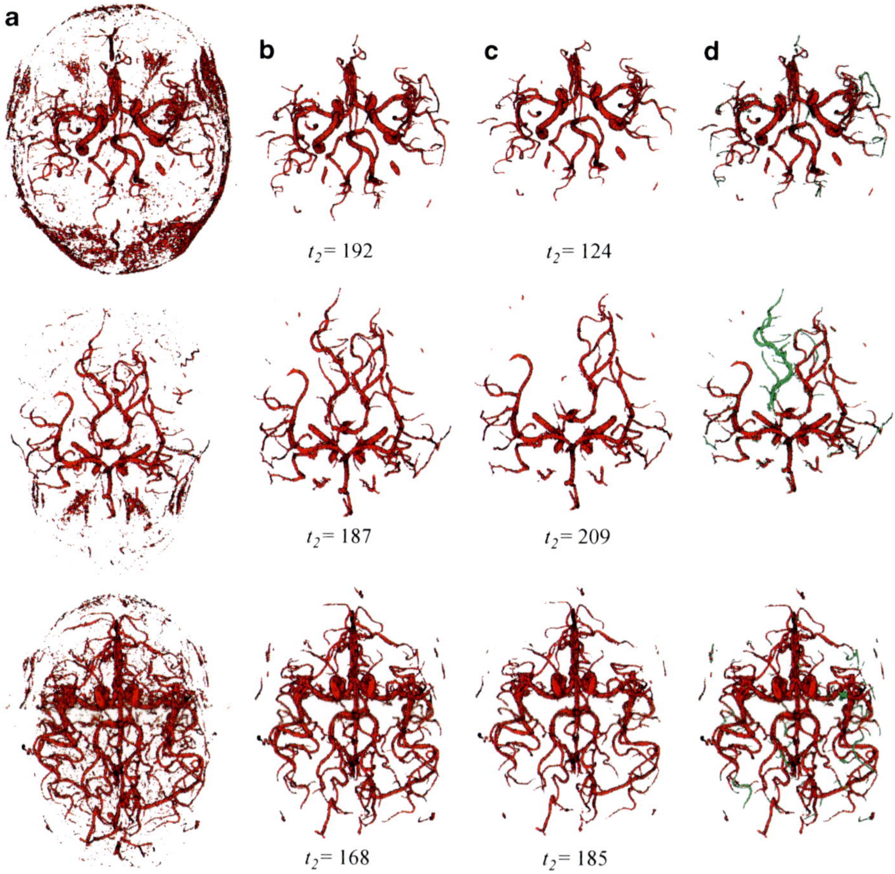

Fig. 14.5 Each *row* relates to one patient: our segmentation before (**a**) and after (**b**) noise and small fat voxels are eliminated with the connectivity filter, the Wilson–Noble's segmentation (**c**) after the connectivity filter, and the differences (**d**) between both the approaches highlighted in *green*

our approach. In the opposite cases, such as in the bottom row of Fig. 14.6, the Chung–Noble's thresholding adds some fat tissues to the vascular trees, whereas our approach correctly separates the former.

14.3.2 *Validating the Segmentation Accuracy with Special Phantoms*

It is very difficult to get accurate "ground truth" data to evaluate the segmentation performance by manually segmenting complete vasculatures. Although qualitative visual analysis by expert radiologists confirms the advantages of our approach, its

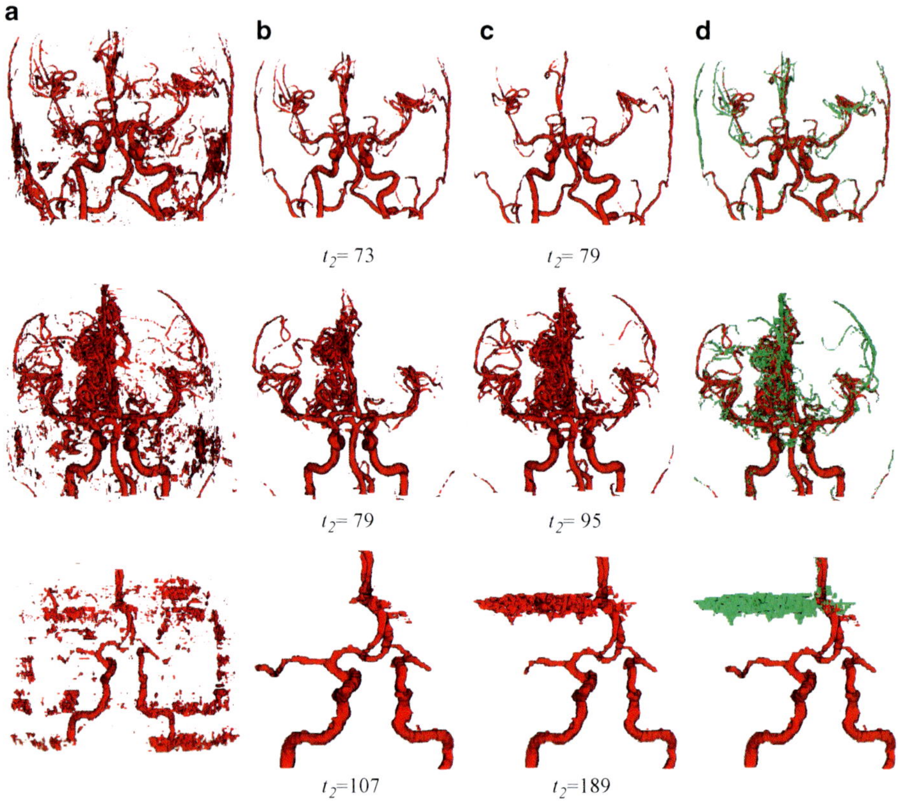

Fig. 14.6 Each *row* relates to one patient: our segmentation before (**a**) and after (**b**) noise and small fat voxels are eliminated with the connectivity filter, the Chung–Noble's segmentation (**c**) after the connectivity filter, and the differences (**d**) between both the approaches highlighted in *green*

quantitative validation is of prime importance. Thus, we have constructed three wooden phantoms in Fig. 14.7 to imitate geometric features of blood vessels typical for any vascular system including different sizes, bifurcations, and zero and high curvature. Each set of 2D slices obtained by scanning the phantoms with a CT scanner is manually segmented to produce "ground truth" region maps. Then, synthetic TOF- or PC-MRA signals are generated with inverse mapping methods according to their marginal probability distributions $p(q\,|\,3)$ ("blood vessels") and mixed $p(q\,|\,1)$, $p(q\,|\,2)$ ("background") in Fig.14.3b, d, respectively. The resulting gray level distributions for each slice are similar to those in Fig. 14.1b, d.

Figure 14.7 compares the results of our, the Wilson–Noble's, and the Chung–Noble's segmentation, the errors being in terms of the numbers of wrong (i.e., missed or extra) voxels relative to the total voxels number in the manually segmented 3D phantoms. In total, our approach produces 0.18–1.34% erroneous voxels compared to 3.97–9.52% for the Wilson–Noble's approach on the synthetic TOFMRA data and 0.14–0.79% of erroneous voxels compared to 2.12–4.01% for the Chung–Noble's

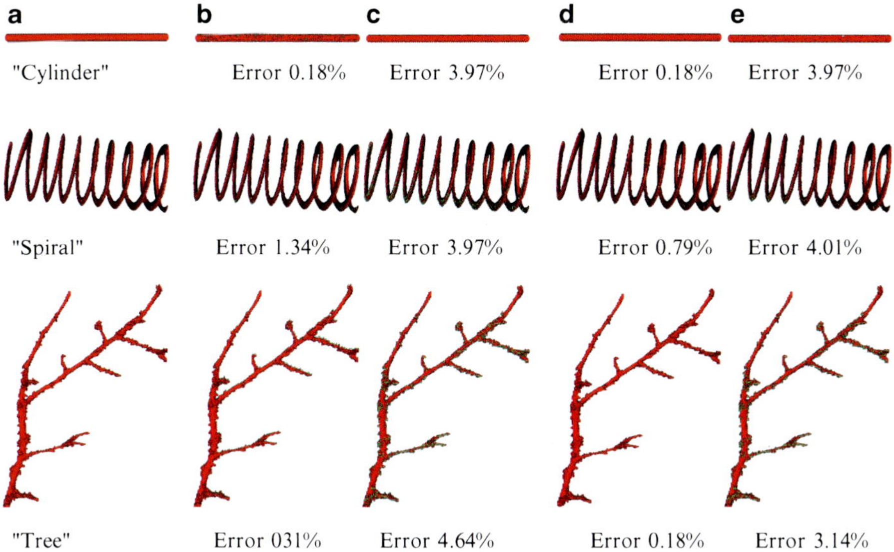

Fig. 14.7 True 3D geometrical phantoms; our (**a**) and the Wilson–Noble's (**b**) segmentation of their synthetic TOF-MRA 3D images, and our (**c**) and Chung–Noble's (**d**) segmentation of their synthetic PC-MRA images

approach on the synthetic PC-MRA data. The error constituents per each 2D slice of the three phantoms for all the approaches and data types are plotted in Fig. 14.8.

Table 14.1 combines the error statistics for all the 440 synthetic TOF- or PC-MRA slices in these three phantoms segmented with our, the Wilson–Noble's or Chung–Noble's, and three other segmentation algorithms.

14.4 Conclusion

These and other experiments confirm high accuracy of the proposed LCDG-based extraction of blood vessels from the TOF- and PC-MRA images. Our present implementation on a single 2.4 GHZ Pentium 4 CPU with 512 MB RAM using C++ programming language takes about 5 s for segmenting one TOF-MRA 3D data set with 93 2D slices of size 512×512 pixels each and 3 s for one PC-MRA 3D data set with 123 2D slices of size 256×256 pixels each. The proposed segmentation is not limited to only MRA but suitable also for computer tomographic angiography (CTA). The algorithm's code, sample data, and segmentation results for the TOF-MRA, PC-MRA, and CTA images will be provided on our web site.[2]

[2] http://louisville.edu/speed/bioengineering/faculty/bioengineering-full/dr-aymanel-baz/elbazlab.html.

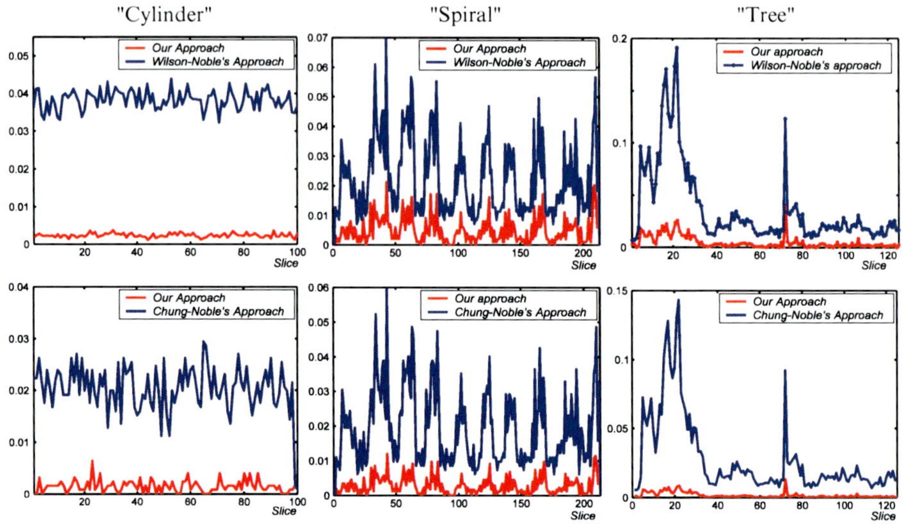

Fig. 14.8 Total errors per slice in each 3D geometrical phantom for our and the Wilson–Noble's segmentation (*top*) and for our and the Chung–Noble's segmentation (*bottom*)

Table 14.1 The minimum ε_n, maximum ε_x, mean $\bar{\varepsilon}$ segmentation error, and the standard deviation σ of errors on the TOF-MRA and PC-MRA phantoms for our approach (OA) and the Wilson–Noble's (WN) or Chung–Noble's (CN) one, respectively, as well as for the other algorithms using the iterative thresholding (IT) [37], the gradient-based deformable model (DMG) [38], and the deformable model based on the gradient vector flow (GVF) [39]

	TOF-MRA phantoms					PC-MRA phantoms				
	OA	WN	IT	DMG	GVF	OA	WN	IT	DMG	GVF
ε_n,%	0.09	0.110	4.81	10.1	2.45	0.02	0.08	3.71	9.80	1.96
ε_x,%	2.10	12.1	33.1	21.8	13.6	1.25	7.90	29.1	20.8	12.1
$\bar{\varepsilon}$,%	0.61	6.20	18.8	11.9	5.96	0.37	2.90	10.9	9.80	3.12
σ,%	0.93	7.40	8.41	3.79	2.79	0.62	4.30	6.22	2.10	2.06

14.5 Appendix: EM-Based Precise LCDG-Approximation of a Probability Distribution

14.5.1 Sequential EM-Based Initialization

The initial LCDG model closely approximating a given marginal gray level distribution F is built using the conventional EM algorithm [21, 35, 36] adapted to the DGs. The approximation involves the following steps:

1. The distribution F is approximated with a mixture P_K of K positive DGs relating each to a dominant mode.
2. Deviations between F and P_K are approximated with the alternating "subordinate" components of the LCDG as follows.

   (a) The positive and the negative deviations are separated and scaled up to form two seemingly "probability distributions" D^{p} and D^{n}.
   (b) The same conventional EM algorithm is used iteratively to find a subordinate mixture of positive or negative DGs that approximates best D^{p} or D^{n} respectively (i.e., the sizes $C_{\mathrm{p}} - K$ and C_{n} of the mixtures are found by minimizing sequentially the total absolute error between each "distribution" D^{p} or D^{n} and its mixture model by the number of the components).
   (c) The obtained positive and negative subordinate mixtures are scaled down and then added to the dominant mixture yielding the initial LCDG model of the size $C = C_{\mathrm{p}} + C_{\mathrm{n}}$.
   (d) The resulting initial LCDG has K dominant weights, say, $w_{\mathrm{p},1}, \ldots, w_{\mathrm{p},K}$ such that $\sum_{r=1}^{K} w_{\mathrm{p},r} = 1$, and a number of subordinate weights of smaller values such that $\sum_{r=K+1}^{C_{\mathrm{p}}} w_{\mathrm{p},r} - \sum_{l=1}^{C_{\mathrm{n}}} w_{\mathrm{n},l} = 0$.

14.5.2 *Modified EM Algorithm for Refining LCDGs*

The initial LCDG is refined by approaching the local maximum of the log-likelihood in (14.3) with the EM process adapting that in [32] to the DGs. The latter extends in turn the conventional EM-process in [35] onto the alternating components.

Let

$$p_{w,\Theta}^{[m]}(q) = \sum_{r=1}^{C_{\mathrm{p}}} w_{\mathrm{p},r}^{[m]} \psi(q \mid \theta_{\mathrm{p},r}^{[m]}) - \sum_{l=1}^{C_{\mathrm{n}}} w_{\mathrm{n},l}^{[m]} \psi(q \mid \theta_{\mathrm{n},l}^{[m]})$$

denote the current LCDG at iteration m. Relative contributions of each signal $q \in Q$ to each positive and negative DG at iteration m are specified by the respective conditional weights

$$\pi_{\mathrm{p}}^{[m]}(r \mid q) = \frac{w_{\mathrm{p},r}^{[m]} \psi(q \mid \theta_{\mathrm{p},r}^{[m]})}{p_{w,\Theta}^{[m]}(q)}; \quad \pi_{\mathrm{n}}^{[m]}(l \mid q) = \frac{w_{\mathrm{n},l}^{[m]} \psi(q \mid \theta_{\mathrm{n},l}^{[m]})}{p_{w,\Theta}^{[m]}(q)}, \tag{14.4}$$

such that the following constraints hold:

$$\sum_{r=1}^{C_{\mathrm{p}}} \pi_{\mathrm{p}}^{[m]}(r \mid q) - \sum_{l=1}^{C_{\mathrm{n}}} \pi_{\mathrm{n}}^{[m]}(l \mid q) = 1; \quad q = 1, \ldots, Q-1. \tag{14.5}$$

The following two steps iterate until the log-likelihood is increasing and its changes become small:

$E-step^{[m]}$: Find the weights of (14.4) under the fixed parameters $w^{[m-1]}$, $\Theta^{[m-1]}$ from the previous iteration $m-1$, and

$M-step^{[m]}$: Find conditional MLEs $w^{[m]}$, $\Theta^{[m]}$ by maximizing $L(w,\Theta)$ under the fixed weights of (14.4).

Considerations closely similar to those in [21, 35, 36] show this process converges to a local log-likelihood maximum. The further evidence in [32] demonstrates it is actually a block relaxation MM-process (in a very general way, this is also shown in [36]). Let the log-likelihood of (14.3) be rewritten in the equivalent form with the constraints of (14.5) as unit factors:

$$L(w^{[m]},\Theta^{[m]}) = \sum_{q=0}^{Q} f(q)\left[\sum_{r=1}^{C_\mathrm{p}} \pi_\mathrm{p}^{[m]}(r \mid q)\log p^{[m]}(q) - \sum_{l=1}^{C_\mathrm{n}} \pi_\mathrm{n}^{[m]}(l \mid q)\log p^{[m]}(q)\right]. \quad (14.6)$$

Let the terms $\log p^{[m]}(q)$ in the first and second brackets be replaced with the equal terms

$$\log w^{[m]}_{\mathrm{p},r} + \log \psi(q \mid \theta^{[m]}_{\mathrm{p},r}) - \pi_\mathrm{p}^{[m]}(r \mid q)$$

and

$$\log w^{[m]}_{\mathrm{n},l} + \log \psi(q \mid \theta^{[m]}_{\mathrm{n},l}) - \pi_\mathrm{n}^{[m]}(l \mid q)$$

respectively, which follow from (14.4). At the E-step, the conditional Lagrange maximization of the log-likelihood of (14.6) under the Q restrictions of (14.5) results just in the weights $\pi_\mathrm{p}^{[m+1]}(r \mid q)$ and $\pi_\mathrm{n}^{[m+1]}(l \mid q)$ of (14.4) for all $r = 1,\ldots,C_\mathrm{p}$;, $l = 1,\ldots,C_\mathrm{n}$; and $q \in Q$. At the M-step, the DG weights

$$w^{[m+1]}_{\mathrm{p},r} = \sum_{q \in Q} f(q)\pi_\mathrm{p}^{[m+1]}(r \mid q)$$

and

$$w^{[m+1]}_{\mathrm{n},l} = \sum_{q \in Q} f(q)\,\pi_n^{[m+1]}(l \mid q)$$

follow from the conditional Lagrange maximization of the log-likelihood in (14.6) under the restriction of (14.2) and the fixed conditional weights of (14.4). Under the latter, the conventional MLEs of the parameters of each DG stem from maximizing the log-likelihood after each difference of the cumulative Gaussians is replaced with its close approximation with the Gaussian density (below "c" stands for "p" or "n", respectively):

$$\mu_{c,r}^{[m+1]} = \frac{1}{w_{c,r}^{[m+1]}} \sum_{q \in Q} q f(q) \pi_c^{[m+1]}(r \mid q),$$

$$(\sigma_{c,r}^{[m+1]})^2 = \frac{1}{w_{c,r}^{[m+1]}} \sum_{q \in Q} (q - \mu_{c,r}^{[m+1]})^2 f(q) \pi_c^{[m+1]}(r \mid q).$$

This modified EM-algorithm is valid until the weights w are strictly positive. The iterations should be terminated when the log-likelihood of (14.3) almost does not change or begins to decrease due to accumulation of rounding errors.

The final mixed LCDG-model $p_C(q)$ is partitioned into the K LCDG-submodels $P_{[k]} = [p(q \mid k) : q \in Q]$, one per class $k = 1, \ldots, K$, by associating the subordinate DGs with the dominant terms so that the misclassification rate is minimal.

References

1. Y. Sato, S. Nakajima, N. Shiraga, H. Atsumi, and S. Yoshida, "Three-dimensional multi-scale line filter for segmentation and visualization of curvilinear structures in medical images," Med Image Anal, vol. 2, No. 2, pp.143–168, 1998.
2. K. Krissian, G. Malandain, and N. Ayache, Model based detection of tubular structures in 3D images, INRIA Technical Report 373, INRIA, 1999.
3. K. Krissian, G. Malandain, and N. Ayache, "Directional anisotropic diffusion applied to segmentation of vessels in 3D images," Proceedings of the International Conference on Scale-Space Theory in Computer Vision, Utrecht, The Netherlands (Lect. Notes Comp. Sci.), Springer, 1997, pp. 345–348.
4. K. Krissian, G. Malandain, N. Ayache, R. Vaillant, and Y. Trousset, "Model based multiscale detection of 3D vessels," Proceedings of the IEEE International Conference on Computer Vision Pattern Recognition, Santa Barbara, CA, USA, 1998, pp. 722–727.
5. A. F. Frangi, W. J. Niessen, K. L. Vincken, and M. A. Viergever, "Multiscale vessel enhancement filtering," Proceedings of the International Conference on Medical Image Computing Computer-Assisted Intervention (MICCAI'98), Cambridge, MA, USA (Lect. Notes Comp. Sci. 1496), Springer, 1998, pp. 130–137.
6. C. Lorenz, I.-C. Carlsen, T. M. Buzug, C. Fassnacht, and J. Weese, "A multi-scale line filter with automatic scale selection based on the Hessian matrix for medical image segmentation," Proceedings of the International Conference on Scale-Space Theories in Computer Vision, Chalana, Kim (Lect. Notes Comp. Sci. 1252), Springer, 1997, pp. 152–163.
7. P. Perona and J. Malik, "Scale-space and edge detection using anisotropic diffusion," IEEE Trans Pattern Anal Mach Intell, vol. 12, pp. 629–639, 1990.
8. F. Catte, T. Coll, P.L. Lions, and J. M. Morel, "Image selective smoothing and edge detection by nonlinear diffusion," SIAM J Numer Anal, vol. 29, pp. 182–193, 1992.
9. S. Aylward, S.M. Pizer, E. Bullitt, and D. Eberly, "Intensity ridge and widths for 3D object segmentation and description," Proceedings of the Workshop on Mathematical Methods in Biomedical Image Analysis (MMBIA '96), IEEE Computer Society, Washington, DC, USA, 1996, pp. 131–138.
10. E. Bullitt, A. Liu, S. Aylward, and S. M. Pizer, "Reconstruction of the intracerebral vasculature from MRA and a pair of projection views," Proceedings of the International Conference on Information Processing Medical Imaging, Poultney, VT, USA (Lect. Notes Comp. Sci. 1230), Springer, 1997, pp. 537–542.

11. V. Prinet, O. Monga, C. Ge, S. L. Xie, and S. D. MA, "Thin network extraction in 3D images: application to medical angiograms", Proceedings of the IAPR International Conference on Pattern Recognition, Vienna, Austria, August 1996, vol. 3, IEEE CS Press, 1996, pp. 386–390.
12. V. Caselles, R. Kimmel, and G. Sapiro, "Geodesic active contours," Int J Comput Vis, vol. 22, pp. 61–79, 1997.
13. T. McInerney and D. Terzopoulos, "Medical image segmentation using topologically adaptable surface," Proceedings of the LNCS Series 1205, (CVRMED-MRCAS'97), 1997, pp. 23–32.
14. L. M. Lorigo, O. D. Faugeras, W. E. L. Grimson, and R. Keriven, "Curves: curve evolution for vessel segmentation," Med Image Anal, vol. 5, pp. 195–206, 2001.
15. O. Wink, W. J. Niessen, and M. A. Viergever, "Fast delineation and visualization of vessels in 3-D angiographic images," IEEE Trans Med Imaging, vol. 19, pp. 337–346, 2000.
16. T. Deschamps and L. D. Cohen, "Fast extraction of tubular and tree 3D surfaces with front propoagation methods," Proceedings of the 16th IAPR International Conference on Pattern Recognition, Quebec City, Canada, August 2002, IEEE CS Press, vol.1, 2002, pp. 731–734.
17. R. Manniesing and W. Niessen, "Local speed functions in level set based vessel segmentation," Proceedings of the International Conference on Medical Image Computing Computer-Assisted Intervention (MICCAI'04), Rennes, Saint-Malo, France, 2004, pp. 475–482.
18. M. Descotesux, L. Collins, and K. Siddiqi, "Geometric flows for segmenting vasulature in MRI: theory and validation," Proceedings of the International Conference on Medical Image Computing Computer-Assisted Intervention (MICCAI'04), Rennes, Saint-Malo, France, 2004, pp. 500–507.
19. D. L. Wilson and J. A. Noble, "An adaptive segmentation algorithm for time-of-flight MRA data," IEEE Trans Med Imaging, vol. 18, pp. 938–945, 1999.
20. A. C. S. Chung and J. A. Noble, "Statistical 3D vessel segmentation using a Rician distribution," Proceedings of the International Conference on Medical Image Computing Computer-Assisted Intervention (MICCAI'99), Cambridge, England, 1999, pp. 82–89.
21. A. Webb, Statistical pattern recognition, Wiley, New York, 2002.
22. D. Nain, A. Yezzi, and G. Turk,"Vessels segmentation using a shape driven flow," Proceedings of the International Conference on Medical Image Computing Computer-Assisted Intervention (MICCAI'04), Rennes, Saint-Malo, France, 2004, pp. 51–59.
23. R. Gan, A. C. S. Chung, W. C. K. Wong, and S. C. H. Yu, "Vascular segmentation in three-dimensional rotational angiography based on maximum intensity projections," Proceedings of the IEEE International Symposium on Biomedical Imaging (ISBI, 2004), Arlington, VA, USA, 2004, pp. 133–136.
24. P. Dl, C. Be, R. Ja, and A. Al, "Enhanced image detail using continuity in the MIP Z-buffer: applications to magnetic resonance angiography," J Magn Reson Imaging, vol. 11, pp. 378–388, 2000.
25. P. Reuze, J. L. Coatrieux, L. M. Luo, and J. L. Dillenseger, "A 3-D moment based approach to blood vessels detection and quantification in MRA," Technol Health Care, vol. 1, pp. 181–188, 2000.
26. M. Hernandez-Hoyos, M. Orkisz, J. P. Roux, and P. Douek, "Inertia based vessel axis extraction and stenosis quantification in 3D MRA images," in Proceedings Computer Assisted Radiology and Surgery (CARS), 1999, pp. 180–193.
27. B. Verdonck, I. Bloch, H. Maitre, D. Vandermeulen, and G. Marchal, "A new system for blood vessel segmentation and visualization in 3D MR and spiral CT angiography," in Proceedings Computer Assisted Radiology and Surgery (CARS), 1999, pp. 177–182.
28. P. E. Summers, A. H. Bhalerao, and D. J. Hawkes, "Multiresolution, model-based segmentation of MR angiograms," J Magn Reson Imaging, vol. 7, pp. 950–957, 1997.
29. Y. Masutani, T. Schiemann, K. H. Hhne, "Vascular shape segmentation and structure extraction using a shape a shape-based region growing model," Proceedings of the International Conference on Medical Image Computing Computer-Assisted Intervention (MICCAI'98), Massachusetts Institute of Technology, Cambridge MA, USA, 1998, pp. 1242–1249.
30. T. Chen and D. N. Metaxas, "Image segmentation based on the integration of Markov random fields and deformable models," Proceedings of the International Conference on Medical Image Computing Computer-Assisted Intervention (MICCAI'2000), Pittsburgh, PA, USA, 2000, pp. 256–265.

31. T. Chen and D. N. Metaxas, "Gibbs prior models, marching cubes, and deformable models: a hybrid framework for 3D medical image segmentation," Proceedings of the International Conference on Medical Image Computing Computer-Assisted Intervention (MICCAI'03), Montreal, Quebec, Canada, vol. 2, 2003, pp. 703–710.
32. G. Gimel'farb, A. A. Farag, and A. El-Baz, "Expectation-maximization for a linear combination of Gaussians," Proceedings of the 17th IAPR International Conference on Pattern Recognition, Cambridge, UK, 23–26 August 2004, IEEE CS Press, vol. 3, 2004, pp. 422–425.
33. M. Sabry, Ch. B. Sites, A. A. Farag, S. Hushek, and T. Moriarty, "A fast automatic method for 3D volume segmentation of the human cerebrovascular," Proceedings of the 13th International Conference on Computer Assisted Radiology Surgery (CARS'02), Paris, France, 26–29 June 2002, 2002, pp. 382–387.
34. J. W. Lamperti, Probability, Wiley, New York, 1996.
35. M. I. Schlesinger and V. Hlavac, Ten lectures on statistical and structural pattern recognition, Kluwer Academic, Dordrecht, 2002.
36. T. Hastie, R. Tibshirani, and J. Friedman, The elements of statistical learning, Springer, New York, 2001.
37. S. Hu and E. A. Hoffman, "Automatic lung segmentation for accurate quantization of volumetric X-ray CT images," IEEE Trans Med Imaging, vol. 20, pp. 490–498, 2001.
38. M. Kass, A. Witkin, and D. Terzopoulos, "Snakes: active contour models," Int J Comput Vis, vol. 1, pp. 321–331, 1987.
39. C. Xu and J. L. Prince, "Snakes, shapes, and gbooktitleadient vector flow," IEEE Trans Image Process, vol. 7, pp. 359–369, 1998.

Biographies

Ayman S. El-Baz received the BSc and MS degrees in electrical engineering from Mansoura University, Egypt, in 1997 and 2000, respectively and the PhD degree in electrical engineering from University of Louisville, Louisville, KY. He joined the Bioengineering department, University of Louisville, in August 2006. His current research includes developing new computer-assisted diagnosis systems, image modeling, image segmentation, 2D and 3D registration, visualization, and surgical simulation including finite element analysis, about which he has authored or coauthored more than 70 technical articles.

Georgy Gimel'farb graduated from Kiev Polytechnic Institute, Kiev, Ukraine and received the PhD degree in engineering cybernetics from the Institute of Cybernetics, Academy of Sciences of the Ukraine, and the DSc (Eng) degree in control in engineering from the Higher Certifying Commission of the USSR, Moscow, Russia. After a long time with the Institute of Cybernetics, he joined the

University of Auckland, Auckland, New Zealand, in July 1997, where he is currently an Associate Professor of computer science. His research is focused on image analysis, computer vision, and statistical pattern recognition, with main contributions in the areas of computational stereovision and probabilistic texture modeling and analysis.

Ahmed A. Elnakib was born on 21, January, 1982 in Mansoursa, Egypt. Mr. Elnakib received his BSc and MSc degrees in Electronics and Communication Engineering from Mansoura University. Mr. Elnakib joined the BioImaging Laboratory as a research assistant in September 2008. His current research includes developing new computer-assisted diagnosis systems, image modeling, image segmentation, 2D and 3D registration, and shape analysis of detected lung nodules. In 2003, Mr. Elnakib was a recipient of the Dakahlia Governorate award for outstanding students.

Dr. Falk is the chief of Medical Imaging at Jewish Hospital in Louisville, Kentucky. He completed his training at the Medical College of Wisconsin in 1987. He is a fellowship-trained neuroradiologist with additional interest and expertise in body imaging, orthopedic imaging, and advanced 3D image manipulation. Dr. Falk is the former president of the Jewish Hospital medical staff and is on the Board of Trustees. He also serves on the Board of Directors of The Physicians Incorporated, a not-for-profit, independent practice association, which has approximately 2,500 members in the Louisville area.

Dr. Mohamed Abou El-Ghar was born on 14, August, 1966 in the Nile Delta town of Mansoura-Egypt. He received a MBBch in Medicine from Mansoura University in 1989, followed by 4 years as a radiodiagnosis resident. In 1997, he received a MS degree in radiodiagnosis from Tanta university, and in 2002 he received M.D. degree from Al-Azhar university. Now, he is a consultant of radiodiagnosis in Mansoura urology and nephrology center.

Chapter 15
Noninvasive Imaging for Risk Prediction in Carotid Atherosclerotic Disease

Holger Poppert, R. Feurer, L. Esposito, T. Saam, and Dirk Sander

Abstract Carotid artery stenosis is a leading, treatable risk factor for stroke. Until recently, risk prediction has, for the main part, been based on the degree of lumen narrowing which does not reflect our actual insight into the etiology of arterio-embolic stroke. The risk of carotid artery stenosis is in fact defined by the texture and morphology of the atherosclerotic lesion. Both B-mode ultrasound and MRI can depict these features and have considerable potential for detection of vulnerable plaques.

Keywords Atherosclerotic plaque • Risk prediction • Ultrasound imaging • MRI • Doppler imaging

15.1 Introduction

About two thirds of ischemic strokes are of embolic origin, namely, arterio-embolic and cardiogenic. The majority of arterio-embolic strokes arise from atherosclerotic stenosis of the large brain-supplying vessels, mainly due to stenosis of the internal carotid artery (ICA). These strokes would be potentially avoidable by preventive endarterectomy or angioplasty.

However, according to data from the Framingham study, the prevalence of a stenosis of at least 50% is 5–7% in women and 7–10% in men aged older than 65 years. In the Asymptomatic Carotid Surgery Trial (ACST), the to-date largest interventional study in asymptomatic ICA-stenosis, 3,120 patients with carotid artery stenosis of at least 60% were randomized to either operational or nonsurgical

H. Poppert (✉)
Department of Neurology, Klinikum Rechts der Isar, Technische Universitaet Muenchen, Ismaningerstr. 22, 81675, Muenchen, Germany
e-mail: poppert@neurovasc.de

Jasjit S. Suri et al. (eds.), *Atherosclerosis Disease Management*, DOI 10.1007/978-1-4419-7222-4_15, © Springer Science+Business Media, LLC 2011 treatment [1]. The annual risk of stroke due to the recognized stenosis was 0.8% in the group with nonsurgical treatment, whereas the risk of stroke from any cause was twice as high. Thus, most subjects would probably never suffer from stroke due to carotid artery stenosis. The periprocedural risk of stroke and death in the group with surgical treatment was 2.8%. During a period of 5 years, there was a slight but significant benefit due to surgical treatment regarding the all-inclusive rate of strokes. These figures clarify the need for noninvasive diagnostic parameters for individual stratification of risk to allow for a risk- and resource-efficient treatment.

15.2 Carotid Plaque

The progression of atherosclerosis is a complex and lifelong process, driven by inflammation and the subsequent immune response producing a continuous effect on the structural morphology and stability of the plaque. Accordingly, researchers and clinicians increasingly evaluate morphological and functional characteristics of the vessel wall.

A number of characteristic features that increase the risk of rupture can be identified and classified from histological studies of coronary and carotid plaques: a large lipid core; thinning of the fibrous cap; active inflammation with infiltration of the cap by macrophages and T cells releasing cytokines and proteinases, thus stimulating breakdown of cap collagen and smooth-muscle cell apoptosis; intraplaque hemorrhage and outward remodeling; and finally, erosion or fissures of the plaque surface and thrombus superimposition [7–9]. Novel applications of imaging techniques enable the investigation of the plaque material in vivo with increasing precision and reliability.

15.3 B-Mode Ultrasound

B-mode ultrasound enables the examiner to directly image arterial walls and atherosclerotic plaques. The technique is noninvasive, safe, economical, and widely available.

Echogenicity and echotexture (Fig. 15.1) have been reported to correlate with plaque composition and potential disposition to rupture, although histological evaluation showed only poor concordance in some studies [2–5]. Echolucent lesions are thought to consist of a high proportion of lipid and possibly hemorrhage, whereas echo-opacity is associated with calcification and fibrous tissue. Surface irregularities and inhomogeneity have been proposed as features of unstable plaques, though this is difficult to quantify objectively. Despite numerous studies attempting to identify the pathological components of atheroma ultrasonographically or to correlate the appearance of carotid atheroma on ultrasound with symptoms or brain imaging findings, no clear message has yet emerged from this work. The separate detection of distinctive features such as hemorrhage, necrosis, ulceration, and other surface characteristics has been suggested to be unreliable [6, 7].

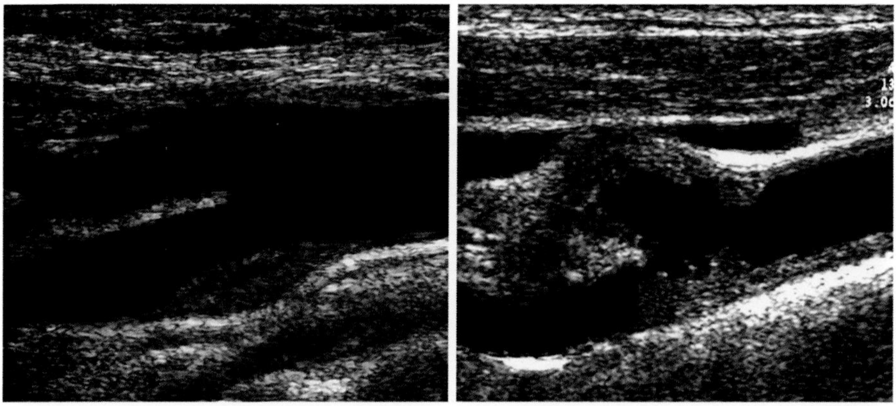

Fig. 15.1 Plaques with different morphologies. *Left*: homogeneous, regular, intact surface. *Right*: inhomogeneous irregular, ulcerated surface

A number of longitudinal studies have investigated the correlation of these sonographic features with the incidence of stroke, with initially promising but conflicting results [1, 8–11]. In particular, the ACST, which assessed plaque echogenicity in 1,543 asymptomatic patients, did not reveal any association between plaque echogenicity and 5-year risk of ipsilateral stroke [1].

In conclusion, although B-mode ultrasound has considerable potential for detection of vulnerable plaques, it is currently limited by the fact that results are highly observer-dependent. Techniques to overcome this limitation by improvement of B-mode image quality and development of dedicated software for image normalization and interpretation are under investigation. A further promising approach consists in the visualization of intraplaque neovascularization by microbubble-based ultrasound contrast medium [12]. In a retrospective study with 147 subjects, Staub et al. showed that the presence and degree of neovascularization was directly associated with cardiovascular risk factors and cardiovascular disease and events [13]. Subjects with intraplaque neovascularization had a hazard ratio (HR) of 4.0 (95% CI 1.3–12.6).

15.4 Intima–Media Thickness

Ultrasound measurements of intima–media thickness (IMT) assess the extent and severity of atherosclerosis. IMT is defined as the distance between the lumen–intima interface and the media–adventitia interface and can be measured at several sites along the vessel wall (Fig. 15.2, Normal (left) and pathologically increased (right) common carotid IMT Fig. 15.3). Generally, IMT measurements >1 mm are considered abnormal. Today, carotid IMT, measured with high-resolution B-mode ultrasound, is the standard for noninvasive surrogate measurements of carotid atherosclerosis (Fig. 15.4) [14].

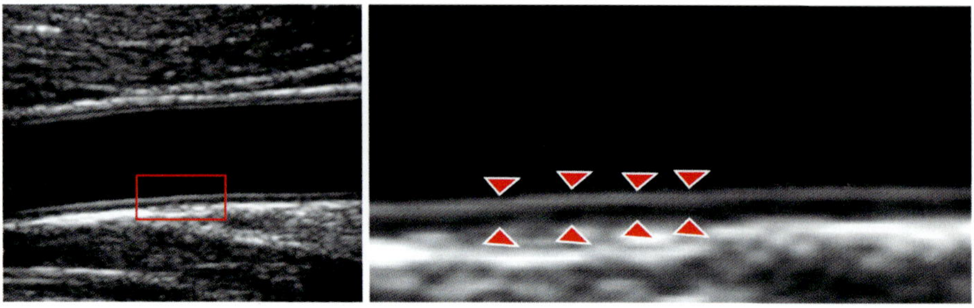

Fig. 15.2 Measurement of common carotid artery IMT according to the leading edge method

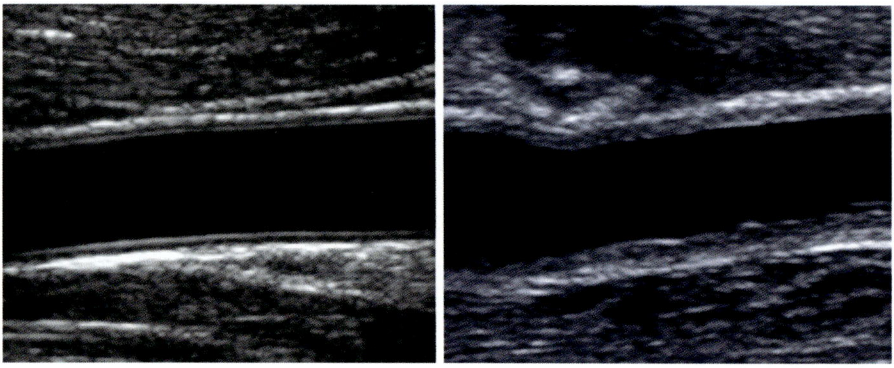

Fig. 15.3 IMT: example. Normal (*left*) and pathologically increased (*right*) common carotid IMT

There is a direct relation between the number of risk factors and IMT, as well as between carotid IMT and clinical cardiovascular disease [15]. Reduction of IMT progression and regression has been shown to improve patient cardiovascular outcomes [16]. IMT is used as an exposure variable to predict cardiovascular disease, as an outcome variable to study determinants of atherosclerosis, and to establish the effect of risk factor-modifying therapy. The use of carotid IMT is beneficial in predicting elevated risk in patients without documented coronary artery disease (CAD) and in guiding intensity of treatment if the IMT continues to progress. Increases in the IMT of the carotid artery have been found to be directly associated with elevated risk of stroke in older patients without a history of cardiovascular disease. In a systematic meta-analysis, data from 37,197 subjects who were followed up for a mean of 5.5 years were reviewed. The analysis provides data on the use of carotid IMT to predict myocardial infarction (MI) and stroke in the general population. For an absolute carotid IMT difference of 0.1 mm, the future risk of MI increases by 10–15%, and the stroke risk increases by 13–18% [16]. The risk for both end points decreased with the number of adjustments for risk factors.

Since the publication of this meta-analysis, several new studies have analyzed the impact of IMT for vascular risk prediction, particularly focusing on stroke risk. An updated overview is given in Table 15.1. The current evidence, however, is insufficient for a general recommendation to use IMT progression as a surrogate for vascular risk

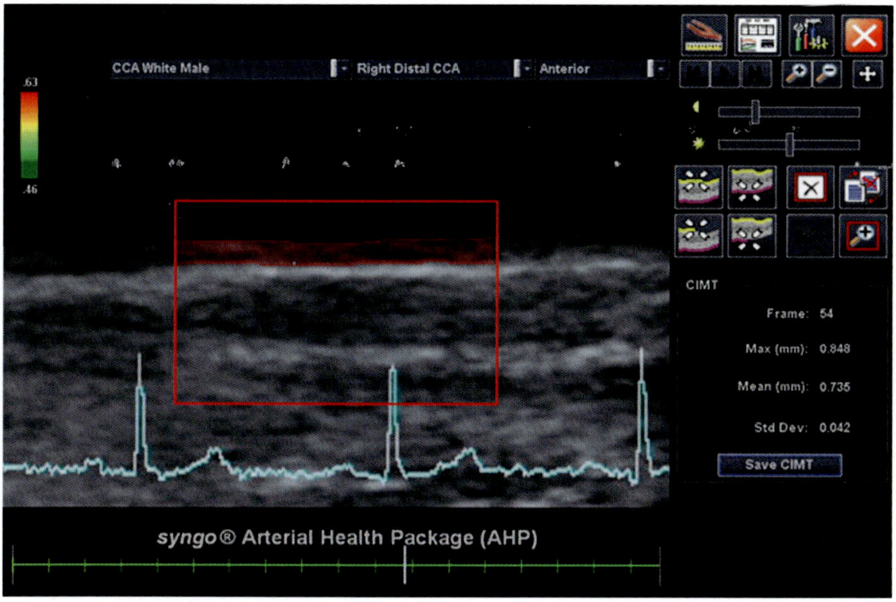

Fig. 15.4 Computer-assisted automatic measurement of common carotid artery IMT. Image analysis system built into the duplex machine

in pathophysiological studies or clinical trials. Two issues have to be addressed before this recommendation can be given: First, quantitative estimates of the association between IMT progression and event risk for different populations and interventions are required. Second, the link between the treatment effects on IMT change and those on clinical end points have to be shown in more than one trial and for more than one type of intervention. Therefore, principal investigators of large IMT studies are cooperating to generate a database containing individual participant data of carotid ultrasound, risk factors, and vascular end points to investigate whether individual IMT progression can be employed as a surrogate for vascular risk (individual progression of carotid IMT as a surrogate for vascular risk [PROG-IMT] project) [17].

Vascular events are rare in young individuals, which makes IMT particularly attractive as an end point in epidemiological and treatment studies in young populations. The nonlinear relationship has major implications for the design of IMT studies in young populations, as it limits the transferability of the results from older samples.

15.5 Degree of Stenosis

Until now, risk prediction has, for the main part, been based on the degree of lumen narrowing. Two different measurement methods have been established: The so-called local degree of stenosis is calculated from the relation between the narrowest traversed residual lumen and the assumed original lumen at the identical position. Whereas both the plaque as well as the vessel wall can be shown with

Table 15.1 Association between IMT and cerebrovascular events in prospective population-based studies

Study	Participants (number)	IMT	Follow-up	No. of events	Risk
Prati et al. [36]	1,348	CCA	Mean: 12.7 years	115 (ischemic stroke, TIA, vascular death)	IMT > 1 mm vs. <1 mm: 10.4% (6.4–17.1) vs. 5.6% (3.2–10.1) HR: 5.6 (3.5–11)
Folsom et al. [15]	6,698	Composite CCA and ICA	Median: 3.9 years	59 (Stroke)	HR per 1 mm SD change in IMT:1.4 (1.2–1.8)
Chien et al. [37]	2,190	CCA	Median: 10.5 years	94 (Stroke)	HR per 1 mm SD change in IMT:1.47 (1.28–1.69)

CCA common carotid artery; *ICA* internal carotid artery; *HR* hazard ratio; *SD* standard deviation. All data adjusted for several conventional risk factors duplex ultrasonography, the assumed original vascular diameter has to be estimated by interpolation between pre- and poststenotic lumen diameters. The so-called distal degree of stenosis is calculated from the relation between the narrowest traversed residual lumen and the lumen distal of the stenosis.

The most important studies so far comparing thromboendarterectomy with nonsurgical treatment have used different approaches: In the European Carotid Surgery Trial (ECST), the local degree of stenosis was used (Fig. 15.5), whereas the North American Symptomatic Carotid Endarterectomy Trial (NASCET) used the distal degree of stenosis [18, 19]. The measured degrees of stenosis show an approximately linear relationship: a local degree of stenosis of 80% corresponds to a distal degree of stenosis of 70%. Mainly, the NASCET classification is used internationally; for pooled analyses, the ECST data were recalculated to NASCET data. In the following lines, the percentage values refer to the NASCET classification, if not otherwise specified.

It should additionally be appreciated that the common two-dimensional display of the normally eccentrically configured stenosis in the longitudinal diameter results in considerable inaccuracies from the use of an arbitrary insonation angle. The angle-corrected systolic maximal velocity is considered as reliable parameter for a degree of stenosis over 60%.

15.6 Origin of Stroke

The majority of strokes related to stenosis of the carotid artery are thromboembolic in origin; hemodynamic strokes due to preceding atherosclerotic stenosis or occlusion can be considered exceptions. Indeed, large multicenter studies show a clear

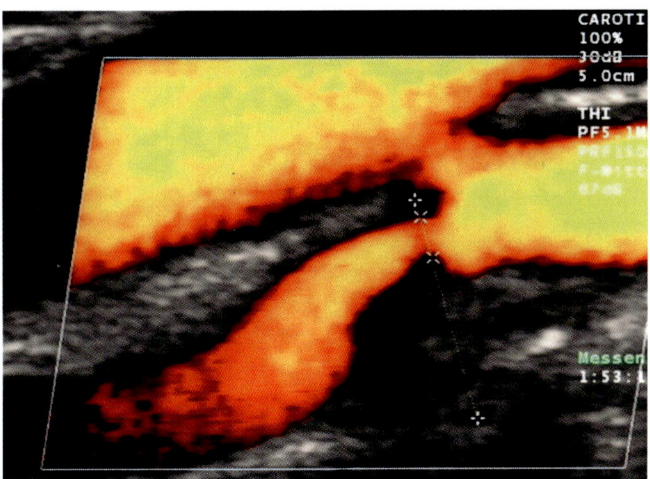

Fig. 15.5 High-grade internal carotid stenosis. Color-coded duplex sonography

statistical correlation between the degree of lumen narrowing of a recently symptomatic carotid artery stenosis and the risk of recurrent stroke. But both the progress in degree of stenosis as well as the embolic event is a consequence of a progressive inflammatory process. The correlation between stroke risk and degree of carotid artery stenosis is, therefore, considered to be coincidental and is not a direct consequence of the reduced lumen. In the ACST study mentioned above, patients without clinical symptoms did not show any correlation between the degree of stenosis and the incidence of stroke.

Consequently, a stratification of risk that is based only on the residual vascular lumen is of low prognostic value and does not reflect our actual insight into the etiology of arterio-embolic stroke.

Finally, the underlying atherosclerosis has to be seen as a complex, lifelong process that is supported by inflammatory processes with continuously varying effects on the structural morphology and functional state of the vessel wall. The risk of carotid artery stenosis is defined by the texture and morphology of the athero-sclerotic lesion. Such features can be clearly verified by histology and can be classified according to established standards.

15.7 Morphology and Texture of Plaques

15.7.1 Histology

Well-established histological criteria for classification of atherosclerotic plaques considering texture and structure have been created by Stary et al. on behalf of the American Heart association (AHA) (Table 15.2) [20]. Due to frequently seen inhomogeneity of morphology and predominant consistence of a lesion over the longitudinal axis, various slices have to be considered for classification. The region that is most advanced or classified as clinically significant determines the overall classification.

15.7.2 Stages of Atherosclerosis

The early stages of atherosclerosis from the first decade of life onward are characterized by an increasing accumulation of lipids in the tunica intima. Initially, mainly foam cells are to be found (AHA classes I and II); later, there are also disseminated extracellular lipid depositions (AHA class III). In these stages, the development proceeds of MI increase by 10–15%, and the stroke risk increases by 13–18% [16], consistently and predictably in qualitative terms. Afterward, several characteristic lesion types and clinical syndromes can emerge. The "advanced lesion" (AHA class IV-V) is characterized by the extended deposition of lipids, cells, and components of the cellular matrix with structural breakdown of the vessel wall in addition to

Table 15.2 AHA plaque classification

Class	Properties
0	No intimal thickening
I	Isolated macrophage foam cells
II	Multiple foam cells layers; smooth muscle cells contain lipid droplets; fatty streaks may be visible by gross inspection
III	Preatheroma; isolated extracellular lipid pools added, increase of smooth muscle cells
IV	Atheroma; singular, confluent, clearly delineated extracellular lipid core, covered by a proteoglycan-containing layer, foam cells, and smooth muscle cells
V	(Originally also known as Va) fibroatheroma; fibrin-containing cap and possibly small calcifications
VIa	(Mostly Class IV or V) + surface rupture
VIb	(Mostly Class IV or V) + hemorrhage or hematoma
VIc	(Mostly Class IV or V) + thrombus
VIabc	Class IV or V + surface rupture + hemorrhage + thrombus
VII	(Originally also known as Vb) calcification predominates
VIII	(Originally also known as Vc) fibrous plaque; the intima is replaced by fibrous connective tissue, no lipid core

Adapted from ref. [20]

signs of prolonged processes of repair and alteration. In the course of this, the lipid core evolves, embedded by smooth-muscle cells and clearly definable by histology, with foam cells, monocytes, macrophages, and T cells, and with consecutive thickening of the vessel wall but initially with only slight or without any lumen narrowing. In immediate proximity of the lipid core are found single capillaries. The swelling of the intima is partially compensated by external expansion. According to definition, neither a defect in the surface nor apposition of a thrombus can be found. The lesion is confined to the intima.

The region between the lipid core and the vascular lumen contains abundant proteoglycans, foam cells, and a few fibers of collagen. Lesions of class IV type are, therefore, considered as exposed to fissuring. Next, the separating layer between lipid core and vascular lumen is converted by connective tissue into the fibrous cap. The increase in volume is no longer determined solely by the growth of the lipid core but also by additional deposition of freshly formed plaque matrix. Whereas histological changes of AHA classes I–IV concern the intima, smooth-muscle cells of the tunica media can also be reduced in lesions of AHA stage V. Media and tunica adventitia may contain deposits of lymphocytes, macrophages, and foam cells. The complex interaction of involved cells ultimately influences the development of the plaque and thereby the extent of progression and potential destabilization. This process is based on the balance of cellular migration and proliferation as well as construction and degradation of cellular matrix.

There is histological evidence for the infiltration of macrophages and T cells secreting cytokines and proteases into the cap. These cause, among other things, the degradation of collagen and the apoptosis of smooth-muscle cells.

Simultaneously, the ongoing accumulation of lipids, cells, and extracellular matrix promotes the buildup of the lipid-rich necrotic core, which is separated from the vascular lumen by the fibrotic cap. The maximally tolerable mechanical stress due to shear force and traction is defined by the thickness of the cap at its thinnest position.

A thin or fissured cap, a large lipid-rich necrotic core, and especially a disadvantageous ratio of these components are considered reliable features of increased plaque vulnerability. Finally, there is danger of erosion and rupture with the liberation of atheromatous material and apposition of thrombus. Evidence of such additional "complications" of AHA class VI is defined in special subgroups, namely, rupture of surface (AHA class VIa), plaque hemorrhage (AHA class VIb), apposition of thrombus (AHA class VIc), or all three features (AHA class VIabc). Usually, this development is based on a lesion classified as AHA IV or V. After the healing of the relevant attributes, the lesion can "regress" into a lesion of AHA class V. This "switch of classes" can occur repetitively with increasing progress of stenosis.

Indeed, the rupture of the surface commonly has its origin at the side of the lesion facing the vascular lumen, but there is also evidence that primary subendothelial hemorrhage can result in rupture of the lesion.

In advanced stages, hemorrhage into the plaque is found regularly – probably due to the rupture of finest fragile, newly formed vessels serving the plaque.

The neoformation of vessels is itself considered an indicator for substantial remodeling and indicates "per se" the existence of an inflammatory unstable plaque. Besides this, the liberated membranes of the erythrocytes themselves are abundant in cholesterol and further trigger atherogenesis. The previous fissuring of plaques with apposition of thrombus and healing through incorporation of the thrombus is hypothesized as another cause for intimal hematoma.

15.7.3 "Stable Plaque"

Through increasing calcification and fibrous remodeling, lesions of AHA classes IV or V can also develop directly into stable plaques with small risk of rupture. AHA class VII describes the calcified lesion; AHA class VIII is characterized by a largely or completely degraded lipid core with remodeling of the thickened intima into connective tissue.

Safe noninvasive differentiation of these plaques considered as stable (AHA classes VII and VIII) from the AHA classes IV–VI would be of great value in making individual decisions for appropriate therapy.

Some of the histologically verified features of unstable plaques are also promising candidates as parameters in noninvasive, decision-guiding imaging techniques. These include irregularity of the surface, plaque vascularization, plaque ulceration, thinning of the fibrous cap, infiltration of the fibrous cap with macrophages and T cells, intimal hemorrhage, and apposition of thrombus-facing vascular lumen.

Reliable, noninvasive in vivo detection of these features was impossible for a long time. Neither conventional angiography nor the predominantly used imaging of the vascular lumen by magnetic resonance imaging (MRI) or computed tomography permits the consideration of such attributes of unstable plaques. High-resolution multicontrast MRI-based tissue characterization represents the most promising modality on that score.

15.8 High-Resolution Magnetic Resonance Imaging

High-resolution multicontrast MRI is the first noninvasive imaging technique that allows characterization of the composition of human carotid atherosclerotic plaque, such as fibrous tissue, lipid and necrotic core, calcium, hemorrhage, thrombus, and the status of the fibrous cap. This was demonstrated for the first time in 1996 by Toussaint et al., using a clinical imager with routinely available pulse sequences in comparison with histological evaluation after surgical endarterectomy that was performed <4 days later [21]. These results were soon after confirmed by other research groups.

However, the establishment of MRI-based plaque imaging in research and clinical routine required a standardized multisequence imaging protocol and reproducible classification system for assessment of plaque composition. This was provided by Cai and colleagues in 2002 [22]. They created a widely accepted, detailed classification scheme for MRI-based plaque imaging, allowing categorization of carotid plaques into distinct lesion types based on the American Heart Association (AHA) histological criteria modified specifically for MRI use (Table 15.3) [22].

The classification is based on a standardized multicontrast imaging protocol including three-dimensional time-of-flight (TOF) MR angiography, T1-weighted (T1W), T2-weighted (T2W), and proton-density-weighted (PDW) studies with black blood imaging technique and ECG gating. Usually, after an initial coronal localizer sequence, TOF is performed and used to identify the location of the carotid bifurcation and the region of maximal stenosis on one or both sides. Axial images are acquired through the common carotid artery >10 mm below the carotid bifurcation to a point >10 mm distal to the extent of the stenosis identified on the time of flight sequence to ensure that the entire carotid plaque is imaged.

Table 15.3 Classification of carotid plaque lesion types according to the modified AHA criteria

Class	Histological classification	Modified MRI classification
I	Foam cells	Nearly normal wall thickness
II	Fatty streaks	
III	Preatheroma	Diffuse intimal thickening or small eccentric plaque without calcification
IV	Atheroma	Plaque with a lipid or necrotic core surrounded by fibrous tissue with possible calcification
V	Fibroatheroma	
VI	Surface defect and hemorrhage	Surface defect and hemorrhage
VII	Calcified plaque	Calcified plaque
VIII	Fibrotic plaque	Fibrotic plaque

Meanwhile, contrast medium is often used to improve the diagnostic accuracy and reproducibility to detect and quantify morphological and compositional features.

By this means, the atherosclerotic plaque can be assigned to one of the six classification classes. The histological lesion classes I and II were combined in class I-II, as it was not possible to distinguish discrete foam cells from multiple foam cell layers. This class shows near-normal wall thickness without calcification.

Class III was defined as diffuse intimal wall thickening or small eccentric plaque without calcification.

In class IV lesions, the cap still constitutes only preexisting intima according to Stary, whereas class V lesions are defined as those in which major parts of the fibromuscular cap represent replacement of tissue disrupted by accumulated lipid and hematoma or organized thrombotic deposits. However, discrimination between proteoglycan composition of class IV and the dense collagen of class V is not possible by MRI, which is why Cai et al. combined both in lesion class IV-V. These lesions are characterized by a lipid or necrotic core surrounded by fibrous tissue with possible calcification. Class VI shows a complex plaque with possible surface defect, hemorrhage, or thrombus. Class VII represents a calcified lesion. Class VIII is characterized by a fibrotic plaque without a lipid core and with possible small calcifications. A good agreement of classification obtained by MR imaging and the histological AHA classification could be repeatedly proved (Fig. 15.6).

On the basis of histological studies, the lesion types IV–VI are regarded as high-risk, unstable plaques [22]. In contrast, lesion classes VII and VIII are regarded as stable plaques. Based on MRI examinations of 50 patients, symptomatic stenoses were found to be caused mainly by class IV-V and VI plaques in one study. In contrast, lesion types VII and VIII and are regarded as stable plaques [23]. Among the asymptomatic stenoses, these classes were found significantly less frequently. The mean degree of stenosis was similar in both groups.

MRI plaque imaging was also shown to predict the risk for periprocedural cerebral ischemia in patients who undergo invasive therapy of carotid artery stenosis [24]. Patients with MRI-detected unstable plaques were found to be at a significantly higher risk of developing periprocedural diffusion-weighted imaging (DWI) lesions and neurological symptoms.

Apart from this descriptive classification, distinct features that are assumed to represent peculiar markers of plaque instability can be detected. Since histological studies have shown that especially plaques containing a lipid-rich necrotic core or intraplaque hemorrhage are especially prone to rupture and lead to cerebral ischemia, MRI studies have concentrated in particular on the visualization of these plaque components.

15.8.1 Fibrous Cap Status and Lipid Core

MRI can not only detect the presence of a fibrous cap but can also visualize differences in cap classes, such as thinning or rupture of the fibrous cap, which

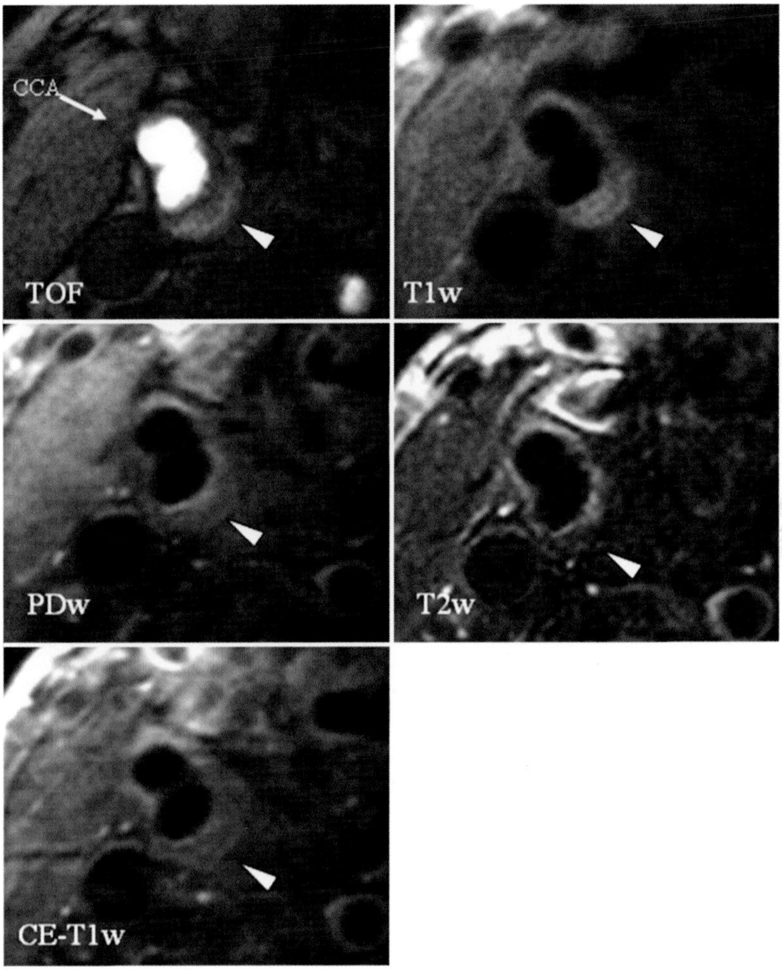

Fig. 15.6 Axial MR images of the right internal carotid artery of an 85-year-old patient with hypertension and a positive family history for cardiovascular disease. The patient suffered a right hemispheric stroke 8 days before the MRI scan. Images show a large lipid/necrotic core with intraplaque hemorrhage, which is hyperintense on TOF and T1W images (chevron). The lumen surface is slightly irregular and the fibrous cap is not visible, suggesting a thin or ruptured fibrous cap. The plaque is consistent with a complicated American Heart Association Lesion type VI with intraplaque hemorrhage and a thin or ruptured fibrous cap

is believed to be the critical event leading to thromboembolic complications due to resultant exposure of thrombogenic subendothelial plaque constituents (Fig. 15.7). The potential of MR imaging to accurately define the relative thickness of the fibrous cap and lipid core has been shown by direct comparison with histology-derived measurements [25, 26]. In 2000, Hatsukami et al. reported the use of a 3D TOF for identifying ruptured fibrous caps with a 1.5-T whole-body scanner. On these sequences, the fibrous cap can be identified as a hypointense

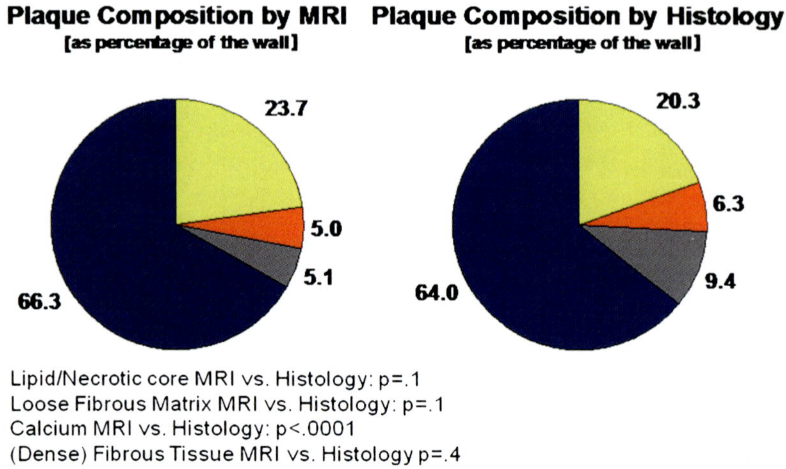

Fig. 15.7 Plaque composition determined by MRI and histology. This figure illustrates that in vivo MRI is able to quantify the major components of the carotid atherosclerotic plaque with good correlation with histopathology; plaque composition was calculated as percentage of the vessel wall for MRI and histology (the paired t-test was used for statistical comparison). (With permission of Saam T et al. [35]

inner rim. Dense calcification located adjacent to the lumen would also appear dark in this sequence but can be identified on T1-, T2-, and PD images as dark structure, whereas fibrous tissue will appear gray. By using preoperative images of 22 consecutive endarterectomy patients, the in vivo state of the fibrous cap was categorized as an intact, thick, an intact, thin, or a ruptured fibrous cap on MRI, gross, and histological sections. They found a high level of agreement between the MR findings and the histological examination of excised endarterectomy specimens, with Cohen $\kappa = 0.83$ (95% confidence interval = 0.67, 1.00) [27]. Discrimination of the FC from the necrotic core can further be improved by using gadolinium-based, contrast-enhanced T1W sequences. Contrary to the underlying necrotic core, the fibrous tissue shows moderate to strong contrast enhancement. Use of a gadolinium contrast agent not only facilitates the identification of the fibrous cap and necrotic core but also allows precise morphological measurements of both components and might again improve identifying vulnerable plaques: Aside from a thin or ruptured fibrous cap, also the underlying lipid-rich, necrotic core represents per se a feature of ongoing inflammation and plaque vulnerability. Both components can be described separately or, as it is increasingly applied, as area ratio of the FC and lipid core.

Yuan et al. were able to show that MRI identification of fibrous cap rupture was highly associated with a recent history of Transient Ischemic Attack (TIA) or stroke in 28 symptomatic and 25 asymptomatic subjects [28]. When compared with patients with a thick fibrous cap, patients with a ruptured cap were 23 times more likely to have had a recent cerebral ischemia. Prospective studies further highlight the ability of MRI to detect vulnerable plaque and underline the potential clinical impact of fibrous cap imaging: Takaya et al. investigated different MRI-derived plaque characteristics in a prospective study [29]. They found a significant association between a thin or ruptured fibrous cap and cerebral ischemia. The relative size of the lipid-rich necrotic core (%LRNC), defined as the proportion of the vessel wall at the most affected cross-sectional location, was another feature associated with future strokes. A study investigating the effects of statins on carotid plaque measured by MRI showed a significant reduction in %LRNC under therapy with high doses of rosuvastatin for 24 months [30]. The overall plaque volume remained unchanged. This finding underlines the potential of plaque MRI for monitoring the morphological and compositional response to therapy in carotid plaques.

15.8.2 Hemorrhage

High-resolution MRI allows accurate detection of intraplaque hemorrhage of different ages with good sensitivity and moderate-to-good specificity by the use of modified criteria for cerebral hemorrhage [31, 32]. Fresh hemorrhage (<1 week) produces a hyperintense signal on T1W and TOF images and iso- to hypointense signal on T2W and PD images. Recent hemorrhage (1–6 weeks) produces a hyperintense signal on T1W, T2W, TOF, and PD images (Fig. 15.8) [31].

Takaya et al. prospectively compared 14 subjects with intraplaque hemorrhage detected by multicontrast MRI and 15 controls with comparably sized carotid plaques but without hemorrhage [33]. Carotid lesions with evidence of early or recent intraplaque hemorrhage at baseline were significantly more likely to increase in overall wall volume, lumen stenosis, and lipid-rich necrotic core volume over an 18-month period.

These findings suggest that intraplaque hemorrhage is a contributory factor to more rapid progression of atherosclerosis and thus represents a marker of plaque instability detectable by MRI. Takaya et al. found intraplaque hemorrhage to correlate with subsequent cerebrovascular events (HR 5.2). This has recently been confirmed Singh et al. [34].

The comparative weighting of single features according to their predictive value for validation of plaque vulnerability still needs to be defined. Saam et al. showed significant differences between symptomatic and asymptomatic plaques in the same patient of all abovementioned parameters. Compared with asymptomatic plaques, symptomatic plaques had a higher incidence of fibrous cap rupture ($P=0.007$), juxtaluminal hemorrhage or thrombus ($P=0.039$), class I hemorrhage ($P=0.021$), and complicated AHA class VI lesions ($P=0.004$), and a lower incidence of uncomplicated AHA class IV and V lesions ($P=0.005$) [35]. Symptomatic plaques also had larger hemorrhage ($P=0.003$) and loose matrix ($P=0.014$) areas and a smaller lumen area ($P=0.008$). Large prospective studies are still needed.

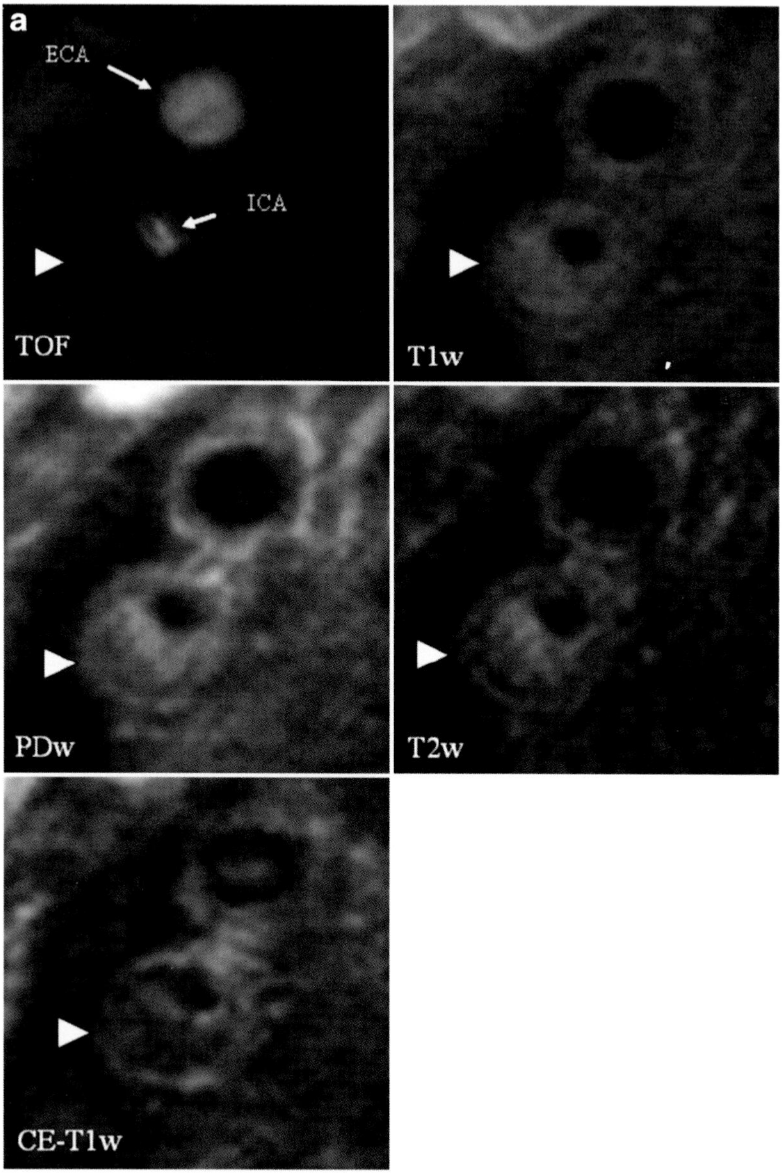

Fig. 15.8 (**a**) Axial MR images of a 74-year-old asymptomatic man with a 90% carotid stenosis in his right internal carotid artery. Cross-sectional MR images show a large eccentric plaque in the right ICA with a large necrotic core without intraplaque hemorrhage (arrow), which is covered by a thick layer of dense and loose fibrous tissue that can be depicted on T1W-, PD- and T2W- images, as compared with the moderately hyperintense area of plaque near the lumen surface. The plaque is consistent with a thick-cap fibroatheroma (American Heart Association lesion type IV-V). (ICA; ECA=external carotid artery) (**b**) Axial MR images of a 59-year-old woman with hypercholesterolemia and hypertension and no other cardiovascular risk factors. The chevron points to a very large lipid/necrotic core in the right common carotid artery (CCA) with a maximum thickness of 8 mm. Despite the marked increase of the vessel wall area, duplex

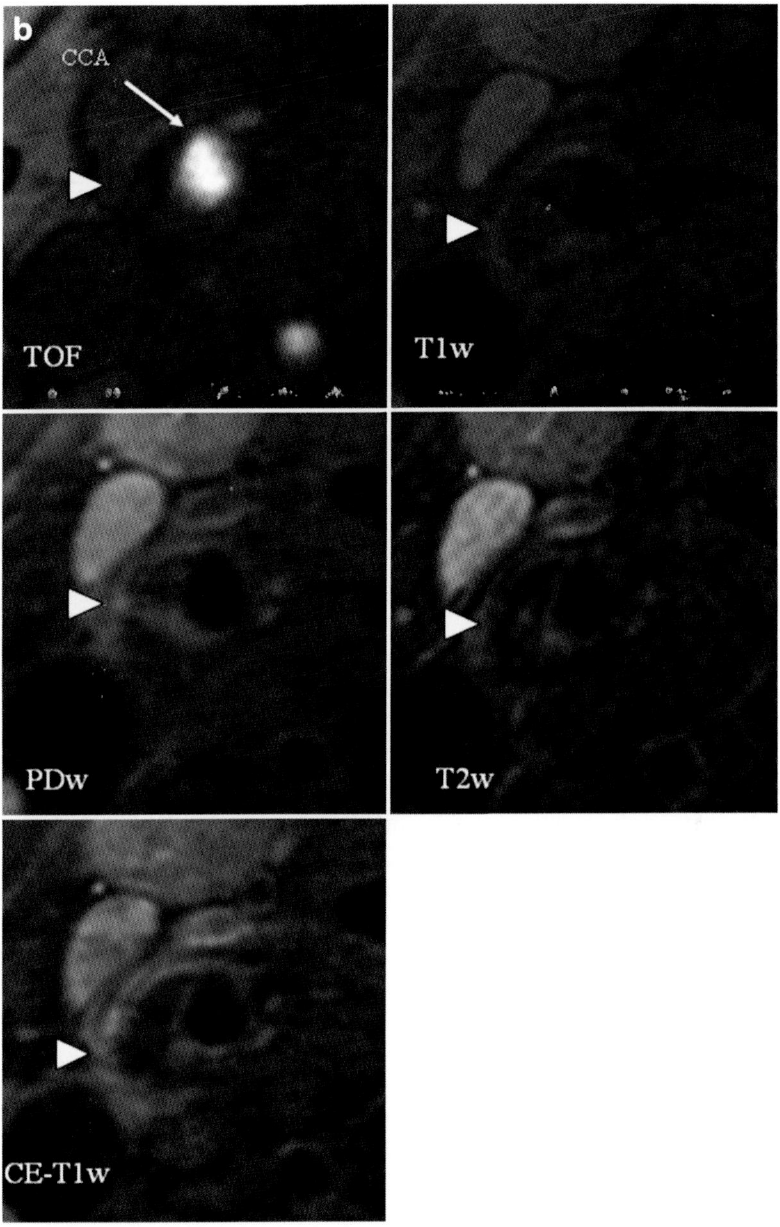

Fig. 15.8 (continued)

sonography revealed no luminal stenosis. The lumen surface is smooth and regular, the fibrous cap is intact and thin. Of note, the fibrous cap and the lipid/necrotic core can best be seen on the contrast-enhanced T1W images; these clearly delineate the enhancing fibrous cap from the underlying necrotic core, which does not show any contrast uptake. The plaque is consistent with a thin-cap fibroatheroma (American Heart Association lesion type IV-V)

15.8.3 Perspective

Further developments might lead to an improvement in image resolution by high matrix size imaging, modification of imaging parameters, and improved, faster hardware. Further shortening of image acquisition times could reduce artifacts and improve imaging quality in a clinical setting with less compliant subjects.

Until recently, most of the MR plaque imaging studies have been performed with 1.5T scanners; 3T MR scanners are now becoming common in research and clinical practice. These scanners provide improvement in the signal-to-noise ratio, contrast-to-noise ratio, and spatial resolution and allow reducing scan time.

Moreover, further specific techniques are under evaluation. Dynamic contrast enhancement using repeated fast image acquisition to measure the contrast uptake property of the plaque might help to detect and quantify neovasculature or increased endothelial permeability. This might provide important information about plaque inflammation.

One further approach to evaluate inflammation processes in the plaque tissue is Ultrasmall super paramagnetic particles of iron oxide (USPIO)-enhanced MRI imaging. The iron particles accumulate predominantly in activated macrophages and cause T2* susceptibility effect, visible on T2*-weighted sequences as magnetic susceptibility artifacts or signal voids. The degree of so detected USPIO uptake is thought to correlate with the extent of inflammation and thus represent a surrogate measure of plaque vulnerability. However, further research is needed for interpretation of the imaging results as large longitudinal studies investigating a correlation between the degree of USPIO uptake and the development of symptoms over time have not yet been performed. Furthermore, the technique is very time-consuming as images have to be acquired before and about 36 h after USPIO infusion, thus degrading future practicability in clinical routine.

A fundamental disadvantage of the "manual" findings is that the assessment can only be based on two-dimensional cross-section images. A total volumetric quantification of different plaque components and the assessment of additional risk factors such as the minimum thickness of the fibrous cap between the vascular lumen and necrotic core would be desirable. This appears theoretically feasible given the achievable submillimeter resolution of the carotid lumens, vessel walls, and atherosclerotic plaques even in 1.5T MRI (Fig. 15.9). However, it is time-consuming and remains imprecise when attempted "by hand." A computer-based three-dimensional analysis could overcome these limitations and, in addition, provide objective and largely independent results from the expertise of the individual evaluator. This kind of automatic classification, which extracts a stenosis class from four MR data sets, can basically be achieved in two ways: either through an algorithmically reproduced rule set or a learning-based approach. The rule set has the advantage in that the algorithm is very intuitive and easy to understand. A learning-based method, in contrast, can allow recognition of the even previously unknown intensity correlations by the algorithm and their use for classification. Both methods are currently being developed and evaluated in various research groups.

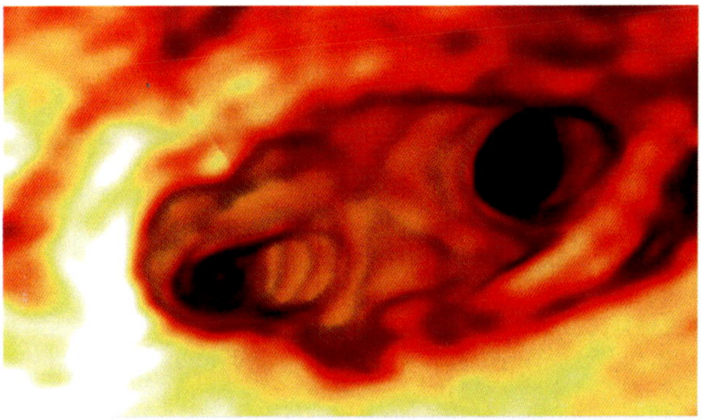

Fig. 15.9 MRI 3D Surface rendering. View from common carotid artery into internal (*left*) and external (*right*) carotid artery

15.8.4 Limitations of MRI-Based Plaque Imaging

MRI suffers from a number of disadvantages such as the expense and time involved in capturing and analyzing images and the lack of wide availability. Difficulties with MRI interpretation may arise from motion artifacts due to patient movement, swallowing, and artery-wall pulsation. Positioning of the subject on a vacuum pillow or use of a (sometimes custom-made) head holder can help maintaining subjects in a comfortable and stable position and avoid head–neck region movement during the MRI scan.

Additionally, MRI is hampered by several contraindications such as claustrophobia and the presence of pacemakers or metallic implants.

However, MRI is noninvasive, does not involve radiation exposure, and is highly reproducible and will therefore certainly have important applications in natural history studies and in clinical trials.

References

1. Halliday, A., A. Mansfield, J. Marro, et al. (2004) Prevention of disabling and fatal strokes by successful carotid endarterectomy in patients without recent neurological symptoms: randomised controlled trial. Lancet, **363**(9420): pp. 1491–502.
2. AbuRahma, A.F., J.T. Wulu, Jr., and B. Crotty (2002) Carotid plaque ultrasonic heterogeneity and severity of stenosis. Stroke, **33**(7): pp. 1772–5.
3. Droste, D.W., M. Karl, R.M. Bohle, et al. (1997) Comparison of ultrasonic and histopathological features of carotid artery stenosis. Neurol Res, **19**(4): pp. 380–4.
4. Hatsukami, T.S., B.D. Thackray, J.F. Primozich, et al. (1994) Echolucent regions in carotid plaque: preliminary analysis comparing three-dimensional histologic reconstructions to sonographic findings. Ultrasound Med Biol, **20**(8): pp. 743–9.
5. Goes, E., W. Janssens, B. Maillet, et al. (1990) Tissue characterization of atheromatous plaques: correlation between ultrasound image and histological findings. J Clin Ultrasound, **18**(8): pp. 611–7.

6. Lammie, G.A., J. Wardlaw, P. Allan, et al. (2000) What pathological components indicate carotid atheroma activity and can these be identified reliably using ultrasound? Eur J Ultrasound, **11**(2): pp. 77–86.
7. Tegos, T.J., K.J. Kalomiris, M.M. Sabetai, et al. (2001) Significance of sonographic tissue and surface characteristics of carotid plaques. AJNR Am J Neuroradiol, **22**(8): pp. 1605–12.
8. Polak, J.F., L. Shemanski, D.H. O'Leary, et al. (1998) Hypoechoic plaque at US of the carotid artery: an independent risk factor for incident stroke in adults aged 65 years or older. Cardiovascular Health Study. Radiology, **208**(3): pp. 649–54.
9. Mathiesen, E.B., K.H. Bonaa, and O. Joakimsen (2001) Echolucent plaques are associated with high risk of ischemic cerebrovascular events in carotid stenosis: the tromso study. Circulation, **103**(17): pp. 2171–5.
10. Prabhakaran, S., T. Rundek, R. Ramas, et al. (2006) Carotid plaque surface irregularity predicts ischemic stroke: the northern Manhattan study. Stroke, **37**(11): pp. 2696–701.
11. Gronholdt, M.L., B.G. Nordestgaard, T.V. Schroeder, et al. (2001) Ultrasonic echolucent carotid plaques predict future strokes. Circulation, **104**(1): pp. 68–73.
12. Coli, S., M. Magnoni, G. Sangiorgi, et al. (2008) Contrast-enhanced ultrasound imaging of intraplaque neovascularization in carotid arteries: correlation with histology and plaque echogenicity. J Am Coll Cardiol, **52**(3): pp. 223–30.
13. Staub, D., M.B. Patel, A. Tibrewala, et al. (2010) Vasa vasorum and plaque neovascularization on contrast-enhanced carotid ultrasound imaging correlates with cardiovascular disease and past cardiovascular events. Stroke, **41**(1): pp. 41–7.
14. Touboul, P.J., M.G. Hennerici, S. Meairs, et al. (2007) Mannheim carotid intima-media thickness consensus (2004–2006). An update on behalf of the Advisory Board of the 3rd and 4th Watching the Risk Symposium, 13th and 15th European Stroke Conferences, Mannheim, Germany, 2004, and Brussels, Belgium, 2006. Cerebrovasc Dis, **23**(1): pp. 75–80.
15. Folsom, A.R., R.A. Kronmal, R.C. Detrano, et al. (2008) Coronary artery calcification compared with carotid intima-media thickness in the prediction of cardiovascular disease incidence: the Multi-Ethnic Study of Atherosclerosis (MESA). Arch Intern Med, **168**(12): pp. 1333–9.
16. Lorenz, M.W., H.S. Markus, M.L. Bots, et al. (2007) Prediction of clinical cardiovascular events with carotid intima-media thickness: a systematic review and meta-analysis. Circulation, **115**(4): pp. 459–67.
17. Lorenz, M.W., H. Bickel, M.L. Bots, et al. (2010) Individual progression of carotid intima media thickness as a surrogate for vascular risk (PROG-IMT) – rationale and design of a meta-analysis project. Am Heart J, **159**(5): pp. 730–6.e2.
18. (1998) Randomised trial of endarterectomy for recently symptomatic carotid stenosis: final results of the MRC European Carotid Surgery Trial (ECST). Lancet, **351**(9113): pp. 1379–87.
19. (1991) Beneficial effect of carotid endarterectomy in symptomatic patients with high-grade carotid stenosis. North American Symptomatic Carotid Endarterectomy Trial Collaborators. N Engl J Med, **325**(7): pp. 445–53.
20. Stary, H.C. (2000) Natural history and histological classification of atherosclerotic lesions: an update. Arterioscler Thromb Vasc Biol, **20**(5): pp. 1177–8.
21. Toussaint, J.F., G.M. LaMuraglia, J.F. Southern, et al. (1996) Magnetic resonance images lipid, fibrous, calcified, hemorrhagic, and thrombotic components of human atherosclerosis in vivo. Circulation, **94**(5): pp. 932–8.
22. Cai, J.M., T.S. Hatsukami, M.S. Ferguson, et al. (2002) Classification of human carotid atherosclerotic lesions with in vivo multicontrast magnetic resonance imaging. Circulation, **106**(11): pp. 1368–73.
23. Esposito, L., M. Sievers, D. Sander, et al. (2007) Detection of unstable carotid artery stenosis using MRI. J Neurol, **254**(12): pp. 1714–22.
24. Poppert, H., L. Esposito, R. Feurer, et al. (2009) MRI plaque Imaging identifies high risk patients for therapy of carotid artery stenosis. Cerebrovasc Dis, **27**(Suppl 6): p. 61.

25. Trivedi, R.A., J. U-King-Im, M.J. Graves, et al. (2004) Multi-sequence in vivo MRI can quantify fibrous cap and lipid core components in human carotid atherosclerotic plaques. Eur J Vasc Endovasc Surg, **28**(2): pp. 207–13.
26. Yuan, C., L.M. Mitsumori, M.S. Ferguson, et al. (2001) In vivo accuracy of multispectral magnetic resonance imaging for identifying lipid-rich necrotic cores and intraplaque hemorrhage in advanced human carotid plaques. Circulation, **104**(17): pp. 2051–6.
27. Hatsukami, T.S., R. Ross, N.L. Polissar, et al. (2000) Visualization of fibrous cap thickness and rupture in human atherosclerotic carotid plaque in vivo with high-resolution magnetic resonance imaging. Circulation, **102**(9): pp. 959–64.
28. Yuan, C., S.X. Zhang, N.L. Polissar, et al. (2002) Identification of fibrous cap rupture with magnetic resonance imaging is highly associated with recent transient ischemic attack or stroke. Circulation, **105**(2): pp. 181–5.
29. Takaya, N., C. Yuan, B. Chu, et al. (2006) Association between carotid plaque characteristics and subsequent ischemic cerebrovascular events: a prospective assessment with MRI – initial results. Stroke, **37**(3): pp. 818–23.
30. Underhill, H.R., C. Yuan, X.Q. Zhao, et al. (2008) Effect of rosuvastatin therapy on carotid plaque morphology and composition in moderately hypercholesterolemic patients: a high-resolution magnetic resonance imaging trial. Am Heart J, **155**(3): pp. 584.e1–8.
31. Chu, B., A. Kampschulte, M.S. Ferguson, et al. (2004) Hemorrhage in the atherosclerotic carotid plaque: a high-resolution MRI study. Stroke, **35**(5): pp. 1079–84.
32. Cappendijk, V.C., K.B. Cleutjens, S. Heeneman, et al. (2004) In vivo detection of hemorrhage in human atherosclerotic plaques with magnetic resonance imaging. J Magn Reson Imaging, **20**(1): pp. 105–10.
33. Takaya, N., C. Yuan, B. Chu, et al. (2005) Presence of intraplaque hemorrhage stimulates progression of carotid atherosclerotic plaques: a high-resolution magnetic resonance imaging study. Circulation, **111**(21): pp. 2768–75.
34. Singh, N., A.R. Moody, D.J. Gladstone, et al. (2009) Moderate carotid artery stenosis: MR imaging-depicted intraplaque hemorrhage predicts risk of cerebrovascular ischemic events in asymptomatic men. Radiology, **252**(2): pp. 502–8.
35. Saam, T., M.S. Ferguson, V.L. Yarnykh, et al. (2005) Quantitative evaluation of carotid plaque composition by in vivo MRI. Arterioscler Thromb Vasc Biol, **25**(1): pp. 234–9.
36. Prati, P., A. Tosetto, D. Vanuzzo, et al. (2008) Carotid intima media thickness and plaques can predict the occurrence of ischemic cerebrovascular events. Stroke, **39**(9): pp. 2470–6.
37. Chien, K.L., T.C. Su, J.S. Jeng, et al. (2008) Carotid artery intima-media thickness, carotid plaque and coronary heart disease and stroke in Chinese. PLoS One, **3**(10): p. e3435.

Biographies

Holger Poppert is assistant professor at the Department of Neurology at Technische Universität München, Munich, Germany. He is heading a cerebrovascular research group and the Diagnostic Neurological Sonography Lab in his institution. His major interests are in the fields of embolic stroke mechanisms and noninvasive plaque imaging.

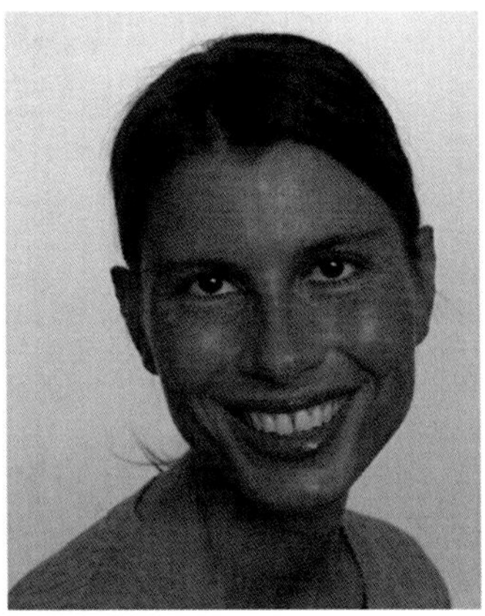

Regina Feurer finished her medical degree at the Technical University of Munich in June 2006. Since June 2007 she works as an intern at the department for neurology of the "Klinikum rechts der Isar der Technischen Universität München". Ms Feurer joined the neurovascular research group of Dr. H. Poppert and had a focus on the question about the relevance of Patent Foramen Ovale in view of stroke recurrence.

Dr. Lorena Esposito graduated from Heidelberg University, Germany, and received her M.D. degree in April 2004. She started her residency in neurology at the hospital Klinikum rechts der isar of the Technische Universität in Munich, Germany, in July 2004. There she joined the Cerebrovascular Resarch Group of PD Dr. Holger Poppert. Her research is focused on plaque imaging of carotid artery stenosis using high resolution MRI. She is currently working on classification systems for detection of vulnerable carotid plaques in follow-up cohorts.

Priv.-Doz. Dr. med. Tobias Saam was born on June 24th 1973 in Heidelberg. He graduated from the medical University of Heidelberg in 2001 and worked 1.5 years as a resident in the radiology department in Heidelberg. From 2003 to 2006 he worked as a senior research fellow in the Vascular Imaging lab in Seattle, Washington, USA. Since 2006 he is working at the Ludwig-Maximilians-University in Munich as a radiology fellow. His area of expertise is vessel wall imaging with a focus on imaging the atherosclerotic "vulnerable" plaque. He received the Young Investigator Award at the 3rd International Symposium "Integrated Biomarkers in Cardiovascular Diseases 2008" in Seattle, WA, USA and the Best Scientific Paper Award at ECR 2007. He was recipient of the Novartis stipend for therapeutic research in 2007. He received the Coolidge Award from General Electric 2010 for his "Habilitation" entitled "Methodological Development and Clinical Evaluation of high-resolution Magnetic Resonance Imaging of Atherosclerotic Carotid Plaques". He has authored or co-authored more than 40 articles in peer-reviewed journals.

Dirk Sander, MD, is Head of the Department of Neurology & Neurologic Rehabilitation, Benedictus Hospital Tutzing and Professor of Neurology at the Faculty of Medicine, Technische Universität of Munich, Germany.

Professor Sander graduated from the Medical School at the University of Göttingen, Germany, and subsequently obtained specialist qualifications in neurology and intensive care. Professor Sander is currently Head of the Department of Neurology at the Benedictus Hospital in Tutzing and Professor of Neurology at the Technische Universität of Munich. He has received the Scientific Prize of the Deutsche Schlaganfallhilfe and the Award for Excellency in Stroke Research from the European Stroke Council. He has been published more than 200 articles in peer-reviewed journals such as Stroke, Lancet, Lancet Neurology, Neurology, Archives of Neurology, Circulation and Cerebrovascular Diseases.

Chapter 16
Noninvasive Targeting of Vulnerable Carotid Plaques for Therapeutic Interventions

Karol P. Budohoski, Victoria E.L. Young, Tjun Y. Tang, Jonathan H. Gillard, Peter J. Kirkpatrick, and Rikin A. Trivedi

Abstract Patients with carotid atherosclerosis develop symptoms most commonly due to plaque progression or rupture. Many factors influence the risk of a lesion rupturing. The degree of stenosis as seen on digital subtraction angiography is no longer a sufficient means of stratifying patients' risk and thus guiding therapy. The term "vulnerable plaque" is used to describe plaques that have a high risk of causing symptoms. Such plaques typically have active inflammation and a thin fibrous cap overlying a large lipid-rich, necrotic core, often with intraplaque hemorrhage. The endothelial surface of such plaques may be irregular, with endothelial ulceration or fibrous cap rupture. A plaque may be characterized as "vulnerable" based on morphology without necessarily causing a significant stenosis. Various means of depicting the characteristics of plaque vulnerability are available. The following chapter aims to describe the established, as well as some of the emerging noninvasive techniques used to image vulnerable carotid plaques. A brief description of novel agents, targeted to specific molecules involved in the atherogenic process, is presented. Furthermore, a comparison of the various techniques in the context of their ability to depict the many characteristics of high risk atherosclerotic lesions is made.

Keywords Vulnerable plaque • Atherosclerotic plaque • Atheroma • Carotid atherosclerotic disease • Noninvasive imaging

16.1 Introduction

Ever since the work of C. Miller Fisher in 1951, who first recognized the connection between an occluded carotid artery and cerebral infarction, carotid atherosclerotic disease has been recognized as a thromboembolic source that may cause a stroke [1].

R.A. Trivedi (✉)
Academic Neurosurgery Unit, University of Cambridge, Cambridge, UK
and
Department of Neurosurgery, Box 166, Addenbrooke's Hospital,
Hills Road, CB2 0QQ Cambridge UK
e-mail: rikin.trivedi@addenbrookes.nhs.uk

Jasjit S. Suri et al. (eds.), *Atherosclerosis Disease Management*,
DOI 10.1007/978-1-4419-7222-4_16,

Carotid endarterectomy (CEA) has become a widely accepted treatment method for this, often focal, pathology. A series of multicenter, randomized, controlled trials (RCTs) have since evaluated the benefit of surgical vs. medical treatment, as well as stratified the risk of stroke for symptomatic and asymptomatic patients with various degrees of luminal stenosis [2–8]. The results of these studies showed that the risk of stroke increased with narrowing of the arterial lumen, making measurements of luminal stenosis the gold standard in stratifying patients' risk and identifying patients suitable for surgical treatment. Furthermore, Rothwell et al. [2], in an analysis of pooled data from the three major RCTs of CEA for symptomatic patients, European Carotid Surgery Trial (ECST), North American Symptomatic Carotid Endarterectomy Trial (NASCET), and Veterans' Affairs Trial, demonstrated that a significant benefit from surgery can be achieved only in patients with severe stenosis (70–99%), with an absolute risk reduction (ARR) of ipsilateral stroke within the next five years of 16%. A marginal benefit was observed in symptomatic patients with moderate stenosis (50–69%) with an ARR of 4.6% [2]. For asymptomatic patients, combined data from the Asymptomatic Carotid Atherosclerosis Study (ACAS) and the Asymptomatic Carotid Surgery Trial (ACST) demonstrated that patients who benefited from surgical treatment were those who had a significant narrowing of ≥70%. The decrease of 5-year stroke risk was shown to be from 11.5 to 6% [5, 8]. In a subgroup analysis performed in the ACST study it was shown that the benefit was significant only for males of <75 years old, with no benefit demonstrated for females [5].

It is important to note that the major trials of carotid endarterectomy for symptomatic patients have used digital subtraction angiography (DSA) to determine the degree of luminal narrowing, thus making it the "gold standard" in guiding therapy for these patients [3, 4, 6, 7]. However, the trials used different methods for quantifying stenosis, causing difficulty in comparing results. Furthermore, DSA has its disadvantages. It is expensive, time consuming, and associated with a relatively high morbidity (0.5–1.3% permanent neurological complications and 0.4–1.3% of transient neurological complications [9–11]). Because of the risks and low availability, DSA is not usually used for asymptomatic patients, where ultrasonography is preferred [5, 8]. Also, the adaptive mechanism of arterial remodeling that may lead to the formation of large plaques without significant luminal narrowing cannot be properly assessed using DSA alone [12].

Histological studies have been used to characterize plaque morphology and identify vulnerable atheromatous disease [13, 14]. As understanding of the morphology has increased, there has been a shift of interest toward identifying characteristics of the carotid arterial wall, which may play a part in the clinical risk assessment, especially for asymptomatic patients.

16.2 The Vulnerable Plaque

Since the pioneering works of Moor et al. [15] in 1968 and Imparato et al. [16] in 1979, the histological properties of carotid atherosclerotic plaques have been defined based on excised carotid specimens from patients who have suffered a stroke or transient ischemic attack (TIA) (Fig. 16.1). Those plaques were termed "culprit plaques," while plaques

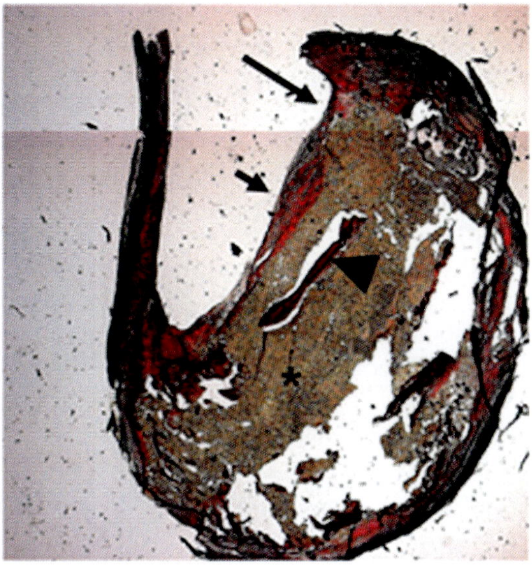

Fig. 16.1 Histological section of a carotid atherosclerotic plaque using elastic van Gieson (EVG) stain for connective tissue. Characteristics of plaque vulnerability: fibrous cap (*short arrow*), thinning of fibrous cap in the shoulder region (*long arrow*), large lipid-rich necrotic core (*asterisk*), intraplaque hemorrhage (*arrowhead*)

Table 16.1 Criteria for defining vulnerable plaques and diagnostic imaging methods used [17]

Type of criteria	Criterion	Imaging
Major	Active inflammation	CE-MRI, USPIO-MRI, FDG-PET
	Thin FC with large LRNC	MRI, CE-MRI, CT
	FC disruption	MRI, CE-MRI, SPECT
	Severe stenosis	MRA, CTA, B-mode ultrasonography
Minor	IPH	MRI
	Expansive remodeling	MRI
	Superficial calcified nodules	MRI, CT
	Yellow coloring on angioscopy	–
	Endothelial dysfunction	–

MRI magnetic resonance imaging, *CE-MRI* contrast-enhanced MRI, *USPIO-MRI* ultrasmall superparamagnetic iron oxide-enhanced MRI, *FDG-PET* fluorodeoxyglucose positron emission tomography, *CT* computed tomography, *SPECT* single photon emission tomography

exhibiting these characteristics were termed "vulnerable." A vulnerable plaque, as proposed in the consensus document [17, 18], is a plaque prone to thrombosis and a plaque with which there is associated a high probability that it will undergo rapid progression and become a culprit plaque, that is, cause an ischemic stroke or an acute cardiovascular event. The characteristics of a vulnerable or high-risk plaque have been divided into major and minor criteria and are summarized in Table 16.1 [17, 18].

These characteristics, if present in a plaque, suggest a high risk of experiencing ischemic events, independently of luminal stenosis. Takaya et al. [19] reported a significant correlation between the presence of a thin or ruptured fibrous cap (FC) as seen on magnetic resonance imaging (MRI) and an ipsilateral ischemic stroke.

Various underlying mechanisms that can cause mechanical failure of the FC, leading to plaque rupture, have been suggested, including biomechanical stress [20–22], hemorrhage within the plaque [13, 23, 24], inflammation within the plaque with an influx of macrophages [25–27], destruction of the FC by proteolytic enzymes such as matrix metalloproteinases (MMPs) [28], and apoptotic cell death of FC fibroblasts [25–27].

In the following sections, we aim to discuss which noninvasive diagnostic imaging modalities are best used to visualize in vivo the particular characteristics of vulnerable plaques. We discuss the role of imaging modalities such as B-mode ultrasonography and contrast-enhanced ultrasonography (CES); high-resolution MRI, contrast-enhanced MRI (CE-MRI), dynamic contrast-enhanced MRI (DCE-MRI) and USPIO-enhanced MRI, [18-F]Fluorodeoxyglucose positron emission tomography (FDG-PET), and single photon emission computed tomography (SPECT).

16.3 B-Mode Ultrasonography

B-mode ultrasonography (US) is a technique with high resolution; it is noninvasive, rapidly applicable, readily available, and the costs involved are far lower than in any other technique used for evaluation of carotid atherosclerotic disease. B-mode US is currently the primary screening tool for evaluation of patients with carotid artery disease. It has been successfully used in many studies trying to identify US characteristics of high-risk patients with atherosclerosis. Recent investigations are focusing on targeting specific sites through implementation of specific contrasting agents, similarly to MRI and nuclear imaging [29]. The limitations of B-mode US include high operator variability, variability between centers, artifacts arising from calcifications, and difficulty in distinguishing subtotal occlusion from total occlusion [30].

The main indicators of plaque vulnerability using B-mode US are as follows:

- Intima/media thickness (IMT) [31–33]
- Plaque echogenicity: echolucent plaques are associated with IPH and lipids; echogenic plaques are associated with fibrous tissue [5, 34–36]
- Plaque irregularity [37, 38]
- Neovascularization and inflammation on molecular contrast enhanced ultrasonography (CEU) [29, 37–39]

16.3.1 Intima/Media Thickness

One of the first studies performed by Salonen et al. on 1,288 Finnish men [40] to determine whether ultrasonographic characteristics of vulnerable carotid plaques can be connected to the risk of subsequent stroke did not confirm intima/media thickness (IMT) to be a predictor of stroke and myocardial infarction. However, a subsequent study performed on 4,476 patients who had no prior history of cardiovascular disease, with a mean follow-up period of 6.2 years found that an increase in IMT in those patients was significantly associated with a higher risk of cardiovascular events and stroke [33], suggesting that, indeed, IMT assessment may be beneficial in risk stratification of asymptomatic patients.

16.3.2 Plaque Echogenicity

Echolucent plaques on B-mode ultrasonography are thought to have a higher component of soft tissue, such as lipid or hemorrhage. Echogenic plaques, on the other hand, are thought to be composed primarily of fibrous tissue [34]. Thus, echolucency is used as a surrogate marker for the presence of LRNC in atherosclerotic plaques and subsequently as one of the characteristics of vulnerability.

The Cardiovascular Health Study examined the correlation of plaque echogenicity and risk of incidental stroke [35] on a cohort of 4,886 patients, >65 years old, with a mean follow-up of 3.3 years. None of the subjects had any prior history or symptoms of cardiovascular disease. It was shown that an echolucent internal carotid artery plaque was associated with a risk of ipsilateral, nonfatal stroke. The age- and sex-adjusted odds ratios (OR) for incidence of stroke were OR: 2.53 (95% CI: 1.42, 4.53) with a relative risk (RR) independent of the severity of stenosis of 2.78 (95% CI: 1.36, 5.69). However, it has to be noted that the Asymptomatic Carotid Surgery Trial, which included patients based on the severity of stenosis measured using ultrasound, did not confirm that echolucency of plaques had any influence on the risk of subsequent stroke [5].

Poor reproducibility of carotid artery plaque imaging with ultrasonography is one of the major limitations of this method [41] and one of the areas of ongoing research. Sabetai et al. [36] have proposed a computer-based system to measure and compare the echolucency of plaques. All B-mode US images were normalized by applying a gray scale value predefined for blood and adventitia and subsequently compared using a gray scale median (GSM). They reported a good interobserver agreement; however, the method described used a mean value of grayscale for the entire plaque, which was a significant limitation of the study disallowing evaluation of any regional instabilities within an overall highly echogenic plaque. Another method used a pixel distribution analysis, which took into account the possible variation of echogenicity within a plaque [42]. The authors used this technique to compare carotid plaques between symptomatic ($n=18$) and asymptomatic patients ($n=27$). More IPH and lipids and less calcifications were observed in the group of symptomatic patients compared with controls. Furthermore, LRNC of symptomatic patients was larger ($p=0.005$) and the lumen narrower ($p=0.01$). These characteristics of plaque vulnerability were consistent with histology.

16.3.3 Plaque Irregularity

Plaque surface morphology was extensively studied using angiograms of patients taken from the European Carotid Surgery Trial [7]. Evaluation of 3,007 digital subtraction angiograms suggested that plaque irregularities, such as ulcerations, are an independent predictor for risk of ischemic stroke following surgical management [43, 44].

Prabhakaran et al. [38] proposed a study to assess whether plaque irregularities could be studied in patients with carotid artery disease noninvasively, using B-mode ultrasonography. One thousand nine hundred and thirty-nine patients were recruited. Baseline characteristics were assessed, including B-mode US of carotid arteries. One thousand and ninety-one (56.4%) subjects had been demonstrated to harbor a carotid plaque. The lesions were than characterized by location of plaque, number of plaques, degree of stenosis, maximal carotid plaque thickness, and surface irregularities. Plaque surface irregularity was shown to be the predictive value for ischemic stroke with a hazard ratio (HR) of 4.0 (95% CI: 1.7, 9.4). However, the analysis was performed with disregard to laterality (all strokes were taken into consideration, not just those ipsilateral to the analyzed carotid plaque). In view of this, plaque irregularities were associated with generalized atherosclerosis, rather than being deemed a direct source of embolic material. These findings were later confirmed by Kitamura et al. [37] in a study of 1,358 Japanese men without prior history of cardiovascular disease. The authors showed that persons with a plaque with an irregular surface shown on B-mode US have an age-adjusted RR of experiencing ischemic stroke of 7.7 (95% CI: 2.0, 30.0), suggesting that plaque surface irregularities in fact are markers of plaque vulnerability.

16.3.4 Molecular Contrast-Enhanced Ultrasonography

The introduction of contrasting media designed for ultrasonography has allowed for broadening the scope of ultrasonographic imaging of carotid atherosclerotic disease [39]. These contrast media rely on an array of microbubbles filled with gas, which produce an acoustic field detectable with ultrasonography. Microbubbles are designed to act in blood in a similar way as red blood cells do, thus they have the ability to penetrate into microvasculature and provide information on capillary blood flow through an area of interest. Furthermore, these microbubbles have now been designed so that they can specifically bind to various molecules, therefore, allowing functional, molecular imaging. Perfusion imaging using microbubbles as contrast agents for US has been made feasible through developments in the ability to detect harmonics and overtones produced by the bubbles over tissue noise. An additional development in microbubble effectiveness was the ability to control their size and stability to gain the best signal without increasing the size above that of the capillary vessels [45]. The first microbubble contrasting agent approved by the FDA was Albunex (1994) [46]. Microbubble contrast media have evolved since their introduction, providing improved contrast on images, better stability, and standardization of the size. The most common use of contrast-enhanced ultrasonography (CEU) so far is to visualize the contours of heart chambers during echocardiography. However, the directions of CEU developments include imaging of microvascular perfusion and inflammation, both of which have a potential role in management of atherosclerosis.

Staub et al. [47] have studied the possibilities of CEU to depict neovascularization and vasa vasorum in carotid atherosclerotic lesions in 147 human subjects. CEU findings were defined as: plaque neovascularization (absent or present) and adventitial vasa vasorum (absent or present). A correlation between the presence of both neovascularization and vasa vasorum as seen on CEU with a history of cardiovascular disease and cardiovascular events was performed. The findings showed that significantly more patients with either neovascularization or vasa vasorum present had a history of cardiovascular disease than those with absent neovascularization or vasa vasorum (38% vs. 20%, $p=0.031$and 73% vs. 54%, $p=0.029$ for neovascularization and vasa vasorum, respectively).

CEU using microbubbles has been also used to target molecular markers of atherosclerosis. These studies, however, are currently focusing on animal models and are discussed briefly in Sect. 16.7.

16.4 Magnetic Resonance Imaging

MRI is well suited for imaging carotid disease as a noninvasive method, which does not include ionizing radiation and hence can be performed sequentially to track disease progression or response to medications. MRI, however, has also its limitations, namely the long image acquisition times, that are especially difficult for patients with an impaired neurological status. Furthermore, MRI is a relatively expensive method, which requires skilled staff, trained in imaging carotid vessels and specialist equipment. However, as MRI imaging evolves and MRI machines become more widely available the potential for it to become a core modality in tailoring treatment strategies for patients with carotid atherosclerotic disease seems highly probable.

MRI-derived measurements have been shown to be reproducible and can be used for classifying plaques according to the American Heart Association criteria [48]. It has been shown that imaging plaques in vivo using MRI corresponds well to postoperative histological examination of the excised plaques [49, 50].

Conventional MRI with three-dimensional (3D) data acquisition has extremely good soft tissue resolution and can be readily used to gather multiple parameters describing plaque morphology including FC thickness, size of lipid-rich necrotic core (LRNC) [51, 52], fibrous cap disruption (which combines desquamation of the FC as well as fissured/ruptured plaque) [53, 54], and intraplaque hemorrhage (IPH) [55, 56]. Furthermore, in conjunction with morphological features biomechanical stress modeling can be performed [20–22, 57]. Contrast enhancement allows to better distinguish between FC and LRNC [50, 52], while dynamic contrast enhancement allows assessment of plaque neovascularization [58–60]. A novel method, relying on ultrasmall superparamagnetic iron oxide (USPIO) particles that are taken up by macrophages can be used for direct visualization of the extent of inflammation within the plaque [21, 61–65].

16.4.1 Morphological Parameters

MRI morphological characteristics that are considered most important in deeming the plaque either stable or vulnerable include thickness of the fibrous cap, size of lipid-rich necrotic core, fibrous cap disruption (encompassing fibrous cap rupture and changes in the endothelial lining of the fibrous cap, often with platelet aggregation), eccentricity of the plaque, and intraplaque hemorrhage.

16.4.1.1 Fibrous Cap and Lipid Rich-Necrotic Core

Fibrous cap (FC) thickness and lipid-rich necrotic core (LRNC) size are two most readily available parameters that potentially may give additional information regarding the stability of a carotid atheromatous plaque. Mostly histological studies have shown that one of the most common features of plaques prone to rupture is thin FC overlying a large LRNC [13, 18, 49, 66].

Sequences used to image the FC and LRNC rely on the ability of cholesterol esters, which form the predominate part of LRNC to shorten T2 relaxation time. The result was a hypointense area on T2-weigthed (T_2W) images representing LRNC. Touze et al. [67] have studied the reproducibility of high-resolution MRI of carotid atheromatous plaques. They analyzed imaging of 85 patients both symptomatic and asymptomatic for intra- and interobserver agreement. Substantial intraobserver agreement was noted for the identification of LRNC ($\kappa=0.69$; $\kappa>0.7$ indicates good agreement between two tests) and moderate for FC ($\kappa=0.58$). Interobserver agreement was moderate for identification of LRNC ($\kappa=0.58$) and fair for FC ($\kappa=0.28$). Intraobserver reproducibility for quantitative measures of plaque, lipids and fibrous component was good with intraclass correlation coefficient (ICC) of 0.90, 0.72, and 0.77, respectively. Furthermore, the imaging data was validated against histology with good results. Hatsukami et al. [53] performed an analysis of the ability of high-resolution MRI to distinguish between differences in FC thickness and rupture. They used 3D multiple overlapping thin slab angiography (MOTSA) sequence to identify thick, thin, and ruptured plaques. Results showed an 89% agreement with histological assessment.

Trivedi et al. [51] in a study of carotid plaque composition with emphasis on quantifying LRNC and FC using 2D, blood-suppressed, fast-spin echo, T_2W MRI sequences reported a high interobserver agreement. The size of LRNC was measured at the point of the largest diameter, and the thickness of FC was measured directly above the thickest part of the LRNC. The ratio of FC thickness to LRNC was calculated. The study, however, excluded all plaques with IPH. Cai et al. [52] used CE-MRI to measure the dimensions of FC and LRNC in intact plaques. Sequences included double inversion recovery (DIR) T1-weighted (T_1W), time-of-flight (TOF), and proton density-weighted (PDw). A good correlation of both measures FC and LRNC with histology was observed ($r=0.80$, $p<0.001$; $r=0.84$, $p<0.001$, respectively). Furthermore, a significantly better ($p<0.001$) correlation of the size of FC on CE-MRI with histology than on T_2W MRI was observed [52]. The factor that may have had an influence on the observed difference was the presence of intraplaque hemorrhage, as when present the difference between T_2W and CE-MRI measurements was 19.6%±30.7%, while when analyzing only plaques without hemorrhage the difference was 2.3%±26.7%. On T_2W, hemorrhage within the LRNC imaging increases signal intensity, thereby obscuring the boundary between LRNC and FC. This is not the case with CE-MRI, where hemorrhage causes only slight enhancement and does not substantially influence differentiation from the surrounding enhanced tissue [68].

From the above results it can be concluded that identification and measurements of plaque size and LRNC are reliable using high-resolution MRI. However, due to limitations in spatial resolution, identification of FC poses a greater challenge and requires more sophisticated methodology and often administration of contrast.

16.4.1.2 Fibrous Cap Disruption and Platelet Aggregation

Histopathological studies have shown that fibrous cap disruption is more frequently found in patients with a history of transient ischemic attack or stroke [69, 70]. Intact plaques can be differentiated from injured or ruptured ones using high-resolution MRI with multicontrast protocols [52]. Spagnoli et al. [70] have demonstrated that superficial thrombus formation on a plaque corresponds to an inflammatory process within the plaque. Furthermore, studies on case series have shown that it is possible to identify fibrous cap disruption of carotid plaques using multisequence, cross-sectional MRI with black and bright blood sequences [71] (Fig. 16.2). It has to be noted, though, that the histological definition of a thin FC (<250 μm [53]) is often

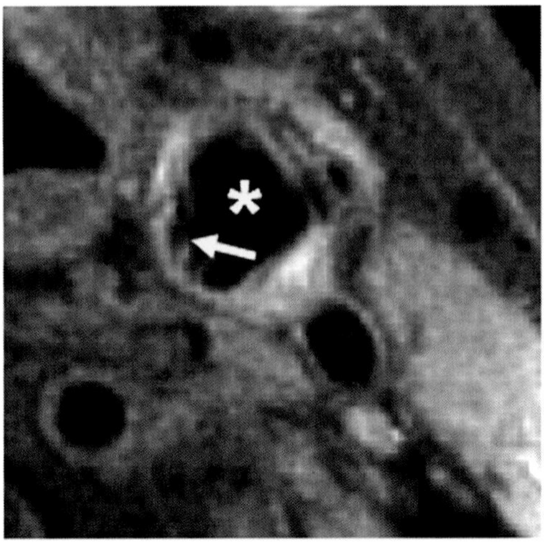

Fig. 16.2 MR image showing FC disruption (*arrow*) in a symptomatic patient with moderate stenosis. *Asterisk* indicates arterial lumen

below the spatial resolution of conventional MRI scanners and sometimes cannot be identified. In such cases, the plaque is classified as follows [24]:

- If fibrous cap is clearly visible on CE-T_1W as being between the LRNC and the vessel lumen – classified as thick FC
- If fibrous cap not evident on CE-T_1W as being between the LRNC and the vessel lumen – classified as thin FC
- If fibrous cap absent on all imaging sequences – classified as ruptured FC

Undoubtedly the spatial resolution is one of the limitations of MRI usefulness in imaging atherosclerosis, especially in small-diameter vessels, and this is one area where improvements are still needed.

16.4.1.3 Severity of Stenosis

The severity of stenosis is the major determinant of future risk of stroke in patients with carotid stenosis, second only to the neurological status. Many of the major randomized trials of carotid endarterectomy including NASCET, ECST, and Veterans' Affairs Trial [3, 6, 7] have used DSA (the accepted gold standard) to measure the extent of luminal narrowing and thus, establishing it as the basis of risk stratification for these patients and in guiding further therapy. The advent of novel noninvasive imaging techniques as well as the cost, time, and morbidity of DSA has lead to change in the preferences in evaluation of carotid atheroma patients. Contrast-enhanced MRI has been validated against DSA with a good sensitivity and specificity [72, 73]. The use of MRI in assessing the degree of stenosis, however, is beyond the scope of this chapter.

16.4.1.4 Intraplaque Hemorrhage

Intraplaque hemorrhage (IPH) is a frequent finding in atherosclerotic plaques of the carotid arteries. Microvessels in the area of an atherosclerotic plaque have a tendency to be fragile. Without proper smooth muscle cell support and with a leaky endothelium it is common for them to rupture and cause a hemorrhage [55, 74]. There have been suggestions that subsequent hemorrhages can act as an atherogenic factor, leading to progression of the plaque by contributing to the supply of inflammatory cells and LDLs. This was confirmed in a histopathological study performed on rabbits where induced lesions with IPH had significantly greater lipid content on histological examination than plaques of control animals [75]. Additionally, in a study performed by Takaya et al. [55], on 29 patients with baseline MRI classified as either containing signs of IPH (hemorrhage group; $n=14$) or not (controls; $n=15$), the presence of IPH was associated with a 23% greater increase in the size of LRNC, as well as a 7% greater decrease in the lumen diameter compared with controls over the course of 18 months. In the same study, using a serial multi contrast-weighted MRI protocol (T_1W, T_2W, PDw, 3D TOF), the authors confirmed

Table 16.2 Stages of IPH, localization of methemoglobin, and MRI characteristics

	Characteristics		
		MRI	
Stage of hemorrhage	Histology	T_1W/TOF	T_2W/PDw
Fresh (Type I)	Intact RBC with intracellular methemoglobin	Hyperintense	Isointense
Recent (Type II)	Lytic RBC with extracellular methemoglobin	Hyperintense	Hyperintense
Old	Amorphemosid	Hypointense	Hypointense

RBC red blood cells, T_1W T1 weighted, *TOF* time-of-flight, T_2W T2 weighted, *PDw* proton density weighted. Criteria adapted from Chu et al. [56] and modified according to Takaya et al. [55]

that plaques which had a documented IPH at baseline had a 43% greater risk of a secondary IPH at 18 months.

Moody et al. [54] demonstrated that magnetic resonance direct thrombus imaging (MR-DTI) using a 3D T_1W, gradient echo sequence can be reliably used to detect methemoglobin within plaque hemorrhage (specificity and sensitivity of 84% with a positive predictive value of 93% and negative predictive value of 70%). Later works by Chu et al. [56] have focused not only on identifying the presence of the thrombus but also the age of the thrombus as an important factor in plaque vulnerability assessment. The investigators have used imaging methodology previously defined for grading the age of intracerebral hemorrhage [76, 77]. Three stages of IPH have been distinguished: fresh hemorrhage (≤1 week old), recent hemorrhage (between 1 and 6 weeks old), and old hemorrhage (≥6 weeks old). The terminology, however, was modified so that fresh and recent hemorrhage became Type I and Type II hemorrhage, respectively [55]. This was due to an observation that over a period of 18 months IPH evolution on MRI showed a different pattern than that in intracerebral hemorrhages (94% of IPH found at baseline did not demonstrate a change of signal intensity after 18 months [55]). Table 16.2 summarizes the current understanding of different stages of intraplaque hemorrhages.

16.4.1.5 Expansive Remodeling

Arterial expansive remodeling is a process whereby growth of the plaque is not reflected in the narrowing of the arterial lumen [12]. This process has been implicated in contributing to plaque vulnerability by disguising large lesions [78, 79]. Such lesions may not manifest themselves in angiography or B-mode ultrasonography, despite extensive atherosclerotic changes. Patients harboring plaques exhibiting substantial expansive remodeling often present with a major cerebrovascular event as the first manifestation of the disease. MRI is potentially an ideal method for assessment of the extent of this phenomenon as it allows for simultaneous imaging of both the arterial lumen and wall [80].

The most frequently used sequences are black-blood fast spin-echo sequences [78, 80–82]. Corti et al. [81, 83] studied the effects of simvastatin lipid-lowering therapy on human atherosclerotic lesions (aortic and carotid lesions) demonstrated by changes in vessel wall thickness (VWT) and vessel wall area (VWA), which are often used as surrogate markers of expansive remodeling. Changes in both parameters were statistically significant at 12, 18, and 24 months after initiation of treatment regime. At 24 months the VWA in carotid arteries decreased from 46.5 ± 2 mm^2 to 38.0 ± 2 mm^2 ($p<0.0001$), the maximal VWT from 2.65 ± 0.9 mm^2 to 2.14 ± 0.6 mm^2 ($p<0.001$), while the minimal VWT remained unchanged. During the follow-up period the lumen area increased only marginally from 32.6 ± 2 mm^2 to 34.2 ± 2 mm^2. Similar finding were observed by Yonemura et al. [82], who compared the effects of using high- vs. low-dose atorvastatin in aortic plaques. This suggests that a therapeutic effect can be achieved without a substantial decrease in the degree of stenosis; hence, imaging of the plaque itself may be useful in treatment follow-up.

16.4.1.6 Superficial Calcified Nodules

Superficial calcified nodules have been first reported in coronary arteries and subsequently in carotid arteries as well [66]. In the classification of atherosclerotic lesions proposed by the American Heart Association (AHA), plaques with superficial calcifications are classified as lesions Type Vb [53]. Later in a modification to the AHA classification [66], these lesions have been classified as a "calcified nodule" and described as an eruptive nodular calcification with underlying fibrocalcific plaque. The description implies that the clinically significant calcified nodules are close to the vessel lumen and are associated with a disrupted FC and superficial thrombi, in contrast to calcifications that are present deep within the plaque and which are considered benign [66].

An in vivo MRI study on 123 patients was able to visualize, using fast-spin echo multicontrast imaging, calcifications with a sensitivity of 76%, specificity of 86%, and an intraobserver agreement of $\kappa=0.75$ [49]. However, this analysis was performed without regard to the location of calcified nodules. Juxtaluminal nodules, which are associated with a greater risk of rupture however, are more difficult to visualize than those located deeper within the plaque because they appear as a hypointense signal on T_1W, T_2W, and TOF images and are not easily distinguishable from the lumen. This limitation, however, can be overcome by an addition of bright-blood sequences to the imaging, making the hypointense calcification clearly different from the hyperintense signal of blood on TOF [68].

The rapid development of new MRI sequences as well as technological advances of the MRI machines allowing for better spatial resolution, shorter imaging times, and the development of novel, highly specific contrast agents is undoubtedly making MRI the imaging modality of choice in the evaluation of patients with atherosclerosis. Recently Underhill et al. [24] proposed a risk assessment system for patients with carotid atheroma based solely on MRI characteristics. The system was designed to aid in objective and uniform, noninvasive stratification of patients' risk arising from their plaque burden. Additionally, the authors aimed to define features of carotid plaques, which may precede the FC disruption and IPH. The classification system was named the Carotid Atherosclerosis Score (CAS) (Fig. 16.3). It was

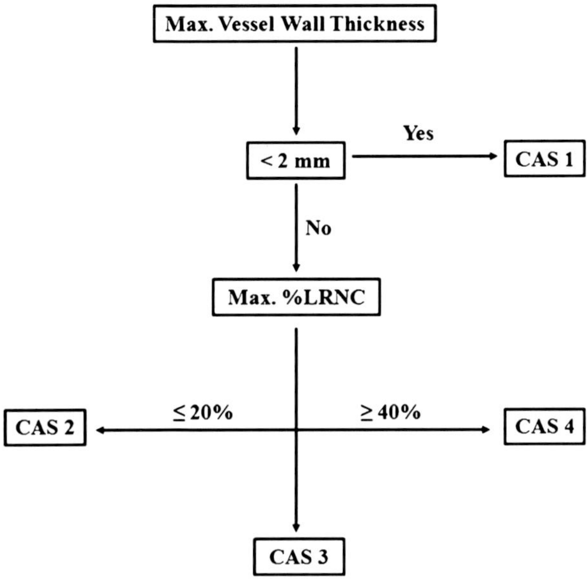

Fig. 16.3 Carotid atherosclerosis score. A schematic illustration of the Carotid Atherosclerosis Score (CAS). CAS 1: low risk; CAS 2: medium–low risk; CAS 3: low–high risk; CAS 4: high risk plaques. Max %LRNC-maximum percentage of plaque wall occupied by LRNC. Adapted from Underhill et al. [24]

determined that the strongest predictors of both FC disruption and IPH were the maximal vessel wall thickness and the percentage of wall occupied by the LRNC. The presence of IPH and calcifications were secondary discriminators. Despite being histologically validated, CAS was based on cross-sectional data and will need to undergo a series of prospective studies for comparison of its predictive value against the degree of stenosis seen on DSA. Nevertheless, it is a step forward in developing largely noninvasive risk assessment methods to guide therapy in patients with carotid atheroma.

The main MRI sequences used and the appearance of various plaque components are summarized in Table 16.3.

16.4.2 Flow Modeling with Shear Stress Estimation

Plaque rupture is a result of structural failure of the inner wall of a diseased artery. One of the proposed mechanisms is biomechanical stress exerted upon the endothelium in the lesion site combined with the diminishing ability of withstanding such stress. Furthermore, it was proposed that tensile wall stress and wall shear stress can be responsible for progression as well as rupture of atherosclerotic lesions [84]. Stress can be measured in carotid arteries with the aid of finite elements analysis (FEA) [85]. Stress modeling using FEA has been validated against histological findings using

Table 16.3 MRI characteristics of carotid plaques

		Appearance of plaque components				
Technique	RT/TE (ms)	LRNC	FC	Thrombus	IPH	Calcifications
SE T2w	1 RR/20/55	Hypointense	Hyperintense	Hyperintense	Hypointense	Dark
FSE T2w	2 RR/30/50	Hypointense	Hyperintense	–	Hypointense	Dark
FSE PDw	2 RR/20	Hypointense	Isointense	–	Isointense	Dark
DIR FSE T1w	570-800/9.3-14	Isointense	Hyperintense	–	Hyperintense	Dark
DIR FSE T2w	3RR/40	Hyperintense/Isointense	Isointense	–	Hyperintense	Dark
DIR FSE PDw	3RR/20	Hyperintense/Isointense	Hyperintense	–	Hyperintense	Dark
3D TOF	23/3.8	Isointense	Hypointense	–	Hyperintense	Dark
3D EPI T1w with IR	10.3/4.0	Hypointense	Isointense	–	Hyperintense	Dark

TR repetition time, *TE* echo time, *LRNC* lipid-rich necrotic core, *FC* fibrous cap, *IPH* intraplaque hemorrhage, *T1w* T1 weighted, *T2w* T2 weighted, *PDw* proton density weighted, *SE* spin echo, *FSE* fast spin echo, *DIR* double inversion recovery, *TOF* time-of-flight, *EPI* echo planar imaging, *IR* inversion recovery, *RR* R-peak (ECG)

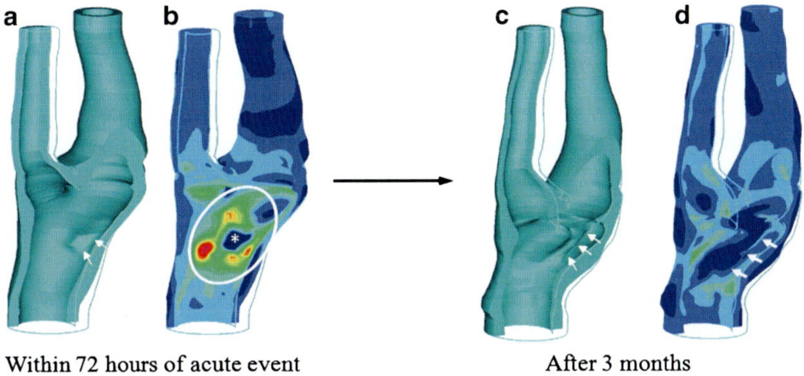

Fig. 16.4 Tensile stress evolution after an acute event caused by plaque rupture. All images are of the same region, obtained from the same patients at different time points. (**a**) 3D reconstruction showing plaque rupture (*white arrows*) located below the carotid bifurcation. (**b**) Wall tensile stress map (*light gray* indicates high stress *dark gray* and *black* indicate low stress) showing high stress (*light gray* area within *white ellipse*) surrounding rupture site (*asterisk*). (**c**) 3D reconstruction obtained 3 months after acute event. Healing process visible in the area of plaque rupture (*white arrows*). (**d**) Wall tensile stress map showing an enlargement of low stress area (*white arrows*)

excised atherosclerotic plaques ex vivo [57]. U-King-Im et al. [57] confirmed that high-resolution MR-derived 2D geometrical arterial models can be used to perform FEA. Trivedi et al. [22] have expanded on this work and performed an MRI-based FEA to predict the differences in plaque tensile stress for symptomatic and asymptomatic patients with carotid atherosclerosis. Results showed a substantial difference in principal tensile stress between the symptomatic and asymptomatic patients (627.6 kPa compared with 370.2 kPa, $p=0.05$, respectively). Figure 16.4 demonstrates changes of tensile stress of arterial wall following an acute event.

16.4.3 Active Inflammation

Inflammation plays a major role in all aspects of atherosclerosis, that is, plaque initiation, plaque progression, and plaque rupture. It has become a target of emerging therapies for atherosclerosis. Systemic markers of inflammation, such as C-reactive protein (CRP), have been correlated with the severity of atherosclerotic disease and risk of cardiovascular events [27]. These circulating markers of inflammation are likely to be a reaction to the local inflammatory process within the plaque itself. The exact mechanisms of inflammation within a plaque are beyond the scope of this chapter; however, there are two aspects of this process, which are potential targets of diagnostic imaging of vulnerable plaques. They are macrophage infiltration of the plaque as part of the inflammatory process and neovascularization. These processes can be visualized using USPIO-enhanced MRI and DCE-MRI [50, 59, 63, 65], respectively. The extent of these processes can be a useful tool in assessment of patients' risk of future stroke and in monitoring therapy.

16.4.3.1 Dynamic Contrast-Enhanced MRI and Neovascularization

Contrast-enhanced MRI (CE-MRI) of atherosclerosis is a developing area. Most commonly gadolinium-based contrast media are used for their T1 shortening ability. Accumulation of the media leads to signal increase on T_1W sequences. By using gadolinium chelates it is possible to investigate late-phase enhancement, which is a surrogate of contrast diffusion to the extracellular compartment, and observe the diffusion of contrast over time. For both CE-MRI and DCE-MRI the choice of image acquisition is essential. Care needs to be taken to sufficiently suppress the blood signal to allow clear visualization of the vessel wall. This can be achieved by using black blood T_1W sequences with either double inversion-recovery (DIR) [86] or more recently proposed, quadruple inversion-recovery (QIR) [87]. For dynamic contrast-enhanced MRI, on the other hand, a sequence that does not suppress blood is sometimes used, e.g., spoiled gradient-recalled echo (SPGR) T_1W.

The deposition of lipid within the vessel leads to the development of microvasculature triggered by the infiltration of inflammatory cells. Previous studies have shown that the extent of microvasculature within a plaque corresponds to plaque rupture and the incidence of intraplaque hemorrhage [88, 89]. One MRI technique for analyzing plaque neovascularization is dynamic contrast-enhanced MRI (DCE-MRI). This technique was originally developed for analysis of tumor neovasculature. It uses sequencing that can be incorporated into routine CE-MRI. The measured parameters may not have very good spatial resolution, however, this method gives priority to temporal resolution. Furthermore, it uses widely available contrast (licensed in humans) to assess tissue microvasculature and endothelial permeability [90]. First observations of contrast enhancement of atherosclerotic plaques were performed by Lin et al. [91] in pigs and subsequently by Aoki et al. in humans [92]. These first accounts report a bright rim around the imaged artery, which was thought to be due to the presence of vasa vasorum in the adventitia of the artery. Later works noted enhancement within the plaque especially in the FC [50, 93].

The principal of DCE-MRI image analysis today is to use kinetic modeling of dynamic contrast enhancement seen on MRI typically in T_1W [58, 59, 94]. The applied kinetic model is based on an assumption that tissues have two compartments, the intravascular space and an extravascular space, which determine the concentration of contrast as well as a third compartment which has no uptake of contrast properties. Two parameters that are considered for analysis are fractional plasma volume (V_p) and transfer constant of the contrast agent (K^{trans}) and represent the two compartments: intravascular space and extravascular, extracellular space. V_p is associated with the intravascular space and in consequence represents the actual microvascular volume [59] (Fig. 16.5). V_p was found to be a good marker of plaque neovasculature ($r=0.68$, $p<0.001$) [58, 59]. K^{trans} represents the extravascular, extracellular space and in turn is used to estimate the permeability of the microvasculature in the region of interest by measuring only extravasated contrast. K^{trans} in a study of 27 patients who subsequently underwent CEA was found to correlate with plaque macrophage content ($r=0.75$, $p<0.001$) and amount of microvessels ($r=0.71$, $p<0.001$) [59]. This showed that neovascularization measured using MRI can be used as a marker of inflammation within the plaque.

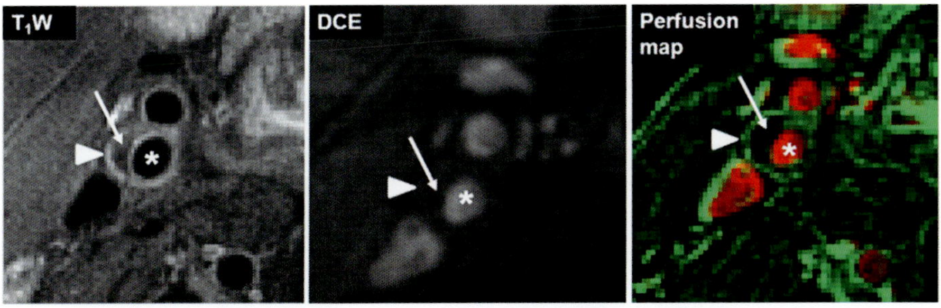

Fig. 16.5 Right internal carotid artery with atherosclerotic plaque (asymptomatic patient). T_1W – T1 weighted sequence; DCE – dynamic contrast-enhanced fat saturated, *black blood*, gradient echo sequence. Perfusion map shows enhancement of adventitia vasa vasorum (*arrow head*) and in the lipid core (*arrow*). $K^{trans}=0.115\ min^{-1}$ in keeping with literature for moderate severity of disease [60]

Further development of MRI contrast is now being driven by designing agents that would bind to specific targets in the atherogenic process on the cellular/molecular level. This could overcome the major limitation of all CE-MRIs, which is nonspecific enhancement of surrounding tissue. Contrast media that bind to thrombus, specific inflammatory mediators, and apoptotic cells or factors that promote angiogenesis have been under investigation and have the potential to act similarly as the contrast agents used in nuclear medicine. Many of these new techniques use gadolinium as a basis for the agent to provide the contrast on MRI.

16.4.3.2 USPIO-Enhanced MRI and Macrophage Content

One agent, that is not gadolinium based, which is used to target inflammation at a cellular level is ultrasmall superparamagnetic iron oxide (USPIO) particles. USPIO consists of a microcrystalline magnetite core, within dextran coating. Typically the diameter of USPIO is around 30 nm [95]. The particles accumulate in the reticuloendothelial system (RES) of the liver and spleen after injection into the circulation. The half life of USPIO particles in circulation is much longer than that of larger superparamagnetic iron oxides with a diameter of approximately 150 nm. This allows USPIO particles to be taken up by macrophages and transported into loci of inflammation, such as the atherosclerotic plaques where they can be detected on imaging [96, 97].

USPIO acts as an enhancing media in MRI by creating a large dipolar magnetic field gradient acting on water molecules that are in close proximity, thus reducing the T2 relaxation times [98]. USPIO has a predominantly T2/T2* shortening effect and a negative contrast is observed on T_2W. The T2 effect creates a hypointense area on conventional spin-echo (SE) MR sequences (Fig. 16.6). Because of a T_2* sensitivity of gradient echo (GE) MRI sequences they are more sensitive to USPIO than spin echo sequences but may be subject to poorer image quality [98].

Studies have shown that USPIO particles are taken up by macrophages and transported to atherosclerotic lesions where they cause a T2*-weighted (T_2*W) signal [96]. Later the same was confirmed in atherosclerotic plaques in the carotid arteries

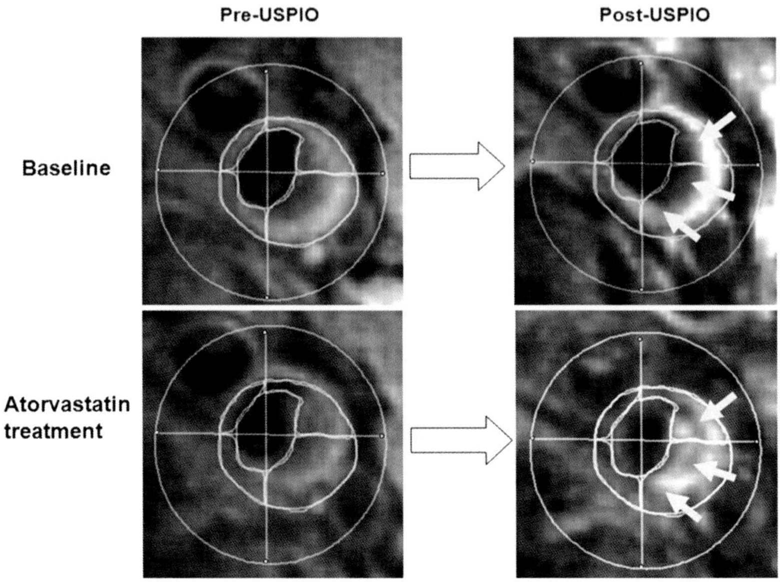

Fig. 16.6 T2* effect of USPIO particles in carotid plaque (*white arrows*). *Top row* – baseline images. *Bottom row* – images after 12 weeks of treatment with atorvastatin (80 mg/day)

of humans [64, 65, 97, 99]. Furthermore, Kooi et al. used excised plaques from ten patients who prior to CEA underwent USPIO-enhanced MRI of the carotid plaque (mean time between administration of USPIO and surgery 4.7 ± 3.0 days). This showed a signal loss on USPIO-enhanced T_2*W GE MRI, which was correlated with the presence of iron on histological examination of the excised plaques (Perls staining and electron microscopy) [65]. Later Trivedi et al. [64] determined a temporal relationship of the intensity of signal loss on double-inversion recovery, blood-suppression T_2*W USPIO-enhanced MRI (Fig. 16.7). A time interval for greatest signal loss was demonstrated between 24 and 36 h after injection of USPIO, thus establishing the optimal window for postinfusion imaging. In later studies USPIO-enhanced MRI has been correlated against the degree of stenosis and biomechanical stress on the plaque [21, 62]. In one study 71 patients with asymptomatic carotid atheroma of ≥40% stenosis underwent multi sequence USPIO-enhanced MRI. No correlation was found between the degree of stenosis and plaque inflammation measured with USPIO, suggesting that inflammation may be an independent risk factor for plaque vulnerability [62]. This supports earlier works suggesting that risk stratification of patients with carotid atheroma should not be made on the basis of lumen narrowing, especially for those for whom the evidence behind surgical treatment is not as strong [17–19, 100]. Furthermore, a study undertaken on 18 patients with angiographically proven carotid atherosclerosis has shown a highly significant correlation for USPIO-enhanced signal change and maximal biomechanical stress ($p < 0.009$) [21].

Another aspect of noninvasive imaging of atheroma inflammation is its potential to validate new therapies. So far only one study has been preformed where the outcome measure was assessed using USPIO-enhanced MRI. In the ATHEROMA

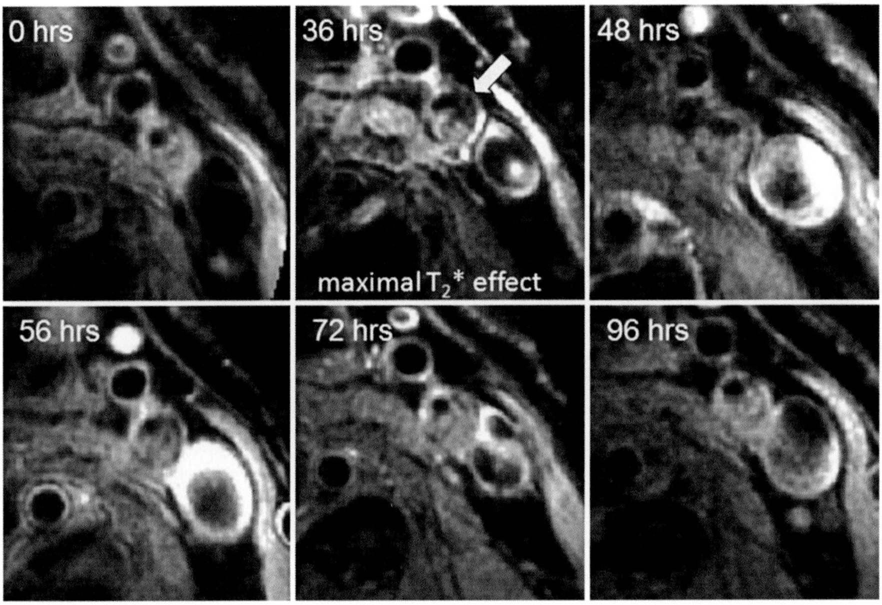

Fig. 16.7 Temporal resolution of USPIO-enhanced MRI. Maximal T2* effect (*white arrow*) was seen 36 h post USPIO administration

trial [61], 47 patients were randomized to either low-dose (10 mg) or high-dose (80 mg) atorvastatin for 12 weeks. Aggressive lipid lowering with high-dose atorvastatin in the trial was associated with a significant reduction of inflammation as seen on USPIO-enhanced MRI after 3 months of therapy. The observed effects of statin therapy at 3 months were in clear contrast to the observed effects of previous interventional studies [13, 78, 81, 83, 101], where only morphological effects were analyzed. Corti et al. [78, 81, 83] were able to demonstrate a reduction in vessel wall area and vessel wall thickness at 12 months but not 6 months as there is an inevitable delay in gross morphological changes. These results imply a potential role of inflammation imaging in future evaluation of anti-inflammatory medications for the treatment of atherosclerosis, allowing for early detection of treatment effect.

16.5 Positron Emission Tomography and Single Photon Emission Computed Tomography

Nuclear imaging techniques are not suitable for morphological imaging due to low spatial resolution. Nuclear imaging techniques are very sensitive in detecting radioactive tracers designed to accumulate in specific tissues. This being the reason why PET and SPECT are mainly being used to assess plaque inflammation [102, 103], plaque apoptotic activity [104], and plaque proteolytic enzyme activity [104], rather than morphological parameters.

16.5.1 Inflammation

In recent years, PET has been used in studies of carotid atherosclerotic plaque inflammation. Fluorine-18-labeled 2-deoxy-D-glucose (FDG) is a glucose analog that accumulates in cells proportionally to their metabolic activity. Labeling with radioactive fluorine-18 allows for detection with nuclear imaging [105]. A study conducted on individuals known to harbor various infections has shown that FDG accumulates in loci of inflammation [102] and can be measured using PET (FDG-PET).

The first study of FDG-PET in carotid atherosclerosis, performed by Rudd et al. [106], tested the hypothesis that plaque inflammation could be successfully identified by FDG accumulation as measured using FDG-PET (Fig. 16.8). The study was performed on eight patients with unilaterally symptomatic carotid atheroma due to undergo CEA. Coregistration of the lesion area with CT was performed. A difference in uptake was shown between the symptomatic and asymptomatic side. Histological analysis using carotid plaque autoradiography showed that uptake and accumulation of FDG was predominant on the border between FC and LRNC in plaques from the symptomatic sides; however, no specific identification of cells was performed. In a subsequent study, Tawakol et al. [107] assessed the localization of FDG accumulation in atherosclerotic carotid plaques using staining designed to identify macrophages (anti-CD68 antibodies). A correlation was found between mean FDG uptake as seen of FDG-PET and plaque macrophage content on histology ($r=0.85$; $p<0.0001$) [107]. FDG-PET has been assessed for reproducibility with encouraging results: with high interobserver agreement: ICC values from 0.71 to 0.97 depending on the artery analyzed (0.71 for aortic arch) and intraobserver agreement: ICC values between 0.93 and 0.98 [108]. Additional confirmation of good specificity of FDG-PET in imaging

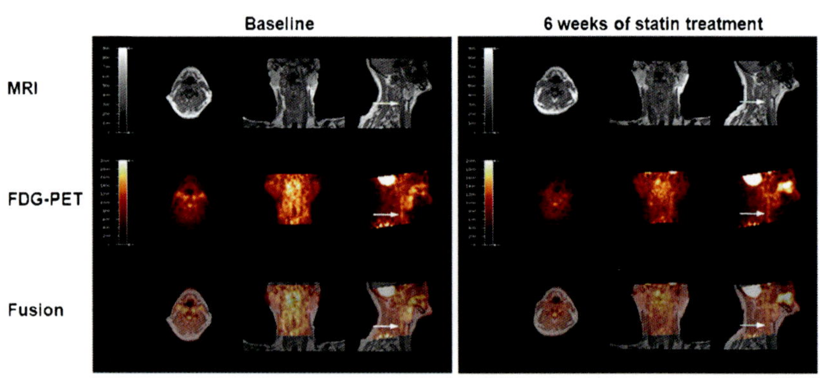

Fig. 16.8 FDG-PET as imaging biomarker in proof of principle trials: example from ongoing proof of principle randomized controlled trial of patient with recent TIA imaged with FDG-PET and MRI at baseline and 6 weeks after commencing either 10 or 80 mg Atorvastatin (blinded dataset). Mean and maximum SUV values in the region of left carotid artery plaque (*white arrows*) have decreased by 14.5 and 7.4%, respectively (image courtesy of Dr, Elizabeth Warburton, Department of Clinical Neurosciences, University of Cambridge, Addenbrooke's Hospital, Cambridge, UK)

inflammation of atherosclerosis came from a subsequent study by Rudd et al. [109]. The investigators aimed to determine the relationship between FDG accumulation, cardiovascular risk factors, and systemic markers of inflammation. FDG uptake was found to be significantly higher in plaques of patients with a history of cardiovascular disease and patients who had higher levels of serum inflammatory biomarkers, such as IL-18 and CRP. Additionally, an atheroprotective biomarker adiponectin was negatively correlated with FDG uptake by plaques.

Inflammation was postulated to be a surrogate marker for identifying plaques with a large LRNC [110]. This, however, has been contradicted by other studies where a weak correlation between plaque composition on MRI and plaque inflammation on FDG-PET was observed [111].

Despite the vast array of evidence that FDG-PET can image active inflammation within atherosclerotic plaques, only few prospective interventional studies, using this modality to assess the effectiveness of antiatherosclerotic therapies have been conducted. Tahara et al. [112] were the first to use FDG-PET as an outcome measure in an interventional study of statins. They assessed 43 subjects for plaque uptake of FDG before and after 3 months of simvastatin treatment or diet alone. Simvastatin attenuated the standardized uptake values (SUVs) of FDG compared with diet alone ($p<0.001$). Another prospective study was conducted by Lee et al. [113] who assessed changes in uptake of FDG as an effect of lifestyle changes in 60 healthy patients. They analyzed 60 subjects and found a total of 352 FDG-positive lesions at baseline (mean 5.9 ± 5.0 lesions/subject). The total amount of FDG-positive lesions was significantly reduced after a mean of 17.1 months of lifestyle changes (mean 2.1 ± 2.2 lesions/subject; $p<0.0001$).

FDG-PET is limited by its spatial resolution (order of 4–5 mm). The advent of integrated CT-PET technology should provide automatic coregistration of areas of high signal on PET with the structural information provided by CT. Another disadvantage of PET-FDG imaging is the associated radiation burden, which limits the number of PET studies that patients can have (two in total), reducing its potential contribution in assessing therapeutic interventions. It is also not formally licensed to be used in diabetics by the FDA in the USA, excluding an important group of vasculopathies from imaging and potential risk stratification.

16.6 The Vulnerable Plaque in Clinical Trials

It is now well recognized that harboring a vulnerable plaque, whether or not it results in severe stenosis, is associated with an increased risk of future TIA or stroke [19, 78]. While the management of patients who have clear indications for surgical or endovascular treatment is established, it is the patients who do not meet those criteria who are most problematic: symptomatic patients with mild stenosis and asymptomatic patients with severe stenosis. Those are the patients who potentially need the closest and most frequent follow-up. Those are also the patients who could benefit most from pharmacological therapy directed at the reduction of plaque vulnerability and postponing surgery. Recently, a number of prospective clinical trials have been conducted to address the problem of high-risk plaques (Table 16.4).

The benefits arising from statin treatment of cardiovascular diseases have been established [114]. Experimental studies have also indicated that statin therapy has a positive effect on carotid atherosclerotic plaques. In a randomized study that included 24 patients with severe, symptomatic carotid atheroma (11 randomized to 40 mg of pravastatin and 13 to no lipid-lowering therapy), it was demonstrated that pravastatin had a plaque-stabilizing effect. After 3 months of treatment, patients underwent CEA and the excised plaques were subjected to histological analysis. Specimens obtained from pravastatin-treated individuals, compared with the control group, demonstrated a decrease in lipid content (8.2% ± 8.4% vs. 23.9% ± 21.1%; $p<0.05$), oxidized-LDL immunoreactivity (13.3% ± 3.6% vs. 22.0% ± 6.5%; $p<0.001$), macrophage content (15.0% ± 10.2% vs. 25.3% ± 12.5%; $p<0.05$), MMP-2 immunoreactivity (3.6% ± 3.9% vs. 8.4% ± 5.3%; $p<0.05$), and apoptotic cell death (17.7% ± 7.8% vs. 32.0% ± 12.3%; $p<0.05$) [115]. New imaging modalities, including MRI, PET, SPECT, and B-mode US have the ability to measure these characteristic features of vulnerable plaques in vivo. Their additional advantage of using these methods is that they can be performed sequentially to monitor the progression of the disease and the effectiveness of therapy.

Corti et al. [81, 83] have conducted a prospective longitudinal study that included 18 hypercholesterolemic subjects with documented carotid atherosclerotic lesions. The patients were treated throughout the duration of the study with simvastatin. Imaging using MRI was performed at 6 weeks, 6, 12, 18, and 24 months. The measured parameters included lumen area (LA), vessel wall thickness (VWT), and vessel wall area (VWA). They have reported that after 12 months of treatment there is significant reduction in both VWA and VWT (from 49 ± 11 mm^2 to 42 ± 11 mm^2 and from 2.7 ± 0.5 mm^2 to 2.4 ± 0.5 mm^2, respectively) without accompanying changes in the LA.

Follow-up analysis revealed a further decrease in VWA and VWT at 18 and 24 months. The late changes were accompanied by a slight increase in LA. A later study, designed to compare the effects of aggressive and conventional lipid-lowering therapies (80 mg/day vs. 20 mg/day of simvastatin) on aortic and carotid plaque vulnerability, also used high-resolution MRI. The study included 51 patients with asymptomatic aortic and/or carotid atherosclerotic disease and used the same MRI protocol for plaque characterization. Similarly at least 12 months were required to detect changes in carotid plaque size as seen on MRI. Carotid VWA decreased by 14 and 18% at 12 and 24 months, respectively, while VWT decreased by 10 and 17%, respectively [80].

The ORION study [101] used 2D T_1W, T_2W, and PDw black-blood sequences and 3D TOF bright-blood angiography to evaluate the effect of high- and low-dose rosuvastatin on carotid plaque volume and composition. 43 patients with 16–79% stenosis evaluated by B-mode US were included in this study. The morphological parameters included LA, VWA, and normalized wall index (NWI=WA/[LA=WA]). Plaque composition was assessed as percentage of WA and included percentage of LRNC, percentage of calcification, and percentage of FC. The authors did not

Table 16.4 MRI characteristics of vulnerable plaques in clinical trials

Authors	Design	*n*	Intervention	Time	Imaging	Results
Corti et al. [81, 83]	Long.	18	Simvastatin	24 months	DIR, FSE PDw, T2w	VWA: decrease of 15 and 18% at 12 and 24 months; VWT: decrease of 11 and 19% at 12 and 24 months; LA: increase of 5% at 24 months
Lima et al. [80]	RCT	27	20 vs. 80 mg/day simvastatin	6 months	DIR, FSE PDw, T2w	PV: decrease of 12% from baseline; PA: decrease of 12% from baseline; no difference between groups
Corti et al. [78]	RCT	51	20 vs. 80 mg/day simvastatin	24 months	DIR, FSE PDw, T2w	VWA: decrease of 14 and 18% at 12 and 24 months, respectively; VWT: decrease of 10 and 17% at 12 and 24 months, respectively; LA: increase of 4 and 5% at 12 and 24 months, respectively; no differences between groups
Tahara et al. [112]	RCT	43	Statin vs. diet	3 months	FDG-PET	SUV: significant reduction only in statin group at 3 months
Crouse et al. [116]	RCT	984	40 mg/day rosuvastatin vs. placebo	24 months 12 weeks	B-mode US	CIMT: progression rate at 24 months decreased in rosuvastatin group −0.0014 mm/year (95% CI: −0.0041, 0.0014) vs. placebo 0.0131 mm/year (95% CI: 0.0087, 0.0174)
Tang et al. [61]	RCT	47	10 vs. 80 mg/day atorvastatin		USPIO-enhanced	ΔSI: 0.203 (95% CI: 0.065, 0.198) reduction from baseline at 6 weeks (high dose); difference of 0.240 (95% CI: 0.134, 0.347) between groups at 12 weeks

Long. longitudinal, *USPIO* ultrasmall superparamagnetic iron oxide, *FDG-PET* fluorodeoxyglucose positron emission tomography, *B-mode US* B-mode ultrasonography, *VWT* vessels wall thickness, *VWA* vessel wall area, *LA* lumen area, *PA* plaque area, *PV* plaque volume, *ΔSI* signal intensity change, *SUV* standardized uptake value, *CIMT* carotid intima/media thickness observe a change in any of the morphological parameters at 24 months. However, a reduction in percentage of LRNC of 41.4%±9.6% ($p=0.005$) during the treatment period was noted for those patients who at baseline had an identified LRNC. In the subgroup analysis the decrease was significant only in the high-dose statin group (37.0%±9.3%, $p=0.014$). Furthermore, from the patients without a LRNC at baseline none developed such during the study duration. Percentage of calcifications and percentage of FC did not show any difference both between baseline and 24 months and between high- and low-dose groups.

A different method of imaging atherosclerotic lesion changes was adapted for the ATHEROMA study [61]. Similarly, patients were randomized to high- (80 mg/day atorvastatin) and low-dose (10 mg/day atorvastatin) therapy; however, imaging of carotid arteries was performed using USPIO-enhanced MRI. The following fat-suppressed fast spin-echo DIR sequences were used: T1w, STIR, intermediate, T_2W, and T2*W with a multishot spiral acquisition. Changes in plaque inflammation between the two groups were assessed at 6 and 12 weeks of allocated treatment. USPIO signal intensity changes (ΔSI) were calculated in a way that a positive ΔSI indicated a decrease in macrophage activity. At 6 weeks, both groups showed a positive ΔSI; however, at 12 weeks the macrophage activity in the low-dose group increased ($\Delta SI=-0.038$) while in the high-dose group decreased further ($\Delta SI=0.203$). The difference between the two groups at 12 weeks was $\Delta SI=0.240$ ($p<0.0001$), demonstrating that a better effect of decreasing plaque inflammation can be achieved with 80 mg/day than 10 mg/day atorvastatin. This first implementation of a novel method, USPIO-enhanced MRI, into a clinical trial setting allowed the investigators to show beneficial effects of aggressive statin therapy as early as 6 weeks after initiation of treatment (Fig. 16.6), which was a vast improvement compared with the previous studies using only morphological parameters [78, 80, 81, 83, 101].

The usefulness of FDG-PET in evaluation of carotid plaque vulnerability has so far been tested in one randomized controlled trial. Tahara et al. [112] randomized 43 asymptomatic patients harboring atherosclerotic lesions to either simvastatin (40 mg/day). At 3 months, patients receiving simvastatin demonstrated a reduction in SUV in their atherosclerotic plaques compared with the diet group ($p<0.01$). Despite the early changes detected in the study, FDG-PET has not been readily used in clinical trials, due to both the radioactive exposure connected and the cost of performing nuclear imaging.

A number of studies have focused on the morphological parameters as seen using B-mode US and the risk of stroke for patients with carotid atherosclerotic disease. The parameters evaluated were IMT [33, 40], plaque echolucency [35, 36], and plaque irregularity [43]. The METEOR study randomized 984 low-risk (Framingham Risk Score <10%) individuals to either 40 mg/day rosuvastatin or placebo and measured changes of IMT at 24 months. Results showed a reduction in the rate of progression of IMT in patients receiving statins compared with those receiving placebo (−0.0014 mm/year vs. 0.0131 mm/year for rosuvastatin and placebo groups, respectively; $p<0.001$), however, the study did not demonstrate any regression in plaque burden measured by IMT [116].

The availability and relatively low cost of B-mode US is an advantage for implementation in prospective studies. However, conventional US imaging allows only for the evaluation of gross morphological changes and requires a long follow-up period to depict significant changes. The usefulness of novel CEU techniques [39, 46, 47, 117], which potentially may provide additional and earlier information about the vulnerability of atherosclerotic lesions, remains to be tested in prospective clinical trials.

16.7 Future Directions

Imaging of the morphology and composition of carotid plaques is becoming a part of clinical practice and the design of clinical trials. With the aid of targeted contrast media, it is now possible to depict early changes in the functional state of carotid plaques. The usefulness of USPIO has been demonstrated in the ATHEROMA study, which detected changes in inflammatory processes within the plaque as early as 6 weeks posttreatment [61]. Furthermore, FDG-PET has also been used to detect early changes at 3 months after statin administration [112]. Both these studies demonstrated changes within the plaque up to 9 months before any morphological manifestations could be visualized.

New contrast media for MRI, nuclear imaging, and ultrasonography are being developed to target processes of atherosclerosis formation at early stages and with high specificity to predetermined molecule- or tissue-specific targets.

Molecular MRI contrasts have been used mostly in animal models to specifically target lipids, e.g., HDL [118], oxidized-LDL [119]; various adhesion molecules, e.g., VCAM-1 [120]; inflammatory cells, e.g., macrophage scavenger receptors (MSR) [121]; proteolytic enzymes, e.g., matrix metalloproteinases [122]; and thrombus, e.g., fibrin, which has since been studied in humans [123].

A study by Lancelot et al. [122], performed in ApoE knockout mice, has focused on matrix metalloproteinases (MMPs), enzymes that are implicated to play a part in atherosclerotic plaque rupture, due to their collagen-degrading properties [124]. MR images obtained showed a markedly better contrast enhancement of aortic plaques using MMP-specific media (P947), compared with standard contrasts. Furthermore, observations showed that the highest contrast intensities were observed in the FC and shoulders of the plaque, but not the LRNC, which is consistent with histological reports regarding MMPs localization within an atherosclerotic lesion [125, 126].

Another study conducted in ApoE knockout mice by Briley-Saebo et al. [119] focused on in vivo MRI of oxidized-LDL. Oxidized-LDLs are suspected to play a part in the initiation and progression of the inflammatory process within a plaque, as they are involved in the initiation of a T-cell mediated autoimmune response [127]. Micelles containing gadolinium with either murine (MDA2 and E06) or human (IK17) IgG antibodies against oxidation-specific epitopes were used. ApoE knockout mice showed maximal uptake of micelles at 72 h (MDA2 and IK17) and

96 h (E06) after injection. This was manifested by 125–231% enhancement compared to adjacent muscle.

So far, the only molecular MRI media that was used in humans for imaging atherosclerosis was a fibrin-specific contrast EP-2104R. Thrombus demonstrated high signal amplification compared with adjacent muscle and surrounding blood [123].

As in the case of MRI, developments in nuclear molecular imaging are heading in the direction of devising highly specific, targeted contrasting agents that would have affinity to single molecules. This would allow for imaging of the functional status of an atherosclerotic lesion at a molecular level. With the aid of nuclear molecular imaging, studies are focusing on depicting markers of inflammation within a plaque, e.g., alfa(v)beta3-integrin [128], IL-2 [103]; markers of active recruitment of inflammatory cells into the lesion, e.g., VCAM-1 [129]; markers of apoptotic cell death, which are implied in plaque destabilization, are being targeted, e.g., annexin A5 (AA5) [104]; and activity of proteolytic enzymes, which could be contributing to plaque destabilization, such as MMPs [104].

Molecular-targeted media are also being studied in CEU. However, due to their character, they differ from those used for MRI or nuclear imaging. They are composed of microbubbles covered with molecules with affinity toward specific targets, e.g., antibodies against specific epitopes [29, 39]. These microbubbles cannot be directed at elements of atherosclerotic lesions residing within the neointima such macrophages, oxidized-LDLs, apoptotic cells, and proteolytic enzymes, as they are not diffusible [29, 39]. One of the most common ways of using microbubbles in CEU is to image microvessels [29, 47]. The use of these media for targeting molecular markers of inflammation relies mostly on endothelial and vascular cell adhesion molecules (ECAMs and VCAMs). These are molecules expressed on the luminal endothelial surface, which play a part in the adhesion and recruitment of inflammatory cells (lymphocytes and monocytes) into the lesion area and thus may be responsible for the progression and destabilization of atherosclerotic plaques [25, 26]. The ability of microbubbles to target VCAM-1 expressed on the arterial endothelium has been shown in ApoE knockout mice [117]. However, one of the main limitations of CES is the fact that by relying on adhesion molecules within vessels for targeted binding, microbubbles are subject to shear stress during blood flow. In large vessels with atherosclerotic plaques the shear stress tends to be high, especially during systole. The main objective during development of microbubbles is to facilitate adhesion during diastole, when shear is low and generate sufficiently strong adhesion that will be able to withstand high-shear systole. So far, most microbubble contrasting media rely on monoclonal antibodies. Because of the potentially small amount of reactions between target molecules and monoclonal antibodies there is a risk of insufficient adhesion to withstand high-velocity and high-shear blood flow [39].

Importantly, for implementation of any one method into clinical practice substantial technological improvements in the image processing protocols as well as imaging machines themselves are needed. Which method, whether MRI, nuclear imaging, or ultrasonography will prevail in the arena of molecular imaging remains to be seen; however, the availability and relative cost of the methods will surely play a role.

Table 16.5 summarizes the various diagnostic imaging modalities that use molecular imaging to target atherosclerosis, including MRI, nuclear imaging, and contrast-enhanced ultrasonography.

16.8 Summary

Carotid atherosclerosis imaging is an important aspect of patients' risk stratification and management. Various characteristics that are not detectable using DSA are factors in determining actual plaque burden for patients. The imaging modalities most readily used in evaluation of carotid plaques are B-mode ultrasonography, MRI, and nuclear imaging.

B-mode ultrasonography is currently one of the most frequently used screening tools for carotid atherosclerotic disease. Despite being routinely used for assessment of the degree of stenosis, there are conflicting results regarding the predictive value of various ultrasonographic characteristics of plaque vulnerability. This aspect requires validation in prospective interventional trials [5, 35, 41]. The emergence of CES with targeted contrast agents allows for functional assessment of carotid plaques rather than morphological [46, 47, 117]. Validation of this method is needed; however, due to the wide availability and relatively low cost, CEU may become the method of choice in future research and disease management.

MRI is useful in morphological characterization of lesions due to high tissue resolution. It can be used to distinguish FC [51–53, 67, 132], LRNC [51, 52, 67, 132], FC disruption [53, 69, 133], and IPH [56, 75]. Furthermore, it can be used to evaluate plaque neovascularization by using DCE-MRI [58–60, 89, 91] and inflammation with USPI-enhanced MRI [21, 62, 65, 96, 97, 99] and to calculate the shear and tensile stress [20, 22, 57, 85]. MRI has been used in clinical trials [61, 78, 80, 83, 101, 115] of emerging pharmacological therapies for carotid atheroma and has been evaluated against histology [24, 51, 61].

A new and developing field is molecular and targeted MRI [119, 122, 130, 131]. With the aid of these methods, MRI is able to depict the functional state of the atheromatous plaque at the molecular level and guide medical interventions. Molecular MRI can also be used to study carotid disease pathophysiology in vivo. Unfortunately, it relies on highly sophisticated technology both in contrast media development as in the MR image processing techniques. That is one of the reasons why, despite the great potential, molecular MRI techniques are at present still in the experimental stage and surely need further validation before they could be applicable in humans.

PET is a promising technique used for functional evaluation of atherosclerotic lesions. A vast amount of research is currently underway to develop new, highly specific contrasting agents with affinity to inflammatory molecules [105, 128, 129]. So far, most studies have focused on using fluorine-18-labeled 2-deoxy-D-glucose (FDG) as a marker of inflammation [106–113]. FDG-PET has been evaluated in various experimental and clinical settings. Although the method has been shown to

Table 16.5 Summary of studies using molecular imaging of atherosclerosis

Authors	Technique	Target	*n*/population	Results
Rudd et al. [106]	FDG-PET	Inflammation/macrophages	8/Humans (symptomatic stenosis with TIA)	FDG uptake: greater for symptomatic patients
Tahara et al. [112]	FDG-PET	Inflammation/macrophages	43/Humans (cancer screening)	FDG uptake: greater for the diet alone group than simvastatin group
Kwee et al. [111]	FDG-PET	Inflammation/macrophages	50/Humans (symptomatic carotid stenosis)	Weak correlation between MRI/CT and FDG-PET
Laitinen et al. [128]	PET	Alfa(v)beta3-integrin	Mice	Signal intensity was associated with macrophage content and FDG uptake
Nahrendotf et al. [129]	PET	VCAM-1	Mice (ApoE–/–): normal vs. statin-enriched diet	Signal intensity lower in the atorvastatin group
Haider et al. [104]	SPECT	MPPs, Phosphatydylserine (apoptosis)	12/Rabbits (6-high chol. diet; 6-control)	MPI and annexin A5 uptake greater for high cholesterol group than control
Annovazzi et al. [103]	SPECT	IL-2R	14/Humans	Enhancement correlated with IL-2R+ cells on histology
Kelly et al. [130]	MRI	VCAM-1	3/Mice	Signal intensity: greater before administration of contrast
Lipinski et al. [131]	MRI	MSR A t. I and II	Ex vivo	Signal intensity: greater when incubated with contrast agent than without
Briley-Saebo et al. [119]	MRI	Oxy-LDL	33/Mice (ApoE–/–)	Signal intensity maximum at 72–96 h

Lancelot et al. [122]	MRI	MMPs	Ex vivo (rabbit and human aorta specimens)	MMP-rich MRI images correspond to MMP activity on histology
Spuentrup et al. [123]	MRI	Fibrin	11/Humans	High signal amplification of thrombus
Kaufmann et al. [117]	CES	VCAM-1	23/Mice (ApoE–/– and wild-type) and ex vivo	Signal intensity: greater for ApoE–/–; microbubbles attach to endothelial cells

NIRF near infrared fluoroscopy, *MMP* matrix metalloproteinase, *FDG* fluorine-18-labeled 2-deoxy-D-glucose, *FDG-PET* fluorine-18-labeled 2-deoxy-D-glucose-labeled positron emission tomography, *SPECT* single photon emission computed tomography, *MPI* matrix metalloproteinase inhibitor, *VCAM-1* vascular cell adhesion molecule-1, *MSR A* macrophage scavenger receptor A, *oxy-LDL* oxidized-low density lipoproteins, *CES* contrast-enhanced ultrasonography be able to depict inflammation in a highly specific manner [107, 109], has good correlation with histology [106], and good reproducibility [8, 108], controversies still remain as regards the ability of FDG-PET to discriminate between different components of a plaque [110, 111].

The main limitation of positron emission tomography is high exposure to radiation. This makes PET not suitable for continuous monitoring of disease progression in patients with atherosclerosis. Other limitations include costs and availability of this technique. PET ranks among the most expensive methods used in diagnostic imaging.

References

1. Fisher M (1951) Occlusion of the internal carotid artery. AMA Arch Neurol Psychiatry 65: 346–377.
2. Rothwell PM, Eliasziw M, Gutnikov SA, et al. (2003) Analysis of pooled data from the randomised controlled trials of endarterectomy for symptomatic carotid stenosis. Lancet 361: 107–116.
3. Mayberg MR, Wilson SE, Yatsu F, et al. (1991) Carotid endarterectomy and prevention of cerebral ischemia in symptomatic carotid stenosis. Veterans Affairs Cooperative Studies Program 309 Trialist Group. JAMA 266: 3289–3294.
4. Hobson RW, 2nd, Weiss DG, Fields WS, et al. (1993) Efficacy of carotid endarterectomy for asymptomatic carotid stenosis. The Veterans Affairs Cooperative Study Group. N Engl J Med 328: 221–227.
5. Halliday A, Mansfield A, Marro J, et al. (2004) Prevention of disabling and fatal strokes by successful carotid endarterectomy in patients without recent neurological symptoms: randomised controlled trial. Lancet 363: 1491–1502.
6. Barnett HJ, Taylor DW, Eliasziw M, et al. (1998) Benefit of carotid endarterectomy in patients with symptomatic moderate or severe stenosis. North American Symptomatic Carotid Endarterectomy Trial Collaborators. N Engl J Med 339: 1415–1425.
7. Randomised trial of endarterectomy for recently symptomatic carotid stenosis: final results of the MRC European Carotid Surgery Trial (ECST). (1998) Lancet 351: 1379–1387.
8. Endarterectomy for asymptomatic carotid artery stenosis. Executive Committee for the Asymptomatic Carotid Atherosclerosis Study. (1995) JAMA 273: 1421–1428.
9. Willinsky RA, Taylor SM, TerBrugge K, et al. (2003) Neurologic complications of cerebral angiography: prospective analysis of 2,899 procedures and review of the literature. Radiology 227: 522–528.
10. Johnston DC, Chapman KM, and Goldstein LB (2001) Low rate of complications of cerebral angiography in routine clinical practice. Neurology 57: 2012–2014.
11. Hankey GJ, Warlow CP, and Molyneux AJ (1990) Complications of cerebral angiography for patients with mild carotid territory ischaemia being considered for carotid endarterectomy. J Neurol Neurosurg Psychiatry 53: 542–548.
12. Glagov S, Weisenberg E, Zarins CK, et al. (1987) Compensatory enlargement of human atherosclerotic coronary arteries. N Engl J Med 316: 1371–1375.
13. Stary HC, Chandler AB, Dinsmore RE, et al. (1995) A definition of advanced types of atherosclerotic lesions and a histological classification of atherosclerosis. A report from the Committee on Vascular Lesions of the Council on Arteriosclerosis, American Heart Association. Circulation 92: 1355–1374.
14. Stary HC, Chandler AB, Glagov S, et al. (1994) A definition of initial, fatty streak, and intermediate lesions of atherosclerosis. A report from the Committee on Vascular Lesions of the Council on Arteriosclerosis, American Heart Association. Circulation 89: 2462–2478.

15. Moore WS, and Hall AD (1968) Ulcerated atheroma of the carotid artery. A cause of transient cerebral ischemia. Am J Surg 116: 237–242.
16. Imparato AM, Riles TS, and Gorstein F (1979) The carotid bifurcation plaque: pathologic findings associated with cerebral ischemia. Stroke 10: 238–245.
17. Naghavi M, Libby P, Falk E, et al. (2003) From vulnerable plaque to vulnerable patient: a call for new definitions and risk assessment strategies: Part I. Circulation 108: 1664–1672.
18. Naghavi M, Libby P, Falk E, et al. (2003) From vulnerable plaque to vulnerable patient: a call for new definitions and risk assessment strategies: Part II. Circulation 108: 1772–1778.
19. Takaya N, Yuan C, Chu B, et al. (2006) Association between carotid plaque characteristics and subsequent ischemic cerebrovascular events: a prospective assessment with MRI – initial results. Stroke 37: 818–823.
20. Tang D, Yang C, Kobayashi S, et al. (2001) Steady flow and wall compression in stenotic arteries: a three-dimensional thick-wall model with fluid-wall interactions. J Biomech Eng 123: 548–557.
21. Tang TY, Howarth SP, Li ZY, et al. (2008) Correlation of carotid atheromatous plaque inflammation with biomechanical stress: utility of USPIO enhanced MR imaging and finite element analysis. Atherosclerosis 196: 879–887.
22. Trivedi RA, Li ZY, U-King-Im JM, et al. (2007) Identifying vulnerable carotid plaques in vivo using high resolution magnetic resonance imaging-based finite element analysis. J Neurosurg 107: 536–542.
23. Nighoghossian N, Derex L, and Douek P (2005) The vulnerable carotid artery plaque: current imaging methods and new perspectives. Stroke 36: 2764–2772.
24. Underhill HR, Hatsukami TS, Cai J, et al. (2010) A noninvasive imaging approach to assess plaque severity: the carotid atherosclerosis score. AJNR Am J Neuroradiol 31(6):1068–1075.
25. Garcia de Tena J (2005) Inflammation, atherosclerosis, and coronary artery disease. N Engl J Med 353: 429–430; author reply 429–430.
26. Hansson GK (2005) Inflammation, atherosclerosis, and coronary artery disease. N Engl J Med 352: 1685–1695.
27. Kriszbacher I, Koppan M, and Bodis J (2005) Inflammation, atherosclerosis, and coronary artery disease. N Engl J Med 353: 429–430; author reply 429–430.
28. Newby AC (2005) Dual role of matrix metalloproteinases (matrixins) in intimal thickening and atherosclerotic plaque rupture. Physiol Rev 85: 1–31.
29. Lindner JR (2009) Contrast ultrasound molecular imaging of inflammation in cardiovascular disease. Cardiovasc Res 84: 182–189.
30. Perkins JM, Galland RB, Simmons MJ, et al. (2000) Carotid duplex imaging: variation and validation. Br J Surg 87: 320–322.
31. Mannami T, Konishi M, Baba S, et al. (1997) Prevalence of asymptomatic carotid atherosclerotic lesions detected by high-resolution ultrasonography and its relation to cardiovascular risk factors in the general population of a Japanese city: the Suita study. Stroke 28: 518–525.
32. O'Leary DH, Polak JF, Kronmal RA, et al. (1992) Distribution and correlates of sonographically detected carotid artery disease in the Cardiovascular Health Study. The CHS Collaborative Research Group. Stroke 23: 1752–1760.
33. O'Leary DH, Polak JF, Kronmal RA, et al. (1999) Carotid-artery intima and media thickness as a risk factor for myocardial infarction and stroke in older adults. Cardiovascular Health Study Collaborative Research Group. N Engl J Med 340: 14–22.
34. Gronholdt ML, Wiebe BM, Laursen H, et al. (1997) Lipid-rich carotid artery plaques appear echolucent on ultrasound B-mode images and may be associated with intraplaque haemorrhage. Eur J Vasc Endovasc Surg 14: 439–445.
35. Polak JF, Shemanski L, O'Leary DH, et al. (1998) Hypoechoic plaque at US of the carotid artery: an independent risk factor for incident stroke in adults aged 65 years or older. Cardiovascular Health Study. Radiology 208: 649–654.
36. Sabetai MM, Tegos TJ, Nicolaides AN, et al. (2000) Reproducibility of computer-quantified carotid plaque echogenicity: can we overcome the subjectivity? Stroke 31: 2189–2196.

37. Kitamura A, Iso H, Imano H, et al. (2004) Carotid intima-media thickness and plaque characteristics as a risk factor for stroke in Japanese elderly men. Stroke 35: 2788–2794.
38. Prabhakaran S, Rundek T, Ramas R, et al. (2006) Carotid plaque surface irregularity predicts ischemic stroke: the northern Manhattan study. Stroke 37: 2696–2701.
39. Lindner JR (2004) Microbubbles in medical imaging: current applications and future directions. Nat Rev Drug Discov 3: 527–532.
40. Salonen JT and Salonen R (1991) Ultrasonographically assessed carotid morphology and the risk of coronary heart disease. Arterioscler Thromb 11: 1245–1249.
41. de Bray JM, Baud JM, Delanoy P, et al. (1998) Reproducibility in ultrasonic characterization of carotid plaques. Cerebrovasc Dis 8: 273–277.
42. Lal BK, Hobson RW, 2nd, Hameed M, et al. (2006) Noninvasive identification of the unstable carotid plaque. Ann Vasc Surg 20: 167–174.
43. Eliasziw M, Streifler JY, Fox AJ, et al. (1994) Significance of plaque ulceration in symptomatic patients with high-grade carotid stenosis. North American Symptomatic Carotid Endarterectomy Trial. Stroke 25: 304–308.
44. Rothwell PM, Gibson R, and Warlow CP (2000) Interrelation between plaque surface morphology and degree of stenosis on carotid angiograms and the risk of ischemic stroke in patients with symptomatic carotid stenosis. On behalf of the European Carotid Surgery Trialists' Collaborative Group. Stroke 31: 615–621.
45. Lindner JR (2010) Microbubbles in medical imaging: current applications and future directions. Nat Rev Drug Discov 3: 527–532.
46. Christiansen C, Kryvi H, Sontum PC, et al. (1994) Physical and biochemical characterization of Albunex, a new ultrasound contrast agent consisting of air-filled albumin microspheres suspended in a solution of human albumin. Biotechnol Appl Biochem 19 (Pt 3): 307–320.
47. Staub D, Patel MB, Tibrewala A, et al. (2010) Vasa vasorum and plaque neovascularization on contrast-enhanced carotid ultrasound imaging correlates with cardiovascular disease and past cardiovascular events. Stroke 41: 41–47.
48. Cai JM, Hatsukami TS, Ferguson MS, et al. (2002) Classification of human carotid atherosclerotic lesions with in vivo multicontrast magnetic resonance imaging. Circulation 106: 1368–1373.
49. Saam T, Ferguson MS, Yarnykh VL, et al. (2005) Quantitative evaluation of carotid plaque composition by in vivo MRI. Arterioscler Thromb Vasc Biol 25: 234–239.
50. Yuan C, Kerwin WS, Ferguson MS, et al. (2002) Contrast-enhanced high resolution MRI for atherosclerotic carotid artery tissue characterization. J Magn Reson Imaging 15: 62–67.
51. Trivedi RA, U-King-Im JM, Graves MJ, et al. (2004) MRI-derived measurements of fibrous-cap and lipid-core thickness: the potential for identifying vulnerable carotid plaques in vivo. Neuroradiology 46: 738–743.
52. Cai J, Hatsukami TS, Ferguson MS, et al. (2005) In vivo quantitative measurement of intact fibrous cap and lipid-rich necrotic core size in atherosclerotic carotid plaque: comparison of high-resolution, contrast-enhanced magnetic resonance imaging and histology. Circulation 112: 3437–3444.
53. Hatsukami TS, Ross R, Polissar NL, et al. (2000) Visualization of fibrous cap thickness and rupture in human atherosclerotic carotid plaque in vivo with high-resolution magnetic resonance imaging. Circulation 102: 959–964.
54. Moody AR, Murphy RE, Morgan PS, et al. (2003) Characterization of complicated carotid plaque with magnetic resonance direct thrombus imaging in patients with cerebral ischemia. Circulation 107: 3047–3052.
55. Takaya N, Yuan C, Chu B, et al. (2005) Presence of intraplaque hemorrhage stimulates progression of carotid atherosclerotic plaques: a high-resolution magnetic resonance imaging study. Circulation 111: 2768–2775.
56. Chu B, Kampschulte A, Ferguson MS, et al. (2004) Hemorrhage in the atherosclerotic carotid plaque: a high-resolution MRI study. Stroke 35: 1079–1084.
57. U-King-Im JM, Li ZY, Trivedi RA, et al. (2006) Correlation of shear stress with carotid plaque rupture using MRI and finite element analysis. J Neurol 253: 379–381.

58. Kerwin W, Hooker A, Spilker M, et al. (2003) Quantitative magnetic resonance imaging analysis of neovasculature volume in carotid atherosclerotic plaque. Circulation 107: 851–856.
59. Kerwin WS, O'Brien KD, Ferguson MS, et al. (2006) Inflammation in carotid atherosclerotic plaque: a dynamic contrast-enhanced MR imaging study. Radiology 241: 459–468.
60. Kerwin WS, Oikawa M, Yuan C, et al. (2008) MR imaging of adventitial vasa vasorum in carotid atherosclerosis. Magn Reson Med 59: 507–514.
61. Tang TY, Howarth SP, Miller SR, et al. (2009) The ATHEROMA (Atorvastatin Therapy: Effects on Reduction of Macrophage Activity) Study. Evaluation using ultrasmall superparamagnetic iron oxide-enhanced magnetic resonance imaging in carotid disease. J Am Coll Cardiol 53: 2039–2050.
62. Tang TY, Howarth SP, Miller SR, et al. (2008) Correlation of carotid atheromatous plaque inflammation using USPIO-enhanced MR imaging with degree of luminal stenosis. Stroke 39: 2144–2147.
63. Trivedi RA, U-King-Im JM, Graves MJ, et al. (2004) Noninvasive imaging of carotid plaque inflammation. Neurology 63: 187–188.
64. Trivedi RA, U-King-Im JM, Graves MJ, et al. (2004) In vivo detection of macrophages in human carotid atheroma: temporal dependence of ultrasmall superparamagnetic particles of iron oxide-enhanced MRI. Stroke 35: 1631–1635.
65. Kooi ME, Cappendijk VC, Cleutjens KB, et al. (2003) Accumulation of ultrasmall superparamagnetic particles of iron oxide in human atherosclerotic plaques can be detected by in vivo magnetic resonance imaging. Circulation 107: 2453–2458.
66. Virmani R, Kolodgie FD, Burke AP, et al. (2000) Lessons from sudden coronary death: a comprehensive morphological classification scheme for atherosclerotic lesions. Arterioscler Thromb Vasc Biol 20: 1262–1275.
67. Touze E, Toussaint JF, Coste J, et al. (2007) Reproducibility of high-resolution MRI for the identification and the quantification of carotid atherosclerotic plaque components: consequences for prognosis studies and therapeutic trials. Stroke 38: 1812–1819.
68. Kampschulte A, Ferguson MS, Kerwin WS, et al. (2004) Differentiation of intraplaque versus juxtaluminal hemorrhage/thrombus in advanced human carotid atherosclerotic lesions by in vivo magnetic resonance imaging. Circulation 110: 3239–3244.
69. Fisher M, Paganini-Hill A, Martin A, et al. (2005) Carotid plaque pathology: thrombosis, ulceration, and stroke pathogenesis. Stroke 36: 253–257.
70. Spagnoli LG, Mauriello A, Sangiorgi G, et al. (2004) Extracranial thrombotically active carotid plaque as a risk factor for ischemic stroke. JAMA 292: 1845–1852.
71. Chu B, Ferguson MS, Chen H, et al. (2009) Magnetic [corrected] resonance imaging [corrected] features of the disruption-prone and the disrupted carotid plaque. JACC Cardiovasc Imaging 2: 883–896.
72. Anzalone N, Scomazzoni F, Castellano R, et al. (2005) Carotid artery stenosis: intraindividual correlations of 3D time-of-flight MR angiography, contrast-enhanced MR angiography, conventional DSA, and rotational angiography for detection and grading. Radiology 236: 204–213.
73. Anzidei M, Napoli A, Geiger D, et al. (2010) Preliminary experience with MRA in evaluating the degree of carotid stenosis and plaque morphology using high-resolution sequences after gadofosveset trisodium (Vasovist) administration: comparison with CTA and DSA. Radiol Med 115(4): 634–647.
74. Virmani R, Narula J, and Farb A (1998) When neoangiogenesis ricochets. Am Heart J 136: 937–939.
75. Kolodgie FD, Gold HK, Burke AP, et al. (2003) Intraplaque hemorrhage and progression of coronary atheroma. N Engl J Med 349: 2316–2325.
76. Gomori JM, Grossman RI, Goldberg HI, et al. (1985) Intracranial hematomas: imaging by high-field MR. Radiology 157: 87–93.
77. Gomori JM, Grossman RI, Hackney DB, et al. (1988) Variable appearances of subacute intracranial hematomas on high-field spin-echo MR. AJR Am J Roentgenol 150: 171–178.
78. Corti R, Fuster V, Fayad ZA, et al. (2005) Effects of aggressive versus conventional lipid-lowering therapy by simvastatin on human atherosclerotic lesions: a prospective, randomized,

double-blind trial with high-resolution magnetic resonance imaging. J Am Coll Cardiol 46: 106–112.
79. Saam T, Hatsukami TS, Takaya N, et al. (2007) The vulnerable, or high-risk, atherosclerotic plaque: noninvasive MR imaging for characterization and assessment. Radiology 244: 64–77.
80. Lima JA, Desai MY, Steen H, et al. (2004) Statin-induced cholesterol lowering and plaque regression after 6 months of magnetic resonance imaging-monitored therapy. Circulation 110: 2336–2341.
81. Corti R, Fayad ZA, Fuster V, et al. (2001) Effects of lipid-lowering by simvastatin on human atherosclerotic lesions: a longitudinal study by high-resolution, noninvasive magnetic resonance imaging. Circulation 104: 249–252.
82. Yonemura A, Momiyama Y, Fayad ZA, et al. (2005) Effect of lipid-lowering therapy with atorvastatin on atherosclerotic aortic plaques detected by noninvasive magnetic resonance imaging. J Am Coll Cardiol 45: 733–742.
83. Corti R, Fuster V, Fayad ZA, et al. (2002) Lipid lowering by simvastatin induces regression of human atherosclerotic lesions: two years' follow-up by high-resolution noninvasive magnetic resonance imaging. Circulation 106: 2884–2887.
84. Kaazempur-Mofrad MR, Isasi AG, Younis HF, et al. (2004) Characterization of the atherosclerotic carotid bifurcation using MRI, finite element modeling, and histology. Ann Biomed Eng 32: 932–946.
85. Baldewsing RA, de Korte CL, Schaar JA, et al. (2004) Finite element modeling and intravascular ultrasound elastography of vulnerable plaques: parameter variation. Ultrasonics 42: 723–729.
86. Simonetti OP, Finn JP, White RD, et al. (1996) "Black blood" T2-weighted inversion-recovery MR imaging of the heart. Radiology 199: 49–57.
87. Yarnykh VL, and Yuan C (2002) T1-insensitive flow suppression using quadruple inversion-recovery. Magn Reson Med 48: 899–905.
88. Mofidi R, Crotty TB, McCarthy P, et al. (2001) Association between plaque instability, angiogenesis and symptomatic carotid occlusive disease. Br J Surg 88: 945–950.
89. Moreno PR, Purushothaman KR, Fuster V, et al. (2004) Plaque neovascularization is increased in ruptured atherosclerotic lesions of human aorta: implications for plaque vulnerability. Circulation 110: 2032–2038.
90. Jackson A, O'Connor JP, Parker GJ, et al. (2007) Imaging tumor vascular heterogeneity and angiogenesis using dynamic contrast-enhanced magnetic resonance imaging. Clin Cancer Res 13: 3449–3459.
91. Lin W, Abendschein DR, and Haacke EM (1997) Contrast-enhanced magnetic resonance angiography of carotid arterial wall in pigs. J Magn Reson Imaging 7: 183–190.
92. Aoki S, Aoki K, Ohsawa S, et al. (1999) Dynamic MR imaging of the carotid wall. J Magn Reson Imaging 9: 420–427.
93. Wasserman BA, Smith WI, Trout HH, 3rd, et al. (2002) Carotid artery atherosclerosis: in vivo morphologic characterization with gadolinium-enhanced double-oblique MR imaging initial results. Radiology 223: 566–573.
94. Akella NS, Twieg DB, Mikkelsen T, et al. (2004) Assessment of brain tumor angiogenesis inhibitors using perfusion magnetic resonance imaging: quality and analysis results of a phase I trial. J Magn Reson Imaging 20: 913–922.
95. EMA (2008) International Nonproprietary Name (INN): superparamagnetic iron oxide nanoparticles stabilised with dextran and sodium citrate. Procedure No. EMEA/H/C/801.
96. Schmitz SA, Coupland SE, Gust R, et al. (2000) Superparamagnetic iron oxide-enhanced MRI of atherosclerotic plaques in Watanabe hereditable hyperlipidemic rabbits. Invest Radiol 35: 460–471.
97. Schmitz SA, Taupitz M, Wagner S, et al. (2001) Magnetic resonance imaging of atherosclerotic plaques using superparamagnetic iron oxide particles. J Magn Reson Imaging 14: 355–361.
98. Bulte JW, Brooks RA, Moskowitz BM, et al. (1999) Relaxometry and magnetometry of the MR contrast agent MION-46L. Magn Reson Med 42: 379–384.
99. Trivedi R, U-King-Im JM, and Gillard J (2003) Accumulation of ultrasmall superparamagnetic particles of iron oxide in human atherosclerotic plaque. Circulation 108: e140; author reply e140.

100. Trivedi RA, U-King-Im JM, Graves MJ, et al. (2005) Non-stenotic ruptured atherosclerotic plaque causing thrombo-embolic stroke. Cerebrovasc Dis 20: 53–55.
101. Underhill HR, Yuan C, Zhao XQ, et al. (2008) Effect of rosuvastatin therapy on carotid plaque morphology and composition in moderately hypercholesterolemic patients: a high-resolution magnetic resonance imaging trial. Am Heart J 155: 584 e581–e588.
102. Sugawara Y, Braun DK, Kison PV, et al. (1998) Rapid detection of human infections with fluorine-18 fluorodeoxyglucose and positron emission tomography: preliminary results. Eur J Nucl Med 25: 1238–1243.
103. Annovazzi A, Bonanno E, Arca M, et al. (2006) 99mTc-interleukin-2 scintigraphy for the in vivo imaging of vulnerable atherosclerotic plaques. Eur J Nucl Med Mol Imaging 33: 117–126.
104. Haider N, Hartung D, Fujimoto S, et al. (2009) Dual molecular imaging for targeting metalloproteinase activity and apoptosis in atherosclerosis: molecular imaging facilitates understanding of pathogenesis. J Nucl Cardiol 16: 753–762.
105. Sheikine Y, and Akram K (2010) FDG-PET imaging of atherosclerosis: Do we know what we see? Atherosclerosis 211(2): 371–380.
106. Rudd JH, Warburton EA, Fryer TD, et al. (2002) Imaging atherosclerotic plaque inflammation with [18F]-fluorodeoxyglucose positron emission tomography. Circulation 105: 2708–2711.
107. Tawakol A, Migrino RQ, Bashian GG, et al. (2006) In vivo 18F-fluorodeoxyglucose positron emission tomography imaging provides a noninvasive measure of carotid plaque inflammation in patients. J Am Coll Cardiol 48: 1818–1824.
108. Rudd JH, Myers KS, Bansilal S, et al. (2007) (18)Fluorodeoxyglucose positron emission tomography imaging of atherosclerotic plaque inflammation is highly reproducible: implications for atherosclerosis therapy trials. J Am Coll Cardiol 50: 892–896.
109. Rudd JH, Myers KS, Bansilal S, et al. (2009) Relationships among regional arterial inflammation, calcification, risk factors, and biomarkers: a prospective fluorodeoxyglucose positron-emission tomography/computed tomography imaging study. Circ Cardiovasc Imaging 2: 107–115.
110. Silvera SS, Aidi HE, Rudd JH, et al. (2009) Multimodality imaging of atherosclerotic plaque activity and composition using FDG-PET/CT and MRI in carotid and femoral arteries. Atherosclerosis 207: 139–143.
111. Kwee RM, Teule GJ, van Oostenbrugge RJ, et al. (2009) Multimodality imaging of carotid artery plaques: 18F-fluoro-2-deoxyglucose positron emission tomography, computed tomography, and magnetic resonance imaging. Stroke 40: 3718–3724.
112. Tahara N, Kai H, Ishibashi M, et al. (2006) Simvastatin attenuates plaque inflammation: evaluation by fluorodeoxyglucose positron emission tomography. J Am Coll Cardiol 48: 1825–1831.
113. Lee SJ, On YK, Lee EJ, et al. (2008) Reversal of vascular 18F-FDG uptake with plasma high-density lipoprotein elevation by atherogenic risk reduction. J Nucl Med 49: 1277–1282.
114. Amarenco P, Labreuche J, Lavallee P, et al. (2004) Statins in stroke prevention and carotid atherosclerosis: systematic review and up-to-date meta-analysis. Stroke 35: 2902–2909.
115. Crisby M, Nordin-Fredriksson G, Shah PK, et al. (2001) Pravastatin treatment increases collagen content and decreases lipid content, inflammation, metalloproteinases, and cell death in human carotid plaques: implications for plaque stabilization. Circulation 103: 926–933.
116. Crouse JR, 3rd, Raichlen JS, Riley WA, et al. (2007) Effect of rosuvastatin on progression of carotid intima-media thickness in low-risk individuals with subclinical atherosclerosis: the METEOR Trial. JAMA 297: 1344–1353.
117. Kaufmann BA, Sanders JM, Davis C, et al. (2007) Molecular imaging of inflammation in atherosclerosis with targeted ultrasound detection of vascular cell adhesion molecule-1. Circulation 116: 276–284.
118. Chen W, Vucic E, Leupold E, et al. (2008) Incorporation of an apoE-derived lipopeptide in high-density lipoprotein MRI contrast agents for enhanced imaging of macrophages in atherosclerosis. Contrast Media Mol Imaging 3: 233–242.
119. Briley-Saebo KC, Shaw PX, Mulder WJ, et al. (2008) Targeted molecular probes for imaging atherosclerotic lesions with magnetic resonance using antibodies that recognize oxidation-specific epitopes. Circulation 117: 3206–3215.

120. Nahrendorf M, Jaffer FA, Kelly KA, et al. (2006) Noninvasive vascular cell adhesion molecule-1 imaging identifies inflammatory activation of cells in atherosclerosis. Circulation 114: 1504–1511.
121. Mulder WJ, Strijkers GJ, Briley-Saboe KC, et al. (2007) Molecular imaging of macrophages in atherosclerotic plaques using bimodal PEG-micelles. Magn Reson Med 58: 1164–1170.
122. Lancelot E, Amirbekian V, Brigger I, et al. (2008) Evaluation of matrix metalloproteinases in atherosclerosis using a novel noninvasive imaging approach. Arterioscler Thromb Vasc Biol 28: 425–432.
123. Spuentrup E, Botnar RM, Wiethoff AJ, et al. (2008) MR imaging of thrombi using EP-2104R, a fibrin-specific contrast agent: initial results in patients. Eur Radiol 18: 1995–2005.
124. Galis ZS, and Khatri JJ (2002) Matrix metalloproteinases in vascular remodeling and atherogenesis: the good, the bad, and the ugly. Circ Res 90: 251–262.
125. Schafers M, Riemann B, Kopka K, et al. (2004) Scintigraphic imaging of matrix metalloproteinase activity in the arterial wall in vivo. Circulation 109: 2554–2559.
126. Amirbekian V, Aguinaldo JG, Amirbekian S, et al. (2009) Atherosclerosis and matrix metalloproteinases: experimental molecular MR imaging in vivo. Radiology 251: 429–438.
127. Stemme S, Faber B, Holm J, et al. (1995) T lymphocytes from human atherosclerotic plaques recognize oxidized low density lipoprotein. Proc Natl Acad Sci USA 92: 3893–3897.
128. Laitinen I, Saraste A, Weidl E, et al. (2009) Evaluation of alphavbeta3 integrin-targeted positron emission tomography tracer 18F-galacto-RGD for imaging of vascular inflammation in atherosclerotic mice. Circ Cardiovasc Imaging 2: 331–338.
129. Nahrendorf M, Keliher E, Panizzi P, et al. (2009) 18F-4V for PET-CT imaging of VCAM-1 expression in atherosclerosis. JACC Cardiovasc Imaging 2: 1213–1222.
130. Kelly KA, Allport JR, Tsourkas A, et al. (2005) Detection of vascular adhesion molecule-1 expression using a novel multimodal nanoparticle. Circ Res 96: 327–336.
131. Lipinski MJ, Amirbekian V, Frias JC, et al. (2006) MRI to detect atherosclerosis with gadolinium-containing immunomicelles targeting the macrophage scavenger receptor. Magn Reson Med 56: 601–610.
132. Toussaint JF, LaMuraglia GM, Southern JF, et al. (1996) Magnetic resonance images lipid, fibrous, calcified, hemorrhagic, and thrombotic components of human atherosclerosis in vivo. Circulation 94: 932–938.
133. Yuan C, Zhang SX, Polissar NL, et al. (2002) Identification of fibrous cap rupture with magnetic resonance imaging is highly associated with recent transient ischemic attack or stroke. Circulation 105: 181–185.

Biographies

Karol P. Budohoski, MD is currently enrolled as a PhD student at the University of Cambridge, UK. His thesis focuses on clinical studies of cerebrovascular disease.

Victoria E.L. Young, MPhil, MRCS is currently completing a PhD on MR imaging of carotid disease at the University of Cambridge, UK.

Tjun Y. Tang, MD, MRCS has completed a MD thesis on imaging carotid plaque inflammation using contrast-enhanced MRI and is currently focusing on clinical training in vascular surgery

Jonathan H. Gillard, MD, FRCR MBA is Professor of Neuroradiology at the University of Cambridge with academic interests in imaging of cerebrovascular disease.

Peter J. Kirkpatrick, FRCS (SN), FMedSci practices as a neurosurgeon and is the head of cerebrovascular surgery at the University of Cambridge.

Rikin A. Trivedi, PhD, FRCS (SN) practices as a neurosurgeon with academic interests in cerebrovascular disorders.

Chapter 17
Noninvasive Imaging of Carotid Atherosclerosis

Robert M. Kwee, Robert J. van Oostenbrugge, Leo Hofstra, Jos M.A. van Engelshoven, Werner H. Mess, Joachim E. Wildberger, and M. Eline Kooi

Abstract Carotid atherosclerosis is an important cause of stroke. Because stroke results in considerable morbidity, mortality, and costs, prevention is pivotal. Patient symptomatology and degree of luminal stenosis are currently the main grounds to perform carotid endarterectomy (CEA). However, many patients undergo CEA with its attendant risks without taking advantage, whereas in others CEA is probably incorrectly withheld. Noninvasive imaging of carotid plaque characteristics may be used to improve risk stratification for stroke. Histopathologically, vulnerable plaques (i.e., plaques that have a high tendency to cause future thromboembolic events) are characterized by the presence of a large lipid-rich necrotic core with a thin overlying fibrous cap, neovasculature growth, macrophage infiltration, intraplaque hemorrhage, cell death, ulceration, and thrombogenicity. Several imaging modalities, including ultrasonography, transcranial Doppler, magnetic resonance imaging, multidetector-row computed tomography, and nuclear imaging techniques may be used to characterize one or more of these plaque features in vivo. For each technique, accuracy and reproducibility, (dis)advantages and limitations, and clinical potential will be outlined.

Keywords Carotid atherosclerosis • Stroke • Plaque • Imaging

17.1 Introduction

Stroke results in considerable morbidity and mortality. Of the 15 million people worldwide who suffer a stroke annually, 5 million die and another 5 million are left permanently disabled, placing a considerable burden on family and community [1].

R.M. Kwee (✉)
Department of Radiology, Maastricht University Medical Center, Maastricht, the Netherlands
and
Cardiovascular Research Institute Maastricht, Maastricht University Medical Center, Maastricht, the Netherlands
e-mail: rmkwee@gmail.com

Jasjit S. Suri et al. (eds.), *Atherosclerosis Disease Management*,
DOI 10.1007/978-1-4419-7222-4_17,

Given these dismal statistics and the high treatment costs of stroke, prevention of the disease is of particular importance.

Carotid atherosclerosis is an important cause of stroke [2]. Currently, degree of luminal stenosis and symptomatology are the main grounds to perform carotid endarterectomy (CEA). However, for symptomatic patients with 70–99% stenosis, 6.3 patients need to be operated to prevent one ipsilateral ischemic stroke in 5 years [3]. For patients with asymptomatic 60–99% stenosis and for symptomatic patients with 50–69% stenosis, the numbers needed to treat are even as high as 16.9 [4] and 21.7 [3], respectively. Thus, risk stratification for stroke needs to be improved.

It is currently believed that plaque properties other than degree of obstruction alone are important to identify patients at high risk of stroke. The so-called "vulnerable plaque" is a plaque that has a high tendency to cause ischemic events due to cerebral emboli originating from a platelet-rich thrombus on the plaque surface or from plaque rupture. Histological analysis of CEA specimens suggests that vulnerable plaques are characterized by a large lipid-rich necrotic core with a thin overlying fibrous cap, neovasculature growth, macrophage infiltration, intraplaque hemorrhage, cell death, ulceration, and thrombogenicity [5–11]. Early recognition of these plaque features may identify high-risk patients who might particularly benefit from aggressive interventions.

Several noninvasive imaging modalities may be used to identify vulnerable carotid plaque features in humans in vivo (see Table 17.1). In this chapter, we provide an overview of these modalities. For each technique, accuracy and reproducibility (dis)advantages and limitations, and clinical potential will be outlined.

Table 17.1 Noninvasive imaging modalities for characterization of carotid plaques

Imaging technique	Potential parameters for vulnerable plaque identification
US	
Conventional B-mode US	Plaque echogenicity
Contrast-enhanced US	Neovascularization, ulcerations
TCD	Plaque thrombogenicity (microemboli from plaque surface)
MDCT	LRNC (including hemorrhage), ulcerations
MRI	
Conventional MRI	LRNC, fibrous cap status, intraplaque hemorrhage, ulcerations
Dynamic contrast-enhanced MRI	Neovascularization and macrophage accumulation
USPIO-enhanced MRI	Macrophage accumulation
Nuclear imaging techniques	
^{18}F-FDG PET	Macrophage accumulation
Annexin A5 scintigraphy	Cell death

17.2 Ultrasonography

This high-resolution, noninvasive technique is a very useful method to visualize carotid plaques because it is readily available, rapidly applicable, and can be performed at relatively low cost. It is usually performed as an initial screening modality in TIA/stroke patients to assess for carotid stenosis.

17.2.1 Conventional B-Mode US

Studies comparing in vivo B-mode ultrasonic plaque morphology to histological analysis of CEA specimens have shown that echolucent plaques contain significantly more soft tissue (LRNC and hemorrhage) than echo-rich plaques, which are primarily composed of fibrous tissue and calcifications [12, 13]. Thus, echolucent plaques (Fig. 17.1) are thought to be more vulnerable than echo-rich plaques (Fig. 17.2). Several large prospective studies have investigated whether assessment of carotid plaque echogenicity by standard B-mode ultrasonography (US) can predict the occurrence of stroke. These studies [14–16] found conflicting results. The Cardiovascular Health Study found that asymptomatic elderly patients with a hypoechoic plaque (visually assessed) have a relative risk (RR) of ipsilateral ischemic stroke of 2.78 (95% CI, 1.4–5.7), independent of stenosis grade and other cardiovascular risk factors [14]. Grønholdt et al. [15] performed gray-scale median (GSM) measurement of plaques of 111 asymptomatic and 135 symptomatic patients

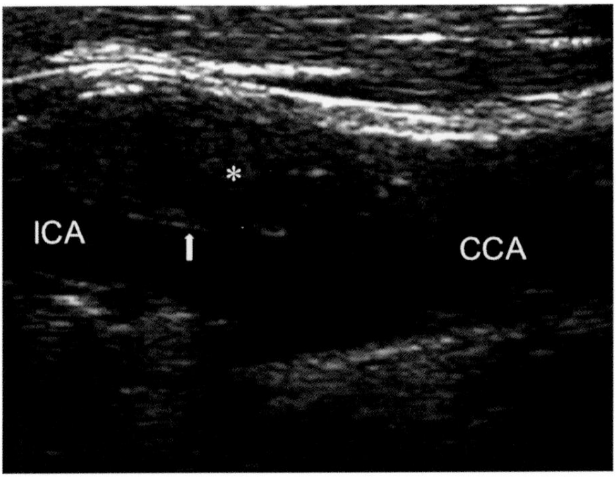

Fig. 17.1 Standard B-mode US. Posterolateral view of the carotid bulb showing a large echolucent plaque (*asterisk*) at the origin of the internal carotid artery (ICA). The fibrous cap is just visible as a *thin line* (*arrow*) slightly more echointense than the body of the plaque (CCA: common carotid artery). Reproduced with permission from Kwee et al. [102]

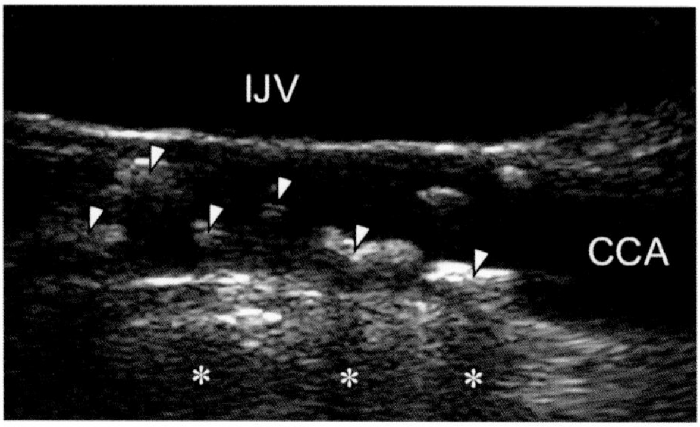

Fig. 17.2 Standard B-mode US. Anterolateral view of the common carotid artery (CCA) showing a very inhomogeneous plaque. Multiple large echorich areas (*arrowheads*), which partly give rise to acoustic shadowing (*asterisks*), are depicted. These areas are presumably calcifications (IJV: internal jugular vein). Reproduced with permission from Kwee et al. [102]

with ≥50% carotid stenosis. Patients were followed up for a mean period of 4.4 years. In symptomatic patients, RR of ipsilateral ischemic stroke for echolucent (GSM < 74) versus echo-rich (GSM ≥ 74) plaques was 3.1 (95% CI, 1.3–7.3), independent of other cardiovascular risk factors. However, no elevated risk was found among asymptomatic patients. In 1,543 asymptomatic patients with 60–99% carotid stenosis, the Asymptomatic Carotid Surgery trial [16] also did not find an association between plaque echogenicity (plaque echolucency <25% versus ≥25) and 5-year risk of ipsilateral stroke. However, no adjustment for age, sex, and other cardiovascular risk factors was made.

Plaque echogenicity can be assessed either visually [14, 16] or objectively by means of a computer-assisted GSM measurement [15]. Studies investigating the reproducibility of visual assessment of plaque echogenicity by standard B-mode US found fair-to-good interobserver agreement (kappa coefficient [κ] = 0.38–0.74) [17–19]. An objective characterization of plaque echogenicity may be more reliable and less observer dependent [20]. However, even computer-assisted GSM measurement only assesses the median brightness of the entire plaque; regional instability, such as hemorrhage, may exist within a plaque even with a high GSM value. This may explain why there is no consensus yet as to which GSM threshold value is most sensitive to distinguish vulnerable from stable plaques. A stratified gray-scale measurement of carotid plaque echogenicity [21] or pixel segmentation with tissue mapping [22] may be better methods to characterize plaque composition. Another limitation of standard B-mode ultrasound is that accurate interpretation of images may be hampered by artifacts, e.g., acoustic shadowing caused by calcifications. This can be minimized by applying real-time compound ultrasound imaging, which uses multiple scanning angles to improve image quality (Fig. 17.3). Indeed, compared to standard B-mode US, real-time compound US has shown to improve visualization of plaque texture and surface [19], and it achieves higher interobserver agreement for classification of plaque echogenicity (κ=0.83 vs. κ=0.74)[19].

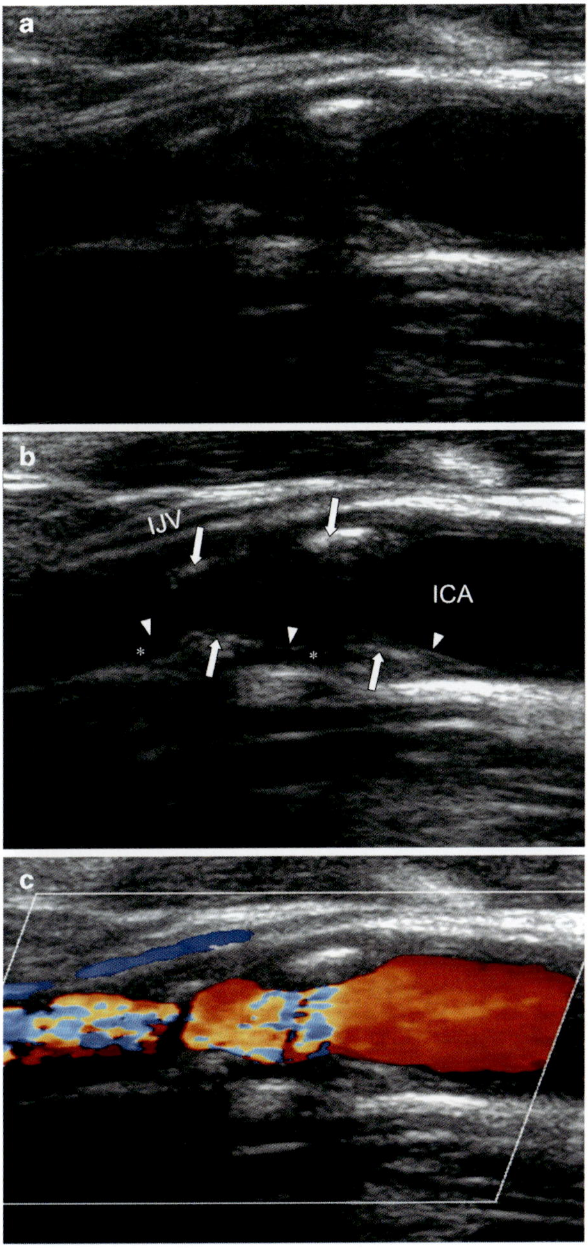

Fig. 17.3 Standard B-mode US versus compound US. Anterolateral view of the internal carotid artery (ICA) shows a heterogeneous plaque (**a**–**c**). Compared to conventional ultrasound (**a**), compound ultrasound (**b**) improves visualization of plaque echogenicity, plaque surface, and vessel wall demarcation. Echorich (*arrows*) and echolucent (*asterisks*) areas are depicted. The fibrous cap is visible as a *thin echointense line* (*arrowheads*). Duplex ultrasound (**c**) demonstrates demarcation between lumen and plaque surface and vessel wall. *IJV* internal jugular vein. Reproduced with permission from Kwee et al. [102]

17.2.2 Contrast-Enhanced US

At present, imaging of neovascular development in atherosclerotic lesions with conventional US is not feasible, since the vessel diameter is well below the resolution capacity of currently available US systems. With the help of a US contrast agent, it may be possible to depict neovascularization in a carotid plaque [23]. US contrast agents are gas-filled microbubbles, which behave as pure intravascular tracers. They have a good safety profile [24]. Since microbubbles are smaller than erythrocytes, they will also enter the microcirculation. Microbubbles have a high degree of echogenicity: a strong signal will be evoked even by a small number of air bubbles as compared to the signal from the surrounding tissue. Thus, the intensity increase of the US signal from the carotid plaque after administration of microbubbles is thought to reflect the amount of neovascularization. Because microbubbles also improve the delineation between lumen and vessel wall, contrast-enhanced (CE) US can also reveal plaque surface irregularities and ulcers which are not detected by standard B-mode US [23, 25, 26].

A few studies have validated CE US findings with histology: Shah et al. [27] found a strong correlation between visual assessment of degree of plaque enhancement at CE US (scored on a 4-point scale) and histological degree of neovascularization (Spearman $\rho=0.68$). In addition, Coli et al. [28] found significant larger neovessel density in plaques with higher enhancement compared to plaques with lower enhancement at CE US. Both studies did not report the reproducibility of the method. There are also no published studies yet comparing an objective measure of degree of plaque enhancement (such as a kinetic approach with a model-based evaluation of time-intensity curves) to histology. An objective measure may be more accurate and reproducible than visual assessment.

A few cross-sectional studies have assessed the relation between degree of plaque enhancement at CE US and patient symptomatology: Giannoni et al. [29] found that microbubble contrast enhancement at the base of the plaque was significantly more frequently present in plaques of symptomatic patients than in plaques of clinically asymptomatic patients. Accordingly, another study [30] also found that plaques ipsilateral to the symptomatic side had more intense contrast agent enhancement than plaques of asymptomatic patients.

17.3 Transcranial Doppler

Transcranial Doppler (TCD) monitoring of the middle cerebral artery through the temporal bone can reveal microemboli originating from carotid plaques (Fig. 17.4). Microemboli are defined as signals with a duration of less than 300 ms and an amplitude of 3 dB above the background blood-flow signal [31]. Most microemboli are easily recognized since they produce a characteristic sound, a microembolic signal (MES), stemming from a short-lasting signal intensity increase. Sitzer et al. [11] showed that MESs are significantly associated with plaque ulceration and intraluminal

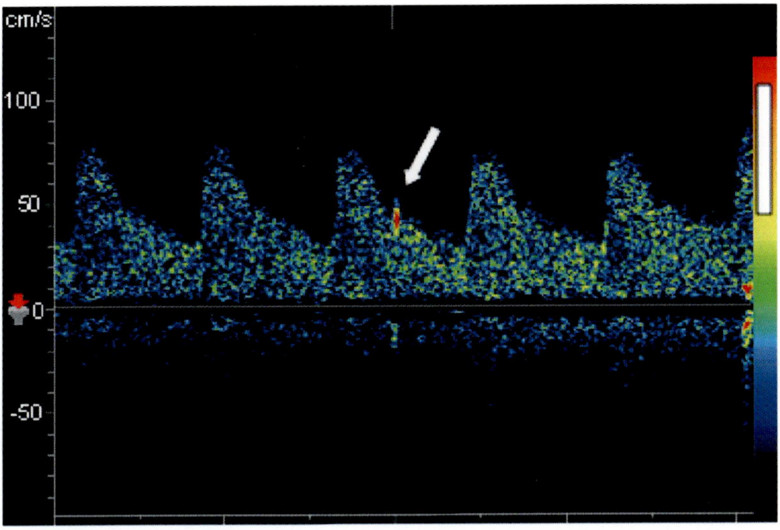

Fig. 17.4 TCD monitoring of the left middle cerebral artery reveals an asymptomatic microembolus (*arrow*) in a patient with a recent history of TIA. Reproduced with permission from Kwee et al. [102]

thrombosis. An individual MES does not cause neurological symptoms but may represent an early warning sign of a neurological event in the near future, since it indicates an active process of the plaque, possibly leading to a clinically manifest thromboembolism. TCD is relatively inexpensive, can be performed at bedside, and allows monitoring for prolonged periods with a high temporal resolution, with excellent reproducibility (κ=0.89–0.99) [32, 33]. Six to eight hour lasting examinations have been shown to be feasible and to produce reliable results in patients with a very low incidence of MESs [34]. A limitation of TCD monitoring of the middle cerebral artery is that it cannot determine the exact origin of MESs. Except from carotid atherosclerosis, microemboli may also arise from aortic arch atherosclerosis or cardiac disorders. In addition, the manual method of MES detection is very labor-intensive and time-consuming. This may be overcome by automated MES analysis [35]. Another limitation of transtemporal TCD is that MES detectability mainly depends on the thickness and homogeneity of the temporal bone. In daily clinical practice, TCD examinations cannot be performed in up to 18% of TIA/stroke patients due to an insufficient temporal bone window [36]. Furthermore, when studying symptomatic patients, the time since last symptoms must be considered, as the prevalence of MESs decreases significantly over time [37].

A number of relatively small prospective studies suggested that MESs are predictors of combined TIA and stroke risk in symptomatic [38–40] and asymptomatic carotid stenosis [40, 41]. The results of three larger prospective studies [42–44] further support the hypothesis that MES detection by TCD can select high-risk patients; Markus and MacKinnon [42] performed 1-h TCD registration in 200 symptomatic patients with >50% carotid stenosis, with a mean time between last symptoms and TCD of 31 days. Patients were followed up until surgical intervention, stroke, or study end at 90 days.

They found that the presence of MESs predicted stroke, and the combined end point of ipsilateral stroke and TIA with an odds ratio of 4.7 (95% CI, 2.0–11.0) after Cox regression to control for age, sex, smoking, hypertension, time from last symptoms and stenosis grade. The absence of MESs identified a group at low risk for stroke alone and stroke and TIA: 0 and 7.5%, respectively, versus 3.5 and 15.5% in all 200 subjects [42]. Abbott et al. [43] prospectively investigated 202 patients with 231 asymptomatic 60–99% carotid stenoses for a mean period of 34 months and found that TIA or stroke occurred more commonly ipsilateral to MES-positive arteries. However, the number of events that occurred in their study was too small to achieve statistical significance [43]. Spence et al. [44] investigated 319 patients with asymptomatic ≥60% carotid stenosis by TCD. Patients with MESs were found to have much more likely a stroke during the first year of follow-up than patients without MESs (15.6%, 95% CI, 4.1–79; versus1%, 95% CI, 1.01–1.36; $P<0.0001$). However, after adjustment for age, sex, cholesterol and smoking, this difference was not significant ($P=0.38$) [44].

17.4 Multidetector-Row Computed Tomography

Single-slice computed tomography has shown to be limited in characterizing carotid plaque morphology [45]. Multidetector-row computed tomography (MDCT), on the other hand, allows evaluating carotid atherosclerosis with thinner slices (0.5–1.0 mm) and less partial volume averaging, making a detailed analysis of plaque composition possible. An advantage of MDCT is that scanning time is very short. A disadvantage is that patients are exposed to ionizing radiation (the effective dose of a carotid MDCT examination is approximately 2.3 mSv). In addition, the use of iodinated contrast, which is necessary for accurate plaque assessment, should be administered with caution in patients with preexisting renal insufficiency [46].

de Weert et al. [47] investigated whether MDCT could accurately classify carotid atherosclerotic plaque component, based on differences in HU values (Fig. 17.5). They found that MDCT measurements of calcifications and fibrous tissue correlated well with histology (Pearson $\rho=0.86$ and 0.87, respectively). LRNC quantification only correlated well with histology in mildly calcified plaques (Pearson $\rho=0.88$). In severely calcified plaques, the correlation was much lower (Pearson $\rho=0.48$) because the blooming effect of calcifications overshadows the LRNC [47]. Vessel wall and plaque component volumes can be assessed with good interobserver agreement (intraclass correlation coefficient [ICC]=0.76–0.99) [48]. Other investigators showed that MDCT has high sensitivity (93.8%) and specificity (98.6%) in detecting plaque ulcerations. [49]. Using multiplanar reformatting (MPR), plaque ulcerations can be detected with excellent interobserver agreement ($\kappa=1.00$) [50]. Due to limited soft-tissue contrast, MDCT cannot reliably distinguish intraplaque hemorrhage from LRNC [47], and fibrous cap status cannot be assessed. Observer variability in the assessment of vessel wall and plaque component is also mainly caused by limited soft-tissue contrast, making it sometimes

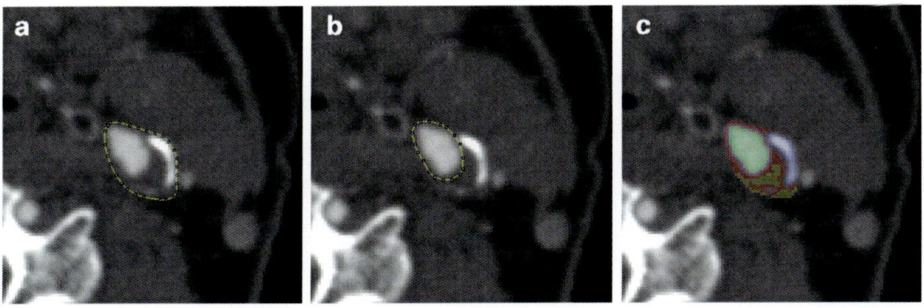

Fig. 17.5 MDCT images of a transverse section of a carotid plaque at the level of the bifurcation. ROI encompassing the plaque and arterial lumen has been drawn on the CT image (**a**). To differentiate lumen area from the atherosclerotic plaque area and from calcified tissue, a second ROI has been drawn (**b**). This second ROI should include the attenuated lumen area, but no calcifications. After the input of the cutoff values that differentiate the plaque components and the lumen, a pixel map based on HU values was obtained (*green* arterial lumen; *blue/white* calcifications; *yellow* lipid; *magenta/red* fibrous tissue) (**c**). This method of MDCT analysis has been previously described by de Weert et al. [47, 48]

difficult to differentiate normal vessel wall from plaque and to differentiate vessel wall from perivascular tissue. The use of automatic software may improve reproducibility of the method, and it can save time in image review. Wintermark et al. [51] demonstrated the potential of using an automated classification computer algorithm to analyze MDCT images. However, although the software worked excellent in classifying calcifications, it worked less well in differentiating LRNC/hemorrhage from fibrous tissue. Further improvement in automatic measurement software is needed. The application of dual-energy technology may provide more accurate measurements of soft-tissue components of the plaque, since it allows the removal of calcifications [52].

Several cross-sectional MDCT studies have been performed, demonstrating that plaques which are more calcified are less often associated with ischemic symptoms [53–55]. An explanation may be that calcified plaques, which contain relatively less lipid, are mechanically more stable and less prone to rupture [56]. It has also been demonstrated that MDCT-assessed wall volume, number of lipid clusters, and lipid clusters closer to the lumen are associated with ischemic symptoms [55].

17.5 Magnetic Resonance Imaging

Magnetic resonance imaging (MRI) is well suited for evaluating carotid plaques; it is widely available, provides excellent soft-tissue contrast and multiplanar imaging capability, and is free of ionizing radiation. A limitation of MRI is that it is contraindicated in patients with claustrophobia, pacemakers, defibrillators, or other implanted electronic devices.

17.5.1 Conventional MRI

The ability of MRI to distinguish and quantify several carotid plaque components in vivo has been extensively validated with histology. A multisequence non-CE MRI protocol, consisting of time-of-flight (TOF)-, T1-, proton density-, and T2-weighted turbo spin-echo (TSE) images may be sufficient to assess plaque carotid plaque components (Fig. 17.6). Using aforementioned protocol and predefined MRI tissue classification criteria, Saam et al. [57] found moderate-to-strong correlations with histological measurements (for LRNC, hemorrhage, calcifications, loose matrix, and dense fibrous tissue, Pearson $\rho=0.75$, 0.66, 0.74, 0.70, and 0.55, respectively) and good-to-excellent interobserver agreement (ICC = 0.73–0.98) [57]. However, other investigators [58] showed that interobserver agreement in identifying FCs by non-CE MRI (TOF or T2-weighted TSE images) is insufficient ($\kappa=0.33$–0.58). The use of gadolinium-based contrast agents improves contrast between the fibrous cap and LRNC [59] (Fig. 17.7), allowing more precise assessments of these components. Cai et al. [60], who compared CE

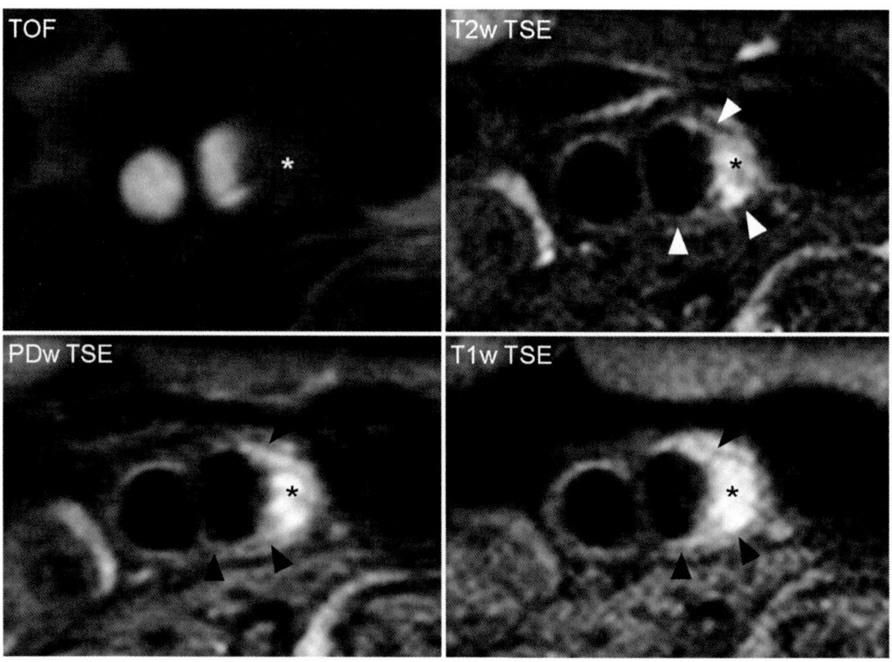

Fig. 17.6 Multisequence non-CE MRI: co-registered time-of-flight (A), T2-weighted turbo spin-echo (TSE), proton-density-weighted TSE, and T1-weighted TSE images of a transverse section of a plaque in the internal carotid artery, just above the carotid bifurcation. A LRNC with recent hemorrhage is identified as a hyperintense signal area in all sequences (*asterisk*). Calcifications, located at the outer rim of the plaque, are identified as areas of hypointense signals on T2-weighted TSE, proton density-weighted TSE, and T1-weighted TSE images (*arrowheads*). The remaining tissue is defined as fibrous tissue

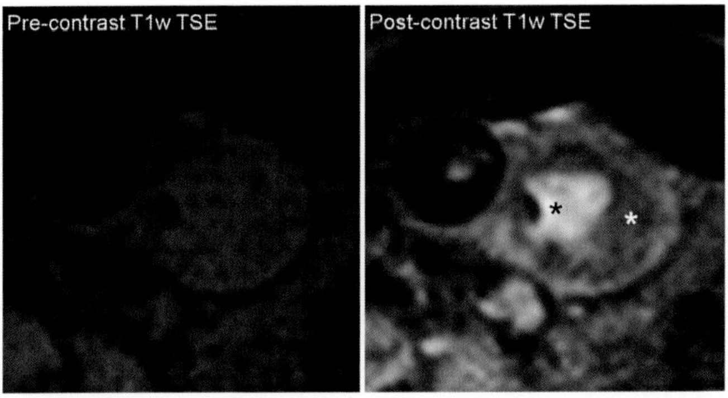

Fig. 17.7 Pre and postcontrast T1-weighted TSE images of a transverse section of a plaque in the internal carotid artery. On the precontrast image, the fibrous cap and LRNC cannot be distinguished. After intravenous administration of a gadolinium-based contrast agent (gadopentate dimeglumine), the fibrous cap enhances strongly (*black asterisk*), whereas the LRNC shows no/slight enhancement (*white asterisk*)

MRI to histology, found strong correlations for FC length and area (Pearson $\rho=0.73$ and 0.80, respectively) and a very strong correlation for LRNC area (Pearson $\rho=0.84$) [60]. Using CE MRI, interobserver agreement in differentiating between fibrotic and/or calcified plaques, plaques with a LRNC and an intact and thick FC, and plaques with a LRNC and a thin and/or ruptured FC is good ($\kappa=0.64$–0.78) [61]. A disadvantage of the use of gadolinium-based contrast is that it is contraindicated in patients with renal failure because of the risk of nephrogenic systemic fibrosis [62]. A disadvantage of the use of a multisequence protocol is that total scan time is rather lengthy (up to 45 min). A single MRI pulse sequence may be more useful in clinical practice. Moody et al. [63] demonstrated that a single T1-weighted magnetization-prepared three-dimensional (3D) gradient echo sequence is able to detect intraplaque hemorrhage, a hallmark of complicated plaque [64], with high sensitivity and specificity (both 84%) and good interobserver agreement ($\kappa=0.75$). Advantages of this 3D sequence are its short scan time (approximately 3.5 min), the possibility of MPR, and that images are easy to interpret (Fig. 17.8). The high accuracy and good interobserver agreement of this technique has been validated by other investigators [65, 66]. Cappendijk et al. [67] demonstrated that the size of the LRNC can be semiquantitatively analyzed using a single T1-weighted turbo field-echo sequence. MRI has also shown to be able to detect plaque ulcerations [68]. However, there are no published studies assessing its accuracy and reproducibility in a large series of patients. A high-resolution 3D sequence, enabling MPR, may be needed to reliably detect ulcerations.

Most research has been performed at MR scanners with a field strength of 1.5 T. MRI at 3.0 T has several advantages over 1.5 T: increased signal-to-noise ratio and/or improved spatial resolution, and potential reductions of imaging time. Underhill et al. [69] showed that there is an excellent agreement between both field strengths for the assessment of all plaque components (ICC=0.88–0.96).

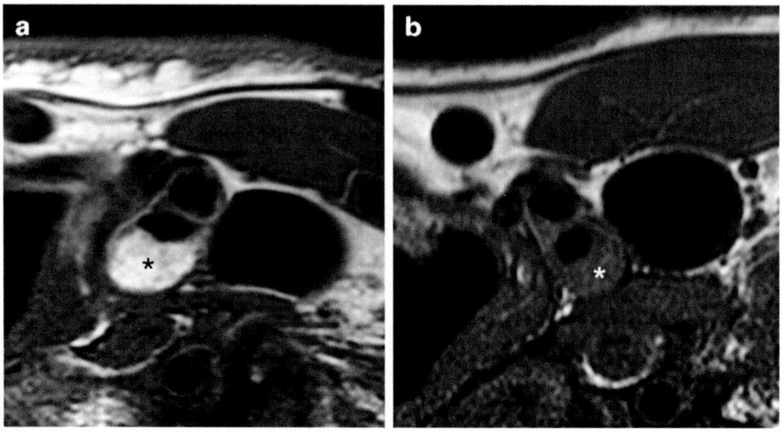

Fig. 17.8 T1-weighted TFE images of transverse sections of plaques in the internal carotid artery of two different patients (**a** and **b**). In patient 1 (**a**), high signal intensity is identified within the plaque (*black asterisk*), consistent with the presence of intraplaque hemorrhage. In patient 2 (**b**), no high signal intensity is identified within the plaque (*white asterisk*), consistent with the absence of intraplaque hemorrhage

However, hemorrhages and calcifications appeared larger at 3.0 T due to increased magnetic susceptibility at higher field strength [69]. Plaque component quantification by automatic software has shown to be feasible [70, 71]. The use of such software may further improve the reproducibility of MRI, and it can save time in image review. The development of new coil technologies and new pulse sequences may also improve the performance of MRI.

Several prospective longitudinal studies have been performed, both in asymptomatic and symptomatic patients: Takaya et al. [72] studied 154 asymptomatic patients with 50–79% carotid stenosis. Multisequence non-CE MRI was performed at baseline, and patients were clinically followed up for an average period of 38 months. They found that plaques with a thin or ruptured fibrous cap, intraplaque hemorrhage, larger maximum %LRNC, and larger maximum wall thickness were associated with the occurrence of subsequent ipsilateral TIA or stroke ($n = 12$) [72]. Singh et al. [73] studied 75 asymptomatic men with 50–70% carotid stenosis (98 eligible arteries). A 3D T1-weighted fat-suppressed spoiled gradient-echo sequence was performed at baseline, to detect intraplaque hemorrhage, and patients were clinically followed up for an average period of 25 months. Sensitivity and specificity for the prediction of ipsilateral TIA or stroke were 17 and 100%, respectively [73]. Altaf et al. followed up 64 symptomatic patients with 30–69% carotid stenosis for a median period of 28 months [74] and 66 symptomatic patients with 60–99% carotid stenosis for a median period of 33.5 days [75]. At baseline, a 3D T1-weighted fat-suppressed spoiled gradient-echo sequence was used to detect intraplaque hemorrhage. Sensitivity and specificity of MR-depicted intraplaque hemorrhage for the prediction of ipsilateral TIA or stroke varied between 33–34 and 91–96%, respectively [74, 75].

17.5.2 Dynamic CE MRI

After the injection of gadolinium-based contrast agents, temporal enhancement of the vessel wall can be observed (Fig. 17.9), which is thought to be caused by increased vascularity of the adventitial vasa vasorum feeding plaque neovasculature [76]. Kerwin et al. [77] compared dynamic CE MRI findings with histology. Using a 2D dynamic CE T1-weighted fast field echo (FFE) sequence and kinetic modeling, they assessed time-varying signal intensities within the plaques to estimate the fractional plasma volume (v_p) and transfer constant (K^{trans}) of contrast material into the extracellular space. Correlations between K^{trans} and v_p measurements, and neovasculature content within the plaque were found to be strong (Pearson $\rho=0.71$ and 0.68, respectively). In addition, there were also strong and moderate correlations between both parameters and plaque macrophage content (Pearson $\rho=0.75$ and 0.54), respectively. Thus, dynamic CE MRI can be used to noninvasively assess degree of plaque neovascularization and inflammation, which are closely related to each other.

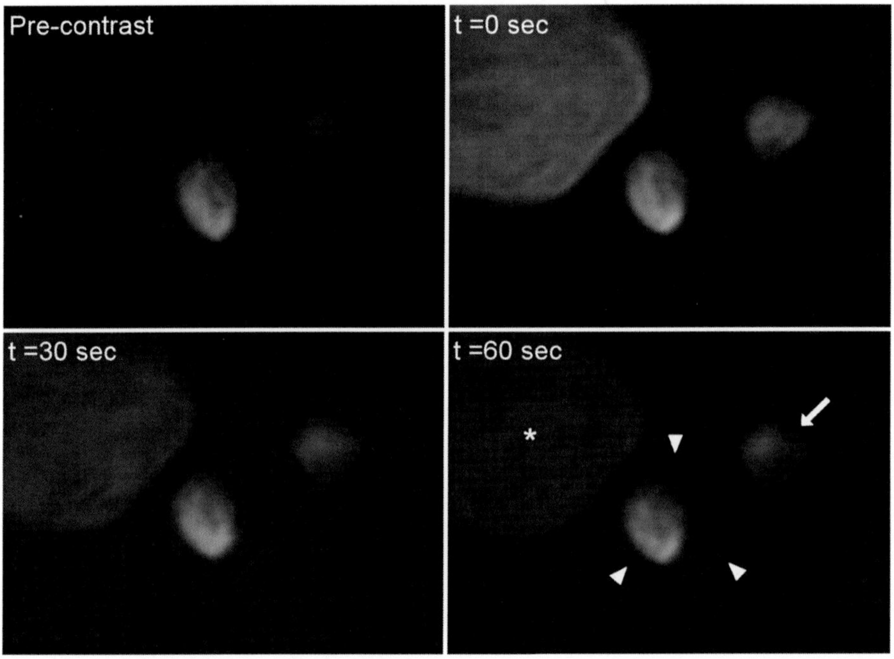

Fig. 17.9 Precontrast (*upper left*) and dynamic contrast-enhanced T1-weighted FFE images at bolus arrival (*upper right*), 30 ((*lower left*), and 75 s later (*lower right*). A saturation slab has been positioned above the slice, so that the internal jugular vein appears *dark* on the precontrast image. After administration of a gadolinium-based contrast agent (gadopentate dimeglumine), strong enhancement of the internal jugular vein (*asterisk* in the *lower right image*) can be observed. There is also enhancement of the adventitia of the internal carotid artery (*arrowheads* in the *lower right image*), which is thought to be caused by increased vascularity of the adventitial vasa vasorum feeding plaque neovasculature. *Arrow* in the *lower right image* points at the external carotid artery

17.5.3 Ultrasmall Particles of Iron Oxide-Enhanced MRI

Ultrasmall particles of iron oxide (USPIO)-enhanced MRI provides a means to visualize macrophage accumulation. After intravenous injection, USPIOs are phagocytosed by cells of the monocyte-macrophage system throughout the whole body, including those within atheromatous plaques. The uptake of USPIO nanoparticles causes localized magnetic field inhomogeneities within the vessel wall, resulting in signal loss on T2*-weighted FFE images (Fig. 17.10) [78]. USPIO particles also reduce the T1 relaxation time, which can lead to a signal increase. The resulting effect of the competing T1 signal increase and T2* signal decrease is

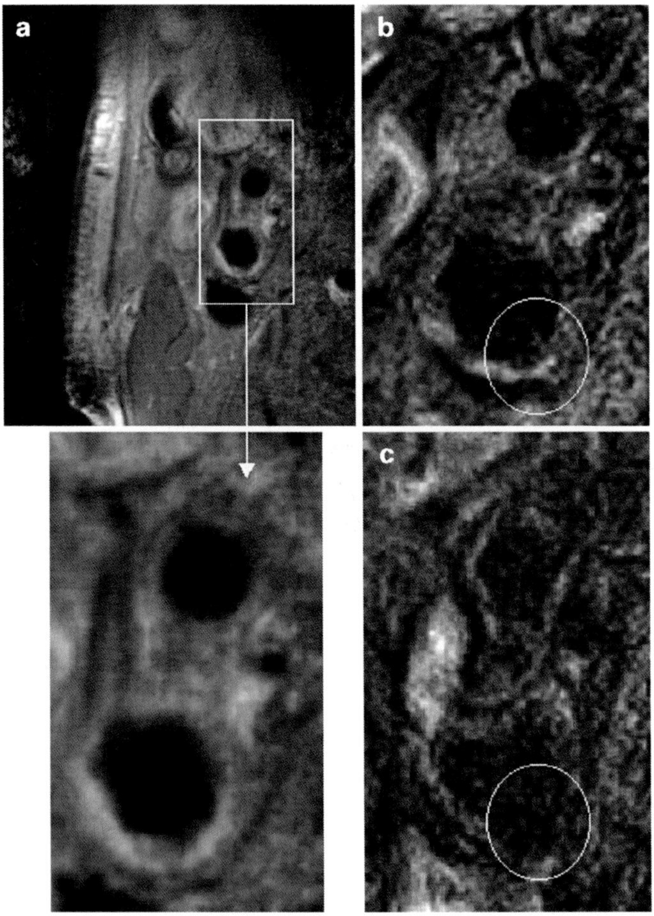

Fig. 17.10 Proton density-weighted TSE (**a**) pre- (**b**) and postcontrast (**c**) T2*-weighted FFE images of a transverse section of the external (above) and internal (below) carotid artery. Twenty-four hours after intravenous administration of ferumoxtran-10, a signal decrease can be observed on the postcontrast T2*-weighted FFE image (*circle*), consistent with ferumoxtran-10 uptake by macrophages. Reproduced with permission from Kooi et al. [78]

dependent upon the USPIO concentration and distribution within the plaque (T1 effect dominates at low concentrations, while the T2* susceptibility effect dominates at higher concentrations of clustered USPIOs), as well as the MRI pulse sequence used [79]. The dextran-coated USPIO ferumoxtran-10 is most suited for carotid plaque imaging because it has a much longer plasma half-life (24–36 h) than other human-approved USPIOs (3–6 h). All published studies on humans have used this type of USPIO. Kooi et al. [78] initially validated the use of the technique with histology; they showed that ferumoxtran-10 predominantly accumulates in macrophages in ruptured and rupture-prone plaques. Trivedi et al. [80] investigated a larger series of patients. In plaques with visually focal signal loss, they found a moderate-to-strong correlation between ferumoxtran-10-induced signal decrease and plaque macrophage content (Spearman ρ=0.49–0.60). However, in plaques with visually diffuse signal loss, they occasionally measured an increase in signal intensity, and no significant correlation with macrophage content was found [80]. Trivedi et al. [80] suggested that these plaques may mainly contain extracellular ferumoxtran-10, and the increase in signal intensity may be explained by the T1 effect of ferumoxtran-10. They also concluded that it is probable that only a focal area of signal loss represents ferumoxtran-10 accumulation in macrophages [80]. The same investigators [81] investigated the reproducibility of the technique: inter-observer agreement on identification of areas of signal loss and on the size of these areas proved to be excellent (κ=0.88) and good (ICC=0.75), respectively.

An advantage of USPIO-enhanced MRI is that it can be performed simultaneously with standard high-resolution MRI of the carotid artery wall. A disadvantage of USPIO-enhanced MRI is that images must be obtained twice: before and after administration of USPIOs. The optimum imaging time window to detect maximal signal change post-USPIO infusion is between 36 and 48 h [82]. In addition, accurate assessment of signal loss after USPIO administration may be difficult, especially when signal loss is only modest and obscured by calcifications and blood degradation products, which also have strong susceptibility effects. New MRI pulse sequences, providing positive contrast, may facilitate measurements of plaque USPIO uptake [83].

Several cross-sectional studies have been performed, demonstrating that degree of ferumoxtran-10 uptake of carotid plaques is related to patient symptomatology: plaques ipsilateral to the symptomatic side were shown to have significantly more vessel quadrants of signal loss compared to contralateral asymptomatic plaques [79] and asymptomatic plaques of asymptomatic patients [84]. Furthermore, contralateral asymptomatic plaques of symptomatic patients were also shown to have significantly more vessel quadrants of signal loss than asymptomatic plaques of asymptomatic patients [85].

17.6 Nuclear Imaging Techniques

Nuclear imaging techniques are very sensitive in detecting radioactive tracers targeted at carotid plaques. Single photon emission computed tomography (SPECT), which uses single-headed or multi-headed rotating gamma cameras, produces multiple cross-sectional images from which a 3D dataset can be reconstructed. Positron emission tomography (PET) is another tomographic nuclear technique, using positron-emitting tracers, and is also capable of producing 3D images. Disadvantages of nuclear imaging techniques are the lack of detailed anatomic information in the area of tracer uptake and exposure of the patient to ionizing radiation. Nuclear imaging techniques are generally costly, labor-intensive, and time-consuming.

17.6.1 Fluorine-18-Fluorodeoxyglucose PET

Fluorine-18-fluorodeoxyglucose (^{18}F-FDG) is a glucose analog that competes with glucose for uptake into metabolically active cells, where it accumulates in proportion to metabolic activity. ^{18}F-FDG PET imaging can be used to identify the increased glycolytic activity of inflammatory cells [86], including those in an inflamed plaque. An initial ^{18}F-FDG PET study by Rudd et al. [87], who performed autoradiography of excised plaques, confirmed ^{18}F-FDG accumulation in macrophage-rich areas of the plaque. A subsequent study by Tawakol et al. [88] validated in vivo ^{18}F-FDG PET findings with histology; they found a strong correlation between mean plaque ^{18}F-FDG uptake and mean percentage of plaque macrophages (Spearman $\rho = 0.85$). Coregistration of PET images with CT or MRI is necessary for good anatomical localization of focal ^{18}F-FDG uptake [87–89]. Degree of plaque inflammation at ^{18}F-FDG PET can be measured as mean and maximum blood-normalized standardized uptake value (SUV), also known as the target-to-background ratio. Both parameters are normalized for blood ^{18}F-FDG activity by dividing them through the mean SUV of blood (as measured in the jugular vein) [89]. Using this method, CT-guided ^{18}F-FDG PET analysis achieves excellent interobserver agreement (ICC = 0.94–0.96) [89]. MRI-guided ^{18}F-FDG PET has also shown to be a highly reproducible technique, and it has been shown that the excellent anatomic detail provided by MR can be used to facilitate partial volume correction of the PET images [90]. A disadvantage of ^{18}F-FDG PET is that patients need to fast before the examination, to minimize competition between uptake of physiologic glucose and ^{18}F-FDG. In addition, PET imaging should be performed some time after ^{18}F-FDG injection; after approximately 3 h, the contrast between the ^{18}FDG concentration in plaque and blood is optimal [87]. During the period between ^{18}F-FDG injection and imaging, patients need to rest on a bed to reduce ^{18}F-FDG uptake in neck muscles.

Several studies relating degree of plaque ^{18}FDG uptake to clinical ischemic symptoms, have been performed: Davies et al. [91] and Arauz et al. [92] found that in the majority of symptomatic patients who were scheduled for CEA, high plaque ^{18}F-FDG uptake could be observed in a vascular location compatible with the clinical presentation. In addition, Rudd et al. [87] found 27% higher ^{18}F-FDG uptake of plaques on the symptomatic side compared to contralateral asymptomatic plaques.

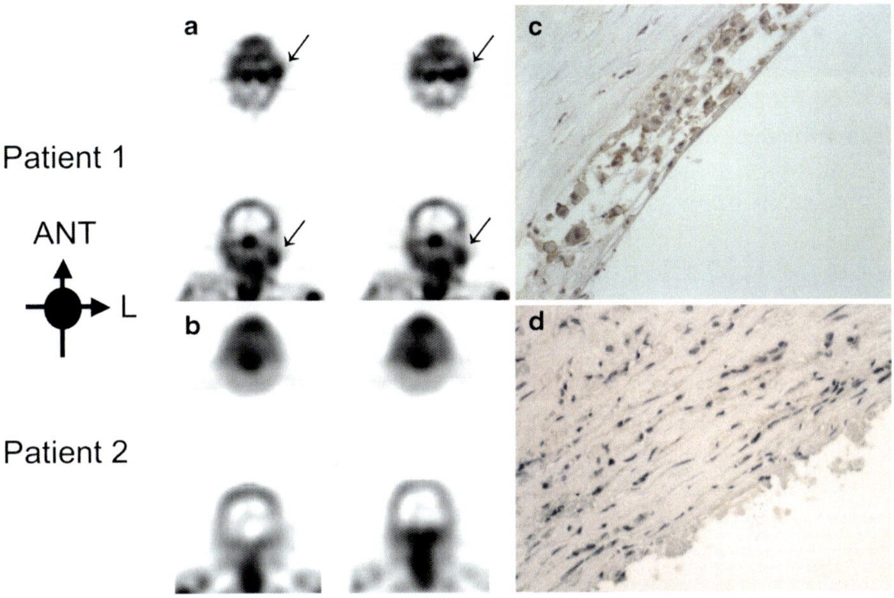

Fig. 17.11 Transverse and coronal SPECT images of carotid artery plaques of two different patients obtained with ^{99m}Tc-annexin A5 (**a** and **c**). Patient 1 experienced a left-sided TIA 3 days before imaging and had >70% carotid stenosis on both sides. Uptake of ^{99m}Tc-annexin A5 is evident only in the left carotid artery region (*arrows*) (**a**). Histopathological analysis of the corresponding left CEA specimen (polyclonal rabbit anti-annexin A5 antibody, ×400) shows substantial infiltration of macrophages into the neointima, with extensive binding of annexin A5 (*brown*) (**b**). In contrast, SPECT images of patient 2 (**c**), who had had a right-sided TIA 3 months before imaging, do not show evidence of ^{99m}Tc-annexin A5 uptake in the carotid artery region on either side. Doppler US revealed >70% carotid stenosis on the affected side. Histopathological analysis of the right CEA specimen from patient 2 (polyclonal rabbit anti-annexin A5 antibody, ×400) shows a lesion rich in smooth-muscle cells, with negligible binding of annexin A5 (**d**). *ANT* anterior, *L* left. Modified with permission from Kietselaer et al. [94].

17.6.2 Annexin A5 Scintigraphy

Cell death is a recognized feature of advanced atherosclerotic disease; symptomatic carotid atherosclerotic plaques display apoptosis and oncosis [9]. A high rate of cell death in atherosclerotic plaques may contribute to destabilization of the fibrous cap, and it may increase the risk of plaque rupture and thrombosis. Annexin A5 is a plasma protein, which binds with high affinity to phosphatidylserine, which is expressed on the cell membrane of apoptotic cells [93]. A pilot study [94] demonstrated the feasibility of in vivo imaging of apoptosis in carotid plaques with radiolabeled (^{99m}Tc-) annexin A5. In four patients who were scheduled for CEA, SPECT imaging was performed 6 h after the infusion of ^{99m}Tc-annexin A5. In the two patients that had suffered a TIA more than 3 months before imaging and who had been treated with statins and antiplatelet agents, no ^{99m}Tc-annexin A5 uptake was observed. Histological analysis confirmed stable, smooth muscle, and collagen-rich lesions, with negligible macrophage infiltration and no intraplaque hemorrhage. By contrast, the two patients that had experienced a TIA within a week before undergoing ^{99m}Tc-annexin A5 SPECT demonstrated significant ^{99m}Tc-annexin A5 uptake in the culprit carotid vessel; contralateral arteries, although similarly stenotic, did not demonstrate ^{99m}Tc-annexin A5 uptake. Annexin A5 binding was histologically traced to apoptotic macrophages (Fig. 17.11).

17.7 Summary and Future Research Directions

Large randomized controlled trials have demonstrated a benefit of CEA in patients with significant stenosis [3, 4]. The clinical practice guidelines of the Society for Vascular Surgery recommend CEA in symptomatic patients with ≥50% carotid stenosis and in asymptomatic patients with ≥60% stenosis and low perioperative risk [95]. However, the numbers of patients needed to undergo CEA to prevent one ipsilateral stroke in 5 years are 11.8 [96] and 16.9 [4], respectively. Thus, in clinical practice, many patients undergo surgery with its attendant risks without taking advantage of it. In others, CEA is probably incorrectly withheld on basis of degree of stenosis alone. There is, therefore, an urgent need for new adjuncts to differentiate high-risk, vulnerable plaques from stable plaques. In this view, multiple noninvasive imaging modalities, capable of assessing vulnerable plaque features in vivo, have undergone investigation (Table 17.1). Large prospective studies using standard B-mode US found conflicting results on the association between plaque echogenicity and the occurrence of stroke, which casts doubts as to its utility. The use of compound US in combination with other methods to characterize plaque composition may have more potential in assessing plaque vulnerability. CE US seems to be a promising technique to detect plaque neovasularization, and it may also be used to detect plaque ulcerations. However, there still is limited experience with this technique. The accuracy of CE US should be further investigated, and standardized acquisition and evaluation methods should be developed before applying this technique in prospective studies. MES-positive symptomatic patients, assessed by TCD, have been shown to be at high risk for developing ipsilateral stroke; these results should be further validated. Whether MES-positive asymptomatic patients are also at increased risk still has to be determined; the number of ipsilateral strokes that occurred in existing studies was too small to achieve statistical significance. Future studies in asymptomatic patients with 60–99% carotid stenosis, who have an annual ipsilateral ischemic stroke risk of 2% [4], need approximately 820 years of patient follow-up to detect an RR even as large as 4.0. New techniques for automated analysis should be further developed to make MES assessment by TCD more practical in clinical use. MDCT has shown to be able to identify and quantify several carotid plaque components in vivo, and it can reliably be used to indentify plaque ulcerations. However, at present, MDCT cannot distinguish LRNC from hemorrhage, and fibrous cap status cannot be assessed. In addition, accuracy of LRNC quantification is low in severely calcified plaques. Large prospective longitudinal studies are needed to determine whether MDCT-assessed carotid plaque parameters can predict the risk of ipsilateral stroke. Multisequence conventional MRI, in combination with dynamic CE MRI, is able to assess the majority of plaque features that are thought to be associated with plaque vulnerability. In patients with asymptomatic 50–79% carotid stenosis, it has been demonstrated that, using multisequence non-CE MRI, plaques with thin or ruptured fibrous caps, intraplaque hemorrhage, larger maximum %LRNC, and larger maximum wall thickness are associated with increased risk of ipsilateral TIA/stroke. A T1-weighted magnetization prepared 3D gradient echo sequence is a highly sensitive and specific technique to detect intraplaque hemorrhage. Using this sequence, several prospective studies have suggested that the absence of intraplaque hemorrhage may have a high negative predictive value for the occurrence of future ipsilateral TIA and stroke, both in symptomatic and asymptomatic patients. Adequately powered studies are needed to investigate whether carotid MRI is able to identify patients at risk for the clinically important end point of ipsilateral stroke alone. USPIO-enhanced MRI is able to quantify macrophage accumulation within carotid plaques. Although USPIO-enhanced MRI may be helpful in identifying vulnerable plaques, a disadvantage of this method is that MRI should be performed twice: before and after USPIO administration, with an interval of approximately 2 days. More importantly, USPIO-based contrast agents are currently not available in the market. ^{18}F-FDG PET is also able to assess macrophage accumulation within carotid plaques. Cross-sectional studies showed that degree of plaque ^{18}F-FDG uptake is related to symptomatology. Whether ^{18}F-FDG PET is useful for improving risk stratification for stroke still needs to be investigated by prospective studies. It has been demonstrated that Annexin A5 scintigraphy is able to visualize apoptosis of macrophages in carotid plaques. However, there is limited clinical experience in the use of Annexin A5 scintigraphy for carotid plaque imaging. Since the technique is limitedly available, costly, uses ionizing radiation, and a long time interval between radiolabeled Annexin A5 injection and imaging is needed, it may be less appropriate as a screening technique in patients with carotid atherosclerosis.

Future studies should investigate whether one imaging modality or a multimodality approach is most effective in improving risk stratification of stroke. One should especially focus on the group of symptomatic patients with 50–69% carotid stenosis since the benefit of CEA in this group is questionable, particularly if CEA is not performed within 2 weeks of last symptoms [96]. The other group of patients on whom investigators should focus is the group of asymptomatic patients with ≥60% stenosis, where the balance between benefits and risks of CEA is small [4]. If noninvasive imaging can identify subgroups of patients who are at high risk of developing ipsilateral ischemic stroke, a randomized controlled trial is warranted to weigh the risks of surgery versus nonsurgery in these subgroups.

Noninvasive imaging may also be used as a means in choosing the type of surgical intervention (i.e., CEA or carotid artery stenting). For instance, the Imaging in Carotid Angioplasty and Risk of Stroke Study has already shown that patients with echolucent plaque (GSM≤25) on standard B-mode US have an increased risk of stroke in carotid artery stenting [97]. Whether patients with echolucent plaques are better off with CEA or nonsurgery than with CAS needs to be determined. Last, noninvasive imaging may be used to investigate the natural course of carotid plaques and to monitor the effects of drug therapy on carotid atherosclerosis. For instance, MES detection by TCD has shown to be a feasible method to evaluate the efficacy of antiplatelet therapy [98], and it has been shown that multisequence MRI can be used to assess the effect of statin therapy on plaque volume and composition [99], whereas USPIO-enhanced MRI [100] and ^{18}F-FDG PET [101] can be used to assess the effect of statin therapy on plaque inflammation. Future studies should assess whether the effects of drug therapy, as monitored by noninvasive imaging, also result in a reduced risk of developing ipsilateral ischemic stroke.

References

1. Mackay J, Mensah GA (2004) The atlas of heart disease and stroke. World Health Organization, Geneva, Switzerland
2. Albers GW, Amarenco P, Easton JD, et al; American College of Chest Physicians. Antithrombotic and thrombolytic therapy for ischemic stroke (2008) American College of Chest Physicians Evidence-Based Clinical Practice Guidelines (8th Edition). Chest 133 (Suppl 6):630–669
3. Rothwell PM, Eliasziw M, Gutnikov SA, et al; Carotid Endarterectomy Trialists' Collaboration (2003) Analysis of pooled data from the randomised controlled trials of endarterectomy for symptomatic carotid stenosis. Lancet 361:107–116
4. Executive Committee for the Asymptomatic Carotid Atherosclerosis Study (1995) Endarterectomy for asymptomatic carotid artery stenosis. JAMA 273:1421–1428
5. Spagnoli LG, Mauriello A, Sangiorgi G, et al (2004) Extracranial thrombotically active carotid plaque as a risk factor for ischemic stroke. JAMA 292:1845–1852
6. Redgrave JN, Lovett JK, Gallagher PJ, et al (2006). Histological assessment of 526 symptomatic carotid plaques in relation to the nature and timing of ischemic symptoms: the Oxford plaque study. Circulation 113:2320–2328.
7. Virmani R, Kolodgie FD, Burke AP, et al (2005) Atherosclerotic plaque progression and vulnerability to rupture: angiogenesis as a source of intraplaque hemorrhage. Arterioscler Thromb Vasc Biol 25:2054–2061
8. Gao P, Chen ZQ, Bao YH, et al (2007) Correlation between carotid intraplaque hemorrhage and clinical symptoms: systematic review of observational studies. Stroke 38:2382–2390
9. Crisby M, Kallin B, Thyberg J, et al (1997) Cell death in human atherosclerotic plaques involves both oncosis and apoptosis. Atherosclerosis 130:17–27
10. Park AE, McCarthy WJ, Pearce WH, et al (1998) Carotid plaque morphology correlates with presenting symptomatology. J Vasc Surg 27:872–878
11. Sitzer M, Müller W, Siebler M, et al (1995) Plaque ulceration and lumen thrombus are the main sources of cerebral microemboli in high-grade internal carotid artery stenosis. Stroke 26:1231–1233
12. European Carotid Plaque Study Group (1995) Carotid artery plaque composition – relationship to clinical presentation and ultrasound B-mode imaging. Eur J Vasc Endovasc Surg 10:23–30
13. Grønholdt ML, Wiebe BM, Laursen H, et al (1997) Lipid-rich carotid artery plaques appear echolucent on ultrasound B-mode images and may be associated with intraplaque haemorrhage. Eur J Vasc Endovasc Surg 14:439–445
14. Polak JF, Shemanski L, O'Leary DH, et al (1998). Hypoechoic plaque at US of the carotid artery: an independent risk factor for incident stroke in adults aged 65 years or older. Cardiovascular Health Study. Radiology 208:649–654

15. Grønholdt ML, Nordestgaard BG, Schroeder TV, et al (2001) Ultrasonic echolucent carotid plaques predict future strokes. Circulation 104:68–73
16. Halliday A, Mansfield A, Marro J, et al; MRC Asymptomatic Carotid Surgery Trial (ACST) Collaborative Group (2004). Prevention of disabling and fatal strokes by successful carotid endarterectomy in patients without recent neurological symptoms: randomised controlled trial. Lancet 363:1491–1502
17. Montauban van Swijndregt AD, Elbers HR, Moll FL, et al (1998) Ultrasonographic characterization of carotid plaques. Ultrasound Med Biol 24:489–493
18. de Bray JM, Baud JM, Delanoy P, et al (1998) Reproducibility in ultrasonic characterization of carotid plaques. Cerebrovasc Dis 8:273–277
19. Kern R, Szabo K, Hennerici M, et al (2004) Characterization of carotid artery plaques using real-time compound B-mode ultrasound. Stroke 35:870–875
20. Sabetai MM, Tegos TJ, Nicolaides AN, et al (2000). Reproducibility of computer-quantified carotid plaque echogenicity: can we overcome the subjectivity? Stroke 31:2189–2196
21. Sztajzel R, Momjian S, Momjian-Mayor I, et al (2005) Stratified gray-scale median analysis and color mapping of the carotid plaque: correlation with endarterectomy specimen histology of 28 patients. Stroke 36:741–745
22. Lal BK, Hobson RW 2nd, Pappas PJ, et al (2000) Pixel distribution analysis of B-mode ultrasound scan images predicts histologic features of atherosclerotic carotid plaques. J Vasc Surg 35:1210–1217
23. Feinstein SB (2006) Contrast ultrasound imaging of the carotid artery vasa vasorum and atherosclerotic plaque neovascularization. J Am Coll Cardiol 48:236–243
24. Piscaglia F, Bolondi L; Italian Society for Ultrasound in Medicine and Biology (SIUMB) Study Group on Ultrasound Contrast Agents (2006). The safety of Sonovue in abdominal applications: retrospective analysis of 23188 investigations. Ultrasound Med Biol 32:1369–1375
25. Kono Y, Pinnell SP, Sirlin CB, et al (2004) Carotid arteries: contrast-enhanced US angiography – preliminary clinical experience. Radiology 230:561–568
26. Magnoni M, Coli S, Cianflone D (2009) A surprise behind the dark. Eur J Echocardiogr 10:887–888
27. Shah F, Balan P, Weinberg M, et al (2007) Contrast-enhanced ultrasound imaging of atherosclerotic carotid plaque neovascularization: a new surrogate marker of atherosclerosis? Vasc Med 12:291–297
28. Coli S, Magnoni M, Sangiorgi G, et al (2008). Contrast-enhanced ultrasound imaging of intraplaque neovascularization in carotid arteries: correlation with histology and plaque echogenicity. J Am Coll Cardiol 52:223–230
29. Giannoni MF, Vicenzini E, Citone M, et al (2009) Contrast carotid ultrasound for the detection of unstable plaques with neoangiogenesis: a pilot study. Eur J Vasc Endovasc Surg 37:722–727
30. Xiong L, Deng YB, Zhu Y, et al (2009) Correlation of carotid plaque neovascularization detected by using contrast-enhanced US with clinical symptoms. Radiology 251:583–589
31. Consensus Committee of the Ninth International Cerebral Hemodynamic Symposium (1995) Basic identification criteria of Doppler microembolic signals. Stroke 26:1123
32. Markus H, Bland JM, Rose G, et al (1996) How good is intercenter agreement in the identification of embolic signals in carotid artery disease? Stroke 27:1249–1252
33. Van Zuilen EV, Mess WH, Jansen C, et al (1996) Automatic embolus detection compared with human experts. A Doppler ultrasound study. Stroke 27:1840–1843
34. Mackinnon AD, Aaslid R, Markus HS (2004) Long-term ambulatory monitoring for cerebral emboli using transcranial Doppler ultrasound. Stroke 35:73–78
35. Cullinane M, Reid G, Dittrich R, et al (2000) Evaluation of new online automated embolic signal detection algorithm, including comparison with panel of international experts. Stroke 31:1335–1341
36. Wijnhoud AD, Franckena M, van der Lugt A, et al (2008) Inadequate acoustical temporal bone window in patients with a transient ischemic attack or minor stroke: role of skull thickness and bone density. Ultrasound Med Biol 34:923–929
37. Sliwka U, Lingnau A, Stohlmann WD, et al (1997) Prevalence and time course of microembolic signals in patients with acute stroke. A prospective study. Stroke 28:358–363

38. Censori B, Partziguian T, Casto L, et al (2000) Doppler microembolic signals predict ischemic recurrences in symptomatic carotid stenosis. Acta Neurol Scand 101:327–331
39. Valton L, Larrue V, Le Traon AP, et al (1998) Microembolic signals and risk of early recurrence in patients with stroke or transient ischemic attack. Stroke 29:2125–2128
40. Molloy J, Markus HS (1999) Asymptomatic embolization predicts stroke and TIA risk in patients with carotid artery stenosis. Stroke 30:1440–1443
41. Siebler M, Nachtmann A, Sitzer M, et al (1995) Cerebral microembolism and the risk of ischemia in asymptomatic high-grade internal carotid artery stenosis. Stroke 26:2184–2186
42. Markus HS, MacKinnon A (2005) Asymptomatic embolization detected by Doppler ultrasound predicts stroke risk in symptomatic carotid artery stenosis. Stroke 36:971–975
43. Abbott AL, Chambers BR, Stork JL, et al (2005) Embolic signals and prediction of ipsilateral stroke or transient ischemic attack in asymptomatic carotid stenosis: a multicenter prospective cohort study. Stroke 36:1128–1133
44. Spence JD, Tamayo A, Lownie SP, et al (2005) Absence of microemboli on transcranial Doppler identifies low-risk patients with asymptomatic carotid stenosis. Stroke 36:2373–2378.
45. Walker LJ, Ismail A, McMeekin W, et al (2002) Computed tomography angiography for the evaluation of carotid atherosclerotic plaque: correlation with histopathology of endarterectomy specimens. Stroke 33:977–981
46. Ellis JH, Cohan RH (2009) Reducing the risk of contrast-induced nephropathy: a perspective on the controversies. AJR Am J Roentgenol 192:1544–1549
47. de Weert TT, Ouhlous M, Meijering E, et al (2006) In vivo characterization and quantification of atherosclerotic carotid plaque components with multidetector computed tomography and histopathological correlation. Arterioscler Thromb Vasc Biol 26:2366–2372.
48. de Weert TT, de Monyé C, Meijering E, et al (2008) Assessment of atherosclerotic carotid plaque volume with multidetector computed tomography angiography. Int J Cardiovasc Imaging 24:751–759
49. Saba L, Caddeo G, Sanfilippo R (2007) CT and ultrasound in the study of ulcerated carotid plaque compared with surgical results: potentialities and advantages of multidetector row CT angiography. AJNR Am J Neuroradiol 28:1061–1066
50. de Weert TT, Cretier S, Groen HC, et al (2009). Atherosclerotic plaque surface morphology in the carotid bifurcation assessed with multidetector computed tomography angiography. Stroke 40:1334–1340
51. Wintermark M, Jawadi SS, Rapp JH, et al (2008). High-resolution CT imaging of carotid artery atherosclerotic plaques. AJNR Am J Neuroradiol 29:875–882
52. Uotani K, Watanabe Y, Higashi M, et al (2009) Dual-energy CT head bone and hard plaque removal for quantification of calcified carotid stenosis: utility and comparison with digital subtraction angiography. Eur Radiol 19:2060–2065
53. Miralles M, Merino J, Busto M, et al (2006) Quantification and characterization of carotid calcium with multi-detector CT-angiography. Eur J Vasc Endovasc Surg 32:561–567
54. Nandalur KR, Hardie AD, Raghavan P, et al (2007) Composition of the stable carotid plaque: insights from a multidetector computed tomography study of plaque volume. Stroke 38:935–940
55. Wintermark M, Arora S, Tong E, et al (2008) Carotid plaque computed tomography imaging in stroke and nonstroke patients. Ann Neurol 64(2):149–57.
56. Abedin M, Tintut Y, Demer LL (2004) Vascular calcification: mechanisms and clinical ramifications. Arterioscler Thromb Vasc Biol 24:1161–1170
57. Saam T, Ferguson MS, Yarnykh VL, et al (2005) Quantitative evaluation of carotid plaque composition by in vivo MRI. Arterioscler Thromb Vasc Biol 25:234–239
58. Touzé E, Toussaint JF, Coste J, et al; HIgh-Resolution magnetic resonance Imaging in atherosclerotic Stenosis of the Carotid artery (HIRISC) study group (2007) Reproducibility of high-resolution MRI for the identification and the quantification of carotid atherosclerotic plaque components: consequences for prognosis studies and therapeutic trials. Stroke 38:1812–1819
59. Wasserman BA, Smith WI, Trout HH 3rd, et al (2002) Carotid artery atherosclerosis: in vivo morphologic characterization with gadolinium-enhanced double-oblique MR imaging initial results. Radiology 223:566–573

60. Cai J, Hatsukami TS, Ferguson MS, et al (2005) In vivo quantitative measurement of intact fibrous cap and lipid-rich necrotic core size in atherosclerotic carotid plaque: comparison of high-resolution, contrast-enhanced magnetic resonance imaging and histology. Circulation 112:3437–3444
61. Kwee RM, van Engelshoven JM, Mess WH, et al (2009) Reproducibility of fibrous cap status assessment of carotid artery plaques by contrast-enhanced MRI. Stroke 40:3017–3021
62. Perez-Rodriguez J, Lai S, Ehst BD, et al (2009) Nephrogenic systemic fibrosis: incidence, associations, and effect of risk factor assessment – report of 33 cases. Radiology 250:371–377
63. Moody AR, Murphy RE, Morgan PS, et al (2003) Characterization of complicated carotid plaque with magnetic resonance direct thrombus imaging in patients with cerebral ischemia. Circulation 107:3047–3052
64. Stary HC, Chandler AB, Dinsmore RE, et al (1995) A definition of advanced types of atherosclerotic lesions and a histological classification of atherosclerosis. A report from the Committee on Vascular Lesions of the Council on Arteriosclerosis, American Heart Association. Circulation 92:1355–1374
65. Cappendijk VC, Cleutjens KB, Heeneman S, et al (2004) In vivo detection of hemorrhage in human atherosclerotic plaques with magnetic resonance imaging. J Magn Reson Imaging 20:105–110
66. Bitar R, Moody AR, Leung G, et al (2008) In vivo 3D high-spatial-resolution MR imaging of intraplaque hemorrhage. Radiology 249:259–267
67. Cappendijk VC, Heeneman S, Kessels AG, et al (2008) Comparison of single-sequence T1w TFE MRI with multisequence MRI for the quantification of lipid-rich necrotic core in atherosclerotic plaque. J Magn Reson Imaging 27:1347–1355
68. Chu B, Ferguson MS, Underhill H, et al (2005) Images in cardiovascular medicine. Detection of carotid atherosclerotic plaque ulceration, calcification, and thrombosis by multicontrast weighted magnetic resonance imaging. Circulation 112:3–4
69. Underhill HR, Yarnykh VL, Hatsukami TS, et al (2008) Carotid plaque morphology and composition: initial comparison between 1.5- and 3.0-T magnetic field strengths. Radiology 248:550–560
70. Hofman JM, Branderhorst WJ, ten Eikelder HM, et al (2006) Quantification of atherosclerotic plaque components using in vivo MRI and supervised classifiers. Magn Reson Med 55:790–799
71. Liu F, Xu D, Ferguson MS, et al (2006) Automated in vivo segmentation of carotid plaque MRI with Morphology-Enhanced probability maps. Magn Reson Med 55:659–668
72. Takaya N, Yuan C, Chu B, et al (2006) Association between carotid plaque characteristics and subsequent ischemic cerebrovascular events: a prospective assessment with MRI – initial results. Stroke 37:818–823
73. Singh N, Moody AR, Gladstone DJ, et al (2009) Moderate carotid artery stenosis: MR imaging-depicted intraplaque hemorrhage predicts risk of cerebrovascular ischemic events in asymptomatic men. Radiology 252:502–508
74. Altaf N, Daniels L, Morgan PS, et al (2008) Detection of intraplaque hemorrhage by magnetic resonance imaging in symptomatic patients with mild to moderate carotid stenosis predicts recurrent neurological events. J Vasc Surg 47:337–342
75. Altaf N, MacSweeney ST, Gladman J, et al (2007) Carotid intraplaque hemorrhage predicts recurrent symptoms in patients with high-grade carotid stenosis. Stroke 38:1633–1635
76. Aoki S, Aoki K, Ohsawa S, et al (1999) Dynamic MR imaging of the carotid wall. J Magn Reson Imaging 9:420–427
77. Kerwin WS, O'Brien KD, Ferguson MS, et al (2006) Inflammation in carotid atherosclerotic plaque: a dynamic contrast-enhanced MR imaging study. Radiology 241:459–468
78. Kooi ME, Cappendijk VC, Cleutjens KB, et al (2003) Accumulation of ultrasmall superparamagnetic particles of iron oxide in human atherosclerotic plaques can be detected by in vivo magnetic resonance imaging. Circulation 107:2453–2458
79. Tang T, Howarth SP, Miller SR, et al (2006) Assessment of inflammatory burden contralateral to the symptomatic carotid stenosis using high-resolution ultrasmall, superparamagnetic iron oxide-enhanced MRI. Stroke 37:2266–2270

80. Trivedi RA, Mallawarachi C, U-King-Im JM, et al (2006) Identifying inflamed carotid plaques using in vivo USPIO-enhanced MR imaging to label plaque macrophages. Arterioscler Thromb Vasc Biol 26:1601–1606
81. Trivedi RA, U-King-Im JM, Graves MJ, et al (2004) In vivo detection of macrophages in human carotid atheroma: temporal dependence of ultrasmall superparamagnetic particles of iron oxide-enhanced MRI. Stroke 35:1631–1635
82. Tang TY, Patterson AJ, Miller SR, et al (2009) Temporal dependence of in vivo USPIO-enhanced MRI signal changes in human carotid atheromatous plaques. Neuroradiology 2009 51:457–465
83. Howarth SP, Li ZY, Tang TY, et al (2008) In vivo positive contrast IRON sequence and quantitative T(2)* measurement confirms inflammatory burden in a patient with asymptomatic carotid atheroma after USPIO-enhanced MR imaging. J Vasc Interv Radiol 19:446–448
84. Howarth SP, Tang TY, Trivedi R, et al (2009) Utility of USPIO-enhanced MR imaging to identify inflammation and the fibrous cap: a comparison of symptomatic and asymptomatic individuals. Eur J Radiol 70:555–560
85. Tang TY, Howarth SP, Miller SR, et al (2007) Comparison of the inflammatory burden of truly asymptomatic carotid atheroma with atherosclerotic plaques contralateral to symptomatic carotid stenosis: an ultra small superparamagnetic iron oxide enhanced magnetic resonance study. J Neurol Neurosurg Psychiatry 78:1337–1343
86. Love C, Tomas MB, Tronco GG, et al (2005) FDG PET of infection and inflammation. Radiographics 25:1357–1368
87. Rudd JH, Warburton EA, Fryer TD, et al (2002) Imaging atherosclerotic plaque inflammation with [18F]-fluorodeoxyglucose positron emission tomography. Circulation 105:2708–2711
88. Tawakol A, Migrino RQ, Bashian GG, et al (2006) In vivo 18F-fluorodeoxyglucose positron emission tomography imaging provides a noninvasive measure of carotid plaque inflammation in patients. J Am Coll Cardiol 48:1818–1824
89. Rudd JH, Myers KS, Bansilal S, et al (2008) Atherosclerosis inflammation imaging with 18F-FDG PET: carotid, iliac, and femoral uptake reproducibility, quantification methods, and recommendations. J Nucl Med 49:871–878
90. Izquierdo-Garcia D, Davies JR, Graves MJ, et al (2009) Comparison of methods for magnetic resonance-guided [18-F]fluorodeoxyglucose positron emission tomography in human carotid arteries: reproducibility, partial volume correction, and correlation between methods. Stroke 40:86–93
91. Davies JR, Rudd JH, Fryer TD, et al (2005) Identification of culprit lesions after transient ischemic attack by combined 18F fluorodeoxyglucose positron-emission tomography and high-resolution magnetic resonance imaging. Stroke 36:2642–2647
92. Arauz A, Hoyos L, Zenteno M, et al (2007) Carotid plaque inflammation detected by 18F-fluorodeoxyglucose-positron emission tomography. Pilot study. Clin Neurol Neurosurg 109:409–412.
93. Koopman G, Reutelingsperger CP, Kuijten GA, et al (1994) Annexin V for flow cytometric detection of phosphatidylserine expression on B cells undergoing apoptosis. Blood 84:1415–1420
94. Kietselaer BL, Reutelingsperger CP, Heidendal GA, et al (2004) Noninvasive detection of plaque instability with use of radiolabeled annexin A5 in patients with carotid-artery atherosclerosis. N Engl J Med 350:1472–1473
95. Hobson RW 2nd, Mackey WC, Ascher E, et al (2008) Society for Vascular Surgery. Management of atherosclerotic carotid artery disease: clinical practice guidelines of the Society for Vascular Surgery. J Vasc Surg 48:480–486
96. Rothwell PM, Eliasziw M, Gutnikov SA, et al; Carotid Endarterectomy Trialists Collaboration (2004) Endarterectomy for symptomatic carotid stenosis in relation to clinical subgroups and timing of surgery. Lancet 363:915–924
97. Biasi GM, Froio A, Diethrich EB, et al (2004) Carotid plaque echolucency increases the risk of stroke in carotid stenting: the Imaging in Carotid Angioplasty and Risk of Stroke (ICAROS) study. Circulation 110:756–762

98. Markus HS, Droste DW, Kaps M, et al (2005). Dual antiplatelet therapy with clopidogrel and aspirin in symptomatic carotid stenosis evaluated using Doppler embolic signal detection: the Clopidogrel and Aspirin for Reduction of Emboli in Symptomatic Carotid Stenosis (CARESS) trial. Circulation 111:2233–2240
99. Underhill HR, Yuan C, Zhao XQ, et al (2008) Effect of rosuvastatin therapy on carotid plaque morphology and composition in moderately hypercholesterolemic patients: a high-resolution magnetic resonance imaging trial. Am Heart J 155:584
100. Tang TY, Howarth SP, Miller SR, et al (2009) The ATHEROMA (Atorvastatin Therapy: Effects on Reduction of Macrophage Activity) Study. Evaluation using ultrasmall superparamagnetic iron oxide-enhanced magnetic resonance imaging in carotid disease. J Am Coll Cardiol 53:2039–2050
101. Tahara N, Kai H, Ishibashi M, et al (2006) Simvastatin attenuates plaque inflammation: evaluation by fluorodeoxyglucose positron emission tomography. J Am Coll Cardiol 48:1825–1831
102. Kwee RM, van Oostenbrugge RJ, Hofstra L, et al (2008) Identifying vulnerable carotid plaques by noninvasive imaging. Neurology 70:2401–2409

Biographies

Dr. Robert M. Kwee obtained his Medical Doctor Degree at the University of Utrecht in 2006. In 2007, he started his PhD program on noninvasive imaging of carotid atherosclerosis at the Department of Radiology of the Maastricht University Medical Center. In 2010, he started his residency in Radiology at the Maastricht University Medical Center and he will obtain his PhD.

R.J. van Oostenbrugge, MD, PhD, born 27-12-1965, studied Medicine at the University of Maastricht, The Netherlands. After his graduation in 1990 he was trained in Neurology at the University Hospital Maastricht (later: Maastricht University Medical Center), where he was also involved in the neuro-oncology research program. In 1999 he obtained his PhD at the University of Maastricht for his thesis "Interphase cytogenetics in the cytodiagnosis of leptomeningeal metastases". This PhD thesis was rewarded the Dr. Pelerin Award. Since 2000 he is a staff member of the Department of Neurology at the Maastricht University Medical Center, where he is engaged in clinical research of the causes and consequences of cerebrovascular disease in general and in molecular biological and genetic studies in cerebrovascular disease with a emphasis on cerebral small vessel disease.

Jos van Engelshoven was born on 18 June, 1945 in Maastricht, the Netherlands. He studied medicine at the University of Nijmegen in the Netherlands (1963–1971) and received his PhD degree in 1983. After registration as radiologist (1978) he joined the staff of the department of radiology of the St. Annadal Hospital in Maastricht and between 1986 and 2008 he was professor of radiology and chairman of the radiology department of the academic hospital of Maastricht. His main fields of research are cardiovascular and MR imaging.

Werner H. Mess received his medical degree in 1989 at the University of Düsseldorf, whereupon he was a resident in Neurology in Mannheim/Heidelberg and Düsseldorf until 1994. During that period he also completed a thesis on the natural course of patients with an asymptomatic carotid artery stenosis. In 1995 he went to the Netherlands and received his PhD in medicine at the University of Maastricht in 2003 on the technical aspects of emboli detection with transcranial Doppler sonography. Since 2007 he is a professor of clinical neurophysiology. His main research interest is the evaluation of functional and anatomical properties of the carotid artery plaque, especially in terms of emboligenicity applying ultrasound technology.

Joachim Ernst Wildberger received his M.D. in 1994 at the Rhenian-Westphalian Technical University in Aachen. After an internship in cardiology in Mönchenglad bach and in Diagnostic Radiology in Leipzig, he started his residency in Diagnostic Radiology at the Technical University in Aachen, Germany. He was board certified for Diagnostic Radiology in December 1998 and became a fellow/staff member at this department thereafter. He received his Ph.D. in Radiology in 2002, was appointed as vice-chairman of the department in 2006 and received a professorship at the Technical University Aachen in 2007. In July 2007 he became Head of Department, Diagnostic Radiology, HELIOS Klinikum Berlin-Buch, Charité Campus Buch. In July 2008 he was appointed as Professor and Chairman of the Department of Radiology at the Maastricht University Medical Center (MUMC+).

His main research topics are technical developments and functional imaging in multislice spiral-CT as well as image-guided interventions. He received the Research and Education Fund of the European Congress of Radiology in 2003 and 2005, the Hanns-Langendorff Award of the German Society of Medical Radiation Protection in 2004 and the Eugenie-und-Felix-Wachsmann Award of the German Academy of Education and Training in Radiology in 2008.

Dr. M. Eline Kooi obtained a Master's Degree in Experimental Physics (cum laude) at the University of Amsterdam. She obtained her PhD in 1999 in Molecular Physics at the same university. In 1999, she joined the department of Radiology of the Maastricht University Medical Center. Since 2001, she has a permanent position at this department. She is a registered Medical Physicist in Radiology since 2004. Her research is focused on the development and validation of new applications of cardiovascular imaging.

Part IV
Treatment and Monitoring of Atherosclerosis

Chapter 18
Treatment of Carotid Stenosis: Carotid Endarterectomy and Carotid Angioplasty and Stenting

Franco Nessi, Michelangelo Ferri, Emanuele Ferrero, and Andrea Viazzo

Abstract Surgical treatment for steno-obstructive pathology of the extracranial carotid artery is today the gold standard for the prevention of cerebral ischemic disease associated to the manifestation of atherosclerosis. During the 1990s, a number of randomised trials incontrovertibly demonstrated the advantages which are derived from carotid endarterectomy with respect to medical therapy alone, in patients with symptomatic or asymptomatic stenosis. As a whole, these trials indicated a perioperative rate of major adverse events (stroke, mortality and MACE) of about 9%. Many projects have since been conducted in the field of carotid surgery, and today the reported complication rates are lower than 3%, with excellent long-term results (North American Symptomatic Carotid Endarterectomy Trial Collaborators, N Engl J Med 325(7):445–534, 1991; Mayberg et al., JAMA 266(23):3289–945, 1991). The results of new arising techniques, such as carotid stenting, must be compared with these complication rates in order to have a clear benchmark for a more objective future dissemination.

Keywords Carotid stenosis • Carotid endarterectomy • Carotid artery stenting

18.1 Introduction

Stroke is the third cause of death in the USA and in the Western World [1, 2], accounting for approximately 1 out of 15 deaths. The morbidity rate of the acute event survivors is in all cases extremely high; stroke is the main cause of disability in the Western World. Of long-term surviving patients, 48% have hemiparesis, 22% cannot walk unassisted, 24–53% are either completely or partially dependent for their normal everyday activities, 12–18% are aphasic and 32% suffer from post-traumatic depression. In total, the costs for treating stroke patients amount to

F. Nessi (✉)
Vascular and Encovascular Surgery Unit, Mauriziano Umberto I Hospital, Turin, Italy
e-mail: fnessi@mauriziano.it

Jasjit S. Suri et al. (eds.), *Atherosclerosis Disease Management*,
DOI 10.1007/978-1-4419-7222-4_18, © Springer Science+Business Media, LLC 2011

10 billion dollars a year in the USA. The main purpose of extracranial carotid atherosclerotic pathology surgery is stroke prevention. The first extracranial carotid reconstruction was performed in 1951, and the first endarterectomy (CEA) was performed by DeBakey in 1953. The most convincing proof of the correlation between carotid bifurcation stenosis and ipsilateral cerebral ischemic pathology was supplied by the operation performed in 1954 by Eastcott, Pickering and Robb, in which a patient with relapsing transient ischemic attacks (TIA) was healed from symptoms by bifurcation resection followed by end-to-end anastomosis between the internal carotid artery (ICA) and common carotid artery (CCA) [3]. During the 1990s, a number of randomised trials incontrovertibly demonstrated the benefits of carotid endarterectomy (CEA) compared to medical therapy alone, in patients with symptomatic or asymptomatic stenosis [4, 5]. These trials indicated a perioperative rate of major adverse events (stroke, mortality and MACE) of about 9% [6]. Three major clinical trials were published during the early 1990s: North American Symptomatic Carotid Endarterectomy Trial (NASCET) [4], European Carotid Surgery Trial (ECST) [7] and Asymptomatic Carotid Atherosclerosis Study (ACAS) [8], from which most of the current indications for surgical treatment of carotid obstruction pathology were derived and reported complication rates were lower than 3%, with excellent long-term results [9, 10]. The results of new arising techniques, such as carotid stenting, must be compared with these complication rates, in order to have a clear benchmark for a more objective future dissemination. The prevention of transient ischemic attacks is today considered a secondary objective, because it is common opinion that transient ischemic attacks do not cause irreversible brain damage, but are rather associated to a high level of subsequent stroke (from 5–10% during the first week, and risk from 10–20% during the following 3 months [11, 12]).

18.1.1 Diagnostic Tests

18.1.1.1 Duplex Scan

Doppler Ultrasound (duplex scan) is a non-invasive test that uses ultrasound waves to reconstruct an image of the carotid arteries and the blood flow through the arteries. With duplex scan, it is possible to carry out the characteristics of the carotid plaque, in particular of the echolucent degree. Grogan et al. have demonstrated that symptomatic plaques are more echolucent and less calcific than asymptomatic plaques and are associated to a higher degree of interplaque necrosis to histopathology. Echolucency is thus indicative of a greater degree of instability and thus the B-mode analysis features of the plaque must be taken into consideration in the pre-operative decision-making process. Echogenicity is evaluated in terms of ultrasound reflection ratio and is maximum (hyperechoic) for calcific plaques (Fig. 18.1) and minimum for hemorrhagic plaques (anechoic) (Fig. 18.2). Some anatomo-pathologies display by more or less intermediate

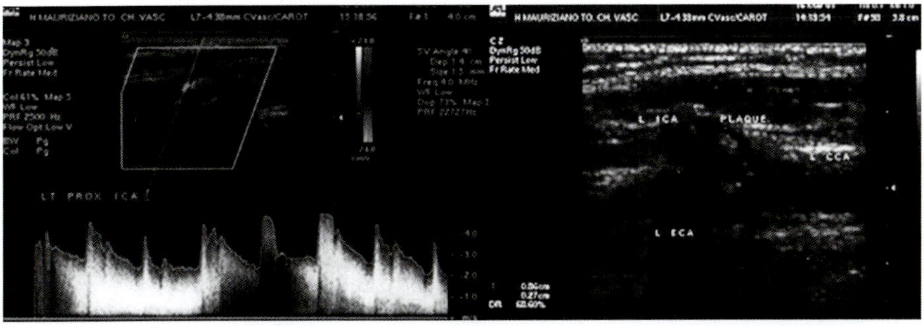

Fig. 18.1 Calcific plaque detected by duplex scan

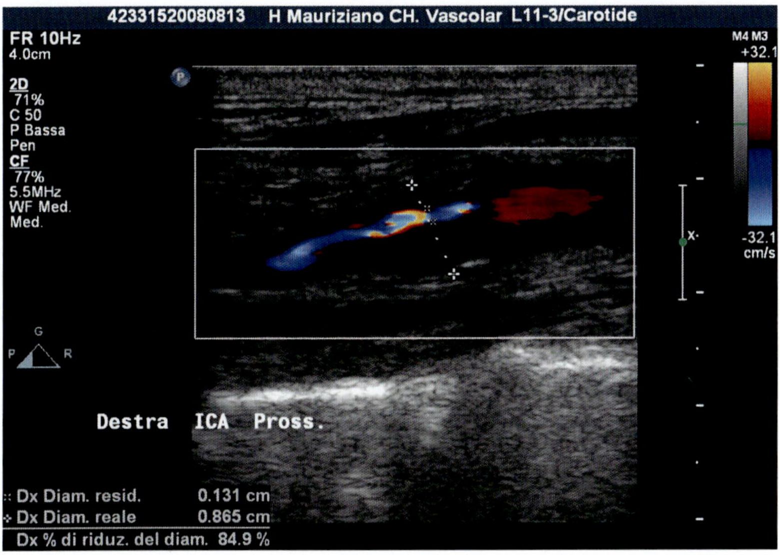

Fig. 18.2 Anechoic plaque detected by duplex scan

echogenicity (hypoechoic), such as the case of fibrolipid or cholesterinic plaques (Fig. 18.3). Furthermore, plaques can be classified according to their composition (homogenous or mixed). Therefore, they may be defined as dishomogeneous or heterogeneous. Such features allow to accurately define the endoluminal surface of the plaque which may be smooth, irregular or ulcerated (Fig. 18.4). The features of the plaque are certainly a predictive index of the risk of stroke. Duplex scan presents 87% sensitivity and 75% specificity in identifying severe internal carotid artery stenosis [13, 14]. The advantages of duplex scan are non-invasive, relatively inexpensive and widely available; the disadvantages are: it is "operator dependant" compared to other imaging modalities; it is difficult to

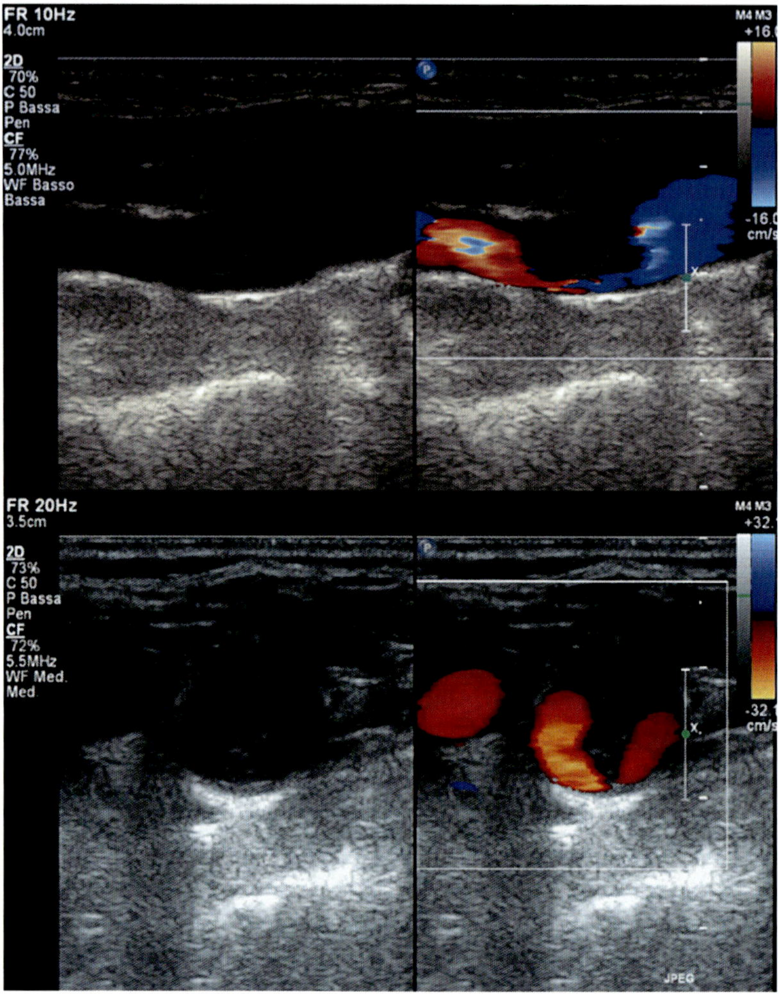

Fig. 18.3 Hypoechoic plaque detected by duplex scan

distinguish between "trickle flow" seen in severe stenosis, and complete occlusion.

18.1.1.2 Magnetic Resonance Imaging

In extracranial carotid artery disease contrast-enhanced magnetic resonance (MR) angiography has become a non-invasive imaging alternative for catheter angiography. The major task of carotid MR is to assess carotid artery stenosis. According to two current meta-analyses, carotid MR has a high sensitivity of about 94% and a high specificity of about 92% for the diagnosis of severe carotid artery stenosis [15, 16].

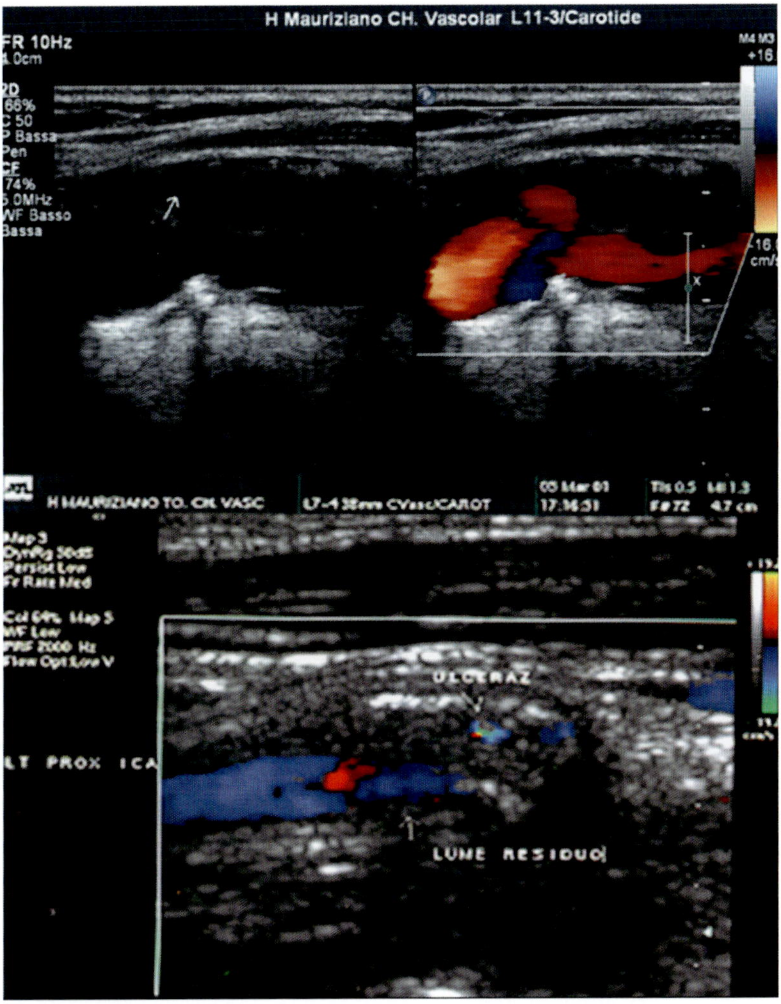

Fig. 18.4 Irregular or ulcerated plaque detected by duplex scan

18.1.1.3 Computed Tomography Angiography

New emerging modality (multi-slice helical imaging), with high sensitivity and specificity, for detection of severe internal carotid artery stenosis [17, 18], may supplant the role of MRA in the evaluation of carotid artery stenosis.

18.1.1.4 Angiography

Angiography scan is currently considered the "gold standard," but involves some risk. Pictures are taken of the blood vessel while a dye is injected. The disadvantages are that it is expensive, invasive procedure with potential risks and complications (0.5–2.0% risk of stroke).

18.1.2 Symptomatic and Asymptomatic Carotid Stenosis

A first major distinction within obstructive carotid pathology is between symptomatic and asymptomatic stenosis. In literature, extracranial carotid stenosis is considered symptomatic in patients who have experienced clinical ischemic symptom of the cerebral hemisphere or of the ipsilateral retina 6 months prior. The following specific clinical symptoms are considered: TIA, defined as sudden appearance of signs and/or symptoms referable to cerebral or visual focal deficit attributed to insufficient blood flow lasting for less than 24 h [19] (typically motor/sensory dysfunctions on one side, aphasia, dysarthria, amaurosis fugax) and stroke (minor or major). The 6-month interval was conventionally established in the main surgical trials (NASCET and ECST). Consequently, asymptomatic is any carotid stenosis which is not included in the definition of symptomatic, and therefore also if the aforesaid symptom occurred before 6 months prior to diagnosis or in case of non-specific symptoms (typically vertigo, dizziness or loss of consciousness). Such a 6-month interval has not yet been adopted in all guidelines and clinical trials. Recent reviews have indicated to considerably reduce the conventional time of 6 months (to not more than 3 months) since the last cerebral ischemic or ipsilateral ocular episode in order to consider a carotid stenosis as symptomatic. In a recent review, Rothwell showed the benefit of surgery and that the risk of symptomatic stenosis is maximum within the first 2 weeks after the symptom, while it is minimum 12 weeks after the symptom [19]. According to data from NASCET and ECST, the benefits of carotid endarterectomy are greater than those of medical therapy in symptomatic patients in terms of ipsilateral stroke and death reduction, if the perioperative death and invalidating stroke rate (at 1 month) is <6%. Such a benefit is modest for stenosis from 50 to 69% (calculated using the NASCET method) (NNT 22), and high for stenosis from 70 to 99% (NNT 6 and 14, respectively), providing in absence of near occlusion/string sign. If we consider patients with asymptomatic carotid stenosis, the data obtained from ACAS and ACST [20] indicate that CEA is beneficial for patients with carotid stenosis >70%, calculated using the NASCET method, aged from 40 to 75 years, with a life expectancy of at least 5 years and performed at centres with a perioperative complication rate lower than 3%. Diagnostics using a non-invasive method must be carried out in all suspected cases of carotid stenosis. The examination of choice is duplex scan (Fig. 18.5) because this provides highly specific information on artery morphology and plaques, on haemodynamic and on the contralateral carotid artery. The demonstrated reliability of this examination is used to indicate suitably of surgical approach [21]. Invasive second level examinations (angiography, spiral CT, angioMR; Fig. 18.6) may be indicated if the stenosis is suspected by means duplex scan but not surely proved. The plaque must always, when possible, be analysed considering

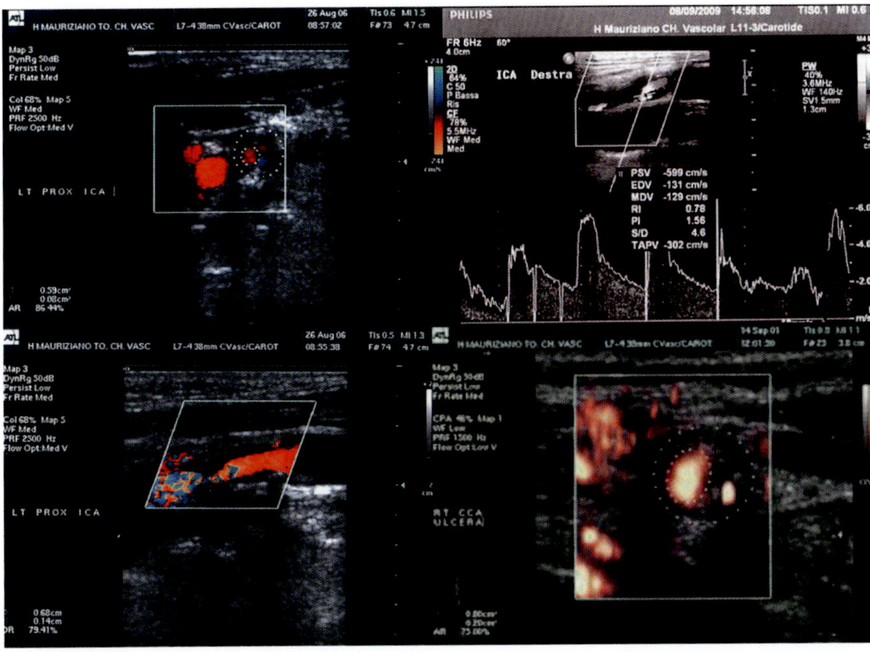

Fig. 18.5 Carotid stenosis detected by duplex scan

the clinical/symptomological case history determined by carotid stenosis of patients, divided into symptomatic and asymptomatic groups for the sake of simplicity. A carotid plaque is defined at risk when it is symptomatic, haemodynamically significant, with cholesterinic or haemorrhagic component and with ulcerated surface [22, 23] (Fig. 18.7).

18.1.3 Anaesthesiological Technique

CEA can be performed under general, local, locoregional under duplex scan with regional deep cervical plexus block and superficial infiltration anaesthesia, mainly depending on the surgeon and anaesthesiologist's experience and preference. For CEA performed with regional anaesthesia, the assessment of mental status (conversing with the patient) and the contralateral motor and sensory responses during carotid cross clamping are adequate. If CEA is performed under general anaesthesia, a number of special methods have been utilised for intra-operative neurologic monitoring, although those practises are not mandated as standard care. The special techniques include: (1) Electroencephalography (EEG), somatosensory evoked potential and Neurotrack (4-channel processed EEG). The 16-channel raw EEG read by a neurologist on site is considered as a gold standard and a focal

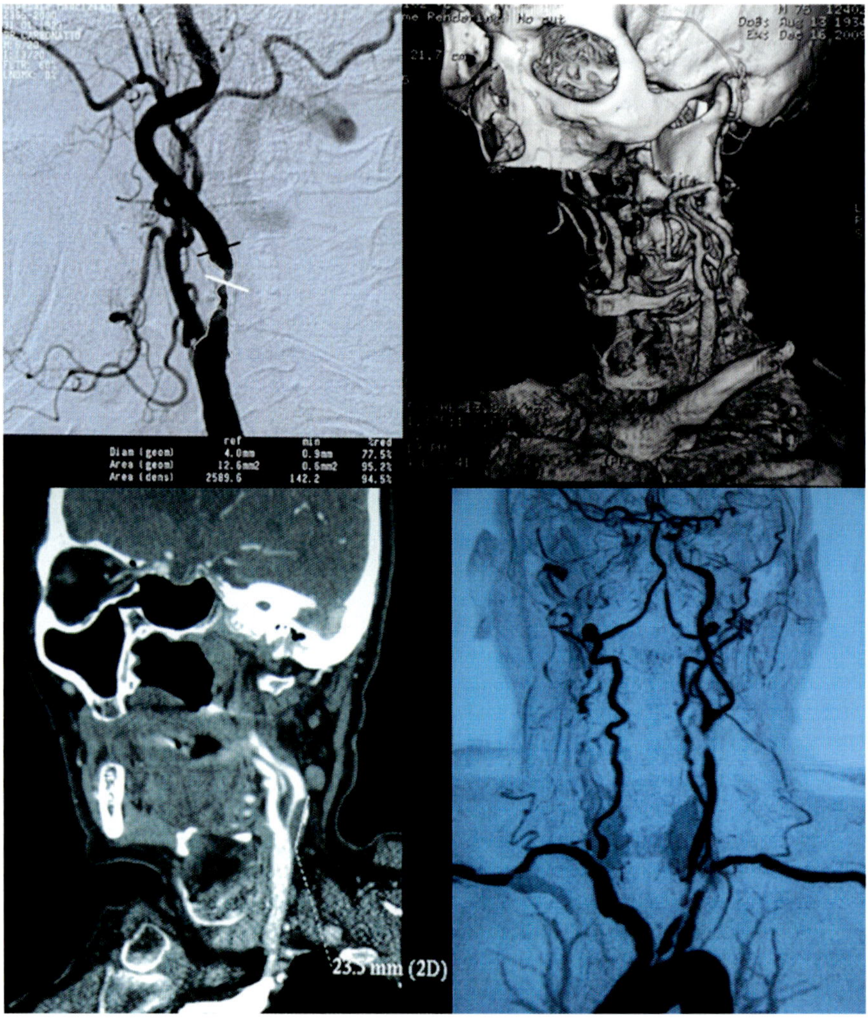

Fig. 18.6 Carotid stenosis detected by: angiography, spiral CT and angioMR scan

decrease in waveform frequency and/or amplitude is the signs for ischemia; (2) Transcranial Doppler. It measures cerebral blood flow velocity and is helpful in detecting emboli; (3) Cerebral perfusion pressure including MAP and stump pressure; (4) Cerebral oximetry or near-infrared spectroscopy (NIRS). It non-invasively measures regional cerebral O_2 saturation (rSo_2); the monitor rSo_2 tends but lack specificity. Studies have shown that 20% of reduction in rSo_2 had high false-positive value (66%) but 97.4–100% negative predictive value, i.e., if no rSo_2 decrease, cerebral ischemia is unlikely, but a decrease does not always indicate ischemia; and (5) Jugular mixed venous O_2 saturation. This is an invasive and non-specific technique. A reading reduces to <50% suggesting possible ischemia.

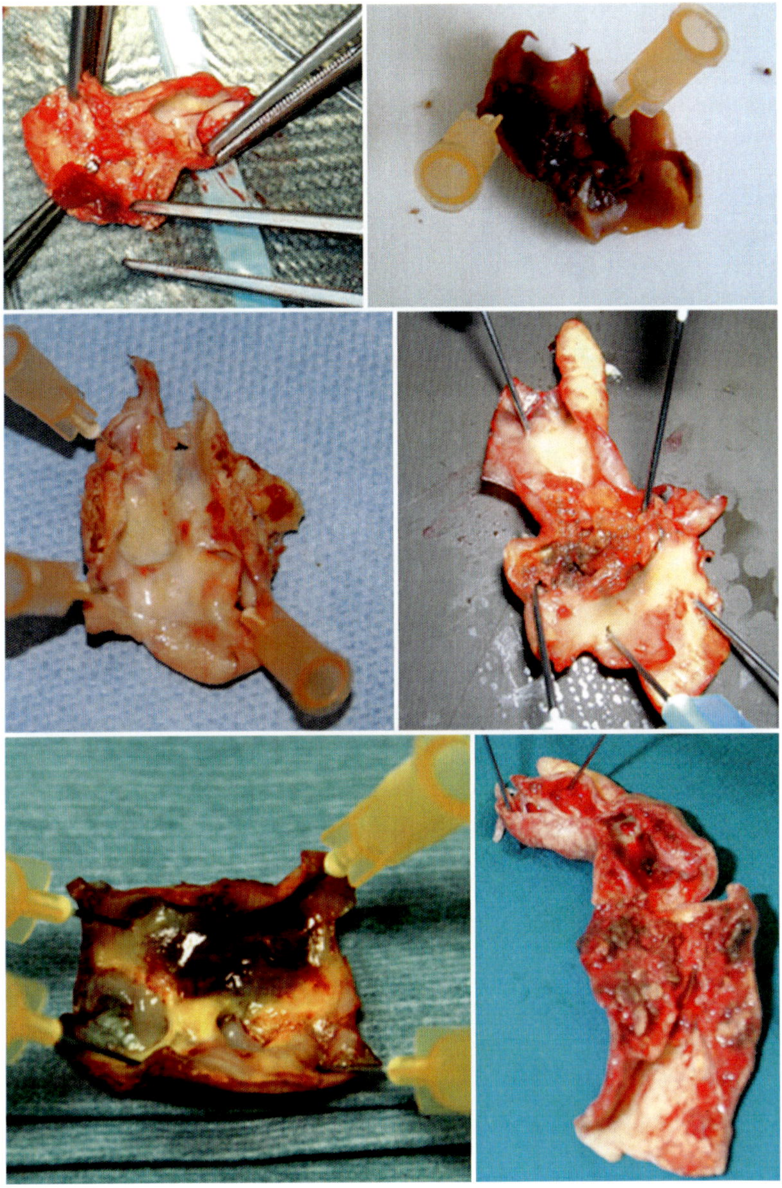

Fig. 18.7 Different type of carotid plaque

Currently no single method has been infallible and shown to improve neurologic outcome. With regard to the anaesthesiological management of CEA procedures, most recent studies indicate only a C degree recommendation in favour of locoregional anaesthesia. However, the only randomised trial conducted until today (GALA Trial [24]) has demonstrated that there is no statistic significance in results in terms of TIA/stroke/mortality between general anaesthesia or local anaesthesia in carotid surgery. Our approach consists in using locoregional anaesthesia in over 90% of cases, with selective block on C2–C3–C4 level or by means of infiltrations by planes of local anaesthetic, on laterocervical level along the SCM (Fig. 18.8). Local/locoregional anaesthesia allows remarkable advantages, such as direct monitoring of cerebral functions, greater haemodynamic stability and good post-operative pain control. Cerebral function is monitored during carotid clamping substantially by means of careful evaluation by the anaesthetist and surgeon, asking the patient to make simple movements of the contralateral hand so as to evaluate degree of consciousness and possible focal deficits. If general anaesthesia is used, cerebral activity is instead monitored by means of electroencephalogram (EEG), thus requiring the presence of a neurologist in the operating theatre or by measuring stump pressure (Fig. 18.9), which today is still a reliable predictor of cerebral perfusion during CEA under general anaesthesia [25]. Use of heparin IV at a dosage of 50 U.I./kg associated to a close monitoring of activated clotting time (ACT) values is very important in our experience. Systemic heparinisation is considered effective when ACT increases to more than 200 s. During haemostasis and closure, we indicate administering a dose of Protamine capable of antagonising half of the administered heparin, when ACT is still higher than 200 s. It has been demonstrated that intra-operative ACT monitoring allows to substantially reduce the risk of cerebral ischemia and laterocervical haematoma [26].

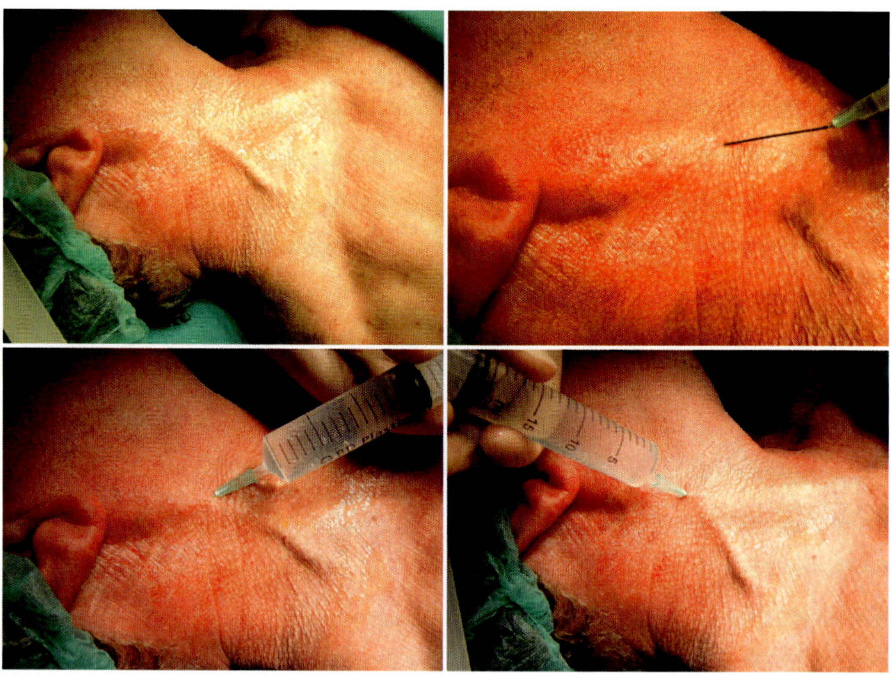

Fig. 18.8 Local anaesthesia procedure

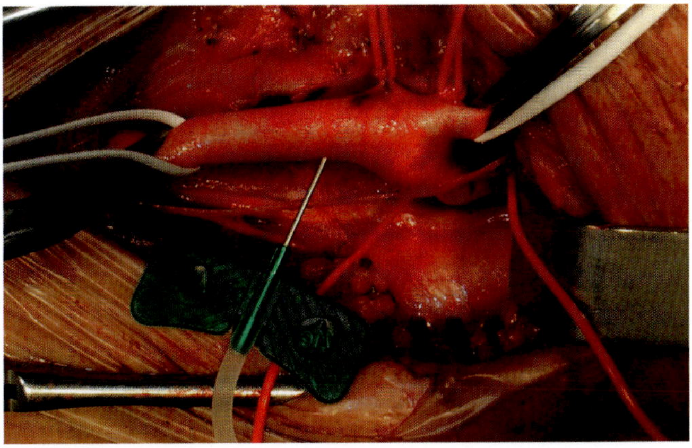

Fig. 18.9 Intraoperative stump pressure measure

18.1.4 Surgical Technique

The patient is put in supine position on the operating table with head turned by 45° towards the opposite side and neck well-extended. An oblique incision is made along the front edge of the sternocleidomastoid muscle (SCM), centred on the carotid bifurcation located by means of perioperative duplex scan examination. The incision must be slightly inclined towards the rear along its distal segment to avoid the parotid: an excessively frontal incision would injure the cervico-facial branch of the facial nerve. The limits of the operating range are: laterally, the internal jugular vein; superiorly, the belly of the digastric muscle posterior; and inferiorly, the omohyoid muscle. The internal jugular vein is exposed, located deeply with respect to SCM; the common carotid artery (CCA) is in posteromedial position and thus located. Now it can be performed through two types of access: the direct access and the retrojugular access. With the direct access the tireolinguofacial venous trunk, with high anatomic variability, is sectioned between ligatures. The carotid bifurcation and, higher up, the external carotid artery (ECA) and internal carotid artery (ICA) are then exposed.

With retrojugular route the access to the ICA necessitates median retraction and often collapse of the internal jugular vein (IJV). In order to distally prepare ICA where it is healthy and without atherosclerotic lesions, the occipital artery must be bound and cut to prevent medial upward mobilisation of the hypoglossal nerve which crosses both ECA and ICA (Fig. 18.10). After having exposed the bifurcation, the mobilisation manoeuvres are carried out after systematic heparinisation and with the CCA clamped, to prevent potential cerebral embolisms from instable plaques. The vessels are taped to control the superior thyroid artery, thus avoiding damage to the superior laryngeal nerve which often crosses the first portion of the artery, and then ECA and ICA. Systemic heparinisation is obtained by means of heparin sodium (usually dosage 50 U.I./kg) administered some minutes before

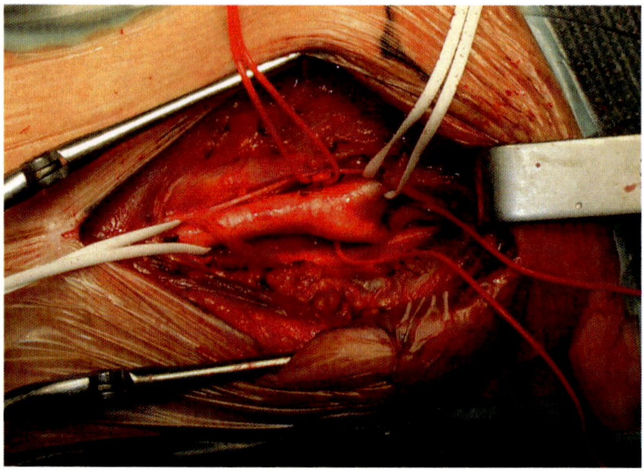

Fig. 18.10 Carotid biforcation

clamping. Clamping is performed after having checked the patient's haemodynamic stability and taken arterial pressure to sufficient values to ensure compensation activation via the contralateral carotid and vertebrobasilar vessels. CCA, ECA and superior thyroid artery are clamped in sequence and distal ICA is clamped last. During mobilisation of the carotid bifurcation, care must be devoted to avoid injury to the vagus nerve, which is often tenaciously adherent to carotid bifurcation and ICA. An accurate balance of the four elements on which the choice of the carotid thromboendarterectomy depends can now be determined:

- Proximal and distal extension of atherosclerotic lesions
- Morphology of the carotid sinus
- Diameter of ICA
- Possible presence of carotid kinking or excessive length of ICA.

18.1.4.1 Standard CEA (Direct Suture or with Patch)

Endarterectomy (CEA) by longitudinal arteriotomy is performed on carotid bifurcation and ICA at the bulb level on the posterolateral margin. The plaque is removed on CCA, ECA and superior thyroid artery, and ICA using an endarterectomy spatula. CEA is ended with a "pseudo-eversion" of the internal carotid or by extending the arteriotomy on ICA to end of plaque. The arteriotomy is finally closed by direct suture of the vessel. Direct suture is normally used for small arteriotomies (Fig. 18.11). The results of NASCET trial suggest that it is better to close using a patch in order to avoid restenosis. The patch can be made of either autologous material (vein) or alloplastic material (Dacron or PTFE). Use of patch for closing the endarterectomy is indicated for: small sized arteries, laceration of the artery wall or extended arteriotomies (Fig. 18.12).

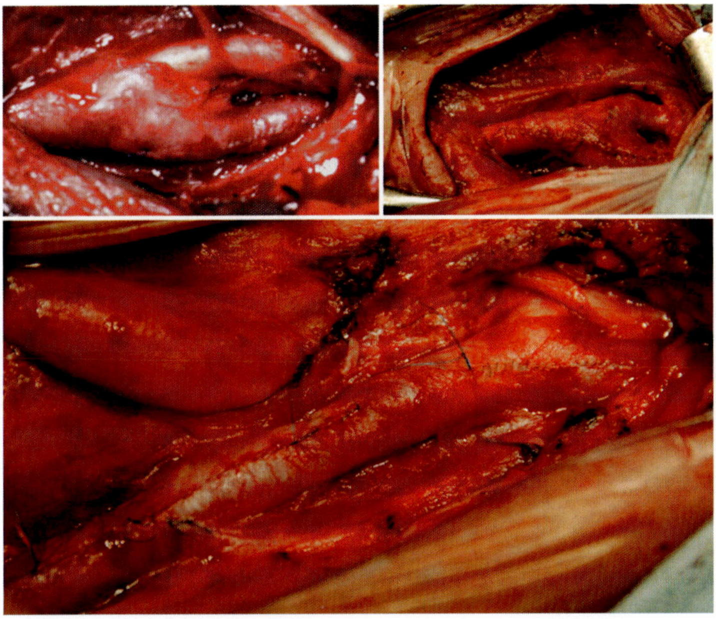

Fig. 18.11 Standard CEA (direct suture)

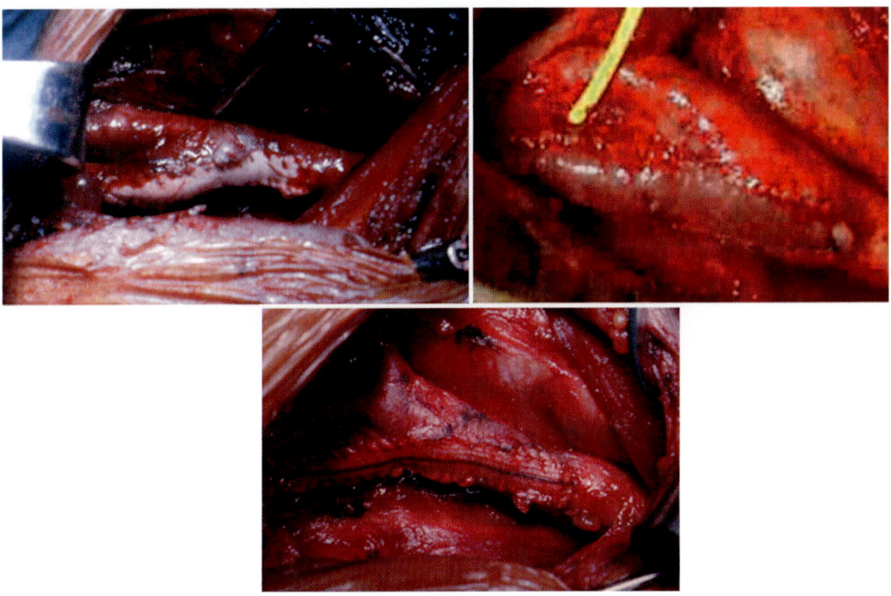

Fig. 18.12 Standard CEA (with patch)

18.1.4.2 Eversion Technique

This technique involves a bevelled sectioning of ICA on bifurcation level. ICA's endarterectomy is performed by undermining along the plane of cleavage in circular sense, everting the vessel wall by turning it back like a glove finger. Dissection is thus continued in cranial sense to reach the distal limit of the plaque to its detachment, which is normally clean. After longitudinal arteriotomy of CCA, endarterectomy is

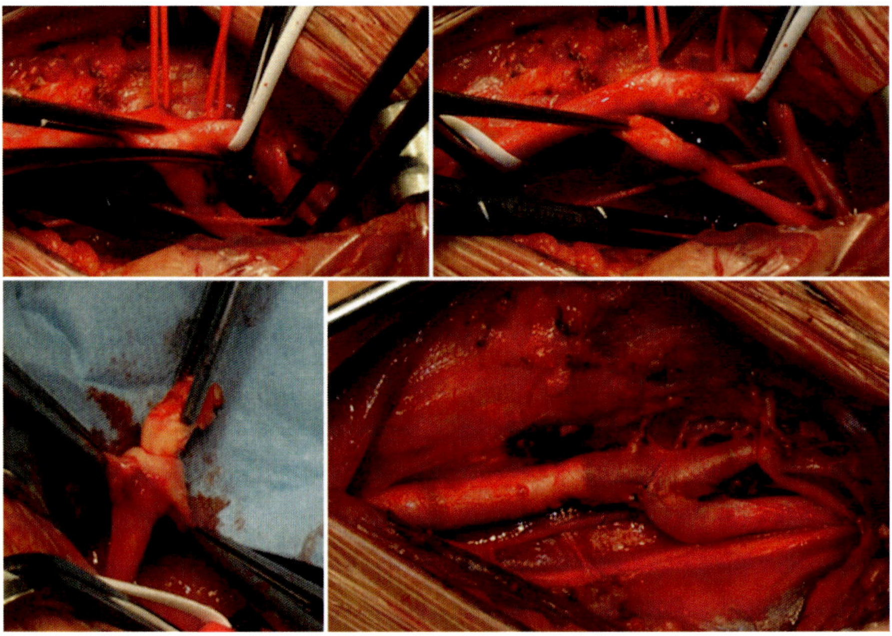

Fig. 18.13 CEA (eversion technique)

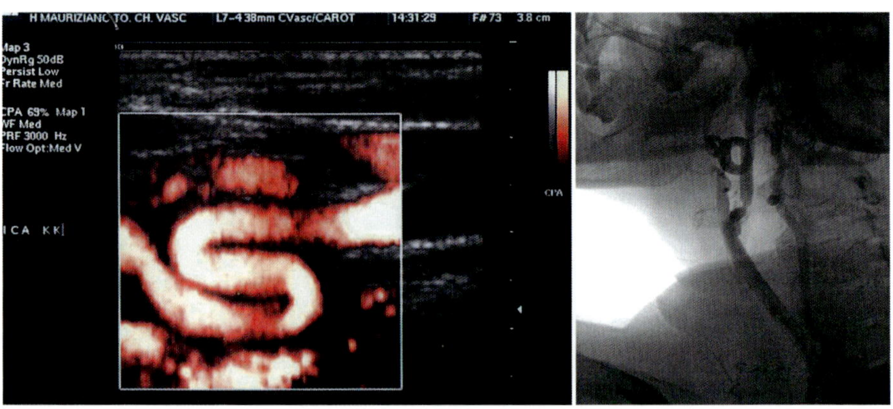

Fig. 18.14 Kinking detected by duplex scan; coiling detected by angiography scan

ended on level of the latter and of ECA. ICA reimplantation is carried out along the arteriotomy on CCA, by means of anastomosis prepared with due hemisutures using 6–7/0 non-absorbable monofilament thread (Fig. 18.13). This CEA technique is indicated in case of kinking/coiling on ICA level (Fig. 18.14) or presence of small sized carotid vessels.

18.2 Special Issues

18.2.1 Shunt

The placement of shunt is to maintain cerebral blood flow (CBF) during carotid artery cross-clamping. It seems to provide maximal neuroprotection. In case of clamping intolerance, we position a Pruitt-Inahara shunt (Fig. 18.15) and carry out carotid endarterectomy by means of standard technique, direct suture or with patch, or by means of eversion technique. Routine shunting is not indicated in carotid surgery because a higher incidence of stroke has been demonstrated with respect to selective shunting, although the data available in literature are excessively limited to either supporting or rejecting the routine use of shunts; large-scale randomised trials have been requested; until today, no monitoring method during shunting has been able to produce better results [27]. The risks of insertion of shunt are associated with dislodging emboli and intimal wall dissection.

18.2.2 Quality Check

At this end of the procedure, it is our practise to carry out a routine check of CEA by means of intra-operative duplex scan (Fig. 18.16) to evaluate the distal endpoint

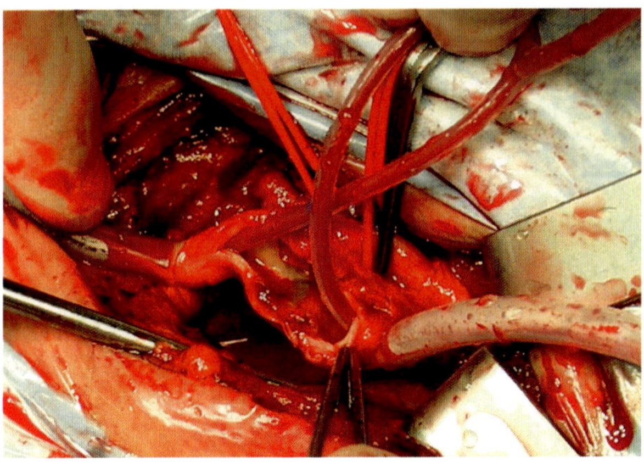

Fig. 18.15 Pruitt-Inahara shunt

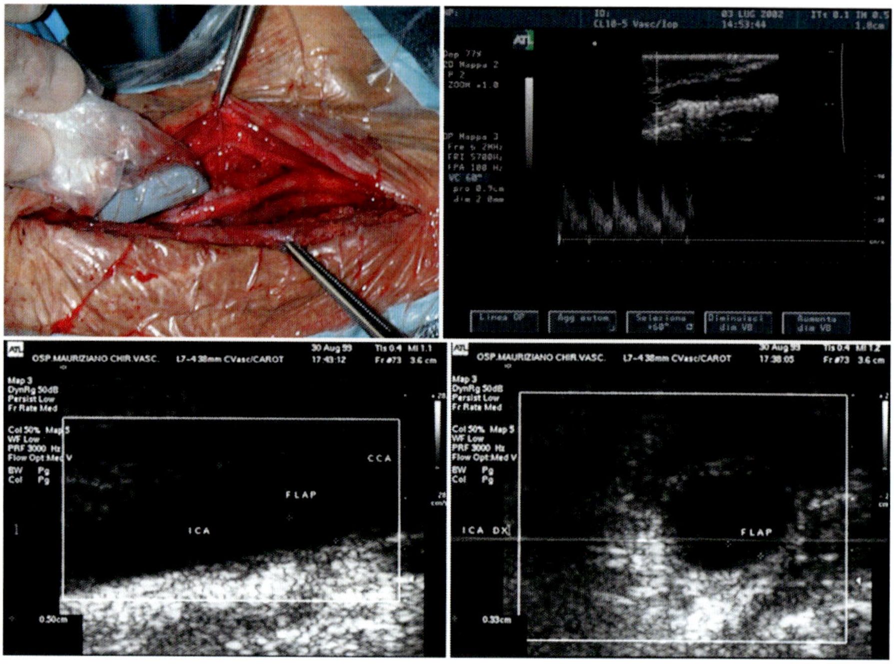

Fig. 18.16 Intra-operative duplex scan

and the presence of intimal residual flaps. Problems if found and correlated to the surgical procedure can be immediately corrected. Finally, patients are discharged after an average of 2.5 days of post-operative hospitalisation, obviously in absence of complications (laterocervical haematoma, neck tissue oedema, fever) and with neurological and haemodynamic stability, with ECG unchanged with respect to pre-operative examination.

18.2.3 Urgent Surgery

In cases with suspected cerebral lesion caused by symptomatic carotid artery stenosis patients must receive pre-operative cerebral imaging to determine a functional pre-operative evaluation. Cerebral imagining is carried out by means of cerebral CT scan (Fig. 18.17) or by means of cerebral MR scan (Fig. 18.18) or diffusion-weighted MR (DWI; Fig. 18.19). Patients with impaired consciousness or an infarct larger than 2.5 cm on CT or MR scans or both were excluded from surgery. Patients with stupor or coma (NIHSS >22) were excluded from CEA.

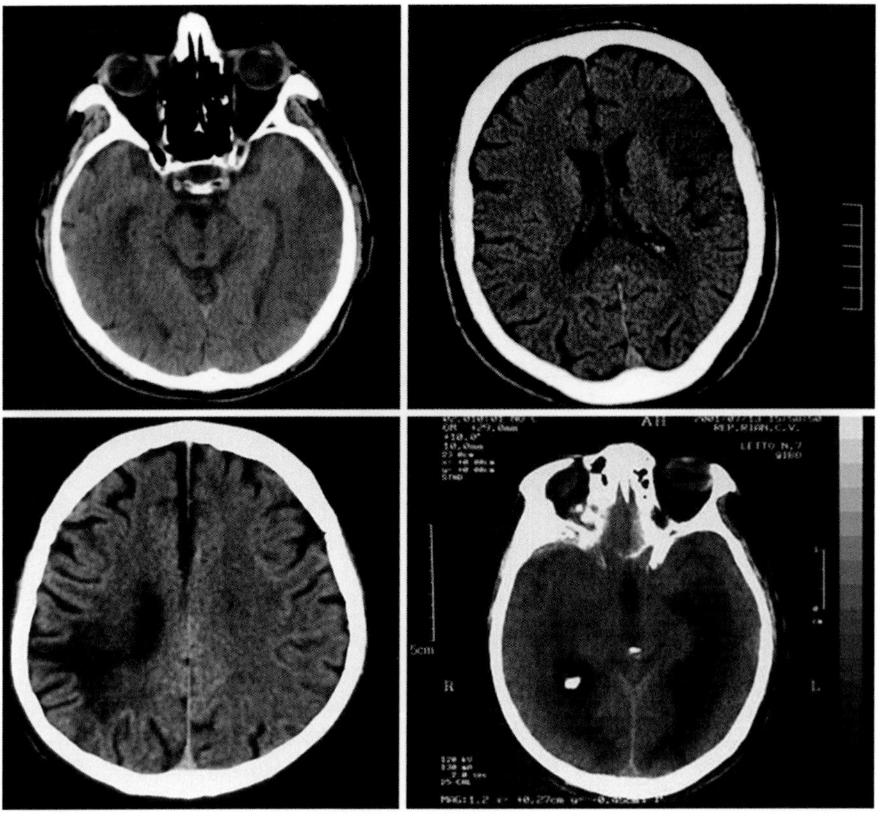

Fig. 18.17 Cerebral CT scan

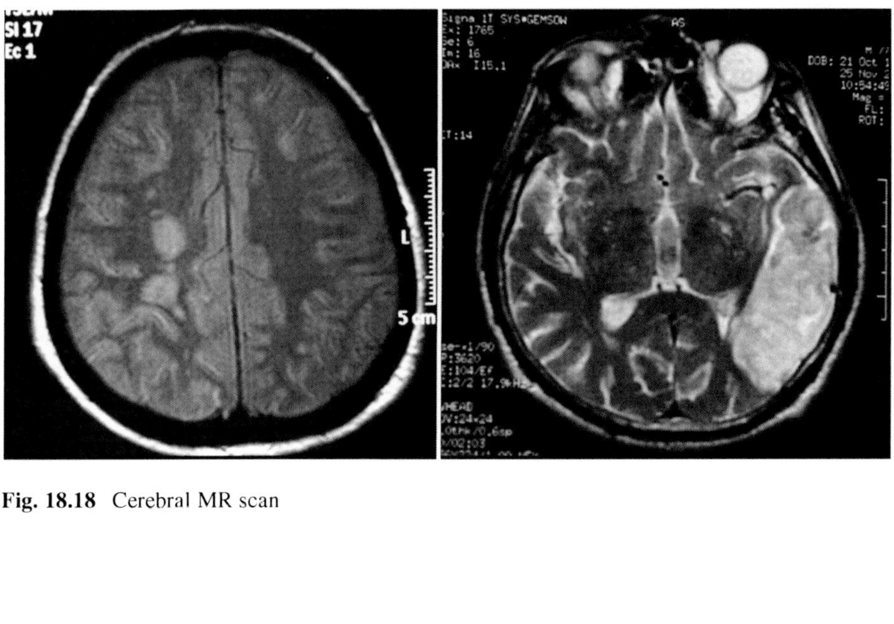

Fig. 18.18 Cerebral MR scan

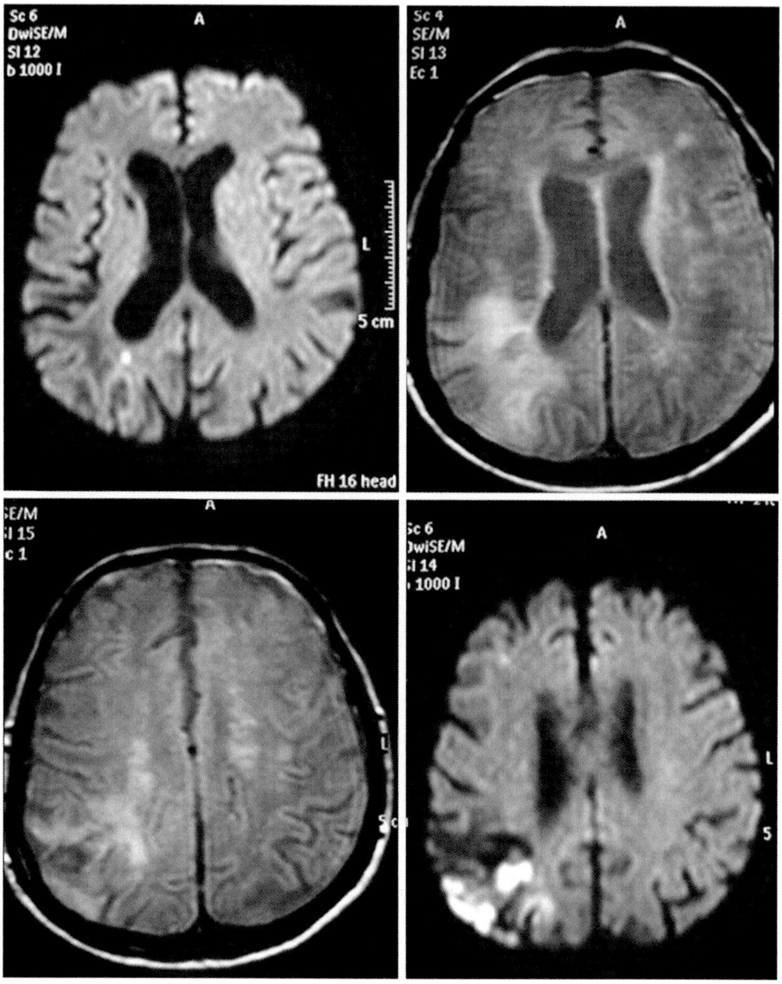

Fig. 18.19 Cerebral diffusion-weighted MR scan

18.2.4 CEA/CABG: Staged, Combined, Reversed Approach

The patients requiring CEA commonly have multiple co-existing diseases. The significant ones are coronary artery disease with decompensated congestive heart failure and unstable angina, hypertension (diastolic BP > 110 mmHg), diabetic mellitus (glucose > 200 mg/dl). If not well-controlled, these co-morbidities increase post-operative morbidity and mortality although most post-operative neurologic deficits appear to be related to surgical technique. Because myocardial infarction is the main perioperative cause of death in patients receiving CEA, it is important to carry out an evaluation of the patient's cardiology history. If needed, these patients receive myocardial revascularisation, including surgical coronary artery bypass grafting (CABG); in this case, timing with carotid endarterectomy is decided according to which of the two pathologies is clinically the most severe. In case of patients receiving CABG, the two procedures can be carried out either at the same or at different times, and today it is not clear which approach gives the best results. Literature indicates that approximately 12% of patients receiving CABG have carotid stenosis ≥80% (sec. ECST) [28], while 50% of CEA candidate patients have coronary disease. Patients with surgical indication for both districts may be treated by means of three approaches described in literature: "staged approach" (CEA and/or CAS first and then CABG) [30–34], characterised by a lower neurological morbidity–mortality; "reversed staged approach" (CABG first and then CEA), with lower cardiac morbidity–mortality and higher neurological morbidity–mortality; "combined approach" (CEA/CABG), with intermediate neurological and cardiac risks with respect to the previous two. In patients receiving CEA, myocardial infarction (IMA) is one of the first causes of perioperative risk and of long-term mortality [35]. The debate is still open in literature on the opportunity for carrying out such intervention in one step, because some writers believe that the combined procedure may cause a higher incidence of adverse post-operative events (mortality/major stroke/minor stroke/cardiomyopathologies) [36–41]. In our centre, both pathologies are treated by using the combined approach (CEA/CABG) when patients have high degree carotid stenosis (>85%) or are symptomatic, with results in terms of stroke and mortality (3.6 and 10.9%, respectively) in line with those in literature.

18.3 The Results of Surgery

Randomised trials have incontrovertibly demonstrated the better results of surgery with respect to medical therapy, although these results – above all with regards to the incidence of complications – are starting to be dated. NASCET and ECST were published 17 years ago, while ACAS was published approximately 13 years ago. As a whole, these trials showed an average incidence of greater adverse events (stroke, TIA, death, major adverse cardiovascular events) in the arm receiving carotid endarterectomy reaching an average of 9.8%. Some major centres have recently revised their case histories demonstrating that the results of carotid surgery are a far in terms of complications and adverse events from those shown in the trials, reaching and in some cases even exceeding the 1.5% risk described in ACAS, which at the time of its publication was least to say optimistic. LaMuraglia et al. report the result of 2,236 cases, of which 36% symptomatic, carried out between 1997 and 2004. Within case studies, there was a stroke rate of 0.8%, a myocardial infarction rate of 0.5% and a mortality rate of 0.6%. A total incidence of 2.5% is obtained by adding all cases of stroke, TIA, major adverse cardiovascular events and death [42]. Similarly, Long et al. analysed the outcome of 1,927 cases of endarterectomy (of which 37% on symptomatic patients), carried out from 1999 to 2003: incidence of stroke 0.8%, incidence of myocardial infarction 0.5%, mortality rate of 0.6%; with a total rate of stroke, TIA, major adverse cardiovascular events and death of approximately 3% [10]. Feasby et al. instead analysed 3,283 cases in 2000–2001, of which 62% symptomatic, reporting a stroke rate of 2.9%, a myocardial infarction rate of 1.2%, a mortality rate of 0.8% and an extraordinary total rate of stroke, TIA, major adverse cardiovascular events and death of 1.6% [9]. A recent review, supported by the Italian Registry for Vascular System Activities, collected data from 89 surgical vascular centres (nearly all the centres existing in Italy) during 2007; on a total of 5,962 CEA, perioperative mortality and stroke rote were in total 1.1% (stroke 0.9%; mortality 0.2%) [29]. Considering the total complication rate of these four registers, an incidence in the range from 1.1 to 3% is obtained, which is very far from the 9.8% rate of the first trial. Obviously, these results remain statically non-comparable, because register data is being compared to randomised trial data. However, the low rate of complications indicated by the large case histories is emblematic of the progress made by carotid surgery in approximately 15 years and at the same time is the real benchmark for carotid pathology endovascular treatment.

18.3.1 Personal Experience

From our point of view, 3,507 carotid endarterectomy procedures were carried out at the Vascular Surgery Unit of Umberto I Hospital, Turin, Italy, from 1998 to 2008, with a minor/major stroke/ TIA incidence of 2.7% (94/3,507), of which 1.6% TIA, 0.5% minor stroke and 0.6% major stroke; the perioperative mortality rate was 0.14% (5/3,507) and the main cause was myocardial infarction. With regards to the incidence of other complications, our series indicate 4.6% (161/3,507) of cranial nerve lesions (in 80% of cases to the inferior laryngeal nerve), 4.3% (150/3,507) of cervical bleeding, 0.26% cases of infection (9/3,507), 0.08% cases (3/3,507) all fatal of brain haemorrhage/hemorrhagic stroke and 0.05% (2/3,507) of incidence of pseudo-aneurism. These results are absolutely not comparable, and actually superior, to those shown in literature. It is our opinion that the improved results depend on the learning curve, as demonstrated by the fact that from 1998 to today there has been a gradual reduction of surgical times, and of mortality and morbidity rate, and on stricter pre-operative assessments to decide indication for surgical approach.

18.4 Endovascular Technique

18.4.1 Carotid Stenting Technique

For carrying out the first angioplasty and carotid stent, radiologist Klaus Mathias from Dortmund used techniques driving from his experience in peripheral interventions [43]. The 0.035 in. guides were initially used to directly cross the lesion, which was later treated with a balloon and stent in absence of direct angiographic vision using the vertebral column as anatomic landmark. Important changes to the technique were introduced in the 1990s also capitalising on the experience accumulated in coronary interventions. Merit of having considerably improved the carotid stenting technique goes to Cardiologist Gary Roubin, who worked first in Alabama and now in New York, with the use of long inserters advanced into the common carotid for viewing the stenosis with contrast means and lower profile guides and balloons used for coronary interventions [44].

18.4.2 Vascular Access

We prefer access from the femoral artery which allows an easier cannulation of the common carotid. We use brachial access only in case of femoral artery occlusion or unsuccessful cannulation of the common carotid from the femoral artery. In this case, we use brachial artery for treating the contralateral carotid. Use of the radial artery was described.

18.4.3 Diagnostic Catheters

Selective cannulation of the common carotid by means of diagnostic catheter is needed both to acquire adequate angiographic information and for advancing the supporting guides. Diagnostic caterer sizes range from 4 to 6 French. A good quality carotid angiography may even be obtained using a 4 French caterer selectively positioned in the carotid. Carotid angiography is an integral part of CAS intervention. An intracranial angiography is recommended before the intervention. This will provide information on possible intracranial stenotic lesions and a basic image of intracranial vascularisation which may be very useful for solving embolism complications.

18.4.4 Common Carotid Artery Access

The most decisive factor for the success of a CAS procedure technique is the possibility of gaining access to the common carotid artery. In all published experiences, the first cause of procedure failure is the impossibility of advancing into the common carotid artery. The long 6 or 7 French introducer sheath is the preferred instrument for cannulation of the common carotid by the Roubin school. This approach is needed to position a relatively distal diagnostic caterer in the common carotid. The push and pull technique of the caterer on the floppy guide 0.035 in. is used to feed the caterer into the vessel. A long 0.035 in. guide (220–260 cm) is then positioned in the external carotid. We generally use stiff-type or stiff-Amplatz high supporting guides. In order to facilitate the passage of the guide in the external carotid, and avoid crossing the lesion of the internal carotid causing possible dislocation of material, angiographic road-mapping application may be used to visualise the vessels even without further injection of contrast medium. After having positioned the guide in the external carotid, the diagnostic caterer is removed and the introducer sheath is slowly advanced with its internal persuader. The guiding catheter (diameter 7 or 8 French) is the second, most cost-effective access system to the common carotid. In general, they provide sufficient support for angioplasty by using low profile stents and protection systems available also with supporting guides. Very soft specific guide caterers with 5 cm distal tip and segment, which allow a more distal advancement of the caterer in the common carotid, are currently available. If access to the common carotid is problematic, we recommending suspending attempts after 30 min and reassessing surgical approach, because in our experience and according to that of other centres most complications occurs indeed after prolonged manoeuvres with caterers in the aortic arch.

18.4.5 Protection Systems

Trials with transcranial-Doppler have demonstrated that carotid stenting is associated to a higher rate of embolisation of fragments which respect to surgical endarterectomy [45, 46]. Various cerebral protection systems have been suggested in order to reduce these embolisations, which can cause perioperative neurological complications [47]. The first system, a distal occlusion balloon, was developed and used for the first time in 1990 by Theron [48]. Three different cerebral protection systems are currently used: two distal protection systems, such as distal occlusive balloons and filters, and proximal protection of the common and external carotids. Distal occlusive balloons system was the first protection systems to be used on a wide scale [49, 50]. It consists of a 0.014 in. guide carrying a balloon in distal portion, which may be inflated and deflated through a very thin channel. The stenotic lesion is crossed by the guide and the balloon is positioned distally with respect to the stenosis, where it is inflated to block off the flow of blood into the internal carotid. Subsequently, angioplasty and stenting is carried out. Finally, an aspiration catheter is advanced on the guide and the column of blood contained in the internal carotid upstream of the balloon is aspirated. Possible debris dislocated during stenting is thus eliminated. The balloon is then deflated and the guide is retracted. The advantages of this system are low profile (2.2 F) and good system torqueability. Possible disadvantages are that the occlusion is not tolerated by 6–10% of patients [49] and that visualisation by means of contrast is not possible during inflation. Protection filters consist of a metallic structure (or skeleton) covered by a polyethylene membrane or close knit Nitinol wire mesh, with holes varying in diameter from 80 to 200 μm [51–54]. The filters are generally fixed to the distal portion of a 0.014 in. guide. During the procedure, the filters are wound within a release catheter, by means of which they are carried distally to the stenosis. In presence of tight stenosis with calcific plaques or very fibrous stenosis, the passage of the closed filter may be impossible. Generally, after a delicate dilatation (balloons of diameter 2–2.5 mm), the stenosis can be passed by the release catheter with the filter. At the end of the procedure the filter is closed with a retraction catheter and removed from the carotid. Possible complications related to the filter are vasospasm due to the presence of the guide inside the vessel, usually reversible by infracarotid administration of nitrates (Fig. 18.20). Adequate anticoagulation is however recommended to prevent clotting in the filter. Proximal protection systems instead allow to implement cerebral protection before passing any device through the stenosis. These systems consist of a long introducer sheath with a balloon on the distal end which is inflated in the common carotid. A second balloon inflated in the external carotid guarantees the total blockage of the antegrade flow in the internal carotid. Proximal protection systems exploit the cerebral vascular connection of the circle of Willis. By means of this circle, a back pressure is applied in the internal carotid after occlusion of the common and external carotids. After positioning the stent and

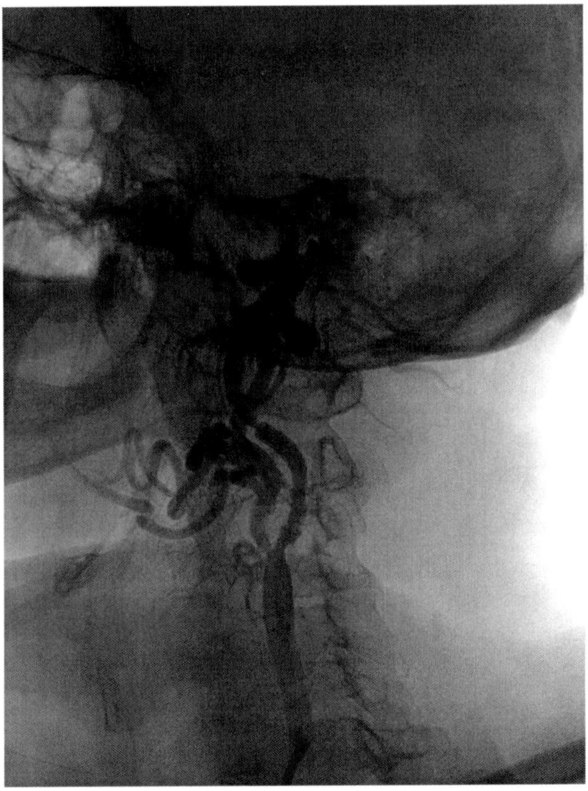

Fig. 18.20 ICA vasospasm complication post-CAS

before deflating the balloons in the common and external carotid, the blood present in the internal carotid containing possible dislocated debris is aspirated. The advantage of proximal protection is that the entire stenting procedure is protected and, if applied correctly, does not imply any type of embolisation. The disadvantages are that it is not tolerated by all patients.

18.4.6 Stent Implantation

All current carotid angioplasty procedures include elective stent implantation. Excellent immediate and long-term results are obtained with the stent, higher than angioplasty with simple balloon. Pre-dilatation with coronary balloons of diameter 3.5–4.0 mm is used only in the case of severe calcified stenosis which prevents cross lesion by the device. A stent of diameter from 6 to 9 mm is commonly used, and the diameter of the distal common carotid artery is employed as reference. Relatively long stents allowing to cover the entire lesion are used. The stent is positioned as less distally as possible, while guarantying coverage of the entire stenosis (Fig. 18.21). Self-expanding stents are now exclusively used in the carotid because of their lowest risk of deformation or breakage with respect to balloon-expandable stents. The most commonly used stent in the carotid artery is Carotid Wallstent (Boston Scientific). This stent, known as a "mesh-wire" stent, is available with a very small profile (5.5 French), with flexible shaft and rapid exchange, allowing the use of short guides. A half released stent can be reclosed, allowing exact positioning of the distal end of the stent. Nitinol self-expanding stents have more recently been introduced. These are characterised by a higher radial force and a greater adaptability to the tortuosity of the vessel and to differences of size between internal carotid and common carotid. Nitinol stents, because of the design formed by crowns connected by short bridges, do not allow repositioning the stent after having retracted the release sheath. Some Nitinol stents are conical and have a smaller diameter on the distal portion to be positioned in the internal carotid and a larger diameter of the proximal portion, to be positioned in the common carotid. Today, we cannot say which stent design and material provides the best long-term results because no trials comparing the various types of stent have been carried out. Furthermore, choosing the most suitable stent depends on the positioning ease and the lower risk of acute complications. After implantation, post-expansion of the stent with balloon is needed to obtain an acceptable angiographic result in nearly 100% of the cases. This part of the procedure implies a major risk of embolisation of material and transcranial ultrasonography has shown that the highest number of signals is indeed during post-expansion. Because of the risk of embolisation despite the use of protection system, we recommend using an undersized balloon with respect to the diameter of the vessel and inflation pressures not higher than ten atmospheres, after intravenous administration of atropine. A brief inflation is performed to reduce the risk of embolisation. In carotid stenting, unlike coronary stenting, there is not need to obtain a residual stenosis close to 0%. Angiography

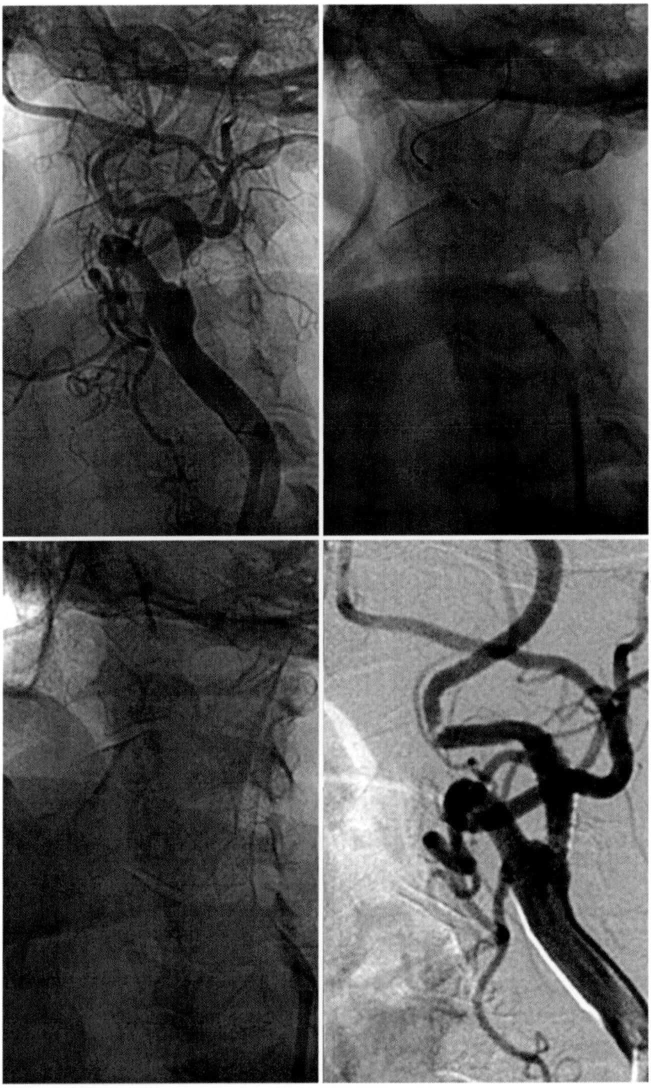

Fig. 18.21 Carotid artery stenting

with presence to residual stenosis of up to 40% (Fig. 18.22), obtained without incurring in excessive embolisation risk, guarantee excellent clinical and ultrasonography results both immediate and in the long-term.

18.4.7 Pharmaceutical Protocol

Before carotid stenting we administer ASA 100–325 mg and Ticlopodine (250 mg twice a day, starting at least 3 days before the procedure) or Clopidogrel

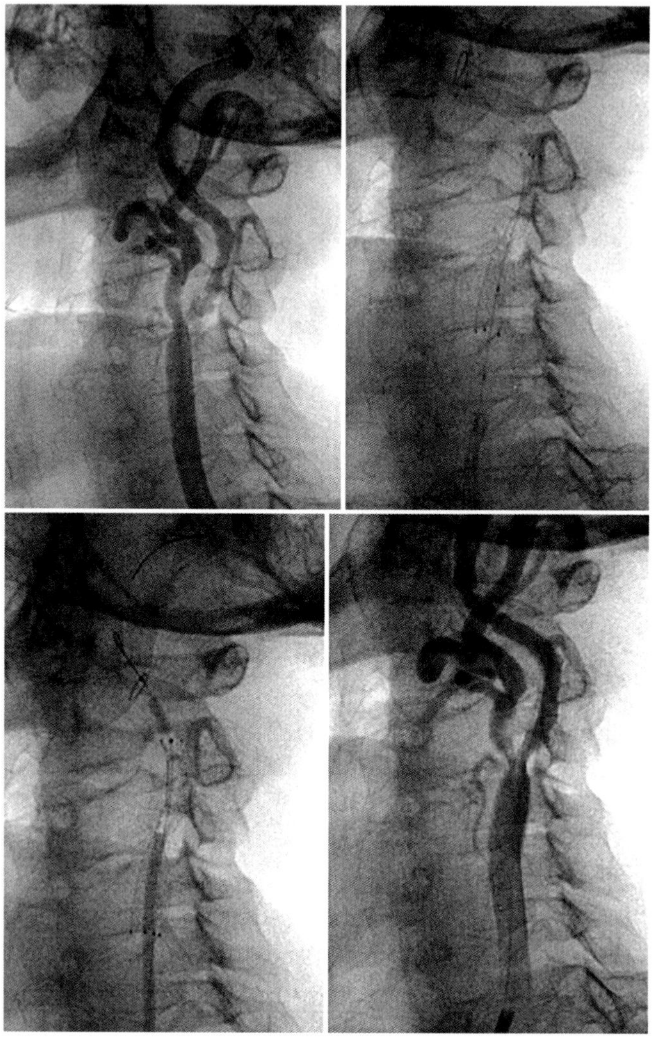

Fig. 18.22 Residual ICA stenosis at angiography control post-CAS

(300 mg the day before the procedure). Heparin 70–100 U.I./kg is administered during the procedure, maintaining ACT between 250 and 350 s. Repeating ACT is advisable at the end of the procedure. In case of values >250 s, it is advisable to neutralise heparin with protamine sulphate due to the risk of infracranial haemorrhages. Shortly prior to post-expansion, we recommend administering to all patient 1 mg of atropine in vein to prevent or attenuate bradicardia or asystole. During the procedure, we keep ready an infusion ready in pump or drip of dopamine for use in case of prolonged hypotension. After the procedure, therapy is continued with ASA for undermined time and Ticlopodine (250 mg twice a day) or Clopidogrel (75 mg) for 1 month. In the increasingly more frequent case

of CAS proceeding aortic-coronary bypass, we have reduced the associated antiplatelet therapy (Ticlopodine or Clopidogrel plus ASA) to a few days, maintaining only aspirin without observing any problems. The use of glycoprotein IIb/IIIa inhibitors is not recommended during carotid stenting procedures [55, 56].

18.4.8 Interdisciplinary Collaboration

The close collaboration with neurologists is important for discussing indications and solving possible complications. A carotid-duplex scan's expert is needed for reliable functional evaluation and patient follow-up in a carotid stenting programme. Collective collaboration, which includes discussing indications and possible reciprocal referral of patients, is welcome but not always easy. Unquestionably, there is a certain conflict of interests in the treatment of carotid stenosis. In the patient's interest, the procedure should be carried out by the specialist who carries out the procedure with the lowest incidence of complications.

18.4.9 Clinical Results of Carotid Stenting

Surgical endarterectomy (CEA) has been carried out with excellent results for many years and must be considered the gold standard for treating carotid stenosis [4, 57]. In order to demonstrate the equality or superiority of an alternative technique, the incidence of perioperative complications and the long-term efficacy of the new therapeutic approach must be critically evaluated. Clinical efficiency is the true objective of the treatment and evaluated in years of survival without stroke. In order to evaluate procedural safety, the only value parameter is the endpoint combined with the incidence of stroke/death during the first 30 days. Strokes include minor events (not disabling or permanent) and major events (disabling and permanent), but not TIA. Death may result from any cause. Another important definition is that of symptomatic and asymptomatic patients. Only the patients who had a TIA, a stroke or a amaurosis fugax episode during the 6 months prior to the procedure and which is clearly correlated to the lesion to be treated are defined symptomatic. All other patients completely asymptomatic or with non-specific symptoms (e.g. syncope, vertigo or contralateral stroke/TIA) must be considered as truly asymptomatic patients [4, 57, 58].

18.4.10 Perioperative Complications

The indication to carry out any type of revascularisation of carotid arteries depends on the incidence of complications associated to the suggested procedure [59]. Procedural safety criteria have been defined for CEA by the American Heart

Association and by the Society of Vascular Surgeons [59, 60]. According to these guidelines, the treatment of carotid stenosis must only be carried out if the risk of stroke/death during the first 30 days is <6% in patients with symptomatic lesions and <3% in asymptomatic patients. These limits are based on two major trials on CEA. In NASCET (North American Symptomatic Carotid Endarterectomy Trial) [4–57, 60], the stroke/death incidence after 30 days in symptomatic patients was 6.7%. The 3% limit in asymptomatic patients is based on ACAS (Asymptomatic Carotid Artery Study) [61], in which the perioperative incidence of stroke/death after CEA in low-risk patients was 2.3%. Other studies on less carefully selected asymptomatic patients obtained less favourable results. In VACS (Veterans Affairs Cooperative Study) [62], the incidence of permanent stroke and death was 4.7%. Until now, three randomised trials comparing CEA and CAS are available, although with a relatively low number of patient [63, 64]. In the three trials, the incidence of stroke/death of the two techniques was not significantly different. These results are limited to evaluating perioperative complications and suggest clinical equivalence of the two procedures. It must further be noted that the largest study (CAVATAS), which included 500 patients per arm [63], the incidence of stroke/death was higher than 9% in patients treated with angioplasty (in CAVATAS stent was only used for 26% of the cases), and in patients treated with CEA. The randomised SAPPHIRE study, presented at the annual AHA congress in 2002, contained more encouraging results [64]. One hundred and fifty six patents were randomised for carotid stenting with cerebral protection and 151 patents for endarterectomy. Symptomatic patients were 32% of the stent group and 29% of the surgical group. The incidence of stroke/death in the group of patients treated with CAS was 4.5% and in the group treated with CEA was 6.6% (p=NS). The combined stroke/death/infarction end-point was significantly lower in the CAS group (5.8% vs. 12.6%; p=0.47). Despite the result of stroke and death, these results exceed the limits set in AHA guidelines [59, 60] based on NASCET and ACAS [4, 61]. A possible explanation that patients with contralateral occlusion of the carotid, restenosis after CEA, age >80 years and presence of concomitant cardiac or pulmonary pathologies, where exclusive from the last two trials, while these higher risk patients were treated in the CAS versus CEA randomised trials. An important advantage of the less invasive therapeutic approach of CAS versus CAE is the absence of lesions to cranial nerves and problems related to healing of the surgical wound. On the other hand, the incidence of complications related to percutaneous vascular access is low, by virtue of the small diameter of the catheters used.

18.5 Conclusions

Nowadays, CEA is the gold standard for the treatment of extracranial carotid steno-occlusive pathology, with long-term operative results which have been incontrovertibly demonstrated in major randomised trials. However, the complication rates shown in these trials do not reflect the results obtained today in major centres, with a high volume of procedures. In particular, new treatments – such as carotid stenting – must be evaluated on the basis of these results and not of those of previous trials. CEA, in expert hands, is the best treatment for stroke prevention, with total complication rates greatly lower than 3%. Carotid stenting is emerging as a low invasive therapeutic alternative for the treatment of carotid stenosis. The results of the first randomised trials and CAS with cerebral protection register indicated results similar to those obtained in the best surgical endarterectomy case histories. The carotid stenting procedure appears preferable in high operating risk patients, above all those with symptomatic carotid stenosis, in patients with associated coronary disease and in those older than 80 years.

References

1. Miniño AM, Heron MP, Smith BL. Death Statistical Reports: preliminary data 2004. Natl Vital Stat Rep 2006;54(19):1–49.
2. American Heart Association. Heart Disease and Stroke Statistics 2006 "Update". Dallas, TX: American Heart Association; 2006.
3. Thompson JE. Carotid surgery: the past is prologue. The John Homans lecture. J Vasc Surg 1997;25(1):131–40.
4. North American Symptomatic Carotid Endarterectomy Trial Collaborators. Beneficial effect of carotid endarterectomy in symptomatic patients with high-grade carotid stenosis. N Engl J Med 1991;325(7):445–53.
5. Mayberg MR, Wilson SE, Yatsu F, et al. Carotid endarterectomy and prevention of cerebral ischemia in symptomatic carotid stenosis. Veterans Affairs Cooperative Studies Program 309 Trialist Group. JAMA 1991;266(23):3289–94.
6. Bates ER, Babb JD, Casey DE Jr., et al. ACCF/SCAI/SVMB/SIR/ASITN 2007 Clinical Expert Consensus Document on carotid stenting. Vasc Med 2007;12(1):35–83.
7. European Carotid Surgery Trialists' Collaborative Group. MRC European Carotid Surgery Trial: interim results for symptomatic patients with severe (70–99%) or with mild (0–29%) carotid stenosis. Lancet 1991;337(8752):1235–43.
8. Executive Committee for the Asymptomatic Carotid Atherosclerosis Study. Endarterectomy for asymptomatic carotid artery stenosis. JAMA 1995;273(18):1421–8.
9. Feasby TE, Kennedy J, Quan H, et al. Real-world replication of randomized controlled trial results for carotid endarterectomy. Arch Neurol 2007;64(10):1496–500.
10. Long GW, Nuthakki V, Bove PG, et al. Contemporary outcomes for carotid endarterectomy at a large community-based academic health center. Ann Vasc Surg 2007;21(3):321–7.
11. Eliasziw M, Kennedy J, Hill MD, Buchan AM, Barnett HJ. Early risk of stroke after a transient ischemic attack in patients with internal carotid artery disease. CMAJ 2004;170(7):1105–9.
12. Lisabeth LD, Ireland JK, Risser JM, Brown DL, Smith MA, Garcia NM, Morgenstern LB. Stroke risk after transient ischemic attack in a population-based setting. Stroke 2004;35:1842–46.
13. Nederkoorn PJ, Mali WPTM, Eikelboom BC, et al. Preoperative diagnosis of carotid artery stenosis: accuracy of noninvasive testing. Stroke 2002;33:2003–8 (Level II evidence).
14. Blakeley DD, Oddone EZ, Hasselblad V, et al. Noninvasive carotid artery testing: a metaanalytic review. Ann Intern Med 1995;122:360–7 (Level II evidence).
15. Wardlaw JM, Chappell FM, Best JJ, Wartolowska K, Berry E. Non-invasive imaging compared with intra-arterial angiography in the diagnosis of symptomatic carotid stenosis: a meta-analysis. Lancet 2006;367:1503–12.

16. Debrey SM, Yu H, Lynch JK, Lövblad KO, Wright VL, Janket SJ, Baird AE. Diagnostic accuracy of magnetic resonance angiography for internal carotid artery disease: a systematic review and meta-analysis. Stroke 2008;39:2237–48.
17. Radoux B, Marro B, Koskas F, et al. Carotid artery stenosis: prospective comparison of CT, three-dimensional gadolinium-enhanced MR, and conventional angiography. Radiology 2001;220:179–85.
18. Anderson GB, Ashforth R, Steinke DE, et al. CT angiography for the detection and characterization of carotid artery bifurcation disease. Stroke 2000;31:2168–73.
19. Rothwell PM, Eliasziw M, Gutnikov SA, et al. Endarterectomy for symptomatic carotid stenosis in relation to clinical subgroups and timing of surgery. Lancet 2004;363(9413): 915–24.
20. Halliday AW, Thomas D, Mansfield A. The Asymptomatic Carotid Surgery Trial (ACST). Rationale and design. Steering Committee. Eur J Vasc Surg 1994;8(6):703–10.
21. Johnston DC, Goldstein LB. Clinical carotid endarterectomy decision making: non invasive vascular imaging versus angiography. Neurology 2001;56:990–1.
22. Van Damme H, Vivario M, Boniver J, et al. Histologic characterization of carotid plaques. Cardiovasc Pathol 1994;3:9–17.
23. Imparato AM, Riles TS, Mintzer K, Bauman F. The importance of hemorrhage in the relationship between gross morphologic characteristics and cerebral symptoms. Ann Sure 1983;197:195.
24. GALA Trial Collaborative Group, Lewis SC, Warlow CP, Bodenham AR, Colam B, Rothwell PM, Torgerson D, Dellagrammaticas D, Horrocks M, Liapis C, Banning AP, Gough M, Gough MJ. General anaesthesia versus local anaesthesia for carotid surgery (GALA): a multicentre, randomised controlled trial. Lancet 2008;372(9656):2132–42.
25. Jacob T, Hingorani A, Ascher E. Carotid Artery Stump Pressure (CASP) in 1135 consecutive endarterectomies under general anesthesia: an old method that survived the test of times. J Cardiovasc Surg (Torino) 2007;48(6):677–81.
26. de Sousa AA, Dellaretti MA, Faglioni W Jr., Carvalho GT. Monitoring of activated coagulation time in carotid endarterectomy. Surg Neurol 2005;64 Suppl 1:S1.6–9.
27. Rerkasem K, Rothwell PM. Routine or selective carotid artery shunting for carotid endarterectomy (and different methods of monitoring in selective shunting). Cochrane Database Syst Rev 2009;7(4):CD000190.
28. Brown KR, KresovikTF, Chin MH, Kresovik RA, Grund Sherry L, Hendel ME. Multistate population-based outcomes of combined carotid endarterectomy and coronary artery bypass. J Vasc Surg 2003;37:32–9.
29. Palombo D, Lucertini G, Mambrini S, Spinella G, Pane B. Carotid endarterectomy: results of the Italian Vascular Registry. J Cardiovasc Surg (Torino) 2009;50(2):183–7.
30. Bandyk DF, Back MR, Johnson BL, Shames ML. Carotid intervention prior to or during coronary artery bypass grafting. When is it necessary? J Cardiovasc Surg (Torino) 2003;44(3):401–5.
31. Yoon YS, Shim WH, Kim SM, Park KJ, Cho SY. Carotid artery stenting in patients with symptomatic coronary artery disease. Yonsei Med J 2000;41(1):89–97.
32. Kovacic JC, Roy PR, Baron DW, Muller DW. Staged carotid artery stenting and coronary artery bypass graft surgery: initial results from a single center. Catheter Cardiovasc Interv 2006;67(1):142–8.
33. Randall MS, McKevitt FM, Cleveland TJ, Gaines PA, Venables GS. Is there any benefit from staged carotid and coronary revascularization using carotid stents? A single-center experience highlights the need for a randomized controlled trial. Stroke 2006;37(2):435–9.
34. Dubinsky RM, Lai SM. Mortality from combined carotid endarterectomy and coronary artery bypass surgery in the US. Neurology 2007;68(3):195–7.
35. Mackey WC, O'Donnell TF, Callow AD. Cardiac risk in patients undergoing carotid endarterectomy: impact on perioperative and long-term mortality. J Vasc Surg 1990;11:226–34.
36. De Feo M, Renzulli A, Onorati F, Marmo J, Galdieri N, De Santo LS, Della Corte A, Cotrufo M. The risk of stroke following CABG: one possible strategy to reduce it? Int J Cardiol 2005;98(2):261–6.
37. Giangola G, Migaly J, Riles TS, Lamparello PJ, Adelman MA, Grossi E, Colvin B, Pasternak PF, Galloway A, Culliford AT, Esposito R, Ribacove G, Crawford BK, Glassman L, Baumann

FG, Spencer FC. Perioperative morbidity and mortality in combined vs. staged approaches to carotid and coronary revascularization. Ann Vasc Surg 1996;10(2):138–42.

38. Ghosh J, Murray D, Khwaja N, Murphy MO, Walker MG. The influence of asymptomatic significant carotid disease on mortality and morbidity in patients undergoing coronary artery bypass surgery. Eur J Vasc Endovasc Surg 2005;29(1):88–90.
39. Osaka M, Nakata H, Sakamoto H. Clinical outcome for coronary artery bypass grafting in patients with severe carotid occlusive disease. J Cardiol 2001;38(6):303–9.
40. Antunes PE, Anacleto G, de Oliveira JM, Eugenio L, Antunes MJ. Staged carotid and coronary surgery for concomitant carotid and coronary artery disease. Eur J Cardiothorac Surg 2002;21(2):181–6.
41. Naylor AR, Cuffe RL, Rothwell PM, Bell PR. A systematic review of outcomes following staged and synchronous carotid endarterectomy and coronary artery bypass. Eur J Vasc Endovasc Surg 2003;25(5):380–9.
42. LaMuraglia GM, Brewster DC, Moncure AC, et al. Carotid endarterectomy at the millennium: what interventional therapy must match. Ann Surg 2004;240(3):535–44; discussion 544–6.
43. Mathias K. Perkutane transluminale Katheterbehandlung supraaortaler Arterienobstruktionen. Angio 1981;3:47–50.
44. Roubin SG, New G, Iyer SS, et al. Immediate and late clinical outcomes of carotid artery stenting in patients with symptomatic and asymptomatic carotid artery stenosis. A 5-year prospective analysis. Circulation 2001;103:532–7.
45. Jordan WD, Voellinger DC, Doblar DD, et al. Microemboli detected by transcranial Doppler monitoring in patients during carotid angioplasty versus carotid endarterectomy. Cardiovasc Surg 1999;7:33–38.
46. Markus HS, Clifton A, Buckenham T, et al. Carotid angioplasty: detection of embolic signals during and after the procedure. Stroke 1994;25:2403–6.
47. Ackerstaff RG, Moons KG, van de Vlasakker CJ, et al. Association of intraaoperative transcranial Doppler monitoring variables with stroke from carotid endarterectomy. Stroke 2000;31:1817–23
48. Theron JG, Payelle GG, Coskun O, et al. Carotid artry stenosis: treatment with protected balloon angioplasty and stent placement. Radiology 1996;201:627–36.
49. Tübler T, Schlüter M, Dirsch O, et al. Balloon-protected carotid artery stenting: relationship of periprocedural neurological complications with the size of particulate debris. Circulation 2001;104:2791–96.
50. Martin JB, Pache JC, Treggiari-Venzi M, et al. Role of the distal balloon protection technique in the prevention of cerebral embolic events during carotid stent placement. Stroke 2001;32:479–84.
51. Reimers B, Corvaja N, Moshiri S, et al. Cerebral protection with filter devices during carotid artery stenting. Circulation 2001;104:12–5.
52. Williams DO. Carotid filters: new to the Interventionalist's toolbox. Circulation 2001;104:2–3.
53. Al-Mubarak N, Colombo A, Gaines PA, et al. Multicenter evaluation of carotid artery stenting with a filter protection system. J Am Coll Cardiol 2002;39:841–6.
54. Cremonesi A, Castriota F. Efficacy of a nitinol filter device in the prevention of embolic events during carotid interventions. J Endovasc Ther 2002;9:155–9.
55. Qureshi AI, Saad M, Zaidat OO, et al. Intracerebral hemorrhages associated with neurointerventional procedures using a combination of antithrombotic agents including abciximab. Stroke 2002;33:1916–9.
56. Hofmann R, Kerschner K, Steinwender C, et al. Abciximab bolus injection does not reduce cerebral ischemic complications of elective carotid artery stenting: a randomized study. Stroke 2002;33:725–7.
57. MRC European Carotid Surgery Trial. Randomised trial of endarterectomy for recently symptomatic carotid stenosis: final results of the MRC European Carotid Surgery Trial (ECST). Lancet 1998;351:1379–1387.
58. Barnett HJ, Taylor DW, Eliasziw M, et al. Benefit of carotid endarterectomy in patients with symptomatic moderate or severe stenosis: North American Symptomatic Carotid Endarterectomy Trial Collaborators. N Engl J Med 1998;339:1415–25.

59. Moore WS, Barnett HJM, Beebe HG, et al. Guidelines for carotid endarterectomy: a multidisciplinary consensus statement from the Ad Hoc committee, American Heart Association. Stroke 1995;26:188–201.
60. Biller J, Feinberg WM, Castaldo JE, et al. Guidelines for carotid endarterectomy: a statement for healthcare professionals from a special writing group of the stroke council, American Heart Association. Circulation 1998;97:501–509.
61. Executive committee for the Asymptomatic Carotid Atherosclerosis Study. Endarterectomy for asymptomatic carotid artery stenosis. JAMA 1995;273:1421–8.
62. Hobson RW, Weiss DG, Fields WS, et al. Efficacy of carotid endarterectomy for asymptomatic carotid stenosis: the Veterans' Affairs Cooperative Study group. N Engl J Med 1993;328:221–7.
63. Carotid and Vertebral Artery Transluminal Angioplasty Study. Endovascular versus surgical treatment in patients with carotid stenosis in the carotid and vertebral artery transluminal angioplasty study (CAVATAS): a randomised trial. Lancet 2001;357:1729–91.
64. Yadav J, et al. SAPPHIRE: Stenting and angioplasty with protection in patients at high risk for endarterectomy. Scientific communication. 75th Scientific Session of the American Heart Association. Chicago, 2002.

Biographies

Franco Nessi, graduated in medicine and surgery from Turin University, Italy. Received the specialization in General Surgery from Turin University, Italy and Vascular Surgery from Pavia University, Italy. He directed the division of Vascular Surgery and Kidney Transplantation at Hospital "Maggiore della Carità" of Novara, Italy, from 1997 to 2003. From 2003 is the director of Vascular and Endovascular Surgery Hospital Mauriziano Umberto I of Turin, Italy. He is lecturer at the School of Specialisation in Vascular Surgery, University of Turin, Italy. He worked for major international trials: ECST, ILAILL, GALA TRIAL. Now he is a collaborator for the ACST-2 trial.

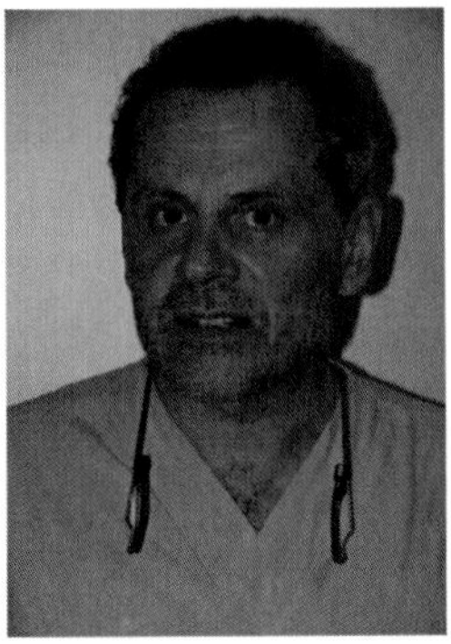

Michelangelo Ferri, graduated in medicine and surgery from Perugia University, Italy. Received the PhD degree in General Surgery, the specialization in Vascular Surgery and in General Surgery from Perugia University, Italy. From 1999 is consultant in Vascular and Endovascular Surgery Hospital Mauriziano Umberto I of Turin. He worked for major international trials: ECST, EVEREST, EUROCAST, RITA, ACST, GALA TRIAL. Now he is a collaborator for the ACST-2 trial.

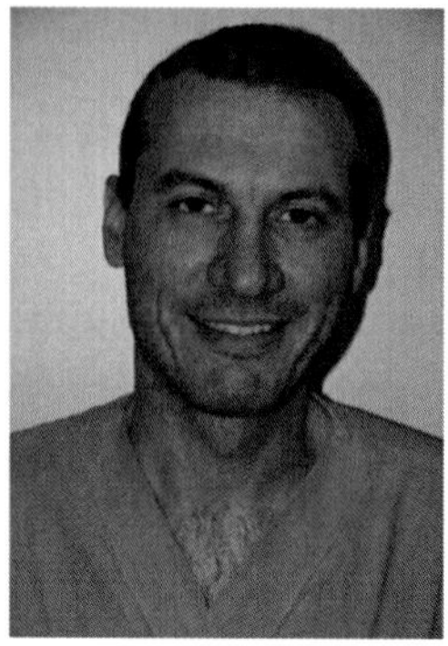

Emanuele Ferrero, graduated in medicine and surgery from Genova University, Italy. Received the specialization in Vascular Surgery from Siena University, Italy. From 2006 is consultant in Vascular and Endovascular Surgery Hospital Mauriziano Umberto I of Turin. He worked for major international trials: ILAILL, GALA TRIAL. Now he is a collaborator for the ACST-2 trial. He is author of many international publication of vascular surgery. He is reviewer of "Journal of the Royal Society of Medicine".

Andrea Viazzo, graduated in medicine and surgery from Amedeo Avogadro Piemonte Orientale Novara University, Italy. Received the specialization in Vascular Surgery from Turin University, Italy. From 2007 is consultant in Vascular and Endovascular Surgery Hospital Mauriziano Umberto I of Turin. He worked for a major international trial: GALA TRIAL. Now he is a collaborator for the ACST-2 trial.

Chapter 19
Drug Therapy and Follow-Up

Mario Eandi

Abstract Atherosclerosis is a systemic disease of the arteries and is considered the leading cause of cardiovascular diseases (CVDs), coronary artery disease (CAD), peripheral artery disease (PAD), and stroke. Atherogenesis is dramatically accelerated by the concomitant presence of major cardiovascular risk factors such as dyslipidemia, hypertension, smoking, and visceral obesity. Atherothrombosis and aneurismal disease related to atherosclerosis underlie the majority of acute cardiovascular events in developed countries.

Different pharmacological classes of drugs have been developed to treat a single specific risk factor of CVD: hipolipemic and antihypertensive agents are landmarks of *antiatherogenic* therapy from several years. In recent times, the development of modern imaging techniques offers new tools for documenting the effect of drugs on progression of atherosclerosis lesions. The introduction of these technologies highlights the role of surrogate end-point and biomarkers as predictors of clinical outcomes in CVDs. The potential value to public health and industries in accelerating the discovery and development processes for cardiovascular therapeutics through smaller, shorter studies, using validated endpoints other than mortality and irreversible morbidity is substantial and very attractive.

Keywords Atherogenesis • Surrogate biomarkers • Antiatherogenic therapy • Hipolipidemic drugs • Antihypertensive drugs

19.1 Introduction

Atherosclerosis is a systemic disease of the arteries with important sequelae in many regional circulations, including the heart, brain, kidneys, mesentery, and limbs. Artery diseases are classified into atherosclerotic occlusive disorders, nonatherosclerotic

M. Eandi (✉)
Department Anatomy, Pharmacology and Forensic Sciences, University of Turin
Via Pietro Giuria, 13 - 10126 Torino, Italy
e-mail: mario.eandi@unito.it

Jasjit S. Suri et al. (eds.), *Atherosclerosis Disease Management*,
DOI 10.1007/978-1-4419-7222-4_19,

occlusive disorders, and aneurysms. Atherosclerotic diseases and nonatherosclerotic arterial occlusive disorders are subdivided into coronary, cerebral, peripheral, renal, and mesenteric disorders. Similarly, aneurysms are classified according to the arterial bed affected (Fig. 19.1) [1].

Atherosclerosis lesions tend to occur focally, preferentially affecting the proximal left anterior descending coronary artery, the proximal portions of the renal arteries, and the carotid bifurcation in the extracranial circulation. Indeed, atherosclerotic lesions often form at branching points of arteries, regions of disturbed blood flow. Postmortem observations and intravascular ultrasound studies have shown that, in addition to focal flow-limiting stenoses, nonocclusive intimal atherosclerosis also occurs diffusely in affected arteries.

Atherosclerosis is considered the leading cause of cardiovascular diseases (CVDs), mainly coronary artery disease (CAD), peripheral artery disease (PAD), and stroke. The onset of atherosclerotic disease in one vascular bed increases the risk of disease in others ("cross-risk"): patients with one ischemic event have an increased likelihood of experiencing another event in the future [2–4].

Atherosclerosis is the major cause of cardiovascular morbidity and mortality in developed countries. Moreover, current predictions estimate that by the year 2020, CVD, notably atherosclerosis, will become the leading global cause of total disease burden. Today, according to the World Health Organization, about 972 million

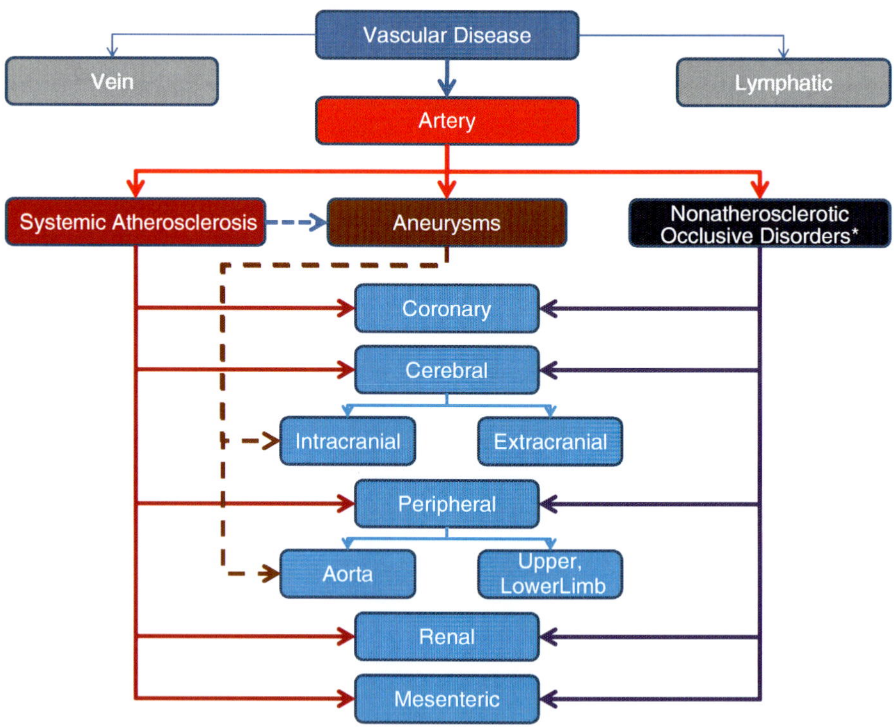

Fig. 19.1 Major classification of vascular diseases. *Asterisk* includes inflammatory, artery dysplasias, congenital, traumatic, and infections [1]

people are hypertensive patients worldwide and 300 million are in areas at high risk of CVD. It is estimated that in the 2020 about 1.5 billion people will be hypertensive and about 600 million at high risk to experience major cardiovascular events. CVD annually yields 28% of all deaths in the developed world, despite major progress in diagnosis and treatment over the past 25 years.

Despite a decline in the incidence of coronary heart disease (CHD) in Europe, data from the World Health Organization Global Burden of Disease report 1990–2020 indicate a marked projected increase in the burden of atherosclerotic CVD. Demographic trends partly account for this increase: worldwide the proportion of people older than 60 years is growing faster than any other age group, contributing to increase cardiovascular event rates in elderly patients. Moreover, with improvement in clinical management, more patients survive their initial acute coronary event [5].

Several epidemiological and experimental studies have identified the major risk factors associated with the development of atherosclerosis [6, 7]. In particular, the prospective, community-based Framingham Heart Study provided rigorous and robust support for the concept that hypercholesterolemia, hypertension, and other factors are correlated with cardiovascular risk.

Figure 19.2 shows the risk factors recognized by the current National Cholesterol Education Project Adult Treatment Panel III (ATP III) [8]. The weight of evidence supporting various risk factors differs. Hypercholesterolemia and hypertension certainly predict coronary risk, but other "nontraditional" risk factors, such as levels

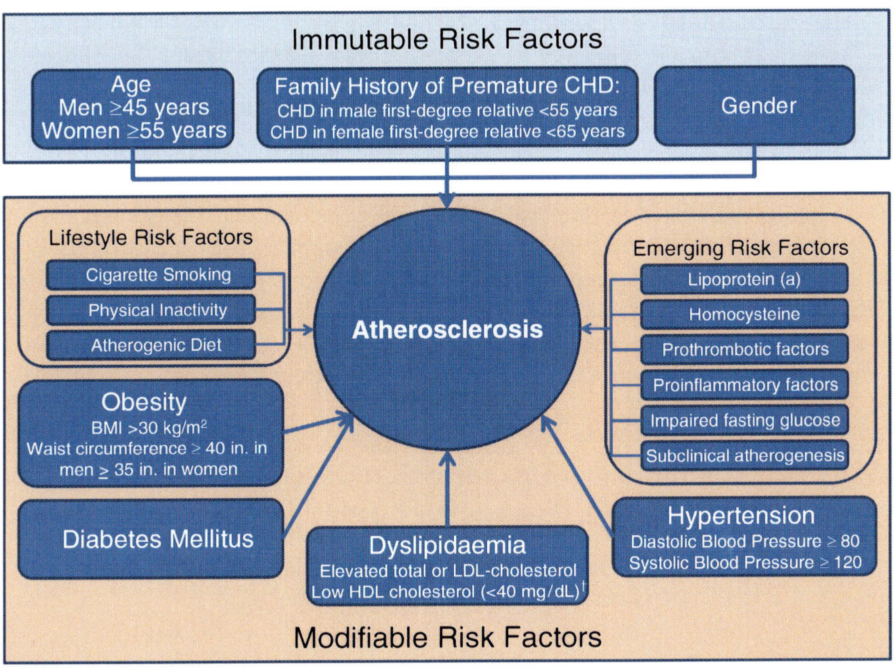

Fig. 19.2 Risk factors of atherosclerosis recognized by the current National Cholesterol Education Project Adult Treatment Panel III (ATPIII) [8]

of homocysteine, lipoprotein(a) [Lp(a)], or infection, remain controversial. Moreover, the causality of some biomarkers that predict cardiovascular risk, such as C-reactive protein (CRP), remains uncertain.

Considering the clinical perspective, cardiovascular risk factors are classified into two categories: those nonmodifiable, such as age and gender, and those modifiable by lifestyle and/or pharmacotherapy, such as stress, physical inactivity, cigarette smoking, hypercholesterolemia, hypertension, and diabetes mellitus.

For many decades, research and development of new drugs for the treatment of atherosclerotic diseases were oriented to the control of major risk factors. Therefore, several pharmacological classes of drugs have been developed to treat a single specific risk factor of CVD, such as hypercholesterolemia, hypertension, or diabetes mellitus. More recently, the importance of comprehensive risk factor management in each individual cardiovascular patient has been acknowledged and introduced in the current treatment guidelines [9–16].

The results obtained by the increasing amount of studies on the molecular mechanisms underlying the pathogenesis of atherosclerosis and atherothrombosis are opening new interesting perspective for the prevention and cure of CVD. Nowadays, new drugs for atherosclerosis are designed assuming a specific molecular target among the several mechanism of atherogenesis.

Various pharmacological therapies have been designed to reduce the risk, development, and progression of the atherosclerotic plaque. In addition, different approaches have been pursued for the prevention of the complications arising after rupturing of atherosclerotic plaques.

Finally, the results of experimental studies conducted on animal models and the outcomes of long-term clinical studies indicate that sufficiently drastic changes in the plaque environment induced by drug treatment can stabilize and cause regression of even advanced lesions.

This chapter will address a brief clinical and pharmacological description of hypocholesterolemic drugs and antihypertensive agents, the two main classes of drugs traditionally used in primary and secondary prevention of major atherosclerosis risk factors. Antiplatelets agents and antidiabetic drugs are also used in secondary prevention of major CV events (strokes and myocardial infarction), but will not be considered in this work. A bigger attention is focused on the biomarkers and surrogate end-point issues, specifically regarding the employment of measures obtained from imaging devices on atherosclerotical lesions.

19.2 Physiopathology of Atherosclerosis as Target of Drug Action

19.2.1 Atherogenesis and Atherothrombosis

The pathology of atherosclerosis is a complex process that consists of the generation of "atheroma" lesions in the intima layer of the arteries. These lesions progressively evolve from isolated fatty streak lesions to form complex "atherosclerotic plaques" which deform the arterial wall, disturb the blood flow, cause stenosis with ischemia of the underlying tissues (Fig. 19.3) [17].

Atherosclerotic plaques are asymmetric focal thickenings of the intimal (or innermost) layer of the arterial wall, and consist of diverse cell types, connective tissue components, lipids, and cellular debris. Inflammatory and immune cells derived from circulating blood are key constituents of the plaque, together with vascular endothelial and smooth muscle cells [18, 19].

Erosion or rupture of vulnerable, lipid-rich atherosclerotic plaque triggers the formation of a platelet-rich thrombus that may partially or completely occlude the artery. Platelets are key players in all phases of atherothrombosis, including the initial steps of atherogenesis, the progression of fatty streaks to atherosclerotic lesions, and the resulting thrombotic complications. Under the high shear flow of a ruptured plaque, platelets may adhere directly to von Willebrand factor and the activated endothelium, initiating the process of platelet activation. After initial activation, potent amplification mechanisms, such as platelet-to-platelet aggregation and fibrin formation ensue, leading to a growing thrombus at the site of plaque rupture [5].

Atherothrombosis, which results from direct interaction between atherosclerotic plaque and arterial thrombosis, underlies the majority of acute cardiovascular events, independently of the specific vascular bed in which they occur.

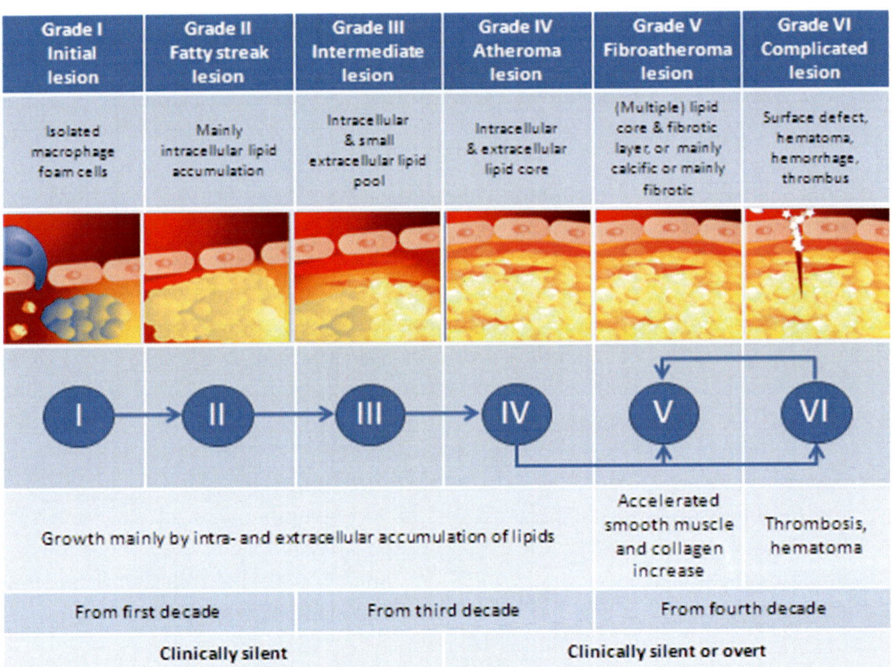

Fig. 19.3 The phases of atherothrombosis (modified from: [17])

Moreover, the atherosclerotic lesions may induce ectasia and cause development of aneurismal disease. This event frequently occurs in the aorta.

Atherogenesis in humans usually occurs over a period of many decades. The initiation, progression, and ultimate rupture of a vulnerable atherosclerotic plaque with an ensuing thrombo-embolic event typically evolves over several decades. Growth of atherosclerotic plaques probably does not occur in a linear fashion, but rather discontinuously, with periods of relative quiescence punctuated by periods of rapid evolution. After a generally prolonged "silent" period, atherosclerosis may become clinically manifest. However, most plaque ruptures remain clinically silent, as the fibrous cap of the plaque is constantly undergoing remodeling, rupture, thrombosis, and healing [18].

The process of atherogenesis mainly occurs in large- and medium-sized arteries and involves endothelial dysfunction, inflammation, oxidative stress, cholesterol accumulation, apoptosis, and extracellular matrix degradation. Indeed, atherosclerosis and atherothrombosis are currently viewed as inflammatory disorders [20–22]. Furthermore, atherogenesis is dramatically accelerated by the concomitant presence of major cardiovascular risk factors such as dyslipidemia, hypertension, smoking, and visceral obesity, which act at the endothelium of the arterial wall to favor atherosclerotic plaque formation [23, 24].

19.2.2 Endothelial Dysfunction

The vascular endothelium is an active paracrine, endocrine, and autocrine organ that is indispensable for the regulation of vascular tone and the maintenance of vascular homeostasis [25]. In the last two decades, endothelial dysfunction has been shown to represent a key early step in the development of atherosclerosis. Moreover, deleterious alterations of endothelium functions have also been involved in plaque progression and correlated with the occurrence of atherosclerotic complications [26, 27].

Under normal homeostatic conditions, the endothelium maintains normal vascular tone and blood fluidity, with no or little expression of proinflammatory factors. Shear stress, the frictional force induced at the endothelial surface by blood flow, is a key factor in the modulation of endothelial cell function and morphology. Laminar blood flow induces a positive mean shear stress that assures normal endothelial cell function, thereby exerting an atheroprotective effect. In contrast, perturbations of laminar blood flow, such as at the carotid bifurcation, are associated with endothelial dysfunction resulting from low mean shear stress and a flow reversal phenomenon [28].

Conventional cardiovascular risk factors including aging, hypercholesterolemia, hypertension, hyperglycemia, smoking, and a family history of premature atherosclerotic disease are all associated with alteration in endothelial function (Fig. 19.4) [29]. Endothelial dysfunction has also been associated with obesity, elevated CRP, and chronic systemic infection [30–36].

Cardiovascular risk factors with hemodynamic factors induce endothelial dysfunction via a number of mechanisms including activation of expression of

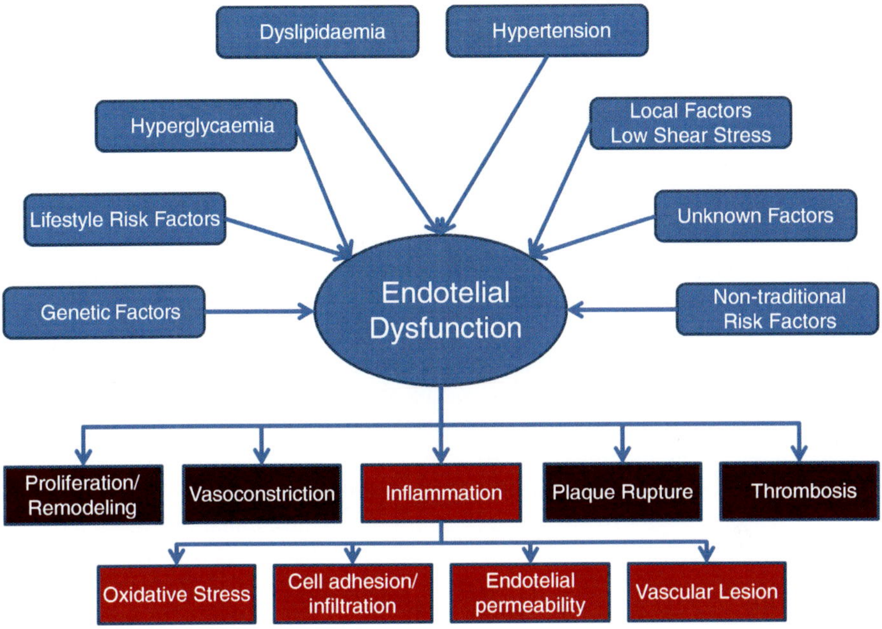

Fig. 19.4 Risk factors affecting the endothelium and the consequences of endothelial dysfunction on atherogenesis and atherothrombosis

genes regulating vasoconstriction, oxidative stress, adhesion and/or infiltration of inflammatory cells such as monocytes, proliferation of smooth cells, and increase in endothelial permeability [5].

Endothelial dysfunction is characterized by a reduction of the bioavailability of vasodilators, particularly nitric oxide (NO), and/or an increase in endothelium-derived contracting factors. The resulting imbalance leads to an impairment of endothelium-dependent vasodilation, which is the functional characteristic of endothelial dysfunction. Moreover, endothelial dysfunction also induces a specific state of endothelial activation, characterized by a proinflammatory, proliferative, and procoagulatory states that favor all stages of atherogenesis (Fig. 19.4). Endothelial dysfunction may lead to enhanced permeability to lipoproteins, enhanced lipid influx, and cholesterol accumulation in the arterial wall. The resulting chronic inflammatory process is accompanied by a loss of antithrombotic factors and an increase in vasoconstrictor and prothrombotic products [37].

Reactive oxygen species (ROS) are generated at sites of inflammation and injury. Further, oxidative stress increases vascular endothelial permeability and promotes leukocyte adhesion, which is coupled with alterations in endothelial signal transduction and redox-regulated transcription factors [38].

Therefore, the vascular endothelium, which regulates the passage of macromolecules and circulating cells from blood to tissues, is a major target of oxidative stress, playing a critical role in the pathophysiology of atherosclerosis.

The status of endothelial function may reflect the propensity of an individual to develop atherosclerotic disease, and thus may serve as a marker of an unfavorable cardiovascular prognosis.

The gold standard test for the evaluation of coronary endothelial function requires invasive quantitative coronary angiography (QCA) to examine the changes in diameter in response to intracoronary infusions of endothelium-dependent vasodilators such as acetylcholine. Endothelial function of the coronary microvasculature can be assessed with intracoronary Doppler techniques to measure coronary blood flow in response to pharmacological or physiological stimuli [39].

Noninvasive tests for the assessment of coronary endothelial function that have been described include Doppler echocardiography, positron emission tomography, and phase-contrast magnetic resonance imaging (MRI). Further, brachial artery ultrasound is a widely used noninvasive measure of endothelial function. Importantly, endothelial dysfunction assessed by this technique correlates with measures of coronary endothelial dysfunction [26].

Finally, also strain-gage venous impedance plethysmography, and measures of arterial compliance and waveform morphology provide biomarkers of endothelial function [39].

19.2.3 Lipoprotein Cholesterol Retention in the Arterial Intima

In human and animal models, hypercholesterolemia and other forms of atherogenic dyslipidemia favor focal activation of the endothelium in large- and medium-sized arteries.

Cholesterol, a fat-like substance (lipid), is a structural and functional component of cell membranes and is a precursor of bile acids and steroid hormones. The total amount of cholesterol in an organism at a given time represents the balance between the sum of the rate of exogenous (dietary) and cholesterol absorption and the rate of endogenous cholesterol synthesis and the elimination, mainly through metabolism (Fig. 19.5).

Cholesterol is transported in the blood by three major classes of lipoproteins: low-density lipoproteins (LDL), very low-density lipoproteins (VLDL), and high-density lipoproteins (HDL). Another lipoprotein class, intermediate density lipoprotein (IDL), resides between VLDL and LDL, and in clinical practice it is included in the LDL measurement. Chylomicrons, the triglyceride-rich lipoproteins (TGRLP) formed in the intestine from dietary fat, are a class of lipoproteins similar to VLDL that transport the cholesterol absorbed through the intestine.

LDL cholesterol contains a single apolipoprotein, namely apo B-100 (apo B) and makes up 60–70% of the total serum cholesterol. The VLDL are TGRLP and contain 10–15% of the total serum cholesterol. The major apolipoproteins of VLDL are apo B-100, apo Cs (C-I, C-II, and C-III), and apo E. VLDL are produced by the liver and are precursors of LDL.

The apolipoproteins of chylomicrons are the same as for VLDL except that apo B-48 is present instead of apo B-100.

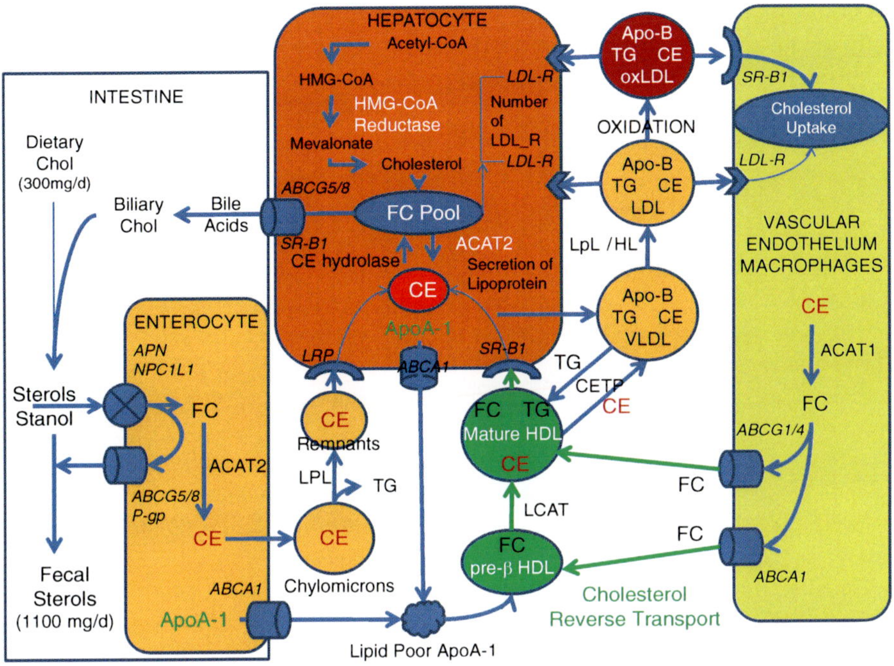

Fig. 19.5 Overview of cholesterol transport and metabolism. *ABCA1* the adenosine triphosphate binding cassette (ABC) transporter A1 to lipid-poor apolipoprotein (apo) A-I (ApoA-I), *ABCG1/G4* the ABC transporters G1 and G4 to mature, spherical HDL, *ANP* aminopeptidase N, *ACAT* acyl-CoA: cholesterol acyltransferase (ACAT), *APO* apolipoprotein, *CE* cholesteryl ester, *CETP* cholesteryl ester transfer protein, *HDL* high-density lipoprotein, *LCAT* lecithin:cholesterol acyl-transferase, *LDL* low-density lipoprotein, *LDL-R* LDL receptor, *LPL* lipoprotein lipase, *oxLDL* oxidized LDL, *NPC1L1* Niemann–Pick C1-like 1 protein, *SR-B1* the scavenger receptor, *TG* triglyceride, *VLDL* very low-density lipoprotein

HDL contains apo A-I (70%) and apo A-II, and normally makes up 20–30% of the total serum cholesterol.

LDL is the major atherogenic lipoprotein which has been identified as the primary target of cholesterol-lowering therapy. VLDL remnants, a form of partially degraded VLDL relatively enriched in cholesterol ester, appear to promote atherosclerosis, similar to LDL. Partially degraded chylomicrons, called chylomicron remnants, probably carry some atherogenic potential.

The accumulation of LDL and other atherogenic lipoproteins (VLDL and VLDL remnants) in the subendothelial matrix is a primary initiating event in atherosclerosis [5].

The net difference between the rates of LDL influx and efflux from the arterial intima, which corresponds to the amount of LDL that is retained (trapped) as a result of binding to cellular and matrix components or degradation, is a critical element in atherosclerotic lesion formation.

Evidence to date indicates that intimal LDL uptake, concentration, and residence time are all significantly increased at plaque sites, and furthermore, are enhanced by a hypertensive state. Moreover, higher circulating LDL levels lead to greater LDL accumulation and greater risk of CVD [5].

The correlation between LDL-C level and atherosclerosis is documented by a wide variety of observational and experimental evidence accumulated over several decades from animal, pathological, clinical, genetic, and different types of population studies.

A direct relationship between the levels of LDL-C (or total cholesterol) and the rate of new-onset CHD in men and women who were initially free of CHD has been found in the Framingham Heart Study, the Multiple Risk Factor Intervention Trial (MRFIT), and the Lipid Research Clinics (LRC) trial [40–42].

The same correlation holds for recurrent coronary events in people with established CHD [43].

The power of elevated LDL to cause CHD is shown most clearly in persons with genetic forms of hypercholesterolemia. In these persons, advanced coronary atherosclerosis and premature CHD occur commonly even in the complete absence of other risk factors.

Finally, a causal role for LDL has been corroborated by the controlled clinical trials of LDL lowering; recent trials especially have revealed a striking reduction in the incidence of CHD [44].

The association between LDL-C levels and CHD risk is continuous, but risk rises more steeply with increasing LDL-C concentrations. This results in a curvilinear relationship that becomes linear when the CHD risk is plotted on a log scale (log-linear relationship). This relationship is consistent with a large body of epidemiological data and with data available from clinical trials of LDL-lowering therapy. Moreover, the results of HPS study provided new evidence to support the log-linear relationship between LDL-C levels and CHD risk, even at low LDL-C concentrations. In fact, HPS results suggest that reducing serum LDL-C from any baseline level further lowers the risk in high-risk patients. Overall, these data suggest that for every 30-mg/dL change in LDL-C, the relative risk for CHD is changed in proportion by about 30%; the relative risk is set at 1.0 for LDL-C=40 mg/dL.

As a consequence, first, when persons with low LDL-C have the same absolute risk (because of other risk factors) as those with high LDL-C, the same absolute benefit is attained for a given milligram-per-deciliter lowering of LDL-C, and second, when persons with low LDL-C have a lower absolute risk than those with higher LDL-C, less absolute benefit is attained for a given LDL-C lowering in the low LDL-C group [45].

Elevated LDL-C plays a role in the development of the mature coronary plaque and contributes to plaque instability as well. Conversely, LDL cholesterol lowering stabilizes plaques and reduces the risk of acute coronary syndromes.

LDL lowering earlier in life slows atherosclerotic plaque development and reduces the cardiovascular risk. LDL-lowering therapy in patients with advanced coronary atherosclerosis (short-term risk reduction) can stabilize plaques and contribute to prevent acute coronary syndromes [46].

Strong epidemiological evidence have shown that decreased level of high-density lipoprotein (HDL) cholesterol (C) is an independent risk factor for atherosclerosis in general and CAD in particular. HDL-C levels are inversely correlated with CHD morbidity and mortality even in patients with normal levels of LDL-C. High HDL-cholesterol levels conversely convey reduced risk. It is estimated that CAD risk increases by 1–3% for every 1% reduction in HDL-C concentration [47].

A low HDL level often reflects the presence of other atherogenic factors. In many persons, a low HDL level correlates with elevations of serum triglycerides and remnant lipoproteins, and shows linkage with small, dense LDL particles. The tight association among low HDL, small LDL particles, and elevated triglycerides has evoked the term *lipid triad*. Moreover, a low HDL level can be a sign of insulin resistance and its associated metabolic risk factors [48].

Several experimental and epidemiological data indicate that HDL directly participates in the atherogenic process. In vitro, HDL promotes efflux of cholesterol from foam cells in atherosclerotic lesions (reverse cholesterol transport). Other studies indicate that the antioxidant and anti-inflammatory properties of HDL inhibit atherogenesis [49, 50].

19.2.4 Proinflammatory Oxidized LDL and the Role of Monocytes–Macrophages

LDL-C can become oxidized in the presence of oxidative stress and form a lipoprotein species that is particularly atherogenic. Oxidized LDL (oxLDL) readily accumulates within the arterial wall, it can cause endothelial dysfunction and it is proinflammatory. Several factors may influence the susceptibility of LDL to oxidation, including its size and composition, type 2 diabetes or the metabolic syndrome, hypertension, and the presence of endogenous antioxidant compounds, such as α-tocopherol. Individuals with high levels of oxidative stress are at an increased risk for cardiovascular events. Antioxidant vitamins have been tested in experimental models of atherosclerosis as a potential approach to prevent CHD. However, clinical trials of antioxidants failed to demonstrate consistent beneficial effects on cardiovascular outcomes.

The formation of oxidized, biologically active lipids and oxysterols is a key factor in the initiation of the inflammatory response in the artery wall. Oxidized bioactive lipids are also potent modulators of nuclear transcription factors and gene expression in target cells. Moreover, oxLDL can activate endothelial cells to express adhesion molecules for monocytes such as ICAM-1 and VCAM-1. As a result, circulating monocytes adhere to the endothelium, and subsequently migrate by diapedesis through endothelial cell junctions to the subendothelial space [51].

Macrophage colony-stimulating factor, a growth factor produced in the inflamed intima, induces monocytes entering the plaque to differentiate into macrophages. This step is critical for the development of atherosclerosis and is associated with upregulation of both scavenger receptors and toll-like receptors.

Extensively modified or highly oxidized LDL are avidly taken up by monocyte–macrophages via scavenger receptors to form foam cells characterized by the presence of numerous cytoplasmic droplets of cholesteryl esters derived from LDL cholesterol.

Excess accumulation of cholesterol uptake in macrophages, in the absence of efficient cholesterol efflux to HDL, leads to foam cell formation.

Foam cell macrophages also secrete a spectrum of lipid and protein mediators of inflammation including platelet-derived activating factor, eicosanoids, interleukins (IL), phospholipases, and metalloproteases. Cytokines and growth factors secreted by macrophages and T cells are important for the migration and proliferation of smooth muscle cells and for extracellular matrix production. Interestingly, scavenger receptor-mediated uptake of oxidized LDL by macrophages results in induction of the expression of nuclear peroxisome proliferator-activated receptor γ (PPARγ) and leads to transcriptional induction of the CD36 receptor. High levels of expression of PPARγ have been detected in lesion foam cells, thereby suggesting that these nuclear receptors play a central role in the control of macrophage lipid metabolism and pathogenesis of atherosclerosis [52].

Macrophages, smooth muscle cells, and endothelial cells equally contribute to the inflammatory reaction via the CD40/CD40 L mechanism, as these cells express CD40 at their surface.

Activation of the CD40 receptor via ligand binding results in production of inflammatory cytokines, metalloproteases, and adhesion proteins. In addition, the CD40/CD40 L interaction is essential for major immune reactions involving T and B lymphocytes; indeed, recent studies in this area suggest that lymphocytes may play a major role in atherogenesis [53, 54].

T-cell infiltrates are typically present in atherosclerotic lesions. The atherosclerotic lesion contains cytokines that promote a Th1 rather than Th2 response. T-cell cytokines stimulate the production of proinflammatory factors at a later stage of the cytokine cascade. As a result, elevated levels of IL-6 and CRP may be detected in the peripheral circulation.

Thus, the dynamic disequilibrium between inflammatory and anti-inflammatory activity is the major factor in enhancing the progression of atherosclerosis [55].

Metabolic factors such as dyslipidemia may affect this process in several ways. In particular, dyslipidemia contributes to lipid deposition in the artery, initiating new rounds of immune cell recruitment.

19.2.5 Apoptosis, Plaque Rupture, and Thrombus Formation

The atherosclerotic plaque is a site of intense apoptotic activity that primarily involves macrophages and T lymphocytes [56]. However, proinflammatory cytokines may induce apoptosis in all plaque cell types, partly as a result of excessive production of NO. Expression of inducible NO synthase is directly correlated with the degree of apoptosis in human plaques. Complex plaques become unstable or vulnerable in response to both local and systemic factors; thrombus formation at such lesions may also be accelerated by the same factors. Apoptosis plays an important role in thrombus formation in acute coronary syndromes. Active inflammation is a key feature of vulnerable plaque. Rupture occurs preferentially at shoulder sites where the cap is thin, and where activated immune cells and macrophages are abundant. These cells produce numerous proinflammatory agents and proteolytic enzymes (such as metalloproteases) that weaken the cap and activate cells in the core region, thereby exacerbating intraplaque inflammation. The stable plaque is thereby transformed into a vulnerable, fragile structure, which may rupture under mechanical stress, with induction of thrombus formation and triggering of an acute coronary syndrome. The disrupted plaque is therefore a stimulus to both thrombosis and coagulation [18]. Alternatively, plaque erosion may occur at the endothelial surface, and the thrombus appears to be superimposed on a de-endothelialized but intact plaque [23].

Amounts of coronary calcium, as detected by electron beam computed tomography (EBCT) or spiral CT, correlate positively with coronary plaque burden [57–59].

Several studies indicate that, in persons with multiple risk factors, a concomitantly high coronary calcium score places persons in the range of a CHD risk equivalent [59–61]. A report by the American College of Cardiology/American Heart Association (ACC/AHA) acknowledged the potential power of coronary calcium to predict major coronary events [62]. However, ATP III does not recommend EBCT for indiscriminate screening for coronary calcium in asymptomatic persons, particularly in persons without multiple risk factors. Measurement of coronary calcium is promising for older persons in whom the traditional risk factors lose some of their predictive power, and a high coronary calcium score could be used to decide the use of LDL-lowering drugs for primary prevention in older persons [63].

19.3 Biomarkers and Surrogate Endpoints

The clinical development of drugs for atherosclerosis requires that clinical efficacy and safety are demonstrated against individual therapeutic indication, such as hypercholesterolemia, or essential hypertension. To this end, it is sufficient to obtain the evidence, on a substantial number of patients, of the drug's effectiveness in changing specific intermediate or surrogate endpoints, e.g., reduction of diastolic and systolic blood pressure (SBP), LDL-C levels, glycemia. However, to get the authorization to use the medicinal product in primary or secondary prevention of atherosclerosis or overall cardiovascular risk, the pharmaceutical company needs to provide the clinical evidence that the long-term use of the drug significantly reduces the rate of cardiovascular morbidity and mortality. Therefore, the "gold standard" for measuring clinical cardiovascular efficacy in drug development is the morbidity and mortality trial. However, such trials are very expensive, may require the inclusion of 10,000–15,000 subjects, followed for at least 5 years, to demonstrate a significant incremental benefit of a novel drug over and above that provided by currently available therapies.

The potential value to public health in accelerating the discovery and development processes for cardiovascular therapeutics through smaller, shorter studies, using validated endpoints other than mortality and irreversible morbidity is substantial and very attractive.

Surrogate endpoints are biomarkers that are predictors of clinical outcomes and that can therefore be used to assess the efficacy or safety of treatments. A biomarker is a characteristic that, objectively measured, is evaluated as an indicator of physiologic or pathologic processes or as an indicator of response to a therapeutic intervention. Typical surrogate endpoints used to assess the clinical efficacy of cardiovascular drugs include levels of LDL-C, HDL-C, triglycerides, CRP, and blood pressure.

The need for more rapid drug development highlights the role that surrogate markers may play in establishing the efficacy of drugs for managing CVD, and more specifically, atherosclerosis.

At present time, there is no common agreed-upon standard regarding the body of evidence needed for a biomarker to be considered a surrogate marker of clinical efficacy. Boissel et al. proposed a first framework for the validation of biomarkers [64]. Subsequently, Espeland et al. modified the terminology of Boissel et al. and described the clinical and statistical characteristics that a biomarkers should have to be considered a surrogate marker of efficacy in atherosclerotic disease [65]. Efficiency, linkage, and congruence are the clinical criteria outlined for validating surrogate markers (Table 19.1).

The association of vascular imaging and soluble markers of disease activity can provide valuable information to pharmaceutical companies and regulatory agencies during the development of novel treatments for atherosclerotic disease.

Table 19.1 Framework for the validation of new surrogate biomarkers of clinical efficacy, proposed by Boissel et al. and subsequently adapted by Espeland et al. [64, 65]

Clinical criteria	
Efficiency	• The surrogate marker should be relatively easy to evaluate, preferably by noninvasive means, and more readily available than the gold standard • The time course of changes in the surrogate marker should precede that of the endpoints so that disease and/or disease progression may be established more quickly via the surrogate • Clinical trials based on surrogates should require fewer resources, less participant burden, and a shorter time frame
Linkage	• The quantitative and qualitative relationship between the surrogate marker and the clinical endpoint should be established on the basis of epidemiologic and clinical studies • The nature of this relationship may be understood in terms of its pathophysiology or in terms of an expression of joint risk
Congruence	• The surrogate should produce parallel estimates of risk and benefits that are related to the target disease process as endpoints • Individuals with and without vascular disease should exhibit differences in surrogate marker measurements • In intervention studies, anticipated clinical benefits should be deducible from the observed changes in the surrogate marker

Many vascular imaging technologies are available to investigate and collect information on vascular structure and on the development of atherosclerosis. Three of these vascular imaging technologies: QCA, assessment of carotid intima–media thickness (cIMT) by ultrasound, and the determination of "plaque volume" using intravascular (or intracoronary) ultrasound (IVUS), meet or are at least close to meeting the established criteria for surrogacy [64, 65]. Recently, MRI has also emerged as one of the most promising noninvasive technique for longitudinal in vivo study of large atherosclerotic arteries.

These methods are appropriate for detecting atherosclerosis in specific vascular beds and for predicting clinical risk across populations. Moreover, considering atherosclerosis as a systemic arterial disorder and the "cross-risk," such methods can be used to support a clinical diagnosis of systemic atherosclerosis and overall cardiovascular risk [66–68]. Therefore, with convenient standardization in application and analysis, these techniques have been successfully used in clinical research to assess the safety and efficacy of new pharmaceutical therapies. Indeed, vascular imaging data have already been accepted in support of regulatory approval for supplemental indications for statins to slow the progression of atherosclerosis [69].

19.3.1 Quantitative Coronary Angiography

The QCA utilizes an image-analysis techniques, largely independent on the subjective determinations of cardiologists, to measure the coronary lumen diameter and to calculate the percent stenosis on images acquired by classic X-ray contrast angiography performed during coronary catheterization.

The introduction of QCA in the 1980s allowed to study the correlation between the extent of coronary stenosis and established risk factors such as hypercholesterolemia and hypertension. Thus, it became possible to measure the rates of coronary atherosclerotic progression in human subjects and to evaluate the effect of therapies.

Several placebo-controlled, long-term clinical trials and the meta-analysis of Rossouw aimed to study the correlation of antiatherosclerotic effects of different therapies, including the hypocholesterolemic drugs, and the QCA measures demonstrated a robust parallelism among favorable changes in lipid levels, favorable changes in angiographic atherosclerosis, and reduction in the incidence of cardiovascular events [70].

However, a next meta-analysis of angiographic studies in patients with CAD and MI showed that a consistent percentage of subjects who experienced an MI had coronary artery stenoses of <50% luminal narrowing. Thus, the QCA, which better detects coronary artery stenoses >70%, under-represents the patients with acute MI [71].

Further, because atherosclerosis is primarily a disease of the vessel wall and not the vessel lumen and because the latter is the only part of the vessel visualized using contrast angiography, it is now established that QCA provides a very limited look at atherosclerosis burden. Thus, although QCA was once the gold standard for assessing the progression of atherosclerosis in clinical trials, its use has now reverted to clinical diagnosis alone [69].

19.3.2 Carotid B-Mode Ultrasound

The development of high-resolution ultrasound techniques made it possible to measure the vessel wall intima–media thickness and the lumen diameter along the axis of the ultrasound beam. Other advantages of medical ultrasound are the noninvasive nature, permitting serial measures of vessel structure, and the lack of risk of vascular injury or ionizing radiation.

Recently, measurement of cIMT by B-mode ultrasound came out as a gold standard of the quantitative research tool in the noninvasive study of atherosclerosis [72].

Several epidemiologic and prospective observational studies showed that quantitative measures of cIMT correlate with established cardiac risk factors, with cardiovascular and cerebrovascular events, and with the risk of developing cardiovascular events in asymptomatic subjects [69].

Several clinical trials have been conducted using cIMT as an endpoint to assess the efficacy of antiatherosclerotic therapies, some of which also included measures of cardiovascular outcome.

Recently, Espeland et al. and Duivenvoorden et al. reviewed the measure of cIMT as a surrogate of CVD [65, 73]. They could confirm that the decrease of carotid atherosclerosis progression caused by statins is consistent with the decrease in cardiovascular events as observed in clinical trials. This results support the use of atherosclerosis imaging, specifically the measure of cIMT by B-mode ultrasound, as a benchmarking tool for a well-informed decision whether to proceed to large morbidity and mortality studies in the assessment of a novel therapeutic strategy. Indeed, if a novel cardiovascular compound is added to a statin and does not show any beneficial effect or even promotes atherosclerosis progression, it is highly unlikely, if not impossible, that this compound will have beneficial effects on clinical endpoints and it will not be worthwhile embarking on large clinical outcome trials in a phase III cardiovascular program. However, when the novel compound shows clinically relevant regression of arterial wall thickness, benefit on clinical endpoints is likely, but still large clinical endpoint trials are required to definitely confirm the safety and effects on morbidity and mortality [69].

19.3.3 Coronary Intravascular Ultrasound

Coronary intravascular ultrasound (IVUS) is an innovative imaging modality that allows the acquisition of intravascular images of the vessel wall, using a miniaturized transducer attached to the tip of a catheter. The data obtained, analyzed either manually or semiautomatically, outline the intimal lining of the vessel lumen and the external elastic membrane that separates the media from the adventitia. The difference between the cross-sectional area (CSA) bordered by the external elastic membrane and the CSA of the vessel lumen represents the vessel wall or atheroma CSA. The atheroma volume can be calculated by summing the multiple vessel wall CSA slices along a vessel segment.

Because coronary IVUS is an invasive procedure performed during a cardiac catheterization, it involves a risk of complications, including vascular injury, thrombosis, dissection, cardiac arrhythmia, and infection. Furthermore, for ethical and practical reasons, it cannot be applied too frequently on the same patients.

The clinical relevance of IVUS imaging measures have been documented by correlations with data obtained by in vitro histopathology, by epidemiologic and prospective observational studies, and by pharmacological clinical trials [74, 75].

A prospective follow-up study of 56 individuals with known CVD in whom serial coronary IVUS studies were performed, showed a relationship between coronary risk factors and IVUS-documented progression of atherosclerosis in the left main coronary artery [76]. The 18 subjects who had adverse cardiovascular events during the follow-up had an average annual plaque progression greater than that of the other subjects ($25.2 \pm 19.5\%$ vs. $5.9 \pm 15.6\%$, $P < 0.001$). Moreover, the investigators demonstrated a positive linear relationships between plaque progression and the CVD risk ($r = 0.41$–0.60; $P < 0.002$–0.0001) calculated using three different cardiovascular risk scores to subjects, i.e., Framingham, PROCAM, and SCORE.

Although not as abundant as the cIMT data, accumulating evidence supports the use of coronary IVUS in the assessment of drugs developed to slow or revert the progression of atherosclerosis.

Several studies have used coronary IVUS to assess the efficacy of statins or other class of lipid-lowering drugs. The most recent studies have used the nominal change in the percent atheroma volume as the primary endpoint. Coronary IVUS, as a surrogate marker, is becoming a gold standard for documenting the effect of drugs on progression of atherosclerosis.

The discussions are now ongoing among academia, industry, and regulatory agency, and highlight the need for the application of new tools in drug development. Recently, the European Agency for the Evaluation of Medicinal Products (EMEA), in collaboration with the European Society of Cardiology, has developed a biomarker task force to address this issue, recognizing the opportunity to redesign the process of drug development [77].

Likewise, the US Food and Drug Administration (FDA) has put forward a critical path initiative to discuss the introduction of "innovation" path "directed toward improving the product development process itself by establishing new evaluation tools" (http://www.fda.gov/oc/initiatives/criticalpath/whitepaper.html).

19.3.4 Magnetic Resonance Imaging

MRI has been shown to be a useful tool for the noninvasive evaluation of atherosclerotic plaques in the aorta and carotid arteries. In humans, the MRI evaluation of the thoracic aorta proved to closely correlate with transesophageal echocardiography findings [78–80].

Lima et al. studied the reproducibility of plaque size measurements by the combined surface and transesophageal technique in ten patients with documented atherosclerosis who repeated MRI within 7 days of the initial study [81]. Cross-sectional plaque area measurements (2D) were closely correlated between studies 1 and 2, with an intraclass correlation coefficient of $R=0.91$ and a coefficient of variation of 23.9%. However, the most reproducible index of plaque size was 3D plaque volume, with an intraclass correlation coefficient of $R=0.97$ and a coefficient of variation of 4.8%. Intra-observer and inter-observer concordances were 0.91 and 0.81, respectively, for plaque volume. On the basis of these reproducibility studies, it has been calculated that changes in aortic plaque volume >4.6% can be considered as accurately measured by this MRI method [81].

19.4 Hypolipidemic Drugs

Hypolipidemic drugs are used to decrease the serum levels of LDL-C, and to increase the levels of HDL-C, thus to prevent two independent risk factors associated to atherosclerosis and cardiovascular mortality. To decrease the high levels of serum triglycerides may be also an objective of hypolipidemic drugs.

Actually, the major classes of hypolipidemic drugs introduced in the clinical practice are the HMG CoA reductase inhibitors (statins), the fibric acid derivatives (fibrates), the nicotinic acid, the bile acid sequestrants, and the selective inhibitors of cholesterol absorption from the intestine. Table 19.2 summarizes the major characteristics of statins, fibrates, nicotinic acid, and bile acid sequestrants.

The clinical use of hypolipidemic drugs is well established and should follow the recommendation of international guidelines periodically upgraded [8, 10, 82–85].

19.4.1 Statins

Statins are selective inhibitors of HMG CoA reductase, the rate-limiting step in cholesterol biosynthesis (Fig. 19.5) [86]. Inhibition of cholesterol synthesis reduces hepatic cholesterol content and upregulates LDL receptors, which lowers serum LDL-cholesterol levels. LDL receptors also remove small LDL particles, as well as IDL and VLDL remnants [87, 88].

The latter effect contributes to lowering of TGRLP by statins [88, 89].

Moreover, statins also apparently reduce the hepatic release of lipoproteins into the circulation, at least in part enhancing the removal of lipoproteins by LDL receptors within hepatocytes or in the space of Disse [90, 91].

Reductions in LDL cholesterol of 18–55% are observed, depending on the statin and the dose administered. The dose-effect correlation of each statin is log-linear; therefore, LDL-cholesterol levels fall by about 6% with each doubling of the dose [92, 93].

Statins usually increase the HDL-C levels by 5–10%, but greater increases can occur in persons with low HDL and elevated triglycerides. Conversely, triglyceride

Table 19.2 Major characteristics of currently available drugs affecting lipoprotein metabolism [8]

	Statins	Fibric acids	Nicotinic acid	Bile acid sequestrants
Agents (daily doses)	Lovastatin (20–80 mg) Pravastatin (20–40 mg) Simvastatin (20–80 mg) Fluvastatin (20–80 mg) Atorvastatin (10–80 mg) Rosuvastatin (5–40 mg)	Gemfibrozil (600 mg BID) Fenofibrate (200 mg) Clofibrate (1,000 mg BID)	Immediate release (1.5–3 g) Extended-sustained release (1–2 g)	Cholestyramine (4–16 g) Colestipol (5–20 g) Colesevelam (2.6–3.8 g)
Lipid/lipoprotein effects				
LDL	18–55%	5–20% *(may be increased in patients with high TG)*	5–25%	15–30%
HDL	5–15%	10–20%	15–35%	3–5%
TG	7--30%	20–50%	20–50%	No change or increase
Clinical trial				
Major outcomes	Reduced major coronary events, CHD deaths, need for coronary procedures, stroke, and total mortality	Reduced major coronary events	Reduced major coronary events, and possibly total mortality	Reduced major coronary events and CHD deaths
Side effects	Myopathy Increased liver enzymes	Dyspepsia Gallstones Myopathy Unexplained non-CHD deaths in WHO study	Flushing Hyperglycemia Hyperuricemia (or gout) Upper GI distress Hepatotoxicity	Gastrointestinal distress Constipation Decreased absorption of other drugs
Contraindications				
Absolute	Active or chronic liver disease	Severe renal disease	Chronic liver disease Severe gout	Dysbetalipoproteinemia TG > 400 mg/dL
Relative	Concomitant use of certain drugs[a]	Severe hepatic disease	Diabetes Hyperuricemia Peptic ulcer disease	TG > 200 mg/dL

LDL low-density lipoprotein, *HDL* high-density lipoprotein, *TG* triglyceride

[a]Cyclosporine, macrolide antibiotics, various antifungal agents, and cytochrome P-450 inhibitors (fibrates and niacin should be used with appropriate caution)

levels generally decrease by 7–30% with the statins. When triglyceride levels are >200 mg/dL, triglycerides fall in direct proportion to LDL-cholesterol lowering. However, with triglyceride levels of <150 mg/dL, triglyceride responses are inconsistent, and, with very high triglyceride levels the LDL-C lowering is less than the proportional [93–99].

The statins are the most effective and practical class of drugs for reducing LDL-C concentrations and cardiovascular risk. Clinical trials in patients with various LDL-C serum levels and with and without CHD have shown consistently that statins reduce the relative risk of major coronary events by ≈30% and produce a greater absolute benefit in patients with higher baseline risk [100].

Results from five long-term clinical trials have documented a decrease in CHD and total mortality, reductions in myocardial infarctions, revascularization procedures, stroke, and peripheral vascular disease (Table 19.3) [94–99]. These trials documented benefits in men and women, in middle-aged and older persons, and in primary and secondary prevention.

19.4.1.1 Secondary Prevention Studies

The Scandinavian Simvastatin Survival Study (4S) [97], The Cholesterol And Recurrent Events (CARE) study [94], and The Long-term Intervention with Pravastatin in Ischemic Disease (LIPID) study [95] showed significant reductions in recurrent myocardial infarction and coronary death, coronary artery procedures, and stroke. Moreover, two of the trials (4S and LIPID) reported a reduction in total mortality with statin therapy.

4S study showed the reduction in coronary events in both men and women, in individuals younger and older than 60 years of age, and in subjects with other risk factors, including smoking, hypertension, and diabetes [101]. In addition, a 30% reduction in cerebrovascular events was reported in 4S.

CARE study extended the findings of 4S to individuals with average cholesterol levels. However, post hoc analysis revealed no reduction in coronary events among patients in the treatment group with baseline LDL levels below 125 mg/dL. Diabetic subjects treated with pravastatin ($n=5{,}282$) had a 25% reduction in major coronary events compared with placebo. LIPID study also included subjects with unstable angina and showed a 24% reduction of CHD deaths in the pravastatin group.

Several other types of clinical trials with statin therapy also showed favorable results [102, 103].

Statin therapy reduces the risk of essentially every clinical manifestation of the atherosclerotic process.

Several clinical trials have examined the effect of treatment to lower LDL-cholesterol goals and earlier treatment of patients. No single trial conclusively confirms a specific LDL-cholesterol goal lower than 100 mg/dL. However, several studies (MARS [104]; LAARS [105]; AVERT [106, 107]; MIRACL [108]) showed a clinical benefit in the treatment group with on-treatment LDL-C from 72 to

Table 19.3 Clinical outcome studies using statins

Study	Patients (% F)	Duration	Drug (dose/day)	Baseline LDL (mg/dL)	LDL-C Change (%)	Major coronary Events (%)	CABG and PTCA (%)	Coronary mortality (%)	Total mortality (%)	Stroke (%)	NNT
Secondary prevention trials											
4S [97]	4,444 (19)	5.4 years	Simvastatin 20–40 mg/day	188	−35*	−35*	−37*	−42*	−30*	−27*	15
CARE [94]	4,159 (14)	5 years	Pravastatin 40 mg/day	139	−27*	−25*	−27*	−24*	−9	−31*	33
LIPID [95]	9,014 (17)	5 years	Pravastatin 40 mg/day	150	−25*	−29*	−24*	−24*	−22*	−19*	28
Primary prevention trials											
WOSCOPS [99]	6,595 (0)	4.9 years	Pravastatin 40 mg/day	192	−26*	−31*	−37*	−33	−22*	–	42
AFCAPS/ TexCAPS [96]	6,605 (15)	5 years	Lovastatin 20–40 mg/day	150	−25*	−37*	−33	NS	NS	–	24

NNT indicates number needed to treat to prevent one major coronary event (100/absolute risk reduction)
*Statistically significant changes at $P<0.05$ or lower

98 mg/dL. All these studies indicate that further benefit accrues in patients treated to an LDL-C level below 100 mg/dL.

AVERT [106] and MIRACL [108] trials, measuring both clinical endpoints and other endpoints, such as vascular function, showed an early (1 week–3 months) benefit of statin treatment for patients with atherosclerosis or acute coronary syndromes. MIRACL study, including patients with non-Q MI or unstable angina, demonstrated that statin treatment initiated in hospital was associated with a 16% relative risk reduction at 16 weeks and was safe. Accordingly, a very large observational study from Sweden also supports the concept of early in-hospital initiation of statin treatment that has been shown to be associated with an adjusted 25% lowering of total mortality at 1 year [109].

19.4.1.2 Primary Prevention Studies

The availability of statins made it possible to definitively test whether LDL lowering would reduce CHD risk. Both the WOSCOPS and the AFCAPS/TexCAPS trials showed that statin therapy significantly reduced the relative risk for major coronary events (Table 19.3) [96, 99].

WOSCOPS found a significant reduction in the primary end point of coronary death and nonfatal myocardial infarction after 5 years [99]. Because of the lower baseline risk in the WOSCOPS population, the number needed to treat (NNT) to prevent one major coronary event was higher (NNT=42) than that found in 4S (NNT=15), CARE (NNT=33), or LIPID (NNT=28). However, considering the subgroup of high-risk individuals (0.2% event rate per year), including those younger than 55 years of age and with vascular disease, smoking, or minor ECG abnormalities, or older hypercholesterolemic individuals with any additional risk factor, the NNT would have been reduced from 42 to 17 [110].

Because of the very small number of deaths in both placebo and treatment groups, AFCAPS/TexCAPS could not conclude about the effects of cholesterol-lowering therapy on total mortality. AFCAPS/TexCAPS found that lovastatin prevented first acute major coronary events in men and women with average LDL levels and low HDL levels [96]. Individuals with HDL<40 mg/dL benefited the most. This study raises the question of whether the National Cholesterol Education Program (NCEP) should recommend statin treatment in patients whose risk profile includes HDL<40 mg/dL.

19.4.1.3 Angiographic Trials

Statins slow the progression and induce the regression of coronary atherosclerosis, reduce the formation of new lesions, and reduce the incidence of coronary events [111].

Several prevention trials included both angiographic outcomes and clinical endpoints. The absolute change in arterial narrowing was relatively small, but the frequency of cardiovascular events observed in the treated group in a period of only

2 years decreased substantially in most of these studies. This apparent disparity between the small degree of angiographic change and the relatively large differences in clinical event rates led to the concept of plaque stabilization [112].

Three angiographic studies are particularly noteworthy. POSCH (using surgery) and FATS (using nicotinic acid and a statin or sequestrant) studies achieved LDL levels near 100 mg/dL and showed favorable changes in coronary lesions [113, 114]. The Post-CABG trial, by examining the benefits of moderate vs. aggressive LDL lowering on progression of atherosclerosis in saphenous vein grafts, tested the concept that a lower LDL is better [103]. Using a statin and a sequestrant when needed, the moderate treatment group was treated to maintain LDL levels between 130 and 140 mg/dL, while the aggressive treatment group was titrated to a target LDL of <95 mg/dL. The aggressively treated group had less progression, fewer new lesions, and needed less revascularization.

19.4.1.4 Ultrasound Trials: cIMT Biomarkers

Trials that used serial B-mode carotid ultrasound in patients with and without CHD showed that statins slow progression and induce regression of carotid atherosclerosis [115, 116].

Espeland et al. reviewed the measure of cIMT as a surrogate of CVD, and conducted a meta-analysis that included seven statin trials [65, 115, 117–122]. The main results are summarized in Table 19.4. In this analysis, a −0.012 mm/year change in cIMT (95% CI, −0.015 to −0.007) was associated with an odds ratio of 0.48 (95% CI, 0.30–0.78) for cardiovascular events.

Serial cIMT measurements, as a marker of CVD progression and/or regression, were used to study the effect of various interventions on CVD, as summarized in Table 19.5 [123].

19.4.1.5 Coronary Intravascular Ultrasound Studies

The main results of representative coronary IVUS studies assessing therapies directed at slowing the progression of atherosclerosis are summarized in Table 19.6. A consistent relationship between the on-treatment level of LDL-C, an accepted surrogate marker of cardiovascular risk, and the nominal change in percent atheroma volume as measured by IVUS has been shown by these studies.

Most of the IVUS studies on statins have used active controls, and their dimension was tailored to detect treatment differences in the progression of atherosclerosis but not in the frequency of cardiovascular endpoints. Even so, there is evidence that ultrasound measures of atherosclerosis, based on imaging of the arterial wall and quantification of either vessel wall thickness (VWT) or cross-sectional vessel wall area (VWA), have meaningful clinical relevance. A cross-study analysis of three clinical trials in subjects with known CAD comparing the same therapies, i.e., pravastatin (40 mg/day) and atorvastatin (80 mg/day), provides additional and

Table 19.4 Clinical trials involving statins using both cIMT and cardiovascular event outcomes (adapted from [65])

Clinical trial	No. of patients[a]	Statin	Relative impact on IMT progression of primary outcome[b] (mm/year)	Relative impact on reported cardiovascular endpoints	
				Abstracted CVD events	Odds ratio (95% CI)
ACAPS [115]	919	Lovastatin	–0.015 (–0.023 to –0.007) ($P=0.001$)	CVD death, MI, stroke	0.34 (0.12–0.69)
KAPS [118]	447	Pravastatin	–0.014 (–0.022 to –0.006) ($P=0.005$)	CVD death, MI, stroke	0.57 (0.22–1.47)
PLAC-II [117]	151	Pravastatin	–0.009 (–0.031 to –0.013) ($P=0.44$)	Clinical coronary events	0.37 (0.11–1.24)
CAIUS [119]	305	Pravastatin	–0.014 (–0.021 to –0.005) ($P=0.0007$)	CVD death, MI, stroke	1.02 (0.14–7.33)
REGRESS [120]	255	Pravastatin	–0.030 (–0.056 to –0.004) ($P=0.002$)	Clinical events	0.51 (0.24–1.07)
BCAPS [121]	793	Fluvastatin	–0.008 (–0.013 to –0.003) ($P=0.002$)	CVD death, MI, stroke	0.64 (–0.24 to 1.66)
FAST [122]	164	Pravastatin	Significant benefit ($P<0.001$)	CVD death, MI	0.32 (0.10–1.06)
Pooled estimate			–0.012 (–0.016 to –0.007)[c]		0.48 (0.30–0.78)

[a] Arms used in meta-analyses
[b] Data are means (95% CI) (reported P-value)
[c] Excludes FAST

Table 19.5 Baseline and maximum cIMT change in various studies evaluating cIMT as end point [123]

Study	Intervention	Duration	No. of patients	Mean SDBaseline cIMT (mm)	Δ cIMT (% of baseline)	Study quality[a]
Nolting et al. [124]	Simvastatin (80 mg)	2 years	153	0.92	−5.4*	1
Kastelein et al. [125]	Simvastatin/ezetimibe (80/10 mg) vs. simvastatin (80 mg)	24 months	720	0.7 ± 0.2	+1.2	5
Smilde et al. [126]	Atorvastatin (40 mg) vs. simvastatin (20 mg)	2 years	330	0.93 ± 0.2	−3.3e*	5
Taylor et al. [127]	Atorvastatin (80 mg) vs. pravastatin (40 mg)	12 months	138	0.94 ± 0.65	−14.7*	4
Crouse et al. [128]	Rosuvastatin (40 mg) vs. placebo	24 months	738	1.16 ± 0.2	−0.24*	4

*Significant change ($P<0.05$) from baseline

[a] Quality of studies assessed using the criteria outlined by Jadad et al. [129]

Table 19.6 Representative coronary IVUS studies assessing statin therapies directed at slowing the progression of atherosclerosis

Trial	No	Years	Test/control	% Δ LDL-C (T)	% Δ HDL-C (T)	IVUS endpoint	Δ IVUS T/C (*P*)	Events T/C
ASTEROID [130]	507	2	Rosuvastatin 40 mg/ baseline	–53	+15	Δ PAV	–6.8/baseline	N.A.
REVERSAL [131]	502	1.5	Atorvastatin 80 mg/ pravastatin 40 mg	–46	+2.9	% Δ TAV	–0.4/+2.7 ($P=0.02$)	6/9
GAIN	131	1	Atorvastatin to target/usual care	–42	+9	% Δ TAV	+2.5/+11.8 ($P=0.138$)	2/2
ESTABLISH [132]	70	0.5	Atorvastatin 20 mg/usual care	–42	+2.4	% Δ TAV	–13.1/+8.7 ($P=0.0001$)	0/0
PTCA-Pravastatin [133]	25	3	Pravastatin 10/diet	–26	+29	% Δ PI	–7/+27 ($P=0.0005$)	N.A.

PAV percent atheroma volume, *TAV* total atheroma volume, *PI* plaque index, *N.A.* not applicable, *N.S.* not significant, *PTCA* percutaneous transluminal coronary angioplasty, *Apo A* apolipoprotein A, *IVUS* intravascular ultrasound, *T* test, *C* control substantial support for this conclusion. ARBITER [127] and REVERSAL [131] studies, using cIMT and coronary IVUS, respectively, to assess the progression of coronary atherosclerosis, showed a significant greater effect in patients treated with atorvastatin (80 mg/day), compared with patients treated with pravastatin (40 mg/day). Accordingly, PROVE-IT TIMI-22 trial, a morbidity and mortality study in patients with acute coronary syndrome, showed a significant reduction in coronary events in the patients treated with the more potent dose of statin [134].

19.4.1.6 Magnetic Resonance Imaging Studies

Using MRI, several authors demonstrated the plaque regression in human aortas in response to lipid-lowering therapy [135, 136].

Corti et al. studied the effects of simvastatin on a total of 44 aortic and 32 carotid artery plaques detected in 21 asymptomatic hypercholesterolemic patients at baseline. The effects of statin on these atherosclerotic lesions were evaluated as changes vs. baseline in lumen area (LA), VWT, and VWA by MRI [137].

Maximal reduction of plasma total and LDL cholesterol by simvastatin (23 and 38%, respectively; $P<0.01$ vs. baseline) was achieved after about 6 weeks of therapy and maintained thereafter throughout the study.

Significant reductions in maximal VWT and VWA (10 and 11% for aortic and 8 and 11% for carotid plaques, respectively, $P<0.01$), without changes in LA, have been reported at 12 months. Further decreases in VWT and VWA ranging from 12 to 20% were observed at 18 and 24 months.

A slight but significant increase (ranging from 4 to 6%) in LA was seen in both carotid and aortic lesions at these later time points. The study showed that long-term lipid-lowering therapy with simvastatin is associated with sustained vascular remodeling and significant regression of established atherosclerotic lesions in humans [137].

Similarly, by using combined surface/transesophageal MRI, Lima et al. detected significant atherosclerotic plaque regression and reverse remodeling in the thoracic aorta of 27 patients after only 6 months of simvastatin treatment (20–80 mg daily). The MRI outcomes were strongly associated with LDL-cholesterol reduction [81].

Recently, Yonemura et al. investigated the long-term effect of 20 vs. 5-mg atorvastatin on thoracic and abdominal plaques and the association between plaque progression and on-treatment LDL-cholesterol levels in 36 hypercholesterolemic patients [138]. MRI was performed at baseline and 1 and 2 years of treatment.

The 20-mg dose of atorvastatin markedly reduced LDL-cholesterol levels (–47%) vs. 5-mg (–35%) dose. After 2 years of treatment, regression of thoracic plaques was found in the 20-mg group (–15% VWA reduction), but not in the 5-mg group (+7%). Moreover, although the 20-mg dose induced plaque regression (–14%) from baseline to 1 year, no further significant regression was seen from 1 to 2 years of treatment (–1%). Progression of abdominal plaques was found in the 5-mg group (+10%), but not in the 20-mg group (+2%). Plaque progression in the 5-mg group was found from baseline to 1 year (+8%), but not from 1 to 2 years (+2%).

The degree of thoracic plaque regression correlated with LDL-cholesterol reduction ($r=0.61$), whereas thoracic plaque change from 1 to 2 years correlated with on-treatment LDL-cholesterol levels ($r=0.64$) [138].

19.4.1.7 Pleiotropic Effects of Statins

Recent experimental and clinical studies suggest that statins may exert vascular protective effects beyond cholesterol reduction. The cholesterol-independent or pleiotropic effects of statins appear to impact on several areas of the atherogenic process, including endothelial function, oxidative stress, inflammation, plaque stability, and thrombogenic response [139].

Statins inhibit cholesterol synthesis at an early metabolic step limiting the availability of mevalonate, therefore affecting not only synthesis of cholesterol, bile acids, steroid hormones, and vitamin D, but also the isoprene pathway downstream from mevalonate.

The cholesterol-independent or "pleiotropic effects" of statins usually are correlated to biochemical mechanisms modulated through the isoprene pathway, which is important for the synthesis of dolichol (glycosylation), ubiquinone (mitochondrial respiration), heme-A (oxygen transport), transfer RNA, and for protein prenylation. A large proportion of the pleiotropic effects of statins appears to be due to the inhibition of the synthesis of isoprenoids such as farnesylpyrophosphate and geranylgeranylpyrophosphate, which are important post-translational lipid attachments for intracellular signaling molecules (Fig. 19.6) [140, 141].

Focusing on the isoprene pathway has revealed that the multi-system effects of statins may go through common pathways such as Rho and Rac prenylation. Decrease in Rho GTPase responses as a consequence of statin treatment increases the production and bioavailability of endothelium-derived NO. The mechanism involves, in part, Rho/Rho-kinase (ROCK)-mediated changes in the actin cytoskeleton, which leads to decreases in eNOS mRNA stability. The regulation of eNOS by Rho GTPases may be an important mechanism underlying the cardiovascular protective effect of statins [142].

Early cardiovascular effects of statins, especially those mediated by endothelial NO release and resulting in improved endothelial function, may be dissociated from the LDL effect and appear more and more to show pleiotropic properties [143]. The discovery of anti-inflammatory properties correlated to the nonpharmacophore moiety of statins which directly interacts with an integrin site revealed that this part of the structure is far from being inert and has contributed to modify the notion that all pleiotropic effects had to be linked with cholesterol synthesis inhibition [144]. The emergence of new biomarkers of CAD risk, such as CRP, and the evidence that some of them may predict outcome without correlating with LDL levels has enticed researchers to explore new atherogenic mechanism and their relationship with the pleiotropic effects of statins [145].

Exploratory clinical research in areas where statins were not anticipated to be effective have revealed novel efficacy on CVD such as arrhythmias [146, 147],

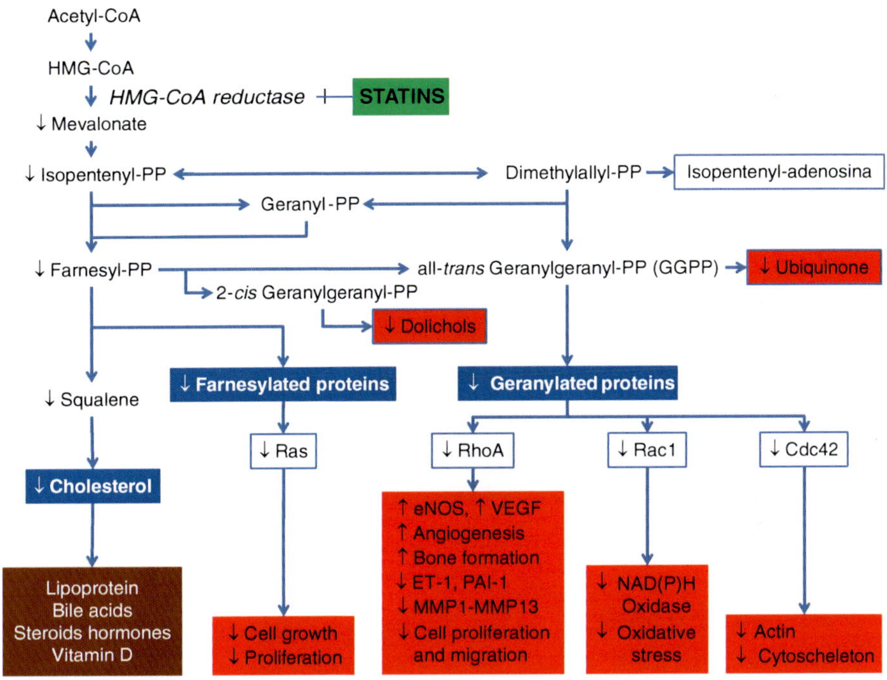

Fig. 19.6 Cholesterol biosynthesis pathway. Inhibition of hydroxy methylglutaryl coenzyme A (HMG CoA) reductase by statins reduces the synthesis of mevalonate, cholesterol, and isoprenoids. Reducing these metabolites with statins leads to subsequent changes in bioactive proteins, including the activation of the GTP-binding proteins rho and rac. *Bmp-2* bone morphogenetic protein-2, *eNOS* endothelial nitric oxide synthase, *t-PA* tissue-type plasminogen activator, *ET-1* endothelin-1, *PAI-1* plasminogen activator inhibitor-1 (modified from [139])

congestive heart failure [148, 149], stroke [150, 151], and diabetes [152], therefore stimulating further investigation into pleiotropic effects. Moreover, many new and unanticipated effects of statins have emerged which may provide new and exciting applications beyond the cardiovascular field in conditions as diverse as Alzheimer's disease [153, 154], arthritis [155], bone repair [156], antiphospholipid antibodies syndrome [157], multiple myeloma [158], and multiple sclerosis [159].

The mechanisms of the statin pleiotropic effects and their potential benefits on atherogenesis have been extensively reviewed and are summarized in Table 19.7. The knowledge regarding pleiotropic effects of statins and atherosclerosis risk reduction has dramatically improved. Nevertheless, there is still much research to be done in humans to assess the relative magnitude of what is accounted for by a reduction in cholesterol synthesis and what is due to alteration of the isoprene pathway. The full impact of statin therapy on each of the atherogenesis processes is not fully understood, and the existence of pleiotropic effects related to statin therapy has become a topic of much debate, as lipid lowering itself improves many aspects of the atherogenic process. Indeed, the meta-analysis conducted by Gould et al. showed

Table 19.7 Summary of the pleiotropic effects and potential benefits of statin therapy [139]

Effect	Benefit
Increased eNOS expression and synthesis of nitric oxide Decreased synthesis of endothelin-1 Inhibition of LDL-C oxidation Reduced oxygen free-radical scavenging Increased DAF expression on ECs Increased PI 3-kinase/Akt activity	Improvement of endothelial dysfunction, antioxidant effect, and vascular cytoprotection
Reduced number and activity of inflammatory cells Reduced leukocyte-EC adhesion Reduced levels of C-reactive protein Reduced natural killer and T-cell activation Reduced monocyte activation Increased γ-interferon production Decreased inhibition of LFA-1 TH1-to-TH2 shift	Reduced inflammatory response and immunomodulation
Reduced macrophage cholesterol accumulation Reduced production of metalloproteinases Increased collagen synthesis Increased smooth muscle content	Stabilization of atherosclerotic plaques
Inhibition of platelet adhesion/aggregation Reduced fibrinogen concentration Reduced blood viscosity	Reduced thrombogenic response

DAF decay-accelerating factor, *EC* endothelial cell, *LFA-1* leukocyte function-associated antigen-1, *TH* T helper

that the slope of the relationship between cholesterol reduction and mortality risk reduction was the same for statins and nonstatins, while the mortalitory risk reduction correlated to statin long-term treatment was found to be a consequence of cholesterol reduction alone [160].

Although these data suggest statins may not provide benefit beyond lipid lowering, results from other studies support the existence of pleiotropic effects. Indeed, in WOSCOPS, 4S, and HPS trials, the clinical benefit associated with statin therapy was independent of baseline LDL-C level [139].

Evidence in support of the pleiotropic effects of statins also comes from their effects on stroke incidence. Indeed, while a review of 45 prospective observational cohorts involving 450,000 individuals failed to find an association between stroke and blood cholesterol level [161], a meta-analyses of clinical trials showed a clear association between statin therapy and reduced cerebrovascular events [139, 150]. Moreover, also the reduction in stroke incidence reported during HPS was independent of baseline cholesterol level. Thus, statins may reduce stroke incidence by mechanisms independent of cholesterol lowering. It has been suggested that statins may positively influence blood pressure, reducing a significant risk factor for the development of stroke. Statins have been shown to positively ameliorate angiotensin II-induced hypertension, left ventricular hypertrophy, fibrosis, and macrophage infiltration in rats [162]. Statins also downregulate angiotensin II type 1 receptor expression in platelets and vascular smooth muscle and endothelial cells. Moreover, several small clinical trials have shown that statins may have blood pressure-lowering effect [163]. However, in these studies, patients with both essential hypertension and hypercholesterolemia were enrolled, and several population studies have shown a significant association between hypertension and dyslipidemia. Thus, it is uncertain whether the antihypertensive action of statins must be ascribed to pleiotropic mechanisms.

In summary, current developments in pharmacogenomics, bioinformatics, signal transduction analysis, and imaging technologies coupled with the explosion of information on biological markers of atherosclerosis are enlightening the relevance of pleiotropic effects of statins, whether beneficial, neutral, or adverse [164].

19.4.2 Fibrates

The fact that treatment with statins does not prevent nearly 70% of cardiovascular events despite their consistent and significant effect on reducing the LDL-C levels, and the epidemiologic evidence that HDL-C levels are inversely correlated to cardiovascular risk, has made therapies to increase HDL-C levels (fibrates and nicotinic acid), an increasingly important option to combat CAD [165].

Gemfibrozil, fenofibrate, and clofibrate are the three most important fibrates currently available.

The mechanism of action of the fibrates is complex and not fully explained. Recent findings showed that fibrates are agonists for the nuclear transcription factor *peroxisome proliferator-activated receptor-alpha (PPARα).* Through this mechanism, fibrates downregulate the apolipoprotein C-III gene and upregulate genes for apolipoprotein A-I, fatty acid transport protein, fatty acid oxidation, and possibly lipoprotein lipase. The effects of fibrates on lipoprotein lipase and apolipoprotein C-III (an inhibitor of lipoprotein lipase) enhance the catabolism of TGRLP, whereas increased fatty acid oxidation reduces the formation of VLDL triglycerides. These effects account for serum triglyceride lowering, which is the major action of fibrates. Serum triglyceride lowering combined with increased synthesis of apolipoprotein A-I and A-II tend to raise HDL-cholesterol levels. Triglyceride lowering also transforms small, dense LDL into normal-sized LDL. The effect of PPAR activity on other atherogenic mechanisms is now being evaluated [166–171].

The use of fibrates typically reduces triglyceride levels by 25–50% and raises HDL-C by 10–15%; the greater effects generally occur in persons with very high triglyceride levels and very low HDL-cholesterol levels. The LDL-cholesterol-lowering effects of gemfibrozil and clofibrate are generally in the range of 10% or less in persons with primary hypercholesterolemia. Fenofibrate is more effective and frequently reduces LDL-cholesterol levels by 15–20% when triglycerides are not elevated. Fibrate therapy induces only slight changes in LDL cholesterol in persons with combined hyperlipidemia, while generally rises LDL-cholesterol levels in persons with hypertriglyceridemia [172–174].

Thus, fibrates primarily target atherogenic dyslipidemia. In addition, the ability of fibrates to lower triglycerides has led to their wide usage in persons having very high triglyceride levels and chylomicronemia. The purpose of fibrate therapy in such persons is to reduce the risk for acute pancreatitis. Their value for this purpose is well recognized. Finally, fibrates are highly effective for reducing β-VLDL concentrations in persons with dysbetalipoproteinemia [173].

The major primary and secondary prevention trials with fibrates are summarized in Table 19.8.

Long-term treatment with clofibrate and gemfibrozil reduced the risk of fatal and nonfatal myocardial infarction in two large primary prevention trials (WHO Clofibrate trial and Helsinki Heart Study) [175, 176], and gemfibrozil therapy reduced CHD death and nonfatal myocardial infarction and stroke in a recently reported secondary prevention trial (Veteran Administration–HDL Intervention Trial) [181]. However, this beneficial effect on cardiovascular outcomes has not been observed in all large fibrate trials [177].

WHO Clofibrate Study showed a 25% reduction of nonfatal myocardial infarction, but a significantly higher total mortality in the clofibrate group, due to an increase in non-CHD deaths [175].

Therefore, the use of clofibrate in general medical practice decreased markedly after this study.

Luckily, the safety profile of gemfibrozil ensued better. The Helsinki Heart Study demonstrated no change in total mortality and a 37% reduction in fatal and nonfatal myocardial infarctions during the course of the study. After 8.5–10 years of follow-up, the increase in noncardiac deaths and all-cause mortality observed in gemfibrozil group was not statistically significant [183]. Moreover, after 10 years of follow-up, no difference in cancer rates was observed between patients who had received gemfibrozil or placebo.

Early secondary prevention trials with clofibrate therapy gave suggestive evidence of CHD risk reduction. However, in the Coronary Drug Project, clofibrate therapy failed to significantly reduce the risk for CHD [177], and, likewise, in the BIP trial, bezafibrate therapy did not significantly reduced recurrent major coronary events in persons with established CHD [182].

In contrast, gemfibrozil therapy in the VA-HIT 48 trial reduced the risk for CHD death and nonfatal myocardial infarction by 22%, and also significantly reduced the stroke rates [181]. In both the VA-HIT and BIP studies, no increased risk of non-CHD mortality has been reported.

Thus, taken as a whole, clinical trials of fibrate therapy strongly suggest a reduction in CHD incidence, although results are less robust than with statin therapy. Furthermore, a reduction in total mortality has not been demonstrated with fibrate therapy.

The VA-HIT trial revealed that modification of other lipid risk factors could reduce the risk for CHD when LDL cholesterol is in the range of 100–129 mg/dL. Gemfibrozil therapy, which raised HDL and lowered triglyceride, significantly reduced the primary endpoint of fatal and nonfatal myocardial infarction without lowering LDL-cholesterol levels. This study thus raises the possibility of efficacy from optional use of nonstatin drugs when LDL-cholesterol levels in CHD patients are in the range of 100–129 mg/dL [181].

Table 19.8 Prevention clinical trials with CHD endpoints using fibrates that modify triglyceride-rich lipoproteins

Study/drug	Patients (sex)	Duration	Lipid and lipoprotein values (baseline or placebo and on-treatment) Group	TC	TG	Non-HDL-C	HDL-C	% Change in coronary event rate (drug vs. placebo)	*P*
Primary prevention									
WHO trial [175]Clofibrate	15,745 (M) (subset=4,935)	5 years	Placebo	257	210	–	–	–20	=0.05
			On-treatment	229	160	–	–		
Helsinki Heart Study [176]Gemfibrozil	4,081 (M)	5 years	Baseline	289	175	242	47	–34	<0.02
			On-treatment	247	115	196	51		
Secondary prevention									
Coronary Drug Project [177]Clofibrate	1,103 (M)+2,789 (M) placebo	5 years	Baseline	250	177	–	–	–5	NS
			On-treatment	234	149	–	–		
Newcastle Trial [178]Clofibrate	400 (M)97 (F)	5 years	Baseline	245	337	–	–	–49	<0.01
			On-treatment	217	–	–	–		
			Baseline	270					
			On-treatment	229	–				
Scottish Trial [179] Clofibrate	593 (M)124 (F)	6 years	Baseline	264	–	–	–	–44	NS
			On-treatment	229	–	–	–		
			Baseline	280	–	–	–		
			On-treatment	228	–	–	–		
Stockholm Study [180] Clofibrate+Nicotinic ac.	219 (M)60 (F)	5 years	Baseline	251	208	203	48	–35	<0.01
			On-treatment	218	166	–	–		
VA-HIT Trial [181]Gemfibrozil	2,531 (M)	5 years	Baseline	175	161	143	32	–22	<0.006
			On-treatment	170	115	136	34		
BIP [182]Bezafibrate	2,825 (M)265 (F)		Baseline	212	145	177	35	–9.4	=0.26
			On-treatment	202	115	161	41		

The fibrates are generally well-tolerated in most persons. Gastrointestinal complaints are the most common side effects. All fibrates appear to increase the lithogenicity of bile, increasing the likelihood of cholesterol gallstones. The combination of a fibrate with a statin increases the risk for myopathy, which can lead to rhabdomyolysis. None of well-established side effects can account for the increased total mortality observed in the WHO clofibrate study.

19.4.3 Nicotinic Acid

Nicotinic acid or niacin given in pharmacological doses lowers serum total and LDL-cholesterol and triglyceride levels and raises HDL-cholesterol levels.

The mechanism of action of nicotinic acid is complex and multifactorial. New concepts regarding its mechanism of action have emerged only relatively recently with the discovery and identification of cellular receptors for niacin and their potential involvement in pharmacologic actions of niacin [184].

Current evidence indicates that niacin acts on multiple tissues and targets to beneficially modulate lipid/lipoprotein profile, anti-inflammatory processes, and adverse flushing reactions (Table 19.9).

Pharmacologic doses of niacin reduce total cholesterol, triglycerides, VLDL, LDL, and Lp(a), and increase HDL levels. In addition, niacin increases larger HDL2 subfractions and decreases atherogenic small, dense LDL particles. It decreases apolipoprotein (apo) B-containing particles while elevating apo A-I particles. Furthermore, niacin selectively increases apoA-I-containing HDL particles, a cardioprotective subfraction and efficient mediator of the reverse cholesterol transport pathway.

The liver appears to be the major target organ of niacin to increase HDL–apoA-I and decrease TGs and VLDL/LDL particles.

In addition to its lipid-regulating role, new data are emerging to indicate vascular anti-inflammatory and antioxidative properties of niacin that can shed light on the pleiotropic role of this drug in reducing atherosclerosis.

New evidence indicates that niacin acts on GPR109A (HM74A) and GPR109B (HM74) receptors expressed in adipocytes and immune cells. In adipocytes, GPR109A activation reduces triglyceride (TG) lipolysis, resulting in decreased free fatty acid (FFA) mobilization to the liver. However, this mechanism has yet to be confirmed in humans because the plasma FFA decrease is transient and is followed by a rebound increase in FFA levels. Furthermore, recently niacin has been shown to inhibit diacylglycerol acyltransferase 2 (DGAT2) isolated from human hepatocytes, resulting in accelerated hepatic apolipoprotein (apo)B degradation and decreased apoB secretion, thus explaining reductions in VLDL and LDL. Kinetic and in vitro studies indicate that niacin retards the hepatic catabolism of apoA-I but not liver scavenger receptor B1-mediated cholesterol esters, suggesting that niacin inhibits hepatic holoparticle HDL removal. Indeed, recent preliminary evidence suggests that niacin decreases surface expression of hepatic β-chain of adenosine triphosphate synthase, which has been implicated in apoA-I/HDL holoparticle catabolism.

Table 19.9 Nicotinic acid (niacin) affects multiple tissue enzymes and receptors [184]

Organ/tissue	Cell	Target enzyme/receptor	Primary effect	Clinical effect
Adipose tissue	Adipocyte	↑ GPR109A (HM74A)	↓ Lipolysis ↓ FFA mobilization	↓ VLDL–TG
Liver	Hepatocyte	↓ DGAT2	↓ TG synthesis ↓ ApoB secretion	↓ VLDL–TG ↓ ApoB ↑ LDL size ↓ Lp(a)
		↓ β-Chain ATP synthase	↓ HDL (apoA-I) catabolism	↑ Lp–A-1, HDL2
Artery	Endothelium	↑ NAD(P)H ↓ Redox-sensitive genes	↓ LDL oxidation ↓ MCP-1 ↓ VCAM-1	↓ Vascular inflammation
	Macrophage	Unknown	↑ PGJ2 ↑ PPARγ, ABCA1	↑ Cholesterol efflux ↑ HDL-C
Skin	Langerhans cell/ macrophage	GPR109A	↑ PGD2 and PGE2, acting on DP1, EP2, and EP4 receptors	Flushing

ABCA1 ATP-binding cassette A1, *Apo* apolipoprotein, *ATP* adenosine triphosphate, *DGAT2* diacylglycerol acyltransferase 2, *FFA* free fatty acid, *HDL* high-density lipoprotein, *HDL-C* HDL cholesterol, *LDL* low-density lipoprotein, *Lp(a)* lipoprotein(a), *MCP-1* monocyte chemotactic protein-1, *NAD(P)H* reduced nicotinamide adenine dinucleotide phosphate, *PG* prostaglandin, *PPAR* peroxisome proliferator-activated receptor, *TG* triglyceride, *VCAM-1* vascular cell adhesion molecule-1, *VLDL* very low-density lipoprotein, ↑ increased, ↓ decreased

Niacin is the most potent available lipid-regulating agent to increase HDL levels [185]. The increases in HDL cholesterol are generally in the range of 15–30%, but increases of 40% have been noted with very high doses [186, 187].

Nicotinic acid typically reduces triglyceride levels by 20–35%, but reductions of 50% have been noted with high doses in hypertriglyceridemic persons.

The changes in HDL cholesterol and triglyceride concentrations tend to be curvilinear (log-linear); thus, smaller doses of nicotinic acid still produce significant increases in HDL or reductions in triglyceride with fewer side effects.

Although smaller doses often increase HDL-cholesterol levels, doses of 2–3 g/day are generally required to produce LDL-cholesterol reductions of 15% or greater. Indeed, in one carefully controlled study in patients with hypercholesterolemia, reductions in LDL cholesterol of 5, 16, and 23% were observed with daily doses of 1.5, 3.0, and 4.5 g, respectively [188].

Nicotinic acid can also lower Lp(a) up to 30% with high doses. Whether Lp(a) lowering by nicotinic acid therapy reduces the risk for CHD is not known.

Among lipid-lowering agents, nicotinic acid appears to be most effective for favorably modifying all the lipoprotein abnormalities associated with atherogenic dyslipidemia.

Several clinical trials (secondary prevention and angiographic studies) indicate that treatment with niacin, alone or in combination with other lipid-lowering agents, significantly reduces total mortality and coronary events, retards the progression, and induces regression of coronary atherosclerosis.

The Coronary Drug Project trial showed that nicotinic acid reduces the risk of recurrent myocardial infarction [177]. Total mortality was decreased in a 15-year follow-up of the persons originally treated with nicotinic acid [189].

Decreased rates of atherosclerotic progression were also observed in three quantitative angiographic trials: FATS, HATS, and CLAS [190–192]. In all these trials, nicotinic acid was combined with other LDL-lowering drugs and effects were compared with placebo.

The HATS trial demonstrated that the combined use of niacin and simvastatin dramatically improved angiographic end points and led to a statistically significant reduction in clinical cardiovascular events (>80% decrease) and significant coronary stenosis regression compared with placebo.

The ARBITER 2 trial showed that the combined therapy with extended release of niacin 1,000 mg and a statin slowed the progression of atherosclerosis, which was measured by a decrease in cIMT in individuals with moderately low HDL-cholesterol levels [184].

Despite such broad-ranging beneficial effects on lipids and atherosclerosis, the clinical use of niacin is limited by the major side effects such as flushing and hepatotoxicity. Indeed, the tolerability and the safety profile of nicotinic acid are problematic. Flushing of the skin is common and is intolerable for many persons. However, most persons develop tolerance to the flushing after more prolonged use of the drug. GPR109A-mediated production of prostaglandin D2 (PGD2) in macrophages and Langerhan cells causes skin capillary vasodilation and explains, at least in part, niacin's effect on flushing.

Importantly, although these flushing effects are mediated by the nicotinic acid receptor, they appear to be independent of the beneficial lipid-altering effects of nicotinic acid.

Although the mechanism by which nicotinic acid induces flushing is not completely understood, observations suggest that blockade of the PGD2 receptor, specifically the subtype 1 (DP1), with laropiprant, a selective antagonist, may suppress the flushing symptoms associated with nicotinic acid in the human. Recently, a fixed dose combination tablet of prolonged release nicotinic acid with laropiprant has been developed and approved for the treatment of dyslipidemia, particularly in patients with combined mixed dyslipidemia and in patients with primary hypercholesterolemia. Adding laropiprant to nicotinic acid reduced the symptoms of flushing caused by nicotinic acid (EPHAR, Tredaptive 2008, http://www.emea.europa.eu).

A variety of gastrointestinal symptoms, including nausea, dyspepsia, flatulence, vomiting, diarrhea, and activation of peptic ulcer, may occur.

Hepatotoxicity, hyperuricemia and gout, and hyperglycemia are the three major adverse effects of nicotinic acid. The risk of all three is increased with higher doses. Nicotinic acid reduces insulin sensitivity, and doses >3 g/day often worsen hyperglycemia in persons with type 2 diabetes. Other adverse effects include conjunctivitis, nasal stuffiness, acanthosis nigricans, ichthyosis, and retinal edema (toxic amblyopia).

Due to low tolerability, nicotinic acid is typically not used primarily to lower LDL levels. Instead, it is generally used in combination with other drugs, especially the statins.

Furthermore, although nicotinic acid can favorably modify the lipoprotein profile, especially in patients with atherogenic dyslipidemia, its long-term use is limited for many patients by side effects. For this reason, the drug is generally reserved for patients at higher short-term risk, i.e., for those with CHD, CHD risk equivalents, or multiple (2+) risk factors with 10-year risk for CHD of 10–20%. Its use for long-term prevention of CHD in persons with 10-year risk <10% is not well established.

19.4.4 Bile Acid Sequestrants

The sequestrants (cholestyramine, colestipol, and colesevelam) are anion exchangers that bind bile acids in the intestine. Therefore, the enterohepatic recirculation of bile acids is reduced and the conversion of cholesterol to bile acids in the liver is increased through feedback regulation. The resulting decrease in hepatocyte cholesterol content upregulates LDL-receptor expression, which in turn lowers serum LDL-cholesterol concentrations.

Therefore, the principal action of bile acid sequestrants is to lower LDL cholesterol.

However, in some persons, they increase hepatic VLDL production, thereby raising serum triglyceride levels. As a consequence, they are contraindicated as monotherapy in persons with triglycerides levels of >400 mg/dL and in familial dysbetalipoproteinemia. Moreover, they usually should be used as monotherapy only in persons with triglyceride levels of <200 mg/dL.

Bile acid sequestrants induce LDL-lowering synergic effect when combined with other cholesterol-lowering drugs. Specifically, sequestrants are useful in combined drug therapy with statins. Although doubling the dose of a statin produces only a 6% further reduction in LDL cholesterol, adding a moderate dose of a sequestrant to a statin can further lower LDL cholesterol by 12–16%. Furthermore, sequestrants combined with plant stanol esters apparently enhance LDL lowering. Thus, sequestrants in combination with dietary options for lowering LDL cholesterol (plant stanols/sterols and viscous fiber) should enable many persons to achieve their LDL-cholesterol goal without the need for an agent that is systemically absorbed.

The Lipid Research Clinics Coronary Primary Prevention Trial showed a significant reduction of the CHD risk in patients treated with cholestyramine [193, 194].

Beneficial outcomes also have been demonstrated in other clinical trials in which sequestrants were combined with other lipid-modifying drugs [190, 191].

Because they remain unabsorbed in their passage through the gastrointestinal tract, bile acids sequestrants lack systemic toxicity, but cause various gastrointestinal symptoms, including constipation, abdominal pain, bloating, fullness, nausea, and flatulence. Moreover, they can decrease the absorption of a number of drugs that are administered concomitantly. The general recommendation is that other drugs should be taken either an hour before or 4 h after administration of the sequestrant.

Finally, because of their bulk, they lack convenience of administration. Eight to 10 g/day cholestyramine or 10–20 g/day colestipol reduce LDL-cholesterol concentrations by 10–20%. Smaller doses of sequestrants (8–10 g/day) generally are well-tolerated; higher doses (16–20 g/day) are less well-tolerated. Colesevelam is a much more potent bile acid sequestrant, which has been primarily evaluated at doses of 2.6–3.8 g/day inducing reductions in LDL cholesterol of 12–18%.

19.4.5 Cholesterol Absorption Inhibitors

Total cholesterol pool depends on the two main functions: the rate of dietary cholesterol absorption through the intestinal wall and the rate of endogenous cholesterol synthesis. New approaches to lower TC and LDL-C consist of inhibiting the exogenous cholesterol pathway responsible for absorption of cholesterol from the diet and bile. In the intestine, free cholesterol is solubilized into micelles, absorbed into enterocytes, esterified to cholesteryl ester, packaged into chylomicrons, and released into lymph and blood [6]. Inhibitors of cholesterol absorption are used to disrupt this pathway [195].

Ezetimibe is the first of a new class of highly selective cholesterol absorption inhibitors. Although the mechanism is not yet fully elucidated, ezetimibe appears to block the protein transporter called Niemann–Pick C1-like 1 protein (NPC1L1) that is located at the apical membrane of the small intestine enterocytes. It was shown that labeled ezetimibe glucuronide binds specifically to a single site in cells expressing NPC1L1, but does not bind to membranes from animals lacking the gene encoding this protein. Once ingested, ezetimibe undergoes extensive metabolism to its glucuronide metabolite in the intestine wall This metabolite is transported to the liver and excreted in the bile before accumulating in intestinal villi. Unlike other agents such as lipase inhibitors (e.g., orlistat) and bile acid sequestrants (e.g., cholestyramine), ezetimibe does not affect the absorption of other substances such as triglyceride, fat-soluble vitamins, and bile acids.

The result of ezetimibe's action on NPC1L1 in the enterocytes consists of a decreased absorption of dietary and biliary cholesterol in the small intestine and subsequently a decreased delivery of LDL to the liver. Increased clearance of plasma LDL through the liver ensues through upregulation of LDL receptors on the surface of hepatocytes.

NPC1L1 is also expressed in human hepatocytes and is similarly blocked by ezetimibe. However, the clinical effects and possible off-target effects of the interaction between ezetimibe and hepatic NPC1L1 are unclear [123].

Compared with placebo, ezetimibe significantly reduces LDL-C levels by 17.2–22.3% when used alone [196, 197]. In combination with different statins, ezetimibe induces an additive reduction in LDL-C levels by 6–20% [198]. Ezetimibe has also been shown to favorably modify other lipid subtypes in addition to LDL-C. Indeed, the administration of ezetimibe to hyperlipidemic patients reduces the serum levels of oxidized cholesterol, tryglycerides, apolipoprotein B-100, and increases HDL-C levels.

The cholesterol lowering of ezetimibe is synergic with the statin effects. Indeed, statins reduce serum LDL levels by reducing hepatic cholesterol production and upregulating the LDL receptors in the liver. Emerging evidence shows that ezetimibe might be useful when combined with a fibrate or niacin.

Several trials have compared the effects of combining ezetimibe with a low-dose statin with the standard approach of switching to a higher dose statin. The combination therapy 10 mg of pravastatin and 10 mg of ezetimibe was more effective than 40 mg of pravastatin alone on various measures of lipids, as well as insulin resistance and high-sensitivity CRP.

To investigate the potential role for ezetimibe being used in preference to a highly potent statin, 2,959 patients were randomized to receive an ezetimibe–simvastatin combination or rosuvastatin. Those patients randomized to the combination had significantly lower LDL-C levels by the end of the trial, and a significantly higher chance of attaining a LDL-C target of less than 70 mg/dL.

Several clinical trials have documented a significant improvements in LDL-C in specific patient populations such as intolerance to statins, Black, white, Hispanic and African American, familial hypercholesterolemia, type 2 diabetes mellitus, metabolic syndrome, CAD, HIV infection, and renal transplantation [123, 195].

Recently, the ENHANCE (Ezetimibe and Simvastatin in Hypercholesterolemia Enhances Atherosclerosis Regression) study that used cIMT measurements to assess the impact of ezetimibe therapy on atherosclerosis produced as many questions as answers [125]. This study investigated the effect of ezetimibe–simvastatin on cIMT in persons with familial hypercholesterolemia who were randomized to receive simvastatin, 80 mg, and either ezetimibe, 10 mg, or placebo. The primary end point was a change in cIMT after 24 months of treatment. At the conclusion of the study, the simvastatin–ezetimibe group had significantly reduced LDL-C (−39.1 vs. 55.6%; $P<0.01$) and hs-CRP (−49.2 vs. −23.5%; $P<0.01$) levels compared with the simvastatin group. However, the primary outcome, cIMT change, did not differ between the treatment groups ($P=0.15$) (Table 19.5). Indeed, 81% of the study participants were already receiving statin therapy before the start of the study. This fact may have affected cIMT stabilization and/or regression for study participants before the study began, thus dampening the potential impact of ezetimibe on cIMT change during the study.

19.5 Antihypertensive Drugs

19.5.1 Hypertension and Atherogenesis

Hypertension is an established cardiovascular risk factor correlated to CAD, heart failure, renal failure, and stroke. Data from epidemiological studies involving more than one million individuals have indicated that death from both CHD and stroke increases progressively and linearly from levels as low as 115 mmHg SBP and 75 mmHg diastolic blood pressure (DBP) upward [199]. The increased risks are present in individuals ranging from 40 to 89 years of age. For every 20 mmHg systolic or 10 mmHg diastolic increase in BP, there is a doubling of mortality from both CHD and stroke. Actually, high SBP and DBP are considered two independent risk factors of CVD.

Hypertension is a state of sustained hemodynamic stress that is sensed by the endothelium. Over the long term, the sensing mechanism becomes maladaptive and perpetuates a vicious cycle of endothelial dysfunction, oxidative stress, and cellular injury.

Endothelial dysfunction and oxidative stress are early and common events in the pathogenesis of both atherosclerotic and hypertensive disease. ROS directly injure the endothelial lining and potentiate the effect of vasoconstrictive factors in the vessel wall as well as in leukocytes and platelets [200].

An increase in Ca^{2+} influx has been associated with both hypertension and early atherosclerotic lesions [201]. Hypercholesterolemia has been recognized as an important factor in altering SMC permeability of Ca^{2+} by activating the voltage-operated Ca^{2+} channels. The significant increase in SMC Ca^{2+} concentration has been shown to cause SMC proliferation and leads to atherosclerotic plaque progression [202].

Altered renin-angiotensin system (RAS) activity has a central role in the genesis of both hypertension and atherosclerosis. Indeed, the RAS has an important role in many types of CVD, including also myocardial infarction and diabetic nephropathy.

The RAS induces vasoconstriction and increases intravascular volume, both of which increase myocardial workload. Thus, initially, the RAS has been described as a circulating hormonal system, affecting CVD through hemodynamic and endocrine factors. More recently, however, the studies on the mechanism of the RAS focused on tissue-based cellular effects occurring in the arteries and myocardium that are reparative responses to tissue injury. Therefore, it became evident that the

RAS directly contributes to coronary ischemic events via atherosclerosis, altered postinfarct remodeling, and reduced fibrinolysis. Moreover, an important role of the RAS in the pathophysiology of chronic kidney diseases (CKD) has also been documented.

The RAS, mainly through the action of angiotensin II, activates intracellular signaling pathways that promote atherothrombosis through inflammation, endothelial dysfunction, growth, altered fibrinolysis, and potentiation of LDL oxidation (Fig. 19.7) [203, 204].

Angiotensin II induces nicotinamide adenine dinucleotide/nicotinamide adenine dinucleotide phosphate oxidase in all vascular components and causes a reduction in endothelial NOS activity. Moreover, a positive feedback loop among RAS activation, angiotensin II release, oxidative stress, and endothelial dysfunction has been described. In addition, aldosterone also induces oxidative stress and promotes proinflammatory/fibrogenic activity at vascular and nonvascular sites of injury. Finally, a significant cross-talk is observed among RAS, ROS, and dyslipidemia mediated via LOX-1, a novel lectin-like receptor for oxidated LDL, that is overexpressed in blood vessels of hypertensive animals, and LOX-1 activation is associated with the generation of ROS [163].

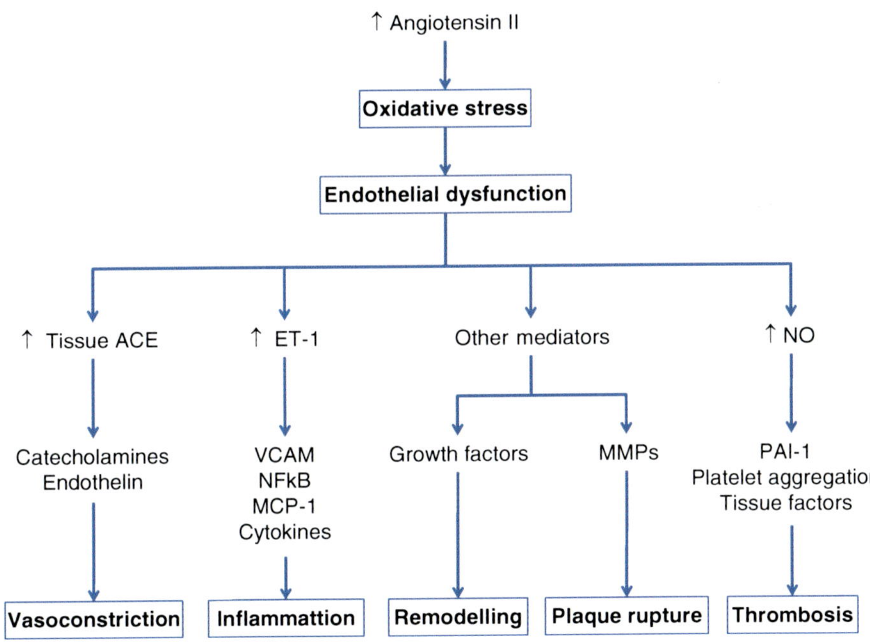

Fig. 19.7 The multiple effects of increased tissue production of angiotensin II in the pathogenesis of hypertension and atherosclerosis (modified from [203]). *ACE* angiotensin-converting enzyme, *ET-1* endothelin-1, *MCP-1* monocyte chemoattractant protein-1, *MMP* matrix metalloproteinase, *NF-kB* nuclear factor-kB, *NO* nitric oxide, *PAI-1* plasminogen activator type 1, *VCAM* vascular cell adhesion molecule

Detailed reviews of the role of the RAS in atherosclerosis, hypertension, and other CVD have recently been published [200, 205–207].

19.5.2 The Renin-Angiotensin System as a Target of Antiatherosclerotic Drugs

The antiatherosclerotic activity of drugs that inhibit the RAS has been shown in animal models of atherosclerosis and also documented in clinical trials in humans [200].

It is now well established that many of the effects of the RAS are mediated by angiotensin II (A-II), an oligopeptide produced by two enzymatic cleavages of angiotensinogen (Fig. 19.8).

Two distinct similar systems generate angiotensin II: the circulating RAS mainly responsible for short-term endocrine functions and the tissue RAS mainly regulating long-term autocrine and paracrine changes within all the major organs, including the heart, blood vessels, and kidneys.

In the circulating RAS, renin, which is released into circulation by the juxtaglomerular apparatus of the kidney in response to decreased glomerular perfusion, catalyzes

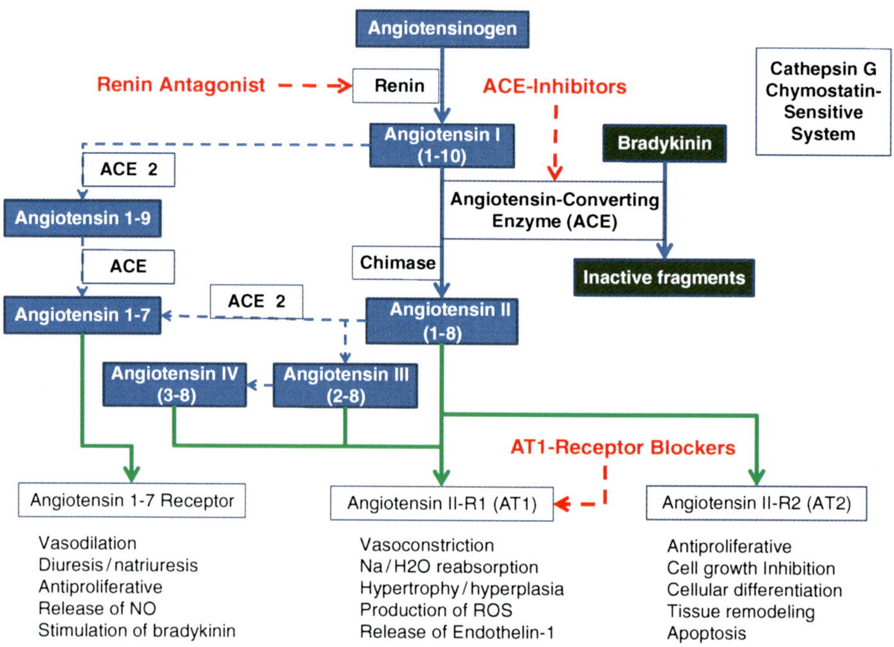

Fig. 19.8 Overview of the renin-angiotensin system (RAS), its complex network of endocrine, autocrine, and oparacrine effects involved in the pathogenesis of hypertension and atherosclerosis, and the site of action of RAS-inhibiting drugs. Alternative enzymatic pathways (*dashed lines*). *ACE* angiotensin-converting enzyme, *NO* nitric oxide, *ROS* reactive oxygen species

the cleavage of angiotensinogen to angiotensin I. Angiotensin I, a decapeptide, is cleaved by lung angiotensin-converting enzyme (ACE), which also degrades bradykinin, into the circulating octapeptide angiotensin II.

In the tissue RAS, all the components necessary to generate angiotensin II are present within the tissue and blood vessels. Thus, tissue renin generates tissue angiotensin I, subsequently catalyzed by tissue ACE into tissue angiotensin II. A large proportion of total body ACE exists on endothelial cells, throughout atherosclerotic plaques, and in the parenchymal cells of certain tissues.

Alternative enzyme pathways also generate angiotensin II in the tissue RAS (Fig. 19.8). The relative importance of alternate pathways is uncertain. Moreover, in addition to angiotensin II, a number of other biologically active peptides can be generated through the action of various peptidases.

Angiotensin II may be cleaved into angiotensin III and angiotensin IV [206].

In the heart, kidneys, and testes, an ACE-related carboxypeptidase (ACE2) converts angiotensin I to angiotensin 1–9, subsequently cleaved to angiotensin 1–7 by ACE. ACE2 also converts angiotensin II to angiotensin 1–7.

Angiotensin II, and at least in part other active peptides of the RAS, function through two receptors, type 1 (AT1) and type 2 (AT2). Other specific receptors have been identified for angiotensin IV and angiotensin 1–7 [206].

The AT1 is ubiquitously and abundantly distributed in adult tissues, including blood vessels, heart, kidney, adrenal gland, liver, brain, and lungs. The AT2 in the adult is limited mainly to the myocardium, vascular epithelium, uterus, ovary, brain, pancreas, and adrenal medulla.

The AT1 activation promotes cell growth and regulates the expression of bioactive substances such as aldosterone, vasoconstrictive hormones, growth factors, cytokines, and extracellular matrix components. The AT1 also initiates several autoregulatory feedback loops of the RAS. Through positive loops, angiotensin II stimulates the expression of its precursor, angiotensinogen, and also enhances ACE activity. However, the AT1 also engages in a negative feedback loop by inhibiting the secretion of renin.

Recently, Yasuda et al. found that mechanical stress can activate AT1 receptor independently of angiotensin II [208]. Indeed, they showed that mechanical stress not only activates extracellular signal-regulated kinases in vitro, but also induces cardiac hypertrophy in vivo, without the involvement of angiotensin II. Moreover, all these events have been shown to be inhibited by candesartan as an inverse agonist for AT1 receptor. The innovative concepts that AT1 receptor directly mediates mechanical stress-induced cellular responses, and the evidence of the inverse-agonistic activity of AT1-receptor blockers, open novel perspectives for developing drugs to prevent atherosclerosis and organ damage in CVD.

The AT2 often counterbalances the effects of the AT1 by favoring apoptosis and inhibiting the growth of vascular smooth muscle and cardiac myocytes. However, it may also have a role in cell growth and in the progression of inflammation and fibrosis. Indeed, in the human heart, it is predominantly located in interstitial fibroblasts. Moreover, the AT2 is upregulated with atherosclerosis, vascular injury, MI, and heart failure.

It is now currently accepted that the effects of angiotensin II may be correlated to the ratio of AT1:AT2 expression. However, although angiotensin II seems to be the ligand responsible for most signal transduction through the AT1 and the AT2, newly described components of the RAS indicate that it is not the only ligand, and the AT1 and the AT2 are not its only receptors. Indeed, angiotensin III may stimulate aldosterone synthesis and inflammation, and angiotensin IV may induce vasoconstriction and promote the plasminogen activator inhibitor production by activating the AT1 and the AT4, respectively [200].

Finally, the angiotensin 1–7, via a specific receptor, seems to be involved in modulating most of the effects mediated by bradykinin.

The main effects of angiotensin II and bradykinin in the cardiovascular system and their potential role on the atherosclerotic disease are summarized in Table 19.10 [207].

Table 19.10 Effects of angiotensin II and bradykinin in the cardiovascular system and their potential role on the atherosclerotic disease (modified from [207])

Effect	Angiotensin II	Bradykinin
Vascular tone	*Vasoconstriction* Direct effect (AT1 receptor stimulation) Activates sympathetic nervous system Augments peripheral nonadrenergic activity Increases endothelin-1 Reduces NO	*Vasodilation* Increases nitric oxide production Increases prostacyclin and prostaglandin E2 (prostaglandins)
Sodium and water	*Sodium and water retention* Increases aldosterone synthesis and secretion Increases antidiuretic hormone Decreases renal blood flow	*Natriuresis and diuresis* Direct tubular effect
Fibrinolysis thrombosis	*Prothrombotic* Increases plasma activator inhibitor type 1 (PAI-1) Increases platelet aggregation and adhesion Increases tissue factor	*Enhances fibrinolytic function* Increases tissue-type plasminogen activator (tPA)
Endothelial function	*Promotes endothelial dysfunction* Increases superoxide radicals that deplete NO Reduces NO Increases endothelin-1 Inhibits endothelial cell migration Promotes uptake of oxLDL-C Increases reactive oxygen species that deplete NO, oxidizes LDL-C, expresses chemoattractant proteins, increases cytokine production, and expresses proinflammatory markers	*Improves endothelial function* Increases NO production

(continued)

Table 19.10 (Continued)

Effect	Angiotensin II	Bradykinin
Atherosclerosis, inflammation, and cellular effects	*Promotes atherosclerosis and vascular damage* Vascular smooth muscle activation, hypertrophy, and migration Stimulates production of extracellular matrix, matrix glycoprotein, and metalloproteinases (type IV collagenases and elastases) Collagen synthesis in vascular smooth muscle cells Increases adhesion molecules vascular cell adhesion molecule-1 (allows monocytes, after rolling, to adhere and transmigrate through the endothelium and invade the subendothelial space to form atherosclerotic foam cells) and intracellular adhesion molecule Activates mitogen-activated protein kinases Induces growth-factor expression (platelet-derived, basic fibroblast, insulin-like, and transforming growth factor-β) Generates superoxide anions (increasing oxidative stress) Increases interleukin-6 Increase inflammatory mediators Activates transcription factor nuclear factor-kappa B (causes expression of molecules for cell adhesion and proliferation) Increases monocyte chemoattractant protein-1 (recruits monocytes to the vessel wall) Monocyte and macrophage activation (monocyte colony-stimulating factor)	*Antiatherosclerotic, vasculoprotective*Inhibits cell growth

19.5.3 *First-Line Antihypertensive Drugs*

The first-line agents recommended for antihypertensive patients belong to six pharmacologic classes: diuretics, β-blockers (BBs), calcium-channel blockers (CCBs), α1-blockers, angiotensin-converting enzyme inhibitors (ACEIs), and angiotensin II type 1 receptor blockers (ARBs). Other old and new drugs are also available to treat hypertension in specific clinical conditions. Several international guidelines on clinical management of hypertensive patients were published and are periodically revisited.

Actually, four different classes of drugs are available to inhibit the RAS, control the hypertension, decrease the progression or revert the process of atherogenesis, and prevent the risk of CVD: the renin antagonists, the ACEIs, the ARBs, and the aldosterone antagonists. The ACEIs and the ARBs are among the most important agents developed to control the risk of hypertension, reduce the progression of heart failure and of CKD.

Studies in patients with hypertension showed that treatment with ACE inhibitors and ARBs reverses endothelial dysfunction and reduces oxidant stress and inflammation, suggesting that inhibition of RAS improves endothelial function that is mediated in part by a reduction of ROS. Inhibition of RAS by ARBs also reduces the oxidizability of LDL. There is accumulating evidence that antihypertensive regimens that inhibit the RAS may provide incremental end-organ protection [200, 205, 207, 209–212].

The HOPE (Heart Outcome Prevention Evaluation) trial assessed the role of an ACE inhibitor, ramipril, in patients who were at high risk for cardiovascular events but who did not have left ventricular dysfunction or heart failure. Ramipril significantly reduced the rate of death, myocardial infarction, and stroke in a broad range of high-risk patients who did not have a low ejection fraction or heart failure, thus suggesting that the use of ACE inhibitors may prevent the progression of initially clinically silent atherosclerosis [213].

A recent paper overviewed 11 clinical trials supporting the use of RAS inhibitors (ACEIs and ARBs) as monotherapy or combination therapy based on the known role of the RAS in BP regulation and the vascular response to injury [210]. Combining ACE inhibitors and ARBs to provide more extensive RAS inhibition may provide greater antihypertensive efficacy and end-organ protection than the use of either class alone.

The role of Ca^{2+} channel blockers in the treatment of atherosclerosis still has not been clearly established [202]. The antiatherosclerotic effect of Ca^{2+} channel blockers has been shown to be limited to the attenuation of the initial stages of atherosclerosis including fatty streak formation without having any effect on pre-existing lesions [214]. Nevertheless, lacidipine, a dihydropiridine CCB, reduced progression of carotid atherosclerosis independent of the reduction in clinic and ambulatory blood pressure [215, 216].

A recent clinical studies showed that amlodipine, a third generation highly lipophilic dihydropiridine CCB with antioxidant properties, prevented the development and progression of atherosclerosis [217].

Moreover, it has been shown that amlodipine in combination with statins induce a synergistic antiatherosclerotic effect, inhibiting new lesion formation and the progression of established coronary atherosclerosis [214].

The antiatherosclerostic effect of amlodipine is associated with an increase in the resistance of LDL to oxidative modification, prevention of SMC membrane remodeling, inhibition of SMC proliferation and migration, attenuation of TNF-α-induced endothelial apoptosis, modulation of vascular cell gene expression, and extracellular matrix formation [218].

Other antihypertensive agents, such as some β-blockers, reverse endothelial dysfunction, reduce the generation of ROS, attenuate vascular inflammation, and achieve reductions in systolic and DBP.

In summary, almost all therapeutic strategies for hypertension improve endothelial function, including dietary modifications, exercise, and pharmacotherapy.

19.5.4 Health Outcomes Associated with First-Line Antihypertensive Agents

Until recently, the international guidelines on hypertension recommended low-dose diuretics and β-blockers as first-line treatment for patients with uncomplicated hypertension [219, 220]. Several early clinical trials conducted in patients with the highest level of blood pressure provided a wealth of evidence about the health benefits of low-dose diuretics and β-blockers and conditioned this recommendation, precluding further long-term placebo-controlled trials [221].

Thus, the next large long-term trials have compared one active treatment against another active treatment, including not only diuretics and β-blockers, but also α-blockers, CCBs, ACEIs, and ARBs [222–224]. These comparative trials in hypertension have provided a patchwork of evidence about the health benefits of antihypertensive agents, entangling the question of which first-line treatment regimen is optimal.

Recently, several meta-analyses have been published on this topic. Psaty et al. used the method of network meta-analysis to summarize the available direct and indirect clinical evidence concerning the efficacy and safety of various antihypertensive drugs used as first-line agents [225]. The primary aim of this meta-analysis was to compare low-dose diuretics with each of the other five active first-line therapies evaluated in large long-term trials in terms of major CVD end points and all-cause mortality.

In this network meta-analysis, the authors combined the data from 42 clinical studies, published until December 2002 from the USA, Europe, Scandinavia, Australia, Japan, and China, and included 192,478 patients followed up for an average of 3–4 years and randomized to seven major treatment strategies, including placebo or no-treatment.

In a preliminary standard meta-analysis, any active treatment showed a significant reduction in the risk of all major outcomes in comparison with placebo, an untreated control group, or usual care. However, combining all types of active treatments in this analysis involved the unevaluated assumption that all treatments had a similar effect. The significant heterogeneity observed for the outcomes of stroke and major cardiovascular events may be the result of important differences between drug classes.

The network meta-analysis compared low-dose diuretics with no-active treatment or any other class of antihypertensive drugs, incorporating both the direct and indirect comparisons between treatments. Yet, the direct and indirect comparisons were similar. Changes in systolic and diastolic blood pressure were similar between comparison treatments.

For all outcomes, the network meta-analysis confirmed that low-dose diuretics were superior to placebo. Several other treatment strategies were significantly better than placebo for some end points, but none of the other first-line treatment strategies was significantly better than low-dose diuretics for any major CVD outcome.

On the contrary, compared with α-blockers, CCBs, or ACE inhibitors, low-dose diuretics were associated with reduced risks of CHF. Compared with ACE inhibitors, low-dose diuretics were associated with a reduced risk of stroke. Compared with β-blockers, α-blockers, CCBs, or ACE inhibitors, low-dose diuretics were associated with reduced risks of CVD events.

This network meta-analysis provided compelling evidence that low-dose diuretics are the most effective first-line treatment for preventing the occurrence of CVD morbidity and mortality. Current clinical practice and treatment guidelines reflect this evidence, at least in part.

Recently, the role of β-blockers as first-line therapy for hypertension relative to each of the other major classes of antihypertensive drugs has been questioned.

Wiysonge on behalf of the Cochrane Hypertension Group conducted a meta-analysis to quantify the effectiveness and safety of β-blockers on morbidity and mortality endpoints in adults with hypertension [226].

The results of this meta-analysis showed that the risk of all-cause mortality was not different between first-line β-blockers and placebo, diuretics or RAS inhibitors, but was higher for β-blockers compared with CCBs. The risk of cardiovascular mortality was not significantly different between β-blockers and placebo or each class of antihypertensive drugs.

Analogously, CHD risk was not significantly different between β-blockers and placebo or each of the other major classes of antihypertensive drugs. The risk of stroke was lower for first-line β-blockers compared with placebo, but higher compared with CCBs and RAS inhibitors, while not significantly different compared with diuretics.

Overall, the effect of β-blockers on total CVD was better than that of placebo, but significantly worse than that of CCBs and not significantly different from that of diuretics or RAS inhibitors. This results are correlated to the relative effectiveness on stroke.

Finally, patients on β-blockers were more likely to discontinue treatment due to side effects than those on diuretics and RAS inhibitors.

The results of this meta-analysis induced the authors to conclude that the available evidence does not support the use of β-blockers as first-line drugs in the treatment of hypertension.

The relatively weak effect of β-blockers to reduce stroke, the absence of an effect on CHD when compared with placebo or no-treatment, and the trend toward worse outcomes in comparison with CCBs, RAS inhibitors, and thiazide diuretics are the main reasons that justify this conclusion.

However, it is not known at present whether there are differences between the different subtypes of β-blockers or whether β-blockers have differential effects on younger and elderly patients.

19.5.5 Role of Blood Pressure Lowering

In two meta-regression analysis, Staessen et al. demonstrated that BP lowering was the major determinant of the benefits of antihypertensive treatment on all-cause and cause-specific cardiovascular outcomes [227, 228]. The BPLTTC reported that the reduction in CHD and stroke produced by antihypertensive treatment increased with lower BP targets, and that in other trials, it was proportional to the differences in the achieved systolic BP between randomized groups [229].

The same group investigated whether, for the same degree of BP lowering, prevention of stroke was superior to the protection against CHD. They found that a 10 mmHg decrease in systolic BP antihypertensive treatment prevented CHD and stroke to a similar relative extent. The absolute benefit (i.e., the number of patients to be treated to prevent one event) depends on the rate of CHD or stroke in the population to which the present findings might be extrapolated.

Recently, Verdecchia et al. investigated whether protection from CHD and stroke conferred by ACEIs and CCBs in hypertensive or high-risk patients may be explained by the specific drug regimen [230].

They reviewed 28 outcome trials that compared either ACEIs or CCBs with diuretics, β-blockers, or placebo, and extracted summary statistics regarding 9,509 incident cases of CHD (myocardial infarction and coronary death) and 5,971 cases of stroke from a total of 179,122 patients.

In placebo-controlled trials, ACEIs decreased the risk of CHD ($P<0.001$), and CCBs reduced stroke incidence ($P<0.001$). There were no significant differences in CHD risk between regimens based on diuretics/β-blockers and regimens based on ACEIs ($P=0.46$) or CCBs ($P=0.52$). The risk of stroke was reduced by CCBs ($P=0.041$) but not by ACEIs ($P=0.15$) compared with diuretics/β-blockers.

Prevention of CHD was explained by systolic BP reduction ($P<0.001$) and use of ACEIs ($P=0.028$), whereas prevention of stroke was explained by systolic BP reduction ($P=0.001$) and use of CCBs ($P=0.042$).

This quantitative overview confirms that ACEIs and CCBs protect against CHD and stroke mainly by reducing BP. However, over and beyond BP reduction, ACEIs appear superior to CCBs for the prevention of CHD, whereas CCBs appear superior to ACEIs for the prevention of stroke.

These data have relevant clinical implications in suggesting that the ancillary properties of these drug classes might provide specific contributions to the prevention of CHD and stroke, respectively. Indeed, other potential modifiers or confounders of the outcome results, including the patients' age and sex distribution and year of publication of the trials, did not contribute to the variance explained by our meta-regression models.

Despite the results of many randomized trials, and the widespread use of blood pressure-lowering drugs, uncertainty remains about which drugs to use and who to treat. Previous meta-analysis did not solve all the uncertainties.

Therefore, Law et al. conducted the largest meta-analysis of epidemiological cohort studies ever published, aiming to determine the quantitative efficacy of different classes of blood pressure-lowering drugs in preventing CHD and stroke, and who should receive treatment [231].

They reviewed 147 randomized trials of blood pressure-lowering drugs recording CHD events and strokes: 108 "blood pressure difference" trials studied differences in blood pressure between study drug and placebo or control group not receiving the study drug, and 46 "drug comparison" trials compared drugs. Seven trials with three randomized groups fell into both categories.

The results of this recent meta-analysis are complex, innovative, at least in part, and also provocative, compared with those of other previously published meta-analyses.

In the blood pressure difference trials, β-blockers showed a special effect, over and above that due to blood pressure reduction, in preventing recurrent CHD events in people with a history of CHD: risk reduction 29% (95% CI, 22–34%) compared with 15% (11–19%) in trials of other drugs. The extra effect was limited to a few years after myocardial infarction, with a risk reduction of 31% compared with 13% in people with CHD with no recent infarct ($P=0.04$).

In the other blood pressure difference trials, after exclusion of CHD events in trials of β-blockers in people with CHD, there was a 22% reduction in CHD events (17–27%) and a 41% (33–48%) reduction in stroke for a blood pressure reduction of 10 mmHg systolic or 5 mmHg diastolic. This results indicate that the benefit is explained by blood pressure reduction itself.

The five main classes of blood pressure-lowering drugs (thiazides, β-blockers, ACEIs, ARBs, and CCBs) showed equivalent effectiveness in preventing CHD events and strokes, with the exception that CCBs had a greater preventive effect on stroke (RR=0.92, 95% CI, 0.85–0.98).

Moreover, the percentage reductions in CHD events and stroke were similar in people with and without CVD and regardless of blood pressure before treatment, down to 110 mmHg SBP and 70 mmHg DBP.

The authors also estimated that, in people aged 60–69 with a DBP before the treatment of 90 mmHg, three drugs at half-standard dose in combination could reduced the risk of CHD by 46% and of stroke by 62%; one drug at standard dose had about half this effect.

The meta-analysis also showed that drugs other than CCBs (with the exception of noncardioselective β-blockers) reduced the incidence of heart failure by 24% (19–28%) and CCBs by 19% (6–31%).

In conclusion, this meta-analysis demonstrated that, with the exception of the extra protective effect of β-blockers given shortly after a myocardial infarction and the minor additional effect of CCBs in preventing stroke, all the classes of blood pressure-lowering drugs have a similar effect in reducing CHD events and stroke for a given reduction in blood pressure. These results seem to exclude the pleiotropic effects of antihypertensive drugs.

Moreover, the proportional reduction in CVD events was the same or similar regardless of pretreatment blood pressure and the presence or absence of existing CVD. These results induced the authors to suggest a simplification of the actual guidelines on the use of blood pressure-lowering drugs: these agents, eventually as a poly-pill, should be offered to people with all levels of blood pressure. Indeed, the results indicate the importance of lowering blood pressure in everyone over a certain age, rather than measuring it in everyone and treating it in some. However, this provocative proposal will encounter a lot of unsurmontable difficulties because they tend to increase the already high economic impact of cardiovascular drugs.

19.5.6 Antihypertensive Treatment and Carotid Intima–Media Thickness

Hypertension promotes carotid intima–media thickening. Several trials tested the effects of antihypertensive drugs on cIMT [121, 216, 232–251]. Recently, Wang et al. reviewed the randomized controlled trials that evaluated the effects of an antihypertensive drug vs. placebo or another antihypertensive agent of a different class on cIMT (Table 19.11) [252]. Conducting a meta-analysis on 22 trials published before 2005, they investigated whether antihypertensive treatment reduces cIMT, whether new antihypertensive drugs are more effective than old agents, diuretics and β-blockers, in the prevention of carotid intima–media thickening, and whether ACEIs and CCBs are equally effective in this regard. Moreover, they also studied the relevance of using cIMT as an intermediate outcome measure for the prevention of mortality and cardiovascular events.

Eight trials comparing active treatment with placebo or no-treatment included 3,329 patients with diabetes or CHD. A random-effects model showed that active antihypertensive treatment, initiated with an ACEI, a β-blocker, or a CCB, significantly reduced the yearly increase in cIMT by 7 μm/year (95% CI, −14 to −2; $P=0.01$). The combined results of the β-blocker trials were not statistically significant with borderline significant heterogeneity among individual trials ($P=0.05$).

In nine trials including 4,564 hypertensive patients, CCBs, ACEIs, an ARB, or an α-blocker, compared with diuretics or β-blockers, in the presence of similar blood pressure reductions, decreased intima–media thickening by 3 μm/year (95% CI, −5 to −0.3; $P=0.01$). The overall beneficial effect of the newer over older drugs was largely attributable to the decrease of intima–media thickening by 5 μm/year ($P=0.007$) in four trials of CCBs involving 3,619 patients, because overall ACEIs were not different from old drug classes ($P=0.19$).

Five trials including 287 patients with hypertension or diabetes, compared CCBs with ACEIs. CCBs did not differentially affect blood pressure, but significantly reduce the yearly increase in cIMT by 23 μm/year (95% CI, −42 to −4; $P=0.02$). This differential effect of CCBs might contribute to their superior protection against stroke.

The meta-regression analysis showed that the BP difference during follow-up did not predict the treatment-induced difference in the yearly changes of carotid IMT. However, the treatment-induced differences in the yearly changes of cIMT correlated weakly and inversely with the differences in lumen diameter of the common carotid artery during follow-up ($P=0.02$).

In conclusion, this meta-analysis showed that CCBs are more effective than ACEIs, which in turn are more effective than placebo or no-treatment, but not more active than diuretics/β-blockers in the prevention of carotid intima–media thickening. Whether these findings are attributable to functional changes through vasodilation without a structural decrease in CSA and whether these findings have implications for the long-term prevention of cardiovascular complications such as stroke, remain to be proven [252].

Table 19.11 Meta-analytic effects of antihypertensive treatment on changes in carotid intima–media thickness (cIMT) in trials comparing active treatment with placebo or no-treatment, old with new drugs, and angiotensin-converting enzyme inhibitors (ACEIs) with calcium entry blockers (CCBs) (adapted from [252])

Characteristics of randomized controlled trials				Results of meta-analysis				
Trial	Treatments (test/ref)	Follow-up (years)	Basal mean SBP/DBP (mmHg)	Patients Ref:test (No.)	Baseline cIMT Ref:test (μm)	Change/year (Ref:Test μm/year)	Difference (95% CI) (μm/year)	*P*
Active treatment vs. placebo or no-treatment								
Migdalis [232]	Fosinopril/No-Treat	1.0	168/94	20:20	680:701	103:30		
PART-2 [233]	Ramipril/Plac	4.7	133/79	309:308	780:800	5:8		
SECURE [234]	Ramipril 2.5 mg/Plac	4.5	132/76	227/232	1,146:1,148	22:14		
	Ramipril 10 mg/Plac			227/234	1,146:1,160	22:18		
Hosomi [235]	Enalapril/No-Treat	2.0	140/76	50:48	700:700	20:10		
PREVEND [236]	Fosinopril[a]/Plac	4.0	130/76	323:319	770:770	11:8		
All ACEIs				*929:1,161*			*−6 (−12 to 0.4)*	*0.07*
BCAPS [121]	Metoprolol[a]/Plac	3.0	139/85	390:393	893:912	8:7		
ELVA [237]	Metoprolol/Plac	3.0	138/81	44:35	897:894	12:−12		
All BBs				*434:428*			*−10 (−33 to 13)*	*0.41*
PREVENT [238]	Amlodipine/Plac	3.0	129/79	186:191	1,258:1,259	11:−4		
All active treatment vs. placebo or no-treatment trials				*1,549:1,780*			*−7 (−12 to −2)*	*0.01*
New vs. old drugs								
MIDAS [239]	Isradipine SR/HCTZ	3.0	150/97	441:442	1,170:1,170	50:40		
VHAS [240]	Verapamil/Chlorthalidone	4.0	168/102	191:186	908:902	16:15		
INSIGHT [241]	Nifedipine GITSHCTZ/A	4.0	161/94	164:160	660:668	5:−1		
ELSA [216]	Lacidipine/Atenolol	3.75	164/101	1,012:1,023	1,162:1,159	15:13		
All CCBs				*1,808:1,811*			*−5 (−9 to −1)*	*0.007*
CELIMENE [242]	Enalapril/Celiprolol	0.75	159/99	40:42	592:575	−52:−32		
Roman [243]	Ramipril/HCTZ	0.5	150/94	28:22	750:770	160:100		
PHYLLIS [244]	Fosinopril/HCTZ	2.6	160/98	127:127	1,220:1,220	10:−2		
	Fosinopril[a]/HCTZ			126:128	1,210:1,200	−2:−2		

All ACEIs				*321:319*			*−1 (−5 to 2)*	*0.52*
LAARS [245]	Losartan/Atenolol	2.0	160/101	111:114	821:833	−37:−38		
DAPHNE [246]	Doxazosin/HCTZ	3.0	163/100	39:41	1,430:1,390	−60:−50		
All trials of new vs. old drugs				*2,279:2,285*			*−3 (−5 to −0.3)*	*0.03*
CCBs vs. ACEIs								
Koshiyama [247]	Amlodipine/ACEIs	0.5	–	11:11	–	22:−104		
Topouchian [248]	Verapamil/Trendalapril	0.5	158/99	18:21	680:720	−80:−40		
Pontremoli [249]	Nifedipine GITS/Lisinopril	0.5	161/104	16:15	820:840	−65:−110		
Stanton [250]	Amlodipine/Lisinopril	1.0	165/100	34:35	792:763	−27:−48		
ELVERA [251]	Amlodipine/Lisinopril	2.0	175/93	63:63	1,057:1,019	0:−17		
All CCBs vs. ACEIs Trials				*142:145*			*−23 (−42 to −4)*	*0.02*

ACEI angiotensin-converting enzyme inhibitor, *BB* β-blocker, *CCB* calcium-channel blocker, *GITS* gastrointestinal therapeutic system, *HCTZ* hydrochlorothiazide, *HCTZ/A* hydrochlorothiazide/amiloride, *SBP/DBP* systolic and diastolic blood pressure

[a]Possible association with pravastatin (40 mg daily) or fluvastatin (40 mg daily) in a 2 x 2 factorial design

19.6 Conclusions

As a brief comment on this work, it could be useful to highlight some major points of interest in recent evolutions of atherosclerosis management.

In order to provide wide awareness about pharmacological treatment of atherosclerosis, it is remarkable to recall that in the last two decades, therapies formerly focused on lowering cholesterol blood levels have considerably changed their target: hypolipemic drugs as statins, fibrates, niacin, bile acid sequestrants, and selective inhibitors of cholesterol absorption have been supported by antihypertensive agents with the aim to treat atherosclerosis not only as a proved disease, but also in its preliminary stage.

Pharmacological treatment of hypertension restrains the mechanism of shear stress and high blood pressure in promoting atherosclerosis development and progression. The association of hypolipemic drugs and antihypertensive agents leads to a better control of atherosclerosis since its initial lesions and reveals a substantially different approach of atherosclerosis treatment: more comprehensive antiatherogenic therapy is now based on earlier treatment of causes that may enhance atherosclerosis progression rather than secondary prevention of complications arising after ruptures of atherosclerotic plaques.

Besides the use of classic X-ray contrast angiography to evaluate anatomic progression of atherosclerotic lesions, the arise of modern imaging techniques as high-resolution ultrasound techniques and MRI opened new horizons in documenting the effect of drugs on progression of atherosclerosis. Combined with the evaluation of soluble markers of disease (levels of LDL-C, HDL-C, triglycerides, CRP) and blood pressure monitoring, these methods offer a big potential value in accelerating the discovery and development of cardiovascular therapeutics: the identification of surrogate end-point or biomarkers, predictors of clinical outcomes in CVD, can be a powerful tool to assess the clinical efficacy of antiatherosclerotic therapy and CVD management.

Extensive discussions are now ongoing among academia, industry, and regulatory agency, focused on the need for the application of new tools in drug development.

References

1. Hiatt, W.R., Goldstone, J., Smith, S.C., Jr., McDermott, M., Moneta, G., Oka, R., Newman, A.B., Pearce, W.H.: Atherosclerotic peripheral vascular disease symposium II: nomenclature for vascular diseases. Circulation **118** (2008) 2826–2829
2. Adams, R.J., Chimowitz, M.I., Alpert, J.S., Awad, I.A., Cerqueria, M.D., Fayad, P., Taubert, K.A.: Coronary risk evaluation in patients with transient ischemic attack and ischemic stroke: a scientific statement for healthcare professionals from the Stroke Council and the Council on Clinical Cardiology of the American Heart Association/American Stroke Association. Stroke **34** (2003) 2310–2322
3. Kannel, W.B.: Risk factors for atherosclerotic cardiovascular outcomes in different arterial territories. J Cardiovasc Risk **1** (1994) 333–339

4. Haffner, S.M., Lehto, S., Ronnemaa, T., Pyorala, K., Laakso, M.: Mortality from coronary heart disease in subjects with type 2 diabetes and in nondiabetic subjects with and without prior myocardial infarction. N Engl J Med **339** (1998) 229–234
5. Chapman, M.J.: From pathophysiology to targeted therapy for atherothrombosis: a role for the combination of statin and aspirin in secondary prevention. Pharmacol Ther **113** (2007) 184–196
6. Hopkins, P.N., Williams, R.R.: A survey of 246 suggested coronary risk factors. Atherosclerosis **40** (1981) 1–52
7. Kannel, W.B., Wilson, P.W.: An update on coronary risk factors. Med Clin North Am **79** (1995) 951–971
8. Third Report of the National Cholesterol Education Program (NCEP): Expert panel on detection, evaluation, and treatment of high blood cholesterol in adults (adult treatment panel III) final report. Circulation **106** (2002) 3143–3421
9. Conti, C.R.: Evolution of NCEP guidelines: ATP1–ATPIII risk estimation for coronary heart disease in 2002. National Cholesterol Education Program. Clin Cardiol **25** (2002) 89–90
10. Kuhar, M.B.: Update on managing hypercholesterolemia. The new NCEP guidelines. AAOHN J **50** (2002) 360–364
11. Smith, S.C., Jr., Allen, J., Blair, S.N., Bonow, R.O., Brass, L.M., Fonarow, G.C., Grundy, S.M., Hiratzka, L., Jones, D., Krumholz, H.M., Mosca, L., Pearson, T., Pfeffer, M.A., Taubert, K.A.: AHA/ACC guidelines for secondary prevention for patients with coronary and other atherosclerotic vascular disease: 2006 update endorsed by the National Heart, Lung, and Blood Institute. J Am Coll Cardiol **47** (2006) 2130–2139
12. Segura, J., Ruilope, L.M.: New guidelines of the European Society of Hypertension. Curr Hypertens Rep **10** (2008) 337–338
13. Marzolo, M., Verlato, F., Zotta, L., Guadagnin, M.L., Borgese, L., Cravatari, M., Nardi, M., Camporese, G., Andreozzi, G.M.: Occult atherosclerosis and physical vascular examination: a simple strategy to avoid inadequate cardiovascular prevention and under-use of diagnostic vascular guidelines in outpatients. A multicenter study by Angiology Care Units in North-Eastern Italy. Int Angiol **27** (2008) 426–432
14. Gidding, S.S., Lichtenstein, A.H., Faith, M.S., Karpyn, A., Mennella, J.A., Popkin, B., Rowe, J., Van Horn, L., Whitsel, L.: Implementing American Heart Association pediatric and adult nutrition guidelines: a scientific statement from the American Heart Association Nutrition Committee of the Council on Nutrition, Physical Activity and Metabolism, Council on Cardiovascular Disease in the Young, Council on Arteriosclerosis, Thrombosis and Vascular Biology, Council on Cardiovascular Nursing, Council on Epidemiology and Prevention, and Council for High Blood Pressure Research. Circulation **119** (2009) 1161–1175
15. de la Sierra, A., Zamorano, J.L., Ruilope, L.M.: Application of hypertension guidelines in clinical practice: implementation of the 2007 ESH/ESC European practice Guidelines in Spain. J Hypertens **27 Suppl 3** (2009) S27–S32
16. Smith, S.C., Jr., Jackson, R., Pearson, T.A., Fuster, V., Yusuf, S., Faergeman, O., Wood, D.A., Alderman, M., Horgan, J., Home, P., Hunn, M., Grundy, S.M.: Principles for national and regional guidelines on cardiovascular disease prevention: a scientific statement from the World Heart and Stroke Forum. Circulation **109** (2004) 3112–3121
17. Munger, M.A., Hawkins, D.W.: Atherothrombosis: epidemiology, pathophysiology, and prevention. J Am Pharm Assoc (2003) **44** (2004) S5–S12; quiz S12–S13
18. Libby, P., Theroux, P.: Pathophysiology of coronary artery disease. Circulation **111** (2005) 3481–3488
19. Libby, P.: Atherosclerosis: disease biology affecting the coronary vasculature. Am J Cardiol **98** (2006) 3Q–9Q
20. Libby, P.: Inflammation and cardiovascular disease mechanisms. Am J Clin Nutr **83** (2006) 456S–460S
21. Libby, P.: Inflammation in atherosclerosis. Nature **420** (2002) 868–874
22. Libby, P., Ridker, P.M., Hansson, G.K.: Inflammation in atherosclerosis: from pathophysiology to practice. J Am Coll Cardiol **54** (2009) 2129–2138

23. Naghavi, M., Libby, P., Falk, E., Casscells, S.W., Litovsky, S., Rumberger, J., Badimon, J.J., Stefanadis, C., Moreno, P., Pasterkamp, G., Fayad, Z., Stone, P.H., Waxman, S., Raggi, P., Madjid, M., Zarrabi, A., Burke, A., Yuan, C., Fitzgerald, P.J., Siscovick, D.S., de Korte, C.L., Aikawa, M., Juhani, K.E., Assmann, G., Becker, C.R., Chesebro, J.H., Farb, A., Galis, Z.S., Jackson, C., Jang, I.K., Koenig, W., Lodder, R.A., March, K., Demirovic, J., Navab, M., Priori, S.G., Rekhter, M.D., Bahr, R., Grundy, S.M., Mehran, R., Colombo, A., Boerwinkle, E., Ballantyne, C., Insull, W., Jr., Schwartz, R.S., Vogel, R., Serruys, P.W., Hansson, G.K., Faxon, D.P., Kaul, S., Drexler, H., Greenland, P., Muller, J.E., Virmani, R., Ridker, P.M., Zipes, D.P., Shah, P.K., Willerson, J.T.: From vulnerable plaque to vulnerable patient: a call for new definitions and risk assessment strategies: part I. Circulation **108** (2003) 1664–1672
24. Naghavi, M., Libby, P., Falk, E., Casscells, S.W., Litovsky, S., Rumberger, J., Badimon, J.J., Stefanadis, C., Moreno, P., Pasterkamp, G., Fayad, Z., Stone, P.H., Waxman, S., Raggi, P., Madjid, M., Zarrabi, A., Burke, A., Yuan, C., Fitzgerald, P.J., Siscovick, D.S., de Korte, C.L., Aikawa, M., Airaksinen, K.E., Assmann, G., Becker, C.R., Chesebro, J.H., Farb, A., Galis, Z.S., Jackson, C., Jang, I.K., Koenig, W., Lodder, R.A., March, K., Demirovic, J., Navab, M., Priori, S.G., Rekhter, M.D., Bahr, R., Grundy, S.M., Mehran, R., Colombo, A., Boerwinkle, E., Ballantyne, C., Insull, W., Jr., Schwartz, R.S., Vogel, R., Serruys, P.W., Hansson, G.K., Faxon, D.P., Kaul, S., Drexler, H., Greenland, P., Muller, J.E., Virmani, R., Ridker, P.M., Zipes, D.P., Shah, P.K., Willerson, J.T.: From vulnerable plaque to vulnerable patient: a call for new definitions and risk assessment strategies: part II. Circulation **108** (2003) 1772–1778
25. Hadi, H.A., Carr, C.S., Al Suwaidi, J.: Endothelial dysfunction: cardiovascular risk factors, therapy, and outcome. Vasc Health Risk Manag **1** (2005) 183–198
26. Anderson, T.J., Gerhard, M.D., Meredith, I.T., Charbonneau, F., Delagrange, D., Creager, M.A., Selwyn, A.P., Ganz, P.: Systemic nature of endothelial dysfunction in atherosclerosis. Am J Cardiol **75** (1995) 71B–74B
27. Kinlay, S., Ganz, P.: Role of endothelial dysfunction in coronary artery disease and implications for therapy. Am J Cardiol **80** (1997) 11I–16I
28. Traub, O., Berk, B.C.: Laminar shear stress: mechanisms by which endothelial cells transduce an atheroprotective force. Arterioscler Thromb Vasc Biol **18** (1998) 677–685
29. Libby, P., Ridker, P.M., Maseri, A.: Inflammation and atherosclerosis. Circulation **105** (2002) 1135–1143
30. Celermajer, D.S., Sorensen, K.E., Gooch, V.M., Spiegelhalter, D.J., Miller, O.I., Sullivan, I.D., Lloyd, J.K., Deanfield, J.E.: Non-invasive detection of endothelial dysfunction in children and adults at risk of atherosclerosis. Lancet **340** (1992) 1111–1115
31. Prasad, A., Zhu, J., Halcox, J.P., Waclawiw, M.A., Epstein, S.E., Quyyumi, A.A.: Predisposition to atherosclerosis by infections: role of endothelial dysfunction. Circulation **106** (2002) 184–190
32. Steinberg, H.O., Chaker, H., Leaming, R., Johnson, A., Brechtel, G., Baron, A.D.: Obesity/insulin resistance is associated with endothelial dysfunction. Implications for the syndrome of insulin resistance. J Clin Invest **97** (1996) 2601–2610
33. Thogersen, A.M., Jansson, J.H., Boman, K., Nilsson, T.K., Weinehall, L., Huhtasaari, F., Hallmans, G.: High plasminogen activator inhibitor and tissue plasminogen activator levels in plasma precede a first acute myocardial infarction in both men and women: evidence for the fibrinolytic system as an independent primary risk factor. Circulation **98** (1998) 2241–2247
34. Al Suwaidi, J., Higano, S.T., Holmes, D.R., Jr., Lennon, R., Lerman, A.: Obesity is independently associated with coronary endothelial dysfunction in patients with normal or mildly diseased coronary arteries. J Am Coll Cardiol **37** (2001) 1523–1528
35. Fichtlscherer, S., Zeiher, A.M.: Endothelial dysfunction in acute coronary syndromes: association with elevated C-reactive protein levels. Ann Med **32** (2000) 515–518
36. Fichtlscherer, S., Rosenberger, G., Walter, D.H., Breuer, S., Dimmeler, S., Zeiher, A.M.: Elevated C-reactive protein levels and impaired endothelial vasoreactivity in patients with coronary artery disease. Circulation **102** (2000) 1000–1006
37. Bonetti, P.O., Lerman, L.O., Lerman, A.: Endothelial dysfunction: a marker of atherosclerotic risk. Arterioscler Thromb Vasc Biol **23** (2003) 168–175

38. Lum, H., Roebuck, K.A.: Oxidant stress and endothelial cell dysfunction. Am J Physiol Cell Physiol **280** (2001) C719–C741
39. Deanfield, J., Donald, A., Ferri, C., Giannattasio, C., Halcox, J., Halligan, S., Lerman, A., Mancia, G., Oliver, J.J., Pessina, A.C., Rizzoni, D., Rossi, G.P., Salvetti, A., Schiffrin, E.L., Taddei, S., Webb, D.J.: Endothelial function and dysfunction. Part I: methodological issues for assessment in the different vascular beds: a statement by the Working Group on endothelin and endothelial factors of the European Society of Hypertension. J Hypertens **23** (2005) 7–17
40. Wilson, P.W., D'Agostino, R.B., Levy, D., Belanger, A.M., Silbershatz, H., Kannel, W.B.: Prediction of coronary heart disease using risk factor categories. Circulation **97** (1998) 1837–1847
41. Stamler, J., Wentworth, D., Neaton, J.D.: Prevalence and prognostic significance of hypercholesterolemia in men with hypertension. Prospective data on the primary screenees of the Multiple Risk Factor Intervention Trial. Am J Med **80** (1986) 33–39
42. Rifkind, B.M.: Lipid Research Clinics Coronary Primary Prevention Trial: results and implications. Am J Cardiol **54** (1984) 30C–34C
43. Rossouw, J.E., Lewis, B., Rifkind, B.M.: The value of lowering cholesterol after myocardial infarction. N Engl J Med **323** (1990) 1112–1119
44. Law, M.R., Wald, N.J., Rudnicka, A.R.: Quantifying effect of statins on low density lipoprotein cholesterol, ischaemic heart disease, and stroke: systematic review and meta-analysis. BMJ **326** (2003) 1423
45. Grundy, S.M., Cleeman, J.I., Merz, C.N., Brewer, H.B., Jr., Clark, L.T., Hunninghake, D.B., Pasternak, R.C., Smith, S.C., Jr., Stone, N.J.: Implications of recent clinical trials for the National Cholesterol Education Program Adult Treatment Panel III guidelines. Circulation **110** (2004) 227–239
46. Brown, W.V.: Cholesterol lowering in atherosclerosis. Am J Cardiol **86** (2000) 29H–32H
47. Forrester, J.S., Shah, P.K.: Emerging strategies for increasing high-density lipoprotein. Am J Cardiol **98** (2006) 1542–1549
48. Vega, G.L., Grundy, S.M.: Hypoalphalipoproteinemia (low high density lipoprotein) as a risk factor for coronary heart disease. Curr Opin Lipidol **7** (1996) 209–216
49. Tall, A.R.: An overview of reverse cholesterol transport. Eur Heart J **19 Suppl A** (1998) A31–A35
50. Tall, A.R.: Cholesterol efflux pathways and other potential mechanisms involved in the athero-protective effect of high density lipoproteins. J Intern Med **263** (2008) 256–273
51. Tedgui, A., Mallat, Z.: Anti-inflammatory mechanisms in the vascular wall. Circ Res **88** (2001) 877–887
52. Tontonoz, P., Nagy, L., Alvarez, J.G., Thomazy, V.A., Evans, R.M.: PPARgamma promotes monocyte/macrophage differentiation and uptake of oxidized LDL. Cell **93** (1998) 241–252
53. Tedgui, A., Mallat, Z.: Apoptosis as a determinant of atherothrombosis. Thromb Haemost **86** (2001) 420–426
54. Hansson, G.K.: Atherosclerosis – an immune disease: the Anitschkov lecture 2007. Atherosclerosis **202** (2009) 2–10
55. Sposito, A.C., Chapman, M.J.: Statin therapy in acute coronary syndromes: mechanistic insight into clinical benefit. Arterioscler Thromb Vasc Biol **22** (2002) 1524–1534
56. Tedgui, A., Mallat, Z.: Apoptosis, a major determinant of atherothrombosis. Arch Mal Coeur Vaiss **96** (2003) 671–675
57. Greenland, P., Abrams, J., Aurigemma, G.P., Bond, M.G., Clark, L.T., Criqui, M.H., Crouse, J.R., 3rd, Friedman, L., Fuster, V., Herrington, D.M., Kuller, L.H., Ridker, P.M., Roberts, W.C., Stanford, W., Stone, N., Swan, H.J., Taubert, K.A., Wexler, L.: Prevention Conference V: Beyond secondary prevention: identifying the high-risk patient for primary prevention: noninvasive tests of atherosclerotic burden: Writing Group III. Circulation **101** (2000) E16–E22
58. Detrano, R.C.: Coronary artery scanning using electron beam computed tomography. Am J Card Imaging **10** (1996) 97–100

59. Detrano, R.C., Wong, N.D., Doherty, T.M., Shavelle, R.M., Tang, W., Ginzton, L.E., Budoff, M.J., Narahara, K.A.: Coronary calcium does not accurately predict near-term future coronary events in high-risk adults. Circulation **99** (1999) 2633–2638
60. Raggi, P.: Regression of calcified coronary artery plaque assessed by electron beam computed tomography. Z Kardiol **89 Suppl 2** (2000) 135–139
61. O'Malley, P.G., Taylor, A.J., Jackson, J.L., Doherty, T.M., Detrano, R.C.: Prognostic value of coronary electron-beam computed tomography for coronary heart disease events in asymptomatic populations. Am J Cardiol **85** (2000) 945–948
62. O'Rourke, R.A., Brundage, B.H., Froelicher, V.F., Greenland, P., Grundy, S.M., Hachamovitch, R., Pohost, G.M., Shaw, L.J., Weintraub, W.S., Winters, W.L., Jr., Forrester, J.S., Douglas, P.S., Faxon, D.P., Fisher, J.D., Gregoratos, G., Hochman, J.S., Hutter, A.M., Jr., Kaul, S., Wolk, M.J.: American College of Cardiology/American Heart Association Expert Consensus document on electron-beam computed tomography for the diagnosis and prognosis of coronary artery disease. Circulation **102** (2000) 126–140
63. Grundy, S.M.: Age as a risk factor: you are as old as your arteries. Am J Cardiol **83** (1999) 1455–1457; A1457
64. Boissel, J.P., Collet, J.P., Moleur, P., Haugh, M.: Surrogate endpoints: a basis for a rational approach. Eur J Clin Pharmacol **43** (1992) 235–244
65. Espeland, M.A., O'Leary D, H., Terry, J.G., Morgan, T., Evans, G., Mudra, H.: Carotid intimal-media thickness as a surrogate for cardiovascular disease events in trials of HMG-CoA reductase inhibitors. Curr Control Trials Cardiovasc Med **6** (2005) 3
66. Mitchell, J.R., Schwartz, C.J.: Relationship between arterial disease in different sites. A study of the aorta and coronary, carotid, and iliac arteries. Br Med J **1** (1962) 1293–1301
67. Wofford, J.L., Kahl, F.R., Howard, G.R., McKinney, W.M., Toole, J.F., Crouse, J.R., 3rd: Relation of extent of extracranial carotid artery atherosclerosis as measured by B-mode ultrasound to the extent of coronary atherosclerosis. Arterioscler Thromb **11** (1991) 1786–1794
68. Dormandy, J., Heeck, L., Vig, S.: Lower-extremity arteriosclerosis as a reflection of a systemic process: implications for concomitant coronary and carotid disease. Semin Vasc Surg **12** (1999) 118–122
69. Revkin, J.H., Shear, C.L., Pouleur, H.G., Ryder, S.W., Orloff, D.G.: Biomarkers in the prevention and treatment of atherosclerosis: need, validation, and future. Pharmacol Rev **59** (2007) 40–53
70. Rossouw, J.E.: Lipid-lowering interventions in angiographic trials. Am J Cardiol **76** (1995) 86C–92C
71. Smith, S.C., Jr., Blair, S.N., Criqui, M.H., Fletcher, G.F., Fuster, V., Gersh, B.J., Gotto, A.M., Gould, K.L., Greenland, P., Grundy, S.M., Hill, M.N., Hlatky, M.A., Houston-Miller, N., Krauss, R.M., LaRosa, J., Ockene, I.S., Oparil, S., Pearson, T.A., Rapaport, E., Starke, R.D.: Preventing heart attack and death in patients with coronary disease. Endorsed by the board of trustees of the American College of Cardiology. Cardiovasc Nurs **32** (1996) 26–28
72. Simon, A., Gariepy, J., Chironi, G., Megnien, J.L., Levenson, J.: Intima-media thickness: a new tool for diagnosis and treatment of cardiovascular risk. J Hypertens **20** (2002) 159–169
73. Duivenvoorden, R., Nederveen, A.J., de Groot, E., Kastelein, J.J.: Atherosclerosis imaging as a benchmark in the development of novel cardiovasular drugs. Curr Opin Lipidol **18** (2007) 613–621
74. Prati, F., Arbustini, E., Labellarte, A., Dal Bello, B., Sommariva, L., Mallus, M.T., Pagano, A., Boccanelli, A.: Correlation between high frequency intravascular ultrasound and histomorphology in human coronary arteries. Heart **85** (2001) 567–570
75. Tuzcu, E.M., Kapadia, S.R., Tutar, E., Ziada, K.M., Hobbs, R.E., McCarthy, P.M., Young, J.B., Nissen, S.E.: High prevalence of coronary atherosclerosis in asymptomatic teenagers and young adults: evidence from intravascular ultrasound. Circulation **103** (2001) 2705–2710
76. von Birgelen, C., Hartmann, M., Mintz, G.S., van Houwelingen, K.G., Deppermann, N., Schmermund, A., Bose, D., Eggebrecht, H., Neumann, T., Gossl, M., Wieneke, H., Erbel, R.: Relationship between cardiovascular risk as predicted by established risk scores versus plaque progression as measured by serial intravascular ultrasound in left main coronary arteries. Circulation **110** (2004) 1579–1585

77. Hertog, M.G.: Workshop on CV prevention with a focus on metabolic syndrome and surrogate markers. Taskforce 1 of the Cardiovascular Round Table, European Society of Cardiology: Executive Summary Report. Eur Heart J **27** (2006) 114–115
78. Fayad, Z.A., Nahar, T., Fallon, J.T., Goldman, M., Aguinaldo, J.G., Badimon, J.J., Shinnar, M., Chesebro, J.H., Fuster, V.: In vivo magnetic resonance evaluation of atherosclerotic plaques in the human thoracic aorta: a comparison with transesophageal echocardiography. Circulation **101** (2000) 2503–2509
79. Fayad, Z.A., Fuster, V.: Characterization of atherosclerotic plaques by magnetic resonance imaging. Ann N Y Acad Sci **902** (2000) 173–186
80. Momiyama, Y., Kato, R., Fayad, Z.A., Tanaka, N., Taniguchi, H., Ohmori, R., Kihara, T., Kameyama, A., Miyazaki, K., Kimura, K., Arakawa, K., Kusuhara, M., Nagata, M., Nakamura, H., Ohsuzu, F.: A possible association between coronary plaque instability and complex plaques in abdominal aorta. Arterioscler Thromb Vasc Biol **26** (2006) 903–909
81. Lima, J.A., Desai, M.Y., Steen, H., Warren, W.P., Gautam, S., Lai, S.: Statin-induced cholesterol lowering and plaque regression after 6 months of magnetic resonance imaging-monitored therapy. Circulation **110** (2004) 2336–2341
82. Wiviott, S.D., Cannon, C.P.: Update on lipid-lowering therapy and LDL-cholesterol targets. Nat Clin Pract Cardiovasc Med **3** (2006) 424–436
83. Smith, E.E., Abdullah, A.R., Amirfarzan, H., Schwamm, L.H.: Serum lipid profile on admission for ischemic stroke: failure to meet National Cholesterol Education Program Adult Treatment Panel (NCEP-ATPIII) guidelines. Neurology **68** (2007) 660–665
84. Singh, V., Deedwania, P.: Reducing morbidity and mortality in high risk patients with statins. Vasc Health Risk Manag **5** (2009) 495–507
85. Kent, S.M., Coyle, L.C., Flaherty, P.J., Markwood, T.T., Taylor, A.J.: Marked low-density lipoprotein cholesterol reduction below current national cholesterol education program targets provides the greatest reduction in carotid atherosclerosis. Clin Cardiol **27** (2004) 17–21
86. Endo, A.: The discovery and development of HMG-CoA reductase inhibitors. J Lipid Res **33** (1992) 1569–1582
87. Vega, G.L., Grundy, S.M.: Management of primary mixed hyperlipidemia with lovastatin. Arch Intern Med **150** (1990) 1313–1319
88. Broyles, F.E., Walden, C.E., Hunninghake, D.B., Hill-Williams, D., Knopp, R.H.: Effect of fluvastatin on intermediate density lipoprotein (remnants) and other lipoprotein levels in hypercholesterolemia. Am J Cardiol **76** (1995) 129A–135A
89. Bakker-Arkema, R.G., Davidson, M.H., Goldstein, R.J., Davignon, J., Isaacsohn, J.L., Weiss, S.R., Keilson, L.M., Brown, W.V., Miller, V.T., Shurzinske, L.J., Black, D.M.: Efficacy and safety of a new HMG-CoA reductase inhibitor, atorvastatin, in patients with hypertriglyceridemia. JAMA **275** (1996) 128–133
90. Arad, Y., Ramakrishnan, R., Ginsberg, H.N.: Effects of lovastatin therapy on very-low-density lipoprotein triglyceride metabolism in subjects with combined hyperlipidemia: evidence for reduced assembly and secretion of triglyceride-rich lipoproteins. Metabolism **41** (1992) 487–493
91. Twisk, J., Gillian-Daniel, D.L., Tebon, A., Wang, L., Barrett, P.H., Attie, A.D.: The role of the LDL receptor in apolipoprotein B secretion. J Clin Invest **105** (2000) 521–532
92. Stein, E.A., Lane, M., Laskarzewski, P.: Comparison of statins in hypertriglyceridemia. Am J Cardiol **81** (1998) 66B–69B
93. Jones, P., Kafonek, S., Laurora, I., Hunninghake, D.: Comparative dose efficacy study of atorvastatin versus simvastatin, pravastatin, lovastatin, and fluvastatin in patients with hypercholesterolemia (the CURVES study). Am J Cardiol **81** (1998) 582–587
94. Sacks, F.M., Pfeffer, M.A., Moye, L.A., Rouleau, J.L., Rutherford, J.D., Cole, T.G., Brown, L., Warnica, J.W., Arnold, J.M., Wun, C.C., Davis, B.R., Braunwald, E.: The effect of pravastatin on coronary events after myocardial infarction in patients with average cholesterol levels. Cholesterol and Recurrent Events Trial investigators. N Engl J Med **335** (1996) 1001–1009
95. Prevention of cardiovascular events and death with pravastatin in patients with coronary heart disease and a broad range of initial cholesterol levels. The Long-Term Intervention with Pravastatin in Ischaemic Disease (LIPID) Study Group. N Engl J Med **339** (1998) 1349–1357

96. Downs, J.R., Clearfield, M., Weis, S., Whitney, E., Shapiro, D.R., Beere, P.A., Langendorfer, A., Stein, E.A., Kruyer, W., Gotto, A.M., Jr.: Primary prevention of acute coronary events with lovastatin in men and women with average cholesterol levels: results of AFCAPS/TexCAPS. Air Force/Texas Coronary Atherosclerosis Prevention Study. JAMA **279** (1998) 1615–1622
97. Randomised trial of cholesterol lowering in 4444 patients with coronary heart disease: the Scandinavian Simvastatin Survival Study (4S). Lancet **344** (1994) 1383–1389
98. LaRosa, J.C., He, J., Vupputuri, S.: Effect of statins on risk of coronary disease: a meta-analysis of randomized controlled trials. JAMA **282** (1999) 2340–2346
99. Shepherd, J., Cobbe, S.M., Ford, I., Isles, C.G., Lorimer, A.R., MacFarlane, P.W., McKillop, J.H., Packard, C.J.: Prevention of coronary heart disease with pravastatin in men with hypercholesterolemia. West of Scotland Coronary Prevention Study Group. N Engl J Med **333** (1995) 1301–1307
100. Maron, D.J., Fazio, S., Linton, M.F.: Current perspectives on statins. Circulation **101** (2000) 207–213
101. Pyorala, K., Pedersen, T.R., Kjekshus, J., Faergeman, O., Olsson, A.G., Thorgeirsson, G.: Cholesterol lowering with simvastatin improves prognosis of diabetic patients with coronary heart disease. A subgroup analysis of the Scandinavian Simvastatin Survival Study (4S). Diabetes Care **20** (1997) 614–620
102. The effect of aggressive lowering of low-density lipoprotein cholesterol levels and low-dose anticoagulation on obstructive changes in saphenous-vein coronary-artery bypass grafts. The Post Coronary Artery Bypass Graft Trial Investigators. N Engl J Med **336** (1997) 153–162
103. Knatterud, G.L., Rosenberg, Y., Campeau, L., Geller, N.L., Hunninghake, D.B., Forman, S.A., Forrester, J.S., Gobel, F.L., Herd, J.A., Hickey, A., Hoogwerf, B.J., Terrin, M.L., White, C.: Long-term effects on clinical outcomes of aggressive lowering of low-density lipoprotein cholesterol levels and low-dose anticoagulation in the post coronary artery bypass graft trial. Post CABG Investigators. Circulation **102** (2000) 157–165
104. Blankenhorn, D.H., Azen, S.P., Kramsch, D.M., Mack, W.J., Cashin-Hemphill, L., Hodis, H.N., DeBoer, L.W., Mahrer, P.R., Masteller, M.J., Vailas, L.I., Alaupovic, P., Hirsch, L.J.: Coronary angiographic changes with lovastatin therapy. The Monitored Atherosclerosis Regression Study (MARS). Ann Intern Med **119** (1993) 969–976
105. Kroon, A.A., Aengevaeren, W.R., van der Werf, T., Uijen, G.J., Reiber, J.H., Bruschke, A.V., Stalenhoef, A.F.: LDL-Apheresis Atherosclerosis Regression Study (LAARS). Effect of aggressive versus conventional lipid lowering treatment on coronary atherosclerosis. Circulation **93** (1996) 1826–1835
106. Pitt, B., Waters, D., Brown, W.V., van Boven, A.J., Schwartz, L., Title, L.M., Eisenberg, D., Shurzinske, L., McCormick, L.S.: Aggressive lipid-lowering therapy compared with angioplasty in stable coronary artery disease. Atorvastatin versus Revascularization Treatment Investigators. N Engl J Med **341** (1999) 70–76
107. Brown, W.V.: The benefit of aggressive lipid lowering. Atheroscler Suppl **1** (2000) 15–19
108. Schwartz, G.G., Olsson, A.G., Ezekowitz, M.D., Ganz, P., Oliver, M.F., Waters, D., Zeiher, A., Chaitman, B.R., Leslie, S., Stern, T.: Effects of atorvastatin on early recurrent ischemic events in acute coronary syndromes: the MIRACL study: a randomized controlled trial. JAMA **285** (2001) 1711–1718
109. Stenestrand, U., Wallentin, L.: Early statin treatment following acute myocardial infarction and 1-year survival. JAMA **285** (2001) 430–436
110. West of Scotland Coronary Prevention Study: identification of high-risk groups and comparison with other cardiovascular intervention trials. Lancet **348** (1996) 1339–1342
111. Ballantyne, C.M., Herd, J.A., Dunn, J.K., Jones, P.H., Farmer, J.A., Gotto, A.M., Jr.: Effects of lipid lowering therapy on progression of coronary and carotid artery disease. Curr Opin Lipidol **8** (1997) 354–361
112. Brown, B.G., Zhao, X.Q.: Importance of endothelial function in mediating the benefits of lipid-lowering therapy. Am J Cardiol **82** (1998) 49T–52T

113. Buchwald, H., Varco, R.L., Matts, J.P., Long, J.M., Fitch, L.L., Campbell, G.S., Pearce, M.B., Yellin, A.E., Edmiston, W.A., Smink, R.D., Jr., et al.: Effect of partial ileal bypass surgery on mortality and morbidity from coronary heart disease in patients with hypercholesterolemia. Report of the Program on the Surgical Control of the Hyperlipidemias (POSCH). N Engl J Med **323** (1990) 946–955
114. Brown, B.G., Zhao, X.Q., Sacco, D.E., Albers, J.J.: Arteriographic view of treatment to achieve regression of coronary atherosclerosis and to prevent plaque disruption and clinical cardiovascular events. Br Heart J **69** (1993) S48–S53
115. Furberg, C.D., Adams, H.P., Jr., Applegate, W.B., Byington, R.P., Espeland, M.A., Hartwell, T., Hunninghake, D.B., Lefkowitz, D.S., Probstfield, J., Riley, W.A., et al.: Effect of lovastatin on early carotid atherosclerosis and cardiovascular events. Asymptomatic Carotid Artery Progression Study (ACAPS) Research Group. Circulation **90** (1994) 1679–1687
116. Salonen, R., Nyyssonen, K., Porkkala, E., Rummukainen, J., Belder, R., Park, J.S., Salonen, J.T.: Kuopio Atherosclerosis Prevention Study (KAPS). A population-based primary preventive trial of the effect of LDL lowering on atherosclerotic progression in carotid and femoral arteries. Circulation **92** (1995) 1758–1764
117. Crouse, J.R., 3rd, Byington, R.P., Bond, M.G., Espeland, M.A., Craven, T.E., Sprinkle, J.W., McGovern, M.E., Furberg, C.D.: Pravastatin, lipids, and atherosclerosis in the carotid arteries (PLAC-II). Am J Cardiol **75** (1995) 455–459
118. Salonen, R., Nyssonen, K., Porkkala-Sarataho, E., Salonen, J.T.: The Kuopio Atherosclerosis Prevention Study (KAPS): effect of pravastatin treatment on lipids, oxidation resistance of lipoproteins, and atherosclerotic progression. Am J Cardiol **76** (1995) 34C–39C
119. Mercuri, M., Bond, M.G., Sirtori, C.R., Veglia, F., Crepaldi, G., Feruglio, F.S., Descovich, G., Ricci, G., Rubba, P., Mancini, M., Gallus, G., Bianchi, G., D'Alo, G., Ventura, A.: Pravastatin reduces carotid intima-media thickness progression in an asymptomatic hypercholesterolemic mediterranean population: the Carotid Atherosclerosis Italian Ultrasound Study. Am J Med **101** (1996) 627–634
120. de Groot, E., Jukema, J.W., Montauban van Swijndregt, A.D., Zwinderman, A.H., Ackerstaff, R.G., van der Steen, A.F., Bom, N., Lie, K.I., Bruschke, A.V.: B-mode ultrasound assessment of pravastatin treatment effect on carotid and femoral artery walls and its correlations with coronary arteriographic findings: a report of the Regression Growth Evaluation Statin Study (REGRESS). J Am Coll Cardiol **31** (1998) 1561–1567
121. Hedblad, B., Wikstrand, J., Janzon, L., Wedel, H., Berglund, G.: Low-dose metoprolol CR/XL and fluvastatin slow progression of carotid intima-media thickness: Main results from the Beta-Blocker Cholesterol-Lowering Asymptomatic Plaque Study (BCAPS). Circulation **103** (2001) 1721–1726
122. Sawayama, Y., Shimizu, C., Maeda, N., Tatsukawa, M., Kinukawa, N., Koyanagi, S., Kashiwagi, S., Hayashi, J.: Effects of probucol and pravastatin on common carotid atherosclerosis in patients with asymptomatic hypercholesterolemia. Fukuoka Atherosclerosis Trial (FAST). J Am Coll Cardiol **39** (2002) 610–616
123. Al Badarin, F.J., Kullo, I.J., Kopecky, S.L., Thomas, R.J.: Impact of ezetimibe on atherosclerosis: is the jury still out? Mayo Clin Proc **84** (2009) 353–361
124. Nolting, P.R., de Groot, E., Zwinderman, A.H., Buirma, R.J., Trip, M.D., Kastelein, J.J.: Regression of carotid and femoral artery intima-media thickness in familial hypercholesterolemia: treatment with simvastatin. Arch Intern Med **163** (2003) 1837–1841
125. Kastelein, J.J., Akdim, F., Stroes, E.S., Zwinderman, A.H., Bots, M.L., Stalenhoef, A.F., Visseren, F.L., Sijbrands, E.J., Trip, M.D., Stein, E.A., Gaudet, D., Duivenvoorden, R., Veltri, E.P., Marais, A.D., de Groot, E.: Simvastatin with or without ezetimibe in familial hypercholesterolemia. N Engl J Med **358** (2008) 1431–1443
126. Smilde, T.J., van Wissen, S., Wollersheim, H., Trip, M.D., Kastelein, J.J., Stalenhoef, A.F.: Effect of aggressive versus conventional lipid lowering on atherosclerosis progression in familial hypercholesterolaemia (ASAP): a prospective, randomised, double-blind trial. Lancet **357** (2001) 577–581
127. Taylor, A.J., Kent, S.M., Flaherty, P.J., Coyle, L.C., Markwood, T.T., Vernalis, M.N.: ARBITER: arterial biology for the investigation of the treatment effects of reducing

cholesterol: a randomized trial comparing the effects of atorvastatin and pravastatin on carotid intima medial thickness. Circulation **106** (2002) 2055–2060

128. Crouse, J.R., 3rd, Raichlen, J.S., Riley, W.A., Evans, G.W., Palmer, M.K., O'Leary, D.H., Grobbee, D.E., Bots, M.L.: Effect of rosuvastatin on progression of carotid intima-media thickness in low-risk individuals with subclinical atherosclerosis: the METEOR Trial. JAMA **297** (2007) 1344–1353
129. Jadad, A.R., Moore, R.A., Carroll, D., Jenkinson, C., Reynolds, D.J., Gavaghan, D.J., McQuay, H.J.: Assessing the quality of reports of randomized clinical trials: is blinding necessary? Control Clin Trials **17** (1996) 1–12
130. Nissen, S.E., Nicholls, S.J., Sipahi, I., Libby, P., Raichlen, J.S., Ballantyne, C.M., Davignon, J., Erbel, R., Fruchart, J.C., Tardif, J.C., Schoenhagen, P., Crowe, T., Cain, V., Wolski, K., Goormastic, M., Tuzcu, E.M.: Effect of very high-intensity statin therapy on regression of coronary atherosclerosis: the ASTEROID trial. JAMA **295** (2006) 1556–1565
131. Nissen, S.E., Tuzcu, E.M., Schoenhagen, P., Brown, B.G., Ganz, P., Vogel, R.A., Crowe, T., Howard, G., Cooper, C.J., Brodie, B., Grines, C.L., DeMaria, A.N.: Effect of intensive compared with moderate lipid-lowering therapy on progression of coronary atherosclerosis: a randomized controlled trial. JAMA **291** (2004) 1071–1080
132. Okazaki, S., Yokoyama, T., Miyauchi, K., Shimada, K., Kurata, T., Sato, H., Daida, H.: Early statin treatment in patients with acute coronary syndrome: demonstration of the beneficial effect on atherosclerotic lesions by serial volumetric intravascular ultrasound analysis during half a year after coronary event: the ESTABLISH Study. Circulation **110** (2004) 1061–1068
133. Takagi, T., Yoshida, K., Akasaka, T., Hozumi, T., Morioka, S., Yoshikawa, J.: Intravascular ultrasound analysis of reduction in progression of coronary narrowing by treatment with pravastatin. Am J Cardiol **79** (1997) 1673–1676
134. Cannon, C.P., Braunwald, E., McCabe, C.H., Rader, D.J., Rouleau, J.L., Belder, R., Joyal, S.V., Hill, K.A., Pfeffer, M.A., Skene, A.M.: Intensive versus moderate lipid lowering with statins after acute coronary syndromes. N Engl J Med **350** (2004) 1495–1504
135. Yonemura, A., Momiyama, Y., Fayad, Z.A., Ayaori, M., Ohmori, R., Higashi, K., Kihara, T., Sawada, S., Iwamoto, N., Ogura, M., Taniguchi, H., Kusuhara, M., Nagata, M., Nakamura, H., Tamai, S., Ohsuzu, F.: Effect of lipid-lowering therapy with atorvastatin on atherosclerotic aortic plaques detected by noninvasive magnetic resonance imaging. J Am Coll Cardiol **45** (2005) 733–742
136. Ayaori, M., Momiyama, Y., Fayad, Z.A., Yonemura, A., Ohmori, R., Kihara, T., Tanaka, N., Nakaya, K., Ogura, M., Sawada, S., Taniguchi, H., Kusuhara, M., Nagata, M., Nakamura, H., Ohsuzu, F.: Effect of bezafibrate therapy on atherosclerotic aortic plaques detected by MRI in dyslipidemic patients with hypertriglyceridemia. Atherosclerosis **196** (2008) 425–433
137. Corti, R., Fuster, V., Fayad, Z.A., Worthley, S.G., Helft, G., Smith, D., Weinberger, J., Wentzel, J., Mizsei, G., Mercuri, M., Badimon, J.J.: Lipid lowering by simvastatin induces regression of human atherosclerotic lesions: two years' follow-up by high-resolution noninvasive magnetic resonance imaging. Circulation **106** (2002) 2884–2887
138. Yonemura, A., Momiyama, Y., Fayad, Z.A., Ayaori, M., Ohmori, R., Kihara, T., Tanaka, N., Nakaya, K., Ogura, M., Taniguchi, H., Kusuhara, M., Nagata, M., Nakamura, H., Tamai, S., Ohsuzu, F.: Effect of lipid-lowering therapy with atorvastatin on atherosclerotic aortic plaques: a 2-year follow-up by noninvasive MRI. Eur J Cardiovasc Prev Rehabil **16** (2009) 222–228
139. Liao, J.K.: Beyond lipid lowering: the role of statins in vascular protection. Int J Cardiol **86** (2002) 5–18
140. Nilsson, J.: Nobel seminar: molecular medicine – cell biology in disease. The mevalonate pathway – a link between cellular cholesterol homeostasis and growth regulation. J Intern Med **232** (1992) 10–13
141. Takemoto, M., Liao, J.K.: Pleiotropic effects of 3-hydroxy-3-methylglutaryl coenzyme a reductase inhibitors. Arterioscler Thromb Vasc Biol **21** (2001) 1712–1719
142. Rikitake, Y., Liao, J.K.: Rho GTPases, statins, and nitric oxide. Circ Res **97** (2005) 1232–1235

143. Laufs, U., Wassmann, S., Hilgers, S., Ribaudo, N., Bohm, M., Nickenig, G.: Rapid effects on vascular function after initiation and withdrawal of atorvastatin in healthy, normocholesterolemic men. Am J Cardiol **88** (2001) 1306–1307
144. Weitz-Schmidt, G., Welzenbach, K., Brinkmann, V., Kamata, T., Kallen, J., Bruns, C., Cottens, S., Takada, Y., Hommel, U.: Statins selectively inhibit leukocyte function antigen-1 by binding to a novel regulatory integrin site. Nat Med **7** (2001) 687–692
145. Ridker, P.M., Rifai, N., Rose, L., Buring, J.E., Cook, N.R.: Comparison of C-reactive protein and low-density lipoprotein cholesterol levels in the prediction of first cardiovascular events. N Engl J Med **347** (2002) 1557–1565
146. Kayikcioglu, M., Can, L., Evrengul, H., Payzin, S., Kultursay, H.: The effect of statin therapy on ventricular late potentials in acute myocardial infarction. Int J Cardiol **90** (2003) 63–72
147. Young-Xu, Y., Jabbour, S., Goldberg, R., Blatt, C.M., Graboys, T., Bilchik, B., Ravid, S.: Usefulness of statin drugs in protecting against atrial fibrillation in patients with coronary artery disease. Am J Cardiol **92** (2003) 1379–1383
148. Aronow, W.S., Ahn, C.: Frequency of congestive heart failure in older persons with prior myocardial infarction and serum low-density lipoprotein cholesterol ≥125 mg/dl treated with statins versus no lipid-lowering drug. Am J Cardiol **90** (2002) 147–149
149. Node, K., Fujita, M., Kitakaze, M., Hori, M., Liao, J.K.: Short-term statin therapy improves cardiac function and symptoms in patients with idiopathic dilated cardiomyopathy. Circulation **108** (2003) 839–843
150. Hebert, P.R., Gaziano, J.M., Chan, K.S., Hennekens, C.H.: Cholesterol lowering with statin drugs, risk of stroke, and total mortality. An overview of randomized trials. JAMA **278** (1997) 313–321
151. Corvol, J.C., Bouzamondo, A., Sirol, M., Hulot, J.S., Sanchez, P., Lechat, P.: Differential effects of lipid-lowering therapies on stroke prevention: a meta-analysis of randomized trials. Arch Intern Med **163** (2003) 669–676
152. Freeman, D.J., Norrie, J., Sattar, N., Neely, R.D., Cobbe, S.M., Ford, I., Isles, C., Lorimer, A.R., Macfarlane, P.W., McKillop, J.H., Packard, C.J., Shepherd, J., Gaw, A.: Pravastatin and the development of diabetes mellitus: evidence for a protective treatment effect in the West of Scotland Coronary Prevention Study. Circulation **103** (2001) 357–362
153. Crisby, M., Carlson, L.A., Winblad, B.: Statins in the prevention and treatment of Alzheimer disease. Alzheimer Dis Assoc Disord **16** (2002) 131–136
154. Kirsch, C., Eckert, G.P., Mueller, W.E.: Statin effects on cholesterol micro-domains in brain plasma membranes. Biochem Pharmacol **65** (2003) 843–856
155. Leung, B.P., Sattar, N., Crilly, A., Prach, M., McCarey, D.W., Payne, H., Madhok, R., Campbell, C., Gracie, J.A., Liew, F.Y., McInnes, I.B.: A novel anti-inflammatory role for simvastatin in inflammatory arthritis. J Immunol **170** (2003) 1524–1530
156. Wong, R.W., Rabie, A.B.: Statin collagen grafts used to repair defects in the parietal bone of rabbits. Br J Oral Maxillofac Surg **41** (2003) 244–248
157. Meroni, P.L., Raschi, E., Testoni, C., Tincani, A., Balestrieri, G.: Antiphospholipid antibodies and the endothelium. Rheum Dis Clin North Am **27** (2001) 587–602
158. van de Donk, N.W., Kamphuis, M.M., Lokhorst, H.M., Bloem, A.C.: The cholesterol lowering drug lovastatin induces cell death in myeloma plasma cells. Leukemia **16** (2002) 1362–1371
159. Neuhaus, O., Archelos, J.J., Hartung, H.P.: Statins in multiple sclerosis: a new therapeutic option? Mult Scler **9** (2003) 429–430
160. Gould, A.L., Cariski, A.T.: Meta-analysis by the Program on the Surgical Control of the Hyperlipidemias (POSCH) investigators. Ann Surg **227** (1998) 314–316
161. Cholesterol, diastolic blood pressure, and stroke: 13,000 strokes in 450,000 people in 45 prospective cohorts. Prospective studies collaboration. Lancet **346** (1995) 1647–1653
162. Dechend, R., Fiebler, A., Lindschau, C., Bischoff, H., Muller, D., Park, J.K., Dietz, R., Haller, H., Luft, F.C.: Modulating angiotensin II-induced inflammation by HMG Co-A reductase inhibition. Am J Hypertens **14** (2001) 55S–61S
163. Sarkar, K., Sinha, A.K., Mehta, J.L.: The role of statins in endothelial dysfunction in hypertension. Curr Opin Cardiol **21** (2006) 316–321

164. Davignon, J.: Emphasis on pleiotropic effects, a new paradigm shift? [Review]. Coron Artery Dis **15** (2004) 223–225
165. Venkatesh, P.K.M., Caskey, D.M., Reddy, P.C.M.: Therapies to increase high-density lipoprotein cholesterol and their effect on cardiovascular outcomes and regression of atherosclerosis. Am J Med Sci **336** (2008) 64–68
166. Eisenberg, S., Gavish, D., Oschry, Y., Fainaru, M., Deckelbaum, R.J.: Abnormalities in very low, low and high density lipoproteins in hypertriglyceridemia. Reversal toward normal with bezafibrate treatment. J Clin Invest **74** (1984) 470–482
167. Schoonjans, K., Staels, B., Auwerx, J.: Role of the peroxisome proliferator-activated receptor (PPAR) in mediating the effects of fibrates and fatty acids on gene expression. J Lipid Res **37** (1996) 907–925
168. Fruchart, J.C., Brewer, H.B., Jr., Leitersdorf, E.: Consensus for the use of fibrates in the treatment of dyslipoproteinemia and coronary heart disease. Fibrate Consensus Group. Am J Cardiol **81** (1998) 912–917
169. Vu-Dac, N., Schoonjans, K., Kosykh, V., Dallongeville, J., Fruchart, J.C., Staels, B., Auwerx, J.: Fibrates increase human apolipoprotein A-II expression through activation of the peroxisome proliferator-activated receptor. J Clin Invest **96** (1995) 741–750
170. Neve, B.P., Fruchart, J.C., Staels, B.: Role of the peroxisome proliferator-activated receptors (PPAR) in atherosclerosis. Biochem Pharmacol **60** (2000) 1245–1250
171. Pineda Torra, I., Gervois, P., Staels, B.: Peroxisome proliferator-activated receptor alpha in metabolic disease, inflammation, atherosclerosis and aging. Curr Opin Lipidol **10** (1999) 151–159
172. Pauciullo, P., Marotta, G., Rubba, P., Cortese, C., Caruso, M.G., Gnasso, A., Fischetti, A., Motti, C., Mancini, M.: Serum lipoproteins, apolipoproteins and very low density lipoprotein subfractions during 6-month fibrate treatment in primary hypertriglyceridaemia. J Intern Med **228** (1990) 425–430
173. Leaf, D.A., Connor, W.E., Illingworth, D.R., Bacon, S.P., Sexton, G.: The hypolipidemic effects of gemfibrozil in type V hyperlipidemia. A double-blind, crossover study. JAMA **262** (1989) 3154–3160
174. Kornitzer, M., Dramaix, M., Vandenbroek, M.D., Everaert, L., Gerlinger, C.: Efficacy and tolerance of 200 mg micronised fenofibrate administered over a 6-month period in hyperlipidaemic patients: an open Belgian multicenter study. Atherosclerosis **110 Suppl** (1994) S49–S54
175. A co-operative trial in the primary prevention of ischaemic heart disease using clofibrate. Report from the Committee of Principal Investigators. Br Heart J **40** (1978) 1069–1118
176. Frick, M.H., Elo, O., Haapa, K., Heinonen, O.P., Heinsalmi, P., Helo, P., Huttunen, J.K., Kaitaniemi, P., Koskinen, P., Manninen, V., et al.: Helsinki Heart Study: primary-prevention trial with gemfibrozil in middle-aged men with dyslipidemia. Safety of treatment, changes in risk factors, and incidence of coronary heart disease. N Engl J Med **317** (1987) 1237–1245
177. Clofibrate and niacin in coronary heart disease. JAMA **231** (1975) 360–381
178. Trial of clofibrate in the treatment of ischaemic heart disease. Five-year study by a group of physicians of the Newcastle upon Tyne region. Br Med J **4** (1971) 767–775
179. Ischaemic heart disease: a secondary prevention trial using clofibrate. Report by a research committee of the Scottish Society of Physicians. Br Med J **4** (1971) 775–784
180. Carlson, L.A., Rosenhamer, G.: Reduction of mortality in the Stockholm Ischaemic Heart Disease Secondary Prevention Study by combined treatment with clofibrate and nicotinic acid. Acta Med Scand **223** (1988) 405–418
181. Rubins, H.B., Robins, S.J., Collins, D., Fye, C.L., Anderson, J.W., Elam, M.B., Faas, F.H., Linares, E., Schaefer, E.J., Schectman, G., Wilt, T.J., Wittes, J.: Gemfibrozil for the secondary prevention of coronary heart disease in men with low levels of high-density lipoprotein cholesterol. Veterans Affairs High-Density Lipoprotein Cholesterol Intervention Trial Study Group. N Engl J Med **341** (1999) 410–418
182. Secondary prevention by raising HDL cholesterol and reducing triglycerides in patients with coronary artery disease: the Bezafibrate Infarction Prevention (BIP) study. Circulation **102** (2000) 21–27

183. Huttunen, J.K., Heinonen, O.P., Manninen, V., Koskinen, P., Hakulinen, T., Teppo, L., Manttari, M., Frick, M.H.: The Helsinki Heart Study: an 8.5-year safety and mortality follow-up. J Intern Med **235** (1994) 31–39
184. Kamanna, V.S., Kashyap, M.L.: Nicotinic acid (niacin) receptor agonists: will they be useful therapeutic agents? Am J Cardiol **100** (2007) S53–S61
185. Vega, G.L., Grundy, S.M.: Lipoprotein responses to treatment with lovastatin, gemfibrozil, and nicotinic acid in normolipidemic patients with hypoalphalipoproteinemia. Arch Intern Med **154** (1994) 73–82
186. Luria, M.H.: Effect of low-dose niacin on high-density lipoprotein cholesterol and total cholesterol/high-density lipoprotein cholesterol ratio. Arch Intern Med **148** (1988) 2493–2495
187. McKenney, J.M., Proctor, J.D., Harris, S., Chinchili, V.M.: A comparison of the efficacy and toxic effects of sustained- vs. immediate-release niacin in hypercholesterolemic patients. JAMA **271** (1994) 672–677
188. Illingworth, D.R., Stein, E.A., Mitchel, Y.B., Dujovne, C.A., Frost, P.H., Knopp, R.H., Tun, P., Zupkis, R.V., Greguski, R.A.: Comparative effects of lovastatin and niacin in primary hypercholesterolemia. A prospective trial. Arch Intern Med **154** (1994) 1586–1595
189. Canner, P.L., Berge, K.G., Wenger, N.K., Stamler, J., Friedman, L., Prineas, R.J., Friedewald, W.: Fifteen year mortality in Coronary Drug Project patients: long-term benefit with niacin. J Am Coll Cardiol **8** (1986) 1245–1255
190. Blankenhorn, D.H., Nessim, S.A., Johnson, R.L., Sanmarco, M.E., Azen, S.P., Cashin-Hemphill, L.: Beneficial effects of combined colestipol-niacin therapy on coronary atherosclerosis and coronary venous bypass grafts. JAMA **257** (1987) 3233–3240
191. Brown, G., Albers, J.J., Fisher, L.D., Schaefer, S.M., Lin, J.T., Kaplan, C., Zhao, X.Q., Bisson, B.D., Fitzpatrick, V.F., Dodge, H.T.: Regression of coronary artery disease as a result of intensive lipid-lowering therapy in men with high levels of apolipoprotein B. N Engl J Med **323** (1990) 1289–1298
192. Brown, B.G., Zhao, X.Q., Chait, A., Fisher, L.D., Cheung, M.C., Morse, J.S., Dowdy, A.A., Marino, E.K., Bolson, E.L., Alaupovic, P., Frohlich, J., Albers, J.J.: Simvastatin and niacin, antioxidant vitamins, or the combination for the prevention of coronary disease. N Engl J Med **345** (2001) 1583–1592
193. The Lipid Research Clinics Coronary Primary Prevention Trial results. I. Reduction in incidence of coronary heart disease. JAMA **251** (1984) 351–364
194. The Lipid Research Clinics Coronary Primary Prevention Trial results. II. The relationship of reduction in incidence of coronary heart disease to cholesterol lowering. JAMA **251** (1984) 365–374
195. Hamilton, P.: Role of ezetimibe in the management of patients with atherosclerosis. Coron Artery Dis **20** (2009) 169–174
196. Bays, H.E., Moore, P.B., Drehobl, M.A., Rosenblatt, S., Toth, P.D., Dujovne, C.A., Knopp, R.H., Lipka, L.J., Lebeaut, A.P., Yang, B., Mellars, L.E., Cuffie-Jackson, C., Veltri, E.P.: Effectiveness and tolerability of ezetimibe in patients with primary hypercholesterolemia: pooled analysis of two phase II studies. Clin Ther **23** (2001) 1209–1230
197. Knopp, R.H., Dujovne, C.A., Le Beaut, A., Lipka, L.J., Suresh, R., Veltri, E.P.: Evaluation of the efficacy, safety, and tolerability of ezetimibe in primary hypercholesterolaemia: a pooled analysis from two controlled phase III clinical studies. Int J Clin Pract **57** (2003) 363–368
198. Melani, L., Mills, R., Hassman, D., Lipetz, R., Lipka, L., LeBeaut, A., Suresh, R., Mukhopadhyay, P., Veltri, E.: Efficacy and safety of ezetimibe coadministered with pravastatin in patients with primary hypercholesterolemia: a prospective, randomized, double-blind trial. Eur Heart J **24** (2003) 717–728
199. Lewington, S., Clarke, R., Qizilbash, N., Peto, R., Collins, R.: Age-specific relevance of usual blood pressure to vascular mortality: a meta-analysis of individual data for one million adults in 61 prospective studies. Lancet **360** (2002) 1903–1913
200. Jacoby, D.S., Rader, D.J.: Renin-angiotensin system and atherothrombotic disease: from genes to treatment. Arch Intern Med **163** (2003) 1155–1164
201. Schmitz, G., Hankowitz, J., Kovacs, E.M.: Cellular processes in atherogenesis: potential targets of Ca2+ channel blockers. Atherosclerosis **88** (1991) 109–132

202. Tulenko, T.N., Laury-Kleintop, L., Walter, M.F., Mason, R.P.: Cholesterol, calcium and atherosclerosis: is there a role for calcium channel blockers in atheroprotection? Int J Cardiol **62 Suppl 2** (1997) S55–S66
203. Ruilope, L.M., Rosei, E.A., Bakris, G.L., Mancia, G., Poulter, N.R., Taddei, S., Unger, T., Volpe, M., Waeber, B., Zannad, F.: Angiotensin receptor blockers: Therapeutic targets and cardiovascular protection. Blood Press **14** (2005) 196–209
204. Schmidt-Ott, K.M., Kagiyama, S., Phillips, M.I.: The multiple actions of angiotensin II in atherosclerosis. Regul Pept **93** (2000) 65–77
205. Lonn, E.: Angiotensin-converting enzyme inhibitors and angiotensin receptor blockers in atherosclerosis. Curr Atheroscler Rep **4** (2002) 363–372
206. Krum, H., Gilbert, R.E.: Novel therapies blocking the renin-angiotensin-aldosterone system in the management of hypertension and related disorders. J Hypertens **25** (2007) 25–35
207. Higgins, J.P.: Can angiotensin-converting enzyme inhibitors reverse atherosclerosis? South Med J **96** (2003) 569–579
208. Yasuda, N., Akazawa, H., Qin, Y., Zou, Y., Komuro, I.: A novel mechanism of mechanical stress-induced angiotensin II type 1-receptor activation without the involvement of angiotensin II. Naunyn Schmiedebergs Arch Pharmacol **377** (2008) 393–399
209. O'Keefe, J.H., Wetzel, M., Moe, R.R., Bronsnahan, K., Lavie, C.J.: Should an angiotensin-converting enzyme inhibitor be standard therapy for patients with atherosclerotic disease? J Am Coll Cardiol **37** (2001) 1–8
210. Weir, M.R.: Effects of renin-angiotensin system inhibition on end-organ protection: can we do better? Clin Ther **29** (2007) 1803–1824
211. Ruilope, L.M., Redon, J., Schmieder, R.: Cardiovascular risk reduction by reversing endothelial dysfunction: ARBs, ACE inhibitors, or both? Expectations from the ONTARGET Trial Programme. Vasc Health Risk Manag **3** (2007) 1–9
212. Yusuf, S., Teo, K.K., Pogue, J., Dyal, L., Copland, I., Schumacher, H., Dagenais, G., Sleight, P., Anderson, C.: Telmisartan, ramipril, or both in patients at high risk for vascular events. N Engl J Med **358** (2008) 1547–1559
213. Dagenais, G.R., Yusuf, S., Bourassa, M.G., Yi, Q., Bosch, J., Lonn, E.M., Kouz, S., Grover, J.: Effects of ramipril on coronary events in high-risk persons: results of the Heart Outcomes Prevention Evaluation Study. Circulation **104** (2001) 522–526
214. Jukema, J.W., van der Hoorn, J.W.: Amlodipine and atorvastatin in atherosclerosis: a review of the potential of combination therapy. Expert Opin Pharmacother **5** (2004) 459–468
215. Zanchetti, A., Hennig, M., Baurecht, H., Tang, R., Cuspidi, C., Carugo, S., Mancia, G.: Prevalence and incidence of the metabolic syndrome in the European Lacidipine Study on Atherosclerosis (ELSA) and its relation with carotid intima-media thickness. J Hypertens **25** (2007) 2463–2470
216. Zanchetti, A., Bond, M.G., Hennig, M., Neiss, A., Mancia, G., Dal Palu, C., Hansson, L., Magnani, B., Rahn, K.H., Reid, J.L., Rodicio, J., Safar, M., Eckes, L., Rizzini, P.: Calcium antagonist lacidipine slows down progression of asymptomatic carotid atherosclerosis: principal results of the European Lacidipine Study on Atherosclerosis (ELSA), a randomized, double-blind, long-term trial. Circulation **106** (2002) 2422–2427
217. Mason, R.P., Marche, P., Hintze, T.H.: Novel vascular biology of third-generation L-type calcium channel antagonists: ancillary actions of amlodipine. Arterioscler Thromb Vasc Biol **23** (2003) 2155–2163
218. Mason, R.P.: Mechanisms of atherosclerotic plaque stabilization for a lipophilic calcium antagonist amlodipine. Am J Cardiol **88** (2001) 2M–6M
219. The sixth report of the Joint National Committee on prevention, detection, evaluation, and treatment of high blood pressure. Arch Intern Med **157** (1997) 2413–2446
220. The fifth report of the Joint National Committee on detection, evaluation, and treatment of high blood pressure (JNC V). Arch Intern Med **153** (1993) 154–183
221. Psaty, B.M., Smith, N.L., Siscovick, D.S., Koepsell, T.D., Weiss, N.S., Heckbert, S.R., Lemaitre, R.N., Wagner, E.H., Furberg, C.D.: Health outcomes associated with antihypertensive therapies used as first-line agents. A systematic review and meta-analysis. JAMA **277** (1997) 739–745

222. Brown, M.J., Palmer, C.R., Castaigne, A., de Leeuw, P.W., Mancia, G., Rosenthal, T., Ruilope, L.M.: Morbidity and mortality in patients randomised to double-blind treatment with a long-acting calcium-channel blocker or diuretic in the International Nifedipine GITS study: Intervention as a Goal in Hypertension Treatment (INSIGHT). Lancet **356** (2000) 366–372
223. Dahlof, B., Devereux, R.B., Kjeldsen, S.E., Julius, S., Beevers, G., de Faire, U., Fyhrquist, F., Ibsen, H., Kristiansson, K., Lederballe-Pedersen, O., Lindholm, L.H., Nieminen, M.S., Omvik, P., Oparil, S., Wedel, H.: Cardiovascular morbidity and mortality in the Losartan Intervention For Endpoint reduction in hypertension study (LIFE): a randomised trial against atenolol. Lancet **359** (2002) 995–1003
224. Major outcomes in high-risk hypertensive patients randomized to angiotensin-converting enzyme inhibitor or calcium channel blocker vs. diuretic: the Antihypertensive and Lipid-Lowering Treatment to Prevent Heart Attack Trial (ALLHAT). JAMA **288** (2002) 2981–2997
225. Psaty, B.M., Lumley, T., Furberg, C.D., Schellenbaum, G., Pahor, M., Alderman, M.H., Weiss, N.S.: Health outcomes associated with various antihypertensive therapies used as first-line agents: a network meta-analysis. JAMA **289** (2003) 2534–2544
226. Wiysonge, C.S., Bradley, H., Mayosi, B.M., Maroney, R., Mbewu, A., Opie, L.H., Volmink, J.: Beta-blockers for hypertension. Cochrane Database Syst Rev (2007) CD002003
227. Staessen, J.A., Wang, J.G., Thijs, L.: Cardiovascular protection and blood pressure reduction: a meta-analysis. Lancet **358** (2001) 1305–1315
228. Staessen, J.A., Wang, J.G., Thijs, L.: Cardiovascular prevention and blood pressure reduction: a quantitative overview updated until 1 March 2003. J Hypertens **21** (2003) 1055–1076
229. Turnbull, F.: Effects of different blood-pressure-lowering regimens on major cardiovascular events: results of prospectively-designed overviews of randomised trials. Lancet **362** (2003) 1527–1535
230. Verdecchia, P., Reboldi, G., Angeli, F., Gattobigio, R., Bentivoglio, M., Thijs, L., Staessen, J.A., Porcellati, C.: Angiotensin-converting enzyme inhibitors and calcium channel blockers for coronary heart disease and stroke prevention. Hypertension **46** (2005) 386–392
231. Law, M.R., Morris, J.K., Wald, N.J.: Use of blood pressure lowering drugs in the prevention of cardiovascular disease: meta-analysis of 147 randomised trials in the context of expectations from prospective epidemiological studies. BMJ **338** (2009) b1665
232. Migdalis, I.N., Gerolimou, B., Kozanidou, G., Hatzigakis, S.M., Karmaniolas, K.D.: Effect of fosinopril sodium on early carotid atherosclerosis in diabetic patients with hypertension. J Med **28** (1997) 371–380
233. MacMahon, S., Sharpe, N., Gamble, G., Clague, A., Mhurchu, C.N., Clark, T., Hart, H., Scott, J., White, H.: Randomized, placebo-controlled trial of the angiotensin-converting enzyme inhibitor, ramipril, in patients with coronary or other occlusive arterial disease. PART-2 Collaborative Research Group. Prevention of atherosclerosis with ramipril. J Am Coll Cardiol **36** (2000) 438–443
234. Lonn, E., Yusuf, S., Dzavik, V., Doris, C., Yi, Q., Smith, S., Moore-Cox, A., Bosch, J., Riley, W., Teo, K.: Effects of ramipril and vitamin E on atherosclerosis: the study to evaluate carotid ultrasound changes in patients treated with ramipril and vitamin E (SECURE). Circulation **103** (2001) 919–925
235. Hosomi, N., Mizushige, K., Ohyama, H., Takahashi, T., Kitadai, M., Hatanaka, Y., Matsuo, H., Kohno, M., Koziol, J.A.: Angiotensin-converting enzyme inhibition with enalapril slows progressive intima-media thickening of the common carotid artery in patients with non-insulin-dependent diabetes mellitus. Stroke **32** (2001) 1539–1545
236. Asselbergs, F.W., van Roon, A.M., Hillege, H.L., de Jong, P.E., Gans, R.O., Smit, A.J., van Gilst, W.H.: Effects of fosinopril and pravastatin on carotid intima-media thickness in subjects with increased albuminuria. Stroke **36** (2005) 649–653
237. Wiklund, O., Hulthe, J., Wikstrand, J., Schmidt, C., Olofsson, S.O., Bondjers, G.: Effect of controlled release/extended release metoprolol on carotid intima-media thickness in patients with hypercholesterolemia: a 3-year randomized study. Stroke **33** (2002) 572–577
238. Pitt, B., Byington, R.P., Furberg, C.D., Hunninghake, D.B., Mancini, G.B., Miller, M.E., Riley, W.: Effect of amlodipine on the progression of atherosclerosis and the occurrence of clinical events. PREVENT Investigators. Circulation **102** (2000) 1503–1510

239. Borhani, N.O., Mercuri, M., Borhani, P.A., Buckalew, V.M., Canossa-Terris, M., Carr, A.A., Kappagoda, T., Rocco, M.V., Schnaper, H.W., Sowers, J.R., Bond, M.G.: Final outcome results of the Multicenter Isradipine Diuretic Atherosclerosis Study (MIDAS). A randomized controlled trial. JAMA **276** (1996) 785–791
240. Zanchetti, A., Rosei, E.A., Palu, C., Leonetti, G., Magnani, B., Pessina, A.: The Verapamil in Hypertension and Atherosclerosis Study (VHAS): results of long-term randomized treatment with either verapamil or chlorthalidone on carotid intima-media thickness. J Hypertens **16** (1998) 1667–1676
241. Simon, A., Gariepy, J., Moyse, D., Levenson, J.: Differential effects of nifedipine and co-amilozide on the progression of early carotid wall changes. Circulation **103** (2001) 2949–2954
242. Boutouyrie, P., Bussy, C., Hayoz, D., Hengstler, J., Dartois, N., Laloux, B., Brunner, H., Laurent, S.: Local pulse pressure and regression of arterial wall hypertrophy during long-term antihypertensive treatment. Circulation **101** (2000) 2601–2606
243. Roman, M.J., Alderman, M.H., Pickering, T.G., Pini, R., Keating, J.O., Sealey, J.E., Devereux, R.B.: Differential effects of angiotensin converting enzyme inhibition and diuretic therapy on reductions in ambulatory blood pressure, left ventricular mass, and vascular hypertrophy. Am J Hypertens **11** (1998) 387–396
244. Zanchetti, A., Crepaldi, G., Bond, M.G., Gallus, G., Veglia, F., Mancia, G., Ventura, A., Baggio, G., Sampieri, L., Rubba, P., Sperti, G., Magni, A.: Different effects of antihypertensive regimens based on fosinopril or hydrochlorothiazide with or without lipid lowering by pravastatin on progression of asymptomatic carotid atherosclerosis: principal results of PHYLLIS – a randomized double-blind trial. Stroke **35** (2004) 2807–2812
245. Ludwig, M., Stapff, M., Ribeiro, A., Fritschka, E., Tholl, U., Smith, R.D., Stumpe, K.O.: Comparison of the effects of losartan and atenolol on common carotid artery intima-media thickness in patients with hypertension: results of a 2-year, double-blind, randomized, controlled study. Clin Ther **24** (2002) 1175–1193
246. Hoogerbrugge, N., de Groot, E., de Heide, L.H., de Ridder, M.A., Birkenhageri, J.C., Stijnen, T., Jansen, H.: Doxazosin and hydrochlorothiazide equally affect arterial wall thickness in hypertensive males with hypercholesterolaemia (the DAPHNE study). Doxazosin Atherosclerosis Progression Study in Hypertensives in the Netherlands. Neth J Med **60** (2002) 354–361
247. Koshiyama, H., Tanaka, S., Minamikawa, J.: Effect of calcium channel blocker amlodipine on the intimal-medial thickness of carotid arterial wall in type 2 diabetes. J Cardiovasc Pharmacol **33** (1999) 894–896
248. Topouchian, J., Asmar, R., Sayegh, F., Rudnicki, A., Benetos, A., Bacri, A.M., Safar, M.E.: Changes in arterial structure and function under trandolapril-verapamil combination in hypertension. Stroke **30** (1999) 1056–1064
249. Pontremoli, R., Viazzi, F., Ravera, M., Leoncini, G., Berruti, V., Bezante, G.P., Del Sette, M., Deferrari, G.: Long term effect of nifedipine GITS and lisinopril on subclinical organ damage in patients with essential hypertension. J Nephrol **14** (2001) 19–26
250. Stanton, A.V., Chapman, J.N., Mayet, J., Sever, P.S., Poulter, N.R., Hughes, A.D., Thom, S.A.: Effects of blood pressure lowering with amlodipine or lisinopril on vascular structure of the common carotid artery. Clin Sci (Lond) **101** (2001) 455–464
251. Terpstra, W.F., May, J.F., Smit, A.J., Graeff, P.A., Meyboom-de Jong, B., Crijns, H.J.: Effects of amlodipine and lisinopril on intima-media thickness in previously untreated, elderly hypertensive patients (the ELVERA trial). J Hypertens **22** (2004) 1309–1316
252. Wang, J.-G.M.D.P., Staessen, J.A.M.D.P., Li, Y.M.D.P., Van Bortel, L.M.M.D.P., Nawrot, T.P., Fagard, R.M.D.P., Messerli, F.H.M.D., Safar, M.M.D.: Carotid intima-media thickness and antihypertensive treatment: a meta-analysis of randomized controlled trials. Stroke **37** (2006) 1933–1940

Biography

Mario Eandi, MD, is a Full Professor of Clinical Pharmacology at the School of Medicine and Surgery, University of Turin, Italy. He was a member of the Permanent Committee for the Revision of Italian Pharmacopeia and a member of the Technical-Scientific Committee at the Italian Agency for Drug (AIFA), and he has been appointed President of the Regional Conference for the Ethic Committees (Piemonte-Italy).He has authored numerous papers, articles, book chapters, and books on pharmacokinetics, pharmacodynamics, pharmacoeconomics, and mathematical modeling and simulation. He is editor-in-chief for the italian journal "Farmecocomia e Percorsi Terapeutici".

Chapter 20
Control of Inflammation with Complement Control Agents to Prevent Atherosclerosis

Perla Thorbjornsdottir, Gudmundur Thorgeirsson, Girish J. Kotwal, and Gudmundur Johann Arason

Abstract Atherosclerosis is characterized by progressive accumulation of lipids, macrophages, and cell debris in the *tunica intima* of vascular walls, in a process consisting of and driven by chronic inflammation. Complement is a potent inducer of inflammation, and a number of studies have implicated complement in the pathogenesis of atherosclerosis, identifying its components in lesions from patients as well as experimental animals. The key question has been whether complement activation serves a role in lesion initiation or progression or whether it simply serves as an enhancing factor for processes already in play. This enigma has been recently solved by the demonstration that diet-induced atherosclerosis in a mouse model (C57BL) can be significantly (50%) reduced by regular injections with the complement inhibitor, VCP (vaccinia virus complement control protein). Since VCP has also been shown to protect against reperfusion injury following myocardial infarction, it shows unique promise in the fight against atherosclerosis as well as its most significant acute events.

Keywords Complement inhibitors • Vaccinia virus • Atherosclerosis • Inflammation • Pathways

20.1 Atherosclerosis

20.1.1 Pathogenesis of Atherosclerosis

Atherosclerosis is the most common disease of modern times; with prevalence values estimated as 85% at age 50 and accounting for more than 30% of global mortalities, it is the leading cause of death in all regions of the world except

G.J. Kotwal (✉)
InflaMed Inc, Louisville, KY, USA
e-mail: GJKOTW01@gmail.com

Jasjit S. Suri et al. (eds.), *Atherosclerosis Disease Management*,
DOI 10.1007/978-1-4419-7222-4_20,

Sub-Saharan Africa [30, 224]. The disease is characterized by the formation of atheromatous (fat-containing) lesions in large- and medium-sized arteries, often culminating in occlusion of an artery supplying the heart (myocardial infarction, MI), the brain (ischemic stroke), or extremities (peripheral artery disease, PAD). In coronary arteries, atherosclerosis may also be diagnosed as angina pectoris (AP), a clinically milder form of the disease, caused by narrowing of the artery and a transient (usually exercise-related) ischemia, felt as chest pain. This condition may precede MI but may also persist for years or even decades as stable AP. Atherosclerosis of coronary arteries (angina pectoris and/or the acute myocardial syndromes (ACS) myocardial infarction, unstable angina, or sudden cardiac death) is variably known as coronary artery disease (CAD), atherosclerotic cardiovascular disease (CVD), ischemic heart disease (IHD), or coronary heart disease (CHD).

The pathogenesis of the disease is complex and not yet fully understood. Atherosclerosis is characterized by progressive accumulation of lipids, macrophages, cell fragments, and connective tissue material in the tunica intima of vascular walls, in a process consisting of and driven by chronic inflammation [92, 108, 111]. Temporally, the process of atherosclerosis may be divided into four stages: lesion initiation, lesion progression, plaque rupture (with ensuing thrombosis and infarction), and thrombolysis (leading to reperfusion). The processes governing each of these stages are incompletely understood, but it is becoming increasingly clear that inflammation is the driving force behind the early atheromatous stages as well as plaque rupture, and it is also responsible for a major part of the damage occurring after reperfusion of the artery. Atherosclerosis is a slow process, with the early lesions (fatty streaks) developing already in childhood. They consist of subendothelial accumulation of cholesterol-containing macrophages. Fatty streak lesions can be seen in the aorta in the first decade of life, the coronary arteries in the teens and carotid artery in the twenties [111, 201]. The disease is chronic, with slow progress during childhood and adolescence, and then it accelerates in adult life to result in plaque erosion and rupture when the lesions become more developed [18, 19, 70]. The lesions of atherosclerosis occur primarily in large- and medium-sized elastic and muscular arteries such as the iliac, coronary and carotid arteries, and the aorta. The process can lead to narrowing of the arterial lumen, or infarction, and consequently transient or more prolonged ischemia of the tissue nourished by the artery [170]. Lesions are more likely to form at branching points where the blood flow is turbulent and therefore slower, which allows components of the blood a closer contact with the endothelium.

Fatty streaks are asymptomatic and are precursors of more advanced lesions, which are packed with lipid-rich necrotic debris and often have a thin fibrotic cap, prone to rupture. Advanced lesions grow into the lumen of the artery, and if large enough they can block the blood flow. During times of added oxygen demands, e.g., upon a physical exertion, this condition can be felt as chest pain (angina pectoris). If the fibrotic cap over the lesion ruptures, it can lead to acute occlusion due to the formation of a thrombus. This can lead to myocardial infarction, stroke, or peripheral artery disease [111]. For further information about atherosclerotic disease, refer [54, 67, 70, 92, 108, 111, 210].

20.1.2 Stages of Lesion Development

Atherosclerotic lesions usually arise in the tunica intima layer of the artery. They have been categorized into three stages: the fatty streak, the fibrotic plaque, and the complicated lesion [67]. The fatty streak is the first stage, it is an asymptomatic lesion, which does not narrow the arterial lumen. The lesions are made mainly of lipids and macrophages, with only few smooth muscle cells. The fibrotic plaque is a more developed form of atherosclerotic lesion and is thought to be the most common precursor to occlusive lesions. It grows into the arterial lumen and causes narrowing of it. Usually fibrotic plaques are found at the same locations as the fatty streak precursors. They are the major determinants of clinically significant disease. They consist of a layer of smooth muscle cells covered with a fibrotic cap, and beneath the fibrous material are the lipids, foam cells (macrophages and an increasing number of smooth muscle cells), and a necrotic core [108, 111]. Over time, the fatty streak develops through the fibrotic plaque to the third stage, the complicated lesion. Virtually all occlusive lesions are of this stage. Complicated lesions arise because of necrosis, calcification, ulcerification on the surface and bleeding from small vessels that grow into the lesion from the tunica media of the arterial wall [67, 111]. Although advanced lesions may be big enough to block the blood flow, the most important complication is the rupture of the lesion leading to the formation of a thrombus and subsequently infarction of the tissue supplied by an artery in the heart (myocardial infarction, MI), the brain (ischemic stroke), or the extremities (peripheral artery disease, PAD). Perfusion may be ultimately restored by the activation of the thrombolytic system or by medical intervention, but the tissue damage incurred during the ischemic period may be substantial, and processes activated during reperfusion may add to the damage.

Ross and Glomset first proposed in 1976 that the initiating event in atherogenesis is injury to the endothelium [56, 77, 171, 172]. This is currently recognized as endothelial dysfunction [6], which may have many causes, including elevated levels of modified low-density lipoprotein (LDL), free radicals caused by cigarette smoke, hypertension, diabetes mellitus, genetic alteration, elevated plasma homocysteine concentrations, infectious microorganisms, and other factors. Endothelial injury leads to platelet adhesion, monocyte adhesion and infiltration, and smooth muscle migration and proliferation.

20.1.3 Animal Models of Atherosclerosis

The etiology of atherosclerosis is very complex and many factors contribute to the disease course. Bearing in mind the complexity of the disease and the problems of studying its causes on humans, researchers have realized the strong need of finding a suitable animal model for studying the etiology and pathogenesis of the disease. The tendency to develop atherosclerosis differs between animal species and strains.

Mice do not spontaneously develop atherosclerotic lesions; they need to be fed with high-fat diet that promotes hyperlipidemia [147]. Beverly Paigen surveyed in the mid-1980s a large number of inbred strains to find the most susceptible mouse strain for developing atherosclerotic lesions [146, 147]. The susceptibility to develop fatty streaks is rare among inbred strains, but a few strains were discovered, in particular the SWR and the C57 and C58 families [146], with the C57BL/6 mouse strain being the most susceptible. This mouse strain became the mouse model of choice for diet-induced atherosclerosis, mainly because of its wide availability, reproducible mating patterns, and the susceptibility for lesion formation. The lesions are normally restricted to the fatty streak stage [146, 148, 194], but after prolonged feeding some investigators have noted more advanced lesions containing cellular debris and collagen [204]. The lesions are restricted to the aortic root and most commonly characterized by deposition of lipids and foam cells. It is interesting that unlike humans, female mice are more predisposed to develop fatty streaks than the males [145].

The need for a more extensive and faster form of the disease has driven the development of genetically modified mouse strains. Today there are two gene-targeted mouse strains that are widely used, both based on the C57BL/6 strain; LDL receptor knockout mice (LDLR–/–) and ApoE-deficient mice (ApoE–/–). The lesion formation is accelerated in both mouse strains [48, 232], and only the former mouse strain is dependent on high-fat diet for fatty streak development. The latter mouse strain develops atherosclerotic lesions without high-fat diet [250], though the disease course is more accelerated with the Paigen diet [158]. ApoE–/– mice also develop lesions beyond the fatty streak stage, with intermediate complexity of macrophage foam cells, necrotic cores, and fibrotic caps [161].

20.1.4 Initiation and Progression of Atherosclerotic Lesions

The initiating event in the fatty streak formation is the adherence of monocytes to the activated endothelium. The first step in adhesion is a rolling of leukocytes along the endothelial surface, mediated by transient binding of endothelial selectins (E- and P-selectin) to carbohydrate ligands on leukocytes or L-selectin of lymphocytes to carbohydrates on endothelium. The firm adhesion of monocytes and T cells to the endothelium can be mediated by the integrin VLA-4, interacting with VCAM-1 on the endothelium [28]. The monocytes then migrate over the endothelium into the subendothelial space in response to chemoattractant molecules produced there. In the intima, the monocytes differentiate into macrophages and start engulfing LDL particles. Their attempt to eliminate LDL particles, however, ultimately fails, and the LDL-laden macrophages are referred to as foam cells. These cells have increased expression of scavenger receptors and increased internalization of modified lipoproteins [49, 109]. M-CSF is thought to be a candidate activator of several steps that stimulate the change of monocyte to the foam cell macrophage. M-CSF augments scavenger receptor expression and increases the production of cytokines and growth factors by these cells. Both human and experimental plaques overexpress M-CSF [39, 168]. Mice lacking M-CSF show retarded lesion development with markedly reduced macrophage accumulation [195].

The fatty streak can evolve into a more complex lesion, accelerated by risk factors such as hypertension, hyperlipidemia, and hyperglycemia. The fibrotic plaque is characterized by a growing mass of extracellular lipid and the accumulation of smooth muscle cells from the tunica media, which proliferate and lay down extracellular matrix. As the macrophages, the smooth muscle cells may take up lipids in the lesion and become lipid laden foam cells [12, 159]. The foam cells may ultimately die either by necrosis or by apoptosis, liberating their contained lipid and producing a necrotic core with extracellular lipid. Continued inflammation leads to increased numbers of macrophages and lymphocytes within the lesion. Cytokines and growth factors secreted by these cells are important for the smooth muscle cell migration and proliferation and extracellular matrix production. Activation of these cells leads to release of hydrolytic enzymes, chemokines, and growth factors, which can lead to further damage and ultimately to focal necrosis. Thus the continuous accumulation of leukocytes, migration and proliferation of smooth muscle cells, and formation of fibrotic tissue lead to enlargement of the lesion. In the beginning, the fibrotic plaque grows towards the tunica media, but as the lesion size increases it starts to bulge into the arterial lumen and causes narrowing of it.

With increasing evolvement of the lesion, it eventually becomes covered with a fibrotic cap that overlies a lipid core and necrotic tissue. This eventually develops into a complicated lesion [170]. Stable lesions usually have thick and stable fibrous caps; in contrast, activation of macrophages in the lesion especially in the shoulder region (which is the place of entry) may lead to release of proteases that cause thinning of the cap. This leads to an unstable plaque, which can result in critical stenosis or thrombosis, and consequently clinical symptoms [69]. In less severe cases, these may consist of transient ischemia of the perfused tissue with no or only minor tissue destruction (stable or unstable angina pectoris), but thrombosis leads to infarction and more substantial tissue damage (myocardial infarction).

20.1.5 Myocardial Infarction

Myocardial infarction occurs as a result of erosion or uneven thinning and rupture of the fibrous cap of a lesion in the right, the left anterior descendent (LAD) or the circumflexa coronary artery, or one of their tributaries. Inflammation is thought to be the underlying cause of the plaque rupture. The degradation of the fibrous cap may result from elaboration of metalloproteinases produced by macrophages [170]. T cells may stimulate the production of metalloproteinases in the lesion, leading to plaque instability [187]. Vulnerable plaques are often nonocclusive and active inflammation is evident, the fibrotic cap is thin and macrophages are accumulated at the sites of plaque rupture. The stability of atherosclerotic lesions may be influenced by calcification and neovascularization, which is often seen in advanced lesions. The small vessels may provide a conduct for entry of inflammatory cells [111, 170].

Occlusion of a coronary artery leads to myocardial damage not only through hypoxia of the ischemic area supplied by the artery but, paradoxically, also in part through processes initiated as the blood clot is removed by natural thrombolysis or medical intervention. Ischemia/reperfusion injury occurs when the tissue lacks blood supply for an extended period and then the blood supply is restored. This process triggers an intense inflammatory response, which can be quite deleterious to the tissue. Ischemia/reperfusion injury can occur in various parts of the body under varying circumstances, but the current thesis focuses on the damage incurred after myocardial infarction. Reperfusion of the ischemic myocardium is necessary to salvage the tissue from death and for the limitation of the infarct size. However, reperfusion after ischemia is involved in complex pathological changes, which can accelerate the process initiated during ischemia, and it can also lead to new pathological changes initiated after the reperfusion [14, 162].

Ischemia/reperfusion injury causes attraction, activation, adhesion, and migration of neutrophils at the site of tissue injury, which causes tissue damage through the release of numerous inflammatory mediators. Once activated, neutrophils can release free radicals, along with the free radicals formed during the ischemia, which can cause severe coronary endothelial and cardiac myocyte injury, as free radicals inactivate nitric oxide (a chemical mediator with anti-atherogenic properties, e.g., vasorelaxation), stimulate the expression of adhesion molecules, and activate neutrophils.

20.1.6 Atherosclerotic Risk Factors

Atherosclerosis, the underlying cause of CVD, is a complex process and both environmental and genetic factors play a key role. The etiology is still unknown, but some of the risk factors have been identified [60]. Hypercholesterolemia (elevated levels of serum cholesterol) is special among the risk factors in being sufficient and necessary for the development of atherosclerotic lesions in humans and in experimental animals, even in the absence of other risk factors. Statistical correlation has also been established in epidemiological studies with smoking, hypertension, diabetes, and obesity. Male gender and age are also important [60]. Family history is a significant risk factor for CVD and it has been difficult to find the specific genetic factors that lead to atherosclerotic susceptibility, because most cases of atherosclerosis are associated with polygenic factors (and to complicate the issue even further, some of these factors may relate to the underlying risk factors). The atherosclerosis that results from the monogenic disorder familial hypercholesterolemia, which is caused by a mutation in the LDL receptor gene, is an exception from the general rule [60].

Up to a quarter of patients with CVD do not have traditional risk factors, and new nontraditional risk factors, "conditional risk factors" [149, 151, 198] have been identified as also important. From these, it is evident that inflammation plays a central role along with cholesterol in the pathogenesis of atherosclerosis. Among the conditional risk factors are triglycerides, small dense LDL, lipoprotein(a) [45, 110], homocysteine [13], fibrinogen [53, 75], and inflammation markers such as

C-reactive protein (CRP) [46]. This list may in future also include fasting plasma insulin and C3 [128, 175].

Obesity can be an underlying risk factor for cardiovascular disease (CVD), and the majority of obese persons who develop CVD have typically a clustering of a few major risk factors, which is called the metabolic syndrome. These factors are strongly intercorrelated, and diagnosis of metabolic syndrome is based on the co-occurrence of at least three of the following five risk factors: hypertension (>130/85 mmHg), elevated glucose (>6.1 mmol/L) associated with insulin resistance, elevated triglycerides (>1.69 mmol/L), reduced high-density lipoprotein (HDL) cholesterol (<1.04 mmol/L in men and 1.29 mmol/L in women), and obesity, especially increased waist circumference (>88 cm in women and 102 cm in men) [74]. Risk factors are known to be synergistic in precipitating CVD, i.e., the presence of more than one risk factor confers a risk, which is higher than the sum of the individual risk factors. Diagnosis of the metabolic syndrome constitutes a particularly strong risk for CVD.

The overall contribution of all the currently known risk factors has been recently reviewed in a world-wide case–control study, which identified the following main risk factors:[1] current smoking (OR f: 2.86, m: 3.05), diabetes (OR f: 4.26, m: 2.67), hypertension (OR f: 2.95, m: 2.32), abdominal obesity (OR f: 2.26, m: 2.24), ApoB/ApoA1 ratio (OR f: 4.42, m: 3.76), and psychosocial index (OR f: 3.49, m: 2.58). In addition, three protective factors were recognized; fruit/vegetable intake (OR f: 0.58, m: 0.74), exercise (OR f: 0.48, m: 0.77), and alcohol consumption (OR f: 0.41, m: 0.88) [248].

20.1.7 Low-Density Lipoprotein and Lipid Transport

Lipids are vital for various functions of the body, but as they are insoluble in plasma they need to be transported as part of lipoprotein particles. The main lipoprotein particles in plasma are VLDL (very low-density lipoprotein), LDL (low-density lipoprotein), and HDL (high-density lipoprotein). VLDL, synthesized by the liver, is a triglyceride-rich lipoprotein particle with the main carrier protein being apolipoprotein B100 (apoB). With delipidation of its fatty material and loss of triglycerides, the particle becomes smaller, resulting in the cholesterol-rich LDL particle. HDL is also a cholesterol-rich particle with apolipoprotein A1 as the main carrier. Its role is to reverse the cholesterol transport, taking cholesterol from peripheral cells, and it serves as reservoir for apolipoproteins. Both VLDL and HDL have an additional apolipoprotein, apoE, which has a role in delivering triglycerides to tissues and to distribute cholesterol to cells [222].

Cholesterol is a necessary component of cell membranes. LDL is the main carrier of cholesterol in the blood. LDL is secreted by the liver, and functions in the transport of cholesterol by the blood to peripheral tissues where it provides cells with the cholesterol they need. LDL leaves the blood stream by crossing the endothelium of capillaries by transcytosis or by diffusing passively through the junction between the

[1] OR = odds ratio, f = females, m = males. The psychosocial index is a model-dependent index combining positive exposure to depression, perceived stress at home or work (general stress), low locus of control, and major life events.

endothelial cells [93, 229]. The LDL then binds to LDL receptors on tissue cells via the apoB particles. The receptor-mediated uptake is tightly regulated, and when the cell has enough cholesterol, it shuts down the LDL receptor synthesis and thus prevents intracellular accumulation of cholesterol [32]. If the uptake of LDL by cells is blocked, it accumulates in the blood and when the levels are raised, the transport over arterial endothelium and retention in tunica intima is increased. This increase is probably because the arterial tissue is without a lymphatic system, which normally eliminates extra LDL [73]. It is retained in the tunica intima most likely because of interaction between the apoB part of the LDL particle and the matrix proteoglycans of the extracellular matrix [31, 192, 244]. The first event in atherosclerosis is the accumulation of LDL in the subendothelial matrix in the intimal layer of the artery.

LDL particles have to be modified to be atherogenic. Although the most extensive modification of LDL takes place in the arterial intima, it is possible that LDL circulating in the bloodstream is already mildly modified, which can enhance the modification in the intima. There is evidence for the presence of mildly oxidized or desialylated LDL circulating in the blood plasma [15, 214]. During retention in the extracellular spaces, the LDL particles become exposed to many agents secreted by intimal cells that can modify them even further, for example it is well known that LDL undergoes oxidation [78, 203] and enzymatic modification [22, 23]. The modification involves reactive oxygen species produced by endothelial cells and macrophages, and several enzymes in the intima are also thought to be involved.

Shortly after the beginning of lipoprotein accumulation and retention in the intima layer of the artery, monocytes can be seen adhering to the surface of the endothelium. They transmigrate over the endothelial layer, where they proliferate and mature into macrophages [186]. The macrophages start taking up the lipoproteins, but do not succeed in eliminating them and eventually become large lipid-filled cells, often termed foam cells. This process is fuelled by the continuing process of LDL modification; the degree of modification can vary greatly [51, 72, 133], but LDL must be extensively modified before it can be taken up by macrophages rapidly enough to form foam cells. At later stages, LDL is also taken up by smooth muscle cells, migrating into the lesion from the tunica media, and this also leads to foam cell formation [12, 159]. Unlike macrophages, which according to immunofluorescence microscopy may both enter and leave the atheroma [65, 66], smooth muscle cells are trapped in the intima. The sustained presence of atherogenic LDL, stimulation of endothelial cells, and recruitment of foam cell precursors lead to a process of chronic inflammation [111].

Oxidized LDL (oxLDL) is not bound by the LDL receptor, instead the uptake of oxLDL leading to foam cell formation is mediated by scavenger receptors expressed on macrophages and smooth muscle cells. This uptake is not regulated by ingested cholesterol and therefore allows the macrophages and smooth muscle cells to become filled with LDL, i.e., converted to foam cells. The differentiation of monocytes to macrophages causes elevated expression of scavenger receptors, which under normal circumstances recognize and ingest pathogens and apoptotic cells. Scavenger receptors also recognize altered molecular patterns as present on oxLDL [49, 109].

Atherosclerotic lesions become more developed with the formation of a necrotic core, where apoptotic cells play an important role [64]. The lipid-laden macrophages and smooth muscle cells undergo programmed cell death, which at first may lead to regression of the atherosclerotic lesion. Accumulation of apoptotic cells in the plaque may be due to inefficient removal of the dead cells and some apoptotic cells may remain "mummified" rather than being phagocytized. Many agents in the lesion are able to promote apoptosis, including oxysterols in the oxLDL particle. The ultimate release of oxidized and insoluble lipids from necrotic cells can contribute to the formation of advanced lesion and influence plaque stability.

Recruitment of monocytes to the lesion site and their differentiation into macrophages is at first protective. It serves to remove oxLDL, which is both cytotoxic and proinflammatory. However, progressive accumulation of macrophages and the uptake of oxLDL but unsuccessful elimination ultimately lead to the development of atherosclerotic lesions through a process of chronic inflammation.

20.1.8 Inflammation in Atherosclerosis

Inflammation is the response of the body against irritation or injury to eliminate the harmful agents involved, such as microbes or necrotic cells. In inflammation, phagocytic cells accumulate in the inflamed tissue to ingest the agent causing the inflammation. The inflamed tissue may also attract other cells, such as T cells, to assist the phagocytes. If the inflammatory cells fail to eliminate the harmful matter, the inflammation becomes chronic, and this can cause tissue damage due to the destructive action of activated complement as well as enzymes, cytokines, and other mediators produced by the inflammatory cells.

Atherosclerosis remains the leading cause of death in the world despite life style changes and new drug therapies to lower blood cholesterol [169, 170, 202]. Until the past decade, the disease was considered to be caused mainly by lipid accumulation in the arterial wall. It came to light in the past decade that there must be other factors involved. Today it is accepted that inflammation plays a key role in the pathogenesis of the disease and atherosclerosis is now considered a chronic inflammatory disease [70, 107, 108, 111]. Both the innate and the adaptive immune system have been suspected to contribute to the development of the lesion, with oxLDL [123] and heat shock proteins (HSP) [76, 108, 245] identified as potential autoantigens; however, research has shown that the absence of an adaptive immune response does not protect animals from developing atherosclerosis in the presence of high levels of plasma cholesterol over long periods [47], leaving inflammation as the main immunological culprit of the disease. In addition to the lipids, inflammatory cells including macrophages and T cells play a central role [76, 108], and are observed during all stages of atherosclerotic disease. Neutrophils are rarely seen in atherosclerotic lesions. As in many inflammatory diseases, the complement system has been suspected of mediating some of the tissue damage seen in atherosclerosis [23, 26, 135, 138, 144, 217, 220]. This notion has been confirmed and reinforced by recent studies in which complement has in fact been identified as a rate-limiting step in the initiation and progression of lesions. These studies will be reviewed in more detail in Sect. 20.3, but let us first look at some basic features of the complement system.

20.2 The Complement System

20.2.1 The Complement Cascade

The complement system is an important part of the innate immune system [239, 240]. It has many functions including initiation of inflammation, recruitment of leukocytes, clearance of immune complexes, neutralization of pathogens, regulation of antibody responses, and disruption of cell membranes. In addition, it is a major effector in immunopathological diseases [122, 243].

The complement system was first discovered in 1889 as a bactericidal principle of serum. This bacteriolytic activity was shown to be dependent on two different factors: a heat-stable factor and a heat-labile factor. The heat-stable factor was found to bind specifically to the pathogen and was termed antibody. After this binding (sensitization), any further steps in bacteriolysis were mediated by the nonspecific heat-labile factor, termed complement [174]. The complement system was later found to consist of about 30 proteins, some of which act in a cascade-like reaction sequence, while others serve as control proteins or as cellular receptors. The key components are present in blood in precursor forms and need to be activated. The complement proteins constitute nearly 10% of the total serum proteins and form one of the major defense systems of the body [174, 239, 240, 243]. The proteins are mainly produced by the liver [174, 243], but other cells, such as monocytes, T cells, macrophages, adipocytes, granulocytes, skin fibroblasts, and keratinocytes can also produce the complement components locally, in particular C3. The importance of the complement system is best seen by increased susceptibility to infection and increased tendency for immune complex disease in individuals lacking particular complement components. The main effectors of the complement system are the third component (C3) and the terminal components C5b-9. When activated, C3 binds covalently to pathogen surfaces, making them susceptible to destruction by receptor-mediated phagocytosis. C5b-9 forms a membrane attack complex, which inserts pores in pathogen membranes, causing lysis of the target cell.

There are three pathways of complement activation, all leading to the activation of the C3 component. These are the classical pathway (C1 → C4 → C2 → C3), which is initiated by antibody bound to antigen, the lectin pathway (MBL/ficolin/MASP → C4 → C2 → C3), activated by carbohydrates, and the alternative pathway (C3b → fB → C3), activated in the presence of microbial pathogens. The activation by the enzyme cascade leads to a huge amplification of the initial stimulus (Figs. 20.1 and 20.2).

The terminology of the complement proteins can be repelling for those unfamiliar to the names used. It is thus necessary to introduce the names in brief. All the proteins belonging to the classical and the terminal pathways (MAC) are termed "components" and thus designed with the letter C (component) and a number,

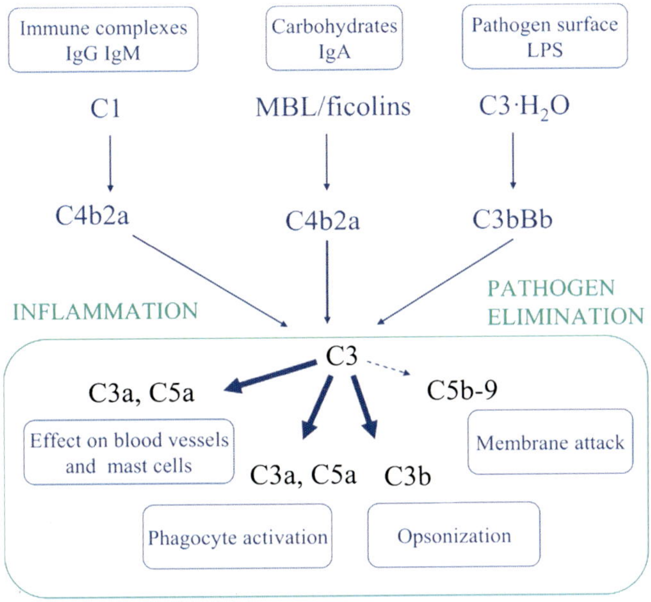

Fig. 20.1 Physiological aspects of the complement system. IgG or IgM bound with immune complexes or on pathogen surface lead to the activation of C3 via the classical pathway. Certain carbohydrates on IgA antibodies or carbohydrate-rich antigens lead to the activation of C3 via the lectin pathway. C3 may also be activated through the alternative pathway; this is initiated by LPS or by complex surface structures of pathogens. C3 is centrally positioned in the complement cascade, and its activation leads to pathogen elimination due to phagocytosis (by opsonization and phagocyte activation) or by direct cytolysis; more primitively (and even in the absence of pathogens), it may lead to inflammation through an effect on blood vessels and mast cells. Reprinted with permission from ref. [7] (Copyright 2008 The Icelandic Medical Student)

e.g., C1. Proteins of the alternative pathway and the first regulatory proteins discovered were designed with the term "factor" and capital letter, e.g., factor B. Factor A turned out to be identical to C3 and thus that term was dropped. The lectin pathway was discovered much later and a formal terminology has not yet been adopted. In that pathway, the first component is either mannose-binding lectin (MBL) or ficolin, and the first enzyme to be activated is called MBL-associated serine protease (MASP), after that the pathway is the same as the classical pathway.

20.2.2 Complement Activation Pathways

20.2.2.1 The Classical Pathway

The classical pathway has a role both in innate and in adaptive immunity and is primarily activated by antibody bound to antigens. The activation is through the binding of C1 to pathogen surfaces. The C1 molecular complex consists of C1q and two molecules of each zymogen C1r and C1s. C1q can bind to pathogen surfaces

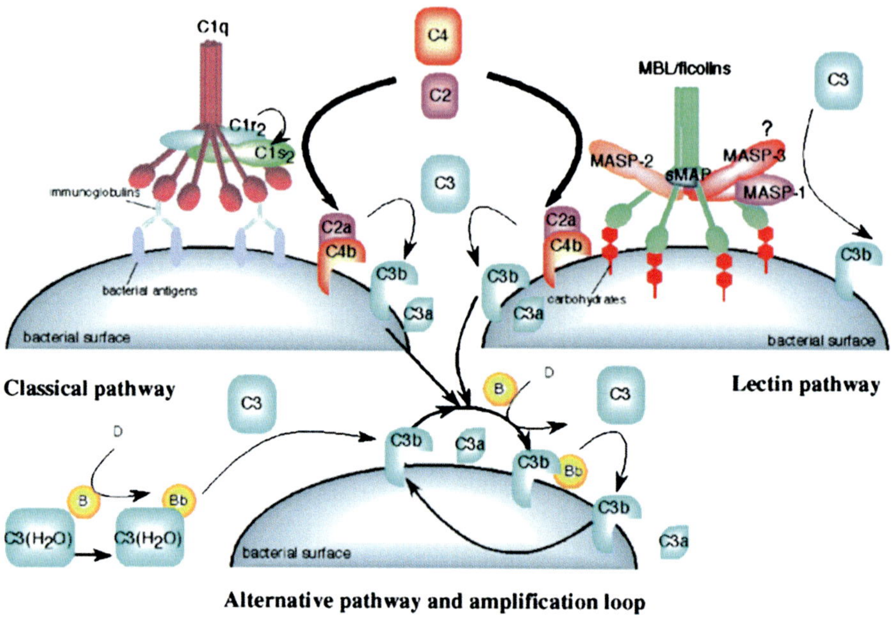

Fig. 20.2 Activation of the classical, lectin, and alternative pathways. The classical pathway is initiated by the binding of the C1 complex to antibodies bound to antigen on the surface of bacteria. The C1 complex consists of C1q and two molecules of C1r and C1s. The binding of the recognition subcomponent C1q to the Fc portion of immunoglobulins results in autoactivation of the serine protease C1r. C1r then cleaves and activates C1s, the enzyme that translates the activation of the C1 complex into complement activation through the cleavage of C4 and C2 to form a C4bC2a enzyme complex. C4bC2a acts as a C3 convertase and cleaves C3, resulting in products that bind to and result in the destruction of invading bacteria. The lectin pathway is initiated by binding of either of MBL or ficolin, associated with MASP-1, MASP-2, MASP-3, and sMAP to an array of carbohydrate groups on the surface of a bacterial cell. As with C1s, MASP-2 is responsible for the C4 and C2 activation, leading to the generation of the same C3 convertase as the classical pathway. MASP-1 is able to cleave C3 directly. The alternative pathway is initiated by the low-grade activation of C3 by hydrolyzed C3 [C3 (H_2O)] and activated factor B (Bb). The activated C3b binds factor B (B), which is cleaved into Bb by factor D (D) to form the alternative pathway C3 convertase, C3bBb. Once C3b is attached to the surface, the amplification loop consisting of the alternative pathway components is activated and the C3 convertase enzymes cleave many molecules of C3 to C3b, which bind covalently around the site of complement activation. Reprinted with permission from Nature Reviews Immunology [55] (copyright (2002) Macmillan Magazines Ltd)

in three different ways: it can bind to domains of IgM or IgG complexed with antigen, it can bind directly to some bacterial surfaces, or it can bind to C-reactive protein (CRP), an acute phase protein. The binding of C1q causes a conformational change in the C1 complex, leading to the activation of C1r molecules, which in turn activate the serine protease C1s. The activation leads to the production of the C3 cleaving enzyme, C3 convertase, by cleaving C4 and C2. The cleavage generates four fragments: two large fragments C4b and C2b, which together form the C3 convertase, and two small fragments C4a and C2a. The most important activity of the C3 convertase is the production of large number of C3b, which coat pathogen surfaces, marking them for phagocytosis after binding to C3b receptors on the phagocytes. The cleavage of the C3 molecule also leads to the formation of C3a, which serves as a chemoattractant for leukocytes [174, 239, 243].

20.2.2.2 The Lectin Pathway

The activation through the lectin pathway starts with binding of L and H ficolins to *N*-acetylglycosamine (GlcNAc) or mannose binding lectin (MBL) to carbohydrates containing mannose and certain other sugars on pathogen surfaces (bacteria and viruses). The access to these sugars is blocked by sialic acid in vertebrate cells. MBL is a serum protein, which is normally present in low concentration in blood, but during an acute phase response the production in the liver increases. The MBL molecule and the M-, L-, and H-ficolin molecules are very similar in function to the classical pathway C1q [55, 62, 199]. Like C1q, the lectins (MBL and the ficolins) form a complex with two molecules of the protease zymogen, MASP (MBL-associated serine protease), which are activated through the binding of the lectins to carbohydrates. There are three types of MASP molecules. MASP-1 is capable of directly cleaving C3 [61, 114] as well as C2. MASP-2 is homologous to C1r and C1s of the classical pathway, which cleaves C4 and C2 after binding of lectins to pathogen surfaces and generates the C3 convertase [215]. MASP-1 is not necessary for activation through the lectin pathway, but it appears to play a role as an amplifier of complement activation by cooperating with MASP2 [121]. MASP-3 seems to inhibit MASP-2 activity [55, 61].

20.2.2.3 The Alternative Pathway

This pathway differs from the two other pathways by being independent of special binding proteins for activation. The alternative pathway is always activated by spontaneous hydrolysis of C3 in plasma. Spontaneous hydrolysis of C3 leads to altered conformation of the molecule and production of $C3{\cdot}H_2O$, which leads to the spontaneous cleavage of C3 and production of C3b. The hydrolyzed C3 molecule allows binding to the plasma protein, factor B, followed by cleavage of factor B by factor D to generate the fragment Bb. This leads to the formation of the convertase C3bBb, which is in fluid phase and can cleave many molecules of C3. C3b is highly reactive and quickly binds to either pathogen surfaces or molecules of water. C3b molecules in serum are thus quickly inactivated by hydrolysis, but C3b molecules forming near pathogen surfaces may bind to the surface. Regulatory proteins prevent complement activation by C3b deposited on host cells, and factor I can also inactivate C3b by cleaving it to the inactive form, iC3b [174, 239, 243].

Nascent C3b molecules can bind to the C3 convertase and generate the C5 convertase (C4bC2bC3b). Cleavage of C5 is the next step in the complement cascade, leading to the formation of C5a, which is a very strong anaphylatoxin, and C5b, the first molecule in the membrane attack complex (MAC).

20.2.2.4 The Membrane Attack Complex

The MAC is one of the major steps in the complement activation process. It lyses target cells and can activate neighbor cells to secrete proinflammatory molecules. The assembly of MAC starts with the cleavage of C5 by C5 convertase, generating C5a and C5b. C5b is the initiating factor in the assembly of MAC, binding to one of each molecule C6, C7, and C8. The C7 and C8 undergo a conformational change, allowing the complex to become inserted into cell membranes and initiating polymerization of C9 molecules. This leads to pore formation and up to 16 molecules of C9 can be added. The pore leads to disruption of the cell membrane by destroying the proton gradient across the membrane, leading to cell death [174, 239, 243].

The importance of the early complement components in fighting infections is especially apparent from studies on individuals with homozygous genetic defects in the activation of C3 and in the C3 protein. These defects are associated with a wide range of pyogenic infections and the individuals have substantially increased frequency of infections, reflecting the importance of C3 for opsonization of a wide range of pathogenic bacteria. Defects in the components of the MAC do not have as adverse consequences, but individuals with MAC defect do have increased susceptibility to Neisserial infections, emphasizing an important role for MAC in defense against this pathogen [243].

20.2.3 Complement Inhibitors

The complement system is very important in host defense against infections, but uncontrolled activation can lead to host cell damage. Host tissues express a number of membrane and fluid phase regulatory proteins, preventing autologous damage of host tissues. The regulation occurs predominantly at two steps in the process of complement activation; at the level of convertase enzymes (C3 and C5 convertase) and in the formation of the MAC [120, 223]. In addition, one regulator, C1-inhibitor (C1-INH) operates at the level of C1.

Regulators of complement activation (RCA) consist of a family of proteins, which control the assembly and stability of the convertase enzymes [85]. The membrane bound complement regulators decay accelerating factor (DAF) [134], membrane cofactor protein (MCP) [16], and complement receptor 1 (CR1) [1] protect cells against damage by inhibiting C3 cleavage and deposition [86], while CD59 inhibits the assembly of MAC and lysis of host cells [167]. Two plasma proteins, factor H [233, 252] and C4-binding protein (C4BP) [234], inhibit complement activation by binding to the active complement fragments C4b and C3b, preventing the formation of C3 convertase and promoting the dissociation (decay) of the convertases already formed. C4BP, factor H, CR1, and MCP can in addition act as cofactors for the cleavage of C3b by factor I.

20.2.4 The Complement Inhibitor Vaccinia Virus Complement Control Protein

Vaccinia virus complement control protein (VCP) is a strong inhibitor of the classical, lectin, and alternative pathways of complement, acting on both C4 and C3 [89]. Identified in 1988 as a product of the vaccinia virus, VCP is a 35 kDa soluble protein with structural [99] and functional [97] resemblance to human C4 binding protein (C4BP) as well as other proteins of the regulators of complement activation (RCA) family. In vitro studies have shown that VCP is bound by C4b and C3b and serves as a cofactor with factor I in cleaving these two molecules. VCP has been shown to inhibit the formation and accelerate the decay of the classical pathway C3 convertase (C4b2a) as well as the alternative pathway convertase (C3bBb) [97, 116, 181]. Like its soluble mammalian RCA counterparts C4BP and factor H, but unlike the surface-bound RCA molecules DAF, MCP, and CR1, VCP displays heparin-binding capabilities [196]. This suggests an in vivo role in connection with heparan sulfate proteoglycans lining the endothelial cell layer [127, 197]. By blocking complement activation at multiple sites, VCP downregulates proinflammatory chemotactic factors (C3a, C4a, and C5a) resulting in reduced cellular influx and inflammation. Animal studies have shown that VCP significantly enhances functional recovery following traumatic brain injury [79, 94, 154, 155], modulates inflammation, and improves spinal cord integrity following spinal cord injury [163, 164] and has both therapeutic and prophylactic activities in Alzheimer's disease [44, 98, 102, 156]. Others have shown its beneficial effect in reducing hyperacute xenorejection following transplantation [2, 4, 5]. VCP has proven safe in all animal model studies tested to date (mouse, rat, rabbit, pig, and baboon), and through these experiments, it has become increasingly clear that VCP may hold great promise as a potential therapeutic molecule in inflammatory conditions [89].

20.2.5 Complement and Inflammation

The complement system is one of the key components in inflammation. The anaphylatoxins C3a, C4a, and C5a have proinflammatory properties and play a major role in the inflammatory process by acting on specific receptors on myeloid cells. They are involved in smooth muscle contraction and increased vascular permeability [27]. C3a and C5a are also involved in increased induction of adhesion molecules on the surface of vascular endothelium and the release of histamine and TNF-α from mast cells [88, 91, 205]. The activity of C3a and C5a leads to clearance of pathogens by influx of antibodies, complement and phagocytic cells to the infected site. The C5a molecule is considered to have the highest biological activity of the anaphylatoxins, by acting as a chemoattractant for neutrophils and monocytes, leading to their recruitment to sites of infection [87]. In addition, cleavage of C3a, C4a, and C5a by the anaphylatoxin inactivator carboxypeptidase N removes the C-terminal arginine from the anaphylatoxins to form C3a des arg, C4 des arg, and C5 des arg, respectively. This destroys the activity of the C3a and C4a, while C5a des arg retains a considerable degree of activity after cleavage [174, 243].

Activity of MAC may be important in inflammation even without cytolytic activity. Sublytic doses of C5b-9 can induce the release of proinflammatory mediators and the synthesis of IL-8 and MCP-1 [96], increased expression of adhesion molecules [213] and entry into cell cycle [178], leading to increase in DNA synthesis and cell proliferation [135]. Intense complement activation can lead to cell lysis and subsequently to necrotic cell death, leading to an inflammatory reaction in the surrounding tissues. Apoptotic cells do not under normal circumstances lead to inflammation, because the complement system helps to eliminate these cells from the tissues [212]. If the elimination of this waste material fails, the apoptotic cells could accumulate and evoke an autoimmune response.

20.2.6 The Complement System and Myocardial Infarction

In addition to a presumed role in atherogenesis, activation of complement components is a major contributor to the tissue damage occurring during and after myocardial infarction [14, 124]. Hill and Ward, over 30 years ago, first proposed a role for complement activation in myocardial reperfusion injury [80] and subsequent studies have shown that local activation of the complement system is a major component in reperfusion injury following myocardial ischemia [117, 184, 226]. Today it is evident that C1q [173], C3, C4, and C5 [117, 157], and MAC [104, 184] are deposited in the infarcted tissue, and the activation takes place through all three activation pathways of the complement system, the classical pathway, the alternative pathway, and the lectin pathway [33, 40, 200]. Recent studies indicate that the lectin pathway is particularly important in this context [241]. Complement activation occurs early in the process; in a rat model, it is initiated after 2 h of ischemia [227]. This leads to production of chemotactic factors such as the anaphylatoxins (C3a and C5a), which can attract inflammatory cells to the ischemic tissue, and formation of the MAC (C5b-9), which can activate several steps in the inflammatory cascade (see p. 14). The presence of MAC in the necrotic tissue has been shown to be associated with loss of the regulatory protein CD59 [225]. The complement system is also involved in the tissue damage occurring during the ischemia itself [41]. In addition, complement has been strongly linked to the major risk factors of myocardial infarction, in particular C3, C4, and MBL [10, 29, 101, 128–131, 180].

20.3 The Role of Complement in Atherogenesis

20.3.1 Historical Notes

The initial decades after the discovery of the complement system were focused mainly on the biochemical characterization of its many components and factors [174]. A physiological role was initially mainly suspected and confirmed to relate to pathogen elimination, but studies of genetically deficient subjects led investigators to suspect a role for the classical activation pathway in safe elimination of immune complexes [103], and this notion was later confirmed [9, 11]. Atherosclerosis was for a long time considered to be a disease of lipid metabolism, and a connection between complement and atherosclerosis was not suspected until relatively recently. The first studies suggesting a role for complement in atherosclerosis were made by in Hungary in 1975–83 by George Füst and his group [57–59, 208, 209]. These results showed that the complement system was activated and levels of C3 and immune complexes raised in MI. Independently, Geertinger and Sorensen [63] made the important discovery that deficiency in complement component C6 protected against diet-induced atherosclerosis in rabbits. This finding was not followed up but other groups entered the field, notably the groups of Horea Rus in Cluj-Napoca, Romania (since 1983) [136, 139–141, 176, 177, 179, 236–238] and Sucharit Bhakdi in Mainz, Germany (since 1985) [136, 183, 184, 188, 206]. These studies showed that complement is activated immediately after LDL-influx over the endothelium and that this leads to formation of foam cells and typical changes in the forming plaque. The combined results of these studies have yielded sound results that implicate the complement system in the pathogenesis of atherosclerosis [23, 138, 144, 217, 220]. In spite of these advances, the field did not move fast in the first 20 years, and it is only relatively recently that the field has attracted the interest that it deserves. The interest of the present authors was raised by the observation that the frequency of C4B*Q0 was significantly raised in Hungarian patients with myocardial infarction [101] and was correspondingly decreased in the general population of middle age individuals [100]. This led to a collaborative study between Icelandic and Hungarian scientists focusing on the relationship between C4B*Q0 and cardiovascular disease (CVD) and its risk factors [8, 10, 11, 29]. These studies showed that C4B*Q0 interacts with smoking to promote angina pectoris (AP) (OR 30.07) and myocardial infarction (MI) (OR 22.66) and also leads to a significantly increased 1-year mortality after MI (hazard ratio 3.50). These results firmly establish a role for complement in the pathogenesis of cardiovascular disease, and notably, the interaction of smoking and C4B*Q0 is observed even in the general population of middle age smokers in Iceland as a significant decrease in the carrier frequency of C4B*Q0 [8].

20.3.2 Complement and Cardiovascular Disease: Initial Findings

Prior to 2005, a number of studies had implicated the activation of complement in the pathogenesis of atherosclerosis, showing complement components to be deposited in atherosclerotic lesions from patients as well as experimental animals [137, 138, 236, 237], with higher concentrations in more developed lesions [237]. Activation of complement was shown to take place through both the classical and the alternative pathways, but not much information was gained about activation through the lectin pathway. Complement components C3 and C4, complement regulators, terminal complexes (C5b-9), and immunoglobulins were all shown to immunolocalize in human atherosclerotic plaques [137, 138, 237] and increased mRNA levels of complement components were seen in fibrotic plaques compared to normal arteries [246]. This indicated local activation and production of the complement components in the arterial wall. Furthermore, the activation of complement was shown to be the first visible sign of inflammation in the arterial wall in an animal model of atherosclerosis, suggesting that the activation of complement formed a link between the deposition of LDL and monocyte recruitment [188]. The early products of the complement components, C3d, and the late components, C5b-9, were localized in different areas in atherosclerotic lesions, C3d mainly in the superficial intima and C5b-9 in the deeper intima layers in association with smooth muscle cells, cell debris, and extracellular lipids [176, 217]. C5b-9 deposition was found to correlate with the severity of the lesion [237], as concentrations were greater in developed lesions, such as fibrotic plaques, than in fatty streaks and healthy intima [136].

Despite these advances, and the indication that complement seemed involved in all stages of cardiovascular disease [23, 138, 144, 217, 220], the information in 2005 was incomplete about the extent of the involvement at each stage. Moreover, the importance of complement in lesion development was a matter of dispute, since in C6-deficient rabbits fed with high-fat diet, complement appeared to play a obligatory and rate limiting role in the development of lesions [63, 185] as C6 deficiency suppressed the development of atherosclerotic lesions without affecting the serum levels of lipids. In marked contrast to these findings, three studies on complement deficient mouse models with a defect in a key element in lipid metabolism (LDLR–/– and/or ApoE–/–) reported that the development of atherosclerotic lesions was not severely affected by the complement deficiency [35, 150, 153]. In experiments with C6-deficient rabbits, complement seemed to play an obligatory and rate-limiting role in lesion development. There was extensive difference in lesion size between C6-deficient and C6-sufficient animals [63, 185] with similar levels of plasma cholesterol and lipoproteins in both groups [185]. Staining for C5b-9 was positive in all lesions of C6-competent rabbits and negative in all C6-deficient rabbits and moreover, the C5b-9 staining was never seen outside the atherosclerotic lesions [185]. In the discordant studies, deficiency of C5 in the ApoE–/– mice did not lead to any visible changes [150], while the deficiency of C3 in LDLR–/– or LDLR–/–ApoE–/– mice led to increase in lesion size [35, 153] with increased buildup of LDL and macrophages and slower transition from fatty streaks to fibrotic plaques in the former mouse strain [35]. In the C3–/–LDLR–/– mouse model, there was a slower transition from fatty streaks to fibrotic plaques (increased buildup of LDL and macrophages and slower influx of smooth muscle cells and deposition of collagen).

The discordance between the studies described above could theoretically be explained by a difference between diet-induced and genetically driven atherosclerotic disease. Although lesions of genetically modified mouse strains are more similar to lesions in the human by being more extensive and progressing beyond the fatty streak stage, it should be noted that the course of the disease in these mice is different from that in human patients, as in human patients the disease is generally diet-induced and only very rarely the result of genetic defects in the major lipid transport proteins (such defects are known as familial hypercholesterolemia). The atherogenic drive of ApoE–/– and LDLR–/– mice is thus much stronger than that of the human, and the suitability of these mouse strains for revealing the atherogenic effect of complement has been questioned [23]. Another possible confounding factor is the fact that the mice in the discrepant studies were deficient from birth in both complement and key components in lipid transport, thus opening possibilities for the defects to also act on ontogenic development. It appeared possible to resolve the controversy between these two lines of evidence by differences between genetically driven and diet-induced disease [23], as it appeared possible that the strong atherogenic drive in genetically induced disease could mask the importance of complement in the disease development. This hypothesis was tested by using the inbred mouse strain C57BL/6, which is the background strain of ApoE- and LDLR-deficient mice [35, 150, 153] but without genetical modifications in the lipid profile; the importance of complement in the disease was studied by using a complement inhibitor instead of gene knockout

20.3.3 The Effect of VCP on Diet-Induced Atherosclerosis in C57BL Mice

During the first decades of research into the connection between complement and atherosclerosis, results had been obtained that indicated that complement activation takes place in the initiation and/or progression of atherosclerotic lesions. However, this left out the key question of whether complement activation serves a role in the initiation and/or progression of lesions or whether it simply serves as an enhancing factor for processes already in play. To discern between these alternative, we injected the complement inhibitor VCP into female C57BL/6J mice fed with atherogenic diet. Remarkably, it turned out that the development of fatty streaks in this animal model of diet-induced atherosclerotic disease was significantly retarded by the injection of VCP [216]. These results uniquely identify complement as an obligatory and rate-limiting step in the initiation and initial progress of atherosclerosis. More specifically, evaluation of lesions in mice fed on atherogenic diet for 15 weeks and injected at weekly intervals in the last 7 weeks with either VCP or saline indicated a significant reduction in lesion size in the VCP-injected mice (Fig. 20.3), compared to the saline-injected mice ($p=0{,}004$, Fig. 20.4). No lesions were seen in the control mice fed with normal diet.

After 15 weeks on atherogenic diet lipids and macrophages were abundant in aortic walls but lesions did not progress beyond the fatty streak stage. This is consistent with previous observations on the C57BL/6 mouse model [146, 148, 194]. Normal histology showed no smooth muscle cells in the intimal layer of the aorta. The observation that lesions do not progress beyond the foam cell stage and

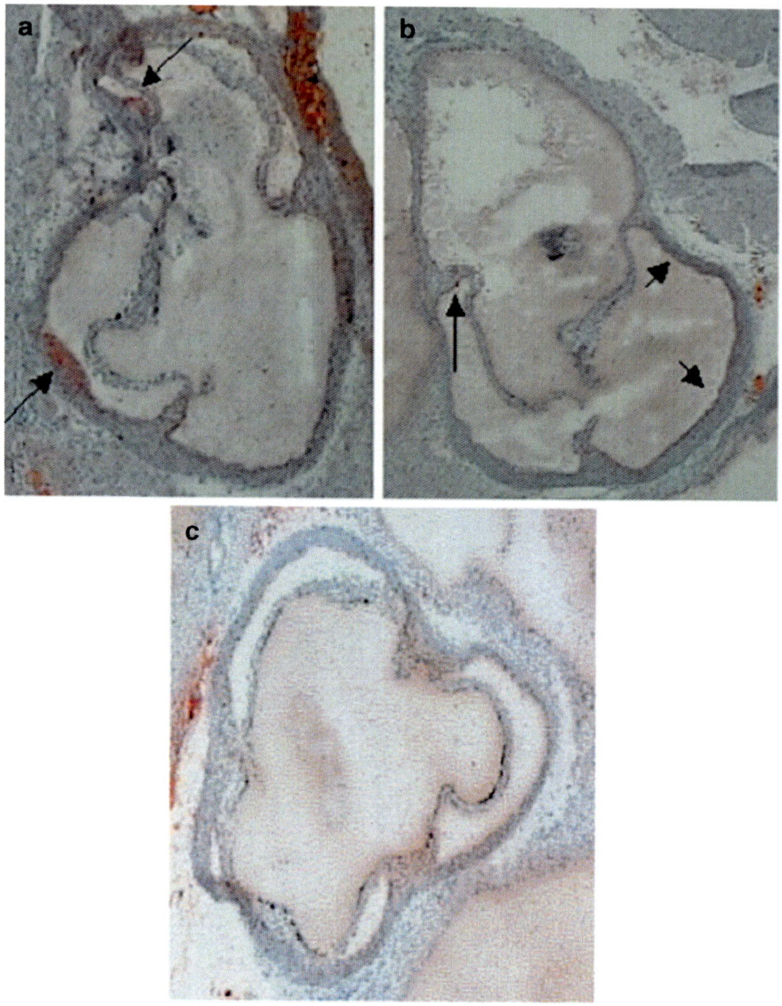

Fig. 20.3 Development of fatty streaks in mice fed with high-fat diet for 15 weeks and injected with saline (**a**) or 20 mg/kg VCP (**b**) in week 8–15. Note the lipid plaques (*arrows*) in the intima (**b**) or intima and media (**a**). No lipid staining is evident in a mouse fed with chow diet and injected with saline (**c**). Oil red O staining. Reprinted with permission from *Annals of the New York Academy of Sciences* [copyright (2005) by the New York Academy of Sciences] [216]

that lesions were significantly smaller in VCP-injected mice supports earlier findings implicating complement in the initial stages of lesion formation in diet-induced atherosclerosis [63, 185, 188]. It should be pointed out that our experimental design did not explore the maximal protection attainable by VCP as its inhibitory effect disappeared from serum less than 4 days after injection [216] but we had only sufficient VCP for injecting once per week. In addition, we had to confine our experiment to the period of maximal fatty streak formation in week

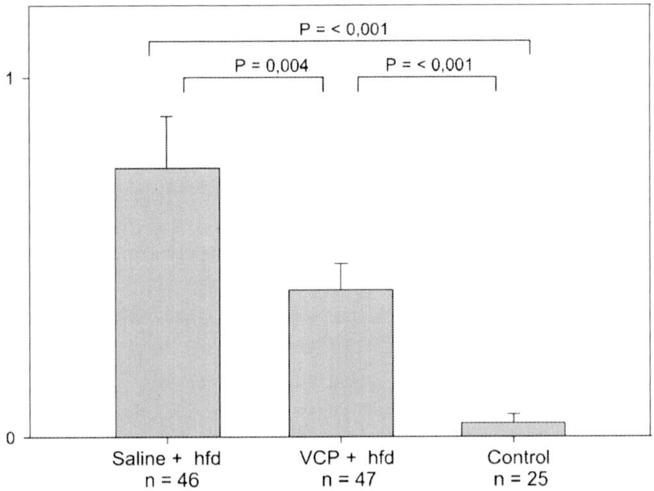

Fig. 20.4 Percent lesion area (calculated by Leica qwin computer analysis) in aortic sections from mice (**a**) fed a high-fat diet and injected with saline, (**b**) fed a high-fat diet and injected with VCP (20 mg/kg), and (**c**) control mice. The lesion size is significantly lower in the VCP-injected mice compared with the saline-treated mice fed with high-fat diet, $p=0.004$. Reprinted with permission from *Annals of the New York Academy of Sciences* [copyright (2005) by the New York Academy of Sciences] [216]

8–15 allowing for some lipid deposition before treatment. Due to its heparin binding sites [127, 196, 197] VCP may be sequestered in the body for periods exceeding its half-life in serum [90], and this may explain the relatively high level of protection attained even by weekly injections. It is likely that the 50% protection observed is a compromise of a much higher level of protection (even 100%) immediately after injection and no or lower protection towards the end of the week. In the light of the results with C6-deficient rabbits [185] showing a complete inhibition of fatty streak formation, a much higher level of protection may be expected with changes in the injection regime.

These results have strong implications from a practical as well as a scientific point of view, as they raise hope that VCP, or indeed other complement inhibitors, may lead to future development of drugs with prophylactic potential. This is important given that previous attempts to assess the usefulness of a complement inhibitor had not yielded convincing results [119, 182]. Complement inhibitor injected animals had been estimated to have smaller lesions, but no statistical evaluation was shown, and the findings were not published in scientifically acclaimed periodicals. In line with our results, previous studies on complement inhibitors had suggested a beneficial effect in xenotransplantation [4, 5], Alzheimer's disease [44, 98, 102, 156], brain and spinal cord injury [79, 94, 154, 155, 163, 164], and reperfusion injury [34, 71, 83, 105, 112, 228, 249], but the focus had been mainly on therapeutic use. From a wider point of view, our results shed new light on the role of complement in atherosclerosis, suggesting that it may be a rate-limiting step in diet-induced disease, although in disease driven by defects in lipid metabolism it may be redundant.

20.3.4 Complement in Atherogenesis: Mode of Action

The results described in Sect. 20.3.3 show that the complement system plays an obligatory and rate-limiting role in diet-induced atherogenesis. Existing theories on how inflammatory processes are initiated focused mainly on modified forms of LDL. Modified LDL induces the expression of adhesion molecules and cytokines in endothelial cells in vitro [20, 42], is chemotactic for monocytes, macrophages, and T cells [82, 160], and can upregulate genes for macrophage colony stimulating factor (M-CSF) [160] and monocyte chemotactic protein (MCP-1) [52] derived from endothelial cells [170]. The results of Sect. 20.3.3, however, suggest that these direct effects of modified LDL may not be sufficient to evoke an inflammatory response in vivo in the absence of complement. The presence of oxidized LDL particles in the intima was known to stimulate the endothelial cells (EC) in the vascular wall to start to produce inflammatory molecules, including adhesion molecules (ICAM-1, VCAM-1), growth factors (M-CSF), and chemokines (MCP-1), leading to adhesion of blood monocytes and lymphocytes to the endothelium and their recruitment to the intima [69, 111]. Activation of complement would permit or accentuate this process, as C3a and C5a cause increased induction of adhesion molecules on the surface of vascular endothelium and the release of histamine and TNF-α from mast cells [88, 91, 205] and sublytic doses of C5b-9 induce the release of proinflammatory mediators, increased expression of adhesion molecules and the synthesis of IL-8 and MCP-1 [96, 213]. Activation of complement may provide the missing link between LDL accumulation and proinflammatory cytokines such as IL-1β or TNF-α in human atherosclerotic lesions, which had been poorly understood [107, 108]; these cytokines induce VCAM-1 expression by endothelial cells [43] and thus binding of leukocytes such as monocytes and T lymphocytes. In addition to LDL, hemodynamic forces, homocysteine levels, sex hormones, and infection [111] have been regarded as candidates for inducing the inflammatory response in the disease, but the results of Sect. 20.3.3 render these possibilities unlikely. Antigen–antibody complexes (immune complexes) constitute a largely overlooked possibility [132, 235]; they are strong activators of complement and thus deserve attention in future studies.

Acute phase proteins and systemic mediators such as CRP and IL-6 are involved in atherosclerosis. Foam cell formation leads to the secretion of IL-6, which stimulates the liver to produce the acute phase proteins CRP, fibrinogen, and plasminogen activator inhibitor (PAI). These proteins mediate inflammation and can possibly accelerate the evolution of atherosclerotic lesions by promoting T-cell activation and infiltration into the lesion [69]. Early clinical studies have found elevated levels of CRP in patients with myocardial infarction and prospective studies have found plasma CRP levels to be an independent indicator of both future coronary risk and outcome in patients with acute coronary syndromes, even in the absence of myocardial damage. Macrophage accumulation in the lesion may be associated with increased plasma concentration of both fibrinogen and CRP [46, 81, 115, 207], which are markers of inflammation and thought to be early signs of atherosclerosis. These findings acquire a new meaning when combined with the results in Sect. 20.3.3. Enzymatically modified LDL is an activator of complement [21–23] and complement is a potent inducer of all the main events in inflammation. Complement may thus provide the missing link between cholesterol trapping and upregulation of leukocyte homing receptors and act in a vicious circle with the acute phase proteins.

The present study provides no clue to what in the atherosclerotic lesion activates complement, but several candidates have been proposed [138, 144, 217]. Complement can be activated in the forming plaque by modified LDL, in particular enzymatically digested LDL (Fig. 20.5) [22, 188, 189, 217]. It could also be activated by immunoglobulins, immune complexes or C-reactive protein (CRP); IgM and IgG are retained in the lesion [237, 238] and can form immune complexes with oxLDL [247] or with other candidate antigens. CRP deposition has also been observed in human atherosclerotic lesions and found to be consistent with the severity of the lesion (see above) [165, 219, 237]. CRP can bind to oxLDL [35, 38, 185] and enzymatically degraded LDL [24] and can activate complement through both the classical and the alternative pathways. CRP may modify the outcome by limiting the activation to the C3 level by preventing formation of MAC and favoring opsonization [25, 68]. The presence of C5b-9 (MAC) in the deeper layers of human lesions [143] suggests that the effect of CRP may be confined to the upper layers where the activation takes place and that CRP and/or MAC inhibitors are less abundant in the deeper layers.

Complement may modulate lesion development in various ways (Fig. 20.5). The current study and studies with C6-deficient rabbits [63, 185] fed with high-fat diet suggest that complement inhibition leads to reduced accumulation of lipids and

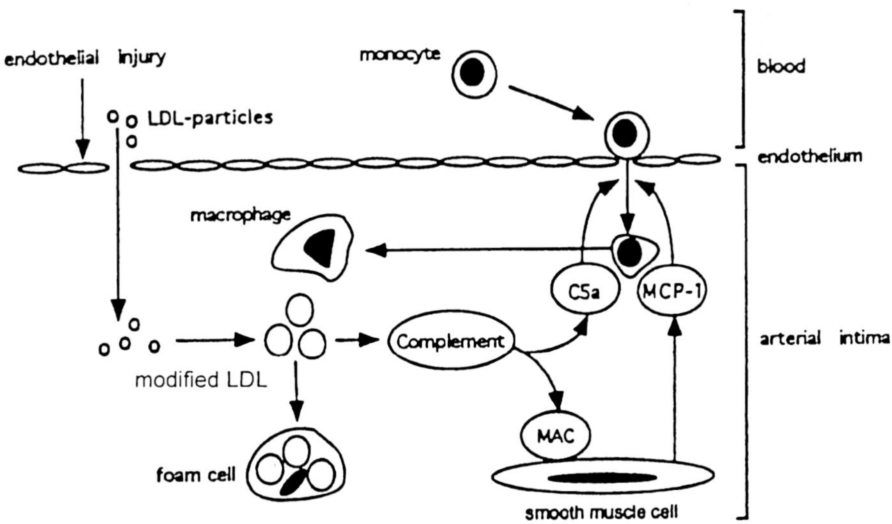

Fig. 20.5 Schematic diagram of cellular interactions in atherogenesis induced by complement. Reprinted from Torzewski et al. [217], copyright (1997), with permission from Elsevier

macrophages in the lesion. Complement may have direct effect on the endothelial integrity. The endothelial cells are exposed to reactive complement components, which are activated in the blood or at the endothelial surface, and are thus dependent upon soluble or membrane-bound complement inhibitors to prevent cell damage. A study by Venneker et al. [230] showed that endothelium of lesional and nonlesional skin from patients with systemic sclerosis had reduced levels of expression of the MCP (CD46) and decay-accelerating factor (DAF, CD55), which control the complement system at the C3 convertase level. This may contribute to vascular damage and therefore make the endothelium more permeable to macromolecular influx. The anaphylatoxins can recruit leukocytes from the circulation [217]. Furthermore, the generation of C5b-9 in the intima of the artery may lead to injury to vascular cells and subsequent release of growth factors and cytokines from endothelial cells, macrophages, and smooth muscle cells [138]. Of these, smooth muscle cells are the most likely target for C5b-9 generation, because they do not express CD59 and are therefore poorly protected for activation of the late complement components [190]. Assembly of terminal complement components (C5b-9) on the smooth muscle cells leads to MCP-1 synthesis, which could explain the initial monocyte recruitment into the arterial wall [217, 218]. In addition, it could explain the smooth muscle cell proliferation, as C5b-9 is mitogenic for smooth muscle cells [135]. Complement activation has been associated with apoptosis. Binding of oxLDL to endothelial cells makes them apoptotic [142], and endothelial cells undergoing apoptosis trigger complement activation through the alternative pathway [221]. C5b-9 deposits have been found localized not only on intact smooth muscle cells and cell debris [176], but also on apoptotic cells [137]. The above results indicate that the activation of complement by apoptotic cells may contribute to the lesion development, as complement activation by apoptotic cells favors removal of apoptotic cells and monocyte recruitment into the lesion [118]. The complement system may thus play a role both in fatty streak formation and the chronic inflammatory process that drives lesion progression.

20.3.5 The Effect of Complement Inhibitors on Reperfusion Injury

Although complement inhibitors have not prior to this study been used with advantage in animal models of atherosclerotic development, their use in models of myocardial infarction has given very promising results. In fact, studies of this kind provide the best evidence so far gained for the importance of complement in reperfusion damage. Previous experimental studies on animals have shown that inhibition of complement activation, either at the time of coronary artery occlusion [41, 113, 157, 242] or just before reperfusion [3, 34, 83, 126, 152, 166, 193, 228, 249] reduces infarct size and neutrophil infiltration. Therefore, inhibiting complement activation in order to reduce ischemic reperfusion injury has been regarded promising.

The first evidence that complement activation had deleterious effect on tissue integrity during ischemia/reperfusion was presented by Maroko et al. [113] in 1978. They showed that by depleting complement with cobra venom factor (CVF), the infarct size in an animal model was significantly reduced. This drove researchers to develop more suitable inhibitors to prevent complement activation, as CVF is not a good inhibitor in animals and humans because it leads to biologically active complement products. Today, complement inhibitors that have specific actions on the complement cascade and are poorly immunogenic have been developed. Various recombinant human complement inhibitors have been developed, as well as monoclonal antibodies, synthetic peptides, and peptidomimetics, which block activation of certain complement components, neutralize an activation fragment or antagonize complement receptors [125].

Modern studies on the involvement of complement in reperfusion injury are based on the use of genetically deficient animal strains, such as C6-deficient rabbits in which ischemia/reperfusion injury was reduced compared to complement sufficient controls [95] or the use of complement inhibitors in wild type strains. The inhibitors such as soluble CR1 (sCR1) [34, 105, 106, 193, 249], C1 esterase inhibitor (C1-INH) [83], C5a monoclonal antibody [3, 228], and C5a receptor antagonist [152, 166] have all been shown to reduce ischemic/reperfusion injury by reducing infiltration of neutrophils and reducing the inflammatory response. Despite promising results in experimental models, clinical trials in humans have not been as promising. The use of sCR1 in man was trialed in patients during cardiopulmonary bypass, the agent proved safe but there were no clinically important differences between the patients who received sCR1 and those receiving the placebo. Further development of this agent was terminated (further information, site: http://www.avantimmune.com) [251]. C1-INH initially showed promising results in clinical trials [17, 36, 37, 50], but a study involving its application to reduce capillary leak during open-heart surgery in 13 neonates lead to nine deaths due to venous thrombosis [50]. It has now been shown that C1-INH has cardioprotective effect at low doses (20–40 IU/kg), but at higher doses the detrimental effects come to light [84]. Human studies using the novel C5 monoclonal antibody, pexelizumab, either in thrombolytic therapy [112] or in angioplasty [71] after MI showed no significant effect of the C5 antibody in the reduction of tissue damage. However, in the latter study, the 90-day mortality was somewhat reduced in those receiving the antibody [71]. In addition, a large clinical study on patients undergoing cardiopulmonary bypass receiving pexelizumab showed no statistically significant difference on the primary endpoint between those receiving the drug and those receiving placebo, but reduction in mortalities or subsequent MI was found [191].

In the light of a still ongoing search for a suitable complement inhibitor in a model of reperfusion injury, we tested the possibility that VCP could be a candidate for future developments in this field, using a model of experimentally induced MI in rats. There are no animal models available which spontaneously develop myocardial infarction. For this reason, research on myocardial infarction is generally confined to surgical ligation of a coronary artery. As surgical skills are becoming better, the use of mice in such studies is on the increase [211, 231]. However, the rat is still the most widely used animal for this kind of research due to low maintenance cost and convenient size. Our results on Sprague-Dawley rats are published in detail elsewhere (Thorbjörnsdottir P, MS Thesis, University of Iceland 2005; Thorbjörnsdottir et al. unpublished). Briefly, to determine the protective effect of VCP on myocardial damage and to find the appropriate dose of VCP, i.v. injections of VCP (1, 4, and 20 mg/kg) or saline were made after 25 min of ischemia (5 min prior to reperfusion). The results showed that 43% protection against myocardial damage could be attained by using 20 mg/kg VCP ($p=0.007$). By extrapolarization, 8–12 mg/kg were calculated to offer maximal protection, and this was verified by repeating the experiments using 8.5 mg/kg. The protection against myocardial damage was 44% ($p<0.001$), and a significant reduction (14%, $p=0.017$) was also observed at a lower dose (4 mg/kg). Thus it is evident that VCP injection reduces the size of the infarcted area in rat myocardium after ischemia reperfusion in a dose-dependent manner. This means that VCP is by now unique among complement inhibitors in being able to interfere with atherosclerosis and its acute effects in two different ways, and unique among drugs with potential in atherosclerosis by interfering not only with atherosclerosis risk factors but also the process of atherosclerosis (Fig. 20.6).

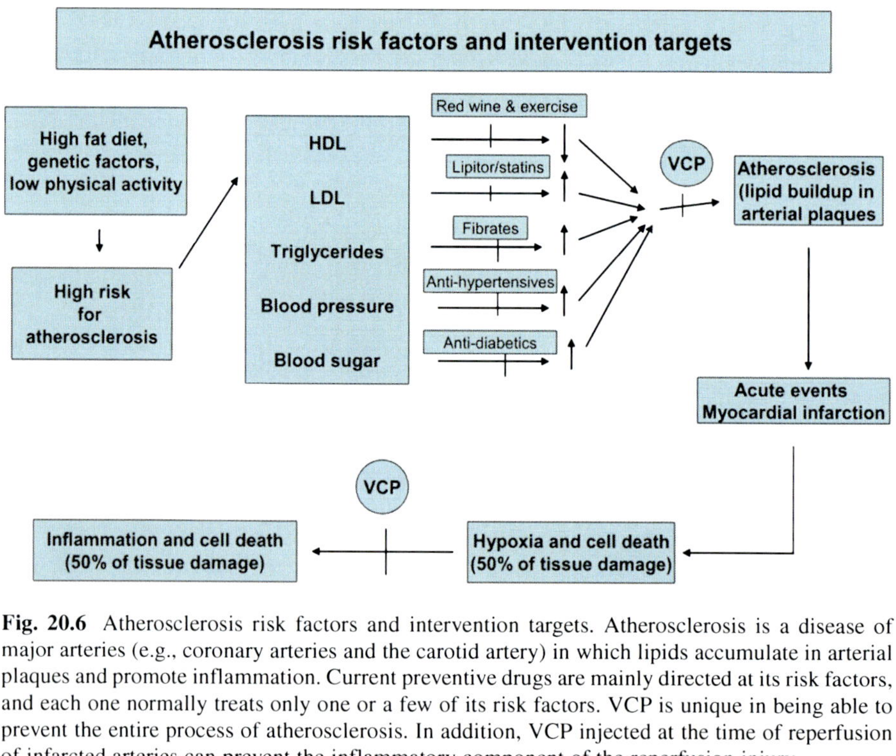

Fig. 20.6 Atherosclerosis risk factors and intervention targets. Atherosclerosis is a disease of major arteries (e.g., coronary arteries and the carotid artery) in which lipids accumulate in arterial plaques and promote inflammation. Current preventive drugs are mainly directed at its risk factors, and each one normally treats only one or a few of its risk factors. VCP is unique in being able to prevent the entire process of atherosclerosis. In addition, VCP injected at the time of reperfusion of infarcted arteries can prevent the inflammatory component of the reperfusion injury

20.3.6 Conclusions

Atherosclerosis is a chronic inflammatory disease. Complement is a major inducer of inflammation. Several studies have indicated that complement could be involved in the pathophysiology of atherosclerosis, with components being immunolocalized in atherosclerotic lesions from patients as well as experimental animals. However, until recently, the extent of the involvement was not known, and in particular, it remained unresolved whether complement activation in atherosclerosis served a role in the initiation and/or progression of lesions, or whether it simply served as an enhancing factor for a process already in play. The controversy over the relative importance of complement in this scenario was made even more poignant because in C6-deficient rabbits fed with high-fat diet, complement appeared to play an obligatory and rate-limiting role, but in genetically modified mice (LDLR–/– and/or ApoE–/–), deficiency of complement (C3 or C5) did not retard lesion progression. It seemed possible to explain the discordance by a difference between diet-induced and genetically driven atherosclerotic disease. To solve this controversy, a study was designed using the inbred mouse strain C57BL/6, which is the background strain of ApoE- and LDLR-deficient mice, but without genetical modifications in the lipid profile; the importance of complement in the disease was studied by using a complement inhibitor instead of gene knockout

Remarkably, the results demonstrate, for the first time, that the development of atherosclerotic lesions in diet-induced atherosclerotic model can be significantly reduced by weekly injections of a complement inhibitor, VCP. The present results combined with previous data [185] support that the complement system plays an obligatory, rate-limiting role in lesion formation in diet-induced atherosclerotic disease. These results increase the understanding of the disease course and raise hope that the progression of atherosclerosis and its main events may in the future be prevented or retarded by the use of complement inhibitors. In another study, VCP administered as a bolus during reperfusion of an infarcted coronary artery also proved to reduce reperfusion damage by 44%. The combined results uniquely identify the complement inhibitor VCP as a protein with a potential role in future development of disease modulators in atherosclerosis and its acute events.

References

1. Ahearn, J. M. & Fearon, D. T. 1989 Structure and function of the complement receptors, CR1 (CD35) and CR2 (CD21). Adv Immunol 46, 183–219.
2. Al-Mohanna, F., Parhar, R. & Kotwal, G. J. 2001 Vaccinia virus complement control protein is capable of protecting xenoendothelial cells from antibody binding and killing by human complement and cytotoxic cells. Transplantation 71, 796–801.
3. Amsterdam, E. A., Stahl, G. L., Pan, H. L., Rendig, S. V., Fletcher, M. P. & Longhurst, J. C. 1995 Limitation of reperfusion injury by a monoclonal antibody to C5a during myocardial infarction in pigs. Am J Physiol 268, H448–57.

4. Anderson, J. B., Smith, S. A., van Wijk, R., Chien, S. & Kotwal, G. J. 2002 Vaccinia virus complement control protein ameliorates hyperacute xenorejection by inhibiting xenoantibody binding. Transplant Proc 34, 3277–81.
5. Anderson, J. B., Smith, S. A., van Wijk, R., Chien, S. & Kotwal, G. J. 2003 Vaccinia virus complement control protein inhibits hyperacute xenorejection in a guinea pig-to-rat heterotopic cervical cardiac xenograft model by blocking both xenoantibody binding and complement pathway activation. Transpl Immunol 11, 129–35.
6. Anderson, T. J. 1999 Assessment and treatment of endothelial dysfunction in humans. J Am Coll Cardiol 34, 631–8.
7. Arason, G. J. 2008 The complement system. Icelandic Med Stud 59, 68–70.
8. Arason, G. J., Bodvarsson, S., Sigurdarson, S. T., Sigurdsson, G., Thorgeirsson, G., Gudmundsson, S., Kramer, J. & Füst, G. 2003 An age-associated decrease in the frequency of C4B*Q0 indicates that null alleles of complement may affect health or survival. Ann N Y Acad Sci 1010, 496–9.
9. Arason, G. J., Geirsson, A. J., Kolka, R., Vikingsdottir, T. & Valdimarsson, H. 2002 Deficiency of complement-dependent prevention of immune precipitation in systemic sclerosis. Ann Rheum Dis 61, 257–60.
10. Arason, G. J., Kramer, J., Blaskó, B., Kolka, R., Thorbjornsdottir, P., Einarsdottir, K., Sigfusdottir, A., Sigurdarson, S. T., Sigurdsson, G., Rónai, Z., Prohászka, Z., Sásvari-Szekely, M., Bodvarsson, S., Thorgeirsson, G. & Füst, G. 2007 Smoking and a complement gene polymorphism interact in promoting cardiovascular disease morbidity and mortality. Clin Exp Immunol 149, 132–8.
11. Arason, G. J., Steinsson, K., Kolka, R., Vikingsdottir, T., D'Ambrogio, M. S. & Valdimarsson, H. 2004 Patients with systemic lupus erythematosus are deficient in complement-dependent prevention of immune precipitation. Rheumatology (Oxford) 43, 783–9.
12. Argmann, C. A., Sawyez, C. G., Li, S., Nong, Z., Hegele, R. A., Pickering, J. G. & Huff, M. W. 2004 Human smooth muscle cell subpopulations differentially accumulate cholesteryl ester when exposed to native and oxidized lipoproteins. Arterioscler Thromb Vasc Biol 24, 1290–6.
13. Arnesen, E., Refsum, H., Bonaa, K. H., Ueland, P. M., Forde, O. H. & Nordrehaug, J. E. 1995 Serum total homocysteine and coronary heart disease. Int J Epidemiol 24, 704–9.
14. Arumugam, T. V., Shiels, I. A., Woodruff, T. M., Granger, D. N. & Taylor, S. M. 2004 The role of the complement system in ischemia-reperfusion injury. Shock 21, 401–9.
15. Avogaro, P., Bon, G. B. & Cazzolato, G. 1988 Presence of a modified low density lipoprotein in humans. Arteriosclerosis 8, 79–87.
16. Ballard, L. L., Bora, N. S., Yu, G. H. & Atkinson, J. P. 1988 Biochemical characterization of membrane cofactor protein of the complement system. J Immunol 141, 3923–9.
17. Bauernschmitt, R., Bohrer, H. & Hagl, S. 1998 Rescue therapy with C1-esterase inhibitor concentrate after emergency coronary surgery for failed PTCA. Intensive Care Med 24, 635–8.
18. Berenson, G. S., Srinivasan, S. R., Bao, W., Newman, W. P., 3rd, Tracy, R. E. & Wattigney, W. A. 1998 Association between multiple cardiovascular risk factors and atherosclerosis in children and young adults. The Bogalusa Heart Study. N Engl J Med 338, 1650–6.
19. Berenson, G. S., Wattigney, W. A., Tracy, R. E., Newman, W. P., 3rd, Srinivasan, S. R., Webber, L. S., Dalferes, E. R., Jr. & Strong, J. P. 1992 Atherosclerosis of the aorta and coronary arteries and cardiovascular risk factors in persons aged 6 to 30 years and studied at necropsy (The Bogalusa Heart Study). Am J Cardiol 70, 851–8.
20. Berliner, J. A., Territo, M. C., Sevanian, A., Ramin, S., Kim, J. A., Bamshad, B., Esterson, M. & Fogelman, A. M. 1990 Minimally modified low density lipoprotein stimulates monocyte endothelial interactions. J Clin Invest 85, 1260–6.
21. Bhakdi, S. 2003 An hypothesis for the immunopathogenesis of atherosclerosis. Clin Nephrol 60 Suppl 1, S49–52.
22. Bhakdi, S., Dorweiler, B., Kirchmann, R., Torzewski, J., Weise, E., Tranum-Jensen, J., Walev, I. & Wieland, E. 1995 On the pathogenesis of atherosclerosis: enzymatic transformation of human low density lipoprotein to an atherogenic moiety. J Exp Med 182, 1959–71.

23. Bhakdi, S., Lackner, K. J., Han, S. R., Torzewski, M. & Husmann, M. 2004 Beyond cholesterol: the enigma of atherosclerosis revisited. Thromb Haemost 91, 639–45.
24. Bhakdi, S., Torzewski, M., Klouche, M. & Hemmes, M. 1999 Complement and atherogenesis: binding of CRP to degraded, nonoxidized LDL enhances complement activation. Arterioscler Thromb Vasc Biol 19, 2348–54.
25. Bhakdi, S., Torzewski, M., Paprotka, K., Schmitt, S., Barsoom, H., Suriyaphol, P., Han, S. R., Lackner, K. J. & Husmann, M. 2004 Possible protective role for C-reactive protein in atherogenesis: complement activation by modified lipoproteins halts before detrimental terminal sequence. Circulation 109, 1870–6.
26. Binder, C. J., Chang, M. K., Shaw, P. X., Miller, Y. I., Hartvigsen, K., Dewan, A. & Witztum, J. L. 2002 Innate and acquired immunity in atherogenesis. Nat Med 8, 1218–26.
27. Björk, J., Hugli, T. E. & Smedegard, G. 1985 Microvascular effects of anaphylatoxins C3a and C5a. J Immunol 134, 1115–9.
28. Blankenberg, S., Barbaux, S. & Tiret, L. 2003 Adhesion molecules and atherosclerosis. Atherosclerosis 170, 191–203.
29. Blaskó, B., Kolka, R., Thorbjornsdottir, P., Sigurdarson, S. T., Sigurdsson, G., Rónai, Z., Sásvari-Szekely, M., Bodvarsson, S., Thorgeirsson, G., Prohászka, Z., Kovacs, M., Füst, G. & Arason, G. J. 2008 Low complement C4B gene copy number predicts short-term mortality after acute myocardial infarction. Int Immunol 20, 31–7.
30. Bonow, R. O., Smaha, L. A., Smith, S. C., Jr., Mensah, G. A. & Lenfant, C. 2002 World Heart Day 2002: the international burden of cardiovascular disease: responding to the emerging global epidemic. Circulation 106, 1602–5.
31. Boren, J., Olin, K., Lee, I., Chait, A., Wight, T. N. & Innerarity, T. L. 1998 Identification of the principal proteoglycan-binding site in LDL. A single-point mutation in apo-B100 severely affects proteoglycan interaction without affecting LDL receptor binding. J Clin Invest 101, 2658–64.
32. Brown, M. S. & Goldstein, J. L. 1986 A receptor-mediated pathway for cholesterol homeostasis. Science 232, 34–47.
33. Buerke, M., Prufer, D., Dahm, M., Oelert, H., Meyer, J. & Darius, H. 1998 Blocking of classical complement pathway inhibits endothelial adhesion molecule expression and preserves ischemic myocardium from reperfusion injury. J Pharmacol Exp Ther 286, 429–38.
34. Buerke, M., Schwertz, H., Seitz, W., Meyer, J. & Darius, H. 2001 Novel small molecule inhibitor of C1s exerts cardioprotective effects in ischemia-reperfusion injury in rabbits. J Immunol 167, 5375–80.
35. Buono, C., Come, C. E., Witztum, J. L., Maguire, G. F., Connelly, P. W., Carroll, M. & Lichtman, A. H. 2002 Influence of C3 deficiency on atherosclerosis. Circulation 105, 3025–31.
36. Caliezi, C., Wuillemin, W. A., Zeerleder, S., Redondo, M., Eisele, B. & Hack, C. E. 2000 C1-Esterase inhibitor: an anti-inflammatory agent and its potential use in the treatment of diseases other than hereditary angioedema. Pharmacol Rev 52, 91–112.
37. Caliezi, C., Zeerleder, S., Redondo, M., Regli, B., Rothen, H. U., Zurcher-Zenklusen, R., Rieben, R., Devay, J., Hack, C. E., Lammle, B. & Wuillemin, W. A. 2002 C1-inhibitor in patients with severe sepsis and septic shock: beneficial effect on renal dysfunction. Crit Care Med 30, 1722–8.
38. Chang, M. K., Binder, C. J., Torzewski, M. & Witztum, J. L. 2002 C-reactive protein binds to both oxidized LDL and apoptotic cells through recognition of a common ligand: Phosphorylcholine of oxidized phospholipids. Proc Natl Acad Sci USA 99, 13043–8.
39. Clinton, S. K., Underwood, R., Hayes, L., Sherman, M. L., Kufe, D. W. & Libby, P. 1992 Macrophage colony-stimulating factor gene expression in vascular cells and in experimental and human atherosclerosis. Am J Pathol 140, 301–16.
40. Collard, C. D., Vakeva, A., Morrissey, M. A., Agah, A., Rollins, S. A., Reenstra, W. R., Buras, J. A., Meri, S. & Stahl, G. L. 2000 Complement activation after oxidative stress: role of the lectin complement pathway. Am J Pathol 156, 1549–56.
41. Crawford, M. H., Grover, F. L., Kolb, W. P., McMahan, C. A., O'Rourke, R. A., McManus, L. M. & Pinckard, R. N. 1988 Complement and neutrophil activation in the pathogenesis of ischemic myocardial injury. Circulation 78, 1449–58.

42. Cushing, S. D., Berliner, J. A., Valente, A. J., Territo, M. C., Navab, M., Parhami, F., Gerrity, R., Schwartz, C. J. & Fogelman, A. M. 1990 Minimally modified low density lipoprotein induces monocyte chemotactic protein 1 in human endothelial cells and smooth muscle cells. Proc Natl Acad Sci USA 87, 5134–8.
43. Cybulsky, M. I. & Gimbrone, M. A., Jr. 1991 Endothelial expression of a mononuclear leukocyte adhesion molecule during atherogenesis. Science 251, 788–91.
44. Daly, J. & Kotwal, G. J. 1998 Pro-inflammatory complement activation by the A beta peptide of Alzheimer's disease is biologically significant and can be blocked by vaccinia virus complement control protein. Neurobiol Aging 19, 619–27.
45. Danesh, J., Collins, R. & Peto, R. 2000 Lipoprotein(a) and coronary heart disease. Meta-analysis of prospective studies. Circulation 102, 1082–5.
46. Danesh, J., Wheeler, J. G., Hirschfield, G. M., Eda, S., Eiriksdottir, G., Rumley, A., Lowe, G. D., Pepys, M. B. & Gudnason, V. 2004 C-reactive protein and other circulating markers of inflammation in the prediction of coronary heart disease. N Engl J Med 350, 1387–97.
47. Dansky, H. M., Charlton, S. A., Harper, M. M. & Smith, J. D. 1997 T and B lymphocytes play a minor role in atherosclerotic plaque formation in the apolipoprotein E-deficient mouse. Proc Natl Acad Sci USA 94, 4642–6.
48. Daugherty, A. 2002 Mouse models of atherosclerosis. Am J Med Sci 323, 3–10.
49. de Villiers, W. J. & Smart, E. J. 1999 Macrophage scavenger receptors and foam cell formation. J Leukoc Biol 66, 740–6.
50. de Zwaan, C., Kleine, A. H., Diris, J. H., Glatz, J. F., Wellens, H. J., Strengers, P. F., Tissing, M., Hack, C. E., van Dieijen-Visser, M. P. & Hermens, W. T. 2002 Continuous 48-h C1-inhibitor treatment, following reperfusion therapy, in patients with acute myocardial infarction. Eur Heart J 23, 1670–7.
51. Diaz, M. N., Frei, B., Vita, J. A. & Keaney, J. F., Jr. 1997 Antioxidants and atherosclerotic heart disease. N Engl J Med 337, 408–16.
52. Dwivedi, A., Anggard, E. E. & Carrier, M. J. 2001 Oxidized LDL-mediated monocyte adhesion to endothelial cells does not involve NFkappaB. Biochem Biophys Res Commun 284, 239–44.
53. Ernst, E. & Resch, K. L. 1993 Fibrinogen as a cardiovascular risk factor: a meta-analysis and review of the literature. Ann Intern Med 118, 956–63.
54. Faxon, D. P., Fuster, V., Libby, P., Beckman, J. A., Hiatt, W. R., Thompson, R. W., Topper, J. N., Annex, B. H., Rundback, J. H., Fabunmi, R. P., Robertson, R. M. & Loscalzo, J. 2004 Atherosclerotic Vascular Disease Conference: Writing Group III: pathophysiology. Circulation 109, 2617–25.
55. Fujita, T., Matsushita, M. & Endo, Y. 2004 The lectin-complement pathway – its role in innate immunity and evolution. Immunol Rev 198, 185–202.
56. Furchgott, R. F. & Zawadzki, J. V. 1980 The obligatory role of endothelial cells in the relaxation of arterial smooth muscle by acetylcholine. Nature 288, 373–6.
57. Füst, G., Surjan, M., Braun, P. & Varkonyi, S. 1975 Serum beta1A/C-globulin level in early myocardial infarction. Acta Med Acad Sci Hung 32, 49–54.
58. Füst, G., Szekely, J. & Gero, S. 1977 Circulating immune complexes in vascular diseases. Lancet 1, 193–4.
59. Füst, G., Szondy, E., Szekely, J., Nanai, I. & Gero, S. 1978 Studies on the occurrence of circulating immune complexes in vascular diseases. Atherosclerosis 29, 181–90.
60. Fuster, V., Gotto, A. M., Libby, P., Loscalzo, J. & McGill, H. C. 1996 27th Bethesda Conference: matching the intensity of risk factor management with the hazard for coronary disease events. Task Force 1. Pathogenesis of coronary disease: the biologic role of risk factors. J Am Coll Cardiol 27, 964–76.
61. Gadjeva, M., Thiel, S. & Jensenius, J. C. 2001 The mannan-binding-lectin pathway of the innate immune response. Curr Opin Immunol 13, 74–8.
62. Garred, P., Honoré, C., Ma, Y. J., Munthe-Fog, L. & Hummelshøj, T. 2009 MBL2, FCN1, FCN2 and FCN3-The genes behind the initiation of the lectin pathway of complement. Mol Immunol 46, 2737–44.

63. Geertinger, P. & Sørensen, H. 1977 On the reduced atherogenic effect of cholesterol feeding in rabbits with congenital complement (C6) deficiency. Artery 1, 177–184.
64. Geng, Y. J. & Libby, P. 2002 Progression of atheroma: a struggle between death and procreation. Arterioscler Thromb Vasc Biol 22, 1370–80.
65. Gerrity, R. G. 1981 The role of the monocyte in atherogenesis: I. Transition of blood-borne monocytes into foam cells in fatty lesions. Am J Pathol 103, 181–90.
66. Gerrity, R. G. 1981 The role of the monocyte in atherogenesis: II. Migration of foam cells from atherosclerotic lesions. Am J Pathol 103, 191–200.
67. Gerrity, R. G. & Antonov, A. S. 1997 The pathogenesis of atherosclerosis. Diabetologia 40 Suppl 2, S108–10.
68. Gershov, D., Kim, S., Brot, N. & Elkon, K. B. 2000 C-Reactive protein binds to apoptotic cells, protects the cells from assembly of the terminal complement components, and sustains an antiinflammatory innate immune response: implications for systemic autoimmunity. J Exp Med 192, 1353–64.
69. Getz, G. S. 2005 Thematic review series: the immune system and atherogenesis. Immune function in atherogenesis. J Lipid Res 46, 1–10.
70. Glass, C. K. & Witztum, J. L. 2001 Atherosclerosis. the road ahead. Cell 104, 503–16.
71. Granger, C. B., Mahaffey, K. W., Weaver, W. D., Theroux, P., Hochman, J. S., Filloon, T. G., Rollins, S., Todaro, T. G., Nicolau, J. C., Ruzyllo, W. & Armstrong, P. W. 2003 Pexelizumab, an anti-C5 complement antibody, as adjunctive therapy to primary percutaneous coronary intervention in acute myocardial infarction: the COMplement inhibition in Myocardial infarction treated with Angioplasty (COMMA) trial. Circulation 108, 1184–90.
72. Griendling, K. K. & Alexander, R. W. 1997 Oxidative stress and cardiovascular disease. Circulation 96, 3264–5.
73. Groszek, E. & Grundy, S. M. 1980 The possible role of the arterial microcirculation in the pathogenesis of atherosclerosis. J Chronic Dis 33, 679–84.
74. Grundy, S. M. 2004 Obesity, metabolic syndrome, and cardiovascular disease. J Clin Endocrinol Metab 89, 2595–600.
75. Hamsten, A., de Faire, U., Walldius, G., Dahlen, G., Szamosi, A., Landou, C., Blomback, M. & Wiman, B. 1987 Plasminogen activator inhibitor in plasma: risk factor for recurrent myocardial infarction. Lancet 2, 3–9.
76. Hansson, G. K. 2001 Immune mechanisms in atherosclerosis. Arterioscler Thromb Vasc Biol 21, 1876–90.
77. Harker, L. A., Ross, R. & Glomset, J. A. 1978 The role of endothelial cell injury and platelet response in atherogenesis. Thromb Haemost 39, 312–21.
78. Hessler, J. R., Morel, D. W., Lewis, L. J. & Chisolm, G. M. 1983 Lipoprotein oxidation and lipoprotein-induced cytotoxicity. Arteriosclerosis 3, 215–22.
79. Hicks, R. R., Keeling, K. L., Yang, M. Y., Smith, S. A., Simons, A. M. & Kotwal, G. J. 2002 Vaccinia virus complement control protein enhances functional recovery after traumatic brain injury. J Neurotrauma 19, 705–14.
80. Hill, J. H. & Ward, P. A. 1971 The phlogistic role of C3 leukotactic fragments in myocardial infarcts of rats. J Exp Med 133, 885–900.
81. Hirschfield, G. M. & Pepys, M. B. 2003 C-reactive protein and cardiovascular disease: new insights from an old molecule. QJM 96, 793–807.
82. Horkko, S., Binder, C. J., Shaw, P. X., Chang, M. K., Silverman, G., Palinski, W. & Witztum, J. L. 2000 Immunological responses to oxidized LDL. Free Radic Biol Med 28, 1771–9.
83. Horstick, G., Berg, O., Heimann, A., Gotze, O., Loos, M., Hafner, G., Bierbach, B., Petersen, S., Bhakdi, S., Darius, H., Horstick, M., Meyer, J. & Kempski, O. 2001 Application of C1-esterase inhibitor during reperfusion of ischemic myocardium: dose-related beneficial versus detrimental effects. Circulation 104, 3125–31.
84. Horstick, G., Kempf, T., Lauterbach, M., Bhakdi, S., Kopacz, L., Heimann, A., Malzahn, M., Horstick, M., Meyer, J. & Kempski, O. 2001 C1-esterase-inhibitor treatment at early reperfusion of hemorrhagic shock reduces mesentery leukocyte adhesion and rolling. Microcirculation 8, 427–33.

85. Hourcade, D., Holers, V. M. & Atkinson, J. P. 1989 The regulators of complement activation (RCA) gene cluster. Adv Immunol 45, 381–416.
86. Hourcade, D., Liszewski, M. K., Krych-Goldberg, M. & Atkinson, J. P. 2000 Functional domains, structural variations and pathogen interactions of MCP, DAF and CR1. Immunopharmacology 49, 103–16.
87. Hugli, T. E. 1986 Biochemistry and biology of anaphylatoxins. Complement 3, 111–27.
88. Jagels, M. A., Chambers, J. D., Arfors, K. E. & Hugli, T. E. 1995 C5a- and tumor necrosis factor-alpha-induced leukocytosis occurs independently of beta 2 integrins and L-selectin: differential effects on neutrophil adhesion molecule expression in vivo. Blood 85, 2900–9.
89. Jha, P. & Kotwal, G. J. 2003 Vaccinia complement control protein: multi-functional protein and a potential wonder drug. J Biosci 28, 265–71.
90. Jha, P., Smith, S. A., Justus, D. E. & Kotwal, G. J. 2003 Prolonged retention of vaccinia virus complement control protein following IP injection: implications in blocking xenorejection. Transplant Proc 35, 3160–2.
91. Johnson, A. R., Hugli, T. E. & Müller-Eberhard, H. J. 1975 Release of histamine from rat mast cells by the complement peptides C3a and C5a. Immunology 28, 1067–80.
92. Johnson, J. L. & Newby, A. C. 2009 Macrophage heterogeneity in atherosclerotic plaques. Curr Opin Lipidol 20, 370–8.
93. Kao, C. H., Chen, J. K., Kuo, J. S. & Yang, V. C. 1995 Visualization of the transport pathways of low density lipoproteins across the endothelial cells in the branched regions of rat arteries. Atherosclerosis 116, 27–41.
94. Keeling, K. L., Hicks, R. R., Mahesh, J., Billings, B. B. & Kotwal, G. J. 2000 Local neutrophil influx following lateral fluid-percussion brain injury in rats is associated with accumulation of complement activation fragments of the third component (C3) of the complement system. J Neuroimmunol 105, 20–30.
95. Kilgore, K. S., Park, J. L., Tanhehco, E. J., Booth, E. A., Marks, R. M. & Lucchesi, B. R. 1998 Attenuation of interleukin-8 expression in C6-deficient rabbits after myocardial ischemia/reperfusion. J Mol Cell Cardiol 30, 75–85.
96. Kilgore, K. S., Schmid, E., Shanley, T. P., Flory, C. M., Maheswari, V., Tramontini, N. L., Cohen, H., Ward, P. A., Friedl, H. P. & Warren, J. S. 1997 Sublytic concentrations of the membrane attack complex of complement induce endothelial interleukin-8 and monocyte chemoattractant protein-1 through nuclear factor-kappa B activation. Am J Pathol 150, 2019–31.
97. Kotwal, G. J., Isaacs, S. N., McKenzie, R., Frank, M. M. & Moss, B. 1990 Inhibition of the complement cascade by the major secretory protein of vaccinia virus. Science 250, 827–30.
98. Kotwal, G. J., Lahiri, D. K. & Hicks, R. 2002 Potential intervention by vaccinia virus complement control protein of the signals contributing to the progression of central nervous system injury to Alzheimer's disease. Ann N Y Acad Sci 973, 317–22.
99. Kotwal, G. J. & Moss, B. 1988 Vaccinia virus encodes a secretory polypeptide structurally related to complement control proteins. Nature 335, 176–8.
100. Kramer, J., Fulöp, T., Rajczy, K., Nguyen, A. T. & Füst, G. 1991 A marked drop in the incidence of the null allele of the B gene of the fourth component of complement (C4B*Q0) in elderly subjects: C4B*Q0 as a probable negative selection factor for survival. Hum Genet 86, 595–8.
101. Kramer, J., Rajczy, K., Hegyi, L., Fulöp, T., Mohacsi, A., Mezei, Z., Keltai, M., Blaskó, G., Ferenczy, E., Anh-Tuan, N., et al. 1994 C4B*Q0 allotype as risk factor for myocardial infarction. BMJ 309, 313–4.
102. Kulkarni, A. P., Pillay, N. S., Kellaway, L. A. & Kotwal, G. J. 2008 Intracranial administration of vaccinia virus complement control protein in Mo/Hu APPswe PS1dE9 transgenic mice at an early age shows enhanced performance at a later age using a cheese board maze test. Biogerontology 9, 405–20.
103. Lachmann, P. J. & Walport, M. J. 1987 Deficiency of the effector mechanisms of the immune response and autoimmunity. Ciba Found Symp 129, 149–71.
104. Langlois, P. F. & Gawryl, M. S. 1988 Detection of the terminal complement complex in patient plasma following acute myocardial infarction. Atherosclerosis 70, 95–105.

105. Lazar, H. L., Bao, Y., Gaudiani, J., Rivers, S. & Marsh, H. 1999 Total complement inhibition: an effective strategy to limit ischemic injury during coronary revascularization on cardiopulmonary bypass. Circulation 100, 1438–42.
106. Lazar, H. L., Hamasaki, T., Bao, Y., Rivers, S., Bernard, S. A. & Shemin, R. J. 1998 Soluble complement receptor type I limits damage during revascularization of ischemic myocardium. Ann Thorac Surg 65, 973–7.
107. Libby, P. 2002 Inflammation in atherosclerosis. Nature 420, 868–74.
108. Libby, P., Ridker, P. M. & Hansson, G. K. 2009 Inflammation in atherosclerosis: from pathophysiology to practice. J Am Coll Cardiol 54, 2129–38.
109. Linton, M. F. & Fazio, S. 2001 Class A scavenger receptors, macrophages, and atherosclerosis. Curr Opin Lipidol 12, 489–95.
110. Luc, G., Bard, J. M., Arveiler, D., Ferrieres, J., Evans, A., Amouyel, P., Fruchart, J. C. & Ducimetiere, P. 2002 Lipoprotein (a) as a predictor of coronary heart disease: the PRIME Study. Atherosclerosis 163, 377–84.
111. Lusis, A. J. 2000 Atherosclerosis. Nature 407, 233–41.
112. Mahaffey, K. W., Granger, C. B., Nicolau, J. C., Ruzyllo, W., Weaver, W. D., Theroux, P., Hochman, J. S., Filloon, T. G., Mojcik, C. F., Todaro, T. G. & Armstrong, P. W. 2003 Effect of pexelizumab, an anti-C5 complement antibody, as adjunctive therapy to fibrinolysis in acute myocardial infarction: the COMPlement inhibition in myocardial infarction treated with thromboLYtics (COMPLY) trial. Circulation 108, 1176–83.
113. Maroko, P. R., Carpenter, C. B., Chiariello, M., Fishbein, M. C., Radvany, P., Knostman, J. D. & Hale, S. L. 1978 Reduction by cobra venom factor of myocardial necrosis after coronary artery occlusion. J Clin Invest 61, 661–70.
114. Matsushita, M. & Fujita, T. 1995 Cleavage of the third component of complement (C3) by mannose-binding protein-associated serine protease (MASP) with subsequent complement activation. Immunobiology 194, 443–48.
115. Mazer, S. P. & Rabbani, L. E. 2004 Evidence for C-reactive protein's role in (CRP) vascular disease: atherothrombosis, immuno-regulation and CRP. J Thromb Thrombolysis 17, 95–105.
116. McKenzie, R., Kotwal, G. J., Moss, B., Hammer, C. H. & Frank, M. M. 1992 Regulation of complement activity by vaccinia virus complement-control protein. J Infect Dis 166, 1245–50.
117. McManus, L. M., Kolb, W. P., Crawford, M. H., O'Rourke, R. A., Grover, F. L. & Pinckard, R. N. 1983 Complement localization in ischemic baboon myocardium. Lab Invest 48, 436–47.
118. Mevorach, D., Mascarenhas, J. O., Gershov, D. & Elkon, K. B. 1998 Complement-dependent clearance of apoptotic cells by human macrophages. J Exp Med 188, 2313–20.
119. Mitsui, E., Fujioka, T. & Yasugi, T. 1990 Effect of anticomplementary agent (K-76 COONa) on experimental atherosclerosis in rabbits. J Nihon Univ Med Assoc 49, 357–366.
120. Miwa, T. & Song, W. C. 2001 Membrane complement regulatory proteins: insight from animal studies and relevance to human diseases. Int Immunopharmacol 1, 445–59.
121. Moller-Kristensen, M., Thiel, S., Sjoholm, A., Matsushita, M. & Jensenius, J. C. 2007 Cooperation between MASP-1 and MASP-2 in the generation of C3 convertase through the MBL pathway. Int Immunol 19, 141–9.
122. Mollnes, T. E., Song, W. C. & Lambris, J. D. 2002 Complement in inflammatory tissue damage and disease. Trends Immunol 23, 61–4.
123. Monaco, C., Crea, F., Niccoli, G., Summaria, F., Cianflone, D., Bordone, R., Bellomo, G. & Maseri, A. 2001 Autoantibodies against oxidized low density lipoproteins in patients with stable angina, unstable angina or peripheral vascular disease; pathophysiological implications. Eur Heart J 22, 1572–7.
124. Monsinjon, T., Richard, V. & Fontaine, M. 2001 Complement and its implications in cardiac ischemia/reperfusion: strategies to inhibit complement. Fundam Clin Pharmacol 15, 293–306.
125. Morgan, B. P. & Harris, C. L. 2003 Complement therapeutics; history and current progress. Mol Immunol 40, 159–70.
126. Mulligan, M. S., Warner, R. L., Rittershaus, C. W., Thomas, L. J., Ryan, U. S., Foreman, K. E., Crouch, L. D., Till, G. O. & Ward, P. A. 1999 Endothelial targeting and enhanced antiinflammatory effects of complement inhibitors possessing sialyl Lewisx moieties. J Immunol 162, 4952–9.

127. Murthy, K. H., Smith, S. A., Ganesh, V. K., Judge, K. W., Mullin, N., Barlow, P. N., Ogata, C. M. & Kotwal, G. J. 2001 Crystal structure of a complement control protein that regulates both pathways of complement activation and binds heparan sulfate proteoglycans. Cell 104, 301–11.
128. Muscari, A., Bozzoli, C., Puddu, G. M., Sangiorgi, Z., Dormi, A., Rovinetti, C., Descovich, G. C. & Puddu, P. 1995 Association of serum C3 levels with the risk of myocardial infarction. Am J Med 98, 357–64.
129. Muscari, A., Massarelli, G., Bastagli, L., Poggiopollini, G., Tomassetti, V., Drago, G., Martignani, C., Pacilli, P., Boni, P. & Puddu, P. 2000 Relationship of serum C3 to fasting insulin, risk factors and previous ischaemic events in middle-aged men. Eur Heart J 21, 1081–90.
130. Muscari, A., Massarelli, G., Bastagli, L., Poggiopollini, G., Tomassetti, V., Volta, U., Puddu, G. M. & Puddu, P. 1998 Relationship between serum C3 levels and traditional risk factors for myocardial infarction. Acta Cardiol 53, 345–54.
131. Muscari, A., Mele, M. C., Bastagli, L., Poggiopollini, G., Tomassetti, V., Drago, G., Legnani, C., Barini, A., Cappelletti, O., Boni, P. & Puddu, P. 2001 Conditional risk factors in men with previous myocardial infarction: relevance of C3 and homocysteine. Acta Cardiol 56, 303–11.
132. Mustafa, A., Nityanand, S., Berglund, L., Lithell, H. & Lefvert, A. K. 2000 Circulating immune complexes in 50-year-old men as a strong and independent risk factor for myocardial infarction. Circulation 102, 2576–81.
133. Navab, M., Berliner, J. A., Watson, A. D., Hama, S. Y., Territo, M. C., Lusis, A. J., Shih, D. M., Van Lenten, B. J., Frank, J. S., Demer, L. L., Edwards, P. A. & Fogelman, A. M. 1996 The Yin and Yang of oxidation in the development of the fatty streak. A review based on the 1994 George Lyman Duff Memorial Lecture. Arterioscler Thromb Vasc Biol 16, 831–42.
134. Nicholson-Weller, A., Burge, J., Fearon, D. T., Weller, P. F. & Austen, K. F. 1982 Isolation of a human erythrocyte membrane glycoprotein with decay-accelerating activity for C3 convertases of the complement system. J Immunol 129, 184–9.
135. Niculescu, F., Badea, T. & Rus, H. 1999 Sublytic C5b-9 induces proliferation of human aortic smooth muscle cells: role of mitogen activated protein kinase and phosphatidylinositol 3-kinase. Atherosclerosis 142, 47–56.
136. Niculescu, F., Hugo, F., Rus, H. G., Vlaicu, R. & Bhakdi, S. 1987 Quantitative evaluation of the terminal C5b-9 complement complex by ELISA in human atherosclerotic arteries. Clin Exp Immunol 69, 477–83.
137. Niculescu, F., Niculescu, T. & Rus, H. 2004 C5b-9 terminal complement complex assembly on apoptotic cells in human arterial wall with atherosclerosis. Exp Mol Pathol 76, 17–23.
138. Niculescu, F. & Rus, H. 1999 Complement activation and atherosclerosis. Mol Immunol 36, 949–55.
139. Niculescu, F., Rus, H., Cristea, A. & Vlaicu, R. 1985 Localization of the terminal C5b-9 complement complex in the human aortic atherosclerotic wall. Immunol Lett 10, 109–14.
140. Niculescu, F., Rus, H. G. & Vlaicu, R. 1987 Activation of the human terminal complement pathway in atherosclerosis. Clin Immunol Immunopathol 45, 147–55.
141. Niculescu, F., Rus, H. G. & Vlaicu, R. 1987 Immunohistochemical localization of C5b-9, S-protein, C3d and apolipoprotein B in human arterial tissues with atherosclerosis. Atherosclerosis 65, 1–11.
142. Norata, G. D., Tonti, L., Roma, P. & Catapano, A. L. 2002 Apoptosis and proliferation of endothelial cells in early atherosclerotic lesions: possible role of oxidised LDL. Nutr Metab Cardiovasc Dis 12, 297–305.
143. Oksjoki, R., Jarva, H., Kovanen, P. T., Laine, P., Meri, S. & Pentikainen, M. O. 2003 Association between complement factor H and proteoglycans in early human coronary atherosclerotic lesions: implications for local regulation of complement activation. Arterioscler Thromb Vasc Biol 23, 630–6.
144. Oksjoki, R., Kovanen, P. T. & Pentikainen, M. O. 2003 Role of complement activation in atherosclerosis. Curr Opin Lipidol 14, 477–82.
145. Paigen, B., Holmes, P. A., Mitchell, D. & Albee, D. 1987 Comparison of atherosclerotic lesions and HDL-lipid levels in male, female, and testosterone-treated female mice from strains C57BL/6, BALB/c, and C3H. Atherosclerosis 64, 215–21.

146. Paigen, B., Ishida, B. Y., Verstuyft, J., Winters, R. B. & Albee, D. 1990 Atherosclerosis susceptibility differences among progenitors of recombinant inbred strains of mice. Arteriosclerosis 10, 316–23.
147. Paigen, B., Morrow, A., Brandon, C., Mitchell, D. & Holmes, P. 1985 Variation in susceptibility to atherosclerosis among inbred strains of mice. Atherosclerosis 57, 65–73.
148. Paigen, B., Morrow, A., Holmes, P. A., Mitchell, D. & Williams, R. A. 1987 Quantitative assessment of atherosclerotic lesions in mice. Atherosclerosis 68, 231–40.
149. Pasternak, R. C., Grundy, S. M., Levy, D. & Thompson, P. D. 1996 27th Bethesda Conference: matching the intensity of risk factor management with the hazard for coronary disease events. Task Force 3. Spectrum of risk factors for coronary heart disease. J Am Coll Cardiol 27, 978–90.
150. Patel, S., Thelander, E. M., Hernandez, M., Montenegro, J., Hassing, H., Burton, C., Mundt, S., Hermanowski-Vosatka, A., Wright, S. D., Chao, Y. S. & Detmers, P. A. 2001 ApoE(-/-) mice develop atherosclerosis in the absence of complement component C5. Biochem Biophys Res Commun 286, 164–70.
151. Pearson, T. A., Blair, S. N., Daniels, S. R., Eckel, R. H., Fair, J. M., Fortmann, S. P., Franklin, B. A., Goldstein, L. B., Greenland, P., Grundy, S. M., Hong, Y., Miller, N. H., Lauer, R. M., Ockene, I. S., Sacco, R. L., Sallis, J. F., Jr., Smith, S. C., Jr., Stone, N. J. & Taubert, K. A. 2002 AHA Guidelines for Primary Prevention of Cardiovascular Disease and Stroke: 2002 Update: Consensus Panel Guide to Comprehensive Risk Reduction for Adult Patients Without Coronary or Other Atherosclerotic Vascular Diseases. American Heart Association Science Advisory and Coordinating Committee. Circulation 106, 388–91.
152. Pellas, T. C., Boyar, W., van Oostrum, J., Wasvary, J., Fryer, L. R., Pastor, G., Sills, M., Braunwalder, A., Yarwood, D. R., Kramer, R., Kimble, E., Hadala, J., Haston, W., Moreira-Ludewig, R., Uziel-Fusi, S., Peters, P., Bill, K. & Wennogle, L. P. 1998 Novel C5a receptor antagonists regulate neutrophil functions in vitro and in vivo. J Immunol 160, 5616–21.
153. Persson, L., Boren, J., Robertson, A. K., Wallenius, V., Hansson, G. K. & Pekna, M. 2004 Lack of complement factor C3, but not factor B, increases hyperlipidemia and atherosclerosis in apolipoprotein E-/- low-density lipoprotein receptor-/- mice. Arterioscler Thromb Vasc Biol 24, 1062–7.
154. Pillay, N. S., Kellaway, L. A. & Kotwal, G. J. 2005 Administration of vaccinia virus complement control protein shows significant cognitive improvement in a mild injury model. Ann N Y Acad Sci 1056, 450–61.
155. Pillay, N. S., Kellaway, L. A. & Kotwal, G. J. 2007 Vaccinia virus complement control protein significantly improves sensorimotor function recovery after severe head trauma. Brain Res 1153, 158–65.
156. Pillay, N. S., Kellaway, L. A. & Kotwal, G. J. 2008 Early detection of memory deficits and memory improvement with vaccinia virus complement control protein in an Alzheimer's disease model. Behav Brain Res 192, 173–7.
157. Pinckard, R. N., O'Rourke, R. A., Crawford, M. H., Grover, F. S., McManus, L. M., Ghidoni, J. J., Storrs, S. B. & Olson, M. S. 1980 Complement localization and mediation of ischemic injury in baboon myocardium. J Clin Invest 66, 1050–6.
158. Plump, A. S., Smith, J. D., Hayek, T., Aalto-Setala, K., Walsh, A., Verstuyft, J. G., Rubin, E. M. & Breslow, J. L. 1992 Severe hypercholesterolemia and atherosclerosis in apolipoprotein E-deficient mice created by homologous recombination in ES cells. Cell 71, 343–53.
159. Raines, E. W. & Ross, R. 1993 Smooth muscle cells and the pathogenesis of the lesions of atherosclerosis. Br Heart J 69, S30–7.
160. Rajavashisth, T. B., Andalibi, A., Territo, M. C., Berliner, J. A., Navab, M., Fogelman, A. M. & Lusis, A. J. 1990 Induction of endothelial cell expression of granulocyte and macrophage colony-stimulating factors by modified low-density lipoproteins. Nature 344, 254–7.
161. Reddick, R. L., Zhang, S. H. & Maeda, N. 1994 Atherosclerosis in mice lacking apo E. Evaluation of lesional development and progression. Arterioscler Thromb 14, 141–7.
162. Ren, G., Dewald, O. & Frangogiannis, N. G. 2003 Inflammatory mechanisms in myocardial infarction. Curr Drug Targets Inflamm Allergy 2, 242–56.

163. Reynolds, D. N., Smith, S. A., Zhang, Y. P., Lahiri, D. K., Morassutti, D. J., Shields, C. B. & Kotwal, G. J. 2003 Vaccinia virus complement control protein modulates inflammation following spinal cord injury. Ann N Y Acad Sci 1010, 534–9.
164. Reynolds, D. N., Smith, S. A., Zhang, Y. P., Mengsheng, Q., Lahiri, D. K., Morassutti, D. J., Shields, C. B. & Kotwal, G. J. 2004 Vaccinia virus complement control protein reduces inflammation and improves spinal cord integrity following spinal cord injury. Ann N Y Acad Sci 1035, 165–78.
165. Reynolds, G. D. & Vance, R. P. 1987 C-reactive protein immunohistochemical localization in normal and atherosclerotic human aortas. Arch Pathol Lab Med 111, 265–9.
166. Riley, R. D., Sato, H., Zhao, Z. Q., Thourani, V. H., Jordan, J. E., Fernandez, A. X., Ma, X. L., Hite, D. R., Rigel, D. F., Pellas, T. C., Peppard, J., Bill, K. A., Lappe, R. W. & Vinten-Johansen, J. 2000 Recombinant human complement C5a receptor antagonist reduces infarct size after surgical revascularization. J Thorac Cardiovasc Surg 120, 350–8.
167. Rollins, S. A. & Sims, P. J. 1990 The complement-inhibitory activity of CD59 resides in its capacity to block incorporation of C9 into membrane C5b-9. J Immunol 144, 3478–83.
168. Rosenfeld, M. E., Yla-Herttuala, S., Lipton, B. A., Ord, V. A., Witztum, J. L. & Steinberg, D. 1992 Macrophage colony-stimulating factor mRNA and protein in atherosclerotic lesions of rabbits and humans. Am J Pathol 140, 291–300.
169. Ross, R. 1993 Rous-Whipple Award Lecture. Atherosclerosis: a defense mechanism gone awry. Am J Pathol 143, 987–1002.
170. Ross, R. 1999 Atherosclerosis – an inflammatory disease. N Engl J Med 340, 115–26.
171. Ross, R. & Glomset, J. A. 1976 The pathogenesis of atherosclerosis (first of two parts). N Engl J Med 295, 369–77.
172. Ross, R. & Glomset, J. A. 1976 The pathogenesis of atherosclerosis (second of two parts). N Engl J Med 295, 420–5.
173. Rossen, R. D., Swain, J. L., Michael, L. H., Weakley, S., Giannini, E. & Entman, M. L. 1985 Selective accumulation of the first component of complement and leukocytes in ischemic canine heart muscle. A possible initiator of an extra myocardial mechanism of ischemic injury. Circ Res 57, 119–30.
174. Rother, K. & Till, G. 1988 The complement system: Springer, Berlin.
175. Ruige, J. B., Assendelft, W. J., Dekker, J. M., Kostense, P. J., Heine, R. J. & Bouter, L. M. 1998 Insulin and risk of cardiovascular disease: a meta-analysis. Circulation 97, 996–1001.
176. Rus, H. G., Niculescu, F., Constantinescu, E., Cristea, A. & Vlaicu, R. 1986 Immunoelectron-microscopic localization of the terminal C5b-9 complement complex in human atherosclerotic fibrous plaque. Atherosclerosis 61, 35–42.
177. Rus, H. G., Niculescu, F., Nanulescu, M., Cristea, A. & Florescu, P. 1986 Immunohistochemical detection of the terminal C5b-9 complement complex in children with glomerular diseases. Clin Exp Immunol 65, 66–72.
178. Rus, H. G., Niculescu, F. & Shin, M. L. 1996 Sublytic complement attack induces cell cycle in oligodendrocytes. J Immunol 156, 4892–900.
179. Rus, H. G., Niculescu, F. & Vlaicu, R. 1987 Presence of C5b-9 complement complex and S-protein in human myocardial areas with necrosis and sclerosis. Immunol Lett 16, 15–20.
180. Saevarsdottir, S., Oskarsson, O. O., Aspelund, T., Eiriksdottir, G., Vikingsdottir, T., Gudnason, V. & Valdimarsson, H. 2005 Mannan binding lectin as an adjunct to risk assessment for myocardial infarction in individuals with enhanced risk. J Exp Med 201, 117–25.
181. Sahu, A., Isaacs, S. N., Soulika, A. M. & Lambris, J. D. 1998 Interaction of vaccinia virus complement control protein with human complement proteins: factor I-mediated degradation of C3b to iC3b1 inactivates the alternative complement pathway. J Immunol 160, 5596–604.
182. Saito, E., Mitsui, E., Fujioka, T., Hara, Y., Tochihara, T., et al. 1990 Effects of K-76 COONa, an anticomplementary agent, on experimental atherosclerosis in rabbits. J Drug Dev 3, 147–154.
183. Salama, A., Hugo, F., Heinrich, D., Hoge, R., Muller, R., Kiefel, V., Mueller-Eckhardt, C. & Bhakdi, S. 1988 Deposition of terminal C5b-9 complement complexes on erythrocytes and leukocytes during cardiopulmonary bypass. N Engl J Med 318, 408–14.

184. Schäfer, H., Mathey, D., Hugo, F. & Bhakdi, S. 1986 Deposition of the terminal C5b-9 complement complex in infarcted areas of human myocardium. J Immunol 137, 1945–9.
185. Schmiedt, W., Kinscherf, R., Deigner, H. P., Kamencic, H., Nauen, O., Kilo, J., Oelert, H., Metz, J. & Bhakdi, S. 1998 Complement C6 deficiency protects against diet-induced atherosclerosis in rabbits. Arterioscler Thromb Vasc Biol 18, 1790–5.
186. Schwartz, C. J., Valente, A. J., Sprague, E. A., Kelley, J. L., Suenram, C. A. & Rozek, M. M. 1985 Atherosclerosis as an inflammatory process. The roles of the monocyte-macrophage. Ann N Y Acad Sci 454, 115–20.
187. Schönbeck, U., Mach, F., Sukhova, G. K., Murphy, C., Bonnefoy, J. Y., Fabunmi, R. P. & Libby, P. 1997 Regulation of matrix metalloproteinase expression in human vascular smooth muscle cells by T lymphocytes: a role for CD40 signaling in plaque rupture? Circ Res 81, 448–54.
188. Seifert, P. S., Hugo, F., Hansson, G. K. & Bhakdi, S. 1989 Prelesional complement activation in experimental atherosclerosis. Terminal C5b-9 complement deposition coincides with cholesterol accumulation in the aortic intima of hypercholesterolemic rabbits. Lab Invest 60, 747–54.
189. Seifert, P. S., Hugo, F., Tranum-Jensen, J., Zahringer, U., Muhly, M. & Bhakdi, S. 1990 Isolation and characterization of a complement-activating lipid extracted from human atherosclerotic lesions. J Exp Med 172, 547–57.
190. Seifert, P. S., Roth, I., Schmiedt, W., Oelert, H., Okada, N., Okada, H. & Bhakdi, S. 1992 CD59 (homologous restriction factor 20), a plasma membrane protein that protects against complement C5b-9 attack, in human atherosclerotic lesions. Atherosclerosis 96, 135–45.
191. Shernan, S. K., Fitch, J. C., Nussmeier, N. A., Chen, J. C., Rollins, S. A., Mojcik, C. F., Malloy, K. J., Todaro, T. G., Filloon, T., Boyce, S. W., Gangahar, D. M., Goldberg, M., Saidman, L. J. & Mangano, D. T. 2004 Impact of pexelizumab, an anti-C5 complement antibody, on total mortality and adverse cardiovascular outcomes in cardiac surgical patients undergoing cardiopulmonary bypass. Ann Thorac Surg 77, 942–9; discussion 949–50.
192. Skalen, K., Gustafsson, M., Rydberg, E. K., Hulten, L. M., Wiklund, O., Innerarity, T. L. & Boren, J. 2002 Subendothelial retention of atherogenic lipoproteins in early atherosclerosis. Nature 417, 750–4.
193. Smith, E. F., 3rd, Griswold, D. E., Egan, J. W., Hillegass, L. M., Smith, R. A., Hibbs, M. J. & Gagnon, R. C. 1993 Reduction of myocardial reperfusion injury with human soluble complement receptor type 1 (BRL 55730). Eur J Pharmacol 236, 477–81.
194. Smith, J. D. 1998 Mouse models of atherosclerosis. Lab Anim Sci 48, 573–9.
195. Smith, J. D., Trogan, E., Ginsberg, M., Grigaux, C., Tian, J. & Miyata, M. 1995 Decreased atherosclerosis in mice deficient in both macrophage colony-stimulating factor (op) and apolipoprotein E. Proc Natl Acad Sci USA 92, 8264–8.
196. Smith, S. A., Mullin, N. P., Parkinson, J., Shchelkunov, S. N., Totmenin, A. V., Loparev, V. N., Srisatjaluk, R., Reynolds, D. N., Keeling, K. L., Justus, D. E., Barlow, P. N. & Kotwal, G. J. 2000 Conserved surface-exposed K/R-X-K/R motifs and net positive charge on poxvirus complement control proteins serve as putative heparin binding sites and contribute to inhibition of molecular interactions with human endothelial cells: a novel mechanism for evasion of host defense. J Virol 74, 5659–66.
197. Smith, S. A., Sreenivasan, R., Krishnasamy, G., Judge, K. W., Murthy, K. H., Arjunwadkar, S. J., Pugh, D. R. & Kotwal, G. J. 2003 Mapping of regions within the vaccinia virus complement control protein involved in dose-dependent binding to key complement components and heparin using surface plasmon resonance. Biochim Biophys Acta 1650, 30–9.
198. Smith, S. C., Jr., Greenland, P. & Grundy, S. M. 2000 AHA Conference Proceedings. Prevention conference V: Beyond secondary prevention: Identifying the high-risk patient for primary prevention: executive summary. American Heart Association. Circulation 101, 111–6.
199. Sorensen, R., Thiel, S. & Jensenius, J. C. 2005 Mannan-binding-lectin-associated serine proteases, characteristics and disease associations. Springer Semin Immunopathol 27, 299–319.
200. Stahl, G. L., Xu, Y., Hao, L., Miller, M., Buras, J. A., Fung, M. & Zhao, H. 2003 Role for the alternative complement pathway in ischemia/reperfusion injury. Am J Pathol 162, 449–55.
201. Stary, H. C. 1994 Changes in components and structure of atherosclerotic lesions developing from childhood to middle age in coronary arteries. Basic Res Cardiol 89 Suppl 1, 17–32.

202. Steinberg, D., Glass, C. K. & Witztum, J. L. 2008 Evidence mandating earlier and more aggressive treatment of hypercholesterolemia. Circulation 118, 672–7.
203. Steinberg, D., Parthasarathy, S., Carew, T. E., Khoo, J. C. & Witztum, J. L. 1989 Beyond cholesterol. Modifications of low-density lipoprotein that increase its atherogenicity. N Engl J Med 320, 915–24.
204. Stewart-Phillips, J. L. & Lough, J. 1991 Pathology of atherosclerosis in cholesterol-fed, susceptible mice. Atherosclerosis 90, 211–8.
205. Stimler, N. P., Brocklehurst, W. E., Bloor, C. M. & Hugli, T. E. 1981 Anaphylatoxin-mediated contraction of guinea pig lung strips: a nonhistamine tissue response. J Immunol 126, 2258–61.
206. Suttorp, N. & Bhakdi, S. 1989 Terminal complement complex and endothelial cells. Z Kardiol 78 Suppl 6, 140–2.
207. Szaba, F. M. & Smiley, S. T. 2002 Roles for thrombin and fibrin(ogen) in cytokine/chemokine production and macrophage adhesion in vivo. Blood 99, 1053–9.
208. Szondy, E., Horvath, M., Mezey, Z., Szekely, J., Lengyel, E., Füst, G. & Gero, S. 1983 Free and complexed anti-lipoprotein antibodies in vascular diseases. Atherosclerosis 49, 69–77.
209. Szondy, E., Mezey, Z., Füst, G., Szekely, J. & Gero, S. 1981 Serial measurement of circulating immune complexes in myocardial infarction. Br Heart J 46, 93–8.
210. Tabas, I. 2010 Macrophage death and defective inflammation resolution in atherosclerosis. Nat Rev Immunol 10, 36–46.
211. Tarnavski, O., McMullen, J. R., Schinke, M., Nie, Q., Kong, S. & Izumo, S. 2004 Mouse cardiac surgery: comprehensive techniques for the generation of mouse models of human diseases and their application for genomic studies. Physiol Genomics 16, 349–60.
212. Taylor, P. R., Carugati, A., Fadok, V. A., Cook, H. T., Andrews, M., Carroll, M. C., Savill, J. S., Henson, P. M., Botto, M. & Walport, M. J. 2000 A hierarchical role for classical pathway complement proteins in the clearance of apoptotic cells in vivo. J Exp Med 192, 359–66.
213. Tedesco, F., Pausa, M., Nardon, E., Introna, M., Mantovani, A. & Dobrina, A. 1997 The cytolytically inactive terminal complement complex activates endothelial cells to express adhesion molecules and tissue factor procoagulant activity. J Exp Med 185, 1619–27.
214. Tertov, V. V., Kaplun, V. V., Sobenin, I. A. & Orekhov, A. N. 1998 Low-density lipoprotein modification occurring in human plasma possible mechanism of in vivo lipoprotein desialylation as a primary step of atherogenic modification. Atherosclerosis 138, 183–95.
215. Thiel, S., Vorup-Jensen, T., Stover, C. M., Schwaeble, W., Laursen, S. B., Poulsen, K., Willis, A. C., Eggleton, P., Hansen, S., Holmskov, U., Reid, K. B. & Jensenius, J. C. 1997 A second serine protease associated with mannan-binding lectin that activates complement. Nature 386, 506–10.
216. Thorbjornsdottir, P., Kolka, R., Gunnarsson, E., Bambir, S. H., Thorgeirsson, G., Kotwal, G. J. & Arason, G. J. 2005 Vaccinia virus complement control protein diminishes formation of atherosclerotic lesions: complement is centrally involved in atherosclerotic disease. Ann N Y Acad Sci 1056, 1–15.
217. Torzewski, J., Bowyer, D. E., Waltenberger, J. & Fitzsimmons, C. 1997 Processes in atherogenesis: complement activation. Atherosclerosis 132, 131–8.
218. Torzewski, J., Oldroyd, R., Lachmann, P., Fitzsimmons, C., Proudfoot, D. & Bowyer, D. 1996 Complement-induced release of monocyte chemotactic protein-1 from human smooth muscle cells. A possible initiating event in atherosclerotic lesion formation. Arterioscler Thromb Vasc Biol 16, 673–7.
219. Torzewski, J., Torzewski, M., Bowyer, D. E., Frohlich, M., Koenig, W., Waltenberger, J., Fitzsimmons, C. & Hombach, V. 1998 C-reactive protein frequently colocalizes with the terminal complement complex in the intima of early atherosclerotic lesions of human coronary arteries. Arterioscler Thromb Vasc Biol 18, 1386–92.
220. Torzewski, M., Torzewski, J., Bowyer, D. E., Waltenberger, J., Fitzsimmons, C., Hombach, V. & Gabbert, H. E. 1997 Immunohistochemical colocalization of the terminal complex of human complement and smooth muscle cell alpha-actin in early atherosclerotic lesions. Arterioscler Thromb Vasc Biol 17, 2448–52.

221. Tsuji, S., Kaji, K. & Nagasawa, S. 1994 Activation of the alternative pathway of human complement by apoptotic human umbilical vein endothelial cells. J Biochem 116, 794–800.
222. Tulenko, T. N. & Sumner, A. E. 2002 The physiology of lipoproteins. J Nucl Cardiol 9, 638–49.
223. Turnberg, D. & Botto, M. 2003 The regulation of the complement system: insights from genetically-engineered mice. Mol Immunol 40, 145–53.
224. Tuzcu, E. M., Kapadia, S. R., Tutar, E., Ziada, K. M., Hobbs, R. E., McCarthy, P. M., Young, J. B. & Nissen, S. E. 2001 High prevalence of coronary atherosclerosis in asymptomatic teenagers and young adults: evidence from intravascular ultrasound. Circulation 103, 2705–10.
225. Vakeva, A., Laurila, P. & Meri, S. 1992 Loss of expression of protectin (CD59) is associated with complement membrane attack complex deposition in myocardial infarction. Lab Invest 67, 608–16.
226. Vakeva, A., Laurila, P. & Meri, S. 1993 Regulation of complement membrane attack complex formation in myocardial infarction. Am J Pathol 143, 65–75.
227. Vakeva, A., Morgan, B. P., Tikkanen, I., Helin, K., Laurila, P. & Meri, S. 1994 Time course of complement activation and inhibitor expression after ischemic injury of rat myocardium. Am J Pathol 144, 1357–68.
228. Vakeva, A. P., Agah, A., Rollins, S. A., Matis, L. A., Li, L. & Stahl, G. L. 1998 Myocardial infarction and apoptosis after myocardial ischemia and reperfusion: role of the terminal complement components and inhibition by anti-C5 therapy. Circulation 97, 2259–67.
229. Vasile, E., Simionescu, M. & Simionescu, N. 1983 Visualization of the binding, endocytosis, and transcytosis of low-density lipoprotein in the arterial endothelium in situ. J Cell Biol 96, 1677–89.
230. Venneker, G. T., van den Hoogen, F. H., Boerbooms, A. M., Bos, J. D. & Asghar, S. S. 1994 Aberrant expression of membrane cofactor protein and decay-accelerating factor in the endothelium of patients with systemic sclerosis. A possible mechanism of vascular damage. Lab Invest 70, 830–5.
231. Verdouw, P. D., van den Doel, M. A., de Zeeuw, S. & Duncker, D. J. 1998 Animal models in the study of myocardial ischaemia and ischaemic syndromes. Cardiovasc Res 39, 121–35.
232. Véniant, M. M., Withycombe, S. & Young, S. G. 2001 Lipoprotein size and atherosclerosis susceptibility in Apoe(-/-) and Ldlr(-/-) mice. Arterioscler Thromb Vasc Biol 21, 1567–70.
233. Vik, D. P., Keeney, J. B., Munoz-Canoves, P., Chaplin, D. D. & Tack, B. F. 1988 Structure of the murine complement factor H gene. J Biol Chem 263, 16720–4.
234. Villiers, M. B., Thielens, N. M., Reboul, A. & Colomb, M. G. 1982 A study of a covalent-like interaction between soluble nascent C4b and C4-binding protein. Biochim Biophys Acta 704, 197–203.
235. Virella, G., Atchley, D., Koskinen, S., Zheng, D. & Lopes-Virella, M. F. 2002 Proatherogenic and proinflammatory properties of immune complexes prepared with purified human oxLDL antibodies and human oxLDL. Clin Immunol 105, 81–92.
236. Vlaicu, R., Niculescu, F., Rus, H. G. & Cristea, A. 1983 Immune deposits in human aortic atherosclerotic wall. Med Interne 21, 3–8.
237. Vlaicu, R., Niculescu, F., Rus, H. G. & Cristea, A. 1985 Immunohistochemical localization of the terminal C5b-9 complement complex in human aortic fibrous plaque. Atherosclerosis 57, 163–77.
238. Vlaicu, R., Rus, H. G., Niculescu, F. & Cristea, A. 1985 Quantitative determinations of immunoglobulins and complement components in human aortic atherosclerotic wall. Med Interne 23, 29–35.
239. Walport, M. J. 2001 Complement. First of two parts. N Engl J Med 344, 1058–66.
240. Walport, M. J. 2001 Complement. Second of two parts. N Engl J Med 344, 1140–4.
241. Walsh, M. C., Bourcier, T., Takahashi, K., Shi, L., Busche, M. N., Rother, R. P., Solomon, S. D., Ezekowitz, R. A. & Stahl, G. L. 2005 Mannose-binding lectin is a regulator of inflammation that accompanies myocardial ischemia and reperfusion injury. J Immunol 175, 541–6.

242. Weisman, H. F., Bartow, T., Leppo, M. K., Marsh, H. C., Jr., Carson, G. R., Concino, M. F., Boyle, M. P., Roux, K. H., Weisfeldt, M. L. & Fearon, D. T. 1990 Soluble human complement receptor type 1: in vivo inhibitor of complement suppressing post-ischemic myocardial inflammation and necrosis. Science 249, 146–51.
243. Whaley, K. (ed.) 1987 Complement in health and disease: MTP Press, Lancaster.
244. Williams, K. J. & Tabas, I. 1995 The response-to-retention hypothesis of early atherogenesis. Arterioscler Thromb Vasc Biol 15, 551–61.
245. Xu, Q., Kleindienst, R., Waitz, W., Dietrich, H. & Wick, G. 1993 Increased expression of heat shock protein 65 coincides with a population of infiltrating T lymphocytes in atherosclerotic lesions of rabbits specifically responding to heat shock protein 65. J Clin Invest 91, 2693–702.
246. Yasojima, K., Schwab, C., McGeer, E. G. & McGeer, P. L. 2001 Complement components, but not complement inhibitors, are upregulated in atherosclerotic plaques. Arterioscler Thromb Vasc Biol 21, 1214–9.
247. Ylä-Herttuala, S., Palinski, W., Butler, S. W., Picard, S., Steinberg, D. & Witztum, J. L. 1994 Rabbit and human atherosclerotic lesions contain IgG that recognizes epitopes of oxidized LDL. Arterioscler Thromb 14, 32–40.
248. Yusuf, S., Hawken, S., Ôunpuu, S., Dans, T., Avezum, A., Lanas, F., McQueen, M., Budaj, A., Pais, P., Varigos, J. & Lisheng, L. 2004 Effect of potentially modifiable risk factors associated with myocardial infarction in 52 countries (the INTERHEART study): case-control study. Lancet 364, 937–52.
249. Zacharowski, K., Otto, M., Hafner, G., Marsh, H. C., Jr. & Thiemermann, C. 1999 Reduction of myocardial infarct size with sCR1sLe(x), an alternatively glycosylated form of human soluble complement receptor type 1 (sCR1), possessing sialyl Lewis x. Br J Pharmacol 128, 945–52.
250. Zhang, S. H., Reddick, R. L., Piedrahita, J. A. & Maeda, N. 1992 Spontaneous hypercholesterolemia and arterial lesions in mice lacking apolipoprotein E. Science 258, 468–71.
251. Zimmerman, J. L., Dellinger, R. P., Straube, R. C. & Levin, J. L. 2000 Phase I trial of the recombinant soluble complement receptor 1 in acute lung injury and acute respiratory distress syndrome. Crit Care Med 28, 3149–54.
252. Zipfel, P. F., Hallstrom, T., Hammerschmidt, S. & Skerka, C. 2008 The complement fitness factor H: role in human diseases and for immune escape of pathogens, like pneumococci. Vaccine 26 Suppl 8, I67–74.

Biographies

Perla Thorbjornsdottir received her M.S. from The University of Iceland (Faculty of Medicine). By using a diet-induced mouse model of atherosclerosis, she demonstrated, for the first time, that atherogenesis may be prevented by using the complement inhibitor VCP (vaccinia virus complement control protein). Complement activation is thus a crucial and rate-limiting step in atherogenesis. Her results also showed that VCP can prevent 44% of the cardiac injury that takes place when a coronary artery is reperfused after myocardial infarction.

Gudmundur Thorgeirsson received his MD degree at the University of Iceland in Reykjavik and his PhD degree in experimental pathology at Case Western Reserve University in Cleveland, Ohio. He was a resident in pathology and internal medicine and subsequently a fellow in cardiology at the University Hospitals of Cleveland from 1974–1982. Since 1982 he has been a cardiologist at Landspitali, University Hospital of Iceland and concurrently a faculty member in pharmacology and internal medicine at the University of Iceland. Currently he is a professor of medicine and the dean of the Faculty of Medicine, University of Iceland. His research interests include endothelial cell biology, cardiovascular epidemiology and cardiovascular genetics.

Girish J. Kotwal received his Ph.D. from McMaster University, Hamilton, Canada. During his post doctoral years at the National Institutes of Health, Bethesda, MD, he developed an interest in poxviral regulation of complement mediated by vaccinia virus complement control protein (VCP). As an independent investigator in Universities across two continents and in collaboration with several research groups around the world he was involved in developing functional models to evaluate the preclinical role of VCP in complement mediated inflammation damage in disease states including Atherosclerosis, Arthritis, Alzheimer's Disease, CNS traumatic injury, ischaemia reperfusion injury, infection, multiple organ dysfunction syndrome, macular degeneration, myocardial infarction, sepsis and xenotransplantation. He is a co-inventor on several patents related to the application of VCP in disease.

Gudmundur Johann Arason received his Ph.D. from Bedford College, University of London, London, UK. During his post doctoral years at the Landspitali University Hospital, Reykjavík, Iceland, he developed an interest in the physiology of complement and its role in immune complex disease and in chronic inflammatory disease. He discovered a defect in immune complex handling in SLE which reflects titers of autoantibodies to C1q. Turning to atherosclerosis in 1995, a collaboration with Hungarian scientists yielded results showing that C4B*Q0 interacts with smoking to promote angina pectoris (OR 30.1) and myocardial infarction (OR 22.7) and also leads to increased one-year mortalities after MI (hazard ratio 3.5). Since 1999, his studies with Girish Kotwal have shown that inhibition of complement by VCP (vaccinia virus complement control protein) can inhibit reperfusion injury by 44% and prevent atherogenesis by 50%. He is a co-inventor on several patents related to the application of VCP in atherosclerosis and its acute events.

Part V
Molecular and Emerging Technologies

Chapter 21
Vibro-Acoustography of Arteries

Cristina Pislaru, James F. Greenleaf, Birgit Kantor, and Mostafa Fatemi

Abstract Vibro-acoustography can reconstruct a map of the object by measuring the acoustic field resulting from object vibration at a specified low frequency (low kilohertz range). This method has been successfully utilized to image arterial calcifications in vitro and in vivo. Recent achievements in transducer technology and software implementation in existing commercial ultrasound machines should facilitate the exploration of the full potential of this unique imaging technique.

Keywords Atherosclerotic plaques • Arterial calcifications • Ultrasound Imaging • Carotid imaging • Vibro-acoustography • Vibrometry • Ultrasound radiation force

21.1 Introduction

Several emerging ultrasound methods are proposed to probe arterial elasticity. These methods can be generally categorized into three major types: (a) transient methods, where an impulse radiation force is used and the response (deformation) of the tissue is measured, from which the elasticity is estimated [1, 2–4]; (b) shear-wave methods, where the viscoelasticity is estimated from the velocity of propagation of shear waves induced by an impulse radiation force [5–11]; and (c) vibro-acoustography, where the object is imaged by applying a localized oscillating force acting on the object and the acoustic signal generated by the vibrating object is measured by a hydrophone [12]. Magnetic resonance elastography has also been developed [13, 14]; this method uses a mechanical actuator to vibrate the body from the surface and then measures the strain waves with a phase-sensitive magnetic resonance imaging technique.

Most of the techniques currently proposed employ the use of ultrasound both to apply the force and to measure the response of the tissue. The use of ultrasound radiation force has several advantages: (1) it is noninvasive, (2) current ultrasound systems can be easily

C. Pislaru (✉)
Ultrasound Research Lab, Department of Physiology and Biomedical Engineering, Mayo Clinic College of Medicine, 200 First Street SW, Rochester, MN 55905, USA
e-mail: pislaru.cristina@mayo.edu

Jasjit S. Suri et al. (eds.), *Atherosclerosis Disease Management*,
DOI 10.1007/978-1-4419-7222-4_21, © Springer Science+Business Media, LLC 2011

modified for the purpose of probing arterial elasticity, (3) the ultrasound radiation force can be precisely focused on the target object, and (4) the radiation force can be produced in a wide range of frequencies. Review articles of some of the methods proposed and their applications in different types of tissue are presented in refs. [9, 11, 15, 16, 17].

This chapter focuses on the potential of vibro-acoustography technique. We will present the current status of this imaging technique for vascular application and the supporting evidence of its potential for carotid imaging.

21.2 Principle of Vibro-Acoustography

Vibro-acoustography is a unique ultrasound method capable of detecting hard inclusions in softer material [12, 15, 18]. The technique uses acoustic radiation pressure produced by ultrasound to induce a vibration in the object from a distance. The image of the object is then reconstructed based on the amplitude (and phase) of the acoustic emission signal that is generated by the vibrating object and detected by a hydrophone. The amplitude of this signal depends on the viscoelastic properties of the object as well as its acoustical properties (absorption, scattering, and reflection) and the surrounding medium [16, 18]. By scanning point by point in a plane or volume, a genuine map of the object can be obtained (Fig. 21.1).

A common approach is to use two unmodulated ultrasound beams, at slightly different frequencies ($f1$ and $f1 + \Delta f$), propagating along the path, and crossing each other at the focal point. The resulting ultrasound field generates an oscillating force at the focal point, producing a vibration of the object at a frequency equal to the difference frequency (Δf), which is set in the low kilohertz range (5–50 kHz). An alternative is to use a single amplitude-modulated ultrasound beam, at a desired frequency to induce the oscillating force. The acoustic emission signal generated

Fig. 21.1 Principle of vibro-acoustographic imaging. The two elements of the confocal transducer generate separate ultrasound signals in the megahertz range, at slightly different frequencies. The ultrasound field in the focal plane produces an oscillating force, acting on the object at a frequency equal to the difference frequency (Δf). This force vibrates the object, which in turn generates a sound field that can be recorded by an audio hydrophone placed near the object. To form an image, the object is scanned point by point, and the acoustic emission is recorded from each point [19]

by the vibrational object is not omnidirectional; thus, the position of the hydrophone is sometimes critical. Either tone-burst or continuous wave ultrasound can be used. The spatial resolution is determined by the ultrasound beam width at the focal point, which is of the order of the incident ultrasound wavelength. The biological noise spectrum is generally below 1 kHz and can be easily filtered out; however, the technique requires a quiet environment for accurate signal detection.

This method has been proven useful for several applications, including the detection of calcifications in excised tissue specimens [19–21] and large vessels in vivo [22].

21.3 Detection of Arterial Calcifications: Experimental Results

(a) *Excised human carotid arteries.* Vibro-acoustography has been proven useful for the detection of calcified atherosclerotic plaques in excised human carotid arteries [19] and breast tissue [23]. In the latter study, it was found that calcified arteries as small as 1 mm in diameter can be detected and the resulting images faithfully resemble X-ray images of the specimen [20, 21, 23].

The technique is also capable of imaging large atherosclerotic plaques [19]. Figure 21.2 shows examples of vibro-acoustographic images of human carotid arteries obtained from human cadavers. Carotid arteries of a young and an old subject were scanned in this example. For these in vitro tests, specimens are secured on a thin sheet of latex, which is almost transparent to the ultrasound and X-ray, and immersed in a water tank for scanning. Vibro-acoustography was performed at 7 kHz difference frequency (Δf) using continuous wave ultrasound from a 3 MHz confocal transducer [19]. Figure 21.2 shows that the normal carotid artery obtained from a young person appears on vibro-acoustographic scan without calcification, while the carotid artery of an old person with atherosclerosis showed the presence of a large calcified plaque (Fig. 21.2).

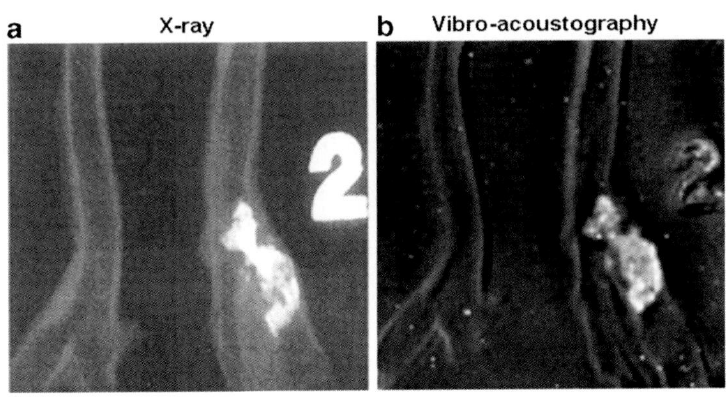

Fig. 21.2 Vibro-acoustography of excised human carotid arteries: (**a**) X-ray image showing a normal (*left*) and a calcified (*right*) carotid artery; (**b**) vibro-acoustographic image at 7 kHz vibration frequency. The calcification (and lead marker number '2', used for identification) is seen as a bright region in both X-ray and vibro-acoustographic images [19]

The hard plaque appears as a bright spot, faithfully resembling the X-ray image of the specimen. The increased brightness of the plaque on vibro-acoustographic image is due to the fact that the hard plaque is an efficient acoustic radiator, producing a strong acoustic emission signal when exposed to the radiation force. This strong signal is easily detected by the sensitive hydrophone placed in the vicinity. Hence, the hard plaque stands out as compared to the soft background. Generally, the contrast in the vibro-acoustographic image is facilitated by some characteristics of this unique technique, such as the absence of speckle, the high spatial resolution conferred by the high frequency transducer used (0.7 mm spatial resolution in the example shown in Fig. 21.2), and the high signal-to-noise ratio with this imaging technique [18]. These in vitro examples have demonstrated the utility of this technique in delineating a stiff inclusion against a soft background.

(b) *In vivo imaging of normal arteries*. The utility of vibro-acoustography for vascular imaging was further tested in vivo in live anesthetized swine [22]. In those studies, femoral arteries were scanned because they were readily accessible using the available setup for vibro-acoustography at that time. Considering current technological advancements, the method may be applicable for carotid imaging using hand-held linear transducers [24].

Figure 21.3 shows examples of in vivo vibro-acoustographic scans obtained from femoral arteries of swine. These images were obtained by using a dual-beam confocal transducer, programmed to send tone bursts of ultrasound to produce 45 kHz vibrations resulting from the intersection of two ultrasound beams (2.980 and 3.025 MHz). The focal point of this transducer was set at the level of the artery. The acoustic emission signal was detected by a hydrophone placed in the vicinity, in direct contact with animal's body. The amplitude and phase of the acoustic emission signal were used to reconstruct the images of the arteries scanned. Raster scans of 5×5 cm were obtained in a coronal plane passing through the artery. The brightness of each pixel in the image is normalized to the maximum acoustic emission signal recorded. The maximum spatial sampling interval set in these examples was 0.25 mm.

These in vivo images obtained from femoral arteries were high-resolution, high-contrast, and speckle-free (Fig. 21.3). Considerable anatomic detail can be observed, including the course and boundaries of the artery and its branching anatomy within the scanning plane. Normal artery walls appear straight, relatively thin, and smooth. The brightness of the walls varies regionally along the length of the artery, with some parts being brighter than other. The lumen appears dark or dense and with intensity similar to or lower than the artery walls and the surrounding muscles. A characteristic effect in vibro-acoustography is that points closer to the hydrophone position have a stronger signal than points further away; this effect may be partially compensated for, if desired, by applying a spatial low-frequency band-pass filter to these images.

Presence of catheters and guide-wires inserted intravascularly and placed within the scanning plane can also be nicely seen in these in vivo images (Fig. 21.3). Guiding catheters (polymers) and metal wires clearly differentiate from the vessel, without producing shadowing effects or strong reflections as in conventional ultrasound B-mode imaging. Some examples are shown in Fig. 21.3c–d. In these examples, data

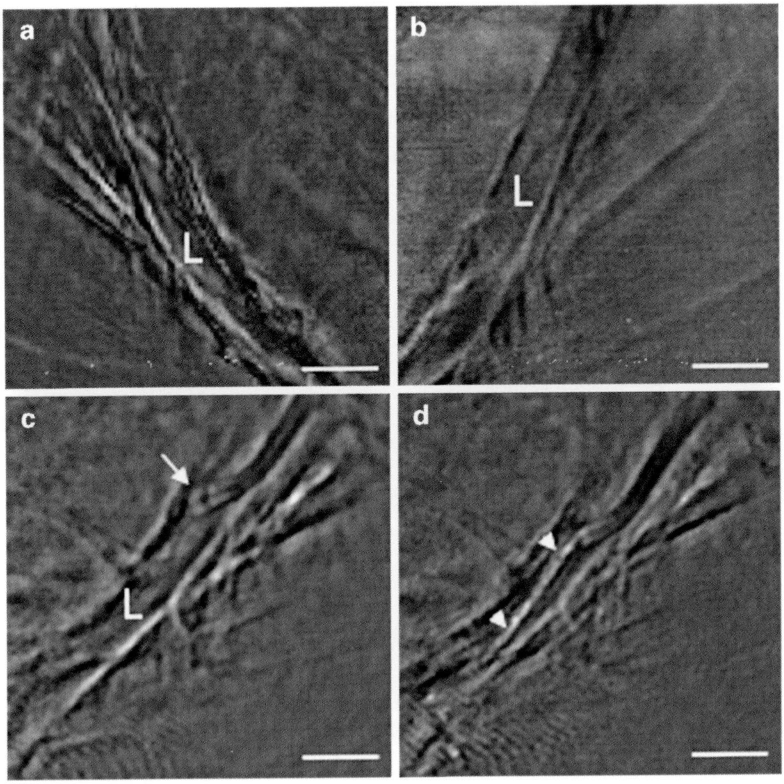

Fig. 21.3 In vivo vibro-acoustographic images of normal femoral arteries from three pigs (**a** and **b** images were taken from two animals and **c**, **d** images from a third animal). Images are reconstructed on the basis of the amplitude of the acoustic emission signal at 45 kHz vibrations. Scans were obtained in a coronal plane passing through the artery and perpendicular to the ultrasound beams. *L* indicates the artery lumen. *Arrows* indicate the tip of the guiding catheter and *arrowheads*, the guide-wire. Scale bar: 1 cm (adapted from ref. [22])

acquisition took about 3–6 min for a 5×5 cm image. However, current advancements now allow up to ten times faster data acquisition (within seconds). The in vivo and other in vitro results [25–27] suggest that vibro-acoustography may have potential to guide the result of interventions. Fast scanning may be achieved by array transducers and electronic scanning [24].

The reproducibility of in vivo vibro-acoustography imaging is good. Figure 21.4 shows vibro-acoustographic scans obtained 1 week apart in the same animals. Internal and external factors that could introduce additional noise and errors, such as breathing artifacts or different imaging setup, do not seem to be critical. Electrocardiographic gating has not been used for these images, even though it would be desirable in a clinical setting.

The reproducibility of morphological measurements from in vivo vibro-acoustographic images is also high. Measurements performed showed a very good agreement between the vibro-acoustography, conventional ultrasound, and X-ray [22].

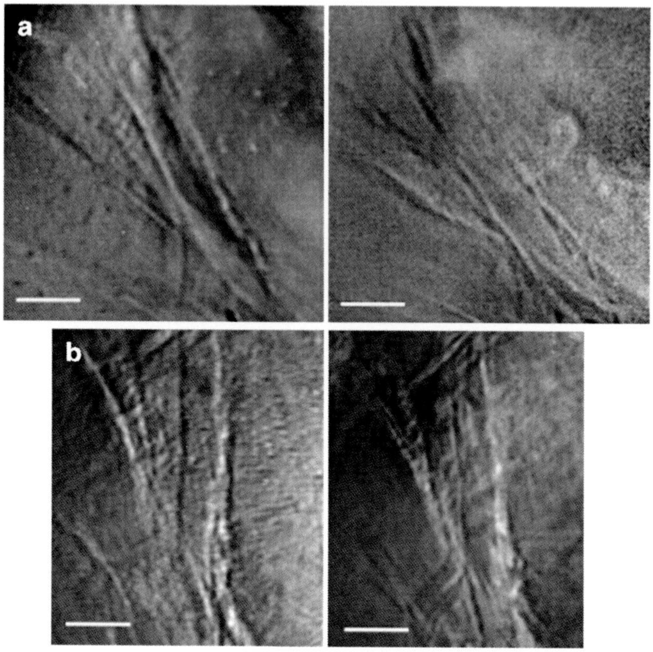

Fig. 21.4 In vivo vibro-acoustographic images of femoral arteries from two animals (**a**, **b**). Vibro-acoustographic scans were obtained at a 1-week interval, demonstrating the good reproducibility of this imaging method between serial scans. Scale bar: 1 cm (adapted from ref. [22])

For instance, the mean difference for the measurement of artery diameter when compared with conventional ultrasound was 5% in average (or ~0.25 mm, bordering on the spatial resolution of these images).

(c) *In vivo detection of calcified arterial plaques*. Further in vivo animal studies have demonstrated that vibro-acoustography is capable to detect calcified arterial plaques in vivo. For the purpose of testing this new technique, a swine model was employed that allowed rapid and reproducible formation of calcified plaques within days after an injection of hydroxyapatite solution [22, 28]. The inflammatory reaction that follows the injection causes partial destruction of structural constituents of the wall, including the external and internal elastic lamina, smooth muscle proliferation, neointimal formation, and calcium deposition, roughly mimicking a calcified human atherosclerotic plaque.

Figures 21.5 and 21.6 show examples of in vivo vibro-acoustographic images of femoral arteries of pigs with experimentally induced arterial calcifications. Calcified plaques were consistently detected and they appear as irregularities along a thickened artery. Most of these plaques show up on vibro-acoustographic images as bright spots, as in previous in vitro tests (Fig. 21.2). However, large plaques appear as dark defects. This "defect" could be due to either a phase difference in the acoustic emission signal and/or to a reduction in the acoustic emission signal due to increased stiffness of the plaque or rigid body motion.

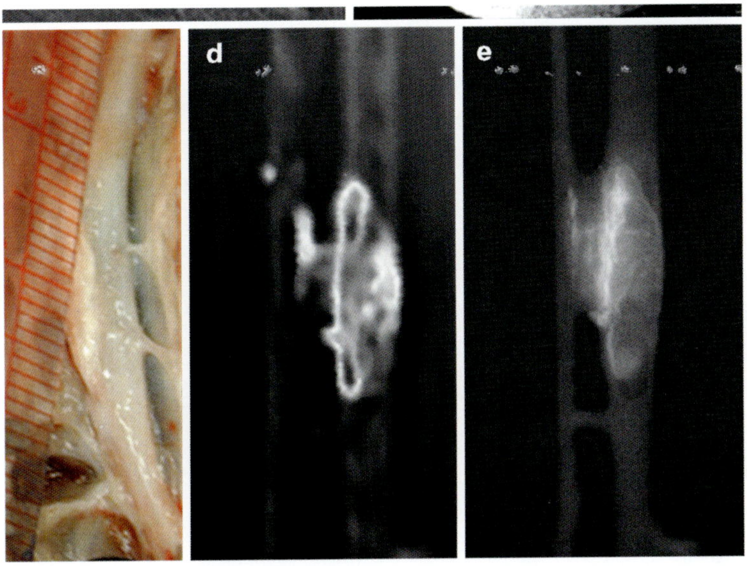

Fig. 21.5 Examples of images obtained from one animal with a calcified plaque. (**a**) In vivo vibro-acoustographic imaging was performed at 7 days after calcium hydroxyapatite injection. (**b**) Contrast angiography at 14 days showed no notable stenosis at the location of injection (at the tip of the instrument). (**c**) Photograph of femoral artery in situ showing the plaque. (**d**, **e**) Vibro-acoustography and X-ray images of the excised specimen [22]

The location and extent of these plaques matched the location and extent of calcifications seen on ex vivo vibro-acoustographic scans and X-ray radiographs of excised arteries. Morphological measurements of plaque area and extent showed very close agreements between the in vivo and in vitro vibro-acoustographic scans and between the vibro-acoustography and X-ray images of excised specimens [22]. Presence of plaques was detected in every case, from small to large plaques, regardless of the degree of luminal narrowing. Figure 21.5 shows an example of an in vivo vibro-acoustographic scan from an animal with a calcified plaque without luminal stenosis, as proven by angiography.

Vibro-acoustography has also potential to image the atherosclerotic plaque in its 3-D configuration, by compiling scans acquired at different depths. This level of detail is necessary for a complete plaque characterization. Figure 21.7 illustrates examples of sequential in vivo vibro-acoustographic scans acquired at different depths from the body surface. There are distinct differences in the appearance of the plaque at different depths. Vibro-acoustography is now implemented in a research ultrasound scanner using linear- and phased-array transducers, which should allow faster scanning and improved spatial (axial) resolution required for 3-D reconstruction of the plaque within minutes, suitable for clinical settings.

The reproducibility of in vivo measurements of plaque length and extent from vibro-acoustographic images is also high. For instance, measurement of plaque extent showed a good agreement between the vibro-acoustography, conventional ultrasound, and X-ray radiography of the excised specimen, with reasonable limits of agreement between these techniques [22]. The sensitivity and specificity for

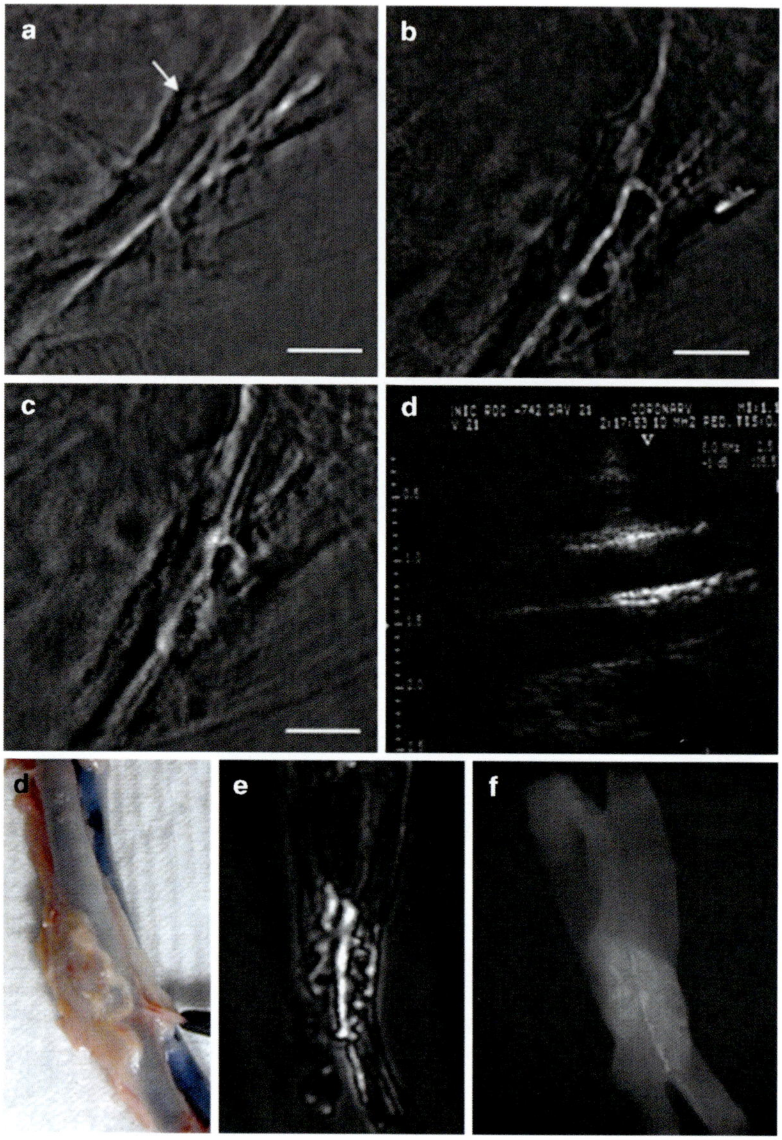

Fig. 21.6 Vibro-acoustographic images obtained from a right femoral artery with a calcified plaque. In vivo vibro-acoustography was performed at baseline (**a**), 7 days (**b**), and 14 days (**c**) after calcium hydroxyapatite injection. B-mode ultrasound at 14 days after injection (**d**). Photographic (**e**), vibro-acoustography (**f**), and X-ray (**g**) images of excised artery containing the plaque are shown for comparison. Scale bar: 1 cm [22]

detecting calcifications in this first in vivo study were 100 and 86%, respectively, and the positive and negative predictive values were 77 and 100%, respectively. Overall, these first in vivo tests demonstrate that vibro-acoustography provides accurate and reproducible measurements of femoral arteries as well as vascular calcifications in living animals, suggesting that this technique has clinical potential.

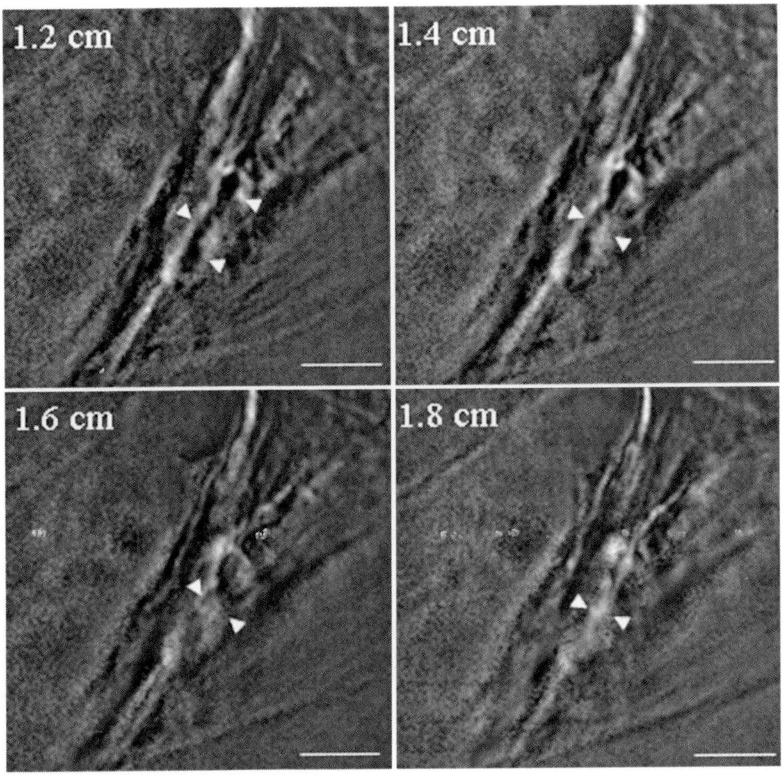

Fig. 21.7 In vivo vibro-acoustographic scans of a femoral artery with a calcified plaque. Images were obtained at four depth levels (in 2-mm increments) and show the variation in plaque area (*arrows*) with depth. Scale bar: 1 cm (adapted from ref. [22])

Currently, vibro-acoustography has only been tested for the detection of large calcified plaques in vivo. No attempt has been done yet to test the potential of this technique for plaque characterization. This task would need high image resolution and fast scanning. Nevertheless, there is evidence to suggest that vibro-acoustography may be suitable for this purpose, given its ability to image plaques three-dimensionally, its theoretical ability to detect regional heterogeneities in tissue stiffness, and recent technological advancements of this technology. The ability to detect regional heterogeneity is important because plaque composition and spatial distribution of its components are more important than the absolute quantity of calcium in the atherosclerotic plaque [29, 30]. In particular, a scattered and spotty calcification pattern in individual fibro-fatty plaques is typical for culprit lesions and may identify plaques prone to rupture [31]. As this imaging technique progresses, future studies should explore this potential.

(d) *Contrast enhanced vibro-acoustography*. Air-filled microbubbles act as strong reflectors for the ultrasound beams. Echocardiographic contrast agent can be used to increase the amplitude of the acoustic emission signal in the vibro-acoustography image. Figure 21.8 shows an example of in vivo vibro-acoustography scans before and during contrast injection (Optison). In this example, the contrast agent was used to confirm the course and boundaries of the artery and the patency of the vessel [22].

Presence of gas-filled microbubbles produces a signal much stronger than the surrounding structures, such as artery walls and muscles. This is an expected finding. Bubbles are highly nonlinear and the two ultrasound frequencies mix (multiply) efficiently in the bubbles, making an especially loud difference frequency signal. In the example shown in Fig. 21.8, small bubbles or clusters of microbubbles can be seen entering the branches coming off the main artery and running parallel to the imaging plane. In vitro studies have shown that the amplitude of the acoustic emission signal is proportional to the concentration of microbubbles [32] and that small microbubbles have strong acoustic emission at twice the difference frequency (harmonic acoustic emission) [33]. In vivo quantification of blood flow in the tissue has not yet been reported.

21.4 Other Applications of This Technique

Besides calcified plaques, the technique has also been used for the detection of breast tumors in vitro and in vivo [21, 23, 24], to image brachytherapy seeds position in excised prostate tissue [25–27, 35], to quantify flow in vitro [32], and to monitor the result of thermal modifications of tissue proteins [36, 37].

21.5 Detection Sensitivity

Vibro-acoustography has very high detection sensitivity, which is an advantage. Higher ultrasound intensity would be needed if detection sensitivity is poor. Ultrasound Doppler methods, at conventional frequencies, can detect displacements of the order of

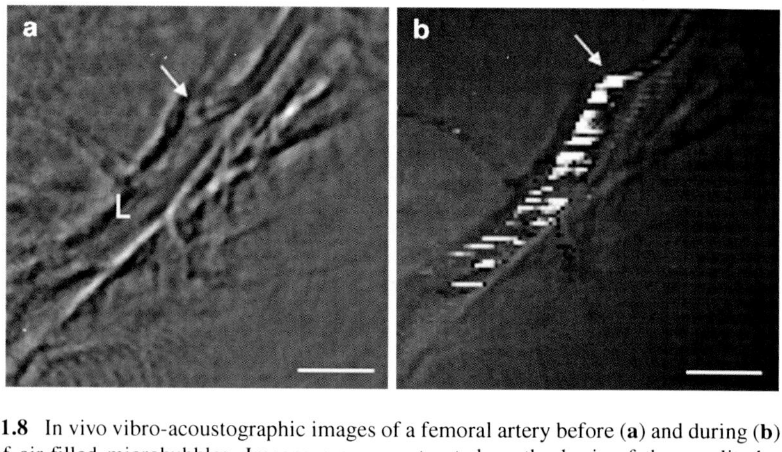

Fig. 21.8 In vivo vibro-acoustographic images of a femoral artery before (**a**) and during (**b**) injection of air-filled microbubbles. Images are reconstructed on the basis of the amplitude of the acoustic emission signal at 45 kHz vibrations. *L* indicates the artery lumen, and *arrows*, the tip of the guiding catheter. Scale bar: 1 cm (adapted from ref. [22])

2a few micrometers. Magnetic resonance imaging has detection sensitivity of the order of 100 nm [13]. Vibro-acoustography can detect motion as small as a few nanometers [12]. This high sensitivity is due to the fact that small vibrations of the object produce an acoustic emission pressure field that is easily detectable by a sensitive hydrophone, while using a low transmittance power to vibrate the object. The use of a hydrophone makes this technique simple to use in clinical settings; however, it requires an acoustically quiet environment for accurate signal detection.

21.6 Image Resolution

The spatial resolution in vibro-acoustographic images depends on the frequency of the incident ultrasound beams (Mega Hertz range) as well as on the sampling interval chosen by the operator. In the examples shown in Fig. 21.2, a 3 MHz transducer was used but the maximum spatial sampling interval chosen in that study was 0.25 mm. Higher image resolution can be obtained with a higher frequency transducer available for vascular imaging. Technologic advances have improved the speed of data acquisition and the spatial resolution [24, 38], allowing imaging of more complex pathology than currently explored. New phased-array transducer designs offer a higher image spatial resolution (i.e., depth resolution, 7 versus 10 mm) and higher frame rate (up to one frame per second for a 100×100-pixel image at a 50 kHz difference frequency) [38]. For instance, for a 7-MHz linear transducer, the spatial resolution in vibro-acoustography would be about 0.4 mm in azimuth and 1.5 mm in elevation. Current advancements allow vibro-acoustographic scanning using a modified commercial scanner [24].

An important advantage in vibro-acoustographic images is the reduction/absence of speckles, which are characteristic with the conventional ultrasound. Speckle is a random interference pattern in an image, formed in a medium containing numerous subresolution scatterers. Speckles reduce spatial resolution and contrast quality in the ultrasound images, decreasing the level of detail displayed and the diagnostic accuracy of conventional ultrasound. In Fig. 21.3, one can notice the near complete elimination of the speckle (noise) in the vibro-acoustographic images. As a result, image clarity increases and spatial resolution and contrast improve. The lack of strong reflections and shadowing effects may help with the visualization of a plaque containing calciums.

21.7 Quantitative Measurements

A limitation of vibro-acoustography is that the amplitude of the acoustic emission signal is not a direct measure of the object hardness (stiffness). Rather, local differences in stiffness can be detected from the relative displacement values, without the need to measure the applied force. The acoustic emission field is a function of several physical parameters that relate to the hardness of the target object (artery) as well as to the surrounding medium. However, it is theoretically possible to measure one property of the target object if one has knowledge of the other variables [18]. A method for quantitative estimation of tissue elasticity using inverse problem approach and vibro-acoustic data has been presented in ref. [39]. Phase images may also be helpful for disclosing subtle differences between the structures. More direct quantitative measurements of the complex elasticity of vessel walls may be achieved using shear-wave methods [5, 6, 8, 40, 41]. However, as opposed to other methods, the acoustic emission resulting from particle vibration has low tissue attenuation within the frequency range used, thus the signal can be detected by the hydrophone from a large distance (tens of centimeters).

Precise morphological measurements that are clinically relevant in carotid imaging can be made from vibro-acoustographic images. The artery diameter, plaque extent, plaque area, wall thickness, and other features can be readily measured, with very high reproducibility [22]. The lack of speckle, reverberations, and shadowing effects improve contrast quality in vibro-acoustographic scans. Potentially, this imaging technique can be combined with other quantitative nonimaging techniques to perform "biopsy-like" measurement of tissue stiffness [6, 40], at locations suspected for abnormalities.

21.8 Exposure Safety

Vibro-acoustographic images can be obtained at intensities below the US Food and Drug Administration (FDA)-recommended safe limit for ultrasound imaging in vivo ($I_{\mathrm{SPTA}} \leq 720$ mW/cm^2). In the animal study reported in ref. [22], the calculated spatial peak temporal average intensity was within this recommended limit, at the same time maintaining good image quality. The mechanical index was less than 1. The thermal safety of vibro-acoustographic scanning has been studied by Chen et al. [7], who showed that heating during a single vibro-acoustographic scan to be below 0.05°C in soft tissue at relevant attenuation levels (up to 0.7 dB/cm/MHz). For other in vivo applications such as breast imaging [34], ultrasound exposures were also in compliance with the FDA-recommended limit.

21.9 Limitations of Vibro-Acoustography

One limitation, as mentioned above, is that vibro-acoustography cannot directly quantify plaque hardness (stiffness) or the amount of calcium in the plaque. A recent method for stiffness estimation from vibro-acoustic data has been presented in ref. [39]. Nevertheless, detection of local inhomogeneities can also be useful. Both mechanical and acoustical properties of tissue can affect the acoustic emission signal: an increase in the absorption of the ultrasound signal in the tissue leads to an increase of the acoustic emission, while an increase in tissue stiffness can lead to an increase or a decrease in the acoustic emission. However, a frequency shift in the spectrum of the acoustic emission has been shown to be linked to mechanical properties [36, 42], as in the resonance frequency-based method used in biomechanics [43]. Other challenges may arise from scanning structures very deep into the body as this may reduce the clarity of vibro-acoustographic images due to increased signal absorption and attenuation at higher depths. In in vivo breast imaging, vibro-acoustographic scans were readily obtained up to 4–5 cm deep into the body [34]. Although focused ultrasound is used in this technique, interference from the intervening structures in the path of each of the two ultrasound beams may affect image clarity. Random changes in sound speed along these paths can produce phase aberrations and defocusing, resulting in blurring of images. However, the hand-held transducers that are currently available allow quick adjustment of transducer position and image optimization. Electrocardiographically triggered imaging is desirable; however, this will increase the scanning time. The in vivo results that are presented here were obtained using a 3 MHz transducer, as a trade-off between the magnitude of the acoustic emission obtained and the image resolution [22]. However, vibro-acoustographic scans can be obtained using higher frequency linear transducers (for example, 7 MHz) that are currently available.

21.10 Clinical Potential

Arterial calcifications occur during the development of atherosclerosis and are almost exclusively seen in the advanced types of atherosclerotic plaques and are absent in normal vessels [44, 45]. Although calcifications do not necessarily localize to vulnerable plaques or flow-limiting stenotic lesions, presence of coronary artery calcification (CAC) is associated with increased risk of major adverse cardiovascular events beyond what can be inferred from traditional risk factors. Testing for CACs can assist with cardiovascular risk prediction in asymptomatic patients [46–51]. Similarly, detection of arterial calcification in peripheral vessels (such as carotid arteries) is predictive of coronary artery atherosclerosis and CACs [52, 53]. Thus, the presence of arterial calcification justifies early and aggressive risk factor modification to lower the incidence of cardiac events in these patients [48].

Presence of peripheral vascular disease also has implications in patients presenting with acute coronary syndromes. Atherosclerosis of noncoronary arteries is an independent predictor of clinical events and a risk factor for cardiovascular death and nonfatal myocardial infarction in these patients [54, 55]. Furthermore, presence of unstable atherosclerotic plaques in any vascular segment is associated with an increased risk for clinical complications [56–60]. Given the considerable morbidity and mortality associated with acute vascular events, modern imaging methods are focused on identification of vulnerable plaques that may require intervention [61–66]. The ability of vibro-acoustography for plaque characterization has yet to be explored.

Vibro-acoustography and other quantitative methods [3, 5, 6, 8, 40, 41] that allow estimation of the complex elasticity of the artery wall may become a valuable complement to conventional ultrasound imaging. These techniques are currently in preclinical testing.

21.11 Summary

Vibro-acoustography can reconstruct a map of the object by measuring the acoustic field resulting from object vibration at a specified low frequency (low kilohertz range). This method has been successfully utilized to image arterial calcifications in vitro and in vivo. Recent achievements in transducer technology and software implementation in existing commercial ultrasound machines should facilitate the exploration of the full potential of this unique imaging technique.

Acknowledgments The work was supported in part by grants EB002640, CA127235, and CA91956 from the National Institutes of Health. Disclosure of Conflict of Interest: Mayo Clinic and some of the authors (M.F. and J.F.G.) have a financial interest associated with the technology presented here. The technology has been licensed in part for limited use to Industry. Mayo Clinic has received royalties of greater than the federal threshold for significant financial interest.

References

1. Catheline S, Wu F, Fink M (1999) A solution to diffraction biases in sonoelasticity: the acoustic impulse technique. J Acoust Soc Am 105:2941–2950
2. Dahl JJ, Dumont DM, Allen JD et al (2009) Acoustic radiation force impulse imaging for noninvasive characterization of carotid artery atherosclerotic plaques: a feasibility study. Ultrasound Med Biol 35(5):707–716
3. Nightingale K, Palmeri M, Nightingale R, Trahey G (2001) On the feasibility of remote palpation using acoustic radiation force. J Acoust Soc Am 110:625–634
4. Sandrin L, Catheline S, Tanter M et al (1992) 2D time resolved pulsed elastography with ultrafast ultrasonic imaging. Ultrason Imaging 21(4):259–272
5. Bercoff J, Tanter M, Fink M (2004) Supersonic shear imaging: a new technique for soft tissue elasticity mapping. IEEE Trans Ultrason Ferroelectr Freq Control 51:396–409
6. Chen S and Greenleaf JF (2008) Characterization of tissue viscoelasticity from shear wave speed dispersion. In: Al-Jumaily A and Fatemi M, eds. Biomedical Applications of Vibration and Acoustics for Imaging and Characterizations. ASME Press, New York, NY, pp. 239–249
7. Chen S, MW Urban, C Pislaru et al (2009) Shearwave dispersion ultrasound vibrometry (SDUV) for measuring tissue elasticity and viscosity. IEEE Trans Ultrason Ferroelectr Freq Control 56(1):55–62
8. Nightingale K, McAleavey S, Trahey G (2003) Shear-wave generation using acoustic radiation force : in vivo and ex vivo results. Ultrasound Med Biol 29:1715–1723
9. Rudenko V, Sarvazyan AP, Emelianov SY (1996). Acoustic radiation force and streaming induced by focused nonlinear ultrasound in a dissipative medium. J Acoust Soc Am 99:2791–2798
10. Sarvazyan AP, Rudenko OV, Swanson SD et al (1998) Shear wave elasticity imaging: a new ultrasonic technology of medical diagnostics. Ultrasound Med Biol 24(9):1419–1435
11. Sarvazyan AP and Egorov V (2009) Mechanical imaging in medical applications. Conf Proc IEEE Eng Med Biol Soc 2009:1975–1978
12. Fatemi M and Greenleaf JF (1998) Ultrasound-stimulated vibro-acoustic spectrography. Science 280:82–85

13. Muthupillai R, Lomas D, Rossman P et al (1995) Magnetic resonance elastography by direct visualization of propagating acoustic strain waves. Science 269:1854–1857
14. Woodrum DA, Romano AJ, Lerman A et al (2006) Vascular wall elasticity measurement by magnetic resonance imaging. Magn Res Med 56:593–600
15. Greenleaf JF, Ehman RL, Fatemi M, Muthupillai R (1998). Imaging elastic properties of tissue. In: Duck FA, Baker AC, Staritt HC, eds. Ultrasound in Medicine (Medical Science Series). Bristol England: Institute of Physics Publishing, pp. 263–277
16. Greenleaf JF, Fatemi M, Insana M (2003) Selected methods for imaging elastic properties of biological tissues. Annu Rev Biomed Eng 5:57–78
17. Manduca A, Oliphant TE, Dresner MA et al (2001) Magnetic resonance elastography: non-invasive mapping of tissue elasticity. Med Image Anal 5(4):237–254
18. Fatemi M and Greenleaf JF (1999) Vibro-acoustography: an imaging modality based on ultrasound-stimulated acoustic emission. Proc Natl Acad Sci USA 96:6603–6608
19. Fatemi M and Greenleaf FG (2000). Probing the dynamics of tissue at low frequencies with the radiation force of ultrasound. Phys Med Biol 45:1449–1464
20. Alizad A, Fatemi M, Whaley DH, Greenleaf JF (2004) Application of vibro-acoustography for detection of calcified arteries in breast tissue. J Ultrasound Med 23(2):267–273
21. Alizad A, Fatemi M, Wold LE, Greenleaf JF (2004) Performance of vibro-acoustography in detecting of microcalcifications in excised human breast tissue: a study on 74 breast tissue samples. IEEE Trans Med Imaging 23:307–311
22. Pislaru C, Kantor B, Kinnick RR et al (2008) In vivo vibroacoustography of large peripheral arteries. Invest Radiol 43:243–252
23. Fatemi M, Wold LE, Alizad A, Greenleaf JF (2002) Vibro-acoustic tissue mammography. IEEE Trans Med Imaging 21:1–8
24. Urban MW, Chalek C, Kinnick RR et al (2010). Implementation of vibro-acoustography on a clinical ultrasound system. IEEE Trans Ultrason Ferroelectr Freq Control. In Print
25. Mitri FG, Trompette P, Chapelon JY (2004) Improving the use of vibro-acoustography for brachytherapy metal seed imaging: a feasibility study. IEEE Trans Med Imaging 23(1):1–6
26. Mitri FG, Davis BJ, Alizad A et al (2008) Prostate cryotherapy monitoring using vibroacoustography: preliminary results of an ex vivo study and technical feasibility. IEEE Trans Biomed Eng 55(11):2584–2592
27. Mitri FG, Davis BJ, Urban MW et al (2009) Vibro-acoustography imaging of permanent prostate brachytherapy seeds in an excised human prostate – preliminary results and technical feasibility. Ultrasonics 49(3):389–394
28. Ashai K, Kantor B, Keelan P et al (1999) Novel model of arterial calcification by direct intramural hydroxyapatite injection. J Am Coll Cardiol 33 (Suppl. A):64A
29. Burke AP, Kolodgie FD, Farb A et al (2002) Morphological predictors of arterial remodeling in coronary atherosclerosis. Circulation 105:297–303
30. Pham PH, Rao DS, Vasunilashorn F et al (2006) Computed tomography calcium quantification as a measure of atherosclerotic plaque morphology and stability. Invest Radiol 41:674–680
31. Ehara S, Kobayashi Y, Yoshiyama M et al (2004) Spotty calcification typifies the culprit plaque in patients with acute myocardial infarction. An intravascular ultrasound study. Circulation 110:3424–3429
32. Belohlavek M, Asanuma T, Kinnick RR, Greenleaf JF (2001) Vibro-acoustography: quantification of flow with highly-localized low-frequency acoustic force. Ultrason Imaging 23(4):249–256
33. Chen S, Kinnick R, Greenleaf JF, Fatemi M (2006) Difference frequency and its harmonic emitted by microbubbles under dual frequency excitation. Ultrasonics 44:e123–e126
34. Alizad A, Whaley DH, Greenleaf JF, Fatemi M (2006) Critical issues in breast imaging by vibro-acoustography. Ultrasonics 44(1):e217–e220
35. Fatemi M and Mitri FG (2009) Vibro-acoustography imaging applications for the prostate. Conf Proc IEEE Eng Med Biol Soc 9:4415–4419

36. Konofagou EE and Hynynen K (2003) Localized harmonic motion imaging: theory, simulations and experiments. Ultrasound Med Biol 29(10):1405–1413
37. Konofagou E, Thierman J, Hynynen K (2003) The use of ultrasound-stimulated acoustic emission in the monitoring of modulus changes with temperature. Ultrasonics 41:337–345
38. Silva GT, Greenleaf JF, Fatemi M (2004) Linear arrays for vibro-acoustography: a numerical simulation study. Ultrason Imaging 26:1–17
39. Brigham JC, Aquino W, Mitri FG et al (2007) Inverse estimation of viscoelastic material properties for solids immersed in fluids using vibroacoustic techniques – art. no. 023509. J Appl Phys 101(2):23509
40. Chen S, Aquino W, Alizad A et al (2010) Thermal safety of vibro-acoustography using a confocal transducer. Ultrasound Med Biol 36(2):343–349
41. Deffieux T, Montaldo G, Tanter M, Fink M (2009) Shear wave spectroscopy for in vivo quantification of human soft tissues visco-elasticity. IEEE Trans Med Imaging 28:313–322
42. Fatemi M and Greenleaf JF (1999). Application of radiation force in noncontact measurement of the elastic parameters. Ultrason Imaging 21:147–154
43. Rao SS (1995) Mechanical Vibrations, 3rd ed., Addison-Wessley, Reading, MA, pp. 43–46
44. Davies MJ (1997) The composition of coronary-artery plaques. N Engl J Med 336(18):1312–1314
45. Stary HC, Chandler AB, Dinsmore RE et al (1995) A definition of advanced types of atherosclerotic lesions and a histological classification of atherosclerosis. A report from the Committee on Vascular Lesions of the Council on Arteriosclerosis, American Heart Association. Circulation 92(5):1355–1374
46. ACCF/AHA (2007) Clinical expert consensus document on coronary artery calcium scoring by computed tomography in global cardiovascular risk assessment and in evaluation of patients with chest pain: a report of the American College of Cardiology Foundation Clinical Expert Consensus Task Force (ACCF/AHA Writing Committee to Update the 2000 Expert Consensus Document on Electron Beam Computed Tomography). Circulation 115:402–426
47. Arad Y, Goodman KJ, Roth M et al (2005) Coronary calcification, coronary disease risk factors, C-reactive protein, and atherosclerotic cardiovascular disease events: the St. Francis Heart Study. J Am Coll Cardiol 46:158–165
48. Greenland P, LaBree L, Azen SP et al (2004) Coronary artery calcium score combined with Framingham score for risk prediction in asymptomatic individuals. JAMA 291:210–215
49. LaMonte MJ, FitzGerald SJ, Church TS et al (2005) Coronary artery calcium score and coronary heart disease events in a large cohort of asymptomatic men and women. Am J Epidemiol 162:421–429
50. Shaw LJ, Raggi P, Schisterman E et al (2004) Prognostic value of cardiac risk factors and coronary artery calcium screening for all-cause mortality. Radiology 228:826–833
51. Taylor AJ, Binderman J, Taylor AJ et al (2005) Coronary calcium independently predicts incident premature coronary heart disease over measured cardiovascular risk factors: mean three-year outcomes in the Prospective Army Coronary Calcium (PACC) project. J Am Coll Cardiol 46:807–814
52. Megnien JL, Sene V, Jeannin S et al (1992) Coronary calcification and its relation to extracoronary atherosclerosis in asymptomatic hypercholesterolemic men. The PCV METRA Group. Circulation 85:1799–1807
53. van Popele NM, Mattace-Raso FU, Vliegenthart R et al (2006) Aortic stiffness is associated with atherosclerosis of the coronary arteries in older adults: the Rotterdam Study. J Hypertens 24:2371–2376
54. CAPRIE Steering Committee (1996) A randomized, blinded trial of clopidogrel versus aspirin in patients at risk of ischaemic events (CAPRIE). Lancet 348:1329–1339
55. Komorovsky R, Desideri A, Coscarelli AS et al (2005) Predictive value of associations between carotid and coronary artery disease in patients with acute coronary syndromes. Am J Cardiol 95:116–119
56. Lusis AJ (2000) Atherosclerosis. Nature 407(6801):233–241

57. Rundek T, Arif H, Boden-Albala B et al (2008) Carotid plaque, a subclinical precursor of vascular events: the Northern Manhattan Study. Neurology 70:1200–1207
58. Christian BA, Kirzeder DJ, Boyd J et al (2009) Showered calcific emboli to the brain, the 'salted pretzel' sign, originating from the ipsilateral internal carotid artery causing acute cerebral infarction. Stroke 40(5):e319–e321
59. Mathiesen EB, Bonaa KH, Joakimsen O (2001) Echolucent plaques are associated with high risk of ischemic cerebrovascular events in carotid stenosis: the Tromso Study. Circulation 103:2171
60. Prabhakaran S, Rundek T, Ramas R (2006) Carotid plaque surface irregularity predicts ischemic stroke: the northern Manhattan study. Stroke 37:2696–2701
61. Chen JW and Wasserman BA (2005). Vulnerable plaque imaging. Neuroimaging Clin N Am 15: 609–621
62. Kwee RM, van Oostenbrugge RJ, Hofstra L et al (2008) Identifying vulnerable carotid plaques by noninvasive imaging. Neurology 70(24 Pt 2):2401–2409
63. Touze E, Toussaint JF, Coste J et al (2007) HIgh-Resolution magnetic resonance Imaging in atherosclerotic Stenosis of the Carotid artery (HIRISC) study group. Reproducibility of high-resolution MRI for the identification and the quantification of carotid atherosclerotic plaque components: consequences for prognosis studies and therapeutic trials. Stroke 38(6):1812–1819
64. Trivedi RA, Gillard JH, Kirkpatrick PJ (2008) Modern methods for imaging carotid atheroma. Br J Neurosurg 22(3):350–359
65. U-King-Im JM, Young V, Gillard JH (2009) Carotid-artery imaging in the diagnosis and management of patients at risk of stroke. Lancet Neurol 8(6):569–580
66. Waxman S, Ishibashi F, Muller JE (2006) Detection and treatment of vulnerable plaques and vulnerable patients: novel approaches to prevention of coronary events. Circulation 114:2390–2411

Biographies

Cristina Pislaru received the M.D. degree from the Carol Davila University of Medicine and Pharmacy, Bucharest, Romania. After few years of research at the Catholic University of Leuven, Leuven, Belgium, she joined the Physiology & Biomedical Engineering Department, Mayo Clinic College of Medicine, in April 1999. She currently holds the academic rank of Assistant Professor of Biomedical Engineering, Mayo Clinic College of Medicine. Her current research interests include studying cardiac mechanics, with focus on myocardial ischemia and diastolic dysfunction, development of noninvasive quantitative methods for assessment of mechanical properties of heart muscle and blood vessels, finite element modeling of fluid-structure interaction in the heart, and intracardiac ultrasound.

Birgit Kantor earned her M.D. at RWTH University Aachen in Germany and her Ph.D. at the University of Dusseldorf in Germany. She is board certified in Internal

Medicine and Cardiology and is a faculty member of the Division of Cardiovascular Diseases at Mayo Clinic, Rochester, Minnesota. She currently holds the academic rank of Associate Professor of Medicine at Mayo Clinic College of Medicine. Dr Kantor's NIH-funded research lies at the interface of vascular biology of atherosclerosis and non-invasive, cutting-edge imaging methods. Her latest NIH grant is entitled "Non-invasive Detection of Vulnerable Plaque". She is a member of a national study section, and she oversees all intramural funding mechanisms at Mayo Clinic's Center for Translational Science Activities (CTSA).

James F. Greenleaf (IEEE/M'73) received the B.S. degree in Electrical Engineering from the University of Utah, Salt Lake City, in 1964, the M.S. degree in Engineering Science from Purdue University, Lafayette, IN, in 1968, and the Ph.D. degree in Engineering Science from the Mayo Graduate School of Medicine, Rochester, MN, and Purdue University in 1970. He is currently Professor of Biomedical Engineering and Associate Professor of Medicine, Mayo Graduate School, and Consultant, Department of Physiology and Biomedical Engineering, and Internal Medicine, Division of Cardiovascular Diseases, Mayo Clinic Rochester. He has served on the IEEE Technical Committee for the Ultrasonics Symposium for 5 years. He served on the IEEE Ultrasonics, Ferroelectrics, and Frequency Control Society (UFFC-S) Subcommittee on Ultrasonics in Medicine/IEEE Measurement Guide Editors, and on the IEEE Medical Ultrasound Committee. Doctor Greenleaf was President of the UFFC-S in 1992 and 1993 and is currently Vice President for Ultrasonics. Doctor Greenleaf has 12 patents and is recipient of the 1986 J. Holmes Pioneer Award and the 1998 William J. Fry Memorial Lecture Award from the American Institute of Ultrasound in Medicine and is a Fellow of IEEE, American Institute of Ultrasound in Medicine, and American Institute for Medical and Biological Engineering. Doctor Greenleaf was the Distinguished Lecturer for IEEE Ultrasonics, Ferroelectrics, and Frequency Control Society (1990/1991) and recipient of the Rayleigh Award (2004). His special field of interest is ultrasonic biomedical science, and he has published more than 327 articles and edited or authored five books in the field.

Mostafa Fatemi received Ph.D. degree in Electrical Engineering from Purdue University. He is currently a Professor of Biomedical Engineering, Department of Physiology and Biomedical Engineering, Mayo Clinic College of Medicine, Rochester, MN. His current research interests include applications acoustic radiation force for biomedical imaging and tissue characterization. Dr. Fatemi has published extensively in the field of medical ultrasound. He is currently an associate editor of IEEE Transactions on Medical Imaging. Dr. Fatemi holds seven patents including three on vibro-acoustography techniques. Dr. Fatemi is a Fellow of Acoustical Society of America and a Fellow of American Institute of Medical and Biological Engineering and a senior member of American Institute of Ultrasound in Medicine.

Chapter 22
Metabonomics in Patients with Atherosclerotic Artery Disease

Filippo Molinari, Pierangela Giustetto, William Liboni, Franco Nessi, Michelangelo Ferri, Emanuele Ferrero, Andrea Viazzo, and Jasjit S. Suri

Abstract Atherosclerosis can be thought of as a complex process involving many aspects of the patient's life, ranging from sex, age, and genetics to lifestyle and nutrition. In the last 10 years, there has been a wide expansion of the "-omics" sciences. Such sciences have proven very effective and accurate in the analysis of complex systems, where the analysis of many factors might help a better understanding of the system itself. Among all the "-omics" sciences, metabolomic and metabonomic are gaining increasing interest. Metabonomics quantitatively measures living systems undergoing the effects of diseases. Unlike genomics and proteomics, metabonomics focuses on the multiparameter evaluation of a living complex system by studying its overall physiological profile. Metabonomics can be thought of as a multiparameter profiling technique of each individual.

We applied metabonomic techniques to the analysis of hematochemical data relative to a population of atherosclerotic patients. Being atherosclerosis a complex disease, we aimed at finding specific correlates of the atherosclerotic outcome by investigating the patients' metabolic variables. We coupled hematochemical data to instrumental variables, in order to gain a deeper comprehension of the atherosclerotic process. We considered the plaque type and the surgical treatment as factors and investigated their correlations with the variables in the database. Results showed that by using less than ten variables it is possible to cluster the patients on the basis of their respective factors.

Keywords Metabonomics • Carotid artery • Stenting • Endarterectomy • Hematochemical data • Plaque • Atherosclerosis • Principal component analysis • Discriminant analysis • Partial least squares

F. Molinari (✉)
Biolab, Department of Electronics, Politecnico di Torino, Corso Duca degli Abruzzi, 24, 10129 Torino, Italy
e-mail: filippo.molinari@polito.it

Jasjit S. Suri et al. (eds.), *Atherosclerosis Disease Management*,
DOI 10.1007/978-1-4419-7222-4_22,

22.1 Atherosclerosis and Metabonomics

Atherosclerosis is a diffuse pathological process that causes the deposition of blood-borne elements in the arterial wall [1]. Almost all territories can be affected by atherosclerosis. The atherosclerotic process that takes place in the carotid arteries has been widely studied in the last 20 years. Carotid wall lesions, in fact, have been correlated with different pathologies such as coronary and cerebrovascular diseases [2], poststroke cognitive impairments [3], platelet aggregability [4], aortic valve damage [5], and diabetes [6]. Large multicentric population and cohort studies have been devoted to the screening of carotid diseases as predictors of severe pathologies [7–10].

Atherosclerosis can be a silent and relatively long degenerative process. Many patients remain asymptomatic for long time. Arterial plaque is probably the most evident final effect of the atherosclerotic process. Cholesterol accumulation, in fact, plays a central role in atherogenesis. When there is an imbalance between cholesterol influx and efflux, the atherosclerosis process continues silently, with little or no clinical signs [1].

The atherosclerotic process can be easily monitored by in vivo procedures. The most used techniques are CT imaging [11, 12] and MRI [13, 14]. Experimental studies investigated the possibility of characterizing arterial plaques using near-infrared radiation [15] or optical coherence tomography [16]. Ultrasound techniques remain the most widely used methodology to assess plagued vessels. Plaque characterization studies were carried out using intravascular ultrasound techniques (IVUS) [17], traditional 2D scans [18], or 3D volume data [19, 20].

The guidelines for the therapeutic treatment of atherosclerotic patients were proposed by large international multicenter studies: the NASCET [21, 22] and the ECST [22]. Essentially, those studies recommend considering the patient symptoms and the stenosis degree as principal variable for deciding the surgical treatment. By following such recommendations, the side effects dropped below few percents with a very good long-term prognosis.

Despite the standardization of the protocols and the international guidelines, the management of atherosclerotic lesions remains a difficult task because of two main factors:

1. The plaque composition and nature can be very different. There are plaques that are defined as "hard", which contain several calcifications. Others can be "soft" and with a prevalent lipid content. "Mixed" plaques can be even more complicated and couple a soft to a hard part. Clearly, the plaque composition influences the therapeutic/pharmacological approach.
2. As already mentioned, some patients may remain asymptomatic for long periods. The degree of stenosis itself is insufficient to explain the possible patient symptoms in full. Therefore, it is now believed that other characteristics should be considered when approaching the atherosclerotic patient.

In fact, recent studies based on ultrasound plaque imaging have demonstrated that symptomatic plaques are more echolucent and less calcific than asymptomatic plaques and are associated to a higher degree of interplaque necrosis to histopathology [23]. Also, it has been shown that the atherosclerotic plaque composition rather than the degree of arterial stenosis can be the determinant of rupture (which is the cause of the most severe and invalidating accidents). This has led to the search for new imaging techniques that can provide information about plaque composition. Several studies demonstrated that a careful characterization of the plaque is extremely important in the management of the atherosclerotic disease [17, 24–26].

To summarize, atherosclerosis can be thought of as a complex process involving many aspects of the patient's life, ranging from sex, age, and genetics to lifestyle and nutrition. In the last 10 years, there has been a wide expansion of the "-omics" sciences. Such sciences have proven very effective and accurate in the analysis of complex systems, where the analysis of many factors might help a better understanding of the system itself. Among all the "-omics" sciences, metabolomics and metabonomics are gaining increasing interest. Metabonomics quantitatively measures living systems undergoing the effects of diseases. Unlike genomics and proteomics, metabonomics focuses on the multiparameter evaluation of a complex living system by studying its overall physiological profile. Metabonomics can be thought of as a multiparameter profiling technique of each individual. From a nomenclature point of view, the term metabolomic and metabonomic can be used interchangeably. However, the term metabonomic is used to describe multiple metabolic changes caused by a biological perturbation, whereas the term metabolomic is focused on a comprehensive metabolic profiling.

In this pilot study, we used metabonomic techniques to analyze the hematochemical data of a population of atherosclerotic patients. Being atherosclerosis a complex disease, we aimed at finding specific correlates of the atherosclerotic outcome by investigating the patients' metabolic variables. We coupled hematochemical data to instrumental variables, in order to gain a deeper comprehension of the atherosclerotic process. We considered the plaque type and the surgical treatment as factors and investigated their correlations with the variables in the database. Results showed that by using less than ten variables it is possible to cluster the patients on the basis of their respective factors.

Even if still preliminary, this study further enforces the need for a better comprehension of the atherosclerotic process and shows that traditionally considered variables (i.e. the stenosis degree and the symptoms) should be coupled to a metabolic profiling of the subject in order to further improve therapy.

22.2 Database and Subject Population Description

In this section, we describe the database we considered and the patient population.

22.2.1 Patient Population

In this study, we enrolled 40 consecutive patients of the Vascular and Endovascular Surgery Unit of the "Umberto I" Hospital of Torino, Italy. All the subjects were

Table 22.1 Treatment types of the 40 patients

Treatment	Number
CEA	3
STENT	17
TEA	9
No treatment	11

CEA carotid endarterectomy, *TEA* thromboendoarterectomy

Table 22.2 Population-related variables

Variable	Unit
Age	years
Treatment type	a.u.
Symptomatic/asymptomatic	Boolean 1/0
Hypertension	Boolean 1/0

instructed about the experiment before being enrolled in the study. All signed an informed consent. This study received the approval by the Ethical Committee of the "Umberto I" Hospital of Torino, Italy.

Eleven subjects were females. The age of the subjects was equal to 73.6±7.9 years, ranging from 52–89. All the subjects were referred to the Vascular and Endovascular Surgery Unit for possible surgical treatment of an extracranic carotid plaque. Eleven patients were symptomatic, of whom five were females. Of the symptomatic patients, two patients were aphonic (one female), two had a small intracranial hemorrhage (one female), four had a transient ischemic attack (TIA) (two females), one presented iposthenia of the right upper limb, one had temporary blindness (female), and one an occlusive stroke with global amnesia.

Table 22.1 reports the treatment type which the subjects underwent to. Three patients underwent carotid endarterectomy (CEA), seventeen were stented, nine were treated by thromboendoarterectomy, and eleven were not surgically treated. This variable was inserted into the database after numerical coding. We used the following code: 1=CEA; 2=STENT; 3=TEA; and 4=No treatment. The variables that are numerically coded have the indication (a.u.) as measurement unit.

Table 22.2 summarizes the population-related variables we considered in our database.

22.2.2 Instrumental Data

All the patients underwent the following diagnostic protocol:

- Echo Color Doppler ultrasound examination of the supra-aortic vessels
- CT angiography
- Contrast MRI plaque imaging

CT and MRI imaging data were used in the presurgical planning. Ultrasound data were more quantitative and were inserted into the database. According to the

Table 22.3 Instrumental variables

Instrumental variable	Unit
Stenosis	%
Controlateral stenosis	%
PSV	cm/s
Plaque type	a.u.

NASCET criterion, we measured the percentage of stenosis of the affected and of the controlateral carotid. Also, we inserted the peak systolic velocity (PSV) into the database. Average values for the percentage of stenosis were 78.1 ± 9.7%, range 55–90%. Average values for the PSV were 274.4 ± 129.7 cm/s, range 70–520 cm/s.

On the basis of their echographic appearance, plaques were assigned one of the following five classes:

1. Calcified
2. Fibro-calcified
3. Fibrous
4. Fibro-soft
5. Ulcerated

The surgically removed plaques were sent to histology and the echographic scoring was validated. We did not find any discrepancy between the ultrasound-based plaque assignment and histology. The plaque type was inserted into the database.

Table 22.3 reports the instrumental variables we considered.

22.2.3 Hematochemical Variables

The same laboratory made all the hematological analyses. The subjects underwent blood analysis 2 days before the decision for surgical treatment. As our purpose was to develop a methodology based on data that are usually acquired during clinical protocols, we kept the same list of blood analysis that is routinely used in the Surgery Unit (where all the patients were treated). Table 22.4 reports the complete list of the 39 variables we considered.

22.3 Analysis Architecture: Combination of ANOVA and Metabonomic Techniques

In this section, we introduce the statistical and metabonomic techniques we applied.

Our analysis strategy was a combination of correlation analysis, ANOVA, and unsupervised and supervised classification techniques. The analysis technique consisted of the following steps:

1. Correlation analysis: the elimination of collinearity among variables is essential when working on large data sets in order to avoid system overfitting. In pilot

Table 22.4 Hematochemical variables

Hematochemical variable	Unit
Glucose	mg/dl
Creatinine	mg/dl
Sodium	mEq/l
Potassium	mEq/l
Chloride	mEq/l
Glutamic oxaloacetic transaminase (GOT or AST)	U/l
Alanine aminotransferase (ALT)	U/l
Gamma-glutamyl transferase (GGT)	U/l
Alkaline phosphatase	U/l
Cholinesterase	U/l
Cholinesterase inhibited	U/l
Dibucaine number	%
Lactate dehydrogenase (LDH)	U/l
Creatinine kinase (CK)	U/l
Muscle-brain creatinine kinase (CK_MB)	%
Prothrombin	%
Prothrombin international normalized ratio (INR)	
Partial thromboplastine time (PPT)	s
PPT/prothrombin ratio	
Fibrinogen	mg/dl
Antithrombin III	%
D-dimer	μg/ml
White blood cells (WBC)	$\times 10^3$ c/μl
Neutrophils	%
Lymphoctyes	%
Monocytes	%
Eosinophils	%
Basophils	%
Large uncolored cells (LUC)	%
Red blood cells (RBC)	$\times 10^6$ c/μl
Hemoglobin (HGB)	g/dl
Hematocrit (HCT)	%
Mean corpuscular volume (MCV)	fl
Mean corpuscular hemoglobin (MCH)	pg
Mean corpuscular hemoglobin concentration (MCHC)	g/dl
Red cell distribution width (RDW)	%
Hemoglobin distribution width (HDW)	g/dl
Platelets	$\times 10^3$ c/μl
Mean platelet volume (MPV)	fl

tests (results not reported in this chapter), we found that the presence of collinear variables in the database produced overperformance of the final classifiers. Therefore, we first eliminated from the database all the variables with a certain degree of linear correlation.

2. Multiway ANOVA: the ANOVA analysis evidenced the variables correlated with our dependent variable. We chose the stenosis degree as dependent variables and selected, among the remaining variables, the ones that explained the higher amount of variance. Only the variables that were most significant for the stenosis degree were used in the subsequent supervised and unsupervised analyses.
3. Combination of unsupervised and supervised analysis techniques: this step enabled the clustering of our subjects on the basis of their plaque typology or on the basis of the surgical intervention they underwent.

The first step of our analysis was the correlation analysis of the variables to avoid collinearity. Then, we used ANOVA to detect the most significant variables that correlated to the dependent variables. Jointly, these first two steps helped us in reducing the problem complexity and representation. Subsequent unsupervised and supervised techniques were conducted on a reduced set of variables, in order to better detail the inner correlation structure of the data set.

Finally, to build the classifier, the raw data set was split into two parts. This procedure is known as cross-validation. We decided to use it in order to avoid possible overfitting of the data set given by the relatively small number of patients.

All the statistical analyses were carried out by using StatGraphics Centurion XV.

In the following, we briefly explain the analysis techniques we used in this research.

22.3.1 Correlation Analysis

The raw data set was a matrix having as rows (statistical units) 40 patients and as columns (variables) the 47 values relative to the hematochemical (39), instrumental (4), and population-related quantities (4).

The entire database was first analyzed in order to detect collinearity among the variables. When correlated variables were found, we deleted one of the two. The criteria for the selection of which variable to eliminate in the pair of correlated variables were the following:

- If one variable was particularly critical or expensive to measure, it was eliminated. Also, we eliminated the variable that could be affected by the higher measurement noise. This was done in order to keep the most reliable and easy variables to measure. As an example, we found correlation between Potassium and the Dibucaine number (correlation coefficient $=0.60$ and $P=10^{-7}$). Because potassium can be easily measured by any hematochemical laboratory and it is included in almost all the hematologic tests that can be prescribed to an atherosclerosis sufferer, we eliminated the Dibucaine number.
- If it was reasonable to state a relationship of cause–effect between two variables, we eliminated the variable that acted as effect. As an example, we found a strong correlation (correlation coefficient $=0.62$ and $P=10^{-5}$) between the degree of

stenosis and the peak systolic value. This correlation was clearly expected as the lumen stenosis perturbs the velocity regimen of the vessel. In this case, it is obvious that stenosis is the cause and the PSV is the effect; therefore, the PSV was eliminated.

All the choices we made were a posteriori validated. We eliminated some collinear variables and used the remaining for the ANOVA analysis, the unsupervised and supervised analyses, and the discriminant analysis (DA). We evaluated performance and then repeated the same procedure but with the dual variables (i.e. by considering the variables we originally eliminated). Results showed that the discriminant and clustering performance slightly decreased.

22.3.2 Multidimensional ANOVA

The remaining variables were analyzed by using ANOVA, in order to extract the correlation structure that constituted the inner texture of the data set. The degree of stenosis was considered as the dependent variable. We considered the plaque type and the treatment type as principal factors and all the other variables (those that were not eliminated by the correlation analysis) as covariables.

Prior of performing unsupervised and supervised clustering, we performed a multivariate analysis of the raw data. Specifically, we aimed at evidencing the variables that are linked to plaque type and treatment type in atherosclerosis. Multidimensional ANOVA tests were performed in order to extract the variables that influenced the factors. This lead to a reduced set of variables that could be easily and effectively processed by metabonomic techniques.

22.3.3 Principal Component Analysis

Principal component analysis (PCA) is an unsupervised technique that effectively represents the information embedded in multidimensional data sets [23]. The raw data are represented in a transformed domain with lower dimensionality. The correlation structure of the raw data generates few latent variables (or principal components – PCs). PCs form an orthogonal basis and are sorted in order of decreasing explained data variance. Each PC can be expressed as $\mathrm{PC} = aX_1 + bX_2 + cX_3 + \cdots$, where X_1, X_2, X_3, ... represent the measured features and a, b, c, ... represent numerical weights. Each statistical unit is assigned a score relative to each extracted component, whereas the correlation coefficients between the original variables and the extracted components (loadings) furnish the significance of the specific PC.

PCA is very effective in representing the inner structure of the data set. In this study, it was used to represent the subjects with respect to the plaque type.

22.3.4 Partial Least Squares

Partial least square (PLS) is a supervised technique that aims at finding the better linear combination explaining experimental data. With reference to our study, we can define $\mathbf{X}$ as the matrix containing the measured variables and $\mathbf{Y}$ the matrix of the intervention type. Hence, $\mathbf{X}$ contains the independent and $\mathbf{Y}$ the dependent variables. The goal of PLS is to find the linear combination of $\mathbf{X}$ that better explains $\mathbf{Y}$. The initial linear model can be expressed as:

$$\mathbf{Y} = \mathbf{XB} + \mathbf{E} \tag{22.1}$$

where $\mathbf{B}$ is the regression coefficient matrix and **E** is the error matrix (same dimensions as **Y**). Given the difference in the numerical values of our data, we centered **X** and **Y** by subtracting their mean value and by normalizing with respect to their standard deviations.

In the presence of many collinear variables, PCA may fail in properly extracting the covariance structure between the predictors. Conversely, PLS only extracts the covariance structure between the predictors and the response variables, thus leading to a more reliable system description. The PLS defines a matrix **T**, so that $\mathbf{T}=\mathbf{XW}$; where **W** is an appropriate weight matrix and **T** is the factor score matrix. Then, the considered linear regression model becomes $\mathbf{Y}=\mathbf{TQ}+\mathbf{E}$; where **Q** is the matrix of regression coefficients (loadings) for **T**, **E** is the error matrix. Once **Q** is calculated, the overall system is equivalent to the one in (22.1), where $\mathbf{B}=\mathbf{WQ}$. This can now be used as a linear predictive regression model. Roughly, PLS can be thought as a PCA applied to the **X** matrix, a PCA applied to the **Y** matrix and the correlation analysis of the two sets of obtained PCs.

Using the standard NIPALS algorithm, we performed all PLS numerical computations.

22.3.5 Discriminant Analysis

Discriminant function analysis (DA) is used to determine the variables that better discriminate between two or more occurring groups in the sample population. In our study, we considered the following groups:

- If the dependent variable was the plaque type, we had five different groups, each one corresponding to a specific plaque type (i.e. calcified, fibro-calcified, fibrous, fibro-soft, and ulcerated).
- If the dependent variable was the intervention type, we had four different groups, each one related to the surgical treatment (i.e. CEA, stenting, TEA, and no intervention).

Computationally, DA determines a set of weight multiplying the **X** variables, so that the assignment error of each statistical unit to the correct **Y** class is minimized.

We used PLS–DA to build a supervised model that could predict plaque type or the intervention type on the basis of patients' instrumental and hematochemical data.

To build the classifier, the raw data set was split into two parts. Cross-validation was carried out it in order to avoid possible over fitting of the data set given by the relatively small number of patients. We randomly selected 30 patients to build the classifier and we used the remaining 10 subjects for validation. We tested 50 different possible combinations of 30 (classifier set) and 10 (validation set) subjects, with the condition that in the validation subset there was at least one subject for each considered group (the group referred either to a plaque type or to an intervention type). We enforced this condition as a preliminary study (results not reported in this paper) revealed that the classifier did not perform correctly when built on a heterogeneous subset and validated on a homogeneous one. Therefore, we only considered the random combinations leading to subsets in which all the classes were present.

22.4 Database Reduction

Among the variables (hematochemical data), we started removing all the variables with a correlation coefficient greater than 0.6. However, subsequent validations revealed that the system was still affected by the presence of collinear variables. Therefore, we iteratively eliminated all the variables with a correlation coefficient higher than 0.5 and, then, 0.4. We found that 0.4 was the optimal choice for our data set. This threshold was then confirmed by the final DA.

- The variables that were eliminated when considering a correlation coefficient greater than 0.6 were PSV, chloride, cholinesterase inhibited, dibucaine number, lymphocytes, PTT, Hgb, and HCT. It is important to note that the PSV was eliminated as it was collinear to the percentage of stenosis. This is not surprising, since, as already pointed out in Sect. 22.3.1, clearly the degree of stenosis influence the velocity regimen into the vessel. Chloride was eliminated since collinear to sodium, dibucaine number was collinear to potassium, cholinesterase inhibited to prothrombin, lymphocytes to neutrophils, and HGB and HCT to LDH.
- The variables that were eliminated while considering a correlation coefficient greater than 0.5 were ALT, alkaline phosphatase, fibrinogen, and MCHC as they were collinear to age; potassium as it was collinear to eosinophils; RBC as it was collinear to LDH; INR as it was collinear to PTT; and neutrophils as they were collinear to LUC.
- When considering a threshold of 0.4, the following variables were further eliminated: eosinophils and basophils since collinear to age; AST since collinear to CK; GTT since collinear to antithrombin III; creatinine, HDW, and monocytes since collinear to PTT; MCV since collinear to platelets; MCH since collinear to RDW.

As explained, the variable elimination was gradual. We started by eliminating the variables with a collinearity degree greater than 0.6 and then observed the correlation structure of the remaining data set. We found that, with a threshold of 0.4, our data set was made of uncorrelated variables.

Overall, the correlation analysis eliminated 25 variables. The database was thus reduced to 22 variables. We performed multiway ANOVA on this reduced data set. Table 22.5 reports the results of the ANOVA analysis, showing that only seven variables proved statistically significant for the dependent variable. Such variables were as follows:

1. Hypertension
2. The presence of symptoms
3. LDH
4. D-dimer
5. PTT
6. RDW
7. Platelets

The ANOVA results were extremely interesting. First of all, hypertension resulted as a significant factor for our sample population. This confirms the importance of hypertension as risk factor for atherosclerosis. Also, the result reveals how hypertension

Table 22.5 ANOVA results on the reduced data set of 22 variables

Covariables	*F* ratio	*P*-value
Age	2.25	0.1575
Hypertension	7.04	0.0199
Symptomatic/asymptomatic	6.32	0.0259
Controlateral stenosis	1.69	0.2162
Glucose	0.04	0.8413
Sodium	0.51	0.4880
Cholinesterase	1.03	0.3291
LDH	7.02	0.0200
CK_MB	2.51	0.1369
D-Dimer	5.77	0.0319
PTT	17.84	0.0010
Antithrombin III	1.47	0.2470
Prothrombin	4.40	0.0561
WBC	5.45	0.0562
Neutrophils	0.00	0.9693
RDW	16.69	0.0013
MPV	4.58	0.0518
Platelets	8.34	0.0127
CK	2.22	0.1602
Principal effects		
Plaque type	5.11	0.0107
Treatment type	10.28	0.0010

The stenosis degree was considered as the dependent variable, whereas the plaque type and the intervention type were the factors. Significance level was 95%

might have an influence on the degree of stenosis, which, in turn, might influence the treatment and the plaque type.

Second, the symptoms of the patients are crucial. In fact, symptomatic patients might follow a completely different therapeutic trial with respect to asymptomatic. The presence of symptoms remained a critical variable in our data set.

LDH is related to the amount of tissue turnover and hemolysis. Also, LDH levels are correlated to exudates and transudates, thus to the amount of passively secreted fluids. Therefore, the importance of this metabolite in atherosclerosis might be linked to the increased tendency of elements deposition in the arterial wall.

D-dimer and PTT are linked to the hemolytic process. Therefore, such variables impact on thrombolysis and might be helpful in directing the cause of onset atherosclerotic symptoms.

RDW and platelets, finally, are hematological quantities providing a snapshot of the blood characteristics of the patient. Again, platelets are linked to coagulations and, therefore, to the risk of thrombosis and/or vascular occlusions.

After ANOVA analysis, the data set was reduced to seven hematochemical variables, two factors (plaque type and treatment type), and an independent variable (the degree of stenosis). This further reduced data set was processed by PCA, PLS, and PLS–DA for obtaining patient clustering and discrimination.

22.5 Analysis of the Patients in Function of the Surgical Treatment

PCA was performed on a data set consisting of the 40 subjects (rows) and the following eight observations: stenosis degree, hypertension, symptoms, PTT, D-dimer, LDH, RDW, platelets. All the variables were standardized. We extracted four PCs; their combination explained 71.1% of the total variance of the data. Figure 22.1 reports the graph of the seven eigenvalues. The eigenvalues are proportional to the percent of variability in the data they explain. We chose the eigenvalues greater than 1 (horizontal line in Fig. 22.1), since eigenvalues lower than 1 explain the same variance of the original variables, so they are not useful for data analysis in a reduced domain.

Table 22.6 reports the relative weights of the original variables in the four components. The symptoms have a high weight in the first component, the stenosis percentage in the second, the D-dimer in the second, and the RDW in the fourth. Also, hypertension, LDH, D-dimer, and RDW have negative weights in the second, third, fourth, and first components, respectively.

Figure 22.2 depicts the subjects' distribution on the hyperplane formed by components 1 and 2. Red circles represent the subjects who underwent stenting, green circles TEA, yellow circles CEA, and blue circles the absence of surgical intervention. The black continuous lines represent the projection of the original variables on the hyperplanes. The first component separates the subjects who were stented from those who did not have any intervention. The second component is discriminant on

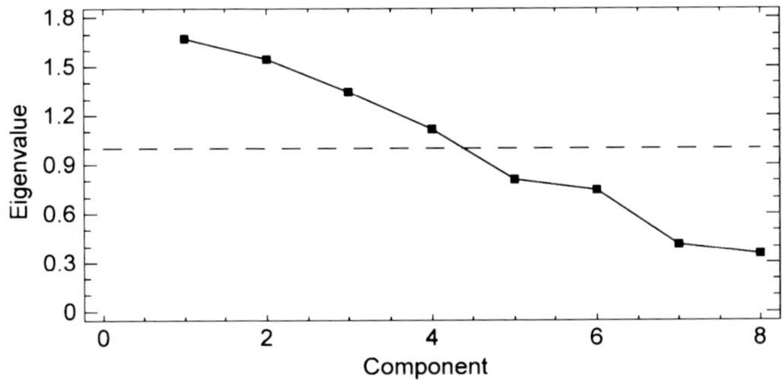

Fig. 22.1 Eigenvalues of the PCA applied to the data set consisting of eight variables (stenosis degree, hypertension, symptoms, PTT, D-dimer, LDH, RDW, and platelets). The first four eigenvalues are higher than 1 (*horizontal dashed line*). Therefore, these four eigenvalues were used to represent the subjects in a reduced domain of four principal components

Table 22.6 PCA components and original variables considering the treatment type as dependent variable

Variable	Component 1	Component 2	Component 3	Component 4
Hypertension	0.29	−0.41	0.50	0.17
Symptoms	0.48	−0.15	−0.24	−0.23
Stenosis percentage	0.35	0.46	0.25	0.43
LDH	0.28	0.16	−0.57	0.43
PTT	−0.38	−0.35	−0.47	0.11
D-dimer	−0.30	0.51	0.12	−0.44
RDW	−0.45	0.21	0.08	0.54
Platelets	0.25	0.39	−0.26	−0.21

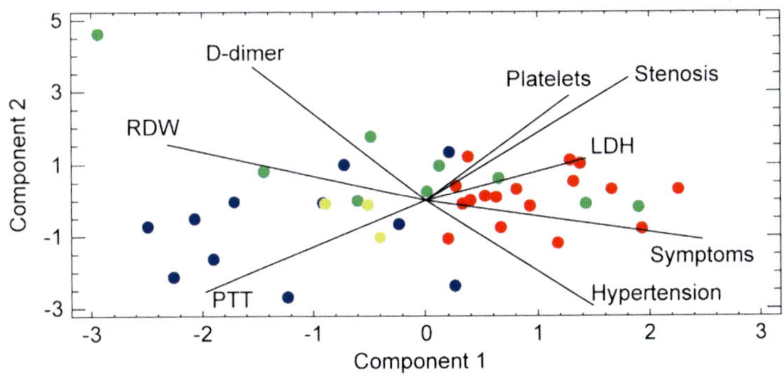

Fig. 22.2 PCA representation of the subjects in the hyperplane of PC1 and PC2. *Yellow circles*=CEA; *Green circles*=TEA; *Red circles*=stenting; and *Blue circles*=no treatment

the TEA versus the CEA interventions. The length of the black line for each original variable is proportional to the importance of that variable for the PC. Hence, it can be noticed how the stenosis degree and the symptoms are correlated to the subjects

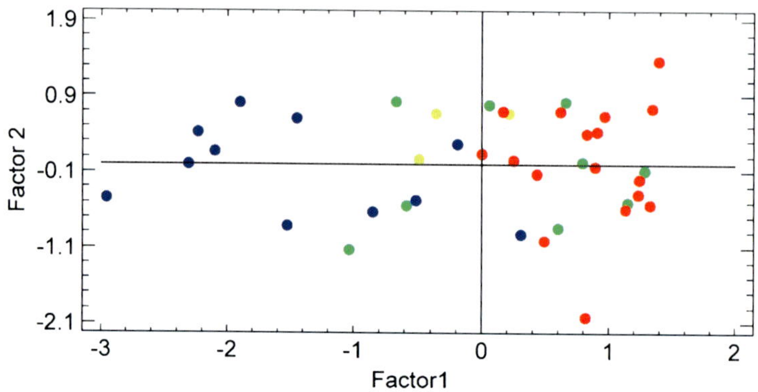

Fig. 22.3 Sample PLS representation of the subjects in the hyperplane defined by Factor 1 and Factor 2. *Yellow circles* = CEA; *Green circles* = TEA; *Red circles* = stenting; and *Blue circles* = no treatment

that were stented, whereas the RDW and the PTT to those who did not receive surgical treatment.

The analysis of the subjects with respect to surgical intervention was refined by using PLS as supervised approach. Independent variables were same as PCA, the dependent variable was the surgical treatment. Figure 22.3 reports the PLS discrimination of the subjects in the hyperplane (factor 1, factor 2). Factor 1, in particular, is discriminant for the subjects who underwent stenting against those who did not received surgical treatment. In particular, the percent of stenosis is higher in the subjects who underwent stenting compared with those who did not receive surgical treatment (t test, $P<0.001$), and the symptomatic patients are also prevalent among those who were stented (t test, $P<0.002$).

The supervised PLS–DA classifier was built by coding the dummy variable **Y** by a number (1 = CEA; 2 = STENT; 3 = TEA; and 4 = No treatment). To build the PLS–DA classifier, we relied on the previously performed ANOVA analysis and removed all the variables with a correlation coefficient higher than 0.4. Figure 22.4 shows a sample of the discrimination of the classifier performed by the first two functions. The black "+" (plus sign) marks the centroids of the clusters. In this specific case.

The PLS–DA classifier was then used to test the 50 trials made up of 40 subjects (corpus set) and 10 subjects (validation set). The corpus and validation sets were randomly selected. We tested 50 combinations, which led to corpus and validation sets containing at least one subject for each group. Hence, we classified 500 subjects, 50 of whom underwent CEA; 100, TEA; 220, stenting; and 130 received no treatment. Each element of the validation set was assigned to a group on the basis of its distance from the centroids. For all the considered classifiers, the discriminant function P-value was always lower than 0.01. The classification performance was as follows:

- 84.6% sensitivity, 94.4% specificity, and 95.1% efficiency for the stenting treatment
- 85.7% sensitivity, 92.7% specificity, and 91.5% efficiency for the TEA treatment

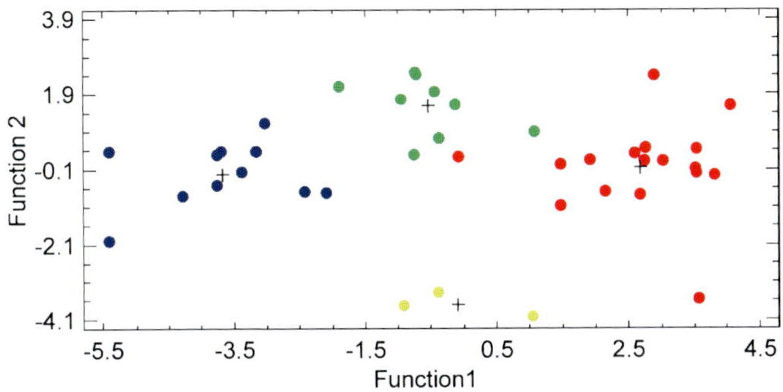

Fig. 22.4 PLS–DA clustering of the subjects in the hyperplane formed by the first two discriminant functions. *Yellow circles* = CEA; *Green circles* = TEA; *Red circles* = stenting; and *Blue circles* = no treatment

- 63.6% sensitivity, 99.2% specificity, and 96.2% efficiency for the CEA treatment
- 92.5% sensitivity, 90.0% specificity, and 90.8% efficiency for the subjects who did not receive surgical treatment

22.6 Patients and Plaque Type

We repeated the same analysis protocol of Sects. 22.3 and 22.4 to the plaque clustering. We carried out the same analysis reported by Sect. 22.5, changing the dependent variable to the plaque typology. We performed PCA, then PLS, and finally, PLS–DA to test how much the variables explained the plaque typology that affected each subject (Figs. 22.5–22.7).

22.7 Discussions and Future Perspectives

The application of metabonomic techniques to the analysis of the metabolic profile of atherosclerosis sufferers lead to very promising results. In the following, we discuss the major results of our pilot study.

- We showed that it is possible to classify a patient on the basis of the metabolic pattern. Classification can be done either with respect to the plaque type or the treatment the patient was subjected to. This result is very important since it demonstrates the existence of a strong correlation between the metabolic subjects' profile, their atherosclerosis expression, and the surgical treatment. In a future perspective, this study could be widened by considering a larger database, incorporating more instrumental variables and maybe genetic data. A further investigation

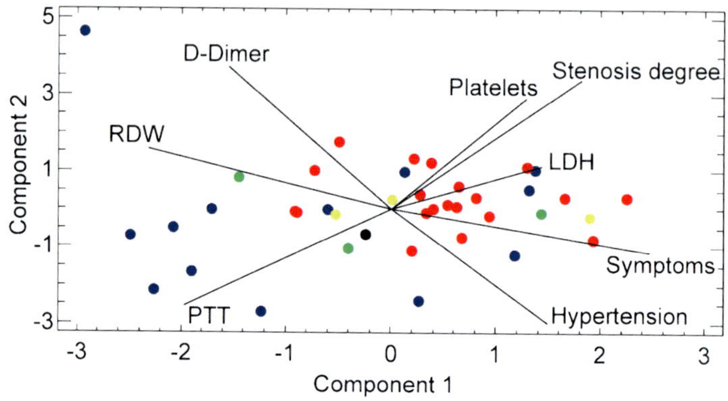

Fig. 22.5 PCA representation of the subjects in the hyperplane of PC1 and PC2. *Yellow circles* = CEA; *Green circles* = TEA; *Red circles* = stenting; and *Blue circles* = no treatment

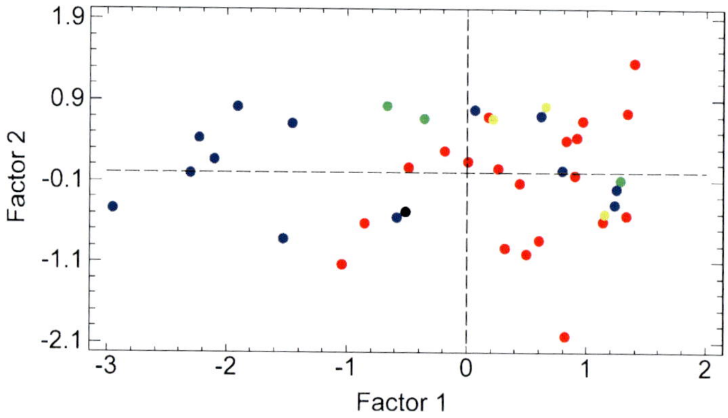

Fig. 22.6 PLS representation of the subjects in the hyperplane of the first two factors. *Yellow circles* = CEA; *Green circles* = TEA; *Red circles* = stenting; and *Blue circles* = no treatment

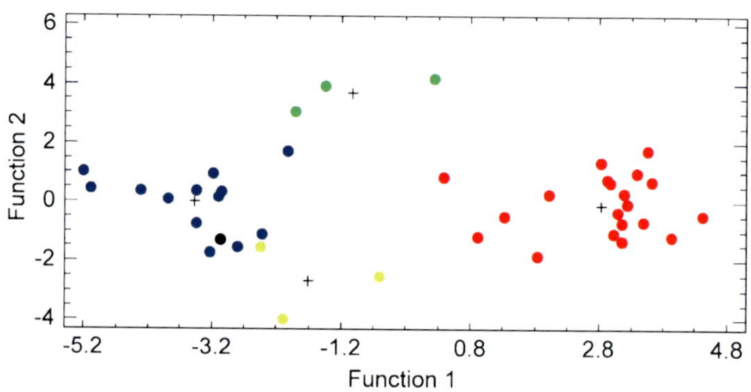

Fig. 22.7 PLS–DA representation of the subjects in the hyperplane of the first two discriminant functions. *Yellow circles* = CEA; *Green circles* = TEA; *Red circles* = stenting; and *Blue circles* = no treatment

of the existing relationships between the atherosclerotic process outcome and the patients profile could be very useful for gaining a better comprehension of the vascular degeneration of each specific patient.

- To the best of our knowledge, this study is the first incorporating metabolic, instrumental, and personal data of atherosclerosis patients for classification purposes. Despite the small number of subjects tested (40) and of the considered variables (47), this study showed that clear correlations between different variables exist. Therefore, one of the noteworthy results of this study was the evidencing that atherosclerosis is better described by a holistic approach, rather than by a reductionist one.
- Correlation analysis and ANOVA showed that the presence of symptoms and of hypertension is correlated to the patients' therapeutic line. Presently, international accepted guidelines consider the presence of symptoms and the degree of stenosis as the unique criteria for the therapeutic indication. Therefore, our results are in accordance to these international guidelines, but they also point out that symptoms and stenosis alone might be insufficient to a full comprehension of the atherosclerotic process.
- The metabonomic profiling of atherosclerotic patients could be very useful in assessing the individual aspects of the disease evolution. A significant scope of improvement of this work could be the development of evolutionary models. Such models should consider a metabolic snapshot of the patient to derive possible evolutionary scenarios. This could be very important for both the monitoring of the patients and the assessment of the efficacy of drug therapy. Also, an evolutionary model could be used for the definition of personalized therapeutic protocols.
- From a methodological point of view, we showed that, by properly combining statistical, supervised, and unsupervised techniques, it is also possible to describe and analyze complex and quite big databases consisting of inhomogeneous variables (metabolic and instrumental data).

This pilot study adapting metabonomic techniques to the profiling of pathologic subjects could be a basis for the development of a clinically applicable methodology devoted to the profiling, classification, and evolutionary modeling in atherosclerosis.

References

1. J.J. Badimon, B. Ibanez, and G. Cimmino, Genesis and dynamics of atherosclerotic lesions: implications for early detection, Cerebrovasc Dis, 27(Suppl 1), (2009), 38–47.
2. S.H. Johnsen and E.B. Mathiesen, Carotid plaque compared with intima-media thickness as a predictor of coronary and cerebrovascular disease, Curr Cardiol Rep, 11(1), (2009), 21–7.
3. Y.H. Lee and S.J. Yeh, Correlation of common carotid artery intima media thickness, intracranial arterial stenosis and post-stroke cognitive impairment, Acta Neurol Taiwan, 16(4), (2007), 207–13.
4. Y. Fusegawa, H. Hashizume, T. Okumura, Y. Deguchi, Y. Shina, Y. Ikari, and T. Tanabe, Hypertensive patients with carotid artery plaque exhibit increased platelet aggregability, Thromb Res, 117(6), (2006), 615–22.

5. M. Soylu, A.D. Demir, O. Ozdemir, Y. Uzun, and S. Korkmaz, Relationship between plaque morphology of carotid artery and aortic valve calcification, Angiology, 54(6), (2003), 637–40.
6. L.A. Lange, D.W. Bowden, C.D. Langefeld, L.E. Wagenknecht, J.J. Carr, S.S. Rich, W.A. Riley, and B.I. Freedman, Heritability of carotid artery intima-medial thickness in type 2 diabetes, Stroke, 33(7), (2002), 1876–81.
7. P.J. Touboul, R. Hernandez-Hernandez, S. Kucukoglu, K.S. Woo, E. Vicaut, J. Labreuche, C. Migom, H. Silva, and R. Vinueza, Carotid artery intima media thickness, plaque and Framingham cardiovascular score in Asia, Africa/Middle East and Latin America: the PARC-AALA study, Int J Cardiovasc Imaging, 23(5), (2007), 557–67.
8. A.H. Thakore, C.Y. Guo, M.G. Larson, D. Corey, T.J. Wang, R.S. Vasan, R.B. D'Agostino, Sr., I. Lipinska, J.F. Keaney, Jr., E.J. Benjamin, and C.J. O'Donnell, Association of multiple inflammatory markers with carotid intimal medial thickness and stenosis (from the Framingham Heart Study), Am J Cardiol, 99(11), (2007), 1598–602.
9. R. Elosua, L.A. Cupples, C.S. Fox, J.F. Polak, R.A. D'Agostino, Sr., P.A. Wolf, C.J. O'Donnell, and J.M. Ordovas, Association between well-characterized lipoprotein-related genetic variants and carotid intimal medial thickness and stenosis: The Framingham Heart Study, Atherosclerosis, 189(1), (2006), 222–8.
10. P.J. Touboul, J. Labreuche, E. Vicaut, and P. Amarenco, Carotid intima-media thickness, plaques, and Framingham risk score as independent determinants of stroke risk, Stroke, 36(8), (2005), 1741–5.
11. J.A. Goldstein, Coronary plaque characterization by computed tomographic angiography: present promise and future hope, JACC Cardiovasc Imaging, 2(2), (2009), 161–3.
12. S. Motoyama, T. Kondo, H. Anno, A. Sugiura, Y. Ito, K. Mori, J. Ishii, T. Sato, K. Inoue, M. Sarai, H. Hishida, and J. Narula, Atherosclerotic plaque characterization by 0.5-mm-slice multislice computed tomographic imaging, Circ J, 71(3), (2007), 363–6.
13. R.M. Botnar, Coronary plaque characterization by T(1)-weighted cardiac magnetic resonance, JACC Cardiovasc Imaging, 2(6), (2009), 729–30.
14. B. Sun, D.P. Giddens, R. Long, Jr., W.R. Taylor, D. Weiss, G. Joseph, D. Vega, and J.N. Oshinski, Automatic plaque characterization employing quantitative and multicontrast MRI, Magn Reson Med, 59(1), (2008), 174–80.
15. S. Waxman, Near-infrared spectroscopy for plaque characterization, J Interv Cardiol, 21(6), (2008), 452–8.
16. I.K. Jang, G.J. Tearney, B. MacNeill, M. Takano, F. Moselewski, N. Iftima, M. Shishkov, S. Houser, H.T. Aretz, E.F. Halpern, and B.E. Bouma, In vivo characterization of coronary atherosclerotic plaque by use of optical coherence tomography, Circulation, 111(12), (2005), 1551–5.
17. J. Qian, A. Maehara, G.S. Mintz, M.P. Margolis, A. Lerman, J. Rogers, S. Banai, S. Kazziha, C. Castellanos, L. Dani, M. Fahy, G.W. Stone, and M.B. Leon, Impact of gender and age on in vivo virtual histology-intravascular ultrasound imaging plaque characterization (from the global Virtual Histology Intravascular Ultrasound [VH-IVUS] registry), Am J Cardiol, 103(9), (2009), 1210–4.
18. L.A. Baroncini, A. Pazin Filho, L.O. Murta Junior, A.R. Martins, S.G. Ramos, J. Cherri, and C.E. Piccinato, Ultrasonic tissue characterization of vulnerable carotid plaque: correlation between videodensitometric method and histological examination, Cardiovasc Ultrasound, 4, (2006), 32.
19. B. Chiu, M. Egger, J.D. Spence, G. Parraga, and A. Fenster, Quantification of carotid vessel wall and plaque thickness change using 3D ultrasound images, Med Phys, 35(8), (2008), 3691–710.
20. M. Egger, J.D. Spence, A. Fenster, and G. Parraga, Validation of 3D ultrasound vessel wall volume: an imaging phenotype of carotid atherosclerosis, Ultrasound Med Biol, 33(6), (2007), 905–14.
21. M. Fisher, A. Martin, M. Cosgrove, and J.W. Norris, The NASCET-ACAS plaque project North American Symptomatic Carotid Endarterectomy Trial Asymptomatic Carotid Atherosclerosis Study, Stroke, 24(12 Suppl), (1993), I24–5; discussion I31–2.

22. L. Saba and G. Mallarini, A comparison between NASCET and ECST methods in the study of carotids Evaluation using Multi-Detector-Row CT angiography, Eur J Radiol, (2009), (epub ahead of print).
23. J.K. Grogan, W.E. Shaalan, H. Cheng, B. Gewertz, T. Desai, G. Schwarze, S. Glagov, L. Lozanski, A. Griffin, M. Castilla, and H.S. Bassiouny, B-mode ultrasonographic characterization of carotid atherosclerotic plaques in symptomatic and asymptomatic patients, J Vasc Surg, 42(3), (2005), 435–41.
24. B.G. Rubin, Impact of plaque characterization on carotid interventions, Perspect Vasc Surg Endovasc Ther, 18(4), (2006), 312–5.
25. P. Schoenhagen, M. Barreto, and S.S. Halliburton, Quantitative plaque characterization with coronary CT angiography (CTA): current challenges and future application in atherosclerosis trials and clinical risk assessment, Int J Cardiovasc Imaging, 24(3), (2008), 313–6.
26. T.T. de Weert, M. Ouhlous, P.E. Zondervan, J.M. Hendriks, D.W. Dippel, M.R. van Sambeek, and A. van der Lugt, In vitro characterization of atherosclerotic carotid plaque with multidetector computed tomography and histopathological correlation, Eur Radiol, 15(9), (2005), 1906–14.

Biographies

Dr. Filippo Molinari received the Italian Laurea and the Ph.D. in electrical engineering from the Politecnico di Torino, Torino, Italy, in 1997 and 2000, respectively. He is leader in ultrasound imaging focused towards tissue characterization, vascular quantification for diagnostics and therapeutics. Currently, he is Assistant Professor at Politecnico di Torino, Italy – Department of Electronics.

Dr. Pierangela Giustetto received the Italian Laurea in Radiology techniques for medical images and radiotherapy from University di Torino, Italy, in 1987. In 2006, she received the Italian first level Laurea in Physics from the University of Torino, Italy. She is now completing her M.S. degree in Physics working in the field of advanced ultrasound technologies and interactions with human tissues and nanomolecules.

Dr. William Liboni received the Italian Laurea in Medicine in 1969 from the Università degli Studi di Torino and then specialized in radiology and nuclear medicine in 1975 and 1993, respectively. Since 1994 he directed the Neurology Division at the Gradenigo Hospital of Torino, Italy. His research focuses on the functional assessment of neurologically impaired subjects and on the non-invasive monitoring of chronic pathologies.

Franco Nessi, graduated in medicine and surgery from Turin University, Italy. Received the specialization in General Surgery from Turin University, Italy and Vascular Surgery from Pavia University, Italy. He directed the division of Vascular Surgery and Kidney Transplantation at Hospital "Maggiore della Carità" of Novara, Italy, from 1997 to 2003. From 2003 is the director of Vascular and Endovascular Surgery Hospital Mauriziano Umberto I of Turin, Italy. He is lecturer at the School of Specialisation in Vascular Surgery, University of Turin, Italy. He worked for major international trials: ECST, ILAILL, GALA TRIAL. Now he is a collaborator for the ACST-2 trial.

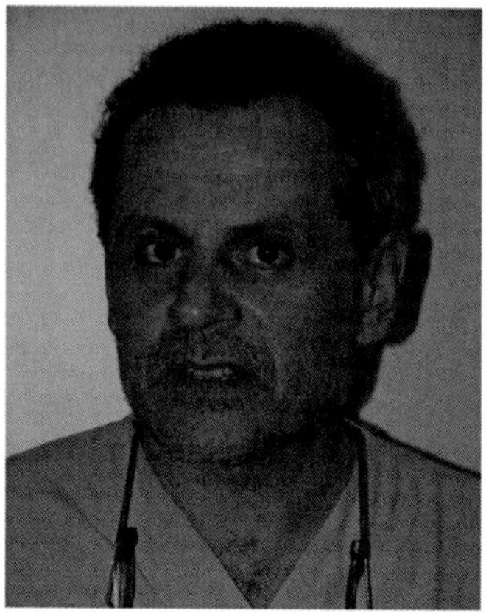

Dr. Michelangelo Ferri, MD, is specialist in Vascular Surgery and currently consultant at the Vascular Surgery Division of the Hospital Mauriziano "Umberto I" of Torino, Italy.

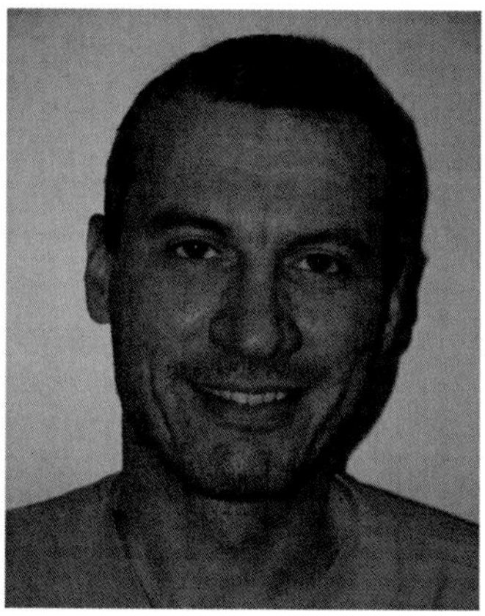

Dr. Emanuele Ferrero, MD, is specialist in Vascular Surgery and currently consultant at the Vascular Surgery Division of the Hospital Mauriziano "Umberto I" of Torino, Italy.

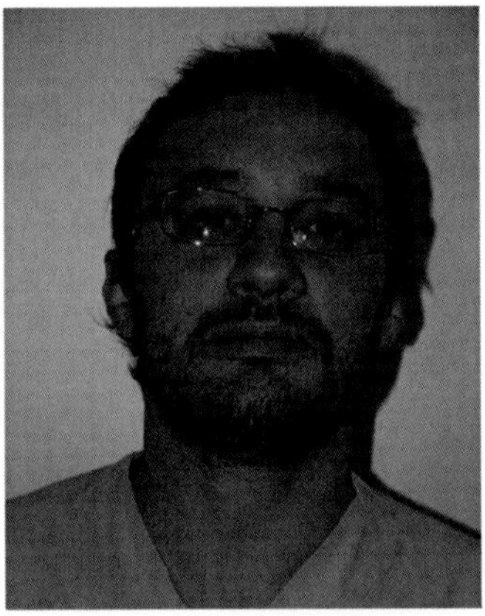

Dr. Andrea Viazzo, MD, is specialist in Vascular Surgery and currently consultant at the Vascular Surgery Division of the Hospital Mauriziano "Umberto I" of Torino, Italy.

Dr. Jasjit S. Suri is an innovator, scientist, a visionary, an industrialist and an internationally known world leader in Biomedical Engineering. Dr. Suri has spent over 20 years in the field of biomedical engineering/devices and its management. He received his Doctrate from University of Washington, Seattle and Business Management Sciences from Weatherhead, Case Western Reserve University, Cleveland, Ohio. Dr. Suri was crowned with President's Gold medal in 1980 and the Fellow of American Institute of Medical and Biological Engineering for his outstanding contributions.

Chapter 23
Molecular Imaging of Atherosclerosis

Patrick Kee and Wouter Driessen

Abstract Atherosclerotic cardiovascular disease is an insidious condition that develops over an extended period of time. By the time the disease is apparent, the arterial wall has already undergone a substantial amount of remodeling. Current diagnostic tools are based on risk factor profiling, biomarker measurements, and lesion detection. Although relatively effective in large cohorts of subjects, risk factors and biomarkers are poor predictors of future risk in individual patients, because of the high occurrence of such factors. Conventional imaging modalities can detect only flow-limiting lesions. By the time advanced disease is diagnosed, mechanical and pharmacological interventions have only modest impact on disease progression. There is clearly a need to devise better imaging tools to measure plaque burden and disease activities within the arterial wall, to diagnose this disease at an earlier stage and to predict the short-term risks of developing complications. Given that molecular changes at a cellular level precede gross anatomic changes in the affected organ, application of techniques for detection of molecular changes in the arterial wall in living subjects might be feasible. Molecular imaging has the potential to (1) screen for early stages of atherosclerosis at which interventions are most effective, (2) follow the course of the disease and aid in the titration of therapy, and (3) detect the presence of vulnerable plaque that requires aggressive intervention. This review article, an overview of the current state of the art of molecular imaging, covers essential components that are vital to the success of the imaging strategy and highlights the potential challenges intrinsic to each component.

Keywords Atherosclerosis • Molecular imaging • Biomarkers • Contrast agents

---

P. Kee (✉)
Division of Cardiology, University of Texas Health Sciences, 6431 Fannin, MSB 1.247, Houston, TX 77030, USA
e-mail: patrick.kee@uth.tmc.edu

Jasjit S. Suri et al. (eds.), *Atherosclerosis Disease Management*,
DOI 10.1007/978-1-4419-7222-4_23, © Springer Science+Business Media, LLC 2011

23.1 Introduction

The principle of molecular imaging is based on the principle that molecular changes at a cellular level precede gross anatomic changes in the affected organ. This imaging technique is particularly valuable in disease conditions in which early detection could have an impact on the management and prognosis of the disease. Although considered the first choice in the diagnosis of obstructive vascular disease, detection of luminal narrowing by contrast angiography is an incomplete representation of the disease process in the arterial wall during the development of atherosclerosis. As described by Glagov et al., the artery can accommodate an increasing atherosclerotic burden by outward expansion and thickening of the arterial wall (positive remodeling) during which the size of the arterial lumen for adequate blood flow is maintained [31]. When compensatory mechanisms fail in advanced disease, luminal narrowing (or negative remodeling) occurs. Molecular pathways responsible for the development of various stages of atherosclerosis are currently under study; research focuses on inflammatory pathways that play an important role in the recruitment of inflammatory cells, lipid deposition, plaque destabilization, and neovascularization. Molecular markers in association with these pathways can be detected by immunohistochemistry of tissue sections with the translation of this technique into a living organism in which molecular expression can be visualized in real-time, molecular imaging represents a powerful technique for the monitoring and diagnosing of disease processes.

23.2 Molecular Markers

Among the components required for molecular imaging, the identification of suitable and meaningful molecular markers in atheroma is probably the most important element for the success of molecular imaging in atherosclerosis. Without identification of appropriate molecular markers, the imaging results would be clinically irrelevant. Suitable molecular markers for imaging of atheroma should have a number of characteristics.

The molecular markers should be widespread, abundant (i.e., expressed at high levels on the target tissue), specific for atheroma, and preferably expressed in specific vascular territories. For instance, the inflammatory component of atherosclerosis is considered an early marker in pathogenesis, and the expression of adhesion molecules in the atheroma promotes the adhesion of circulating monocytes to the inflamed endothelium. This process, however, is not specific to atheroma. Because infection and other inflammatory conditions can induce the expression of adhesion molecules, the use of such molecular markers can be confounded by other disease processes.

The molecular markers should be specific for the disease stage. Management of atherosclerosis can occur at different stages of the disease. Thus, detection of early disease markers could be useful for the screening of patients with cardiac risk factors or a strong family history of premature coronary artery disease. On the other hand, late disease markers could be useful for the identification of vulnerable plaque that might lead to short-term risks in patients with established disease. The levels and pattern of expression of these molecular markers in different stages are useful for "fingerprinting" the stage of disease. Changes in the levels and pattern of expression in an individual would be valuable for assessment of disease progression and efficacy of therapy.

The molecular markers should have prognostic importance. Achieving this goal would distinguish this imaging technique from conventional imaging tools. The ability to detect molecular markers that predict disease progression or future clinical events will guide management and might reveal surrogate markers for evaluation of new therapeutic interventions.

Currently, a number of molecular markers have been evaluated for molecular imaging of atherosclerosis. These markers are known components of atheroma in animal models with advanced atherosclerosis (Table 23.1). Although advanced atheroma in these animal models share some characteristics with human atheroma, the animals have a lipoprotein profile different from that of humans and plaque rupture is rarely observed [14]. Nonetheless, these animal models are useful for

Table 23.1 Early and later markers of atherosclerosis

Class	Molecules	References
Early markers		
Selectins	P-selectin, E-selectin	[27, 74, 84]
Integrins	LFA-1 (CD11a/CD18), Mac-1 (CD11b/CD18), VLA4 ($\alpha_4\beta_1$ integrin)	[25, 47, 64]
Immunoglobulin superfamily	ICAM-1, VCAM-1	[1, 8, 11, 38, 46, 48, 63, 65]
Late markers		
Nonspecific markers	Subendothelial matrix/absence of glycocalyx, collagen, fibronectin	[16, 36, 62]
Enzyme	Tissue factors, cysteine protease, cathepsin K, matrix metalloproteinase	[2, 3, 29, 37, 57, 66, 94]
Cellular components	Oxidized LDL, smooth muscle cells, macrophages, scavenger receptors on foam cells, activated leukocytes	[5, 7, 10, 17, 41, 51, 56, 67]
Apoptosis markers	Phosphatidylserine	[37, 44, 50, 76]
Angiogenesis, neovessel formation	$\alpha_v\beta_3$ integrin (RGD-containing peptides), nonspecific uptake of FDG	[4, 13, 15, 59, 95, 98]
Thrombogenesis	Fibrin, glycoprotein IIb/IIIa	[26, 38, 80]
Other novel markers	Endothelial markers identified by high-throughput technologies such as in vivo phage display (see above)	[40, 61, 73, 85]

the evaluation of various molecular imaging strategies prior to their translation to applications in humans. The detection of factors that are associated with plaque destabilization can in part predict plaque rupture and short-term risks. Such factors include macrophage infiltration, protease production, apoptosis, neovascularization, and intraplaque hemorrhage. Many of these factors can be detected by molecular imaging (Table 23.1).

Our laboratory has taken a different approach for the identification of molecular markers of atherosclerosis, an approach motivated by our desire to identify molecular markers that are stage-specific and potentially useful for screening and prognostic applications. Our screening strategy is based on the hypothesis that the vascular endothelium of each different organ expresses a unique set of surface markers – termed a vascular ZIP code. During pathologic processes such as atherosclerosis, the pattern of surface marker expression is likely to be altered. Genomic approaches, such as high-throughput sequencing and gene arrays, to identify prominent markers in atherosclerosis generally do not allow for the molecular heterogeneity intrinsic to the microanatomic and pathophysiological context of the atherosclerotic process. For example, many genes implicated in atherosclerosis might be expressed in restricted – but perhaps highly specific and accessible – cellular locations such as the neovasculature of evolving plaques.

Over the past decade we have used a nonbiased combinatorial screening method – in vivo phage display – to uncover vascular ZIP codes in animal models and in patients [6, 54, 70–72]. Phage display is a highly versatile technology that involves genetic manipulation of bacteriophage such that peptides and proteins can be expressed on their surface [78, 81]. Although it was initially developed to map epitope binding sites of immunoglobulins, the range of applications for this technique has broadened considerably over the past decades, such as in the construction of phage-displayed peptide libraries and the development of screening methods in which the libraries are used to isolate peptide ligands. Specifically, an in vivo selection method has been developed in which peptides that home to specific vascular beds are selected after intravenous administration of a phage display random peptide library [6, 55, 71]. This strategy has revealed a vascular address system that allows tissue-specific targeting of normal blood vessels and targeting of pathophysiological conditions with a vascular component (e.g., angiogenesis in tumors and atherosclerosis). In the in vivo phage display procedure, phage capable of homing to certain organs or disease states after intravenous injection is recovered from the library. A typical scheme for identification of phage homing to atheroma consists of: (1) injecting a phage library into an atherosclerotic animal, (2) recovering and expanding phage particles bound to atheroma, and (3) reinjecting the subset of bound phage into another atherosclerotic animal. This process is repeated until only a few bound phage with the highest affinity to the atheroma are enriched, a technique termed biopanning. The amino acid sequence of the recovered peptides is determined by sequencing of the DNA corresponding to the insert in the phage genome. Sequences are analyzed to monitor enriched peptide motifs compared to the unselected library. Statistically significant enriched sequences are further analyzed with ClustalW and BLAST software to identify proteins, or ligands, with which the motifs share sequence homology, and are subsequently cross-referred with data from proteomic or microarray studies. Ultimately, this approach allows circulating homing peptides to be detected in an unbiased functional assay, without preconceived notions about the nature of their target. Aside from their carrier function for targeted drug delivery, the peptides themselves can be used as drug discovery leads for peptidomimetic drugs or for therapeutic modulation of their corresponding receptor(s). The genetic approach for receptor identification entails comparison of the peptide sequences of the bound phage with protein libraries. Peptide sequence similarities between phage peptides and known proteins are next used to identify the endothelial marker bound by the phage peptide(s). This approach is relatively quick but is restricted to known molecular markers of uncertain clinical significance and is likely to miss molecular markers that are not included in the protein libraries. The biochemical approach for receptor identification includes affinity chromatography on columns coupled with control and targeting peptides. In brief, proteins binding to the targeting peptide are purified from cell- or tissue-extracts. Proteins specifically bound to the peptide columns are eluted by excess peptide, and eluted fractions are evaluated by two-dimensional electrophoresis. Unique bands observed only in fractions eluted from the column with the targeting peptide are analyzed by mass spectrometry. For this approach, the homing peptide and control needs to be chemically synthesized to purify and elute the targeted protein; therefore, this method is more expensive and not as efficient as genomic or in silico approaches [23, 24, 69, 71].

We have injected phage into various animal models of atherosclerosis to identify phage with the highest affinity to atheroma. There are few examples of the successful use of in vivo phage display in these models. The Weissleder group has screened a phage-display random peptide library in the ApoE$^{-/-}$ model to identify peptides targeting atheroma in vivo. By sequence-homology searches, several homologs implicated in atherosclerosis (e.g., transferrin, leukemia inhibitory factor, and VLA-4) were identified. The VLA-4 homologous peptide, prioritized for further validation, led to the development of multimodal nanoparticles which could be used for the early detection of atherosclerosis [49]. In another example, Robert et al. screened a phage-displayed monoclonal single-chain antibody (Fv) library in an adult male Bourgogne brown rabbit with complex plaques containing intramural thrombi induced by denudation of the aorta and an 8-month atherogenic diet. The authors identified several monoclonal single-chain antibodies recognizing the atherosclerotic aorta after a single round of in vivo biopanning. The protein recognized by the single-chain antibody, of approximately 56 kDa, was affinity-purified and identified by mass spectrometry as vitronectin [73].

23.3 Potential Molecular Targets of Atherosclerosis for Molecular Imaging

A pattern of expression of molecular markers in atherosclerosis is described in Fig. 23.1. Prior to the appearance of fatty streaks, endothelial dysfunction is considered the earliest evidence of vasculopathy. Adhesion molecules such as selectins

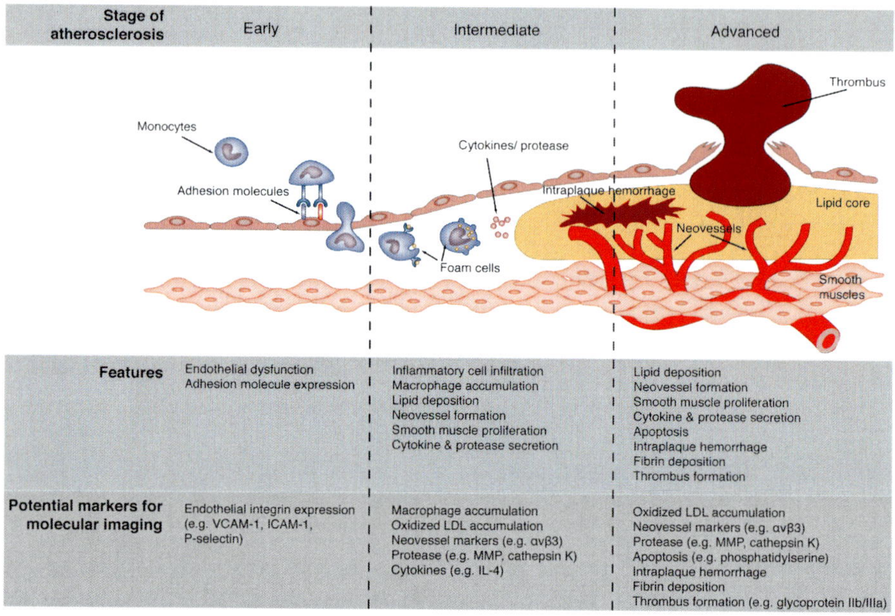

Fig. 23.1 Features of different stages of atherosclerosis and potential markers for their identification by molecular imaging

and integrins are established early markers of atherosclerosis. Intermediate stage atherosclerosis is highlighted by infiltration of oxidized low-density lipoprotein (LDL) and foam cell formation. Advanced and vulnerable lesions usually are characterized by a well-developed lipid core, a thin fibrous cap, and neovascularization. Advanced lesions that contain high levels of tissue factor and proteolytic enzymes further destabilize the lesion. Finally, disruption of the thin fibrous cap results in fibrin deposition and thrombus formation. Many of these molecular markers have been evaluated as potential targets for molecular imaging (Table 23.1).

23.3.1 Adhesion Molecules

The presence of inflammatory cells in atherosclerotic plaque in the arterial wall indicates that factors modulating their recruitment and transmigration into the arterial wall play an important role in the development of atherosclerosis. The attachment of inflammatory cells to activated endothelium is mediated by a number of cellular adhesion molecules, such as P-selectin, E-selectin, intercellular adhesion molecule (ICAM)-1, and vascular cell adhesion molecule (VCAM)-1. The selectins are responsible for the initial tethering and rolling of the leukocytes on the vessel wall.

Among the adhesion molecules, VCAM-1 has been most preferentially studied for molecular imaging. VCAM-1 is expressed more specifically in lesions and lesion-prone regions in the aorta, whereas ICAM-1 is present beyond the lesions [42, 68]. Expression of VCAM-1 in atheroma has been detected by ultrasound- [38, 46], near-infrared [48], nuclear [11], and paramagnetic contrast agents [63].

23.3.2 *Oxidized LDL and Foam Cells*

Entrapped LDL particles undergo oxidation in the arterial wall and are potent mediators of inflammation. Resident macrophages engulf oxidized LDL and become foam cells that populate the lipid core. Thus, the detection of oxidized LDL and the accumulation of macrophages in the arterial wall are attractive targets for imaging. Passive tracking of oxidized lipid deposition in the arterial wall is possible by detection of the interaction of lipophilic gadolinium-based agent with the LDL receptor [60]. Targeted radiolabeled or gadolinium-based contrast agents, conjugated to a murine monoclonal antibody that recognizes malondialdehyde–lysine epitopes on modified LDL, detected the deposition of oxidized LDL in the arterial wall of atherosclerotic mice [10, 79, 88]. Nonspecific uptake of particulate gadolinium, iron oxide, and iodinated particles by macrophages allows their detection in the arterial wall by MR [5] and CT imaging [41]. Another approach involves exogenous labeling of monocytes prior to their injection into animals for tracking of monocytes to atherosclerotic lesions [51, 83].

23.3.3 *Neovessel Formation*

Normal arteries derive a fair portion of their supply of oxygen and nutrients from passive diffusion. In atherosclerosis, the arterial wall becomes thickened as a result of positive remodeling and outward expansion. Therefore, passive diffusion alone is not sufficient for the supply of oxygen to the deeper layers of the arterial wall; consequently, hypoxia stimulates the growth of neovessels into the arterial wall from the vasa vasorum to meet the metabolic demands [97]. The presence of neovasculature is a definitive marker of disease progression. These vessels also facilitate the infiltration of inflammatory cells that further destabilize the atheroma. Neovessels are fragile, and their rupture leads to intraplaque hemorrhage in the arterial wall. Hemosiderin released from erythrocytes is proinflammatory, and cholesterol released from the red cell membrane adds to the lipid pool in the plaque.

Disease activity in atherosclerosis correlates with neovessel-density in the arterial wall [35, 39], a property rendering them a suitable target for molecular imaging. The density of plaque neovascularization in the carotid artery was semiquantified by grading of the enhancement of the plaque by an ultrasound contrast agent [21]. Plaques with higher enhancement of contrast agent showed more extensive neovascularization by histology. Plaque vascularization visualized by ultrasound contrast agent was prominent in the fibrous and fibro-fatty tissue but not in the calcific or the necrotic and hemorrhagic tissue in human carotid arteries [93]. Compared with transvascular ultrasound, intravascular ultrasound produces images of the coronary arterial wall at superior spatial resolution. Prototypic intravascular ultrasound devices using harmonic or subharmonic imaging have been developed to resolve bubble-specific signals in the adventitia, independent of the background B-mode image of the arterial structures [32, 33].

Other than perfusion imaging, potentially specific targets for neovessel imaging in atheroma include integrin expression and angiogenic factors. The integrin $\alpha_v\beta_3$ is highly expressed in atherosclerotic plaques by both medial and intimal smooth muscle cells and by endothelial cells of angiogenic microvessels. A peptidomimetic of Arg-Gly-Asp (RGD) conjugated to gadolinium chloride produced strong enhancement of the external structure of the aortic wall in apoE-deficient mice by MR imaging [13]. Another targeted MR contrast agent was prepared by conjugation of a peptidomimetic vitronectin antagonist against integrin $\alpha_v\beta_3$ to paramagnetic nanoparticulate emulsion. Two hours after injection into hypercholesterolemic rabbits, the targeted contrast agent increased the MRI signal in the adventitia by about 50% [98]. Imaging tracers can also be modified as drug delivery vehicles, also known as "theranostic" agents. For example, the antiangiogenic drug fumagillin, incorporated into the same targeted MR contrast agent used for imaging, effectively reduced microvessel development in treated animals, with associated reduction in MRI enhancement in the arterial wall [99]. Another potential target for neovessel imaging is an angiogenic molecule, the extradomain B of fibronectin, which is inserted into the fibronectin gene by alternative splicing during angiogenesis, but is virtually undetectable in normal tissue. An antibody against extradomain B, labeled with radioiodine or infrared fluorophores, was injected into atherosclerotic apoE-deficient mice. By radiographic and fluorescent imaging, areas with extradomain B were shown to colocalize with lipid areas of atheroma [62].

23.3.4 Proteolytic Enzymes

Macrophages are the main source of proteolytic enzymes in atherosclerosis. Proteolytic enzymes such as matrix metalloproteinases (MMPs), cysteine protease, and cathepsin K are found in fibrous caps of human atherosclerotic plaques, predominantly at the macrophage-rich shoulder regions and have been implicated in the destabilization of atheroma [30].

Direct detection of MMPs has been targeted by a peptide-based MMP inhibitor conjugated to gadolinium complex [57], or a broad-spectrum MMP-inhibiting macrocyclic compound conjugated to a technetium-99m tracer [29].

However, detection of the proteinases has been focused on the harnessing of their enzymatic activities to improve imaging sensitivity. Local expression of proteolytic enzymes in atheroma offers a unique opportunity for the design of activatable probes that are quenched in the uncleaved form and become strongly fluorescent after proteolysis. This approach has the potential to improve signal-to-noise ratio in molecular imaging. Activatable near-infrared probes have been designed for the detection of the expression of cysteine protease [19], MMP [22] and cathepsin K [43] and their imaging applications have been successfully demonstrated in murine models of atherosclerosis. To improve sensitivity and anatomic localization of signal [66], combined fluorescence molecular tomography with CT imaging in their studies of murine atherosclerosis.

23.3.5 *Apoptosis*

Programmed cell-death or apoptosis is observed in advanced atheroma and might predict the vulnerability of plaque to rupture and thrombosis. One of the earliest events in apoptosis is the exposure of phosphatidylserine on the cell plasma membrane; its detection therefore appears to be an attractive target for diagnosis of the vulnerable plaque.

The most commonly used targeting agent is annexin A5, an endogenous ligand of phosphatidylserine. Annexin A5 radiolabeled with technetium 99m was used to perform noninvasive imaging of apoptosis in porcine coronary atherosclerosis [44]. Most of the atherosclerotic lesions were American Heart Association class II, and 60% of the injured coronary arteries were positive for focal uptake of annexin A5 label. Subsequently, another study was carried out by the same group to evaluate whether annexin A5 labeling was correlated with MMP-9 production and apoptosis [37]. Two imaging contrast agents targeting MMP and phosphatidylserine were injected into an atherosclerotic rabbit with aortic denudation and 4 months of cholesterol feeding. There was significant uptake of the tracers in the injured segments. Interventions such as statin therapy, withdrawal of atherogenic diet [37], or caspase inhibition [76], effectively reduced uptake of annexin A5, which was correlated with pathologically verified apoptosis, macrophages, and MMP-9-positive areas in the atherosclerotic lesions. This study demonstrated a pathologic relationship between MMP production and apoptosis [37].

In a pilot study, four human subjects with a recent history of transient ischemic attack were injected with radiolabeled annexin A5 before carotid endarterectomy. Uptake of annexin A5 was observed in the ultrasound-verified carotid artery lesion [50]. Pathologically, those lesions had unstable features including macrophage infiltration and intraplaque hemorrhage. Immunohistochemical analysis demonstrated specific binding of annexin A5 to macrophage plasma membranes.

Homing peptides against phosphatidylserine identified via phage display methodology, have been prepared and conjugated to gadolinium chloride for MR imaging. This contrast agent highlighted atheroma in apoE-deficient mice 60 min after intravenous injection [12].

23.3.6 *Fibrin Deposition and Thrombus Formation*

Fibrin deposition on atheroma or atherosclerotic ulcers does not necessarily lead to occlusive thrombus formation but may herald the imminent risk of further complications of arterial occlusion and end-organ infarction. Noninvasive detection of fibrin with targeted ultrasound contrast agent has been investigated by our group and others. Fibrin targeting can be achieved via a multistage technique by first pretargeting fibrin with anti-fibrin antibodies and avidin in situ, followed by administration of biotinylated perfluorocarbon nanoemulsion. This technique was successfully highlighted arterial thrombi in vivo [58]. Platelet aggregation can also be detected by the target of platelet glycoprotein IIb/IIIa receptor. One potential candidate is the linear hexapeptide (lysine-glutamine-alanine-glycine-aspartate-valine) that is coupled to the lipid shell via a PEG spacer in MRX-408 (ImaRX Pharmaceutical Corp, Tucson, Arizona) [92]. Preliminary studies demonstrated MRX-408 specifically bound and enhanced the detection of blood clot by ultrasound in vitro and in vivo, and has the potential to enhance thrombolysis when activated by ultrasound [77, 100]. Our laboratory has studied the binding characteristics in vitro and targeting in vivo of anti-fibrinogen-conjugated echogenic liposomes (ELIP) to thrombus. We have demonstrated the binding of anti-fibrinogen-conjugated ELIP to thrombotic components of atheroma and have provided acoustic enhancement for intravascular and transvascular ultrasound imaging in a Yucatan miniswine model of atherosclerosis [38]. Recently, our laboratory has utilized the intrinsic fibrin-binding capacity of tissue plasminogen activator in the preparation of a dual function ELIP formulation that effectively targeted thrombus in vivo as well as facilitated ultrasound-enhanced thrombolysis in vivo [86, 87].

Various fibrin-targeting MR imaging agents have also been developed to detect fibrin deposition in association with atheroma. These include gadolinium-based nanoparticles conjugated to a monoclonal antibody against fibrin [28], fibrin-binding peptide [9], or fibrin-targeting small molecules [80], all of which targeted and enhanced the visualization of fibrin clots in various animal models.

23.4 Homing Ligands

Homing ligands serve as a vital link that allows interaction between the contrast agent and molecular marker. Ideally, the initial attachment of the ligand to the endothelial cell receptor must be rapid (rapid on-rate), and the dissociation of the ligand from the receptor should be slow (slow off-rate) for efficient accumulation of contrast agent at the sites for molecular imaging. Ligand–receptor interaction appears most efficient at sites in the circulation where wall shear stress is low. For instance, leukocyte adhesion occurs at sites such as curvatures and branch points at which average and oscillatory shear stresses are low. Coincidentally, these sites are favorable for atheroma formation. Most studies showed that adhesion of antibody-coated microbubbles to solid-state ligands was optimal at shear stresses ranging from 0.02 to 0.5 Pa, far below the mean value of 1.2 Pa in the human common carotid artery [75]. The substantially higher wall shear stress in animals smaller than humans poses technical challenges for optimal targeting in such animal models [20].

Modifications such as increasing ligand surface density might enhance the adhesion rate of targeted contrast agents [53, 74, 84].

A number of molecules have been evaluated as homing ligands for molecular imaging. Antibodies against various molecular markers are most commonly used as homing ligands because of their availability, relatively high affinity for molecular targets, and ease of conjugation to imaging agents. However, their size and nonhuman origin are potential sources of immunogenicity in humans. Humanized antibodies and truncated forms are suitable alternatives and should overcome some of these disadvantages. Antibodies as homing ligands, however, exhibit poor on-rates. For instance, monoclonal antibodies, such as anti-P-selectin, have frequently been used for targeting atheroma. Microbubbles coated with anti-P-selectin monoclonal antibody display poor capture efficiency under both static and steady-state, low shear conditions [74, 84]. In contrast, carbohydrates such as sialyl Lewis^x [53] and glycosulfopeptides such as PSGL-1 [74] exhibit higher capture efficiency and stronger adhesion to P-selectin, in comparison to anti-P-selectin antibodies. The inclusion of two or more targeting agents on a contrast agent might be a potential solution for the improvement of binding efficiency. Dual-targeting microbubbles coated with anti-VCAM-1 antibody and sialyl Lewis^x demonstrated twice the binding efficiency of single-targeted microbubbles at 6 dyn/cm^2 in a flow chamber [27]. Another dual-targeting microparticle of iron oxide coated with antibodies against VCAM-1 and P-selectin was five- to sevenfold more efficient than were single-targeted agents in binding to aortic root atherosclerosis in apoE-deficient mice [63]. Finally, dual-targeting microbubbles coated with antibodies against VCAM-1 and P-selectin was used successfully to stage age-dependent atheroma in a murine model of atherosclerosis (Kaufmann et al. date missing).

Short peptides are the next commonly used homing ligands in molecular imaging. Short peptides can form noncovalent linkages with protein targets. Whereas their relative affinity for their targets is weaker than that of antibodies, their small size potentially reduces their immunogenicity. The other important advantage of peptides as homing ligands is related to their derivation from phage display technology. When discovered by in vivo biopanning in appropriate animal models, homing peptides can identify new molecular markers responsible for disease in a stage-specific manner. This advantage provides not only new targets for stage-specific diagnosis with molecular imaging, but also new and previously unidentified molecular pathways for further investigation. In the process of in vivo biopanning, there is an additional selection resulting in peptide sequences that will most likely evade endogenous clearance mechanisms and will bind to intended targets with high affinity and specificity [69]. Known molecular targets can also be used to prepare the corresponding high-affinity peptide sequence. With a biopanning approach that was specifically designed to identify peptide sequences that bind VCAM-1, Kelly et al. have identified cyclic VCAM-1-targeted peptide sequences (VINP-28) that were successful in ex vivo imaging of excised aorta [48]. Subsequently, VINP-28, attached in a multivalent fashion to magnetofluorescent nanoparticles, appeared to facilitate cellular internalization of the contrast agent [65].

More recently, the ability to screen random libraries of single-stranded oligonucleotides (RNA or DNA) has given rise to a new class of targeting ligands. In a process termed SELEX (Systematic Evolution of Ligands by Exponential Enrichment) [34] or in vitro selection [18], these oligonucleotide structures (nucleic acid aptamers) can be screened for adhesion to immobilized targets, cells, or tissues. In brief, the randomized libraries of oligonucleotides are incubated with the target of interest; after a series of washes and/or elutions, specifically bound aptamers are enhanced by PCR. This process can be repeated several times, to enrich specific target-binding aptamers.

In general, peptides and aptamers have relatively lower affinities than those characteristic of antibodies. However, their lower affinity is compensated for by their specificity. Peptides identified by in vivo biopanning have already undergone in vivo selection, and those with the optimal affinity and specificity are identified. Furthermore, affinity of homing ligands can be further enhanced by combination of multiple homing ligands on to a single contrast agent [27] or by augmentation of the number of homing ligands displayed on a scaffold – a process known as multivalency [4, 96].

The homing ligands need to be conjugated to molecules or contrast agents to allow their visualization in vivo, but this reaction can alter the chemical characteristics of the ligands. The chemical modification can result in either altered affinity or altered metabolism in vivo. The density of homing ligands on the surface of structures such as liposomes or nanoparticles could also result in steric hindrance among the homing ligands themselves and affect their affinity for the targets. Attempts have been made to predict the behavior of the homing ligands in vivo from in vitro and ex vivo, experiments, either alone or in association with the contrast agents. Characterization and optimization of various homing ligands in vivo can be performed by the use of flow chamber simulation, BiaCore analysis, and atomic force microscopy.

Competitive binding assays are important to demonstrate the specificity of the targeting agent for the intended target. This procedure usually involves preinjection of the animal with unlabeled homing ligands to occupy the binding sites on the target receptor, prior to the injection of targeted contrast agent.

23.5 Contrast Agents

For the detection of molecular markers by immunohistochemistry, images can be obtained by chemical reactions that generate color, fluorescence, or radioactivity. In living subjects, the signal is generally obscured by anatomic barriers, and strategies are therefore needed to allow detection of the signal by external detectors, for example, by the detection of signal attenuation caused by the contrast agent as an energy pulse passes through the body. Agents such as CT contrast agents containing high molecular weight compounds, or gas-containing ultrasound contrast agents, fall into this category. Alternatively, it is possible to detect high-energy signal emitted from the contrast agents by external sensors. Radioactive or fluorescent compounds can emit high-energy photons, which can, to some extent, pass through anatomic barriers. On the other hand, gas-containing contrast agents not only attenuate ultrasound but also have inherent signal amplification. Signal emitted from contrast agents can be enhanced further by the packaging of a high concentration of the molecules into liposomes or nanoparticles, a technology achieved by encapsulation CT or MRI contrast agents or perfluorocarbon gases inside liposomal or polymer preparations.

A major consideration in molecular imaging is the achievement of a high signal-to-noise ratio. The signal from the targeted contrast agents is related to the capacity of the targeted contrast agent to remain associated with the intended target and to be distinguished from the surrounding tissues. Noise is related to the clearance of unbound contrast agents from the circulation at the time of imaging as well as to signal with similar intensity as the contrast agent in the surrounding tissues. The optimization of both parameters can be challenging because both signal and noise tend to peak early after injection and fall precipitously over time. An ideal targeted contrast agent should remain bound to the target at a high concentration while unbound contrast agents are cleared rapidly from the circulation. Contrast agents become apparent if they emit strong signal toward the external detectors (e.g., radioactive agents), emit unique signal due to the lack of endogenous counterpart (e.g., ^{18}F), or exist at sufficiently high concentrations that can be distinguished from the surrounding structures (e.g., radio-opaque contrast and paramagnetic agents). Radioactive agents for SPECT or PET imaging generally have higher sensitivity for detection than CT or MRI contrast agents. The other solution is via suppression of the signal in the surrounding tissues as well as signal from unbound contrast agents. In ultrasound imaging, signal suppression in the surrounding tissues can be achieved by nonlinear imaging that eliminates tissue signal and enhances contrast visualization. The unbound contrast agents can be cleared by a high mechanical index ultrasound pulse. Another potential solution can be achieved by activatable probes that are "silent" in the unbound state and become detectable when bound to the intended targets. These activatable probes, which have been designed for fluorescence and MRI imaging, self-quench in the inactivated state. When the molecular structure is altered by a chemical process, e.g., in the presence of cysteine protease [19], MMP [22], or cathepsin K [43], the fluorescence tags are no longer quenched and fluorescence signal is emitted.

Due to the hemodynamics of the circulation and the variability in the affinity of homing ligands, a number of modifications are needed to prolong the circulation time of the contrast agents to enable their interrogation of endothelial surfaces. Large contrast agents such as liposomes are prone to clearance mechanisms because of their size. Spacer arms such as polyethylene glycol are used to decorate the surface of liposomes to enhance their circulation time [52] and to allow more flexible display of the homing ligands.

The size of the contrast agents can affect their accessibility to molecular targets. Large contrast agents such as liposomes are intravascular tracers that are useful for interrogation of the endothelial surface. Large contrast agents can occasionally have access to deeper structures in the arterial wall when the endothelial surface is disrupted by pathologic processes; they can also travel through new channels created by neovascularization in advanced atherosclerosis. Small contrast agents might be capable of accessing deeper arterial structures. Some compounds such as USPIO can be taken up by resident macrophages and accumulate at sufficiently high concentrations for detection in atheroma.

23.6 Imaging Modalities

The density of any particular molecular marker in the arterial wall is expected to be low, in the range of nanomolar or femtomolar concentrations. The sparsity of these molecular markers will require imaging tracers and imaging modalities with high sensitivity for their detection. Other challenges of imaging the cardiovascular system in humans are related to the deep locality, highly mobile nature, and small size of the vascular structures affected by atherosclerosis. Imaging modalities that have been applied successfully to small animal models need to be reengineered for humans.

Clearly, validation of various components of molecular imaging needs to be performed in small animal models prior their translation to the clinical arena. Small animal imaging tools will remain the workhorse in early phases of research. Optical and ultrasound imaging are examples of high-throughput and inexpensive imaging modalities.

The ideal imaging modality for molecular imaging should be capable of visualizing deep-seated structures with high spatial and temporal resolution. Unfortunately, none of the currently available imaging tools can achieve those goals, and there are significant tradeoffs that need to be considered. The commonly used imaging modalities for molecular imaging include ultrasound, optical, SPECT, PET, CT, and MRI, all of which differ in their spatial and temporal resolution as well as their sensitivity (Table 23.2).

Ultrasound imaging has good theoretical temporal and spatial resolution. When conjugated to appropriate homing ligands, targeted ultrasound contrast agents are suitable for molecular imaging. Our laboratory has used both transvascular and intravascular ultrasound to detect adhesion molecules and fibrin in atheroma with targeted echogenic liposomes [38]. Both ultrasound contrast agents and tissue backscattering appear echogenic in B-mode ultrasound imaging. Thus, specialized imaging algorithms such as nonlinear imaging [21, 82, 93] and subharmonic imaging [32] should be considered for differentiation of bound contrast agents from underlying tissues. Kaufmann et al. have prepared and injected anti-VCAM-1-conjugated microbubbles into wild-type and apoE-deficient mice [46]. To create animals with different stages of atherosclerosis, these investigators divided wild-type and apoE-deficient mice into two groups, to each of which was allocated a normal or hypercholesterolemic chow. A graded response in the expression of VCAM-1 in the arterial wall was observed among the animal groups. Retention of microbubbles in

Table 23.2 Imaging characteristics of various imaging modalities for molecular imaging

Imaging techniques	Spatial resolution	Imaging depth	Temporal resolution	Sensitivity	Molecular probe used
PET	1–2 mm	No limit	10 s to min	10^{-11}–10^{-12} mol/L	Nanograms
SPECT	1–2 mm	No limit	Minutes	10^{-10}–10^{-11} mol/L	Nanograms
Optical bioluminescence	3–5 mm	1–2 cm	Seconds to minutes	10^{-15}–10^{-17} mol/L	Micrograms to milligrams
Optical fluorescence	2–3 mm	<1 cm	Seconds to minutes	10^{-9}–10^{-12} mol/L	Micrograms to milligrams
MRI	25–100 μm	No limit	Minutes to hours	10^{-3}–10^{-5} mol/L	Micrograms to milligrams
CT	50–200 μm	No limit	Minutes	10^{-3}–10^{-5} mol/L	Micrograms to milligrams
Ultrasound	50–500 μm	mm to cm	Seconds to minutes	10^{-3}–10^{-5} mol/L	Micrograms to milligrams

the aortic arch was specifically visualized by Siemens Contrast Pulse Sequencing, and the level of VCAM-1 was successfully quantified by this technique. However, spatial resolution tends to suffer when such imaging algorithms are used that limits their utility. The lifespan of ultrasound contrast agents in the circulation is short-lived due to their entrapment and destruction. Thus, appropriate timing is crucial for signal quantification. Proper controls with an isotype IgG-conjugated ultrasound contrast agent are needed to determine nonspecific binding. Finally, most imaging protocols require a destructive pulse to distinguish the specific signal of the bound from the freely circulating contrast agents. The main limitations of this imaging modality are the failure to visualize deeper structures that are obscured by air, tissues, or bony structures, and less than ideal spatial resolution in practice due to significant background noise.

Optical imaging with fluorescence and near-infrared is a high-throughput and highly sensitive imaging tool for small animals. Near-infrared spectrum has the added advantage of elimination of background tissue autofluorescence, a step resulting in an excellent signal-to-noise ratio. A number of fluorescence and NIR tracers are available that can be readily conjugated to homing ligands. Unbound tracers are almost completely cleared from the animals, and bound agents are thereby left at the targets. This imaging modality has been successfully employed in tumor imaging, in which solid tumors are relatively large and superficially located. When imaging atherosclerosis in small animals, only ex vivo imaging has been reported, which also requires the removal of intervening organs to demonstrate signal enhancement [62]. Tissue penetration is an obvious limitation of this imaging technique. The relatively lower density of molecular markers expressed in atheroma vs. solid tumors also results in a lower local concentration of bound tracers. The use of FMT might have an advantage over reflectant fluorescence imaging in the quantification of fluorescence signal from deep tissues. Modern FMT equipment can coregister quantitative tomographic data with CT, MR, and PET imaging. Nahrendorf et al. performed coregistration of fluorescence molecular tomography with CT to localize and quantify protease activity in the aortic root of apoE deficient mice [66]. In addition, they measured the reduction in protease activity in the aortic root after the animals were treated with atorvastatin.

Nuclear imaging with SPECT or PET is the imaging modality of choice for molecular imaging because of the sensitivity of probe detection. Nuclear imaging is useful for the detection of organ perfusion defects, tumor metastases, infection, and metabolic anomalies in humans. Targeted nuclear tracers have been evaluated in atheroma for detection of macrophage accumulation [17, 51, 67], oxidized LDL [88–91], inflammation [11, 48], and apoptosis [37, 44, 50, 76]. The efficiency of isotope counting in PET is superior to that in SPECT due to the inherent collimation created by annihilation of positrons. Despite the relatively poor spatial resolution of nuclear imaging, quantitative data can be coregistered with CT or MR imaging for anatomical localization. Obviously, nuclear tracers have short half-lives and could affect the preparation of targeted agents. Radioactivity in the lungs, liver, and spleen can obscure visualization of atheroma-prone regions such as the coronary artery, aortic arch, and thoracic aorta.

CT and MR imaging offer excellent tissue penetration, anatomical definition, and tissue characterization. With untargeted contrast agents, features of plaque vulnerability such as soft plaque, lipid core, thin fibrous cap, thrombus formation, and neovascularization can be identified with CT and MR imaging. With the advent of multidetector CT, the speed of acquisition has improved and noninvasive visualization of the coronary artery has become technically feasible. Untargeted MR contrast agents such as Gadolinium and iron-oxide particles are apparently taken up readily by macrophages and facilitate their detection in the atheroma [5, 41]. In general, the sensitivity of detection of CT and MR contrast agents is poor, and high local concentrations of contrast agents are required to create sufficient contrast from the surrounding structures. However, at high concentrations, CT contrast agents can create artifacts such as beam hardening that impair proper visualization. Attempts have been made to increase the number of X-ray-attenuating molecules in the chemical structure of contrast agents or to encapsulate large amounts of contrast agents in liposomes or nanoparticles to improve their sensitivity for detection. In MR imaging, techniques to improve signal-to-noise ratio include the use of higher magnetic field strengths or hyperpolarized isotopes, decreasing the acquisition time by a reduction in k-space sampling, and the use of molecules that exist in low quantities in vivo (e.g., fluorine).

23.7 Image Analysis

To differentiate among levels of molecular markers and to quantify disease activities and plaque burden in the arterial wall are important considerations in the development of molecular imaging as a diagnostic tool. Early animal studies focused on proof of concept of the imaging technique by comparison of the increase in arterial wall mass between normal animals and those with advanced disease. Recently, attempts have been made to determine whether the imaging technique could grade the lesions in animal models with different stages of atherosclerosis, from the age of the atherosclerosis-susceptible animals, duration of atherogenic diet, and regression of atherosclerosis by withdrawal of the atherogenic diet or lipid-lowering therapy.

Haider et al. studied lesion regression in cholesterol-fed animals with lipid-lowering therapy and withdrawal of the atherogenic diet [37]. By targeting macrophage apoptosis and MMP activity with radioactive contrast agents, they demonstrated a reduction in radionuclide uptake after diet or statin intervention. In another study in which, apoE-deficient mice were given an atherogenic diet with or without admixed atorvastatin, Nahrendorf et al. showed a consistent reduction of VCAM-1 in the aortic root of treated animals studied with VCAM-1-targeted MR and fluorescent reflectance imaging [65]. Kaufmann et al. attempted to quantify age-dependent change in signal enhancement in the aortic arch of mice lacking the LDL receptor and the Apobec-1 editing peptide [45]. They prepared dual-targeting microbubbles coated with antibodies against P-selectin and VCAM-1 for ultrasound-based molecular imaging and found that it was possible to detect a difference in the signal intensity in the lesions in a graded manner as a function of age.

Care needs to taken to ensure that longitudinal analysis is performed at identical sites or segments in the animals. Fiducial landmarks such as the visualization of all three aortic valve cusps in the aortic root, spinal vertebral levels, and major arterial branches such as the renal arteries and iliac bifurcation, are suitable reference structures. With multimodality imaging, these landmarks are crucial for fusion of images. Distribution of atheroma is patchy. An arterial segment rather than a single cross section should always be used to quantify average lesion enhancement and overall lesion volume.

Different imaging modalities require specific quantitative analysis. In targeted ultrasound imaging, the objective is to quantify the amount of bound targeted microbubbles in the region of interest. This region can be difficult to define. Although high-frequency ultrasound (i.e., >40 MHz) allows clear delineation of the layers of the arterial wall in small animals, microbubbles are generally not seen well at frequencies higher than 20 MHz. Deep tissue penetration is also not possible with high-frequency ultrasound. For enhancement of contrast-to-noise ratio, nonlinear imaging is used to visualize the microbubbles independent of the surrounding tissues, which are equally reflective under B-mode imaging. When a nonlinear algorithm is used for imaging, ultrasound frequency generally drops to a relatively low level (i.e., <20 MHz). As a result, both the near and far walls of the artery and its lumen will be encompassed by the beam width in small animals; it becomes impossible to delineate the finer details of the artery [46]. The region of interest for analysis covers the entire artery in nonlinear ultrasound imaging. To differentiate between bound and circulating microbubbles, one applies a destructive pulse after a period of contrast circulation. Obviously, the period of circulation prior to the destructive pulse must be tailored to each animal model and contrast agent. The difference in average signals in pre- and postdestructive contrast frames corresponds to the extent of binding of the targeted microbubbles.

In optical imaging such as fluorescence and near-infrared fluorescence imaging, the contrast agent is allowed to circulate until the background signal has dropped to a minimum. The suitable targeted fluorochrome should remain bound to the plaque. Plaque target-to-background ratio is usually derived by:

$$\frac{\text{Plaque signal} - \text{Normal artery signal}}{\text{Adjacent aortic background signal}}.$$

Fluorescence images can be coregistered with the anatomic white-light images in ex vivo macroscopic fluorescence reflectance imaging [62]. A new fluorescence-mediated tomography imaging technique uses temporally and spatially resolved illuminators and detectors to detect near-infrared fluorochromes noninvasively and deep within the body [19].

In PET imaging, the maximum standardized uptake value (SUV) within a region of interest is measured. The placement of the region of interest can be problematic in control subjects when a nonspecific tracer such as ^{18}F-FDG is used, because its uptake may be low in delayed images. To overcome that problem, the investigator chooses random sites in the vessel wall. Standardized uptake value is usually defined as:

$$\frac{\text{Tissue activity / Volume of tissue}}{\text{Body activity / Body weight}}.$$

In MR imaging, contrast agents that produce a positive signal can be quantified based on the contrast-to-noise ratio:

$$\frac{\text{Blood pool signal} - \text{Plaque signal}}{\text{Standard deviation of the noise}}.$$

Image intensity can be normalized relative to a fiduciary marker such as a test tube with gadolinium–DTPA in saline placed within the field of view. Signal enhancement in the aortic wall can be measured in all segmented slices. An integrated measurement of contrast enhancement in a segment of the arterial wall is a better estimation of contrast uptake than that from a single section. The quantification of negative MR contrast such as USPIO can be problematic. It is uncertain how best to quantify the T2*-weighted signal and whether the degree of signal loss is in any way proportional to the inflammatory load within the plaque. For calculation of the relative signal intensity, the signal change is referenced to the adjacent muscle, with the assumption that the muscle does not take up USPIO. Furthermore, the signal reduction is dependent on the reproducibility of coil positioning and on image coregistration before and after infusion of USPIO.

23.8 Conclusions

Molecular imaging of atherosclerosis represents the new frontier of personalized medicine. This imaging technique has the potential to measure disease activity and plaque burden in the arterial wall in an individual with great accuracy. If successful, the implication of this technology will be profound and will change how atherosclerosis is diagnosed and treated. Molecular markers with prognostic significance will be excellent surrogate markers that will accelerate the development of drugs and devices. Before such promises are realized, there are some major technologic hurdles to overcome. Discovery of appropriate molecular markers that specifically identify and stage atheroma is vital. This goal can be achieved with new screening tools and with animal models that produce human-like lesions. Ultimately, the relevance of those molecular markers in human atheroma will need to be validated. The other components of molecular imaging also need further refinement. Homing ligands need to be rendered less immunogenic for repeated administration, and their on- and off-rates must also be improved. The contrast agents and the imaging modalities are closely related. Ideally, a highly sensitive imaging technique with high spatial resolution and tissue penetration is needed. At this point of development, no single imaging modality can fulfill these criteria. Future imaging tools will likely be multimodal in nature to achieve both high sensitivity and high spatial resolution. Finally, signal quantification that accurately represents disease activity and progression is requisite for the evolution of this technology into useful clinical tools.

References

1. Abe Y, Sugisaki K, Dannenberg AM. (1996) Rabbit vascular endothelial adhesion molecules: ELAM-1 is most elevated in acute inflammation, whereas VCAM-1 and ICAM-1 predominate in chronic inflammation. J Leukoc Biol 60:692–703.
2. Aikawa M, Rabkin E, Okada Y et al. (1998) Lipid lowering by diet reduces matrix metalloproteinase activity and increases collagen content of rabbit atheroma: a potential mechanism of lesion stabilization. Circulation 97:2433–44.
3. Aikawa M, Voglic SJ, Sugiyama S et al. (1999) Dietary lipid lowering reduces tissue factor expression in rabbit atheroma. Circulation 100:1215–22.
4. Almutairi A, Rossin R, Shokeen M et al. (2009) Biodegradable dendritic positron-emitting nanoprobes for the noninvasive imaging of angiogenesis. Proc Natl Acad Sci USA 106: 685–90.
5. Amirbekian V, Lipinski MJ, Briley-Saebo KC et al. (2007) Detecting and assessing macrophages in vivo to evaluate atherosclerosis noninvasively using molecular MRI. Proc Natl Acad Sci USA 104:961–6.
6. Arap W, Kolonin MG, Trepel M et al. (2002) Steps toward mapping the human vasculature by phage display. Nat Med 8:121–7.
7. Babaev VR, Gleaves LA, Carter KJ et al. (2000) Reduced atherosclerotic lesions in mice deficient for total or macrophage-specific expression of scavenger receptor-A. Arterioscler Thromb 20:2593–9.
8. Behm C, Kaufmann B, Carr C et al. (2008) Molecular imaging of endothelial vascular cell adhesion molecule-1 expression and inflammatory cell recruitment during vasculogenesis and ischemia-mediated arteriogenesis. Circulation 117(22):2902–11.
9. Botnar RM, Perez AS, Witte S et al. (2004) In vivo molecular imaging of acute and subacute thrombosis using a fibrin-binding magnetic resonance imaging contrast agent. Circulation 109:2023–9.
10. Briley-Saebo K, Shaw P, Mulder W et al. (2008) Targeted molecular probes for imaging atherosclerotic lesions with magnetic resonance using antibodies that recognize oxidation-specific epitopes. Circulation 117(25):3206–15.
11. Broisat A, Riou LM, Ardisson V et al. (2007) Molecular imaging of vascular cell adhesion molecule-1 expression in experimental atherosclerotic plaques with radiolabelled B2702-p. Eur J Nucl Med Mol Imaging 34:830–40.
12. Burtea C, Laurent S, Lancelot E et al. (2009) Peptidic targeting of phosphatidylserine for the MRI detection of apoptosis in atherosclerotic plaques. Mol Pharm 6:1903–19.
13. Burtea C, Laurent S, Murariu O et al. (2008) Molecular imaging of alpha v beta3 integrin expression in atherosclerotic plaques with a mimetic of RGD peptide grafted to Gd-DTPA. Cardiovasc Res 78:148–57.
14. Calara F, Silvestre M, Casanada F et al. (2001) Spontaneous plaque rupture and secondary thrombosis in apolipoprotein E-deficient and LDL receptor-deficient mice. J Pathol 195:257–63.
15. Calcagno C, Cornily J, Hyafil F et al. (2008) Detection of neovessels in atherosclerotic plaques of rabbits using dynamic contrast enhanced MRI and 18F-FDG PET. Arterioscler Thromb Vasc Biol 28(7):1311–7.
16. Caravan P, Das B, Dumas S et al. (2007) Collagen-targeted MRI contrast agent for molecular imaging of fibrosis. Angew Chem Int Ed Engl 46:8171–3.
17. Carrió I, Pieri PL, Narula J et al. (1998) Noninvasive localization of human atherosclerotic lesions with indium 111-labeled monoclonal Z2D3 antibody specific for proliferating smooth muscle cells. J Nucl Cardiol 5:551–7.
18. Chapman KB, Szostak JW. (1994) In vitro selection of catalytic RNAs. Curr Opin Struct Biol 4:618–22.
19. Chen J, Tung CH, Mahmood U et al. (2002) In vivo imaging of proteolytic activity in atherosclerosis. Circulation 105:2766–71.

20. Cheng C, Helderman F, Tempel D et al. (2007) Large variations in absolute wall shear stress levels within one species and between species. Atherosclerosis 195:225–35.
21. Coli S, Magnoni M, Sangiorgi G et al. (2008) Contrast-enhanced ultrasound imaging of intraplaque neovascularization in carotid arteries: correlation with histology and plaque echogenicity. J Am Coll Cardiol 52:223–30.
22. Deguchi J-o, Aikawa M, Tung C-H et al. (2006) Inflammation in atherosclerosis: visualizing matrix metalloproteinase action in macrophages in vivo. Circulation 114:55–62.
23. Dias-Neto E, Nunes DN, Giordano RJ et al. (2009) Next-generation phage display: integrating and comparing available molecular tools to enable cost-effective high-throughput analysis. PLoS One 4:e8338.
24. Driessen WH, Ozawa MG, Arap W et al. (2009) Ligand-directed cancer gene therapy to angiogenic vasculature. Adv Genet 67:103–21.
25. Elitok S, Brodsky SV, Patschan D et al. (2006) Cyclic arginine-glycine-aspartic acid peptide inhibits macrophage infiltration of the kidney and carotid artery lesions in apo-E-deficient mice. Am J Physiol Renal Physiol 290(1):F159–66.
26. Elmaleh DR, Fischman AJ, Tawakol A et al. (2006) Detection of inflamed atherosclerotic lesions with diadenosine-5′,5‴-P1,P4-tetraphosphate (Ap4A) and positron-emission tomography. Proc Natl Acad Sci USA 103:15992–6.
27. Ferrante EA, Pickard JE, Rychak J et al. (2009) Dual targeting improves microbubble contrast agent adhesion to VCAM-1 and P-selectin under flow. J Control Release 140:100–7.
28. Flacke S, Fischer S, Scott MJ et al. (2001) Novel MRI contrast agent for molecular imaging of fibrin: implications for detecting vulnerable plaques. Circulation 104:1280–5.
29. Fujimoto S, Hartung D, Ohshima S et al. (2008) Molecular imaging of matrix metalloproteinase in atherosclerotic lesions: resolution with dietary modification and statin therapy. J Am Coll Cardiol 52:1847–57.
30. Galis ZS, Sukhova GK, Lark MW et al. (1994) Increased expression of matrix metalloproteinases and matrix degrading activity in vulnerable regions of human atherosclerotic plaques. J Clin Invest 94:2493–503.
31. Glagov S, Weisenberg E, Zarins CK et al. (1987) Compensatory enlargement of human atherosclerotic coronary arteries. N Engl J Med 316:1371–5.
32. Goertz D, Frijlink M, Tempel D et al. (2007) Subharmonic contrast intravascular ultrasound for vasa vasorum imaging. Ultrasound Med Biol 33(12):1859–72.
33. Goertz DE, Frijlink ME, Tempel D et al. (2006) Contrast harmonic intravascular ultrasound: a feasibility study for vasa vasorum imaging. Invest Radiol 41:631–8.
34. Gold L, Polisky B, Uhlenbeck O et al. (1995) Diversity of oligonucleotide functions. Annu Rev Biochem 64:763–97.
35. Gössl M, Versari D, Mannheim D et al. (2006) Increased spatial vasa vasorum density in the proximal LAD in hypercholesterolemia-implications for vulnerable plaque-development. Atherosclerosis 192(2):246–52.
36. Graf K, Dietrich T, Tachezy M et al. (2008) Monitoring therapeutical intervention with ezetimibe using targeted near-infrared fluorescence imaging in experimental atherosclerosis. Mol Imaging 7:68–76.
37. Haider N, Hartung D, Fujimoto S et al. (2009) Dual molecular imaging for targeting metalloproteinase activity and apoptosis in atherosclerosis: molecular imaging facilitates understanding of pathogenesis. J Nucl Cardiol 16:753–62.
38. Hamilton AJ, Huang S-L, Warnick D et al. (2004) Intravascular ultrasound molecular imaging of atheroma components in vivo. J Am Coll Cardiol 43:453–60.
39. Herrmann J, Lerman LO, Rodriguez-Porcel M et al. (2001) Coronary vasa vasorum neovascularization precedes epicardial endothelial dysfunction in experimental hypercholesterolemia. Cardiovasc Res 51:762–6.
40. Hong HY, Lee HY, Kwak W et al. (2007) Phage display selection of peptides that home to atherosclerotic plaques: IL-4 receptor as a candidate target in atherosclerosis. J Cell Mol Med 12(5B):2003–14.

41. Hyafil F, Cornily JC, Feig JE et al. (2007) Noninvasive detection of macrophages using a nanoparticulate contrast agent for computed tomography. Nat Med 13:636–41.
42. Iiyama K, Hajra L, Iiyama M et al. (1999) Patterns of vascular cell adhesion molecule-1 and intercellular adhesion molecule-1 expression in rabbit and mouse atherosclerotic lesions and at sites predisposed to lesion formation. Circ Res 85:199–207.
43. Jaffer FA, Kim DE, Quinti L et al. (2007) Optical visualization of cathepsin K activity in atherosclerosis with a novel, protease-activatable fluorescence sensor. Circulation 115:2292–8.
44. Johnson LL, Schofield L, Donahay T et al. (2005) 99mTc-annexin V imaging for in vivo detection of atherosclerotic lesions in porcine coronary arteries. J Nucl Med 46:1186–93.
45. Kaufmann BA, Carr CL, Belcik JT et al. (2010) Molecular imaging of the initial inflammatory response in atherosclerosis: implications for early detection of disease. Arterioscler Thromb Vasc Biol 30:54–9.
46. Kaufmann BA, Sanders JM, Davis C et al. (2007) Molecular imaging of inflammation in atherosclerosis with targeted ultrasound detection of vascular cell adhesion molecule-1. Circulation 116:276–84.
47. Kawamura A, Miura S, Murayama T et al. (2004) Increased expression of monocyte CD11a and intracellular adhesion molecule-1 in patients with initial atherosclerotic coronary stenosis. Circ J 68(1):6–10.
48. Kelly KA, Allport JR, Tsourkas A et al. (2005) Detection of vascular adhesion molecule-1 expression using a novel multimodal nanoparticle. Circ Res 96:327–36.
49. Kelly KA, Nahrendorf M, Yu AM et al. (2006) In vivo phage display selection yields atherosclerotic plaque targeted peptides for imaging. Mol Imaging Biol 8:201–7.
50. Kietselaer BL, Reutelingsperger CP, Heidendal GA et al. (2004) Noninvasive detection of plaque instability with use of radiolabeled annexin A5 in patients with carotid-artery atherosclerosis. N Engl J Med 350:1472–3.
51. Kircher M, Grimm J, Swirski F et al. (2008) Noninvasive in vivo imaging of monocyte trafficking to atherosclerotic lesions. Circulation 117(3):388–95.
52. Klibanov AL, Maruyama K, Torchilin VP et al. (1990) Amphipathic polyethyleneglycols effectively prolong the circulation time of liposomes. FEBS Lett 268:235–7.
53. Klibanov AL, Rychak JJ, Yang WC et al. (2006) Targeted ultrasound contrast agent for molecular imaging of inflammation in high-shear flow. Contrast Media Mol Imaging 1:259–66.
54. Koivunen E, Arap W, Rajotte D et al. (1999) Identification of receptor ligands with phage display peptide libraries. J Nucl Med 40:883–8.
55. Kolonin MG, Sun J, Do KA et al. (2006) Synchronous selection of homing peptides for multiple tissues by in vivo phage display. FASEB J 20:979–81.
56. Korosoglou G, Weiss RG, Kedziorek DA et al. (2008) Noninvasive detection of macrophage-rich atherosclerotic plaque in hyperlipidemic rabbits using "positive contrast" magnetic resonance imaging. J Am Coll Cardiol 52:483–91.
57. Lancelot E, Amirbekian V, Brigger I et al. (2008) Evaluation of matrix metalloproteinases in atherosclerosis using a novel noninvasive imaging approach. Arterioscler Thromb Vasc Biol 28:425–32.
58. Lanza GM, Wallace KD, Scott MJ et al. (1996) A novel site-targeted ultrasonic contrast agent with broad biomedical application. Circulation 94:3334–40.
59. Leong-Poi H, Christiansen J, Klibanov AL et al. (2003) Noninvasive assessment of angiogenesis by ultrasound and microbubbles targeted to alpha(v)-integrins. Circulation 107: 455–60.
60. Li H, Gray BD, Corbin I et al. (2004) MR and fluorescent imaging of low-density lipoprotein receptors. Acad Radiol 11:1251–9.
61. Liu C, Bhattacharjee G, Boisvert W et al. (2003) In vivo interrogation of the molecular display of atherosclerotic lesion surfaces. Am J Pathol 163:1859–71.
62. Matter CM, Schuler PK, Alessi P et al. (2004) Molecular imaging of atherosclerotic plaques using a human antibody against the extra-domain B of fibronectin. Circ Res 95:1225–33.

63. McAteer MA, Schneider JE, Ali ZA et al. (2008) Magnetic resonance imaging of endothelial adhesion molecules in mouse atherosclerosis using dual-targeted microparticles of iron oxide. Arterioscler Thromb Vasc Biol 28:77–83.
64. Montecucco F, Steffens S, Burger F et al. (2008) Tumor necrosis factor-alpha (TNF-alpha) induces integrin CD11b/CD18 (Mac-1) up-regulation and migration to the CC chemokine CCL3 (MIP-1alpha) on human neutrophils through defined signalling pathways. Cell Signal 20(3):557–68.
65. Nahrendorf M, Jaffer FA, Kelly KA et al. (2006) Noninvasive vascular cell adhesion molecule-1 imaging identifies inflammatory activation of cells in atherosclerosis. Circulation 114:1504–11.
66. Nahrendorf M, Waterman P, Thurber G et al. (2009) Hybrid in vivo FMT-CT imaging of protease activity in atherosclerosis with customized nanosensors. Arterioscler Thromb Vasc Biol 29:1444–51.
67. Nahrendorf M, Zhang H, Hembrador S et al. (2007) Nanoparticle PET-CT imaging of macrophages in inflammatory atherosclerosis. Circulation 117(3):379–87.
68. Nakashima Y, Raines EW, Plump AS et al. (1998) Upregulation of VCAM-1 and ICAM-1 at atherosclerosis-prone sites on the endothelium in the ApoE-deficient mouse. Arterioscler Thromb Vasc Biol 18:842–51.
69. Ozawa MG, Zurita AJ, Dias-Neto E et al. (2008) Beyond receptor expression levels: the relevance of target accessibility in ligand-directed pharmacodelivery systems. Trends Cardiovasc Med 18:126–32.
70. Pasqualini R. (1999) Vascular targeting with phage peptide libraries. Q J Nucl Med 43:159–62.
71. Pasqualini R, Ruoslahti E. (1996) Organ targeting in vivo using phage display peptide libraries. Nature 380:364–6.
72. Rajotte D, Arap W, Hagedorn M et al. (1998) Molecular heterogeneity of the vascular endothelium revealed by in vivo phage display. J Clin Invest 102:430–7.
73. Robert R, Jacobin-Valat M-J, Daret D et al. (2006) Identification of human scFvs targeting atherosclerotic lesions: selection by single round in vivo phage display. J Biol Chem 281:40135–43.
74. Rychak JJ, Li B, Acton ST et al. (2006) Selectin ligands promote ultrasound contrast agent adhesion under shear flow. Mol Pharm 3:516–24.
75. Samijo SK, Willigers JM, Barkhuysen R et al. (1998) Wall shear stress in the human common carotid artery as function of age and gender. Cardiovasc Res 39:515–22.
76. Sarai M, Hartung D, Petrov A et al. (2007) Broad and specific caspase inhibitor-induced acute repression of apoptosis in atherosclerotic lesions evaluated by radiolabeled annexin A5 imaging. J Am Coll Cardiol 50:2305–12.
77. Schumann PA, Christiansen JP, Quigley RM et al. (2002) Targeted-microbubble binding selectively to GPIIb IIIa receptors of platelet thrombi. Invest Radiol 37:587–93.
78. Scott JK, Smith GP. (1990) Searching for peptide ligands with an epitope library. Science 249:386–90.
79. Shaw PX, Horkko S, Tsimikas S et al. (2001) Human-derived anti-oxidized LDL autoantibody blocks uptake of oxidized LDL by macrophages and localizes to atherosclerotic lesions in vivo. Arterioscler Thromb Vasc Biol 21:1333–9.
80. Sirol M, Fuster V, Badimon JJ et al. (2005) Chronic thrombus detection with in vivo magnetic resonance imaging and a fibrin-targeted contrast agent. Circulation 112:1594–600.
81. Smith GP, Scott JK. (1993) Libraries of peptides and proteins displayed on filamentous phage. Methods Enzymol 217:228–57.
82. Stieger SM, Dayton PA, Borden MA et al. (2008) Imaging of angiogenesis using Cadence contrast pulse sequencing and targeted contrast agents. Contrast Media Mol Imaging 3:9–18.
83. Swirski FK, Pittet MJ, Kircher MF et al. (2006) Monocyte accumulation in mouse atherogenesis is progressive and proportional to extent of disease. Proc Natl Acad Sci USA 103: 10340–5.
84. Takalkar AM, Klibanov AL, Rychak JJ et al. (2004) Binding and detachment dynamics of microbubbles targeted to P-selectin under controlled shear flow. J Control Release 96:473–82.

85. Thapa N, Hong H-Y, Sangeetha P et al. (2008) Identification of a peptide ligand recognizing dysfunctional endothelial cells for targeting atherosclerosis. J Control Release 131:27–33.
86. Tiukinhoy-Laing SD, Buchanan K, Parikh D et al. (2007) Fibrin targeting of tissue plasminogen activator-loaded echogenic liposomes. J Drug Target 15:109–14.
87. Tiukinhoy-Laing SD, Huang S, Klegerman M et al. (2007) Ultrasound-facilitated thrombolysis using tissue-plasminogen activator-loaded echogenic liposomes. Thromb Res 119:777–84.
88. Torzewski M, Shaw PX, Han KR et al. (2004) Reduced in vivo aortic uptake of radiolabeled oxidation-specific antibodies reflects changes in plaque composition consistent with plaque stabilization. Arterioscler Thromb Vasc Biol 24:2307–12.
89. Tsimikas S. (2002) Noninvasive imaging of oxidized low-density lipoprotein in atherosclerotic plaques with tagged oxidation-specific antibodies. Am J Cardiol 90:22L–7L.
90. Tsimikas S, Palinski W, Halpern SE et al. (1999) Radiolabeled MDA2, an oxidation-specific, monoclonal antibody, identifies native atherosclerotic lesions in vivo. J Nucl Cardiol 6:41–53.
91. Tsimikas S, Shortal BP, Witztum JL et al. (2000) In vivo uptake of radiolabeled MDA2, an oxidation-specific monoclonal antibody, provides an accurate measure of atherosclerotic lesions rich in oxidized LDL and is highly sensitive to their regression. Arterioscler Thromb Vasc Biol 20:689–97.
92. Unger EC, McCreery TP, Sweitzer RH et al. (1998) In vitro studies of a new thrombus-specific ultrasound contrast agent. Am J Cardiol 81:58G–61G.
93. Vicenzini E, Giannoni MF, Puccinelli F et al. (2007) Detection of carotid adventitial vasa vasorum and plaque vascularization with ultrasound cadence contrast pulse sequencing technique and echo-contrast agent. Stroke 38:2841–3.
94. Wagner S, Breyholz H-J, Faust A et al. (2006) Molecular imaging of matrix metalloproteinases in vivo using small molecule inhibitors for SPECT and PET. Curr Med Chem 13:2819–38.
95. Waldeck J, Häger F, Höltke C et al. (2008) Fluorescence reflectance imaging of macrophage-rich atherosclerotic plaques using an alphavbeta3 integrin-targeted fluorochrome. J Nucl Med 49:1845–51.
96. Weissleder R, Kelly K, Sun EY et al. (2005) Cell-specific targeting of nanoparticles by multivalent attachment of small molecules. Nat Biotechnol 23:1418–23.
97. Williams JK, Heistad DD. (1996) Structure and function of vasa vasorum. Trends Cardiovasc Med 6:53–7.
98. Winter PM, Morawski AM, Caruthers SD et al. (2003) Molecular imaging of angiogenesis in early-stage atherosclerosis with alpha(v)beta3-integrin-targeted nanoparticles. Circulation 108:2270–4.
99. Winter PM, Neubauer AM, Caruthers SD et al. (2006) Endothelial alpha(v)beta3 integrin-targeted fumagillin nanoparticles inhibit angiogenesis in atherosclerosis. Arterioscler Thromb 26:2103–9.
100. Wu Y, Unger EC, McCreery TP et al. (1998) Binding and lysing of blood clots using MRX-408. Invest Radiol 33:880–5.

Biographies

Patrick Kee is an assistant professor in the Division of Cardiology at the University of Texas Health Science at Houston. He is a clinical cardiologist with research interests in lipid metabolism and molecular imaging of atherosclerosis. Currently, his research focuses include the discovery of novel molecular markers of atherosclerosis and the validation of those markers in atherosclerotic animal models.

Wouter Driessen is a postdoctoral fellow in the David H. Koch Center at the University of Texas MD. Anderson Cancer Center in Houston. He obtained his Ph.D. in pharmaceutics from the University of Florida in Gainesville with a focus on nanoparticle preparation and characterization. His current research interest is in the area of targeted delivery of drugs, genes, nucleic acids, and imaging agents to angiogenic sites.

Chapter 24
Biologic Nanoparticles and Vascular Disease

Maria K. Schwartz, John C. Lieske, and Virginia M. Miller

Abstract Sustained or repeating chronic infections have been linked to arterial calcification. Biologic nanoparticles (NPs) that propagate in culture have been isolated from calcified soft tissues. When inoculated into experimental animals, NPs appear to exacerbate the arterial response to injury, up to and including medial discontinuity and vascular occlusion. These NPs may be re-assembled proteins and microvesicles derived from mammalian cells and/or microorganisms.

Keywords Arterial calcification • Calcium • Hydroxyapatite • Infection • Inflammation

24.1 Introduction

Calcification of arterial tissue is a common occurrence that is a strong predictor of cardiovascular and all-cause mortality [16, 23]. Increases in circulating low-density lipoproteins (LDL) that lead to abnormal lipid metabolism initiate endothelial cell inflammatory responses characterized by expression of adhesion molecules, infiltration of macrophages, and endothelial dysfunction [37, 53]. However, significant arterial calcification can be observed in individuals whose circulating levels of LDL and cholesterol are below that which would typically require pharmacological intervention [28, 36, 40, 44] and in patients with organ transplants and on renal dialysis [9, 61]. This chapter will explore alternative mechanisms which may contribute to accelerated arterial calcification with particular attention to the contribution of nano-sized biologics as potential stimuli for inflammation and calcific nidi.

V.M. Miller (✉)
Departments of Surgery and Physiology and Biomedical Engineering, Mayo Clinic, 4-62 Medical Science Building, 200 First Street Southwest, Rochester, MN 55905, USA
e-mail: miller.virginia@mayo.edu

Jasjit S. Suri et al. (eds.), *Atherosclerosis Disease Management*,
DOI 10.1007/978-1-4419-7222-4_24, © Springer Science+Business Media, LLC 2011

24.2 Infection and Atherosclerosis

Sustained or repeated episodes of subclinical infection may initiate and maintain enzymatic dysregulation that contributes to accelerated arterial calcification [3, 54]. Indeed, *Chlamydia pneumoniae*, *Helicobacter pylori*, *Mycoplasma pneumonia*, and periodontal disease [47] have been implicated in accelerated arterial disease processes, although the exact mechanisms by which infection might lead to calcification are not clear. It has been suggested that activation of toll-like receptors or pathogen degradation products lead to the generation of specific calcium nucleating sites, called matrix vesicles, on vascular smooth muscle cells [6, 8, 22, 32, 52, 59, 64]. While the exact composition and functional attributes of matrix vesicles are not defined, calcifying, nano-sized particles can be isolated from calcified diseased human tissues including arterial calcifications [46], polycystic kidney disease [25], kidney stones [10, 35], prostatitis [67], and aortic valve calcification [9]. These isolated nanoparticles (NPs) can be sustained in cell culture, allowing for compositional and functional analysis. Whether these particles represent a type of transferrable, infective, inflammatory stimulus that initiates calcific processes, are protein/mineral complexes formed as result of biochemical interactions, or both remains a subject of debate and investigation.

24.3 History of Biologic Nanoparticles

Calcium carbonate spheres ranging from 30 to 300 nm in size were observed in deposits in hot springs of Viterbo, Italy by geologist Dr. Robert Folk and coworkers. Because their structure was similar to bacteria and because of their nanometer size, they were referred to as "nan(n)obacteria" [20]. Similar structures isolated from bovine serum propagate and calcify under cell culture conditions, supporting the idea that these particles might be biological and perhaps represent a new type of life form [30]. Related structures were observed on a Martian meteorite shortly thereafter [43]. In addition to geological sources and mammalian blood [30, 41, 51], similar nanobacteria-like particles, or nanoparticles (NPs) have been identified in and isolated from a variety of diseased tissue, including calcified arteries [46], atherosclerotic plaques [21], calcified aortic valves [9], and kidney stones [10, 35]. Moreover, NPs have been observed in or linked to prostatitis [60], ovarian cancer [58], gallstones [63], and calcium deposition in the placenta [2].

To date, these NPs have largely been considered to be similar not by chemical or biological analysis, but rather due to their similar morphology (Fig. 24.1). Indeed, a definitive lack of biochemical criteria for identification and classification of NPs continues to cause confusion and debate within the field. However, since this is the current state of the art, for the remainder of this chapter the term biologic nanoparticles (NPs) will be used to refer to nano-sized particles derived from biological tissue regardless of their composition.

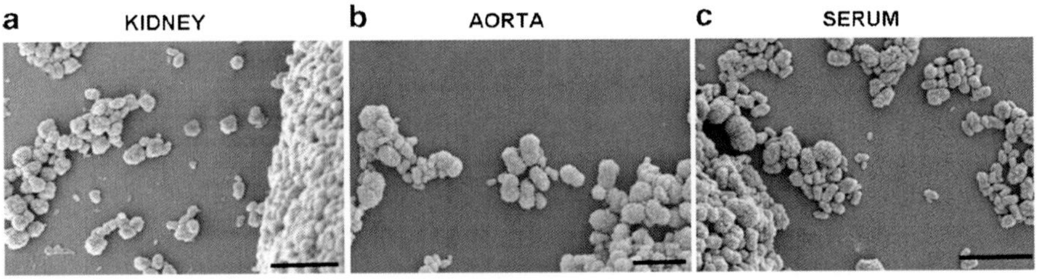

Fig. 24.1 SEM micrographs of NPs derived from human kidney stones (**a**), human aorta (**b**), and bovine serum (**c**). Bar indicates 1 μm. (**b**, **c**) are shown as they appear in a previous publication [57]. Kidney stones were demineralized in HCl, then neutralized, and centrifuged; the resulting NP pellet was rinsed in phosphate-buffered saline (PBS) and seeded into culture flasks containing standard Dulbecco's minimal essential media (DMEM;10% FBS, 1.8 mM Ca^{2+}, pH 7.5, 10% CO_2). Aortic-derived NPs were prepared from calcified aneurysms collected as surgical waste from patients undergoing vascular repair at Mayo Clinic, Rochester, MN. The tissue was homogenized in sterile PBS, then filtered through a 0.2 μm filter before seeding into DMEM-containing flasks. Bovine serum-derived NPs were obtained from NANOBAC OY (Kuopio, Finland). Modified from Fig. 1 of reference [57]

24.4 Biochemical Characterization of Biologic Nanoparticles

Biologic NPs, which are coccoid or spherical in shape, range in size from 50 to 200 nm and propagate and calcify when placed in standard cell culture conditions [30, 45, 62]. These have been derived from three sources: mammalian blood, renal stones, and calcified arterial tissue. Biologic NPs form a hydroxyapatite [HA; $Ca_5(PO_4)_3(OH)$] shell [30, 45, 62] when seeded into standard cell culture media and a large percentage (40.4–42.8%) of total NP culture is inorganic material (calcium and phosphorous contribute 23.4–23.5% and 12.3–14.6%, respectively) [29]. Biologic NPs also contain lipids [66], appear to have a membrane structure [30, 31, 35, 46], stain positive for DNA, uptake [^{3}H]uridine [30, 46], and stain gram negative [30, 31]. It should be noted that the typical preparation of biologic NPs from tissue involves homogenization or pulverization of the diseased tissue or renal stone and the resultant material is forced through a 0.2 μm membranous filter. Therefore, it may be suggested that this process results in cell remnants/fragments which retain biological activity under cell culture conditions. However, not all tissue homogenates or blood plasma produce sustainable cultures of biologic NPs [35, 46, 56]. Therefore, further investigation is necessary to fully define the origin and composition of these NPs.

The presence and characterization of nucleic acids in biologic NP cultures is controversial. A 16S RNA sequence has been isolated from bovine-derived NPs and added to the NCBI nucleotide gene bank (*Nanobacteria sanguineum*, accession no. X98418). However, other groups have been unable to replicate this sequence [14, 35, 41]. Although DNA has been identified in and isolated from NP cultures by many groups [14, 30, 35, 46], it has been suggested that the DNA present in NP cultures is environmental contamination resulting from isolation techniques and not native to the NPs themselves [14]. To date, data are inconclusive, although it is fair to say that neither a DNA nor RNA sequence unique to NPs has yet been reported to settle this debate.

Whatever be the origin of biologic NPs, the density of culture media seeded with NPs (determined by changes in optical density or turbidity) doubles within 1–5 days [30]. Reducing agents such as β-mercaptoethanol (BME) increase NP propagation in culture while metabolic inhibitors such as sodium azide, antimycin A, potassium cyanide [35], and 5-fluorouracil [13] retard it (Table 24.1). Certain antimicrobials, such as tetracycline, ampicillin [13, 35, 56], trimethoprim, nitrofurantoin, and trimethoprim-sulfame-thoxazole [13], also inhibit NP propagation. Interestingly, other antibiotics, such as nalidixic acid [35], penicillin, neomycin, and chloramphenicol [13], do not affect NP propagation, suggesting some degree of specificity. Notably, a pair of antibiotics that do not chelate calcium, gentamicin and vancomycin, also have a negative effect on NP propagation but do not completely inhibit it (Table 24.1) [13, 35, 56]. The difference in efficacy of various antibiotics on reducing

Table 24.1 Effects of various compounds on biologic NP propagation in culture

Function	Compound	Calcium chelator	Effect on NP cultures	Source
Reducing agent	β-Mercaptoethanol	–	Increase	[35]
Metabolic inhibitor	Sodium azide	–	Inhibitory	[35]
	Antimycin A	–	Inhibitory	[35]
	Potassium cyanide	–	Inhibitory	[35]
	5-Fluorouracil	–	Inhibitory	[13]
Antimicrobial	Tetracycline	+	Inhibitory	[13, 35, 56]
	Ampicillin	+	Inhibitory	[13, 35]
	Trimethoprim	–	Inhibitory	[13]
	Nitrofurantoin	–	Inhibitory	[13]
	Trimethoprim-sulfamethoxazole	–	Inhibitory	[13]
	Nalidixic acid		No effect	[35]
	Penicillin	+	No effect	[13]
	Neomycin	–	No effect	[13]
	Chloramphenicol	–	No effect	[13]
	Gentamicin	–	Modulatory	[13, 35, 56]
	Vancomycin	–	Modulatory	[13]
	Erythromycin	–	No effect	[13]
	Ciprofloxacin	+	No effect	[13]
	Doxycycline	+	Initial inhibition, no effect over time	[13]
Osteoclast inhibitor	Bisphosphonates	–	Inhibitory	[13]
Extracellular RNA degrader	RNase	–	No effect	[56]

+ chelator, – non-chelator or unknown

the propagation of NPs may reflect differences in the ability of the drugs to chelate calcium, but other mechanisms also appear to be involved [56].

NPs in culture also incorporate [^{35}S] methionine, suggesting ongoing protein synthesis [30]. A number of mammalian and non-mammalian proteins have been characterized from biologically derived NPs in culture, including EF-Tu, GroEL, pyruvate dehydrogenase, and polyribonucleotide nucleotidyl transferase, which appear to be prokaryotic [35]. However, the presence of these proteins has been disputed; a number of groups argue that the proteins associated with the biologic NPs are serum proteins accumulated from the culture media [41, 51, 66]. Conversely, the protein profiles of biologic NPs differs from those of HA crystals incubated in culture media [35, 45]. Clarification of a protein signature for NPs derived from different sources would lead to a better understanding of NPs as a potential cause or result of disease processes.

Initial experiments involving propagation and characterization of biologic NPs all utilized a biofilm that adhered to culture flasks over time [14]. However, NPs exist both as an adherent biofilm and floating in the culture media, much the same way a standard biofilm would [4, 15]. These floating particles are referred to as planktonic NPs, and are observed both in culture flasks and as the sole state when NPs are cultured in simulated microgravity [11]. While NPs in both states are still biologically derived NPs, the characteristics of NPs in biofilm may differ from those of the planktonic form. For example, biofilms form a matrix containing polysaccharides, proteins, and other polymeric substances [15]. NP biofilm derived from arterial homogenates contains polysaccharides similar to those of more conventional biofilm [56]. Therefore, matrix NPs may have biological effects distinct from planktonic NPs both in vitro and in vivo.

24.5 Are Biologic Nanoparticles Lifeforms?

As noted above, biologic NPs were originally described as nan(n)obacteria [20, 30, 43]. Their apparent "growth" in culture, the susceptibility of those cultures to aerobic inhibitors [13, 35], and the isolation of the 16S RNA sequence appeared to support the hypothesis that biologic NPs were, in fact, a novel organism. However, this claim ignited controversy. First and foremost, these NPs were considered too small to contain the minimal amount of volume deemed necessary to sustain the enzymatic and concentration gradient-mediated processes of an independent life form. Scientists argued that a minimum diameter of at least 100 nm was necessary to maintain the proposed biochemical processes [39, 48]. Indeed, a panel appointed by the National Research Council determined that for life *as we currently know it* to exist, a minimum size of 200 nm was required. However, the same panel also suggested that organisms could be as small as 50 nm if they operate by different "molecular rules" than those of organisms identified and described in 1999 [33]. Indeed, the range in size of biologic NPs overlaps with that of both viruses and

Table 24.2 Size comparisons for biologic and inorganic nanoparticles, microvesicles, typical pathogens, and blood elements

	Comparative sizes
Ribosomes	20 nm
Elemental nanoparticles	5–30 nm
Generally nanotechnology	<90 nm
Exosomes	59–90 nm
Prions	70–150 nm
Biologic nanoparticles	50–200 nm
Viruses	10–300 nm
Microvesicles	100–700 nm
Platelets	2–4 μm
Archaebacteria	0.1–5 μm
Bacteria	0.5–5 μm
Red blood cells	6–8 μm

archaeabacteria (Table 24.2), supporting the idea that the biologic NPs (50–200 nm) are not too small to be an organism.

However, an important component of the debate regarding whether biologic NPs are an organism is the lack of a reliably isolated genomic sequence [14, 35, 41]. It is argued that nucleic acid sequences associated with NPs are a contaminant rather than derived from NPs [14]. On the other hand, recent evidence suggests that current standard methods of nucleic acid sequencing may not be able to detect certain atypical genomes such as small subunit ribosomal RNA (ssRNA) sequences [26]. Thus, it remains conceivable that NPs could represent a biologic life form, especially a symbiotic and/or deficient one.

Another aspect of the debate centers on the origin of biologic NPs. Several groups have argued that NPs result from protein–mineral interactions [41, 51, 65, 66]. Others contend that current experimental evidence suggests enzymatic processes to play a role [13, 35, 42, 62, 46, 56]. However, one thing remains certain. Whatever be their origin, biologic NPs may be a factor in cardiovascular disease processes.

24.6 Biologic Nanoparticles as a Transmissible Cause of Disease

The previous sections acknowledge that the classification of NPs derived from mammalian tissues remains under dispute. However, these NPs have been isolated and propagated from diseased tissues and implicated in calcific disease processes in experimental animal model systems. Cell-derived matrix vesicles and/or cell-derived microparticles/microvesicles may contribute to vascular calcification [1, 18, 28, 32, 34] and human-derived NPs (whatever be their composition – cell-derived debris or hydroxyapatite–protein complexes [41, 51, 66]) may act in the same manner.

Biologically derived NPs from different sources exhibit varying degrees of cytotoxicity. NPs derived from healthy human serum did not show a cytotoxic effect when applied to 3T6 mammalian cells in culture. However, toxicity of NPs derived from fetal bovine serum (FBS) varied between sources, since FBS-derived NPs from one supplier did not have an effect on 3T6 cell survival, while FBS-derived NPs from two other suppliers significantly reduced cell viability, with an increasing effect over time. The NPs seemed to exert the greatest effect between 48 and 72 h after inoculation of cell cultures. The most cytotoxic of the NPs were then introduced to cultures of various mammalian fibroblast cells (B_6MCA, BHK, CHO). The cytotoxic effects (i.e., cell lysis) of NPs were dependent on cell line, time, and dose. In addition, the FBS-derived NPs appeared to nonrandomly adhere and be internalized into 3T6 cells [12]. These results provided justification for the evaluation of other effects of biologic NPs on cellular activity, including the response to vascular cell injury.

When ^{99m}Tc-labled FBS-derived NPs were injected intravenously into healthy animals, they appeared in the urine within 15 min of the inoculation. Other NPs were observed (by single photon emission computed tomography) in the kidneys, heart, spleen, intestine and stomach after 45 h [5]. However, biologic NPs appear to have different effects when injected into animals with vascular injury [55, 57]. Inoculation of NPs derived from human kidney stones or calcified human arteries into rabbits with endothelium removed from one carotid artery resulted in narrowing, occlusion, and calcification of the region of injury. Injured vessels in animals inoculated with saline, lipopolysaccharide (LPS, as a surrogate for sub-clinical infection), or hydroxyapatite crystals (HA, to control for inorganic material in NPs shells) developed similar degrees of intimal hyperplasia but no narrowing or calcification. The injured vessels in animals inoculated with HA exposed to culture conditions or NPs derived from bovine serum did not occlude, but rather developed a discontinuous internal elastic lamina and media layer at the site of damage [55, 57]. This study suggests that bovine-derived NPs have different pathophysiological effects than human-derived NPs.

While the debate over the composition and definition of biologic NPs is likely to continue, one point cannot be debated: biologic NPs have effects that render them relevant to cardiovascular research. Indeed, the pathophysiological consequences of biologic NPs seem to be dependent upon their specific composition. For instance, in the majority of cases, circulating biologic NPs derived from human aneurysm and kidney stones caused vascular occlusion in endothelium-denuded vessels. Moreover, bovine-derived NPs modulate the vascular response to injury in a manner similar to that of hydroxyapatite (HA) incubated in culture media [57]. These differential effects of NPs from different sources and origins provide further rationale for investigating the effects of biologic NPs on physiological systems, as even pure HA particles appear to have pathophysiological effects when proteins are attached. As such, it is evident that nano-sized particles, whether inorganic or cellular in nature, may have long-term cardiovascular effects based solely on their size, as well as their composition.

One can argue that it does not matter if NPs are a life form or not. What should be debated is whether or not NPs are a novel transmissible form of disease-causing agent. The gold standard for demonstrating that a suspected agent causes disease is Koch's Postulates. Initially, biologic NPs were described as nanobacteria. NPs have been identified in and isolated from calcified human arterial tissue and propagated in culture [45], thus fulfilling Koch's first two conditions. Experiments have also shown that when previously unexposed rabbits are inoculated with these NPs, arterial pathology, including occlusion with and without calcification, results [55, 57]. The heterogeneous nature of the responses most likely reflects the heterogeneity of NP isolates, as NPs of bovine origin promoted arterial disorganization rather than occlusion. This could reflect differences in composition between NP isolates. While there were no attempts to isolate NPs from the rabbit tissues to fulfill the fourth and final of Koch's Postulates, studies to date have built an excellent case suggesting that biologic NPs affect vascular response to injury and vascular calcification and could be a transmissible cause of disease.

Koch's Postulates were developed to provide criteria for assessing microbial causation of disease. However, through experiments designed to define the biological nature of the human-derived NPs, it is clear that their composition is heterogeneous and not consistent with a single microbe. Instead, biologic NPs may represent particles derived from commensal bacterial or autologous tissues that modulate physiological processes [56]. Therefore, Koch's Postulates may not be adequate to assess their contribution to pathophysiology. Alternatively, Hill's Criteria [24] were developed to assess the potential role of environmental contaminants such as carbon and coal dust (which may be present in NP size ranges) in disease. These criteria are another way to evaluate whether an association or a causative relationship exists between a stimulus, in this case NPs, and diseases such as arterial calcification.

The first criterion to be evaluated is the *strength* of the apparent association between the presence of NPs and soft tissue calcification. In chronic in vivo rabbit experiments discussed above, the only vessels to calcify were in rabbits inoculated with human arterial-derived NPs [55, 57]. Structures of size similar to NPs have been *consistently* identified, by multiple groups, in soft tissue calcifications [9, 10, 21, 30, 35, 41, 46, 51, 58, 60, 63]. Further, the response to NPs appears to be *specific* – vascular occlusion only occurred in rabbits inoculated with human arterial-derived NPs with no occlusion in the injury controls [55, 57]. There is also *temporality* of the response, that is, vascular occlusion followed inoculation with NPs [55, 57].

A large body of literature already exists to link cardiovascular disease to biochemical factors such as lipoproteins and chronic infection [7, 19, 47, 50]. Therefore, the idea that NPs are biochemically active components of the blood, perhaps derived from cell or bacterial membranes, is *plausible*. In addition, the possibility that NPs promote soft tissue calcification is *coherent*; it does not violate any known biological law. And again, the proliferation of evidence demonstrating a link between blood lipoproteins, cytokines, and chronic infection to cardiovascular disease provides an *analogy* to NPs and cardiovascular disease. As such, seven of Hill's nine criteria are met. The final two, showing a *biological gradient*, and *experiment*, or showing that eliminating NPs from a system removes or lessens the disease process, are still under investigation. Consequently, Hill's Criteria also support the idea that NPs may be environmental or endogenous agents, carried within the blood (internal environment), that can be causative in exaggerating the vascular response to injury.

24.7 Nanoparticles: Toward a Unifying Hypothesis

Many times over the course of development in a field, it is discovered that two parallel lines of research are actually studying the same phenomenon. Microvesicles, or microparticles, with a diameter of <1 μm are released from cell membranes during events such as apoptosis and cell activation [38]. These microparticles can be identified by the surface markers they carry from their cell of origin and are associated with pathophysiology [17, 27]. It is possible that the human-derived NPs described in this chapter are actually the equivalent of microparticles derived from human apoptotic cells or commensal bacteria (Fig. 24.2). If that were the case, it would explain the plethora of conflicting literature surrounding biologic NPs to date. For example, the tendency of NPs to stain gram negative might be explained if they originated from a gram-negative bacteria. Indeed, "nanobacteria" may in fact be smaller fragments of commensal bacteria. The possibility of crossover between these two fields certainly warrants deeper investigation in the future.

24.8 Conclusions

Nanotechnology has emerged as a scientific discipline. However, effects of nano-sized materials on biological systems are, at best, only peripherally understood. Nano-materials have demonstrated unique effects on biological systems depending on a number of factors including their original composition and size. Indeed, materials on a nano level often have different characteristics and effects than the same material at a macro level [49]. NPs that appear to self-propagate and self-calcify in vitro have been derived from human tissues [9, 10, 21, 35, 46]. These NPs contain nucleic acids and proteins that may either be derived from the environment or

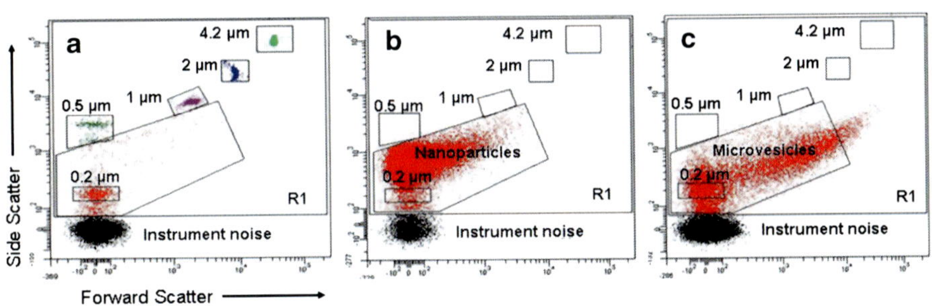

Fig. 24.2 Representative scatter *dot plots* of flow cytometric analysis. (**a**) *Scatter dot plot* derived from twice filtered Hanks'/HEPES buffer pH 7.4 plus size calibration beads. (**b**) *Scatter dot plot* of decalcified NPs cultured from a human kidney stone. (**c**) *Scatter dot plot* of isolated microvesicles isolated from blood of healthy individuals. All plots were obtained using a FACSCanto™ flow cytometry. Most of the signal from the decalcified, cultured NPs fell within <0.5 μm. Although the range of signal from microvesicles isolated from human blood were <1 μm, some fell within the size range of NPs

characteristic of the NPs themselves, and further study of their origins is merited. In particular, the evidence indicating that the formation of NP biofilm is both a chemical and biological process compels further examination of the composition of human-derived NPs.

Previous research has potentially linked biologic NPs to a variety of diseases [2, 9, 10, 25, 35, 45, 58, 60, 63, 67]. Experiments described in this chapter provide evidence that human-derived NPs increase the vascular response to injury, leading to vascular occlusion and even calcified plaque formation [55, 57]. Several of Hill's Criteria [24] and the first three of Koch's Postulates have been fulfilled by these studies. However, further work is required in order to better understand the pathophysiological mechanisms of action of human-derived NPs on vascular systems.

The current body of research relating to biologic NPs draws a number of conflicting conclusions and contradictory implications. However, one potential explanation for these contradictions may be found in another area of research. Microvesicles (often the same size as NPs) can be derived from cell membranes; they carry cell surface markers and can initiate pathophysiological responses [27, 38]. It may be that human-derived NPs are, in fact, cellular or the bacterial equivalent of cell-derived microparticles. It is clear that biologic NPs represent a novel agent that can transmit disease in experimental model systems. Perhaps, it is best not to get caught up in the debate over whether they are "alive" or not, but rather focus on their pathophysiological effects.

References

1. Abedin M, Tintut Y, Demer LL. (2004) Vascular calcification: mechanisms and clinical ramifications. Arterioscler Thromb Vasc Biol 24:1161–1170.
2. Agababov RM, Abashina TN, Suzina NE, et al. (2007) Link between the early calcium deposition in placenta and nanobacterial-like infection. J Biosci 32:1163–1168.
3. Aikawa E, Nahrendorf M, Figueiredo JL, et al. (2007) Osteogenesis associates with inflammation in early-stage atherosclerosis evaluated by molecular imaging in vivo. Circulation 116:2841–2850.
4. Akerman KK, Juronen I, Kajander EO. (1993) Scanning electron microscopy of nanobacteria – novel biofilm producing organisms in blood. Scan Electron Microsc 15:sIII:90–91.
5. Akerman KK, Kuikka JT, Ciftcioglu N, et al. (1997) Radiolabeling and in vivo distribution of nanobacteria in rabbit. Proc SPIE 3111:436–442.
6. Arbour NC, Lorenz E, Schutte BC, et al. (2000) TLR4 mutations are associated with endotoxin hyporesponsiveness in humans. Nat Genet 25:187–191.
7. Beck JD, Offenbacher S. (2001) The association between periodontal diseases and cardiovascular diseases: A state-of-the-science review. Ann Periodontol 6:9–15.
8. Boskey AL, Boyan BD, Schwartz Z. (1997) Matrix vesicles promote mineralization in a gelatin gel. Calcif Tissue Int 60:309–315.
9. Bratos-Perez MA, Sanchez PL, Garcia de Cruz S, et al. (2008) Association between self-replicating calcifying nanoparticles and aortic stenosis: A possible link to valve calcification. Eur Heart J 29:371–376.
10. Ciftcioglu N, Bjorklund M, Kuorikoski K, et al. (1999) Nanobacteria: An infectious cause for kidney stone formation. Kidney Int 56:1893–1898.

11. Ciftcioglu N, Haddad RW, Golden CD, et al. (2005) A potential cause for kidney stone formation during space flights: Enhanced growth of nanobacteria in microgravity. Kidney Int 67:483–491.
12. Ciftcioglu N, Kajander EO. (1998) Interaction of nanobacteria with cultured mammalian cells. Pathophysiology 4:259–270.
13. Ciftcioglu N, Miller-Hjelle MA, Hjelle JT, et al. (2002) Inhibition of nanobacteria by antimicrobial drugs as measured by a modified microdilution method. Antimicrob Agents Chemother 46:2077–2086.
14. Cisar JO, Xu D-Q, Thompson J, et al. (2000) An alternative interpretation of nanobacteria-induced biomineralization. Proc Natl Acad Sci USA. 97:11511–11515.
15. Costerton JW, Stewart PS, Greenberg EP. (1999) Bacterial biofilms: a common cause of persistent infections. Science 284:1318–1322.
16. Detrano R, Guerci AD, Carr JJ, et al. (2008) Coronary calcium as a predictor of coronary events in four racial or ethnic groups. N Engl J Med 358:1336–1345.
17. Diamant M, Tushuizen ME, Sturk A, et al. (2004) Cellular microparticles: new players in the field of vascular disease? Eur J Clin Invest 34:392–401.
18. Doherty TM, Asotra K, Fitzpatrick LA, et al. (2003) Calcification in atherosclerosis: Bone biology and chronic inflammation at the arterial crossroads. Proc Natl Acad Sci USA 100:11201–11206.
19. Epstein SE, Zhu J, Burnett MS, et al. (2000) Infection and atherosclerosis: Potential roles of pathogen burden and molecular mimicry. Arterioscler Thromb Vasc Biol 20:1417–1420.
20. Folk RL. (1992) Bacteria and nannobacteria revealed in hardgrounds, calcite cements, native sulfur, sulfide minerals, and travertines. Geological Society of America Annual Meeting, Abstracts with Programs 24:104
21. Folk RL. (1998) Nannobacteria in the natural environment and in medicine. Alpe Adria Microbiol J 7:87–95.
22. Greenhill NS, Presland MR, Rogers KM, et al. (1985) X-ray microanalysis of mineralized matrix vesicles of experimental saccular aneurysms. Exp Mol Pathol 43:220–232.
23. Hecht HS, Budoff MJ, Berman DS, et al. (2006) Coronary artery calcium scanning: Clinical paradigms for cardiac risk assessment and treatment. Am Heart J 151:1139–1146.
24. Hill AB. (1965) The environment and disease: association or causation? Proc R Soc Med 58:295–300.
25. Hjelle JT, Miller-Hjelle MA, Poxton IR, et al. (2000) Endotoxin and nanobacteria in polycystic kidney disease. Kidney Int 57:2360–2374.
26. Huber H, Hohn MJ, Rachel R, et al. (2002) A new phylum of Archaea represented by a nanosized hyperthermophilic symbiont. Nature 417:63–67.
27. Jayachandran M, Litwiller RD, Owen WG, et al. (2009) Circulating microparticles and endogenous estrogen in newly menopausal women. Climacteric 12:1–8.
28. Jayachandran M, Miller VM, Owen WG. (2008) Blood borne microparticles of age-associated thrombogenicity. Basic Clin Pharmacol Toxicol 102:56.
29. Kajander EO, Ciftcioglu N. (1998) Nanobacteria: An alternative mechanism for pathogenic intra- and extracellular calcification and stone formation. Proc Natl Acad Sci USA 95:8274–8279.
30. Kajander EO, Kuronen I, Akerman KK, et al. (1997) Nanobacteria from blood, the smallest culturable autonomously replicating agent on Earth. Proc SPIE 3111:420–428.
31. Khullar M, Sharma SK, Singh SK, et al. (2004) Morphological and immunological characteristics of nanobacteria from human renal stones of a north Indian population. Urol Res 32:190–195.
32. Kim KM. (1976) Calcification of matrix vesicles in human aortic valve and aortic media. Federation Proc 35:156–162.
33. Knoll A. (1999) Overview size limits of very small microorganisms: Proceedings of a Workshop. National Academy of Sciences, pp. 164.
34. Kockx MM, Muhring J, Bortier H, et al. (1996) Biotin- or digoxigenin-conjugated nucleotides bind to matrix vesicles in atherosclerotic plaques. Am J Pathol 148:1771–1777.
35. Kumar V, Farell G, Yu S, et al. (2006) Cell biology of pathologic renal calcification: contribution of crystal transcytosis, cell-mediated calcification, and nanoparticles. J Investig Med 54:412–424.

36. Lakoski SG, Greenland P, Wong ND, et al. (2007) Coronary artery calcium scores and risk for cardiovascular events in women classified as "low risk" based on framingham risk score: the multi-ethnic study of atherosclerosis (MESA). Arch Intern Med 167:2437–2442.
37. Libby P. (2002) Inflammation in atherosclerosis. Nature 420:868–874.
38. Lynch SF, Ludlam CA. (2007) Plasma microparticles and vascular disorders. Br J Haematol 137:36–48.
39. Maniloff J. (1997) Nannobacteria: Size limits and evidence (letter; comment). Science 276:1776; discussion 1777.
40. Manson J, Allison M, Rossouw JE, et al. (2007) Estrogen therapy and coronary-artery calcification. N Engl J Med 356:2591–2602.
41. Martel J, Ding-E Young J. (2008) Purported nanobacteria in human blood as calcium carbonate nanoparticles. Proc Natl Acad Sci USA 105:5549–5554.
42. Mathew G, McKay DS, Ciftcioglu N. (2008) Do blood-borne calcifying nanoparticles self-propagate? Int J Nanomedicine 3:265–275.
43. McKay DS, Gibson EK, Jr., Thomas-Keprta KL, et al. (1996) Search for past life on Mars: Possible relic biogenic activity in Martian meteorite ALH84001. Science 273:924–930.
44. Miller VM, Black DM, Brinton EA, et al. (2009) Using basic science to design a clinical trial: Baseline Characteristics of Women Enrolled in the Kronos Early Estrogen Prevention Study (KEEPS). J Cardiovasc Transl Res 2:228–239.
45. Miller VM, Hay M. (2004) Principles of sex-based differences in physiology: Advances in molecular and cell biology, vol. 34. London: Elsevier Publishing Company.
46. Miller VM, Rodgers G, Charlesworth JA, et al. (2004) Evidence of nanobacterial-like structures in human calcified arteries and cardiac valves. Am J Physiol Heart Circ Physiol 287:H1115–H1124.
47. Muhlestein JB, Anderson JL. (2003) Chronic infection and coronary artery disease. Cardiol Clin 21:333–362.
48. Nealson KH. (1997) Nannobacteria: Size limits and evidence. Science 276:1776; discussion 1777.
49. Nel A, Xia T, Madler L, et al. (2006) Toxic potential of materials at the nanolevel. Science 311:622–627.
50. Prasad A, Zhu J, Halcox JPJ, et al. (2002) Predisposition to atherosclerosis by infections. Role of endothelial dysfunction. Circulation 106:184–190.
51. Raoult D, Drancourt M, Azza S, et al. (2008) Nanobacteria are mineralo fetuin complexes. PLoS Pathog 4:e41.
52. Rogers KM, Stehbens WE. (1986) The morphology of matrix vesicles produced in experimental arterial aneurysms of rabbits. Pathology 18:64–71.
53. Ross R. (1986) The pathogenesis of atherosclerosis – an update. N Engl J Med 14:488–500.
54. Schoppet M, Shanahan CM. (2008) Role for alkaline phosphatase as an inducer of vascular calcification in renal failure? Kidney Int 73:989–991.
55. Schwartz MA-K, Lieske JC, Kumar V, et al. (2008) Human-derived nanoparticles and vascular responses to injury in rabbit carotid arteries: proof of principle. Int J Nanomedicine 3:243–248.
56. Schwartz MK, Hunter LW, Huebner M, et al. (2009) Characterization of biofilm formed by human-derived nanoparticles. Nanomedicine 4:931–941.
57. Schwartz MK, Lieske JC, Hunter LW, et al. (2009) Systemic injection of planktonic forms of mammalian-derived nanoparticles alters arterial response to injury in rabbits. Am J Physiol Heart Circ Physiol 296:1434–1441.
58. Sedivy R, Battistutti WB. (2003) Nanobacteria promote crystallization of psammoma bodies in ovarian cancer. APMIS 111:951–954.
59. Shao JS, Cai J, Towler DA. (2006) Molecular mechanisms of vascular calcification: Lessons learned from the aorta. Arterioscler Thromb Vasc Biol 26:1423–1430.
60. Shoskes DA, Thomas KD, Gomez E. (2005) Anti-nanobacterial therapy for men with chronic prostatitis/chronic pelvic pain syndrome and prostatic stones: Preliminary experience. J Urol 173:474–477.

61. Sigrist M, Bungay P, Taal MW, et al. (2006) Vascular calcification and cardiovascular function in chronic kidney disease. Nephrol Dial Transplant 21:707–714.
62. Vali H, McKee MD, Cifticioglu N, et al. (2001) Nanoforms: A new type of protein-associated mineralization. Geochim Cosmochim Acta 65:63–74.
63. Wang L, Shen W, Wen J, et al. (2006) An animal model of black pigment gallstones caused by nanobacteria. Dig Dis Sci 51:1126–1132.
64. Yang IA, Holloway JW, Ye S, et al. (2003) TLR4 Asp299Gly polymorphism is not associated with coronary artery stenosis. Atherosclerosis 170:187–190.
65. Young JD, Martel J, Young D, et al. (2009) Characterization of granulations of calcium and apatite in serum as pleomorphic mineralo-protein complexes and as precursors of putative nanobacteria. PLoS One 4:e5421.
66. Young JD, Martel J, Young L, et al. (2009) Putative nanobacteria represent physiological remnants and culture by-products of normal calcium homeostasis. PLoS One 4:e4417.
67. Zhou Z, Hong L, Shen X, et al. (2008) Detection of nanobacteria infection in type III prostatitis. Urology 71:1091–1095.

Biographies

Maria K. Schwartz, Ph.D. is a Postdoctoral Research Fellow at the Mayo Clinic, College of Medicine. Maria is now an Assistant Professor at the College of St. Benedict and St. John's University. She received her training at Mayo Clinic College of Medicine, Mayo Graduate School, Rochester, MN. Her research focuses on the vascular effects of biologically derived nanoparticles and chronic models of asthma.

Virginia M. Miller, MBA, Ph.D. is Professor of Surgery and Physiology at the Mayo Clinic, College of Medicine. She received her training at the University of Missouri, Columbia, Mo and Carlson School of Management, University of Minnesota. Research in her laboratory focuses on soluble factors in the blood which differentially affect development of vascular disease in men and women including hormones and infection. She is currently President of elect for the Organization for the Study of Sex Differences.

John C. Lieske is Professor of Medicine at the Mayo Clinic in Rochester Minnesota. He directs the Kidney Stone Clinic in the Division of Nephrology and Hypertension and is Medical Director of the Renal Testing Laboratory in the Department of Laboratory Medicine and Pathology. He graduated from Northwestern University with a degree in Biomedical Engineering, and then completed medical school at the University of Chicago, an Internal Medicine residency at Emory University, and a Nephrology Fellowship at the University Chicago. His current research interests are in the pathogenesis, epidemiology, and treatment of kidney stone disease, and the development of urinary biomarkers of renal disease.

Chapter 25
(Shear) Strain Imaging Used in Noninvasive Detection of Vulnerable Plaques in the Carotid Arterial Wall

T. Idzenga, H.H.G. Hansen, and C. L. de Korte

Abstract The primary trigger for myocardial infarction and stroke is destabilization of atherosclerotic plaques. The chance of a plaque to rupture is related to its composition and geometry. Ultrasound (shear) strain imaging allows assessment of local tissue mechanics and possible risk assessment of vulnerable plaques. Intravascularly, in coronary arteries using a catheter, strain imaging has been demonstrated to be successful. At different intraluminal pressures, ultrasound data of the artery wall were recorded and local radial strains were estimated using cross-correlation methods. It has been shown in vitro and in vivo that softer lipidic plaques can be distinguished from harder fibrous and calcified plaques on basis of their strain values.

However, plaque rupture often occurs without preceding clinical symptoms. A relatively cheap noninvasive technique would make it possible to screen people before an actual cardiovascular event occurs and possibly give a risk assessment of the plaques present. Given the successful results of intravascular strain imaging, noninvasive versions of the technique are being developed by multiple research centers. These techniques focus on (shear) strain imaging of the carotid artery wall. At the moment, most of these techniques have been shown to give promising results for simulated and experimental data of vessel-like phantoms. Furthermore, the first in vivo results show good correspondence between calcifications and histology. A few studies also show in vivo reproducibility of the technique.

Various methods for noninvasive ultrasound strain imaging have been developed, and the first results demonstrate the potential of the methods to detect vulnerable plaques. Further validation of these methods will open the door for clinical screening of plaques.

Keywords Ultrasound • Strain imaging • Compounding • Shear strain • Vulnerable plaque detection

T. Idzenga (✉)
Department of Pediatrics, Clinical Physics Laboratory, CUKZ 833, Radboud University Nijmegen Medical Center, P.O. Box 9101, 6500HB Nijmegen, The Netherlands
e-mail: t.idzenga@cukz.umcn.nl

Jasjit S. Suri et al. (eds.), *Atherosclerosis Disease Management*,
DOI 10.1007/978-1-4419-7222-4_25,

25.1 Introduction

Atherosclerosis is a systemic disease in which lipid-rich content is deposited in the arterial vessel wall. Myocardial infarction and stroke are two leading causes of death [1, 2]. The primary trigger for these two causes is destabilization of atherosclerotic plaques. In 76% of the myocardial infarctions and strokes, rupture of plaques with superimposed thrombus formation is involved [3]. Although atherosclerosis is a systemic disease, plaques usually develop in the conducting arteries (e.g., the coronary, carotid, and femoral artery). They start off as so-called "fatty streaks" on the intimal layer of the arterial wall, small accumulations of atherogenic lipoproteins, macrophages, white blood cells, and smooth muscle cells in the intima. These fatty streaks can either disappear or develop into advanced stable plaques or into vulnerable plaques that are prone to rupture. A stable plaque has a small lipid pool that contains few macrophages and is separated from the blood by a thick fibrous cap. Vulnerable plaques mostly consist of a large pool of lipids and thrombogenic material covered by a thin fibrous cap [4]. Furthermore, a stable plaque tends to grow slowly into the luminal area, thereby occluding the blood flow and decreasing the blood supply of tissues. This decrease in blood supply can often be recognized from clinical symptoms like angina pectoris and overall fatigue. These symptoms can be used as a warning and give a surgeon time to perform catheterization, stenting, or bypass surgery before an acute event occurs. In the case of a vulnerable plaque, most of the luminal area is maintained due to outward remodeling [5, 6]. Based on the schematic representation of Fig. 25.1, vulnerable plaques seem less dangerous since the luminal area is initially not reduced. However, when the fibrous cap ruptures, the blood comes in contact with the contents of the core, leading to thrombus formation, which may cause a heart attack or stroke by blocking a coronary or cerebral artery. Early identification of vulnerable plaques is, therefore, of crucial importance to prevent morbidity and mortality and a well-addressed topic in literature [7–12]. Upon identification of a vulnerable plaque, it is of importance to classify the condition of the plaque, i.e., whether it is prone to rupture. Furthermore, knowledge on development of vulnerable plaques can help predict which plaques will develop into vulnerable plaques.

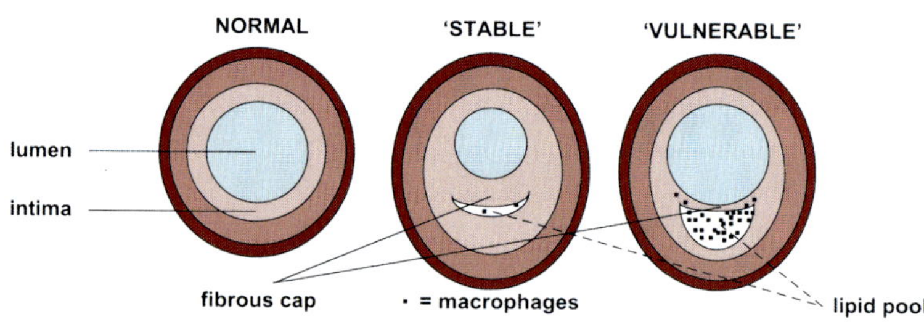

Fig. 25.1 Schematic representations of a healthy vessel, a vessel with a stable plaque, and a vessel with a vulnerable plaque

A promising technique for noninvasive vulnerable plaque detection/classification is ultrasound strain imaging. Ultrasound strain imaging was first described by Ophir et al. [13]. The technique enables assessment of strain in tissue using ultrasound. Movement of tissue is estimated from sequential ultrasound images using cross-correlation techniques. From these displacement estimates, the strain in the tissue can be calculated. At first, the technique was mainly applied to detect hard and soft tumors within breast, prostate, thyroid, and liver tissue. The tissue was strained by application of an external force. Given its power to differentiate between various tissues, the technique has also been applied for atherosclerotic plaque detection. De Korte et al. were the first to apply the technique intravascularly to distinguish stable from vulnerable plaques in coronary arteries [14, 15]. Instead of applying an external force to deform the tissue, the pulsating nature of the blood flow was used as deforming force. Although the intravascular results were convincing, its intravascular nature restricts the technique to be applied only in the cath lab and withholds application as a screening tool. Therefore, noninvasive detection and differentiation of plaques are preferable. At this moment, multiple approaches for noninvasive ultrasound (shear) strain imaging have already been developed [8, 16–21], and preliminary in vivo results have just been published. The availability of noninvasive ultrasound strain imaging promises to enable early screening of risk populations for vulnerable plaques, albeit no longer in coronary arteries, but for superficial arteries like the carotid artery and the femoral artery.

In this chapter, we give a short explanation of ultrasonic strain imaging and a review of existing techniques for estimating strain in the carotid arterial wall. The techniques that will be discussed are: intravascular strain imaging, noninvasive strain imaging (in longitudinal and transverse cross-sections), and noninvasive shear strain imaging.

25.2 Ultrasound Strain Imaging

Strain imaging is based on the principle that soft tissue deforms more than hard tissue when an external force is applied, see Fig. 25.2. To measure strain, a registration of a tissue is made before (the predeformation state) and after (the postdeformation state) applying force to the tissue. From cross-correlation of the pre- and postdeformation registration, local tissue displacements are estimated. The tissue strains are derived from these estimates by first-order spatial derivation. Two types of strain can be defined, normal strain and shear strain. The first type is obtained by derivation of tissue displacement in the same direction as the displacement. The second type of strain is obtained by derivation of tissue displacement in a direction orthogonal to that of displacement. Strain in tissue can, therefore, be represented as a tensor.

J. Ophir and coworkers were first to describe the concept of strain imaging using ultrasound data [13]. Ultrasound data are very suitable for strain imaging, due to the periodicity of the ultrasound wave. Figure 25.3 shows simulated ultrasound

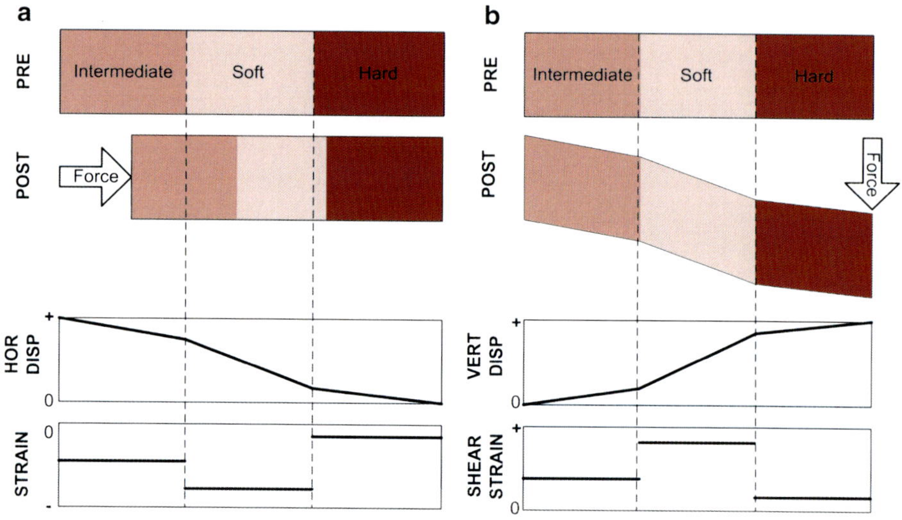

Fig. 25.2 Principle of (**a**) normal strain and (**b**) shear strain imaging: the softer the tissue, the higher the (shear) strain (Note that strain is negative when a tissue is "compressed"). Normal strain and shear strain can be calculated by comparing a registration of a tissue before and after an external force is applied

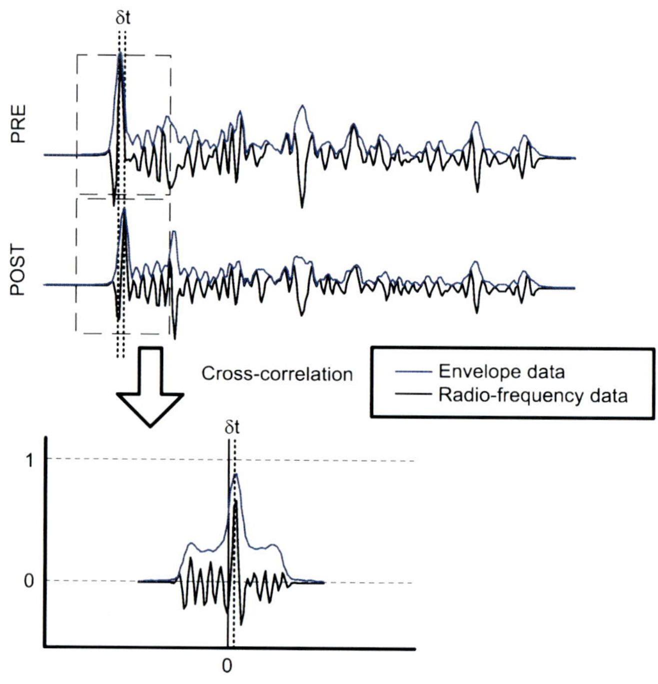

Fig. 25.3 Traditionally, displacements and strains were calculated locally by 1D cross-correlation of pre- and postdeformation ultrasound radiofrequency/envelope data

signals for the tissue of Fig. 25.2a in pre- and postdeformation state. Ophir and coworkers determined the local tissue displacements by cross-correlating one-dimensional windows of pre- and postdeformation radiofrequency data. An example of the cross-correlation function for the marked windows is shown also in Fig. 25.3. The location of the peak of the normalized cross-correlation corresponds to the time shift δt between the two rf signals, caused by the displacement of the tissue. Since time corresponds to depth in ultrasound, the time shift can be translated into a local tissue displacement for the tissue in the window of interest. By repeating this cross-correlation procedure for multiple depths, the displacement field in the direction of the ultrasound beam can be estimated. Consequently, spatial derivation can be applied to obtain strains. Instead of direct point-to-point spatial derivation, often a least-squares strain estimator (LSQSE) is applied [22]. An LSQSE calculates a least-squares linear fit through the displacement values. The slope of this fit corresponds to the strain. This is done to reduce the error in displacement estimates caused by high frequency noise.

25.3 Intravascular Strain Imaging

In the following couple of years, ultrasound strain imaging techniques improved and the range of applications increased. Vulnerable and stable plaques have different composition. The fatty content of a vulnerable plaque and the fibrotic tissue of a stable plaque have different mechanical properties. For detection of the different plaques, ultrasound strain imaging became interesting. The first reports on this application described intravascular ultrasound (IVUS) strain imaging. They came from three different research groups [23–25]. For this technique, an IVUS catheter was inserted in a diseased coronary artery, and rf A-line data were recorded for the full circumference for a low (predeformation) and a high (postdeformation) intraluminal pressure. Next, radial strains were calculated by applying cross-correlation techniques to each separate A-line. The strain image was plotted next to the IVUS echograms (Fig. 25.4). An alternative method for visualization of strain is the palpogram: only the strain values at the lumen to vessel wall boundary (the rupture-prone region) is plotted on top of the B-mode data [26]. This is the most important layer to access risk, since rupture will occur in this layer.

The technique has first been validated in vessel-mimicking phantoms with hard and soft layers; both layers were correctly identified [23]. Next, the technique was tested in vitro on excised coronary and femoral arteries. The obtained radial strain maps were locally compared with corresponding histology. It was shown that strain values were different for fibrous, fatty, and fibro-fatty plaques [14]. Then, the step towards in vivo imaging was taken. IVUS strain imaging also proved to be successful in differentiating between fibrous and fatty materials and between plaques with and without macrophages in vivo [27]. Furthermore, palpography was able to classify vulnerable and stable plaques in vitro [28]. The number of studies that confirm the usefulness of intravascular ultrasound strain imaging of the coronary arteries for

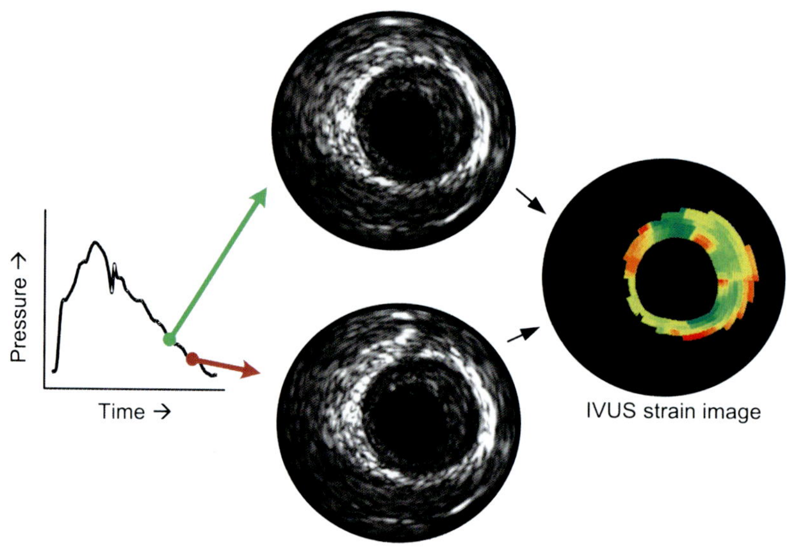

Fig. 25.4 Principle of IVUS Elastography: rf data of a coronary artery are recorded at two different intraluminal pressure levels. From these recordings, a radial strain image is obtained

vulnerable plaque detection is still increasing. However, a major drawback of IVUS elastography that will always remain is its invasive nature. This means that the technique is only applicable in patients that are already undergoing an interventional procedure in the cath lab.

Using an intravascular catheter, Brusseau et al. [29] acquired rf data of a single excised human carotid artery at several intraluminal pressure steps. A 1D cross-correlation-based technique was applied to derive radial strain maps. A low strain spot was observed at all pressure levels which corresponded well to the site of a collagen-rich region observed from histology data. This result supports that, in principle, strain imaging of the carotid arteries instead of the coronary arteries is possible, but they still used an intravascular approach.

Most people who are at risk of having a myocardial infarction or a stroke are asymptomatic [30]. Ultrasound is relatively cheap and harmless compared to other imaging modalities like MRI and CT. Therefore, a noninvasive variant for vascular ultrasound strain imaging would be ideal for preventive screening purposes. The imaging depth of the ultrasound (maintaining an adequate resolution) excludes imaging the coronary arteries. Noninvasive ultrasound strain imaging, therefore, focuses on superficial arteries, like carotid or femoral arteries. Since atherosclerosis is a systemic disease, the local findings in the carotid artery might also serve as a surrogate marker for overall disease in a patient [31].

25.4 Noninvasive Strain Imaging Techniques

A major difference between the invasive intravascular and the noninvasive transcutaneous approach is the placement of the ultrasound probe. This placement imposes two difficulties on the step from intravascular to noninvasive. First, the center frequency of the intravascular ultrasound probe was 30–40 MHz, which has limited imaging depth. For the intravascular approach, the required imaging depth is not large, but for the noninvasive approach, the required depth increases. Noninvasively center frequencies need to be lower due to the frequency-dependent attenuation of the ultrasound signal by skin, fat, and muscle tissue between the probe and the artery. Typically, center frequencies for noninvasive strain imaging are around 7–10 MHz. Consequently, the spatial resolution of the ultrasound signal and the strain images is also lower. The second difficulty with noninvasive radial strain imaging is misalignment between the ultrasound beam and the radial strain (see Fig. 25.5). When the probe is orientated in longitudinal direction with respect to the carotid artery, the ultrasound beam is aligned with the radial strain. It is not possible to image radial strain in the entire transverse cross-section. In this section, methods and results of strain imaging in the longitudinal and transverse cross-section are described and discussed.

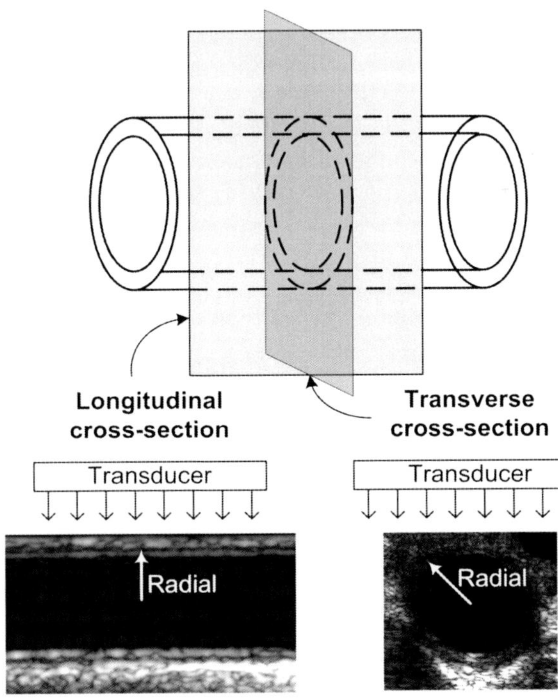

Fig. 25.5 Schematic overview of imaging planes used for ultrasonic imaging of the carotid artery. The orientation of the ultrasound beams with respect to the radial strains in the cross-section are shown

25.4.1 Longitudinal Cross-Section

25.4.1.1 Doppler-Based Methods

The earliest techniques for noninvasive strain imaging method to assess vascular atherosclerosis came from the field of Doppler imaging of blood flow velocity [32]. Bonnefous and coworkers used cross-correlation to noninvasively estimate the radial motion of a human carotid artery from in vivo recorded echo data. The technique was also applied to determine strain in cadaverous human arterial samples. For this, the samples were placed in a temperature-regulated conservation bath and perfused with blood. To prevent blood clot formation, anticoagulation drugs were added. A peristaltic pump and a frequency generator were used to generate a controlled pulsating blood flow through the arterial segments. During pulsations, the intraluminal pressure and echo data in the longitudinal plane were recorded. Radial strain images were derived and compared to histology data. A high correlation was found between the stiffness of the lesions and the level of strain.

Kanai et al. [33] developed another method that originated from the field of Doppler-based blood flow velocity. This phase tracking method that is based on the principle that tissue displacement between a pre- and postdeformation situation induces a phase change in the echo signal. As long as the center frequency of the ultrasound beam is known, the technique can provide very accurate displacement and strain estimates [34]. The method was experimentally validated using a homogeneous vessel phantom. Compared to the theoretical strain profile, the error and the standard deviation in radial strain were 12.0 and 14.1%, respectively. The technique was also applied to in vitro recordings of a femoral artery. Low-strain regions corresponded well with calcified regions, and higher strains were observed for smooth muscle cells and collagen regions. The technique does not take into account lateral motion, i.e., in plane motion in the direction perpendicular to the ultrasound beam. For tissue that has a lot of lateral motion between frames, it can be expected that the technique is not applicable. However, recently [35], the authors have demonstrated a plane wave imaging sequence that enables recording rf data at frame rates of 3,500 Hz. For such high frame rates, the lateral interframe motion is negligible.

25.4.1.2 Registration-Based Method

In 1999, Maurice et al. published a theoretical framework for a registration-based strain imaging method called the Lagrangian speckle motion estimator or LSME [36]. Later, this technique was applied in noninvasive vascular strain imaging [19]. LSME deforms a predeformation image in multiple iterations until it best matches the postdeformation image. The translations and deformations of the precompression image that are required to find the best match correspond to the 2D displacements between pre- and postdeformation situation. From these displacements, the strains can be derived. The method was first tested in simulated rf data of transverse cross-sections of cylindrical tissue. They simulated a homogeneous cylindrical blood vessel and a blood vessel with soft and hard regions. The results demonstrated the potential of the technique to detect differences between hard and soft vascular tissue. Furthermore, it was shown that the technique allowed estimation of all 2D components of the strain tensor, which is useful for deriving strain components in other directions, like the radial and circumferential directions. In a later study, Maurice et al. show that LSME can also be used intravascularly [37]. For various intraluminal pressures, rf data of an excised human carotid artery were acquired, and radial strains were estimated. A low-strain region was found to correspond to a collagen-rich region. Although these publications report strain images for transverse vessel cross-sections, in vivo work of Maurice et al. mainly focuses on longitudinal recordings. The first in vivo results for the LSME method in human subjects were published in 2007 [11]. Axial strains were determined for longitudinal cross-sections of two healthy subjects and two 75-year-old asymptomatic patients with severe carotid stenosis. Schmitt et al. showed that axial strains could be reproducibly measured for the healthy subjects over 5–7 heart cycles using an adapted version of the LSME. Cumulated axial strain curves for an ROI in the plaque region and an ROI in the wall region showed that the hard calcified plaque tissue strained less than the wall tissue. Furthermore, the strain pattern in the stenotic region was much more heterogeneous than that for the vessel walls of the healthy subjects. In the most recent study, Maurice et al. have applied LSME to measure axial strains for healthy female and male subjects in four age categories [38]. Left and right common and internal carotid arteries were imaged longitudinally, and the measurements were performed in duplicate by two different radiologists. For the strain measurements in the common carotid arteries, the authors reported a good correlation between both radiologists, whereas less consistency was observed in the internal common carotid arteries. The correlation between measurements of the left and the right common carotid arteries was also larger than the correlation between the left and right internal carotid artery. Axial strains for male common carotid arteries were less than for female common carotid arteries of the same age. However, it has to be noted that few subjects were investigated in each age group.

25.4.1.3 Cross-Correlation-Based Methods

The next group of methods to be discussed are based on coarse-to-fine or multilevel cross-correlation [21, 39, 40]. With these methods, the pre- and postdeformation data are cross-correlated iteratively with decreasing window sizes instead of fixed window sizes. The benefit of these methods is the ability to estimate large as well as small displacements and strains with high accuracy. The "coarse" displacement estimates are used to guide the algorithm in the following iteration for estimation of smaller displacements. Since envelope-based cross-correlation is more robust when large translational motion occurs, often the first iteration is carried out on the B-mode/envelope signal, whereas in the following iterations rf data is used.

Several research groups have reported results on noninvasive strain imaging of carotid arteries using a coarse-to-fine approach. In 2008, a two-dimensional multistep coarse-to-fine algorithm [41] was applied in a clinical pilot study [42]. Based on echo intensity and percentage of stenosis, 16 atherosclerotic plaques were classified as being soft or calcified. Maximum axial strains were calculated for a certain region within each plaque and cumulated over time for three or more subsequent heartbeats. It was observed that calcified plaques deformed less than soft plaques.

In a work by Kim et al., a multistep strain imaging method was applied to in vivo images of a healthy subject and a diseased subject [43]. Axial strains were shown for two points in the brachial artery wall during 5 s. The two points were separated by 0.2 mm. During the acquisition, the external compression applied with the transducer was increased. A nice cyclic pattern was observed. With increasing external pressure, the variations between the peak systolic and diastolic pressure increased up to a strain difference of 20%. The peak strain rate was reported to be 100%/s during diastole and −250%/s during systole. For the diseased vessel, the strain rate was about three times lower. Recently, the group has applied the method to examine peripheral artery-vein bypass grafts [44]. Strain images were estimated in a 78-year-old subject who had a femoral-to-popliteal artery in situ bypass before and after a stenosis developed. Significantly lower strain values were observed in the stenotic region compared to the normal tissue. Strain values were about four times lower than those measured in a 81-year-old subject who did not develop stenosis after bypass surgery. In another study [45], the technique was applied to evaluate dialysis fistula stenosis in two subjects. Again, strain values in the stenotic regions were lower than in the "normal" regions of fistula.

Figure 25.6 shows the axial/radial strain measured in the adventitial layer of a healthy female volunteer. A coarse-to-fine algorithm [40] was used to derive longitudinal displacement and shear strain.

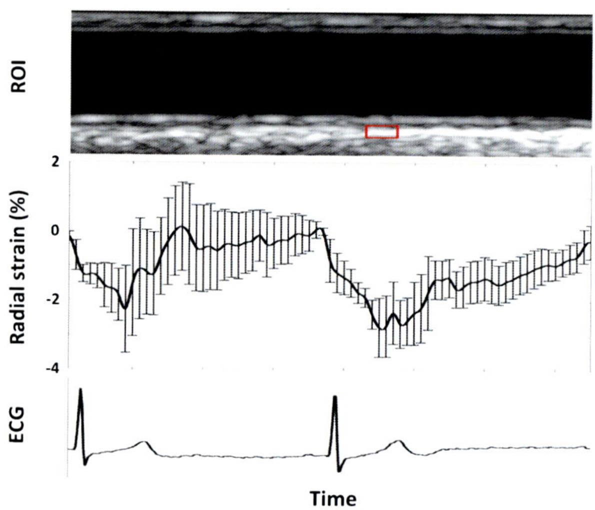

Fig. 25.6 Radial strain (mean ± SD within the selected ROI) estimated in the adventitial layer in a healthy female volunteer. The selected region of interest (~0.5 × 0.5 mm) is indicated in the top part, and the cardiac cycle is indicated by the ECG in the bottom part

25.4.2 *Transverse Cross-Sections*

Analogous to LSME, the cross-correlation-based algorithms can also be extended to two dimensions to derive the full strain tensor. As explained in Sect. 4.1.2, extension to 2D enables the estimation of radial and circumferential strains for recordings of transverse vessel cross-sections. In a study by Ribbers et al. [20], 2D strain data for both longitudinal and transverse cross-sections of a homogeneous vessel phantom with a concentric lumen were presented. The strain data were obtained using a 2D coarse-to-fine algorithm. For the transverse cross-sections also, radial and circumferential strain images were constructed. The constructed radial strain images revealed an inverse quadratic decay outward, which corresponded to the strain profile predicted by theory. Next to the phantom results, axial strain images were derived from in vivo recordings of longitudinal and transverse cross-sections of 12 carotid arteries. A soft and a hard plaque were identified. The authors' conclusion was that noninvasive strain imaging of carotid arteries was feasible, but the quality of radial and circumferential strain images was severely deteriorated by the lateral contribution. Lateral strains cannot be estimated as accurate as axial strains, since no phase information is available in the lateral direction to enhance the cross-correlation procedure.

Dedicated techniques for radial and circumferential strain imaging for transverse cross-sections are scarcely reported. In the next paragraphs, two methods are described that require little or no lateral information for the estimation of radial strains in transverse cross-sections.

25.4.2.1 A-Line Based Beam Steering

As explained earlier, the radial direction does not correspond to the axial direction for transverse strain imaging. Although the phase tracking method of Kanai et al. [46] can only be applied to determine axial displacements, they found a way to circumvent the problem [47]. Adjacent ultrasound beams are all steered through the center of the lumen by changing the time delays of the transducer elements. By using this electronically, steering the ultrasound beams become aligned with the radial strains, and the phase tracking method can be applied to obtain radial strains for small segments of the cross-section. A disadvantage of the technique is that only a partial strain image of the artery can be obtained since it is impossible to steer the ultrasound beam through the center for all 180°. Despite the promising initial results for a rubber tube and in vivo acquisitions of a carotid artery, no further results on this technique were reported.

25.4.2.2 Image-Based Beam Steering and Compounding

Another method dedicated to radial strain imaging for transverse cross-sections of superficial arteries is a method developed by Hansen et al. [8]. This technique also makes use of electronic beam steering. However, in this case, the beam steering is not performed separately for each rf line, but for the entire image plane, as shown in Fig. 25.7. Rf data are acquired at multiple beam steering angles in pre- and post-compression situations. For each angle, there are segments of the cross-section in which the ultrasound beam is (closely) aligned with the radial strain. For these segments, axial strains are estimated using a 2D coarse-to-fine cross-correlation algorithm. These strains are then converted into radial strains by means of projection. The segments with radial strain estimates from the various beam steering angles are then compounded to a radial strain image for the entire cross-section. This technique also enables estimation of circumferential strains by selecting segments for which the circumference is (closely) aligned with the ultrasound beam.

The technique was first applied on a homogeneous gelatin vessel phantom with a concentric lumen. Data were recorded at beam steering angles ranging from −45 to 45°, with an angular increment of 15°. A good similarity was found between the theoretical and the estimated radial and circumferential strain images. Since the projection angles were limited to 30°, lateral information was required to complete the images at 3 and 9 o'clock. For the circumferential strain image, these are the 6 and 12 o'clock regions. This results in noisy strain in these regions.

The method was also partially tested in vivo [48]: recordings of several common carotid arteries were obtained with and without beam steering at an angle of 20°. It was illustrated that radial strains could be estimated reproducibly in segments of the cross-section for several cardiac cycles with and without beam steering. Figure 25.8 shows an example of a B-mode image without a plaque and the mean cumulative axial/radial strain in the selected ROI during the cardiac cycle estimated with a coarse-to-fine cross-correlation-based strain imaging algorithm.

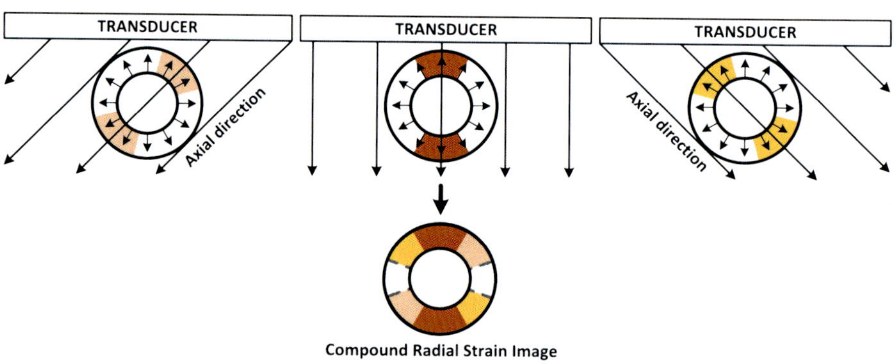

Fig. 25.7 Schematic overview of a beam steering approach for radial strain imaging of transverse cross-sections of a superficial artery. Acquisitions are performed at multiple beam steering angles. For each acquisition, angle radial strains are calculated for segments of the cross-section. The segments are added together to form a compound radial strain image

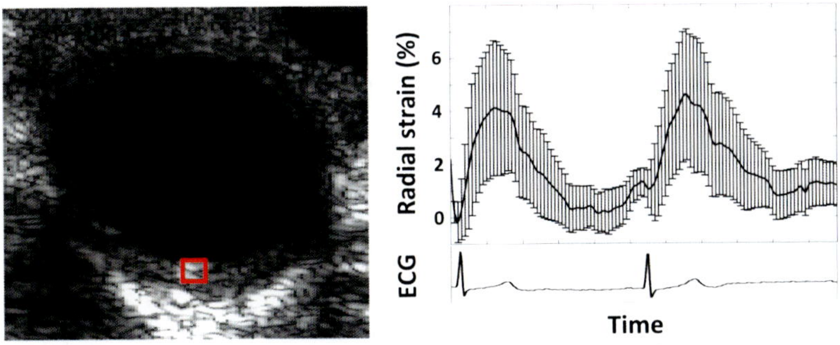

Fig. 25.8 An in vivo B-mode image (*left panel*) of a common carotid artery and a radial strain curve in the selected ROI (*right panel*). The radial strain curve was synchronous with the cardiac cycle

25.5 Noninvasive Shear Strain Imaging Techniques

Shear strain is the part of the strain tensor that handles spatial derivatives of estimated displacements in a direction orthogonal to the direction of displacement (see Fig. 25.2b). Shear strain has first been used in detection and characterization of tumors, e.g., in breast tissue [49]. They have shown in simulations that the shear strain between a tumor and the surrounding tissue could possibly be used to differentiate between benign and malignant tumors. Later, Thitaikumar et al. carried out an in vivo feasibility study on bonding at the boundary of an inclusion using axial shear strain elastography [50]. Recently, shear strain imaging has been used in vascular applications [17, 18, 42]. A hypothesis on the role of shear strain in atherosclerosis has been proposed by our group [51]. This hypothesis would make risk assessment of plaques possible.

25.5.1 Echo-Tracking

Cinthio and coworkers have already presented measurements of longitudinal displacement and shear strain in layers of the carotid arterial wall [18, 52–55]. They used an echo-tracking technique based on block matching to estimate longitudinal displacement. This technique was based on distinct echos in the conventional B-mode images [52]. To estimate shear strain, ROIs that contained distinct anatomical structures were selected in different layers of the arterial wall, e.g., in the adventitia layer and the surrounding tissue. The selected ROIs were tracked based on the B-mode/envelope data, resulting in motion estimation of only the selected anatomical structure. Cinthio et al. used a 5–12-MHz linear array transducer and small ROIs (~0.5×0.5 mm, depending on the speckle size) in their in vivo experiments. In the two selected ROIs, the longitudinal displacement (i.e., the direction of blood flow) was estimated. The shear strain was defined as the difference in displacement between the two layers divided by the radial distance between the two layers. For this method to be applicable, a distinct echo needs to be present during the full cardiac cycle [54]. In a study with ten healthy, normotensive subjects [18], the intima–media complex of the carotid artery showed a distinct antegrade longitudinal movement (in the direction of blood flow) in early systole. This was followed by a retrograde movement later in systole. In diastole, there was another antegrade movement, followed by a gradual return to the original position. The adventitial region showed a similar pattern of significantly smaller amplitude. This resulted in a maximum shear strain of 0.36 rad (SD 0.26, range 0.09–0.62) between the intima–media complex and the adventitial region.

With their technique, Cinthio et al. demonstrated a multiphase bidirectional pattern in longitudinal movement of the carotid artery during the cardiac cycle. Furthermore, they identified the presence of a shear strain between the intima–media complex and the adventitial layer and between the adventitial layer and the surrounding tissue. Drawbacks of this method are that only the shear strain between the two selected ROIs can be calculated and that distinct echos need to be present in both ROIs during the entire cardiac cycle.

25.5.2 *Relative Lateral Shift*

A similar method was used by Shi et al. [42]. They used a two-dimensional multilevel cross-correlation technique to track lateral motion of the arterial tissue [21]. They selected an ROI in the plaque and one in the arterial wall. For both ROIs, they calculated the cumulated lateral displacement and calculated the difference between these two as the relative lateral shift, a rough estimate of the shear strain. Shi et al. applied this method to 16 patients that were scheduled for a carotid endarterectomy. They found a cyclic pattern in their relative lateral shift during the cardiac cycle which was as expected. However, they also found a (assumed linear) increasing/decreasing trend in their data. They corrected for this trend by fitting a straight line through the data and removing the general trend. This resulted in relative lateral shift values ranging from 0 to 3 mm. When comparing their findings to the identifications of the plaque by a radiologist, they found that calcified plaques only showed a small relative lateral shift (<1 mm), whereas soft plaques showed higher relative lateral shift (2–3 mm). However, also some of the soft plaques showed low relative lateral shifts of <1 mm, and they suggested that additional parameters are required to clearly differentiate between the types of plaque.

25.5.3 *Radiofrequency-Based Ultrasound*

Idzenga et al. investigated the estimation of longitudinal shear strain in the carotid arterial wall using a coarse-to-fine strain algorithm [40] based on rf data [17]. They compare these estimates to those based on the B-mode/envelope data as a benchmark. The raw rf data provide besides echo level also phase-information, which the envelope does not. This makes it possible to obtain a more precise estimate of displacements [40, 56]. Furthermore, the use of 2D signal windows in the coarse-to-fine strain algorithm enables a 2D estimation of displacements, and the various strains [40, 57, 58].

The approach Idzenga et al. used to calculate shear strain is similar to the approach used by Cinthio et al., i.e., calculating the shift in two ROIs and divide by the distance between the two. The pre- and postdeformation ultrasound images are combined into a displacement image using normalized cross-correlation. The resolution of this image is determined by the size of the finest window in the coarse-to-fine algorithm and the overlap between windows. In the resulting shear strain image, a region of interest is selected, and in this region, the average shear strain is calculated. In simulations and phantom experiments, they compared this rf-based technique to the envelope-based results using exact knowledge of applied displacement and strain. They also showed two in vivo examples of the rf-based approach in healthy subjects.

In the simulations, there was a linear relation between the shear strain estimates and the applied displacement. The root mean squared error and the variance of the rf-based estimates were significantly smaller than for the envelope-based estimates. This suggests that the rf-based shear strain estimates were closer to the true value and more accurate than the envelope-based estimates. Similar results were found in the phantom experiments. Although, the estimated values for shear strain (rf- as well as envelope-based) underestimated the presumed applied shear strain. In the in vivo examples, they showed that the shear strain in the *adventitia* of the carotid artery wall varies periodically "in phase" with the cardiac cycle (see Fig. 25.9).

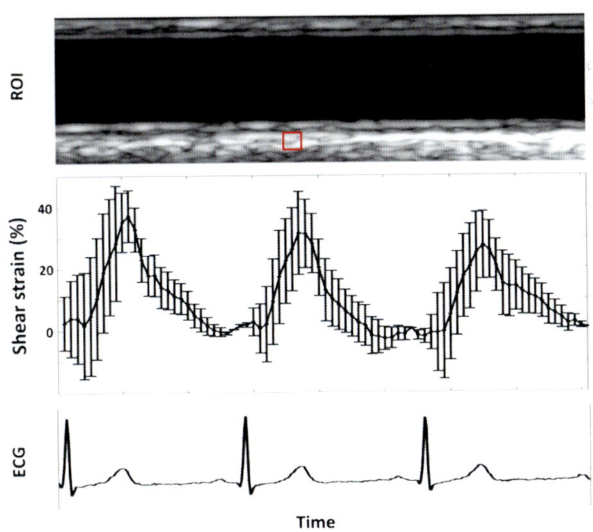

Fig. 25.9 Longitudinal shear strain (mean ± SD) estimated in the adventitial layer in a healthy female volunteer. The selected region of interest is indicated in the *top part*, and the cardiac cycle is indicated by the ECG in the *bottom part*

The cumulated shear strain showed an increase during systole, whereas during the diastole, it showed a decrease returning to the initial value. This is a similar pattern as found by Cinthio et al. [18]. Idzenga et al. have shown that estimation of longitudinal shear strain in the *adventitia* of the carotid artery using ultrasound radiofrequency signals is accurate and feasible.

25.6 Conclusions

Intravascular strain imaging has shown that differentiation between vulnerable plaques and stable plaques is possible. The invasive nature of this technique, however, imposes a threshold on diagnosing plaques in patients without clinical symptoms. To lower this threshold, noninvasive techniques have been developed for strain imaging. Most of these techniques have shown potential for detection of vulnerable plaques in the carotid arteries. The first in vivo pilot studies show that strain imaging enables differentiating between soft and hard tissue. The techniques for shear strain imaging have potential to predict development of vulnerable plaques. Studies in which in vivo acquisitions are compared with histology data are required to validate the developed methods, and patient studies will have to confirm that noninvasive vulnerable plaque detection and prediction are possible.

References

1. Thom T, Haase N, Rosamond W, Howard VJ, Rumsfeld J, Manolio T, et al. Heart disease and stroke statistics – 2006 update: a report from the American Heart Association Statistics Committee and Stroke Statistics Subcommittee. Circulation 2006 Feb 14;113(6):e85–151.
2. Weintraub HS. Identifying the vulnerable patient with rupture-prone plaque. Am J Cardiol 2008 Jun 16;101(12A):3F–10F.
3. Kolodgie FD, Burke AP, Skorija KS, Ladich E, Kutys R, Makuria AT, et al. Lipoprotein-associated phospholipase A2 protein expression in the natural progression of human coronary atherosclerosis. Arterioscler Thromb Vasc Biol 2006 Nov;26(11):2523–9.
4. Schaar JA, Muller JE, Falk E, Virmani R, Fuster V, Serruys PW, et al. Terminology for high-risk and vulnerable coronary artery plaques. Report of a meeting on the vulnerable plaque, June 17 and 18, 2003, Santorini, Greece. Eur Heart J 2004 Jun;25(12):1077–82.
5. Glagov S, Weisenberg E, Zarins CK, Stankunavicius R, Kolettis GJ. Compensatory enlargement of human atherosclerotic coronary arteries. N Engl J Med 1987 May 28;316(22):1371–5.
6. Pasterkamp G, Schoneveld AH, van der Wal AC, Haudenschild CC, Clarijs RJ, Becker AE, et al. Relation of arterial geometry to luminal narrowing and histologic markers for plaque vulnerability: the remodeling paradox. J Am Coll Cardiol 1998;32(3):655–62.
7. Hurks R, Peeters W, Derksen WJ, Hellings WE, Hoefer IE, Moll FL, et al. Biobanks and the search for predictive biomarkers of local and systemic outcome in atherosclerotic disease. Thromb Haemost 2009 Jan;101(1):48–54.
8. Hansen HH, Lopata RG, De Korte CL. Noninvasive carotid strain imaging using angular compounding at large beam steered angles: validation in vessel phantoms. IEEE Trans Med Imaging 2009 Jun;28(6):872–80.

9. Lerakis S, Synetos A, Toutouzas K, Vavuranakis M, Tsiamis E, Stefanadis C. Imaging of the vulnerable plaque: noninvasive and invasive techniques. Am J Med Sci 2008 Oct;336(4):342–8.
10. Young JJ, Phillips HR, Marso SP, Granada JF, McPherson JA, Waksman R, et al. Vulnerable plaque intervention: state of the art. Catheter Cardiovasc Interv 2008 Feb 15;71(3):367–74.
11. Schmitt C, Soulez G, Maurice RL, Giroux MF, Cloutier G. Noninvasive vascular elastography: toward a complementary characterization tool of atherosclerosis in carotid arteries. Ultrasound Med Biol 2007 Dec;33(12):1841–58.
12. Behler RH, Nichols TC, Zhu H, Merricks EP, Gallippi CM. ARFI imaging for noninvasive material characterization of atherosclerosis. Part II: toward in vivo characterization. Ultrasound Med Biol 2009 Feb;35(2):278–95.
13. Ophir J, Céspedes EI, Ponnekanti H, Yazdi Y, Li X. Elastography: a method for imaging the elasticity in biological tissues. Ultrason Imaging 1991;13:111–34.
14. de Korte CL, Pasterkamp G, van der Steen AFW, Woutman HA, Bom N. Characterization of plaque components using intravascular ultrasound elastography in human femoral and coronary arteries in vitro. Circulation 2000;102(6):617–23.
15. de Korte CL, Carlier SG, Mastik F, Doyley MM, van der Steen AFW, Serruys PW, et al. Morphological and mechanical information of coronary arteries obtained with Intravascular elastography: a feasibility study in vivo. Eur Heart J 2002;23(5):405–13.
16. Hasegawa H, Kanai H. Reduction of influence of variation in center frequencies of RF echoes on estimation of artery-wall strain. IEEE Trans Ultrason Ferroelectr Freq Control 2008 Sep;55(9):1921–34.
17. Idzenga T, Hansen HHG, Lopata RGP, De Korte CL. Non-invasive assessment of shear strain in the carotid arterial wall based on ultrasound radiofrequency data. Proc IEEE Ultrason Symp 2009 2010 39:2456–2459.
18. Cinthio M, Ahlgren AR, Bergkvist J, Jansson T, Persson HW, Lindstrom K. Longitudinal movements and resulting shear strain of the arterial wall. Am J Physiol Heart Circ Physiol 2006 Jul;291(1):H394–402.
19. Maurice RL, Ohayon J, Fretigny Y, Bertrand M, Soulez G, Cloutier G. Noninvasive vascular elastography: theoretical framework. IEEE Trans Med Imaging 2004 Feb;23(2):164–80.
20. Ribbers H, Lopata RG, Holewijn S, Pasterkamp G, Blankensteijn JD, De Korte CL. Noninvasive two-dimensional strain imaging of arteries: validation in phantoms and preliminary experience in carotid arteries in vivo. Ultrasound Med Biol 2007 Apr;33(4):530–40.
21. Shi H, Varghese T. Two-dimensional multi-level strain estimation for discontinuous tissue. Phys Med Biol 2007 Jan 21;52(2):389–401.
22. Kallel F, Ophir J. A least-squares strain estimator for elastography. Ultrasonic Imaging 1997 Jul;19(3):195–208.
23. de Korte CL, Céspedes EI, van der Steen AFW, Lancée CT. Intravascular elasticity imaging using ultrasound: feasibility studies in phantoms. Ultrasound Med Biol 1997;23(5):735–46.
24. Ryan LK, Foster FS. Ultrasonic measurement of differential displacement and strain in a vascular model. Ultrasonic Imaging 1997;19:19–38.
25. Shapo BM, Crowe JR, Erkamp R, Emelianov SY, Eberle M, O'Donnell M. Strain imaging of coronary arteries with intraluminal ultrasound: experiments on an inhomogeneous phantom. Ultrasonic Imaging 1996;18:173–91.
26. Céspedes EI, de Korte CL, van der Steen AFW. Intravascular ultrasonic palpation: assessment of local wall compliance. Toronto, Canada, IEEE, 1997 pp. 1079–82.
27. de Korte CL, Sierevogel M, Mastik F, Strijder C, Velema E, Pasterkamp G, et al. Identification of atherosclerotic plaque components with intravascular ultrasound elastography in vivo: a Yucatan pig study. Circulation 2002;105(14):1627–30.
28. Schaar JA, de Korte CL, Mastik F, Strijder C, Pasterkamp G, Serruys P, et al. Vulnerable plaque detection with intravascular elastography: a sensitivity and specificity study. Circulation 2001;104(17):II–459.
29. Brusseau E, Fromageau J, Finet G, Delachartre P, Vray D. Axial strain imaging of intravascular data: results on polyvinyl alcohol cryogel phantoms and carotid artery. Ultrasound Med Biol 2001;27(12):1631–42.

30. Nicolaides A, Sabetai M, Kakkos SK, Dhanjil S, Tegos T, Stevens JM, et al. The asymptomatic carotid stenosis and risk of stroke (ACSRS) study – aims and results of quality control. Int Angiol 2003 Sep;22(3):263–72.
31. O'Leary DH, Polak JF, Kronmal RA. Carotid-artery intima and media thickness as a risk factor for myocardial infarction and stroke – reply. N Engl J Med 1999 Jun 3;340(22):1763.
32. Bonnefous O, Brevannes L, Denis E, Sananes JC, Montaudon M, Laurent FH, et al. New noninvasive echographic technique for arterial wall characterization. Radiology 1996 Nov;201:1129.
33. Kanai H, Sato M, Koiwa Y, Chubachi N. Transcutaneous measurement and spectrum analysis of heart wall vibrations. IEEE Trans Ultrason Ferroelectr Freq Control 1996 Sep;43(5):791–810.
34. Hasegawa H, Kanai H. Reduction of influence of variation in center frequencies of RF echoes on estimation of artery-wall strain. IEEE Trans Ultrason Ferroelectr Freq Control 2008 Sep;55(9):1921–34.
35. Hasegawa H, Kanai H. Simultaneous imaging of artery-wall strain and blood flow by high frame rate acquisition of RF signals. IEEE Trans Ultrason Ferroelectr Freq Control 2008 Dec;55(12):2626–39.
36. Maurice RL, Bertrand M. Lagrangian speckle model and tissue-motion estimation – theory. IEEE Trans Med Imaging 1999 Jul;18(7):593–603.
37. Maurice RL, Brusseau E, Finet G, Cloutier G. On the potential of the Lagrangian speckle model estimator to characterize atherosclerotic plaques in endovascular elastography: in vitro experiments using an excised human carotid artery. Ultrasound Med Biol 2005 Jan;31(1):85–91.
38. Maurice RL, Soulez G, Giroux MF, Cloutier G. Noninvasive vascular elastography for carotid artery characterization on subjects without previous history of atherosclerosis. Med Phys 2008 Aug;35(8):3436–43.
39. Chen H, Shi H, Varghese T. Improvement of elastographic displacement estimation using a two-step cross-correlation method. Ultrasound Med Biol 2007 Jan;33(1):48–56.
40. Lopata RG, Nillesen MM, Hansen HH, Gerrits IH, Thijssen JM, De Korte CL. Performance evaluation of methods for two-dimensional displacement and strain estimation using ultrasound radio frequency data. Ultrasound Med Biol 2009 Mar;35(5):796–812.
41. Shi H, Varghese T. Two-dimensional multi-level strain estimation for discontinuous tissue. Phys Med Biol 2007 Jan 21;52(2):389–401.
42. Shi H, Mitchell CC, McCormick M, Kliewer MA, Dempsey RJ, Varghese T. Preliminary in vivo atherosclerotic carotid plaque characterization using the accumulated axial strain and relative lateral shift strain indices. Phys Med Biol 2008 Nov 21;53(22):6377–94.
43. Kim K, Weitzel WF, Rubin JM, Xie H, Chen XC, O'Donnell M. Vascular intramural strain imaging using arterial pressure equalization. Ultrasound Med Biol 2004 Jun;30(6):761–71.
44. Weitzel WF, Kim K, Henke PK, Rubin JM. High-resolution ultrasound speckle tracking may detect vascular mechanical wall changes in peripheral artery bypass vein grafts. Ann Vasc Surg 2009 Mar;23(2):201–6.
45. Weitzel WF, Kim K, Park DW, Hamilton J, O'Donnell M, Cichonski TJ, et al. High-resolution ultrasound elasticity imaging to evaluate dialysis fistula stenosis. Semin Dial 2009 Jan;22(1):84–9.
46. Kanai H, Sato M, Koiwa Y, Chubachi N. Transcutaneous measurement and spectrum analysis of heart wall vibrations. IEEE Trans Ultrason Ferroelectr Freq Control 1996 Sep;43(5):791–810.
47. Nakagawa N, Hasegawa H, Kanai H. Cross-sectional elasticity imaging of carotid arterial wall in short-axis plane by transcutaneous ultrasound. Jpn J Appl Phys 2004 May;43(5B):3220–6.
48. Hansen HHG, Lopata RGP, Holewijn S, Truijers M, de Korte CL. Non-invasive vascular ultrasound strain imaging: different arteries, different approaches. IFMBE Proc 2009; 22:298–302.
49. Konofagou EE, Harrigan T, Ophir J. Shear strain estimation and lesion mobility assessment in elastography. Ultrasonics 2000 Mar;38(1–8):400–4.
50. Thitaikumar A, Krouskop TA, Garra BS, Ophir J. Visualization of bonding at an inclusion boundary using axial-shear strain elastography: a feasibility study. Phys Med Biol 2007 May 7;52(9):2615–33.

51. Idzenga T, Pasterkamp G, De Korte C. Shear strain in the adventitial layer of the arterial wall facilitates development of vulnerable plaques. Biosci Hypotheses 2009;2(5):339–42.
52. Cinthio M, Ahlgren AR, Jansson T, Eriksson A, Persson HW, Lindstrom K. Evaluation of an ultrasonic echo-tracking method for measurements of arterial wall movements in two dimensions. IEEE Trans Ultrason Ferroelectr Freq Control 2005 Aug;52(8):1300–11.
53. Cinthio M, Jansson T, Persson HW, Lindstrom K, Ahlgren AR. Non-invasive measurements of longitudinal strain of the arterial wall. Proc IEEE Ultrason Symp 2007;570–2.
54. Persson M, Ahlgren AR, Eriksson A, Jansson T, Persson HW, Lindstrom K. Non-invasive measurement of arterial longitudinal movement. Proc IEEE Ultrason Symp 2002;2:1783–6.
55. Persson M, Ahlgren AR, Jansson T, Eriksson A, Persson HW, Lindstrom K. A new non-invasive ultrasonic method for simultaneous measurements of longitudinal and radial arterial wall movements: first in vivo trial. Clin Physiol Funct Imaging 2003 Sep;23(5):247–51.
56. Konofagou E, Ophir J. A new elastographic method for estimation and imaging of lateral displacements, lateral strains, corrected axial strains and Poisson's ratios in tissues. Ultrasound Med Biol 1998 Oct;24(8):1183–99.
57. Chen H, Varghese T. Multilevel hybrid 2D strain imaging algorithm for ultrasound sector/phased arrays. Med Phys 2009 Jun;36(6):2098–106.
58. Varghese T, Bilgen M, Ophir J. Multiresolution imaging in elastography. IEEE Trans Ultrason Ferroelectr Freq Control 1998;45(1):65–75.

Biographies

Tim Idzenga obtained his MSc in Medical Electrical Engineering in 1993 with a project on the acoustical properties of the eye. In 1999, he got his PhD on "Intravascular Ultrasound Elastography." Since 2004, he is the head of the Clinical Physics Laboratory and since 2006 associate professor on Medical Ultrasound Technologies. Dr. de Korte was awarded a VENI in 2000 and a VIDI in 2006 by the Netherlands Organization for Scientific Research (NWO). He has published over 70 peer-reviewed publications and holds three patents. His main research interests are echocardiographic deformation imaging and segmentation, vascular ultrasound strain imaging, echographic muscle dynamics imaging, and quantitative echography for diagnosing fatty liver.

Hendrik Hansen received the MSc degree in applied physics from Twente University, Enschede, The Netherlands, in 2003. In 2008, he received his PhD from the Erasmus Universitypstyle Medical Center, Rotterdam, The Netherlands. As a postdoc at the Clinical Physics Laboratory, Radboud University Nijmegen Medical

Centre, he uses radiofrequency ultrasound to investigate the hypothesis that shear strain in the arterial wall stimulates the development of vulnerable plaques in the carotid arterial wall which are more prone to rupture.

Chris de Korte was born in Roermond, The Netherlands, in 1982. He received the MSc degree in applied physics from the Eindhoven University of Technology, Eindhoven, The Netherlands in 2000. At the moment, he is pursuing his PhD study at the Clinical Physics Laboratory of the Radboud University Nijmegen Medical Centre. The research project that he is working on focuses on noninvasive ultrasound strain imaging of carotid arteries. His current research interests include strain imaging, elastography, ultrasound, and atherosclerosis.

Chapter 26
Intravascular Photoacoustic and Ultrasound Imaging: From Tissue Characterization to Molecular Imaging to Image-Guided Therapy

Bo Wang, Jimmy Su, Andrei Karpiouk, Doug Yeager, and Stanislav Emelianov

Abstract Successful diagnosis and treatment of atherosclerosis demands imaging modalities that can characterize the composition of atherosclerotic plaques, stage the disease, and guide interventional therapy. In this chapter, combined intravascular photoacoustic (IVPA) and intravascular ultrasound (IVUS) imaging is used to address these issues. Based on the difference in optical absorption spectra, lipid-rich tissues can be differentiated using spectroscopic IVPA imaging. Using gold nanoparticles as contrast agent, events happening at the molecular and cellular levels may be visualized in IVPA images. IVPA imaging can also monitor stent deployment during interventional therapy procedures. Design of integrated IVUS/IVPA catheters is discussed. Based on the structural information provided by IVUS images, IVPA images may add further clinically relevant information, helping the management of atherosclerosis.

Keywords Intravascular photoacoustic imaging • Intravascular ultrasound imaging • Plasmonic nanoparticles • Molecular imaging • Spectroscopic imaging • Plasmon resonance coupling • Therapy • Stent

Abbreviations

IVPA	Intravascular photoacoustic
IVUS	Intravascular ultrasound
PVA	Polyvinyl alcohol
Au NPs	Gold nanoparticles
CT	Computational tomography
MRI	Magnetic resonance imaging
LDL	Low-density lipoprotein

S. Emelianov (✉)
Department of Biomedical Engineering, The University of Texas at Austin, Austin, TX, USA
e-mail: emelian@mail.utexas.edu

Jasjit S. Suri et al. (eds.), *Atherosclerosis Disease Management*,
DOI 10.1007/978-1-4419-7222-4_26,

MMP	Matrix metalloproteinase
mPEG-SH	Methoxypolyethylene glycol-thiol
PVDF	Polyvinylidene fluoride
IgG	Immunoglobulin G
3D	Three dimension

26.1 Intravascular Ultrasound and Photoacoustic Imaging

Various imaging modalities can be applied to image atherosclerosis. Among them, angiography is the most widely used modality to detect the systematic distribution and the degree of stenosis of vessels. However, in angiography, the arbitrary projection of vessels onto a 2D plane may misrepresent the true vessel lumen narrowing, therefore leading to the misjudgment of the plaque distribution [1]. To accurately demonstrate the distribution of plaques, 3D noninvasive or invasive imaging of atherosclerosis is required. Compared to the noninvasive imaging modalities, such as computational tomography (CT) and magnetic resonance imaging (MRI), intravascular imaging modalities generally provide better resolution [2]. In this chapter, clinically available intravascular ultrasound (IVUS) imaging and an emerging imaging modality – intravascular photoacoustic (IVPA) imaging are used to examine atherosclerotic plaques. The combination of the two imaging modalities can provide clinically important information such as the vulnerability and the stage of the plaques. Combined IVUS/IVPA imaging may also be used to monitor interventional therapies such as the deployment of coronary stents.

26.1.1 Intravascular Ultrasound Imaging

Intravascular ultrasound (IVUS) imaging is a well-developed technology that has been routinely used in catheterization labs for diagnostic imaging and guidance during interventional procedures [1]. In IVUS imaging of coronary arteries, a signal element or array-based IVUS catheter around 1–1.2 mm in diameter is inserted into the lumen of the artery to image the vessel wall. Using a high-frequency ultrasound transducer located at the tip of the catheter, acoustic pressures are generated in a particular direction, thus forming an ultrasound beam. The transducer is then used to receive the backscattered acoustic waves from the vessel wall. The acoustic beams generated from the catheter are either mechanically (single-element IVUS catheter) or electronically (array-based IVUS catheter) rotated at a speed of around 30 revolutions per second to form real-time, cross-sectional images of the vessel. A volumetric view of the artery can also be generated by pulling the IVUS catheter back along the artery. Therefore, IVUS imaging is a minimally invasive modality that permits real-time, high-resolution imaging of the vessel wall.

IVUS imaging can delineate the thickness and certain structures within the vessel wall. Therefore, IVUS imaging may be used to select the most appropriate option of transcatheter therapy (rotational atherectomy, stents, etc.) [3]. However, histopathological information obtained with IVUS imaging is limited [1]. Angioscopy and histological studies generally report low sensitivity of IVUS imaging in detection of thrombus and lipid-rich lesions [4]. To overcome the limitations of IVUS imaging, several investigational techniques such as integrated backscatterer imaging, IVUS elastography [5], and spectral radiofrequency signal analysis [6] have been introduced, which facilitate the discrimination of atheroma. However, because of the limited acoustic contrast among different types of soft tissues, limited information can be extracted from IVUS images alone.

26.1.2 Intravascular Photoacoustic Imaging

Compared to ultrasound imaging in which contrast is based on acoustic backscattering of tissue, photoacoustic imaging is sensitive to the optical absorption contrast within the tissues. In photoacoustic imaging, a laser pulse with nanosecond duration is emitted onto the tissue. After absorbing the laser energy, the tissue generates broadband photoacoustic (also known as optoacoustic) signals due to its fast thermal expansion [7]. The amplitude of the photoacoustic signal is proportional to the laser fluence, optical absorption coefficient of the tissue, and the temperature-dependent Grüneisen coefficient [8]. Under known laser fluence and constant tissue temperature, photoacoustic imaging can map the optical absorption property of the tissue. Blood vessels are most widely imaged by photoacoustic imaging, because of their abundance and high optical absorption contrast in the visible wavelength region [9, 10]. Functional information, such as the level of blood oxygenation or the velocity of the blood flow, can be extracted using multiwavelength photoacoustic imaging or frequency analysis of the photoacoustic signals [11, 12].

IVPA imaging is an extension of general photoacoustic imaging to intravascular applications. Figure 26.1a, b shows representative setup of the combined IVUS/IVPA imaging system for ex vivo artery imaging. The system is triggered by a tunable OPO laser capable of generating 5-ns duration laser pulses. The laser beam is delivered through an optical fiber. A single-element IVUS catheter is inserted inside the artery lumen and aligned with the laser beam. After the arterial tissue is irradiated by a laser pulse, the IVUS catheter receives the photoacoustic signals generated from the tissue. Then, after a user-defined delay, the IVUS transducer performs ultrasound imaging on the same region of the vessel wall. As the artery is rotated 360° by a stepper motor, co-registered radio frequency IVUS and IVPA signals from one cross-section of the artery are acquired. Currently, the digitized IVPA and IVUS signals are stored and processed off-line.

In Fig. 26.1b, the laser beam is delivered externally onto the vessel wall. This setup guarantees a well-defined light path and laser beam profile, which is suitable for the well controlled ex vivo imaging. Figure 26.1c is another benchtop imaging system in

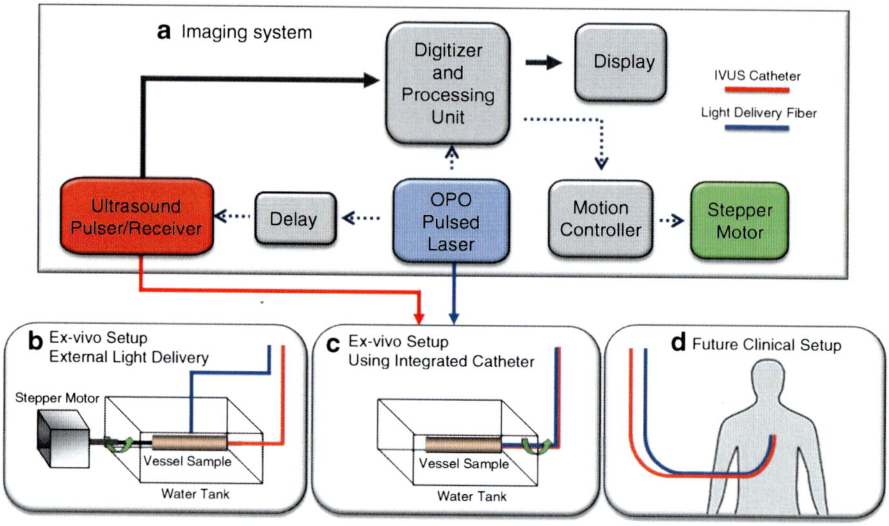

Fig. 26.1 (**a**) The overall configuration of a combined IVUS/IVPA imaging system. The system is capable of acquiring co-registered IVUS and IVPA images of the vessel wall. (**b**, **c**) The setup of two benchtop laboratory IVUS/IVPA imaging systems for ex vivo tissue imaging. In first setup (**b**), the laser beam is delivered onto the vessel sample externally through an optical fiber. In other setup (**c**), the laser beam is delivered into the lumen of the vessel sample through an integrated IVUS/IVPA imaging catheter. (**d**) In future clinical implementation, a cross-sectional scan of vessel can be performed using an integrated IVUS/IVPA imaging catheter

the transition toward in vivo combined IVUS/IVPA imaging, where the laser beam is delivered into the vessel lumen through an integrated catheter. The setups in Fig. 26.1b, c used a single-element IVUS transducer for receiving radio frequency signals. A stepper motor was used to rotate the vessel sample for cross-sectional scan. Finally, in the future, an integrated catheter that is capable of combined IVUS/IVPA imaging of the cross-section of the human artery will be developed (Fig. 26.1d).

Using a commercially available, single-element 40-MHz coronary IVUS catheter (Boston Scientific, Inc.), Sethuraman et al. developed the first benchtop combined IVUS/IVPA imaging system to demonstrate the feasibility of IVPA imaging on tissue-mimicking phantoms and ex vivo vessels [14]. The axial resolution of the IVPA image is around 40 μm, which is comparable with the IVUS image. Meanwhile, because of the lower acoustical than optical attenuation inside the tissue, the imaging depth of IVPA imaging is several millimeters, which is larger than the intravascular optical coherence tomography (OCT) imaging. Figure 26.2 shows the IVUS, IVPA, and combined IVPA/IVUS images of the cross-sectional view (Fig. 26.2a–c) and the 3D volumetric view of a vessel-mimicking phantom (Fig. 26.2d–f). The phantom was imaged at 680 nm wavelength using a system setup shown in Fig. 26.1b. The vessel-mimicking phantom was made out of 8% polyvinyl alcohol (PVA) by weight. Silica powder (0.4% by weight) was added as acoustic scatters. An inclusion made out of PVA and 0.1% graphite powder was embedded spirally inside the PVA phantom. IVUS (Fig. 26.2a, d) clearly images the

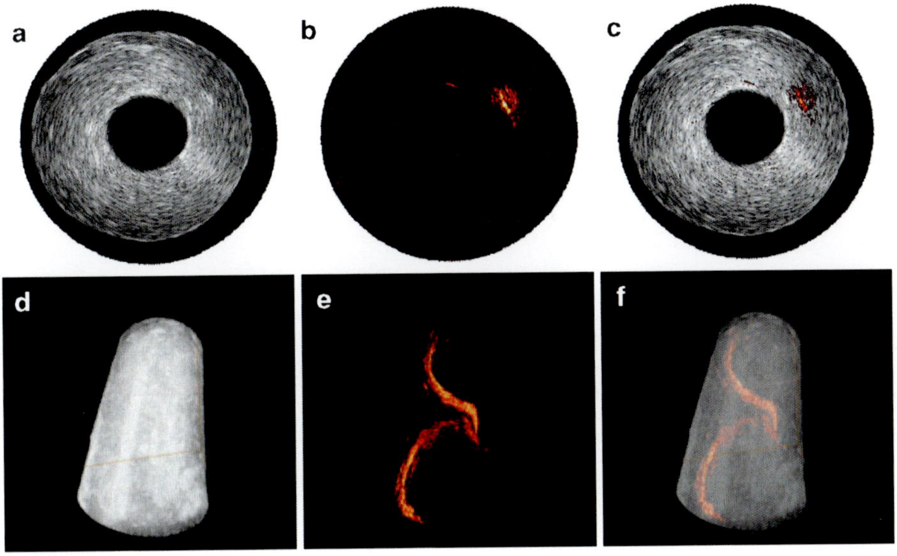

Fig. 26.2 (**a–c**) Cross-sectional and (**d–f**) three-dimensional IVUS images (**a**, **d**), IVPA images (**b**, **e**) and combined IVUS/IVPA images (**c**, **f**) of a tissue-mimicking phantom with a spiral inclusion embedded in the phantom wall (reproduced from [13], with permission)

structure of the phantom. However, due to the limited acoustic contrast between the inclusion and the vessel wall, IVUS imaging cannot detect the inclusion. On the other hand, the inclusion can be clearly visualized from IVPA images (Fig. 26.2b, e) because of the high optical absorption of the graphite powder. As the IVPA and IVUS images are co-registered, the combined IVUS/IVPA image successfully displays the relative location of the inclusion in the context of the surrounding phantom.

It is important to display the IVPA image together with its co-registered IVUS image. IVPA and IVUS imaging are complementary to each other. While IVPA provides the tissue composition related optical absorption information, IVUS provides the structural information of the vessel wall [15]. Moreover, combining the two modalities may utilize the hardware resources more efficiently. Many components used for IVPA imaging, such as the transducer, radio frequency signal receiver, and the analog-to-digital convertor, can easily be shared with IVUS imaging.

26.2 Arterial Tissue Characterization Using Spectroscopic IVPA Imaging

The vulnerability of atherosclerotic plaques depends more on the composition of the plaques than on the degree of stenosis within the blood vessel [16]. Therefore, an imaging technique that can differentiate tissue composition in the arterial wall is

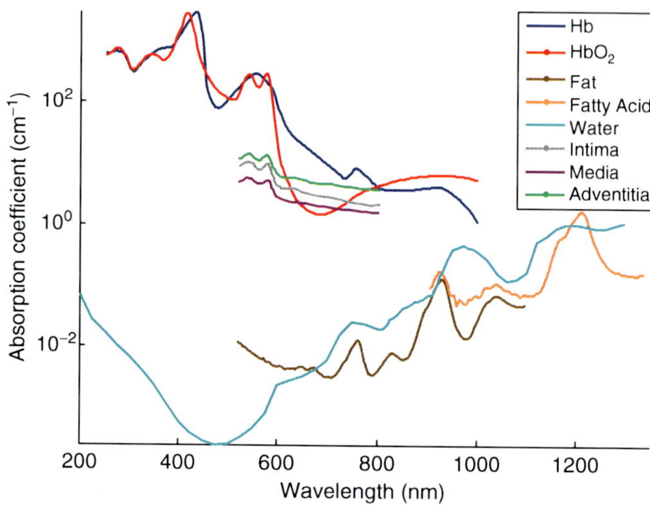

Fig. 26.3 Optical absorption spectra of representative tissue types in the artery [18, 19]

required. For example, the ability to image the distribution of the lipid and the thickness of the fibrous cap covering the lipid region of the plaques will help to identify the classical rupture-prone plaques [17]. Spectroscopic IVPA imaging may be used to characterize the arterial tissue based on the differences in the optical absorption spectra among various tissue types.

The optical absorption spectra of the arterial tissue can be reconstructed through multiwavelength IVPA imaging with controlled laser fluence and tissue temperature. Based on reconstructed optical absorption spectra and the known absorption spectra from various tissue types (Fig. 26.3), the composition of plaques may be identified. In order to image the arterial wall and achieve larger imaging depths, spectroscopic IVPA imaging is conducted in the "optical window" (typically, 650–1,200 nm wavelength) where the attenuation due to blood absorption is relatively low.

Multiwavelength IVPA imaging from 680 to 900 nm wavelength was performed using a benchtop IVPA imaging system (Fig. 26.1b) on both atherosclerotic and healthy New Zealand white rabbit aortas [20]. The atherosclerotic lesions were created by feeding a rabbit with 10 months of 0.15% cholesterol diet. After laser fluence compensation, to demonstrate the wavelength-dependent photoacoustic amplitude change, the first derivative (i.e., slope) was calculated between IVPA images acquired at 680 and 900 nm:

$$\frac{\mathrm{d}S}{\mathrm{d}\lambda} = \frac{S(\lambda_i) - S(\lambda_j)}{\lambda_i - \lambda_j}, \tag{26.1}$$

where S is the photoacoustic signal amplitude, and λ is the wavelength (i.e., $\lambda_i = 900$ nm and $\lambda_j = 680$ nm.). The resulting first derivative (i.e., spectroscopic IVPA) images of the diseased and control aortas are shown in Fig. 26.4a, b, respectively. Compared to the

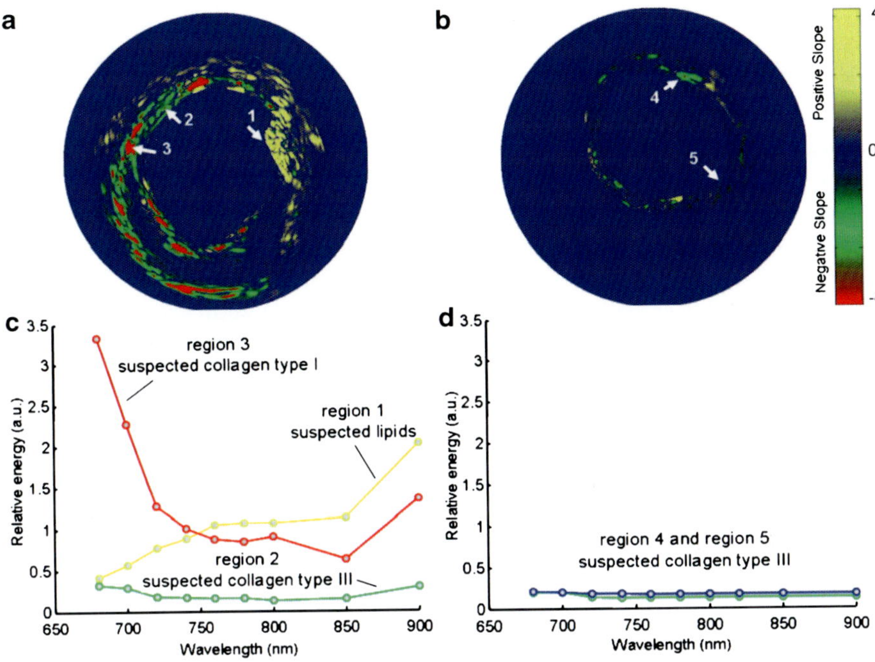

Fig. 26.4 The spectroscopic (first derivative) IVPA images of (**a**) the atherosclerotic and (**b**) normal aorta calculated at 900 nm using a finite difference approach. The reference image for evaluating the first derivative was obtained at 680 nm. Images (**c**, **d**) showing the variations of the relative energy in the photoacoustic responses with wavelength observed in (**a**) atherosclerotic and (**b**) control aorta (reproduced from [20], with permission)

normal aorta, the diseased aorta shows higher variations in the photoacoustic signal amplitude between 680 and 900 nm wavelength. The multiwavelength photoacoustic response of representative regions in the diseased and control aortas are plotted in Fig. 26.4c, d. In the diseased aorta, signal amplitudes in region 1 increase at longer wavelengths. This trend is similar to the increased optical absorption of fatty acid (Fig. 26.3). Moreover, the yellow regions (positive slopes) in Fig. 26.4a are mainly located close to the vessel lumen, which also correlates to the location of lipid deposits in this type of lipid-rich plaque. Region 2 in Fig. 26.4a has similar constant photoacoustic responses as regions 3 and 4. This may indicate the presence of collagen type III, the main composition in a healthy aorta wall. On the other hand, the negative slopes such as region 3 may indicate the presence of collagen type I in the atherosclerotic aorta. The prominent differences of the wavelength-dependent photoacoustic signal amplitude between diseased and normal rabbit aortas indicate that spectroscopic IVPA imaging within the 680–900 nm wavelength range may identify atherosclerotic lesions in the artery.

The specific composition of the lesions may be further identified by comparing the behavior of photoacoustic signals versus wavelength with the optical absorption spectrum of certain tissue type. Spectroscopic IVPA imaging at longer wavelength, around 1,200 nm, can be used to differentiate lipid in atherosclerotic plaques. In this wavelength range, the spectra of most chromophores in tissue are similar to water. Moreover, low optical scattering in this wavelength range will further increase the imaging depth. Compared to the relatively flat absorption spectrum of water from 1,200 to 1,230 nm, the optical absorption spectrum of fatty acid has a sharp peak around 1,210 nm. This difference in optical absorption spectra between fatty acid and water may be used to distinguish lipid-rich tissue from other water-based tissues in atherosclerotic plaques [21].

Multiwavelength IVPA imaging was performed on two aorta samples: a rabbit aorta containing lipid-rich plaques and a normal rabbit aorta. The aorta samples were imaged from a 1,200 to 1,230 nm wavelength with a step size of 10 nm, using a benchtop imaging system shown in Fig. 26.1b. The differential approach that had been used to process images in Fig. 26.4 was applied on IVPA images acquired at 1,200 and 1,230 nm. Regions showing decreasing photoacoustic amplitude from 1,200 to 1,230 nm, similar to the absorbance spectrum of fatty acid, were identified and color-coded. The spectroscopic IVPA image of the diseased rabbit aorta showed extensive scattered lipid deposits in the thickened intimal layer (Fig. 26.5a). The distribution of lipid was confirmed by both the oil red O stain and the H&E histology stain (Fig. 26.5b, c). On the other hand, the spectroscopic IVPA image of the control aorta did not show lipid presence in the arterial wall (Fig. 26.5d).

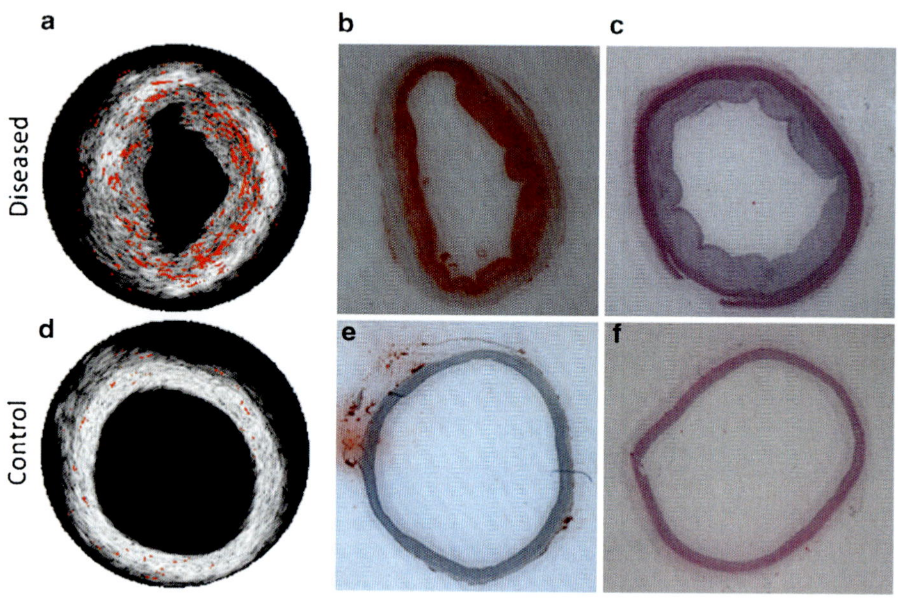

Fig. 26.5 Lipid regions (*orange color*) were demonstrated on top of the IVUS images for diseased (**a**) and control (**d**) rabbit aorta. (**b**, **e**) Oil red O stain for lipid and (**c**, **f**) H&E stain close to the imaged cross-section of diseased and control rabbit aortas (reproduced from [22], with permission)

Besides the finite difference approach applied here for processing multiwavelength IVPA data, other correlation-based approaches, such as the intraclass correlation method, may be used to differentiate chromophores with characteristic optical absorption peak [23]. However, correlation-based methods usually require data points at more wavelengths. Due to the limitation of the available wavelengths of the laser used in the experiments (680–950 nm, 1,200–2,400 nm), additional data covering the whole peak of the optical absorption of lipid around 1,210 nm could not be captured. If the aorta is imaged from 1,150 to 1,230 nm, then correlation-based methods may provide higher sensitivity than the finite difference approach.

Decreased accuracy of the spectroscopic IVPA imaging may occur due to the wavelength-dependent optical attenuation [24]. The effect of such optical attenuation on spectroscopic imaging may be reduced or removed by imaging in a narrower wavelength range or by using computational compensation such as a model-based inversion scheme [25].

It should be noted that arterial tissue characterization using spectroscopic IVPA imaging requires no external contrast agent. Based on the known tissue optical absorption spectra, spectroscopic IVPA imaging is capable of differentiating multiple tissue types relevant to the pathology of atherosclerosis. After post processing, spectroscopic IVPA imaging has a relatively high resolution, around 80 μm. Moreover, the high penetration depth of light in the near infrared region enables visualization and characterization of the arterial tissue at up to 1 cm from the lumen. The high imaging depth makes IVPA imaging a potential tool for imaging vasa vasorum or adventitial fat, which are important indicators of the progression of atherosclerosis [26, 27].

26.3 Molecular and Cellular-Specific IVPA Imaging

During the development of atherosclerosis, complex interactions take place at the molecular and cellular level, with various atherosclerosis-related biomarkers present at different stages of the disease progression [28]. The ability to detect and localize certain biomarkers within atherosclerotic plaques would help to understand the pathology of atherosclerosis, facilitate the diagnosis of the disease while it is asymptomatic, identify the stage and vulnerability of plaques, and further determine the proper therapeutic procedures. Because of the limited imaging contrast from most of these biomarkers, external contrast agents that target certain biomarkers are introduced for different imaging modalities [29–31]. IVPA imaging requires such contrast agents to have high optical absorption in the NIR wavelength region. Eligible candidates include but are not limited to dyes [32], carbon nanotubes [29], or metallic nanoparticles [33, 34]. Gold nanoparticles have outperformed other contrast agents because they are not subjected to photobleaching and exhibit good biocompatibility, high volumetric optical absorption, and tunable optical absorption peak(s).

Molecular and cellular-specific IVPA imaging can be realized using gold nanoparticles (Au NPs) as contrast agents. These nanometer-sized contrast agents have the unique property of being able to penetrate the plaques embedded in the vessel wall. The surface of the Au NPs can be conjugated with antibodies targeting various biomarkers for molecular or cellular-specific imaging [35]. Moreover, once the Au NPs reach their targeted biomarkers, they aggregate and form high local concentration regions. A single spherical Au NP has a surface resonance peak at around 530 nm wavelength. Once the distance between the Au NPs is comparable to their diameter, the surface plasma resonance will be coupled. As a result, the peak optical absorption of the aggregated Au NPs is broadened and red-shifted [36]. The change of optical absorption spectrum due to plasmon resonance coupling has been utilized to image cancer cells targeted by Au NPs [35]. In this section, plasmon resonance coupling in aggregated Au NPs is utilized to detect macrophages – a critical type of cells involved in the progression of atherosclerosis.

26.3.1 IVPA Imaging of Macrophages Labeled With Au NPs: Cell Study

Macrophages in atherosclerosis originate from monocytes circulating in the blood stream, which enter into the injured arterial wall and differentiate into macrophages. Macrophages internalize low-density lipoprotein (LDL) particles and turn into foam cells, which may later contribute to the formation of lipid pools in the plaques [28, 37]. Macrophages that infiltrate into the fibrous cap of plaques also accelerate the progression of disease by releasing matrix metalloproteinases (MMPs). MMPs can weaken the fibrous cap, making the plaques prone to rupture [38].

Because of the nonspecific uptake property of macrophages, Au NPs with a coating of methoxypolyethylene glycol-thiol (mPEG-SH) may be used in IVPA imaging to detect macrophages. After incubating murine macrophages (J774A.1) with PEGylated 50 nm sphere Au NPs overnight, macrophages were loaded with particles. Compared to the control sample (Fig. 26.6a), the dark field image of incubated cells showed that the cells had changed into a golden color (Fig. 26.6b). The optical absorption spectrum of the cells loaded with Au NPs was also different from the spectrum of single Au NPs: the peak absorption was red-shifted; the absorption was increased in the NIR wavelength region (Fig. 26.6c). These changes indicated plasmon resonance coupling among the aggregated nanoparticles after they were internalized by macrophages.

To test whether IVPA imaging can detect macrophages loaded with Au NPs, macrophages labeled with 50 nm spherical Au NPs and Au NPs only were suspended in gelatin gel and microinjected into a vessel-mimicking phantom made out of PVA (Fig. 26.7). The phantom was imaged by the benchtop combined IVUS/IVPA imaging system shown in Fig. 26.1b. When imaged at 532 nm, both macrophages loaded with Au NPs and Au NPs generate photoacoustic signals. No photoacoustic signals are

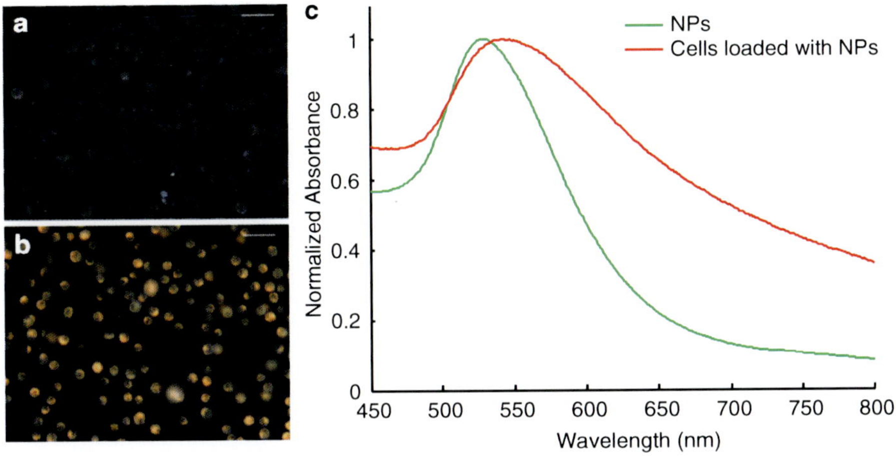

Fig. 26.6 Dark field images of the (**a**) murine macrophages and (**b**) macrophages loaded with 50 nm spherical Au NPs. The scale bars are 50 μm. (**c**) Normalized absorbance spectra of Au NPs and cells loaded with particles (adapted from [39], with permission)

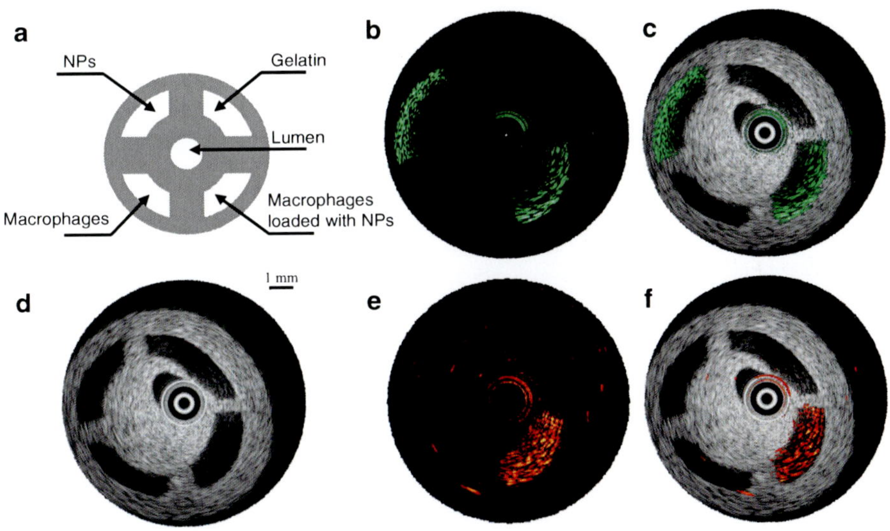

Fig. 26.7 (**a**) The diagram and (**d**) IVUS image of the tissue-mimicking vessel phantom with four compartments. The IVPA images of the same cross-section of the phantom were taken at 532 nm (**b**) and 680 nm wavelength (**e**). The combined IVUS and IVPA images of the phantom at (**c**) 532 nm wavelength and (**f**) 680 nm wavelength, indicating the origin of the photoacoustic response in IVPA images (reproduced from [39], with permission)

detected from the macrophages and gelatin gels (Fig. 26.7b, c). As the laser beam propagates from the outer boundary of the phantom towards the center of the lumen, photoacoustic signal amplitude exponentially decreases due to the attenuation of laser fluence. This attenuation of fluence is primarily caused by strong absorbing Au NP at 532 nm wavelength. When imaged at 680 nm, photoacoustic signals are only generated from macrophages loaded with Au NPs (lower right compartment in Fig. 26.7f), and not from the nonaggregated nanoparticle suspension (upper left compartment in Fig. 26.7f). This is because the optical absorption of macrophages labeled with Au NPs increases in the NIR wavelength region due to plasmon resonance coupling, which occurs as a result of the aggregation of particles within the macrophages following endocytosis (Fig. 26.6c). From 680 to 750 nm, the photoacoustic amplitude of macrophages labeled with Au NPs showed a monotonic decline (Fig. 26.8), similar to the normalized absorbance shown in Fig. 26.6c. The background signal amplitude generated from gelatin gel and the PVA phantom itself was low and remained relatively constant in the 680–730 nm wavelength range.

The cell phantom experiment demonstrated that macrophages loaded with Au NPs can be detected using IVPA imaging. More importantly, in the NIR wavelength region, only macrophages labeled with Au NPs can be detected, but not free-floating nanoparticles. This guarantees that IVPA imaging in the NIR wavelength does not directly image Au NPs, rather, it images only the macrophages that are labeled with particles. Single Au NPs that have not reached the targeted cells, such as Au NPs circulating in the blood stream, will not generate detectable photoacoustic signal in this wavelength range. Therefore, IVPA imaging is capable of monitoring the particle uptake by the cells immediately following the injection of contrast agent.

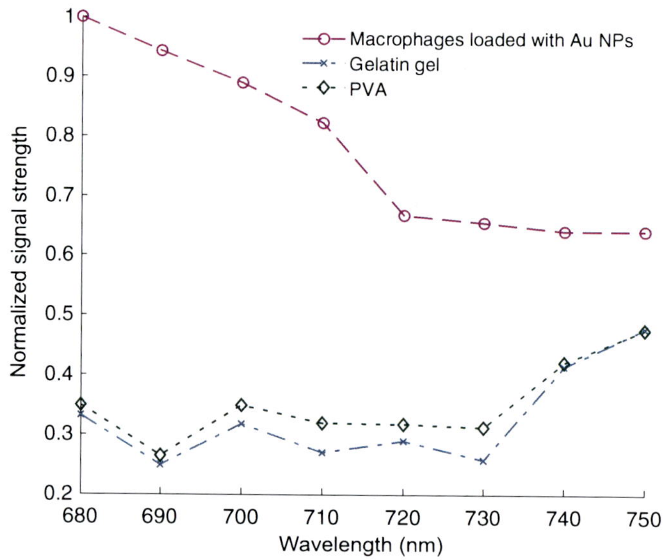

Fig. 26.8 The normalized photoacoustic strength of macrophages loaded with Au NPs, gelatin gel and PVA in the wavelength region from 680 to 750 nm (reproduced from [39], with permission)

26.3.2 *Spectroscopic IVPA Imaging of Macrophages in an Animal Model of Atherosclerosis*

Cell phantom experiments have shown that IVPA in the NIR region is capable of detecting aggregated Au NPs in macrophages. However, other tissue types will also generate photoacoustic signal in this wavelength region (Fig. 26.4). Spectroscopic IVPA imaging was used to differentiate photoacoustic signals generated by Au NPs from signals generated by arterial tissues.

In an experiment, Au NPs were delivered in vivo into an animal model of atherosclerosis. A New Zealand white rabbit was maintained on a high-cholesterol diet (0.5%) for 2 weeks, and received balloon angioplasty in the aorta. The rabbit was then continued on the high-cholesterol diet for an additional 3 months. To achieve the maximum contrast agent delivery, 100 mg of 20 nm spherical gold nanoparticles were conjugated with anti-rabbit IgG antibodies and injected into the rabbit through the ear vein. The rabbit was sacrificed 24-h post-injection. The excised aorta was imaged using the imaging system shown in Fig. 26.1b along with a control rabbit that had received a high-cholesterol diet, but no Au NPs injection.

Multiwavelength IVPA imaging from 700 to 780 nm was performed on the excised aortas. The phantom experiment shows that photoacoustic signal amplitude from macrophages labeled with Au NPs monotonically declines in the NIR wavelengths (Fig. 26.8). Therefore, to distinguish Au NPs from multiwavelength IVPA images, photoacoustic signals having negative slopes similar to the macrophages labeled with Au NPs were selected and considered as regions containing Au NPs. Moreover, because higher concentrations of NPs per cell yield a relatively flat absorption spectrum in the NIR wavelengths, the slopes of the photoacoustic amplitude in the spectroscopic image were color-coded to qualitatively indicate the concentration of NPs per cell.

IVUS and combined IVUS/IVPA (700 nm) of the control and the Au NPs injected aorta are shown in Fig. 26.9. The tissue disconnection at 3 o'clock in the IVUS image was caused by rotation artifact. High photoacoustic signals can be observed at the outer boundary of the aorta due to high laser fluence (Fig. 26.9b, e). The spectroscopic IVPA images were overlaid onto the IVUS images to demonstrate the location of Au NPs. The red color indicates higher concentration of Au NPs. In the control aorta, no Au NPs were shown on the artery (Fig. 26.9c). In the Au NPs-injected rabbit, the combined IVUS and spectroscopic IVPA image shows Au NPs in the intima and the adventitia layers of the aorta (Fig. 26.9f).

The distribution of the Au NPs in the injected rabbit aorta can be better visualized in volumetric 3D images. Figure 26.10 presents 3D images reconstructed from 20 frames of cross-sectional scans (step size 0.4 mm) on the aorta extracted from a rabbit injected with Au NPs. The combined spectroscopic IVPA image and the IVUS image showed one layer of Au NPs located in the intima layer of the aorta. Meanwhile, particles are also present in the adventitia layer of the aorta. The distribution of Au NPs was confirmed by histology stains in the imaged region of the particle-injected aorta. Black regions in the silver stained slides indicate the location

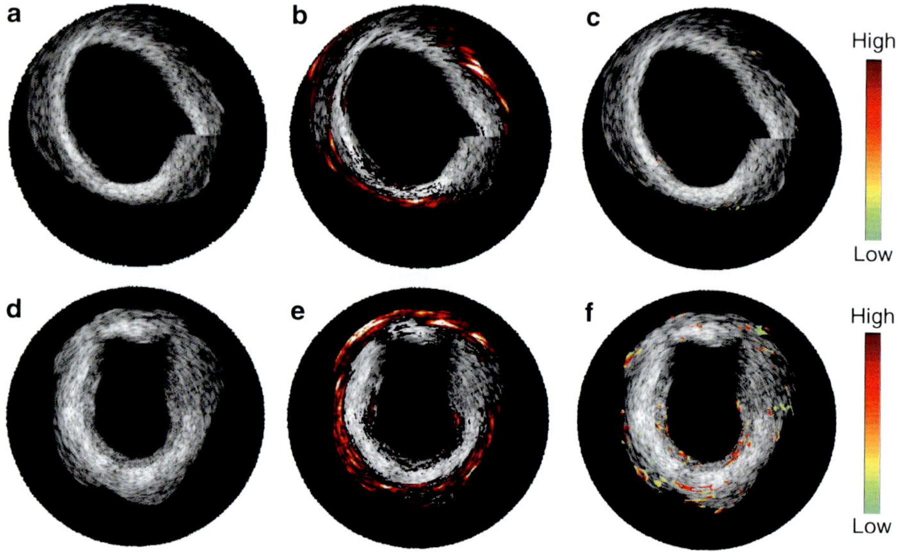

Fig. 26.9 (**a**, **d**) IVUS, (**b**, **e**) combined IVUS and IVPA image at 700 nm, and (**c**, **f**) combined IVUS and spectroscopic IVPA images of aortas from the control (*first row*) and Au NP injected (*second row*) rabbits (reproduced from [40], with permission)

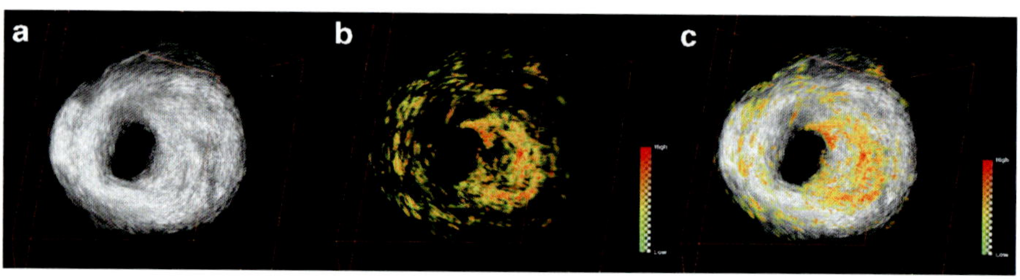

Fig. 26.10 (**a**) 3D IVUS, (**b**) spectroscopic IVPA, and combined (**c**) IVUS and spectroscopic IVPA images of the aorta from a rabbit injected with Au NPs (reproduced from [40], with permission)

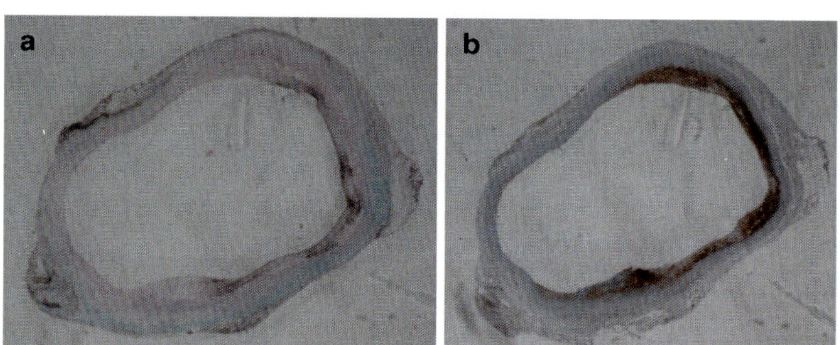

Fig. 26.11 (**a**) Silver stain for Au NPs and (**b**) RAM 11 stain for macrophages were performed on the cross-section of aorta that had been imaged by the combined IVUS/IVPA imaging (reproduced from [40], with permission)

of Au NPs (Fig. 26.11a). In the silver stain image, Au NPs located in the intima layer from 12 to 6 o'clock, correlate with the location of the layer of Au NPs in the inner boundary of the aorta shown in the combined IVUS and spectroscopic IVPA image (Fig. 26.10c). Similar to the spectroscopic IVPA image, scattered Au NPs are present in the adventitia layer of the aorta in the silver stain. Particles in the intima layer can be delivered through the leaky endothelium, while the particles in the adventitia layer may be delivered through vasa vasorum – neovasculature penetrating plaques through the adventitia. RAM11 stain for macrophages on the same cross-section as the silver stain showed the activity of macrophages in the corresponding regions that contain Au NPs.

By utilizing the coupled plasmon resonance of spherical Au NPs, IVPA in the NIR wavelength region may directly image targeted biomarkers. Here, plasmon resonance coupling was achieved through internalization of the contrast agent into macrophages. Similarly, coupled resonance will occur when the Au NPs target receptors on the cell surface [41, 42]. Spectroscopic IVPA imaging can successfully differentiate Au NPs from other tissue types by analyzing the slope of signal amplitude versus the wavelength. Moreover, the value of the slope may also qualitatively reflect the local concentration of Au NP per cell. Using Au NPs as a contrast agent, IVPA may image the clinical relevant biomarkers at the molecular and cellular level with high resolution and adequate imaging depth.

26.4 IVPA Monitoring of Stent Deployment

Coronary stents are currently the most widely used coronary intervention for atherosclerosis in the United States. Over 500,000 stenting procedures are annually performed in the United States alone [43]. While the procedure is more than 90% successful overall [44], stents have brought along several unique issues, including restenosis, hyperplasia, and stent drift [45–47]. The ability to visualize stents both during the stenting procedure and during post-surgery follow-up is important to correctly assess the stent with respect to the plaques and the vessel [48]. In post-surgery, monitoring the stent's position within the vessel wall and subsequent restenosis is required as current 6-month restenosis rates range from 10 to 50% [44]. Immediately following a stenting procedure, it is important to determine whether the stent was properly expanded and that the stent is deployed in contact with the lumen wall [49]. Malapposition can occur resulting in the stent being detached from the vessel wall. Detached stent struts can cause turbulent eddies to form in the vessel which can lead to thrombosis in the area of the stent. Post surgery follow-up procedures are important to determine stent integrity and to monitor how much restenosis has occurred around the stent struts. The distance between the stent and the vessel lumen must be determined to assess stent viability.

A wide variety of imaging modalities have been proposed to image stents in vivo. Currently, the most common method for assessing stent position is x-ray coronary angiography/fluoroscopy [50]. Though widely used, this procedure is problematic due to its use of ionizing radiation, and possible complications in using iodinated contrast agents. Furthermore, x-ray fluoroscopy can only depict a two-dimensional projection, which can lead to an underestimation of the lumen diameter and the stent strut apposition within the lumen.

Other imaging modalities such as MRI and CT have problems with metallic susceptibility artifacts, which prevent the vessel lumen from being accurately visualized relative to the stent [43, 51]. CT imaging is also limited by radiation exposure, which prevents the use of long-scan times. The use of OCT directly competes with these disadvantages with a resolution of 10–20 μm and a lack of metallic susceptibility artifacts which allows individual stent struts to be visualized but has severe depth limitations, allowing only a penetration depth of about 2 mm [49, 52]. The presence of blood flowing through the vessel limits this depth even further, requiring clinicians to flush the vessel during the imaging procedure [53, 54]. Furthermore, in OCT the tissue behind the stent strut becomes hidden due to scattering shadows, preventing complete diagnosis of the stent's relation to the vessel lumen [53]. IVUS has also been used to assess stent patency and monitor vessel restenosis. IVUS can suffer from low contrast between the stent and surrounding tissue depending on orientation of the transducer and metal surface [55, 56]. Comet-tail-like artifacts are also present in the IVUS image due to the acoustic reverberation of the ultrasound pulse within the metal [57].

To counteract these disadvantages, IVPA can be used in conjunction with IVUS to image stents deployed in vivo. Like IVUS imaging, IVPA has sufficient depth penetration and resolution to visualize the stent and surrounding tissue. Moreover, the bulk metal in stents is highly optically absorbing which provides high IVPA contrast of the stent relative to the background vessel. The imaging modality is tomographic which allows for accurate 3D reconstructions of the stent and vessel combined.

To demonstrate the feasibility of the IVPA system to image stents, a study was performed using a commercially available 5.0-mm stent (Cordis BX Velocity™) embedded within a tissue-mimicking phantom. A 10-mm outer diameter vessel was created with three different interior regions where the inner diameter varied along the stent (Fig. 26.12). These three regions were constructed so that the aforementioned stent was embedded approximately 1.0 mm inside the vessel wall, deployed adjacent to the vessel wall, and malapposed approximately 1.0 mm from the lumen wall, respectively. The malapposed region created a gap between the stent and the vessel wall. Imaging experiments were conducted to determine whether sufficient axial resolution could determine the relation between the stent and the vessel walls.

Imaging of the vessel and stent was performed using a prototype IVUS/IVPA benchtop system similar to Fig. 26.1b. Laser light irradiated the vessel from the vessel exterior at 800-nm wavelength. This wavelength was chosen for sufficient depth penetration and laser energy levels. For pullback-based 3D imaging, a 1D axis was placed under the water tank to move the sample along the longitudinal axis.

The cross-sectional ultrasound, photoacoustic and combined images of different parts of the vessel phantom with the stent are shown (Fig. 26.13). With the ultrasound signal displayed at 40 dB, the photoacoustic images at 15 dB show high contrast between the stent and the background.

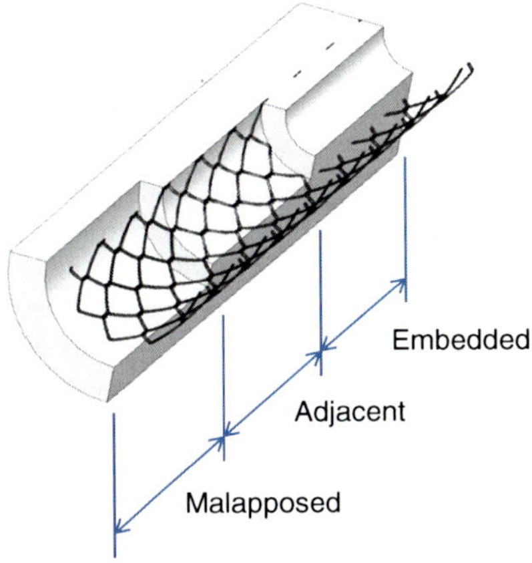

Fig. 26.12 Cutaway diagram of the vessel phantom, which consisted of three regions of varying distances between the stent and the vessel wall to model: Stent embedded (within the vessel wall), deployed (adjacent to the vessel wall), and malapposed (separate from the vessel wall) (reproduced from [58], with permission)

Indeed, the metal stent expresses a high photoacoustic signal due to the high optical absorption of the metal compared to that of the vessel phantom, which has little photoacoustic response at this wavelength. This high optical absorption allowed for high contrast of the stent to the background tissue. The ultrasound image, however, was able to visualize the complete vessel, including the structure and thickness of the vessel wall. Therefore, the co-registered images showed the location of the stent (IVPA image) in relationship to the vessel structure (IVUS image). By scanning along the length of the vessel, the total radial distance between the stent and the lumen wall at all points along the stent was assessed. In the region where the stent was embedded within the vessel (Fig. 26.13a), the stent struts were measured to be embedded 0.7–1.0 mm within the vessel wall. In the region where the stent was merely adjacent to the vessel wall (Fig. 26.13b), the image gave good qualitative agreement showing the correct position of the stent with respect to the vessel. Finally, the malapposed section could also be quantitatively measured, and the images showed that the stent was malapposed away from the lumen wall by 0.8–1.1 mm (Fig. 26.13c).

By combining a set of 80 cross-sectional images, a 3D image of the entire vessel and stent was reconstructed (Fig. 26.14). The structure of the stent was clearly seen in the context of the vessel structure. The display transparency (alpha value) of the ultrasound image was modified so that only the photoacoustic signal could be seen, leaving only the structure of the stent (Fig. 26.14b). The shape and position of the stent within the vessel could then easily be assessed and visualized (Fig. 26.14c). The photoacoustic image also allowed the inner diameter of the stent to be correctly measured at 5.0 mm, which was the manufacturer-reported diameter of the stent.

Figure 26.15 demonstrates that a stent deployed into an excised sample of rabbit aorta could also be visualized in the IVPA image. The rabbit aorta modeled multiple plaques

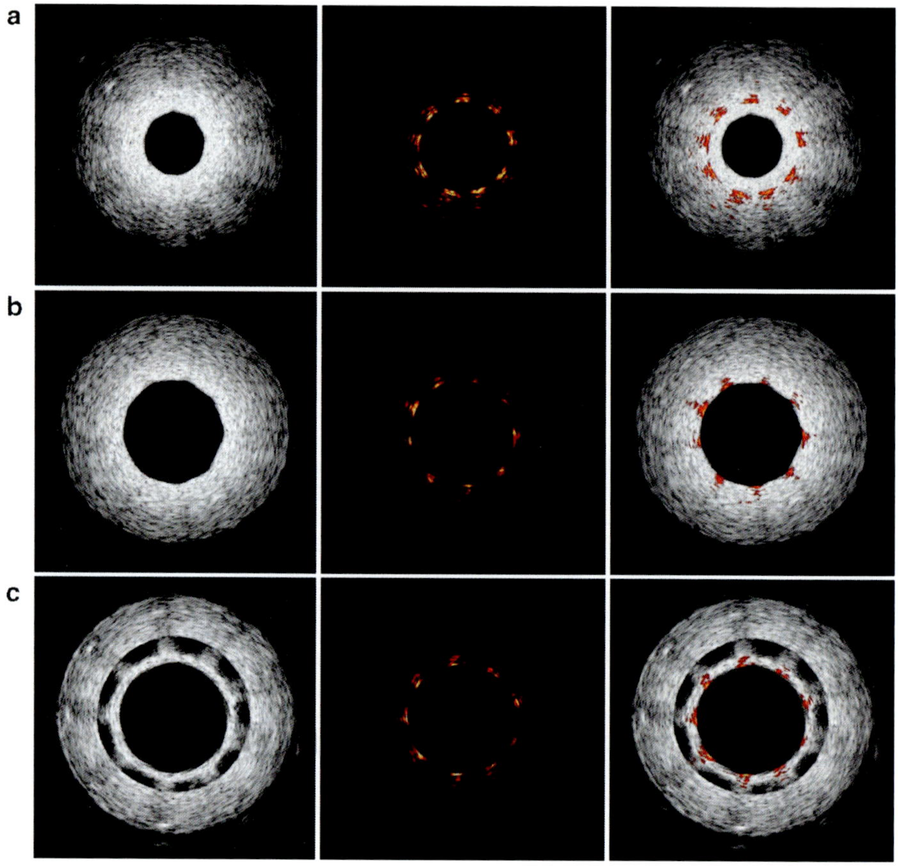

Fig. 26.13 IVUS (*left column*), IVPA (*middle column*) and combined IVUS/IVPA images (*right column*) from the three different stent regions in the vessel. (**a**) Stent embedded within the vessel. (**b**) Stent adjacent to lumen wall. (**c**) Stent detached from lumen wall. Due to the fabrication of this section of the PVA phantom, a thin layer of PVA was formed on the surface of the stent. This PVA film is the source of the additional inner ultrasound ring in panel (**c**) where the stent is located (reproduced from [58], with permission)

due to a high-cholesterol diet. Even though the tissue was irradiated externally by the laser, sufficient laser fluence was able to penetrate the vessel wall to generate photoacoustic signals with enough contrast from the stent. Thus, IVPA imaging may be able to image stents embedded deep inside the arterial wall due to restenosis.

These coronary arterial stents are well visualized using combined IVUS and IVPA imaging. As demonstrated, ultrasound imaging can provide useful structural information of the vessel wall. Photoacoustic imaging utilizes the difference in optical absorption of laser energy and offers high contrast in viewing the metal stent relative to the surrounding vessel. In the combined IVPA/IVUS images the full structure of the phantom is visible and not obscured behind the stent struts, thus allowing one to see the position of the stent within the vessel wall, regardless of the depth within the vessel wall at which the stent is located.

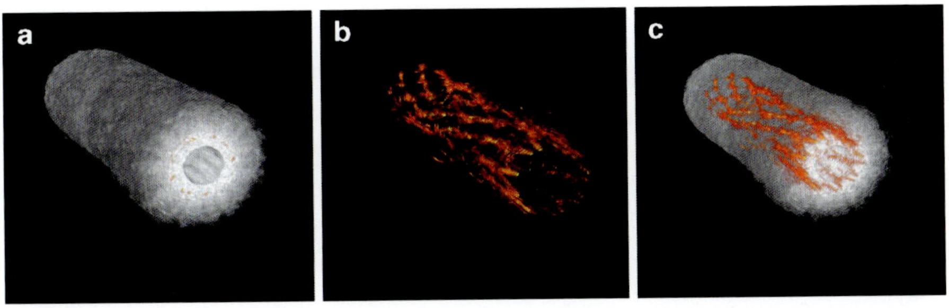

Fig. 26.14 Three-dimensional reconstructed (**a**) IVUS, (**b**) IVPA and (**c**) combined images of trisectional phantom. Individual cross-sections can show the position of the stent within the vessel. Photoacoustic signal alone can assess the shape of the stent to determine the condition of the stent (reproduced from [58], with permission)

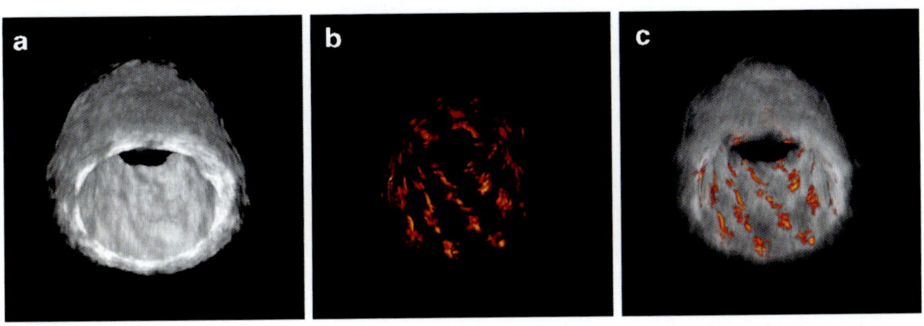

Fig. 26.15 Reconstructed (**a**) IVUS, (**b**) IVPA and (**c**) combined images of a stent deployed within an excised section of an atherosclerotic rabbit aorta. Stent is visible as adjacent to lumen wall

The use of IVUS/IVPA imaging to image stents is a natural progression in cardiovascular disease treatment as stents are commonly used to treat blood vessels that have narrowed due to atherosclerosis. Since recent studies have shown that stents can drift over time, there is a need to detect stent position and viability with respect to the regions of atherosclerosis, while also being able to analyze the progression of plaque vulnerability.

26.5 Design of an Integrated Catheter for Combined Coronary IVUS/IVPA Imaging

Unlike the benchtop imaging system shown in Fig. 26.1b, c, clinical implementation of the combined IVUS/IVPA imaging requires an integrated catheter that can deliver laser light into the vessel lumen. The integrated catheter should be composed of both ultrasound and optical parts, allowing both IVUS and IVPA imaging.

To perform IVPA imaging, short laser pulses should be delivered into the vessel lumen using a light delivery system. The light delivery system is usually fiberoptic-based. The tip of the optical fiber should be equipped with a unit to redirect the light sideways, i.e., normal to the longitudinal axis of the catheter direction. Microoptics such as mirrors, prisms, etc. or special assemblies of the fiber's tips such as a side fire fiber or an axicone can be utilized. The laser beam can also be delivered through optical fibers assembled in a bundle.

Several types of optical fiber-based photoacoustic probes have been reported. Beard et al. designed a probe suitable for photoacoustic and photothermal imaging. The probe was based on a multimode optical fiber coupled with both pulsed and continuous wave lasers. The tip of the fiber was covered by an optically transparent Fabre-Perot polymer film sensor. The laser pulse that propagated through the film would initiate photoacoustic transients in the imaging object. These transients were detected by the sensor at the tip of the probe [59]. In the design by Viator et al., the laser beam was delivered through an optical fiber that was polished 45° at the distal end and sealed with a glass tube. The resulting side fire fiber could deliver light onto the vessel wall. A polyvinylidene fluoride (PVDF) transducer was used to acquire photoacoustic signals [60]. Yang et al. designed an endoscopic photoacoustic probe. The probe delivered light to the distal end of the probe using an optical fiber, where it was then reflected at a nearly right angle by a scanning mirror. Photoacoustic transients were reflected by the mirror and detected by a single-element ultrasound transducer. To obtain the image from the whole cross-section, the mirror is rotated inside the probe using a geared micromotor and a magnetic shaft [61]. However, none of these photoacoustic catheters/probes have been used for combined ultrasound and photoacoustic imaging. Because ultrasound imaging can provide structural information that photoacoustic imaging may lack, the ability to acquire co-registered IVUS and IVPA image of the vessel is critical.

Two prototypes of integrated IVUS/IVPA catheters that are capable of both IVUS and IVPA imaging are presented here. The prototypes are designed to be mechanically rotated into a lumen. Both designs were tested using a 40-MHz single-element ultrasound transducer and a single 600-μm-core multimode optical fiber [62].

In the first design, a laser beam was directed sideways by a micromirror mounted at the tip of the integrated catheter. A microphotograph of the catheter and its schematic diagram depicting the alignment of an ultrasound and light beams are shown in Fig. 26.16. The IVUS imaging transducer was fixed facing away from fiber in the position resulting in the maximum overlap of the ultrasound and light beams and also avoiding a direct interaction of light with the ultrasound transducer. As shown in Fig. 26.16b, light emitted from the fiber was redirected at a near right angle using a custom-made silver micromirror. The rigid part of the catheter's distal end shown in Fig. 26.16 was measured to be 14-mm long and 4-mm thick.

This prototype of the integrated IVUS/IVPA imaging catheter was tested in phantom studies. The point target phantom used in the experiments is shown in Fig. 26.17a. The phantom was made with 12 graphite rods embedded in a tissue-mimicking environment. To mimic ultrasound and optical properties of soft tissues, 8% gelatin gel was mixed with 40-μm silica particles and 20% low-fat milk [63, 64]. In experiments, the phantom was fixed inside a water tank while the prototype of the imaging catheter was rotated inside the lumen of the phantom Fig. 26.1c.

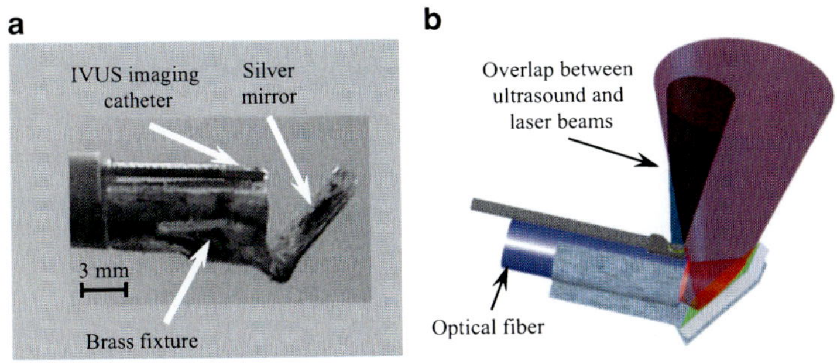

Fig. 26.16 (**a**) Microphotograph of the integrated IVUS/IVPA imaging catheter based on a single-element IVUS imaging catheter and a single optical fiber equipped by a micromirror to irradiate an area of an artery imaged by the IVUS imaging catheter. (**b**) Schematic diagram of the catheter showing the alignment of light and ultrasound beams (reproduced from [62], with permission)

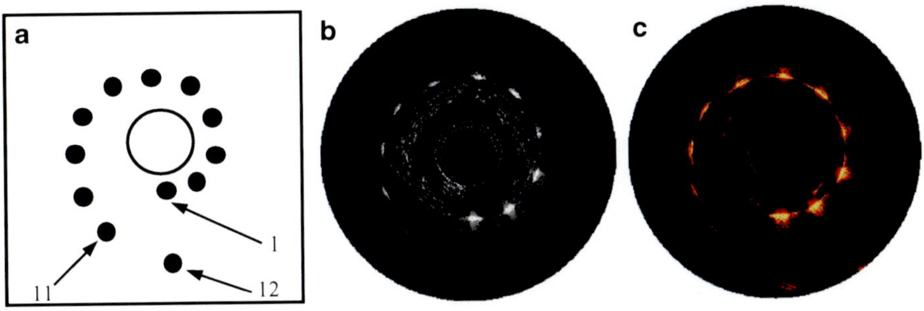

Fig. 26.17 (**a**) Schematic diagram of the phantom consisting of 12-point targets distributed spirally away from the center of the lumen and of 12th point target located separately. (**b**) IVUS image of the phantom embedded in the tissue-mimicking environment. (**c**) IVPA image of the phantom (reproduced from [62], with permission)

The ultrasound and photoacoustic images of the phantom are shown in Fig. 26.17b, c respectively. Lumen, tissue-mimicking background and point targets are imaged in the IVUS image. The brightness of the point targets decreases with increasing distance from the catheter. Due to the attenuation of high-frequency ultrasound in the tissue-mimicking environment, only 11-point targets are detected in the IVUS image (Fig. 26.17b). The co-registered IVPA image of the phantom is shown in Fig. 26.17c, where all point targets are clearly visualized. Note that the point target embedded deepest inside the phantom (point target #12 in Fig. 26.17a) can still be detected by IVPA imaging.

In another design of an integrated catheter, the laser beam is redirected by a side fire fiber. The light delivery system is based on an optical fiber polished with a 33° angle to utilize the effect of total internal reflection on the angled surface of the fiber. Since a gas must be kept next to this surface, the distal end of the fiber was covered by a sealed 1-mm-thick quartz pipe to keep air near the tip (Fig. 26.18a). The IVUS imaging transducer was fixed on the fiber such that the ultrasound and optical beams were aligned and overlapped as shown in Fig. 26.18b.

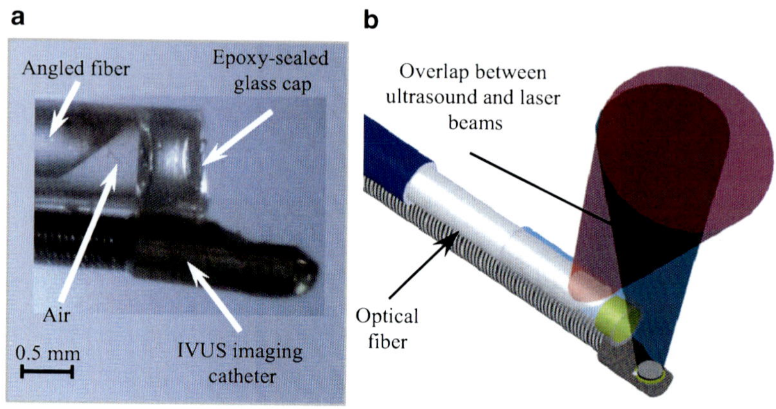

Fig. 26.18 (**a**) Microphotograph of the integrated IVUS/IVPA imaging catheter based on a single-element IVUS imaging catheter and a side fire optical fiber equipped with gas-trapping cap. (**b**) Schematic diagram of the catheter showing the alignment of light and ultrasound beams (reproduced from [62], with permission)

The performance of this prototype integrated IVUS/IVPA imaging catheter was tested using the same phantom as shown in Fig. 26.17a. The IVUS and IVPA images of the phantom are shown in Fig. 26.18a, b respectively. Similarly, IVUS image outlines the structure of the phantom, while the IVPA image clearly identifies the high optical absorbing point targets. These images differ from Fig. 26.17b, c because of slightly different cross-section of the same phantom was imaged.

The design of an integrated catheter for combined IVUS/IVPA imaging is not limited to those designs mentioned above. Different types of the ultrasound units and light delivery systems can be mounted into a single device to perform combined IVUS/IVPA imaging of a transverse cross-section of an artery. The ultrasound units can be made out of commonly used PVDF. Light delivery systems that are based on either a single optical fiber or a fiber bundle can be integrated with either single element-based or array-based IVUS catheters.

A diagram of the integrated IVUS/IVPA imaging catheter consisting of a single-element IVUS imaging catheter and a fiber bundle-based light delivery system is shown in Fig. 26.20a. The fiber bundle consists of multiple side fire optical fibers. The distal end of each side fire fiber is polished at an angle and equipped with a gas-trapping cap to keep air around the tip as shown in Fig. 26.20b. Each side fire fiber is used to irradiate a sector of the imaging cross-section of the artery while the ultrasound transducer is mechanically rotated inside the bundle.

Another type of the integrated IVUS/IVPA imaging catheter based on an array IVUS catheter and a single side fire fiber is shown in Fig. 26.21a. In this design, the light delivery system is rotated mechanically inside the ultrasound array. To perform the combined IVUS/IVPA imaging, the electronically rotated ultrasound beam should be synchronized with the mechanical rotation of the side fire fiber to irradiate the area of the artery. To improve the light illumination of the integrated IVUS/IVPA imaging catheter, the tip of the optical fiber can also be polished as an

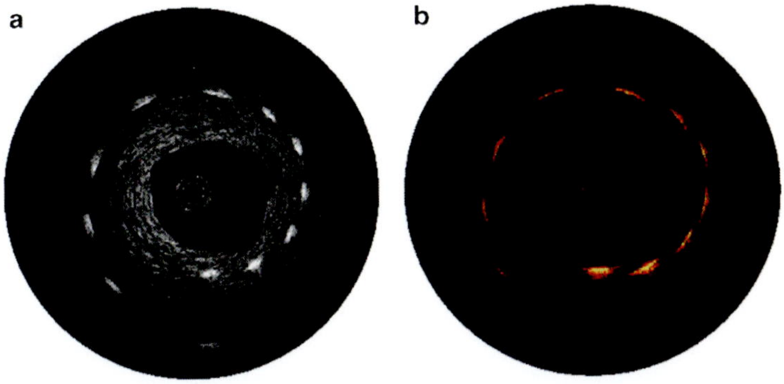

AU1] **Fig. 26.19** (**a**) IVUS and (**b**) IVPA images of the phantom schematically shown in Fig. 26.17a (reproduced from [62], with permission)

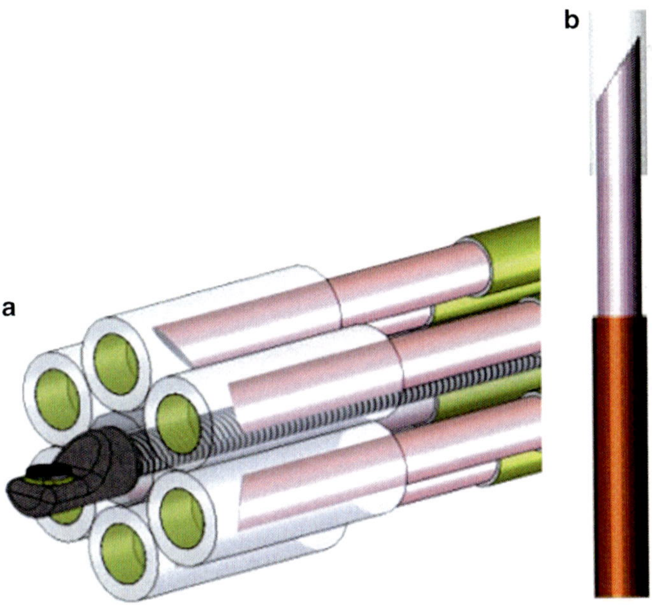

Fig. 26.20 (**a**) Diagram of the integrated IVUS/IVPA imaging catheter based of a single-element IVUS imaging catheter and on optical bundle consisting of six single side fire fibers. (**b**) Diagram of the side fire fiber with a gas-trapping cap

inversed axicon and equipped with a gas-trapping cap; the diagram of the distal tip of such an optical fiber is shown in Fig. 26.21b. With this configuration, the optical fiber does not need to be rotated, and all the ultrasound elements could record photoacoustic transients simultaneously. Therefore, IVPA image of the whole cross-section of an artery can be acquired with one laser pulse.

Regardless of which form is implemented for the integrated IVUS/IVPA imaging catheters, the design of the catheter faces some common challenges. For example,

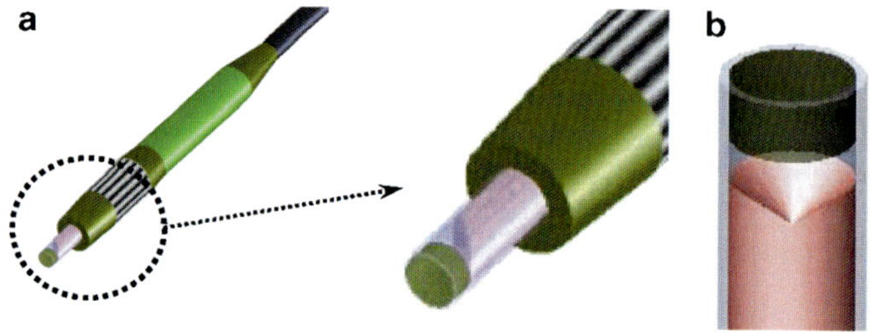

Fig. 26.21 (**a**) Diagram of the integrated IVUS/IVPA imaging catheter based on an ultrasound array and a single side fire fiber. (**b**) Axicon-like distal tip of optical fiber

the mechanical properties of the catheter have to be suitable for the coronary imaging application. The overall diameter of the integrated catheter should be comparable to the available clinical IVUS catheters, i.e., 1.0–1.2 mm. The integrated catheter has to be flexible enough to easily glide inside the coronary arteries along the guide wire. Moreover, the optical delivery system has to be able to sustain high laser fluence so that a desirable signal-to-noise ratio of the in IVPA images may be achieved.

In this chapter, combined IVUS and IVPA imaging has been introduced. While the IVUS image demonstrates the structure of the vessel, addition of a co-registered IVPA image shows the optical absorption information of the same region that has been imaged by IVUS. Combined IVUS/IVPA imaging has broad applications, from detecting vulnerable plaques, to molecular and cellular imaging of plaques, to monitoring the stent deployment. With the development of a new generation of integrated IVUS/IVPA imaging catheters this technology has a promising future in clinical practice.

Acknowledgments The authors would like to acknowledge Dr. Konstantin Sokolov for his help with bioconjugated gold nanoparticles and the helpful discussions on animal experiments, James Amirian and Dr. Richard Smalling for their assisstance in the animal experiments, and Dr. Silvio Litovsky for his help with the histology analysis. Partial support from National Institutes of Health under grants HL 096981 and HL 084076 is acknowledged. We also would like to acknowledge the technical support from Boston Scientific, Inc.

References

1. S Nissen, P Yock (2001), Intravascular ultrasound novel pathophysiological insights and current clinical applications, Circulation 103: 604–616
2. BD MacNeill, HC Lowe, M Takano et al (2003), intravascular modalities for detection of vulnerable plaque: current status, Arterioscler. Thromb. Vasc. Biol. 23: 1333–1342
3. PJ Fitzgerald, PG Yock (1993), Mechanisms and outcomes of angioplasty and atherectomy assessed by intravascular ultrasound imaging, J. Clin. Ultrasound 21: 579–588
4. F Damian, S Udo, WH Hans (1998), Comparison of angioscopic, intravascular ultrasonic, and angiographic detection of thrombus in coronary stenosis, Am. J. Cardiol. 82: 1273–1275

5. CL de Korte, G Pasterkamp, AF van der Steen et al (2000), Characterization of plaque components with intravascular ultrasound elastography in human femoral and coronary arteries in vitro, Circulation 102: 617–623
6. A König, MP Margolis, R Virmani et al (2008), Technology Insight: in vivo coronary plaque classification by intravascular ultrasonography radiofrequency analysis, Nat. Clin. Pract. Cardiovasc. Med. 5: 219–229
7. AA Oraevsky, AA Karabutov (2003), Optoacoustic Tomography, CRC Press, Boca Raton
8. V Gusev, A Karabutov (1991), Laser Optoacoustics, American Institute of Physics, New York
9. X Wang, Y Pang, G Ku et al (2003), Noninvasive laser-induced photoacoustic tomography for structural and functional in vivo imaging of the brain, Nat. Biotechnol. 21: 803–806
10. HF Zhang, K Maslov, G Stoica et al (2006), Functional photoacoustic microscopy for high-resolution and noninvasive in vivo imaging, Nat. Biotechnol. 24: 848–851
11. HP Brecht, DS Prough, YY Petrov et al (2007), In vivo monitoring of blood oxygenation in large veins with a triple-wavelength optoacoustic system, Opt. Express 15: 16261–16269
12. H Fang, K Maslov, LV Wang (2007), Photoacoustic Doppler effect from flowing small light-absorbing particles, Phys. Rev. Lett. 99: 184501
13. B Wang, AB Karpiouk, SY Emelianov (2008), Design of catheter for combined intravascular photoacoustic and ultrasound imaging, Proceedings of the 2008 IEEE Ultrasonics Symposium: 1150–1153
14. S Sethuraman, SR Aglyamov, JH Amirian et al (2007), Intravascular photoacoustic imaging using an IVUS imaging catheter, IEEE Trans. Ultrason. Ferroelectr. Freq. Control. 54: 978–986
15. S Emelianov, S Aglyamov, J Shah et al (2004), Combined ultrasound, optoacoustic, and elasticity imaging, Proc. SPIE 5320: 101–112
16. M Naghavi, P Libby, E Falk et al (2003), From vulnerable plaque to vulnerable patient a call for new definitions and risk assessment strategies: Part I, Circulation 108: 1664–1672
17. ME Brezinski, GJ Tearney, BE Bouma et al (1996), Optical coherence tomography for optical biopsy: properties and demonstration of vascular pathology, Circulation 93: 1206–1213
18. SA Prahl (2001), Optical properties spectra, retrieved 2009, http://omlc.ogi.edu/spectra/.
19. CL Tsai, JC Chen, WJ Wang (2001), Near-infrared absorption property of biological soft tissue constituents, J. Med. Biol. Eng. 21: 7–14
20. S Sethuraman, JH Amirian, SH Litovsky et al (2008), Spectroscopic intravascular photoacoustic imaging to differentiate atherosclerotic plaques, Opt. Express 16: 3362–3367
21. TJ Allen, PC Beard (2009), Photoacoustic characterisation of vascular tissue at NIR wavelengths, Proc. SPIE 7177: 71770A
22. B Wang, JL Su, et al (2010). Detection of lipid in atherosclerotic vessels using ultrasound-guided spectroscopic intravascular photoacoustic imaging. Opt. Express 18(5): 4889–4897
23. S Mallidi, T Larson, J Tam et al (2009), Multiwavelength photoacoustic imaging and plasmon resonance coupling of gold nanoparticles for selective detection of cancer, Nano Lett. 9: 2825–2831
24. M Sivaramakrishnan, K Maslov, HF Zhang et al (2007), Limitations of quantitative photoacoustic measurements of blood oxygenation in small vessels, Phys. Med. Biol. 52: 1349–1362
25. J Laufer, D Delpy, C Elwell et al (2007), Quantitative spatially resolved measurement of tissue chromophore concentrations, Phys. Med. Biol. 52: 141–168
26. HM Kwon, G Sangiorgi, EL Ritman et al (1998), Adventitial vasa vasorum in balloon-injured coronary arteries: visualization and quantitation by a microscopic three-dimensional computed tomography technique, J. Am. Coll. Cardiol. 32: 2072–2079
27. D Vela, L Buja, M Madjid et al (2007), The role of periadventitial fat in atherosclerosis, Arch. Pathol. Lab. Med. 131: 481
28. J Sanz, ZA Fayad (2008), Imaging of atherosclerotic cardiovascular disease, Nature 451: 953–957
29. A De La Zerda, C Zavaleta, S Keren et al (2008), Carbon nanotubes as photoacoustic molecular imaging agents in living mice, Nat. Nanotechnol. 3: 557–562

30. K Briley-Saebo, W Mulder, V Mani et al (2007), Magnetic resonance imaging of vulnerable atherosclerotic plaques: current imaging strategies and molecular imaging probes, J. Magn. Reson. Imaging 26: 460–479
31. A Conjusteau, SA Ermilov, D Lapotko et al (2006), Metallic nanoparticles as optoacoustic contrast agents for medical imaging, Proc. SPIE 6086: 1605–7422
32. D Razansky, M Distel, C Vinegoni et al (2009), Multispectral opto-acoustic tomography of deep-seated fluorescent proteins in vivo, Nat. Photon. 3: 412–417
33. M Eghtedari, A Oraevsky, JA Copland et al (2007), High sensitivity of in vivo detection of gold nanorods using a laser optoacoustic imaging system, Nano Lett. 7: 1914–1918
34. A Doiron, K Homan, S Emelianov et al (2009), Poly(lactic-co-glycolic) acid as a carrier for imaging contrast agents, Pharm. Res. 26: 674–682
35. JS Aaron, J Oh, TA Larson et al (2006), Increased optical contrast in imaging of epidermal growth factor receptor using magnetically actuated hybrid gold/iron oxide nanoparticles, Opt. Express 14: 12930–12943
36. W Rechberger, A Hohenau, A Leitner et al (2003), Optical properties of two interacting gold nanoparticles, Opt. Commun. 220: 137–141
37. E Falk (2006), Pathogenesis of atherosclerosis, J. Am. Coll. Cardiol. 47: C7–12
38. ZS Galis, GK Sukhova, R Kranzhofer et al (1995), Macrophage foam cells from experimental atheroma constitutively produce matrix-degrading proteinases, Proc. Natl. Acad. Sci. U.S.A. 92: 402
39. B Wang, E Yantsen, T Larson et al (2009), Plasmonic intravascular photoacoustic imaging for detection of macrophages in atherosclerotic plaques, Nano Lett. 9: 2212–2217
40. B Wang, P Joshi, V Sapozhnikova et al (2010), Intravascular photoacoustic imaging of macrophages using molecularly targeted gold nanoparticles, Proc. SPIE 7564: 75640A
41. S Mallidi, T Larson, J Aaron et al (2007), Molecular specific optoacoustic imaging with plasmonic nanoparticles, Opt. Express 15: 6583–6588
42. J Aaron, N Nitin, K Travis et al (2007), Plasmon resonance coupling of metal nanoparticles for molecular imaging of carcinogenesis in vivo, J. Biomed. Opt. 12: 034007
43. D Maintz, RM Botnar, R Fischbach et al (2002), Coronary magnetic resonance angiography for assessment of the stent lumen: a phantom study, J. Cardiovasc. Magn. Reson. 4: 359–367
44. SC Smith Jr, TE Feldman, JW Hirshfeld Jr et al (2006), ACC/AHA/SCAI 2005 Guideline Update for Percutaneous Coronary Intervention – summary article: a report of the American College of Cardiology/American Heart Association Task Force on Practice Guidelines (ACC/AHA/SCAI Writing Committee to Update the 2001 Guidelines for Percutaneous Coronary Intervention), Circulation 113: 156–175
45. M Joner, AV Finn, A Farb et al (2006), Pathology of drug-eluting stents in humans: delayed healing and late thrombotic risk, J. Am. Coll. Cardiol. 48: 193–202
46. EL Eisenstein, KJ Anstrom, DF Kong et al (2007), Clopidogrel use and long-term clinical outcomes after drug-eluting stent implantation, JAMA 297: 159–168
47. DE Kandzari, MB Leon, JJ Popma et al (2006), Comparison of zotarolimus-eluting and sirolimus-eluting stents in patients with native coronary artery disease: a randomized controlled trial, J. Am. Coll. Cardiol. 48: 2440–2447
48. D Maintz, M Grude, EM Fallenberg et al (2003), Assessment of coronary arterial stents by multislice-CT angiography, Acta. Radiol. 44: 597–603
49. P Barlis, K Dimopoulos, J Tanigawa et al (2010), Quantitative analysis of intracoronary optical coherence tomography measurements of stent strut apposition and tissue coverage, Int. J. Cardiol. 141:151–156
50. DR Elgort, CM Hillenbrand, S Zhang et al (2006), Image-guided and -monitored renal artery stenting using only MRI, J. Magn. Reson. Imaging 23: 619–627
51. J Hug, E Nagel, A Bornstedt et al (2000), Coronary arterial stents: safety and artifacts during MR imaging, Radiology 216: 781–787
52. TL Slottow, R Pakala, T Okabe et al (2008), Optical coherence tomography and intravascular ultrasound imaging of bioabsorbable magnesium stent degradation in porcine coronary arteries, Cardiovasc. Revasc. Med. 9: 248–254

53. Y Kawase, K Hoshino, R Yoneyama et al (2005), In vivo volumetric analysis of coronary stent using optical coherence tomography with a novel balloon occlusion-flushing catheter: a comparison with intravascular ultrasound, Ultrasound Med. Biol. 31: 1343–1349
54. IK Jang, BE Bouma, DH Kang et al (2002), Visualization of coronary atherosclerotic plaques in patients using optical coherence tomography: comparison with intravascular ultrasound, J. Am. Coll. Cardiol. 39: 604–609
55. G Finet, C Cachard, P Delachartre et al (1998), Artifacts in intravascular ultrasound imaging during coronary artery stent implantation, Ultrasound Med. Biol. 24: 793–802
56. A Gronningsaeter, T Lie, K Bolz et al (1995), Ultrasonographic stent-imaging artifacts, J. Vasc. Invest. 3: 140–149
57. MC Ziskin, DI Thickman, NJ Goldenberg et al (1982), The comet tail artifact, J. Ultrasound Med. 1: 1–7
58. JL-S Su, B Wang, SY Emelianov (2009), Photoacoustic imaging of coronary artery stents, Opt. Express 17: 19894–19901
59. PC Beard, F Perennes, E Draguioti et al (1998), Optical fiber photoacoustic-photothermal probe, Opt. Lett. 23: 1235–1237
60. J Viator, G Paltauf, S Jacques et al (2001), Design and testing of an endoscopic photoacoustic probe for determination of treatment depth after photodynamic therapy, Proc. SPIE 4256
61. J-M Yang, K Maslov, H-C Yang et al (2009), Photoacoustic endoscopy, Opt. Lett. 34: 1591–1593
62. AB Karpiouk, B Wang, SY Emelianov (2010), Development of a catheter for combined intravascular ultrasound and photoacoustic imaging, Rev. Sci. Instrum. 81: 014901
63. EL Madsen, JA Zagrebski, MC MacDonald et al (1991), Ultrasound focal lesion detectability phantoms, Med. Phys. 18: 1171–1181
64. MD Waterworth, BJ Tarte, AJ Joblin et al (1995), Optical transmission properties of homogenised milk used as a phantom material in visible wavelength imaging, Australas. Phys. Eng. Sci. Med. 18: 39–44

Biographies

Bo Wang received the BS and MS degrees in Control Science and Engineering in 2004 and 2006, respectively, from Harbin Institute of Technology, Harbin, P. R. China. In 2006, she joined the Ultrasound Imaging and Therapeutics Research Laboratory at the University of Texas at Austin. She is currently a PhD candidate at the University of Texas at Austin. Her research interests are in the areas of using combined intravascular photoacoustic and ultrasound imaging to detect and differentiate vulnerable plaques, using gold nanoparticles as a contrast agent for intravascular photoacoustic imaging.

Jimmy Su received the BS degree in Biomedical Engineering from Johns Hopkins University, Baltimore, Maryland in 2002. From 2002 to 2003, he worked as an electrical engineer, developing biological agent detection systems in Austin, Texas. In 2006, he obtained the MS degree in Biomedical Engineering from the University of Texas, studying the stability of human walking. In 2007, he joined the Ultrasound Imaging and Therapeutics Research Laboratory at the University of Texas as a PhD student. His research interests are in the areas clinical imaging, using ultrasound and photoacoustic imaging for the detection of foreign objects in vivo, and image processing.

Andrei B. Karpiouk received his BS, MS and PhD degrees in Laser Physics from the Moscow Engineering Physics Institute (Technical University), Moscow, Russia in 1992, 1994, and 2002, respectively. In 2003–2004, he was a postdoctoral fellow in the Laser Center at the UTMB, Galveston, Texas. Now, Dr. Karpiouk holds Research Associate III position in the Department of Biomedical Engineering, the UT at Austin, Texas. His research interests are in combined medical imaging, biomedical engineering, research of laser–tissue interaction processes, measurements of biomechanical properties of soft tissues, etc.

Doug Yeager has served as a graduate research assistant for the Ultrasound Imaging and Therapeutics Lab as well as the Biomedical Optics and Nanodiagnostics Lab at the University of Texas at Austin since 2009, with research focusing on combined intravascular ultrasound and photoacoustic imaging of atherosclerotic plaques. He holds a BS degree in Biomedical Engineering from Texas A&M University, College Station.

Stanislav Emelianov received the BS and MS degrees in physics and acoustics in 1986 and 1989, respectively, from the Moscow State University, and the PhD degree in physics in 1993 from Moscow State University and the Institute of Mathematical Problems of Biology of the Russian Academy of Sciences, Russia. In 1989, he joined the Institute of Mathematical Problems of Biology, where he was engaged in both mathematical modeling of soft tissue biomechanics and experimental studies of noninvasive visualization of tissue mechanical properties. Following his graduate work, he moved to the University of Michigan, Ann Arbor, as a postdoctoral fellow in the Bioengineering Program, Electrical Engineering and Computer Science Department. From 1996 to 2002, Dr. Emelianov was a Research Scientist at the Biomedical Ultrasonics Laboratory at the University of Michigan. During his tenure at Michigan, Dr. Emelianov was involved primarily in the theoretical and practical aspects of elasticity imaging. In 2002, Dr. Emelianov joined the University of Texas at Austin where he formed the Ultrasound Imaging and Therapeutics Research Laboratory. Dr. Emelianov is currently an Associate Professor of Biomedical Engineering. His research interests are in the areas of medical imaging for therapeutics and diagnostic applications including molecular/cellular imaging, hybrid nanoparticles acting as imaging contrast and therapeutic agents, functional imaging, photoacoustic imaging, elasticity imaging, image-guided therapy, and ultrasound biomicroscopy.

Chapter 27
Evaluation Criteria of Carotid Artery Atherosclerosis: Noninvasive Multimodal Imaging and Molecular Imaging

Rakesh Sharma and Jose Katz

Abstract Carotid artery stenosis is a disease developed due to cardiovascular incapacity and cerebral infarction. Carotid artery stenosis is related with deep, subcortical, or cortical infarctions. In presurgery evaluation, asymptomatic dyslipidemia or symptomatic carotid artery stenosis are evaluated by imaging. Cardiovascular ischemia is occasionally interpreted as active and silent infarcts. In advanced atherosclerosis, better information is extracted out from presurgery clinical symptoms combined with dyslipidemia evaluation and associated information from cerebral angiography, carotid duplex ultrasound, computer-assisted topographic angiography (CTA) and magnetic resonance angiography (MRA). In present chapter, the central idea is that carotid artery disease is manifestation of structural and or molecular changes visible in the carotid artery wall and physical characteristics of flowing blood. To evaluate the carotid artery disease burden and plaque type, a new criterion of presurgery evaluation was proposed in this chapter by imaging atherosclerosis followed by postsurgery plaque characterization using biomarkers in endarterectomy samples (changes in tissue expression of mRNA-encoded inflammation modulatory proteins, oxidation, lipid transport, calcification, proteolysis, or hemorrhage, oligonucleotide microarray analysis, and high in situ hybridization – GenePaint and immunohistochemistry – ProteinPaint) with or without statin treatment of carotid artery disease. Present time, new multimodal molecular imaging techniques are emerging to give better new insights of plaque staging by molecular events in carotid artery disease progress and its evaluation.

R. Sharma (✉)
Department of Medicine, Columbia University, New York 10033, NY, USA
and
Center of Nanobiotechnology, Florida State University and Tallahassee Memorial Hospital, Tallahassee 32304, FL, USA
and
Innovations and Solutions Inc, USA, 3945 West Pensacola Street, Tallahassee 32304, FL, USA

J.Katz
Dr katz's Cardiology Centers, Medison Ave. New York, NY, USA

Jasjit S. Suri et al. (eds.), *Atherosclerosis Disease Management*,
DOI 10.1007/978-1-4419-7222-4_27,

Keywords Pre-surgery evaluation • Carotid artery disease • Atherosclerosis • Metabolic & molecular geography • NMR-biochemical signature • Molecular imaging • Endarterectomy • Molecular-metabolic paint

27.1 Introduction

Carotid artery disease is called "stenosis" or narrowing of carotid arteries or blockage to blood flow to brain. If narrowing of artery is severe is called "stroke" or interruption of blood flow to brain. The term was invented by Chiari in 1905 and later Hunt, Moniz, and Hultquist recently defined the term carotid artery disease [1]. Every year, more than 750,000 of people suffer from stroke due to irregular heart rhythm or atrial fibrillation caused by blood clots, and blockage of carotid arteries due to cholesterol deposits (Atherosclerosis) [1]. These cholesterol deposits form plaques and blood clots stop blood supply to brain though carotid artery. Both blood clots and atherosclerotic plaques increase the risk of stroke. At the first clinical visit, a careful presurgical evaluation of the carotid artery disease is necessary to measure risk of stroke and development of stenosis. The presurgery evaluation includes both measuring structural changes in the carotid artery wall and in biochemical contents of flowing blood. If evaluation warrants for advanced stage of stenosis, removal of blockage, or carotid artery wall deposits "endarterectomy" is performed. Later, postsurgery evaluation is continued with follow-up. Post surgery evaluation includes detailed examination of excised tissue.

In this chapter, we introduce the readers to atherosclerotic lesions in the cardiovascular system followed by presurgical and post surgery evaluation of disease in patients on lipid-lowering drugs mainly statins. The lesion composition ranges widely in severity from the limited fatty streak to the extensive complicated lesion. In between these two extremes there are other lesion types, some of which are especially vulnerable to disruption at the lesion surface, often leading to thrombosis and vessel occlusion. After initial plaque classification proposed by Stary et al. [2], continuous efforts of NIH/ACC/AHA continued for classification of atherosclerosis plaques as recommended in year 2002 [3, 4] applicable to multimodal plaque imaging [5]. In the following sections, we highlight that disease with or without lipid-lowering statin therapy causes detectable changes in the clinical symptoms, serum lipids, blood biomarkers, in vivo diagnostic US/MRI/CT/PET/SPECT imaging, and contrast-enhanced molecular imaging (for plaque volume and wall thickening) to determine the stage of disease before surgery by presurgery evaluation. In the next section, endarterectomy procedure is described with surgical criteria of endarterectomy procedure. In the next section, postsurgery evaluation by endarterectomy specimen histopathology and ex vivo MRI/NMR, ex vivo molecular staining is described for characterization of different plaque components for disease burden and follow-up of treatment response. A comparable approach of plaque typing and classification is presented based on biomarkers for specific enzyme matrix metalloprotease activities, and specific gene expressions in characterization of carotid atherosclerotic lesions. In the next section, newly coming up molecular imaging techniques are reviewed for their clinical use with introduction of new possibilities of multimodal imaging and molecular geography.

What Is the Goal of Evaluation of Carotid Artery Disease?

The major approach is (1) to find out the location and stage of disease progress; (2) to observe the efficacy of drug therapy and progress of disease to decide the surgical treatment and follow-up of disease recovery. Overall purpose is to get insight of disease progress and its molecular events in carotid artery if it needs removal of artery (endarterectomy) and to determine the course of treatment follow-up.

In the following section, clinical symptoms of carotid atherosclerosis are described to evaluate the possibility of disease burden and the plaque development.

27.1.1 Clinical Symptoms of Carotid Artery Disease

The patients usually remain asymptomatic or symptomatic associated with difficulty in speaking or slurring of speech; difficulty in finding words or expressing themselves; weakness or numbness in a limb; facial numbness or droopiness; difficulty with balance or walking indicate the chances of "Transient Ischemic Accident" [1]. Myocardial infarction is the associated disease. Severe stenosis of the internal carotid artery (ICA) (>70%) in patients with transient ischemic attacks (TIA) is considered asymptomatic (Table 27.1).

With these symptoms, the patient is advised for treatment of hypertension, hypercholesterolemia, diabetes, low fat diets, blood thinning, and smoking/alcohol prevention measures followed by presurgical evaluation if need arises. Narrowing of carotid artery causes "swooshing or bruits obserbed by stethoscope".

What Surgeon Needs to Evaluate in Carotid Artery Disease?
The main approach[1] involves the following.

27.1.1.1 Presurgery Evaluation

1. To investigate the cause of clinical symptoms by blood tests in lab for elevated lipids, proteins, enzymes, inflammatory markers [4].

Table 27.1 A presurgery evaluation of carotid artery stenosis by different imaging modalities and clinical symptoms

Clinical symptoms	Presurgery	Postsurgery
Difficulty in speaking or slurring of speech	++	–
Difficulty in finding words or expressing yourself	+++	+
Weakness or numbness in a limb	+++	–
Facial numbness or droopiness	+	–
Difficulty with balance or transient ischemic accident	+++	–

Sign "+" indicates prominent symptom and sign "–" indicates no symptom

[1]The first author participated in atherosclerosis program at SCORE labs in department of Medicine, Columbia University, New York under supervison of Dr. JK Katz and his cardiology labs.

2. To locate the disease on in vivo images to visualize artery wall contents, intima–media thickness, % stenosis, lumen area and plaque volume (by combined US/MRI/CT and PET/molecular multimodal imaging) to help in measuring the dimensions of endarterectomy tissue size [5].
3. To monitor the wall pressure, evaluation of wall shear forces and carotid artery microfluidics.
4. Poor drug response in lipid lowering with continuous deterioration, may need endarterectomy procedure.

27.1.1.2 Postsurgery Evaluation

1. After endarterectomy, to get cellular and tissue details of plaque development by ex vivo plaque pathology with ex vivo plaque imaging for diagnostic accuracy
2. To get ex vivo molecular staining of plaque for insight of cellular and molecular basis of plaque development and better follow-up of drug or postsurgery treatment
3. To detect postsurgery complications with possible local therapy by using new in vivo cellular and tissue targeting nanotechnology-based techniques in development

The following section describes the presurgery evaluation and use of different techniques.

27.2 Presurgery Evaluation

It includes serum lipids, blood biomarkers, in vivo imaging and drug treatment response.

Dyslipidemia is the disorder of lipids in blood. Recently, we reported the following modified criteria[2] of dyslipidemia in carotid artery disease suitable for presurgery evaluation [4, 6, 7]. The lipid dysfunction includes mainly hypercholesterolemia, hypertriglyceridemia, low HDL-cholesterol and apolipoprotein changes. The target lipid levels (LDL-C <2.5 mmol/L, <3.5 mmol/L, <4.5 mmol/L, and total cholesterol/HDL-C ratio <6.0, <5.0, <4.0) indicate the high risk, moderate risk, and low risk of dyslipidemia. The triglycerides >1.7 mmol/L with HDL-C <1.0–1.3 mmol/L indicate metabolic syndrome.

Blood biomarkers molecular are indicators of inflammation, angiogenesis, phagocytosis, and endothelial dysfunction including: (1) cell adhesion, (2) cell proliferation, (3) extracellular lipid transport, (4) intracellular lipid metabolism and trafficking, (5) inflammation, (6) apoptosis, (7) proteolysis, and (8) angiogenesis, and (9) calcification (Table 27.2).

[2]http://www.cmaj.ca/cgi/content/full/169/9/921/DC1/

Table 27.2 Serum lipids, inflammation markers, enzymes, and proteins to evaluate the disease burden by clinical chemistry laboratory tests. Reproduced with permission from [6]

Laboratory tests	Normal	Asymptomatic (presurgery)	Symptomatic (surgery needed)
Serum lipids:			
Triglycerides	<1 mmol/L	>1–1.7 mmol/L	>1.7 mmol/L
Total Cholesterol	<4.14 mmol/L	4.15–7.2 mmol/L	>7.21 mmol/L
HDL	>1.4 mmol/L	1.2 mmol/L	<1.1 mmol/L
LDL-C	<2.5 mmol/L	2.5–3.5 mmol/L	<4.5 mmol/L
Cholesterol:HDL	<4.0	<5.0	<5.0–6.0
Apoprotein-A(Lp-A)	<24 mg/dL	>24–30 mg/dL	>30 mg/dL
C-reactive Protein	<1 mg/mL	1.0–3.0 mg/mL	>10 mg/mL
VCAM	<445 ng/mL	445–460 ng/mL	>460 ng/mL
ICAM	<250 ng/mL	255–280 ng/mL	>280 ng/mL
MMP-9/-8/-2/	41/3–6/631 ng/mL	93/7.2/785 ng/mL	154/10,794 ng/mL
Serum proteins[a]			

SMC a-smooth muscle actin, *anti-CD-68* monocytel macrophages, *anti-CD-3* T-lymphocytes, *PE-CAM* endothelial cells, *FLK-2* fetal liver kinase-2, *VEGF* Vascular endothelial growth factor, *VCAM* vascular cell adhesion molecule; *TIMP* tissue inhibitor of metalloproteinases

[a]For information, specific proteins mediate, modulate, or suppress different atherogenic processes as follows:

Cell adhesion	E-selectin, P-selectin, L-selectin
Cell proliferation	TGF-8, cyclin A
Extracellular lipid transport	ApoB.100, Apo[a]
Intracellular cholesterol metabolism and transport	ACAT, CEH, LDL-R HMG-CoA reductase, caveolin, Lp-PLA$_2$
Inflammation	CD-68, CD-3, HLA-DR, IL-2Rec, TNF-α, IL-1β, IF-γ, IL-6, ELAM-1, MCP1, HSP-65, E1, E2 antibodies, C-reactive protein
Apoptosis	CD95, caspase-3, -8
Angiogenesis	VEGF-1, FLK-1,-2, angiopoietin-1,-2
Proteolysis	MMP-1,-2,-3,-9,-12; TIMP-1,-2; GPx-1, MPO, Phospholipase A$_2$II
Calcification	Osteopontin, osteocalcin, SMAD-4

How Presurgery Evaluation is Important?

With disease progress, carotid artery wall gets thicker and carotid artery wall layers get deposits of lipids, collagen, calcium, thrombi as pockets (so-called plaque) and the process is called "atherosclerosis" as a result of carotid artery "stenosis" [7]. The plaque composition, size, and biology (plaque typing) with other plaque characteristics such as lipids, surface ulceration, subintimal hemorrhage, and plaque composition may play a significant role in staging the atherosclerosis [8]. The ischemic neurologic symptoms suggest the burden of carotid atherosclerosis and possible myocardial infarction [9, 10]. In other words, plaque typing, carotid artery surface characteristics, and plaque size (degree of stenosis) may also predict the development of thrombus-mediated acute coronary syndromes and clinical risk. Percutaneous interventions by carotid angioplasty may be used as a balloon catheter in CAD evaluation before surgery to make plaque flat against the artery wall surface and a pipe like stent is placed to make clear wider passage for blood flow in the artery. In severe blockage, procedure of surgical endarterectomy is performed and patient is evaluated by postsurgery plaque examination for further course of disease progress. The postsurgery diagnostic criteria for vulnerable plaques include three important details: (1) the plaque's fibrous cap is thinnest at the marginal shoulders of the lesion, (2) the core rich in extracellular lipid located deep in the intima comprises >30% of the lesion's cross-sectional area with media–intima thickness <2 mm, and (3) the distribution of monocyte–macrophage foam cells is concentrated primarily in the microanatomic regions adjacent to the lipid-rich core and the thinning fibrous cap. These atherosclerotic lesion features are intimal xanthoma, intimal thickening, pathological intimal thickening, fibrous cap atheroma, thin fibrous cap atheroma, calcified nodule, and fibrocalcific plaque [11]. Under this nomenclature, the atheroma with a thin fibrous cap (<65 μm) equates to the "vulnerable plaque" by virtue of its presence in >70% of cases with plaque rupture. The noninvasive in vivo imaging technologies help to characterize plaque morphology and composition to determine the degree of stenosis.

How Does Plaque Develop?

Carotid atherogenesis is now believed to be a combination of several different processes including injury to the vessel endothelium, accumulation of lipoproteins in the subendothelial space, smooth muscle cell (SMC) proliferation and migration, and monocyte entry into the intimal layer with subsequent transformation into macrophages. As a result plaques get vulnerable and rupture to cause vascular blockage. The whole process is completed in different following stages in the order of foam cells→lipid core →calcification→fiber cap→ulceration→ hemorrhage→thrombus [9].

First stage: Excessive uptake of modified lipoproteins by macrophages or SMCs leads to the formation of foam cells. The accumulation of foam cells and extracellular lipid in the intimal layer leads to the formation of the fatty streak.

Second stage: At this stage, atherosclerosis does not cause significant luminal stenosis. The progression of a fatty streak to a fibrous plaque occurs when smooth muscles cells produce a fibrous cap around the lipid core. This lipid core consists of extracellular cholesteryl esters derived from lipid-laden foam cells that eventually become necrotic and lyse. Endothelium dysfunction progress with inflammation and SMCs undergo phagocytosis and apoptosis. Fibrous plaques may remain as stable lesions (even though they may grow larger) or proceed to become complicated plaques rendered unstable by ulceration, hemorrhage, thrombosis, and/or calcification.

Third stage: This complicated plaque is the form of atherosclerosis frequently associated with a clinically acute, occlusive event. The distribution of various biophysical events of atherosclerotic plaques in the vascular system follow a common pattern, typically forming deposits along bifurcations and bends, suggesting that plaque localization is partly determined by the dynamics of blood flow. The relationship between flow dynamics, plaque formation, and arterial wall thickness is closely linked to tensile stress and shear stress. In the carotid artery the size of the bulb, the bifurcation angle, and wall curvature have profound effects on plaque development [9].

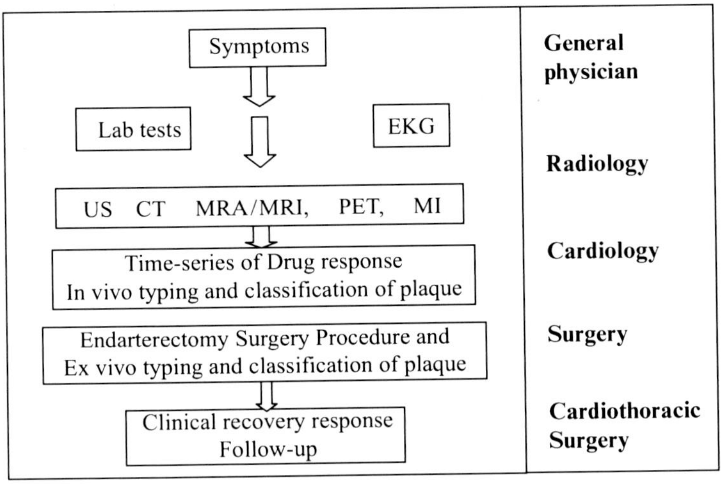

Fig. 27.1 Algorithm of carotid artery disease evaluation

Different plaque characteristics can be used to develop plaque typing and classification of stages to define the carotid artery burden by imaging, pathology, and molecular staining techniques as useful tools in surgical evaluation as shown in Fig. 27.1. Plaque typing and classification is described in detail for imaging and pathology in the following section.

27.2.1 Presurgery Evaluation by Imaging of Carotid Artery Disease

After physical examination, next step is the evaluation of carotid arteries and associated internal and external branches by in vivo procedures of oculoplethysmography, arteriography, Doppler ultrasound, magnetic resonance angiography (MRA) and CT as major choices of imaging modalities. However each imaging technique has merits and disadvantages. In the next section, imaging modalities are described with their performance in evaluation of CAD and plaque typing as shown in Table 27.3.

27.2.1.1 Available In Vivo Imaging Techniques

(a) MR angiography allows visual inspection of the entire cerebrovascular system including carotid artery system. It provides the information about tandem atherosclerotic disease, plaque morphology with dimensionality, and collateral circulation which may affect disease management (see Fig. 27.2). The quality of the angiogram depends on selective catheterization of the carotid artery with at least two views after injections in aortic arch. However, suboptimal studies

Table 27.3 Components of presurgery and postsurgery evaluation

Presurgery evaluation of CAD

Symptoms	Lab tests	In vivo imaging
Slurred speech	TG/VLDL	Ultrasound
Poor expression	LDL	MR angiography
Weak/numb limbs	HDL	Computer-assisted tomography
Face numb or droopiness	Cholesterol	Positron emission tomography
↓	ApoB	Single photon emission computed tomography
	Lipo A	Postgadlinium MRI
	Enzymes	
	ICAM/VCAM	
	Proteins	
Excised plaque ↓		

Plaque contents important in postsurgery evaluation of CAD and ex vivo imaging of atherosclerotic plaque

Extent of CAD burden	Ex vivo imaging	Ex vivo pathology	Ex vivo molecular/ nanotechnology
Normal/no plaque			
Type I	No calcification	Foam cells	NIRF
Type II	Wall thickening	Fatty streaks, SMC	NIRF, FIT
Type III	Thick intima	Preatheroma	Ferritin
Type IV	Lipid core	Lipid/fiber core	Collagen
Type Va	Fibroatheroma Calcification	Lipid/fiber core	Proteoglycan
Type Vb	Hemorrhage	Hemorrhage	Integrin
Type Vc	Thrombus	Thrombus	
Type VI	Calcification	Calcification	Ratiometric
Type VII	Fibrotic core	Fibrotic core	Immunostaining
Type VIII	No lipid/Calc	Calcification/no lipid	Ratiometric

can lead to misinterpretations as an irregular stenosis can be either under- or overestimated in a single projection. The disadvantages of angiography include its invasive nature, high cost and risk of morbidity and mortality in approximately 4% patients with the risk of serious neurologic complications or death approximately 1% (range 0–5.7%). Cerebrovascular symptoms, advanced age, diabetes, hypertension, elevated serum creatinine and peripheral vascular disease further increase the risk of morbidity. The size of the catheter, amount of contrast and procedure duration also affect the risk of complications.

(b) Carotid duplex ultrasonography (CDUS) is a noninvasive, safe, and relatively inexpensive technique for evaluation the carotid arteries. CDUS uses B-mode ultrasound imaging and Doppler ultrasound to detect focal increases in blood flow velocity indicative of high grade carotid stenosis. The peak systolic velocity is the gauge to measure the severity of the stenosis. In addition, the end diastolic velocity, spectral configuration, and internal/common carotid artery ratio provide better information. Color Doppler flow technique is less accurate but more efficient technique. Doppler ultrasound [Transcranial Dopler (TCD)] images major intracerebral arteries through the orbit and at the base of the brain. It provides additional information of intracranial hemodynamic consequences as a result of development of collateral flow patterns in the circle of Willis and progressing stenosis. TCD may be used to improve the accuracy of CDUS in identifying surgical carotid disease. Transcranial Doppler is often used in conjunction with CDUS to evaluate the hemodynamic significance of ICA stenosis.

(c) The MRA techniques most often used for evaluating the extracranial carotid arteries by two-dimensional time-of-flight (2D TOF) or MOTSA. MRA produces a reproducible three-dimensional image of the carotid bifurcation with good sensitivity for detecting high grade carotid stenosis. However, slow, turbulent, and absent blood flow are poorly distinguished by MRA; therefore, the degree of stenosis is consistently overestimated. Even normal or very mildly stenotic arteries may appear diseased on MRA as nonlaminar blood flow in the carotid bulb contributes to loss of signal intensity. Irregular stenoses which disturb flow are particularly susceptible to over-interpretation. Turbulent and complex flow patterns in the carotid artery proximal and distal locations to the stenosis. It also leads to overestimation of the length of the stenotic segment. MRA is more sensitive technique and it is specific for detecting high grade carotid stenosis along the wall [11]. The majority of angiographic-MRA correlations are performed using the 2D TOF MRA technique typically known as "bright-blood" or dephased "black-blood" angiography [12, 13]. These techniques with multiple contrast T1/T2/PD-w magnetic resonance imaging (MRI) offer the measurement of stenosis. In general, the degree of carotid artery stenosis overestimates for all ranges of stenosis using formula shown in (27.1):

$$\%\ \text{Stenosis} = 1 - \left(D_{\text{stenosed}} - D_{\text{normal}}\right) \times 100, \qquad (27.1)$$

where D is diameter measured at maximum wall thickness location and normal wall location.

In different studies on presurgery studies, sensitivity for 2D TOF MRA predicted 50–99% angiographic stenosis varied between 73 and 100% and specificity varied between 59 and 99%, respectively [12]. However, the presurgery evaluation has demerits. The MRA technique cannot distinguish very severe stenosis from vascular occlusion. It cannot distinguish ulceration, atheroma. Both false positive and false negatives may occur. Combined techniques of 3D TOF with 2D TOF improve the performance of MRA but are not routinely performed due to high cost and long scan time [12]. The MRA is less operator-dependent and produces an image of the carotid artery. However, it is more expensive and time-consuming. It is less readily available and may not be performed if the patient is critically ill, unable to lie supine, or has claustrophobia, a pacemaker or ferromagnetic implants. Due to these difficulties, 17% of MRA studies remain incomplete because of the patient intolerance or patient could not lie still during imaging.

(d) Spiral CT provides an anatomic depiction of the carotid artery lumen and images of adjacent soft tissue and bony structures. The three-dimensional carotid artery reconstruction (sum of piled up image slices) accurately measures the residual lumen diameter with distinct calcification. CT angiography requires a contrast bolus comparable to that administered during a conventional angiogram. So, impaired renal function is a relative contraindication for contrast bolus use [5].

27.2.1.2 Choice of Imaging Modalities in Presurgery Evaluation

1. MRA has been considered the gold standard. However, angiography is a risky procedure and majority of patients with carotid ischemic symptoms do not have severe carotid stenosis. Therefore, patients are generally selected for angiography using one of the noninvasive tests earlier done. Other two black blood MRA and carotid artery ultrasound (CDUS) techniques have better sensitivity than specificity in the detection of surgical carotid artery disease.
2. Both carotid ultrasound and MRA provide better information than conventional angiography in the presurgical assessment of patients with carotid artery disease. Both ultrasound and MRA are cost effective with less overall error rate comparable to the interobserver reliability. However, bypassing angiography before surgery requires any other noninvasive imaging be highly specific and sensitive technique to evaluate CAD (see Fig. 27.2).
3. TCD ultrasonography may be beneficial in clinical setting, increasing the specificity to carotid duplex ultrasound in detecting a <1.5-mm residual lumen diameter (Table 27.4).

27.2.2 Plaque Classification and Plaque Typing

Several documents have been reported from American Heart Association, American College of Cardiology, and other regulatory agencies on plaque histopathology classification and plaque typing useful in in vivo imaging [3, 4, 13–15].

The following section describes the importance of different imaging techniques in classification of plaques.

27.2.2.1 Human Atherosclerotic Lesions and ACC/ACR Classification on Imaging

The imaging modalities provide rough estimation of wall thickness and lumen area in different plaque types. In this direction, several investigators and American

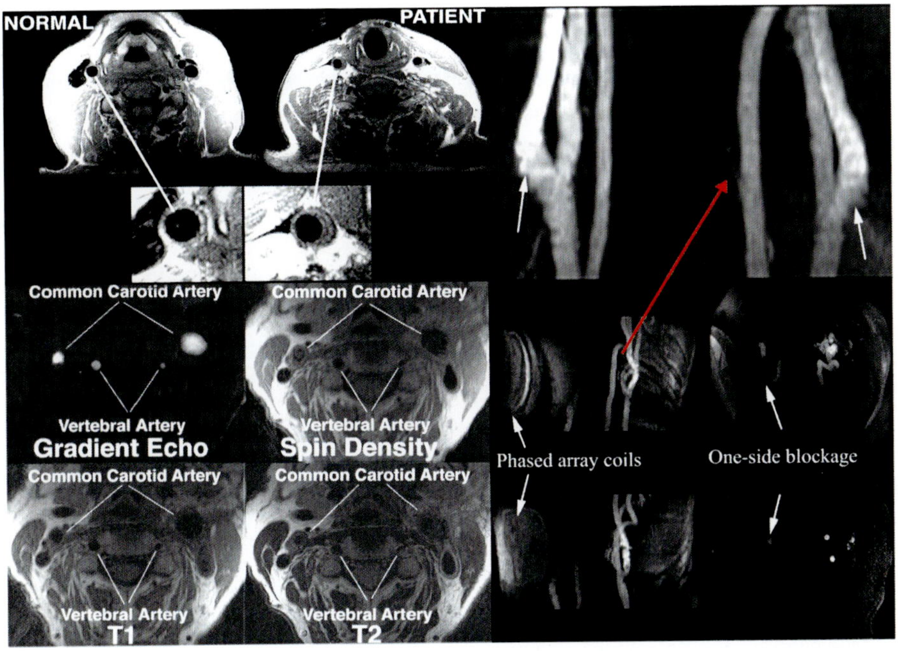

Fig. 27.2 *Top left*: A comparison is shown between normal carotid artery wall with atherosclerotic thickened wall by *arrows*. *Top right*: The bright-blood MR angiograph shows the location of bifurcation of common artery into internal and external artery branches on both sides of neck. *Bottom left*: A multicontrast imaging approach is shown to distinguish common carotid artery on gradient echo, proton spin density, T1-weighted and T2-weighted images. The images were acquired at scan parameters of TR = 100 ms (for GE); TR = 1,000–2,500 ms (for SE) to generate T1 or T2 weighting; slice thickness = 1 mm, matrix 256 × 256; FOV = 1.8 × 4.0 cm^2. *Bottom right*: The one side carotid artery block is shown on *right hand side* well detected by using a phased array coil on lateral sides as shown with *arrow*

Table 27.4 Different imaging modalities are shown to compare their sensitivity to visualize morphology, vasculature, and flow in the artery

	% Sensitivity		
Imaging modality	Morphology	Vasculature	Flow
Angiography	+	++	++
US	+	++++	++++
MRA	+++	+++	
CT	++	+	–
PET/SPECT	+	+	–
Optical	+	++	+++
Molecular imaging	++++	++++	+

Sign "+" indicates the relative benefit of technique, sign "–" indicates none

Colleges of Cardiology and Radiology have classified plaques using following common principles.

A system for classifying atherosclerotic lesions based on *imaging* is not established yet. The following criteria is a tentative guideline [3] that provides a concise, convenient way to designate lesion types as shown in Fig. 27.3.

Type I lesions contain enough atherogenic lipoproteins to elicit an increase in macrophages and formation of scattered foam cells.

Type II lesions consist primarily the layers of macrophage foam cells and lipid-laden SMCs and include lesions grossly designated as fatty streaks.

Type III lesions are intermediate between type II and IV; they contain scattered extracellular lipid droplets and particles that disrupt the coherence of some intimal SMC. This extracellular lipid is the immediate precursor of the larger, confluent and more disruptive core of extracellular lipid that characterizes type IV lesions (atheroma that is potentially symptom producing).

Type IV and Va lesions are thought to become the vulnerable plaque.

Type Va lesions have a lipid core and contain thick layers of fibrous connective tissue.

Type Vb lesions are possibly calcified.

Type Vc lesions contain mainly fibrous connective tissue with little or no calcium or lipid. This plaque type may abruptly provoke sufficient thrombosis or stenosis to cause a clinical event, such as myocardial infarction, stroke, or peripheral insufficiency.

Type VI lesions, when these develop a fissure, hematoma, or thrombus.

Type VII lesions are highly calcified.

Type VIII lesions are fibrotic plaque without lipid core and with possible small calcifications.

Two types of vulnerable plaques occur: those that rupture and provoke thrombosis and those in which the endothelial/intimal surface becomes denuded or ulcerated and provoke thrombosis. Combined imaging modalities display better plaque details as shown in Table 27.5.

Other forthcoming clinical evaluation techniques of carotid artery stenosis are as follows:

1. Plaque microfluidics is graded by contours
2. Blood flow dynamics offers a quick visualization of blood flow projectiles

These techniques are in development or clinical evaluation phase but show great promise. Basically, graphical analyses showed the shape of the trajectory to determine when clinically significant regression was achieved and whether subgroups such as race and gender have different trajectories. The carotid artery [external carotid artery (ECA)] flow, waveform, and occlusion geometry as hemodynamic wall parameters are associated with intimal hyperplasia and atherosclerosis. Two major parameters of wall shear stress and oscillatory shear properties predict the flow and velocity of blood. Any location of artery occlusion (distal, proximal, stump, smooth) is presumed to have distinct effect on carotid hemodynamics

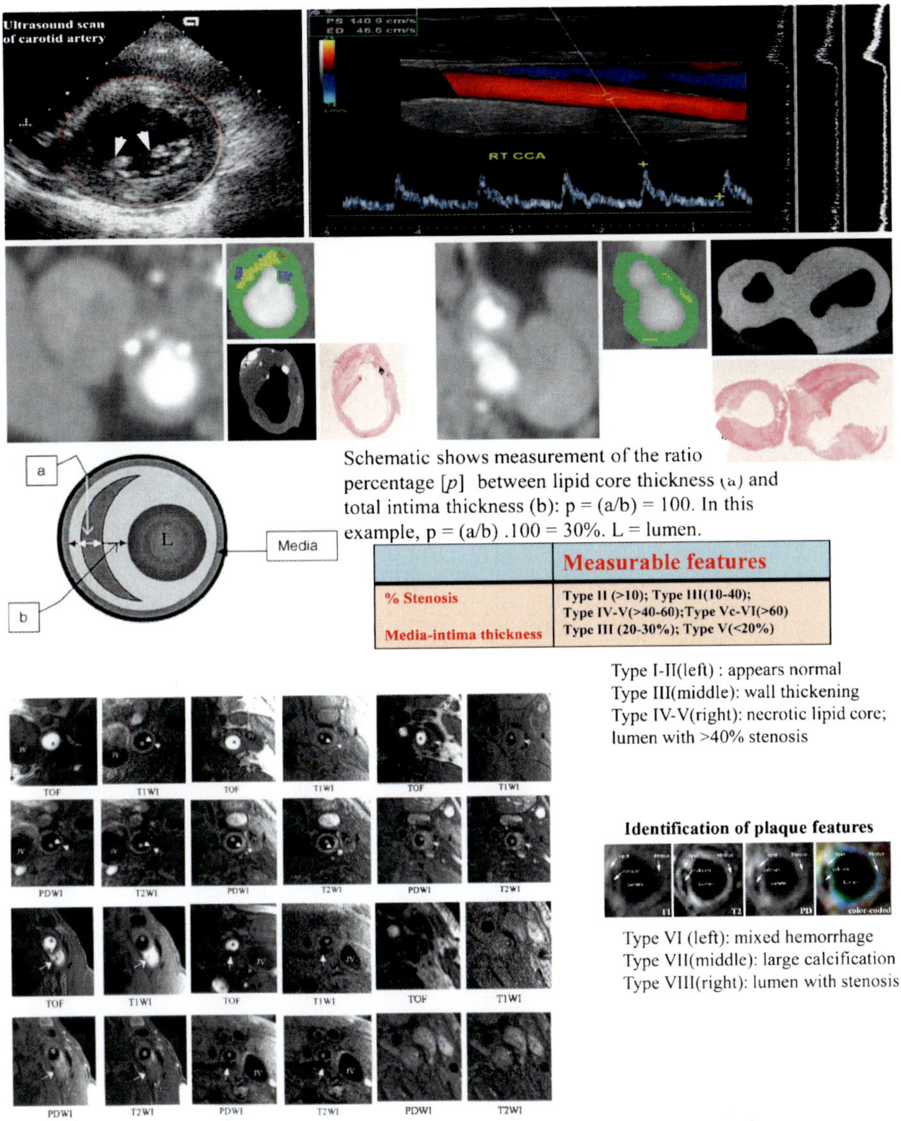

Fig. 27.3 Plaque classification by US, MRI, and CT imaging modalities. *Top*: Ultrasound reflective echoes with flow information as signal from different carotid artery structures are shown. *Middle rows*: CT images are shown with bright signal from calcium. Notice the accuracy of wall deposits detection shown with *arrows*. *Bottom*: Different plaques are shown on MRI multiple contrast. Notice the plaque and lumen are distinct on each of the T1-w/T2-w/proton density-weighted images and indicate the stage of plaque types (shown as type I through VIII). Reproduced with permission from [3]

Table 27.5 Plaque classification by in vivo imaging and injectable bioimaging techniques

Plaque	MRI	CT	US	PET	Nanotechnology (Ferridex®)	Molecular (labeling)
Type 1					Foam cells	Foam cells
Type 2	++ T1-w				?	Fatty streaks
Type 3	+ T2-w		+		?	+ Intima–media, lipids
Type 4	– PD-w/T2-w	– Ca		+	Necrotic core	+ Fibrous cap/ lipid core
Type 5	– Ca;+T1-w	++ Ca	+++	+	+ Ca, proteoglycan	++ Ca, fibroatheroma
Ulceration	+ T2-w				?	
Thrombus	+ T1-w				?	
Hemorrhage	+ T1-w				?	Hemorrhage
Type 6			++	++	?	
Type 7	– T1/T2/PD-w	+++ Ca	+++		++Ca	+++ Ca
Type 8	+/– T1/T2-w			+++		+++ fibrous core

Sign "+" indicates the enhancement and sign "–" indicates loss of signal on image

applicable to carotid sinus [16–20]. For details of these clinical imaging techniques, readers are referred to original reports (Fig. 27.4).

In following section, we describe experimental design and in vivo and ex vivo evaluation of carotid artery disease. Our approach was to quantitate the disease burden or % stenosis using in vivo and ex vivo imaging of carotid artery wall, plaque size and propose a diagnostic accuracy scale of histopathology-MR comparison with possibility of molecular imaging.

27.2.2.2 A Proposed Carotid Artery Stenosis Evaluation of Plaque Characterization

As described in previous sections, carotid artery disease is a lipid disorder causing biophysical changes in blood flow, vascular wall morphology, and molecular changes in both cells and vascular tissue. The carotid artery atherosclerosis is managed by lipid-lowering statin therapy because drugs inhibit cholesterol biosynthesis and exert pleiotropic effects on carotid artery wall. In advanced stages of disease with poor drug response, the following criterion determined the need of surgical intervention of carotid endarterectomy (CEA) and follow-up:

1. The first approach was if short-term drug therapy alters the lipid composition or dimensions of the plaque as measured in vivo MRI and ex vivo by high resolution 3 T MRI and postprocessed feature space analysis.
2. A second approach was if short-term drug therapy causes changes in ex vivo carotid artery wall tissue expression of mRNA that encodes proteins that modulate inflammation, oxidation, lipid transport, calcification, proteolysis, or hemorrhage.

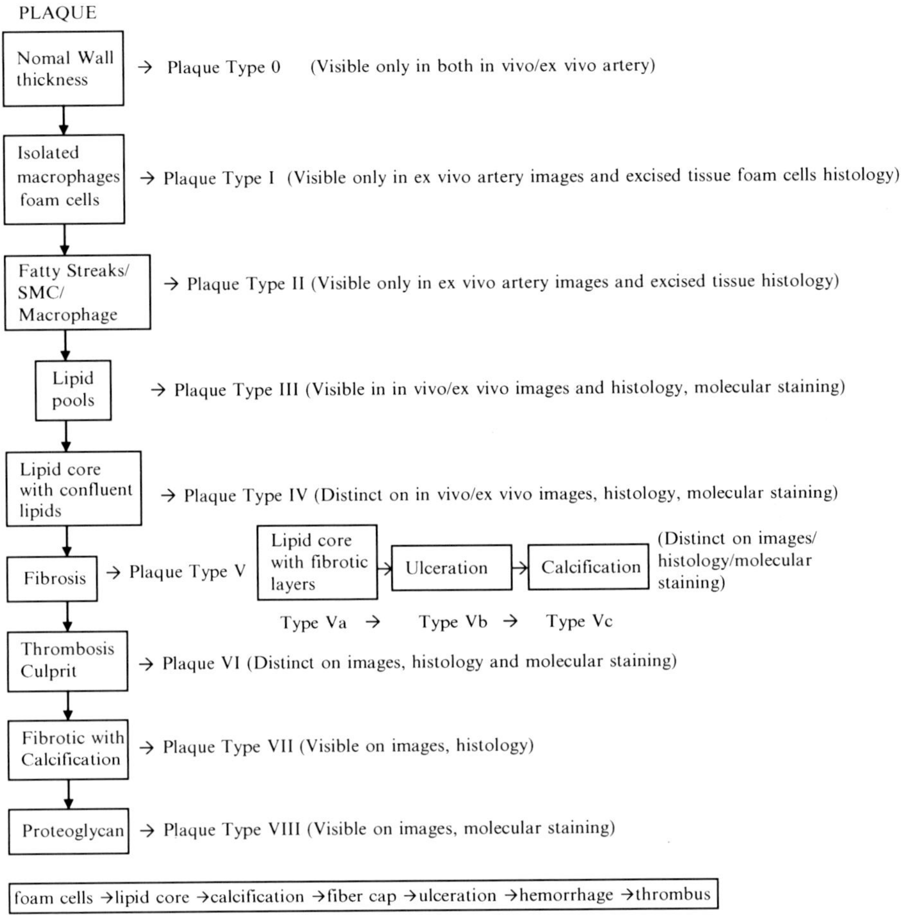

Fig. 27.4 AHA plaque classification of and relationship between human atherosclerotic plaques [2]

Laser capture microdissection, oligonucleotide microarray analysis, and high throughput in situ hybridization (GenePaint), and immunohistochemistry (ProteinPaint) are being used to examine CEA tissues that have and have not been exposed to drug treatment.

3. A third approach was to determine in candidates for bilateral CEA if short-term drug therapy produces in CEA tissue and/or plasma levels of proteins involved in processes leading to atherosclerosis. This approach is being pursued using mass spectroscopy imaging and high sensitivity protein microarray analysis (PMA).
4. A fourth approach was to develop nanoparticles suitable to image the molecular events in artery. Different molecular imaging techniques have specific advantage and accuracy to map out the chemical and biophysical events.

Criteria of Patient Selection

Men and women >18 years old ($n=60$) were recruited to the study having bilateral atherosclerotic carotid artery disease requiring endarterectomy on only one side or already had endarterectomy on one side and have atherosclerosis in the contralateral carotid artery. These patients had LDL-C >100 and <200 mg/dL on current care with or without well-controlled conditions such as diabetes, hypothyroidism, hypertension, congestive heart failure. Exceptions were patients with any malignancy in the past 5 years other than treated squamous cell or basal cell skin cancer, claustrophobic in an MRI magnet, hypersensitive to lipid-lowering drugs. Subjects were recruited from candidates for CEA[3] presenting to the surgical service and patients at Dr Jose Katz's Cardiology Clinic with the instructions of the tissue resulting from their surgery for consent. The potential risks of undergoing the MRI scan were minimal. The main discomfort was the knocking sound of the MRI magnet gradients during image acquisition. No additional discomfort or risk to patient beyond the scheduled surgery was incurred by donating carotid plaque tissue.

Lab tests included clinical lipid chemistry and immunology. Serum was used for lipid profile, proteins, inflammatory markers, and enzymes as described in previous section in Tables 27.1 and 27.2 on lab tests in presurgery evaluation. The risk associated with taking the lipid-lowering statin included symptomatic side effects and abnormal lab tests. Symptomatic effects observed include headache, dizziness, sore throat, diarrhea, upset stomach, muscle weakness, constipation, flatulence, abdominal pain, and rash. Laboratory abnormalities included temporary mild increases in levels of liver transaminases generally <3 × ULN.

Patients were given statin treatment under supervision of a physician during the course of 45 min MRI study. The risks associated with taking a statin were rather low. The benefits are considerable: about 43% lowering of LDL-cholesterol, representing a significant reduction in risk for cardiovascular disease. The treatment of dyslipidemias included therapeutic agents in Phase II and III clinical trials of safety and efficacy including eight statins, three bile acid sequestrants, and three fibrates.

Statistical Methods Used in Diagnostic Accuracy in In Vivo MRI Images

We developed a carotid artery algorithm to standardize the visibility of different wall and plaque components based on the gross appearance on images by comparing them on different weighting schemes of multiple contrast protocol [3].

In the following section, our focus is on quantitative MRI technique to measure different plaque constituents. The following steps were used in quantitative evaluation:

---

[3]First author participated in his internship at heart and Vascular Surgery, Tallhassee memorial Hospital, Tallahassee, FL 32304 under mentors Drs. J Hurt, Murrah, Khairrallah, and Saint; DrKatz's Cardiology, NY.

1. Images were organized into stacks of contiguous slices.
2. Calibration of image stacks in prescribed field-of-view and slice distance factors.
3. Adjustment of contrast on images using optimization to define plaque and vessel wall tissue density.
4. Segmentation of internal, external, and common carotid arterial lumen.
5. Segmentation of internal, external, and common carotid arterial external wall boundary.
6. Identification of atherosclerostic regions in normal wall and measurement of average normal wall thickness.
7. Calculation of internal, external, and common carotid arterial lumen volume per unit length.
8. Calculation of internal, external, and common carotid arterial total (normal + plaque) wall volume per unit length.
9. Calculation of internal, external, and common carotid arterial plaque volume per unit length.
10. Measurement of maximal total (normal + plaque) internal, external, and common carotid arterial wall area and intima-media thickness.
11. Measurement of minimal internal, external, and common carotid arterial lumen areas.
12. Measurement of molecular biomarkers on plaque images and on slices.

27.2.3 *Presurgery Assessment of Carotid Plaque Magnetic Resonance Imaging and Spectroscopy In Vivo: Where We Are Today?*

A noninvasive, in vivo MR imaging monitored the plaque changes over time to identify, quantify, and characterize carotid plaques before endarterectomy or post-surgery evaluation [21, 22]. Subsequently, assessment was made on changes related to presurgery clinical symptoms and events. High resolution magnetic resonance imaging (MRI) with magnetic resonance spectroscopy (MRS) were used either alone or together to demonstrate the presence of different MR visible lipid components in atherosclerotic lesions before surgery [23]. The MR image intensity of a carotid artery tissue was determined by its relaxation times, T1, T2, and its proton density (PD) so-called multiple contrast by combinations of these three image characteristics were reported to identify a particular tissue component in the plaque: report by NASCET [24, 25]. ESCET clinical studies based on projection MRA demonstrated the need to measure the degree of carotid stenosis in order to determine the presurgery clinical usefulness before CEA in symptomatic patients. Measurements of presurgical % stenosis on projectional MR angiograms was shown to correlate with luminal stenosis but overestimated the degree of luminal stenosis because of signal loss due to a complex flow pattern [26, 27]. To combat flow artifacts, several flow insensitive MRA methods are available insensitive to signal loss from complex flow [12, 28, 29]. The cross-sectional information of vessel lumen diameter available from MRA source images was reported to measure the actual luminal area more accurately than the projection MRA techniques [29]. Still, in vivo MR imaging of the carotid arterial wall and atherosclerotic plaque remains a challenge due to (1) the artery's small size, (2) its pulsatile motion, (3) blood flowing in its lumen, (4) demand of high resolution, small field-of-view (FOV) images, and (5) suppression of artifacts caused by vessel wall motion and blood flow. Recent techniques using chemical shift imaging and lipid suppression techniques have provided better fat–water contrast information about plaque morphology [30]. Good news is, in contrast to ultrasonography, the utility of MRI images of plaques is not limited by calcium deposits in plaques (Table 27.4).

27.2.3.1 Magnetic Resonance Macroimaging at 1.5 T In Vivo of Carotid Arteries by Multiple Contrast

Magnetic resonance macroimaging (MRM) of excised plaque tissue using a standard clinical scanner operating at 1.5 T (GE Horizon LX) demonstrates the potential of characterizing the components of carotid plaques under clinical conditions. Several coils are placed together in array manner, so-called phase array around the neck covering carotid arteries to enhance the MRI signal from plaque. Investigators have used such custom made phased array surface coil to study the calcification deposits in carotid arteries of a patient with significant stenosis 1 day before endarterectomy, and the plaque was resected at surgery [30, 31]. Carotid artery phase array coil has two rectangular elements, each with dimensions $\Delta x \times \Delta z = 3 \times 8$ cm, overlapping 1 cm in the x-direction. The overall dimensions of the coil pair were $\Delta x \times \Delta z = 5 \times 8$ cm at optimizing the SNR at the depth of $y = -30$ to $y = -40$ mm from the surface over the required coverage area. Differentiation of the plaque from the rest of the vessel wall was based on multiple contrast approach optimal imaging protocol for visualizing and quantifying carotid plaques by in vivo T1-weighted, T2-weighted, and proton density (PD)-weighted imaging sequences. In next section, our focus is to describe the microimaging to get microscopy details on 1.5 Tesla MRI imaging of Carotid plaques. For in vivo studies, GE Horizon LX system was equipped with a bilateral two-phased array coil (6 cm, Ultra Image). Axial PD-w, Tl-w, and T2-w images were acquired centered at the level of the carotid bifurcation. Inferior and superior spatial saturation bands were applied to decrease the high signal from intravascular flowing protons. An oblique spatial saturation band was used at tangent to the skin surface to decrease the strong fat signal. Spectrally selective fat saturation T1-w and PD-w images were acquired. Peripheral cardiac gating with a digital oximetry sensor was used to limit motion artifacts related to vessel pulsatility during the image acquisition. To eliminate ghosting due to patient motion, a special molded rubber head rest was used to minimize lateral head movement. Images are downloaded directly from the MR scanner with the eFILM software (DICOM protocol) to a PC workstation and processed using several different software packages including the Virtual Tool Kit (VTK), NIH ImageJ, and IDL as reported elsewhere [20, 23].

27.2.3.2 MRI at 1.5 T of Carotid Arteries In Vivo: Multiple Contrast Technique and Quantitative Analysis

The noninvasive MRI method characterized and quantified the atherosclerotic plaques in vivo in longitudinal study involving serial measurements over time. These measures evaluated and monitored the therapeutic drug treatment. In vivo imaging of the carotid arterial wall and its atherosclerotic plaque imaging suffers from (1) the small dimensions of the plaque components (<1 mm), (2) the pulsatile motion of whole plaque structure to the heart beat, and (3) blood flowing through the lumen. Thus, an effective MR protocol must provide (1) high resolution together with sufficient SNR, (2) compensation for cardiac motion, and (3) suppression of flow artifacts. We have overcome these obstacles by using (1) a custom head–neck phased array surface coil (11 cm × 6 cm, Ultra Image) to achieve high sensitivity and to minimize noise from surrounding tissue, (2) peripheral gating to compensate for the pulsatile arterial motion, and (3) standard saturation techniques to suppress flow artifacts. The phase array surface coil with its bilateral construction enabled the imaging of both carotid arteries simultaneously. We developed a locator MR protocol to precisely localize the carotid bifurcation which we used as the fiduciary landmark. In the next section, MR-imaging procedure is described. First, a sagittal scout scan was performed to check for correct position of the coil relative to the carotid bifurcations in the neck of the subject. Second, axial bright-blood MR-TOF slices (1.5 mm) were acquired covering a distance of about 3–4 cm above and below the tentative location of the carotid bifurcation as determined from the scout scan. These images locate both carotid bifurcations with an accuracy of 0.75 mm (half the slice thickness). Third, the position of each carotid bifurcation was confirmed in lateral TOF projection images constructed from the axial slices. Fourth, 6–8 axial slices (3 mm) of T1-w, T2-w, and PD-w images were acquired above and below the carotid bifurcation. The total imaging time required for the entire procedure was about 40–45 min [21–23].

27.2.3.3 Multiple Contrast Technique

Multiple MRI sequences and image combination of three MR weightings for T1-w, T2-w, and proton density-w images offer complete atherosclerosis plaque characterization with high sensitivity and specificity in vivo as shown in Fig. 27. 2 [30, 32]. The idea was to classify plaques based on identification of calcification, lipid-rich necrotic core, fibrous core and recent hemorrhage using a simple logistic regression model of cut-off signal intensities by means of ROC curves [32, 33].

27.2.4 *Quantitative MRI Analysis*

27.2.4.1 Segmentation[4]

In vivo MRI has limited precision to visualize, separate out (segment out) and quantitate different components of carotid lesions based on their MR images acquired in vivo and ex vivo. The segmentation of plaque features was based on color-coded feature maps of lumen, plaque, and artery wall. The advanced carotid lesions are usually a complex mixture of components, including a lipid-rich core, regions of extensive calcification and inflammation, as well as fibrotic areas enriched in extracellular matrix, thrombus, and thickened intima. These regions differ in molecular composition and dynamics, proton chemical shifts, spin densities, and relaxation properties. MRI distinguishes these different components to quantify them. We developed an approach to perform segmentation using (1) identification of TR/TE combinations that optimize contrast between atheroma and fibrous cap components; (2) parametric imaging to segment and quantitate different artery structures and lesion components based on their T1 or T2 relaxation constants; (3) feature space analysis for separation and delineation of different lesion components by multiple contrast of TI-w, T2-w, and PD-w. (4) Advanced edge detection techniques for boundaries between plaque components of images as described earlier [21]. The following sections describe the approach of segmentation mentioned above.

27.2.4.2 Selection of TE and TR for Optimizing Contrast Between Specific Atheroma Core vs. Fibrous Cap Plaque Components

Important chemical molecules in plaque are lipids, elastin/collagen, calcium, and thrombus heme protein with distinct structures of atheromatous core and the fibrous cap. Chemical shift techniques are not much useful while spin echo sequences for T2-weighted images provide better structural contrast and molecular contrast than T1-weighted images. The lipid-rich core (dark) and fibrocelluar fissure (light) by T2-w images offer better contrast than Tl-w images to distinguish atheromatous core and fibrous tissue [34, 35]. In our study, T2-w sequences provided good contrast between atheroma and fibrous cap with T2-w but suffered from background noise. To achieving best contrast between these components with least noise, a broad range of TR and TE values were used to optimize contrast. Advantage of T2-w images was better contrast achievable at minimal demand on MRI hardware. T1-w sequences provided a much higher MR signal than T2-w sequences with less background noise than T2-w sequences. From MRI physics standpoint, the time between excitation and data collection of T1-w sequences is much shorter than data

[4]The experiments were conducted with data analysis by first author as postdoc. scientist trainee at Atherosclerosis and Lipoprotein Training Center, Baylor College of Medicine, Houston, TX under supervision of Dr. Joel D Morrisett.

collection time for T2-w contrast sequences. Hence, T1-w sequences are much less subject to motion artifacts than T2-w sequences. 3D display of TE/TR showed heavy T1-w sequences with extremely short echo times provide better contrast between atheroma and fibrous tissue than T2-w sequences. The frequency selective sequence with an extremely short TE showed better contrast between atheroma and fibrous tissue.

Let us discuss the chemical nature of plaque tissue and content on images. Based on image contrast, three chemical plaque components are distinct as intraplaque hemorrhage, fibrous tissue, and atheroma. Hemorrhage appears hyperintense to fibrous tissue on TI-w and hypointense on T2-w. The optimal TE and TR parameters distinguish hemorrhage from other plaque constituents using in vivo MR images due to different relaxation constants of old, intermediate, and fresh thrombus arising from the different spin states of heme iron in the clot [21]. Other approach was to identify calcified regions of plaque by MRI. Calcified regions have very low water content (and consequently few hydrogen atoms) and appear hypointense on all imaging sequences. Normal arterial media can be differentiated from plaque components in most instances by its location and by the configuration of the arterial wall at the location of interest [3, 21, 23]. However, different signal intensity characteristics facilitate the automated segmentation. The media–intima thickness and adventitia may or may not appear on in vivo images. The optimization of MRI offers an assessment of both normal carotid arterial tissue and atherosclerotic carotid arterial plaque [36]. In our previous study, tissue contrast was optimized by performing a comprehensive evaluation of tissue signal intensity behavior in response to a broad range of pulse sequence parameters. Other study reported experimentally derived maps between matrices of input factors (TR, TE, flip angle, etc.) and tissue signal intensity. In our previous study, optimal MRI sequence settings were conclusive to decide the course of T1/T2/PD-weighting in MRI [21, 37].

27.2.4.3 Parametric Imaging for Segmentation and Quantitation of Lesion Components Based on Their T1 and T2 Values

Parametric imaging used to do "imaging microdissection" to separate out components with different T1 and T2 constants. Plaque calcification with short T2 (2–20 ms) appear dark in highly calcified areas of the tissue (red); while lipid components with short T2 (20–26 ms) appear hypointense in areas rich in lipids (yellow); fibrocellular components with long T2 (26–34 ms) appear isointense in fibrocellular lipid-rich areas (green); and fibrous components with long TE (34–52 ms) appear brighter in fibrous areas (blue) corresponding to integrated values of 23, 34, 26, and 17%, respectively [21, 23]. The remainder area corresponds to open lumen. Parametric imaging is ideal for segmentation and quantitation of different tissue components but segmentation takes long measurement time. A series of multislice spin echo images at seven different echo times at TR = 4 s. Other approach such as ex vivo diffusion MR imaging was reported to identify thrombus and hemorrhage in excised plaque tissues [34]. Crystallized lipid components appeared dark on diffusion-weighted (DW) images due to lack of lipid diffusion in the solid state of these compounds.

27.2.4.4 Feature Space Analysis for Segmentation and Quantitation of Different Lesion Components Based on Combinations of Their T1, T2, and PD Values

Any of the T1-w image, T2-w image, or PD-w image alone will produce a distinct contrast for each of the plaque components to discriminate them. The approach of combining contrast information contained in three sets of each T1-w image, the T2-w image, or the PD-w images generated better discrimination after subjecting images to feature space analysis for simultaneous plaque content discrimination [23]. We determined the 3D distribution of voxel intensity for T1-w, T2-w, and PD-w images from the same anatomical location [21, 23]. 3D feature space plot were produced, with the voxel intensity of the TI-w images, the T2-w images, and the PD-w images along the x, y, and z axes, respectively.

27.2.4.5 Boundary Detection

In general, 30–40 axial MR images are acquired in one session. Lumen delineation is simplest on digital images. Manual delineation of the areas of lumen, arterial wall, and other different components is laborious and biased susceptible to inter-operator variability. Semi-automated boundary detection algorithms are preferable and rapid method. In this direction, similarity-matched region or "seed growing" is a simple means of automatic tracing of intensity similarities to create boundary [38]. A more sophisticated and robust image processing procedure for determining the subareas of MR data sets by "SNAKES" algorithm was reported as deformable, active contour models that are attracted to image features such as lines, edges, and contours [39]. The SNAKE is an "energy minimizing spline" guided by internal constraint forces and influenced by external forces that will pull it toward prominent image features. SNAKES rely on specified initial conditions to place them near the desired contour either by using user interface, automated method, or other mechanism. Through energy minimization, the "SNAKE" slithers to its equilibrium position. An additional inflational force makes the SNAKE behave like a balloon. As the balloon passes by edges of artery, it is stopped when the change in image intensity is greater and passes over edges when the change is small. Yuan et al. reported area segmentation using the SNAKE algorithm useful in analyzing MR images of the carotid arteries [39]. The segmented features in different sessions demonstrated the molecular events (see Figs. 27.3 and 27.5). Several advanced image processing techniques of image analysis include parametric imaging, feature space analysis, and SNAKES boundary detection algorithm are available to optimize discrimination and quantitation of plaque components.

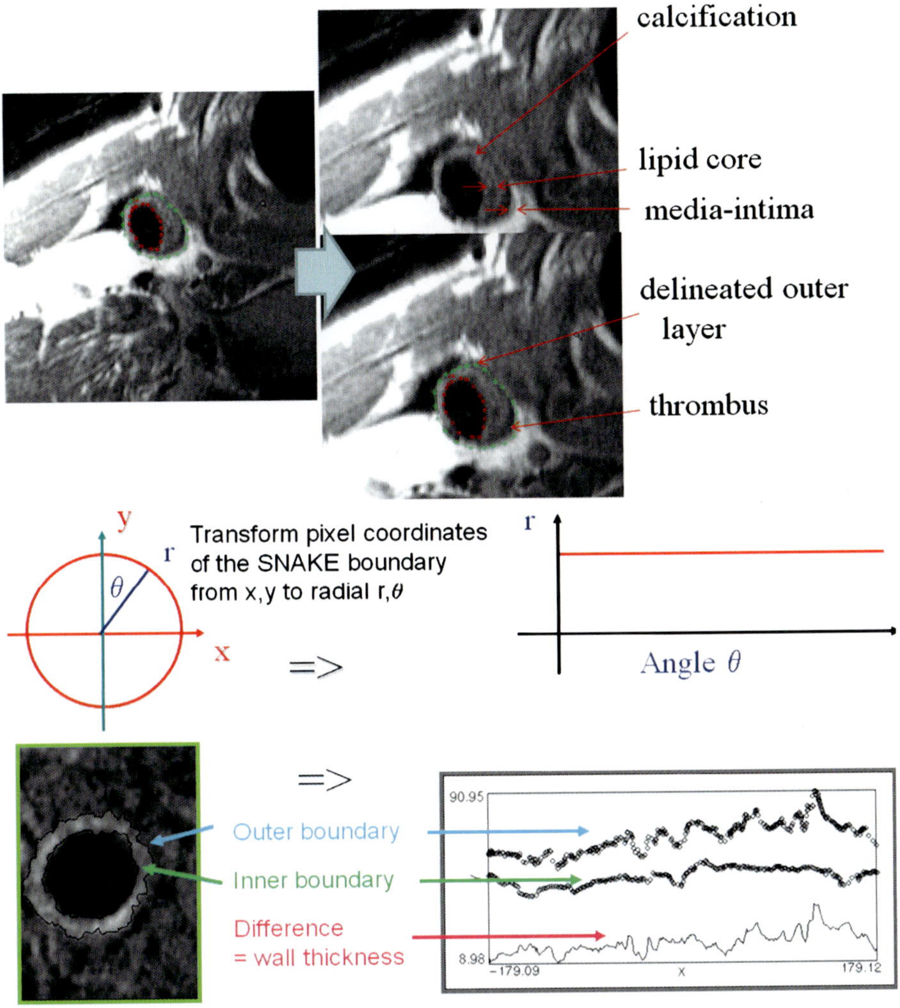

Fig. 27.5 *Top*: The figure shows the boundary detection around inner (*red color*) and outer walls (*green color*) by delineation method to measure wall thickening. *Bottom*: A semiautomated SNAKES algorithm fro boundary detection. The threshold-based edge detection method is shown for semiautomated delineation of outer wall and internal wall of carotid artery with lumen area measurement. The wall demarcation at all angles measures the lumen area and wall area. Reproduced with permission from [24]

27.2.5 *Time Series of Changes in Different Sets of Images from Same Location*

In different sessions, same carotid artery tissue was imaged to compare images of the same anatomical region of carotid artery (the branching or bifurcation) as point

Table 27.6 The mean ± sd values are shown computed from artery wall images

Right common carotid artery	Lumen (mm^2)	Wall thickness (mm^2)	Fibrous core	Lipid core	Hematoma	Calcification
Mean	29.68	1.14	PDW +++	++	+/–	– – – –
Standard deviation	2.47	0.19	T1W +++	++	+/–	– – – –
CV	3.1%	67%	T2W +++	–	+/–	– – – –

Serial images were acquired from two subjects imaged three times over 5 months. Different structures appear different on T1/T2/PD-w multiple contrast images. +++ Hyperintense; ++ isointense; +/– variable; – – – – no signal

of reference by a technique "registration" from MR TOF bright-blood transverse and lateral projection images [26]. This fiduciary origin was defined as the point where the common carotid artery first splits into two distinctly separate branches. It was common observation that image slices from different sessions were not lined up exactly, but even if they are out of register they were only off (shifting) by no more than half the slice thickness or 5° on vertical sides. Corresponding artery segments from each subject were grouped for the three imaging sessions. The following measurements for each of lumen, walls and plaques in these groups of three serial mean ± sd values were computed from artery wall images to determine precision and statistical power. Serial images by T1-w/T2-w/PD-w techniques were acquired from two subjects imaged three times over 5 months showed the results indicated in Table 27.6. Patients presenting significant carotid stenosis, wall thickness, and lumen measurements showed much complicated segmentation, especially near the bifurcation; hence, it needed objective techniques that minimize operator bias especially in studies with many patients (Fig. 27.5).

What information is important in presurgery evaluation?

27.2.6 Presurgery Evaluation of Aggressive Statin Treatment on Carotid Atherosclerotic Lesions by Serial Ultrasound and MRI Measurements

Assessment of the efficacy of drug therapy is done by visualizing disease stabilization and regression of carotid artery plaques. Ultrasonography monitors and measures the plaque and wall changes by lipid-lowering drug therapy. We cite some lead examples of statin drug therapy monitoring based on plaque and wall structural features. In LAARS study, decreased intima–media thickness in carotid artery were measured at baseline, 6 months, 12 months, and 24 months after drug therapy [40]. In multiple regression analysis, change in intima–media thickness by B-mode ultrasonography was significantly correlated with changes in Lp[a] and apoA-I concentrations [41]. Intima–media thickness of the common carotid artery was measured at baseline up to 2 years in MARS study [42]. At 2-year follow-up, annual rate of change in intima–media thickness was –0.038 mm/year in lovastatin-treated patients, indicating regression of carotid lesions, and +0.019 mm/year in placebo patients, indicating progression; 4-year changes were –0.028 and +0.015 mm/year, respectively. Annual rate of change in intima–media thickness was significantly correlated with on-trial concentrations of LDL-C, TG, apoB, apoC-I11, and apoE. In the Italian ultrasound CAIUS study, asymptomatic men and women with LDL-C of 150–250 mg/dL and ultrasound evidence of carotid disease were randomized to pravastatin 40 mg/day or placebo [43]. The primary endpoint (rate of change in intima–media thickness) of the common carotid artery, bifurcation, and ICA indicated less progression with pravastatin (–0.004 mm/year) compared with placebo patients (+0.009 mm/year). In other RIS study [44], hypertensive men counseled on diet, smoking cessation, and lipid-lowering drugs showed decreased intima–media thickness. Other several carotid ultrasound studies reported greater LDL-C lowering by ACAPS study [45], PLAC 11 study [46], and KAPS study [47]. Crisby et al. [48] reported a study of 11 patients with symptomatic carotid artery stenosis received 40 mg/day of pravastatin, while 13 control subjects received no lipid-lowering therapy for 3 months prior to endarterectomy. Lesions in the pravastatin-treated group showed less lipids, less-oxidized LDL, fewer macrophages, fewer T-cells, less MMP-2 immunoreactivity, greater TIMP-I immunoreactivity, higher collagen content, and lower frequency of cell death. Conclusive evidence showed endarterectomy on the unoperated carotid artery vessel. Changes in these variables over 24 months were assessed using statistical multilevel linear models for longitudinal data (hierarchical or random effect models) to interpolate for an endarterectomy on the carotid artery [49–51]. In all these reports, primary outcomes were statin postdrug treatment changes in plaque volume, residual lumen area, and wall thickness.

27.3 Postsurgery CEA

When surgical procedure is needed?

27.3.1 Selection Criteria of Surgical Procedure

Aggressive statin treatment reduces plasma LDL-C to 60–80 mg/dL during 2 years. Statin also affects human carotid atherosclerotic lesions (causing >40% stenosis to lesser stenosis) by significantly decreasing both lesion volume and maximum wall thickness, and increasing stenosed lumen area as measured by MRI. If drug treatment is not sufficient to reduce % stenosis to less than 40% it needs surgical procedure.

Our approach was to develop a diagnostic accuracy criteria based on ex vivo carotid artery tissues obtained at endarterectomy for examination and decision if intima–media thickness changed from occluded to less occluded or less % stenosis (shift from >40% to lesser) by microscopic histopathology examination and micro dissection. The endarterectomy was needed for two purposes: to confirm the plaque staging and to decide the course of follow-up treatment. These visible plaque changes were possible due to the surgical extraction of the entire atherosclerotic area of artery en bloc. The ex vivo endarterectomy procedures of sampling and examining specific plaque components in the lesion tissues enhanced the sensitivity of tissue analysis: (1) by reducing variability among plaque constituents from a single plaque type tissue; (2) by enhancing discrimination among plaque molecular deposits or tissues; and (3) by improving localization of selected molecular lipid, collagen, cholesterol, calcification sites or elastin, fibrous, atheroma tissue cores. The histopathology further confirmed the plaque characteristics (e.g., presence of fatty streaks or fibrous plaques by their lipid contents, or by characterization of inclusions rich in cholesteryl esters).

Statins may have effects upon lipid metabolism, e.g., high level inhibition of cholesterol biosynthesis within 13 h after lovastatin treatment [7]. The statin drugs showed rapid effects in humans during aggressive treatment altering plasma lipids as shown in Table 27.7.

27.3.2 *Evaluation Criteria of Endarterectomy Specimen*

Study Design: Carotid artery plaque tissue images were obtained from patients undergoing CEA. Patients had previous endarterectomy of one carotid artery on record to be eligible. Patients with presurgery imaging were screened for endarterectomy with informed consent into the study. Before surgery, they were screened for bilateral MRI. One day after surgery, they began treatment with a statin prescribed in lowering plasma cholesterol and favorable pleiotropic effects on the arterial wall as well [21]. The unoperated contralateral carotid artery was monitored by serial MRI for wall thickness, lumen area, intima–media thickness, and plaque volume (1 time/year over 2 year) as shown in Fig. 27.5. The calculations for statistical

Table 27.7 A comparative effect of statins is shown on lipid lowering presumably reduces the risk of carotid artery disease and atherogenesis

	Efficacy on			Recommended
Statins	Lipid lowering	Endothelium		daily dose
Atorvastatin	↓L	↓A		10–80 mg
Fluvastatin		↓A		20–80 mg
Pravastatin	↓L	↔A	↑E	10–40 mg
Rosuvastatin		↓A		10–40 mg
Simvastatin	↑L	↓A	↑E	1–80 mg

L lipid-lowering or LDL oxidation resistance, *A* smooth cell proliferation, *E* endothelial dilatation. The *arrows* indicate increase by ↑, decrease by ↓; and no change by ↔

power are based on 16 subjects for the study[5]. MRI detection coil was bilateral. It monitored the effects of statin treatment on the progression, stabilization, or regression of lesion(s) in the unoperated carotid, and the extent of restenosis in the intact artery. The Institutional Research Board (IRB) procedures for Phase II and III clinical trials, clinical monitoring and treatment management for all patients included optimum compliance, medication management by approved medical record keeping programs using tablet vials with caps equipped with microcomputer chips to monitor the times of vial opened for medication delivery [52].

At every MR imaging session, bright-blood TOF images, axial T1-w images, T2-w images, and PD-w images were acquired according to the MR protocol (FOV, 18 cm×18 cm; 512×512 voxels and a MR slice thickness of 1 mm) for imaging carotid arteries in vivo with a clinical MR scanner of 1.5 T (GE Medical System, Milwaukee, WI). These parameters provided a quantitative value of % stenosis and continuous wall thickness by getting sum of areas (see Fig. 27.5) at in-plane resolution of 0.254×8.13 mm. MR slices (1 mm) were reformatted (MRI-reformat) cubic interpolation cubic interpolation to construct 3D slice box. Graphics routine visualization toolkit (VTK, Kitware) was used for transformation to measure areas of the plaque and lumen. Briefly, lumen and plaque areas were outlined manually followed by iterative SNAKES edge detection algorithm applied to determine the optimum boundary of each outlined feature.

27.3.2.1 Results

The plaque total volumes calculated with the procedure shown above may not be sensitive to rotations or motion of the head or neck thereby measurement errors were minimal due to repositioning (see Fig. 27.6). Visual comparison of serial imaging sessions was done by 3D reconstructs of the carotid artery slices (see Fig. 27.7), extending from the common, through the bifurcation, into the internal branch. The change in plaque volume after drug treatment was calculated as follows: absolute change (in A mm^3 per unit time) or relative change (A% per unit time) = total plaque volume before treatment − total plaque volume after treatment. The constants used in procedure (such as the total number of voxels in the MR images and the area per voxel) are not subject to variation. The measurements (such as manual outlining of the plaque area per MRI slice, choice of the FOV, slice thickness of the reformatted MRI slices) were subject to variation. Data predicted the regression of atherosclerosis measured as changes in wall thickness, lumen diameter, and plaque volume determined by serial MRI over 2 years. It evaluated the effects of statin treatment by quantitative serial MRI measurements of atherosclerotic lesions and provided a rational basis of design and plan of clinical studies on drug treatment induced regression of carotid atherosclerosis [21, 24].

[5]The statin evaluation experiment data was analyzed by first author as postdoc. scientist trainee at Atherosclerosis and Lipoprotein Training Center, Baylor College of Medicine, Houston, TX under supervision of Dr. Joel D Morrisett.

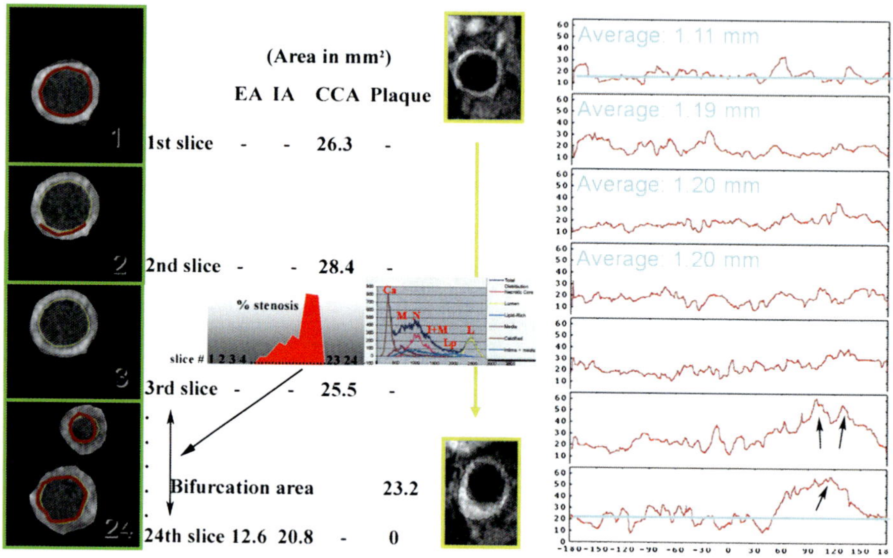

Fig. 27.6 *Left panel*: Illustration of how individual serial slice volumes are summed to obtain the composite lesion volume measurement. *Center panel*: Plot of lumen areas on different MR slice as a function of slice location is shown along the vessel to calculate % stenosis. Other plot shows different wall structures of their contribution. *Right panel*: MR slice through carotid bifurcation illustrating how areas of lumen and plaque are circumscribed and computed. The *arrows* indicate wall thickness

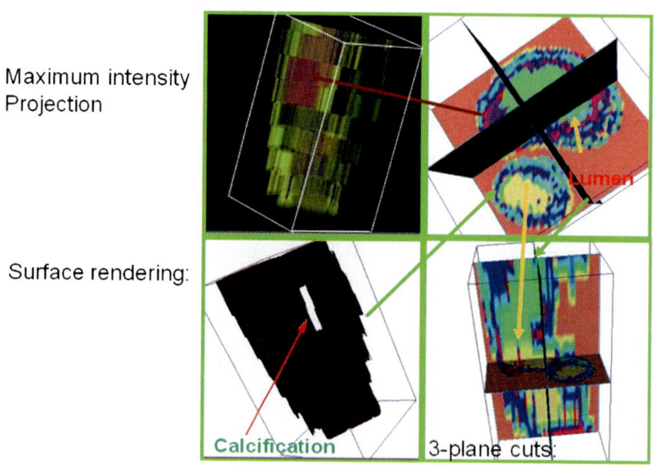

Fig. 27.7 Reconstruction of a carotid artery from MR image slices showing a plaque at the bifurcation, extending into the internal branch. *Top left*: Carotid plaques often have radial gradients of composition, reflecting successive masses of plaque development. *Top right*: Maximum intensity projection technique constructs the distribution of different plaque constituents in three planes. *Bottom left*: Rendering color-coded gradient distinguishes the calcification hypointensity. *Bottom right*: Three-plane cut shows the bifurcation with rendering display in any plane

In the following section, we describe the details of technique to develop criteria of diagnostic accuracy in identification of examination of different plaque contents and molecular deposits. Several regulatory agencies including American Heart Association, American College of Cardiology, and other regulatory bodies strongly suggested endarterectomy for CEA plaque histopathology classification and plaque typing useful in ex vivo imaging and ex vivo molecular staining techniques to compare with histopathology [7].

27.3.3 Plaque Histopathology Classification

A system for classifying atherosclerotic *lesion histopathology* was modified and established to provide a concise, convenient way to designate lesion pathology types in detail [2].

Type I lesions contain enough atherogenic lipoproteins to elicit an increase in macrophages and formation of scattered foam cells.

Type II lesions have primarily the layers of macrophage foam cells and lipid-laden SMCs and include lesions grossly designated as fatty streaks.

Type III lesions are intermediate between type II and type IV; they contain scattered extracellular lipid droplets and particles that disrupt the coherence of some intimal SMC. This extracellular lipid is the immediate precursor of the larger, confluent and more disruptive core of extracellular lipid that characterizes type IV lesions (atheroma that are potentially symptom producing).

Type IV and Va lesions are thought to become the vulnerable plaque.

Type Va lesions have a lipid core and contain thick layers of fibrous connective tissue.

Type Vb lesions are possibly calcified plaques.

Type Vc contain mainly fibrous connective tissue with little or no calcium or lipid. This plaque type may abruptly provoke sufficient thrombosis or stenosis to cause a clinical event, such as myocardial infarction, stroke, or peripheral insufficiency.

Type VI lesions when these lesions develop fissure, hematoma, or thrombus.

Type VII lesions are highly calcified.

Type VIII lesions are fibrotic plaque without lipid core and with possible small calcifications.

27.3.4 Endarterectomy Procedure

CEA operation involves:

1. A longitudinal incision 3–6 cm above and below the bifurcation, identification of the intima–media plane, and “shelling out” the diseased intima intact [53].

The remaining vessel was closed directly to its previous unoccluded dimension or closed with a Dacron patch to expand the diameter of the lumen; Typical tissues are resected for surgically accessible ICA stenoses >70% of the more distal internal carotid lumen diameter if:

- The stenosis is symptomatic, causing TIA or nondisabling stroke.
- There is no worse distal, ipsilateral, carotid arterial disease.
- The patient is in a stable medical condition.
- The rates of major surgical complications among patients of the treating surgeon are <6%.

2. Endarectomy is not done in asymptomatic stenoses of <60%. Symptomatic stenoses of <70% and asymptomatic stenoses of >60% are uncertain indications.
3. Symptomatic stenoses were the operative indications in 39% of the patients, the remaining 61% being symptom free.
4. Eighty percent of the patients had their second procedure within 2 days of the first.
5. There was no operative mortality and the combined stroke mortality was only 1%.

For details, readers may read the original paper [54].

Previous clinical trials have evaluated the efficacy of stroke prevention by CEA in symptomatic patients. The North American Symptomatic Carotid Endarterectomy Trial (NASCET) clearly demonstrated a benefit of surgery in stroke prevention as compared to optimal medical therapy after only 18 months of follow-up. The European Carotid Surgery Trial and VA Cooperative Study produced similar conclusions [55].

27.3.5 Carotid Artery Tissue Processing

Tissues were stored in PBS-50% glycerol buffer at 20°C until imaged or analyzed. Slits in the tissue created at surgery are closed by annealing opposing sides with superglue so as to recover the original spatial relationships within the tissue. The tissue was X-rayed using a FAXITRON analyzer to identify areas of calcification and guide subsequent dissection. After imaging, the tissue was marked with a thin line of India ink along the long axis. Transverse segments 3 mm long (corresponding to the in vivo MRI slice thickness) were cut with a scalpel, starting at the bifurcation which also serves as a point of reference. Segments were inked on the proximal end to preserve correct end-to-end relationship. Segments were then mounted for either frozen or paraffin sectioning. Adjacent sections were stained. Images of histologic sections (5 μm) are projected with a Nikon Microphot FXA microscope using a 1× lens onto a Sony Trinitron Superfine pitch screen. Morphologic features such as the fidl tissue, residual lumen, plaque wall, thrombus, and lipid core are circumscribed and the enclosed areas computed with Optimas Software (v5.2). Means of triplicate measurements of features from histologic slides are subjected to statistical analysis [21, 23].

27.3.6 Development of a Technique for Diagnostic Accuracy and Measurement of Plaque Composition: MRM at 1.5 Tesla Ex Vivo

We designed a holder to image carotid plaques shown in Fig. 27.8. The idea of excised plaque imaging was to simulate the same microimaging technique with matched physical factors used to place patient inside clinical MR suite [21]. Ex vivo plaque microimaging offered better plaque chemical composition and area of different plaque structures as shown in Fig. 27.9.

27.3.7 Carotid Plaque Magnetic Resonance Spectroscopy at 9.4 T Ex Vivo: Molecular Nature of Plaque Type III–IV in Excised Tissues[6]

NMR spectroscopy data of plaque lipid components combined with plaque image features (in voxels with characteristic image intensity) provide molecular signatures of plaque components. Major plaque components were lipids present in three different phases at or near body temperature. Still it is a significant challenge to characterize and quantify the lipids in carotid plaques [24]. Earlier C-13 Magic angle spinning (MAS) NMR detected the anisotropic phases in plaques to determine chemical and physical properties of lipids in liquid, liquid-crystalline, and solid phases [56]. Crystalline cholesterol in a lipid-rich carotid plaque was observed [57]. In a heterogeneous mixture of lipids in a plaque, cross polarization (CP) showed enhanced signals of solid phases relative to nonsolid phases. The CP/MAS NMR spectrum of the plaques showed few resonance signals of lipids other than esterified cholesterol, cholesterol monohydrate and no signal from both liquid phase and crystalline lipid phase. The threshold detection of crystalline cholesterol was ~2 mg. Major findings were as follows:

1. All plaques with crystalline cholesterol showed cholesterol: phospholipids molar ratio of >1:1 ratio at which a typical phospholipids bilayer becomes saturated with cholesterol.
2. Lipids in plaques can also exist in nonsolid phases of isotropic liquids and mesomorphic liquid-crystals.
3. Two major liquid-crystalline phases were present:
   (a) Thermotropic – composed mainly of cholesteryl esters
   (b) Lyotropic – composed of phospholipid bilayers

[6]The NMR experiment data was analyzed by first author as postdoc. scientist trainee at Atherosclerosis and Lipoprotein Training Center, Baylor College of Medicine, Houston, TX under supervision of Dr. Joel D Morrisett.

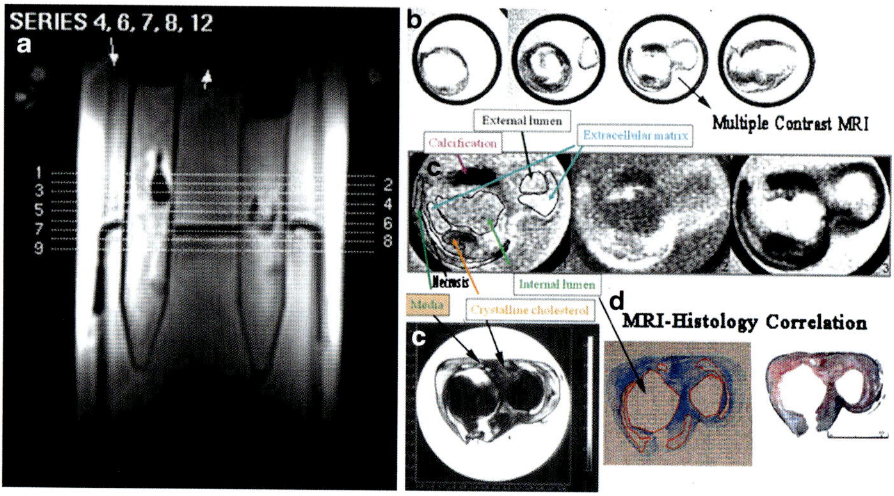

Fig. 27.8 *Left*: Figure shows carotid endarterectomy specimen holder which accommodates four samples in thermostat for MRI microimaging (shown in **a**) to generate transverse proton density-weighted (PDE), T1-weighted (T1-w) and T2-weighted (T2-w) MR images at a location distal to the bifurcation of a typical carotid endarterectomy specimen as shown in (**b**). The spatial and contrast resolution of the MR images permit to distinguish multiple tissue components from each other: calcification(Ca), lipid-rich (LR), lumen (Lu), necrotic area (N), part of the media (M) and media with some intima (I+M) shown in (**c**). Combined information of MR pixel intensities of the T1-w/T2-w/PD-w images distinguish constituents for both the external and the internal branches together with the distributions of the individual components derived by manually circumscribing the areas with different pixel intensities as shown in (**d**)

4. Cholesteryl esters existed in a liquid crystalline phase only in a certain temperature range below body temperature; while phospholipid bilayer existed at temperature range above body temperature.
5. Liquid crystalline phases were anisotropic and thermal motions detected by MAS NMR with or without CP.
6. Mobile isotropic lipid phase were observed without CP. The isotropic phase detected by CP without MAS.
7. Molecular motions of lipids in isotropic and liquid crystalline phases distinguished these phases by NMR.
8. Lipid poor and calcium-rich CEA specimens showed calcium phosphate (hydroxylapatite) detectable and measurable.
9. The P-31 spectrum of a carotid plaque typically showed a strong signal. Delipidation of the plaque sample showed no change in signal intensity on P-31 MAS NMR spectrum. It indicated that the spectrum originated from nonlipid source or phosphorus. Chemical analysis of plaque showed a P:Ca ratio = 1:2, very near the theoretical ratio = 1:2:1 in hydroxyapatite (a crystalline form of calcium mineral) in calcified atherosclerosis plaques.
10. Solid state C-13 and P-31 NMR studies suggested in vivo correlates for micro-imaging experiments on carotid plaques imaged ex vivo.

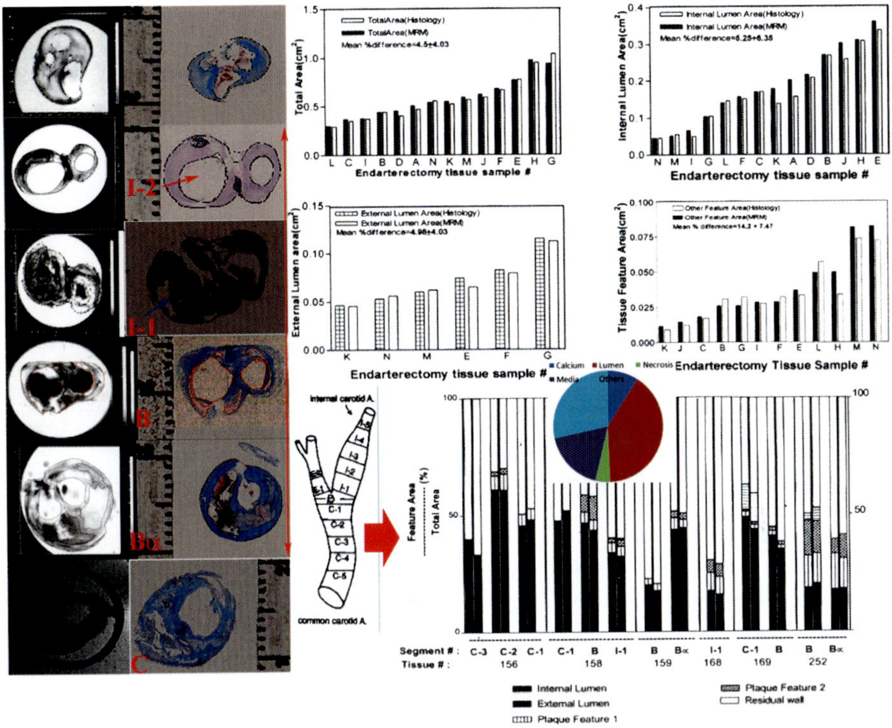

Fig. 27.9 *Left*: Ex vivo 1.5 T images acquired with the sample holder shown in Fig. 27.8. *Left*: Lateral view of sample MRM images with registered histology images are shown with manual delineation of areas (at different levels C→ I2–E2 segments of carotid artery). Notice the region of bifurcation B is distinct with occlusion of lumen. *Top right* : Areas on MRM are compared with histology images shown as *bars* for different segments of artery E–N. *Bottom right*: Plaque composition of one sample is shown at different levels C-3 to I-1 shown with sketch of artery and *bars*. Comparison of areas determined by MRM and by histology for the smaller external carotid lumen and larger internal carotid lumen showed mean percentage differences of 5%. By integrating the areas under these distributions, the percentage composition was calcium: 6.6%; Lumen: 29.7%; Necrosis; 3.0%; Media 3.0%; and other components 20.5% compared to the total area of the internal branch shown as *pie circle*. The external branch contained 54.6% lumen and 45.4% intima+media

27.3.8 Discriminative Analysis of Plaque Constituents

MRM provided digital 2D and 3D distribution of plaque components containing cholesterol crystals, thrombi, calcification, and necrotic core. Previous studies of CEA tissues showed different types of lesions and their components [58]. The common carotid bifurcation into the internal and external branches was the location of extensive calcification and appeared dark on a T1-w image. The ex vivo resolution obtained at 9.4 T was sufficient to visualize components (e.g., lipid-rich domains, fibrous caps, calcium deposits) on contiguous images. By multiplying the plaque feature area with slice thickness (1 mm), and summing up the resulting volumes, it was possible to compute the composite volume of individual plaque components (different colors) and their fractional contribution in the total lesion volume. Contiguous MRM images (1 mm slices) of a carotid plaque (see Fig. 27.9) indicated extensive chemical heterogeneity. It is important to note that MRM does not show the artifacts of tissue structures during processing while histologic sections clearly show them (e.g., splitting, folding, loss of lipid, loss of thrombus, etc.).

27.3.9 Associations Between Histopathology and MRI: Diagnostic Accuracy and Quantitation of Plaque Features

We developed a method to measure carotid wall thickening, lumen area, % stenosis, plaque size, and molecular deposits by quantitation similar to previous reports [19, 59, 60]. A quantitative comparison of carotid artery tissue using MRI and histology digital images was described [20, 58] for areas of necrotic core, residual lumen, calcification, fibrosis, and thrombus as distinguishable features on in vivo and ex vivo MR images. We developed a method to compare the plaque constitution by MRI with histopathology [21–23] distal images and calculating plaque composition. The idea was that distinctive features of atherosclerotic carotid vessel are visible on MR images based on the following criteria:

1. A fat suppression RF pulse may suppress the intensity of the lipid-rich adventitia, making the clear boundary adventitia–media or intima–media easier to see.
2. For full carotid wall thickness, we used conditions of selective partial visualization of the adventitia.

Matching of MR slices and histologic sections (see Fig. 27.9), it was possible to register two digital images to extract out constitution and composition of plaque [23, 59]. The next task was to ensure that ex vivo image slices (1 mm at 9.4 T; and 1 mm at 1.5 T) exist in correct registration with 5 μm histologic section images similar to the report by Yuan et al. [30]. Our approach was to cut 3 mm tissue segments from the end of the tissue, ink the proximal face of each segment to preserve correct end orientation, and mount the segment in optical coherence tomography (OCT) for frozen sectioning. Exact 5 μm histology sections are cut every 100-μm interval, generating 10 sections from 1mm thick tissue [21].

Averaging the ten histologic images simulated the average partial voluming effect on the MR images to get similar features from both digital images. After the frozen sections are cut for immunohistochemical localization, the remaining block was stored at −80°C for molecular staining experiments. Microdissecting by quantitative histology measured different plaque morphologic features such as area of a calcium deposit, lipid-rich region inside lumen, inside the media, and inside the adventitia by circumscribing and integrating segmented areas on histology digital image from all slides [21, 23]. Microdissecting of plaque components such as lipid-rich areas, thrombi, calcium deposits, and fibrotic areas identified plaque components by proton relaxation constants of different plaque components as molecular signatures [21]. It is important to note that plaque size, shape, and intensities of specific plaque structures on MR images with corresponding size, shape, and color or texture features on the histologic images are from a different set of images (MRI image of 1 mm slice is 200 times thicker than histologic section) to get information of thin wall, lumen, highly calcified areas, fat, crystalline cholesterol monohydrate. Crystalline calcium with restricted motion further restrict fast relaxing protons so they appear dark on the all T1-w/T2-w/PD-w MR images. Cholesterol clefts appear as dark regions in the T2-w MR images.

In the following section we describe the molecular protein localization in carotid artery plaques. The principle of digital display of molecular density distribution or "molecular geography" was imbibed by first author using NMR-biochemical correlation to reconstruct molecular patterns in vivo/ex vivo[7]. Now art has grown as molecular staining and 3D reconstructed molecular holography, genePaint, enzymePaint, proteinPaint to visualize regions of metabolic events with color density.

27.4 Ex Vivo Molecular Staining Techniques in Atherosclerosis

27.4.1 Matrix Metalloproteases

In plaques, monocytes and monocyte-derived macrophages play a major role in lipid-filled foam cells in carotid artery. The vulnerable plaque showed enhanced macrophages or foam cells infiltration in the fibrous cap >65 um thick overlying a necrotic core [60]. The extracellular matrix metalloproteinases (MMPs) breakdown the collagen and elastin molecules and carotid artery plaque becomes vulnerable to rupture. Several carotid artery MMPs were reported as MMP-9 (92 Kda gelatinase or gelatinase A), MMP-3 (stromelysin-1), MMP-I (interstitial collagenase), MMP-7 (matrilysin), and MMP-12 (metalloelastase) [60–62]. SMCs secrete MMP's such as MMP-2 (72 kDa gelatinase or gelatinase B), MMP-9, and MMP-3 [62]. In situ hybridization or histochemical techniques can be used to demonstrate that MMP-9, MMP-3, MMP-1, MMP-7, and MMP-12 can all be expressed in various types of atherosclerotic lesions [63, 64]. MMP's are synthesized as inactive zymogens which must be activated, normally by proteolytic cleavage. The antibody staining and in situ hybridization study detected the presence of MMPs in atherosclerotic lesions [65]. In other study, extracts of atherosclerotic lesions run on zymography gels indicated the presence of both active and inactive MMP's [66]. Their experiments clearly indicated that active MMPs exist in the atherosclerotic lesion.

Intracellular expression of MMP-9 in unstable plaque suggested the recent plaque rupture (unstable angina) and use of different MMPs in postsurgery evaluation. The metalloproteases immunostaining digital maps were reconstructed over surface of carotid artery to visualize the locations of high inflammation [61].

[7] Sharma R. The relaxation times of human tissues and MR-biochemical correlation in medicine. Ph.D awarded at india institute of technology, Delhi (1995).

27.4.2 DNA Microarray

Mapping specific genes on the 2D surface of histologic cross-sections of atherosclerotic arteries showed distribution of Apolipoprotein E, CD68, TIMP, Janus kinase 1 (JAK-1), VEGF receptor-2, and phospholipase D macromolecules as shown in Fig. 27.10. These pharmacogenomic and pharmacoproteomic approaches likely discriminate tissues exposed to the pleiotropic effects of drugs and other interventions. High density cDNA-based microarrays can be used to screen human aortic SMCs under conditions of mechanical stress. In previous study, out of an array of 5,000 genes only 5 genes were associated with atherosclerosis: VEGF, COX-1, tenascin-1, PAI-I, BAX, BCL, GAPDH, histone H3 and MP-1to visualize 3D distribution of gene expression [68]. DNA microarray analysis was reported the adipophilin gene (ADFP) using Western blotting [69–71]. However, these techniques of gene profiling on DNA Microarray remain as bench experiments.

27.4.3 The Protein Microarray

Localization of proteins or mapping proteins responsible of carotid artery disease is emerging to identify the myleperoxidase, proteases, CD-68 protein to express metalloproteins as pharmacoproteomic approach. Other examples are CD-45, CD-31 antibodies to immunostain the lesions [67].

27.5 Multimodal Molecular Imaging of Atherosclerosis Plaque Activity: An Emerging Art

Till date, most of the in vivo imaging techniques are being used in extracting information of structural details and localized deposits in carotid artery wall with little success in mapping metabolic and molecular events. Using newly available artificial intelligence, nanotechnology and biophysical mapping techniques have evidence of great hopes of molecular imaging such as labeled nanoparticles, susceptibility and flow imaging, chemical shift imaging, molecular algorithms, and multimodal bioimaging. Molecular imaging techniques are capable of imaging the cellular metabolic events and molecular events associated with release of active molecules out of inflammation, angiogenesis, and pathogenesis. These active molecules generate visible imaging signal from the ligands or radiolabeled contrast agents upon binding with them based on the principles of bioluminescence, fluorescence, optical imaging, and NMR, US, PET, SPECT, autoradiography, or nanoparticle sensitive techniques as described in following sections. These techniques depend on the following concepts:

1. Ruptured plaques, acute infarction, sudden thrombosis, hemorrhage, and vascular obstruction are predictable by molecular changes.
2. Lipid core and local rupture induce the clotting event due to interaction of serum clotting factor with expressed tissue factor.

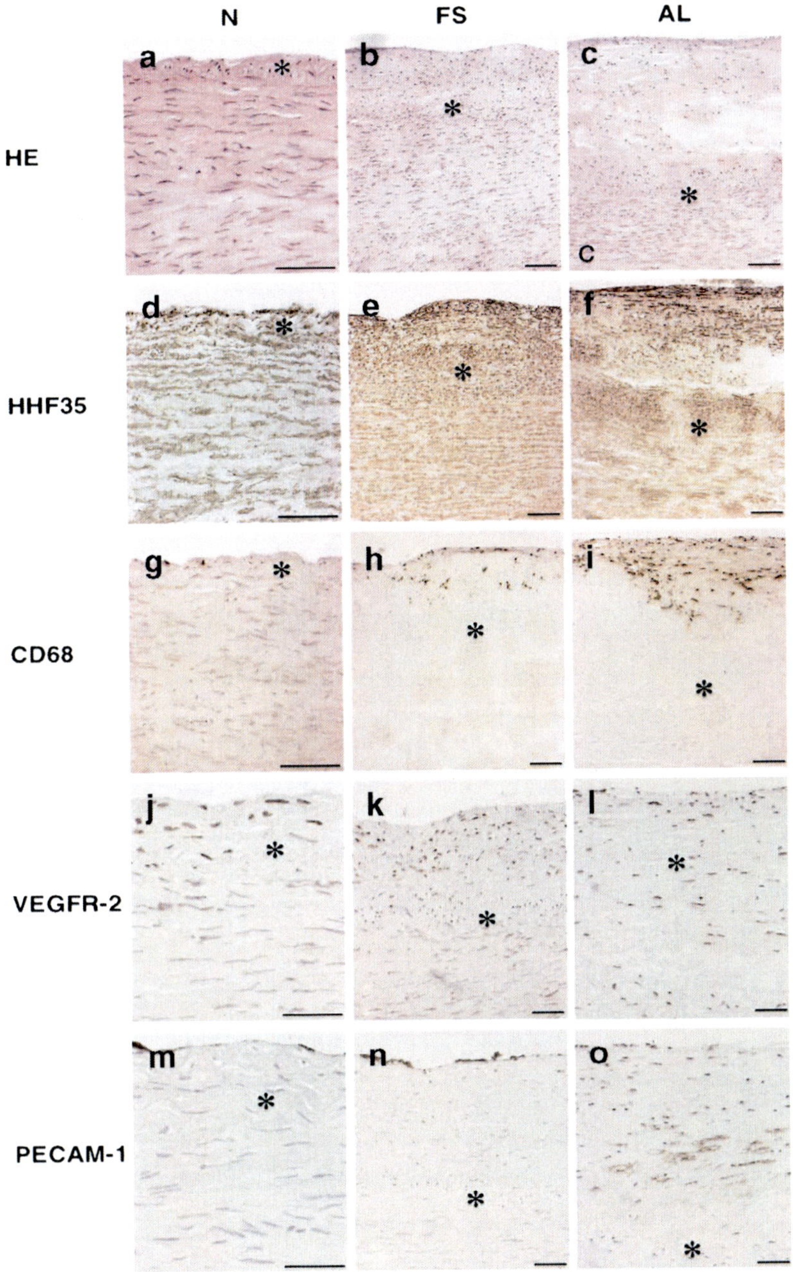

Fig. 27.10 Representative examples of histology of the arterial samples collected for DNA array analysis (organ donor, male, 36 years, serial sections for each type of the lesions). (**a**–**c**) Hematoxylin–eosin stainings; (**d**–**f**) immunostainings for SMC a-actin; (**g**–**i**) immunostainings for macrophages; (**j**–**l**) immunostainings for VEGF receptor-2; (**m**–**o**) immunostainings for PECAM-1. *N* normal, *FS* fatty streak, *AL* advanced lesion. *Asterisk* indicates internal elastic lamina. *Bar*: 100 mm. Reproduced with permission from [67]

3. Fibrin matrix, hemorrhage in plaque mass are detectible with nanoparticles as atheroma targets.
4. Active macrophages around endothelium matrix secrete MMPs that digest extracellular matrix or lipid core or thrombotic core of inflammatory cells.
5. Angiogenesis weakens fibrous cap and promotes plaque rupture due to erosion of extracellular matrix.

Major emphasis of new molecular imaging still is on reducing imaging time and using cheaper, user-friendly, biocompatible and precise methods beyond picomolar scale (technology developed by Visen® for enzymes). FDG-PET/CT and MRI multimodal imaging has emerged as clinical modality for classifying plaques with collagen, lipid core, calcification [5]. The idea was tracking inflammation within plaque when used with tracer [18F]-fluoro-2 deoxy-D-glucose (FDG) deposited in macrophages associated with high lipid-rich areas on MRI. The CT and T2-w/PD-w MR images distinguished the calcification and lipid core. Other component of collagen appeared as iso- to hyperintense on T2-w images and hyperintense on PD-w images. We demonstrated the bright signal intensities of calcification by using susceptibility-weighted T1-w images in carotid artery disease [65]. We developed a technique of sodium MRI/^{18}FDG-PET microimaging to explore the link of intracellular sodium and enhanced glycolysis in cardiovascular tissue [72]. We observed the possible association of ischemia and sodium ion–glucose uptake as manifestation.

The physical principles of molecular imaging techniques are similar to the biochemical or physical methods based on the simulation of color intensities as digital image generated with possibility of measurement of end-product molecules to determine metabolic rates and events during carotid artery disease. Important color mapping techniques are bioluminescence, photoluminescence, fluorescence, acquisition of NMR, PET, US, and SPECT metabolite or molecular images. Significant molecules sensitive to optical near-infrared fluorescence, bioluminescence, photoluminescence, and fluorescence, acquisition of NMR, PET, US, and SPECT molecular imaging are as follows:

- PET-CT imaging radiolabeled precursors such as Vivotag® ^{64}Cu-DTPA for T2-w MRI (sensitivity up to 5 μg Fe/mL) and ^{18}F-deoxyglucose for PET/CT imaging radiolabeled precursors, near infrared fluorochrome (NIRF).
- ^{99m}Tc compounds for SPECT.
- Photo dyes such as Cy 5.5.
- Biosensors such as enzyme sensors, perfluorocarbon sensors.
- Peptides, proteins, antibody fragments, nanoparticles, phage, aptamers.

27.5.1 Multimodal Imaging Principles

In following section, we introduce readers with brief details of newly identified potential molecular imaging techniques.

Optical near-infrared fluorescence is the technique based on biosensor Cyclo [Cys-Arg-Gly-Asp-Cys-]-Gly-Lys [Cy 5.5] that illuminates excitation (625 ± 18 nm)/ emission filter (700 ± 17.5 nm) fluorescence signals produced from infra-red fluorescence reflectance from tissue molecules captured with charged coupled device camera in vivo FX (Kodak MI system). Different biomolecules generate distinct signal from Cy 5.5 emission after fluorescence reflectance from tissue molecules. Bioluminescence is the technique for in vivo imaging with promise of detecting necrotic area in artery. However, clinical utility is still in question. Photonic imaging is the technique based on photoacoustic effect detectable by optical imaging with promise of detecting bioluminescence signals from the cardiovascular tissue. Fluorescence imaging technique is based on the fluorescence emission from fluorescent-labeled compounds injected in the body. The carotid artery plaques have been reported with great promise in tracking molecular targets by helium–cadmium laser (Omnichrome, model 356XMS, Melles Griot). It emits radiation at a wavelength of 325 nm and a continuous wave power of 10 mW to induce endogenous tissue fluorescence after laser energy delivered through a fused silica optical fiber with a 600-μm core diameter. The fluorescence is collected through this optical fiber and directed to a monochromator (model 1235, Seiko-EG&G), to excite and detect fluorescence signals with a single fiber [73]. Optical signals get collected at the entrance slit of the monochromator. The tissue fluorescence was spectrally get dispersed by diffraction grating (150 grooves/mm) and imaged on the detector of an optical multichannel analyzer [73].

As described in previous sections, MR/NMR imaging is the technique based on odd-numbered nuclei spins to generate spin echo and digital resonance signal from the tissue nuclei in the form of simulated 2D slices or 3D images. Different image signal intensities of tissue nuclei generate visual information of chemical shifts of molecules or their magnetic moment characteristics.

PET imaging is the technique based on positron–electron pair production in tissue and release of two photons 180° apart after annihilation from the tissue detectable by germanium or borosilicate crystals as dynamic gray color maps at different angles [73]. Precursor metabolite gets integrated in metabolism and metabolic end-products serve as biomarkers on images such as ^{18}F-deoxyglucose gets in glucose metabolism and generate the image of oxygen-deprived hypoxia locations in the tissue.

Ultrasound imaging is based on the echo formed from return waves after propagation of high frequency sound waves sent from piezoelectric transducer in the body tissue. The difference of Doppler's frequency at the two interfaces in the tissues simulates digital images depending on the motion, density, and flow [44]. Probabilistic Neural Network (PNN) and Support Vector Machine (SVM) classifiers are validated by leave-on-out estimate for identifying echolucent radii to generate gray-colored plaque information [16].

SPECT is the technique of single photon emission computed tomography using ^{99m}Tc radiolabeled contrast agents [74]. This technique is better as radiolabel is washed out in 48 h.

Nanomolar technology to picomolar technology is still in infancy. We reported a simple approach of picotechnology in biological use based on optical principles [75].

Using similar approach with robust software and precise measurement beyond picomolar scale technology was developed by Visen® for enzymes to display 3D picomolar distribution of different metabolites in the digital images of carotid artery tissue [75].

In vivo multimodal imaging including optical near-infrared fluorescence, molecular imaging (peptides, proteins, antibody fragments, nanoparticles, phage, and aptamers) and reporter agents is also emerging to visualize atheroma, calcification, inflammation, and angiogenesis using multifunctional and multimodal nanoparticles to measure fluorescence [76]. The idea was to evaluate stenosis in plaque at less priced microscopy facility using laser scaning, epifluorescence, confocal, multiphoton microscopy with adjunct optoacoustic imaging, fluorescence tomography, optical projection tomography, reflectance imaging. Important advantage of these fluorescence techniques are that these are compatible with angioscopy, bypass graft imaging [77].

27.5.2 Available Multimodal Imaging Techniques

Our lab has acquired various range of imaging probes including the enzyme imaging probes, near-infrared probes, and multimodal nanoparticles usable in imaging macrophages, adhesion molecules, drug delivery, and targeted imaging. Major applications may be possible as in vivo imaging molecular readouts by optical, MRI, and SPECT modalities and in in vivo imaging targeted diagnostics/therapeutics. The main aim of these probes still remains to develop high sensitive, high spatial resolution of reporter probes such as PET/MRI and combined FMT/MRI for common platform to obtain anatomic, chemical, and physiological information of artery. We shall focus on molecular imaging of atherosclerosis and angiogenesis in carotid artery disease. Some of them are established based on active protein localization as follows:

- Matrix metalloproteases MMP-1,-2,-3,-7,-8,-9,-12,-13 and disintegration MMPs, cysteine-rich cathepsins (S, K, B and L), serine proteases (tissue plasminogen activator and urokinase type PA, elastase) enzymes to define plaque expansion, rupture, and atherogenesis [65, 78, 79].
- Preclinical evaluation and phase I clinical trial of a ^{99m}Tc-labeled synthetic polymer used in blood pool imaging [80].
- Protease activable reporters based NIRF imaging fluorochrome agents for carotid wall fluorescence-mediated tomography (FMT) imaging with clinical usable modality. Example. Prosense680 [81].
- Cathepsin B and K inflammatory biomarkers [82].
- MMP-2 and -9 gelinolytic activity on GGPRQITAG peptide substrate by using FMT-CT to detect NIRF signal from carotid atheroma [83].
- Cysteine protease by OCT [84].
- Apoptosis markers by annexin sensors [85].
- Oxidative stress sensors by hypochlorous acid [86].

- Integrisense680 sensor-based reflectance imaging [87].
- Fluorescent deoxyglucose sensors for glycolytic flux in atherosclerosis [88, 89].
- Dextran-coated iron oxide magnetic particles to visualize phagocytosis in macrophages by T2-w low signal and off-resonance positive contrast [90–92].
- NIRF-derived nanoparticles for ^{64}Cu-DTPA-based PET/MRI and VT680 fluorochrome-based NIRF microscopy was reported for macrophage function in plaque [93] with possible clinical use.
- The development of particles for multimodal MRI and NIRF imaging of VCAM-1 expression is in progress [94].
- Apoptosis sensor of NIRF/MRI with possible angiogenic-targeted perfluorocarbon emulsions, microbubbles, micelles, quantum dots, liposomes, lipoproteins are emerging possibilities for MRI/ultrasound/NIRF multimodal imaging of atherosclerosis [95, 96].
- Fumagillin bound $\alpha_V\beta_3$ integrin-targeted particles on perfluorocarbon platform were developed for MRI to detect angiogenesis [97].
- Photosensitizers with NIRF fluorescence capabilities can detect inflamed atheroma and monitor the photodynamic therapy or siRNA molecular therapy [98] (Table 27.8).

Table 27.8 A scheme is shown for multimodal imaging techniques with their targets and use in artery samples

Contrast agent	Modality	Target	Events
Inflammation imaging sensor			
MMP sense	NIRF	MMP	EM
Prosense	NIRF	Cys protease	EM
Fluorescent FAST	NIRF	Cys protease/MMP	EM
Cathepsin K/D/-5 sense	NIRF	Proteinase	N
Urokinase-PA sense	NIRF	Serine proteinase	N
USPIO	MRI	Macrophage	N
CLIO-Cy 5.5/ VT680or750/-Cy7	MRI/NIRF	Macrophage	N
^{64}Cu-CLIO-VT680/VT750	PET/MRI/NIRF	Macrophage	N
VINP-28	MRI/NIRF	VAM-1	
Apoptosis imaging sensor			
Annexin-Cy 5.5	MRI	Phosphtidylserine	EM, I
CLIO-Annexin-Cy 5.5	MRI/NIRF	Phosphtidylserine, VCAM, selectin	EM, I
Angiogenesis imaging sensors			
Liposomes, micelles	US, MRI, NIRF	Hormones	TH
Fumagilin and perfluorocarbon MNP	MRI, NIRF, SPECT, US	Fibrin, collagen, integrin $\alpha_V\beta_3$	TH
RGD peptides	NIRF	Integrin $\alpha_V\beta_3$	N

NIRF near-infrared fluorescence, *MRI* magnetic resonance imaging, *PET* positron emission tomography; *US* ultrasound, *EM* extracellular matrix, *N* nacrosis, *I* ischemia, *TH* thrombosis, *N* necrosis

27.5.3 *Molecular Imaging of Carotid Artery Plaques: How it Works?*

Molecular imaging of atherosclerosis offers in vivo biology insight as well as new clinically translatable strategy to identify and classify high-risk carotid artery plaques. Rapid growth of optical near-infrared fluorescence and molecular imaging is making it possible to get chemical and molecular events in plaque inflammation and angiogenesis with the possibility of clinical intervention such as intravascular catheters, noninvasive tomography as achievable dream of single platform multimodal imaging. The development of multimodal, multifunctional nanoparticles to image carotid artery is a growing science in understanding and treating vascular disease.

How nanoparticles generate contrast?

Suppose two tissues have same T1 and T2 relaxation constants. After nanoparticle injection, only tissue A expresses the molecular epitope that binds the paramagnetic particles. The paramagnetic particles affect the relaxation constants as follows:

$$1/T_{1A} = 1/T_{1B} + r_1[\mathrm{NP}], \qquad (27.2)$$

$$1/T_{2A} = 1/T_{2B} + r_2[\mathrm{NP}], \qquad (27.3)$$

where T_{1B} and T_{2A} are unchanged relaxation constants while $r_1 = 1/T_{1A}$ and $r_2 = 1/T_{2A}$ are observed relaxation constants after nanoparticle binding. The r_1 and r_2 are relativities of nanoparticles calculated by slope from plot r_2 vs. concentration C^* of nanoparticle.

$$r_1 = y = A * \left[1 - \exp^{(-r_1 \times \mathrm{TR})}\right]$$

and

$$r_2 = A + C * \left[\exp^{(-r_2 \times \mathrm{TE})}\right],$$

For SE pulse sequence, MR signal intensities of each tissue will be:

$$S_A = k_1\left(1 - 2\mathrm{e}^{-(\mathrm{TR}-\mathrm{TE}/2)/T_{1A}} + \mathrm{e}^{-\mathrm{TR}/T_{1A}}\right)\mathrm{e}^{-\mathrm{TE}/T_{2A}}, \qquad (27.4)$$

$$S_B = k_2\left(1 - 2\mathrm{e}^{-\mathrm{TR}-\mathrm{TE}/2)/T_{1B}} + \mathrm{e}^{-\mathrm{TR}/T_{1B}}\right)\mathrm{e}^{-\mathrm{TE}/T_{2B}}, \qquad (27.5)$$

where k_1, and k_2 are scan intrinsic properties such as flip angle, coil sensitivity, proton density, etc.

$$\mathrm{CNR} = \left|\frac{(S_A - S_B)}{N}\right| \qquad (27.6)$$

Optimizing TR is suitable technique to generate highest CNR of two tissues at different T1 relaxation constants.

$$\mathrm{TR}_{\mathrm{opt}} = \frac{T_{1\mathrm{A}} T_{1\mathrm{B}}}{\left(T_{1\mathrm{B}} - T_{1\mathrm{A}}\right) \ln\left(k_{\mathrm{A}} T_{1\mathrm{B}} / k_{\mathrm{B}} T_{1\mathrm{A}}\right)} \tag{27.7}$$

Over several years, our lab has got a library of nanoparticles with bioimaging and therapeutic application potentials. For molecular imaging, our strategy was make 10–50 nm-sized nanoparticles for vascular endothelium tissue targeting and functionalization (binding with drug, antibodies, peptides, polysaccharides, avidin–biotin cross-linked with polymers) suited for in vivo targeting. The chemical ligand groups conjugated with nanoparticles and surface ionic properties play significant role in targeting specificity. For vascular imaging applications, we observed potentials of liposome vesicles (50–70 nm) in US [99] and MRI [100], perfluorocarbon core emulsions (200–300 nm) for MRI, US, fluorescence, nuclear, CTI [101], HDL, LDL micelles for MRI [102, 103]. In nanoparticle, we reported polymer hydroxyl acidic core (PGLA, PLA), dendrimers [polyamidoamine (PAMAM), diaminobutane (DAB)] suitable to make superparamagnetic iron oxide (15–60 nm SPIO) particles to cause dephasing and loss of T2* signal intensity due to susceptibility effects as suitable for passive targeted imaging inflammation of cardiovascular tissue[104–106].

Using model predictions, minimum concentration of nanoparticles can be measured needed to generate max CNR [107]. The relative deposited concentration of iron oxide nanoparticles such as Combodex® or Ferridex® in tissues may be measured as relative dephasing signal produced as shown in following Eq. 27.8:

$$[\mathrm{MNP}]\,\alpha - \ln\left[\frac{S(t)}{S_0}\right], \tag{27.8}$$

where S_0 is signal intensity before injection and $S(t)$ is signal intensity in postinjection of contrast agent in tissue at constant TE and short TR.

Our efforts were to measure CNR values, r_1, r_2 values of nanoparticles and their physical characteristics of different multimodal nanoparticles. We identified the macrophage enzyme action as important in uncaging the nanosphere to make free sequestered heavy metal that enhances the proton relaxation or MRI signal. Louie et al. 2000 reported enzyme galactosidase to identify its source as transfected lacZ gene [107]. Other gold, carbon nanotube fullerenes(4 nm), quantum dots cadmium selenide spheres (2–10 nm) metal-based agents are in process of standardization and may be useful in fluorescent imaging [108]. Other investigators have reported possibilities of viral capsid protein cages with gadolinium as potential nanospheres for drug encapsulation and imaging atherosclerosis [109, 110]. We prepared chemical exchange saturation transfer (CEST) chelated lanthanide agents to suppress water signal by selective excitation at 4.57 ppm useful in 21 Tesla microimaging [105]. We tested other potential examples of fluorine moieties on nanoparticles, multivalent maghemite Fe_3O_4-based nanospheres for nuclear SPECT, US, paramagnetic effects based on their fluorescence and superparamagnetic effects [106]. Some technical developments are cited as following:

- Radionuclide-linked antibodies, peptides, annexin-phosphatidyl serine, FDG-PET, and macrophage chemotaxis are possible choices of nuclear/PET imaging.
- Quantum dots, gold nanoshells, carbon nanotubes are ideal choices in autofluorescence in near-infrared wavelengths.
- Lipid emulsions containing radio-opaque iodinated triglycerides may be the choice for CT imaging.
- Microbubbles to target vascular epitopes, liposomes based on compression and reflectance for US.
- Lanthanides to reduce T2* signal due to dephasing and susceptibility effect for MRI
- Iron oxide maghemite-based nanoparticles for off-resonance MRI to achieve bright intensity around nanoparticles [111].
- Perfluorocarbon-based nanospheres to detect fluorine sitting on nanospheres in deposited areas rich in fibrin protein and different moieties over fibrin protein (with distinct electron shielding, J-coupling) are distinct and different to show up at different frequencies of NMR spectrum. Its application may be extended in chemical shift imaging of fat–water contrast in clinical set-up [112].
- At 21 Tesla MR microscopy, major merits are MR sensitivity to concentration of paramagnetic or superparamagnetic nanoparticle agents, fast pulse sequence used, better detection limit, coil sensitivity, epitope prevalence [111].

27.5.4 Present State of Art in Carotid Artery Atherosclerosis and Angiogenesis Imaging

Presurgery plasma biomarkers give first hand information of vasculature. Unstable and ruptured plaques with 40–60% stenosis need attention of imaging to distinguish the plaque type. For it, molecular events offer a window of sequential events and history including fibrin deposition, erosion, hemorrhage, thrombi. In high-grade stenosis with culprit plaques (type VI–VIII) it needs cardiac catheterization. The following choices of molecular imaging are significant in extracting out visible molecular events of disease from carotid artery images:

- Fibrin imaging by US and MRI using antibody specific to avidin–biotin linked with fibrin [113].
- Tissue glycoprotein imaging for in vivo US and MRI [114].
- Echogenic Liposomes for US application [115].
- Intercellular adhesion molecules (ICAM-1) and vascular cell adhesion molecules (VCAM-1), fibrin, and fibrinogen molecules enhance the signal in carotid artery walls [99].
- Perfluorocarbon particles EP-2104R bound with gadolinium chelates amplify the signal of fibrin in artery wall [116]. The modifications in MR pulse sequences such as rapid steady-state free precession (SSFP) 19F MRI demonstrated location of fibrin clots [117].

- Adhesion heterodimer $\alpha_V\beta_3$ integrin molecules in endothelial cells, monocytes, SMC, fibroblasts signify the event of angiogenesis associated with inflammation and growth factor expression to meet high metabolic demand in carotid artery. Molecules $\alpha_V\beta_3$ integrin and collagen III targeted nanoparticles were reported for MRI application [118–120].
- Macrophage imaging to delineate macrophages and foam cells using USPIO particles for MRI imaging of plaques [90, 121, 122].
- Gadolinium-bound HDL and fluorine micelles visualize the lipid core [123].
- Dextran caged iron oxide particles conjugated with a2AP peptide fragment act as "magnetic switch" to detect the myeloperoxidase activity in plaque indicative of inflammation, infarction, and thrombi formation [123]

Recent reports are exciting in the fields of mesenchymal stem cell implantation into necrotic core, cardiovascular regeneration, targeted delivery of drugs encaged in nanoparticles at the site of injury but clinical use is less known. Still hopes are active for rapid developments in genomics, molecular biology, and nanotechnology to make available molecular imaging modalities in clinical use.

27.5.5 3D Molecular Imaging of Biomarkers by Nanoparticles

Several proteins such as $\alpha_V\beta_3$ integrin and collagen III molecules in carotid artery wall predict the status of restenosis and mural injury in carotid artery disease (see Fig. 27.11). Molecular probes offer the potential to characterize biochemical features by targeting of biochemical epitopes such as perfluorocarbon nanoparticles are echogenic for US, usable for MRI, CTI, SPECT imaging of thrombosis and angiogenesis[119, 124].

New approaches of multilabeling techniques are emerging for multimodal imaging techniques. Nanotechnology is the recently applicable mode of multimodal imaging platform by fusing two or three image types. The superoxide myoglobin-based SPIOM particles were prepared with co-precipitation method as previously described in detail [111, 125]. Other significant nanoparticles in atherosclerosis are meso-2,3-dimercaptosuccinic acid bound anti-human Troponin with linker ethyl-3-[3-dimethylaminopropyl] carbodiimide hydrochloride [105, 111, 125]. The general structures and schemes of nanoparticles are shown as micelles (see Fig. 27.12).

The iron oxide-based nanoparticle MION (monocrystalline iron oxide nanoparticle) were reported for MR imaging with other multimodal imaging techniques. The dextran coating around the nanoparticle was reported to be cross-linked with epichlorin hydrin aminated as shown in Fig. 27.12 and labeled with NIRF Vivotag-680 with the following characteristics [93].

- Derivatization with the chelator DTPA to attach with radiotracer ^{64}Cu.
- Iron oxide core provided contrast in MRI (T2, T2*, or steady-state free-precession sequences).
- Fluorochrome for fluorescence imaging, fluorescence microscopy, flow cytometry, and fluorescence-mediated tomography.

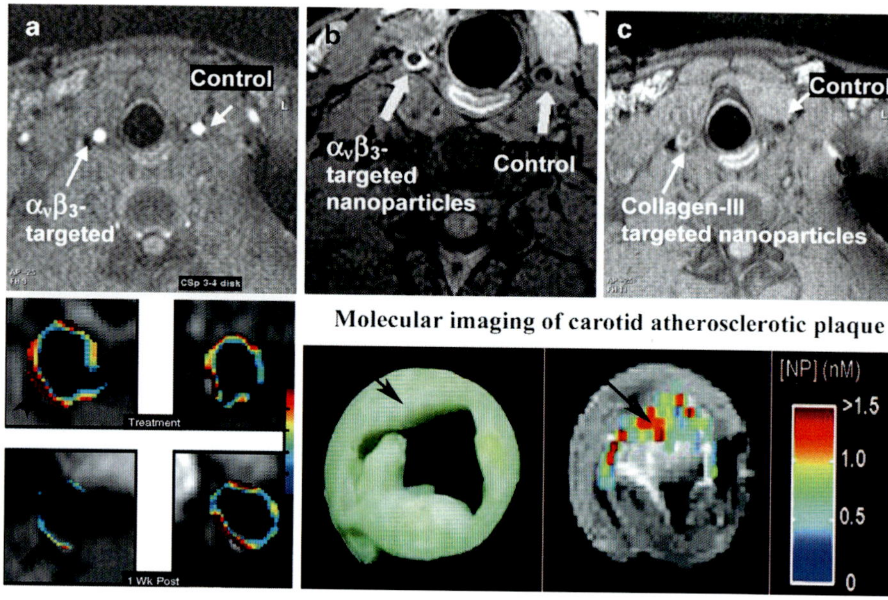

Fig. 27.11 *Top row*: Imaging of alpha V beta3 and collagenase III molecular imaging is shown using perfluorocarbon nanoparticles. The atheroma and angiogenesis can be distinguished by this technique. (*Second row*) *Left*: Color-coded imaging of aorta by alpha V beta3 targeted paramagnetic nanoparticles is shown. Notice the atherosclerosis-related angiogenesis shows reduced activity in the tissue shown by *arrows*. *Right*: First panel shows excised plaque and color-coded plaque region with liposomal nanoparticles shown in next figure. Reproduced with permission from [124]

- Cross-linked aminated polysaccharide coating for biocompatibility, determined blood half-life, and provided linker for attachment of tracers and potentially affinity ligands.
- ^{18}FDG PET for hybrid PET-CT Imaging on X-PET PET-CT system (Mercury Computer Systems, Carlsbad, CA).
- MRI microimaging studies on 7-T horizontal-bore scanner (Bruker Pharmascan, Billerica, MA).
- In vivo fluorescence reflectance imaging, fluorescence microscopy, phosphorimaging autoradiography, flow cytometry using triple fluorescent-labeled images with an upright epifluorescence microscope (Eclipse 80i, Nikon, Melville, NY)
- Histopathology to compare the detectable regions and morphology.

27.5.6 MALDI Imaging Technique

Recently, MALDI, liquid chromatography mass spectroscopy (LCMS) imaging technique was proposed to visualize proteomics by selective mass spectroscopy signal [126] and we proposed the possibility of atherogenic protein as precursor of apoptosis and inducing atherosclerosis. Further, fused MRI images with MALDI

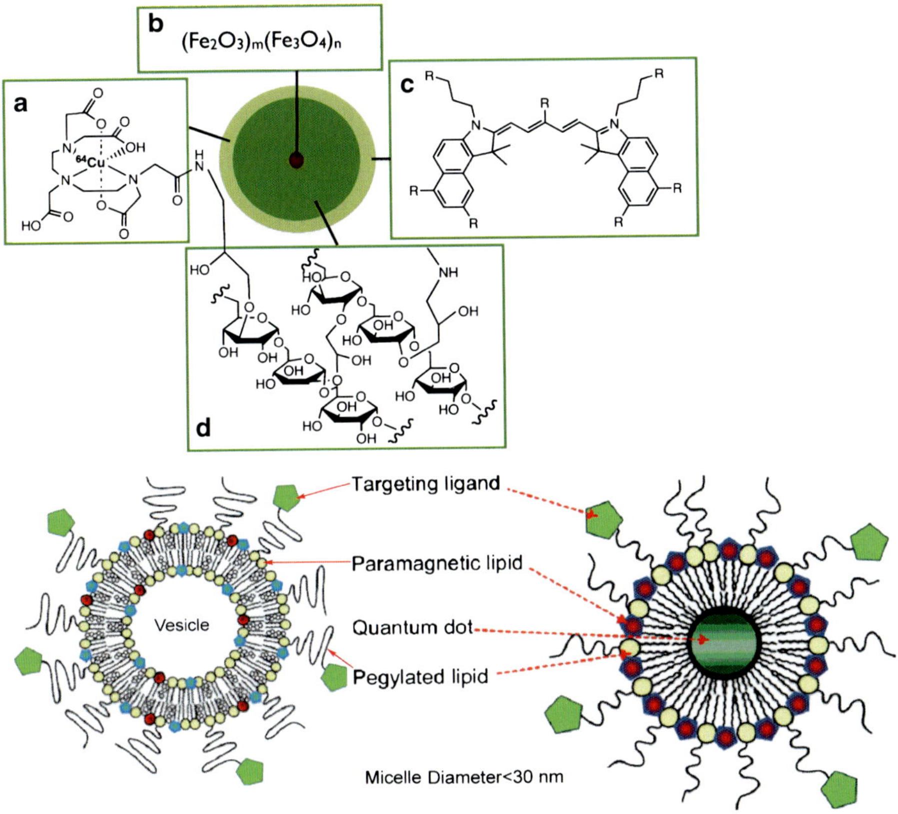

Fig. 27.12 The nanoparticles with their moiety are shown for SPIOM with its chemical branches A, B, C, D (*top*), liposomes (*bottom left*) and quantum dots (*right*) for use in carotid artery imaging. In SPIOM, specific antibody is placed on outer surface of dextran polymer shell containing paramagnetic iron oxide in center as shown in figure on *left*. The quantum dots surrounded by hydrophobic lipid micelles are the source of contrast as shown in figure at *bottom*. Reproduced with permission from [105, 111]

map further presents a guideline for distribution of the atherogenic proteins involved in atherosclerosis mainly lipoproteins. However, the utility of molecular MALDI imaging is not yet ready for clinical use.

We are in the process of developing a method of matrix-assisted laser desorption/ionization (MALDI) of atherosclerostic sample as plaque or serum to identify lipoproteins responsible for atherogenesis. The MALDI signal as mass and charge ratio (*m*/*z*) was digitized as digital distribution appearing image slice. The method was based on iterated manual method to catch narrow range of 2–3 proteins as peak(s) (see Fig. 27.13). By Monte-Carlo simulation, these generate a distribution 3D map of proteomics digits around the carotid artery wall surface. Our idea was

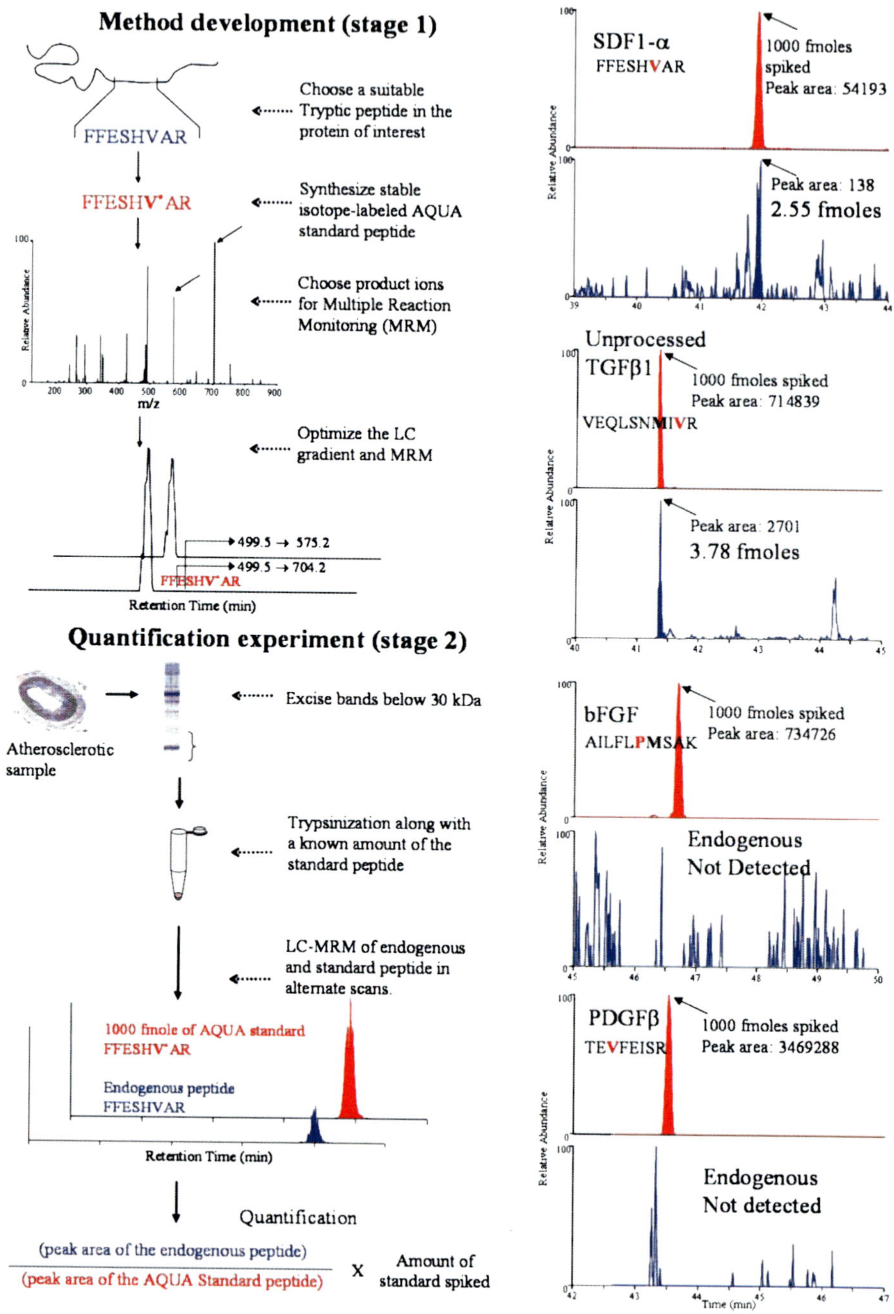

Fig. 27.13 The MALDI imaging offers the distribution of atherogenic proteins at different locations of artery walls. The carotid artery tissue sections are placed on MALDI steel plate and fired for matrix-assisted amplified laser emission desorption as shown in (**a**) on *left*, to record ionization as peaks or *m/z* ratio as shown in (**c**) on *right*. Each peak in protein specific peak has contribution from ionized protons or other selected elements after extracting out specific *m/z*. For quantification of molecules from peaks, trypsinization of plaque content was used followed by peak analysis and area measurement by MUSCOT® software. Reproduced with permission from [126]

to integrating spatially resolved three-dimensional MALDI-IMS image with in vivo MRI image [127]. Visible molecules in plaque seem to be palmitic/palitoleic acid, diacylglycerol in media, HSP-27, vitamin D-binding protein, cytokines/growth factors [126, 127].

What we learn from carotid artery proteomics microimaging?

- Tissue proteomics allows elucidation of the molecular mechanisms of the pathology under study as proteins are effectors of the biological processes taking place in a diseased tissue.
- Although conventional proteomics (2DE) is still the most applied methodology in atherosclerosis research, shotgun proteomics studies are increasing and generating huge novel data (i.e., human cardiac proteins and proteotypic peptides catalogue).
- Isolation of tissue regions/cells by methods like microdissection provides subproteomes that yield more specific information from proteins present in a certain area of the atherome plaque and their levels of expression at this location.
- Aortic stenosis and atherosclerosis are strongly associated since they share clinical risk factors and biological pathways of pathogenesis. In addition, common altered proteins between these diseases have been identified.
- The proteome of human coronary arteries were identified as well as the subproteomes of arterial layers (intima, media, and adventitia) for the first time.
- Novel proteins and proteoglycans involved in atherosclerosis have been identified by proteomic approaches.
- Toll-like receptor-2 ligand-specific activation triggers atherosclerosis, reinforcing the link between innate immunity, inflammation, and atherosclerosis. Pharmacological manipulation of Toll-like receptor-2 pathways has a great therapeutical potential.
- Proteomic profiles (obtained by 2DE or protein microarrays) allow distinguishing unstable from stable atherosclerotic lesions adding new criteria that can complement histological classification.
- The identification of aldehyde dehydrogenase 2(Ada2), as implicated in ischemic preconditioning, points out the value of translational medicine for clinical purposes since the activation of this protein by Ada2 may form the basis of a potential therapeutic approach to prevent or minimize myocardial ischemic injury in the clinical setting [127].
- Most mitochondrial proteins are altered in ischemic hearts. By contrast, in ischemic preconditioning, activation of Protein Kinase (PKC) isozymes preserves mitochondrial function.

27.5.7 *Microfluidics in Carotid Artery Disease*

The external carotid artery (ECA) flow, waveform, and occlusion geometry as hemodynamic wall parameters were reported associated with intimal hyperplasia and atherosclerosis measured with wall shear stress angle gradient (WSSAG) and

Table 27.9 Carotid artery MRA images in different sessions were screened for measurement of wall shear stress, wall shear stress angle gradient and oscillatory shear index. At different branches, the wall shear values change as shown in Fig. 27.13 and in table below

Condition	Wall shear stress (dyne/cm^2)	Wall shear stress angle gradient (radians/cm)	Oscillatory shear index
Normal artery	0	3.0–3.5	0.22–0.49
CCA	4	0.3	0.04
ICA	0.365	0.79	0.037–0.035
ECA	0.35	0.75	0.019–0.017

oscillatory shear index (OSI) shown in Table 27.9 (see also Fig. 27.14). Validated finite volume-based algorithm CFX 4.4 with the SIMPLEC algorithm were applied for pressure correction and algebraic multigrid scheme-based discretization Navier–Stokes equations were applied on structured vascular meshes with speed-up of the iterations for computational simulations. Location of artery occlusion (distal, proximal, stump, smooth) does not affect adverse carotid hemodynamics applicable to carotid sinus but carotid bulb flow is affected by ECA [16–18, 20].

27.5.8 3D Echographic Data Segmentation to Evaluate Carotid Artery Turbulence

By analogy with variational approach in image processing, Doppler imaging regularization was solved in the framework of calculus of variations. Using autoregressive mean curvature flow and PDE numerical schemes of reflection coefficient $\mu_{n=}a_n^{(n)}$ as shown in Eq. 27.9, Doppler imaging displayed the PDE-derived evolution of time spectrum densities as patterns of carotid artery as shown in Fig. 27.14.

$$\partial\mu_n / \partial t = (-i2\pi)^2 n^2 \rho_\upsilon \mu_n + (-i2\pi)_n \mathrm{P}_{n-1} \mathrm{J}_{n-1} \left(A_{n_1} + \mu_n A_{n-1}\right), \tag{27.9}$$

where $\partial\mu_n/\partial t$ is anisotropic heat equation.

The spectrum patterns visualize the curvatures in the artery wall inner surface to evaluate severity of stenosis (see first row in Fig. 27.3).

27.6 Limitations of Techniques in Evaluation of CAD

After first visit of a person with risk of CAD disease, lab tests, imaging modalities are time-consuming and available at different platforms. The measurements may not be conclusive because of limited sensitivity and in vivo measurement limits of EKG, US, MRI, CT, and other molecular imaging techniques. The molecular events inferred from images or serum may not be true or conclusive to define the stage of disease. Molecular imaging still is in infancy to make better diagnostic or therapeutic monitoring. Disease itself is complex due to cardiac incapacity and defined as overlap of several molecular deposits as

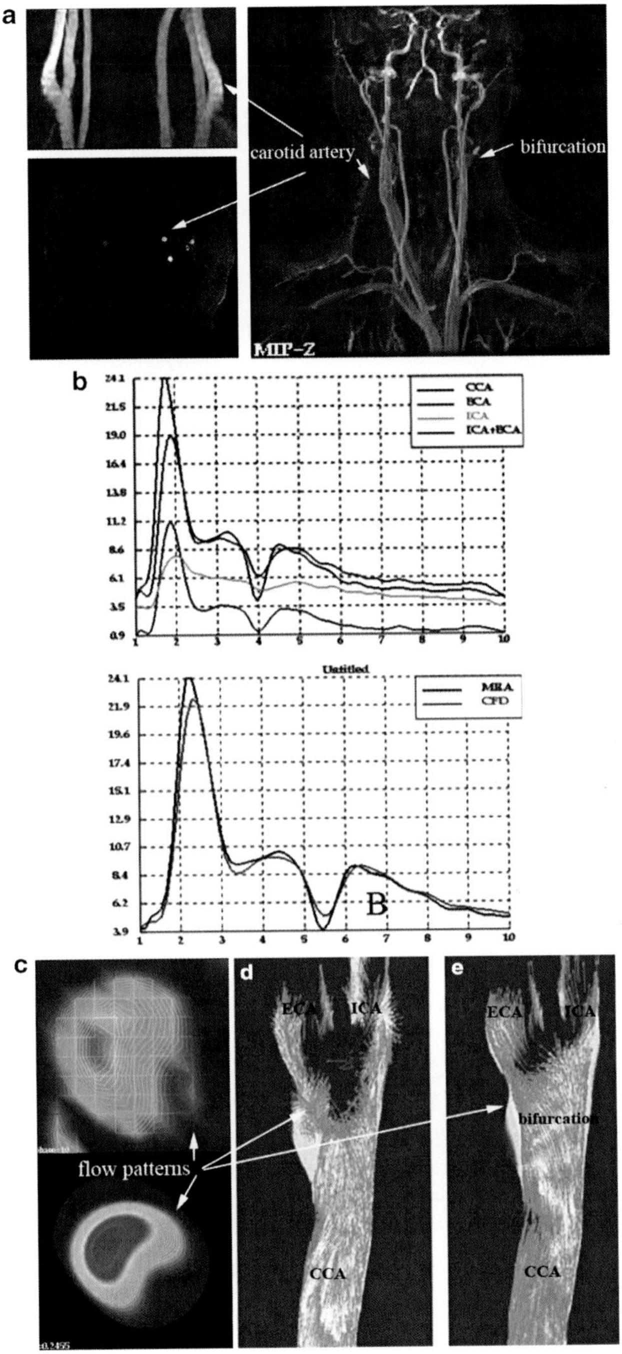

Fig. 27.14 *Left*: (**a**) MR angiogram is shown with carotid artery bright-blood image on Maximum Intensity projection. (**b**) Flow waveform graph of different carotid artery branches were measured with phase contrast MRA (*top*) and compliance of measured with computed waveform (*bottom*). (**c**) Flow waveform digitized velocity profiles measured (*top*) and compliance with computed waveform (*bottom*). (**d–e**) Lateral view of flow in two conditions of occluded (**d**) and normal (**e**) artery. Notice the change in velocity encoding WSSAG and OSI signals in (**d**) vs. (**e**). Modified with permission from [20]

a result of mixed or parallel metabolic imbalances. It all needs real-time evaluation of molecular events and precise measurement of structural changes due to disease.

27.7 Future of Molecular Imaging of Carotid Artery Disease and Nanotechnology

Now trend is developing toward rapid noninvasive multimodal and multifunctional techniques of imaging and mapping molecules in intact tissues to construct molecular geography. Certainly nanotechnology plays significant role in design of imaging contrast agents, execute them and monitor their action as metabolic sensitive nanoparticles to visualize time-series changes due to disease progress or therapeutic drugs. In other direction, efforts are measuring events on images at picomolar levels with three-dimensional spatial information of active molecules in artery such as inflammation marker MMPs and apoptosis marker cathapsin enzymes. Perhaps, nanoscale technology will be replaced by picomolar technology to give insight of metabolite turnover in picomoles across membrane pores, cells barriers.

27.8 Conclusion

Combined physical examination, lab tests, imaging with drug response evaluation overall offers excellent presurgery guideline to the physician if the patient needs surgery and postsurgery follow-up. Our experimental observations on imaging suggested the diagnostic accuracy of plaque characterization, plaque classification, and role of emerging new molecular imaging techniques in extraction of better molecular insight and 3D structural–functional events. A simple criteria of presurgery and postsurgery evaluation of CAD is described to monitor the stage of disease progress or drug-induced regression and disease management using imaging, histopathology, molecular staining. The merging molecular imaging techniques are highlighted as potential diagnostic and therapeutic monitoring tools in evaluation. The chapter presents handful information as source of decision making roadmap illustrated in tables, figures as hands-on-work tips to physicians, nurses, health workers, nurses, and surgery staff.

References

1. Estol CJ. Dr C. Miller Fisher and the history of carotid artery disease, Stroke, 27, (1996), 559–566.
2. Stary HC, Bleakley C, Dinsmore RE, Fuster V, Glagov S, Instill W, Rosenfield ME, Schwartz CJ, Wagner WD, Wissler RW. A definition of advanced types of atherosclerotic lesions and a histological classification of atherosclerosis, Arterioscler Thromb Vasc Biol, 15(9), (1995), 1512–1531.

3. Cai JM, Hatsukami TS, Ferguson MS, Small R, Polissar NL, Yuan C. Classification of human carotid atherosclerosis lesions with in vivo multicontrast magnetic resonance imaging, Circulation, 106, (2002), 1368–1373.
4. Pasternak RC, Smith SC, Bairey-Merz CN, Grundy SM, Cleeman JI, Lenfant C, American College of Cardiology, American Heart Association, National Heart, Lung and Blood Institute. ACC/AHA, NHLBI clinical advisory on the use and safety of statins, J Am Coll Cardiol, 40, (2002), 567–572.
5. Silvera SS, Aidi HE, Rudd JH, Mani V, Yang L, Farkouh M, Fuster V, Fayad ZA. Multimodality imaging of atherosclerotic plaque activity and composition using FDG-PET/CT and MRI in carotid and femoral arteries, Atherosclerosis, 207(1), (2009), 139–143.
6. Turu MM, Krupinski J, Catena E, Rosell A, Montaner J, Rubio F, Sabin JA, Cairols M, Badimon L. Intraplaque MMP-8 levels are increased in asymptomatic patients with carotid plaque progression on ultrasound, Atherosclerosis, 187(1), (2006), 161–169.
7. Koenig W, Khuseyinova N. Biomarkers of atheroclerosis plaque instability and rapture. Atheroscler Thrmob Vascu Biol, 27, (2007), 15–26.
8. Kleinstreuer C, Nazemi M, Archie JP. Hemodynamics analysis of a stenosed carotid bifurcation and its plaque mitigating design, J Biomech Eng, 113 (3), (1991), 330–335.
9. Davies MJ, Richardson PD, Woolf N, Katz DR, Mann J. Risk of thrombosis in human atherosclerotic plaques: role of extracellular lipid, macrophages, and smooth muscle cell content, Br Heart J, 69, (1993), 377–381.
10. Seeger MD, Barratt BS, Lawson GA, Klingman N. The relationship between carotid plaque composition, plaque morphology, and neurologic symptoms, J Surg Res, 58, (1995), 330–336.
11. Bortolani EM, Ghilardi G, Pizzocari P, Coppini P, Longhi F, Trimarchi S. Surgery of the carotid: the morphology of plaque and the clinical correlations, Minerva Cardioangiol, 40(10), (1992), 369–374.
12. Wasserman BA, Haacke EM, Debiao L. Carotid plaque formation and its evaluation with angiography, ultrasound, and MR angiography, J Mag Res Imaging, 4, (1994), 515–527.
13. Naghavi M, Libby P, Falk E, et al. From vulnerable plaque to vulnerable patient: a call for new definitions and risk assessments strategies, Circulation, 108, (2003), 1664–1672.
14. Naghavi M, Libby P, Falk E, et al. From vulnerable plaque to vulnerable patient: a call for new definitions and risk assessments strategies, Circulation, 108, (2003), 1772–1778.
15. Kyriacou E, Pattichis CS, Pattichis MS, Mavrommatis A, Panagiotou S, Christodoulou CI, et al. Classification of atherosclerotic carotid plaques using gray level morphological analysis on ultrasound images. In: Maglogiannis I, Karpouzis K, and Bramer M (Eds) IFIP Internal Federation for Information Processing. V204, Artificial Intelligence Applications and Innovations, 2006, Boston, Springer, pp. 737–744.
16. Hyun S, Kleinstreuer C, Archie JP. Computational analysis of effects of external carotid artery flow and occlusion on adverse carotid bifurcation hemodynamics, J Vasc Surg, 37(6), (2003), 1248–1254.
17. Glor FP, Ariff B, Hughes AD, Crowe LA, Verdonck PR, Barratt DC, McG Thom SA, Firmin DN, Xu XY. Image-based carotid flow reconstruction: a comparison between MRI and ultrasound, Physiol Meas, 25, (2004), 1495–1509.
18. Sui B, Gao P, Lin Y, Gao B, Liu L. Blood flow pattern and wall shear stress in the internal carotid arteries of healthy subjects, Acta Radiol, 49(7), (2008), 806–814.
19. Cebral JR, Yim PJ, Lohner R, Soto O, Choyke PL. Blood flow modeling in carotid arteries with computational fluid dynamics and MR imaging, Acad Radiol, 9, (2002), 1286–1299.
20. Milner JS, Moore JA, Rutt BK, Steinman DA. Hemodynamics of human carotid artery bifurcations: computational studies with models reconstructed from magnetic resonance imaging of normal subjects, J Vasc Surg, 28(1), (1998), 143–156.
21. Sharma, R. MR Imaging in carotid artery atherosclerosis plaque characterization, Magn Reson Med Sci, 1(4), (2002), 217–233.
22. Sharma, R. A device for MR imaging of atherosclerosis plaque in carotid endarterectomy specimens ex vivo, Magn Reson Med Sci, 1(1), (2002), 116–124.

23. Morrisett J, Vick W, Sharma R, Lawrie G, Reardon M, Ezell E, Schwartz J, Hunter G, Gorenstein D. Discrimination of components in atherosclerotic plaques from human carotid endarterectomy specimens by MRI in vivo, J Magn Reson Imaging, 21(5), (2003), 468–474.
24. Sharma R, Singh RB, Gupta RK. A Segmentation Method for Carotid Artery Atherosclerosis Plaque for MRI Contrast and MRI Features, Oxidative Stress Markers in Coronary and Carotid Plaque 16th IEEE Symposium on Computer-Based Medical Systems (CBMS'03) p. 323. http://ieeexplore.ieee.org/iel5/8605/27272/01212809.pdf
25. North American Symptomatic Carotid Endarterectomy Trial (NASCET) Steering Committee. North American Symptomatic Carotid Endarterectomy trial. Methods, patient characteristics, and progress, Stroke, 22, (1991), 711–720.
26. Anderson CM, Saloner D, Lee RE, et al. Assessment of carotid artery stenosis by MR angiography: comparison with X-ray angiography and color-coded Doppler ultrasound, AJNR Am J Neuroradiol, 13, (1992), 989–1003.
27. Masaryk AM, Ross JS, DiCello MC, Modic MT, Paranandi L, Masaryk TJ. 3DFT MR Angiography of the carotid bifurcation, Radiology, 179, (1991), 797–804.
28. Anderson CM, Lee RE, Levin DL, Alonso ST, Saloner D. Measurement of internal carotid artery stenosis from MR angiograms, Radiology, 193(1), (1994), 219–226.
29. Korosec FR, Grist TM, Polzin JA, Weber DM, Mistretta CA. MR angiography using velocity-selective preparation pulses and segmented gradient-echo acquisition, Magn Reson Med, 30, (1993), 704–710.
30. Yuan C, Murakami JW, Hayes CE, Tsuruda JS, Hatsukami TS, Wildy KS, Ferguson MS, Strandness DE. Phased-array magnetic resonance imaging of the carotid artery bifurcation: preliminary results in health volunteers and a patient with atherosclerotic disease, J Mag Res Imaging, 5, (1995), 561–565.
31. Liffers A, Quick HH, Herborn CU, Ermert H, Ladd ME. Geomatrical optimization of a phased array coil for high resolution MR imaging of carotid arteries, Magn Reson Med, 50(2), (2003), 439–443.
32. Wildy KS, Yuan C, Tsuruda JS, Ferguson MS, Wen N, Subramanian DS, Strandess DE. Atherosclerosis of the carotid artery: evaluation by magnetic resonance angiography, J Magn Reson Imaging, 6, (1996), 726–732.
33. Shinnar M, Fallon JT, Wehrli S, Levin M, Dalmacy D, Fayad ZA, Badimon JJ, Harrington E, Fuster V. Diagnostic accuracy of ex vivo MRI for human atherosclerotic plaque characterization, Arterioscler Thromb Vasc Biol, 19, (1999), 2756–2761.
34. Skinner MP, Yuan C, Mitsumori L, Hayes CE, Raines EW, Nelson JA, Ross R. Serial magnetic resonance imaging of experimental atherosclerosis detects lesion fine structure, progression and complications in vivo, Nat Med, 1(1), (1995), 69–73.
35. Meindl S, Jungke M, Bielke G, von Seelan W, Grigat M, Pedrosa P, Higer HP. Using an "Information Manager" as a component of a tissue classification system in NMR tomography. In: Higer HP and Bielke G (Eds) Tissue Characterization in MR Imaging, Berlin, Springer-Verlag, (1990), 139–144.
36. Berr SS, Hurt NS, Ayers CR, Snell JW, Merickel MB. Assessment of the reliability of the determination of carotid artery lumen sizes by quantitative image processing of MR angiograms and images, Magn Reson Imaging, 13, (1995), 827–835.
37. Sharma R, Singh RB. MRI of coronary artery atherosclerosis in rabbits: histopathology MRI correlation and atheroma characterization, Thromb J, 2(1), (2004), 5.
38. Kass M, Witkin A, Terzopoulos D. "SNAKES: Active Contour Models", Proceedings of First International Conference on Computer Vision, 1987, pp. 259–269.
39. Yuan C, Lin E, Hwang JN. Closed contour edge detection of blood vessel lumen and outer wall boundaries in black-blood MR images, Magn Reson Imaging, 17, (1999), 257–266.
40. Kroon AA, van Asten WNJC, Stalenhoef AFH: Effect of apheresis of low-density lipoprotein on peripheral vascular disease in hypercholesterolemic patients with coronary artery disease, Ann Intern Med, 125, (1996), 945–954.
41. Schreiner PJ, Heiss G, Tyroler HA, Morrisett JD, Davis CE, Smith R. Race and gender differences in the association of Lp(a) with carotid artery wall thickness, Arterioscler Thromb Vasc Biol, 16(3), (1996), 471–478.

42. Hodis HN, Mack WJ, LaBree L, Selzer RH, Liu C, Alanpovic P, Kwong_Fu H, Azen SP. Reduction in carotid arterial wall thickness using lovastatin and dietary therapy, Ann Intern Med, 143, (1996), 548–556.
43. Mercuri M, Bond MG, Sintori CR, Veglia F, Crepaldi G, Ferugho FS, Descovich G, Ricci G, Rubba P, Mancini M, et al. Pravastatin reduces carotid intima media thickness progression in asymptomatic hypercholesterolemic mediterannean population: The Carotid Atheroslcerosis Italian Ultrasound Study, Am J Med 101, (1996), 627–634.
44. Suurkla M, Agewall S, Fagerberg B, Wendelhag I, Wikstrand J, for the Risk Intervention Study (RIS) Group. Multiple risk intervention in high risk hypertensive patients: a 3 year ultrasound study of intima-media thickness and plaques in the carotid artery, Arterioscler Thromb Vasc Biol, 16, (1996), 462–470.
45. Furberg CD, Adams HP Jr, Applegate WB, Byington RP, Espeland MA, Hartwell T, Hunninghake DB, Leflkowitz DS, Probstfield J, Riley WA, et al., for the Asymptomatic Carotid Artery Progression Study (ACAPS) Research Group. Effect of lovastatin on early carotid atherosclerosis and cardiovascular events, Circulation, 90, 1994, 1679–1687.
46. Crouse JR III, Byington RP, Bond MG, Espeland MA, Craven TE, Sprinkle JW, McGovern ME, Furberg CD. Pravastatin, lipids, and atherosclerosis in the carotid arteries (PLAC-11), Am J Cardiol, 75, 1995, 455–459.
47. Salonen R, Nyyssönen K, Porkkala E, Rummukainen J, Belder R, Park J-S, Salonen JT. Kuopio Atherosclerosis Prevention Study (KAPS): a population-based primary preventive trial of the effect of LDL lowering on atherosclerotic progression in carotid and femoral arteries, Circulation, 92, (1995), 1758–1764.
48. Crisby M, Fredricksson G, Shah PK, Yano J, Zhu J, Nilsoon J. Pravastatin treatment increases collagen content and decreases lipid content, inflammation, metalloproteinases, and cell death in human carotid plaques: implications for plaque stabilization, Circulation, 103(7), (2001), 926–933.
49. Goldstein H. Multilevel Statistical Models, Second Edition. New York, Wiley Interscience, (1995), 121–136.
50. Huttner HJM, van den Eeden P. The Multilevel Design: A Guide with Annotated Bibliography, 1980–1993. Westort, CO, Greenwood Press, 1995.
51. Kreft I, deLeeuw J. Introducing Multilevel Modeling. Thousand Oaks, CA, Sage Pub, 1998.
52. Insull W Jr. The problem of compliance to cholesterol altering therapy, J Intern Med, 241, (1997), 317–325.
53. Adams GJ, Simoni DM, Bordelon CB, Vick W, Kimball KT, Insull W, Morrisett JD. Bilateral symmetry of human carotid artery atherosclerosis, Stroke, 222, (2003), 2575–2580.
54. King R, Dalquist HD. Method of performing endarterectomy. US patent 4962755, issue date Oct 16, 1990.
55. McCann RL. Surgical management of carotid artery atherosclerotic disease, South Med J, 86(10), 1993, 2S23–2S28.
56. Guo W, Morrisett JD, DeBakey ME, Lawrie GM, Hamilton JA. Quantitation in situ of crystalline cholesterol and calcium phosphate hydroxyapatite in human atherosclerotic plaques by solid state magic angle spinning NMR, Arterioscler Thromb Vasc Biol, 20, (2000), 2682–2688.
57. Guo W, Morrisett JD, Lawrie GM, DeBakey ME, Hamilton JA. Identification of different lipid phases and calcium phosphate deposits in human carotid artery plaques by MAS NMR spectroscopy, Mag Reson Med, 39, (1998), 184–189.
58. Sharma R, Shriniwas BD, Sharma A. Human Carotid Artery: Segmentation Methods in Atherosclerosis Imaging. World Heart Research Book. Edn Halberg F, Singh RB, (2009); V 1, 193–212.
59. Sharma R, Sharma A. Segmentation methods in atherosclerosis vascular imaging, Informa Med Slov, 11(2), (2006), 52–69.
60. Burke AP, Farb A, Malcolm GT, Liang Y-H, Smialek JE, Virmani R. Plaque rupture and sudden death related to exercise in men with coronary artery disease, JAMA 281, (1999), 921–926.
61. Choudhary S, Higgins CL, Chen IY, Reardon M, Lawrie G, Vick GW III, Karmonik C, Via DP, Morrisett JD. Quantitation and localization of matrix metalloproteinases and their inhibitors in human carotid endarterectomy tissues, Arterioscler Thromb Vasc Biol, 26(10), (2006), 2351–2358.

62. Welgus HM, Campbell EJ, Cury JD, Eisen AZ, Senior RM, Wilheim SM, Goldberg GI. Neutral metalloproteinases produced by human mononuclear phagocytes during cellular development: enzyme profile, regulation, and expression, J Clin Invest, 86, (1990), 1496–1502.
63. Ling ZL, Ziekle R, Cheng L, Xiao R, Crow MT, Stetler-Stevenson WG, Froelich J, Lakatta EG. Increased expression of 72kD type IV collagenase (MMP-2) in human atherosclerotic lesions, Am J Pathol, 148, (1996) 121–128.
64. Galis ZS, Sukhova GK, Lark MW, Libby P. Increased expression of matrix metalloproteinases and matrix degrading activity in vulnerable regions of human atherosclerotic plaque, J Clin Invest, 942, (1994), 2493–2503.
65. Sharma R. Katz JK. Preliminary studies on human aldosterone synthase (CYP11B2) gene polymorphisms, metalloprotease-9, apoptosis, and carotid atherosclerosis plaque size by proton MRI, J Renin Angiotensin Aldosterone Sys, 11(3), (2009), 198–204.
66. Nikkari ST, O'Brien KD, Ferguson M, Hatsukami T, Welgus HG, Algers CE, Clowes AW. Interstitial collagenase (MMP-1) expression in human carotid atherosclerosis, Circulation, 92, (1995), 1393–1398.
67. Oksala N, Levula M, Airla N, Pelto-Huikko M, Ortiz RM, Järvinen O, Salenius JP, Ozsait B, Komurcu-Bayrak E, Erginel-Unaltuna N, Huovila AP, Kytömäki L, Soini JT, Kähönen M, Karhunen PJ, Laaksonen R, Lehtimäki T. ADAM-9, ADAM-15, and ADAM-17 are upregulated in macrophages in advanced human atherosclerotic plaques in aorta and carotid and femoral arteries – Tampere vascular study, Ann Med, 41(4), (2009), 279–290.
68. Feng Y, Yang JH, Huang H, Kennedy SP, Turi TG, Thompson JF, Libby P, Lee RT. Transcriptional profile of mechanically induced genes in human vascular smooth muscle cells. Circ Res, 85, (1999), 1118–1123.
69. Nuotio K, Isoviita PM, Saksi J, IJas P, Pitkaniemi J, Sonninen R, Soinne L, Saimanen E, Salonen O, Kovanen PT, Kaste M, Lindsberg PJ. Adipophilin expression is increased in symptomatic carotid atherosclerosis, Stroke, 38, (2007), 1791.
70. Hiltunen MO, Tuomisto TT, Niemi M, Bräsen JH, Rissanen TT, Törönen P, Vajanto I, Ylä-Herttuala S. Changes in gene expression in atherosclerotic plaques analyzed using DNA array, Atherosclerosis, 165(1), (2002), 23–32.
71. Nakai K, Oyanagi M, Hitomi J, Ogasawara K, Inoue T, Kobayashi M, Nakai K, Suwabe A, Habano W, Baba T, Yoshida H, Ogawa A. Screening the single nucleotide polymorphisms in patients with internal carotid artery stenosis by oligonucleotide-based custom DNA array, Bioinf Biol Insights, 1(1), (2007), 63–69.
72. Sharma R, Katz JK. Taxotere chemosensitivity evaluation in mice prostate tumor: validation and diagnostic accuracy of quantitative measurement of tumor characteristics by MRI, PET, and histology of mice tumor, Technol Cancer Res Treat, (7)3, (2008), 155–268.
73. Papazoglou T, Papaioannou T, Arakawa K, Fishbein M, Marmarelis VZ, Grundfest WS. Control of excimer laser aided tissue ablation via laser-induced fluorescence monitoring, Appl Opt, 29, (1990), 4950–4955.
74. Friedlander AH, Freymiller EG. Detection of radiation accelerated atherosclerosis of the carotid artery by panoramic radiography, J Am Dent Assoc, 134(10), (2003), 1361–1365.
75. Shriniwas BD, Sharma R, Sharma A. Extended applications of picotechnology to measure immunoactive biomarkers. Nanotech NSTI 2009 Conference and EXPO Houston, 2009 May 3 WE62.506.
76. Arakawa K, Isoda K, Ito T, Nakajima K, Shibuya T, Ohsuzu F. Fluorescence analysis of biochemical constituent identifies atherosclerostic plaques with a thin fibrous cap, Arterioscler Thromb Vasc Biol, 22, (2002), 1002–1007.
77. Chang K, Jaffer F. Advances in fluorescence imaging of the cardiovascular system, J Nucl Cardiol, 15(3), (2008), 417–428.
78. Lu H, Rateri DL, Daugherty A. Immunostaining of mouse atherosclerotic lesions. In: Sreejayan N and Ren J (Eds) Vascular Biology Protocols, Methods in Mol Med, vol 139, 2007, Torowa, NJ, Humana Press Inc., pp. 77–94.
79. Dollery CM, Owen CA, Sukhova GK, Krettek A, Shapiro SD, Libby P. Neutrophil elastase in human atherosclerotic plaques: production by macrophages, Circulation, 107(22), (2003), 2829–2836.

80. Callahan RJ, Bogdanov A Jr, Fischman AJ, Brady TJ, Weissleder R. Preclinical evaluation and phase I clinical trial of a ^{99m}Tc-labeled synthetic polymer used in blood pool imaging, AJR Am J Roentgenol, 171(1), (1998), 137–143.
81. Jaffer FA, Kim DE, Quinti L, Tung CH, Aikawa E, Pande AN, Kohler RH, Shi GP, Libby P, Weissleder R. Optical visualization of cathepsin K activity in atherosclerosis with a novel, protease-activatable fluorescence sensor, Circulation, 115(17), (2007), 2292–2298.
82. Jaffer FA, Vinegoni C, John MC, Aikawa E, Gold HK, Finn AV, Ntziachristos V, Libby P, Weissleder R. Real-time catheter molecular sensing of inflammation in proteolytically active atherosclerosis, Circulation, 118(18), (2008), 1802–1809.
83. Deguchi JO, Aikawa M, Tung CH, Aikawa E, Kim DE, Ntziachristos V, Weissleder R, Libby P. Inflammation in atherosclerosis: visualizing matrix metalloproteinase action in macrophages in vivo, Circulation, 114(1), (2006), 55–62.
84. Blum G, von Degenfeld G, Merchant MJ, Blau HM, Bogyo M. Noninvasive optical imaging of cysteine protease activity using fluorescently quenched activity-based probes, Nat Chem Biol, 3(10), (2007), 668–677.
85. Ntziachristos V, Schellenberger EA, Ripoll J, Yessayan D, Graves E, Bogdanov A Jr, Josephson L, Weissleder R. Visualization of antitumor treatment by means of fluorescence molecular tomography with an annexin V-Cy5.5 conjugate, Proc Natl Acad Sci USA, 101(33), (2004), 12294–12299.
86. Shepherd J, Hilderbrand SA, Waterman P, Heinecke JW, Weissleder R, Libby P. A fluorescent probe for the detection of myeloperoxidase activity in atherosclerosis-associated macrophages, Chem Biol, 14(11), (2007), 1221–1231.
87. Park K, Hong HY, Moon HJ, Lee BH, Kim IS, Kwon IC, Rhee K. A new atherosclerotic lesion probe based on hydrophobically modified chitosan nanoparticles functionalized by the atherosclerotic plaque targeted peptides, J Control Release, 128(3), (2008), 217–223.
88. Cheng Z, Levi J, Xiong Z, Gheysens O, Keren S, Chen X, Gambhir SS. Near-infrared fluorescent deoxyglucose analogue for tumor optical imaging in cell culture and living mice, Bioconjug Chem, 17(3), (2006), 662–669.
89. Kovar JL, Volcheck W, Sevick-Muraca E, Simpson MA, Olive DM. Characterization and performance of a near-infrared 2-deoxyglucose optical imaging agent for mouse cancer models, Anal Biochem, 384(2), (2009), 254–262.
90. Kooi ME, Cappendijk VC, Cleutjens KB, Kessels AG, Kitslaar PJ, Borgers M, Frederik PM, Daemen MJ, van Engelshoven JM. Accumulation of ultrasmall superparamagnetic particles of iron oxide in human atherosclerotic plaques can be detected by in vivo magnetic resonance imaging, Circulation, 107(19), (2003), 2453–2458.
91. Trivedi RA, Mallawarachi C, U-King-Im JM, Graves MJ, Horsley J, Goddard MJ, Brown A, Wang L, Kirkpatrick PJ, Brown J, Gillard JH. Identifying inflamed carotid plaques using in vivo USPIO-enhanced MR imaging to label plaque macrophages, Arterioscler Thromb Vasc Biol, 26(7), (2006), 1601–1606.
92. Tang TY, Howarth SP, Li ZY, Miller SR, Graves MJ, U-King-Im JM, Trivedi RA, Walsh SR, Brown AP, Kirkpatrick PJ, Gaunt ME, Gillard JH. Comparison of the inflammatory burden of truly asymptomatic carotid atheroma with atherosclerotic plaques contralateral to symptomatic carotid stenosis: an ultra small superparamagnetic iron oxide enhanced magnetic resonance study, J Neurol Neurosurg Psychiatry, 78(12), (2007), 1337–1343.
93. Nahrendorf M, Zhang H, Hembrador S, Panizzi P, Sosnovik DE, Aikawa E, Libby P, Swirski FK, Weissleder R. Nanoparticle PET-CT imaging of macrophages in inflammatory atherosclerosis, Circulation, 117(3), (2008), 379–387.
94. Nahrendorf M, Jaffer FA, Kelly KA, Sosnovik DE, Aikawa E, Libby P, Weissleder R. Noninvasive vascular cell adhesion molecule-1 imaging identifies inflammatory activation of cells in atherosclerosis, Circulation, 114(14), (2006), 1504–1511.
95. McCarthy JR, Patel P, Botnaru I, Haghayeghi P, Weissleder R, Jaffer FA. Multimodal nanoagents for the detection of intravascular thrombi, Bioconjug Chem, 20(6), (2009), 1251–1255.
96. Cormode DP, Skajaa T, Fayad ZA, Mulder WJ. Nanotechnology in medical imaging: probe design and applications, Arterioscler Thromb Vasc Biol, 29(7), (2009), 992–1000.

97. Schmieder AH, Winter PM, Caruthers SD, Harris TD, Williams TA, Allen JS, Lacy EK, Zhang H, Scott MJ, Hu G, Robertson JD, Wickline SA, Lanza GM. Molecular MR imaging of melanoma angiogenesis with alphanubeta3-targeted paramagnetic nanoparticles. Magn Reson Med, 53(3), (2005), 621–627.
98. Medarova Z, Pham W, Farrar C, Petkova V, Moore A. In vivo imaging of siRNA delivery and silencing in tumors, Nat Med., 13(3), (2007), 372–377.
99. Demos SM, Alkan-Onyuksel H, Kane BJ, Ramani K, Nagaraj A, Greene R, Klegerman M, McPherson DD. In vivo targeting of acoustically reflective liposomes for intravascular and transvascular ultrasonic enhancement, J Am Coll Cardiol, 33(3), (1999), 867–875.
100. Flacke S, Fischer S, Scott MJ, Fuhrhop RJ, Allen JS, McLean M, Winter P, Sicard GA, Gaffney PJ, Wickline SA, Lanza GM. Novel MRI contrast agent for molecular imaging of fibrin: implications for detecting vulnerable plaques, Circulation, 104(11), (2001), 1280–1285.
101. Neubauer AM, Sim H, Winter PM, Caruthers SD, Williams TA, Robertson JD, Sept D, Lanza GM, Wickline SA. Nanoparticle pharmacokinetic profiling in vivo using magnetic resonance imaging, Magn Reson Med, 60(6), (2008), 1353–1361.
102. Frias JC, Williams KJ, Fisher EA, Fayad ZA. Recombinant HDL-like nanoparticles: a specific contrast agent for MRI of atherosclerotic plaques, J Am Chem Soc, 126(50), (2004), 16316–16317.
103. Li H, Gray BD, Corbin I, Lebherz C, Choi H, Lund-Katz S, Wilson JM, Glickson JD, Zhou R. MR and fluorescent imaging of low-density lipoprotein receptors, Acad Radiol, 11(11), (2004), 1251–1259.
104. Hayek SS, Sharma R, Kwon S, Sharma A, Chen CJ. Temperature and magnetic resonance characteristics of zinc, manganese, gadolinium, gold, iron magnetic nanoparticles and cytokine synergy in hyperthermia, J Biomed Sci Eng, 1(3), (2008), 147–209.
105. Sharma R, Sharma A, Chen CJ. Temperature and magnetic moment characteristics of MNB$_{0.5}$BZNB$_{0.5}$BGDBXBFEB$_{(2\text{-}X)}$BOB$_4$B magnetic nanoparticles in Hyperthermia, Nanotech Res J, 2(3), (2008), 26–45.
106. Morawski AM, Winter PM, Yu X, Fuhrhop RW, Scott MJ, Hockett F, Robertson JD, Gaffney PJ, Lanza GM, Wickline SA. Quantitative "magnetic resonance immunohistochemistry" with ligand-targeted (19)F nanoparticles, Magn Reson Med, 52(6), (2004), 1255–1262.
107. Louie AY, Hüber MM, Ahrens ET, Rothbächer U, Moats R, Jacobs RE, Fraser SE, Meade TJ. In vivo visualization of gene expression using magnetic resonance imaging, Nat Biotechnol, 18(3), (2000), 321–325.
108. Michalet X, Pinaud FF, Bentolila LA, Tsay JM, Doose S, Li JJ, Sundaresan G, Wu AM, Gambhir SS, Weiss S. Quantum dots for live cells, in vivo imaging, and diagnostics. Science, 307(5709), (2005), 538–544.
109. Allen M, Bulte JW, Liepold L, Basu G, Zywicke HA, Frank JA, Young M, Douglas T. Paramagnetic viral nanoparticles as potential high-relaxivity magnetic resonance contrast agents, Magn Reson Med, 54(4), (2005), 807–812.
110. Winter PM, Cai K, Chen J, Adair CR, Kiefer GE, Athey PS, Gaffney PJ, Buff CE, Robertson JD, Caruthers SD, Wickline SA, Lanza GM. Targeted PARACEST nanoparticle contrast agent for the detection of fibrin, Magn Reson Med, 56(6), (2006), 1384–1388.
111. Sharma R. Nanoparticles that facilitate imaging of biological tissue and methods of forming the same. United States Patent Application 20090220434.
112. Bovey F, Jelinsky L, Mirau P. Nuclear Magnetic Resonance Spectroscopy. San Diego, Academic Press, 1988.
113. Lanza GM, Abendschein DR, Yu X, Winter PM, Karukstis KK, Scott MJ, Fuhrhop RW, Scherrer DE, Wickline SA. Molecular imaging and targeted drug delivery with a novel, ligand-directed paramagnetic nanoparticle technology, Acad Radiol, 9(2), (2002), S330–S331.
114. Oltrona L, Speidel CM, Recchia D, Wickline SA, Eisenberg PR, Abendschein DR. Inhibition of tissue factor-mediated coagulation markedly attenuates stenosis after balloon-induced arterial injury in minipigs, Circulation, 96(2), (1997), 646–652.

115. Hamilton AJ, Huang SL, Warnick D, Rabbat M, Kane B, Nagaraj A, Klegerman M, McPherson DD. Intravascular ultrasound molecular imaging of atheroma components in vivo, J Am Coll Cardiol, 43(3), (2004), 453–460.
116. Botnar RM, Buecker A, Wiethoff AJ, Parsons EC Jr, Katoh M, Katsimaglis G, Weisskoff RM, Lauffer RB, Graham PB, Gunther RW, Manning WJ, Spuentrup E. In vivo magnetic resonance imaging of coronary thrombosis using a fibrin-binding molecular magnetic resonance contrast agent, Circulation, 110(11), (2004), 1463–1466.
117. Caruthers SD, Neubauer AM, Hockett FD, Lamerichs R, Winter PM, Scott MJ, Gaffney PJ, Wickline SA, Lanza GM. In vitro demonstration using 19F magnetic resonance to augment molecular imaging with paramagnetic perfluorocarbon nanoparticles at 1.5 Tesla, Invest Radiol, 41(3), (2006), 305–312.
118. Winter PM, Morawski AM, Caruthers SD, Fuhrhop RW, Zhang H, Williams TA, Allen JS, Lacy EK, Robertson JD, Lanza GM, Wickline SA. Molecular imaging of angiogenesis in early-stage atherosclerosis with alpha(v)beta3-integrin-targeted nanoparticles, Circulation, 108(18), (2003), 2270–2274.
119. Cyrus T, Abendschein DR, Caruthers SD, Harris TD, Glattauer V, Werkmeister JA, Ramshaw JA, Wickline SA, Lanza GM. MR three-dimensional molecular imaging of intramural biomarkers with targeted nanoparticles, J Cardiovasc Magn Reson, 8(3), (2006), 535–541.
120. Corot C, Petry KG, Trivedi R, Saleh A, Jonkmanns C, Le Bas JF, Blezer E, Rausch M, Brochet B, Foster-Gareau P, Balériaux D, Gaillard S, Dousset V. Macrophage imaging in central nervous system and in carotid atherosclerotic plaque using ultrasmall superparamagnetic iron oxide in magnetic resonance imaging, Invest Radiol, 39(10), (2004), 619–625.
121. Trivedi RA, U-King-Im JM, Graves MJ, Cross JJ, Horsley J, Goddard MJ, Skepper JN, Quartey G, Warburton E, Joubert I, Wang L, Kirkpatrick PJ, Brown J, Gillard JH. In vivo detection of macrophages in human carotid atheroma: temporal dependence of ultrasmall superparamagnetic particles of iron oxide-enhanced MRI, Stroke, 35(7), (2004), 1631–1635.
122. Sirol M, Itskovich VV, Mani V, Aguinaldo JG, Fallon JT, Misselwitz B, Weinmann HJ, Fuster V, Toussaint JF, Fayad ZA. Lipid-rich atherosclerotic plaques detected by gadofluorine-enhanced in vivo magnetic resonance imaging, Circulation, 109(23), (2004), 2890–2896.
123. Perez JM, Josephson L, O'Loughlin T, Högemann D, Weissleder R. Magnetic relaxation switches capable of sensing molecular interactions, Nat Biotechnol, 20(8), (2002), 816–820.
124. Jaffer FA, Libby P, Weissleder R. Optical and multimodal molecular imaging insight into atherosclerosis, Arterioscler Thromb Vasc Biol, 29(7), (2009), 1017–1024.
125. Sharma R, Kwon S. New applications of nanoparticles in cardiovascular imaging, J Exp Nanosci, 2(2), (2007), 139–146.
126. Bagnato C, Thumar J, Mayya V, Hwang S, Zebroski H, Claffey KP, Haudenschild C, Eng JK, Lungren DH, Han DK. Proteomics analysis of human coronary atherosclerostic plaque, Mol Cell Proteomics, 6, (2007), 1088–1102.
127. Sharma R. Troponin T. A search of superparamagnetic iron-oxide bound antitroponin nanoparticle for magnetic resonance imaging, Int J Biological Frontiers, 16(2), (2010), 7–24.

Biographies

Rakesh Sharma, did his masters and Ph.D at Indian Institute of Technology, Delhi and second Ph.D in Biochemistry. He was medical college faculty in India for years 1989–1997 and moved to U.S. He was associate professor at Biomedical Engineering, College of Engineering and National High Magnetic Field Lab, Florida. Now he is affliated scientist at Columbia University, New York, and research professor at Center of Nanobiotechnology at Florida State University, Tallahassee, Florida. His research interests are molecular imaging by MRI, PET and nanoparticles design and application development. He is active in the field of bioimaging technology applied to vascular, cancer and drug therapy assessment with 85 peer reviewed publications.

Jose Katz, MD, Ph.D did his Ph.D at University of Illinois, Chicago. Later he served as professor at Free University. He earned his MD and did his cardiology fellowship at Southwestern University Medical Center. He served as senior consultant and professor. Now he is associated faculty professor at Cardiology, Columbia University, New York and CEO of Dr Katz's Cardiology centers in Manhattan, New Jersey tristate areas. His research interests are medicine, mathematical modeling, EKG, MRI and cancer research. Dr Katz has peer reviewed publications over 100 in his credit

Chapter 28
Ultrasound and MRI-Based Technique for Quantifying Hemodynamics in Human Cardiovascular Systems

Fuxing Zhang and Alex J. Barker

Abstract Blood hemodynamics have been shown to correlate with cardiovascular pathologies, such as plaque formation, atherosclerosis, wall remodeling, and aneurysm formation. Thus, quantification of blood hemodynamics in the presence of pathology is of great interest to both researchers and physicians. This chapter reviews currently available as well as 'development-phase' techniques with great promise in future clinical applications. It presents the basic mechanism by which each technique operates, as well as reviews the capabilities and applications of each technique in research and clinical studies of cardiovascular hemodynamics.

Keywords Blood hemodynamics • Ultrasound • Phase-contrast MRI • Echo PIV • Atherosclerosis • Cardiovascular diseases • Wall shear stress

28.1 Introduction

In the USA, approximately 80 million people, or 36% of the population, are living with cardiovascular diseases [1], among which atherosclerosis remains one of the leading causes. Atherosclerosis is a systemic disease process in which fatty deposits, inflammation, cells, and scar tissue build up within the walls of arteries. Several well-defined risk factors, including hypertension, hyperlipidemia, diabetes mellitus, and tobacco smoking, have been implicated in its pathogenesis [2]. Although the causes of atherosclerosis are multifactorial, local hemodynamics, in particular, abnormal wall shear stress (WSS) and oscillatory WSS, play an important role in the pathogenesis and progression of atherosclerotic plaque [3–8]. Plaque has

F. Zhang (✉)
School of Medicine, Department of Pediatrics, Anschutz Medical Campus, University of Colorado at Denver, 1967 S Josephine Street, Apartment 203, Denver, CO 80210, USA
e-mail: zhangf@colorado.edu

Jasjit S. Suri et al. (eds.), *Atherosclerosis Disease Management*,
DOI 10.1007/978-1-4419-7222-4_28,

been demonstrated to occur more frequently in areas of disturbed flow, like at the bifurcation of vessels and in regions of vessel curvature [9, 10]. Areas of low shear stress have also been shown to be more vulnerable to plaque development and rupture [4, 11, 12].

The vulnerability of plaque to rupture is an important clinical question considering that this may precipitate an acute cardiovascular event. For example, carotid artery plaque rupture can precipitate stroke due to thrombo-embolism in the cerebral circulation. However, the detection and quantification of atherosclerosis at its early stage is difficult since atherosclerosis is typically present for decades before the onset of a cardiovascular disease event or symptoms. Therefore, diagnosing subclinical atherosclerosis is of great interest to physicians. Currently, two modalities, ultrafast computed tomography for imaging of artery classifications and B-mode ultrasound for the measurement of carotid intima–media thickness (IMT), are commonly used in clinical studies to help understand the development and progression of subclinical atherosclerosis. These direct imaging-based methods, however, are only effective when plaque has progressed into its middle or late stage due to their limited imaging resolution.

Local hemodynamics have been shown to be closely related with arterial morphology, and small local geometry changes in vasculature might lead to drastic changes in hemodynamics. Thus, accurate quantification of local hemodynamics change might help detect subclinical atherosclerosis, thus to enhance the understanding of its development and progression, as well as its relationship to subclinical events.

In the past decades, many researchers have put their efforts in developing modalities for quantifying blood flow dynamics information, including velocity, flow rate, WSS, oscillatory shear index (OSI), vorticity, etc. This chapter provides the state-of-the-art review of the techniques which have been successfully applied in clinical research studies or applications. The techniques covered in this review include conventional Doppler, color Doppler mapping, vector Doppler, ultrasound speckle tracking, transverse oscillation (TO), echo particle image velocimetry (Echo PIV), and phase-contrast magnetic resonance imaging (PC-MRI). The review introduces the basic principle of each technique, following the research studies or clinical applications. It discusses the advantages and limitations of those techniques as well.

28.2 Ultrasound Doppler

Ultrasound systems can be used to investigate blood flow in human cardiovascular system by the use of Doppler effect. Johann Christian Doppler first described the physical principles of Doppler about 170 years ago. Currently, the ultrasound Doppler technique has been widely employed in the clinical environment to assess blood flow through the heart and relatively large vessels due to its convenience and inexpensiveness. This section introduces several commonly used ultrasound modalities related with Doppler effect, including conventional ultrasound Doppler, color flow mapping (CFM), and vector Doppler.

28.2.1 Conventional Ultrasound Doppler

Continuous wave (CW) Doppler and pulsed wave Doppler are the two main types of ultrasound Doppler systems in common use. Although both take advantage of the Doppler effect, they differ in many aspects, such as transducer design and operation procedures, signal processing, and display of results. They both have their specific applications since each has its advantages and disadvantages.

28.2.1.1 Continuous Wave Doppler

CW Doppler is one of the simplest techniques for blood flow measurements. The basic principle is to detect the frequency shift between the continuously transmitted and received sinusoidal waves. Thus, the transducer in CW Doppler system is composed of two separate crystals for transmitting and receiving signal continuously. The received signals are processed for real-time display of velocity information.

In order to understand the underlying principle of CW Doppler, it is better to first look at the interaction of ultrasound wave and a moving particle. Figure 28.1 shows an ultrasound transducer transmitting a continuous sinusoidal wave and a moving particle with a velocity v, and θ is the angle between the ultrasound insonation direction and the particle moving direction.

Let us say the transmitted signal $e(t)$ is expressed as,

$$e(t) = \cos(2\pi f_0 t), \tag{28.1}$$

where f_0 is the transmitting frequency. Then the received signal $r(t)$ is written as [13, 14]:

$$r(t) = a \cdot e(\alpha(t - t_0)) = a \cdot \cos(2\pi f_0 \alpha(t - t_0)) \tag{28.2}$$

$$\alpha \approx 1 - 2\frac{v_x}{c}, \tag{28.3}$$

$$\alpha t_0 \approx \frac{2d_0}{c}, \tag{28.4}$$

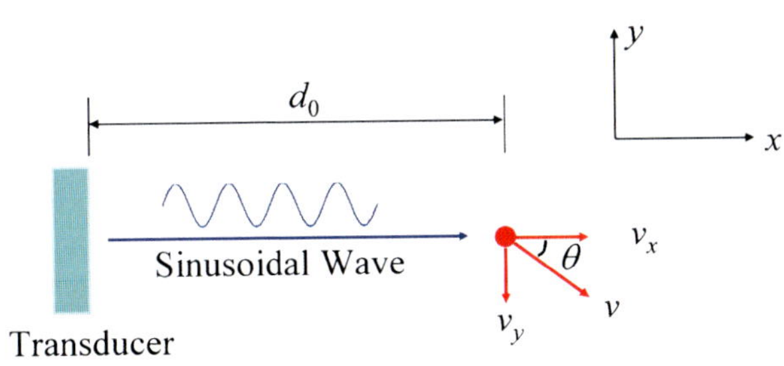

Fig. 28.1 The interaction between an ultrasound wave and a moving particle

where a is a scaling constant, α is the time compression factor, and t_0 is the time from pulse emission to reception.

As is seen, the received signal $r(t)$ is a delayed version of emission signal $e(t)$ with different amplitude and frequency. The scaling factor a, mainly caused by ultrasound wave attenuation, is dependent on many factors such as transmitting frequency, depth of particle, tissue type, etc. The frequency of $r(t)$ is αf_0, related to the particle moving velocity and the attenuation of ultrasound wave. However, the later is not considered in (28.2).

By comparing the frequencies of emitted and received signals, we found there is a frequency shift f_d,

$$f_d = \alpha f_0 - f_0 = -\frac{2v_x}{c} f_0 = -\frac{2v\cos\theta}{c} f_0. \tag{28.5}$$

Equation (28.5) relates the particle moving velocity v and the frequency shift f_d, which is typically called as Doppler shift or Doppler frequency. The negative sign in (28.5) indicates that the received frequency becomes lower when the particle moves away from transducer, and vice versa.

A direct method to detect the Doppler shift from the received signal would be to Fourier transform the signal and get the difference between the obtained frequency spectrum and transmitted frequency f_0. Theoretically, this is an acceptable algorithm; however, it might introduce high percentage of error considering that the Doppler shift is typically in the range of 100–10,000 Hz depending on transmitting frequency, insonation angle, and particle velocity. A more robust implementation is to apply an in-phase quadrature (IQ) demodulation technique on the received signal. A quadrature signal with a frequency f_0 could be expressed as:

$$q(t) = \cos(2\pi f_0 t) + j\sin(2\pi f_0 t). \tag{28.6}$$

Multiplication of $q(t)$ and $r(t)$ will yield [13]

$$\begin{aligned} m(t) = \frac{a}{2}\{&\cos(2\pi f_0[(1-\alpha)t - \alpha t_0]) + \cos(2\pi f_0[(1+\alpha)t - \alpha t_0]) \\ &+ j\sin(2\pi f_0[(1-\alpha)t - \alpha t_0]) + j\sin(2\pi f_0[(1+\alpha)t - \alpha t_0] \end{aligned} \tag{28.7}$$

The resulted signal includes two frequency components, the sum of emitted and received signal frequencies $f_0(1+\alpha)$ and the difference of them $f_0(1-\alpha)$. By applying a bandpass filter on $m(t)$, the summed frequency $f_0(1+\alpha)$ could be removed with only $f_0(1-\alpha)$ component left, which is expressed as [13]:

$$m_f(t) \approx \frac{a}{2}\exp\left(j2\pi f_0 \frac{2v_x}{c} t\right)\exp\left(-j2\pi f_0 \alpha t_0\right). \tag{28.8}$$

Thus, the velocity v_x or v could be obtained by applying Fourier transform on the bandpass signal $m_f(t)$.

It is necessary to note, however, that the derivation of this model only considers the interaction of the ultrasound beam with one moving particle. This is the reason why only one frequency component is present in (28.8). If many particles moving at different velocities are encountered, the obtained signal will consist of a continuum of frequencies.

In modern ultrasound machines, the CW Doppler is typically implemented by digital signal processor (DSP) due to its flexibility and accuracy. The procedures from the received signal to the display of output are shown in Fig. 28.2.

28.2.1.2 Pulsed Wave Doppler

Pulsed wave (PW) Doppler is another commonly used ultrasound modality for quantifying blood flow velocity in cardiovascular system. Different from CW Doppler, PW Doppler does not take advantage of Doppler effect directly, but utilizes the phase change of the received signal.

PW Doppler differs in many aspects from CW Doppler. Most importantly, the pulse sequence is different, as shown in Fig. 28.3. CW Doppler has a continuous sinusoidal pulse sequence, while the PW Doppler sends out pulse trains with a pre-defined pulse repetition frequency (PRF). The cycle number N in each pulse train is determined by the gate size l_g and pulse frequency f_0, which is expressed as:

$$N = \frac{2l_g}{c} f_0, \tag{28.9}$$

where c is the sound velocity. The PRF is limited by the depth of the selected gate d_0,

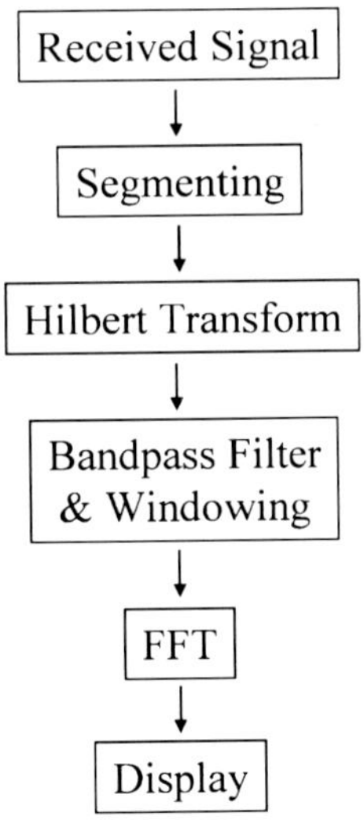

Fig. 28.2 The procedures of implementing continuous wave Doppler: from received signal to output display

$$\mathrm{PRF} \le \frac{c}{2d_0}. \tag{28.10}$$

Secondly the transducer design is different. In CW Doppler, the transmitting and receiving elements are separate; in contrast, the transmitting and receiving elements in PW Doppler are the same. Either single element transducer or array transducer could be used in PW Doppler. Thirdly the CW Doppler continuously receives all reflected data along the ultrasound propagation path, however, the PW Doppler samples data only from the selected gate by controlling the sampling time determined by the gate size and depth. The sample scheme for PW Doppler is shown in Fig. 28.4. Figure 28.4a shows the phase shift between received signals at different time, each of which corresponds to the reflected echo of the

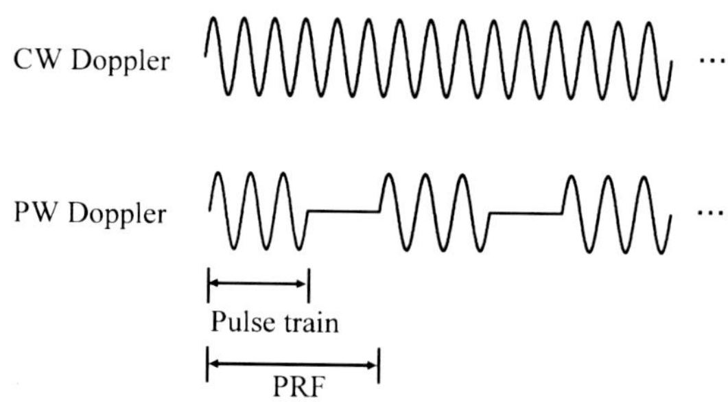

Fig. 28.3 The pulse sequence for CW and PW Doppler

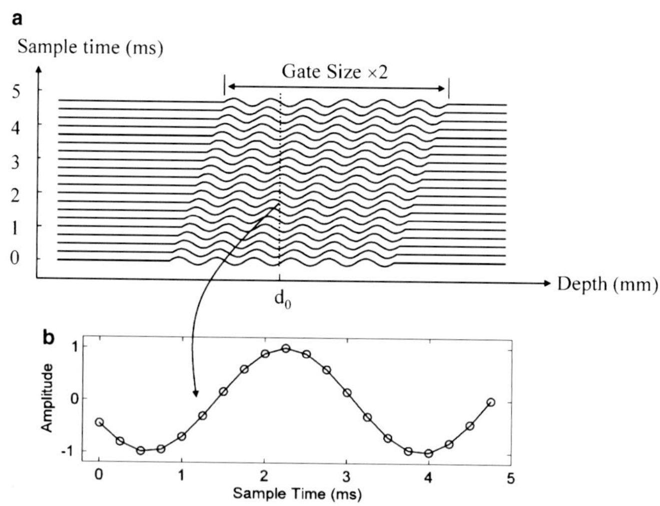

Fig. 28.4 The sampling scheme of PW Doppler: (**a**) the received fast signal at different time; (**b**) the sampled slow signal

transmitted signal (six cycles of sinusoidal pulse). The phase shift is present because the particle is moving toward or away from the transducer. The received signals are sampled at depth d_0 with a sampling frequency of PRF to form a signal with much slower frequency (typically called as slow signal), as shown in Fig. 28.4b.

By using the relationship $t=2d/c$, (28.2) could be transformed from time domain to spatial domain as:

$$r(d) = a \cdot \cos\left(2\pi f_0 \alpha \frac{2(d-d_0)}{c}\right), \tag{28.11}$$

where d is the distance from the transducer. Since the particle is moving with a velocity v_x in the ultrasound wave propagation path, the echoes from the particle will be received at different time, as if the signal is shifted in space with an amount given by $\Delta d = v_x n/\text{PRF}$ [15], where n is the series number of transmitted/received signal. Thus, to model the nth received signal, (28.11) could be modified as [16]:

$$r_n(d) = a \cdot \cos\left(2\pi f_0 \alpha \left(\frac{2(d-d_0)}{c} + n\frac{2v_x}{c\text{PRF}}\right)\right). \tag{28.12}$$

This signal with different n values corresponds to the fast signals in Fig. 28.4a, and sampling this signal at $d=d_0$ yields,

$$r_{d_0}(n) = a \cdot \cos\left(2\pi f_0 \alpha \cdot \frac{2v_x}{c} \frac{n}{\text{PRF}}\right) \tag{28.13}$$

Since $t=n/\text{PRF}$, this signal could be expressed as:

$$r_{d_0}(t) = a \cdot \cos\left(2\pi f_0 \alpha \cdot \frac{2v_x}{c} \cdot t\right) \tag{28.14}$$

And this signal corresponds to the slow signal in Fig. 28.4b. The frequency of this signal is,

$$f_{\text{slow}} = f_0 \alpha \frac{2v_x}{c} = f_0 \alpha \frac{2v\cos\theta}{c}. \tag{28.15}$$

By applying a Fourier transform, the slow signal, f_{slow} is obtained. Thus, the velocity of moving particle is calculated using (28.15), assuming that the ultrasound insonation angle θ and the time compression factor α are known. The angle θ could be estimated by using ultrasound B-mode imaging. From (28.3), it is clear that the compression factor α is also dependent on velocity, which creates problems when calculating the velocity from (28.15). However, since v_x in human application is much smaller than c, α is typically taken as 1 in (28.15).

It is also necessary to note that the sample frequency of the slow signal equals to PRF, therefore the frequency of the slow signal must be smaller than PRF/2 in order to avoid aliasing. This yields a relationship between velocity and PRF:

$$v \leq \frac{c \cdot \mathrm{PRF}}{4 f_0 \cdot \cos\theta} \tag{28.16}$$

The PRF is mainly limited by the depth of measuring gate, as shown in (28.10). In order to increase the threshold of resolvable velocity, the transmitting frequency should be kept as small as possible. On the other hand, the smaller the transmitting frequency, the poorer the spatial resolution is. A tradeoff has to be made between the dynamic range and accuracy of velocity estimation in order to get an optimal f_0, which is typically around 3 MHz in many medical applications.

The implementing procedures, similar to CW Doppler, are shown in Fig. 28.5. Different from CW Doppler, the matched filtering [13] is introduced to improve the signal-to-noise ratio of the received signals. A sampling procedure is necessary to get the slow signal for later processing. The rest of the procedures are the same with those in CW Doppler.

The PW Doppler discussed above only involves one gate where the velocity is estimated. By introducing more gates, a technique called multigated PW Doppler was extended, by which velocity profile along beam direction can be estimated [17, 18].

28.2.2 *Color Flow Mapping*

CW and PW Doppler can only obtain velocity information either in a direction or at a location. Ideally, the goal is to show a mapping of blood velocity in the human body.

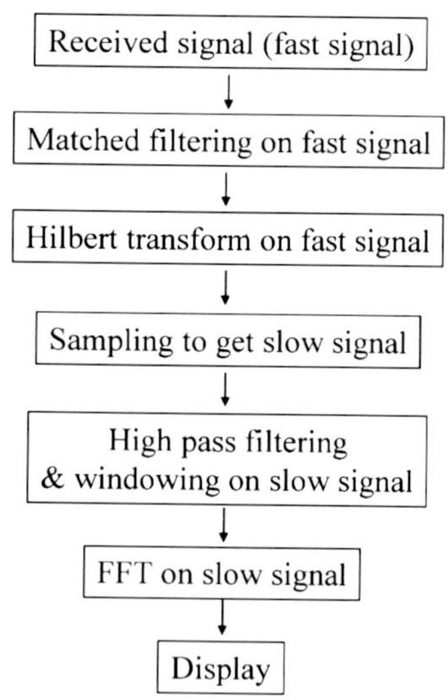

Fig. 28.5 Procedures to implement pulsed wave Doppler in DSP

CFM is proposed to perform this function. Figure 28.6 shows the difference of velocity estimation locations among CW Doppler, PW Doppler, and CFM.

The CFM is an extension of multigated PW Doppler, as the blood velocity is estimated for a number of directions with multiple gates in each direction, as shown in Fig. 28.5c. The pulse sequence is different between the PW Doppler and the CFM. In PW Doppler, the pulse train is sent out continuously at a preselected PRF along one direction, and typically echoes from 128 pulse trains are used to generate one velocity spectrum. In such a case, the sampled slow signal is long enough to estimate the phase shift by using FFT. In CMF, because velocity estimation is made along multiple directions, the number of pulse trains (called packet size) in one direction has to be reduced in order to keep a reasonable frame rate of velocity mapping. The packet size in CMF is variable in different system, and typically ranges from 8 to 16. Such a small packet size would not allow the application of FFT to estimate velocity. A new velocity estimator is thus necessary in CMF. The commonly used method is the autocorrelation approach suggested by Kasai et al. [19]. This estimator is expressed as [13, 19]:

$$v = -\frac{c \cdot PRF}{4\pi f_0 \cdot \cos\theta} \cdot \arctan\left(\frac{\Im\{R(1)\}}{\Re\{R(1)\}}\right)$$
$$-\frac{c \cdot PRF}{4\pi f_0 \cdot \cos\theta} \cdot \arctan\left(\frac{\sum_{i=1}^{N-1}(y(i)x(i-1) - x(i)y(i-1))}{\sum_{i=1}^{N-1}(x(i)x(i-1) + y(i)y(i-1))}\right) \quad (28.17)$$

where R represents the one-lag autocorrelation, $\Im\{\bullet\}$ denotes the imaginary part, and $\Re\{\bullet\}$ the real part, x and y are the real and imaginary parts of the slow signal after Hilbert transform, which is sampled at a particular depth, and N is the packet size (number of pulse trains in one direction). In this way, $x(i)$, $y(i)$, and $x(i-1)$, $y(i-1)$ are the sampled data points from the echoes of two consecutive pulse trains.

Another difference between CMF and conventional PW Doppler is the number of gates in one beam direction. In PW Doppler, only one gate is selected; in contrast, multiple gates are set in CMF to get velocity estimation at different positions along one beam direction. It is necessary to note that using multigate does not increase the

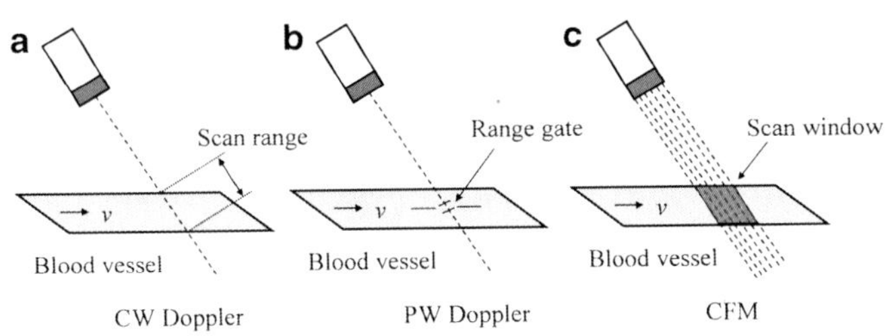

Fig. 28.6 Comparison of velocity estimation locations among CW Doppler, PW Doppler, and CFM

pulsing time, thus does not lower the frame rate. What results is the computational burden as velocity is estimated at many positions.

The procedures to implement CMF, as shown in Fig. 28.7, are similar with those in PW Doppler, except on two aspects: first, one stationary echo canceling filter [13, 20, 21] is used to remove echoes from the vessel walls and boundary tissues; second, the autocorrelation approach, instead of FFT, is employed to estimate velocity.

It is important to mention that the obtained results from autocorrelation estimator are averaged velocities in the range gate in CMF. This is different from the results of FFT in PW Doppler, which is the velocity (frequency) spectrum, including separate velocity information in the range gate. Since velocity in CMF is averaged from a gate, the variance, an indication of the velocity spread within the gate, is also important. The reader is referred to [13] for more details, if interested.

28.2.3 Vector Doppler

All the three techniques discussed above will only provide estimation of one velocity component along the ultrasound beam direction, with the ultimate derivation of the blood velocity requiring an assumption that flow occurs parallel to the vessel wall. This is a reasonable when the vessel wall is linear, and the angle between the ultrasound beam and vessel wall could be obtained by using ultrasound B-mode imaging. However, this assumption fails when the vessel wall is curved, because nonaxial blood flow patterns are present at bifurcation, curves, and obstructions in the vessel [22]. Therefore the obtained velocity might be error prone. Thus, it is necessary to develop a technique which is angle independent and is able to provide two or three components of velocity estimation.

Cross-beam Doppler is one of the commonly studied methods to achieve this purpose. Many researchers proposed different schemes to measure true blood velocity with two or three components [23–28]. A detailed review was provided by

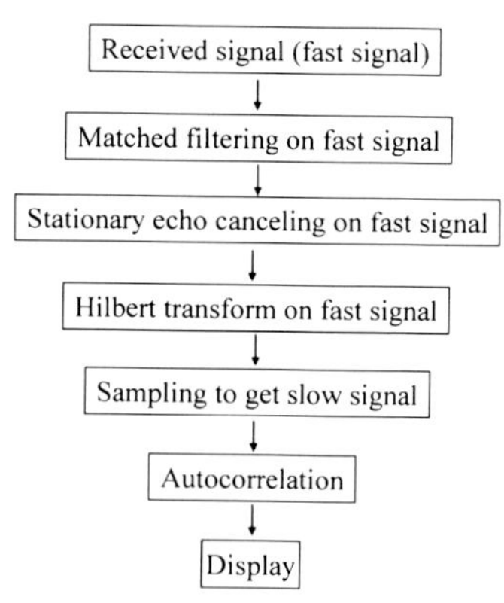

Fig. 28.7 Procedures to implement color flow mapping by DSP

Dunmire et al. [29]. Most of those techniques, in common, share one basic principle that at least two receivers are employed to receive ultrasound echoes from moving particles, from which different velocity components could be derived. In order to do this, the two receivers, either two separate transducers or two apertures in an array transducer, have to be arranged at an angle with respect to the sample volume to be measured. The larger the angle, i.e., the larger the distance between two receivers, the more reliable velocity it could provide.

In this section, vector Doppler, one of the cross-beam methods, is introduced and discussed. Vector Doppler was first proposed by Overbeck et al. in 1992 [30]. The system uses a single transmitter and two receivers to resolve orthogonal velocity vector components, in this way to determine both the magnitude and direction of complex blood velocity. The arrangement of the three transducers is shown in Fig. 28.8. The angle between transmitting beam and blood velocity is θ, the transmitter and the receiver L have an angle of α, and the angle between transmitter and receiver R is β. All angles are measured along the counter-clockwise direction.

From the Doppler (28.5), it is found that the 2cosθ term is introduced due to a delay caused by the round-trip time between emission and reception, where the transmitter and receiver are the same in conventional Doppler system. When the receiver and transmitter are separate in the vector Doppler, as shown in Fig. 28.8, the 2cosθ term could be replaced by $\cos\theta+\cos(\theta-\alpha)$. Thus, (28.5) could be rewritten as:

$$f_{\mathrm{L}} = -f_0\frac{v}{c}\left(\cos\theta + \cos(\theta - \alpha)\right), \tag{28.18}$$

where f_{L} is the Doppler shift detected by the receiver L. A similar relationship could be obtained for the receiver R:

$$f_{\mathrm{R}} = -f_0\frac{v}{c}\left(\cos\theta + \cos(\theta - \beta)\right), \tag{28.19}$$

where f_{R} is the Doppler shift detected by the receiver R. If $\alpha=-\beta$, what results from (28.18) and (28.19) is [30],

$$f_{\mathrm{L}} + f_{\mathrm{R}} = -\frac{2f_0}{c}(1+\cos\alpha)\cdot v\cos\theta, \tag{28.20}$$

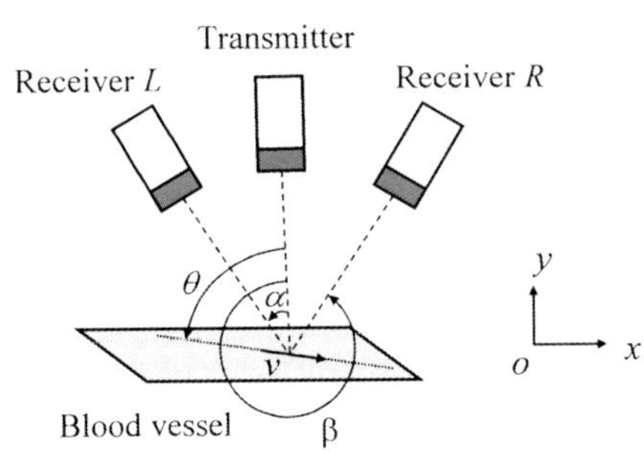

Fig. 28.8 The schematic of vector Doppler for blood velocity measurement

$$f_{\mathrm{L}} - f_{\mathrm{R}} = -\frac{2f_0}{c}\sin\alpha \cdot v\sin\theta. \tag{28.21}$$

Equations (28.20) and (28.21) are the relationships employed in vector Doppler to get two components of velocity, one is along the transmitting beam direction and the other is orthogonal to the beam direction.

Now the question is how to get f_{L} and f_{R} from the signals received by the two receivers. Overbeck et al. proposed a heterodyning technique [30] which utilizes the four quadrature signals obtained from the two receivers to find the sum and difference of the two Doppler shifts. The process is shown in Fig. 28.9. After obtaining the $(f_{\mathrm{L}}+f_{\mathrm{R}})$ and $(f_{\mathrm{L}}-f_{\mathrm{R}})$, the two velocity components could be readily calculated using (28.20) and (28.21). It is necessary to note that a bandpass filter, not shown in Fig. 28.9, should be introduced after the IQ demodulation to remove high-frequency components introduced in demodulation and low-frequency signals due to arterial wall and tissue motion, as discussed in Sect. 28.2.1.1.

This vector Doppler technique was successfully applied by Overbeck on assessing the blood velocity in common carotid artery (CCA) of a healthy volunteer, and both magnitude and angle information were obtained, although no validation or reference results were provided.

After Overbeck's work, many researchers put their efforts on improving the vector Doppler technique. One of the limitations for Overbeck's system is that it could only provide velocity estimation at one voxel in blood flow. Scabia and Capineri proposed and prototyped a system for dynamic vector velocity mapping of blood flow [31, 32]. This system utilizes an array transducer with three apertures which functions as one transmitter at the center as well as two receivers on both sides. The scanning at different voxels in axial direction is achieved by a composition of two different scanning techniques. First, a few large gates are created by adjusting the positions of two receiving apertures, as shown in Fig. 28.10. The scanning of each L_{gate} requires transmitting a series of pulse trains as for in CFM. Second, a number of S_{gate} are achieved by employing the

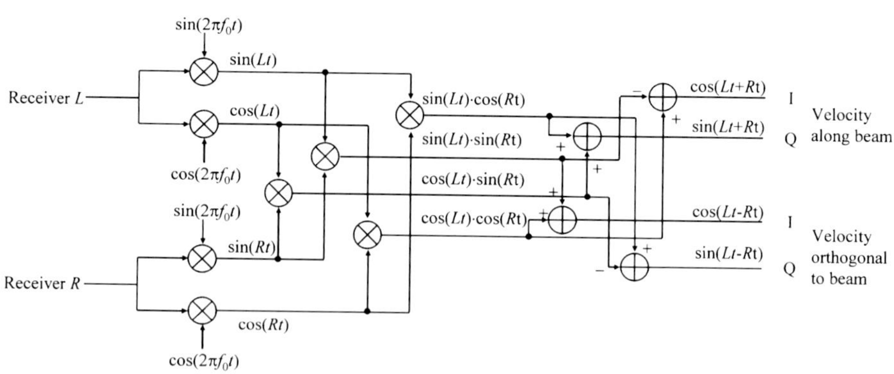

Fig. 28.9 The process to get the sum and difference of two Doppler shifts in vector Doppler

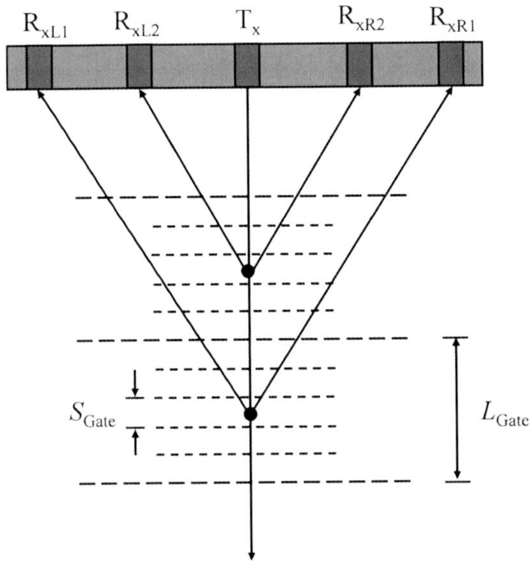

Fig. 28.10 The axial scanning of different voxels for vector velocity mapping by vector Doppler

multigate technique in PW Doppler. The scanning in lateral direction is done by moving the transmitter–receiver group along the array while keeping the distance between them.

One of the criteria to evaluate a 2D vector velocity mapping technique is the frame rate (number of vector maps per second). In Capineri's system, the frame rate could be calculated as:

$$FR = \frac{PRF}{2N \cdot L \cdot M} \tag{28.22}$$

where N is the number of pulse trains to scan a L_{gate}, M is the number of L_{gate}, and L is the number of vector lines in lateral direction. The multiplying factor 2 is due to the fact that only one beamformer is present in this system [32]. Assuming PRF = 10K, N = 32, L = 10, M = 2, the frame rate is estimated to be around 7–8 FPS (frames per second). And this result was obtained without considering the time cost of large amount of signal processing, thus the real frame rate is in general lower than this.

Scabia first applied the 2D vector Doppler on human carotid artery. A 1 × 3 voxel map was obtained with 13.3-kHz PRF and 256 transmit pulses per beam, resulting a vector velocity mapping at 16 FPS. Capineri et al. applied this technique on a phantom flow study and reported that their vector Doppler system could acquire velocity maps with 6 × 9 voxels (interpolated with a factor of 4 to 21 × 23 voxels) at a frame rate of 1 Hz. The accuracy of this technique, however, is not validated in his work.

To the authors' knowledge, no other in vivo study with 2D vector Doppler was found in literatures, indicating more improvement work is expected on this technique before its clinical application.

28.3 Speckle Tracking

One of the main limitations of the vector Doppler technique is its low frame rate in 2D velocity vector estimation. One approach that avoids using multibeam Doppler is the speckle tracking, which was first proposed by Trahey et al. in 1987 [33]. Speckle tracking utilizes a correlation search algorithm to track the displacement of speckle patterns originated from scatters in blood (red blood cells) between two subsequently acquired B-mode images. The searching scheme is shown in Fig. 28.11. First, a cell region of $l \times k$ pixels is selected in the first image, and a search region of $(2M+1)\times(2N+1)$ pixels in the second image is determined. The two regions have the same center position in each image (This is not necessary, but most algorithms employ this scheme). Second, the cell region in the first image is compared with a same size area in the search region of the second image. The compared area in the search region then translates in both directions to generate a series of similarity indexes at different locations. The resulted index map has a size of $(2M+1)\times(2N+1)$, and the highest index corresponds to the position where the cell region matches best with the search region. Thus, the displacement is calculated as the coordinate difference between the highest index position and the center position of the cell region. This procedure could be expressed as [34, 35],

$$\rho_{m,n} = \frac{\sum_{i=1}^{l}\sum_{j=1}^{k}(X_{i,j}-\bar{X})(Y_{i+m,j+n}-\bar{Y})}{\left[\sum_{i=1}^{l}\sum_{j=1}^{k}(X_{i,j}-\bar{X})^2(Y_{i+m,j+n}-\bar{Y})^2\right]^{1/2}}, \tag{28.23}$$

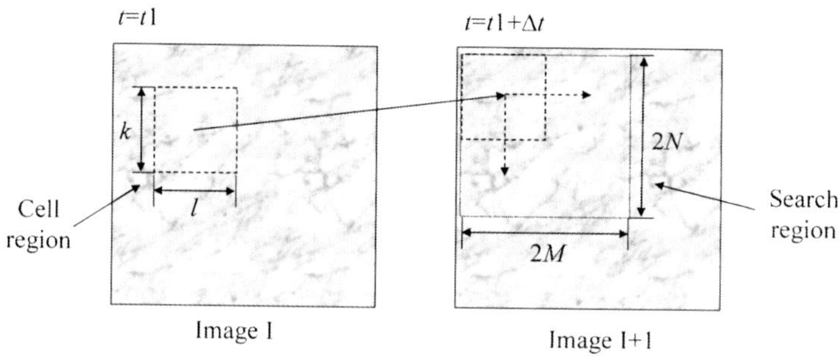

Fig. 28.11 Searching scheme in speckle tracking algorithm

where X and Y are the intensity matrixes of the two B-mode images, and $\bar{X}$ and $\bar{Y}$ are the averaged intensities in the corresponded area. Note that $-M<m<M$ and $-N<n<N$.

By dividing the image into many different cell regions and performing the same operation, a two-dimensional displacement map is thus obtained. Dividing the displacement map by the time interval between the two images, the velocity map is then resulted.

The normalized cross-correlation method using (28.23) requires a large amount of computation, which is not ideal for real-time application. Thus, an algorithm called sum absolute difference (SAD) was proposed to reduce the computation burden [36], and the SAD shows to give similar performance with normalized cross-correlation method. This algorithm is expressed as:

$$\varepsilon_{m,n} = \sum_{i=1}^{l} \sum_{j=1}^{k} \left| X_{i,j} - Y_{i+m,j+n} \right| \tag{28.24}$$

The speckle tracking technique is angle independent, and is able to provide direct estimation of two velocity components. The performance of this technique was first evaluated in test tank experiments under a variety of conditions provided by Trahey et al. [34], and the results showed that it could measure the speckle translations in both directions. It was further applied in vivo to measure the blood velocity in human carotid artery [37] and jugular vein [38], and the results showed that the speckle tracking has superior performance than CFM in some aspects, such as angle independency and no aliasing. However, none of these studies provided validations on the measured velocity information by speckle tracking.

Although speckle tracking seems promising in future clinical application, the speckle decorrelation due to target translation is one of the main factors that contribute to the inaccuracy of this technique, although increasing the cell region size could improve the accuracy in some extent. This limits this technique in detecting relatively unchanged speckle patterns associated with small displacements. This might be the reason why few studies in the past 20 years were reported to assess in vivo blood velocity by speckle tracking. The speckle tracking, however, was successfully applied to quantify tissue motion in many clinical applications [39–42], which, compared with blood motion, causes less speckle decorrelation, thus provides better signal-to-noise ratio.

Another method called plane wave excitation (PWE) was proposed by Udesen [43–45]. This method also applies speckle tracking algorithm to detect 2D speckle displacements between the two ultrasound images. Its difference with previous speckle tracking method lies in the scanning of ultrasound speckle images, which utilizes plane wave instead of focused beam employed in conventional B-mode imaging. In conventional B-mode imaging, the ultrasound pulses from different transducer elements are focused to increase contrast, resolution and SNR. However, this greatly lowers the frame rate since the transmitting procedure has to be repeated to cover the whole field of interest. In PWE method, all the transducer elements

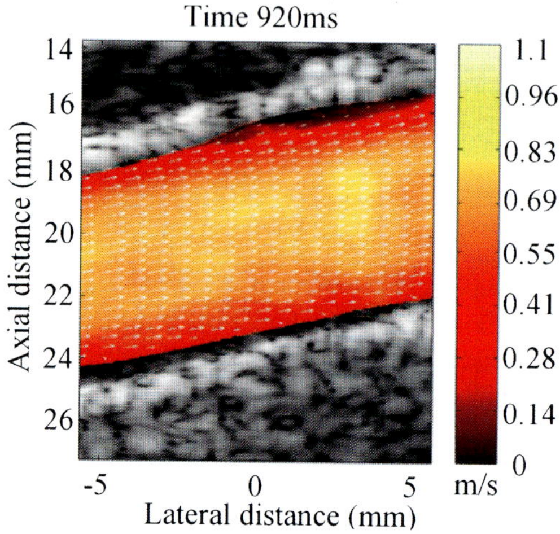

Fig. 28.12 The velocity vector map in human common carotid artery obtained by PWE method [44] (reprinted with permission)

were fired at the same time without any beamforming, thus a pressure wave with nearly plane wave front is generated. The received echoes are then beamformed to generate one B-mode image. The frame rate in the PWE method could be as high as the PRF, which is only limited by the imaging depth. However, this high frame rate comes from a compromise with image contrast. In order to compensate the decrease in SNR, a 13-bit Barker code is used instead of a conventional pulse [46].

The PWE method has been validated against phase-contrast MRI in measuring velocity and flow rate in human carotid artery. Figure 28.12 shows a velocity vector map obtained by the PWE method, which resulted from an average of 40 velocity vectors to reduce the variance and improve the accuracy. This averaging lowers the frame rate to 100 Hz. Therefore, there is a tradeoff between frame rate and the quality of velocity estimation. Figure 28.13 shows the comparison on volumetric flow rate measurements by PWE and MRI. The mean underestimation of the PWE compared with MRI is about 9%.

The PWE method has also been used to study the flow patterns at human carotid bifurcation, as shown in Fig. 28.14. This result is exciting, since it shows that the PWE method may provide a new clinical tool to visualize complex flow patterns as well as quantify some important hemodynamics information, such as vorticity and WSS. This information may be very helpful to study some vascular diseases, including atherosclerosis.

The PWE method shows that applying speckle tracking algorithm on PWE images can provide much better performance on velocity vector detection on conventional B-mode speckle images. This might be explained from a few different aspects. First, the PWE image has better spatial resolution than conventional

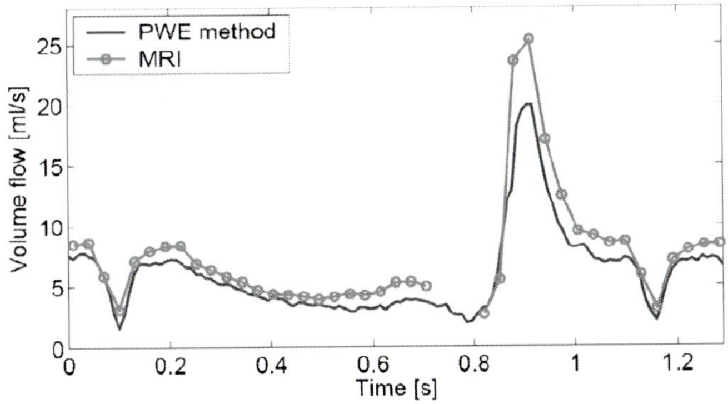

Fig. 28.13 The comparison on volume flow rate in common carotid artery measured by PWE and MRI [44]

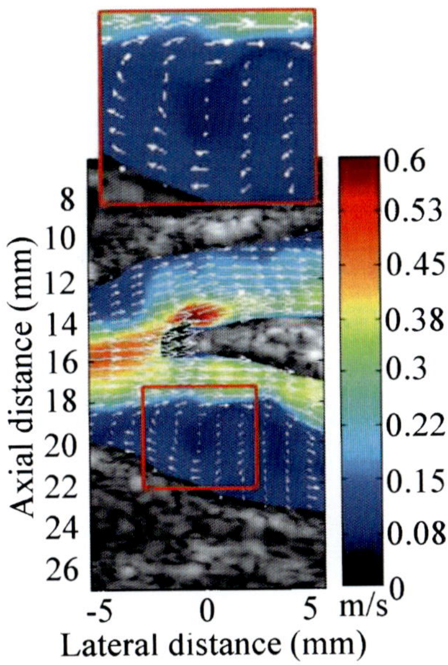

Fig. 28.14 The velocity vector map at carotid bifurcation detected by the PWE method [45] (reproduced with permission)

B-mode image, since a much larger aperture (128 vs. 64) is employed. The better resolution allows better detection of speckle patterns. Second, the frame rate of PWE method is very high (around 4 kHz, or even higher). The speckle decorrelation in such a small time interval might be greatly reduced. Third, taking advantage of high frame rate, the PWE method utilizes a spatial averaging with a kernel size up to 40 frames, which greatly improves the detected velocity vector field.

28.4 Transverse Oscillation

In pulsed Doppler and CFM, only velocity component along ultrasound beam direction could be determined. In pulsed Doppler, the true velocity is angle-corrected by estimating the angle between flow and beam direction using B-mode image, thus the obtained velocity information is quantitative. In CFM, however, the obtained velocity is not angle-corrected. Further, the velocity is estimated using much less pulse trains at each beam direction when compared with pulsed Doppler. Therefore, the CFM technique could only provide qualitative information about blood velocity.

The way by which pulsed Doppler and CFM measure velocity along beam direction is to detect the phase shift in the received echoes from the moving particle. The phase shift is present in a particular direction when two conditions are both met. First, the particle has a velocity component in this direction. Second, the ultrasound pulse oscillates in this direction with a frequency. This simply explains why pulsed Doppler and CFM cannot detect velocity component in transverse beam direction: there is no pulse oscillation in the transverse direction although velocity is present. Based on this theory, Jensen and Munk proposed a novel technique called transverse oscillation (TO) [14, 47–49]. The basic idea in TO method is to create an additional oscillation in the PSF (point spread function of ultrasound pulse) transverse to the beam direction. When transmitting, the TO method uses conventional ultrasound pulses. An apodization function is applied on the received pulses to create an oscillating ultrasound field in transverse beam direction. The simulated PSFs with and without TO were shown in Fig. 28.15.

The detection of the phase shifts in two directions are required to obtain two components of a velocity vector. In order to do this, two ultrasound beams with a distance of $\lambda_x/4$ in lateral (transverse) direction are focused in receiving the echoes. Thus, these two received signals have a 90° phase shift. The two signals are

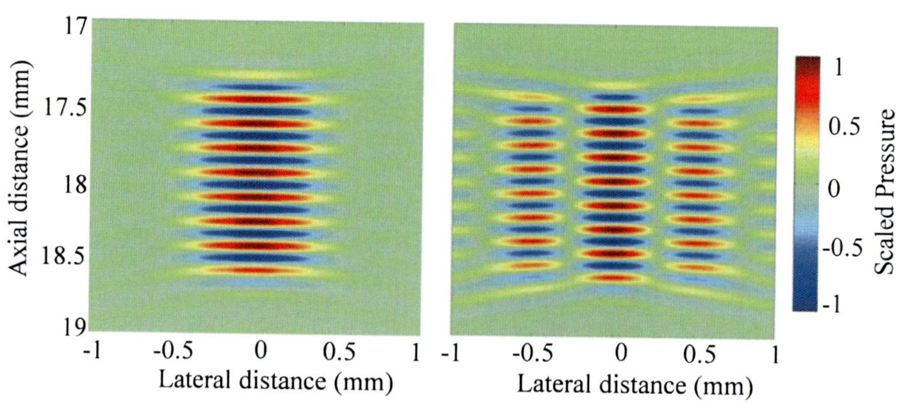

Fig. 28.15 The simulated PSF without (**a**) and with (**b**) transverse oscillation in ultrasound field [50] (reproduced with permission)

thereafter IQ demodulated (Hilbert transform) to get four quadrature signals, which are sampled at interested depth to yield four slow time signals expressed as [16]:

$$\begin{aligned}
r_a(t) &= \cos(At)\cos(Bt), \\
r_b(t) &= \sin(At)\cos(Bt), \\
r_c(t) &= \cos(At)\sin(Bt), \\
r_d(t) &= \sin(At)\sin(Bt),
\end{aligned} \tag{28.25}$$

where $t=n/f_{\text{prf}}$ is sampled time points, $A=2\pi v_x/\lambda_x$, $B=2\pi v_z/\lambda_z$, with v_x as the velocity in lateral direction, and v_z is the velocity in axial direction. λ_x and λ_z are wave lengths in two directions, respectively, and they have relationships with the transmitted pulse wave length λ as,

$$\lambda_x = 2\lambda\frac{z}{L}, \tag{28.26a}$$

$$\lambda_z = \frac{\lambda}{2}, \tag{28.26b}$$

where z is the depth at which velocity is measured and L is the receiving aperture size. From (28.25), two signals could be formed as:

$$r_1(t) = e^{j(B+A)t}, \tag{28.27a}$$

$$r_2(t) = e^{j(B-A)t}. \tag{28.27b}$$

Applying the autocorrelation estimator used in CMF (see (28.17)) on these two signals, their phase shifts could be estimated as:

$$(B+A)\cdot\Delta t = \arctan\frac{\Im\{R_1(1)\}}{\Re\{R_1(1)\}}. \tag{28.28a}$$

$$(B-A)\cdot\Delta t = \arctan\frac{\Im\{R_2(1)\}}{\Re\{R_2(1)\}}. \tag{28.28b}$$

where $\Delta t = 1/f_{\text{prf}}$. Taking B, A, and (28.26b) into (28.28a, 28.28b) yields the two estimators used to determine the two velocity components [16]:

$$v_x = \frac{\lambda_x f_{\text{prf}}}{4\pi}\left[\arctan\frac{\Im\{R_1(1)\}}{\Re\{R_1(1)\}} - \arctan\frac{\Im\{R_2(1)\}}{\Re\{R_2(1)\}}\right], \tag{28.29a}$$

$$v_z = \frac{c f_{\text{prf}}}{8\pi f_0}\left[\arctan\frac{\Im\{R_1(1)\}}{\Re\{R_1(1)\}} + \arctan\frac{\Im\{R_2(1)\}}{\Re\{R_2(1)\}}\right]. \tag{28.29b}$$

Thus, the TO method could be seen as an extension from CFM to provide angle-independent 2D velocity vector mapping of blood flow. This method has been tested in both in vitro phantom flow study and in vivo clinical study to generate 2D velocity vector map, showing great promise in future clinical application. In the phantom study, the measurements by TO method were validated against theoretical profiles in a laminar flow, and the results show that the velocity deviates from the theoretical velocity profile with a relative mean bias of 10.0% and a relative mean standard deviation of 9.8% taken over the entire vessel profile [51]. The initial clinical study on the TO method was carried out on human carotid artery of three volunteers [50]. The 2D velocity vector maps were obtained at both CCA and carotid bifurcation, as shown in Fig. 28.16. The vortex flow at carotid bifurcation was clearly observed from vector maps, which could not be detected by using either pulsed Doppler or CFM. The accuracy of the TO method for in vivo application was evaluated by comparing the volumetric flow detected by the TO method against that by PC-MRI, and a bias of approximately 20% is found. However, this is not a strict validation, since the MRI data was not obtained from the same volunteer, but from literature.

A well-designed validation study was carried out by Hansen et al. to investigate the performance of the TO method for in vivo measurements. The stroke volumes of 11 healthy volunteers were measured by using the TO method and PC-MRI. And a Bland–Altman analysis was performed to show that the flow measurements by the two modalities agree well, where the mean difference was 0.2 ml with the limits of agreement at −1.4 and 1.9 ml.

Although the TO method seems to be promising in clinical application, it also has some limitations. One of its major limitations is its low frame rate of velocity vector map. This is because it typically requires at least 64 pulse trains in each direction to generate acceptable velocity vectors, which is about six times more than CFM. With a 24-kHz PRF, the frame rate is estimated to be around 22 Hz when 16 axial lines of vector are generated [50]. Such a high PRF limits the depth of measurement around 32 mm. The low frame rate of the TO method could not provide enough temporal information which is important in studying pulsatile flow patterns in human cardiovascular system.

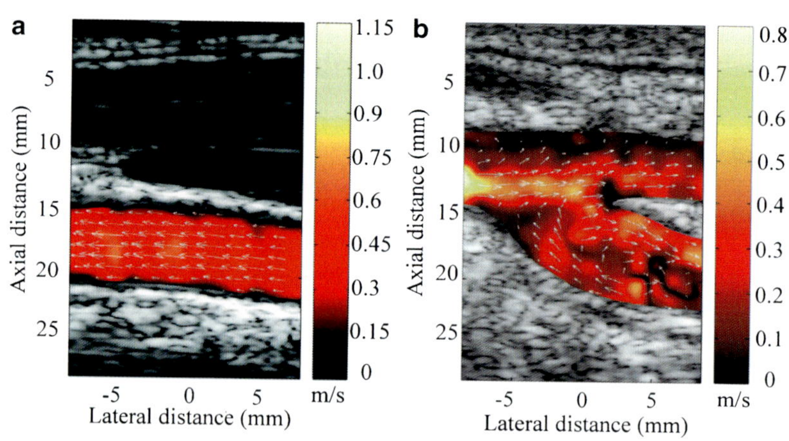

Fig. 28.16 Velocity vector maps obtained by the TO method from common carotid artery (**a**) and carotid bifurcation (**b**) [50] (reproduced with permission)

28.5 Echo Particle Image Velocimetry

Echo PIV is a novel ultrasound-based technique for quantitatively assessing cardiovascular hemodynamics. It was developed by the Cardiovascular Dynamics and Ultrasonics Laboratory at University of Colorado at Boulder. The basic principle and procedure are shown in Fig. 28.17. This technique is based on the synthesis of two techniques: ultrasound B-mode imaging and PIV [52]. Echo PIV consists of detecting and tracking flow tracers within a flow field, and computing localized velocity vectors using a modified PIV algorithm. The flow tracers used in this technique is ultrasound contrast agents, which consists of gas-filled lipid-shelled microbubbles with diameters centered in the range 2–4 μm. Due to the large acoustic impedance mismatch between the bubble and fluid, the bubbles scatter strongly, resulting in a clear B-mode image of particle positions with excellent signal-to-noise ratio.

Similar to speckle tracking, this technique uses cross-correlation as well to track particle movements between the two sequential images. Dividing the particle displacement by the time interval between the two images, the 2D velocity vector field is thus obtained. However, Echo PIV is essentially different from speckle tracking in some aspects. First, Echo PIV tracks the movements of microbubbles, not of speckle patterns in blood vessels. The microbubbles show unique signatures in echoes under ultrasound excitations, thus are easy to detect and track in B-mode images. The SNR from microbubbles is much higher than that of speckle patterns. Second, the decorrelation of microbubble signatures in

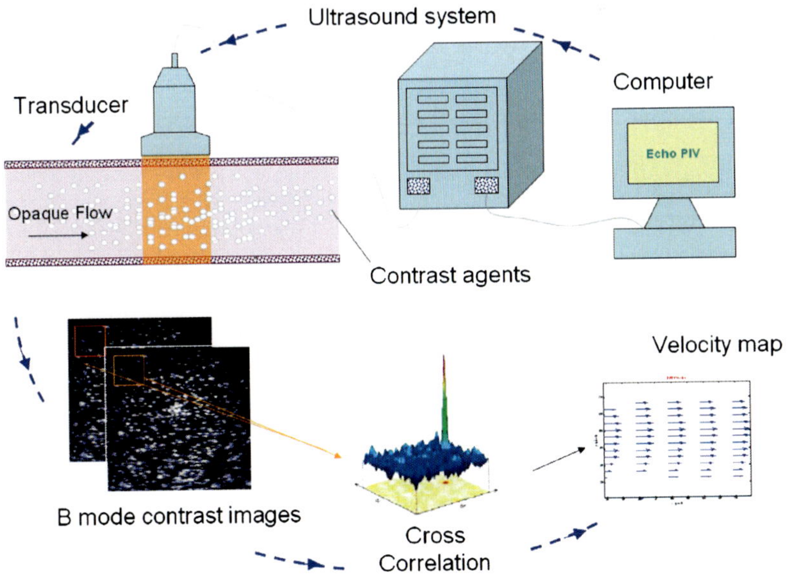

Fig. 28.17 The schematic of basic principle of Echo PIV [52]

two images is much less than that of speckle patterns. The reason is that speckle pattern, arising from a random and complex scattering of ultrasound waves, is only relatively stable as the target is translated within one or at most a few wavelengths [34]. Assuming speckle pattern is stable within two wavelengths (around 300 μm in tissue and blood for a 10 MHz transducer), in order to resolve velocity up to 1.0 m/s, the frame rate of B-mode imaging have to be more than 3,000 FPS. This is not possible for conventional B-mode imaging. In Echo PIV, however, the decorrelation of microbubble pattern is only affected by out-of-plane motion and collapse of microbubbles, not by the in-plane motion of microbubbles. This allows the detection of microbubble pattern within a larger distance, thus lowers the requirement of high frame rate B-mode imaging. Studies show that with conventional B-mode imaging, Echo PIV can capture velocity up to 2 m/s [52–54].

In the past few years, this technique has been improved in both hardware and software, and both in vitro and in vivo studies have been implemented to validate its accuracy and feasibility in quantifying blood flow hemodynamics in human body. In in vitro studies, Echo PIV was validated against analytical solution and pulsed Doppler measurement in laminar flow condition [52, 55], against computational fluid dynamics (CFD) simulation in abdominal aorta aneurysm (AAA) model [53] and against optical PIV measurements in rotating flow condition [52] as well as in a patient-specific compliant model of the carotid bifurcation under physiological condition [54, 56]. In in vivo studies, Echo PIV was validated against PC-MRI for measuring 2D velocity vector field, flow rate, and WSS in human carotid artery [57, 58].

Figure 28.18 shows acquired Echo PIV microbubble images (a) from the in vitro study on an AAA model under the steady flow condition, the resulted velocity vector fields, and streamlines (c and e) against CFD simulations (b and d). The Echo PIV measurements on velocity vector fields agree quite well with CFD simulation, which could be also observed from the streamline fields.

The feasibility of Echo PIV in capturing complex flow fields could be further exemplified from the in vitro study on compliant carotid bifurcation model under pulsatile flow condition, as shown in Fig. 28.19. Figure 28.19a shows the measurements by Echo PIV (b) and optical PIV (a) on time-dependent radial velocity profile, and excellent agreements were found on both amplitude and temporal patterns. The comparison on velocity vector and streamline fields (Fig. 28.19b) indicates that Echo PIV is able to capture localized hemodynamics under complex flow conditions.

Echo PIV has also been validated in a clinical study on human carotid artery [58]. Ten healthy volunteers were involved and the right common carotid artery (rCCA) of each volunteer was scanned by both Echo PIV and PC-MRI. The mean absolute differences (mean ± SD) between the two modalities are 10.0 ± 9.8%, 10.1 ± 8.8%, and 17.0 ± 15.3% for velocity, flow rate, and WSS, respectively. Echo PIV show larger values in WSS measurement when compared with PC-MRI, which might be partially explained by the poor spatial resolution of PC-MRI (1 mm, compared with 0.4–0.5 mm for Echo PIV) in measuring velocity vector field. Figure 28.20 shows the comparison of temporal waveforms of peak velocity, volume flow rate, and

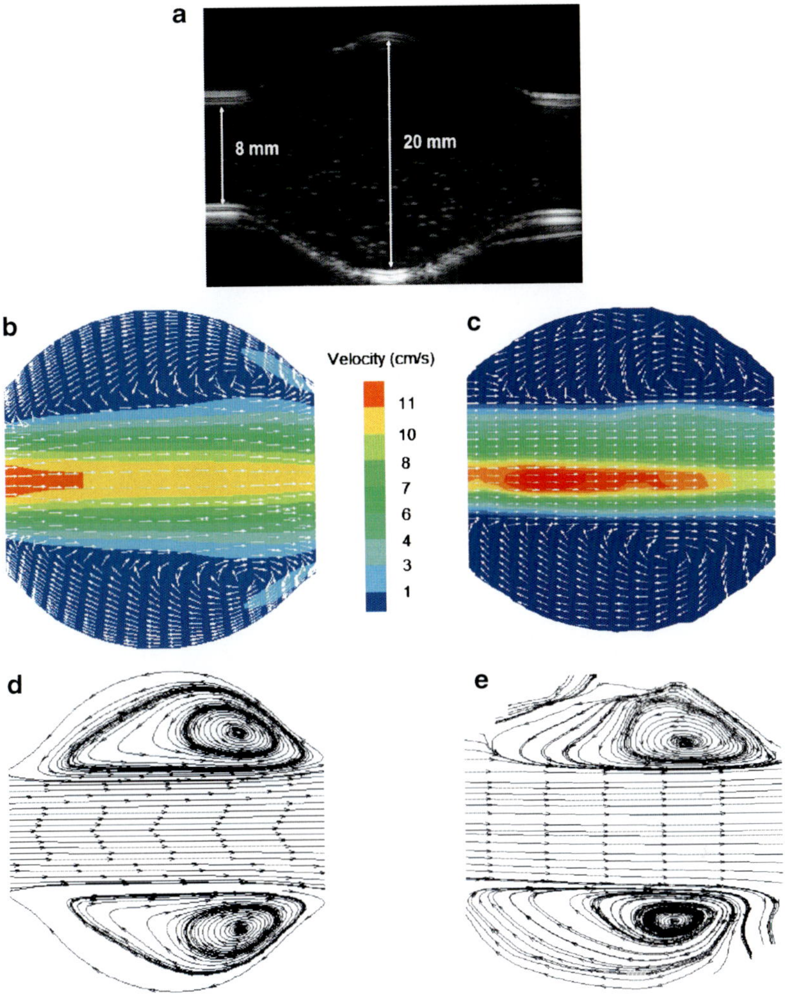

Fig. 28.18 Validation of Echo PIV against CFD simulation in an AAA model: (**a**) microbubble images; (**b**) simulated velocity vector field; (**c**) Echo PIV measured velocity vector field; (**d**) simulated streamline field; (**e**) Echo PIV measured streamline field [53]

WSS measured by Echo PIV and PC-MRI in the rCCA of one example subject. Excellent agreement was observed between the two modalities.

The capability of Echo PIV to quantify localized hemodynamics in carotid bifurcation has also been investigated clinically [59, 60], as shown in Fig. 28.21. As shown in the left panel, the 2D velocity vector field was quantified by Echo PIV at a time point in diastolic phase. The superimposed streamlines show strong recirculation of flow at carotid sinus region, where is one of the most preference locations of atherosclerotic plaques. Echo PIV, able to assess quantitative information of both hemodynamics and morphology of human vasculatures, appears

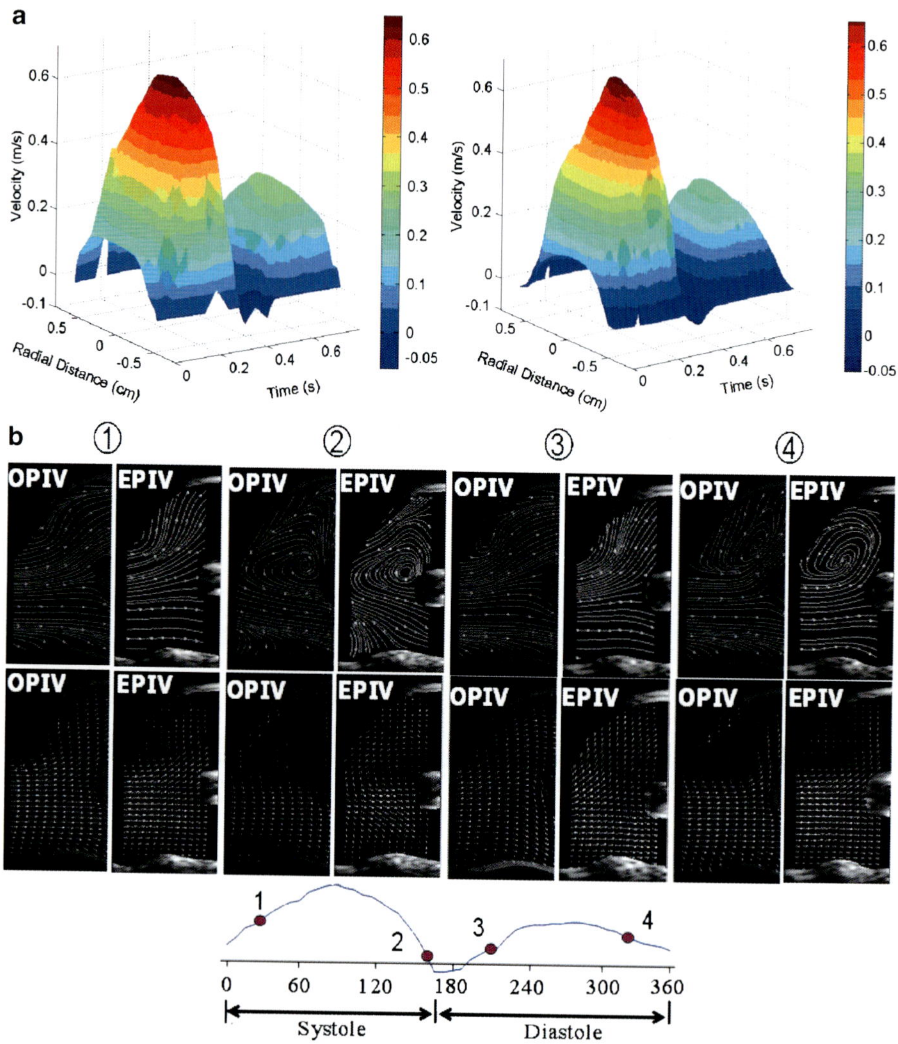

Fig. 28.19 Validation of Echo PIV measurements against optical PIV in a patient-specific human carotid bifurcation model under physiological flow condition: (**A**) radial velocity profile over time, with (**a**) as for optical PIV measurement and (**b**) Echo PIV; (**B**) comparison of streamline and velocity vector fields at four time points over cardiac cycle [54]

to hold promise as a simple-to-use technique for evaluating vascular plaque vulnerability.

In addition to its application in vasculatures, Echo PIV was also clinically applied to investigate blood velocity patterns in human left ventricles (LV). During LV filling, blood flow forms vortices that have specific geometry and anatomical locations. Quantitatively assessing these flow patterns may provide better understanding of LV functions. Hong et al. [61] utilized Echo PIV to quantify

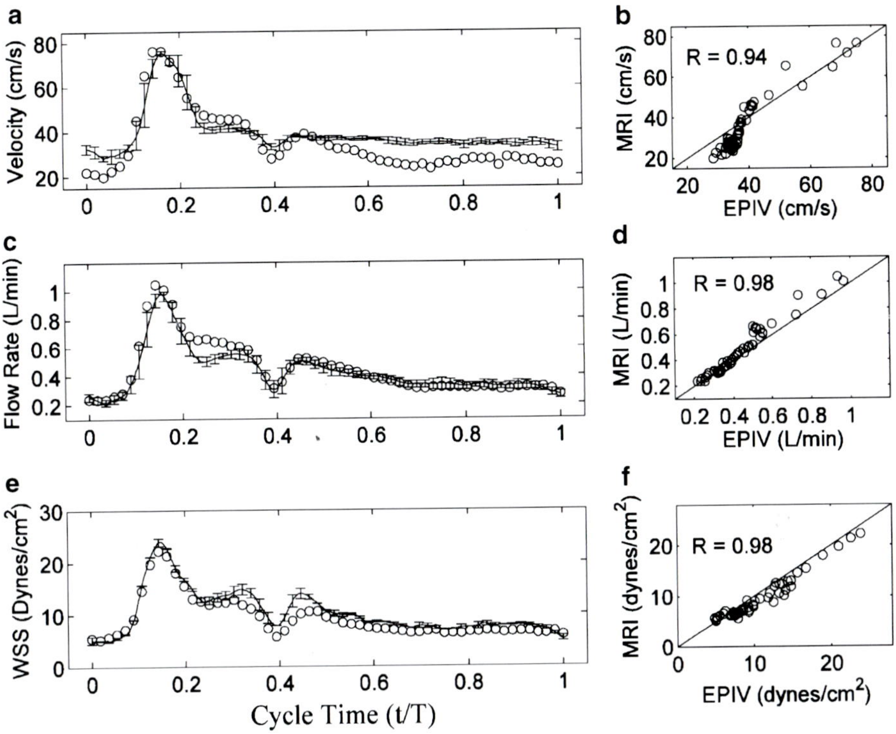

Fig. 28.20 Validation of Echo PIV against phase-contrast MRI in human rCCA [58]

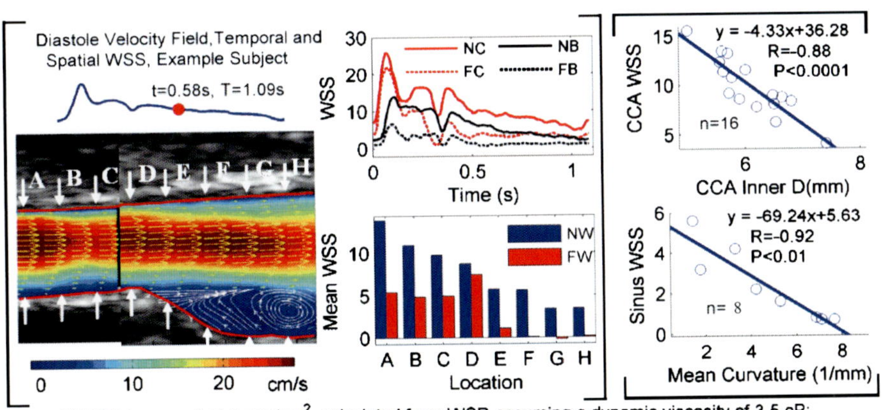

Fig. 28.21 The application of Echo PIV to quantify hemodynamic in both CCA and carotid bifurcation: *left panel*, the velocity vector field with streamline superimposed at carotid sinus region, WSS waveforms over time, and mean WSS vs. locations; *right panel*, the relationship between WSS at CCA and the inner diameter of CCA and the relationship between WSS at carotid sinus and the mean curvature of carotid sinus [59]

vortex parameters in LV of 25 patients (10 normal and 15 patients with abnormal LV systolic function), including vortex depth, transverse position, length, width, and sphericity index. The results showed that the vortex parameters are in the abnormal LV function group are significantly different from the control group, indicating the feasibility to quantify LV vorticity arrangement by Echo PIV. As shown in Fig. 28.22, for example, the vortex in normal subjects showed an elliptical shape (A, Aa, white arrow) and strong pulsatility (Ap, red-colored area), whereas spherical (B, Ba, white arrow) and weak pulsatility (Bp, blue-colored area) vortex was observed in patients with systolic heart failure. This study shows that Echo PIV may serve as a novel approach to depict vortex, the principal quantity to assess the flow structure.

Kheradvar et al. further investigated, both in vitro and in vivo, the validity of Echo PIV for quantitative assessment of the intraventricular blood flow field [62]. In in vitro study, the quantitative comparison against optical PIV showed that Echo PIV measurements were unbiased and consistent with those obtained by optical PIV either in terms of absolute values or in terms of direction. The in vivo study on human LV showed that comparable flow patterns were obtained either from independent

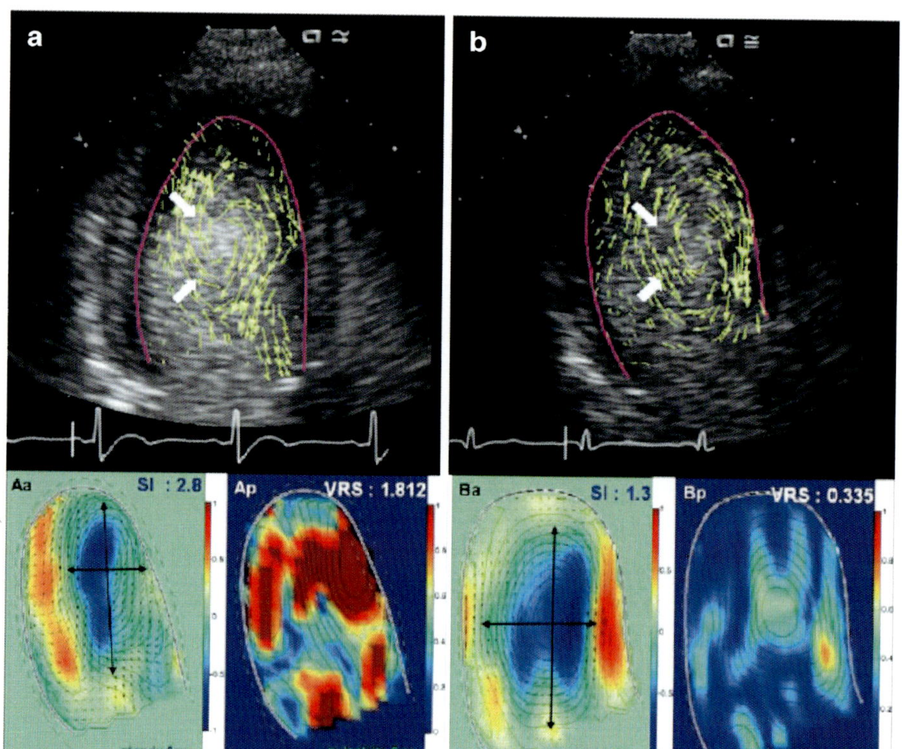

Fig. 28.22 Quantitative vortex flow parameters in normal subjects and lv systolic dysfunction group [61] (reproduced with permission)

examiners or from the same examiner on different attempts. These results indicated good reproducibility of Echo PIV in clinical applications.

Overall, Echo PIV has been showing great promise in both ventricular and vascular applications so far. With further development and improvement, it is believed to become a novel clinical diagnostic tool which may help study and understand many cardiovascular diseases.

28.6 Phase-Contrast MRI

Ultrasound-based techniques have their unique advantages, such as ease of operation, cost-effectiveness, and possibility of real-time application. However, they are limited by low signal-to-noise ratio and low contrast of ultrasound signals, and acoustic window. Flow-sensitive MRI provides alternative routines to quantify blood flow in complex cardiovascular environments as evidenced by the fast increase in both its research and clinical utilization [63–67]. The most common implementation of flow-sensitive MRI is PC-MRI. To its advantage, PC-MRI can resolve difficult-to-interpret flow fields in hard-to-reach image locations. For example, the images and flow data provided by this modality can include full three-vector component velocity fields, unobstructed by acoustic window and tissue depth, with the operator free to choose the imaging plane (2D) or volume (3D) of interest. PC-MRI is a useful and many times the only noninvasive flow imaging method available in situations involving complex, hard-to-interpret blood flow dynamics.

28.6.1 Methodology

The following references provide excellent coverage regarding the fundamentals of MRI, and more specifically PC-MRI. Since this section simply introduces the basic concept for 1D proton velocity encoding, the reader is referred to the books and papers by Pelc, McRobbie, Bushberg, Haacke, Moran, and Young for further details regarding the imaging process, reconstruction fundamentals, advanced velocity encoding, and potential error sources [68–74].

At the most basic level, PC-MRI exploits the change in the phase of transverse spin magnetization when proton spins move along a magnetic field gradient, such as in the case of hydrogen protons in blood flow. The "instantaneous" velocity is obtained as a result of phase shifts encoded by a positionally varying magnetic gradient, over a number of cardiac-gated acquisitions in frequency space, otherwise known as "k-space." In theory, the transverse magnetization, after excitation by an RF pulse, will acquire motion-induced phase shifts during the application of flow-encoding magnetic field gradient pulses. These flow-encoding pulses consist of two lobes, applied in the same direction, but opposite in polarity. Thus, the phase shift as a result of 1D motion in these gradients is written as:

$$\varphi = \gamma(x_0 M_0 + v_x M_1), \tag{28.30}$$

where ϕ is the phase shift, γ is the proton gyromagnetic ratio, x_0 is the original position of the spin, v_x is the velocity of the moving spin, M_0 is the zeroth moment of the x field gradient, and M_1 is the first moment of the x gradient [70, 75]. In the case where a bipolar gradient is used, $M_0 = 0$ and the resultant phase expression is given as follows:

$$\varphi = \gamma(v_x M_1). \tag{28.31}$$

The effect of these flow-encoding gradient pulses is shown in Fig. 28.23 in which a stationary body and a moving body of nuclear spins are influenced by the pulse lobes.

The application of the first gradient pulse causes the strength of the magnetic field to be positionally dependant for the duration of the pulse. Since the proton precessional frequency (or Larmor frequency) is proportional to the strength of the magnetic field, the proton spins will have a locationally proportional resonant frequency associated with their respective position. After the gradient pulse is turned off, will return to a homogenous state (i.e., the excited transverse spin magnetizations resonate at the same frequency), however the net transverse spin field will have acquired a phase shift. This phase shift is determined by the moment of the gradient pulse, and is also a function of the proton spin location in the magnetic field [70].

The phase shift created by the first gradient pulse in Fig. 28.23 can be nulled by a second gradient pulse, having an equivalent moment with an opposite polarity.

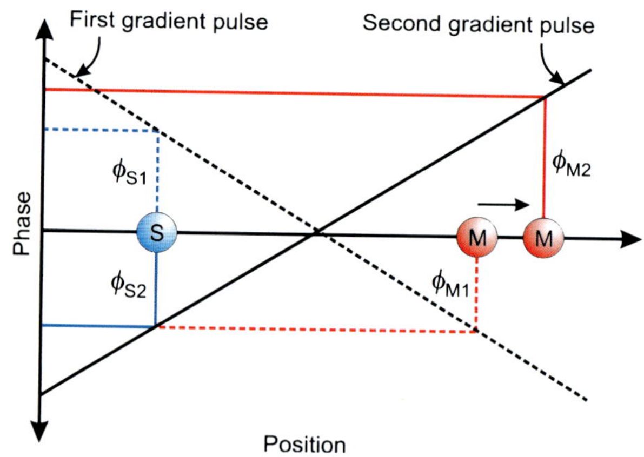

Fig. 28.23 The 1D effect of a bipolar gradient pulse on the phase of a stationary (S) and moving (M) spin magnetization. The first magnetic gradient lobe causes a phase shift encoded by position, x (horizontal axis). The second lobe causes a phase shift opposite in sign, such that the stationary proton exhibits a net-zero phase shift. However, for the moving proton, ϕ_{M1} and ϕ_{M2} are not equal and thus do not cancel. Ultimately, this phase shift is directly related to the proton's velocity

Thus, in the case that the proton spins do not move during the time between the first and second gradient pulses, the resultant phase shifts created by the first gradient pulse will be nulled by the second gradient pulse. However, if the proton spins are moving between the application of the pulses, the phase shift created by the first pulse will not be exactly canceled by the phase shift created by the second pulse. As a result, any remaining phase shift, $\Delta\phi$, will be a function of to the distance the proton spins have moved and, ultimately, directly related to the velocity of the proton spins:

$$\Delta\varphi = v(\gamma \Delta M_1), \tag{28.32}$$

where $\Delta M_1 = 2M_1$. It should be noted that the amount of phase change due to the motion of the proton field can be varied with an adjustment to the gradient pulse. This adjustment is referred to as the velocity encode sensitivity (v_{enc}), which determines the maximum velocity that produces a phase shift of 180°:

$$v_{enc} = \pi / (\gamma \Delta M_1). \tag{28.33}$$

Combining (28.33) and (28.34), the velocity (v) can therefore be determined by the specified v_{enc} and the proton phase shift:

$$v = \left(\frac{v_{enc}}{\pi}\right)\Delta\varphi. \tag{28.34}$$

A typical 2D output from a commercial scanner using this methodology is shown in Fig. 28.24a. The *x*, *y*, and *z* principle directions can be selectively encoded during the imaging sequence, with each additional component summarily contributing to the sequence duration. Depending on the imaging requirements, a reconstruction of the full 3D velocity vector field can be obtained for each voxel, providing a volumetric reconstruction of the hemodynamic characteristics in the region of interest. The results shown in Fig. 28.24b show the systolic flow pathlines in the aorta of a bicuspid aortic valve (BAV) patient. As can be seen, the malformed valve (combined with an enlarged aorta) contribute to a highly helical flow pattern (and therefore altered WSS values) when compared with control patients, Fig. 28.24c [64]. This example demonstrates the capability of this technique to quantify pathologic flow characteristics typically unobservable (or difficult to interpret) using alternative modalities.

28.6.2 *Flow Imaging Capabilities*

28.6.2.1 Global Flow Parameters

Since the publication of the original proposals by Moran to interlace velocity information in common pulse sequences, a number of experimental validations have been published [76–78]. Subsequently, several in vitro and in vivo studies have

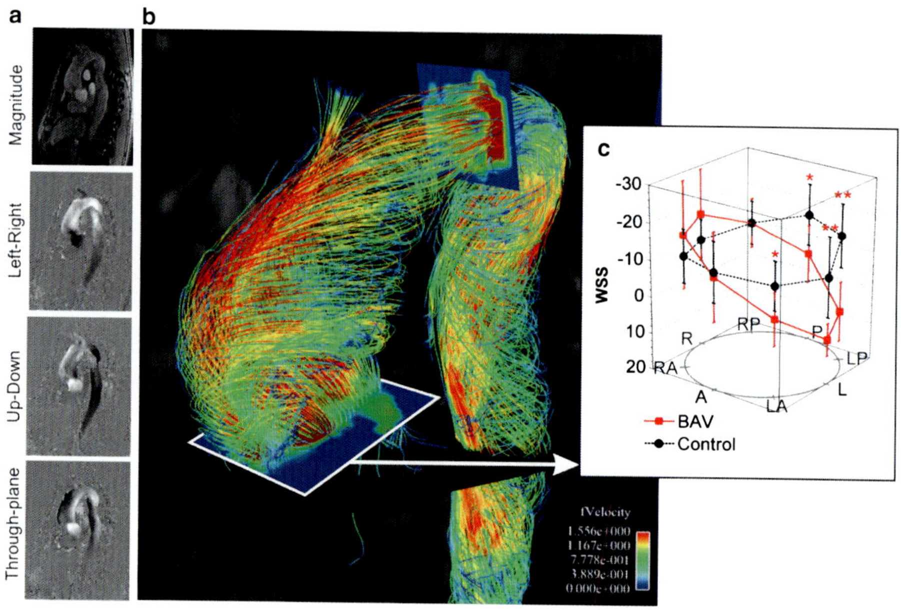

Fig. 28.24 (**a**) 2D, three-direction PC-MRI scan slice images bisecting the aorta longitudinally in a bicuspid aortic valve (BAV) patient. The *vertical column* of images shows the magnitude image, left-right, up-down, and through-plane velocity maps as typically output from the scanner after a three-vector component scan. (**b**) 3D volumetric flow pathline representation of a PC-MRI scan, which highlights exaggerated helical flow commonly observed in BAV patients. (**c**) Average local WSS differences between patients and controls (in dynes/cm^2, $n = 30$, lettering refers to anatomic location) [64]

validated the accuracy of flow measurements made by PC-MRI velocity mapping [79–82]. Studies of steady and pulsatile flow in straight tubes have widely demonstrated that individual velocity measurements are accurate and reproducible to ±5% when flow remains laminar or mildly turbulent [83–85] (most large artery flows remain laminar) [86]. It has been shown by Frayne that time-dependent flow, such as in composite Womersley flow phantoms, can be measured with an accuracy of 2.8% of the time and spatially averaged velocity [87]. Papaharilaou built on the Womersley phantom study to address more clinically relevant geometries, such as that of time-dependent flow in a bypass graft model [88]. In this case, he used CFD to compare the flow from 2D prospectively synchronized contrast enhanced PC-MRI measurements. The selection of the geometry and flow inputs to the model were designed to be physiologically relevant and highly 3D in flow – thereby producing strong secondary flow effects. In this phantom experiment, RMS differences between the experimental and the corresponding CFD velocities were used as an estimate of the random error of the measurement. The RMS accuracy ranged from 7.8 to 11.5% which, as expected, was higher than error measurements of simple, steady, and straight-pipe time-dependent flow. These accuracies are satisfactory when the capability to conceptualize complex flow patterns is considered. As a result, new opportunities have been created regarding hemodynamic structure function studies, the most important of which include the use of local flow parameters.

28.6.2.2 Local Flow Parameters

The ability to correctly and accurately determine spatial velocity gradients is crucial when considering PC-MRI as a tool to quantify local hemodynamic markers known to be concomitant with pathologic activity (such as the relationship between abnormal WSS and atherosclerosis). Frayne and Papaharilaou's validation work demonstrate the feasibility of applying PC-MRI to acquire WSS from the derivative of the velocity field. With this in mind, a number of researchers have presented methods and validations to calculate 2D and 3D WSS vectors [10, 89, 90]. Simulations and phantom experiments show that the ability to accurately determine WSS is dependent on spatial resolution and number of voxels in the cross-section of the flow region. As expected, the experimental measurement of WSS by PC-MRI converges to an analytical solution as the voxel resolution increases [10, 89]. While this implies that one must maximize resolution to obtain accurate WSS values, a drawback is the increase in velocity to noise ratio (VNR). With this in mind, the maximum meaningful voxel resolution of most 1.5 T commercial PC-MRI sequences is approximately $0.5 \times 0.5 \times 5$ mm – which according to simulations, leads to a measured WSS values 80% of the actual value [10]. At 1 mm voxel resolutions, the WSS will be 60% of its actual value. Thus, a common rule-of-thumb is that one must acquire more than ten voxels across the vessel cross-section at the highest resolution possible to be able to interpolate a meaningful (although perhaps, not absolute) WSS quantity. While this leads to an approximation of the original WSS, the relative, regional WSS and OSI patterns can be reliably visualized (such as in the cases of low and oscillating WSS) and co-located to potential problematic regions. With the improvement of pulse sequences and the increase in scanner field strengths, these resolution and accuracy limits are expected to improve.

28.6.3 Clinical and Research Applications

The reader is referred to a number of reviews covering the general purpose implementation of PC-MRI for clinical cardiovascular research, such as the use of flow waveforms and abnormal velocity field patterns to quantify disease progression and presence [71, 72, 91, 92]. In general use, PC-MRI is clinically implemented for the quantification of flow in left-to-right shunts, valvular insufficiencies, coarctations, pulmonary artery anomalies (such as in stenosis, hypoplasia, atrisia, and tertralogy of Fallot), carotid function, and pre- and post-operative vascular function.

In light of the already-available literature, these will not be discussed; however, a few recent developments regarding the use of PC-MRI to evaluate cardiovascular diseases are presented, before MR-based carotid flow imaging is discussed.

28.6.3.1 Recent Developments

Vessel compliance is affected by the presence of pathologic activity (e.g., hypertension), and can be calculated from the deformation and local pressure measured over the cardiac cycle. Unfortunately, the pressure component required to obtain a compliance measurement is a somewhat difficult to measure parameter noninvasively. However, compliance will also affect the speed of blood flow propagation wave along the vessel; thus, velocity sequences at high-temporal resolutions can measure the time of the pulse wave arrival at multiple locations to determine the pulse wave velocity (PWV) [91]. This allows for a measure of vessel compliance on a global scale. Using this technique in combination with PC-MRI, van Der Meer et al. and Grotenhuis et al. found separately that the presence of diabetes and BAV, respectively, increased the PWV [93, 94]. Hardy modified this technique by using a high-temporal resolution pencil excitation (time resolution = 4 ms) to allow PWV velocity calculations in short vessel segments, such as in the carotid artery [95]. These measurements may also prove useful in assessing the diagnoses and prognosis of patients, such as those with systemic or pulmonary hypertension, as well as in the assessment of the atherosclerosis.

As previously discussed, the hemodynamic WSS is a known pathophysiological stimulus leading to gene expression and extracellular matrix remodeling. Thus, the quantification of WSS is important in order to fully understand the progression of disease at the cellular level. As a result, a number of studies measuring WSS in various pathologies using PC-MRI have recently been reported [10, 66, 96–98]. For example, Saloner et al. longitudinally investigated the co-location of WSS and intracranial aneurysm growth, and found that growth was statistically associated with the regions of low WSS [98]. Barker et al. found abnormal WSS in BAV and pulmonary hypertension patients (PHT) [64, 99]. In addition, Markl has reported abnormal WSS in the presence of plaque lesions and in the internal carotid artery (ICA) just distal from the carotid bifurcation [100].

A number of recent studies have also looked at the volumetric 3D arterial flow fields [65–67, 98, 101–109]. These investigations have mostly focused on qualitative flow visualization with streamlines and particle traces in arterial and valvular pathology. However, a number of recent studies have investigated bulk flow quantification, vortex formation, and WSS in arterial pathology. Frydrychowicz et al. found that an obstructing thrombus in the aortic arch cause considerable complex and vertical flow patterns in the proximal descending aorta [103]. Markl et al. and Harloff et al. found significant helical flow in the aorta distal to a sclerotic aortic valve as well as in the ICA [63, 104]. In addition, Harloff et al. detected a stroke mechanism in the descending aorta not previously demonstrable without the use of this technique [110]. Additional efforts to quantitatively analyze the 3D flow fields have also involved flow connectivity mapping, vorticity calculations, and WSS calculations in the pulmonary and systemic arteries [105, 106].

28.6.3.2 Imaging Local Flow in the Carotid

As introduced in Sect. 6.2.2, the resolution limits of commonly available 1.5 T "off-the-shelf" PC-MRI pulse sequences is approximately $0.5\times0.5\times5$ mm. Given that the average adult CCA diameter is 6.5 mm, the vessel size limit is being approached for obtaining accurate local WSS values. Thus, care should be taken in designing studies to locally quantify WSS directly and in some cases, quantification may benefit from patient-specific CFD modeling in combination with MR-derived flow boundary conditions. This coupled CFD method was used by Lee et al. to determine the association of atherosclerotic high-risk regions with carotid morphology [56]. In their study, it was found that bifurcation tortuosity was highly related to regions in the ICA of low and oscillating WSS.

Nonetheless, experimental quantification of WSS at the bifurcation is highly clinically relevant to confirm these findings. Thus, Fig. 28.25 shows the results of a study in which 30 carotid arteries were scanned to obtain three-direction WSS traction vectors at the level of the bifurcation (BIF) and the CCA, focusing on temporally and spatially localized measurements [90]. The local and temporal WSS vectors indicated that PC-MRI measured WSS values correlate with "at-risk" or atheroprone regions at the BIF region of the ICA (when compared with WSS in the CCA).

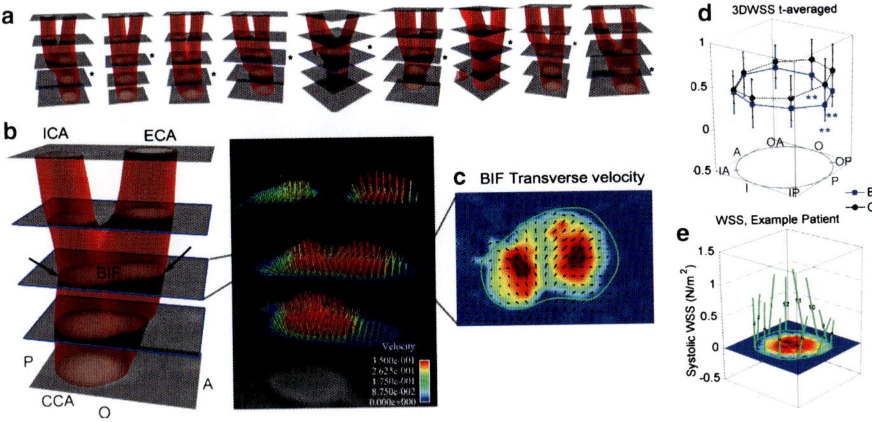

Fig. 28.25 (**a**, **b**) Examples of the MR angiograms of ten (out of 30) carotids analyzed showing the BIF slice position (*). The proximal slice was used for CCA analysis. (**c**) Plot of a BIF slice illustrates the presence of transverse velocities. (**d**) WSS in the BIF as compared to the CCA. Statistically significant values ($p<0.001$) are indicated (**). (**e**) Example systolic WSS vectors demonstrate the WSS significance patterns found in "**d**"

A possible indication of at-risk or disturbed flow is the contribution of velocity components perpendicular to the longitudinal axis of bulk flow. For example, the flow shown in Fig. 28.25c demonstrates the presence of transverse velocity (and thus circumferential WSS) components in the BIF. Supporting this observation, the temporally and spatially averaged circumferential WSS components in the BIF were $25 \pm 28\%$ of the axial WSS and they were $17 \pm 18\%$ of the axial WSS in the CCA. Figure 28.25d shows that this pattern is also locally apparent, with the 3D WSS magnitude (time-averaged over the cardiac cycle) significantly different in the posterior (P) and posterio-lateral (OP and IP) BIF regions. The patient-specific systolic WSS vector plot in Fig. 28.25e illustrates an example case and the areas of low WSS. In addition, it highlights the larger contribution of the circumferential components in these regions. Figure 28.25d highlights the most notable findings in this study – that is, the presence of highly significant ($p < 0.001$) low WSS at the posterio-lateral bifurcation region (as compared to the CCA). This location correlates to the ICA and ICA bulb, regions previously identified the CFD studies to exhibit low WSS (and known regions of focal plaque formation).

Thus, as predicted by the WSS/atherosclerotic risk hypotheses, the local WSS magnitudes were significantly low in atheroprone regions. These locations also correlated well with in silico simulations and provide further support for the utility of MRI in the study of these types of risk hypotheses.

28.7 Summary

The known relationship connecting blood hemodynamics with cardiovascular morphology and pathology emphasizes the need for the flow imaging modalities discussed in this chapter. Thus, a state-of-the-art review was presented, which covers techniques successfully implemented in the research and clinical arenas. Care was taken to review widely employed techniques (such as ultrasound Doppler) as well as emerging techniques (such as Echo PIV and 4D flow sensitive MRI), which show the promise to make major impacts in future structure/function investigations. It is hoped that the implementation of these new emerging techniques will ultimately aid in the detection of subclinical events – such as atherosclerosis – which remains one of the most important new applications for flow imaging research.

References

1. Writing Group M, Lloyd-Jones D, Adams R et al (2009) Heart disease and stroke statistics – 2009 update: a report from the American Heart Association Statistics Committee and Stroke Statistics Subcommittee. Circulation 119(3): e21–e181.
2. Shaaban AM and Duerinckx AJ (2000) Wall shear stress and early atherosclerosis a review. Am J Roentgenol 174(6): 1657–1665.

3. Ku DN, Giddens DP, Zarins CK et al (1985) Pulsatile flow and atherosclerosis in the human carotid bifurcation. Positive correlation between plaque location and low oscillating shear stress. Arterioscler Thromb Vasc Biol 5(3): 293–302.
4. Zarins CK, Giddens DP, Bharadvaj BK et al (1983) Carotid bifurcation atherosclerosis. Quantitative correlation of plaque localization with flow velocity profiles and wall shear stress. Circ Res 53(4): 502–514.
5. Giddens DP, Zarins CK, and Glagov S (1993) The role of fluid-mechanics in the localization and detection of atherosclerosis. J Biomechan Eng: Trans ASME 115(4): 588–594.
6. Gnasso A, Carallo C, Irace C et al (1996) Association between intima-media thickness and wall shear stress in common carotid arteries in healthy male subjects. Circulation 94(12): 3257–3262.
7. Cheng C, Tempel D, van Haperen R et al (2006) Atherosclerotic lesion size and vulnerability are determined by patterns of fluid shear stress. Circulation 113(23): 2744–2753.
8. Stone PH, Coskun AU, Kinlay S et al (2003) Effect of endothelial shear stress on the progression of coronary artery disease, vascular remodeling, and in-stent restenosis in humans – in vivo 6-month follow-up study. Circulation 108(4): 438–444.
9. Lee SW, Antiga L, Spence JD et al (2008) Geometry of the carotid bifurcation predicts its exposure to disturbed flow. Stroke 39(8): 2341–2347.
10. Stalder AF, Russe MF, Frydrychowicz A et al (2008) Quantitative 2D and 3D phase contrast MRI: optimized analysis of blood flow and vessel wall parameters. Magn Reson Med 60(5): 1218–1231.
11. Pedersen EM, Agerbaek M, Kristensen IB et al (1997) Wall shear stress and early atherosclerotic lesions in the abdominal aorta in young adults. Eur J Vasc Endovasc 13(5): 443–451.
12. Motomiya M and Karino T (1984) Flow patterns in the human carotid-artery bifurcation. Stroke 15(1): 50–56.
13. Jensen JA (1996) *Estimation of Blood Velocities Using Ultrasound: A Signal Processing Approach*. New York: Cambridge University Press.
14. Jensen JA (2001) A new estimator for vector velocity estimation. IEEE Trans Ultrason Ferroelectr Freq Control 48(4): 886–894.
15. Munk P (2000) *Estimation of Blood Velocity Vectors Using Ultrasound*. Lyngby: Department of Information Technology, Technical University of Denmark.
16. Udesen J (2006) *2-D Blood Vector Velocity Estimation Using a Phase Shift Estimator*. Lyngby: Department of Electrical Engineering, Technical University of Denmark.
17. Baker DW (1970) Pulsed ultrasonic Doppler blood-flow sensing. IEEE Trans Son Ultrason 17(3): 170.
18. Brandestini M (1978) Topoflow – digital full range Doppler velocity meter. IEEE Trans Son Ultrason 25(5): 287–293.
19. Kasai C, Namekawa K, Koyano A et al (1985) Real-time two-dimensional blood-flow imaging using an auto-correlation technique. IEEE Trans Son Ultrason 32(3): 458–464.
20. Willemetz JC, Nowicki A, Meister JJ et al (1989) Bias and variance in the estimate of the Doppler frequency induced by a wall motion filter. Ultrason Imaging 11(3): 215–225.
21. Hoeks APG, Vandevorst JJW, Dabekaussen A et al (1991) An efficient algorithm to remove low-frequency Doppler signals in digital Doppler systems. Ultrason Imaging 13(2): 135–144.
22. Ku DN, Giddens DP, Phillips DJ et al (1985) Hemodynamics of the normal human carotid bifurcation – in vitro and in vivo studies. Ultrasound Med Biol 11(1): 13–26.
23. Hein IA and Obrien WD (1993) A real-time ultrasound time-domain correlation blood flowmeter .2. Performance and experimental-verification. IEEE Trans Ultrason Ferroelectr Freq Control 40(6): 776–785.
24. Hein IA, Chen JT, Jenkins WK et al (1993) A real-time ultrasound time-domain correlation blood flowmeter .1. Theory and design. IEEE Trans Ultrason Ferroelectr Freq Control 40(6): 768–775.
25. Daigle RE, Miller CW, Histand MB et al (1975) Nontraumatic aortic blood-flow sensing by use of an ultrasonic esophageal probe. J Appl Physiol 38(6): 1153–1160.

26. Hansen PL, Cross G, and Light H (1976) Beam-angle independent Doppler velocity measurement in superficial vessels, in J.P. Woodcock, Editor. *Clinical Blood Flow Measurement*. London: Sector Publishing.
27. Schrank E, Phillips DJ, Moritz WE et al (1990) A triangulation method for the quantitative measurement of arterial blood velocity magnitude and direction in humans. Ultrasound Med Biol 16(5): 499–509.
28. Uematsu S (1981) Determination of volume of arterial blood-flow by an ultrasonic device. J Clin Ultrasound 9(5): 209–216.
29. Dunmire B, Beach KW, Labs KH et al (2000) Cross-beam vector Doppler ultrasound for angle-independent velocity measurements. Ultrasound Med Biol 26(8): 1213–1235.
30. Overbeck JR, Beach KW, and Strandness DE (1992) Vector Doppler – accurate measurement of blood velocity in 2 dimensions. Ultrasound Med Biol 18(1): 19–31.
31. Capineri L, Scabia M, and Masotti L (2002) A Doppler system for dynamic vector velocity maps. Ultrasound Med Biol 28(2): 237–248.
32. Scabia M, Calzolai M, Capineri L et al (2000) A real-time two-dimensional pulsed-wave Doppler system. Ultrasound Med Biol 26(1): 121–131.
33. Trahey GE, Allison JW, and Vonramm OT (1987) Angle independent ultrasonic-detection of blood-flow. IEEE Trans Biomed Eng 34(12): 965–967.
34. Trahey GE, Hubbard SM, and Vonramm OT (1988) Angle independent ultrasonic blood-flow detection by frame-to-frame correlation of B-mode images. Ultrasonics 26(5): 271–276.
35. Trahey GE, Hubbard SM, Fischer TA et al (1988) Angle independent blood-flow mapping using B-mode images. Ultrason Imaging 10(1): 70.
36. Friemel BH, Bohs LN, and Trahey GE (1995) Relative performance of two-dimensional speckle-tracking techniques: normalized correlation, non-normalized correlation and sum–absolute-difference. In IEEE Ultrasonics Symposium, Seatle.
37. Gallippi CM, Bohs LN, Anderson ME et al (2001) Lateral blood velocity measurement in the carotid artery via speckle tracking. In IEEE Ultrasonics Symposium, Atlanta, GA, USA.
38. Bohs LN, Friemel BH, Mcdermott BA et al (1993) Real-time system for angle-independent us of blood-flow in 2 dimensions – initial results. Radiology 186(1): 259–261.
39. Lim P, Mitchell-Heggs L, Buakhamsri A et al (2009) Impact of left ventricular size on tissue Doppler and longitudinal strain by speckle tracking for assessing wall motion and mechanical dyssynchrony in candidates for cardiac resynchronization therapy. J Am Soc Echocardiogr 22(6): 695–701.
40. Kim WJ, Lee BH, Kim YJ et al (2009) Apical rotation assessed by speckle-tracking echocardiography as an index of global left ventricular contractility. Circ Cardiovasc Imaging 2(2): 123–131.
41. Nesser HJ and Winter S (2009) Speckle tracking in the evaluation of left ventricular dyssynchrony. Echocardiogr J Cardiovasc Ultrasound Allied Tech 26(3): 324–336.
42. Ng ACT, Tran DT, Newman M et al (2008) Comparison of left ventricular dyssynchrony by two-dimensional speckle tracking versus tissue Doppler imaging in patients with non-ST-elevation myocardial infarction and preserved left ventricular systolic function. Am J Cardiol 102(9): 1146–1150.
43. Udesen J, Gran F, Hansen KL et al (2007) Fast blood vector velocity imaging: simulations and preliminary in vivo results. In IEEE Ultrasonics Symposium, NYC, NY.
44. Udesen J, Gran F, Hansen KL et al (2008) High frame-rate blood vector velocity imaging using plane waves: simulations and preliminary experiments. IEEE Trans Ultrason Ferroelectr Freq Control 55(8): 1729–1743.
45. Hansen KL, Udesen J, Gran F et al (2009) In-vivo examples of flow patterns with the fast vector velocity ultrasound method. Ultraschall Der Medizin 30(5): 471–477.
46. Gran F, Udesen J, Nielsen MB et al (2008) Coded ultrasound for blood flow estimation using subband processing. IEEE Trans Ultrason Ferroelectr Freq Control 55(10): 2211–2220.
47. Jensen JA and Munk P (1998) A new method for estimation of velocity vectors. IEEE Trans Ultrason Ferroelectr Freq Control 45(3): 837–851.

48. Munk P (1996) *Estimation of the 2-D Flow Vector in Ultrasonic Imaging: a New Approach.* Denmark: Department of Information Technology, Technical University of Denmark.
49. Udesen J and Jensen JA (2006) Investigation of transverse oscillation method. IEEE Trans Ultrason Ferroelectr Freq Control 53(5): 959–971.
50. Udesen J, Nielsen MB, Nielsen KR et al (2007) Examples of in vivo blood vector velocity estimation. Ultrasound Med Biol 33(4): 541–548.
51. Udesen J and Jensen JA (2003) Experimental investigation of transverse flow estimation using transverse oscillation. In IEEE Ultrasonics Symposium, Honolulu, Hawaii.
52. Zheng H, Liu L, Williams L et al (2006) Real time multicomponent echo particle image velocimetry technique for opaque flow imaging. Appl Phys Lett 88: 261915.
53. Liu L, Zheng H, Williams L et al (2008) Development of a custom-designed echo particle image velocimetry system for multi-component hemodynamic measurements: system characterization and initial experimental results. Phys Med Biol 53(5): 1397.
54. Zhang F, Lanning C, Mazzaro L et al (2009) Validation of echo particle image velocimetry in a patient-specific physiologically compliant model of the carotid bifurcation. Ultrasound Med Biol, Unpublished.
55. Kim HB, Hertzberg JR, and Shandas R (2004) Development and validation of echo PIV. Exp Fluids 36(3): 455–462.
56. Zhang F, Lanning C, Mazzaro L et al (2008) Systematic validation of the echo particle image velocimetry technique using a patient specific carotid bifurcation model. In IEEE International Ultrasonics Symposium, Beijing, China.
57. Zhang F (2009) *A Novel Ultrasound-Based Velocimetry Technique (Echo PIV) for Quantitative Assessment of Cardiovascular Hemodynamics: Recent Improvement, In Vitro and In Vivo Validation, and Its Initial Application in Vascular Profiling of Human Carotid Arteries.* Boulder: Mechanical Engineering, University of Colorado.
58. Zhang F, Barker AJ, Gates PE et al (2009) Noninvasive wall shear stress measurements in human carotid artery using echo particle image velocimetry: initial clinical studies. In IEEE Ultrasonics Symposium, Rome, Italy.
59. Zhang FX, Gates PE, Strain WD et al (2009) A novel ultrasound-based wall shear stress biomarker (echo particle image velocimetry) to generate hemodynamic vascular profiles in carotid arteries: clinical studies on 16 healthy subjects. Circulation 120(18): S1037.
60. Gates PE, Zhang F, Strain WD et al (2009) Echo particle image velocimetry (Echo Piv): a novel ultrasound-based method for the in vivo measurement of wall shear stress in human carotid arteries. Hypertension 54(5): 1171.
61. Hong GR, Pedrizzetti G, Tonti G et al (2008) Characterization and quantification of vortex flow in the human left ventricle by contrast echocardiography using vector particle image velocimetry. J Am Coll Cardiol Cardiovasc Imaging 1: 705–717.
62. Kheradvar A, Houle H, Pedrizzetti G et al (2010) Echocardiographic particle image velocimetry: a novel technique for quantification of left ventricular blood vorticity pattern. J Am Soc Echocardiogr 23(1): 86–94.
63. Harloff A, Albrecht F, Spreer J et al (2009) 3D blood flow characteristics in the carotid artery bifurcation assessed by flow-sensitive 4D MRI at 3T. Magn Reson Med 61(1): 65–74.
64. Barker AJ, Lanning C, and Shandas R (2009) Quantification of hemodynamic wall shear stress in patients with bicuspid aortic valve using phase-contrast MRI. Ann Biomed Eng 38: 788–800. DOI: 10.1007/s10439-009-9854-3.
65. Frydrychowicz A, Harloff A, Jung B et al (2007) Time-resolved, 3-dimensional magnetic resonance flow analysis at 3T: visualization of normal and pathological aortic vascular hemodynamics. J Comput Assist Tomogr 31(1): 9–15.
66. Frydrychowicz A, Berger A, Russe MF et al (2008) Time-resolved magnetic resonance angiography and flow-sensitive 4-dimensional magnetic resonance imaging at 3 Tesla for blood flow and wall shear stress analysis. J Thorac Cardiovasc Surg 136(2): 400–407.
67. Hope MD, Meadows AK, Hope TA et al (2008) Evaluation of bicuspid aortic valve and aortic coarctation with 4D flow magnetic resonance imaging. Circulation 117(21): 2818–2819.

68. Moran PR (1982) A flow velocity zeugmatographic interlace for NMR imaging in humans. Magn Reson Med 1(4): 197–203.
69. Haacke EM, Brown RW, Thompson MR et al (1999) *Magnetic Resonance Imaging: Physical Principles and Sequence Design*. New York: Wiley-Liss.
70. Young IA (2000) *Methods in Biomedical Magnetic Resonance Imaging and Spectroscopy*. New York: John Wiley & Sons, Ltd.
71. Pelc NJ, Herfkens RJ, Shimakawa A et al (1991) Phase contrast cine magnetic resonance imaging. Magn Reson Q 7(4): 229.
72. Pelc NJ, Sommer FG, Li KCP et al (1994) Quantitative magnetic-resonance flow imaging. Magn Reson Q 10(3): 125–147.
73. McRobbie DR, Moore EA, Graves MJ et al (2004) *MRI: From Picture to Proton*. Cambridge: Cambridge University Press.
74. Bushberg JT, Seibert JA, Leidholdt EM et al (2002) *The Essential Physics of Medical Imaging*. New York: Lippincott Williams and Wilkins.
75. Taylor CA and Draney MT (2004) Experimental and computational methods in cardiovascular fluid mechanics. Annu Rev Fluid Mech 36: 197–231.
76. Moran PR, Moran RA, and Karstaedt N (1985) Verification and evaluation of internal flow and motion. Radiology 154: 433–441.
77. Bryant DJ, Payne JA, Firmin DN et al (1984) Measurement of flow with NMR using a gradient pulse and phase difference technique. J Comput Assist Tomogr 8: 588–593.
78. van Dijk P (1984) Direct cardiac NMR imaging of the heart wall and blood flow velocity. J Comput Assist Tomo 8: 429–436.
79. Firmin DN, Nayler GL, Klipstein RH et al (1987) In-vivo validation of MR velocity imaging. J Comput Assist Tomogr 11(5): 751–756.
80. Firmin DN, Nayler GL, Kilner PJ et al (1990) The application of phase-shifts in NMR for flow measurement. Magn Reson Med 14(2): 230–241.
81. Oshinski JN, Ku DN, Bohning DE et al (1993) Effects of acceleration on the accuracy of MR phase-velocity measurements. J Magn Reson Imaging 2(6): 665–670.
82. Pettigrew RI, Dannels W, Galloway JR et al (1987) Quantitative phase-flow MR imaging in dogs by using standard sequences – comparison with in-vivo flowmeter measurements. Am J Roentgenol 148(2): 411–414.
83. Oshinski JN, Ku DN, Markou CP et al (1991) Evaluation of MR phase velocity mapping in straight tubes and stenoses. J Magn Reson Imaging 1: 217.
84. Oshinski JN and Ku DN (1991) The accuracy of phase velocity mapping for laminar and turbulent flow. Adv Bioeng ASME BED 20: 21–23.
85. Ku DN, Biancheri CL, Pettigrew RI et al (1990) Evaluation of magnetic-resonance velocimetry for steady flow. J Biomech Eng Trans ASME 112(4): 464–472.
86. Fung YC (1997) *Biomechanics: Circulation*. New York: Springer.
87. Frayne R, Steinman DA, Ethier CR et al (1995) Accuracy of MR phase-contrast velocity-measurements for unsteady-flow. J Magn Reson Imaging 5(4): 428–431.
88. Papaharilaou Y, Doorly DJ, and Sherwin SJ (2001) Assessing the accuracy of two-dimensional phase-contrast MRI measurements of complex unsteady flows. J Magn Reson Imaging 14(6): 714–723.
89. Cheng CP, Parker D, and Taylor CA (2002) Quantification of wall shear stress in large blood vessels using Lagrangian interpolation functions with cine phase-contrast magnetic resonance imaging. Ann Biomed Eng 30(8): 1020–1032.
90. Barker AJ, Zhang F, Gates PE et al (2010) 3-Component phase-contrast MRI WSS vectors in the carotid bifurcation are concurrent with local atherosclerotic Plaque risk hypotheses, Abstract 3701. In 18th Annual Meeting of the International Society of Magnetic Resonance in Medicine, Stockholm, Sweden.
91. Gatehouse PD, Keegan J, Crowe LA et al (2005) Applications of phase-contrast flow and velocity imaging in cardiovascular MRI. Eur Radiol. 15(10): 2172–2184.
92. Weber OM and Higgins CB (2006) MR evaluation of cardiovascular physiology in congenital heart disease: flow and function. J Cardiovasc Magn Reson 8(4): 607–617.

93. van der Meer RW, Diamant M, Westenberg JJM et al (2007) Magnetic resonance assessment of aortic pulse wave velocity, aortic distensibility, and cardiac function in uncomplicated type 2 diabetes mellitus. J Cardiovasc Magn Reson 9(4): 645–651.
94. Grotenhuis HB, Ottenkamp J, Westenberg JJM et al (2007) Reduced aortic elasticity and dilatation are associated with aortic regurgitation and left ventricular hypertrophy in non-stenotic bicuspid aortic valve patients. J Am Coll Cardiol 49(15): 1660–1665.
95. Hardy CJ, Marinelli L, Blezek DJ et al (2008) MRI determination of pulse wave velocity in the carotid arteries. In 16th Annual Meeting of the International Society of Magnetic Resonance in Medicine, Toronto.
96. Oshinski JN, Ku DN, Mukundan S et al (1995) Determination of wall shear stress in the aorta with the use of MR phase velocity mapping. J Magn Reson Imaging 5(6): 640–647.
97. Oshinski JN, Curtin JL, and Loth F (2006) Mean–average wall shear stress measurements in the common carotid artery. J Cardiovasc Magn Reson 8(5): 717–722.
98. Saloner D, Boussel L, Rayz VL et al (2008) CE-MRA and MR velocimetry in the determination of hemodynamic forces in longitudinal studies of intracranial aneurysm growth. In 16th Annual Meeting of the International Society of Magnetic Resonance in Medicine, Toronto, ON.
99. Barker AJ, Lanning C, Ivy D et al (2008) Artery dilatation in pediatric pulmonary hypertension patients decreases hemodynamic wall shear stress. Circulation 118S: 879.
100. Markl M, Zech T, Bauer S et al (2010) Carotid artery wall shear stress: distribution, correlation with geometry and effect of atherosclerosis. In 18th Annual Meeting of the International Society of Magnetic Resonance in Medicine, Stockholm, Sweden.
101. Hope MD, Hope TA, Ordovas K et al (2008) Clinical evaluation of aortic coarctation with 4D flow MR imaging. In 16th Annual Meeting of the International Society of Magnetic Resonance in Medicine, Toronto, ON.
102. Acevedo-Bolton G, Martin A, Rayz V et al (2008) In vivo MR determination of flow fields in patients with intracranial aneurysms using 7D PC-MRI. In 16th Annual Meeting of the International Society of Magnetic Resonance in Medicine, Toronto, ON.
103. Frydrychowicz A, Weigang E, Harloff A et al (2006) Time-resolved 3-dimensional magnetic resonance velocity mapping at 3 T reveals drastic changes in flow patterns in a partially thrombosed aortic arch. Circulation 113(11): E460–E461.
104. Markl M, Harloff A, Foll D et al (2007) Sclerotic aortic valve - Flow-sensitive 4-dimensional magnetic resonance imaging reveals 3 distinct flow-pattern changes. Circulation 116(10): E336–E337.
105. Frydrychowicz A, Arnold R, Harloff A et al (2008) In vivo 3-dimensional flow connectivity mapping after extracardiac total cavopulmonary connection. Circulation 118(2): E16–E17.
106. Reiter G, Reiter U, Kovacs G et al (2008) Magnetic resonance derived 3-dimensional blood flow patterns in the main pulmonary artery as a marker of pulmonary hypertension and a measure of elevated mean pulmonary arterial pressure. Circ Cardiovasc Imaging 1: 23–30.
107. Frydrychowicz A, Berger A, Stalder AF et al (2008) Diameter-dependence of aortic hemodynamics: does size matter? In ISMRM 16th Annual Scientific Meeting, May 3–9, Toronto, ON.
108. Kozerke S, Hasenkam JM, Nygaard H et al (2001) Heart motion-adapted MR velocity mapping of blood velocity distribution downstream of aortic valve prostheses: initial experience. Radiology 218(2): 548–555.
109. Kozerke S, Hasenkam JM, Pedersen EM et al (2001) Visualization of flow patterns distal to aortic valve prostheses in humans using a fast approach for cine 3D velocity mapping. J Magn Reson Imaging 13(5): 690–698.
110. Harloff A, Strecker C, Dudler P et al (2009) Retrograde embolism from the descending aorta visualization by multidirectional 3D velocity mapping in cryptogenic stroke. Stroke 40(4): 1505–1508.

Biographies

Fuxing Zhang was born in Tianjin, China in 1979. He earned his B.S. degree in 2002 and M.S. degree in 2005 from Tsinghua University. He received his Ph.D. degree in 2009 from University of Colorado at Boulder for work on blood flow hemodynamics quantifications. Since then, he has been employed as a research scientist at School of Medicine at University of Colorado, Denver. His research interests include ultrasound signal and image processing, and multidimensional blood velocity vector estimation.

Alex J Barker is a Fulbright postdoctoral research scholar in Medical Physics at the Department of Radiology, University Hospital Freiburg in Germany. The recipient of a F31 NIH Ruth Kirschstein National Research Service Award and Whitaker Scholar Award, Dr. Barker received a PhD in Mechanical Engineering from the University of Colorado, Boulder, subsequently working as a research associate in the Department of Pediatric Cardiology at The Children's Hospital, Denver. He has published a number of articles and 'best paper' proceedings surrounding topics on blood-pool contrast agents, flow sensitive magnetic resonance imaging, and cardiovascular hemodynamics. Dr. Barker can be reached at alexander.barker@freiburg-uniklinik.de.

Editor Biographies

Dr. Jasjit Suri, is an innovator, scientist, a visionary, industrialist and an internationally known world leader in Biomedical Devices and Biomedical Imaging Sciences – applied to Diagnostics and Therapeutics. He worked as Scientist, Manager; Sr. Director, Vice President and Chief Technology Officer (CTO) level positions with several million dollar industries like IBM, Siemens Medical, Phillips Medicals, Fisher and Eigen Inc., companies.

He has written over 300 publications, 60 innovations (patents), 4 FDA clearances, over 20 books in medical imaging and biotechnologies (diagnostic and therapeutic) and has lead as leadership role in releasing products in the men's and women's market applied to the fields of: Cardiology, Neurology (Image Guided Brain Surgery and Spinal Surgery), Urology (Image Guided Prostate Biopsy and HIFU for BPH), Vascular (Atherosclerosis- MR and Ultrasound), Ophthalmology (Thermal Imaging) and Breast Cancer (MR, X-ray-Ultrasound Fusion Guidance) markets.

He received his MS in Neurological MRI from Univ. of Illinois, Chicago, USA, PhD in Cardiac Imaging from University of Washington, Seattle, Washington, USA, and MBA from Ivy League Weatherhead School of Management, Case Western Reserve University, Cleveland, USA. He was crowed with President's Gold Model and Fellow of American Institute of Medical and Biological Engineering by National Academy of Sciences, DC. He has won over 50 awards during his career. Dr. Suri is also Strategic Advisory Board Member of over half a dozen industries and International Journals in Biomedical Imaging and Technologies. He main interests are cancer imaging for diagnosis and therapeutic applications for men's and women's market.

Jasjit S. Suri et al. (eds.), *Atherosclerosis Disease Management*,
DOI 10.1007/978-1-4419-7222-4, © Springer Science+Business Media, LLC 2011

Dr. Chirinjeev Kathuria

Dr. Kathuria holds a Bachelor of Science (B.Sc.) degree and specialized in US Health Care Policy and Administration and a Doctor of Medicine (M.D.) from Brown University. He also holds a Master's in Business Administration (M.B.A.) from Stanford University. Dr. Chirinjeev Kathuria, M.D., M.B.A. has measurable success in building businesses that impact world economies and shift business models. Dr. Kathuria has cofounded and helped build many businesses, which have generated shareholder wealth and jobs. Dr. Kathuria and affiliated companies have been featured in many TV shows and media publications. Dr. Kathuria has extensive experience in the health-care industry and has consulted to a broad range of organizations in the USA, Europe, and Asia. He helped develop Arthur D. Little biotechnology and health-care policy practice in Europe. He conducted a comparative analysis of the European and US biotechnology industries resulting in a paper entitled "Biotechnology in the Uncommon Market" which was published in Biotechnology magazine in December 1992 which helped change at that time the current thinking of biotechnology development. Dr. Kathuria's coauthored papers include "Selectivity Heat Sensitivity of Cancer Cells," "Avascular Cartilage as an Inhibitor to Tumor Invasion," and "Segmentation of aneurysms via connectivity from MRA brain data" the latter was published in the Proceedings of the International Society for Optical Engineering in 1993.

Dr. Filippo Molinari

Dr. Filippo Molinari received the Italian Laurea and the Ph.D. in Electrical Engineering from the Politecnico di Torino, Torino, Italy, in 1997 and 2000, respectively. Since 2002, he has been an assistant professor on the faculty of the Department di Electronics, Politecnico di Torino, where he teaches biomedical signal processing, biomedical image processing, and instrumentation for medical imaging. On March 2009 he was visiting professor at the University of Nagoya, Japan. He is the responsible for the image processing group at the BioLab of the Politecnico di Torino. Dr. Molinari's main research interests include the analysis of strongly nonstationary biomedical signals and the medical imaging applied to the computer-aided diagnosis. Dr. Molinari developed several signal and image processing algorithms, especially in the field of neurology, neurosciences, and in the functional assessment of disabled subjects. Specific interests of Dr. Molinari's research are early diagnosis, therapy, and rehabilitation. In the last 5 years, Dr. Molinari's activity was focused on ultrasound imaging in the field of neurology and cardiology. Dr. Molinari is on the Editorial Board of the Journal of NeuroEngineering and Rehabilitation and acts regularly as reviewer for more than 20 international journals in the field of biomedical engineering and medicine. He has published more than 20 technical papers and has written a collaborative book on advances in diagnostic and therapeutic ultrasound. He is member of the Italian Group of Bioengineering, of the IEEE Engineering in Medicine and Biology Society (EMBS) and of the American Institute of Ultrasound in Medicine (AIUM).

Index

H

I

Q

R

S

T